Solar
Energy
Handbook

McGRAW-HILL SERIES IN MODERN STRUCTURES:
Systems and Management

Thomas C. Kavanagh, Consulting Editor

Baker, Kovalevsky, and Rish *Structural Analysis of Shells*
Coombs and Palmer *Construction Accounting and Financial Management*
Desai and Christian *Numerical Methods in Geotechnical Engineering*
Dubin and Long *Energy Conservation Standards*
Forsyth *Unified Design of Reinforced Concrete Members*
Foster *Construction Estimates from Take-Off to Bid*
Gill *Systems Management Techniques for Builders and Contractors*
Johnson *Deterioration, Maintenance and Repair of Structures*
Johnson and Kavanagh *The Design of Foundations for Buildings*
Kavanagh, Muller, and O'Brien *Construction Management*
Kreider and Kreith *Solar Energy Handbook*
Kreider and Kreith *Solar Heating and Cooling*
Krishna *Cable-Suspended Roofs*
O'Brien *Value Analysis in Design and Construction*
Oppenheimer *Directing Construction for Profit*
Parker *Planning and Estimating Dam Construction*
Parker *Planning and Estimating Underground Construction*
Pulver *Construction Estimates and Costs*
Ramaswamy *Design and Construction of Concrete Shell Roofs*
Schwartz *Civil Engineering for the Plant Engineer*
Tschebotarioff *Foundations, Retaining and Earth Structures*
Tsytovich *The Mechanics of Frozen Ground*
Walker, Walker, and Rohdenburg *Legal Pitfalls in Architecture, Engineering, and Building Construction*
Woodward, Gardner, and Greer *Drilled Pier Foundations*
Xanthakos *Slurry Walls*
Yang *Design of Functional Pavements*
Yu *Cold-formed Steel Structures*

Solar Energy Handbook

JAN F. KREIDER, Ph.D., P.E.   *Editor-in-Chief*

Jan F. Kreider & Associates/Engineers
Boulder, CO

and

Frank Kreith, Dr. de l'Univ., P.E.

Solar Energy Research Institute
Golden, CO

McGRAW-HILL BOOK COMPANY
New York St. Louis San Francisco Auckland Bogotá
Düsseldorf Johannesburg London Madrid Mexico
Montreal New Delhi Panama Paris São Paulo
Singapore Sydney Tokyo Toronto

Library of Congress Cataloging in Publication Data

Main entry under title:

Solar energy handbook.

(McGraw-Hill series in modern structures)
Includes index.
1. Solar energy—Handbooks, manuals, etc.
I. Kreider, Jan F., date   II. Kreith, Frank.
TJ810.H35   621.47   79-22570
ISBN 0-07-035474-X

234567890   DODO   8987654321

The editors for this book were Jeremy Robinson and
Geraldine Fahey, the designer was Elliot Epstein,
and the production supervisor was Sally Fliess.
It was set in Caledonia by Offset Composition Services.

Printed and bound by R.R. Donnelley & Sons Company.

CONTENTS

v

Eyes, though not ours, shall see
Sky high a signal flame,
The sun returned to power above
A world, but not the same
                                    C. D. Lewis

PREFACE

The *Solar Energy Handbook* collects into one volume all the archival data and procedures available for solar system assessment and design, current as of its preparation date. The Handbook differs from a textbook in that it emphasizes applications, not theories, and provides information of a permanent nature which is not expected to change during the life of the book. For this reason proposed solar technologies for which there is no significant field experience are excluded. Nearly two-thirds of the authors, all of whom are preeminent experts, are from the private sector, i.e., excluding academia or government service.

The many forms of solar energy conversion are the principal topic of this Handbook. Sufficient data for the most mature technologies are presented to permit technical and economic analysis and synthesis through the final system design stage. Less detailed design information is given for less mature technologies. Large-scale electric power production by land- or sea-based systems represent the least mature of all technologies covered in this book. Since solar energy has impacts beyond purely technical areas, chapters on macroeconomics, microeconomics, and the law are also included.

The Handbook is separated into six major divisions, each made up of several chapters. The first, Perspective and Basic Principles, treats both the history of solar energy use and basic matters common to all solar systems including solar radiation and optics, heat transfer and thermodynamics, and solar component materials and storage. The second division, Solar-Thermal Collection and Conversion Methods, treats the many methods of producing heat from solar radiation. As such it treats only collectors, not systems, which become the subject of later divisions. Concentrating and nonconcentrating collectors are thoroughly treated in three chapters, and solar ponds are the subject of a fourth chapter.

The third division of the book, Low-Temperature Solar Conversion Systems, describes systems which operate below 100°C. These include the mature technologies of space and water heating using both gas and liquid working fluids as well as the less widely practiced applications of solar cooling and passive heating. System modeling and performance prediction are the subjects of two chapters followed by a chapter on agricultural applications. Finally, ocean temperature gradient power systems are examined.

High-temperature solar systems are treated in the fourth division along with industrial process heat systems. Heat engines and solar-thermal electric power are the subjects of the first two chapters of this division. The fifth division covers Advanced and Indirect Solar Conversion Systems, including wind and biomass energies. Biomass energy conversion is undergoing vigorous development and major new developments are expected in the 1980s. Photovoltaic conversion of sunlight to electricity is included in this division as well.

The final division, Architecture, Economics, the Law, and Solar Energy, treats matters not directly associated with the technical design of solar systems. However, these subjects—microeconomics and macroeconomics, the law, and energy conservation in buildings—are tantamount to the more technical subjects of preceding divisions. These are areas which might not be treated in a handbook devoted strictly to energy, but they are important in the Solar Handbook owing to the far-reaching effects of this energy source.

This Handbook uses hundreds of diverse sources for its material. As many of these as are known have been acknowledged formally in the text. Both the U.S. Customary System and the Système International d'Unites are used bilingually to communicate with the largest possible group of users. Conversion factors and a uniform nomenclature section for the book are contained in the Appendices.

A special acknowledgment is reserved for the small group of solar practitioners, some of whom are contributors to this volume, who persisted in their advancement of the state of the art for decades in the face of nearly universal disinterest. The technologies which are mature today owe their level of refinement in some respects to these persons. The editor-in-chief expresses his appreciation for the support offered by Dottie Lang through the long course of Handbook production. Jeremy Robinson suggested the idea for this Handbook and has been most supportive and helpful throughout.

A final note to the user. The information in this Handbook is current as of its production date. However, solar energy technology is in a state of continuing evolution and some subjects treated are undergoing development and refinement, the details of which can only be followed by review of the current literature. In the course of preparing hundreds of tables and figures and hundreds of thousands of words of text, it is inevitable that a few errors remain. The reader can provide a service to the solar community by reporting errors to the editor for correction of future printings.

Jan F. Kreider
Frank Kreith

Solar
Energy
Handbook

Chapter 1

The History of Terrestrial Uses of Solar Energy*

KEN BUTTI
and
JOHN PERLIN

EARLY UTILIZATION OF SOLAR ENERGY

Solar Architecture in Ancient Greece, Fifth Century B.C. to Third Century B.C.

Space heating and cooling   The first written account of solar energy use comes from ancient Greece. The Greeks did not have any mechanism for cooling their homes in the extreme heat of summer, nor did they have an adequate heating system for the cold winter. They frequently used kitchen stoves fueled by wood and portable braziers fueled by charcoal for heat.

By the fifth century B.C., wood and charcoal had become very scarce. The Greeks, having consumed most of their domestic supply, depended more and more on costly imports from (1) Macedonia and Thrace, (2) the Black Sea region, chiefly Bithynia, (3) the eastern Mediterranean (the coast of Asia Minor, Phoenicia, and Cyprus), and (4) southern Italy.

During this time (the Hellenic period), it became popular to utilize the sun's energy for both heating and cooling buildings. The Greeks developed basic principles of solar architectural design. Aeschylus, Aristole, and Xenophon outlined principles of using the sun's heat in winter. They pointed out that the main rooms of a house should face south and the north side of the building should be sheltered from the cold winds. To minimize solar heat gain in summer, eaves on the south side should provide shade to keep out the hot sun.

Example of solar architecture: the Olynthian house   The most extensively excavated city of ancient Greece, Olynthus, illustrates how solar architectural theory was translated into practice. Other excavations at Colophon, Delos, and Priene suggest that similar techniques prevailed throughout most of the country.

The typical Olynthian home was rectangular in shape (Fig. 1.1). The north wall had few window openings (windows during Greek times were not covered with glass, for transparent glass had not yet been invented), and the main rooms occupied the north wing. They faced an area on the south side of the building called the "pastas." The pastas extended east-west across

*This chapter is abridged from *The Golden Thread* by the authors, Van Nostrand-Reinhold Co., 1980. For their contributions the authors would like to thank Jon Billigmeier, Ph.D., for his French and German translations; Borimir Jordan, Ph.D., Associate Professor of Classics, University of California at Santa Barbara, for his translations and advice on the Greek and Roman sections; and Gloria J. Leitner for editing the entire manuscript.

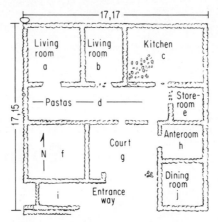

Fig. 1.1 Olynthian house plan. (*From David M. Robinson and J. Walter Graham, "Excavations at Olynthus," Johns-Hopkins, Baltimore, 1938.*)

the entire width of the building and the center section consisted of a colonnade which led into an open-air courtyard.

According to D. Robinson and J. W. Graham, the principal excavators of Olynthus,[12]

The width of the openings of the pastas on the court . . . varies considerably, as for example [house] BVI and AVii6-width between 9.2 m and 3.8 m (30.16 ft and 12.5 ft) respectively. The breadth of the pastas varies from 3 to 4 meters (9.83 ft to 13.1 ft).

Distances between the pillars of the pastas usually varied from 2.75 to 2.2 m (9.02 to 7.2 ft). The court ranged in size from 48 to 14.5 m² (516.1 to 155.1 ft²), or 19.8 to 5.4 percent of the total area of the dwelling.

Olynthus lies at about 40° north latitude. With the noon elevation of the sun at the winter solstice about 26°30', sunlight streamed through the courtyard and the pastas and into the main living rooms. To help retain the solar heat, the floors of the main rooms generally were made of earth. The walls were usually adobe, a poor conductor of heat and therefore a good insulator. Almost all of the walls of the houses at Olynthus measured 0.4 to 0.5 m (1.31 to 1.64 ft) thick.

In summer—with the noon elevation of the sun at the solstice at about 73°30'—eaves shaded the pastas, and the main rooms did not receive the direct rays of the sun. In addition, the adobe walls helped exclude the heat.

Olynthus' orthogonal plan, with avenues running east-west, guaranteed maximum solar exposure for every house, as the excavator observed:[12]

The house plan and the city plan exercised a strong influence upon one another, for the regular rectangular shape [of the houses was] . . . facilitated by the regular city plan.

Other examples of solar architecture in Greece   The Greeks used other types of solar design as well. At Delos, for example, the ground floor of the Maison du Trident and the Maison des Masques was built higher on the north side than on the south side of the building. In this way the south-facing window openings of the main rooms of the house in the north wing could face the sun in winter without obstruction.

In another equally successful technique, the north wing was built two stories high and the south wing only one story high. Thus the upper floor of the north wing was elevated above the south wing and received the winter sunlight.

Solar-Heated Homes, Baths, and Greenhouses in Rome, First Century A.D. to Fifth Century A.D.

Fuel problems   Like Greece, Rome also depleted fuel resources at a voracious pace. For example, at a short distance from Rome, Monte Cimino was a densely forested region up to the Third Century B.C. but by the First Century B.C., Rome had to import pine and other types of wood from as far east as Caucasia. Fuel consumption continued to rise with the standard of living. By the First Century A.D., many of the wealthier citizens enjoyed central heating. Tests done on a reconstructed furnace of the time showed that it would have had to consume

130 kg (287 lb) of wood per hour to heat a large villa adequately, or 16 m³ (565 ft³) every two days.

Public baths required even greater amounts of fuel. Whole tree trunks blazed inside their furnaces, and furnace temperatures reached up to 800°C (1472°F). Such baths could be found throughout Italy. By the Third Century A.D., 800 baths were located in Rome alone. The largest could hold nearly 2000 visitors at a time.

The steady depletion of indigenous supplies of fuel had a natural consequence—a steep rise in the price of wood and charcoal. Energy conservation techniques, including the use of solar energy, became important.

Solar architecture modeled after Greek designs   The three most influential Roman architects—Vitruvius, Palladius, and Faventinus—emphasized proper solar orientation of private villas and public baths. They wrote that rooms primarily used in winter, as well as bathing areas, should face south or southwest. Architectural evidence demonstrates that builders followed these recommendations wherever possible. The most detailed account of the construction and use of villas during the Second Century A.D., *The Letters of Pliny the Younger*, described how all the winter rooms in Pliny's villas faced the winter sun.

The Romans also went beyond techniques of passive solar building design—i.e., simply optimizing the solar energy exposure of a structure. They developed more sophisticated methods of exploiting solar energy including use of glass as a solar heat trap and use of several kinds of solar heat storage.

Use of window glass: the greenhouse effect   The Romans invented window glass in the First Century A.D., but only conjectures about its degree of transmittance have been made. Seneca referred to its transparency and Pliny the Elder differentiated it from opaque glass, stating that window glass allowed the visible spectrum to enter a structure. Other writers of the period remarked that window glass lets in "pure, unfiltered sunlight," that it "admits a clear light." Such comments were, of course, entirely relative since they were based on a comparison with opaque and translucent materials. Glass panes of good quality have been discovered. The largest panes found intact measured 0.33 × 0.55 m (1.08 × 1.8 ft).

We can surmise that the Romans knew about the greenhouse effect. A passage from Palladius indicates that they recognized that clear glass is opaque to low-temperature radiation.

Additional evidence of glass use can be found among the ruins of public baths (Fig. 1.2) built during or after the Imperial period (First Century A.D.). Most of the baths had their south-facing windows glazed. The window frames of the central baths at Pompeii, for example,

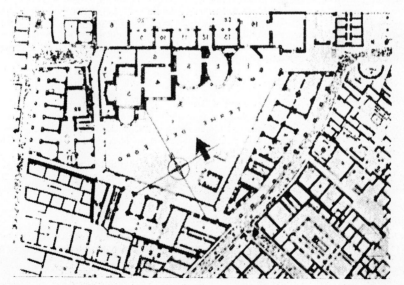

Fig. 1.2   Central baths at Ostia. Arrow points north to the five chambers of the baths. (*From X. Thatcher, "Ostia."*)

measured 1×3 m (6.56×9.84 ft). The accumulation of solar heat inside such baths because of glazing, Seneca noted, allowed the bathers "to broil."

The Romans also seem to have used window glass in their houses. Only if the south-facing windows were glazed could Pliny the Younger have boasted that his favorite winter room not only collected solar heat but increased its intensity. Excavations at Pompeii and Herculaneum, as well as other sites, have verified that by the Imperial period the windows of at least the wealthier homes were made of glass.

The Romans also used glass for greenhouses in which vegetables were grown in winter and exotic plants thrived during inclement weather. Furthermore, they installed glass on the south walls of their winter olive-pressing rooms. Solar heat accumulated inside and prevented congealing of the oil stored in the rooms.

Solar heat storage   Methods of solar heat storage were used by the Romans. For instance, southwesterly facing winter dining rooms had specially made floors that absorbed and stored solar energy. The storage floors were constructed by (1) digging a cavity 0.610 m (2 ft) deep and pounding down the earth inside it; (2) putting rubble or shards of broken earthenware into the hole; (3) gathering cinders and trampling them into a thick mass inside the pit; (4) applying a layer of dark sand mixed with ashes and lime, to a thickness of 15.24 cm (6 in), so that the surface of the floor was black; (5) leveling the floor with a square.

Sun rights   Roman law took into account the sun rights of its citizens. By the Second Century A.D., it had become a civil offense for anyone to place an object in such a way that it obstructed the solar exposure of a structure which required access to the sun's energy. Solar heating must have become a common practice to provoke the enactment of such a law.

Experiments with Solar Hot-Boxes, 1767–1881

Horace de Saussure, 1767   In 1767 Horace de Saussure, a Swiss naturalist, began to test the effectiveness of glass as a solar heat trap. He built five glass boxes, each a cube cut in half parallel to its base. The sides of the first box measured 30.48 cm wide $\times$ 15.24 cm high (12×6 in), the second box measured 25.4 cm wide $\times$ 12.70 cm high (10×5 in), and so forth, to the fifth and smallest box which had sides 10.16×5.08 cm (4×2 in). The bases were cut out so the boxes could be stacked one inside the other. They were attached to a table made of black pearwood.

De Saussure exposed the set of glass boxes to the sun. He found that the temperature of the largest box rose the least, and in each succeeding box the temperature increased. The smallest box reached 87.5°C (189.5°F). Observing that the boxes tended to lose heat by convection, he improved their insulation. The second device he constructed was a box made out of pine 1.27 cm (½ in) thick. It measured 30.48 cm wide $\times$ 22.86 cm high (12×9 in) per side. The interior was lined with black cork 2.54 cm (1 in) thick. Three glass covers were placed 3.81 cm (1.5 in) apart. When de Saussure exposed the device to the sun, he obtained a temperature of 109°C (228.2°F), or 9°C above the boiling point of water. He called it a hot-box, and thus the prototype of the flat-plate solar collector was born (Fig. 1.3).

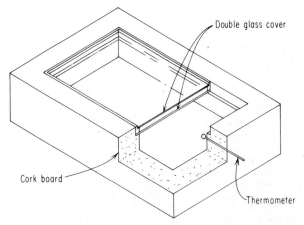

Fig. 1.3   Hot-box.

Despite its superior design, the hot-box lost some heat. To correct this, de Saussure placed the hot-box in a larger container filled with cotton wool as insulation. The container had an open top. Upon exposure to the sun, the temperature inside the hot-box rose to 110°C (230°F), even though the weather was not as favorable as during prior experiments.

As a third measure to eliminate heat loss, he put the hot-box into a glass-covered tin box. As the sun heated the hot-box, de Saussure heated the tin box by conventional means. He was careful always to keep the outer box slightly cooler than the inner one. Using this method, he recorded a temperature of 160°C (320°F) inside the hot-box.

De Saussure did not try to improve the seal on the hot-box by adding more glass covers. He realized that what heat he might have conserved would have been offset by increased absorption, reflection, and dispersion of incoming sunlight by extra layers of glass.

Sir John Herschel, 1837   Several prominent 19th Century scientists substantiated the results of de Saussure's hot-box experiments. Among them was Sir John Herschel, a noted British astronomer. Herschel reported that during an expedition to Cape Town, South Africa, he built a small hot-box out of mahogany with a blackened interior and a single sheet of glass on top. Exposing the box perpendicular to the sun, he observed that a thermometer inside rose to 65°C (149°F). When he improved the insulation of the box by heaping sand around the sides, he found that the temperature reached 80.6°C (177°F).

Herschel also placed the simple hot-box inside a larger container with a glass cover. He piled sand around its outer sides, and the thermometer in the hot-box rose to 115.8°C (240.5°F).

Samuel Pierpoint Langley, 1881   Samuel Pierpoint Langley, an American astrophysicist, also tested the effectiveness of the hot-box. He conducted his experiments in 1881 near Mt. Whitney, California. Langley constructed a copper box with a glass top, measuring 16.5 cm in diameter and 4 cm deep (6.5 × 1.58 in). He placed it in a wooden box which in turn rested inside a larger copper box with a glass cover, 32 cm in diameter and 8 cm deep (12.6 × 3.15 in). Loose cotton packing filled the space between the sides of this copper box and the walls of the wooden box. An outer shell of wood enclosed the walls of the entire apparatus.

On September 9, 1881, Langley recorded that the innermost copper box was 113.3°C (235.9°F)—98.5°C (209.3°F) hotter than the outside shade temperature (Table 1.1). These experiments proved that temperatures exceeding the boiling point of water could be obtained without the use of mirrors or other solar-focusing devices; however, no useful heat was removed from these devices.

History of Solar-Focusing Reflectors, Fourth Century B.C. to 1750s

Antiquity   The ancient Greeks discovered the principle of solar-focusing reflectors. They knew that if a large number of people held plane mirrors in the proper position so that the mirrors focused to the same point, a target at that point could be set on fire. Theophrastus, writing in the Fourth Century B.C., reported that such mirrors were made of polished silver or copper.

Soon the Greeks realized that concave spherical reflectors could produce fire in a similar manner, dispensing with the bulk and large number of people required by the former method. By the Third Century B.C. they were building parabolic mirrors, which were still more efficient. Diocles, a Greek mathematician of the Second Century B.C., gave the first formal geometric proof of the different focal properties of spherical and parabolic concave mirrors (Figs. 1.4 and 1.5). He wrote in his treatise *On Burning Mirrors*,[7]

> [The] property of [the parabolic mirror is] that all rays are reflected to a single point, namely the point whose distance from the surface is equal to one-quarter of the line which [constitutes] the ordinates. . . . The intensity of the burning mirror in this case is greater than that generated from a spherical surface, for from a spherical surface the rays are reflected to a straight line, not to a point.

In antiquity, concave mirrors principally served as a method of kindling fires. Plutarch, for example, noted that the Vestal Virgins used "concave vessels of brass" to reignite their sacred flames. The story of Archimedes burning the fleet at Syracuse by using solar reflectors is mere conjecture and appears to be a legend concocted many years after the supposed event.

Late antiquity and the Middle Ages   Several works of late antiquity and the early Middle Ages discussed, as did Diocles' treatise much earlier, differences in the focal properties and methods of construction of spherical and parabolic concave mirrors. Such tracts included *On the Burning Mirror* by Anthemius of Tralles (500 A.D.) and two works—*A Discourse on the Concave Spherical Mirror* and *A Discourse on the Parabolic Mirror*—by Alhazan of Cairo

TABLE 1.1   Temperature Readings by Langley with Hot-Box, September 9, 1881

Time			Hot-box temperature °C	°F	Temperature in shade °C	°F	Difference, °C
(A.M.)	11h	30m	88.44	191.2			
	11	35	91.22	196.2			
	11	40	94.06	201.3			
	11	45	96.72	206.1			
	11	50	98.88	210.9			
(Noon)	11	55	101.05	213.9			
(P.M.)	12	00	102.61	216.7	15.78	60.4	86.81
	12	05	103.89	219.0	15.78	60.4	88.11
	12	10	105.00	221.0	15.94	60.7	89.06
	12	15	105.94	222.7	16.61	61.9	89.33
	12	20	107.17	224.9	15.67	60.2	91.40
	12	25	108.00	226.4	15.56	60.0	92.44
	12	30	108.94	228.1	15.39	59.7	93.55
	12	35	109.55	229.2	15.39	59.7	94.16
	12	40	110.11	230.2	15.94	60.7	94.17
	12	45	110.67	231.2	17.06	62.7	93.61
	12	50	111.17	232.1	16.78	62.2	94.39
	12	55	111.55	232.8	16.17	61.1	95.38
	1	00	111.83	233.3	15.44	59.8	96.39
	1	05	109.55	229.2	15.56	60.0	93.99
	1	10	111.28	232.3	15.56	59.9	95.78
	1	15	111.83	233.3	16.11	61.0	95.72
	1	20	112.50	234.5	15.94	60.7	96.56
	1	25	112.89	235.2			
	1	30	113.17	235.7			
	1	35	113.33	236.0	16.00	60.8	97.33
	1	40	113.33	236.0	14.83	58.7	98.50
	1	45	112.89	235.2	14.44	58.0	98.45
	1	50	112.11	233.8	14.39	57.9	97.72

SOURCE: S. P. Langley, *Researches in Solar Heat*, U.S. Gov't Printing Office, 1884.

(965–1039 A.D.). Arab contemporaries of Alhazan used spherical concave mirrors made of Damascus steel for distillation purposes.

Europeans had translated Alhazan's works on mirrors into Latin by the late Twelfth or early Thirteenth Century, and they were commonly used as textbooks for several centuries thereafter, stimulating the imagination of early European scientists such as Witelo and Roger Bacon (both of whom lived during the Thirteenth Century). Bacon and Witelo wrote works on mirrors which became standard references. (Bacon dreamed of giant mirrors being used as weapons of war to burn armies and cities.)

The 17th and 18th Centuries   More than 300 years passed before the first experiments with concave mirrors were reported in Europe. One of the first to write about his experiments was Maginus, an Italian astronomer. He constructed a spherical concave mirror in 1619, which had a diameter of 0.509 m (1.67 ft) and could melt lead. Villete, a 17th Century French optician from Lyons, surpassed Maginus' feat by building even larger spherical concave mirrors. The largest of the five he built measured 1.16 m (3.8 ft) in diameter. All were made of an alloy of copper, tin, and tin glass. One of the reflectors, built in 1662, measured 0.763 m (2.5 ft) in diameter and weighed more than 50 kg (110 lb). Its focal line of 1.76 cm (0.66 in) was located at a distance of 0.97 m (3.19 ft) from the center of the mirror. The reflector could be moved to track the sun, and the reflected rays had the ability to pierce through a silver 15-pence piece in 24 s and to melt a small piece of pot iron in 40 s, an iron nail in 30 s, and a steel watch spring in 9 s.

In 1718, two independent researchers, J. Harris and J. T. Desaguliers, tested one of Villete's solar reflectors. Between 9 A.M. and noon on a June morning, they placed various objects at the mirror's focal line and observed that:

1. Copper ore vitrified in 8 s.
2. Iron ore melted in 24 s.

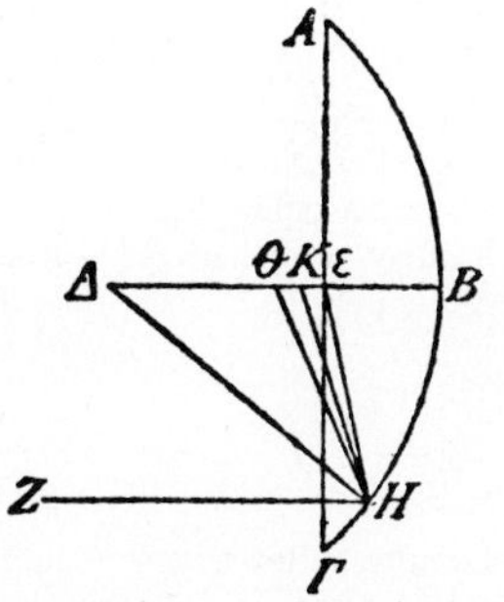

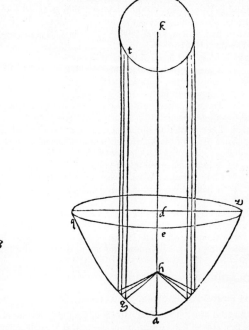

Fig. 1.4 Line focus of spherical mirror. (*From Huxley, "Anthemius de Tralles."*)

Fig. 1.5 Point focus of parabolic mirror. (*From Vitello, "Perspective."*)

3. A silver sixpence melted in 7.5 s.
4. Cast iron melted in 16 s.
5. Tin melted in 3 s.
6. Slate melted in 3 s and was pierced through in 6 s.
7. An emerald melted into a substance resembling turquoise.
8. A diamond weighing 25.92×10^{-5} kg (4 gr) lost 87 percent of its weight.

In 1687 the Baron of Tchirnhausen built the largest and most powerful mirror that had yet been made using a single piece of material. Measuring 1.7 m (5.57 ft) in diameter, the mirror was constructed of plate copper less than twice as thick as an ordinary knife blade. The reflector weighed much less than the ones built by Maginus or Villete, which were made of an alloy. As a result the Baron's reflector was much easier to operate, requiring only one person to move it.

A contemporary of Tchirnhausen, from Dresden, simplified the technique of mirror construction. He made a concave shape out of wood, applied pitch mixed with wax to the inner surface of the concavity, and covered it with gold leaf.

Despite such attempts at improving the construction of mirrors, metallic reflectors built from 1619 to 1755 remained relatively small. This was because alloy mirrors were made by melting the alloy in a concave mold, removing the hardened form, and polishing it—a relatively costly process which put constraints on size. The bulk and weight of alloy reflectors also made them difficult to operate. Mirrors of gilded wood had the advantage of being lighter, but they did not reflect as well as those made of alloy, and they deteriorated rapidly. Hand-built mirrors of plate copper were limited by problems of size and cost. Consequently, during this period scientists used solar-focusing reflectors for spectacular demonstrations rather than practical purposes.

P. Hoesen of Dresden invented a technique that overcame the difficulties of early mirror-building methods. He conducted his work during the late 1750s, although the results were not widely reported until 1769. Hoesen used a concave parabolic shape made out of segments of hardwood and lined the cavity with thin plates of highly polished brass. By adjusting the plates with great precision, he minimized the size of the junctures between the sections so they could hardly be seen. This method allowed him to build comparatively inexpensive, lightweight mirrors with sufficient focal length and relatively large dimensions. For example, one of his devices had a diameter of 3.08 m (10.1 ft) with a focal line of less than 0.013 m (0.043 ft).

The power of Hoesen's mirrors can be attested by the following experiment with one of his smaller devices, 1.55 m (5.08 ft) in diameter:

1. A piece of silver in green talc rock melted in 1 s without any trace of smoke. At the end of a minute, the stone melted.
2. Copper ore melted in 1 s.
3. Tin ore melted in the first second. After a minute of fusion, many grains of tin escaped and the stone changed into a dark glassy substance.
4. Lead melted at the blink of an eye. At the end of 3 s, it flowed in the liquid state.
5. A piece of black striated hematite began to melt after 4 s, without smoke. In 10 min several globules of perfect iron were found adhering to the stone.
6. Asbestos changed into a yellowish-green glass at the end of 3 s.
7. Slate became a black glassy material after 12 s.
8. A half-crown from Saxony began to melt immediately. A hole was pierced through it in 3 s.
9. An iron wheel melted in 3 s and began to flow in 5 s. Large droplets, looking like big peas, fell from the mass, reunited, and appeared to form a green-colored glass.

Unfortunately, although Hoesen's solar reflectors were effective and used by other experimenters, no practical purpose was found for them during this period.

SOLAR-POWERED ENGINES

The First Solar Engines, 1860–1904

August Mouchot, 1860   In 1860, French professor of mathematics August Mouchot began the first scientific investigation of the practical applications of solar power. At this time France was experiencing phenomenal industrial growth but had to import nearly 70 percent of its coal from Belgium and England. With coal prices increasing at nearly 10 percent per year, Mouchot saw vast potential for solar power in France as well as in its colonies.

Drawing from the work of de Saussure, Mouchot conducted his initial experiments with a hot-box. The results disappointed him, and he decided to design his own solar receiver. It consisted of a blackened copper cauldron placed under a glass bell jar mounted on a wooden insulating block. Mouchot placed a number of silvered metal reflectors around the circumference of the apparatus, creating the first primitive solar boiler. It could raise the temperature of 4×10^{-3} m^3 (1.05 gal) of water from 10°C (50°F) to 100°C (212°F) in less than an hour. He later connected a small steam engine to the boiler and demonstrated the first solar-powered engine.

The success of these experiments persuaded the fuel-short French government to sponsor Mouchot's research. Working in Tours, in 1872 he constructed an even larger solar-powered engine (Fig. 1.6). The solar concentrator was a silver-plated copper mirror in the shape of a truncated cone—2.6 m (8.5 ft) in diameter at the aperture and 1 m (3.28 ft) at the base. The boiler was situated in the center of the mirror along its vertical axis. Similar to the first solar boiler Mouchot had made, it consisted of a blackened cauldron set inside a glass bell jar. The mirror focused the direct rays of the sun onto the entire length of the boiler which was 0.79 m (2.6 ft) long. The whole device was mounted equatorially, and rotated at 15°/h to track the sun. The solar-powered boiler could produce 0.14 m^3 (37 gal) of steam per minute at a pressure of 101 kPa (1 atm). A reciprocating engine adapted to the boiler produced 373 W. Later Mouchot used an engine to pump water.

Continuing his research in 1878, Mouchot built a solar engine for the Universal Exposition in Paris (Fig. 1.7). The mirror was nearly 5 times larger than the one he had constructed 6 years earlier, measuring 5.04 m (16.4 ft) in diameter at its widest point and 1 m (3.28 ft) at the narrowest. The reflector was so well balanced that it required a force of only 22 N (5 lb) to rotate it. The boiler had a different design than the first model. It consisted of a series of vertical tubes which were encased under glass and located along the linear focus of the mirror. Under a clear sky the machine could boil 0.07 m^3 (18.5 gal) of water in 30 min, generating a pressure of 608 kPa (6 atm). Mouchot won a gold medal for the device from the Paris exposition.

The following year the French government sent Mouchot to Algeria to study the industrial feasibility of solar engines. He performed a number of experiments there in chemical distillation and irrigation using solar power. At the same time, the government also commissioned two

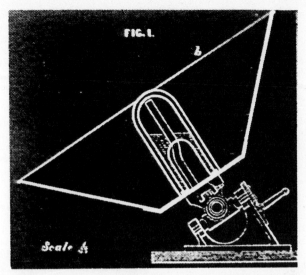

Fig. 1.6 Mouchot's first major sun machine. (*From A. B. Mouchot, "Solar Heat and Its Industrial Applications."*)

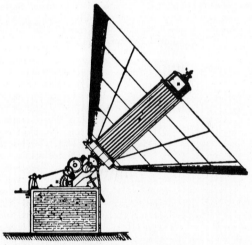

Fig. 1.7 Mouchot's sun motor for the Universal Exposition, 1878. (*From A. B. Mouchot,"Solar Heat and Its Industrial Applications."*)

independent engineers to run a year-long study of Mouchot's invention. After reviewing Mouchot's and the engineers' test results, the government concluded that the cost of constructing and maintaining the silver reflectors was excessive and therefore such solar motors could not meet the demands of commerce. The government withdrew its financial support and Mouchot ended his research.

John Ericsson, 1870   John Ericsson, a prominent American engineer most noted for his design of the Civil War battleship *Monitor*, began solar experiments at about the same time as Mouchot. In the late 1860s he wrote that the potential of solar motors was far beyond calculation, and declared he would dedicate the rest of his life to solving the problems of solar energy application.

Ericsson used an entirely different approach than Mouchot. Disillusioned by the cumbersome cone reflector, he designed a series of parabolic trough-shaped reflectors. His first successful solar engine, built in New York in 1870, had a total collecting area of 3.25 m² (35 ft²). It could drive a small 373-W steam engine, he claimed, with a cylinder diameter of 11.43 cm (4.5 in) and a working speed of 240 r/min. If this was true, it was a remarkable device for it could produce 6 times the horsepower of Mouchot's engine with only one-sixth the collecting area.

In 1872 Ericsson built a second test motor which was a hot-air engine and closely resembled one built by Scottish clergyman Robert Stirling some 56 yrs earlier. The cylinder of the engine lay along the vertical axis of a reflector which had an irregular curved shape. It focused sunlight onto the top of the cylinder, producing a focal line of 20.3 cm (8 in). The engine ran at 400 r/min. By 1876, Ericsson had constructed more than seven test motors, trying to find an alternative to expensive polished metal reflectors. He built the last one in 1883, using a method he had developed of coating window glass with silver so that it could withstand intense solar radiation. By mounting narrow strips of this type of mirror into a steel frame, he could construct a relatively large reflector at modest cost. He built a reflector measuring 3.35 m × 4.8 m (11 ft × 16 ft), with a reflector surface of 15.81 m² (170 ft²). It focused sunlight onto a heat tube 15.87 cm (6.25 in) long (Fig. 1.8). The solar boiler powered a steam engine with a cylinder 15.24 cm (6 in) in diameter and a 20.32-cm (8-in) stroke at 120 r/min. The pressure on the piston was 241.14 kPa (2.38 atm).

Ericsson envisioned the day when Europe's coal mines would be depleted and factories would be forced to move to the sunburnt but solar-energy-rich regions of the world. However he refused to patent, and thus to market, his solar motors until they were perfected. He continued working on solar engines until his death in 1889 but unfortunately revealed few specifics about them to prevent competitors from obtaining patent rights for modifications of his design.

Aubrey Eneas, 1899 A series of crippling coal strikes in the 1890s and widespread discussion of the prospect of a total depletion of coal reserves stimulated Boston engineer Aubrey Eneas to investigate solar power. Over a period of nearly 10 years he built five unsuccessful solar motors before arriving at a design he considered commercially feasible. He formed the Sun Power Co. of Boston with a group of investors from the city.

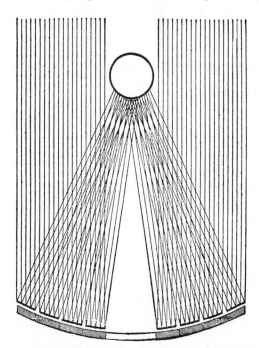

Fig. 1.8 Cross section of Ericsson's parabolic reflector. (*From J. Ericsson, "Contributions to the Centennial."*)

Eneas built his first solar engine in 1900. He patterned the reflector, a large truncated cone, after Mouchot's model. Mounted on two trolley rails, it could be adjusted perpendicular to the sun's rays to accommodate seasonal declination changes (Fig. 1.9). It was also mounted equatorially, and a heavy clock mechanism moved the 3759.9-kg (8300-lb) structure so that it followed the sun's daily path.

Larger than previous reflectors, the cone measured 10.97 m (36 ft) at its mouth and 4.87 m (16 ft) at its base. It consisted of 1786 rectangular glass mirrors, each separately bolted to a steel chassis. With an overall surface area of nearly 93 m² (1000 ft²), the mirror could focus sunlight onto an axial boiler with a concentration ratio of 13.4:1.

The boiler had a unique design. Standing 4.11 m (13.5 ft) high, it was supported by guy wires running from the vertical axis of the cone, similar to the way a bicycle wheel is mounted on an axle. It consisted of two copper tubes fitted one inside the other, which in turn were enclosed in two concentric glass tubes (Fig. 1.10). The surface of the outer tube was coated black. The inner tube received cool water and fed it to the outer tube, where it was heated and turned into steam. The boiler had a water capacity of 0.73 m³ (100 gal) and a steam capacity of 0.22 m³ (58 gal).

The boiler drove an 8206-W (11-hp) compound condensing engine at an average rate of 490 r/min. Within the first hour of operation the engine reached a pressure of 1033.5 kPa (10.2 atm). Under optimum conditions, Eneas reported, it could pump 5.29 m³ (1400 gal) of water per minute. Its average rate was 1.32 m³ (359 gal) per minute.

Eneas could not achieve the kinds of results that Ericsson had claimed. He calculated that his collector had an efficiency of 75 percent, but could only convert 4 percent of the incident solar radiation into mechanical work (the best coal-fired steam plants of the day converted from 10 to 12 percent). Furthermore, it took 13.9 m² (150 ft²) of collector area to produce every 746 W (1 hp), 30 percent more area than Ericsson said his engine required. Nevertheless, Eneas foresaw a bright future for his solar motor in the fuel-short, sun-rich southwest. Hence in 1901 he moved to California, setting up several offices throughout the region during the next 3 yrs. The first person to attempt to sell solar motors commercially, he constructed three large devices and put a price tag of $2160 ($15,552, 1978 dollars) on each. This was equivalent to $196 ($1411, 1978 dollars) per 746 W (1 hp)—some 2 to 5 times more expensive than a conventional steam plant. Because of the high price, and a prevailing attitude of skepticism about solar motors, Eneas was unable to sell more than three solar engines.

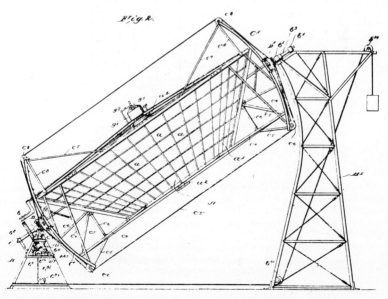

Fig. 1.9 Elevation of Eneas' sun motor. (*From a patent drawing.*)

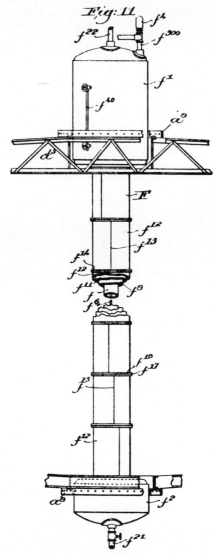

Fig. 1.10   Diagram of Eneas' boiler, f 7, f 8—copper tubes; f 11, f 12—glass sleeves. (*From a patent drawing.*)

A new approach—H. E. Willsie and John Boyle, 1902–1911   In 1902 two American engineers, H. E. Willsie and John Boyle, discovered what they thought was an economical way to harness solar power. Up until 1902, nearly all experimenters had used mirrors and high-temperature engines. Willsie speculated that this stemmed from a fascination with extremely high temperatures. He saw low-temperature, hot-box-type collectors as more practical. They could use both direct and diffuse solar rays, and would be less complex to build and operate.

Willsie obtained favorable results with his initial experiments and filed patent rights for a solar motor in 1903 to form the Willsie Sun Power Co. The following year, in St. Louis, he constructed a low-temperature solar power plant—the first of its kind. The plant was simple in design and operation (Fig. 1.11). The hot-box was made of wood, with a double-layered glass cover and insulation around the sides and bottom of the box. It measured 55.8 m² (600 ft²).

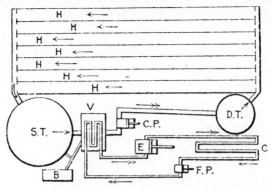

Water from the distributing tank D. T. after flowing through the glass-covered troughs H.H.H. absorbing solar heat is stored in the storage tank S. T. This hot water gives up its heat in the vaporizer V and is sent back by the circulating pump C. P. to the distributing tank. Single-headed arrows indicate the flow of the water. Emergency steam boiler, B, for cloudy periods.

Sulphur dioxide flows in the direction of the double-headed arrows from the vaporizer coils to the engine E. The exhaust vapor goes to the condenser C. The liquid sulphur dioxide is returned by the feed pump F. P. to the vaporizer.

In practice the distributing tank is small; the storage tank S. T. holding both the hot and cold water.

Fig. 1.11 Operational diagram of Willsie's sun motor. (*From H. E. Willsie, "Experiments in the Development of Power from the Sun's Heat."*)

A shallow stream of water flowed through narrow troughs inside the box. The water was heated and ran into a boiler, which in turn heated an ammonia solution (Willsie used ammonia because a high pressure could be obtained at a relatively low temperature). The boiler drove a 4476-W (6-hp) engine.

Willsie claimed a collection efficiency of 85 percent, nearly twice the amount of solar heat collected per square meter than Ericsson's device (the figure seems too high to be taken seriously).

Willsie and his coworker Boyle decided to build a second plant, this one at Needles, CA, in the Mojave Desert. It was an ideal location: the air was dry, nearly 85 percent of the days were sunny, and fuel in the area was expensive. They completed the plant in 1905, modifying the previous design in several major ways. The collector—which occupied 93 m² (1000 ft²)—was divided into two sections, one of which acted as a preheater. Water in the first section could reach 65.55°C (150°F), and 82.22°C (180°F) in the second section.

The heated water flowed through troughs that were placed on an incline to facilitate circulation, and fed into a storage tank. From there the water ran into an open-stack heat exchanger, which was substituted for a conventional boiler. The water temperature dropped an average of 38°C (100°F), creating a pressure of 1378 kPa (13.6 atm).

A vertical, automatic-cutoff, sulfur dioxide engine drove a centrifugal water pump with a power of 1.1×10^4 W (15 hp). The exhaust from the engine cooled in an open-air drip condenser to 17.7°C (64°F).

On several occasions they put the mechanism into operation at night by opening the storage tank. Thus, the Needles solar plant was the first to be able to work when the sun was not shining. The cost-efficiency of the plant was the best ever attained. They estimated the cost of construction of a similar plant at $164 ($1143, 1978 dollars) per 746 W (1 hp) produced. However, while this rate was 16 percent cheaper than Eneas' plant, it was still 2 to 4 times higher than conventional coal-fired steam plants (Table 1.2).

Willsie and Boyle calculated that in sunny but fuel-scarce areas of the country, the plant could pay for itself within 2 yrs of operation. But only 1 yr after he made that estimate, the "producer gas engine" appeared on the market. It reduced the cost of mechanical power even in remote regions of the southwest, and their solar engine had little chance of competing.

TABLE 1.2   Cost of One Electric Horsepower per Hour—Steam vs. Solar

	Steam	Solar
Engineer, 40¢ per hour	0.08	0.08
Fireman, 30¢ per hour	0.06	
Dynamo man, 40¢ per hour	0.08	0.08
Helper, 25¢ per hour	0.05	0.05
Superintendent	0.06	0.06
Coal	1.50	
Interest, maintenance, depreciation—power plant	0.046	0.184
Interest, maintenance, depreciation—electric plant	0.006	0.006
Oil, waste, water (or sulfur dioxide)	0.15	0.15
Total estimated steam plant at $40 per hp	2.078	
Total estimated solar plant at $164 per hp		0.610

SOURCE: H.E. Willsie, "Experiments in the Development of Power from the Sun's Heat," *Engineering News*, New York, May, 1909.

Near-Commercial Success—Frank Shuman, 1906–1917   In 1906 Frank Shuman turned his attention to solar power—the second engineer of national stature in the United States to do so (the first had been John Ericsson). After building a number of working models, in 1907 Shuman constructed his first commercial solar plant near his home in Tacony, a suburb of Philadelphia. Very similar to the solar plants built by Willsie and Boyle, Shuman's plant used a large hot-box to collect solar energy and a low-pressure engine which drove a pump. The collector measured 112 m² (1200 ft²), was well insulated, and was covered by two layers of glass. Black pipe inside the collector functioned as the heat absorber. Ether circulating through the pipes was heated to an average temperature of 93.3°C (200°F) and drove a vertical single-cylinder engine. It generated 2.6×10^3 W (3.5 hp). The plant remained in operation for over 3 yrs.

In 1910 Shuman changed the design drastically to achieve greater collection efficiency:

1. He reduced the collector to approximately 0.94 m² (10.1 ft²), and mounted a number of units in rows angled toward the sun.

2. In place of a steel pipe absorber, he used two sheets of thin copper fused together so that there was a waterway between the sheets.

3. He attached flat mirrors to the top and bottom of each collector.

Test results on a small model measuring 4.6 m² (50 ft²) were favorable, and the project attracted the financial support of British investors interested in building solar-powered irrigation plants in Egypt. Consequently, later in the year Shuman designed a 7.5×10^4 W (100-hp) solar plant.

The plant had 572 collectors, mounted in 26 separate banks, with 478 m² (5148 ft²) of mirrors and 478 m² (5148 ft²) of absorbers. It drove a low-pressure, single-condensing, steam-powered motor operating at an optimum speed of 130 r/min. In tests the collectors operated at 5512 kPa (54 atm) of steam per hour at atmospheric pressure, producing an average of 1.0×10^4 W (14 hp).

While the plant was being readied for shipment to Africa, the eminent British physicist C.V. Boys joined Shuman's project as a consulting engineer. Disappointed with the efficiency of the plant, he pointed out that the collector would have to produce a higher concentration ratio to be economical. Boys recommended reducing the size of the absorbers and placing them at the focus of a parabolic trough reflector so that both sides of the absorber would receive solar radiation. (Boys was unaware of Ericsson's similar design.) In 1912, work began to redesign the solar plant accordingly.

By June of the following year the world's largest solar power plant had been completed and installed at Meadi, Egypt, along the banks of the Nile. It was the culmination of 7 yrs of research and an investment of over $250,000 ($1,711,875 in 1978 dollars).

The plant bore a close resemblance to two of its predecessors: the glass parabolic reflector motor of Ericsson and the low-pressure engine and storage system of Willsie and Boyle. The reflectors occupied nearly 4046 m² (1 acre) and were divided into five sections, each of which measured 62.17 × 4.1 m (204 × 13.5 ft). The collectors were mounted to a crescent-shaped steel frame positioned north-south. The reflectors moved from east to west to track the sun,

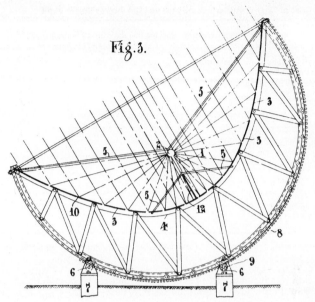

Fig. 1.12 Cross section of Shuman's parabolic reflector. No. 3, glass mirrors; No. 2, boiler. (*From a patent drawing.*)

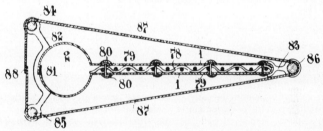

Fig. 1.13 Cross section of Shuman's wedge boiler. No. 1, absorbing surface; No. 2, feed pipe. (*From a patent drawing.*)

and were set 7.26 m (25 ft) apart so as not to shade each other during operation (Fig. 1.12). The mirror surface consisted of a series of long, rectangular sections of silvered window glass, each mounted side by side and angled toward the focal point. The reflectors produced a concentration ratio of 5:1. In place of the type of absorber previously used, Shuman installed a 38.1-cm-high (15 in), wedge-shaped cast-iron receiver which was suspended at the focus of each reflecting trough (Fig. 1.13). On top of the trough, an 8.89-cm-diameter (3.5 in) tube fed water into the passageway below.

The 7579-kPa steam drove a low-pressure engine at an average of 3.7×10^4 W (50 bhp). The engine was designed to operate economically at a steam pressure as low as 27.3 kPa (0.27 atm). Large insulated tanks stored the solar-heated water so that the engine could work at night or during inclement weather.

After extensive tests, Shuman concluded that the plant was mechanically and economically practical. He estimated the cost of constructing a similar plant at $7800 ($48,750, 1978 dollars), compared with $3850 ($24,062, 1978 dollars) to construct an equivalent coal-powered plant in that part of the world. With coal selling in Egypt at $15 for 907 kg ($93.75/ton, 1978 dollars), Shuman calculated that the plant could pay for itself in less than 4 yrs, and after that accumulate a healthy profit (Table 1.3). With these figures in hand, Shuman was able to persuade Field

TABLE 1.3   Cost of Coal-Burning Steam Plant vs. Sun-Power Steam Plant

Sun power steam generator working 365 days for 10 hours per day	
Interest on capital expenditure, $7800 at 5%	$390.00
Wear and tear, depreciation at 5%	390.00
Total	$780.00

Coal-burning steam-generator plant working 365 days for 10 hours per day	
Interest on capital expenditure, $3850 at 5%	$192.50
Wear and tear, depreciation at 5%	192.50
Coal consumption at 2 lb coal per brake horsepower-hour=163 tons at $15 per ton	2445.00
Total	$2830.00
Cost of operating sun-power plant	780.00
Savings	$2050.00

SOURCE: F. Shuman, "Sun-Power Plant," *Scientific American*, Supplement No. 1985, January, 1914.

Marshal Kitchener in Sudan to agree to construct a similar solar plant in his country. He also obtained a 200,000-mark advance from the German Reichstag to build one in German East Africa.

Just 1 yr after the plant had been constructed in Egypt, World War I broke out. Everyone involved in the project returned home to aid in the war effort. This turned out to be more than a temporary postponement, for Shuman died before the war ended. In his absence, and with the growing availability of fuel oils, interest in solar power plants waned and the other projects were never begun.

SOLAR WATER HEATERS

The Solar Water Heater Industry Begins: California, 1890–1909

From the turn of the century to the early 1920s, several factors prevailed in southern California that were conducive to the development of a solar water-heating industry:

1. Conventional methods of water heating were inconvenient (most were not automatic and none had storage).

2. Fuel was scarce in many communities (most rural areas had neither gas nor electricity).

3. Those areas that could obtain fuel had to pay dearly for it (coal cost twice as much as the national average at the time and gas for heating cost more than 10 times what natural gas costs today).

4. Sunshine was amply available.

Bare-tank heaters   The first solar water heaters were simply water tanks placed upright or laid on the ground so that they received maximum sun and minimal shade. By late afternoon, on clear, calm, hot days, a typical 0.113-m³ (30-gal) tank supplied enough hot water for two to three showers at 38.9°C (102°F) (see Fig. 1.14).

Because the tanks were bare and unprotected, water inside readily lost heat at night and cooled to air temperature. A further drawback was that the tanks were slow to heat up during the day because of their poor ratio of area to volume (water temperature did not reach its peak until late afternoon). According to research conducted at the University of California at Davis in 1935, larger-capacity tanks have even poorer ratios of area to volume. The researchers concluded that[3]

> When the quantity of hot water available from a 30 gallon tank [0.113 m³] is not sufficient but the characteristic temperature performance is satisfactory, a larger quantity is obtained by using several tanks in parallel, with all the cold water inlets connected together in one direction and all the hot outlets connected together in the opposite direction.

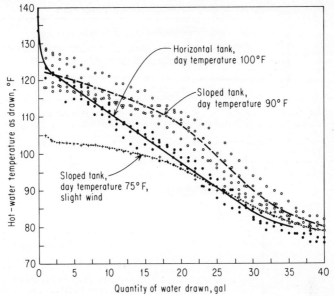

Fig. 1.14 Performance of 30-gal (0.113-m³) bare tank solar heater. (*From F. A. Brooks, "Solar Water Heaters."*)

They also found that tanks inclined at an angle produced 20 percent more hot water than horizontally placed tanks.

Enclosed-tank heaters Bare-tank heaters soon gave way to enclosed tanks for greater efficiency.

Climax, 1891. Clarence M. Kemp, a manufacturer of gas heating equipment, developed an enclosed-tank solar water heater in the 1890s. He used four tanks made of heavy galvanized iron painted dull black. The most common tank size was 0.121 m³ (32 gal).

The tanks were placed side by side and connected with tubing. They were enclosed in a container similar to a hot-box: a wooden box which was insulated with felt paper and covered by a sheet of glass (Fig. 1.15). The most common size of the collector box was 0.915 × 1.37 m (3 × 4.5 ft), and 0.305 m (1 ft) deep. The entire apparatus weighed 56.7 kg (125 lb) and cost $25 ($175, 1978 dollars).

Kemp patented his invention on April 28, 1891, calling it the Climax. It was usually installed on the roof of a house or shed, or secured by brackets to a wall, so that it inclined at an angle, with the four cylindrical tanks lying one above the other. To obtain hot water, one turned on the cold-water valve of the household plumbing system. Cold water entered the lowest tank, forcing hot water out of the upper tank, through pipes along the roof, wall, or window frame, and into the bathtub, sink, or wherever needed.

In another method of installation, the heater connnected to a reservoir system. Opening the hot-water faucet drew hot water from the tanks. Cold water from the reservoir replaced the spent hot water and a float valve in the reservoir opened to refill it with cold water.

The U.C.–Davis experiments conducted during the 1930s tested a similar system—three 0.113-m³ (30-gal) cylindrical tanks housed in a glass-covered wooden box—and found that water temperatures exceeding 48.8°C (120°F) could be obtained by late afternoon. However, at night the water cooled and morning temperatures could never be expected to exceed 43.4°C (110°F). This would be too cool for washing clothes (Fig. 1.16). Nevertheless, Kemp's model had many advantages over other water-heating systems, and over 1600 Climax heaters were sold in southern California by 1900.

Auxiliary Heat Added, 1898. Frank Walker, a Los Angeles contractor and realtor, invented a solar water heater in the spring of 1898 that was superior to the Climax in several ways:

1. Usually consisting of one or two 0.113-m³ (30-gal) water tanks enclosed in a glass-covered box, the heater fit inside the roof so that the glass top was flush with the shingles. This not only

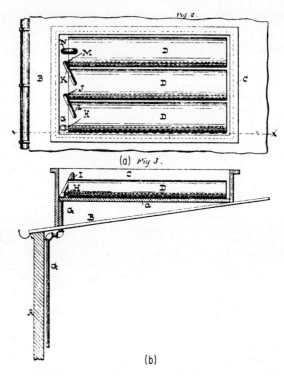

(a) *Fig 3.*

(b)

Fig. 1.15 (*a*) Vertical and (*b*) cross-sectional views of tank-type solar water heater. D, tanks; I, inlet pipe; H, outlet pipe. (*From a patent drawing.*)

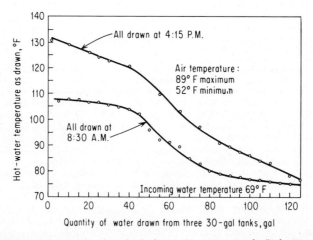

Fig. 1.16 Performance of enclosed triple-tank solar heater. (*From F. A. Brooks, "Solar Water Heaters."*)

looked better than the Climax type of installation, which protruded from the roof, but also provided better insulation for the tanks.

2. A polished metal reflector underneath the tank focused direct sunlight onto it (although later research proved that there was little, if any, additional heat gained).

3. Taking advantage of the difference in density between hot and cold water which prevents them from mixing, the hot water outlet at the top of the tank ensured that only the hottest water would be drawn. The cold-water inlet was located near the bottom of the tank.

4. Walker was the first to connect a solar water heater with a conventional water-heating system. This backup guaranteed hot water at all times.

The Improved Climax, 1905. Charles Haskell, by 1905 owner to the rights of the Climax and Walker solar water heaters, introduced the Improved Climax. The Improved Climax had a flat rectangular tank that was broader and shallower than the old type of cylindrical tank, thereby reducing the depth of water per square meter and producing hotter water faster. Painted black, the tank fitted tightly inside a wooden, glass-covered box. Vertical metal braces supported the top of the tank so that it could withstand the water pressure. Cold water entered through a perforated pipe extending across the bottom of the tank, and hot water left through perforations in the top of the hot-water outlet pipe extending across the upper end of the tank (Fig. 1.17).

The U.C.–Davis experiments found that such a device collected 1.58×10^7 J/m^2 of glass area each day (1400 Btu/ft^2) during mid-September. According to the study, midsummer values would be higher (Fig. 1.18).

The flat tank could not withstand normal water pressure and required an independent reservoir regulated by a float valve. Problems with the float valve became a major irritant to customers. Still, thousands of units were sold between 1904 and 1910.

The Industry Expands: California, 1909–1921

The one unresolved problem shared by all tank heaters was storage: The hot water inside the tanks was not well protected, since only a pane of glass lay between the tanks and the outside air. During chilly, cloudy weather or on cold nights, heat readily escaped and the water cooled.

Day and Night, 1909 William J. Bailey overcame the problem of storage by dividing the solar water heater into two separate units—a solar collector and a hot-water storage tank. Bailey's solar collector is what today we call the flat-plate collector. It consisted of a shallow glass-covered wooden box, 10.16 cm (4 in) deep, lined with felt paper. It contained 1.9-cm-thick (¾ in) galvanized or copper tubing coiled back and forth across a copper sheet to which it was soldered. (Smaller pipe was not used because it increased resistance to flow.)

The separate hot-water storage tank was made of galvanized steel surrounded by 22.86 cm (9 in) of diatomaceous earth as insulation and enclosed inside a box. Water in the tank could stay hot all night and, if necessary, the following day. Bailey guaranteed that when the sun ceased to shine, the water inside the storage tank would not lose over 0.55°C (1°F) per hour. The design relied on a thermosiphon system to circulate the water from the solar collector to the storage tank. The tank was situated higher than the collector, and water flowed up to it as long as the water in the collector was warmer than the water at the bottom of the tank.

The U.C.–Davis studies found that such a thermosiphon system was self-regulating[3]:

> The rise in temperature of the water in the absorber pipe due to the sun's heating depends primarily upon the length of time the water remains in the absorber pipe. These factors act together so that an automatic balance exists between the pipe friction and the force available from the difference in water density due to heating. If the sunshine suddenly becomes more intense, creating an excess in temperature differential, the increased difference in water density provides more force to make the water flow faster; and then it will be in the absorber for a shorter time and will be warmed to a lower degree, thus balancing any temporary temperature excess.

The most common-sized collector measured 1.37×3.81 m (4.5×12.5 ft) and weighed 113.4 kg (250 lb). The average coil length was 45.75 m (150 ft). The collector connected to a 0.227-m^3 (60-gal) tank, measuring 1.52×0.81 m (5×2.67 ft) and weighing 498 kg (1100 lb) installed. Smaller and larger collector and tank sizes were also available.

If a house had a southerly exposure, the collector was installed on the roof. Three strips of 0.63-cm (¼-in) material placed on the bottom of the collector provided airspace between the box and the roof. In houses without a southerly exposure, collectors were installed as awnings, on a nearby shed, or sometimes on the ground. With the exception of ground-level installations, the storage tank was usually placed in the attic. Where there was no room in the attic, a false chimney was built on the roof to house the tank. Usually the solar system connected to an auxiliary conventional heating system such as a wood stove, gas heater, and/or furnace (Fig. 1.19).

C. L. HASKELL.
SOLAR HEATER.
APPLICATION FILED JUNE 28, 1904.

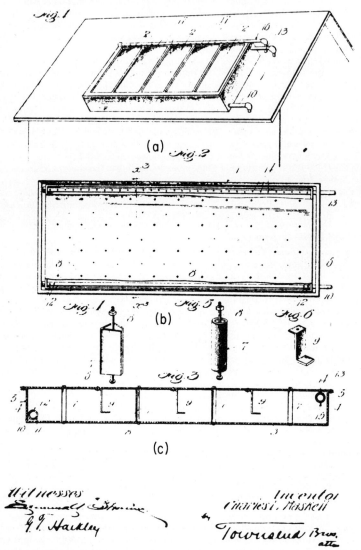

Fig. 1.17 (*a*) Perspective, (*b*) vertical, and (*c*) cross-sectional views of improved Climax solar water heater. (*From a patent drawing.*)

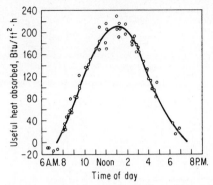

Fig. 1.18 Useful heat output of improved Climax-type solar heater. (*From F. A. Brooks, "Solar Water Heaters."*)

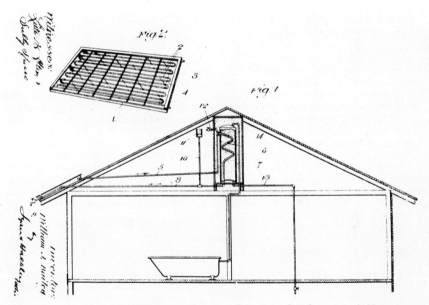

Fig. 1.19 Typical Day and Night solar water heater installation (original nonfreeze system). (*From a patent drawing.*)

Bailey called his product the Day and Night Solar Water Heater to emphasize that it was the only one that could provide steaming-hot, solar-heated water at all hours. In southern and central California it provided water temperatures over 60°C (140°F) at all times during the 9-month solar heating season. The Day and Night supplied 75 percent of a customer's average hot-water needs, at a cost of $33 ($135, 1978 dollars) per square meter of collector installed. Residents all over southern California purchased them (housewives could now do the morning wash with the Day and Night).

Closed-loop system, 1913 In 1913 Bailey's company almost went under because of a disastrous freeze. The water inside the tubing of many Day and Night collectors froze and the tubes split. This spurred Bailey to design a closed-loop system, or "nonfreezing solar water heater," as the Day and Night Company called it. The same collector configuration was connected

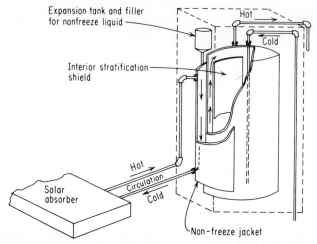

Fig. 1.20 Detail of special Day and Night nonfreeze system. (Note nonfreeze circulation jacket.) (*From Day and Night brochure.*)

to a coil inside the storage tank. A nonfreezing solution, usually water mixed with wood alcohol, circulated through the collector and coil, indirectly heating the water in the storage tank.

This system design provided another advantage: one could also use distilled water in areas where the water was very hard, so that deposits would not accumulate and crust up the tubing. Because of problems with electrolysis, Bailey later modified the closed-loop system. Instead of running through a coil in the storage tank, the nonfreezing solution flowed into a space between the storage tank and a jacket built around it (Fig. 1.20).

With the danger of freezing pipes resolved, Day and Night's business flourished once again. The company expanded its marketing territory to include northern California, the Hawaiian Islands, and Arizona, and even sold some heaters abroad. By the end of World War I, over 4000 models had been sold. Solar water heaters became synonymous with Day and Night's name. The more efficient flat-plate collector and separate storage tank forced the manufacturers of tank solar water heaters out of business.

Solar water heater sales peaked in 1920, with over 1000 sold that year. Then discoveries of natural gas throughout southern California caused an excess in gas production, resulting in low gas prices. In 1900, 28.3 m³ (1000 ft³) of gas with similar heat equivalency to natural gas cost $3.20 ($24, 1978 dollars); by 1927, the same amount of natural gas cost $0.90 ($3.25, 1978 dollars). Networks of new pipelines connected towns and rural areas formerly without gas.

Complementing these new gas discoveries was the concurrent development of the automatic-storage gas water heater—the type in use today. The combination of cheap, accessible supplies of gas and the convenience offered by the new gas heater sent solar sales plummeting. In 1926 Day and Night sold 350 solar water heaters and in 1930, only 40. All told, Day and Night sold over 7000 solar water heaters before production ceased in 1941.

The Industry's Greatest Commercial Success: Florida, 1920s–1950s

The first boom, 1923 Just as the solar water heater was being phased out in California because of the discovery of cheap natural gas, it found a new market in southern Florida. Bailey sold the patent rights to the Day and Night design to H. M. Carruthers, a Florida builder and developer, in 1923. Carruthers formed the Solar Water Heater Company of Miami.

The manufacture of solar water heaters became an established industry in Florida because, although construction boomed, newcomers found that there was no cheap way to heat water by conventional means. Electricity, the main type of energy used for heating water, cost $0.07/kWh ($0.26, 1978 dollars). This meant that an average water-heating bill exceeded $80/yr ($294, 1978 dollars), about half the price of the typical solar water-heating unit. Paying for themselves in 2 yrs, solar water heaters sold well. In less than 2 yrs in business, the Solar

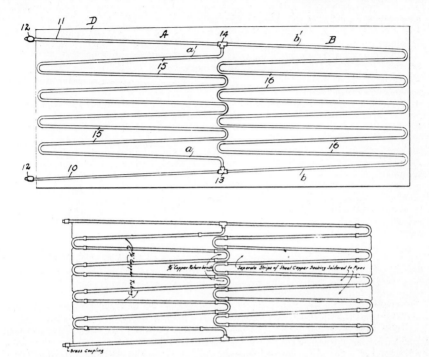

Fig. 1.21 Vertical view of duplex double-coil system. (*From patent and brochure drawings.*)

Water Heater Company became one of the seven largest construction firms in Miami. However, when the building boom collapsed by the summer of 1926, so did the solar water heater business.

1932 to World War II In 1932 the Solar Water Heater Company opened its doors for business once again, but it marketed a modified product. All-metal construction replaced wood, which deteriorated badly in the Miami humidity. Collector boxes were now constructed of galvanized 20-gauge sheet steel and mounted on steel rafters with hot-tipped galvanized J-bolts. The tank was made of 20-gauge galvanized iron and was bolted to the roof (most houses at the time had flat roofs).

The company also found that tilting a collector up slightly more than the latitude (29°20' for Miami) resulted in the best efficiency. For thermosiphon systems in Miami, the minimum advisable slope was 19°30'; the maximum was 49°18' (the collector was normal to the sun at the winter solstice). Interestingly, only a 10 percent difference in efficiency between the various degrees of inclination was reported.

Further tests established that the amount of water heated was directly proportional to the length of the tubing. But adding more tubing in a given area caused undue constriction of flow rate. This brought about an abnormal rise in temperature in the collector, resulting in large heat losses. The obvious solution to the problem of obtaining more hot water was a bigger solar collector. But this would increase the cost of the system, and even if the expense could be absorbed by the customer there might not be enough room on the roof for such a large collector.

The problem was solved with the use of two sets of coils soldered to copper strips, which were placed side by side in a single collector box. Each coil system was shorter than the system in an old-model collector of similar size, but the combined length of the two coil systems was greater. With more tubing in the same size collector box, the company could claim that it was now selling a more compact and efficient collector (Fig. 1.21).

In addition, the company substituted copper tubing for the former galvanized tubing to increase heat conductivity. Regranulated cork placed between the tank and metal box provided extra insulation. A waste product, the cork was low in cost. Since no new housing construction was going on at the time (1932–1934), the Solar Water Heater Company concentrated on the retrofit market. This required determination of the hot-water requirement of a given residence and accurate matching of tank capacity to solar collector size. A properly installed collector produced 1.5 to 1.7 gal of 130°F water per square foot of collector area. Tank size averaged 20 gal per resident.

With the new building boom in the latter part of 1934 [stimulated by the Federal Housing Authority (FHA)], sales of solar water heaters also mushroomed. Because electric rates remained high, at about $0.06/kWh ($0.27, 1978 dollars), solar water heaters made economic sense. For example, a family purchasing a typical solar water heater system on an FHA loan in 1938 would save $33.38 ($148.54, 1978 dollars) in the first year. As sales of solar water heaters increased, new manufacturers emerged. About 10 large firms were in the business by 1938. Unfortunately, with this competition came cost-cutting and soon a reduction in quality. The heaters were being made too small for effective performance. Since most sales were on FHA programs, the government eventually stepped in and established standards for the industry.

By 1941 as many solar water heaters were in use in Miami homes as were electric units. According to a survey of water heating done by the Florida Power and Light Company in 1941, in that year more than twice as many solar units were installed as electric. Estimates of how many solar heaters were installed between 1935–1941 vary from 15,000 to over 30,000 units. More than 5000 solar systems were placed in hotels, apartments, schools, hospitals, factories, and other relatively large buildings. Almost 3000 of these installations used storage tanks of 9.46 m^3 (2500 gal) capacity.

The federal government bought some of the largest systems. The officers' quarters at the giant naval air station at Opalaka (near Miami) used solar water heaters, as did several federal housing projects. The government realized that solar water heaters would hold down operating costs. When war was declared, the government froze the civilian use of copper and the solar water-heater industry came to an abrupt halt.

Postwar After the war, many solar heater firms resumed business. By the early 1950s, however, sales had declined. The increased use of appliances requiring hot water, such as automatic dishwashers and double-cycle washing machines, created a problem—solar collectors sized for prewar hot-water needs were now too small.

Several solar water-heater companies introduced electric boosters to supply extra energy but almost all prewar installations lacked an auxiliary system. Thermostatically controlled electrical resistance heaters were installed in the top third of the storage tank.

Had sufficient quantities of hot water been the only shortcoming of these systems, the introduction of boosters would have placated owners of solar water heaters. At the same time, though, many storage tanks began to burst. Using dissimilar metals in the collector (copper) and tank (steel) brought about electrolytic action, which was exacerbated by water pressure and high temperature. Most tanks were located in the attic, and leaks caused considerable damage. The reputation of solar water heaters as a reliable hot-water source was thus badly shaken.

Moreover, the sharp decrease in electric rates during these years contrasted with a substantial increase in the cost of solar water heaters. By 1958, electricity in Miami cost less than $0.03/kWh ($0.07, 1978 dollars). The price of a solar heater selling for $130 in 1938 ($578, 1978 dollars) now came to $550 ($1202.30, 1978 dollars). See Fig. 1.22 and Table 1.4. Two prime causes for this price rise were the tripled price of copper and the increased wages of workers in the solar market. An unskilled solar employee earning $0.25 to 0.50/h in 1938 ($1.11 to 1.78, 1978 dollars) received about $1.10/h ($4.05, 1978 dollars) in 1958. Since the localized nature of the industry precluded large investments in labor-saving machinery and installations had to be done on site, the increased cost of labor was an important cause of the inflated cost of solar water heaters.

Solar water heaters therefore became priced out of the market. By the late 1950s, the once-thriving industry found itself relegated to a service business—flushing out coils, replacing broken glass, removing damaged tanks, and installing new tanks for those who wished to continue using solar energy.

SOLAR SPACE HEATING

Rediscovery of Solar Architecture: Chicago, 1932–1948

George Fred Keck, 1932 In 1932, Chicago architect George Fred Keck accidentally rediscovered the ancient Greek and Roman principles of solar architecture. Keck had designed an all-glass house for the Chicago Exposition. While supervising its construction, he was amazed

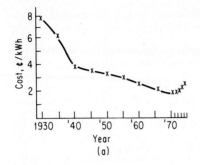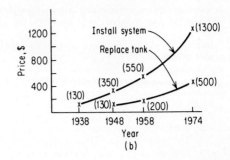

Fig. 1.22 Historical electricity rates and market prices for solar water heaters compared for southern Florida area, 1930–1974. (*a*) Average residential electricity costs in Miami, and (*b*) installed price for 82-gal system with one collector (and price to replace tank). (*From "Solar Water Heating in South Florida, 1923–1974," National Science Foundation, Washington, D.C., 1974.*)

TABLE 1.4 Outcomes of Purchase Decisions in 1938, 1948, and 1958

	Purchase date		
	1938	1948	1958
Installed price of solar unit (loan amount), 82-gal unit with one 4-ft × 12-ft × 8-in collector	$130.00	$350.00	$550.00
Loan interest rate	4%	4½%	$5/$100/year
Term of loan, years	3	3	3
Monthly loan payment	$ 3.84	$ 10.41	$ 17.57
Yearly loan payment	$ 46.07	$124.95	$210.83

	1938 purchase		1948 purchase		1958 purchase	
Year from purchase	Yearly savings	Cumulative savings	Yearly savings	Cumulative savings	Yearly savings	Cumulative savings
1	33.38	33.58	(45.55)	(45.55)	(132.34)	(132.34)
2	18.05	51.43	(51.35)	(96.90)	(133.72)	(266.06
3	16.89	68.32	(49.86)	(146.76)	(134.72)	(400.78)
4	65.37	133.69	76.92	(69.84)	74.72	(326.06)
5	67.80	201.49	77.55	(7.71)	75.02	(251.04)
6	69.18	270.67	77.01	84.72	75.31	(175.73)
7	69.02	339.69	78.05	162.77	73.24	(102.49)
8	69.88	409.57	79.27	242.04	72.63	(29.86)
9	67.48	477.05	81.37	323.41	73.04	43.18
10	66.04	543.09	84.14	407.55	73.42	116.60
11	79.40	622.49	78.49	486.04	73.79	190.39
12	73.60	696.09	77.11	563.15	74.43	264.82
13	75.09	771.18	76.11	639.26	75.87	340.69
14	76.92	848.10	74.72	713.98	81.39	422.08
15	77.55	925.65	75.02	789.00	88.14	510.22

source: J. E. Scott "Solar Water Heating in South Florida, 1923–1974," National Science Foundation, Washington, D.C. 1976.

to find the workmen inside shirtless and perspiring. It was −25°C (−18°F) outside, and the home had no heating plant; yet it was 29°C (85°F) inside because of the accumulation of solar heat.

During the next 8 yrs, Keck studied this phenomenon, discovering that over half of the winter days in the Chicago area and the midwest are sunny and from 5 to 50 percent of the sun's radiation reaches the earth even on cloudy days. He thus began building homes with larger and larger south-facing glass windows. By the early 1940s, he had designed over a dozen solar homes in the Chicago area.

Keck's solar homes shared four features:

1. They were narrow, with most of the important rooms facing south.
2. They were oriented on an east-west axis.
3. The entire south facade was glazed with double-pane glass.
4. Specially placed eaves over the south-facing windows blocked the summer sunlight.

Owners of Keck's solar homes reported fuel savings as high as 38 percent. As a result, the homes generated national interest tempered with skepticism.

The LOF studies: 1942   In 1942 the LOF Glass Co. sponsored a year-long study of a Keck solar home to determine exactly how much energy it conserved. Engineers William C. Knopf and J. C. Peebles of the Illinois Institute of Technology chose a house in the south Chicago suburb of Homewood for the study. It was occupied by its owners, a young couple, who lived in it year-round.

The single-story house had six rooms. Four of them occupied the south side of the house and had large windows, 30.3 m² (326 ft²) in area. On the north side of the house were the utility room and bathroom. The walls and ceilings were insulated with a stitched and creped fibrous blanket. The floor was made of 10.16-cm (4-in) concrete (see Figs. 1.23 and 1.24 for layout and cross section of south-facing wall).

Testing of the house lasted from October 23, 1942, to October 23, 1943. Thermometers recorded temperatures inside the home on the north and south sides at three points: 7.62 cm (3 in) from the floor, 20.32 cm (8 in) from the ceiling, and on the wall. Outside temperatures of the north and south walls were also registered, as well as wind direction and velocity. The researchers calculated the structure's heat gains according to its fuel consumption. Actual heat loss, however, proved difficult to determine. The house was new when the study began. Within the first year, wood around doors and windows started to shrink, resulting in cracks (some large enough to see through) which allowed heat to escape. In addition, the random opening of doors and windows during the most severe weather and above-average losses from high winds due to the exposed location of the house contributed to the problem of making accurate heat loss measurements.

The final test results were somewhat ambiguous. Knopf and Peebles arrived at a heat-loss value of 1.95×10^{11} J (1848 thm) for the heating season. The heating plant of the house added 1.81×10^{11} J (1722 thm). Therefore, they assumed, solar heat had reduced the heating load by 1.3×10^{10} J (127 thm), or 6.8 percent. But they noted that actual heat loss could have put the solar gain anywhere from 4.0×10^{9} J (38 thm) to 4.21×10^{11} J (2288 thm). It was concluded that actual heat gain from the sun was probably closer to 3.48×10^{10} J (330 thm), meaning a 16 percent reduction in the overall heating load.

Because of the unclear results of the test, LOF sponsored a more controlled experiment. It began in 1945 at Purdue University under the direction of Visiting Professor of Mechanical Engineering F. W. Hutchinson. Two full-scale homes were constructed side by side, an "orthodox" house and a "solar" house. They were both 170 m² (1830 ft²) in area and thermally and architecturally identical, except for the amount of glass window area in each. In the orthodox house, 12 percent of the wall area was glass as compared with 22.5 percent in the solar house. The north and west walls of the houses were the same, but the east wall of the solar house had 80 percent more glass than in the orthodox house. In addition, the south wall of the solar house had 10.4 m² (112 ft²) of glass, comprising 51 percent of the wall area, whereas the south wall of the orthodox house had 3.8 m² (41 ft²) of glass, or 18.5 percent of the wall area.

Two nine-week tests took place to determine (1) the degree-per-hour heating load of each house, and (2) the cost of heating the houses to a constant temperature of 21.1°C (70°F).

The first test lasted from December 1, 1945, to February 2, 1946. During this period the sun shone an average of 12.25 percent of the time, and the outside temperature averaged 3.3°C (26°F) (see Table 1.5). The solar home showed a 9 percent lower degree-per-hour heat requirement than the orthodox house. But the houses did not have overall equivalent heat-loss coefficients. This figure, therefore, did not represent a 9 percent reduction in heating costs.

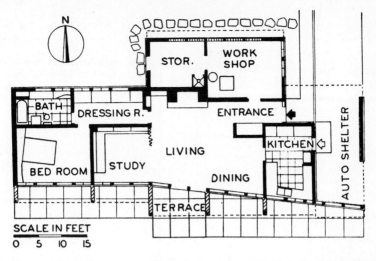

Fig. 1.23 Floor plan of Illinois Institute of Technology test house. (*From W. C. Knopf, "Solar Heating."*)

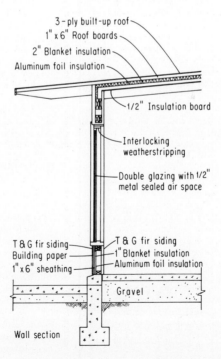

Fig. 1.24 Cross section, south face of Illinois Institute of Technology test home. (*From W. C. Knopf, "Solar Heating."*)

TABLE 1.5 Weekly Averages for the First 9-Week Test Period of the Hutchinson Experiment

Week of	Temperature of air			% reduction in actual degree-hours required, based on *Theoretical Number* from outside air temperature		Hours of sun	% sun	% reduction in degree-hours of solar over orthodox
	Outside	Inside solar	Inside orthodox	Solar	Orthodox			
12/1/45–12/8/45	34.9	40.7	39.1	19.2	13.9	13.0	7.75	6.20
12/8/45–12/15/45	23.2	34.9	33.0	28.0	23.8	18.0	10.70	5.85
12/15/45–12/22/45	8.0	27.6	24.2	34.8	28.3	26.0	15.50	3.35
12/22/45–12/29/45	23.3	29.1	29.0	13.9	18.6	10.0	5.95	.28
12/29/45–1/5/46	29.0	33.9	32.8	18.55	10.5	8.3	4.95	3.42
1/5/46–1/12/46	39.8	41.3	40.8	5.97	3.97	7.0	4.17	1.22
1/12/46–1/19/46	27.0	45.9	36.4	49.8	24.6	47.0	28.00	33.00
1/19/46–1/26/46	21.7	34.8	31.2	30.3	21.8	22.5	18.40	10.65
1/26/46–2/2/46	24.6	36.0	32.0	28.2	18.2	33.5	19.95	12.10
Overall weekly average	25.7	36.0	33.2	24.8	17.6	20.6	12.26	9.02

SOURCE: F. W. Hutchinson, "Solar House," *Heating and Ventilating*, March, 1946.

The second test occurred between December 1, 1946, and February 2, 1947, a period corresponding to the heating season of the first test. Each home was electrically heated to a constant temperature of 21.1°C (70°F). The sun shone an average of 13.2 percent of the time and the average outdoor temperature was about 0°C (32.7°F) (Table 1.6). The orthodox house consumed 9×10^9 J (2514 kWh) and the solar house consumed 1×10^{10} J (2924 kWh). The solar house therefore used 14 percent more electricity because of large heat losses through the unprotected glass windows, especially at night. The solar house also experienced large swings in solar energy input, which often raised interior temperatures above 21°C (70°F) and thereby produced unneeded heat (Fig. 1.25).

Although the Purdue study did little to determine the actual heating performance of a solar home in use, it did allow Hutchinson to gather valuable technical data. In 1946 he coauthored the seminal paper, "A Rational Basis for Solar Heating Analysis." In it he developed a general

TABLE 1.6   Weekly Averages for the Second 9-Week Test Period of the Hutchinson Experiment

		Energy used, kWh		
Week of	Outside air temperature	Solar house	Orthodox house	Hours of sunshine
12/1/46–12/8/46	40.6	214	212	38.0
12/8/46–12/15/46	40.6	249	213	18.2
12/15/46–12/22/46	28.2	331	289	22.4
12/22/46–12/29/46	35.6	296	253	21.0
12/29/46–1/5/47	20.3	456	370	14.4
1/5/47–1/12/47	29.4	373	314	25.9
1/12/47–1/19/47	36.4	296	252	12.2
1/19/47–1/26/47	31.3	284	250	28.0
1/26/47–2/2/47	32.5	425	361	20.2
Overall average or total	32.7	2924	2514	200.3

SOURCE: F. W. Hutchinson, "The Solar House," *Heating and Ventilating*, March, 1947.

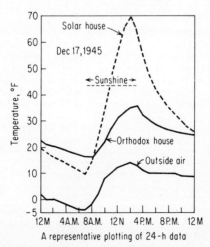

Fig. 1.25   Example of interior temperature variations in the Purdue test homes. (*From F. W. Hutchinson, "Heating and Ventilating."*)

equation which could be used to determined the amount of heat gain or loss during a heating season resulting from the use of large south-facing windows in the place of an opaque wall:

$$Q = F \sum_{n=1}^{n=5088} \Delta_n - 5088 \, (U_o - U_w)(t_i - t'_o) \tag{1.1}$$

where Q = annual saving (based on October 1 to May 1 heating season) in Btu by replacing 1 ft² of nontransmitting south wall with a double-glass solar window which is 100 percent in the shade (because of roof overhang) only at solar noon at June 21 and 100 percent irradiated by noon sun (solar time) only on January 21

t_i = the design inside air temperature

t'_o = the normal outside air temperature during the 7-month heating season from October 1 to May 1 (values given on pp. 258–9 of the ASHVE 1946 Guide)

U_w = overall coefficient of heat transfer of the unit area of wall which is to be replaced by a unit area of window

U_o = overall coefficient of heat transfer of window, including inside and outside air films, two sheets of identical glass and separating air space

5088 = number of hours in the selected 7-month heating season

F = fraction of maximum possible sunshine which occurs during an average year in the 7-month heating season

$\sum_{n=1}^{n=5088} \Delta_n$ = the summation of solar term Δ_n for each of the 5088 h of the heating season

Δ = a solar gain term representing the reduction in excess normal transmission losses, such excess losses being $(U_o - U_w)(t_i - t'_o)$, resulting from solar irradiation and expressed in Btu per hour. The term Δ_n must be numerically evaluated for each hour between sunrise and sunset of each day of the heating season.[11]

In the years that followed, Keck and others continued to build a limited number of solar homes. But doubt lingered over the actual heating value of large south-facing windows. Discussion of the issue continued in both technical and popular journals up until the 1950s. It was, however, generally agreed that by protecting the windows with heavy curtains or shutters at night or during inclement weather, much of the heat loss could be prevented, and in fact result in a modest heat gain from the sun.

Two overriding factors, nevertheless, prevented solar homes from becoming widely accepted: (1) continued reduction in the cost of electricity and heating oils during this period, and (2) as one architect put it, the lack of interest on the part of architects in designing with rationality.

Mechanical House Heating

M.I.T. space-heating experiments, 1938–1961   The most rigorous and extensive scientific study of solar space heating undertaken before the 1970s began in 1938 at the Massachusetts Institute of Technology under the direction of Hoyt Hottel, Professor of Chemical Engineering.

The study had a twofold goal:

1. To conduct a scientific analysis of the performance of the flat-plate collector, which was being used extensively at the time in Florida and elsewhere for domestic water heating.

2. To determine the technical and economic viability of using flate-plate collectors for home space heating.

The systems tested used water as the heat-transfer medium for several reasons. Water has a relatively high capacity for heat absorption and storage, and allows easy determination of the rate of absorption and storage. During the study, which spanned two decades, four test houses were built.

Test House 1: Year-Round Heating, 1939. The goal of the first study was to analyze the performance of the flat-plate collector, ignoring economic considerations. The test house used for the experiment was a laboratory and small office. Its floor area measured 46.5 m² (500 ft²). Fourteen collectors occupied a total of 37.94 m² (408 ft²) of roof space. This portion of the roof inclined 30° southward and was recessed 0.23 m (0.75 ft).

Each collector consisted of:

1. A plate 0.915 × 2.75 m (3 × 9 ft), made of blackened copper sheet 0.05 cm (0.02 in) thick.

2. Six parallel copper tubes soldered to the copper sheet, each 1.45 cm (0.57 in) outside

diameter, spaced 0.15 m (6 in) apart, and soldered at the top and bottom to 1.9-cm (¾-in) outside-diameter copper headers.

3. A 1.9-cm (¾-in) plywood box covered with three layers of double-glazed glass, each 0.91 m² (3 ft²) in area.

4. Insulation lining the bottom of the collector box at a thickness of 13.97 cm (5.5 in).

The collectors were positioned in two groups of seven, one group above the other. Water flowing through three supply lines fed four, four, and six collectors, respectively. A forced circulation system transferred the heated water from the collectors to a storage tank in the basement.

The storage tank, which filled the entire basement, had a capacity of 65.86 m³ (17,400 gal). It was insulated by an effective thickness of 0.6 m (1.96 ft). Such a large and heavily insulated tank was necessary because the study of the performance of the collector required a stable environment (Fig. 1.26). The heat absorbers collected 5.67×10^6 J/m² · day (500 Btu/ft² · day). Collection was year-round and solar energy was stored for use when necessary. There was no backup space-heating system.

In their classic study of flat-plate collector performance, Hoyt Hottel and B. B. Woertz, his collaborator, found it necessary to consider the following variables in evaluating collector design:

1. *Tilt of collector*—Optimum angle of tilt depended on season of use.

2. *Overall transmittance of glass*—Glass high in iron (greenish-tinged) transmitted solar radiation poorly. The preferred glass was water-white, 0.297 cm (0.117 in) thick with a transmittance of 75.5 percent.

3. *Absorptance*—Black demonstrated an absorptance of about 0.98 and excellent adhesion characteristics.

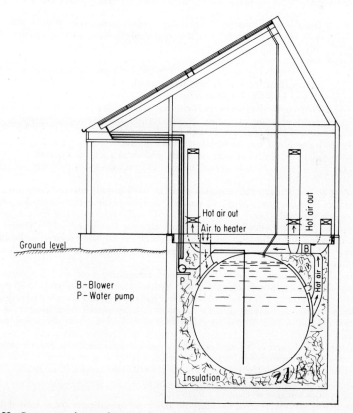

Fig. 1.26 Cross-sectional view of M.I.T.'s first solar house. (*From H. C. Hottel and B. B. Woertz, The Performance of Flat Plate Collectors, "A.S.M.E. Transactions," Feb. 1942.*)

    4. *Heat losses from absorbing plate to the atmosphere*—Five factors were considered:
      (*a*) Temperature of the absorbing plate
      (*b*) Temperature of the outside air and sky
      (*c*) Number of glass covers and their spacing
      (*d*) Tilt of the glass covers
      (*e*) Wind velocity above the glass cover
Factor (*e*) needed to be considered only with collectors having a single glass cover.
    5. *Heat loss through bottom of collector*—Calculated by the coefficient of heat loss of the insulation.
    6. *Effect of dirt on collector*—Found to be quite small. The maximum hindrance was 4.7 percent and the average only 1 percent, even though the test site was in a coal-burning industrial area.
    7. *Shading effect on absorbing plate by walls of collector box*—Collector performance declined by 3.6 percent.
    8. *Effect of heat capacity of collector*—Unusually low useful-heat collection in the morning hours resulted from some of the heat being used to raise covers and bottom insulation to equilibrium temperature.
    9. *Resistance to lateral heat flow in collector plate*—The collector could not be treated as a surface with uniform temperature but could be treated as a surface cooled by tubes 0.15 m (6 in) apart.
    The researchers developed a general equation for the evaluation of collector performance based on the above variables[10]:

$$Q_u/A = F \Sigma \{(q_a/A)(1-D)(1-S) - q_L/A - q_B/A\} - NQ_h/A \qquad (1.2)$$

where Q_u = useful heat collected over a period involving several cycles of heating and cooling, with allowance for dirt on glass, wall shading, heat capacity, and fin effects, in Btu

      A = area

      F = ratio of useful heat collection by a finned surface to that by a surface of the same area maintained at a uniform temperature

      q_a = rate of heat absorption of collector plate

      D = fractional reduction in transmittance of glass plate because of surface dirt

      S = factor allowing for shading of collector plate by walls

      q_L = rate of total heat loss through top of collector

      q_B = rate of heat loss through bottom of collector

      N = number of days or heat-collection periods in a long test period

      Q_h = loss, per collection period, because of heat capacity of collector

    According to the authors, this equation could be used to find unsuspected causes of poor performance. Thus the goal of test 1—a scientific analysis of the performance of the flat-plate collector—was attained.
    Test House 2: Solar Collection, Storage, and Heating Combined, 1947. Hottel and his associates decided that in the second test they would attempt to make solar space heating more economical. They therefore simplified the system by combining solar collection, storage, and heating into one unit.
    The test home consisted of seven cubicles, each comprising a different experiment. In six of the cubicles, the south wall was glazed with figured double-strength glass, 0.55 cm ($\frac{7}{32}$ in) thick, with an airspace of 1.27 cm ($\frac{1}{2}$ in). Cubicle 7 had triple glass of the same type (Fig. 1.27). In all the cubicles except cubicle 1, a stack of water cans completely filled the interior window area. The cans were 3.78×10^{-3} m^3 (1 gal) or 18.9×10^{-3} m^3 (5 gal) in size and were painted flat black.
    In cubicles 2, 4, 6, and 7, the cans served as a radiant water wall. The interior temperature of each cubicle was controlled by raising or lowering a partition separating the water wall from the interior of the house. The partition was thermostatically controlled at 22.2°C (72°F). To inhibit heat loss at night, double-roll aluminized curtains were drawn between the window and the water wall, leaving an airspace of 1.9 cm ($\frac{3}{4}$ in).
    Convective systems were used in cubicles 3 and 5. An insulating board separated the water wall from the interior of the cubicle, of U-0.45 W/m^2 °C. (0.08 Btu/ft^2 · h · °F). A blower measuring 0.15 m (6 in) in diameter which was thermostatically controlled at 22.2°C (72°F)

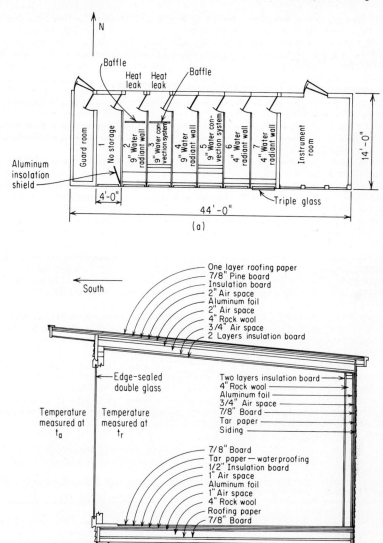

Fig. 1.27 (a) Floor plan and (b) cross section of M.I.T.'s second solar house. (*From Albert G. H. Dietz and Edmund L. Czapek,* "Solar Heating Houses by Vertical South Wall Storage Panels," *Heating, Piping, and Air Conditioning, March 1950.*)

drew air from the room through an opening near the bottom of the insulating board, circulated the air through the water wall, and blew it back out again through the bottom and into the room. In cubicle 1 there was no water wall between the glass and the interior. Each test cell was equipped with a 574-W electrical auxiliary heater which went on whenever the temperature fell below 21.1°C (70°F).

The major problem of the combined collection/storage/heating system was the tremendous loss of heat to the outside. Of all the cubicles, 2 and 3 fared the best. During the December-to-April heating season, they absorbed 8.88×10^8 J/m² (78,400 Btu/ft²). However, they lost 6.27×10^8 J/m² (55,400 Btu/ft²). Hence they registered a net gain of 2.6×10^8 J/m² (23,000 Btu/ft²) or 29.4 percent of the solar energy initially collected (Table 1.7). In all, the cubicles with water walls supplied 37.6 to 48.2 percent of the cells' heat requirements. The researchers concluded that the attractiveness of such a simple system was offset by excessive outward heat losses.

Test House 3: Flat-Plate Collector with Auxiliary System, 1949. Disappointed by the large heat losses incurred in the second test, M.I.T engineers who built test house 3 in 1948 returned to a mechanical solar system resembling that of test house 1—with some notable exceptions. Test house 3 did not rely solely on solar heat, since the researchers realized that an all-solar house would not be economically feasible. Instead, two 3.5-kW immersion-type heaters placed in the solar storage tank provided auxiliary heat when needed.

The roof had 16 collectors, comprising an area of 37.2 m² (400 ft²). They tilted 57° south, favoring winter heat collection. The design of the collectors resembled that of test house 1, except that the collector box covers consisted of only two layers of glass. Fifteen of the collectors measured 0.915×2.74 m (3×9 ft); the sixteenth measured 0.46×2.7 m (1.5×9 ft). To obtain additional solar energy, an overhang reflected sunlight onto the collectors. In addition, large south-facing windows admitted sunlight into the living quarters.

The storage tank occupied a space in the attic. Much smaller than that in house 1, it had a capacity of 4.54 m³ (1200 gal). Insulation consisted of a layer of rock wool between 30.48 and 35.16 cm (12 and 14 in) thick (Fig. 1.28). A pump automatically circulated water from storage to collector whenever the temperature of the collector was 2.78°C (5°F) hotter than the water inside the tank. When the temperature of the water in the collector dropped below the water temperature in the tank, the flow from tank to collector ceased, and air pumped into the tubing prevented it from freezing.

Another pump circulated solar-heated water through 1.9-cm (¾-in) radiant heating coils located in the ceiling, to maintain an interior temperature of 20°C (68°F). The auxiliary heaters began operating when the house temperature dropped below 18.3°C (65°F). The test house, occupied by a family of three, had a living area of 56.5 m² (608 ft²).

The solar absorber collected about 37 percent of the available solar energy. It also collected an additional amount of sunlight reflected from the overhang. On one of its best performance days—March 2—the system stored almost $527,000 \times 10^3$ J (500,000 Btu), providing enough heat for two sunless days when the temperature outside was −1.1°C (30°F). During clear, sunny winter days, solar heat trapped by the south-facing windows was sufficient to heat the house, allowing the collectors to pump heat into the storage tank. In the second winter of operation, test house 3 was destroyed by fire.

TABLE 1.7   Summary of Energy Losses and Gains per Square Foot of Collector Area for M.I.T House 2 (December 1949–April 1950)

Test cell	Solar energy transmitted and absorbed, S'_{VM}, 1000 Btu	Outward losses from collector surface		Solar input to test cells		Heat requirements		Percent gained from sun
		1000 Btu	% of S'_{VM}	1000 Btu	% of S'_{VM}	1000 Btu	% of S'_{VM}	
1	78.4			78.4*	100.0†	116.1†	148.1†	67.5†
		72.6	92.6	5.8	7.4	43.5	55.5	13.3
2.3	78.4	55.4	70.6	23.0	29.4	61.2	78.0	37.6
4	78.4	63.2	80.6	15.2	19.4	32.6	41.6	46.6
5	78.4	65.8	83.9	12.6	16.1	32.6	41.6	38.7
6	78.4	64.6	82.4	13.8	17.6	33.4	42.6	41.4
7	67.7	51.6	76.2	16.1	23.8	33.4	49.4	48.2

*Average of two methods of measuring net solar input.
†Based on total heat loss area (including south window). All other values are based on floor, north wall, and ceiling area only.
SOURCE: A. G. H. Dietz and E. L. Czapek, " Solar Heating of Houses by Vertical, South Wall Storage Panels," *Heating, Piping, and Air Conditioning*, March, 1950.

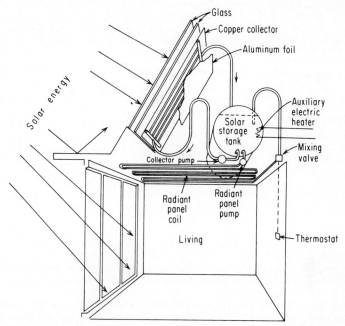

Fig. 1.28 Cross-sectional view of M.I.T's third solar house. (*From "Architectural Record."*)

Test House 4: Prototype for Mass Construction, 1959. The researchers intended test house 4, built in 1959, to serve as a prototype for mass-market production. Therefore they stressed comfort and convenience. The house had two stories of living area, for a total of 134.8 m² (1450 ft²). Special features included double-glazed windows and an underground first floor to minimize heat loss. The south wall of the second story sloped 60° and consisted of 59.5 m² (640 ft²) of collectors.

The absorbing surface of each collector was made of 0.63-cm (¼-in) aluminum. Clamped to the aluminum were six 0.95-cm (⅜-in) outer-diameter copper tubes, situated parallel to each other, 12.7 cm (5 in) apart. Painted flat black, the top of the aluminum sheet and copper tubes had an absorptance of 0.97.

Two plates of low-iron glass covered the collector box. The bottom of the collector, which was integrated as the wall of the second story, was insulated by a layer of aluminum foil at a thickness of 7.62 cm (3 in), an airspace, then another 7.62 cm (3 in) of foil-faced glass-fiber batt. The collectors were connected to a 0.76-m³ (200-gal) expansion tank and a 5.7-m³ (1500-gal) storage tank. Heat losses from the storage tank, which was located in the basement, helped heat the first floor of the house.

A forced-hot-air heating system consisted of an air-to-water heat exchanger and a circulating pump blower. An auxiliary unit consisted of an oil-fired water heater and motorized valve. Two bimetal room thermostats and thermostats in the solar storage tank and auxiliary tank provided temperature control (Fig. 1.29).

During the winter of 1959–1960, a collectible total of 94.86 × 10³ million J (90 million Btu) of solar energy fell on the collectors. Of this amount, the absorbing surface collected 43.1 × 10³ million J (40.9 million Btu), an average efficiency of 45 percent. The solar system supplied 46 percent of the heat load, or 36.26 × 10³ million J (34.4 million Btu) of a total requirement of 78.5 × 10³ million J (74.5 million Btu). Storage-tank heat loss amounted to 6.8 × 10³ million J (6.5 million Btu) (Table 1.8). Researchers observed that the solar system's performance— the low amount of solar heat collected and the low percentage of total heat load provided— stemmed from (1) the severity of the winter and (2) the abnormally low solar incidence.

Performance of the solar heating system during the following winter improved. Although the weather was also more severe than usual, a higher solar incidence enabled the solar system to

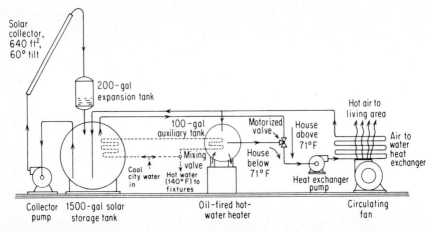

Fig. 1.29   Diagram of the solar heating system in M.I.T's fourth solar house. (*From "A.S.M.E. Journal."*)

take on almost 57 percent of the heating load. Study of test house 4 revealed that the collector wall helped warm the house in winter, but contributed to overheating during the warmer months. In addition, moisture inside the collectors caused problems.

In terms of cost, the study showed that a flat-plate collector in Massachusetts would produce an energy equivalent to 40.7×10^{-3} m^3 of fuel oil per square meter per year (1 gal/ft^2 · year). The expense of such a solar system would have to be reduced by one-fifth to pay for itself within 10 yrs.

Air as the transfer medium and crushed-rock storage

The Boulder House, 1945. George Löf directed the development of the first space-heating system to use air as the heat-transfer medium. He chose an air system because it eliminated the problem of freezing and the possibility of bothersome leaks, and because of its compatibility with forced-air heating systems.

Löf integrated this solar system with an existing conventional system in a house in Boulder, CO. Thus this was also the first retrofit solar space-heating system. The house measured 93 m^2 (1000 ft^2) in area and the collectors occupied 43 m^2 (464 ft^2). The absorber consisted of glass plates measuring 0.74×0.99 m (2.42×3.25 ft). Two-thirds of each plate overlapped, leaving almost 0.305 m (1 ft) of surface exposed. The exposed portion was painted black and the collector cover consisted of several layers of glass. The temperature of the air heated by the absorber increased by as much as 43.3°C (110°F).

Löf's house was also the first to use crushed rock as a storage medium. He chose a pebble bed because it allowed for the maintenance of a high degree of heat stratification. A bin of gravel was located in the crawl space, measuring 5.04 m^3 (180 ft^3) in volume.

Controls allowed for the following heat modes:

1. House cold, solar collector hot: hot air from collector delivered to the house.
2. House hot, solar collector hot: hot air from collector delivered to storage.
3. House cold, solar collector cold, storage hot: hot air from storage delivered to house.
4. House cold, solar collector cold, storage cold: hot air from conventional heater delivered to house.

Solar energy provided about 33 percent of the home's heating load. However, problems occurred with the glass absorbers, which cracked because of thermal stress.

Maria Telkes: Heat of Fusion (Dover House), 1948. Maria Telkes, Professor of Metallurgy at M.I.T, noted that solar heat storage systems based on the specific-heat effect—e.g., systems using water or rock for storage—had drawbacks when used for space heating. Providing more than a 2-day storage capacity when using water as the medium or 1⅓ days when using gravel required a storage area too large to be economically feasible. Telkes substituted a solar storage system based on heat of fusion. She used Glauber's salts ($Na_2SO_4 \cdot 10H_2O$). The salts could store 7.3 times more heat than water, at a temperature range between 26.6°C and 37.8°C (80°F and 100°F), the substance's melting point.

TABLE 1.8   Performance of M.I.T Solar House 4

		Oct.	Nov.	Dec.	Jan.	Feb.	Mar.	Apr.	Total	Av.
Space-heating demand, 10^5 Btu	1959–1960	30.6	77.0	125.1	139.9	188.7	133.3	62.3	687.2	98.2
	1960–1961	36.9	68.0	156.9	184.1	111.8	112.6		670.3	111.7
Space-heating demand supplied by solar heating system, 10^5 Btu	1959–1960	30.4	14.9	34.7	49.7	60.7	84.3	43.4	318.1	45.4
	1960–1961	36.9	54.5	69.7	80.6	68.7	70.5		380.9	63.5
Fraction of space-heating demand by solar heating system, %	1959–1960	98.7	19.4	27.6	35.6	51.2	63.3	69.7		46.3
	1960–1961	100.0	80.7	44.4	43.8	61.1	62.6			56.8
Domestic hot-water-heating demand, 10^5 Btu	1959–1960	15.8	15.0	20.2	21.6	23.7	23.3	19.7	139.3	19.9
	1960–1961	22.9	25.3	29.5	31.0	24.4	34.1		167.1	27.9
Domestic hot-water-heating demand supplied by solar heating system, 10^5 Btu	1959–1960	14.6	8.4	9.8	11.4	12.0	12.1	11.7	80.0	11.4
	1960–1961	21.8	13.7	14.6	13.6	11.3	17.4		92.4	15.4
Fraction of domestic hot-water-heating demand by solar heating system, %	1959–1960	92.3	56.0	48.5	52.8	50.7	51.8	59.4		57.4
	1960–1961	95.3	54.3	49.5	43.8	46.3	55.4			55.3
Total heating demand, 10^5 Btu	1959–1960	46.5	92.0	145.6	161.5	142.4	156.5	82.0	826.5	118.1
	1960–1961	59.8	93.2	186.4	215.1	136.2	146.7		837.4	139.6
Total heating demand supplied by solar heating system	1959–1960	45.0	22.3	44.5	61.1	72.7	96.4	55.4	397.1	59.9
	1960–1961	58.7	68.6	81.3	94.2	80.0	87.9		473.7	79.0
Fraction of gross heating demand supplied by solar heating system, %	1959–1960	96.7	25.4	30.5	37.8	51.1	61.5	67.2		48.2
	1960–1961	98.0	73.6	45.2	43.8	58.7	59.9			56.6

SOURCE: C. D. Engebretson and N. G. Ashar, "Progress in Space Heating with Solar Energy," *ASME Journal*, July, 1960.

In tests conducted at a house in Dover, MA, in 1948, Telkes used 18 collectors. They measured 3.05 × 1.22 m (10 × 4 ft). The collectors stood at a 90° angle behind a double pane of glass on the second floor. A vertical collector can be almost as effective during the winter heating season as a collector tilted at the most favorable angle (in this case, 60°), Telkes found. In addition, there was the advantage that during the summer the collector absorbed only about 43 percent as much heat as a tilted collector. As a result, less heat penetrated the house during hot weather.

The collectors were made of thin iron plate painted black and used air as the transfer medium. This system collected an average of 4533 × 10³ J/m² (400 Btu/ft²), at an efficiency of almost 40 percent.

Heat was stored in bins containing 13.3 m³ (470 ft³) of Glauber's salts. The bins, located on the first floor, also served as radiant-heating panels which supplied sufficient heat when the temperature outside rose above 10°C (40°F). At lower temperatures, air warmed by the heat in the salts was circulated through the house. The solar system supplied 80 percent of the heating load.

However, because such a large quantity of salts was required—over 292.5 kg per square meter of collector area (60 lb/ft²)—the question remains whether the salts functioned by way of heat of fusion or their specific heat.

Conclusions of solar space-heating studies.   Several conclusions can be made from the experiments conducted by researchers at M.I.T and by Löf and Telkes:

1. A solar space-heating system could provide a high percentage of a home's space-heating load.

2. Such a system could be relatively simple.

3. During the period in which the experiments were made (1945–1960), a solar space-heating system could not compete economically with conventional systems for homes.

4. The combination of solar space heating in winter and solar air conditioning would make a solar system financially more attractive.

FINAL REMARKS

The history of solar energy serves as the foundation for present and future work in the field. People in Western society have been making use of the sun intermittently over the past 2500 years, but there appears to be a direct correlation between the use (or nonuse) of solar energy and the price and availability of other energy resources in areas where sufficient amounts of solar energy are available.

REFERENCES

1. Charles Greeley Abbot, *The Sun*, Appleton, 1912.
2. A. S. E. Ackermann, "Utilizing Solar Energy," *Smithsonian Report*, 1916.
3. F. A. Brooks, "Use of Solar Energy for Heating Water," *Smithsonian Report*, 1939.
4. Ken Butti and John Perlin, "Solar Water Heaters in California, 1890–1931," *Soft Tech*, Penguin Books, New York, 1978.
5. Ken Butti and John Perlin, "Solar Water Heaters in Florida, 1923–Present," *CoEvolution Quarterly*, spring 1978.
6. Albert G. H. Dietz and Edmund L. Czapek, "Solar Heating Houses by Vertical South Wall Storage Panels," *Heating, Piping, and Air Conditioning*, March 1950.
7. Diocles, *On Burning Mirrors*, C. J. Toomer (trans.), Springer-Verlag, New York, 1976.
8. C. D. Engebretson and N. G. Ashar, "Progress in Space Heating with Solar Energy," American Society of Mechanical Engineers, Paper 60-WA-88.
9. R. J. Forbes, *Studies in Ancient Technology*, vols. V and VI, E. J. Brill, 1958.
10. H. C. Hottel and B. B. Woertz, "The Performance of Flat Plate Collectors," *ASME Transactions*, February 1942.
11. F. W. Hutchinson and W. P. Chapman, "A Rational Basis for Solar Heating Analysis," *Heating, Piping, and Air Conditioning*, July 1946.
12. David M. Robinson and J. Walter Graham, *Excavations at Olynthus*, pt. VIII, Johns Hopkins Press, Baltimore, 1938.
13. "Solar Water Heating in South Florida, 1923–1974," National Science Foundation report, 1974.

Chapter 2

Fundamentals of Solar Radiation

ELDON C. BOES
Sandia Laboratories, Albuquerque, NM

INTRODUCTION

This introduction describes the contents of Chap. 2, and their importance and uses for solar energy applications.

Motivation for Chap. 2

Solar energy is a variable energy source   Most of the energy conversion systems used by humankind operate with a very manageable energy source in the sense that the amount of source energy consumed by the system is easily controlled to produce a desired result. For example, the amount of natural gas used by a home furnace is controlled by a thermostat so that the result of the energy input, reduced by the system's inefficiencies, is a comfortable internal home temperature. The task of assuring that sufficient input is available to this system is performed by a gas company which provides a sufficient supply of gas and maintains a certain level of pressure in the gas lines. For end-use energy systems consuming electrical energy, the amount of input is controlled by switches; the sufficiency of the input is assured by the electric power utility. For an intermediary energy conversion system such as a coal-fired electrical generator, the amount of energy input is controlled by the amount of coal fed to the boiler.

Solar energy conversion systems differ from other energy conversion systems in this important respect: the amount of energy available to a solar energy system is not easily controlled. For example, it is generally not possible to increase the amount of solar radiation striking a solar collector whenever the collector's output is inadequate for the energy load it is trying to serve.

This distinction of energy conversion systems according to the level of control over the energy input to the system is a very important one for solar energy systems. Obviously, it means that the usual system design problem of specifying a system adequate for the expected load is a far more complicated problem for solar systems than for other familiar energy systems. For solar systems, this design problem has another variable: the energy input.

There is an obvious solution to this problem of designing a solar system adequate to meet a predicted load, and that is to simply design the solar system so large that no matter how reduced the solar radiation input may be, the system is still adequate. Unfortunately, this solution is generally impractical simply because the high cost of solar systems tends to rule out gross overdesign.

These considerations show that an understanding of solar energy as a variable energy resource is essential for proper solar system design.

Factors affecting solar energy availability   There are seven principal factors which strongly affect the amount of solar radiation incident upon a collector.

1. Geographic location. More solar radiation is available for collectors in sunny regions than in regions with cloudy, rainy climates. Solar radiation availability also increases slightly with altitude.

2. Site location of collector. Collectors which are partially shaded by trees or buildings do not intercept as much solar radiation as collectors in full sun.

3. Collector orientation. This factor is extremely important. Collectors which track the sun so as to remain perpendicular to the sun's rays intercept more solar radiation than stationary collectors. In the Northern Hemisphere, stationary collectors which are tilted upward toward the south intercept more solar radiation than horizontal collectors during the winter.

4. Time of day. Obviously, solar collectors intercept less solar radiation at night than during the day. There is generally less solar radiation available to collectors near sunrise and sunset hours than during the middle of the day.

5. Time of the year. At most locations there is more solar energy available to collectors during the summer months than during the winter months just because the days are longer in the summer. At some locations, seasonal weather patterns significantly affect seasonal availabilities of solar energy.

6. Atmospheric conditions. The most important of these atmospheric variables affecting solar radiation are clouds; clouds frequently reduce the amount of solar radiation incident upon a collector by 90 percent. Solar energy availability is also affected by the variations in atmospheric water vapor, dust particles, pollutants, etc.

7. Collector design. A major design distinction is that between flat-plate collectors, which can collect both direct and diffuse solar radiation, and focusing, concentrating collectors, which can only utilize direct solar radiation (see Chap. 8).

Description of Material in Chap. 2

This chapter gives a complete description of solar radiation as the energy source for solar energy systems. The chapter consists of six main sections:

1. Introduction
2. Basic geometric considerations
3. Extraterrestrial solar radiation and atmospheric effects
4. Summaries of solar radiation availabilities
5. Estimation and conversion formulas for solar radiation
6. Selected examples

The remainder of this introduction gives a brief description of the contents of each of these sections.

Basic geometric considerations   An understanding of the sun's virtual motion is necessary for solar collector design and placement. The section on basic geometric considerations provides all of the information and tools required. Formulas for specifying the sun's position for any time, date, and geographic location are given. True solar time is defined and its relationship to clock time is described. Techniques for determining true North are provided. The equations of motion for various types of tracking collectors are provided.

A thorough description of the geometric relationship between the sun's position and a surface of any orientation is given. This includes the angle of incidence of the sun's rays on the surface as well as the equations needed for determining "sunrise" and "sunset" times for the surface and for calculating shading of surfaces by other objects.

Extraterrestrial solar radiation and atmospheric effects   Extraterrestrial solar radiation refers to solar radiation just outside the earth's atmosphere, i.e., in near-earth space. This section describes the spectral distribution of extraterrestrial solar radiation and the annual variation of its intensity due to the annual variation in the distance between the earth and the sun.

The intensity and spectral distribution of solar radiation are considerably altered by the earth's atmosphere. The principal mechanisms causing atmospheric alterations of solar radiation are absorption and scattering, including reflection. Atmospheric scattering of solar radiation has a particularly important consequence: it separates the solar radiation into direct and diffuse components.

All the atmospheric alterations of solar radiation depend upon various parameters such as the atmospheric path length, atmospheric water vapor content, and atmospheric turbidity. The most important atmospheric parameter is the most obvious one—clouds. The nature and magnitudes of the effects of these atmospheric variables are described.

Summaries of solar radiation availabilities   This section basically consists of a complete summary of existing information on the availability of solar radiation for solar energy applications. A brief description of the types of instrumentation used for measuring direct, total, and diffuse radiation is given. The existing solar radiation monitoring networks and data bases for both the United States and the world are described.

A major portion of this section is devoted to tabulations and maps of solar radiation availability. For the United States, these include maps of total radiation available to various orientations of fixed surfaces as well as maps of direct radiation, all on a seasonal basis. Maps showing total radiation on a horizontal surface for the world are also given. Finally, extensive tabulations of solar radiation availabilities to various types of fixed and tracking surfaces are presented.

Estimation and conversion formulas for solar radiation   For many locations in the world, the specific types of solar radiation records needed for system designing do not exist. This section describes formulas for making estimates of the required availabilities and provides guidance concerning the validity and accuracy of these formulas for various situations.

Where solar radiation records do exist, they generally are measurements of total radiation on a horizontal surface. Formulas for converting these into direct and diffuse components, and for converting these to different surface orientations are presented.

For locations with no solar radiation records, interpolation techniques are suggested, and estimation techniques based upon measurements of sunshine duration and cloud cover are presented.

Selected examples   This section presents examples which illustrate the application of the material in Chap. 2 to various solar energy system design and evaluation problems. These examples address such design questions as the following:

- How much solar radiation will be available for a particular solar collector?
- What collector design is best from a solar radiation availability viewpoint?
- What is the best orientation for a collector?
- What is the penalty for using a less-than-optimal collector orientation?
- How much will the actual solar radiation availabilities vary from the average availabilities by day? By month? By year?

BASIC GEOMETRIC CONSIDERATIONS

The Motion of the Earth; Time

There are two major motions of the earth: the revolution of the earth around the sun, and the rotation or spinning of the earth about its axis. Both of these motions are important in solar energy applications.

The earth's orbital motion   The earth revolves in an elliptical orbit with the sun at one focus of the ellipse (Fig. 2.1). The elliptical shape of the orbit is exaggerated in this figure; in fact, the orbit is nearly a circle, having semimajor and semiminor axes of 1.4968×10^8 km and 1.4966×10^8 km, and an eccentricity of $e = 0.0167$. The period of revolution is defined to be one year. Perihelion, the point at which the earth is nearest the sun, occurs on approximately January 2, and aphelion occurs on July 2. As Fig. 2.1 shows, both the earth's orbital motion and its rotation are counterclockwise, as viewed from the North Pole.

The plane containing the earth's elliptical orbit is called the "ecliptic" plane. The "equatorial" plane is the plane containing the earth's equator. The angle between these planes is 23.45°, since the earth's rotational axis intersects the ecliptic plane at an angle of 66.55° (see Fig. 2.1).

The seasons are due to the fact that the earth's axis is inclined with respect to the ecliptic plane. As Fig. 2.1 shows, solar radiation strikes the earth's Northern Hemisphere more directly near aphelion, causing summer in that hemisphere during that portion of the year. At the same time, solar radiation is striking the earth's Southern Hemisphere more obliquely, causing winter there. The equinoxes are dates on which the earth-sun vector lies in the equatorial plane.

Solar declination   The earth-sun vector moves in the ecliptic plane; the angle between the earth-sun vector and the equatorial plane is called the solar declination angle, δ. By convention, δ is considered positive when the earth-sun vector points northward relative to the equatorial plane. The declination δ varies from $-23.45°$ on December 21, the winter* solstice, to $+23.45°$ on June 22, the summer solstice. The solar declination is approximately given by Eq. (2.1):

---

*Throughout this chapter, all seasonally dependent information will refer to seasons relative to the Northern Hemisphere.

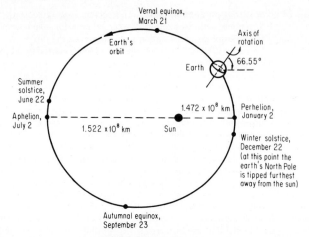

Fig. 2.1 The earth's orbit.

$$\sin \delta = 0.39795 \cos [0.98563 (N - 173)] \tag{2.1}$$

where N is the number of the day. This equation is only accurate to within approximately 1°; accurate tabulations for the solar declination and for the equation of time below are available in Ref. 1.

The Earth's rotation. Solar time   The rotation of the earth about its axis causes the day-night cycle on earth and gives the impression of the sun travelling across the sky each day from east to west. This cycle is the basis for "solar time"; a "solar day" is defined to be the interval of time from the moment the sun crosses the local meridian to the next time it crosses that same meridian. (The "local meridian" at any point is the plane formed by projecting a north-south longitude line through that point out into space from the earth's center.) Because of the earth's forward movement in its orbit during this interval, the time required for one full rotation of the earth is less than a solar day by about 3.95 min.

Solar time vs. standard time; the equation of time   The solar day as defined above varies in length through the year because of (1) the earth's axis is tilted with respect to the ecliptic plane and (2) the angle swept out by the earth-sun vector during a fixed period of time depends upon the earth's position in its elliptic orbit. Moreover, solar noon, the time at which the sun crosses the local meridian, differs for locations of different longitudes. Consequently, standard time, which is a uniform time, and solar time differ. This difference is called the "equation of time"; it varies with date and longitude. The equation of time (EoT) is approximately given by Eq. (2.2):[2]

$$\text{EoT} = 12 + (0.1236 \sin x - 0.0043 \cos x \\ + 0.1538 \sin 2x + 0.0608 \cos 2x) \text{ h} \tag{2.2}$$

where the angle x is a function of the day of the year N:

$$x = 360(N - 1)/365.242° \tag{2.3}$$

Expression (2.2) for the equation of time can be used to convert from local standard time (LST) to solar time:

$$\text{Solar time} = \text{LST} - \text{EoT} - \text{LA} \tag{2.4}$$

where LA is the expression for the adjustment for longitude expressed in hours. Since the conversion from degrees to hours is given by 15° = 1 h (or, 360° = 24 h), the longitude adjustment LA in hours is:

$$\text{LA} = [(\text{local longitude}) - (\text{longitude of local time meridian})]/15 \qquad (2.5)$$

where the longitudes on the right-hand side are in degrees.

Calculating solar noon The local standard time at which solar noon occurs can be approximately determined using the equation of time given in the preceding section. Solar noon can be found more precisely from tables of the sun's position in an almanac or ephemeris; see, for example, Ref. 1.

A very simple technique for obtaining a rough estimate of the local time for solar noon is to simply find the midpoint between the local times for sunrise and sunset. The local times for sunrise and sunset are routinely published in local newspapers.

Hour angle In any time system, it is frequently convenient to express the time in degrees rather than hours. This is especially true for specifying the positions of bodies in space as functions of time. The unit of angular measurement of time is the "hour angle." The basic conversion is that 24 hours equals 360 hour angle degrees.

In solar energy, the solar hour angle h_s is used extensively to express solar time, because it is directly related to the sun's position in the sky. The solar hour angle is measured from solar noon and is positive (negative) before (after) solar noon. For example, at 8 A.M. solar time, the solar hour angle h_s is $+60°$, and at 9 P.M. solar time h_s is $-135°$ in accordance with the right-hand rule.

Since solar time varies with longitude, solar hour angles do also.

The Location of the Sun

The path of the sun across the celestial sphere The "celestial sphere" is an imaginary sphere at an undefined distance from its center, the earth. It is a convenient device for locating celestial objects when their distance from earth is not important. All celestial objects are considered to travel on this sphere, and only the position on the sphere is specified.

The sun travels across the celestial sphere from approximately east to west each day; the sun's path is a circle on the celestial sphere. This daily path depends upon the time of year and upon the latitude of the point of observation on earth. Figure 2.2 shows the sun's paths for a latitude of 34°N for the equinoxes and solstices.

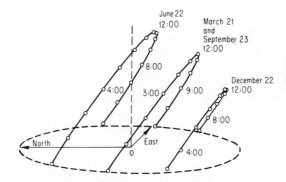

Fig. 2.2 The sun's paths across the sky as seen by an observer at point 0.

Solar azimuth, altitude, and zenith angles The sun's position on the celestial sphere is usually specified in terms of the solar azimuth angle a_s and the solar altitude angle α. These are displayed in Fig. 2.3. The solar altitude angle measures the sun's angular distance from the horizon, and the azimuth angle measures the sun's angular distance from the south. In this handbook, the solar azimuth angle a_s and the wall azimuth angle a_w (defined below) will be considered positive to the east of south and negative to the west of south based on the right-hand rule.

The solar zenith angle z is the sun's angular distance from the zenith, which is the point directly overhead on the celestial sphere. Thus, α and z are complementary angles:

$$\alpha + z = 90° \qquad (2.6)$$

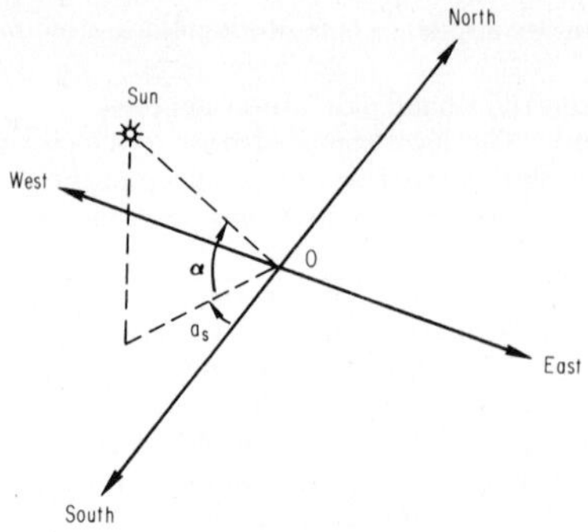

Fig. 2.3   Illustration of the solar altitude angle α and the solar azimuth angle a_s.

The solar altitude and azimuth angles are computed for any time, date, and location using these formulas (where L is the latitude taken positive north of the equator)

$$\sin \alpha = \cos L \, \cos \delta \, \cos h_s + \sin L \, \sin \delta, \tag{2.7}$$

$$\sin a_s = \frac{\cos \delta \, \sin h_s}{\cos \alpha} \tag{2.8}$$

To compute α it is sufficient to apply the inverse sine function in Eq. (2.7). However, in using Eq. (2.8) to find the azimuth angle a_s, it is necessary to distinguish the case where the sun is in the northern half of the sky from the case where the sun is in the southern half of the sky. The complete formula for a_s is

$$a_s = \begin{cases} \sin^{-1}\left(\dfrac{\cos \delta \, \sin h_s}{\cos \alpha}\right), \text{ if } \cos h_s > \dfrac{\tan \delta}{\tan L} \\[4mm] 180° - \sin^{-1}\left(\dfrac{\cos \delta \, \sin h_s}{\cos \alpha}\right), \text{ if } \cos h_s < \dfrac{\tan \delta}{\tan L} \end{cases} \tag{2.9}$$

In the case $\cos h_s = \tan \delta / \tan L$, then the solar azimuth angle is either $-90°$ or $+90°$, depending on whether the hour angle h_s is negative or positive, respectively.

Table 2.1 lists solar altitude and azimuth angles for each hour on the 21st of each month for latitudes of $0°$ through $80°$, in $8°$ steps. Interpolation will give fairly accurate values for times, dates, or latitudes not listed. Although α and a_s are only given for A.M. hours, they can easily be obtained for the afternoon hours by symmetry. See Chap. 16 for graphs of α and a_s called sun path diagrams.

Sunrise and sunset   Sunrise and sunset occur when the solar altitude angle is 0. From Eq. (2.7) we see that the solar hour angle for sunrise is given by:

$$h_{sr} = \cos^{-1}\left(- \tan L \, \tan \delta\right) \tag{2.10}$$

The solar hour angle for sunset is the negative of this:

$$h_{ss} = -h_{sr} \tag{2.11}$$

Solar Incidence Angles

The intensity of solar radiation on a surface depends upon the angle at which the sun's rays strike the surface. The intensity is proportional to the cosine of the angle between the solar rays and the surface normal. This angle i is the "solar incidence angle" for the surface. The computation of i for various fixed and tracking collectors is described in this section.

TABLE 2.1 Solar Altitude and Azimuth Angles, α and a_s, for Various Northern Latitudes

Date and declination	Solar time (A.M.)	0° α	0° a_s	8° α	8° a_s	16° α	16° a_s	24° α	24° a_s
June 21	Noon	67	180	75	180	83	180	90	0
23.5°	11	62	149	69	139	74	120	76	91
	10	53	131	57	122	61	110	63	95
	9	40	122	44	115	47	108	49	99
	8	27	117	31	113	33	110	36	103
	7	14	114	17	112	20	113	22	107
	6	0	114	3	113	6		9	112
	5								
May 21	Noon	70	180	78	180	86	180	86	0
or	11	65	145	71	131	75	108	76	77
July 21	10	54	126	59	116	61	103	62	88
20°	9	42	117	45	110	47	102	48	94
	8	28	113	31	109	33	104	35	98
	7	14	111	17	109	19	106	21	103
	6	0	110	3	110	5	109	8	108
	5								
April 21	Noon	78	180	86	180	86	0	78	0
or	11	71	129	75	106	75	77	71	52
August 21	10	58	113	60	100	61	86	59	72
12°	9	44	107	46	99	46	91	46	82
	8	29	104	34	99	32	94	32	89
	7	15	102	16	100	17	98	18	95
	6	0	102	2	102	3	102	5	101
March 21	Noon	90	90	82	0	74	0	66	0
or	11	75	90	73	63	68	44	62	33
Sept 21	10	60	90	59	76	56	64	52	55
0°	9	45	90	44	82	43	75	40	68
	8	30	90	30	85	29	81	27	77
	7	15	90	15	88	14	86	14	84
	6	0	90	0	90	0	90	0	90
Feb 21	Noon	79	0	71	0	63	0	55	0
or	11	71	53	66	38	59	30	52	24
Oct 21	10	58	69	55	58	50	50	44	43
−11°	9	44	75	41	68	38	62	34	57
	8	29	77	27	73	25	69	22	66
	7	14	79	13	77	11	75	9	74
Jan 21	Noon	70	0	62	0	54	0	46	0
or	11	65	35	58	28	51	23	44	20
Nov 21	10	54	54	49	46	43	40	37	36
−20°	9	42	63	38	57	33	52	28	49
	8	28	67	25	64	21	61	17	58
	7	14	69	11	68	8	66	5	66
Dec 21	Noon	67	0	58	0	51	0	42	0
−23.5°	11	62	31	55	25	48	21	40	18
	10	53	49	47	42	41	37	34	34
	9	40	58	36	53	31	49	25	46
	8	27	63	23	60	19	57	15	55
	7	14	66	10	64	7	63	3	63

TABLE 2.1 Solar Altitude and Azimuth Angles, α and a_s, for Various Northern Latitudes (Continued)

Date and declination	Solar time (A.M.)	32°		40°		48°		56°	
		α	a_s	α	a_s	α	a_s	α	a_s
June 21	Noon	82	0	74	0	66	0	58	0
23.5°	11	74	61	69	42	63	31	56	25
	10	62	80	60	66	56	55	51	46
	9	50	90	49	80	47	72	44	64
	8	37	97	37	91	37	85	36	79
	7	24	103	26	100	27	96	28	92
	6	12	110	15	108	17	106	19	104
	5	1	118	4	117	8	117	11	115
	4							4	127
May 21	Noon	78	0	70	0	62	0	54	0
or	11	72	52	66	37	59	29	52	23
July 21	10	61	73	57	61	53	51	48	44
20°	9	48	85	47	76	44	68	41	62
	8	35	93	35	87	35	82	33	76
	7	23	100	24	97	25	93	25	89
	6	10	107	13	106	15	104	16	102
	5			2	115	5	114	8	113
	4							1	126
April 21	Noon	70	0	62	0	54	0	46	0
or	11	66	38	59	29	52	24	44	21
August 21	10	56	61	51	52	46	45	40	40
12°	9	44	75	42	68	38	62	34	57
	8	32	84	31	80	29	75	26	71
	7	19	93	19	90	19	87	18	84
	6	6	100	8	99	9	98	10	97
	5							2	109
March 21	Noon	58	0	50	0	42	0	34	0
or	11	55	27	48	23	40	20	33	18
Sept 21	10	47	47	42	42	35	38	29	35
0°	9	37	62	33	57	29	53	23	50
	8	25	73	23	70	20	67	16	64
	7	13	82	11	80	10	79	8	77
Feb 21	Noon	47	0	39	0	31	0	23	0
or	11	45	21	37	19	30	17	22	16
Oct 21	10	38	39	32	35	25	33	18	31
−11°	9	29	53	24	50	19	47	13	45
	8	18	64	15	61	11	60	7	59
	7	7	73	4	72	2	72		
Jan 21	Noon	38	0	30	0	22	0	14	0
or	11	36	18	28	16	21	15	13	14
Nov 21	10	31	33	24	31	17	29	10	28
−20°	9	22	46	17	45	11	43	5	42
	8	13	56	8	55	3	55		
	7	1	65						
Dec 21	Noon	34	0	26	0	18	0	10	0
−23.5°	11	33	16	25	15	17	14	9	14
	10	28	31	21	29	14	28	7	27
	9	20	44	14	42	8	41	2	40
	8	10	54	5	53	1	53		

Latitude

TABLE 2.1 Solar Altitude and Azimuth Angles, α and a_s, for Various Northern Latitudes (Continued)

Date and declination	Solar time (A.M.)	Latitude 64° α	64° a_s	72° α	72° a_s	80° α	80° a_s
June 21	Noon	49	0	41	0	33	0
23.5°	11	48	21	41	18	33	16
	10	45	40	39	36	32	33
	9	40	58	35	53	30	49
	8	34	73	31	68	28	64
	7	28	88	27	83	26	79
	6	21	101	22	98	23	94
	5	15	114	18	112	21	109
	4	9	126	14	125	18	123
	3	4	139	10	139	16	138
	2	1	153	8	152	15	152
	1			6	166	14	166
	Midnight			6	180	14	180
May 21	Noon	46	0	38	0	30	0
or	11	45	20	37	18	30	16
July 21	10	42	39	35	35	29	32
20°	9	37	56	32	52	27	48
	8	31	71	28	67	25	64
	7	24	86	24	82	22	79
	6	18	99	19	96	20	94
	5	12	112	14	110	17	108
	4	6	125	10	124	15	123
	3	1	138	7	138	13	137
	2			4	152	11	151
	1			3	166	10	166
	Midnight			2	180	10	180
April 21	Noon	38	0	30	0	22	0
or	11	37	18	29	17	22	16
August 21	10	34	36	27	33	21	31
12°	9	29	53	24	49	19	47
	8	24	68	20	65	17	62
	7	17	82	16	79	14	77
	6	11	95	11	94	12	92
	5	4	109	7	108	9	107
	4			3	122	7	121
	3					5	136
	2					3	151
	1					2	165
	Midnight					2	180
March 21	Noon	26	0	18	0	10	0
or	11	25	17	17	16	10	15
Sept 21	10	22	33	16	31	9	30
0°	9	18	48	13	46	7	45
	8	13	63	9	61	5	60
	7	7	76	5	76	3	75
Feb 21	Noon	15	0	7	0		
or	11	14	15	6	15		
Oct 21	10	12	30	5	30		
−11°	9	8	44	4	44		
	8	3	58	2			

TABLE 2.1   Solar Altitude and Azimuth Angles, α and a_s, for Various Northern Latitudes *(Continued)*

Date and declination	Solar time (A.M.)	Latitude					
		64°		72°		80°	
		α	a_s	α	a_s	α	a_s
Jan 21 or Nov 21 −20°	Noon	6	0				
	11	5	14				
	10	3	28				
Dec 21 23.5°	Noon	3	0				
	11	2	14				

Orientation of fixed surfaces   The orientation of a fixed planar surface such as a wall is specified in terms of its "azimuth angle" a_w and its tilt angle β. These are shown in Fig. 2.4; in this figure, the vector **n** is normal (perpendicular) to the surface. The surface azimuth angle a_w is measured from the south, positive toward the east; the tilt angle β is simply the angle at which the surface is inclined from horizontal and is taken positive for south-facing surfaces.

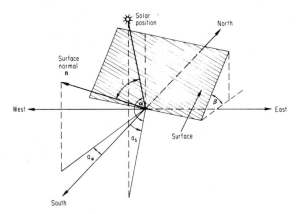

Fig. 2.4   Surface orientation coordinates a_w and β and the solar incidence angle γ.

The solar incidence angle for a fixed surface   The solar incidence angle i for a fixed surface is also shown in Fig. 2.4. It is defined as the angle between the surface's normal **n** and the sun's rays. It is important because the intensity of the direct beam of solar radiation on the surface is proportional to $\cos i$; thus, values of i approaching 90° imply greatly reduced solar radiation intensities available for collection.

The solar incidence angle i can be calculated in terms of the solar position coordinates α and a_s and the surface orientation parameters β and a_w using this equation:

$$\cos i = \cos \alpha \, \cos(a_s - a_w) \, \sin \beta + \sin \alpha \, \cos \beta \tag{2.12}$$

If the right-hand side of (2.12) is negative, the sun's rays will not strike the front side of the surface. The "sunrise" and "sunset" for this surface will occur when $i = 90°$; thus, sunrise and sunset times for a surface can be found by searching for the hour angle h_s at which:

$$\cos (a_s - a_w) = - \tan \alpha \cot \beta \qquad (2.13)$$

If the sunrise hour angle h occurs before sunrise, i.e., $h > h_{sr}$, h_{sr} should be used for h.

Tracking surfaces   Many solar collector designs are "sun-tracking"; that is, they move so as to follow the sun during the day. This tracking usually consists of a rotation of the collector about one, and occasionally two, axes. The principal tracking schemes are described in Table 2.2. The first seven schemes listed are generally better than the last three, as the solar availabilities in Sec. 4 indicate.

The first two tracking schemes listed in Table 2.2 are two-dimensional, "full-tracking" schemes. Both are denoted by N in this chapter since both produce the effect of having the collector oriented normal to the sun's rays. The second of the two rotates at constant speed about a "polar axis," an axis parallel to the earth's axis. Essentially, this rotation is simply undoing the earth's rotation. The declination axis is perpendicular to the polar axis and is horizontal at solar noon. It serves to account for the solar declination angle; in many cases it is only adjusted daily or weekly.

The solar incidence angle for tracking surfaces   Just as for fixed surfaces, the solar incidence angle i for a tracking surface or collector is the angle between the surface's normal n and the sun's rays. Table 2.2 gives the cosine of the solar incidence angle for each of the tracking schemes listed there. When this expression is negative for a particular tracking surface, the incidence angle is greater than 90° and the direct solar rays do not intercept the front of the surface.

Determining True North

The determination of geographic or "true" north or south is important for aligning solar energy collectors, especially for focusing, tracking collectors. There are four relatively simple techniques for determining true north which are described in this section. The first of these is well known; it consists of using a compass, and making appropriate adjustments for the difference between the geographic pole and the magnetic pole. The second consists of using a fairly simple shadow device, the third consists of sighting the sun at true solar noon, and the fourth consists of sighting the North Star (Polaris) at night in the Northern Hemisphere.

Use of a compass   It is a relatively straightforward process to get a reading of magnetic north from a compass. However, in using a compass to determine true geographic north two factors must be considered. The first of these is that an adjustment must be made to correct for the difference in location of the geographic and magnetic north poles. Figure 2.5 indicates the approximate adjustments for the continental United States. Note that in this figure "easterly variation" at a location means that a line to the magnetic pole from that location points east of geographic north. Thus, at such a location, a compass would point to the east of true north. "Westerly variation" has the analogous meaning.

The other factor to be considered in using a compass is the possibility of local deviations in the compass' readings, including effects caused by regional mineral deposits or adjacent steel structures. These can cause serious errors.

A shadow technique for determining north   The primary source of information on this shadow technique is an article by P. L. Harrison.[3] The technique makes use of the fact that if a circular disk is oriented parallel to the earth's equatorial plane, so that the disk's axis is parallel to the earth's axis, then the shadow of the edge of the disk will intersect the disk's axis at a point which remains fixed throughout the day (see Fig. 2.6). The position of this point of intersection depends upon the date; the distance of the point of intersection from the center of the disk is given by

$$y = r \tan \delta \qquad (2.14)$$

where r is the radius of the disk, and δ is the solar declination angle. (Thus, the point of intersection is not strictly fixed, since the solar declination varies slightly through the day.)

This geometrical information can be combined with a simple device to determine true north as follows. The device consists of a semicircular disk inclined so that its diameter is horizontal and its axis is parallel to the earth's axis. Thus, the disk's axis intersects a horizontal plane at an angle equal to the local latitude L, and the disk itself intersects the horizontal at an angle of $90° - L$. A schematic drawing of such a device is shown in Fig. 2.7. If this device is oriented properly in azimuth so that the diameter is on a true east-west line, then the shadow of the edge of the disk will intersect the axis at the correct distance from the center of the disk as

TABLE 2.2 Tracking Schemes, Their Notation, and Solar Incidence Angles

Description	Axis (axes)	Cosine of incidence angle ($\cos i$)	Notation
Movements in altitude and azimuth	Horizontal axis and vertical axis	1	N
Rotation about a polar axis and adjustment in declination	Polar axis and declination axis	1	N
Uniform rotation about a polar axis	Polar axis	$\cos \delta$	P
East-west horizontal	Horizontal, east-west axis	$\sqrt{1 - \cos^2 \alpha \sin^2 a_s}$	EW
North-south horizontal	Horizontal, north-south axis	$\sqrt{1 - \cos^2 \alpha \cos^2 a_s}$	NSH
Rotation about a vertical axis of a surface tilted upward L (latitude) degrees	Vertical axis	$\sin(\alpha + L)$	VL
Rotation about a vertical axis of a surface tilted upward $L + 15°$	Vertical axis	$\sin(\alpha + L + 15)$	VLP
Rotation about a vertical axis of a surface tilted upward $L - 15°$	Vertical axis	$\sin(\alpha + L - 15)$	VLN
Rotation of a horizontal collector about a vertical axis	Vertical axis	$\sin \alpha$	VH
Rotation of a vertical surface about a vertical axis	Vertical axis	$\cos \alpha$	VV
Fixed "tubular" collector	North-South tilted up at angle β	$\sqrt{1 - [\sin(\beta - L)\cos \delta_s \cos h_s + \cos(\beta - L)\sin \delta_s]^2}$	—

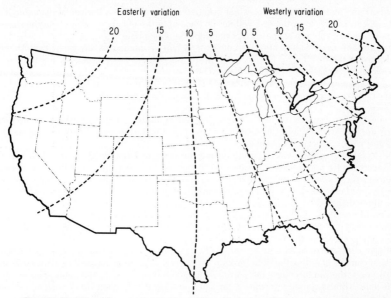

Easterly variation   Westerly variation

20   15   10   5   0   5   10   15   20

Fig. 2.5 Variations of magnetic compass readings, in degrees, from true geographic north.

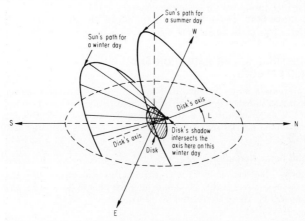

Sun's path for
a summer day

W

Sun's path for
a winter day

Disk's axis

S   L   N

Disk's axis

Disk's shadow
intersects the
axis here on this
winter day

Disk

E

Fig. 2.6 The shadow of a disk parallel to the equatorial plane intersects the disk's axis at a fixed point for a given day.

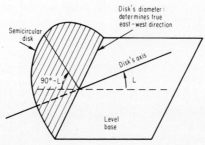

Disk's diameter:
determines true
east–west direction

Semicircular
disk

Disk's axis

$90° - L$ L

Level
base

Fig. 2.7 A device for determining true north with a shadow technique.

described previously. If the device is not oriented so that the disk's horizontal diameter is on a true east-west line, then the disk will not be parallel to the equatorial plane, and the shadow will not intersect the axis at the correct point for the given date.

This device is used to determine a true east-west line by rotating the device in azimuth until the shadow of the disk's edge intersects the disk's axis at the correct distance $y = r \tan \delta$ from the center of the disk for the given date. The disk and its axis must be maintained at the appropriate inclination with respect to horizontal during this process. At that orientation where the shadow-axis intersection is correct, the horizontal diameter of the disk is on a true east-west line.

Of course, the accuracy of this technique depends upon the precision to which the disk and its axis are constructed, and the accuracy with which they are oriented with respect to each other and with respect to horizontal.

The technique is also more accurate during the early or late daylight hours than it is for times nearer solar noon. This is because the device is more sensitive to changes in azimuth nearer sunrise or sunset in the sense that an error in the east-west orientation of the horizontal diameter produces a larger deviation in the shadow-axis intersection from the correct point at these times than it does during the middle of the day.

According to Harrison,[3] this technique is accurate for determining true north to within 1° provided that it is not applied during the middle four hours of the day. More information concerning accuracy of the technique is available in Ref. 3.

Determining north at true solar noon This technique for determining true north is fairly simple; it consists of noting the sun's position at precisely solar noon. According to the definition of solar noon, the sun intersects the north-south longitudinal plane for the point of observation at this time. Thus, a simple recording of the sun's position at solar noon via a transit or a shadow will determine a true north-south line. The precise determination of solar noon should be based upon an ephemeris table combined with an adjustment for local longitude, as described in the section on solar time presented earlier in this chapter.

Use of the North Star In the Northern Hemisphere, the North Star, or Polaris, can be sighted to obtain a true north direction at night which is accurate to within about 1°.

EXTRATERRESTRIAL SOLAR RADIATION AND ATMOSPHERIC EFFECTS

This section presents a basic description of solar energy as a radiant energy source. The characteristics of solar radiation outside the earth's atmosphere are described, and the principal interactions of this radiation with the earth's atmosphere are discussed. This chapter does not include material on the physics and chemistry of the sun itself. A very brief section on radiant energy is included.

Extraterrestrial Solar Radiation

Radiant energy Radiant energy is usually described as a stream of particles, called photons, traveling in transverse waves at the speed of light. Each photon possesses a wavelength λ and an amount of energy E. These are related by the equation

$$E = hc/\lambda \tag{2.15}$$

where h is Planck's constant, $h = 6.6 \times 10^{-34}$ J·s, and c is the velocity of light ($c = 3 \times 10^8$ m/s). The frequency ν of radiation of a given wavelength λ is the number of waves passing a point in a fixed interval of time; it is given by

$$\nu = c/\lambda \tag{2.16}$$

Thus, the equation (2.15) for the energy of a photon of wavelength λ can also be written as

$$E = h\nu \tag{2.17}$$

Radiation exists for widely differing wavelengths. Certain wavelength bands of radiation are given special names. The most common of these is "light" or "visible radiation," which is radiation having wavelengths between approximately 0.35 and 0.75 μm. Radiation having wavelengths from 0.75 to approximately 100 μm is called "infrared radiation"; radiation waves of wavelengths exceeding 100 μm are called "radio waves." Radiation with wavelengths shorter than those of visible light are "ultraviolet," X rays, and gamma rays. Table 2.3 lists these major types of radiation and their corresponding wavelengths and frequency bands.

TABLE 2.3   Types of Radiation and Their Approximate Wavelengths and Frequencies

Radiation type		Approximate wavelength interval (μm)	Approximate frequency band (cycles per second)
Gamma rays		10^{-4}	3×10^{18}
X-rays		5×10^{-2} to 5×10^{-6}	6×10^{19} to 6×10^{15}
Ultraviolet		0.005 to 0.35	9×10^{14} to 6×10^{16}
Visible light		0.35 to 0.75	4×10^{14} to 9×10^{14}
Infrared		0.75 to 300	10^{12} to 4×10^{14}
Radio waves	short	1×10^2 to 1×10^8	3×10^6 to 3×10^{12}
	long	1×10^8	3×10^6

The solar radiation spectrum   Solar radiation is radiant energy issuing from the sun. Although solar radiation covers a fairly wide band of wavelengths, the peak intensity occurs in the interval of visible light. (It is likely that the human eye evolved so as to be most sensitive in the region of peak intensity of solar radiation.)

The approximate distribution of solar radiation with respect to wavelength is shown in Fig. 2.8. This is the "extraterrestrial solar spectrum."

The solar constant   The total energy intensity of extraterrestrial solar radiation, measured just outside the earth's atmosphere and integrated over the entire solar spectrum, is called the "solar constant"; it is denoted I_0. It is to be corrected to the mean earth-sun distance so that the variation of this intensity with the varying earth-sun distance is removed. The term "solar constant" is an historical misnomer since this intensity does vary; however, the variations are not large enough to be of practical concern.

Attempts to determine the value of the solar constant have been made regularly since 1900. The most recent measurements indicate that the value of I_0 is 1.37 kW/m² (Ref. 4). Thus, taking into account the variation in the earth-sun distance through the year, the intensity of extraterrestrial solar radiation just outside the earth's atmosphere varies from 1.32 kW/m² in early July to 1.42 kW/m² in early January. Table 2.4 gives the intensity of extraterrestrial solar radiation for the 21st day of each month.

The total amount of radiant solar energy incident upon a horizontal surface outside the earth's atmosphere for various latitudes for the 21st day of each month is given in Table 2.5. These energy amounts are useful because they clearly represent upper bounds for the amount of solar energy incident upon a horizontal surface on earth for even the clearest of skies.

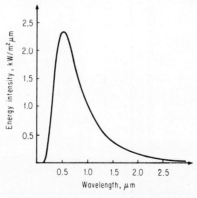

Fig. 2.8   Spectral distribution of extraterrestrial solar radiation.

TABLE 2.4 The Intensity of Extraterrestrial Solar Radiation on the 21st Day of Each Month (kWh/m²)

Date	Jan 21	Feb 21	Mar 21	Apr 21	May 21	Jun 21	Jul 21	Aug 21	Sep 21	Oct 21	Nov 21	Dec 21
Intensity	1.41	1.40	1.38	1.36	1.34	1.33	1.33	1.34	1.36	1.38	1.40	1.41

TABLE 2.5 The Total Amount of Solar Radiation Incident upon a Horizontal Surface Outside the Earth's Atmosphere for the 21st Day of Each Month (kWh/m²)

Latitude	June 21 23.5°	May 21 & July 21 20°	April 21 & Aug 21 12°	March 21 & Sept 21 0°	Feb 21 & Oct 21 −11°	Jan 21 & Nov 21 −20°	Dec 21 −23.5°
0	9.3	9.5	10.1	10.5	10.4	10.1	9.9
8	10.1	10.2	10.5	10.4	9.9	9.2	8.9
16	10.8	10.7	10.6	10.1	9.2	8.2	7.7
24	11.3	11.1	10.6	9.6	8.3	7.0	6.4
32	11.6	11.2	10.4	8.9	7.2	5.7	5.1
40	11.7	11.2	10.0	8.0	6.0	4.4	3.7
48	11.7	11.0	9.4	7.0	4.8	3.0	2.4
56	11.6	10.7	8.7	5.9	3.4	1.7	1.1
64	11.6	10.3	7.9	4.6	2.1	0.5	0.1

Date and Declination

Atmospheric Effects on Solar Radiation

Solar radiation is considerably altered in its passage through the earth's atmosphere. The two principal mechanisms causing these atmospheric alterations are absorption and scattering.

Atmospheric absorption of solar radiation   There are several atmospheric constituents which absorb part of the incoming solar radiation. The first of these is usually called ozone absorption, although the actual absorption is by both oxygen and ozone molecules. This absorption removes nearly all of the ultraviolet solar radiation, so that very little solar radiation with wavelengths less than 0.3 μ reaches the earth's surface.

It is not known how much the quantity of atmospheric ozone varies. However, these variations do not cause variations in the intensity of solar radiation at the earth's surface of sufficient magnitude to be of concern to solar energy system designers.

The other primary absorber of radiation in the atmosphere is water vapor. It absorbs solar radiation in quite specific wavelength bands in the infrared region; consequently, the spectral distribution of terrestrial radiation contains several pronounced dips and peaks in the infrared region. (See Fig. 2.9.)

The amount of water vapor in the atmosphere depends upon the local altitude, climate, and season. The increased solar intensity at higher-altitude locations is partly due to the smaller amounts of atmospheric water vapor at higher altitudes. Since warm air can hold more water vapor without precipitating than can cold air, atmospheric water vapor content are generally much higher in summer than in winter. These variations in atmospheric water vapor content frequently produce 5 to 20 percent variations in the intensity of direct solar radiation at the earth's surface.

In addition to these two primary absorbers, there are other atmospheric constituents which absorb lesser amounts of solar radiation. These include carbon dioxide, oxygen, other gases, and particulate matter.

Atmospheric scattering of solar radiation   The atmospheric constituents primarily responsible for scattering of solar radiation are gas molecules, particulates, and water droplets. This scattering is fairly uniform with respect to direction. However, it is strongly wavelength dependent, and it affects short wavelength radiation most. One consequence is that the scattered radiation eventually reaching the earth's surface from throughout the sky is characteristically blue.

Cloud effects on solar radiation   Clouds are a scattering agent of particular importance. Clouds frequently reduce incoming radiation by as much as 80 to 90 percent by single and multiple scatterings, thus effectively reflecting that amount of radiation back to space.

Because cloud distributions and types are highly variable, their effects on incoming radiation are also highly variable. Thus, clouds produce not only large reductions in the solar radiation available at the earth's surface, but in addition these reductions are quite unpredictable.

Atmospheric Path Length

The amount of atmospheric absorption or scattering affecting incoming solar radiation depends upon the length of the atmospheric path, or the thickness of the atmosphere, through which

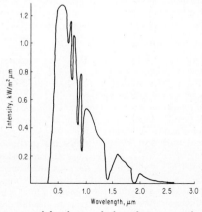

Fig. 2.9   Approximate spectral distribution of solar radiation on earth with an air mass 2 atm.

the radiation travels. Thus, the reduction in intensity of solar radiation at a location on the earth's surface depends not only on the varying composition of the atmosphere at that location, but also upon the location's altitude and upon the radiation angle. For example, solar radiation traveling through the earth's atmosphere at an angle is absorbed and scattered more than solar radiation which comes through the atmosphere from the zenith position directly overhead.

Air mass   The measure of the atmospheric path through which solar radiation must travel is called air mass m. An air mass value of $m = 1$ (dimensionless) is assigned to an atmospheric path directly overhead at sea level. All other air mass values are assigned relative to this unit value.

The two principal factors affecting air mass for an atmospheric path are the direction of the path and the local altitude. The path's direction is simply described in terms of its zenith angle z, which is the angle between the path and the zenith position directly overhead. If the earth were planar, the relationship between m and z would be

$$m = \sec z \tag{2.18}$$

However, for the spherical earth, (2.18) is valid only for zenith angles less than 70°. For larger angles, the relationship is shown in the Bemporad table, Table 2.6.[5]

The adjustment in air mass for local altitude is made in terms of the local atmospheric pressure p. The adjustment is

$$m = (p/p_0)\, m_0 \tag{2.19}$$

where p is the mean local pressure and m_0 and p_0 are the corresponding air mass and pressure at sea level. Spectral air mass effects are described in Chap. 5.

Direct, diffuse and total solar radiation   Solar radiation at the earth's surface consists of two components, direct and diffuse solar radiation. "Direct" radiation refers to that radiation which comes in a beam directly from the sun; it is the type of radiation which casts a distinct shadow.

"Diffuse" solar radiation is the scattered solar radiation which comes from throughout the sky. "Total" solar radiation refers to the combination of these two.

These distinctions are important for solar energy applications. For example, focusing solar collectors can only utilize direct radiation, while flat-plate collectors can capture total radiation. The instruments used to measure the intensities of these different components of solar radiation are also different; they are described in the next section.

SUMMARIES OF SOLAR RADIATION AVAILABILITIES

This section contains information on the availability of solar energy which is essential in solar system designing. The information is based upon actual records of solar radiation measurements.

There are many locations for which solar radiation measurements have never been made. The next section of this chapter provides a variety of suggested approximations and conversion formulas which can be used for estimating solar energy availability under these circumstances.

This section begins with a brief description of the most common types of instruments used for measuring solar radiation intensity, and of the available solar radiation measurement records.

Types of Solar Radiation Instruments

There are dozens of different types of instruments in use for measuring either solar radiation intensity or integrated solar energy over a given time interval. A comprehensive survey of these has been published by the University of Alabama at Huntsville.[6] Only a brief description of the most common types is presented in this handbook.

Most solar radiation instruments measure the intensity of some component of solar radiation, such as the intensity of the direct component or the diffuse component. Records of integrated amounts of solar energy are obtained by summing these intensity readings.

Measurements of direct radiation. Normal incidence pyrheliometers   Normal incidence pyrheliometers are used for measuring the intensity of the direct component of solar radiation on a surface normal to the direction of the rays. Such measurements are directly applicable for focusing collectors which can only utilize the direct component of solar radiation. They are also extremely useful for calculating the amount of total solar radiation available to fixed, tilted collectors when measurements of solar radiation on a tilted surface are not available.

TABLE 2.6 Optical Air Mass Values at Sea Level

zenith angle z	0	30	60	70	80	85	86	87	88	89	90
sec z	1.00	1.15	2.00	2.92	5.76	11.47	14.34	19.10	28.65	57.30	∞
air mass, m	1.00	1.15	2.00	2.92	5.63	10.69	12.87	16.04	20.87	28.35	29.94

Normal incidence pyrheliometers typically consist of (1) a radiation sensor element, mounted inside (2) a collimating tube; this apparatus is mounted on (3) a sun-tracking mechanism. A simplified schematic drawing is shown in Fig. 2.10.

The collimating tube serves to limit the solar energy reaching the radiation sensor to the direct component. In practice, most normal incidence pyrheliometers admit radiation to the sensor from about a 5.7° field of view, even though the solar disk only subtends a 0.5° arc. Thus, measurements of "direct-normal" radiation usually include some diffuse radiation surrounding the solar disk. Under clear sky conditions, this circum-solar diffuse radiation generally represents no more than 5 percent of the instrument reading.

The radiation sensor element is generally either a thermopile, or a photodetector such as a silicon solar cell. Each of these is more appropriate than the other under certain circumstances. The thermopile sensor is a "black receiver" in the sense that it is equally sensitive to radiation of all wavelengths. As such, measurements made with a thermopile type of radiation sensor are most directly applicable for solar energy collectors utilizing nonwavelength-dependent receivers; this includes essentially all thermal solar collectors.

Pyrheliometers with photodetector radiation sensor elements are more appropriate for use with photovoltaic solar collectors. This is especially true if the sensor element and collector are similar photovoltaic devices, that is, devices having the same spectral response. In this case, the pyrheliometer records the amount of solar energy available in the same wavelength region where the collector is sensitive.

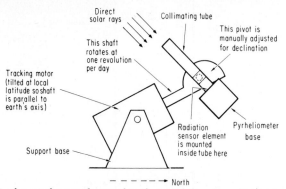

Fig. 2.10   A schematic drawing of a normal incidence pyrheliometer and its sun-tracking mount.

Total or global radiation. Pyranometers   The type of instrument which has been most widely used for making measurements of solar radiation on earth is the pyranometer. It measures the intensity of radiant energy arriving from a hemisphere of directions. Such radiation is called "total" or "global" radiation. In a horizontal position, a pyranometer measures the intensity of total (direct plus diffuse) solar radiation on a horizontal surface. In a tilted position it measures the total radiation intensity on a tilted surface; in this case, the diffuse portion includes reflected radiation from that part of the earth which the pyranometer sees.

The importance of pyranometers lies in the fact that the total radiation which they measure is the same type of radiation which is incident upon flat-plate solar collectors. In a horizontal position, pyranometer measurements also correspond to the radiant heating energy striking the earth's surface, and to the solar energy available to plants.

Pyranometers are similar to pyrheliometers. They consist of a radiation sensor element under a transparent cover, usually a quartz dome (see Fig. 2.11). Just as for a pyrheliometer, the sensor element is usually a thermopile or a photodetector.

Diffuse radiation, shading disks, and shadow bands   There are two commonly used methods for measuring the intensity of diffuse solar radiation. The first of these simply shadows the radiation sensor element of a pyranometer from the direct solar rays with an opaque "shading disk". These are typically hand-held at the end of a slender rod while a measurement is recorded.

A "shadow band" or "shade ring" is a device which can be used to automatically record measurements of diffuse radiation in a routine fashion. The shadow band is mounted over a pyranometer in such a way as to block the direct rays of the sun from the sensor element through the entire day. The shadow band must be reset as necessary to adjust for the changing solar declination.

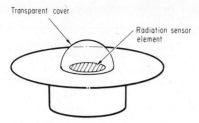

Fig. 2.11 Schematic drawing of a pyranometer used for measuring the intensity of total (direct plus diffuse) solar radiation.

Sunshine duration records There are many locations throughout the world which lack records of solar radiation measurements, but which have records of sunshine duration. Although they are not included in this handbook, these records are quite useful for estimating the availability of solar energy, and estimation formulas based upon sunshine records are described in the next section of this chapter.

Sunshine duration records usually consist of either the number of hours of sunshine for each day, or the percent of the possible number of hours of sunshine by day. The formulas based upon these sunshine duration records usually use monthly means of the daily percent of possible sunshine amounts.

There are several different instruments used for recording sunshine duration. All of these instruments are essentially "on-off" devices; that is, they only record whether or not the intensity of direct radiation exceeds a fixed threshold. They do not have the capability of measuring varying levels of intensity. For this reason, they are often called "sunshine switches." (Of course, the threshold intensity can be selected at any level on most of these instruments.)

Records of Solar Radiation Measurement for the World

The only comprehensive survey of solar radiation monitoring stations on a worldwide basis was conducted in 1966 by the Solar Energy Laboratory of the University of Wisconsin. Essentially all of the information contained in this section is summarized from the report of that survey; see Ref. 7.

The University of Wisconsin report[7] lists about 965 sites around the world having either solar radiation data or sunshine duration data. Because of the sharp increase in interest in application of solar energy, there are now many more stations with such records.

The report described in Reference 7 contains a complete listing of all 965 sites, including site location, length of record, type of data, and instrumentation used. It also lists either the monthly mean daily totals of total radiation on a horizontal surface or the monthly mean daily sunshine duration, or both.

A summary of the number and types of records contained in that report is given in Table 2.7. This table lists the number of stations in each country having records of solar radiation measurements, plus the additional stations which only have sunshine duration records. The table does not list separately all those countries having fewer than five such stations; these are lumped together as the final entry. Of course, many more stations currently exist than are counted in this table.

Records of solar radiation measurements for the United States There are hundreds of locations in the United States with records of solar radiation measurements. Many of these data records are rather short, dating back only to the period of rapidly increasing interest in solar energy utilization of the mid-1970s. However, there are also about 100 U.S. sites which have longer solar radiation measurement records. These records are more valuable because they permit the accurate compilation of long-term average amounts of solar energy available for collection. Solar energy availabilities based upon only a few years data can be misleading because year-to-year variations from average availabilities can be quite large.

The major solar radiation data records for the United States are described in this section. A more complete description of solar radiation data records was published by the University of Alabama in 1976; see Ref. 8.

Solar radiation data records of the National Weather Service The data archiving agency of the National Weather Service (NWS) (formerly the U.S. Weather Bureau) is the National Climatic Center (NCC) of Asheville, NC. All records of the NWS Solar Radiation Network are archived in this one location.

TABLE 2.7   Numbers of Solar Radiation Stations and Additional Sunshine Duration Stations by Country in 1966

| Country | Number of | | |
	Solar radiation stations	Additional sunshine duration stations	Total
Algeria	3	18	21
Angola	5		5
Antarctica	14		14
Argentina	27	15	42
Australia	10	2	12
Austria	18		18
Brazil	1	67	68
Bulgaria	7		7
Canada	25		25
China	1	20	21
Congo	18		18
Czechoslovakia	9		9
Finland	5		5
Formosa	7		7
France	3	40	43
Germany	27		27
Ghana	1	5	6
Hungary	14		14
India	3	40	43
Iran	0	7	7
Italy	47		47
Japan	78		78
Kenya	5		5
Korea	2	7	9
Mauritania	0	5	5
Mexico	5	9	14
New Zealand	6		6
Nigeria	2	13	15
Norway	5	10	15
Poland	12		12
Portugal	10		10
Sudan	6		6
Sweden	12		12
Switzerland	7	1	8
U. of So. Africa	16		16
U.S.S.R.	35		35
United Kingdom	7		7
United States	107		107
Venezuela	18	1	19
Yugoslavia	12		12
All Other Countries, With Less than 5 Stations Each	84	31	115

SOURCE: Ref. 7.

The major solar radiation data sets which are available from the National Climatic Center are summarized in Table 2.8. These data sets are discussed more fully in the following subsections.

The rehabilitated solar data sets for 1950–1975   These are the first two data sets listed in Table 2.8. The data sets are based upon measurements recorded by the Solar Radiation Data Network of the National Weather Service for that time period. This solar data network recorded daily totals of total-horizontal radiation at about 60 sites. Approximately half of these stations used strip-chart recorders which were manually read to give hourly totals.

Because of a general lack of interest in solar radiation measurements during most of this period, the network operation, the data verification, and the instrumentation were to some

TABLE 2.8   A Summary of the Principal Solar Radiation Data Sets Available for the United States from the National Climatic Center, Asheville, NC

Data set name	Period	Number of sites	Description
		Historical Data Sets	
Rehabilitated hourly solar data	1950–1975	26	Hourly values of total-horizontal and direct-normal and other meteorological observations on SOLMET data tapes
Rehabilitated daily solar data	1950–1975	29	Daily totals of total-horizontal radiation, plus some surface meteorological data; printed and on computer tapes
Generated hourly solar data	1950–1975	Approx. 250	Hourly solar data generated with a sunshine and cloud cover model, plus hourly weather data on SOLMET data tapes
Original data records	1950–1975	Approx. 150	Miscellaneous strip chart, circular chart, and manuscript records, plus published data in monthly "Climatological Data National Summary" issues
		Current Data Sets	
Hourly "new network" data tapes	1977– present	38	Hourly measurements of total-horizontal, direct-normal, and weather parameters on SOLMET data tapes.
Monthly solar data publication	1977– present	38	Selected solar data records published by NCC in a special monthly solar radiation report.
Cooperative station data	1977– present	?	Solar data from quality cooperative stations archived by NCC in the above two formats.

extent neglected. Consequently, when interest in solar energy systems accelerated in the mid-1970s, some of this data was rehabilitated in an attempt to remove the various scale-change errors, instrument-drift errors, and calibration errors which were known to exist.

In all, hourly data were rehabilitated for 26 sites, and daily solar radiation data were rehabilitated for an additional 29 sites; the map in Fig. 2.12 displays these 55 locations. In each case, the data were corrected by adjusting all of the recorded measurements so that those measurements which were recorded at noon on clear days would agree with a theoretical model for total-horizontal radiation at noon for that date and location. Data correction factors were selected by interpolation for days in-between these clear solar noon peg points. The theoretical model produced clear, solar-noon values which were dependent upon the date and the location.

The data rehabilitation task also filled the gaps in the solar radiation data records. In some instances these gaps were several years in length. The model used for filling these data gaps was based upon sunshine measurements and/or cloud cover observations. The model is described in the next section of this chapter.

The rehabilitated hourly solar data tapes also have hourly estimates of direct-normal solar radiation. These values are not based upon hourly measurements, since very few direct-normal measurements were recorded during this time period. The procedure used for generating these hourly direct-normal values is given in the next section; it is also described in detail in Ref. 9.

Hourly solar radiation data are available from the National Climatic Center on computer tapes which also have hourly surface weather records. The format is called SOLMET, a standard solar and weather data format adopted by the United States in 1977.

The generated hourly solar data set   The model which was used to fill the gaps in the records of hourly total-horizontal solar radiation as described above was also used to *generate* hourly data for some U.S. sites for which no solar radiation records prior to 1975 exist. Since the model only requires hourly records of percent of possible sunshine and/or opaque cloud

Fig. 2.12 The 55 U.S. sites with rehabilitated solar data for 1950–1975.

Fig. 2.13 Solar radiation network of the U.S. National Weather Service.

cover, records which are available for several hundred locations in the United States, the model made it possible to generate data for numerous additional locations. In all, this generated data is available from the National Climatic Center for about 220 sites in the United States.*

The current U.S. data sets   In 1977 the U.S. National Weather Service established a new 38-station solar data network. The network includes the 34 sites shown for the continental United States in Fig. 2.13, plus one station in each of Alaska, Puerto Rico, Hawaii, and Guam. Every station is equipped with a pyranometer and a pyrheliometer; thus, total-horizontal solar radiation and direct-normal solar radiation are both recorded at 1-min intervals. These data are summed to provide hourly values, and hourly data are available in combination with weather observations on SOLMET computer tapes from the National Climatic Center. This agency also has selected samples of the original 1-min solar data; these are useful as input to simulation models designed to assess the effects of rapid fluctuations in solar intensity on solar systems.

In addition to the SOLMET tapes, certain records of the solar measurements recorded by this new network are available from the National Climatic Center in a monthly publication titled *Monthly Summary Solar Radiation Data*.

*Ref. 20 contains monthly total insolation values for both rehabilitated and generated data sets.

Cooperative station data   Numerous private companies, universities, research laboratories, and individuals installed solar radiation stations in the mid-1970s. The data from some of these independent stations are also archived, disseminated, and reported by the National Climatic Center of Asheville, NC. The primary criteria of NCC for the inclusion of data from these "cooperative stations" in its routine solar data processing and dissemination program are (1) that the data be reliable and (2) that it be useful in the sense that it provide solar radiation availability information for a site not covered by the National Weather Service network. Cooperative station solar data are available from the National Climatic Center in the same SOLMET and monthly report form as the NWS network data.

Research solar radiation stations   In the late 1970s the Division of Solar Energy within the U.S. Department of Energy established several "Research Solar Radiation Data Stations." These stations recorded very careful measurements of a variety of solar radiation parameters. These include direct, diffuse, and total radiation in various spectral wave-bands, circumsolar radiation to several disk diameters around the sun, sky brightness, and the responses of various photovoltaic devices. Simultaneous measurements of surface weather, atmospheric water vapor content, and atmospheric turbidity are recorded.

These research data are particularly useful for various solar radiation correlation and modeling studies. The data are available from the individual stations and from the National Solar Energy Research Institute in Golden, CO.

Other U.S. solar data sources   There are many sources for solar radiation data for the United States in addition to those described above. Some of the major other sources for solar radiation data are:
- Department of Defense, DOD
- Environmental Protection Agency, EPA
- Electric Power Research Institute, EPRI
- Various state and regional networks
- Canada's Solar Radiation Network, described below

Canada's Solar Radiation Network

The Atmospheric Environment Service of Canada operates a solar radiation network consisting of 54 stations. Nearly all of these record hourly totals of total solar radiation on a horizontal surface. At many of the sites hourly values of net (downward minus upward) total radiation and/or diffuse radiation are also recorded. A complete listing of stations, their locations and elevations, and radiation types recorded at each station is given in Table 2.9.

Hourly radiation values for these Canadian stations are published monthly in the *Monthly Radiation Summary*. Information on this publication is available from the Atmospheric Environment Service, 4905 Dufferin Street, Downsview, Ontario, M3H 5T4. The hourly solar radiation values are also available on computer tapes from the Atmospheric Environment Service.

Solar Radiation Availability Maps

This section displays the geographic distributions of several types of solar energy availability on a seasonal basis in the form of isopleth maps. Since the amount of solar radiation available at a location depends strongly on the type of radiation—direct, diffuse, or total—and on the orientation of the surface intercepting the radiation—such as a horizontal surface, a south-facing vertical surface, or a tracking surface normal to the solar rays—these solar energy availability maps are generally only useful for portraying the geographic distributions of the specific radiation types on the specific intercepting surfaces to which the maps refer. Serious errors can be made by directly inferring the availability of one type of solar radiation from a map of another type. An excellent example of this is seen by comparing Figs. 2.29 and 2.21. These show that the total solar radiation available to a south-facing 45° tilted collector in the northern United States is nearly double the total-horizontal solar radiation in the winter. A comparison of Figs. 2.20, 2.21, 2.24, and 2.25 shows that direct-normal solar radiation generally exceeds total-horizontal solar radiation by about 50 percent in the fall and winter.

The next two subsections present solar energy availabilities for the world and the United States, respectively.

Solar radiation availability maps for the world   The worldwide availabilities of total-horizontal solar radiation for the midseason months of January, April, July, and October are given in Figs. 2.14–2.17, respectively. These maps are reprinted from the 1966 report "World

TABLE 2.9   Canada's Solar Radiation Network

Station name	Lat,°	Long,°	Elev., m	Radiation types
Alert, N.W.T.	82.5	62.3	63	Total, net
Bad Lake, SASK.	51.3	108.4	637	Total, net, refl.
Baker Lake, N.W.T.	64.3	96.0	12	Total, net
Beaverlodge, Alta.	55.2	119.4	732	Total
Cambridge Bay, N.W.T.	69.1	105.1	27	Total
Cape St. James, B.C.	51.9	131.0	89	Total
Charlottetown, P.E.I.	46.3	63.1	23	Total, diffuse
Churchill, Man.	58.8	94.1	29	Total, net
Corol Harbor, N.W.T.	64.2	83.4	64	Total
Edmonton Story Plain, Alta.	53.6	114.1	766	Total, diff, net
Elora Research Sta, Ont.	43.7	80.4	376	Total, net
Eureka, N.W.T.	80.0	85.9	10	Total, net
Fort Chimo, Que.	58.1	68.4	37	Total
Fort Nelson, B.C.	58.8	122.6	382	Total
Fort Smith, N.W.T.	60.0	112.0	203	Total
Fredericton, N.B.	45.9	63.6	40	Ttl, net, refl.
Frobisher Bay, N.B.	63.8	68.6	34	Total, net
Goose Bay, NFLD.	53.3	60.4	38	Ttl, diff, net, refl.
Halifax Citadel, N.S.	44.7	63.6	70	Total
Hall Beach, N.W.T.	68.8	81.3	8	Total
Inoucdjowai, Que.	58.5	78.1	5	Total, net
Inwids Bay, N.W.T.	68.3	133.5	103	Total
Isochsen, N.W.T.	78.8	103.5	25	Total, net
Kentville, N.S.	45.1	64.5	49	Total, net
Lethbridge, Alta.	49.6	112.8	929	Net
Montreal, Que.	45.5	73.6	133	Total, diffuse
Moosonee, Ont.	51.3	80.7	10	Total, net
Mould Bay, N.W.T.	76.2	119.3	15	Total, net
Mt. Kobais, B.C.	49.1	119.7	1862	Total
Namarino Departure Bay B.C.	49.2	124.0	8	Total
Nitchequon, Que.	73.2	70.9	536	Total
Normandies, Que.	48.9	72.5	137	Total
Norman Wells, N.W.T.	65.3	126.8	73	Total, net
Ocean Weather St. P.	50.0	145.0	11	Total, net
Ottawa, Ont.	45.5	75.6	98	Total, net
Port Hardy, B.C.	50.7	127.4	22	Total, diffuse
Prince George, B.C.	53.9	122.7	676	Total
Quebec, Que.	46.8	71.4	73	Net
Resolante, N.W.T.	74.7	95.0	67	Ttl.diff.net.refl.
St. John's, NFLD	47.5	52.8	114	Total
Sable Island, N.S.	43.9	60.0	4	Total, net
Sachs Harbour, N.W.T.	72.0	125.3	86	Total
Sandspit, B.C.	53.3	131.8	6	Total
Sept-Isles, Que.	50.2	66.3	53	Total
Suffield, Alta.	50.3	111.2	770	Total
Summerland, B.C.	49.6	119.7	454	Total, net
Swift Current, SASK.	50.3	107.7	825	Total
The Pas, Man.	54.0	101.1	271	Total
Toronto, Ont.	43.7	79.4	107	Total
Toronto Met. Res. Sta	43.8	79.6	194	Ttl.diff,net,refl.
Trout Lake, Ont.	53.8	89.9	220	Total
Vancouver, B.C.	49.3	123.3	87	Total
Whitehorse, Y.T.	60.7	135.1	703	Total, net
Winnipeg, Man.	49.9	97.2	239	Total, diffuse

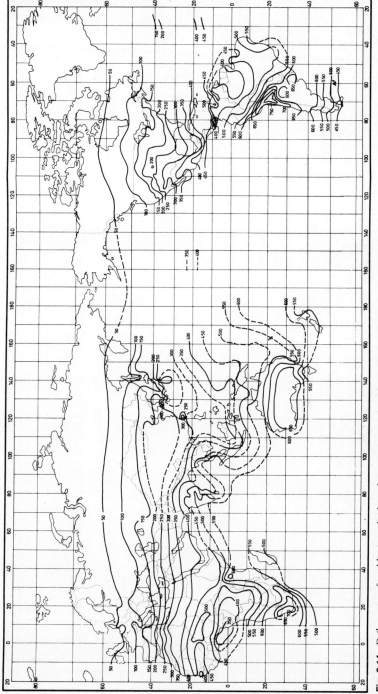

Fig. 2.14 Daily means of total-horizontal solar radiation in January (Langleys). (*From Ref. 7.*)

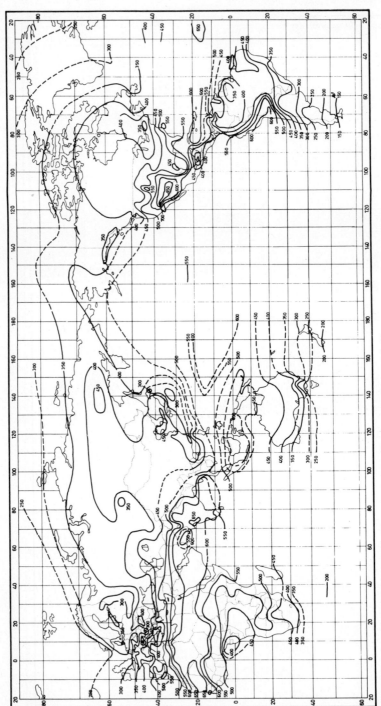

Fig. 2.15 Daily means of total-horizontal solar radiation in April (Langleys). *(From Ref. 7.)*

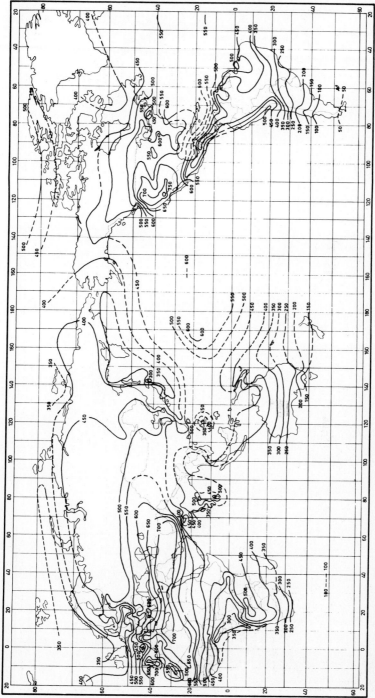

Fig. 2.16 Daily means of total-horizontal solar radiation in July (Langleys). (*From Ref. 7.*)

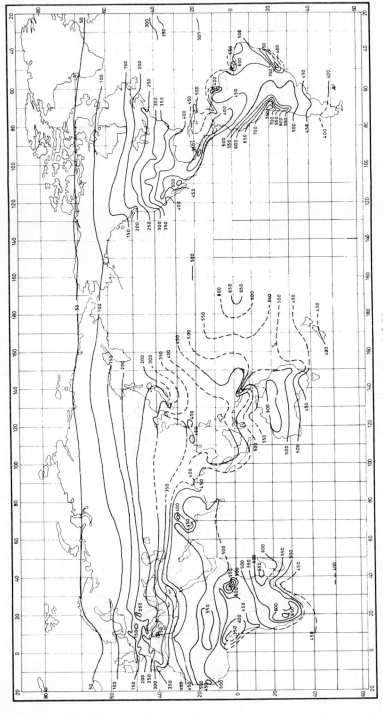

Fig. 2.17 Daily means of total-horizontal solar radiation in October (Langleys). *(From Ref. 7.)*

Distribution of Solar Radiation"; see Ref. 7. (The units are Langleys per day; 1 Langley = 0.0116 kWh/m² = 3.687 Btu/ft².)

The solar energy availability maps for the world shown in Figs. 2.14 to 2.17 refer to total solar radiation on a horizontal surface. The worldwide distributions of direct-normal radiation, or of total radiation on various tilted surfaces, are likely to be significantly different from these. Unfortunately, maps displaying the worldwide distributions of direct-normal radiation or of total radiation on tilted surfaces do not exist; only limited numbers of such maps exist for certain regions of the world.

Solar radiation availability maps for the United States   The maps presented here display the geographic distributions of:

   Total-horizontal radiation, Figs. 2.18–2.21
   Direct-normal radiation, Figs. 2.22–2.25
   Total radiation on a 45° south-facing surface, Figs. 2.26–2.29

All of these distributions are presented on a seasonal basis. The "seasons" used for these maps correspond to the "thermal seasons" for the Northern Hemisphere:

   Spring—March, April, May
   Summer—June, July, August
   Fall—September, October, November
   Winter—December, January, February

All of the maps of solar radiation availability distributions for the United States presented in this section were derived from hourly total-horizontal solar radiation data for 26 sites for the five-year period 1958–1962. The total-horizontal data were converted hour-by-hour to availabilities of direct and diffuse radiation to various other surface orientations. Because data for only 26 sites were available, these maps generally ignore local variations in the availability distributions (maps are revisions of Ref. 10 maps). All of these maps display mean daily amounts in kilowatt hours per square meter.

Maps of total-horizontal radiation   Figures 2.18–2.21 display the seasonal distributions of total-horizontal radiation for the United States. These availabilities represent the amounts of solar energy which are available to horizontal flat-plate collectors, to solar ponds, and to plant life such as agricultural crops. Other maps of the availability of total-horizontal radiation have been drawn[11]; they are in good agreement with those given here.

Maps of direct-normal radiation   Figures 2.22–2.25 display the seasonal availabilities of direct-normal solar radiation for the United States. These availabilities represent the amounts of solar energy which are available to full-tracking, focusing collectors, such as point-focusing parabolic-dish reflector collectors or two-dimensional Fresnel-lens collectors.

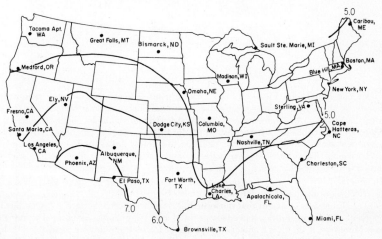

Fig. 2.18   Average daily availabilities of total-horizontal solar radiation for the United States in the spring (kWh/m²).

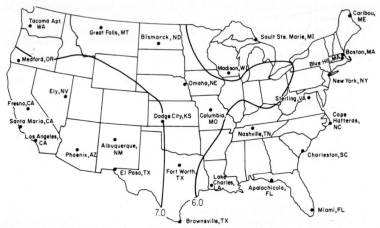

Fig. 2.19 Average daily availabilities of total-horizontal solar radiation for the United States in the summer (kWh/m²).

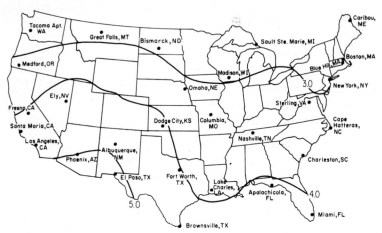

Fig. 2.20 Average daily availabilities of total-horizontal solar radiation for the United States in the fall (kWh/m²).

When contrasted with Figs. 2.19–2.22, these maps indicate that in every season there is generally as much direct-normal solar radiation as total-horizontal solar radiation. In the winter season the direct-normal values exceed the total-horizontal values by about 50 percent.

Maps of total radiation on south-facing, 45° tilted surfaces Figures 2.26–2.29 display the seasonal availabilities of total solar radiation to surfaces facing south and tilted upward 45° from horizontal. These maps represent the amounts of solar energy available for collection by a flat-plate solar collector which is in this orientation.

In general, a 45° tilt is not the optimum tilt for either maximizing solar radiation availability to a collector during the winter heating months or for maximizing collection on an annual basis. However, Table 2.35 indicates that the availabilities on a 45° tilted surface are generally within 10 percent of the availabilities to a collector tilted at the approximately optimal winter-heating angle. Moreover, a 45° tilt is a good tilt-angle choice for such applications as solar water-heating where a reasonably uniform collection rate over the year is desired. The optimal tilt for maximizing annual collection is approximately L-15; such a tilt angle takes advantage of the longer summer days, but it sharply reduces solar energy collection in the winter months.

Fig. 2.21 Average daily availabilities of total-horizontal solar radiation for the United States in the winter (kWh/m²).

Fig. 2.22 Average daily availabilities of direct-normal solar radiation for the United States in the spring (kWh/m²).

Solar Radiation Availability Tables

This section presents a variety of tables of solar energy availability. These tables are useful for determining average amounts of solar radiation which will be available to many different types of solar collectors. Several of the tables also provide direct comparisons of the relative amounts of solar radiation available to different types of collectors; these tables are useful in making such collector design choices as concentrator vs flat-plate, or in selecting the appropriate tracking scheme, or the best tilt angle.

All of the tables in this section present monthly averaged, daily availabilities of solar radiation; availabilities on clear days will be larger. The next section of this chapter provides a simple method for calculating solar availabilities for clear sky conditions. Tables of clear day solar availabilities are also available in the literature; see, for example, Ref. 12.

The first table of this section contains solar energy and clearness information by month for 80 locations in the United States and Canada. The other tables refer only to U.S. locations. The only tabulation of solar radiation availability on a worldwide basis is the lengthy table referred to in Ref. 7. However, numerous national, regional, and continental solar radiation tables are available through national energy agencies and in publications such as *Solar Energy*, the journal of the

Fig. 2.23 Average daily availabilities of direct-normal solar radiation for the United States in the summer (kWh/m²).

Fig. 2.24 Average daily availabilities of direct-normal solar radiation for the United States in the fall (kWh/m²).

International Solar Energy Society. The numerous tables for the United States given here can also be used for many other locations around the world since locations of similar latitude, geography, and weather have similar solar radiation availability characteristics.

Total-horizontal and percent of possible solar radiation for 80 locations in the United States and Canada. Table 2.10 lists the average daily amount of total-horizontal solar radiation by month for 80 different sites in the United States and Canada. This table is reprinted from Ref. 13. In this table $\overline{H}$ represents the monthly average daily totals of total-horizontal radiation.

This table also gives monthly-averaged values of $\overline{K}_T$; $\overline{K}_T$ is the monthly average "percent of possible" solar radiation, or "monthly clearness index." It is defined by

$$\overline{K}_T = \overline{H}/\overline{H}_0 \tag{2.20}$$

where $\overline{H}_0$ is the average daily total amount of extraterrestrial radiation incident upon a horizontal surface during that month. Thus, $\overline{K}_T$ represents the average fraction of radiation which is

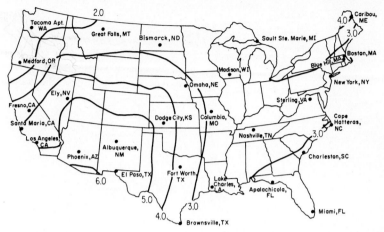

Fig. 2.25 Average daily availabilities of direct-normal solar radiation for the United States in the winter (kWh/m²).

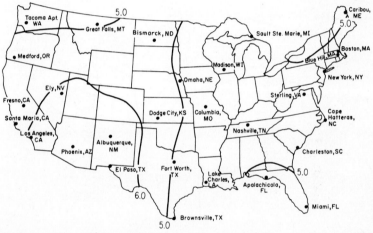

Fig. 2.26 Average daily availabilities of total solar radiation on a south-facing 45° tilted surface for the United States in the spring (kWh/m²).

transmitted either directly or indirectly through the atmosphere for the given month and location. As such, $\overline{K}_T$ is a good parameter for comparing the relative atmospheric clarity for different seasons and at different locations. Note that H_0 can easily be calculated for any date by integrating the expression:

$$I = I_{ext} (cos\,L\,cos\,\delta\,cos\,h_s + sin\,L\,sin\,\delta) \tag{2.21}$$

between the sunrise and sunset hours. The result is

$$H_0 = 2I_{ext}\left[\frac{h_{sr}}{15}\,sin\,L\,\,sin\,\delta + \frac{12}{\pi}\,cos\,L\,\,cos\,\delta\,\,sin\,h_{sr}\right] \tag{2.22}$$

where h_{sr} is the sunrise hour angle, δ is the declination for that day, L is the local latitude, and I_{ext} is the intensity of extraterrestial solar radiation (see Table 2.4).

Fig. 2.27 Average daily availabilities of total solar radiation on a south-facing 45° tilted surface for the United States in the summer (kWh/m²).

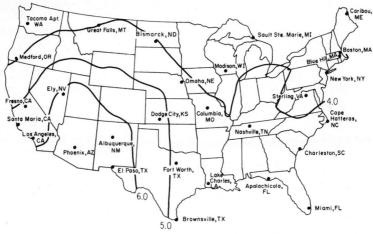

Fig. 2.28 Average daily availabilities of total solar radiation on a south-facing 45° tilted surface for the United States in the fall (kWh/m²).

Table 2.10 contains several other useful pieces of information for each of the 80 sites. These are the site's latitude and elevation above sea level, and the average *daytime* temperature by month. These average daytime temperatures can be used for approximating the average thermal losses from solar collectors operating during the day.

The units used in Table 2.10 are those employed in its original presentation.[13] The energy units, Btu/ft² · day, are converted to kWh/m² per day by multiplying by 0.00315. Temperatures are in degrees Fahrenheit, and elevations are in feet.

Average solar radiation availabilities to different collector types for 26 U.S. locations   This section presents average daily values of both direct and total solar radiation availabilities to a variety of tracking and fixed surfaces for 26 U.S. locations on a seasonal basis. All of these tables are derived from the tables in Ref. 10; however, the values have been modified so as to be in agreement with the rehabilitated hourly solar data sets described earlier. These tables were constructed by converting measured hourly values of total-horizontal radiation to hourly values of direct and total radiation incident on other surfaces; the conversion techniques employed are described in detail in the report.[10]

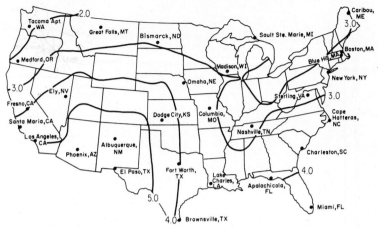

Fig. 2.29 Average daily availabilities of total solar radiation on a south-facing 45° tilted surface for the United States in the winter (kWh/m²).

Solar radiation tables for tracking surfaces   Tables 2.11 through 2.17 list the mean daily availabilities by season of direct and total solar radiation to seven different tracking schemes for 26 U.S. locations. In each table the column labeled D (T) gives the direct (total) radiation availability. Focusing, concentrating collectors would only be able to collect the direct radiation, while flat-plate collectors in these tracking configurations would intercept the amount of total radiation listed.

The tables and their corresponding tracking schemes are as follows:

- Table 2.11—N, a full-tracking surface
- Table 2.12—EW, a surface rotating about a horizontal east-west axis
- Table 2.13—NSH, a surface rotating about a horizontal north-south axis
- Table 2.14—P, a surface rotating about a polar axis
- Table 2.15—VL, a surface tilted $L°$ from horizontal and rotating about a vertical axis
- Table 2.16—VLP, a surface tilted $L + 15$ from horizontal and rotating about a vertical axis
- Table 2.17—VLN, a surface tilted $L - 15°$ from horizontal and rotating about a vertical axis

Note that the solar availabilities listed in these tables refer to incident energy only; they represent the amount of energy available for collection. Since the numbers do not take into account collector optical or thermal loss mechanisms, they do not represent the amounts of energy captured by collectors.

Note also that these availabilities are long-term seasonal averages of daily totals. The average daily values in a particular season could easily differ with these by 10 or 20 percent. Day-to-day variations will be even greater. The total amount of radiation on a clear day could be 30 percent greater than these average values; on a cloudy day the amount could be as low as 20 percent of these tabulated values.

Solar radiation tables for fixed surfaces   Tables 2.18 through 2.32 list the mean daily availabilities by season of direct and total solar radiation to 15 different orientations of fixed surfaces for 26 U.S. locations. Tables 2.18–2.24 list availabilities to south-facing surfaces with tilt angles of 0°, 15°, 30°, 45°, 60°, 75°, and 90°, respectively. Tables 2.25–2.30 list availabilities for 45° tilted surfaces with azimuth angles of ± 15°, ± 30°, ± 45°, ± 60°, ± 75°, and ± 90°, respectively. (Note that a surface with an azimuth angle of a_w generally intercepts about the same amount of solar radiation on the average as does the same surface rotated so that its azimuth is $-a_w$; see Table 2.38.) Finally, Tables 2.31 and 2.32 list average solar availabilities to vertical walls facing east or west $(a_w = ± 90°)$ and southeast or southwest $(a_w = ± 45°)$, respectively.

The values in these tables are long-term average seasonal values. Because of the variability of the weather, values for particular days, and also mean values for particular seasons, can be expected to vary significantly from these tabulated availabilities. Daily values as low as 20

TABLE 2.10 Monthly Means of Total-Horizontal Radiation $\overline{H}$ (Btu/ft² day), Percent Of Possible Radiation K_T, and Daytime Ambient Temperature T_0 (°F)

		Jan	Feb	Mar	Apr	May	Jun	Jul	Aug	Sep	Oct	Nov	Dec
Albuquerque, NM	$\overline{H}$	1150.9	1453.9	1925.4	2343.5	2560.9	2757.5	2561.2	2387.8	2120.3	1639.8	1274.2	1051.6
Lat. 35°03'N	$\overline{K}_T$	0.704	0.691	0.719	0.722	0.713	0.737	0.695	0.708	0.728	0.711	0.684	0.704
El. 5314 ft	T_0	37.3	43.3	50.1	59.6	69.4	79.1	82.8	80.6	73.6	62.1	47.8	39.4
Annette Is., AL	$\overline{H}$	236.2	428.4	883.4	1357.2	1634.7	1638.7	1632.1	1000.4	962	454.6	220.3	152
Lat. 55°02'N	$\overline{K}_T$	0.427	0.415	0.492	0.507	0.484	0.441	0.454	0.427	0.449	0.347	0.304	0.361
El. 110 ft	T_0	35.8	37.5	39.7	44.4	51.0	56.2	58.6	59.8	54.8	48.2	41.9	37.4
Apalachicola, FL	$\overline{H}$	1107	1378.2	1654.2	2040.9	2268.6	2195.9	1978.6	1912.9	1703.3	1544.6	1243.2	982.3
Lat. 29°45'N	$\overline{K}_T$	0.577	0.584	0.576	0.612	0.630	0.594	0.542	0.558	0.559	0.608	0.574	0.543
El. 35 ft	T_0	57.3	59.0	62.9	69.5	76.4	81.8	83.1	83.1	80.6	73.2	63.7	58.55
Astoria, OR	$\overline{H}$	338.4	607	1008.5	1401.5	1838.7	1753.5	2007.7	1721	1322.5	780.4	413.6	295.2
Lat. 46°12'N	$\overline{K}_T$	0.330	0.397	0.454	0.471	0.524	0.466	0.551	0.538	0.526	0.435	0.336	0.332
El. 8 ft	T_0	41.3	44.7	46.9	51.3	55.0	59.3	62.6	63.6	62.2	55.7	48.5	43.9
Atlanta, GA	$\overline{H}$	848	1080.1	1426.9	1807	2618.12	2002.6	2002.9	1898.1	1519.2	1290.8	997.8	751.6
Lat. 33°39'N	$\overline{K}_T$	0.493	0.496	0.522	0.551	0.561	0.564	0.545	0.559	0.515	0.543	0.510	0.474
El. 976 ft	T_0	47.2	49.6	55.9	65.0	73.2	80.9	82.4	81.6	77.4	66.5	54.8	47.7
Barrow, AL	$\overline{H}$	13.3	143.2	713.3	1491.5	1883	2055.3	1602.2	953.5	428.4	152.4	22.9	—
Lat. 71°20'N	$\overline{K}_T$	—	0.776	0.773	0.726	0.553	0.533	0.448	0.377	0.315	0.35	—	—
El. 22 ft	T_0	-13.2	-15.9	-12.7	2.1	20.5	35.4	41.6	40.0	31.7	18.6	2.6	-8.6
Bethel, AL	$\overline{H}$	142.4	404.8	1052.4	1662.3	1711.8	1698.1	1401.8	938.7	755	430.6	164.9	83
Lat. 60°47'N	$\overline{K}_T$	0.536	0.557	0.704	0.675	0.519	0.458	0.398	0.336	0.406	0.432	0.399	0.459
El. 125 ft	T_0	9.2	11.6	14.2	29.4	42.7	55.5	56.9	54.8	47.4	33.7	19.0	9.4
Bismarck, ND	$\overline{H}$	587.4	934.3	1328.4	1668.2	2056.1	2173.8	2305.5	1929.1	1441.3	1018.1	600.4	464.2
Lat. 46°47'N	$\overline{K}_T$	0.594	0.628	0.605	0.565	0.588	0.579	0.634	0.606	0.581	0.584	0.510	0.547
El. 1660 ft	T_0	12.4	15.9	29.7	46.6	58.6	67.9	76.1	73.5	61.6	49.6	31.4	18.4
Blue Hill, MA	$\overline{H}$	555.3	797	1143.9	1438	1776.4	1943.9	1881.5	1622.1	1314	941	592.2	482.3
Lat. 42°13'N	$\overline{K}_T$	0.445	0.458	0.477	0.464	0.501	0.516	0.513	0.495	0.492	0.472	0.406	0.436
El. 629 ft	T_0	28.3	28.3	36.9	46.9	58.5	67.2	72.3	70.6	64.2	54.1	43.3	31.5
Boise, ID	$\overline{H}$	518.8	884.9	1280.4	1814.4	2189.3	2376.7	2500.3	2149.4	1717.7	1128.4	678.6	456.8
Lat. 43°34'N	$\overline{K}_T$	0.446	0.533	0.548	0.594	0.619	0.631	0.684	0.660	0.656	0.588	0.494	0.442
El. 2844 ft	T_0	29.5	36.5	45.0	53.5	62.1	69.3	79.6	77.2	66.7	56.3	42.3	33.1

Location		1	2	3	4	5	6	7	8	9	10	11	12
Boston, MA Lat. 42°22'N El. 29 ft	$\overline{H}$	505.5	738	1067.1	1355	1769	1864	1860.5	1570.1	1267.5	896.7	535.8	442.8
	$\overline{K}_T$	0.410	0.426	0.445	0.438	0.499	0.495	0.507	0.480	0.477	0.453	0.372	0.400
	T_0	31.4	31.4	39.9	49.5	60.4	69.8	74.5	73.8	66.8	57.4	46.6	34.9
Brownsville, TX Lat. 25°55'N El. 20 ft	$\overline{H}$	1105.9	1262.7	1505.9	1714	2092.2	2288.5	2345	2124	1774.9	1536.5	1104.8	982.3
	$\overline{K}_T$	0.517	0.500	0.505	0.509	0.584	0.627	0.650	0.617	0.566	0.570	0.468	0.488
	T_0	63.3	66.7	70.7	76.2	81.4	85.1	86.5	86.9	84.1	78.9	70.7	65.2
Caribou, ME Lat. 46°52'N El. 628 ft	$\overline{H}$	497	861.6	1360.1	1495.9	1779.7	1779.7	1898.1	1675.6	1254.6	793	415.5	398.9
	$\overline{K}_T$	0.504	0.579	0.619	0.507	0.509	0.473	0.522	0.527	0.506	0.455	0.352	0.470
	T_0	11.5	12.8	24.4	37.3	51.8	61.6	67.2	65.0	56.2	44.7	31.3	16.8
Charleston, SC Lat. 32°54'N El. 46 ft	$\overline{H}$	946.1	1152.8	1352.4	1918.8	2063.4	2113.3	1649.4	1933.6	1557.2	1332.1	1073.8	952
	$\overline{K}_T$	0.541	0.521	0.491	0.584	0.574	0.567	0.454	0.569	0.525	0.554	0.539	0.586
	T_0	53.6	55.2	60.6	67.8	74.8	80.9	82.9	82.3	79.1	69.8	59.8	54.0
Cleveland, OH Lat. 41°24'N El. 805 ft	$\overline{H}$	466.8	681.9	1207	1443.9	1928.4	2102.6	2094.4	1840.6	1410.3	997	526.6	427.3
	$\overline{K}_T$	0.361	0.383	0.497	0.464	0.543	0.559	0.571	0.559	0.524	0.491	0.351	0.371
	T_0	30.8	30.9	39.4	50.2	62.4	72.7	77.0	75.1	68.5	57.4	44.0	32.8
Columbia, MO Lat. 38°58'N El. 785 ft	$\overline{H}$	651.3	941.3	1315.8	1631.3	1999.6	2129.1	2148.7	1953.1	1689.6	1202.6	839.5	590.4
	$\overline{K}_T$	0.458	0.492	0.520	0.514	0.559	0.566	0.585	0.588	0.606	0.562	0.510	0.457
	T_0	32.5	36.5	45.9	57.7	66.7	75.9	81.1	79.4	71.9	61.4	46.1	35.8
Columbus, OH Lat. 40°00'N El. 833 ft	$\overline{H}$	486.3	746.5	1112.5	1480.8	1839.1	(2111)	2041.3	1572.7	1189.3	919.5	479	430.2
	$\overline{K}_T$	0.356	0.401	0.447	0.470	0.515	(0.561)	0.555	0.475	0.433	0.441	0.302	0.351
	T_0	32.1	33.7	42.7	53.5	64.4	74.2	78	75.9	70.1	58	44.5	34.0
Davis, CA Lat. 38°33'N El. 51 ft	$\overline{H}$	599.2	945	1504	1959	2368.6	2619.2	2565.6	2287.8	1856.8	1288.5	795.6	550.5
	$\overline{K}_T$	0.416	0.490	0.591	0.617	0.662	0.697	0.697	0.687	0.664	0.598	0.477	0.421
	T_0	47.6	52.1	56.8	63.1	69.6	75.7	81	79.4	76.7	67.8	57	48.7
Dodge City, KS Lat. 37°46'N El. 2592 ft	$\overline{H}$	953.1	1186.3	1565.7	1975.6	2126.5	2459.8	2400.7	2210.7	1841.7	1421	1065.3	873.8
	$\overline{K}_T$	0.639	0.598	0.606	0.618	0.594	0.655	0.652	0.663	0.654	0.650	0.625	0.652
	T_0	33.8	38.7	46.5	57.7	66.7	77.2	83.8	82.4	73.7	61.7	46.5	36.8
East Lansing, MI Lat. 42°44'N El. 856 ft	$\overline{H}$	425.8	739.1	1086	1249.8	1732.8	1914	1884.5	1627.7	1303.3	891.5	473.1	379.7
	$\overline{K}_T$	0.35	0.431	0.456	0.406	0.489	0.508	0.514	0.498	0.493	0.456	0.333	0.349
	T_0	26.0	26.4	35.7	48.4	59.8	70.3	74.5	72.4	65.0	53.5	40.0	29.0
East Wareham, MA Lat. 41°46'N El. 18 ft	$\overline{H}$	504.4	762.4	1132.1	1392.6	1704.8	1958.3	1873.8	1607.4	1363.8	996.7	636.2	521
	$\overline{K}_T$	0.398	0.431	0.469	0.449	0.480	0.520	0.511	0.489	0.508	0.496	0.431	0.461
	T_0	32.2	31.6	39.0	48.3	58.9	67.5	74.1	72.8	65.9	56	46	34.8

TABLE 2.10 Monthly Means of Total-Horizontal Radiation H (Btu/ft² day), Percent of Possible Radiation $\overline{K}_T$, and Daytime Ambient Temperature T_o (°F) (Continued)

		Jan	Feb	Mar	Apr	May	Jun	Jul	Aug	Sep	Oct	Nov	Dec
Edmonton, Alberta	$\overline{H}$	331.7	652.4	1165.3	1541.7	1900.4	1914.4	1964.9	1528	1113.3	704.4	413.6	245
Lat. 53°35'N	$\overline{K}_T$	0.529	0.585	0.624	0.564	0.558	0.514	0.549	0.506	0.506	0.504	0.510	0.492
El. 2219 ft	T_o	10.4	14	26.3	42.9	55.4	61.3	66.6	63.2	54.2	44.1	26.7	14.0
El Paso, TX	$\overline{H}$	1247.6	1612.9	2048.7	2447.2	2673	2731	2391.1	2350.5	2077.5	1704.8	1324.7	1051.6
Lat. 31°48'N	$\overline{K}_T$	0.686	0.714	0.730	0.741	0.743	0.733	0.652	0.669	0.693	0.695	0.647	0.626
El. 3916 ft	T_o	47.1	53.1	58.7	67.3	75.7	84.2	84.9	83.4	78.5	69.0	56.0	48.5
Ely, NV	$\overline{H}$	871.6	1255	1749.8	2103.3	2322.1	2649	2417	2307.7	1935	1473	1078.6	814.8
Lat. 39°17'N	$\overline{K}_T$	0.618	0.660	0.692	0.664	0.649	0.704	0.656	0.695	0.696	0.691	0.658	0.64
El. 6262 ft	T_o	27.3	32.1	39.5	48.3	57.0	65.4	74.5	72.3	63.7	52.1	39.9	31.1
Fairbanks, AL	$\overline{H}$	66	283.4	860.5	1481.2	1806.2	1970.8	1702.9	1247.6	699.6	323.6	104.1	20.3
Lat. 64°49'N	$\overline{K}_T$	0.639	0.556	0.674	0.647	0.546	0.529	0.485	0.463	0.419	0.416	0.47	0.458
El. 436 ft	T_o	-7.0	0.3	13.0	32.2	50.5	62.4	63.8	58.3	47.1	29.6	5.5	-6.6
Fort Worth, TX	$\overline{H}$	936.2	1198.5	1597.8	1829.1	2105.1	2437.6	2293.3	2216.6	1880.8	1476	1147.6	913.6
Lat. 32°50'N	$\overline{K}_T$	0.530	0.541	0.577	0.556	0.585	0.654	0.624	0.653	0.634	0.612	0.576	0.563
El. 544 ft	T_o	48.1	52.3	59.8	68.8	75.9	84.0	87.7	88.6	81.3	71.5	58.8	50.8
Fresno, CA	$\overline{H}$	712.9	1116.6	1652.8	2049.4	2409.2	2641.7	2512.2	2300.7	1897.8	1415.5	906.6	616.6
Lat. 36°46'N	$\overline{K}_T$	0.462	0.551	0.632	0.638	0.672	0.703	0.682	0.686	0.665	0.635	0.512	0.44
El. 331 ft	T_o	47.3	53.9	59.1	65.6	73.5	80.7	87.5	84.9	78.6	68.7	57.3	48.9
Gainesville, FL	$\overline{H}$	1036.9	1324.7	1635	1956.4	1934.7	1960.9	1895.6	1873.8	1615.1	1312.2	1169.7	919.5
Lat. 29°39'N	$\overline{K}_T$	0.535	0.56	0.568	0.587	0.538	0.531	0.519	0.547	0.529	0.515	0.537	0.508
El. 165 ft	T_o	62.1	63.1	67.5	72.8	79.4	83.4	83.8	84.1	82	75.7	67.2	62.4
Glasgow, MT	$\overline{H}$	572.7	965.7	1437.6	1741.3	2127.3	2261.6	2414.7	1984.5	1531	997	574.9	428.4
Lat. 48°13'N	$\overline{K}_T$	0.621	0.678	0.672	0.597	0.611	0.602	0.666	0.630	0.629	0.593	0.516	0.548
El. 2277 ft	T_o	13.3	17.3	31.1	47.8	59.3	67.3	76	73.2	61.2	49.2	31.0	18.6
Grand Junction, CO	$\overline{H}$	848	1210.7	1622.9	2002.2	2300.3	2645.4	2517.7	2157.2	1957.5	1394.8	969.7	793.4
Lat. 39°07'N	$\overline{K}_T$	0.597	0.633	0.643	0.632	0.643	0.704	0.690	0.65	0.705	0.654	0.59	0.621
El. 4849 ft	T_o	26.9	35.0	44.6	55.8	66.3	75.7	82.5	79.6	71.4	58.3	42.0	31.4

Location												
Grand Lake, CO Lat. 40°15′N El. 8389 ft $\bar{H}$	735	1135.4	1579.3	1876.7	1974.9	2369.7	2103.3	1708.5	1715.8	1212.2	775.6	660.5
$\bar{K}_T$	0.541	0.615	0.637	0.597	0.553	0.63	0.572	0.516	0.626	0.583	0.494	0.542
T_o	18.5	23.1	28.5	39.1	48.7	56.6	62.8	61.5	55.5	45.2	30.3	22.6
Great Falls, MT Lat. 47°29′N El. 3664 ft $\bar{H}$	524	869.4	1369.7	1621.4	1970.8	2179.3	2383	1986.3	1536.5	984.9	575.3	420.7
$\bar{K}_T$	0.552	0.596	0.631	0.551	0.565	0.580	0.656	0.627	0.626	0.574	0.503	0.518
T_o	25.4	27.6	35.6	47.7	57.5	64.3	73.8	71.3	60.6	51.4	38.0	29.1
Greensboro, NC Lat. 36°05′N El. 891 ft $\bar{H}$	743.9	1031.7	1323.2	1755.3	1988.5	2111.4	2033.9	1810.3	1517.3	1202.6	908.1	690.8
$\bar{K}_T$	0.469	0.499	0.499	0.543	0.554	0.563	0.552	0.538	0.527	0.531	0.501	0.479
T_o	42.0	44.2	51.7	60.8	69.9	78.0	80.2	78.9	73.9	62.7	51.5	43.2
Griffin, GA Lat. 33°15′N El. 980 ft $\bar{H}$	889.6	1135.8	1450.9	1923.6	2163.1	2176	2064.9	1961.2	1605.9	1352.4	1073.8	781.5
$\bar{K}_T$	0.513	0.517	0.528	0.586	0.601	0.583	0.562	0.578	0.543	0.565	0.545	0.487
T_o	48.9	51.0	59.1	66.7	74.6	81.2	83.0	82.2	78.4	68	57.3	49.4
Hatteras, NC Lat. 35°13′N El. 7 ft $\bar{H}$	891.9	1184.1	1590.4	2128	2376.4	2438	2334.3	2085.6	1758.3	1337.6	1053.5	798.1
$\bar{K}_T$	0.546	0.563	0.583	0.655	0.661	0.652	0.634	0.619	0.605	0.58	0.566	0.535
T_o	49.9	49.5	54.7	61.5	69.9	77.2	80.0	79.8	76.7	67.9	59.1	51.3
Indianapolis, IN Lat. 39°44′N El. 793 ft $\bar{H}$	526.2	797.4	1184.1	1481.2	1828	2042	2039.5	1832.1	1513.3	1094.4	662.4	491.1
$\bar{K}_T$	0.380	0.424	0.472	0.47	0.511	0.543	0.554	0.552	0.549	0.520	0.413	0.391
T_o	31.3	33.9	43.0	54.1	64.9	74.8	79.6	77.4	70.6	59.3	44.2	33.4
Inyokern, CA Lat. 35°39′N El. 2440 ft $\bar{H}$	1148.7	1554.2	2136.9	2594.8	2925.4	3108.8	2908.8	2759.4	2409.2	1819.2	3170.1	1094.4
$\bar{K}_T$	0.716	0.745	0.803	0.8	0.815	0.830	0.790	0.820	0.834	0.795	0.743	0.742
T_o	47.3	53.9	59.1	65.6	73.5	80.7	87.5	84.9	78.6	68.7	57.3	48.9
Ithaca, NY Lat. 42°27′N El. 950 ft $\bar{H}$	434.3	755	1074.9	1322.9	1779.3	2025.8	2031.3	1736.9	1320.3	918.4	466.4	370.8
$\bar{K}_T$	0.351	0.435	0.45	0.428	0.502	0.538	0.554	0.530	0.497	0.465	0.324	0.337
T_o	27.2	26.5	36	48.4	59.6	68.9	73.9	71.9	64.2	53.6	41.5	29.6
Lake Charles, LA Lat. 30°13′N El. 12 ft $\bar{H}$	899.2	1145.7	1487.4	1801.8	2080.4	2213.3	1968.6	1910.3	1678.2	1505.5	1122.1	875.6
$\bar{K}_T$	0.473	0.492	0.521	0.542	0.578	0.597	0.538	0.558	0.553	0.597	0.524	0.494
T_o	55.3	58.7	63.5	70.9	77.4	83.4	84.8	85.0	81.5	73.8	62.6	56.9
Lander, WY Lat. 42°48′N El. 5370 ft $\bar{H}$	786.3	1146.1	1638	1988.5	2114	2492.2	2438.4	2120.6	1712.9	1301.8	837.3	694.8
$\bar{K}_T$	0.65	0.672	0.691	0.647	0.597	0.662	0.665	0.649	0.647	0.666	0.589	0.643
T_o	20.2	26.3	34.7	45.5	56.0	65.4	74.6	72.5	61.4	48.3	33.4	23.8
Las Vegas, NV Lat. 36°05′N El. 2162 ft $\bar{H}$	1035.8	1438	1926.5	2322.8	2629.5	2799.2	2524	2342	2062	1602.6	1190	964.2
$\bar{K}_T$	0.654	0.697	0.728	0.719	0.732	0.746	0.685	0.697	0.716	0.704	0.657	0.668
T_o	47.5	53.9	60.3	69.5	78.3	88.2	95.0	92.9	85.4	71.7	57.8	50.2

TABLE 2.10 Monthly Means of Total-Horizontal Radiation H (Btu/ft² day), Percent of Possible Radiation $\bar{K}_r$, and Daytime Ambient Temperature T_o (°F) (Continued)

		Jan	Feb	Mar	Apr	May	Jun	Jul	Aug	Sep	Oct	Nov	Dec
Lemont, IL	$\bar{H}$	(590)	879	1255.7	1481.5	1866	2041.7	1990.8	1836.9	1469.4	1015.5	(639)	(531)
Lat. 41°40′N	$\bar{K}_r$	(0.464)	0.496	0.520	0.477	0.525	0.542	0.542	0.559	0.547	0.506	(0.433)	(0.467)
El. 595 ft	T_o	28.9	30.3	39.5	49.7	59.2	70.8	75.6	74.3	67.2	57.6	43.0	30.6
Lexington, KY	$\bar{H}$	—	—	—	1834.7	2171.2	—	2246.5	2064.9	1775.6	1315.8	—	681.5
Lat. 38°02′N	$\bar{K}_r$	—	—	—	0.575	0.606	—	0.610	0.619	0.631	0.604	—	0.513
El. 979 ft	T_o	36.5	38.8	47.4	57.8	67.5	76.2	79.8	78.2	72.8	61.2	47.6	38.5
Lincoln, NE	$\bar{H}$	712.5	955.7	1299.6	1587.8	1856.1	2040.6	2011.4	1902.6	1543.5	1215.8	773.4	643.2
Lat. 40°51′N	$\bar{K}_r$	0.542	0.528	0.532	0.507	0.522	0.542	0.547	0.577	0.568	0.596	0.508	0.545
El. 1189 ft	T_o	27.8	32.1	42.4	55.8	65.8	76.0	82.6	80.2	71.5	59.9	43.2	31.8
Little Rock, AR	$\bar{H}$	704.4	974.2	1335.8	1669.4	1960.1	2091.5	2081.2	1938.7	1640.6	1282.6	913.6	701.1
Lat. 34°44′N	$\bar{K}_r$	0.424	0.458	0.496	0.513	0.545	0.559	0.566	0.574	0.561	0.552	0.484	0.463
El. 265 ft	T_o	44.6	48.5	56.0	65.8	73.1	76.7	85.1	84.6	78.3	67.9	54.7	46.7
Los Angeles, CA (WBAS)	$\bar{H}$	930.6	1284.1	1729.5	1948	2196.7	2272.3	2413.6	2155.3	1899.1	1372.7	1082.3	901.1
Lat. 33°56′N	$\bar{K}_r$	0.547	0.596	0.635	0.595	0.610	0.608	0.657	0.635	0.641	0.574	0.551	0.566
El. 99 ft	T_o	56.2	56.9	59.2	61.4	64.2	66.7	69.6	70.2	69.1	66.1	62.6	58.7
Los Angeles, CA (WBO)	$\bar{H}$	911.8	1223.6	1640.9	1866.8	2061.2	2259	2428.4	2198.9	1891.5	1362.3	1053.1	877.8
Lat. 34°03′N	$\bar{K}_r$	0.538	0.568	0.602	0.571	0.573	0.605	0.66	0.648	0.643	0.578	0.548	0.566
El. 99 ft	T_o	57.9	59.2	61.8	64.3	67.6	70.7	75.8	76.1	74.2	69.6	65.4	60.2
Madison, WI	$\bar{H}$	564.6	812.2	1232.1	1455.3	1745.4	2031.7	2046.5	1740.2	1443.9	993	555.7	495.9
Lat. 43°08′N	$\bar{K}_r$	0.49	0.478	0.522	0.474	0.493	0.540	0.559	0.534	0.549	0.510	0.396	0.467
El. 866 ft	T_o	21.8	24.6	35.3	49.0	61.0	70.9	76.8	74.4	65.6	53.7	37.8	25.4
Matanuska, AL	$\bar{H}$	119.2	345	—	1327.6	1628.4	1727.6	1526.9	1169	737.3	373.8	142.8	56.4
Lat. 61°30′N	$\bar{K}_r$	0.513	0.503	—	0.545	0.494	0.466	0.434	0.419	0.401	0.390	0.372	0.364
El. 180 ft	T_o	13.9	21.0	27.4	38.6	50.3	57.6	60.1	58.1	50.2	37.7	22.9	13.9
Medford, OR	$\bar{H}$	435.4	804.4	1259.8	1807.4	2216.2	2440.5	2607.4	2261.6	1672.3	1043.5	558.7	346.5
Lat. 42°23′N	$\bar{K}_r$	0.353	0.464	0.527	0.584	0.625	0.648	0.710	0.689	0.628	0.526	0.384	0.313
El. 1329 ft	T_o	39.4	45.4	50.8	56.3	63.1	69.4	76.9	76.4	69.6	58.7	47.1	40.5

Location		1	2	3	4	5	6	7	8	9	10	11	12
Miami, FL	$\overline{H}$	1292.2	1554.6	1828.8	2020.6	2068.6	1991.5	1992.6	1890.8	1646.8	1436.5	1321	1183.4
Lat. 25°47′N	$\overline{K}_T$	0.604	0.616	0.612	0.600	0.578	0.545	0.552	0.549	0.525	0.534	0.559	0.588
El. 9 ft	T_0	71.6	72.0	73.8	77.0	79.9	82.9	84.1	84.5	83.3	80.2	75.6	72.6
Midland, TX	$\overline{H}$	1066.4	1345.7	1784.8	2036.1	2301.1	2317.7	2301.8	2193	1921.3	1470.8	1244.3	1023.2
Lat. 31°56′N	$\overline{K}_T$	0.587	0.596	0.638	0.617	0.639	0.622	0.628	0.643	0.642	0.600	0.609	0.611
El. 2854 ft	T_0	47.9	52.8	60.0	68.8	77.2	83.9	85.7	85.0	78.9	70.3	56.6	49.1
Nashville, TN	$\overline{H}$	589.7	907	1246.8	1662.3	1997	2149.4	2079.7	1862.7	1600.7	1223.6	823.2	614.4
Lat. 36°07′N	$\overline{K}_T$	0.373	0.440	0.472	0.514	0.556	0.573	0.565	0.554	0.556	0.540	0.454	0.426
El. 605 ft	T_0	42.6	45.1	52.9	63.0	71.4	80.1	83.2	81.9	76.6	65.4	52.3	44.3
New Port, RI	$\overline{H}$	565.7	856.4	1231.7	1484.8	1849	2019.2	1942.8	1687.1	1411.4	1035.4	656.1	527.7
Lat. 41°29′N	$\overline{K}_T$	0.438	0.482	0.507	0.477	0.520	0.536	0.529	0.513	0.524	0.512	0.44	0.460
El. 60 ft	T_0	29.5	32.0	39.6	48.2	58.6	67.0	73.2	72.3	66.7	56.2	46.5	34.4
New York, NY	$\overline{H}$	539.5	790.8	1180.4	1426.2	1738.4	1994.1	1938.7	1605.9	1349.4	977.8	598.1	476
Lat. 40°46′N	$\overline{K}_T$	0.406	0.435	0.480	0.455	0.488	0.53	0.528	0.500	0.486	0.475	0.397	0.403
El. 52 ft	T_0	35.0	34.9	43.1	52.3	63.3	72.2	76.9	75.3	69.5	59.3	48.3	37.7
Oak Ridge, TN	$\overline{H}$	604	895.9	1241.7	1689.6	1942.8	2066.4	1972.3	1795.6	1559.8	1194.8	796.3	610
Lat. 36°01′N	$\overline{K}_T$	0.0382	0.435	0.471	0.524	0.541	0.551	0.536	0.534	0.542	0.527	0.438	0.422
El. 905 ft	T_0	41.9	44.2	51.7	61.4	69.8	77.8	80.2	78.8	74.5	62.7	50.4	42.5
Oklahoma City, OK	$\overline{H}$	938	1192.6	1534.3	1849.4	2005.1	2355	2273.8	2211	1819.2	1409.6	1085.6	897.4
Lat. 35°24′N	$\overline{K}_T$	0.580	0.571	0.576	0.570	0.558	0.629	0.618	0.565	0.628	0.614	0.588	0.608
El. 1304 ft	T_0	40.1	45.0	53.2	63.6	71.2	80.6	85.5	85.4	77.4	66.5	52.2	43.1
Ottawa, Ontario	$\overline{H}$	539.1	852.4	1250.5	1506.6	1857.2	2084.5	2045.4	1752.4	1326.6	826.9	458.7	408.5
Lat. 45°20′N	$\overline{K}_T$	0.499	0.540	0.554	0.502	0.529	0.554	0.560	0.546	0.521	0.450	0.359	0.436
El. 339 ft	T_0	14.6	15.6	27.7	43.3	57.5	67.5	71.9	69.8	61.5	48.9	35	19.6
Phoenix, AZ	$\overline{H}$	1126.6	1514.7	1967.1	2388.2	2709.6	2781.5	2450.5	2299.6	2131.3	1688.9	1290	1040.9
Lat. 33°26′N	$\overline{K}_T$	0.65	0.691	0.716	0.728	0.753	0.745	0.667	0.677	0.722	0.708	0.657	0.652
El. 1112 ft	T_0	54.2	58.8	64.7	72.2	80.8	89.2	94.6	92.5	87.4	75.8	63.6	56.7
Portland, ME	$\overline{H}$	565.7	874.5	1329.5	1528.4	1923.2	2017.3	2095.6	1799.2	1428.8	1035	591.5	507.7
Lat. 43°39′N	$\overline{K}_T$	0.482	0.524	0.569	0.500	0.544	0.536	0.572	0.554	0.546	0.539	0.431	0.491
El. 63 ft	T_0	23.7	24.5	34.4	44.8	55.4	65.1	71.1	69.7	61.9	51.8	40.3	28.0
Rapid City, SD	$\overline{H}$	687.8	1032.5	1503.7	1807	2028	2193.7	2235.8	2019.9	1628	1179.3	763.1	590.4
Lat. 44°09′N	$\overline{K}_T$	0.601	0.627	0.649	0.594	0.574	0.583	0.612	0.622	0.628	0.624	0.566	0.0588
El. 3218 ft	T_0	24.7	27.4	34.7	48.2	58.3	67.3	76.3	75.0	64.7	52.9	38.7	29.2

TABLE 2.10 Monthly Means of Total-Horizontal Radiation H (Btu/ft² day), Percent of Possible Radiation $\overline{K}_T$, and Daytime Ambient Temperature T_o (°F) (*Continued*)

		Jan	Feb	Mar	Apr	May	Jun	Jul	Aug	Sep	Oct	Nov	Dec
Riverside, CA	$\overline{H}$	999.6	1335	1750.5	1943.2	2282.3	2492.6	2443.5	2263.8	1955.3	1509.6	1169	979.7
Lat. 33°57'N	$\overline{K}_T$	0.589	0.617	0.643	0.594	0.635	0.667	0.665	0.668	0.665	0.639	0.606	0.626
El. 1020 ft	T_o	55.3	57.0	60.6	65.0	69.4	74.0	81.0	81.0	78.5	71.0	63.1	57.2
St. Cloud, MN	$\overline{H}$	632.8	976.7	1383	1598.1	1859.4	2003.3	2087.8	1828.4	1369.4	890.4	545.4	463.1
Lat. 45°35'N	$\overline{K}_T$	0.595	0.629	0.614	0.534	0.530	0.533	0.573	0.570	0.539	0.490	0.435	0.504
El. 1034 ft	T_o	13.6	16.9	29.8	46.2	58.8	68.5	74.4	71.9	62.5	50.2	32.1	18.3
Salt Lake City, UT	$\overline{H}$	622.1	986	1301.1	1813.3	—	—	—	—	1689.3	1250.2	—	552.8
Lat. 40°46'N	$\overline{K}_T$	0.468	0.909	0.529	0.579	—	—	—	—	0.621	0.610	—	0.467
El. 4227 ft	T_o	29.4	36.2	44.4	53.9	63.1	71.7	81.3	79.0	68.7	57.0	42.5	34.0
San Antonio, TX	$\overline{H}$	1045	1299.2	1560.1	1664.6	2024.7	814.8	2264.2	2185.2	1844.6	1487.4	1104.4	954.6
Lat. 29°32'N	$\overline{K}_T$	0.541	0.550	0.542	0.500	0.563	0.220	0.647	0.637	0.603	0.584	0.507	0.528
El. 794 ft	T_o	53.7	58.4	65.0	72.2	79.2	85.0	87.4	87.8	82.6	74.7	63.3	56.5
Santa Maria, CA	$\overline{H}$	983.8	1296.3	1805.9	2067.9	2375.6	2599.6	2540.6	2293.3	1965.7	1566.4	1169	943.9
Lat. 34°54'N	$\overline{K}_T$	0.595	0.613	0.671	0.636	0.661	0.695	0.690	0.678	0.674	0.676	0.624	0.627
El. 238 ft	T_o	54.1	55.3	57.6	59.5	61.2	63.5	65.3	65.7	65.9	64.1	60.8	56.1
Sault Ste. Marie, MI	$\overline{H}$	488.6	843.9	1336.5	1559.4	1962.3	2064.2	2149.4	1767.9	1207	809.2	392.2	359.8
Lat. 46°28'N	$\overline{K}_T$	0.490	0.560	0.606	0.526	0.560	0.549	0.590	0.554	0.481	0.457	0.323	0.408
El. 724 ft	T_o	16.3	16.2	25.6	39.5	52.1	61.6	67.3	66.0	57.9	46.8	33.4	21.9
Sayville, NY	$\overline{H}$	602.9	936.2	1259.4	1560.5	1857.2	2123.2	2040.9	1734.7	1446.8	1087.4	697.8	533.9
Lat. 40°30'N	$\overline{K}_T$	0.453	0.511	0.510	0.498	0.522	0.564	0.555	0.525	0.530	0.527	0.450	0.447
El. 20 ft	T_o	35	34.9	43.1	52.3	63.3	72.2	76.9	75.3	69.5	59.3	48.3	37.7
Schenectady, NY	$\overline{H}$	488.2	753.5	1026.6	1272.3	1553.1	1687.8	1662.3	1494.8	1124.7	820.6	436.2	356.8
Lat. 42°50'N	$\overline{K}_T$	0.406	0.441	0.433	0.413	0.438	0.448	0.454	0.458	0.426	0.420	0.309	0.331
El. 217 ft	T_o	24.7	24.6	34.9	48.3	61.7	70.8	76.9	73.7	64.6	53.1	40.1	28.0
Seattle, WA	$\overline{H}$	282.6	520.6	992.2	1507	1881.5	1909.9	2110.7	1688.5	1211.8	702.2	386.3	239.5
Lat. 47°27'N	$\overline{K}_T$	0.296	0.355	0.456	0.510	0.538	0.508	0.581	0.533	0.492	0.407	0.336	0.292
El. 386 ft	T_o	42.1	45.0	48.9	54.1	59.8	64.4	68.4	67.9	63.3	56.3	48.4	44.4
Seattle, WA	$\overline{H}$	252	471.6	917.3	1375.6	1664.9	1724	1805.1	1617	1129.1	638	325.5	218.1
Lat. 47°36'N	$\overline{K}_T$	0.266	0.324	0.423	0.468	0.477	0.459	0.498	0.511	0.459	0.372	0.284	0.269
El. 14 ft	T_o	38.9	42.9	46.9	51.9	58.1	62.8	67.2	66.7	61.6	54.0	45.7	41.5

Location		1	2	3	4	5	6	7	8	9	10	11	12
Seabrook, NJ	$\overline{H}$	591.9	854.2	1195.6	1518.8	1800.7	1964.6	1949.8	1715	1445.7	1971.9	721.8	522.5
Lat. 39°30′N	$\overline{K}_T$	0.426	0.453	0.476	0.481	0.504	0.522	0.530	0.517	0.524	0.508	0.449	0.416
El. 100 ft	T_0	39.5	37.6	43.9	54.7	64.9	74.1	79.8	77.7	69.7	61.2	48.5	39.3
Spokane, WA	$\overline{H}$	446.1	837.6	1200	1864.6	2104.4	2226.5	2479.7	2076	1511	844.6	486.3	279
Lat. 47°40′N	$\overline{K}_T$	0.478	0.579	0.556	0.602	0.603	0.593	0.684	0.656	0.616	0.494	0.428	0.345
El. 1968 ft	T_0	26.5	31.7	40.5	49.2	57.9	64.6	73.4	71.7	62.7	51.5	37.4	30.5
State College, PA	$\overline{H}$	501.8	749.1	1106.6	1399.2	1754.6	2027.6	1968.2	1690	1336.1	1017	580.1	443.9
Lat. 40°48′N	$\overline{K}_T$	0.381	0.413	0.451	0.448	0.493	0.539	0.536	0.512	0.492	0.496	0.379	0.376
El. 1175 ft	T_0	31.3	31.4	39.8	51.3	63.4	71.8	75.8	73.4	66.1	55.6	43.2	32.6
Stillwater, OK	$\overline{H}$	763.8	1081.5	1463.8	1702.6	1879.3	2235.8	2224.3	2039.1	1724.3	1314	991.5	783
Lat. 36°09′N	$\overline{K}_T$	0.484	0.527	0.555	0.528	0.523	0.596	0.604	0.607	0.599	0.581	0.548	0.544
El. 910 ft	T_0	41.2	45.6	53.8	64.2	71.6	81.1	85.9	85.9	77.5	67.6	52.6	43.9
Tampa, FL	$\overline{H}$	1223.6	1461.2	1771.9	2016.2	2228	2146.5	1991.9	1845.4	1687.8	1493.3	1328.4	1119.5
Lat. 27°55′N	$\overline{K}_T$	0.605	0.600	0.606	0.602	0.620	0.583	0.548	0.537	0.546	0.572	0.590	0.589
El. 11 ft	T_0	64.2	65.7	68.8	74.3	79.4	83.0	84.0	84.4	82.9	77.2	69.6	65.5
Toronto, Ontario	$\overline{H}$	451.3	674.5	1088.9	1388.2	1785.2	1941.7	1968.6	1622.5	1284.1	835	458.3	352.8
Lat. 43°41′N	$\overline{K}_T$	0.388	0.406	0.467	0.455	0.506	0.516	0.539	0.500	0.493	0.438	0.336	0.346
El. 379 ft	T_0	26.5	26.0	34.2	46.3	58	68.4	73.8	71.8	64.3	52.6	40.9	30.2
Tucson, AZ	$\overline{H}$	1171.9	1453.8	—	2434.7	—	2601.4	2292.2	2179.7	2122.5	1640.9	1322.1	1132.1
Lat. 32°07′N	$\overline{K}_T$	0.648	0.646	—	0.738	—	0.698	0.625	0.640	0.710	0.672	0.650	0.679
El. 2556 ft	T_0	53.7	57.3	62.3	69.7	78.0	87.0	90.1	87.4	84.0	73.9	62.5	56.1
Upton, NY	$\overline{H}$	583	872.7	1280.4	1609.9	1891.5	2159	2044.6	1789.6	1472.7	1102.6	686.7	551.3
Lat. 40°52′N	$\overline{K}_T$	0.444	0.483	0.522	0.514	0.532	0.574	0.557	0.542	0.542	0.538	0.448	0.467
El. 75 ft	T_0	35.0	34.9	43.1	52.3	63.3	72.2	76.9	75.3	69.5	59.3	48.3	37.7
Washington, DC (WBCO)	$\overline{H}$	632.4	901.5	1255	1600.4	1846.8	2080.8	1929.9	1712.2	1446.1	1083.4	763.5	594.1
Lat. 38°51′N	$\overline{K}_T$	0.445	0.470	0.496	0.504	0.516	0.553	0.524	0.516	0.520	0.506	0.464	0.460
El. 64 ft	T_0	38.4	39.6	48.1	57.5	67.7	76.2	79.9	77.9	72.2	60.9	50.2	40.2
Winnipeg, Man	$\overline{H}$	488.2	835.4	1354.2	1641.3	1904.4	1962	2123.6	1761.2	1190.4	767.5	444.6	345
Lat. 49°54′N	$\overline{K}_T$	0.601	0.636	0.661	0.574	0.550	0.524	0.587	0.567	0.504	0.482	0.436	0.503
El. 786 ft	T_0	3.2	7.1	21.3	40.9	55.9	65.3	71.9	69.4	58.6	45.6	25.2	10.1

SOURCE: Ref. 13, based on unrehabilitated data and solar constant of 1.39 kW/m².

TABLE 2.11   Average Daily Solar Availabilities for a Full-tracking Surface, kWh/m²

	Spring (M,A,M)		Summer (J,J,A)		Fall (S,O,N)		Winter (D,J,F)	
	D	T	D	T	D	T	D	T
Albuquerque	7.6	9.5	8.2	9.9	6.9	8	5.8	6.6
Apalachicola	5	7	4.5	6.8	4.4	5.7	3.3	4.4
Bismarck	4.7	6.6	6.5	8.8	3.9	5	2.7	3.4
Blue Hill	3.6	5.4	4.2	6.5	3.1	4.3	2.2	2.9
Boston	3.6	5.4	4.2	6.5	3.1	4.3	2.2	3
Brownsville	4.2	6.5	5.8	8.2	4.4	6.1	3.2	4.4
Cape Hatteras	4.8	6.8	4.9	7.2	4	5.4	3.1	4
Caribou	5.9	7.2	6	7.6	3.4	4.3	3.9	4.5
Charleston	4.3	6.3	3.9	6.4	3.7	5	3.1	4
Columbia	4.3	6.2	5.8	7.9	4.1	5.2	2.9	3.7
Dodge City	5.9	7.6	7.1	9.2	5.5	6.7	4.6	5.5
El Paso	8	9.7	8.1	9.9	6.9	8.1	6	6.8
Ely	6.9	8.8	8.1	10.3	6.7	7.7	4.8	5.5
Fort Worth	4.5	6.6	5.9	8.1	4.6	6	3.7	4.7
Great Falls	4.5	6.4	6.7	8.9	3.9	5	2.3	2.8
Lake Charles	3.8	6.1	4.3	6.8	3.8	5.3	2.7	3.9
Madison	4.1	6.1	5.1	7.5	3.3	4.5	2.5	3.3
Medford	4.5	6.5	7.5	9.6	3.7	4.8	1.5	2.3
Miami	4.3	6.8	3.5	6	3.5	5.4	3.9	5.3
Nashville	3.9	5.8	4.6	7.2	3.6	4.9	2.4	3.1
New York	3.4	5.3	3.8	6.1	2.8	4	2	2.7
Omaha	4.6	6.6	6	8.2	4.1	5.2	3.2	3.9
Phoenix	7.8	9.7	8	10	6.6	7.7	5.3	6.1
Santa Maria	5.7	7.7	6.9	8.8	5.2	6.5	4.1	4.9
Seattle	4.8	6.3	7.4	9	3.2	4	1.5	2.1
Washington, DC	3.8	5.7	4.3	6.7	3.3	4.5	2.4	3.2

TABLE 2.12   Average Daily Solar Availabilities for a Surface Rotating about a Horizontal East-West Axis, kWh/m²

	Spring (M,A,M)		Summer (J,J,A)		Fall (S,O,N)		Winter (D,J,F)	
	D	T	D	T	D	T	D	T
Albuquerque	5.4	7.6	6	8.1	5.3	6.4	4.8	5.6
Apalachicola	3.8	5.9	3.4	5.9	3.4	4.8	2.8	3.9
Bismarck	3.3	5.4	4.6	6.9	3	4.1	2.3	3.1
Blue Hill	2.6	4.6	3.1	5.5	2.4	3.6	1.9	2.6
Boston	2.6	4.6	3.1	5.4	2.4	3.5	1.9	2.7
Brownsville	3.1	5.6	4.3	6.9	3.3	4.9	2.6	3.9
Cape Hatteras	3.5	5.6	3.6	6	3.1	4.5	2.6	3.5
Caribou	4.2	5.6	4.3	6	2.6	3.5	3.4	3.9
Charleston	3.2	5.3	3	5.6	2.9	4.4	2.6	3.6
Columbia	3.2	5.1	4.2	6.5	3.2	4.5	2.5	3.3
Dodge City	4.3	6.1	5.2	7.3	4.2	5.4	3.9	4.8
El Paso	5.7	7.6	6	8	5.2	6.5	4.9	5.8
Ely	4.9	6.9	5.8	8	5.1	6.2	4	4.8
Fort Worth	3.3	5.5	4.4	6.8	3.5	4.9	3.1	4

TABLE 2.12 Average Daily Solar Availabilities for a Surface Rotating about a Horizontal East-West Axis, kWh/m² (Continued)

	Spring (M,A,M)		Summer (J,J,A)		Fall (S,O,N)		Winter (D,J,F)	
	D	T	D	T	D	T	D	T
Great Falls	3.2	5.4	4.8	7.1	3.1	4.2	2	2.7
Lake Charles	2.8	5.2	3.1	5.9	2.9	4.5	2.2	3.4
Madison	2.9	4.9	3.7	6.1	2.5	3.8	2.1	3
Medford	3.4	5.4	5.5	7.5	2.9	4.1	1.3	2.1
Miami	3.2	5.7	2.6	5.5	2.7	4.7	3.1	4.6
Nashville	2.9	4.9	3.5	6.1	2.8	4.2	2	2.9
New York	2.5	4.6	2.9	5.3	2.2	3.4	1.7	2.5
Omaha	3.2	5.4	4.3	6.6	3.1	4.3	2.6	3.4
Phoenix	5.7	7.7	5.9	8.1	5.1	6.4	4.5	5.3
Santa Maria	4.2	6.2	5.4	7.2	4.2	5.4	3.5	4.4
Seattle	3.5	5	5.3	7	2.5	3.4	1.3	1.9
Washington, DC	2.8	4.9	3.2	5.7	2.6	3.8	2	2.8

TABLE 2.13 Average Daily Solar Availabilities for a Surface Rotating about a Horizontal North-South Axis, kWh/m²

	Spring (M,A,M)		Summer (J,J,A)		Fall (S,O,N)		Winter (D,J,F)	
	D	T	D	T	D	T	D	T
Alburquerque	7.1	9.1	8	9.9	5.7	6.7	4	4.9
Apalachicola	4.8	7	4.4	6.8	3.7	5.1	2.5	3.6
Bismarck	4.1	6.2	6.1	8.4	2.9	4.1	1.5	2.4
Blue Hill	3.3	5.2	4	6.4	2.4	3.6	1.4	2.1
Boston	3.3	5.2	4.1	6.4	2.4	3.7	1.4	2.3
Brownsville	4	6.4	5.7	8.2	3.8	5.6	2.5	3.8
Cape Hatteras	4.5	6.5	4.8	7.2	3.3	4.8	2.1	3.1
Caribou	5.2	6.6	5.7	7.2	2.6	3.4	2.2	2.9
Charleston	4.1	6.1	3.8	6.3	3.1	4.4	2.2	3.2
Columbia	4	5.9	5.6	7.8	3.2	4.5	1.9	2.8
Dodge City	5.5	7.3	6.9	8.8	4.4	5.6	3	3.9
El Paso	7.6	9.4	8	9.8	5.8	7.1	4.4	5.3
Ely	6.4	8.3	7.9	10.1	5.3	6.4	3.1	3.9
Fort Worth	4.3	6.3	5.8	8.1	3.8	5.2	2.6	3.6
Great Falls	4	6	6.3	8.6	2.8	4	1.3	2
Lake Charles	3.7	5.9	4.2	6.9	3.2	4.8	2	3.2
Madison	3.7	5.7	4.9	7.3	2.5	3.8	1.5	2.3
Medford	4.1	6.1	7.2	9.2	2.9	4.1	0.9	1.8
Miami	4.2	6.6	3.4	6.1	3	4.9	3	4.5
Nashville	3.7	5.6	4.5	6.9	2.9	4.3	1.7	2.5
New York	3.1	5	3.6	5.9	2.2	3.5	1.3	2.1
Omaha	4.2	6.3	5.8	8.1	3.2	4.4	2.1	2.8
Phoenix	7.4	9.4	7.9	9.8	5.4	6.7	3.7	4.7
Santa Maria	5.4	7.4	6.8	8.5	4.3	5.5	2.8	3.8
Seattle	4.3	5.8	7	8.6	2.4	3.2	0.9	1.5
Washington, DC	3.5	5.4	4.2	6.6	2.6	3.9	1.6	2.4

TABLE 2.14 Average Daily Solar Availabilities for a Surface Rotating about a Polar Axis, kWh/m²

	Spring (M,A,M)		Summer (J,J,A)		Fall (S,O,N)		Winter (D,J,F)	
	D	T	D	T	D	T	D	T
Aburquerque	7.3	8.9	7.7	9	6.8	7.6	5.4	6.2
Apalachicola	4.8	6.6	4.2	6	4.3	5.5	3.1	4.1
Bismarck	4.6	5.9	6.1	7.5	3.8	4.7	2.5	3.2
Blue Hill	3.5	4.9	4	5.5	3	3.9	2.1	2.6
Boston	3.5	4.9	4	5.5	3	3.9	2.1	2.8
Brownsville	4	6	5.5	7.4	4.3	5.7	3	4.2
Cape Hatteras	4.7	6.1	4.6	6.4	3.9	5.1	2.9	3.8
Caribou	5.7	6.7	5.7	6.7	3.4	4	3.7	4.2
Charleston	4.2	5.9	3.7	5.6	3.6	4.8	2.9	3.8
Columbia	4.1	5.7	5.5	7	4	4.9	2.7	3.4
Dodge City	5.7	7	6.7	8.1	5.3	6.4	4.3	5.1
El Paso	7.7	9.2	7.7	9	6.7	7.8	5.6	6.5
Ely	6.6	8.1	7.7	9.1	6.5	7.4	4.5	5.1
Fort Worth	4.4	5.9	5.6	7.2	4.5	5.6	3.5	4.3
Great Falls	4.4	5.8	6.3	7.7	3.9	4.7	2.2	2.7
Lake Charles	3.7	5.5	4.1	6	3.7	5	2.5	3.6
Madison	4	5.4	4.8	6.4	3.2	4.3	2.3	3.1
Medford	4.3	5.8	7.1	8.4	3.6	4.5	1.4	2.1
Miami	4.2	6.1	3.3	5.3	3.4	5.1	3.7	4.9
Nashville	3.8	5.3	4.4	6.1	3.5	4.7	2.3	3
New York	3.3	4.7	3.6	5.2	2.7	3.8	1.9	2.5
Omaha	4.4	5.9	5.7	7.1	4	4.9	3	3.7
Phoenix	7.5	9.1	7.5	9	6.4	7.4	5	5.7
Santa Maria	5.6	7	6.5	7.8	5.1	6.1	3.9	4.6
Seattle	4.7	5.7	7	8	3.1	3.7	1.5	1.9
Washington, DC	3.7	5.1	4	5.7	3.2	4.2	2.3	3

TABLE 2.15 Average Daily Solar Availabilities for a Surface Tilted L° from Horizontal and Rotating about a Vertical Axis, kWh/m²

	Spring (M,A,M)		Summer (J,J,A)		Fall (S,O,N)		Winter (D,J,F)	
	D	T	D	T	D	T	D	T
Alburquerque	7	8.9	7.7	9.3	6.2	7.3	5	5.8
Apalachicola	4.6	6.6	4.1	6.4	3.9	5.2	2.8	4
Bismarck	4.5	6.1	6.2	8	3.7	4.5	2.4	3.1
Blue Hill	3.4	5.1	4	5.8	2.9	3.9	2	2.7
Boston	3.4	5.1	4.1	6	2.9	3.9	2	2.7
Brownsville	3.8	6.1	5.2	7.7	3.7	5.4	2.6	4
Cape Hatteras	4.5	6.2	4.6	6.7	3.6	5	2.7	3.6
Caribou	5.6	6.8	5.8	7	3.2	3.9	3.5	4
Charleston	4	6	3.7	6.1	3.4	4.7	2.6	3.7
Columbia	4.1	5.9	5.5	7.5	3.8	4.9	2.6	3.4
Dodge City	5.5	7.2	6.7	8.5	5	6.2	4	4.8
El Paso	7.3	9	7.6	9.2	6.1	7.3	5.1	6
Ely	6.5	8.1	7.7	9.5	6.1	7.1	4.2	5

TABLE 2.15 Average Daily Solar Availabilities for a Surface Tilted $L°$ from Horizontal and Rotating about a Vertical Axis, kWh/m² (*Continued*)

	Spring (M,A,M)		Summer (J,J,A)		Fall (S,O,N)		Winter (D,J,F)	
	D	T	D	T	D	T	D	T
Fort Worth	4.2	6.1	5.5	7.6	4.1	5.5	3.2	4.2
Great Falls	4.3	6	6.4	8.2	3.7	4.6	2.1	2.6
Lake Charles	3.4	5.7	3.9	6.4	3.3	4.9	2.3	3.5
Madison	3.8	5.5	4.8	6.9	3.1	4.1	2.2	3
Medford	4.3	6.1	7.2	8.9	3.5	4.5	1.3	2.1
Miami	3.9	6.3	3.1	5.8	3	5	3.2	4.7
Nashville	3.7	5.4	4.4	6.6	3.3	4.6	2.1	2.8
New York	3.3	4.9	3.6	5.6	2.6	3.8	1.8	2.5
Omaha	4.4	6.1	5.7	7.6	3.8	4.8	2.8	3.5
Phoenix	7.2	9.1	7.4	9.3	5.9	7.1	4.7	5.5
Santa Maria	5.4	7.1	6.6	8.2	4.8	5.9	3.6	4.5
Seattle	4.6	5.8	7.1	8.3	3	3.7	1.4	1.9
Washington, DC	3.6	5.4	4.1	6.2	3	4.2	2.1	2.9

TABLE 2.16 Average Daily Solar Availabilities for a Surface Tilted $L + 15°$ from Horizontal and Rotating about a Vertical Axis, kWh/m²

	Spring (M,A,M)		Summer (J,J,A)		Fall (S,O,N)		Winter (D,J,F)	
	D	T	D	T	D	T	D	T
Albuquerque	7.2	8.8	7.6	9.1	6.7	7.6	5.5	6.2
Apalachicola	4.8	6.5	4.2	6.1	4.2	5.4	3.2	4.1
Bismarck	4.5	5.7	6	7.6	3.8	4.5	2.6	3.2
Blue Hill	3.4	4.7	3.8	5.3	3	3.8	2.2	2.7
Boston	3.4	4.8	3.9	5.5	3	3.9	2.1	2.8
Brownsville	3.9	6	5.4	7.4	4.1	5.5	3	4.2
Cape Hatteras	4.6	6	4.6	6.4	3.9	5	3	3.7
Caribou	5.6	6.5	5.6	6.6	3.4	3.9	3.8	4.2
Charleston	4.1	5.8	3.6	5.7	3.6	4.7	2.9	3.8
Columbia	4.1	5.6	5.4	7.1	4	4.9	2.8	3.5
Dodge City	5.6	6.9	6.7	8.2	5.3	6.2	4.4	5.2
El Paso	7.5	9	7.6	9.1	6.6	7.6	5.7	6.4
Ely	6.6	7.9	7.6	9.2	6.5	7.3	4.6	5.2
Fort Worth	4.3	5.9	5.5	7.3	4.4	5.6	3.5	4.4
Great Falls	4.3	5.6	6.2	7.6	3.9	4.5	2.2	2.7
Lake Charles	3.6	5.4	4	6.2	3.6	5	2.5	3.6
Madison	3.9	5.2	4.7	6.5	3.2	4.1	2.4	3
Medford	4.3	5.7	7	8.5	3.6	4.5	1.5	2.1
Miami	4	6.2	3.2	5.6	3.3	4.9	3.6	4.9
Nashville	3.8	5.2	4.3	6.3	3.4	4.6	2.3	3
New York	3.3	4.6	3.6	5.2	2.7	3.7	1.9	2.5
Omaha	4.4	5.8	5.6	7.2	4	4.9	3.1	3.7
Phoenix	7.4	9	7.4	9	6.3	7.3	5.1	5.8
Santa Maria	5.4	7	6.4	7.9	5	6	3.9	4.7
Seattle	4.6	5.5	6.9	7.9	3.1	3.7	1.5	1.9
Washington, DC	3.7	5	4	5.7	3.2	4.1	2.3	2.9

TABLE 2.17 Average Daily Solar Availabilities for a Surface Tilted $L-15°$ from Horizontal and Rotating about a Vertical Axis, kWh/m²

	Spring (M,A,M)		Summer (J,J,A)		Fall (S,O,N)		Winter (D,J,F)	
	D	T	D	T	D	T	D	T
Albuquerque	6.3	8.5	7.2	9	5.4	6.5	4.1	5
Apalachicola	4.2	6.4	3.9	6.2	3.4	4.8	2.3	3.5
Bismarck	4.1	6.1	5.9	8	3.3	4.4	2	2.9
Blue Hill	3.3	5.1	3.9	6.1	2.6	3.8	1.8	2.5
Boston	3.2	5	3.8	6	2.5	3.7	1.7	2.5
Brownsville	3.3	5.8	4.8	7.3	3.2	5	2.2	3.6
Cape Hatteras	4.1	6.1	4.3	6.7	3.1	4.6	2.3	3.2
Caribou	5.2	6.6	5.5	7	2.9	3.7	3	3.6
Charleston	3.6	5.8	3.4	6.1	2.9	4.4	2.2	3.4
Columbia	3.8	5.7	5.1	7.4	3.3	4.5	2.1	3
Dodge City	5	6.9	6.3	8.3	4.3	5.6	3.3	4.3
El Paso	6.6	8.4	7	8.8	5.2	6.5	4.2	5.1
Ely	5.9	7.8	7.2	9.2	5.3	6.4	3.5	4.4
Fort Worth	3.8	5.9	5.1	7.5	3.5	5	2.6	3.6
Great Falls	4.1	5.9	6.1	8.3	3.4	4.4	1.8	2.4
Lake Charles	3.1	5.5	3.6	6.2	2.9	4.5	1.9	3.1
Madison	3.6	5.5	4.6	7	2.7	3.9	1.8	2.8
Medford	4	6	6.8	8.9	3.1	4.2	1.2	2.1
Miami	3.5	5.9	2.8	5.7	2.5	4.6	2.6	4.2
Nashville	3.4	5.4	4.1	6.7	2.9	4.3	1.7	2.7
New York	3	4.9	3.5	5.8	2.3	3.6	1.5	2.4
Omaha	4	6	5.3	7.5	3.3	4.5	2.3	3.1
Phoenix	6.6	8.5	7	9	5.1	6.4	3.8	4.8
Santa Maria	4.9	6.9	6.3	8.1	4.2	5.4	3	3.9
Seattle	4.4	5.8	6.8	8.3	2.7	3.6	1.2	1.8
Washington, DC	3.4	5.3	3.9	6.3	2.6	3.9	1.8	2.7

TABLE 2.18 Average Daily Solar Availabilities on a Horizontal Surface, kWh/m²

	Spring (M,A,M)		Summer (J,J,A)		Fall (S,O,N)		Winter (D,J,F)	
	D	T	D	T	D	T	D	T
Albuquerque	4.9	6.9	5.8	7.8	3.6	4.9	2.6	3.5
Apalachicola	3.5	5.7	3.3	5.8	2.5	4.1	1.7	2.9
Bismarck	2.6	4.7	4	6.4	1.7	2.9	0.9	1.7
Blue Hill	2.2	4.2	2.9	5.3	1.5	2.8	0.8	1.7
Boston	2.2	4.2	2.9	5.3	1.5	2.8	0.8	1.7
Brownsville	2.9	5.4	4.2	6.7	2.6	4.4	1.7	3.1
Cape Hatteras	3.2	5.3	3.5	5.9	2.2	3.7	1.4	2.4
Caribou	3.3	4.8	3.8	5.5	1.5	2.5	1.2	1.9
Charleston	2.9	5.2	2.8	5.5	2.1	3.7	1.4	2.6
Columbia	2.7	4.8	4	6.4	2.1	3.4	1.2	2.1
Dodge City	3.7	5.7	4.9	7	2.8	4.1	1.8	2.8
El Paso	5.2	7.2	5.9	7.8	3.8	5.1	2.8	3.8
Ely	4.2	6.2	5.4	7.6	3.4	4.5	1.9	2.8
Fort Worth	3	5.2	4.2	6.6	2.5	4	1.7	2.8

TABLE 2.18 Average Daily Solar Availabilities on a Horizontal Surface, kWh/m² (*Continued*)

	Spring (M,A,M)		Summer (J,J,A)		Fall (S,O,N)		Winter (D,J,F)	
	D	T	D	T	D	T	D	T
Great Falls	2.6	4.7	4.2	6.7	1.8	2.9	0.7	1.5
Lake Charles	2.6	5	3	5.7	2.1	3.9	1.3	2.6
Madison	2.4	4.5	3.4	5.9	1.6	2.9	0.9	1.8
Medford	2.9	5	5.1	7.2	1.9	3.2	0.6	1.6
Miami	3	5.6	2.6	5.4	2.1	4.1	2	3.6
Nashville	2.6	4.7	3.3	5.9	2	3.4	1.1	2
New York	2.2	4.2	2.7	5.1	1.5	2.8	0.8	1.7
Omaha	2.8	4.9	3.9	6.4	2	3.2	1.2	2.1
Phoenix	5.1	7.2	5.7	7.9	3.7	5	2.5	3.5
Santa Maria	3.8	5.9	5.1	7.1	2.9	4.3	1.8	2.9
Seattle	2.8	4.4	4.7	6.3	1.5	2.4	0.5	1.1
Washington, DC	2.5	4.5	3	5.6	1.7	3.1	0.9	2

TABLE 2.19 Average Daily Solar Availabilities on a South-facing Surface Tilted Upward 15° from Horizontal, kWh/m²

	Spring (M,A,M)		Summer (J,J,A)		Fall (S,O,N)		Winter (D,J,F)	
	D	T	D	T	D	T	D	T
Albuquerque	5.1	7.3	5.8	7.7	4.5	5.6	3.5	4.4
Apalachicola	3.6	5.9	3.2	5.7	3.1	4.5	2.1	3.4
Bismarck	3	5	4.2	6.5	2.2	3.5	1.4	2.3
Blue Hill	2.4	4.5	2.9	5.4	1.9	3.2	1.2	2
Boston	2.4	4.4	2.9	5.4	1.9	3.2	1.2	2.1
Brownsville	3	5.5	4	6.7	3	4.8	2.2	3.6
Cape Hatteras	3.4	5.5	3.5	5.9	2.6	4.2	1.9	2.9
Caribou	3.7	5.2	4	5.7	2	2.9	2	2.7
Charleston	3.1	5.3	2.8	5.5	2.6	4	2	3.1
Columbia	3	5	4.1	6.3	2.6	3.9	1.8	2.6
Dodge City	4	6	4.9	7	3.5	4.8	2.6	3.6
El Paso	5.5	7.5	5.8	7.7	4.5	5.9	3.8	4.7
Ely	4.5	6.7	5.4	7.7	4.2	5.4	2.8	3.6
Fort Worth	3.2	5.3	4.1	6.5	3.1	4.6	2.3	3.4
Great Falls	2.9	5.1	4.5	6.8	2.4	3.5	1.2	1.9
Lake Charles	2.7	5.1	3	5.6	2.6	4.3	1.8	3.1
Madison	2.7	4.8	3.5	6	2	3.3	1.3	2.3
Medford	3.1	5.2	5.2	7.4	2.4	3.6	0.9	1.8
Miami	3.1	5.7	2.4	5.3	2.4	4.4	2.6	4
Nashville	2.7	4.9	3.3	5.9	2.4	3.9	1.4	2.4
New York	2.4	4.4	2.8	5.2	1.8	3.3	1.1	2
Omaha	3	5.1	4	6.5	2.6	3.8	1.8	2.7
Phoenix	5.4	7.6	5.6	7.8	4.4	5.8	3.4	4.3
Santa Maria	4	6	5.1	7.1	3.5	4.9	2.6	3.5
Seattle	3.2	4.7	4.9	6.5	1.9	2.9	0.8	1.4
Washington,DC	2.7	4.7	3.1	5.7	2.1	3.4	1.4	2.3

TABLE 2.20 Average Daily Availabilities on a South-facing Surface Tilted Upward 30° from Horizontal, kWh/m²

	Spring (M,A,M)		Summer (J,J,A)		Fall (S,O,N)		Winter (D,J,F)	
	D	T	D	T	D	T	D	T
Albuquerque	5.1	7.2	5.4	7.2	5	6.2	4.2	5.1
Apalachicola	3.5	5.7	3	5.3	3.3	4.7	2.5	3.7
Bismarck	3.1	5.2	4.1	6.4	2.7	3.8	1.8	2.8
Blue Hill	2.5	4.5	2.9	5.2	2.2	3.5	1.5	2.4
Boston	2.5	4.5	2.9	5.2	2.2	3.5	1.6	2.5
Brownsville	2.8	5.3	3.6	6.1	3.2	4.8	2.4	3.9
Cape Hatteras	3.4	5.4	3.2	5.6	2.9	4.4	2.3	3.3
Caribou	3.9	5.4	4	5.6	2.4	3.3	2.6	3.3
Charleston	3.1	5.2	2.6	5.3	2.8	4.3	2.3	3.4
Columbia	3	5	3.8	6.2	3	4.2	2.1	3
Dodge City	4.1	5.9	4.8	6.7	3.9	5.2	3.3	4.3
El Paso	5.4	7.2	5.3	7.1	5	6.3	4.4	5.4
Ely	4.6	6.6	5.2	7.3	4.8	5.9	3.4	4.3
Forth Worth	3.2	5.2	3.8	6.2	3.3	4.8	2.8	3.8
Great Falls	3.1	5.1	4.4	6.7	2.7	3.9	1.6	2.3
Lake Charles	2.6	5	2.7	5.3	2.8	4.4	2	3.3
Madison	2.7	4.8	3.4	5.9	2.3	3.8	1.7	2.6
Medford	3.2	5.2	5.1	7.1	2.7	3.9	1.1	2
Miami	2.9	5.5	2.2	4.9	2.6	4.6	2.9	4.4
Nashville	2.7	4.7	3.1	5.7	2.6	4.2	1.8	2.6
New York	2.5	4.4	2.7	5	2.1	3.4	1.5	2.3
Omaha	3.1	5.2	3.8	6.2	2.9	4.1	2.2	3.1
Phoenix	5.3	7.4	5.3	7.3	4.9	6.2	4	5
Santa Maria	4	6	4.8	6.6	3.9	5.3	3.1	4
Seattle	3.3	4.8	4.8	6.4	2.2	3.2	1	1.7
Washington, DC	2.7	4.6	3	5.4	2.4	3.7	1.7	2.6

TABLE 2.21 Average Daily Solar Availabilities on a South-facing Surface Tilted Upward 45° from Horizontal, kWh/m².

	Spring (M,A,M)		Summer (J,J,A)		Fall (S,O,N)		Winter (D,J,F)	
	D	T	D	T	D	T	D	T
Albuquerque	4.7	6.7	4.6	6.5	5.2	6.3	4.6	5.6
Apalachicola	3.2	5.2	2.5	4.8	3.4	4.7	2.7	3.9
Bismarck	3	5.1	3.7	6	2.9	4	2.1	2.9
Blue Hill	2.4	4.3	2.5	4.8	2.3	3.5	1.8	2.6
Boston	2.4	4.3	2.6	4.7	2.4	3.5	1.8	2.7
Brownsville	2.5	4.8	3	5.4	3.2	4.8	2.5	3.9
Cape Hatteras	3.1	5	2.8	5.1	3.1	4.4	2.6	3.5
Caribou	3.9	5.3	3.6	5.2	2.5	3.4	3.1	3.7
Charleston	2.8	4.8	2.3	4.6	2.9	4.3	2.5	3.7
Columbia	2.8	4.7	3.4	5.7	3.1	4.4	2.4	3.2
Dodge City	3.8	5.5	4.1	6.1	4.1	5.3	3.6	4.6
El Paso	4.9	6.7	4.5	6.3	5.1	6.4	4.8	5.8
Ely	4.4	6.2	4.5	6.6	5	6.2	3.8	4.6
Fort Worth	2.9	4.9	3.2	5.6	3.4	4.8	3	4

**TABLE 2.21 Average Daily Solar Availabilities on a South-facing Surface Tilted Upward 45°
from Horizontal, kWh/m² (Continued)**

	Spring (M,A,M)		Summer (J,J,A)		Fall (S,O,N)		Winter (D,J,F)	
	D	T	D	T	D	T	D	T
Great Falls	3	5	4.1	6.2	3	4.1	1.9	2.5
Lake Charles	2.3	4.6	2.3	4.8	2.8	4.4	2.2	3.4
Madison	2.7	4.6	3	5.4	2.5	3.7	2	2.9
Medford	3.1	5	4.6	6.5	2.8	4	1.3	2.2
Miami	2.6	4.9	1.8	4.5	2.6	4.5	3	4.5
Nashville	2.6	4.4	2.7	5.1	2.8	4.1	2	2.8
New York	2.3	4.2	2.4	4.6	2.2	3.5	1.6	2.5
Omaha	2.9	4.9	3.5	5.7	3.1	4.2	2.5	3.4
Phoenix	4.9	6.8	4.4	6.4	5	6.2	4.4	5.2
Santa Maria	3.7	5.7	4.1	6	4.1	5.4	3.4	4.3
Seattle	3.2	4.7	4.4	5.9	2.4	3.3	1.2	1.8
Washington, DC	2.6	4.5	2.6	5	2.6	3.7	2	2.8

**TABLE 2.22 Average Daily Solar Availabilities on a South-facing Surface Tilted Upward 60°
from Horizontal, kWh/m²**

	Spring (M,A,M)		Summer (J,J,A)		Fall (S,O,N)		Winter (D,J,F)	
	D	T	D	T	D	T	D	T
Albuquerque	4	5.9	3.5	5.3	5	6	4.8	5.5
Apalachicola	2.7	4.5	1.8	4	3.1	4.5	2.7	3.8
Bismarck	2.8	4.7	3.1	5.3	2.9	4	2.3	3
Blue Hill	2.1	3.9	2.1	4.2	2.3	3.4	1.8	2.6
Boston	2.1	3.9	2.1	4.1	2.3	3.5	1.9	2.7
Brownsville	2.1	4.3	2.1	4.4	2.9	4.4	2.5	3.8
Cape Hatteras	2.7	4.5	2.2	4.3	2.9	4.3	2.6	3.6
Caribou	3.6	4.9	3.1	4.6	2.6	3.4	3.3	3.9
Charleston	2.4	4.3	1.7	4.1	2.7	4.1	2.6	3.6
Columbia	2.5	4.3	2.7	4.9	3	4.3	2.5	3.3
Dodge City	3.3	5	3.3	5.1	4.1	5.2	3.8	4.8
El Paso	4.1	5.8	3.4	5.1	4.8	6	4.8	5.8
Ely	3.8	5.6	3.6	5.6	4.9	6	4	4.8
Forth Worth	2.5	4.3	2.5	4.6	3.3	4.6	3	4
Great Falls	2.8	4.6	3.4	5.5	3	4.1	2	2.5
Lake Charles	2	4.1	1.6	4.1	2.6	4.1	2.2	3.3
Madison	2.3	4.3	2.5	4.7	2.4	3.8	2.1	2.9
Medford	2.7	4.6	3.7	5.5	2.8	3.9	1.3	2.1
Miami	2.1	4.4	1.3	3.7	2.4	4.3	3	4.4
Nashville	2.2	4.1	2.1	4.4	2.6	4	2	2.8
New York	2	3.8	2	4.1	2.1	3.3	1.7	2.4
Omaha	2.6	4.5	2.8	4.9	3.1	4.1	2.6	3.4
Phoenix	4.1	5.9	3.4	5.3	4.8	6	4.4	5.3
Santa Maria	3.2	4.9	3.2	5	3.9	5.2	3.5	4.3
Seattle	2.9	4.3	3.7	5.2	2.4	3.3	1.3	1.9
Washington, DC	2.3	4.2	2.2	4.3	2.5	3.7	2	2.8

TABLE 2.23 Average Daily Solar Availabilities on a South-facing Surface Tilted Upward 75° from Horizontal, kWh/m²

	Spring (M,A,M)		Summer (J,J,A)		Fall (S,O,N)		Winter (D,J,F)	
	D	T	D	T	D	T	D	T
Albuquerque	3.1	4.8	2.3	3.9	4.5	5.4	4.6	5.4
Apalachicola	1.9	3.8	1.1	3	2.8	4	2.5	3.6
Bismarck	2.4	4.1	2.4	4.3	2.7	3.7	2.3	3
Blue Hill	1.7	3.4	1.5	3.5	2.1	3.1	1.8	2.5
Boston	1.7	3.4	1.5	3.4	2.1	3.2	1.9	2.6
Brownsville	1.5	3.4	1.2	3.3	2.4	4	2.4	3.5
Cape Hatteras	2	3.7	1.4	3.5	2.6	4	2.5	3.3
Caribou	3	4.3	2.3	3.7	2.4	3.2	3.3	3.9
Charleston	1.8	3.5	1.1	3.3	2.4	3.6	2.4	3.4
Columbia	2	3.6	1.9	3.9	2.8	3.9	2.4	3.1
Dodge City	2.6	4.2	2.3	4	3.6	4.8	3.7	4.6
El Paso	3.1	4.7	2.1	3.7	4.3	5.5	4.6	5.4
Ely	3	4.8	2.5	4.4	4.5	5.5	3.9	4.6
Fort Worth	1.9	3.5	1.6	3.6	2.9	4.2	2.9	3.7
Great Falls	2.3	4	2.5	4.5	2.8	3.8	2	2.5
Lake Charles	1.4	3.4	1	3.2	2.3	3.7	2.1	3.1
Madison	2	3.7	1.8	3.9	2.3	3.5	2.1	2.9
Medford	2.2	3.9	2.7	4.3	2.5	3.5	1.3	2
Miami	1.5	3.6	0.7	3.1	2.1	3.7	2.8	4.1
Nashville	1.7	3.4	1.4	3.6	2.4	3.5	1.9	2.7
New York	1.7	3.3	1.4	3.4	1.9	3.1	1.7	2.3
Omaha	2	3.9	2	4	2.8	3.8	2.6	3.3
Phoenix	3.1	4.8	2.2	3.9	4.2	5.3	4.1	4.9
Santa Maria	2.4	4.2	2.1	3.7	3.5	4.7	3.3	4.1
Seattle	2.5	3.7	2.8	4.2	2.3	3.1	1.3	1.8
Washington, DC	1.8	3.5	1.5	3.6	2.3	3.3	2	2.8

TABLE 2.24 Average Daily Solar Availabilities on a South-facing Surface Tilted Upward 90° from Horizontal, kWh/m².

	Spring (M,A,M)		Summer (J,J,A)		Fall (S,O,N)		Winter (D,J,F)	
	D	T	D	T	D	T	D	T
Albuquerque	2	3.6	1	2.5	3.6	4.6	4	4.7
Apalachicola	1.1	2.8	0.4	2.3	2.2	3.2	2.2	3.1
Bismarck	1.7	3.4	1.5	3.3	2.4	3.3	2.1	2.7
Blue Hill	1.2	2.7	0.9	2.6	1.8	2.7	1.7	2.3
Boston	1.2	2.7	0.8	2.6	1.8	2.7	1.7	2.3
Brownsville	0.8	2.6	0.3	2.3	1.9	3.2	2	3
Cape Hatteras	1.3	2.8	0.6	2.5	2.1	3.4	2.2	3
Caribou	2.3	3.4	1.5	2.7	2.1	2.8	3.1	3.6
Charleston	1.1	2.8	0.4	2.5	2	3.1	2.1	2.9
Columbia	1.3	2.8	1	2.8	2.3	3.3	2.1	2.9
Dodge City	1.7	3.2	1.2	2.8	3	4.1	3.4	4.1
El Paso	1.9	3.3	0.8	2.2	3.4	4.5	4	4.7
Ely	2.1	3.6	1.2	3	3.8	4.6	3.5	4.2
Fort Worth	1.1	2.8	0.6	2.4	2.3	3.5	2.5	3.3

TABLE 2.24 Average Daily Solar Availabilities on a South-facing Surface Tilted Upward 90° from Horizontal, kWh/m² (Continued)

	Spring (M,A,M)		Summer (J,J,A)		Fall (S,O,N)		Winter (D,J,F)	
	D	T	D	T	D	T	D	T
Great Falls	1.7	3.4	1.6	3.5	2.5	3.4	1.9	2.4
Lake Charles	0.8	2.7	0.3	2.4	1.8	3.1	1.8	2.7
Madison	1.4	2.9	1.1	3	1.9	3	1.9	2.6
Medford	1.5	3.1	1.5	3	2.1	3	1.2	1.8
Miami	0.8	2.8	0.2	2.2	1.6	3.1	2.3	3.6
Nashville	1.1	2.7	0.7	2.6	1.9	3.1	1.7	2.4
New York	1.2	2.6	0.8	2.5	1.6	2.7	1.5	2.2
Omaha	1.4	3	1.2	2.9	2.3	3.3	2.3	3
Phoenix	1.9	3.5	0.8	2.4	3.3	4.4	3.7	4.4
Santa Maria	1.5	3.1	1	2.4	2.8	3.8	2.9	3.7
Seattle	1.8	3	1.7	3	2	2.7	1.2	1.7
Washington, DC	1.3	2.8	0.7	2.7	1.9	2.8	1.8	2.5

TABLE 2.25 Average Daily Solar Availabilities on a 45° Tilted Surface with Azimuth Angle of ± 15°, kWh/m²

	Spring (M,A,M)		Summer (J,J,A)		Fall (S,O,N)		Winter (D,J,F)	
	D	T	D	T	D	T	D	T
Albuquerque	4.7	6.7	4.6	6.4	5.1	6.2	4.6	5.4
Apalachicola	3.2	5.2	2.5	4.8	3.3	4.7	2.6	3.8
Bismarck	3	5.1	3.7	5.9	2.9	4	2	2.9
Blue Hill	2.4	4.3	2.5	4.8	2.3	3.5	1.7	2.5
Boston	2.4	4.3	2.6	4.7	2.3	3.5	1.7	2.6
Brownsville	2.5	4.8	8	5.4	3.1	4.8	2.5	3.9
Cape Hatteras	3.1	5.1	2.8	5.1	3	4.5	2.5	3.4
Caribou	3.9	5.3	3.7	5.2	2.5	3.4	3	3.6
Charleston	2.8	4.9	2.2	4.7	2.8	4.2	2.5	3.5
Columbia	2.8	4.7	3.4	5.6	3.1	4.3	2.4	3.1
Dodge City	3.8	5.5	4.2	6.1	4.1	5.3	3.6	4.5
El Paso	4.9	6.6	4.5	6.2	5	6.3	4.7	5.6
Ely	4.3	6.3	4.5	6.6	4.9	6	3.8	4.5
Fort Worth	2.9	4.9	3.3	5.6	3.4	4.8	3	3.9
Great Falls	3	5	4	6.2	3	4	1.8	2.4
Lake Charles	2.3	4.6	2.3	4.8	2.8	4.4	2.2	3.3
Madison	2.6	4.6	3	5.4	2.4	3.8	1.9	2.8
Medford	3.1	5	4.5	6.5	2.8	3.9	1.2	2.1
Miami	2.6	5.1	1.8	4.4	2.6	4.4	3	4.4
Nashville	2.6	4.4	2.7	5.2	2.7	4.1	1.9	2.8
New York	2.3	4.2	2.4	4.6	2.1	3.4	1.6	2.4
Omaha	2.9	4.9	3.5	5.7	3	4.1	2.5	3.3
Phoenix	4.8	6.8	4.5	6.4	4.9	6.2	4.2	5.2
Santa Maria	3.7	5.6	4.1	6	4	5.3	3.3	4.2
Seattle	3.3	4.7	4.4	5.9	2.4	3.3	1.2	1.8
Washington, DC	2.6	4.5	2.6	5	2.5	3.8	1.9	2.8

TABLE 2.26   Average Daily Solar Availabilities on a 45° Tilted Surface with Azimuth Angle of ± 30°, kWh/m²

	Spring (M,A,M)		Summer (J,J,A)		Fall (S,O,N)		Winter (D,J,F)	
	D	T	D	T	D	T	D	T
Albuquerque	4.7	6.7	4.7	6.5	4.9	5.9	4.3	5.1
Apalachicola	3.1	5.3	2.5	4.8	3.1	4.6	2.5	3.6
Bismarck	2.9	5	3.7	5.9	2.7	3.8	1.9	2.7
Blue Hill	2.3	4.2	2.5	4.8	2.2	3.4	1.6	2.4
Boston	2.3	4.2	2.5	4.8	2.2	3.4	1.7	2.5
Brownsville	2.5	4.8	3	5.5	3	4.7	2.4	3.8
Cape Hatteras	3	5	2.8	5.1	2.8	4.3	2.3	3.3
Caribou	3.8	5.2	3.7	5.2	2.4	3.3	2.8	3.4
Charleston	2.7	4.9	2.3	4.8	2.7	4.1	2.3	3.4
Columbia	2.8	4.6	3.4	5.6	2.9	4.2	2.2	3
Dodge City	3.7	5.5	4.2	6.1	3.9	5.2	3.3	4.3
El Paso	4.9	6.6	4.5	6.4	4.8	6.1	4.4	5.4
Ely	4.2	6.2	4.6	6.7	4.7	5.8	3.5	4.3
Fort Worth	2.9	4.8	3.3	5.6	3.2	4.6	2.8	3.8
Great Falls	2.9	4.9	4	6.2	2.8	3.9	1.6	2.4
Lake Charles	2.3	4.6	2.3	4.8	2.6	4.3	2	3.2
Madison	2.6	4.5	3	5.3	2.3	3.6	1.8	2.7
Medford	3	4.9	4.5	6.5	2.6	3.8	1.1	2
Miami	2.6	5	1.9	4.4	2.4	4.4	2.9	4.3
Nashville	2.5	4.4	2.7	5.2	2.6	4	1.8	2.6
New York	2.3	4.1	2.4	4.6	2	3.3	1.5	2.3
Omaha	2.9	4.8	3.5	5.7	2.9	4	2.3	3.1
Phoenix	4.8	6.8	4.5	6.5	4.7	5.9	4	5
Santa Maria	3.6	5.5	4.1	6	3.8	5.1	3.1	4
Seattle	3.2	4.6	4.4	5.9	2.3	3.2	1.1	1.7
Washington, DC	2.5	4.4	2.6	5	2.4	3.6	1.8	2.7

TABLE 2.27   Average Daily Solar Availabilities on a 45° Tilted Surface with Azimuth Angle of ± 45°, kWh/m²

	Spring (M,A,M)		Summer (J,J,A)		Fall (S,O,N)		Winter (D,J,F)	
	D	T	D	T	D	T	D	T
Albuquerque	4.6	6.6	4.7	6.5	4.5	5.6	3.8	4.7
Apalachicola	3.1	5.2	2.6	4.8	2.9	4.4	2.3	3.5
Bismarck	2.9	4.8	3.7	6	2.5	3.6	1.7	2.5
Blue Hill	2.2	4.2	2.5	4.8	2	3.2	1.4	2.2
Boston	2.2	4.2	2.5	4.7	2	3.2	1.4	2.3
Brownsville	2.5	4.8	3.2	5.6	2.9	4.5	2.2	3.6
Cape Hatteras	3	4.9	2.8	5.2	2.6	4.1	2.1	3.1
Caribou	3.6	5	3.6	5.2	2.2	3.1	2.5	3.1
Charleston	2.7	4.8	2.3	4.7	2.5	3.9	2.1	3.2
Columbia	2.7	4.6	3.4	5.6	2.7	3.9	1.9	2.8
Dodge City	3.6	5.4	4.2	6.1	3.6	4.8	3	4
El Paso	4.8	6.6	4.6	6.4	4.5	5.8	4	4.9
Ely	4.2	6.1	4.6	6.7	4.3	5.5	3.1	3.9
Fort Worth	2.8	4.8	3.4	5.6	3.1	4.4	2.5	3.5

TABLE 2.27 Average Daily Solar Availabilities on a 45° Tilted Surface with Azimuth Angle of ± 45°, kWh/m² (Continued)

	Spring (M,A,M)		Summer (J,J,A)		Fall (S,O,N)		Winter (D,J,F)	
	D	T	D	T	D	T	D	T
Great Falls	2.8	4.8	3.9	6.2	2.5	3.7	1.5	2.1
Lake Charles	2.3	4.5	2.4	4.9	2.5	4.1	1.9	3.1
Madison	2.5	4.5	3	5.4	2.1	3.4	1.5	2.5
Medford	2.9	4.8	4.5	6.4	2.4	3.6	1	1.9
Miami	2.6	5	1.9	4.6	2.3	4.3	2.6	4.1
Nashville	2.4	4.3	2.7	5.2	2.4	3.8	1.6	2.5
New York	2.2	4	2.3	4.6	1.9	3.2	1.3	2.1
Omaha	2.8	4.8	3.5	5.7	2.7	3.8	2.1	2.8
Phoenix	4.7	6.8	4.6	6.6	4.4	5.6	3.6	4.5
Santa Maria	3.5	5.5	4.2	6	3.5	4.8	2.8	3.7
Seattle	3	4.5	4.4	5.9	2.1	3	1	1.6
Washington, DC	2.4	4.3	2.6	5	2.2	3.4	1.6	2.5

TABLE 2.28 Average Daily Solar Availabilities on a 45° Tilted Surface with Azimuth Angle of ± 60°, kWh/m²

	Spring (M,A,M)		Summer (J,J,A)		Fall (S,O,N)		Winter (D,J,F)	
	D	T	D	T	D	T	D	T
Albuquerque	4.4	6.4	4.7	6.5	4.1	5.2	3.3	4.2
Apalachicola	3	5.2	2.6	4.9	2.7	4	2	3.1
Bismarck	2.7	4.6	3.6	5.9	2.2	3.3	1.4	2.3
Blue Hill	2.1	4	2.5	4.8	1.8	3	1.2	2
Boston	2.1	4	2.4	4.8	1.8	3	1.2	2.1
Brownsville	2.4	4.8	3.2	5.6	2.7	4.3	1.9	3.3
Cape Hatteras	2.9	4.8	2.8	5.1	2.4	3.9	1.8	2.8
Caribou	3.4	4.8	3.5	5.1	2	2.8	2.1	2.7
Charleston	2.6	4.7	2.3	4.8	2.3	3.7	1.8	2.9
Columbia	2.6	4.5	3.4	5.6	2.4	3.7	1.6	2.5
Dodge City	3.5	5.2	4.1	6.1	3.2	4.5	2.6	3.5
El Paso	4.7	6.4	4.6	6.4	4.1	5.4	3.5	4.4
Ely	4	5.9	4.6	6.7	3.9	5	2.7	3.4
Fort Worth	2.7	4.7	3.4	5.6	2.7	4.2	2.2	3.1
Great Falls	2.6	4.6	3.9	6.1	2.2	3.3	1.2	1.9
Lake Charles	2.2	4.5	2.4	4.9	2.3	3.9	1.6	2.9
Madison	2.4	4.3	2.9	5.3	1.9	3.2	1.3	2.2
Medford	2.7	4.7	4.4	6.3	2.2	3.3	0.8	1.7
Miami	2.5	4.9	2	4.6	2.1	4.1	2.3	3.8
Nashville	2.4	4.2	2.7	5.2	2.2	3.6	1.4	2.2
New York	2.1	4	2.3	4.5	1.6	3	1.1	1.9
Omaha	2.7	4.6	3.4	5.8	2.4	3.5	1.7	2.6
Phoenix	4.6	6.6	4.6	6.6	4	5.1	3.1	4
Santa Maria	3.4	5.3	4.1	6	3.2	4.4	2.4	3.4
Seattle	2.9	4.3	4.3	5.8	1.9	2.8	0.8	1.4
Washington, DC	2.3	4.2	2.6	4.9	2	3.2	1.4	2.2

TABLE 2.29 Average Daily Solar Availabilities on a 45° Tilted Surface with Azimuth Angle of ± 75°, kWh/m²

	Spring (M,A,M)		Summer (J,J,A)		Fall (S,O,N)		Winter (D,J,F)	
	D	T	D	T	D	T	D	T
Albuquerque	4.2	6.2	4.7	6.5	3.6	4.7	2.8	3.6
Apalachicola	2.9	4.9	2.6	4.9	2.4	3.8	1.7	2.8
Bismarck	2.5	4.4	3.6	5.7	1.8	3	1.1	1.9
Blue Hill	2	3.9	2.4	4.7	1.6	2.7	1	1.8
Boston	2	3.9	2.4	4.6	1.6	2.7	1	1.8
Brownsville	2.4	4.7	3.3	5.7	2.4	4	1.7	3
Cape Hatteras	2.7	4.7	2.8	5.1	2.1	3.6	1.5	2.5
Caribou	3.1	4.5	3.4	4.9	1.7	2.5	1.6	2.3
Charleston	2.5	4.5	2.3	4.8	2	3.4	1.5	2.6
Columbia	2.4	4.3	3.3	5.5	2.1	3.4	1.3	2.2
Dodge City	3.3	5.1	4	6	2.8	4	2	3
El Paso	4.5	6.2	4.6	6.4	3.6	5	3	3.8
Ely	3.7	5.7	4.5	6.7	3.4	4.5	2.1	2.9
Fort Worth	2.6	4.6	3.4	5.6	2.5	3.8	1.8	2.8
Great Falls	2.4	4.4	3.7	5.9	1.9	3	1	1.6
Lake Charles	2.1	4.4	2.4	4.9	2.1	3.7	1.4	2.5
Madison	2.2	4.1	2.9	5.3	1.6	2.9	1.1	2
Medford	2.5	4.5	4.3	6.2	1.9	3.1	0.7	1.6
Miami	2.5	4.9	2	4.5	1.9	3.9	2	3.4
Nashville	2.2	4.1	2.7	5.2	1.9	3.3	1.1	2
New York	1.9	3.8	2.2	4.5	1.4	2.8	0.9	1.7
Omaha	2.5	4.5	3.3	5.6	2.1	3.2	1.4	2.2
Phoenix	4.4	6.3	4.6	6.5	3.5	4.8	2.6	3.5
Santa Maria	3.2	5.2	4.1	5.8	2.8	4.1	2	2.9
Seattle	2.7	4.1	4.1	5.7	1.6	2.5	0.6	1.2
Washington, DC	2.1	4	2.5	4.8	1.7	2.9	1.1	1.9

TABLE 2.30 Average Daily Solar Availabilities on a 45° Tilted Surface with Azimuth Angle of ± 90°, kWh/m²

	Spring (M,A,M)		Summer (J,J,A)		Fall (S,O,N)		Winter (D,J,F)	
	D	T	D	T	D	T	D	T
Albuquerque	4	5.9	4.6	6.4	3.1	4.2	2.2	3
Apalachicola	2.7	4.8	2.6	4.8	2.1	3.4	1.3	2.5
Bismarck	2.2	4.3	3.3	5.6	1.5	2.7	0.8	1.6
Blue Hill	1.8	3.8	2.3	4.5	1.3	2.5	0.7	1.5
Boston	1.8	3.7	2.3	4.5	1.3	2.5	0.7	1.5
Brownsville	2.3	4.7	3.2	5.8	2.1	3.8	1.4	2.8
Cape Hatteras	2.6	4.5	2.7	5.1	1.8	3.3	1.2	2.1
Caribou	2.8	4.2	3.2	4.7	1.4	2.2	1.2	1.8
Charleston	2.3	4.3	2.2	4.7	1.7	3.1	1.2	2.3
Columbia	2.2	4.2	3.1	5.4	1.8	3	1	1.9
Dodge City	3	4.8	3.9	5.9	2.3	3.6	1.6	2.6
El Paso	4.3	5.9	4.6	6.3	3.2	4.4	2.4	3.3
Ely	3.5	5.4	4.4	6.4	2.8	4	1.7	2.5
Fort Worth	2.4	4.4	3.3	5.6	2.1	3.5	1.4	2.4

TABLE 2.30 Average Daily Solar Availabilities on a 45° Tilted Surface with Azimuth Angle of ± 90°, kWh/m² (Continued)

	Spring (M,A,M)		Summer (J,J,A)		Fall (S,O,N)		Winter (D,J,F)	
	D	T	D	T	D	T	D	T
Great Falls	2.2	4.2	3.5	5.7	1.5	2.6	0.7	1.3
Lake Charles	2	4.4	2.4	5	1.8	3.5	1.1	2.3
Madison	2	3.9	2.7	5	1.3	2.6	0.8	1.7
Medford	2.3	4.3	4.1	6	1.6	2.8	0.5	1.4
Miami	2.4	4.8	2	4.6	1.7	3.6	1.7	3.1
Nashville	2.1	3.9	2.6	5.1	1.6	3	0.9	1.7
New York	1.8	3.6	2.1	4.3	1.2	2.4	0.7	1.5
Omaha	2.3	4.3	3.2	5.5	1.7	2.9	1.1	1.8
Phoenix	4.2	6.1	4.5	6.5	3	4.3	2.1	3
Santa Maria	3	5	3.9	5.7	2.4	3.6	1.5	2.5
Seattle	2.4	3.9	3.9	5.4	1.3	2.2	0.5	1.1
Washington, DC	2	3.9	2.4	4.7	1.4	2.7	0.9	1.7

TABLE 2.31 Average Daily Solar Availabilities on East or West Walls, kWh/m²

	Spring (M,A,M)		Summer (J,J,A)		Fall (S,O,N)		Winter (D,J,F)	
	D	T	D	T	D	T	D	T
Albuquerque	2.2	3.8	2.3	3.8	1.9	2.7	1.3	2
Apalachicola	1.4	3	1.2	3.1	1.1	2.3	0.8	1.7
Bismarck	1.4	3	2	3.8	1	1.9	0.6	1.2
Blue Hill	1	2.6	1.2	3	0.8	1.8	0.5	1.2
Boston	1	2.6	1.2	3	0.8	1.8	0.5	1.2
Brownsville	1.1	3	1.6	3.5	1.2	2.5	0.8	1.9
Cape Hatteras	1.4	2.9	1.4	3.3	1	2.3	0.7	1.5
Caribou	1.7	2.9	1.8	3	0.9	1.6	0.9	1.4
Charleston	1.2	2.9	1.1	3	0.9	2.1	0.7	1.6
Columbia	1.2	2.8	1.7	3.5	1	2	0.6	1.3
Dodge City	1.7	3.1	2.1	3.6	1.5	2.4	1.1	1.8
El Paso	2.3	3.7	2.3	3.8	1.8	2.9	1.5	2.2
Ely	2	3.5	2.4	4.1	1.7	2.7	1.1	1.7
Fort Worth	1.3	2.9	1.7	3.5	1.2	2.3	0.9	1.7
Great Falls	1.3	2.9	2	3.8	1	1.9	0.4	1
Lake Charles	1.1	2.9	1.3	3.3	1	2.4	0.7	1.6
Madison	1.2	2.8	1.5	3.4	0.9	2	0.5	1.3
Medford	1.2	2.8	2.1	3.8	0.9	1.9	0.3	1
Miami	1.2	3.1	1	3	0.9	2.5	1	2.2
Nashville	1.1	2.7	1.3	3.3	0.9	2.1	0.5	1.3
New York	1	2.5	1.1	2.8	0.7	1.7	0.4	1.1
Omaha	1.4	3	1.8	3.6	1.1	2	0.7	1.4
Phoenix	2.2	3.8	2.3	3.8	1.7	2.6	1.2	2
Santa Maria	1.6	3.1	1.9	3.3	1.3	2.3	0.9	1.6
Seattle	1.4	2.6	2.2	3.5	0.8	1.6	0.3	0.8
Washington, DC	1.1	2.6	1.2	3.1	0.8	1.8	0.5	1.2

TABLE 2.32   Average Daily Solar Availabilities on Southeast or Southwest Walls, kWh/m²

	Spring (M,A,M)		Summer (J,J,A)		Fall (S,O,N)		Winter (D,J,F)	
	D	T	D	T	D	T	D	T
Albuquerque	2.3	4	1.9	3.4	3.1	3.9	3	3.7
Apalachicola	1.4	3.2	1	2.7	1.8	3	1.7	2.6
Bismarck	1.8	3.4	2	3.7	1.9	2.9	1.5	2.2
Blue Hill	1.2	2.8	1.2	3	1.5	2.4	1.2	1.9
Boston	1.2	2.8	1.2	2.9	1.5	2.4	1.2	1.9
Brownsville	1.1	3	1.2	3.1	1.7	3	1.5	2.6
Cape Hatteras	1.5	3.1	1.2	3.1	1.8	2.9	1.6	2.4
Caribou	2.3	3.4	1.9	3.1	1.7	2.4	2.3	2.8
Charleston	1.3	3	0.9	3	1.6	2.7	1.6	2.4
Columbia	1.4	3	1.5	3.4	1.9	2.8	1.6	2.3
Dodge City	2	3.4	1.8	3.5	2.5	3.5	2.5	3.3
El Paso	2.4	3.8	1.8	3.3	2.9	3.9	3	3.7
Ely	2.3	3.8	2.1	3.9	3.1	3.9	2.6	3.2
Fort Worth	1.4	3	1.4	3.2	2	3.1	1.9	2.7
Great Falls	1.7	3.3	2	3.8	2	2.8	1.3	1.9
Lake Charles	1.1	2.9	0.9	3	1.6	2.8	1.4	2.3
Madison	1.5	3	1.4	3.4	1.6	2.6	1.4	2.1
Medford	1.6	3.1	2.1	3.7	1.7	2.6	0.8	1.5
Miami	1.2	3.1	0.7	2.7	1.4	3	1.8	3
Nashville	1.2	2.9	1.2	3.1	1.6	2.7	1.2	1.9
New York	1.2	2.7	1	2.9	1.3	2.3	1.1	1.7
Omaha	1.6	3.2	1.7	3.5	1.9	2.8	1.7	2.4
Phoenix	2.3	3.9	1.8	3.4	2.8	3.8	2.7	3.4
Santa Maria	1.8	3.4	1.7	3.1	2.3	3.3	2.2	2.9
Seattle	1.8	3	2.2	3.5	1.6	2.3	0.9	1.4
Washington, DC	1.3	2.8	1.1	3	1.5	2.5	1.3	2

percent and as high as 130 percent will occur. Mean daily values for particular seasons will generally be within 20 percent of the listed values.

A table of diffuse solar radiation   Table 2.33 presents a listing of the average daily amounts of diffuse radiation on a horizontal surface by season for 26 U.S. sites. No tabulations of diffuse radiation for other surfaces are given; for a surface which "sees" the ground, the ground's reflectance alters the diffuse radiation incident on the surface. Generally, the sky is brighter than the ground; however, reflected radiation from a snow cover can significantly increase the diffuse radiation incident upon a surface.

Solar Radiation Availability Comparison Tables

This section contains several summary tables which permit direct comparisons of different types of solar radiation availability. These tables are all reprinted from Ref. 10, with some modifications as suggested by the more recently available rehabilitated hourly data.

Relative availabilities for concentrating collectors   Table 2.34 presents a comparison of the amounts of direct solar radiation available to six different one-dimensional tracking schemes for concentrating collectors. This table was derived from the data in Tables 2.11 through 2.17 as follows. For each of the 26 sites and each season the available direct radiation was expressed as a fraction of the corresponding direct-normal radiation available; the numbers printed in Table 2.34 are the averages and ranges of these ratios for the 26 locations.

This table clearly indicates that the two best one-dimensional tracking schemes in terms of direct radiation availability are the polar axis scheme, P, and the latitude +15° tilt/vertical axis scheme, VLP. Both of these intercept more than 95 percent of the direct-normal radiation which is available to a full-tracking concentrator and which represents an upper bound on direct radiation availability.

When seasonal mean daily totals of direct-normal radiation are available, the ratios listed in Table 2.34 provide a means of calculating mean daily availabilities of direct radiation for the other tracking schemes. For example, the table indicates that in the spring the mean daily amount of direct radiation available to a VLP type of tracking collector will be 95 percent of the mean daily amount of direct-normal radiation. For this same example, the range of ratios is from 0.94 to 0.96; thus, using 95 percent should produce a very accurate result. In other instances where the ranges are larger, this technique will be correspondingly less accurate.

TABLE 2.33   Average Daily Diffuse Solar Radiation on a Horizontal Surface, kWh/m²

	Spring (M,A,M)	Summer (J,J,A)	Fall (S,O,N)	Winter (J,F,M)
Albuquerque	2	2	1.3	0.9
Apalachicola	2.2	2.5	1.6	1.3
Bismarck	2.1	2.4	1.2	0.8
Blue Hill	2	2.4	1.3	0.9
Boston	2	2.5	1.3	0.9
Brownsville	2.5	2.5	1.8	1.4
Cape Hatteras	2.1	2.4	1.5	1
Caribou	1.5	1.7	1	0.7
Charleston	2.3	2.7	1.6	1.2
Columbia	2.1	2.4	1.3	0.9
Dodge City	2	2.1	1.3	1
El Paso	2	2	1.3	1
Ely	2	2.2	1.1	0.9
Fort Worth	2.2	2.4	1.5	1.1
Great Falls	2.2	2.5	1.1	0.8
Lake Charles	2.4	2.7	1.8	1.3
Madison	2.1	2.5	1.3	0.9
Medford	2.1	2.1	1.3	1
Miami	2.6	2.8	2	1.6
Nashville	2.2	2.6	1.4	1
New York	2	2.4	1.3	0.9
Omaha	2.1	2.5	1.2	0.9
Phoenix	2.1	2.2	1.3	1
Santa Maria	2.1	2	1.4	1.1
Seattle	1.6	1.6	0.9	0.6
Washington, DC	2	2.6	1.4	1.1

TABLE 2.34   Seasonal Availabilities of Direct Radiation to Various One-dimensional Tracking Schemes Expressed as Fractions of Direct-Normal Radiation

Seasons / Radiation type	Spring Avg.	Spring Range	Summer Avg.	Summer Range	Fall Avg.	Fall Range	Winter Avg.	Winter Range
DP (polar axis)	0.97	0.96–0.98	0.94	0.93–0.95	0.98	0.96–1.0	0.94	0.93–1.0
DNSH (NS horizontal axis)	0.93	0.88–0.97	0.97	0.94–0.99	0.80	0.73–0.87	0.67	0.56–0.79
DEW (EW horizontal axis)	0.73	0.70–0.75	0.74	0.71–0.78	0.77	0.74–0.80	0.85	0.80–0.89
DVL (Latitude tilt, vertical axis)	0.93	0.90–0.96	0.94	0.90–0.97	0.91	0.85–0.95	0.88	0.82–0.93
DVLP (L + 15° tilt, vertical axis)	0.95	0.94–0.96	0.93	0.91–0.94	0.97	0.94–1.00	0.97	0.93–1.00
DVLN (L − 15° tilt, vertical axis)	0.86	0.80–0.92	0.88	0.81–0.92	0.80	0.72–0.86	0.73	0.65–0.81
National average daily values of DN, kWh/m²	6.4		7.1		5.2		4.2	

Relative availabilities for flat-plate vs. concentrating collectors Table 2.35 provides a comparison of direct radiation availabilities to concentrating collectors with the availability of total solar radiation to a fixed surface tilted upward 45° toward the south. For each of seven tracking schemes the ratio of the direct radiation was expressed as a fraction of the total radiation available to a fixed surface oriented as described; this was done for 26 sites and all four seasons. Table 2.35 lists the averages and ranges of these ratios for the 26 sites.

This table indicates that there is almost as much direct radiation available to the better concentrator tracking schemes as there is total radiation to a fixed, 45° tilted, south-facing surface. Overall, the differences are not large. The most important general conclusion is that, on the average, *tracking concentrators intercept about the same amount of direct radiation as total radiation intercepted by reasonably oriented, fixed, flat-plate collectors.*

Winter radiation availabilities for solar heating Table 2.36 presents daily means of total solar radiation available to five different flat-plate collector orientations for 26 U.S. locations for the winter heating season. The five orientations all face south; they include surfaces tilted at 30°, 45°, 60°, and 75°, and a vertical wall.

This table indicates that radiation availabilities during the peak winter heating season are largest for tilt angles of approximately 60°. Solar availabilities for 45° and 75° south-facing tilted surfaces are generally within 10 to 20 percent of these. Furthermore, radiation availabilities on south-vertical walls are only about 10 to 30 percent below the maximum values. Thus, all of these orientations are good for winter heating applications of solar energy.

Availabilities for 15°, 30°, and 45° tilted surfaces Table 2.37 lists the seasonal and the annual availabilities of total radiation to south-facing surfaces tilted upward at 15°, 30°, and 45° at 26 U.S. locations. For nearly all of these locations the 30° tilted surface intercepts the most total radiation on an annual basis. However, this table emphasizes that annual energy availabilities are relatively insensitive to tilt angle; a tilt angle within 15° of the optimum generally reduces annual availability by only a few percent.

Seasonal availabilities on south, east, and west vertical walls Table 2.38 lists average daily availabilities by season on south-, east-, and west-facing vertical walls for 26 U.S. locations. There are two general conclusions to be drawn from this table. The first is that generally east and west walls receive the same amount of solar radiation; the table only contains a few instances where the difference is greater than 10 percent.

The other conclusion is that south walls intercept the most energy in the winter, and east and west walls intercept the most energy in the summer. Consequently rectangular buildings to be solar-heated should be oriented along east-west lines so that solar heat gain is maximized in the winter and minimized in the summer.

TABLE 2.35 Seasonal Availabilities of Direct Radiation to Various Tracking Schemes Expressed as Fractions of the Total Solar Radiation Incident on a Fixed Surface Tilted Upward 45° Toward the South

Seasons / Radiation type	Spring		Summer		Fall		Winter	
	Avg.	Range	Avg.	Range	Avg.	Range	Ave.	Range
DN (full tracking) TN	0.95	0.81–1.19	1.04	0.78–1.28	0.94	0.78–1.1	0.9	0.7–1.05
DP (Polar axis) TP	0.92	0.78–1.15	0.98	0.74–1.22	0.92	0.76–1.08	0.85	0.65–1
DNSH (NS horizontal axis)	0.88	0.74–1.13	1.01	0.76–1.27	0.75	0.63–0.91	0.6	0.42–0.76
DEW (EW horizontal axis)	0.69	0.6–0.85	0.76	0.58–0.96	0.73	0.59–0.85	0.76	0.62–0.92
DVL (latitude tilt, vertical axis)	0.89	0.75–1.09	0.98	0.7 –1.2	0.86	0.67–1	0.79	0.62–0.95
DVLP (L + 15° tilt, vertical axis)	0.9	0.77–1.11	0.96	0.72–1.2	0.91	0.74–1.07	0.87	0.7–1.03
DVLN (L − 15° tilt, vertical axis)	0.82	0.68–0.98	0.92	0.64–1.15	0.75	0.57–0.87	0.66	0.54–0.81

ESTIMATION AND CONVERSION FORMULAS FOR SOLAR RADIATION

This section presents a variety of estimation and conversion formulas for solar radiation availabilities. These formulas provide solar system designers with tools which complement the tabulations of the preceding section.

Clear Sky Solar Radiation Intensities

Extremely accurate formulas are available for computing the intensity of solar radiation under clear sky conditions. However, these require a knowledge of such atmospheric parameters as the total atmospheric water vapor content or the atmospheric turbidity. Measurements of these parameters are not readily available; in fact, measuring the solar radiation intensity directly is often more convenient than using such formulas.

However, there are also a number of fairly simple formulas for estimating solar radiation intensity under clear sky conditions. One such formula for the intensity of the direct component is given in the following subsection. A somewhat more elaborate formula for computing the intensity of direct or total radiation at solar noon on clear days is given in Ref. 14; that reference also contains a tabulation of clear solar-noon values of total-horizontal radiation constructed with the formula.

Direct radiation intensity under clear skies   The intensity of the direct-normal component of solar radiation is given approximately by

$$I_B = I_{\text{ext}} \exp\left[-0.1457(pm_0/1000) - 0.1617(Wm_0)^{0.25}\right] \qquad (2.23)$$

This empirical formula is a slight modification of a formula whose derivation is described in Ref. 15. The parameters are as follows: p is the local atmospheric pressure in millibars, m_0 is "sea level" air mass, W is the atmospheric water vapor content in centimeters, and I_{ext} is the extraterrestial radiation. The value of I_{ext} is given in Table 2.4, and m is given in terms of the solar zenith angle by Table 2.6. Mean monthly values of W for 27 U.S. locations are listed in

TABLE 2.36   Mean Daily Amounts of Total Solar Radiation for Five Different South-facing, Tilted Surfaces in the Winter, kWh/m²

Tilt angle	30°	45°	60°	75°	90°
Albuquerque	5.1	5.6	5.5	5.4	4.7
Apalachicola	3.7	3.9	3.8	3.6	3.1
Bismarck	2.8	2.9	3.0	3	2.7
Blue Hill	2.4	2.6	2.6	2.5	2.3
Boston	2.5	2.7	2.7	2.6	2.3
Brownsville	3.9	3.9	3.8	3.5	3
Cape Hatteras	3.3	3.5	3.6	3.3	3
Caribou	3.3	3.7	4.0	3.9	3.6
Charleston	3.4	3.7	3.6	3.4	2.9
Columbia	3	3.2	3.3	3.1	2.9
Dodge City	4.3	4.6	4.8	4.6	4.1
El Paso	5.4	5.8	5.8	5.4	4.7
Ely	4.3	4.6	4.8	4.6	4.2
Fort Worth	3.8	4	4.0	3.7	3.3
Great Falls	2.3	2.5	2.5	2.5	2.4
Lake Charles	3.3	3.4	3.3	3.1	2.7
Madison	2.6	2.9	2.9	2.9	2.6
Medford	2	2.2	2.1	2	1.8
Miami	4.4	4.5	4.4	4.1	3.6
Nashville	2.6	2.8	2.8	2.7	2.4
New York	2.3	2.5	2.4	2.3	2.2
Omaha	3.1	3.4	3.4	3.3	3
Phoenix	5	5.2	5.3	4.9	4.4
Santa Maria	4	4.3	4.3	4.1	3.7
Seattle	1.7	1.8	1.9	1.8	1.7
Washington, DC	2.6	2.8	2.8	2.8	2.5

TABLE 2.37   Seasonal and Annual Daily Availabilities of Total Solar Radiation Incident on South-facing Surfaces Tilted at 15°, 30°, and 45° from Horizontal, kWh/m²

	Spring (M,A,M)			Summer (J,J,A)			Fall (S,O,N)			Winter (D,J,F)			Annual		
	15°	30°	45°	15°	30°	45°	15°	30°	45	15°	30°	45°	15°	30°	45°
Albuquerque	7.3	7.2	6.7	7.7	7.2	6.5	5.6	6.2	6.3	4.4	5.1	5.6	6.2	6.4	6.3
Apalachicola	5.9	5.7	5.2	5.7	5.3	4.8	4.5	4.7	4.7	3.4	3.7	3.9	4.9	4.9	4.7
Bismarck	5	5.2	5.1	6.5	6.4	6	3.5	3.8	4	2.3	2.8	2.9	4.3	4.5	4.5
Blue Hill	4.5	4.5	4.3	5.4	5.2	4.8	3.2	3.5	3.5	2	2.4	2.6	3.8	3.9	3.8
Boston	4.4	4.5	4.3	5.4	5.2	4.7	3.2	3.5	3.5	2.1	2.5	2.7	3.8	3.9	3.8
Brownsville	5.5	5.3	4.8	6.7	6.1	5.4	4.8	4.8	4.8	3.6	3.9	3.9	5.1	5	4.7
Cape Hatteras	5.5	5.4	5	5.9	5.6	5.1	4.2	4.4	4.4	2.9	3.3	3.5	4.6	4.7	4.5
Caribou	5.2	5.4	5.3	5.7	5.6	5.2	2.9	3.3	3.4	2.7	3.3	3.7	4.1	4.4	4.4
Charleston	5.3	5.2	4.8	5.5	5.3	4.6	4	4.3	4.3	3.1	3.4	3.7	4.5	4.5	4.4
Columbia	5	5	4.7	6.3	6.2	5.7	3.9	4.2	4.4	2.6	3	3.2	4.5	4.6	4.5
Dodge City	6	5.9	5.5	7	6.7	6.1	4.8	5.2	5.3	3.6	4.3	4.6	5.3	5.5	5.4
El Paso	7.5	7.2	6.7	7.7	7.1	6.3	5.9	6.3	6.4	4.7	5.4	5.8	6.4	6.5	6.3
Ely	6.7	6.6	6.2	7.7	7.3	6.6	5.4	5.9	6.2	3.6	4.3	4.6	5.9	6	5.9
Fort Worth	5.3	5.2	4.9	6.5	6.2	5.6	4.6	4.8	4.8	3.4	3.8	4	4.9	5	4.8
Great Falls	5.1	5.1	5	6.8	6.7	6.2	3.5	3.9	4.1	1.9	2.3	2.5	4.3	4.5	4.5
Lake Charles	5.1	5	4.6	5.6	5.3	4.8	4.3	4.4	4.4	3.1	3.3	3.4	4.5	4.5	4.3
Madison	4.8	4.8	4.6	6	5.9	5.4	3.3	3.8	3.7	2.3	2.6	2.9	4.1	4.3	4.2
Medford	5.2	5.2	5	7.4	7.1	6.5	3.6	3.9	4	1.8	2	2.2	4.5	4.5	4.4
Miami	5.7	5.5	4.9	5.3	4.9	4.5	4.4	4.6	4.5	4	4.4	4.5	4.9	4.9	4.6
Nashville	4.9	4.7	4.4	5.9	5.7	5.1	3.9	4.2	4.1	2.4	2.6	2.8	4.3	4.3	4.1
New York	4.4	4.4	4.2	5.2	5	4.6	3.3	3.4	3.5	2	2.3	2.5	3.7	3.8	3.7
Omaha	5.1	5.2	4.9	6.5	6.2	5.7	3.8	4.1	4.2	2.7	3.1	3.4	4.5	4.6	4.6
Phoenix	7.6	7.4	6.8	7.8	7.3	6.4	5.8	6.2	6.2	4.3	5	5.2	6.4	6.4	6.2
Santa Maria	6	6	5.7	7.1	6.6	6	4.9	5.3	5.4	3.5	4	4.3	5.4	5.5	5.3
Seattle	4.7	4.8	4.7	6.5	6.4	5.9	2.9	3.2	3.3	1.4	1.7	1.8	3.9	4	3.9
Washington, DC	4.7	4.6	4.5	5.7	5.4	5	3.4	3.7	3.7	2.3	2.6	2.8	4	4.1	4

Table 2.39; these values include an adjustment factor of 0.81 so as to be directly applicable for clear sky conditions.

In addition to a cloud-free sky (in the vicinity of the sun), this formula assumes that the atmospheric turbidity is low; that is, the atmosphere is relatively free of dust and haze.

Conversion from Total-Horizontal to Direct-Normal

This section presents methods for estimating the amount of direct radiation from measurements of total-horizontal radiation. Formulas for converting from total-horizontal radiation to direct-normal radiation are of interest because records of total-horizontal measurements are far more common than records of direct-normal measurements. Moreover, separating total horizontal radiation measurements into direct and diffuse components is important also for calculating the availability of total radiation on tilted surfaces because this separation is usually the first step employed in converting from horizontal data to tilted-surface data.

In situations where direct-normal radiation data sets or availability tabulations are available, their use is recommended over the application of the conversion formulas present here. Generally, applying these conversion formulas is more difficult and possibly less reliable.

The accuracy of these solar radiation availability conversions depends upon the basic time interval employed. It is generally more reliable to work with hourly data than to work with daily data. This is because the conversion from total radiation to direct radiation requires the separation of the total-horizontal measurement into its direct and diffuse components, and this separation can be performed more reliably for a shorter time interval. Over a longer time interval, such as a day, the separation is complicated by the fact that the direct-horizontal component of the total-horizontal measurement can be reduced by the changing incidence angle or by clouds, and the integrated daily value of total-horizontal radiation cannot distinguish between these cases. For this reason, the use of direct-normal data sets or availability tabulations,

TABLE 2.38 Mean Daily Availabilities of Total Solar Radiation by Season on South, East and West Walls, kWh/m²

	Spring (M,A,M)			Summer (J,J,A)			Fall (S,O,N)			Winter (D,J,F)		
	S	E	W	S	E	W	S	E	W	S	E	W
Albuquerque	3.6	3.6	4	2.5	3.6	4.1	4.6	2.6	2.8	4.7	2	2
Apalachicola	2.8	3	3	2.3	3	3.1	3.2	2.2	2.3	3.1	1.7	1.7
Bismarck	3.4	2.9	3.1	3.3	3.7	3.8	3.3	1.9	2	2.7	1.3	1.2
Blue Hill	2.7	2.5	2.7	2.6	3	3	2.7	1.8	1.9	2.3	1.2	1.2
Boston	2.7	2.5	2.7	2.6	3	3	2.7	1.7	1.8	2.3	1.2	1.1
Brownsville	2.6	3	3	2.3	3.5	3.6	3.2	2.3	2.7	3	1.8	1.9
Cape Hatteras	2.8	3	2.8	2.5	3.3	3.3	3.4	2.3	2.3	3	1.5	1.5
Caribou	3.4	2.7	3	2.7	3	3	2.8	1.6	1.6	3.6	1.3	1.4
Charleston	2.8	2.8	3	2.5	2.9	3.1	3.1	2	2.2	2.9	1.5	1.6
Columbia	2.8	2.8	2.8	2.8	3.6	3.4	3.3	2.1	2	2.9	1.3	1.3
Dodge City	3.2	3.2	3.1	2.8	3.7	3.5	4.1	2.4	2.4	4.1	1.9	1.8
El Paso	3.3	3.6	3.9	2.2	3.6	4	4.5	2.8	3	4.7	2.1	2.2
Ely	3.6	3.3	3.8	3	3.7	4.5	4.6	2.6	2.8	4.2	1.6	1.7
Fort Worth	2.8	3	2.9	2.4	3.5	3.5	3.5	2.4	2.3	3.3	1.7	1.7
Great Falls	3.4	2.8	3	3.5	3.6	4	3.4	1.8	1.9	2.4	1	1
Lake Charles	2.7	2.9	3	2.4	3.2	3.5	3.1	2.3	2.5	2.7	1.6	1.6
Madison	2.9	2.7	2.8	3	3.4	3.5	3	2	2	2.6	1.3	1.3
Medford	3.1	2.9	2.8	3	3.8	3.7	3	1.9	1.8	1.8	1.1	0.9
Miami	2.8	3	3.2	2.2	2.9	3.2	3.1	2.4	2.6	3.6	2.2	2.2
Nashville	2.7	2.7	2.7	2.6	3.3	3.3	3.1	2	2.1	2.4	1.3	1.3
New York	2.6	2.4	2.5	2.5	2.8	2.8	2.7	1.6	1.8	2.2	1.1	1
Omaha	3	2.9	3	2.9	3.7	3.6	3.3	2	2	3	1.4	1.4
Phoenix	3.5	3.7	3.9	2.4	3.8	3.8	4.4	2.6	2.7	4.4	2	1.9
Santa Maria	3.1	3.2	3	2.4	3.8	2.8	3.8	2.5	2.1	3.7	1.6	1.6
Seattle	3	2.7	2.4	3	3.8	3.1	2.7	1.6	1.5	1.7	0.8	0.8
Washington, DC	2.8	2.5	2.7	2.7	3.2	3	2.8	1.9	1.7	2.5	1.3	1.2

which are generally based upon direct measurements or hourly estimates, is much preferred over estimates of direct-normal availabilities which are based solely upon conversion of daily totals of total-horizontal radiation.

Hourly conversions from total to direct radiation Extensive research efforts have been directed toward the problem of estimating hourly values of direct-normal solar radiation based upon hourly measurements of total-horizontal radiation. (Using other weather parameters to estimate direct-normal radiation under general sky conditions generally does not work well because no other weather parameter records the cloud situation in the position of the sun, which is critically important.) The formula given here is the result of these efforts. It is described more fully in Ref. 9.

The formula is described in terms of a functional relationship between k_T, the percent of possible total-horizontal radiation, and k_B the percent of possible direct-normal radiation. These are defined as follows:

$$k_T = I_H/(I_{ext}\sin\alpha)$$
$$k_B = I_H/I_{ext}$$

(2.24)

where I_H is the intensity of total-horizontal radiation, I_{ext} is the intensity of direct-normal radiation outside the earth's atmosphere, α is the solar altitude angle, and I_B is the intensity of direct-normal radiation. Although (2.24) describes these parameters in terms of instantaneous values, they can be applied to hourly integrated values; in this case I_H and I_B represent hourly integrals of radiation, and the expression $(I_{ext}\sin\alpha)$ can be either integrated over the hour, or computed by using the altitude angle at the hour's midpoint. (Of course, I_{ext} is essentially constant over an hour.)

TABLE 2.39   Mean Monthly Values of Total Precipitable Water for Clear Sky Conditions. Values Given in Centimeters.

	Jan.	Feb.	Mar.	Apr.	May	Jun.	Jul.	Aug.	Sep.	Oct.	Nov.	Dec.
Albuquerque, NM	0.47	0.45	0.47	0.52	0.72	1.03	1.73	1.73	1.28	0.84	0.59	0.48
Apalachicola, FL	1.67	1.63	1.77	2.14	2.60	3.35	3.86	3.95	3.58	2.65	1.95	1.77
Bismarck, ND	0.48	0.52	0.59	0.85	1.26	1.83	2.11	2.01	1.48	1.05	0.70	0.56
Brownsville, TX	2.06	2.07	2.17	2.57	3.57	3.39	3.59	3.67	3.75	3.00	2.46	2.21
Boston, MA	0.74	0.70	0.84	1.16	1.63	2.23	2.60	2.51	2.14	1.53	1.16	0.84
Cape Hatteras, NC	1.35	1.18	1.28	1.60	2.19	2.75	3.58	3.58	2.86	2.21	1.53	1.42
Caribou, ME	0.51	0.48	0.60	0.81	1.24	1.79	2.16	2.06	1.66	1.24	0.90	0.60
Charleston, SC	1.43	1.43	1.55	1.89	2.55	3.26	3.79	3.79	3.27	2.34	1.71	1.51
Columbia, MO	0.82	0.73	0.97	1.41	1.79	2.52	2.79	2.75	2.22	1.59	1.18	0.96
Dodge City, KS	0.61	0.60	0.68	0.97	1.40	1.95	2.46	2.35	1.39	1.20	0.83	0.68
El Paso, TX	0.64	0.62	0.68	0.73	0.98	1.51	2.19	2.22	1.87	1.16	0.83	0.71
Ely, NV	0.47	0.45	0.44	0.49	0.68	0.93	1.17	1.25	0.84	0.64	0.57	0.44
Ft. Worth, TX	1.10	1.17	1.32	1.82	2.42	3.02	3.38	3.34	2.94	2.06	1.47	1.22
Fresno, CA	0.88	0.84	0.84	0.93	1.12	1.26	1.63	1.67	1.35	1.07	1.02	0.88
Great Falls, MT	0.51	0.49	0.50	0.61	0.88	1.21	1.30	1.29	1.02	0.75	0.62	0.51
Lake Charles, LA	1.67	1.61	1.75	2.18	2.69	3.32	3.87	3.79	3.42	2.41	1.90	1.79
Madison, WI	0.60	0.60	0.70	1.12	1.58	2.28	2.51	2.42	1.95	1.44	0.88	0.70
Medford, OR	1.05	0.95	0.90	0.93	1.17	1.47	1.53	1.53	1.33	1.18	1.21	0.99
Miami, FL	2.21	2.18	2.31	2.51	3.01	3.74	3.85	3.96	4.03	3.42	2.66	2.32
Nashville, TN	1.02	0.93	1.12	1.60	1.95	2.59	3.04	3.01	2.72	1.76	1.26	1.20
N.Y., NY	0.76	0.75	0.90	1.25	1.74	2.32	2.71	2.65	2.31	1.57	1.24	0.95
N. Omaha, NE	0.62	0.67	0.79	1.13	1.68	2.34	2.69	2.59	1.98	1.42	0.89	0.70
Phoenix, AZ	0.93	0.86	0.85	1.03	1.15	1.52	2.91	2.95	2.10	1.42	0.96	0.91
Raleigh, NC	1.04	1.00	1.12	1.41	2.06	2.69	3.16	3.12	2.54	1.76	1.25	1.05
Santa Maria, CA	1.09	1.09	1.10	1.18	1.39	1.53	1.86	1.81	1.69	1.42	1.26	1.13
Seattle, WA	1.07	1.07	1.02	1.10	1.29	1.62	1.74	1.84	1.67	1.49	1.27	1.14
Washington, DC	0.92	0.87	1.00	1.41	1.98	2.59	3.01	2.95	2.46	1.70	1.24	0.98

SOURCE: Ref. 14.

The relationship between k_T and k_B is a piecewise linear function and is shown in Fig. 2.30 where

$$k_B = A_i k_T + B_i \qquad (2.25)$$

where A_i and B_i are the slope and intercept terms for the different linear segments of the function. The linear segments of the function correspond to different intervals for k_T; those intervals and their corresponding values A_i and B_i are given in Table 2.40.

This formula is applied by first substituting the measured value of I_H and the appropriate altitude angle into Eq. (2.24) to find k_T. Next, a value for k_B is computed using Eq. (2.25) for each hour; finally, Eq. (2.24) can be inverted to compute the hourly value of direct-normal radiation:

$$I_B = k_B I_{ext} \qquad (2.26)$$

The value of I_{ext} is given in Table 2.4.

The relationship (2.25) was derived from direct and total data recorded at five different U.S. sites. These data indicated that the relationship between k_T and k_B is fairly independent of location, season, and time of day. The formula was derived by stratifying the data according to which of the intervals 0.0–0.1, 0.1–0.2, ..., 0.9–1.0 contained the observed value of k_T. For each such interval the mean of observed values of k_B was computed; Eq. (2.25) is simply the piecewise linear function which connects these mean values. The formula is only correct on the average. Actual values of k_B will be distributed around this curve of mean values. An indication of the deviations of actual values of k_T from those predicted by Eq. (2.25) which can be realistically anticipated is given in Fig. 2.31. This is a typical plot of the standard deviations of the distribution of k_B about the mean values of k_B for each of the intervals listed above.

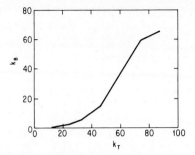

Fig. 2.30 Percent of direct-normal radiation k_B as a function of percent of total horizontal radiation k_T. (*From Ref. 14.*)

TABLE 2.40   Coefficients for the Piecewise Linear Function $k_B = A_i k_T + B_i$

Interval for k_T	A_i	B_i
0.00, 0.05	0.04	0.00
0.05, 0.15	0.01	0.002
0.15, 0.25	0.06	−0.006
0.25, 0.35	0.32	−0.071
0.35, 0.45	0.82	−0.246
0.45, 0.55	1.56	−0.579
0.55, 0.65	1.69	−0.651
0.65, 0.75	1.49	−0.521
0.75, 0.85	0.27	0.395

Mean Daily Conversions from Total-Horizontal to Direct-Normal Radiation A very simple formula for converting from monthly means of daily totals of total-horizontal radiation to monthly means of daily totals of direct-normal radiation is described in Ref. 16. The formula is

$$\overline{B} = -3.17 + 1.31\,\overline{H}\cos z, \qquad \text{kWh/m}^2 \cdot \text{day} \qquad (2.27)$$

where $\overline{B}$ is the monthly mean of the daily total of direct-normal radiation, $\overline{H}$ is the monthly mean of the daily total of total-horizontal radiation, and z is an "average monthly zenith angle." This average zenith angle was taken to be the zenith angle for the 15th day of the month and evaluated at the hour angle equal to one-quarter the sunrise hour angle. This formula represents a linear, least squares fit to a data sample including measurements from Livermore, CA, Raleigh, NC, and Ft. Hood, TX. For this linear regression the square of the correlation coefficient is $R^2 = 0.86$, and the standard deviation about the regression line is $s = 0.68$.

As the regression indicators R^2 and s show, this formula is not very reliable; for instance, the standard deviation about the regression line is about 15 percent of the average of all of the monthly means of daily totals of direct-normal radiation. Consequently, the formula should only be used when no better alternative is available. Although this same formula could be applied to convert an individual daily total, it would be even less reliable.

An alternative to using this formula is to decompose the mean daily total-horizontal value $\overline{H}$ into hourly values using Fig. 2.32, as described below; this can be followed with an application of the hourly conversion formula Eq. (2.25).

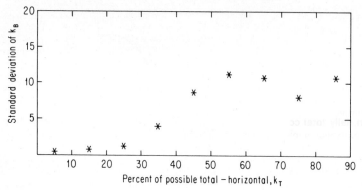

Fig. 2.31 A plot of the standard deviations of k_B values about the line shown in Figure 2.30.

Conversion from Horizontal to Tilted Surfaces

Measurements of total solar radiation on tilted surfaces are quite limited, whereas records of total-horizontal solar radiation measurements are more widely available. Consequently, formulas for converting from total-horizontal radiation data to radiation availabilities on tilted surfaces are frequently employed. Three such conversion formulas are presented in this section.

It is suggested that these formulas only be used when tilted-surface data or availability tables are not available, because use of these formulas is more difficult and probably less reliable.

The first conversion formula assumes that hourly total-horizontal radiation data is available; this is the best of the three techniques. The second formula begins with daily total-horizontal data; it consists of decomposing these daily total values into hourly values, and then using the first formula. The final formula converts from mean daily totals of total-horizontal measurements directly to mean daily totals of total radiation on a south-facing, tilted surface.

Hourly conversions   When hourly values of total-horizontal radiation measurements are available, formulas (2.25) and (2.26) given earlier can be applied to separate the total radiation into its direct beam and diffuse components, I_B and I_D. Those formulas give the direct-normal component I_B; the diffuse component on a horizontal surface is given by the difference

$$I_D = I_H - I_B \tag{2.28}$$

Conversion of these direct and diffuse components to the tilted surface is performed separately. The beam component on the tilted surface is given by

$$I_{BT} = I_B \cos i \tag{2.29}$$

where i is the solar incidence angle for the surface described in Eq. (2.10).

The diffuse component on the tilted surface is given by

$$I_{DT} = 0.5(1 + \cos \beta)I_D + 0.5(1 - \cos \beta)\rho I_T \tag{2.30}$$

The first term on the right-hand side represents the contribution of diffuse sky radiation; it assumes a sky of uniform brightness. The second term represents total radiation reflected onto the surface from the ground. The angle β is the surface tilt angle. The parameter ρ is ground reflectance. Typical values for ρ are 0.1 to 0.2 for terrain free of snow cover, and 0.7 for complete snow cover. A more thorough description of reflectivity values is given in Ref. 17 and Chap. 5.

Finally, the total radiation on the tilted surface is simply the sum of the direct and diffuse components:

$$I_T = I_{BT} + I_{DT} \tag{2.31}$$

Converting from daily to hourly data   This section presents a method for decomposing daily values of total horizontal radiation into hourly values. This decomposition technique is only valid on the average; it does not reproduce the large solar radiation intensity variations which can occur from one hour to the next due to changes in cloud cover.

The technique consists of using the curves in Fig. 2.32 taken from Ref. 12. This figure displays the ratio I_H/H of hourly total-horizontal radiation to daily total-horizontal radiation as a function of the day length and the time from solar noon. The day length is specified by its sunrise hour angle h_{sr}, on the abscissa. The different curves correspond to solar hour midpoints at increasing times from solar noon; for example, the curve labeled 1½ corresponds to the solar hours from 10 to 11 A.M. and from 1 to 2 P.M. According to Ref. 13, these curves are quite accurate in the mean.

Mean daily total conversion from horizontal to a south-facing, tilted surface   This section presents a simple technique for converting from the monthly average daily total radiation on a horizontal surface to the monthly average daily total radiation on a surface tilted upward toward the south. This conversion technique is described in Ref. 18; a brief discussion of its accuracy and an extension of the technique to surfaces of different azimuth are presented in Ref. 19.

The technique consists of approximating the ratio $\overline{R}$ in the equation

$$\overline{H}_T = \overline{R}\,\overline{H} \tag{2.32}$$

where $\overline{H}$ is the given monthly mean of daily total-horizontal radiation, and $\overline{H}_T$ is the desired monthly mean of daily total radiation on the tilted surface. By assuming that the sky and ground are each uniformly bright, the ratio $\overline{R}$ can be expressed as follows:

$$\overline{R} = \left(1 - \frac{\overline{H}_D}{\overline{H}}\right)\overline{R}_B + (1/2)\frac{\overline{H}_D}{\overline{H}}(1 + \cos\beta) + (1/2)\,\rho\,(1 - \cos\beta) \tag{2.33}$$

In this equation β is the surface-tilt angle, ρ is the ground reflectance, $\overline{H}_D$ is the monthly average daily diffuse radiation on a horizontal surface, and $\overline{R}_B$ is the ratio of the monthly average daily direct radiation on the tilted surface to the monthly average daily direct radiation on a horizontal surface. The ratio $\overline{R}_B$ can be approximated by using the trigonometric expression for its value outside the earth's atmosphere:

$$\overline{R}_B = \frac{\cos(L - \beta)\cos\delta\sin h'_{sr} + (\pi/180)h'_{sr}\sin(L - \beta)\sin\delta}{\cos L\cos\delta\sin h_{sr} + (\pi/180)h_{sr}\sin L\sin\delta} \tag{2.34}$$

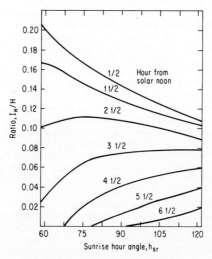

Fig. 2.32   Curves for decomposing daily total-horizontal radiation H into hourly totals I_H.

In this equation, L = latitude, β = surface-tilt angle, δ = solar declination, h_{sr} = sunrise hour angle, and h'_{sr} = surface sunrise hour angle.

The application of Eq. (2.33) requires a value for $\overline{H}_D/\overline{H}$. The approximation given in Ref. 18 is

$$\overline{H}_D/\overline{H} = 1.390 - 4.027\,\overline{K}_T + 5.531\,\overline{K}_T^2 - 3.108\overline{K}_T^3 \qquad (2.35)$$

In this expression, $\overline{K}_T = \overline{H}/\overline{H}_0$, where $\overline{H}_0$ is the average daily extraterrestial solar radiation on a horizontal surface for the month. Note that Eq. (2.35) is based on an early solar constant value of 1.39 W/m² used before satellite data for I_0 were available. When evaluating $\overline{K}_T$ for use in Eq. (2.35) the early solar constant value must be used. Values for $\overline{K}_T$ are given in Table 2.9. An alternative is to calculate H_0 by applying Eq. (2.22) to the middle day for the month.

Estimating Radiation from Cloud Cover and Sunshine Records

There are many locations which lack solar radiation records but which do have records of the amount of sunshine or cloud cover. Numerous formulas for estimating solar radiation values from these records have been developed. Two types of such formulas are discussed here, one for daily data and the other for hourly data. In both cases, the formulas produce estimates of total-horizontal radiation.

Estimating daily total-horizontal radiation from sunshine records   The most frequently used relationship between solar radiation and duration of sunshine is the equation:

$$\overline{H} = \overline{H}_0(a + b\,\mathrm{PP}) \qquad (2.36)$$

In this equation PP represents the percent of possible sunshine, that is, the recorded sunshine duration divided by the maximum possible sunshine duration for the day. The variables $\overline{H}$ and $\overline{H}_0$ are the estimated total-horizontal radiation and the extraterrestrial radiation on a horizontal surface, respectively.

The values of the parameters a and b are derived empirically at locations where both types of data are available. Table 2.41 contains a list of values for these parameters for 18 different locations covering a variety of climate and vegetation types. This table is reprinted from Ref. 7. These 18 different parameter sets permit the computation of daily total-horizontal radiation from recorded daily sunshine duration for essentially any geographic or climatic region.

These models will vary in accuracy; a major source of variation is the basic difference between the capabilities of the total solar radiation instrument, the pyranometer, and the sunshine recorder. The sunshine recorder only measures the duration of sunshine above a preset threshold level, whereas the pyranometer records radiation intensity. Consequently, a formula such as (2.36) can only be expected to be accurate on the average at best. Even this accuracy can suffer if the threshold setting of the sunshine recorder doesn't reasonably match that of the instrument used for the data sample from which the coefficients a and b were derived.

Sunshine and cloud cover models for estimating hourly total-horizontal radiation   Frequently, hourly solar radiation data are desired as input to solar energy systems models at locations for which no such data have been recorded. Consequently, models have been developed for generating hourly total-horizontal solar radiation data from the more commonly available hourly records of sunshine duration or opaque cloud cover. A rather elaborate set of such hourly solar radiation sunshine duration/opaque cloud cover models has been developed by the U.S. National Atmospheric and Oceanographic Administration[14]. These models were employed in preparing the Generated Hourly Solar Data Set described earlier in this chapter (see Table 2.8).

Since extensive hourly solar data sets have already been generated using these models, there is seldom a need for applying these models again. It is generally simpler to obtain and use one of these existing (real or generated) solar data bases. Consequently, these NOAA models are not described here; full details can be found in Ref. 14.

In the situation where a small amount of "typical" hourly solar data is desired, it can be derived using techniques described earlier in this section. The examples of the following section of this chapter indicate some of the possible methods that can be employed.

TABLE 2.41 Linear Regression Coefficients, *a* and *b*, for the Solar Radiation/Sunshine Duration Model [Eq. (2.36)]; from Ref. 7.

Location	Climate*	Veg.†	Sunshine hours in percentage of possible Range	Avg.	*a*	*b*
Albuquerque, NM	BS-BW‡	E‡	68–85	78	0.41	0.37
Atlanta, GA	Cf	M	45–71	59	0.38	0.26
Blue Hill, MA	Df	D	42–60	52	0.22	0.50
Brownsville, TX	BS	GDsp	47–80	62	0.35	0.31
Buenos Aires, Arg.	Cf	G	47–68	59	0.26	0.50
Charleston, SC	Cf	E	60–75	67	0.48	0.09
Darien, Manchuria	Dw	D	55–81	67	0.36	0.23
El Paso, TX	BW	Dsi	78–88	84	0.54	0.20
Ely, NV	BW	Bzi	61–89	77	0.54	0.18
Hamburg, Germany	Cf	D	11–49	36	0.22	0.57
Honolulu, HI	Af	G	57–77	65	0.14	0.73
Madison, WI	Df	M	40–72	58	0.30	0.34
Malange, Angola	Aw-BS	GD	41–84	58	0.34	0.34
Miami, FL	Aw	E-GD	56–71	65	0.42	0.22
Nice, France	Cs	SE	49–76	61	0.17	0.63
Poona, India (Monsoon Dry)	Am	S	25–49	37	0.30	0.51
Stanleyville, Congo	Af	B	65–89	81	0.41	0.34
Tamanrasset, Algeria	BW	Dsp	34–56	48	0.28	0.39
			76–88	83	0.30	0.43

*Climatic classification based on Trewartha's climate map.
†Vegetation classification based on Küchler's map.
‡Definitions of Abbreviations:

Af — Tropical forest climate, constantly moist; rainfall all through the year

Am — Tropical forest climate, monsoon rain; short dry season, but total rainfall sufficient to support rain forest

Aw — Tropical forest climate, dry season in winter

BS — Steppe or semiarid climate

BW — Desert or arid climate

Cf — Mesothermal forest climate. Constantly moist; rainfall all through the year.

Cs — Mesothermal forest climate. Dry season in winter.

Df — Microthermal snow forest climate. Constantly moist; rainfall all through the year.

Dw — Microthermal snow forest climate. Dry season in winter.

B — Broadleaf evergreen trees

Bzi — Broadleaf evergreen, shrubform, minimum height 3 feet, growth singly or in groups or patches.

D — Broadleaf deciduous trees

Dsi — Broadleaf deciduous, shrubform, minimum height 3 feet, plants sufficiently far apart that they frequently do not touch

Dsp — Broadleaf deciduous, shrubform, minimum height 3 feet, growth singly or in groups or patches

E — Needleleaf evergreen trees

G — Grass and other herbaceous plants

GD — Grass and other herbaceous plants; broadleaf deciduous trees

GDsp — Grass and other herbaceous plants; broadleaf deciduous, shrubforms, minimum height 3 feet, growth singly or in groups or patches

M — Mixed: broadleaf deciduous and needleleaf evergreen trees

S — Semideciduous: broadleaf evergreen and broadleaf deciduous trees

SE — Semideciduous: broadleaf evergreen and broadleaf deciduous trees; needleleaf evergreen trees

SELECTED EXAMPLES

These examples illustrate the use of the information in this chapter.

Example 1

For a clear day in January in Phoenix, AZ, find
  1. The conversion from local to solar time
  2. The hourly solar positions
  3. The hourly air mass for the direct solar rays
  4. The intensity of the direct solar beam each hour
  5. The intensity of the direct solar beam on a south-facing 45° tilted surface each hour
  6. The integrated solar energy on this 45° tilted surface for the day

Solution

Because a specific date is needed for specifying the time conversion and the solar position, we assume that the date is January 15. Using Eqs. (2.2) and (2.3), the equation of time (in hours) is

$$Eot = 12 + [0.1236 \sin x - 0.0043 \cos x + 0.1538 \sin 2x + 0.0608 \cos 2x]$$

$$x = 360 (15 - 1)/365.242$$

$$= 13.80°$$

Thus, Eot $= 12.15$ h; since the "12 hour" term merely reflects the fact that solar time is measured from noon, it can be omitted from calculations which refer to the usual 12-hour clock. Since Phoenix's longitude is 112.017°, the conversion from (mountain) standard time (standard meridian is 105°W) to solar time is [from (2.4)]

$$\text{solar time} = \text{standard time} - 0.15 - 0.468.$$

$$= \text{standard time} - 0.618.$$

For example, solar noon occurs at 0.618 h (or 37 m and 5 s) past noon, mountain standard time.
  The rest of the calculations for this example are performed in solar time. The hourly values for solar positions, air mass, and solar intensities are calculated at the midpoint of each solar hour. The calculations for 10:30 A.M. solar time are as follows:

$$h_s \text{ (solar hour angle)} = + 22.5° \qquad \text{(from Sec. 2)}$$

$$\delta \text{ (solar declination)} = \arcsin \{0.39795 \cos [0.98563(15 - 173)]\}$$

$$= -21.27° \qquad \text{[Eq. (2.1)]}$$

$$L \text{ (Phoenix's latitude)} = 33.43°$$

$$\alpha \text{ (solar altitude)} = \arcsin [\cos 33.4° \cos (-21.3°) \cos 22.5°$$

$$+ \sin 33.4° \sin (-21.3°)]$$

$$= 31.2° \qquad \text{[Eq. (2.7)]}$$

$$a_s \text{ (solar azimuth)} = \arcsin \left(\frac{\cos (-21.3°) \sin 22.5°}{\cos 31.2°} \right)$$

$$= 24.6°$$

$$[\text{since } \cos 22.5° > \frac{\tan \delta}{\tan L} \text{ in Eq. (2.9)}]$$

$$z \text{ (solar zenith angle)} = 90° - 31.2°$$

$$= 58.8° \qquad \text{[Eq. (2.6)]}$$

$$m_0 \text{ (air mass)} = \sec 58.8°$$

$$= 1.93 \qquad \text{[Eq. (2.18)]}$$

This is the sea-level air mass for this solar position; since the standard atmospheric pressures at sea level and Phoenix (obtainable from local National Weather Service or Federal Aviation Administration offices) are 1013 and 963 millibars (mbar), respectively, the air mass corrected for Phoenix's altitude is

$$m = (963/1013) \ 1.93$$
$$= 1.83$$

[Eq.(2.19)]

We use Eq. (2.23) to calculate the intensity of direct-normal radiation, with $I_{ext} = 1.41$ kW/m^2 (from Table 2.4) and $W = 0.5$ cm (estimated from values listed in Table 2.39).

$$I_B = 1.41 \exp\{-0.1457 \ (1.83) - 0.1617 \ [(0.5) \ (1.93)]^{0.25}\}$$
$$= 0.92 \text{ kW/m}^2$$

For a south-facing, 45° tilted surface, the solar incidence angle i is given by

$$\cos i = \cos 31.2° \cos 24.6° \sin 45° + \sin 31.2° \cos 45°$$
$$= 0.9162$$

[Eq.(2.12)]

Thus, $i = 23.6°$. The intensity of direct solar radiation on this south-facing, 45° tilted surface is

$$I_{BT} = 0.92 \cos 23.6$$
$$= 0.84 \text{ kW/m}^2$$

[Eq.(2.29)]

The results for the other hours are listed in the following table; note that sunrise and sunset occur at 7:00 A.M. and 5:00 P.M., respectively.

Solar time	h_s	α	a_s	z	m_0	m	I_B	i	I_{BT}
7:30	67.5	5.6	59.9	84.4	9.55	9.08	0.30	65.0	0.13
8:30	52.5	15.9	50.2	74.1	3.65	3.47	0.70	51.0	0.44
9:30	37.5	24.6	38.6	65.4	2.40	2.28	0.85	37.2	0.68
10:30	22.5	31.2	24.6	58.8	1.93	1.83	0.92	23.6	0.84
11:30	7.5	34.8	8.5	55.2	1.75	1.67	0.95	12.1	0.93

The afternoon values are symmetric.

The integrated direct-beam solar radiation on the south-facing, 45° tilted surface is approximately 6.04 kWh/m^2. This value was calculated by simply treating the midhour intensities as hourly average intensities, and summing. This estimate is slightly high. A trapezoidal approximation to this integrated energy gives a lower bound of 5.98 kWh/m^2.

Estimating the integrated total radiation for the day on this tilted surface requires an estimate of the diffuse component contribution. Figure 2.30 indicates that generally under these conditions of high direct-beam intensity ($k_B \sim 0.5$) the diffuse component only contributes about 10 to 20 percent of the total. Thus, a reasonable estimate for a daily total on this tilted surface would be 7 kWh/m^2. A more precise estimate would require a knowledge of the sky conditions (e.g., are there bright clouds?) and the reflectance in the foreground of the surface.

Example 2

What is the best orientation for a flat-plate collector to be used for space heating in the winter in St. Louis, MO? What reductions in solar energy availability will result from changes in the collector's tilt angle or azimuth angle? Answer these same questions for a flat-plate collector to be used for a year-round hot water application.

Solution

Although the solar radiation availability maps indicate that there is slightly less solar energy available at St. Louis than at Columbia, using the tabulated data for Columbia to address the questions at St. Louis should be quite adequate.

For a south-facing collector the winter availabilities of total radiation for various tilt angles in Columbia are as follows (from Table 2.36):

Tilt angle, degrees	30	45	60	75	90
Mean daily energy, kWh/m²	3.0	3.2	3.3	3.1	2.9

Thus, the optimal tilt angle for a south-facing surface in St. Louis in the winter is about 60° from horizontal. Varying from this by 15° results in a 3 to 6 percent reduction in total solar radiation availability; a 30 degree variation from 60° produces a 9 to 12 percent reduction.

From Tables 2.25–2.30, the winter availabilities of total solar radiation for a 45° tilted surface of various azimuth angles in Columbia are:

Azimuth, degrees	0	±15	±30	±45	±60	±75	±90
Mean daily energy, kWh/m²	3.2	3.1	3.0	2.8	2.5	2.2	1.9

The percent reduction for a 60° tilted surface in St. Louis will be similar. Thus, 15° and 30° variations in azimuth from due south will produce approximately 3 and 6 percent reductions in mean daily energy, respectively, and a 60° tilted surface facing southeast or southwest will intercept about 12 percent less solar energy than a south-facing, 60° tilted surface.

The selection of an optimal tilt angle for a year-round hot water application is somewhat design dependent. For example, from Table 2.37 we see that a south-facing surface tilted at 30° intercepts the most solar energy on an annual basis in Columbia. However, this tilt angle would not be optimal for meeting a uniform hot-water requirement over the year. As the following table indicates, surfaces tilted at either 45° or 60° are better in terms of seasonal uniformity of incident solar energy.

Tilt angle	Spring	Summer	Fall	Winter	Annual
30	5.0	6.2	4.2	3.0	4.6
45	4.7	5.7	4.4	3.2	4.5
60	4.3	4.9	4.3	3.3	4.2

The reductions in mean daily incident solar energy due to variations in collector azimuth from south on an annual basis for a 45° tilted surface in Columbia are as follows:

Azimuth angle	0	±15	±30	±45	±60	±75	±90
Mean daily energy, kWh/m²	4.5	4.4	4.3	4.2	4.1	3.9	3.6
Percent reduction	0	2	4	7	9	13	20

Similar reductions would occur at St. Louis. It should be noted that these reductions due to variations in azimuth from south refer to mean daily energy on an annual basis; significantly larger percent reductions occur in the winter, as the calculations above indicate.

Example 3

Which is better for providing domestic hot water, a fixed flat-plate collector, or a tracking concentrating collector? Answer this question for Medford, OR, Albuquerque, NM, Charleston, SC, Boston, MA and Omaha, NE.

Solution

Several factors should be considered in addition to solar radiation availability. These include
  • operating temperature—concentrating collectors out-perform flat-plate collectors above approximately 50°C.

■ collector complexity—flat-plate collectors do not require tracking, but their receiver is more complex.

■ costs—presently these collectors cost about the same per unit area.

■ aesthetics—perhaps the flat-plate collector can be integrated into the architecture more easily.

The solution given here will only compare these two types with respect to incident solar radiation; all other considerations will be ignored.

The only collector orientation and tracking schemes which will be considered will be fixed, south-facing flat-plate collectors tilted at 45° and parabolic troughs tracking about either an east-west horizontal axis or a polar axis. The mean daily availabilities of total radiation for the flat-plate collector will be denoted by $\overline{H}_T$, and the mean daily availabilities of direct radiation to the east-west and polar-axis parabolic troughs will be denoted by $\overline{H}_{D,EW}$ and $\overline{H}_{D,P}$, respectively. The seasonal and annual availabilities for these collectors for the given five locations are as follows:

		Spring	Summer	Fall	Winter	Annual
Albuquerque	$\overline{H}_T$	6.7	6.5	6.3	5.6	6.3
	$\overline{H}_{D,EW}$	5.4	6.0	5.3	4.8	5.4
	$\overline{H}_{D,P}$	7.3	7.7	6.8	5.4	6.8
Boston	$\overline{H}_T$	4.3	4.7	3.5	2.7	3.8
	$\overline{H}_{D,EW}$	2.6	3.1	2.4	1.9	2.5
	$\overline{H}_{D,P}$	3.5	4.0	3.0	2.1	3.2
Charleston	$\overline{H}_T$	4.8	4.6	4.3	3.7	4.4
	$\overline{H}_{D,EW}$	3.2	3.0	2.9	2.6	2.9
	$\overline{H}_{D,P}$	4.2	3.7	3.6	2.9	3.6
Medford	$\overline{H}_T$	5.0	6.5	4.0	2.2	4.4
	$\overline{H}_{D,EW}$	3.4	5.5	2.9	1.3	3.3
	$\overline{H}_{D,P}$	4.3	7.1	3.6	1.4	4.1
Omaha	$\overline{H}_T$	4.9	5.7	4.2	3.4	4.6
	$\overline{H}_{D,EW}$	3.2	4.3	3.1	2.6	3.3
	$\overline{H}_{D,P}$	4.4	5.7	4.0	3.0	4.3

This table indicates that the solar energy available to an east-west parabolic trough is significantly less than the solar energy available to either of the other two schemes. This is because it suffers from rather large incidence angle cosine losses in the morning and evening throughout the year.

The table also shows that the direct radiation available to a parabolic trough tracking about a polar axis is approximately the same as the total radiation incident upon a 45°, south-facing, flat-plate collector. Thus, the selection between these should probably be made in terms of other design factors such as those mentioned above.

Example 4

The average percents of possible sunshine for the midseason months of April, July, October, and January in San Antonio, TX are 56, 77, 67, and 49, respectively. Calculate average daily totals of total-horizontal radiation for these months in San Antonio, and compare the results with both the world maps, Figs. 2.14–2.17, and the U.S. seasonal maps, Figs. 2.18–2.21, and with the values given in Table 2.10.

Solution

The appropriate formula is (2.36) given in Sec. 5 of this chapter:

$$\overline{H} = \overline{H}_0 (a + b \ \text{PP})$$

The most appropriate climate and vegetation classifications pairing for San Antonio for which the coefficients a and b are given in Table 2.41 are those for Brownsville, TX. The resulting equation is

$$\overline{H} = \overline{H}_0 (0.35 + 0.31 \ \text{PP})$$

The latitude of San Antonio is 32.5°; interpolation between the values for latitudes of 24° and 32° gives the values for $\overline{H}_0$ for the 21st day of each month. The calculation of $\overline{H}$ using these values of $\overline{H}_0$ and the given percent of possible sunshine is straightforward. For April,

$$\overline{H} = 10.5\,(0.35 + 0.31\,(.56))$$
$$= 5.5\ \text{kWh/m}^2$$

The results for all the calculations are as follows:

Month	PP	$\overline{H}_0$ (kWh/m²)	$\overline{H}$ (kWh/m²)
April	0.56	10.5	5.5
July	0.77	11.2	6.6
October	0.67	7.5	4.2
January	0.49	6.1	3.1

The inaccuracies introduced by using the values for H_0 for the 21st day of each month instead of the average monthly values of H_0 are not significant when compared to the accuracy of the basic formula employed or the accuracy of the given percent of possible sunshine data.

The following table lists these values for $\overline{H}$ along with values for $\overline{H}$ read from the U.S. and world solar radiation maps of Sec. 4 and the values listed for San Antonio in Table 2.10.

	From this example	From U.S. maps, Figs. 2.18–2.21	From world maps, Figs. 2.14–2.17	From Table 2.10
April (Spring)	5.5	5.7	5.8	5.2
July (Summer)	6.6	6.8	7.3	7.5
October (Fall)	4.2	4.6	4.7	4.7
January (Winter)	3.1	3.1	3.1	3.3

This table indicates that all four of these sources of values for $\overline{H}$ in this example are in general agreement.

Example 5

Estimate the mean daily values of direct-normal radiation and of total radiation on a south-facing, 45° surface for each of the four seasons in San Antonio, TX.

Solution

The easiest way to obtain these estimates is to read the seasonal maps for direct-normal radiation, Figs. 2.22–2.25, and the maps of total radiation on south-facing, 45° tilted surfaces, Figs. 2.26–2.29. The resulting mean daily values are as follows (kWh/m²).

	Spring	Summer	Fall	Winter
Mean daily direct-normal	4.8	6.2	4.8	3.8
Mean daily total on a south-facing, 45° surface	5.1	5.5	5.1	4.3

For comparisons, the mean daily values of total radiation on the south-facing, 45° surface will also be computed using formulas 2.32–2.35. The values of the relevant parameters used in these calculations are as follows:

	April	July	October	January
Date used	21	21	21	21
δ (degrees)	12	20	−11	−20
h_{sr} (degrees)	98	103	83	77
R_B	1.01	0.94	1.33	1.49
K_T (Table 2.10)	0.500	0.647	0.584	0.541
H_D/H	0.37	0.26	0.31	0.34
ρ	0.2	0.2	0.2	0.2
R	0.98	0.95	1.21	1.30
H (U.S. maps)	5.7	6.8	4.6	3.1
H_T	5.6	6.5	5.6	4.0

These values for $\overline{H}_T$, in kWh/m², agree reasonably well with the values read from the maps.

Finally, use of Eq. (2.27) provides another method of computing the mean daily values of direct-normal radiation for San Antonio. The following table summarizes these calculations.

	April	July	October	January
H (from U.S. maps)	5.7	6.8	4.6	3.1
z (degrees)	30	26	48	55
B	5.5	6.7	5.8	3.9

The values of $\overline{B}$ are all within 20 percent of those read from the U.S. maps; for better accuracy, calculations based upon hourly data should be used.

REFERENCES

1. *American Ephemeris and Nautical Almanac*, published annually by the Nautical Almanac Office, U.S. Naval Observatory.
2. H. M. Woolf, "On the Computation of Solar Elevation Angles and the Determination of Sunrise and Sunset Times, NASA-TM-X-1646, p. 23, NASA, September 1968.
3. P. L. Harrison, "A Device for Finding True North," *Solar Energy*, vol. 15, 1974, pp. 303–308.
4. J. Hickey, et al., "Extraterrestrial Solar Irradiance Measurements from the NIMBUS 6 Satellite," Proceedings of Sharing the Sun, Conf. of Amer. Section of ISES, vol. 1, 1976, International Solar Energy Society.
5. N. Robinson, *Solar Radiation*, chap. 3, Elsevier, 1966.
6. E. Carter, et al., *Listing of Solar Radiation Measuring Equipment and Glossary*, published by the Center For Environmental and Energy Studies, U. of Alabama, Huntsville, ERDA/NASA/31293/76/3, 1976.
7. G. Löf, et al., "World Distribution of Solar Radiation," Report No. 21, Solar Energy Laboratory, University of Wisconsin, July 1966.
8. "Solar Radiation Observation Stations with Complete Listing of Data Archived by the National Climatic Center, Asheville, North Carolina and Initial Listing of Data Not Currently Archived," prepared by Center for Environmental and Energy Studies, University of Alabama at Huntsville, Nov. 1966; available from NTIS, U.S. Department of Commerce; DSE/1024-2.
9. C. Randall and M. Whitson, *Hourly Insolation and Meteorological Data Bases Including Improved Direct Insolation Estimator*, Aerospace Corp. Report No. ATR-(7592)-1, El Segundo, CA, December, 1977.
10. E. Boes, et al., "Availability of Direct, Total, and Diffuse Solar Radiation for Fixed and Tracking Collectors in the USA," Sandia Laboratories Report SAND 77-0885; available from NTIS, United States Department of Commerce, Springfield, VA 22161.
11. *Climatic Atlas of the United States*, Environmental Data Service of the Environmental Science Services Administration, U.S. Department of Commerce, 1968.
12. *Applications of Solar Energy for Heating and Cooling of Buildings*, edited by R. Jordan and B. Liu, ASHRAE GRP 170, American Society of Heating, Refrigerating, and Air-Conditioning Engineers, Inc., 1977.
13. R. Jordan and B. Liu, "The Long-Term Average Performance of Flat-Plate Solar Energy Collectors," *Solar Energy*, vol. 7, no. 2, pp. 53–74, 1963.
14. SOLMET: Volume 2. "Rehabilitation of Hourly Solar Radiation Data," National Climatic Center, Asheville, NC, 1977.

15. N. Majumdar, et al, "Prediction of Direct Solar Radiation for Low Atmospheric Turbidity," *Solar Energy*, vol. 13, pp. 383–394, 1972.
16. E. Boes and I. Hall, "Estimating Monthly Means of Daily Totals of Direct Normal Solar Radiation and of Total Solar Radiation on a South-Facing, 45° Tilted Surface," Sandia Laboratories Report SAND 77-0874, 1977, Albuquerque, NM.
17. B. Hunn and D. Calafell, "Determination of Average Ground Reflectivity for Solar Collectors," Proceedings of Sharing the Sun, Conference of the International Solar Energy Society, Amer. Section, vol. 1, Winnipeg, August 1976.
18. B. Liu and R. Jordan, "Daily Insolation on Surfaces Tilted Toward the Equator," *Trans. ASHRAE*, pp. 526–541, 1962.
19. S. Klein, "Calculation of Monthly Average Insolation on Tilted Surfaces," *Solar Energy*, vol. 19, pp. 325–329, 1977.
20. V. Cinquemani, et al., "Input Data for Solar Systems," U.S. Dept. of Commerce, National Climatic Center, 1978.

Chapter **3**

Principles of Optics Applied to Solar Energy Concentrators

W. T. WELFORD
Blackett Laboratory, Imperial College, London, U. K.

and

R. WINSTON
Enrico Fermi Institute and Department of Physics,
University of Chicago, Chicago, IL

BASIC IDEAS IN GEOMETRICAL OPTICS APPLIED TO SOLAR CONVERSION

The Concepts of Geometrical Optics

Geometrical optics is used as the basic tool in designing almost any solar optical system, image-forming or not. We use the intuitive ideas of a ray of light, roughly defined as the path along which light energy travels, together with two types of surfaces which either reflect or transmit the light. When light is reflected from a smooth surface it obeys the well-known law of reflection that the incident and reflected rays make equal angles with the normal to the surface and that both rays and the normal lie in one plane. When light is transmitted the ray direction is changed according to the law of refraction, Snell's law, which states that the sine of the angle between the normal and the incident ray bears a constant-ratio to the sine of the angle between the normal and the refracted ray; again all three directions are coplanar.

A major part of the optical design and analysis of solar devices involves ray tracing, i.e., following the paths of rays through a system of reflecting and refracting surfaces. This is a well-known process in conventional lens design, but the requirements are somewhat different for solar systems so it will be convenient to state and develop the methods *ab initio*. This is because in conventional lens design the reflecting or refracting surfaces involved are almost always portions of spheres and the centers of the spheres lie on one straight line (axisymmetric optical system), so that special methods which take advantage of the simplicity of the forms of the surfaces and the symmetry can be used; nonimaging concentrators, for example, do not in general have spherical surfaces, in fact sometimes there is no explicit analytical form for the surfaces, although usually there is an axis or a plane of symmetry. We shall find it most convenient therefore to develop ray-tracing schemes based on vector formulations, but with the details dealt with in computer programs on an *ad hoc* basis for each different shape.

In geometrical optics we represent the power density across a surface by the density of ray intersections with the surface and the total power by the number of rays. This notion, reminiscent of the useful but outmoded "lines of force" in electrostatics, works as follows: We take N rays spaced uniformly over the entrance aperture of a concentrator at an angle of incidence θ, as in Fig. 3.1. Suppose that after tracing the rays through the system only N' emerge through

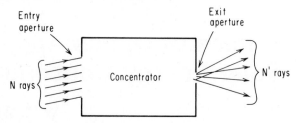

Fig. 3.1   Determining the transmission of a concentrator by ray tracing.

the exit aperture, the dimensions of this latter being determined by the desired concentration ratio; the remaining $(N-N')$ rays are lost by processes which will become clear when we consider some examples. Then the power transmission for the angle θ is N'/N. This can be extended to cover a range of angles θ as required. Clearly N must be taken large enough to ensure that a thorough exploration of possible ray paths is made.

Formulation of the Ray-Tracing Procedure

In order to formulate a ray-tracing procedure suitable for all cases it is convenient to put the laws of reflection and refraction into vector form. Figure 3.2 shows the geometry with unit

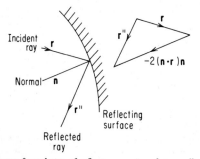

Fig. 3.2   Vector formulation of reflection; $\mathbf{r}$, $\mathbf{r}''$, and $\mathbf{n}$ are all unit vectors.

vectors $\mathbf{r}$ and $\mathbf{r}''$ along the incident and reflected rays and a unit vector $\mathbf{n}$ along the normal pointing into the reflecting surface. Then it is easily verified that the law of reflection is expressed by the vector equation

$$\mathbf{r}'' = \mathbf{r} - 2(\mathbf{n \cdot r})\mathbf{n} \tag{3.1}$$

as in the diagram.

Thus to raytrace "through" a *reflecting* surface we have first to find the point of incidence, a problem of geometry involving the direction of the incoming ray and the known shape of the surface, then we have to find the normal at the point of incidence, again a problem of geometry and finally we have to apply Eq. (3.1) to find the direction of the reflected ray. The process is then repeated if another reflection is to be taken into account. These stages are illustrated in Fig. 3.3. Naturally, in the numerical computation the unit vectors are represented by their components, i.e., the direction cosines of the ray or normal with respect to some Cartesian coordinate system used to define the shape of the reflecting surface.

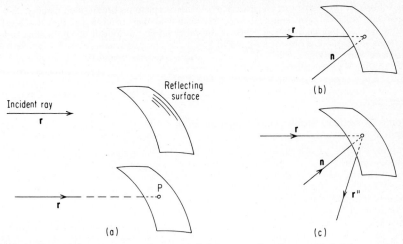

Fig. 3.3   The stages in ray tracing a reflection: (*a*) find the point of incidence *P*; (*b*) find the normal at *P*; (*c*) apply Eq. (3.1) to find the reflected ray **r**″.

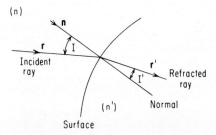

Fig. 3.4   Vector formulation of refraction.

Raytracing through a *refracting* surface is similar but we first have to formulate the law of refraction vectorially. Figure 3.4 shows the relevant unit vectors; it is similar to Fig. 3.2 except that **r**′ is a unit vector along the refracted ray. We denote by n, n′ the refractive indices of the media on either side of the refracting boundary; the refractive index is a parameter of a transparent medium which is related to the speed of light in the medium; specifically, if *c* is the speed of light *in vacuo* the speed in a transparent material medium is *c*/n, where n is the refractive index. For visible light, values of n range from unity to about 3 for usable materials. The law of refraction is usually stated in the form

$$n' \sin I' = n \sin I, \tag{3.2}$$

where *I* and *I*′ are the angles of incidence and refraction as in the figure, and where the coplanarity of the rays and the normal is understood. The vector formulation

$$n'\mathbf{r}' \times \mathbf{n} = n\mathbf{r} \times \mathbf{n} \tag{3.3}$$

contains everything, since the modulus of a vector product of two unit vectors is the sine of the angle between them. This can be put in the form most useful for raytracing by multiplying through vectorially by **n** to give

$$n'\mathbf{r}' = n\mathbf{r} + (n'\mathbf{r}{\cdot}\mathbf{n} - n\mathbf{r}{\cdot}\mathbf{n})\mathbf{n} \tag{3.4}$$

and this is the preferred form for raytracing.* The complete procedure then parallels that for reflection explained by means of Fig. 3.3; i.e., we find the point of incidence, then the direction of the normal, and finally the direction of the refracted ray. Details of the application to lens systems are given, e.g., by Welford (1974).

If a ray travels from a medium of refractive index n towards a boundary with another of index n' < n then it can be seen from Eq. (3.2) that it would be possible to have sin I' greater than unity. Under this condition it is found that the ray is completely reflected at the boundary; this is called "total internal reflection" and we shall find it a useful effect in the design of some solar collectors.

Some Elementary Properties of Image-Forming Optical Systems

In principle, the use of raytracing tells us all there is to know about the geometrical optics of a given optical system, image-forming or not. However, raytracing alone is of little use for inventing new systems having good properties for a given purpose: we need to have ways of describing the properties of optical systems in terms of general performance. In this section we shall introduce some of these concepts.

Consider first a thin converging lens such as would be used as a magnifier or such as has been proposed to increase the flux on the active area of a solar photocell. In section this looks like Fig. 3.5. By the term "thin" we mean that its thickness can be neglected for the purposes

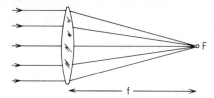

Fig. 3.5 A thin converging lens bringing parallel rays to a focus; since the lens is technically "thin" we do not have to specify the exact plane in the lens from which the focal length f is measured.

under discussion. Elementary experiments show us that if we have rays coming from a point at a great distance to the left, so that they are substantially parallel as in the figure, the rays meet approximately at a point F, the focus. The distance from the lens to F is called the focal length, denoted by f. Elementary experiments also show that if the rays come from an object of finite size at a great distance the rays from each point on the object converge to a separate focal point and we get an image; this is, of course, what happens when a burning glass forms an image of the sun or when the lens in a camera forms an image on the film. It is indicated in Fig. 3.6 where the object subtends the (small) angle 2θ; it is then found that the size of the image is $2f\theta$. This is easily seen by considering the rays through the center of the lens, since these pass through undeviated.

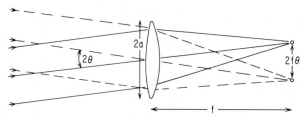

Fig. 3.6 An object at infinity has an angular subtense 2θ; a lens of focal length f forms an image of size $2f\theta$.

*The method of using equation (3.4) numerically is not so obvious as for equation (3.2) since the coefficient of **n** in Eq. (3.4) is actually $n' \cos I' - n \cos I$ and thus it might appear that we have to find **r'** before we can use the equation. The procedure is to find $\cos I'$ via equation (3.2) first and then equation (3.4) is needed to give the complete three-dimensional picture of the refracted ray.

Figure 3.6 contains one of the fundamental concepts we use in solar concentrator theory, the concept of a beam of light of a certain diameter and angular extent. The diameter is that of the lens, $2a$ say, and the angular extent is given by 2θ. These two can be combined as a product, usually without the factor 4, giving θa, a quantity known by various names such as extend, étendue, acceptance, Lagrange invariant, etc. It is, in fact, an invariant through the optical system, provided that there are no obstructions in the light beam and provided we ignore certain losses due to properties of the materials such as absorption and scattering. For example, at the plane of the image the étendue becomes the image height θf multiplied by the convergence angle a/f of the image-forming rays, giving again θa. In discussing three-dimensional (3D) systems, e.g., an ordinary lens such as we have supposed Fig. 3.6 to represent, it is convenient to deal with the square of this quantity, $a^2\theta^2$; this is also sometimes called the étendue but generally it is clear from the context and from dimensional considerations whether the two-dimensional (2D) or 3D form is intended. The 3D form has an interpretation which is fundamental to the theme of this chapter. Suppose we put an aperture of diameter $2f\theta$ at the focus of the lens, as in Fig. 3.7; then this system will only accept rays within the angular range $\pm\theta$ and inside the diameter $2a$; now suppose a flux of radiation B (W/m$^2\cdot$sr) is incident on the lens from the left. The system will actually accept a total flux $\pi^2\theta^2a^2$ W; thus the étendue or acceptance θ^2a^2 is a measure of the power flow which can pass through the system.

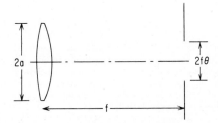

Fig. 3.7   An optical system of acceptance, throughput, or étendue $a^2\theta^2$.

The same discussion shows how the concentration ratio C appears in the context of classical optics: the accepted power $\pi^2\theta^2\alpha^2$ W must flow out of the aperture to the right of the system, if our assumptions above about how the lens forms an image are correct,* and if the aperture has the diameter $2f\theta$. Thus our system is acting as a concentrator of sunlight with concentration ratio $C = (2a/2f\theta)^2 = (a/f\theta)^2$ for the input semiangle θ.

To relate these ideas to practical solar energy collection, we have a distant source which subtends a semiangle of approximately 0.005 rd ($\frac{1}{4}°$), so that this is the given value of θ, the collection angle. Clearly for a given diameter of lens we gain by reducing the focal length as much as possible.

Aberrations of Image-Forming Optical Systems

According to the simplified picture presented in the previous section there is no reason why we could not make a lens system with indefinitely large concentration ratio by simply decreasing the focal length sufficiently. This is, of course, not so, partly because of aberrations in the optical system and partly because of the fundamental limit on concentration prescribed by the second law of thermodynamics.

We can explain the concept of aberrations by reference again to our example of the thin lens of Fig. 3.5. We suggested that the parallel rays shown all converged after passing through the lens to a single point F; in fact, this is only true in the limiting case when the diameter of the lens is taken as indefinitely small. The theory of optical systems under this condition is called paraxial optics or Gaussian optics and it is a very useful approximation for getting at the principal large-scale properties of image-forming systems. If we take a simple lens with a diameter which is a sizeable fraction of the focal length, say $f/4$, we find that the rays from a single point object do not all converge to a single image point. We can show this by raytracing. We first set up a proposed lens design as in Fig. 3.8; the lens has curvatures (reciprocals of radii) c_1 and c_2, center thickness d and refractive index n. If we neglect the central thickness for the moment,

*As we shall see, these assumptions are only valid for limitingly small apertures and objects.

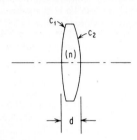

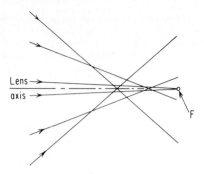

Fig. 3.8 Specification of a singlet lens; the curvature c_1 is positive as shown, c_2 is negative.

Fig. 3.9 Rays near the focus of a lens showing spherical aberration.

then it is shown in specialized treatments (e.g., Welford 1974) that the focal length f is given in paraxial approximation by

$$1/f = (n-1)(c_1 - c_2) \qquad (3.5)$$

and we can use this to get the system to have roughly the required paraxial properties.

Now we can trace rays through the system as specified, using the method outlined above (details of ray-tracing methods for ordinary lens systems are given by Welford [1974]); these will be exact or finite rays, as opposed to the paraxial rays which are implicit in the gaussian optics approximation. The results for the lens in Fig. 3.8 would look as in Fig. 3.9; this shows rays traced from an object point on the axis at infinity, i.e., rays parallel to the axis.

In general for a convex lens the rays from the outer part of the lens aperture meet the axis closer to the lens than the paraxial rays. This effect is known as spherical aberration. (The term is misleading, since the aberration can occur in systems with nonspherical refracting surfaces, but there seems little point in trying to change it at the present late date in the subject.)

Spherical aberration is perhaps the simplest of the different aberration types to describe but it is one of many others. Even if we were to choose the shapes of the lens surfaces to eliminate the spherical aberration or if we eliminated it in some other way we should still find that the rays from object points away from the axis did not form point images, i.e., there would be oblique or off-axis aberrations. Also the refractive index of any material medium changes with the wavelength of the light, and this produces "chromatic aberrations" of various kinds. We do not at this stage need to go into the classification of aberrations very deeply, but this preliminary sketch is necessary to show the relevance of aberrations to the attainable concentration ratio.

The Effect of Aberrations in an Image-Forming System on the Concentration Ratio

Questions of the extent to which it is theoretically possible to eliminate aberrations from an image-forming system have not yet been fully answered; in this chapter we shall attempt to give answers adequate for our purposes. For the moment let us accept that it is possible to eliminate spherical aberration completely but not the off-axis aberrations and we suppose that this has been done for the simple collector of Fig. 3.7. The effect will be that some rays of the beam at the extreme angle θ will fall outside the defining aperture of diameter $2f\theta$. We can see this more clearly by representing the aberration by means of a spot diagram; this is a diagram in the image plane with points plotted to represent the intersections of the various rays in the incoming beam. Such a spot diagram for the extreme angle θ might appear as in Fig. 3.10; the ray through the center of the lens (the "principal" ray in lens theory) meets the rim of the collecting aperture by definition and thus a considerable amount of the flux does not get through. Conversely it can be seen that, in this case at least, some flux from beams at a larger angle than θ will be collected.

We display this information on a graph such as Fig. 3.11; this shows the proportion of light collected at different angles up to the theoretical maximum, θ_{max}. An ideal collector would behave according to the solid line, i.e., collecting all light flux within θ_{max} and none outside.

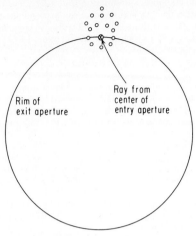

Fig. 3.10 A spot diagram for rays from the beam at the maximum entry angle for an image-forming concentrator; some rays miss the edge of the exit aperture due to aberrations and the étendue is thus less than the theoretical maximum.

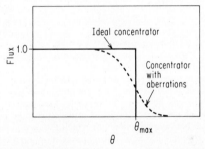

Fig. 3.11 A plot of collection efficiency against angle; the ordinate is the proportion of flux entering the collector aperture at angle θ which emerges from the exit apertures.

At this point it may be objected that all we need to do is to enlarge the collecting aperture slightly to achieve the first requirement and the second requirement doesn't matter; however, we recall that our aim in solar collector design is frequently to achieve maximum concentration because of the requirement for high working temperature so that the collector aperture must not be enlarged beyond $2f\theta$ diameter (see Chap. 8 and 9).

Frequently in discussions of aberrations in textbooks or geometrical optics the impression is given that aberrations are in some sense "small"; this is true in optical systems designed and made to form reasonably good images; e.g., camera lenses, but these systems do not operate with large enough convergence angles (a/f in the notation for Fig. 3.6) to approach the maximum theoretical concentration ratio. If we were to try to use a conventional image-forming system under such conditions we should find the aberrations would be very large and they would severely depress the concentration ratio. Roughly, we can say that this is one limitation which has led to the development of the new, nonimaging solar devices.

The Optical Path Length and Fermat's Principle

There is another way of looking at geometrical optics and the performance of optical systems which we also need to outline.

We noted that the speed of light in a medium of refractive index n is c/n, where c is the speed in vacuum. Thus light travels a distance s in the medium in time $s/v = ns/c$, or, the time taken to travel a distance s in a medium of refractive index n is proportional to ns. The quantity ns is called the "optical path length" corresponding to the length s. Suppose we have a point source O emitting light into an optical system, as in Fig. 3.12; we can trace any number of rays through the system, as outlined above and then we can mark off along these rays points which are all at the same optical path length from O, say P_1, P_2 . . . ; we do this by making the sum of the optical path lengths from O in each medium the same, i.e.,

$$\Sigma \, ns = \text{const.} \tag{3.6}$$

in an obvious notation. These points can be joined to form a surface (we are supposing rays out of the plane of the diagram to be included) and this would be a surface of constant phase of the light waves if we were thinking in terms of the wave theory of light.* We call it a geometrical wavefront, or simply a wavefront, and we can construct wavefronts at all distances along the bundle of rays from O.

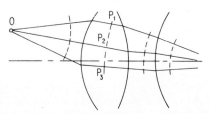

Fig. 3.12 Rays and (in broken line) geometrical wavefronts.

We now introduce a principle which is not so intuitive as the laws of reflection and refraction but which leads to results which are indispensable to our development of the theme of this chapter. It is based on the concept of optical path length and it is a way of predicting the path of a ray through an optical medium. Suppose we have any optical medium, which can have lenses and mirrors and can even have regions of continuously varying refractive index. We wish to predict the path of a light ray between two points in this medium, say A and B in Fig. 3.13. We can propose an infinite number of possible paths, of which three are indicated, but unless A and B happen to be object and image, and we assume they are not, only one or perhaps a small finite number of paths will be physically possible, i.e., paths which rays of light could take according to the laws of geometrical optics. Fermat's principle in the form usually used states that a physically possible ray path is one for which the optical path length along it from A to B is an extremum as compared to neighboring paths. For "extremum" we can often write "minimum," as in Fermat's original statement. It is possible to derive all of geometrical paths from Fermat's principle; it also leads to the result that the geometrical wavefronts are orthogonal to the rays (the theorem of Malus and Dupin), i.e., the rays are normals to the wavefronts. This in turn tells us that if there is no aberration, i.e., all rays meet at one point, then the wavefronts must be portions of spheres; thus also if there is no aberration the optical path length from object point to image point is the same along all rays. We arrive at an alternative way of expressing aberrations, in terms of the departure of wavefronts from the ideal spherical shape. This concept will be useful when we come to discuss the different senses in which an image-forming system can form "perfect" images.

Fig. 3.13 Fermat's principle; it is assumed that in the diagram the medium has a continuously varying refractive index; the full-line path has a stationary optical path length from A to B and is therefore a physically possible ray path.

The Generalized Étendue or Lagrange Invariant and the Phase Space Concept

We now introduce a concept which is essential to the development of the optical principles of all solar collectors. We recall that there is a quantity, $a^2\theta^2$, which is a measure of the power

---

*This construction does not give a surface of constant phase near a focus or near an edge of an opaque obstacle, but this does not affect the present applications.

accepted by the system; here a is the radius of the entrance aperture and θ is the semiangle of the beams accepted, and we found that in paraxial approximation for an axisymmetrical system this is invariant through the optical system. Actually we considered only the regions near the entrance and exit apertures but it is shown in specialist texts on optics that the same quantity can be written down for any region inside a complex optical system; there is one slight complication, that if we are considering a region of refractive index different from unity, say the inside of a lens or a prism, the invariant is written $n^2 a^2 \theta$. The reason for this can be seen from Fig. 3.14, which shows a beam at the extreme angle θ entering a plane parallel plate of glass of refractive index n; inside the glass the angle is $\theta' = \theta/n$, by the law of refraction,* so that the invariant in this region is

$$\text{etendue} = n^2 a^2 \theta^2. \tag{3.7}$$

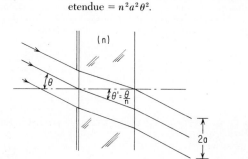

Fig. 3.14 Inside a medium of refractive index n the étendue becomes $n^2 a^2 \theta'^2$.

We might try to use the étendue to obtain an upper limit for the concentration of a system as follows. We suppose we have an axisymmetric optical system of any number of components, i.e., not necessarily the simple system sketched in Fig. 3.7. The system will have an entrance aperture of radius a which may be the rim of the front lens or possibly some limiting aperture inside the system, as in Fig. 3.15; an incoming parallel beam may emerge parallel, as indicated in the figure, or not, and this will not affect the result, but to simplify the argument it is easier to imagine a parallel emerging beam from an aperture of radius a'. The concentration ratio is by definition $(a/a')^2$ and if we use the étendue invariant and assume that the initial and final media are both air or vacuum, i.e., refractive index unity, the concentration ratio becomes $(\theta'/\theta)^2$. Since from obvious geometrical considerations θ' cannot exceed $\pi/2$ this suggests as a theoretical upper limit to the concentration $(\pi/2\theta)^2$.

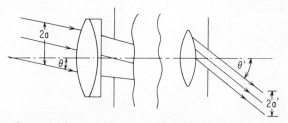

Fig. 3.15 The étendue for a multielement optical system with an internal aperture stop.

Unfortunately this argument is invalid because the étendue as we have defined it is essentially a paraxial quantity and thus it is not necessarily an invariant for angles as large as $\pi/2$; in fact the effect of aberrations in the optical system is to ensure that the paraxial étendue is *not* an invariant outside the paraxial region, so that we have not found the correct upper limit to the concentration.

There is, as it turns out, a suitable generalization of the étendue to rays at finite angles to the axis and we now proceed to explain this. The concept has been known for some time but it has not been used to any extent in classical optical design, so that it is not described in many

*The paraxial approximation is implied, so that $\sin\theta \sim \theta$.

texts. It applies to solar optical systems of any or no symmetry and of any structure, i.e., refracting, reflecting, or continuously varying refractive index.

Let the system be bounded by homogeneous media of refractive indices n and n' as in Fig. 3.16 and suppose we have a ray traced exactly between the points P and P' in the respective input and output media. We wish to consider the effect of small displacements of P and of small changes in direction of the ray segment through P on the emergent ray, so that these changes define a beam of rays of a certain cross section and angular extent. In order to do this we set up a cartesian coordinate system $Oxyz$ in the medium and another, $Ox'y'z'$ in the output medium. The positions of the origins of these coordinate systems and the directions of their axes are quite arbitrary with respect to each other, to the directions of the ray segments and, of course, to the optical system. We specify the input ray segment by the coordinates of P (x,y,z) and the direction cosines of the ray (L,M,N) and similarly for the output segment. We can now represent small displacements of P by increments dx and dy to its x and y coordinates and we can represent small changes in the direction of the ray by increments dL and dM to the direction cosines for the x and y axes. Thus we have generated a beam of area $dxdy$ and angular extent defined by $dLdM$; this is indicated in Fig. 3.17 for the y-section.* Corresponding increments dx', dy', dL' and dM' will occur in the output ray position and direction.

Then the invariant quantity turns out to be $n^2\,dxdydLdM$, i.e., we have

$$n'^2dx'dy'dL'dM' = n^2dxdydLdM \tag{3.8}$$

The proof of this theorem depends on other concepts in geometrical optics which we do not need in this chapter hence we need not give the details here.

The physical meaning of Eq. (3.8) is that it gives the changes in the rays of a beam of a certain size and angular extent as it passes through the system. If there are apertures in the input medium which produce this limited étendue and if there are no apertures elsewhere to cut off the beam then the accepted light power emerges in the output medium, so that the étendue, as defined, is a correct measure of the power transmitted along the beam. It may seem at first remarkable that the choice of origins and directions of the coordinate systems is quite arbitrary†; however, it is not very difficult to show that the generalized étendue or

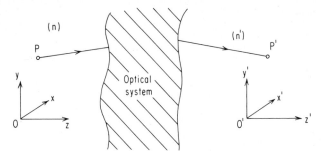

Fig. 3.16   The generalized étendue.

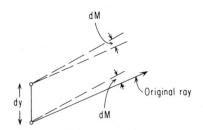

Fig. 3.17   The generalized étendue in the y section.

*It is necessary to note that the increments dL and dM are in direction cosines, not angles; thus in Fig. 3.17 the notation on the figure should not be taken to mean that dM is the angle indicated, merely that it is a measure of this angle.

†This is not quite true; it can be seen from the formulation of the theorem that we cannot choose a direction for the z axis which is perpendicular to the ray direction.

Lagrange invariant as calculated in one medium is independent of coordinate translations and rotations. This, of course, must be so if it is to be a meaningful physical quantity.

The generalized étendue is sometimes written in terms of the optical direction cosines $p = nL$, $q = nM$, when it takes the form

$$dx\,dy\,dp\,dq \tag{3.9}$$

We can now use the étendue invariant to calculate the theoretical maximum concentration ratios of concentrators. Consider first a 2D solar concentrator as in Fig. 3.18; from Eq. (3.8) we have for any ray bundle which traverses the system

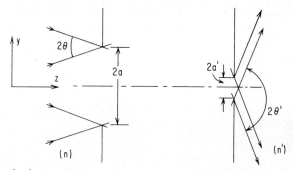

Fig. 3.18 The theoretical maximum concentration ratio for a two-dimensional optical system.

$$n\,dy\,dM = n'\,dy'\,dM' \tag{3.10}$$

and integrating over y and M we obtain

$$4\,na\sin\theta = 4\,n'a'\sin\theta' \tag{3.11}$$

so that the concentration ratio is

$$a/a' = \frac{n'\sin\theta'}{n\sin\theta} \tag{3.12}$$

In this result a' is a dimension of the exit aperture large enough to permit any ray that reaches it to pass and θ' is the largest angle of all the emergent rays. Clearly θ' cannot exceed $\pi/2$ so the theoretical maximum concentration ratio is

$$C_{\max} = \frac{n'}{n\sin\theta} \tag{3.13}$$

Similarly for the 3D case we can show that for an axisymmetric concentrator the theoretical maximum is

$$C_{\max} = \left(\frac{a}{a'}\right)^2 = \left(\frac{n'}{n\sin\theta}\right)^2 \tag{3.14}$$

where again θ is the input semiangle.

The results in Eq. (3.13) and (3.14) are maximum, not necessarily achievable values. Indeed, it is not possible to design a practical 3D concentrator in which the theoretical maximum is attained unless we admit certain impractical designs. We find in practice that if the exit aperture has the diameter given by Eq. (3.14) some of the rays within the incident collecting angle and aperture do not pass it. We also find that in some systems to be described some incident rays are actually turned back by internal reflections and never reach the exit aperture. In addition there are losses due to absorption, imperfect reflection, etc. Thus Eqs. (3.13) and (3.14) give the *theoretical* upper bounds on performance of solar collectors.

The Skew Invariant

There is an invariant associated with the path of a skew ray through an axisymmetric optical system which is useful in collector design. Let S be the shortest distance between the ray and the axis, i.e., the length of the common perpendicular, and let γ be the angle between the ray and the axis. Then the quantity

$$h = n S \sin \gamma \qquad (3.15)$$

is an invariant through the whole system. If the medium has a continuously varying refractive index the invariant for a ray at any coordinate z_1 along the axis is obtained by treating the tangent of the ray at that z value as the ray and using the refractive index value at the point where the ray cuts the transverse plane z_1.

If we use a dynamical analogy then h corresponds to the angular momentum of a particle following the ray path and the skew invariant theorem corresponds to conservation of angular momentum.

Different Versions of the Concentration Ratio

We now have some different definitions of concentration ratio for solar collectors and it is desirable to clarify them by using different names. First, we established upper limits for concentration ratio in 2D and 3D systems, given respectively by Eqs. (3.13) and (3.14). These depend only on the input angle and the input and output refractive indices. Clearly we can call either expression the *theoretical maximum* concentration ratio.

Secondly, an actual system will have entry and exit apertures of dimensions $2a$ and $2a'$; these can be width or diameter for 2D or 3D systems, respectively. The exit aperture may or may not transmit all rays which reach it but in any case the ratios (a/a') or $(a/a')^2$ define a *geometric* concentration ratio.

Thirdly, given an actual system we can trace rays through it and determine the proportion of incident rays within the collecting angle which emerge from the exit aperture. This process will yield an *optical* concentration ratio.

Finally we could make allowances for attenuation in the concentrator by reflection losses, scattering, manufacturing errors, and absorption in calculating the optical concentration ratio. We could call the result the "practical optical concentration ratio."

BASIC DESIGNS OF IMAGE-FORMING SOLAR CONCENTRATORS

In this section we shall study image-forming concentrators of conventional form, e.g., paraboloidal mirrors, short focal length lenses, etc. and estimate their performance. Then we shall show how the departure from ideal performance suggests a principle for the design of nonimaging concentrators, the "edge-ray principle" as we shall call it.

Some General Properties of Ideal Image-Forming Concentrators

The simplest hypothetical image-forming solar concentrator functions as in Fig. 3.19. The rays are coded to indicate that rays from one direction from the sun are brought to a focus at one point in the exit aperture, i.e., the concentrator images the sun (or other source) at the exit aperture. If the exit medium is air, then exit angle θ' must be $\pi/2$ for maximum concentration. Clearly such a concentrator would in practice have to be constructed with glass or another medium of refractive index greater than unity forming the exit surface, as in Fig. 3.20, and the

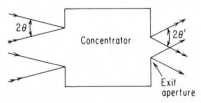

Fig. 3.19 An image-forming concentrator; an image of the source at infinity is formed at the exit aperture of the concentrator.

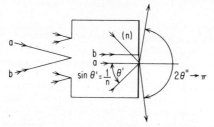

Fig. 3.20 In an image-forming concentrator of maximum theoretical concentration ratio, the final medium in the concentrator would have to have a refractive index n greater than unity; the angle θ' in this medium would be arcsin $1/n$, giving an angle $\pi/2$ in the air outside.

angle θ' in the glass would be such that $\sin \theta' = 1/n$, so that the emergent rays just fill the required $\pi/2$ angle. For typical materials the angle θ' would be about 40°.

Figure 3.20 brings out an important point about the optics of such a concentrator. We have labeled the central or principal ray of the two extreme angle beams a and b respectively and at the exit end these rays have been drawn normal to the exit face; this would be essential if the concentrator were to be used with air as the final medium since if rays a and b were not normal to the exit face some of the extreme angle rays would be totally internally reflected (see above) and thus the concentration ratio would be reduced. In fact the condition that the exit principal rays should be normal to what is in ordinary lens design terms the image plane is not usually fulfilled; such an optical system, called "telecentric," needs to be specially designed and the requirement imposes constraints which would certainly worsen the attainable performance of a concentrator. We shall therefore assume that when a concentrator ends in glass of index n the absorber is placed in optical contact with the glass in such a way as to avoid potential losses through total internal reflection.

An alternative configuration for an image-forming concentrator would be as in Fig. 3.21; the concentrator collects rays over $2\theta_{max}$ as before but the internal optics form an image of the entrance aperture at the exit aperture, as indicated by the arrow coding of the rays. This would be in ordinary optics terminology a telescopic or afocal system. Naturally the same considerations about using glass or a similar material as the final medium hold as for the system of Fig. 3.19 and 3.20 and there is no difference between the systems as far as external behavior is concerned.

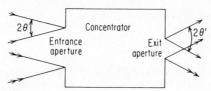

Fig. 3.21 An alternative configuration of an image-forming concentrator; the rays collected from an angle $\pm\theta$ form an image of the entrance aperture at the exit aperture.

If the concentrator terminates in a medium of refractive index n we can gain in maximum concentration ratio by a factor n or n^2, depending on whether it is a 2D or 3D system, as can be seen from Eqs. (3.13) and (3.14). This corresponds to having an extreme angle $\theta' = \pi/2$ in this medium; we then have to reinstate the requirement that the principal rays be normal to the exit aperture and we also have to ensure that the absorber can utilize rays of such extreme angles.

In practice there are problems in utilizing extreme collection angles approaching $\theta' = \pi/2$ whether in air or a higher index medium. There has to be very good matching at the interface between glass and absorber to avoid large reflection losses of grazing incidence rays and irregularities of the interface can cause losses through shadowing. Therefore, in practice it is recommended that values of θ' of say 60° be used; this represents only a small decrease from the theoretical maximum concentration, as can be seen from Eq. (3.13) and (3.14).

Thus in speaking of ideal concentrators we can also regard as *ideal* a system which brings all incident rays within θ_{max} out within θ'_{max} and inside an exit aperture a' given by Eq (3.12), i.e., $a' = na \sin \theta_{max}/n' \sin \theta'_{max}$. Such a concentrator will be ideal but not having the theoretical maximum concentration.

The concentrator sketched in Figs. 3.19 and 3.20 clearly must contain something like a very large aperture (small f-number) photographic objective, or perhaps a high-power microscope objective used in reverse. The speed of a photographic objective is indicated by its f-number or aperture ratio; thus an f/4 objective has a focal length 4 times the diameter of its entrance aperture. This description is not suitable for imaging systems in which the rays form large angles approaching $\pi/2$ with the optical axis for a variety of reasons; it is found that in discussing the resolving power of such systems the most useful measure of performance is the "numerical aperture," or NA, a concept introduced by Ernst Abbe in connection with the resolving power of microscopes. Figure 3.22 shows an optical system with entrance aperture of diameter $2a$. It forms an image of the axial object point at infinity and the semiangle of the cone of extreme rays is α'_{max}. Then the numerical aperture is defined by

$$NA = n' \sin \alpha'_{max} \qquad (3.16)$$

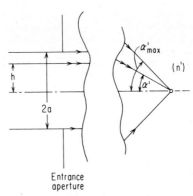

Fig. 3.22   The definition of the numerical aperture of an image-forming system. The N.A. is $n' \sin \alpha$.

where n' is the refractive index of the medium in the image space. We assume that all the rays from the axial object point focus sharply at the image point, i.e., there is no spherical aberration; then Abbe showed that off-axis object points will also be sharply imaged if the condition

$$h = n' \sin \alpha' \times const. \qquad (3.17)$$

is fulfilled for all the axial rays; in this equation h is the distance from the axis of the incoming ray and α' is the angle at which that ray meets the axis in the final medium. Equation (3.17) is a form of the celebrated Abbe sine condition for good image formation; it does not ensure perfect image formation for all off-axis object points but it ensures that aberrations which grow linearly with the off-axis angle are zero. These aberrations are various kinds of "coma" and the condition of freedom from spherical aberration and coma is called aplanatism.

Clearly a necessary condition for our image-forming concentrator to have the theoretical maximum concentration or even for it to be ideal but without theoretical maximum concentration is that the image formation should be aplanatic, although this is not, unfortunately, a sufficient condition.

The constant in Eq. 3.17 has the significance of a focal length; the definition of focal length for optical systems with media of different refractive indices in the object and image spaces is more complicated than for the thin lenses discussed above. In fact it is necessary to define two focal lengths, one for the input space and one for the output space, and their magnitudes are in the ratio of the refractive indices of the two media. In Eq. 3.17 it turns out that the constant is the input side focal length, which we shall denote by f.

From Eq. 3.17 we have for the input semiaperture

$$a = f \cdot NA \qquad (3.18)$$

and also from Eq. (3.12)

$$a' = \frac{a \sin \theta_{max}}{NA} \qquad (3.19)$$

so that by substituting from Eq. 3.18 into Eq 3.19 we have

$$a' = f \sin \theta_{max} \qquad (3.20)$$

where θ_{max} is the input semiangle. To see the significance of this result we recall that we showed that in an aplanatic system the focal length is a constant, independent of the distance h of the ray from the axis used to define it. Thus Eq. 3.20 tells us that in an imaging concentrator with maximum theoretical concentration the diameter of the exit aperture is proportional to the sine of the input angle; this is true even if the concentrator has an exit numerical aperture less than the theoretical maximum n', provided it is ideal in the sense defined above.

From the point of view of conventional lens optics the result of Eq. 3.18 is well known; it is simply another way of saying that the largest aperture aplanatic lens with air as the exit medium is $f/0.5$ since Eq. 3.20 tells us that $a' = f$. The importance of Eq. 3.20 is that it tells us something about one of the shape-imaging aberrations required of the system, namely distortion. A distortion-free lens imaging onto a flat field must obviously have an image height proportional to $\tan \theta$ so that our concentrator lens system is required to have what is usually called barrel distortion; this is illustrated in Fig. 3.23.

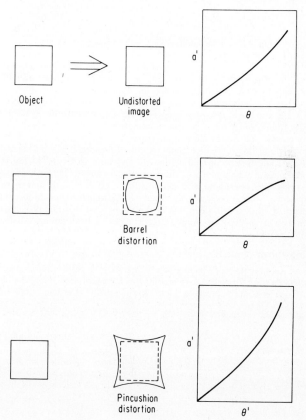

Fig. 3.23 Distortion in image-forming systems; the optical systems are assumed to have an axis of rotation symmetry.

Our picture of an imaging concentrator is gradually taking shape and we can begin to see that certain requirements of conventional imaging can be relaxed. Thus if we can get a sharp image at the edge of the exit aperture and if the diameter of the exit aperture fulfills the requirement of Eqs. (3.18) to (3.20), we do not need perfect image formation for object points at smaller angles than θ_{max}. For example, the image field perhaps could be curved, provided we take the exit aperture in the plane of the circle of image points for the direction θ_{max}, as in Fig. 3.24. Also the inner parts of the field could have point-imaging aberrations, provided these were not so large as to spill rays outside the circle of radius a'. Thus we see that an image-forming concentrator need not, in principle, be so difficult to design as an imaging lens

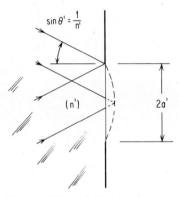

Fig. 3.24   A curved image field with a plane exit aperture.

since the aberrations need to be corrected only at the edge of the field; in practice this relaxation may not be very helpful because the outer part of the field is the most difficult to correct. However, this leads us to a valuable principle for nonimaging concentrators. Not only is it unnecessary to have good aberration correction except at the exit rim, but we do not even need point imaging at the rim itself; it is only necessary that all rays entering at the extreme angle θ_{max} should leave from some point at the rim and that the aberrations inside should not be such as to push rays outside the rim of the exit aperture. The above arguments need only a little modification to apply to the alternative configuration of imaging concentrator in Fig. 3.21, in which the entrance aperture is imaged at the exit aperture. Referring to Fig. 3.25, we can imagine that the optical components of the concentrator are forming an image at the exit aperture of an object at a considerable distance, rather than at infinity, and that this object is the entrance aperture. Alternatively, we can imagine that part of the concentrator is a collimating lens of focal length f, shown in broken line in the figure, and that this projects the entrance aperture to infinity with an angle subtending $2a/f$. The same considerations as before then apply to the aberration corrections.

Image-Forming Mirror Systems

In this section we examine the performance of mirror systems as concentrators. Concave mirrors have, of course, been used for many years as collectors for solar furnaces and the like. Historical material about such systems is given, e.g., by Krenz (1976). However, little seems to have

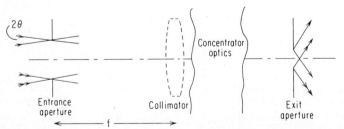

Fig. 3.25   An afocal concentrator shown as two image-forming systems.

been published in the way of angle-transmission curves for such systems. Consider first a simple paraboloidal mirror as in Fig. 3.26. As is well known, this mirror focuses rays parallel to the axis exactly to a point focus, or in our terminology it has no spherical aberration. However, the off-axis beams are badly aberrated; thus in the meridian section (the section of the diagram) it is easily shown by ray tracing that the edge rays at angle θ meet the focal plane further from the axis than the central ray, so that this cannot be an ideal concentrator even for emergent rays at angles much less than $\pi/2$. An elementary geometrical argument [see, e.g., D.A. Harper, *et al.* (1976)] shows how big the exit aperture must be to collect all the rays in the meridian section; referring to Fig. 3.27, we draw a circle passing through the ends of the mirror and the absorber (i.e., exit aperture); then by a well-known property of the circle, if the absorber subtends an angle $4\,\theta_{max}$ at the center of the circle it subtends $2\theta_{max}$ at the ends of the mirror, so that the collecting angle is $2\theta_{max}$. The mirror is not specified to be of any particular shape except that it must reflect all inner rays to the inside of the exit aperture. Then if the mirror subtends 2ϕ at the center of the circle we find

$$\frac{a'}{a} = \frac{\sin 2\theta_{max}}{\sin \phi}$$ (3.21)

and the minimum value of a' is clearly attained when $\phi = \pi/2$. At this point the optical concentration ratio is, allowing for the obstruction caused by the absorber,

$$\left(\frac{a}{a'}\right)^2 - 1 = \frac{1}{4\sin^2\theta_{max}}\left(\frac{\cos^2 2\theta_{max}}{\cos^2\theta_{max}}\right)$$ (3.22)

It can be seen that this is less than 25 percent of the theoretical maximum concentration ratio and less than 50 percent of the ideal for the emergent angle used.

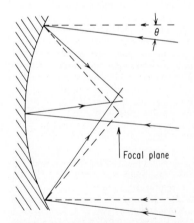

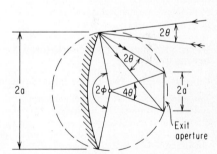

Fig. 3.26 Coma of a paraboloidal mirror. The rays of an axial beam are shown in broken line. The outer rays from the oblique beams at angle θ meet the focal plane further from the axis than the central rays of this beam.

Fig. 3.27 Collecting all the rays from a concave mirror.

If, as is usual, the mirror is paraboloidal, the rays used for this calculation are actually the extreme rays, i.e., the rays outside of the plane of the diagram all fall within the circle of radius a'.

The large loss in concentration at high apertures is basically because the single concave mirror used in this way has large coma, i.e., it does not satisfy Abbe's sine condition [Eq. (3.17)]. The large amount of coma introduced into the image spreads the necessary size of the exit aperture and so lowers the concentration below the ideal value.

There are image-forming systems which satisfy the Abbe sine condition and have large relative apertures. The prototype of these is the Schmidt camera, which has an aspheric plate and a spherical concave mirror, as in Fig. 3.28. The aspheric plate is at the center of curvature of the mirror and thus the mirror must be larger than the collecting aperture. Such a system would have the ideal concentration ratio for a restricted exit angle apart from the central obstruction but these would be practical difficulties in achieving the theoretical maximum. In any case a system of this complexity is clearly not to be considered seriously for solar energy work.

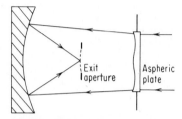

Fig. 3.28   The Schmidt camera. This optical system has no spherical aberration or coma, so that in principle it could be a good concentrator for small collecting angles. However, there are serious practical objections, such as cost and the central obstruction of the aperture.

Conclusions on Image-Forming Concentrators

It must be quite clear by now that, whatever the theoretical possibilities, practical concentrators based on image-forming designs fall a long way short of the ideal.

As to theoretical possibilities, it is certainly possible to have an ideal concentrator of theoretical maximum concentration ratio if we use a spherically symmetric geometry, a continuously varying refractive index, and quite unrealistic material properties (i.e., refractive index between 1 and 2 and no dispersion); this was proved by the example of the Luneburg lens and Luneburg and others (e.g., Morgan, 1958 and Cornbleet, 1976) have shown how designs suitable for perfect imagery for other conjugates can be obtained.

Perfect concentrators cannot be obtained with axial symmetry only and plane apertures if we restrict ourselves to a finite number of elements. However, if we permit unrealistic material properties we can approach indefinitely close to the ideal and in particular we can get within diffraction-limited imagery. Probably it is impossible to have a perfect concentrator with a continuously varying index with axial symmetry and plane apertures, but this has not yet been proven rigorously.

BASIC PRINCIPLES OF NONIMAGING CONCENTRATORS

Generalizing the Two-Dimensional Concentrator

Two-dimensional or troughlike nonimage-forming concentrators for solar energy concentration and similar purposes can be made to have the maximum theoretical concentration ratio. Thus if the concentrator (Fig. 3.29) collects light in an aperture of width $2a$ and over an angle $\pm \theta_i$ all of this light falls on the absorber surface of width $2a' = 2a \sin \theta_i$ at angles of incidence ranging from 0 to $\pi/2$. Concentrators of this kind and several elaborations and developments used for nonplane absorbers have been described by Winston (1974, 1975). If they are assumed to be infinitely long in the direction of the generators or, the practical equivalent, terminated with plane mirrors normal to the generators, they can be regarded as ideal concentrators in the following sense.

a. For the simple system of Fig. 3.29 the étendue or generalized Lagrange invariant of the entering beam appears at the exit aperture, since all rays meet the absorber.

b. A system with a convex absorber as in Fig. 3.30 can be designed to ensure that rays entering at the extreme angle θ_i meet the absorber tangentially if they do not meet its edge, and here also it is found that all rays meet the absorber and that the relation $2a' = 2a \sin \theta_i$ is satisfied, where now $2a'$ is the arc length round the absorber.

c. Let the source be at a finite distance as in Fig. 3.31; here PP' is the source, RR' is the absorber, and QQ' is the entry aperture of the concentrator. Then it can be shown that the étendue of the entering beam is

$$H = PQ' + P'Q - P'Q' - PQ \qquad (3.23)$$

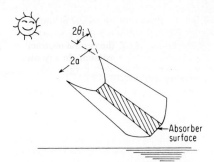

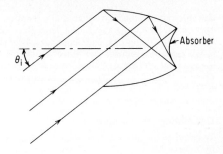

Fig. 3.29 A two-dimensional concentrator.   **Fig. 3.30** A concentrator with a nonplane absorber.

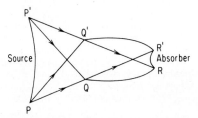

Fig. 3.31 A concentrator for a source at a finite distance.

and the concentrator can be designed to ensure that all of this flux is received by the absorber RR' provided it is convex to the radiation; also it can be arranged that the arc length $RR' = H/2$, so that the system has the maximum theoretical concentration. Equation (3.23) is equivalent to the celebrated Hottel string formula for radiation shape factors (Hottel, 1954).

R. K. Luneburg (1944, 1964) showed that two-dimensional media with continuously varying refractive index could be designed to form perfect images of one plane on another. This suggests the question, what kinds of nonimaging concentrator are possible in two dimensions if we admit media of continuously varying index? In this section we address this question and we obtain a prescription for designing what appears to be the most general form of two-dimensional concentrators.

Statement of the Problem

Let AB and $A'B'$ in Fig. 3.32 represent proposed entry and exit apertures of a concentrator and let the medium between and on either side of these curves have any arbitrary distribution of refractive index. The light source lies to the left of the entry aperture and Σ_α and Σ_β are

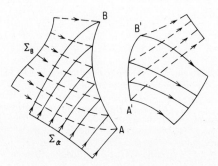

Fig. 3.32 Entry and exit surfaces for a general two-dimensional concentrator.

extreme rays, in the sense that at any point on AB the incoming rays will all be in the angle between the extreme rays*. We have thus defined a (two-dimensional) beam of a certain étendue H, although we have not shown how to calculate it.

We can now propose a set of wavefronts emerging from $A'B'$, of which Σ_α', and Σ_β' shall be extreme wavefronts, and we can postulate that the beam so defined shall have the same étendue H as the entering beam. For this to be possible the optical path length along the arc $A'B'$ must be equal to or greater than $H/2$, but given that condition there is considerable freedom of choice in the pattern of emergent rays.

Calculating the Étendue

To calculate the étendue of a beam in an inhomogeneous medium crossing an aperture defined by a curve AB, as in Fig. 3.33, we use the Hilbert integral adapted to the geometrical optics representation (see Luneburg or Born and Wolf, *loc. cit.*). Let P_1 and P_2 be two points somewhere in a pencil of rays, i.e., the rays from a single wavefront. Then the Hilbert integral from P_1 to P_2 is

$$I(P_1, P_2) = \int_{P_1}^{P_2} n\, \mathbf{k} \cdot \mathbf{ds} \tag{3.24}$$

where n is the local refractive index, $\mathbf{k}$ is a unit vector along the local ray direction and $\mathbf{ds}$ is an element of the path $P_1 P_2$. Thus the Hilbert integral is simply the optical path length along any ray between the wavefronts which pass between P_1 and P_2 and it depends only on the positions of the end points of the path of integration.

Returning to Fig. 3.33, the Hilbert integral from A to B for the rays from Σ_α is

$$I_\alpha(A,B) = \int_A^B n \sin \phi \, ds \tag{3.25}$$

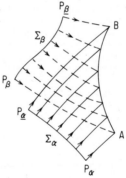

Fig. 3.33   Calculating the étendue by means of the Hilbert integral.

where ϕ is the angle of incidence of a ray on the line element ds. We may write this in the form

$$I_\alpha(A,B) = \langle n \sin \phi \rangle L_{AB} \tag{3.26}$$

where $\langle\,\rangle$ denotes the average and L_{AB} is the length of the curve AB. Thus from the definition of étendue we have for the étendue of the entering beam

$$H = I_\alpha(A,B) - I_\beta(A,B) \tag{3.27}$$

But as remarked above,

$$I_\alpha(A,B) = [P_\alpha B] - [P_\alpha A] \tag{3.28}$$

*This sentence and others like it imply assumptions about the smoothness of the refractive index distribution, but we do not attempt to make them explicit.

where $[P_\alpha B]$ is the optical path length along the ray from P_α to B, similarly for $[P_\alpha A]$, so that we find for the étendue

$$H = [P_\alpha B] + [P_\beta A] - [P_\alpha A] - [P_\beta B] \tag{3.29}$$

This is a straightforward generalization of Eq. (3.23).

A Construction for the Concentrator

In order to have some starting point for the construction we postulate the principle that the optical system between AB and $A'B'$ in Fig. 3.32 must be such that one extreme pencil (i.e., from the wavefront Σ_α) is exactly imaged into one of the emergent extreme pencils and similarly for the other extreme incident pencil from Σ_β. This is a generalization of the edge-ray principle, according to which a system such as that in Fig. 3.29 is designed by making all rays of an extreme pencil emerge at one edge of the exit aperture. If in the present case one of the emergent wavefronts is spherical and the focus is allowed to approach one edge of the exit aperture, we recover the edge-ray principle.

Now suppose we have an optical system of some kind which achieves the transformation of extreme pencils proposed above as a design principle, then in Fig. 3.34, the system takes Σ_α into Σ'_α and Σ_β into Σ'_β and we wish it to do so without loss of étendue. We write down the optical path length from P_α to P'_α and equate it to that from P_α to P'_α and similarly for the other pencil:

$$[P_\alpha B] + [BA']_\alpha + [A'P_\alpha'] = [P_\alpha A] + [AB']_\alpha + [B'P_\alpha']$$
$$[P_\beta A] + [AB']_\beta + [B'P_\beta'] = [P_\beta B] + [BA']_\beta + [A'P_\beta'] \tag{3.30}$$

where $[BA']_\beta$ means the optical path length from B to A' along the ray of the β pencil and similarly for the other symbols. From these we find

$$\left\{ [P_\alpha B] + [P_\beta A] - [P_\alpha A] - [P_\beta B] \right\} - \left\{ [A'P_\beta'] + [B P_\alpha'] - [A'P_\alpha'] - [B'P_\beta'] \right\}$$
$$= [AB']_\alpha - [AB']_\beta + [BA']_\beta - [BA']_\alpha \tag{3.31}$$

If we compare the left side of this equation with Eq. 3.29 we see that it is the difference between the étendues at the entry and exit apertures. We require this difference to vanish so we have to make the right side of Eq. 3.31 vanish. One simple way to do this is to make the optical system such that the α and β ray paths from A to B' coincide exactly and similarly with those from B to A'. We can do this by starting mirrors at A and B in directions that bisect the

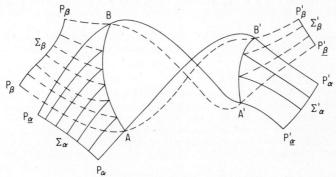

Fig. 3.34 Designing the concentrator.

angles between the incoming α and β rays. We then continue the mirror surfaces in such a way as to make all α rays join up with the corresponding α' rays; i.e., we image the α pencil directly into the α' pencil, and similarly for the other mirror surface. This construction then specifies two mirrors connecting A to A' and B to B'. It completes the construction and uses up all available degrees of freedom in doing so.

Discussion

The design procedure we have described will be successful so long as the reflecting surface is everywhere well defined; this amounts to saying the extreme rays must not form caustics before reaching the reflecting walls. We remark that had we not incorporated reflectors, the vanishing of the right-hand side of Eq. (3.31) would no longer be built into the method of solution but would appear as an additional constraint. this additional constraint is nontrivial in the same sense that the condition $a' = a \sin \theta$ is not automatically satisfied in any given optical system used as a concentrator. We can see this from a simple example: If the concentrator has a plane of symmetry, then the right-hand side of Eq. (3.31) reduces to

$$[AB']_\alpha - [AB']_\beta = 0 \tag{3.32}$$

Thus the system is required to both image the extreme pencils into each other and give equal optical paths to the rays shown in Fig. 3.35.

We can trace in the general design construction, the ideas behind simple concentrators (see, e.g., the various compound parabolic concentrator configurations described in Chap. 8). The results strongly suggest, but do not prove, that it is always necessary to incorporate mirrors in a nonimaging concentrator if it is to have the ideal property of conservation of étendue.

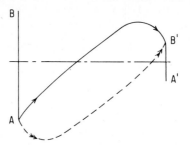

Fig. 3.35   A concentrator with a plane of symmetry.

SOME CONCEPTS OF PHYSICAL OPTICS

The formulation above is entirely phenomenological and it may be of interest to sketch the relationship with the electromagnetic wave representation of light. According to this theory (Born and Wolf, 1975) light can be regarded as waves propagated according to Maxwell's equations with a velocity $c = (\epsilon_0 \mu_0)^{-1/2}$ in vacuum, where ϵ_0 and μ_0 are, respectively, the permittivity and permeability of free space. In a transparent material medium the velocity is found to be $v = (\epsilon \epsilon_0 \mu \mu_0)^{-1/2}$ where ϵ and μ are, respectively, the relative permittivity (dielectric constant) and relative permeability of the medium. In the cases which concern us it is sufficient to take $\mu = 1$ and then we find from the above our definition of refractive index,

$$n = \sqrt{\epsilon} \tag{3.33}$$

While it is possible to develop approximate expressions for ϵ in terms of molecular properties and structure it is found in practice that for solids the theory does not agree well with the experimental results at optical frequencies. Thus we again seem to have a phenomenological treatment, but now the connection with electromagnetic waves makes it possible to include absorption or attenuation in the same representation. Thus in the usual complex number notation a plane wave traveling in the z direction and with y polarization has an electric field given by

$$E_y = E_0 \exp i(knz - \omega t) \tag{3.34}$$

where ω is the angular frequency of the wave motion and $k = 2\pi/\lambda$ is the wave number. It is easily seen that this represents a wave traveling at speed c/n where $c = \omega/k$. Now if we let n take a complex value, say $n - i\kappa$, the above expression acquires a real factor $\exp -\kappa kz$, representing an attenuation or absorption of the wave. The quantity κ/n is variously called the attenuation index or the extinction coefficient; it has the significance that the light is reduced in intensity by the factor $e^{-2\kappa k}$ per unit distance traveled, or, equivalently, the intensity decreases by the factor $e^{-4\pi\kappa}$ per wavelength* traveled. Thus a single parameter, the complex refractive index, represents the optical properties of the material.

Electromagnetic theory also yields quantitative results for the proportions of light reflected and transmitted at interfaces. The relevant formulas, which we give here, are the Fresnel equations for reflection and refraction. They are set out in terms of the electric field strengths of the different beams. For nonzero angle of incidence the formulas depend on the polarization and we use subscripts p and s, respectively, to denote waves polarized with the electric field vector parallel to and perpendicular to the plane of incidence. Then for a wave incident at angle I from a medium of refractive index n the Fresnel formulas for amplitude reflection r and transmission t are

$$\left.\begin{aligned} r_p &= \frac{n'\cos I - n\cos I'}{n'\cos I + n\cos I'} \\[2mm] t_p &= \frac{2n'\cos I}{n'\cos I + n\cos I'} \\[2mm] r_s &= \frac{n'\cos I' - n\cos I}{n'\cos I' + n\cos I} \\[2mm] t_s &= \frac{2n'\cos I'}{n'\cos I' + n\cos I} \end{aligned}\right\} \tag{3.35}$$

These refer, of course, to the complex representation of the electric field. Thus if the second medium is absorbing the appropriate complex refractive index can be used. To calculate the corresponding light intensities we have for the reflectivity

$$\left.\begin{aligned} R_p &= |r_p|^2 \\ R_s &= |r_s|^2 \\ T_p &= 1 - R_p \\ T_s &= 1 - R_s \end{aligned}\right\} \tag{3.36}$$

However, T_p and T_s do not have a simple significance for a heavily absorbing medium.

The Fresnel formulas enable us to calculate the proportions of light reflected at glass or plastic interfaces and also, by putting in the appropriate complex index, at a metal surface.

There are certain special cases to be noted. If the light is incident from, say, glass with index n to air at angle I and if $n \sin I$ is greater than unity then $\cos I'$ is found from the law of refraction (Eq. (3.2)) to be a pure imaginary quantity and from this both R_s and R_p are found to be unity. This is the physical optics formulation of total internal reflection, mentioned above. The reflectance under such conditions is exactly 100 percent apart from small losses due to scatter from dirt and small-scale irregularities on the surface. Thus the phenomenon of total internal reflection is potentially of great utility in applications to optical power transmission and concentration.

Another special case, perhaps not so important for solar energy, occurs when the angle of incidence is

$$\tan I = n'/n \tag{3.37}$$

Under this condition R_p, the reflectance for the parallel polarization, is zero and thus the reflected light is completely polarized in the s direction. The angle of incidence so found is called the Brewster angle.

If Eqs. (3.35) and (3.36) are applied to metal surfaces with appropriate complex refractive indices it is found that a reflectance of unity is never reached. The largest practical value for the visible spectrum is about 92 percent, for aluminum. It is also found that the reflectance of metal surfaces against a dielectric such as glass, rather than air, is even lower (about 85 percent for aluminum). Alternatives to such metal films as reflectors when total internal reflection cannot be used are possible for some applications: these alternatives are multilayers of dielectrics with suitably chosen thicknesses and refractive indices, and metal films overcoated with a few

*Vacuum wavelength, of course.

dielectric layers. Such systems are probably too costly for use in solar energy applications. For descriptions of the theory and applications we refer to Macleod (1969).

The above discussion applies to electromagnetic waves which have substantially plane phasefronts, i.e., the radius of curvature of the phasefront is much larger than the wavelength. If this is not the case, e.g., near a focus, the relationships between electric field and power density or flux are quite complicated. However, keeping to our assumption of large radii of curvature we note that the magnitude and direction of the flux are given by the Poynting vector $\mathbf{P}$, or, for optical frequencies its time average, and we have

$$\mathbf{P} = \mathbf{E} \times \mathbf{H} \tag{3.38}$$

where $\mathbf{E}$ and $\mathbf{H}$ are the electric and magnetic field strengths. Then for nearly plane waves this expression becomes

$$\mathbf{P} = \tfrac{1}{2}\mathbf{k} E_o^2 \sqrt{\frac{\epsilon\epsilon_o}{\mu\mu_o}} \tag{3.39}$$

where $\mathbf{k}$ is a unit vector in the direction of travel of the wave and E_o is the transverse field. Thus from Eq. (3.33) we see that the flux density in a transparent medium of refractive index n is proportional to $nE_o'^2$, where E_o' is the electric field inside the medium, as calculated from the Fresnel formulas, Eqs. (3.35). This result enables us to calculate the flux density after oblique refraction into any given medium.

REFERENCES

Born, M. and E. Wolf, *Principles of Optics*, 5th ed. (Pergamon Press, New York and Oxford, 1975).

Garwin, R. L. *Rev. Sci. Instr.* **23**, pp. 755–757 (1952).

Harper, D. A., R. H. Hildebrand, R. Stiening, and R. Winston, *Appl. Opt.* **15**, pp. 53–60 (1976).

Hottel, H. "Radiant Heat Transmission," in *Heat Transmission*, 3d ed., edited by William H. McAdams (McGraw-Hill Co., New York, 1954).

Krenz, J. H. *Energy Conversion and Utilization* (Allyn & Bacon, Boston, 1976).

Luneburg, R. K. *Mathematical Theory of Optics* (University of California Press, Berkeley and Los Angeles, 1964). This material was originally published in 1944 as loose sheets of mimeographed notes and the book is a word-for-word transcription.

Macleod, H.A., "Thin Film Optical Filters" (Adam Hilger Ltd,, London, 1969).

Marcuse, D., *Light Transmission Optics* (Van Nostrand Reinhold, New York, 1972).

Maxwell, J. C., *Quart. J. Pure Appl. Math.* **2**, pp. 233–247 (1858).

Stavroudis, O. N., *Appl. Opt.* **12**, p. A16 (Oct.). (1973).

Welford, W. T., *Aberrations of the Symmetrical Optical System* (Academic Press, New York, 1974).

Winston, R., *J. Opt. Soc. Amer.* **60**. pp. 245–247 (1970).

Winston, R., *Sol. Energy* **16**, pp. 89–95 (1974).

Winston, R. and H. Hinterberger, *Sol. Energy* **17**, pp. 255–258 (1975).

Chapter **4**

Principles of Thermodynamics and Heat Transfer Applied to Solar Energy*

FRANK KREITH, P.E.
SERI, Golden, CO

and JAN F. KREIDER, P.E.
Jan F. Kreider & Associates, Boulder, CO

The design and analysis of all solar thermal systems requires familiarity with the fundamentals of thermodynamics and heat transfer. This chapter will therefore be devoted to the principles of heat transfer, their relation to thermodynamics, and their application to solar engineering.

Whenever a temperature difference exists, energy may be transferred from the region of higher temperature to the region of lower temperature. According to thermodynamic concepts, the energy which is transferred as a result of a temperature difference is called "heat." Although classical thermodynamics deals with energy and heat transfer, its laws can treat only systems which are in equilibrium. Thermodynamic laws, therefore, can predict the amount of energy required to change a system from one equilibrium state to another, but they cannot predict how fast these changes will occur. The engineering science of heat transfer supplements the first and second laws of classical thermodynamics by providing methods of analysis which can be used to predict rates of energy transfer and temperature distributions.

To illustrate the difference in the kind of information that can be obtained from a thermodynamic and heat-transfer analysis, consider the heating of a metal cylinder placed into a tank of hot water. Thermodynamics can be used to predict the final temperature after the two systems have reached equilibrium, but it cannot tell us what the temperature of the cylinder will be after a given time or how long it will take to obtain equilibrium. A heat transfer analysis, on the other hand, can predict the rate of heat transfer from the water to the cylinder, and from this information we can calculate the temperature of the cylinder as well as the temperature of the water as a function of time.

*Adapted from F. Kreith and J. F. Kreider, "Principles of Solar Engineering," Hemisphere Publishing Corp., Washington, DC, 1978, with permission.

The first section of this chapter deals with the laws of thermodynamics with particular emphasis on the second law. The balance of the chapter covers the application of heat-transfer principles to solar-thermal processes and conversion devices and auxiliary equipment such as heat exchangers.

SOLAR ENERGY AND THE SECOND LAW OF THERMODYNAMICS

One of the unique and technically most attractive features of solar energy from a thermodynamic viewpoint is the feasibility of matching a means of collection to a broad range of tasks which operate over a large range of temperatures (see Chaps. 7 to 9). It is the purpose of this section to quantitatively describe this feature of solar energy and to provide an analytical framework for matching the means of solar energy collection to the task to be performed. A comparison between solar and nonsolar energy resources is made from a second-law viewpoint.

Energy Quantity and Quality

Nearly all spoken and written works on the energy question, whether by technical experts, lay people, or politicians, invariably discuss the conservation of energy. However, the conservation of energy is automatically assured by the first law of thermodynamics. Implicit in this common view is the idea that energy can be saved only by not using it. Therefore, no insight is provided into the optimum means of using energy to perform a given task.

A common measure on energy use efficiency is the first-law efficiency η_1. The first-law efficiency is defined as the ratio of useful output energy of a device to the input energy of the device. For example, a turbine produces work W from an input of heat Q_i. Therefore, the first-law efficiency $\eta_1 = W/Q_i$. For more complex tasks, for example, a distillation column with its waste heat used to operate a bottom-cycle turbine, it is difficult to define first-law efficiency because, the outputs of the device are both mass transfer and work, which cannot be combined easily into a single η_1 value.

Another shortcoming of the first-law efficiency is its very basis—quantities of energy. If an oil-fired boiler operates at a first-law efficiency of 80 percent, little enthusiasm could be expected on the part of a manufacturer in improving its performance. However, use of the same quantity of fuel in an engine-driven heat pump could deliver more heat than the shaft work used to operate the heat pump. Hence, the first-law efficiency $\eta_1 > 100$ percent. Since the first-law efficiency is not bounded above by 100 percent, it cannot be used as a reliable index of possible device performance improvement.

The second law of thermodynamics—one of the outstanding accomplishments of 19th century physics[1]—provides a means of assigning a "quality index" to energy. The concept of "available energy"[2,3]—i.e., energy available to do work, the most valuable form of energy—provides a useful measure of energy quality. With this concept, it is possible to analyze means to minimize the consumption of available energy to perform a given process, thereby ensuring the most efficient possible conversion of energy for the required task. Using the concept of availability, it is possible to define a second-law efficiency η_2 of a process as the ratio of the minimum available energy which *must* be consumed to do a task divided by the actual amount of available energy consumed in performing the task.

The available energy A was first defined by Gibbs[2] as

$$A = E - T_oS + p_oV - D \tag{4.1}$$

where E is the internal energy, S is the entropy, V is the volume, and D is the maximum useful work from diffusion processes (this term does not enter solar analyses and will be dropped hereafter). The environment is characterized by its temperature T_o and its pressure p_o.

The second-law efficiency is a *task* index, not a device index. It is, therefore, much more useful in identifying optimal energy conversion processes than is the first-law efficiency, since it focuses attention on device interactions that transform energy into its two useful types—work (and other ordered forms) and heat. Table 4.1 shows availabilities and both η_1 and η_2 expressions for several common thermal tasks which may be performed by solar energy or by conventional sources. Table 4.1 is based on values of A_{min} calculated for two energy types. The first involves work W:

$$A_{min} = W \tag{4.2}$$

TABLE 4.1   Availabilities and First- and Second-Law Efficiencies for Energy Conversion Systems*

Task	Energy input	
	Input shaft work W_i	Q_r from reservoir at T_r
Produce work W_o	$A = W_i$ $A_{min} = W_o$ $\eta_1 = W_o/W_i$ $\eta_2 = \eta_1$ (Electric motor)	$A = Q_r(1 - T_o/T_r)$ $A_{min} = W_o$ $\eta_1 = W_o/Q_r$ $\eta_2 = \eta_1(1 - T_o/T_r)^{-1}$ (Solar, Rankine cycle)
Add heat Q_a to reservoir at T_a	$A = W_i$ $A_{min} = Q_a(1 - T_o/T_a)$ $\eta_1 = Q_a/W_i$ $\eta_2 = \eta_1(1 - T_o/T_a)$ (Heat pump)	$A = Q_r(1 - T_o/T_r)$ $A_{min} = Q_a(1 - T_o/T_a)$ $\eta_1 = Q_a/Q_r$ $\eta_2 = \eta_1\left(\dfrac{1 - T_o/T_a}{1 - T_o/T_r}\right)$ (Solar water heater)
Extract heat Q_c from cool reservoir at T_c (below ambient)	$A = W_i$ $A_{min} = Q_c(T_o/T_c - 1)$ $\eta_1 = Q_c/W_i$ $\eta_2 = \eta_1(T_o/T_c - 1)$ Vapor (compression air conditioner)	$A = Q_r(1 - T_o/T_r)$ $A_{min} = Q_c(T_o/T_c - 1)$ $\eta_1 = Q_c/Q_r$ $\eta_2 = \eta_1\left(\dfrac{T_o/T_c - 1}{1 - T_o/T_r}\right)$ (Absorption air conditioner)

*The heat reservoirs are considered to be isothermal; T_o is the environmental temperature and processes are reversible; $T_r > T_a > T_o > T_c$.

The second involves heat Q:

$$A_{min} = Q\left(1 - \frac{T_o}{T}\right) \tag{4.3}$$

where $T\,(>T_o)$ is the fixed task and use temperature.

For example, consider the Carnot heat engine cycle. The first-law efficiency $\eta_1 < 1$, implying that a better engine could be built to achieve the supposed $\eta_1 = 1$ maximum. However, the second-law efficiency $\eta_2 = 1$, clearly states that no further improvement is possible. In this as in all other process assessments, η_2 is more useful than η_1 in energy resource use optimization.

Numerical example   Calculate the second-law efficiency of a gas space-heating system designed to maintain a building at 20°C if the environmental temperature is −10°C. The first-law efficiency of the gas furnace is 60 percent. From Table 4.1,

$$\eta_2 = \eta_1 \frac{1 - T_o/T_a}{1 - T_o/T_r}$$

If the flame temperature is 2300 K,

$$\eta_2 = 0.6\left(\frac{1 - 263/293}{1 - 263/2300}\right) = 0.07$$

The second-law efficiency is only 7 percent, indicating an enormous potential for improvement.* The reason for the poor η_2 value is the fundamental mismatch of the low-entropy energy source

*Note that the flame temperature used in this illustrative example does not actually correspond to an isothermal energy source as required for the use of Table 4.1. See Table 4.3 for an expression for η_2 if the reservoir is not isothermal.

to a high-entropy task. If a solar heater operating at 30°C were used to perform the 20°C task, its value of η_2 would be 78 percent.

The Thermodynamic Availability of Sunlight

Coherent light from an equilibrium source has zero entropy and is completely available. However, the sun is not an equilibrium source nor is its radiant energy coherent. Therefore, sunlight in near-earth space has a finite entropy level. In traversing the atmosphere, the entropy level increases further because of direction changes (scattering) and frequency and phase shifts.

Parrot[4] has calculated the availability of beam and diffuse terrestrial solar radiation using Eq. (4.1). For beam radiation the availability A_b is

$$A_b = I \left[1 - \frac{4}{3} \left(\frac{T_o}{T_s} \right) (1 - \cos \theta_s/2)^{1/4} + \frac{1}{3} \left(\frac{T_o}{T_s} \right)^4 \right] \tag{4.4}$$

where T_s is the equivalent surface temperature of the sun (~5800 K), θ_s is the included angle of the sun, and I is the insolation. For $T_a = 300$ K, beam radiation is seen to be 9 percent available, comparable to the availability of the heat of combustion of most fossil fuels.

For diffuse radiation, the availability A_d is

$$A_d = I \left[1 - \frac{4}{3} \left(\frac{T_o}{T_s} \right) + \frac{1}{3} \left(\frac{T_o}{T_s} \right)^4 \right] \tag{4.5}$$

Second-Law Efficiency of Solar-Thermal Processes

If it is assumed that solar devices are only able to deliver heat at the temperature required to perform a task, Table 4.1 can be used to compare the second-law efficiencies of solar and nonsolar energy sources used to perform the same task. For example, both fossil fuel and solar energy can be used to produce industrial process heat at 450 K (~350°F). The former source can be converted to process heat at 80 to 85 percent first-law efficiency. The second-law efficiency is only 30 percent, however, since high combustion temperatures (~2000 K) are used to produce relatively low-temperature heat—an inefficient use of available energy. A solar process-heat system could be designed to collect energy at 500 K to deliver 450 K heat. The second-law efficiency of this process (for an assumed 10 percent parasitic heat loss) is 75 percent, indicating a much closer energy-source-to-task match.

Second-law efficiency values may be assigned to a number of processes by inspection. First-law η_1 values are known for both solar and nonsolar applications.* Likewise, process temperatures are known for many tasks such as water and space heating, cooling, refrigeration, power production, primary metal refining, and the like. Therefore, an η_2 value may be calculated using Table 4.1. This is done in the next section.

Until the mid-1970s, a crucial gap existed in task temperature and energy quantity data for industrial process heat. Process heat is used over a broad temperature range, unlike heat for the applications listed in the preceding paragraph. A comprehensive technical study has provided the requisite data for the industrial heat sector. (See Chap. 21.)

Figure 4.1 summarizes the temperature spectrum of 1972 U.S. process applications. The dashed line depicts process temperature requirements (which may differ from present practice), whereas the solid line includes preheating to the subject temperature. Currently, much heat from fossil sources with high availability levels is used for applications at 300°C or below. The waste of availability is enormous. Table 4.2 indicates second-law efficiencies for the process heat sectors below 500°F (260°C) if fossil fuels are consumed at 80 percent first-law efficiency. Also shown are η_2 values for the same processes if performed by a solar energy system with 10 percent parasitic losses and a 20°C driving force for heat transfer from storage to load device.

---

*In the context of this section, solar energy is assumed to be available at a reservoir storage temperature T_r which depends upon the nature of the collector process. T_r may vary from ambient to 3000 K or more, depending upon the task. Of course, T_r varies continuously over a relatively narrow range in practice, but nominal values are used herein for simplicity.

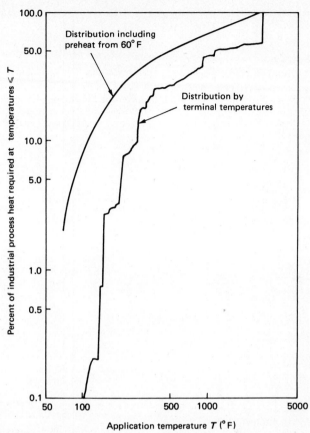

Fig. 4.1 Distribution of industrial process heat temperatures for the United States. (*From F. Kreith and J. F. Kreider, Principles of Solar Engineering, Hemisphere, Washington, DC, 1978.*)

TABLE 4.2 Second-Law Efficiencies for U.S. Industrial Processes below 500°F (260°C)

Temperature, °F	Fraction of U.S. process heat, %	Fossil fuel η_2, %	Solar η_2, %
85	10	< 1	12
120	5	6	52
150	5	10	65
175	5	13	71
210	5	16	72
250	5	20	77
300	5	25	83
370	5	30	85
460	5	35	85
Totals/averages	50	16	61

TABLE 4.3 U.S. Energy Use Efficiency, Current Nonsolar vs. Future Solar

Task	Temperatures, K(°F)	Current η_2, %	Solar† η_2, %	Task	Temperature, K(°F)	Current η_2, %	Solar† η_2, %
Space heating				**Refrigeration**			
Fossil fuels	$T_o = 275$ (35)	4	40	Electric	$T_o = 294$ (70)	3	
($\eta_1 = 0.6$)	$T_a = 294$ (70)			($\eta_1 = 0.9$)	$T_c = 269$ (25)		
Electricity	$T_o = 275$ (35)	2	($T_r = 120°$F)	Gas	$T_o = 294$ (70)	4	28
($\eta_1 = 0.9$)	$T_a = 294$ (70)			($\eta_1 = 0.4$)	$T_c = 269$ (25)		($T_r = 300°$F)
Water heating*				Automobiles,		9	—
Fossil fuels	$T_o = 275$ (35)	10	78	trucks, buses			
($\eta_1 = 0.6$)	$T_a = 333$ (140)			Power production		33	25–35
Electricity	$T_o = 275$ (35)	5	($T_r = 160°$F)	($\eta_1 = 0.33$)‡			
($\eta_1 = 0.9$)	$T_a = 333$ (140)			Low-temperature		2	30
Air conditioning				power cycles,			($T_o = 45°$F)
Electric	$T_o = 308$ (95)	3	—	e.g., OTEC-type			($T_r = 80°$F)
($\eta_1 = 2.0$)	$T_c = 294$ (55)			plant			
				($\eta_1 = 0.02$)			
Absorption	$T_o = 308$ (95)	2	25	Process steam		~30	40–60¶
($\eta_1 = 0.55 \times 0.8$)	$T_c = 294$ (55)		($T_r = 210°$ F)	($\eta_1 = 0.85$)			

*The water heating η_2 values are based on an approximately isothermal hot-water tank. If large temperature excursions are experienced, the following expression for η_2 should be used:

$$\eta_2 = \eta_1 \left[1 - \frac{T_o}{T_a - T_o} \ln T_a/T_o \right]$$

†10% parasitic heat losses.
‡Combined gas turbine, Rankine cycle may achieve $\eta_2 \approx 0.5$ in the future.
¶Depends on temperature; figures shown are a range of expected values.
NOTE: High-temperature direct-heat applications are not included in this table.

The numbers in Table 4.2 may not be precise, nor need they be to demonstrate the potential for improvement. Solar energy could increase the efficiency of energy use by a factor of more than 3 by using available technology. Solar applications above 500°F (260°C) may be somewhat more difficult, but the 50 percent of industrial heat below 500°F (260°C) represents more than 10 percent of U.S. energy consumption. The benefits from the use of solar heat to conserve high-quality fossil fuels for higher-priority uses are obvious.

Potential for Improved Use of Energy in the U.S. Economy by Solar Displacement of Fossil Fuels

Table 4.2 shows the possibility for greatly improving the use of energy in many parts of the U.S. industrial sector. Similar evaluations can be made for other sectors to indicate the potential for improvement of energy use throughout all U.S. energy sectors. Table 4.3 summarizes such a calculation and indicates vividly the areas where solar energy has the most promise for fossil fuel displacement and η_2 improvement: space heating, space cooling, water heating, refrigeration, and process heat. By obvious extension, clothes drying, cooking, agricultural uses, and the like are also suitable candidates. It is to be noted that all of the above uses are high-entropy tasks which match well with commercially available, high-entropy solar collection methods. More difficult solar applications are transportation, power production at high temperature, and direct firing in the primary metals industries.

CONDUCTION HEAT TRANSFER

Conduction is the only heat-transfer mode in opaque solid media. When a temperature gradient exists in such a body, heat will be transferred from the higher to the lower temperature region. The rate at which heat is transferred by conduction, q_k, is proportional to the temperature gradient dT/dx times the area A through which heat is transferred:

$$q_k \propto A \left(\frac{dT}{dx} \right)$$

where $T(x)$ is the temperature and x is the distance in the direction of the heat flow. The actual rate of heat flow depends on the thermal conductivity k, a physical property of the medium. For conduction through a homogeneous medium, the rate equation can therefore be quantitatively expressed as

$$q_k = -k A \left(\frac{dT}{dx} \right) \tag{4.6}$$

The minus sign is a consequence of the second law of thermodynamics, which requires that heat *must* flow in the direction of lower temperature. The temperature gradient, as shown in Fig. 4.2, will be negative if the temperature decreases with increasing values of x. Therefore, if heat transferred in the positive direction x is to be a positive quantity, the negative sign must be inserted in the right-hand side of Eq. (4.6). Equation (4.6) is called Fourier's law of heat conduction and serves to define the thermal conductivity k. If the area is in square meters (m²), the temperature in Kelvin (K), x in meters (m), and the rate of heat flow in watts (W),

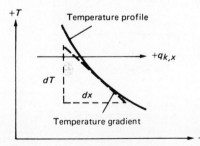

Fig. 4.2 Temperature gradient and direction of heat conduction in a solid. (*From F. Kreith and J. F. Kreider, Principles of Solar Engineering, Hemisphere, Washington, DC, 1978.*)

k has the units of watts per meter per Kelvin (W/m · K). In the English system, the area is in square feet (ft²), x in feet (ft), the temperature in degrees Fahrenheit (°F), and the rate of heat flow in British thermal units per hour (Btu/h). Thus, k has the units of Btu/h · ft · F.

Plane Walls

A direct application of Fourier's law is the case of heat transfer through a wall (Fig. 4.3). When both surfaces of the wall are at uniform temperatures, the heat flow will be in one direction, perpendicular to the wall surfaces. If the thermal conductivity is uniform, integration of Eq. (4.6) gives

$$q_k = -\frac{kA}{\Delta x}(T_2 - T_1) = \left(\frac{kA}{\Delta x}\right)(T_1 - T_2) \tag{4.7}$$

where Δx = thickness of the wall
$\quad T_1$ = temperature at the left surface, where $x = 0$
$\quad T_2$ = temperature at the right surface, where $x = \Delta x$

For many materials, the thermal conductivity is a linear function of temperature:

$$k(T) = k_o(1 + \beta_k T) \tag{4.8}$$

In such cases, integration of Eq. (4.6) gives

$$q_k = \frac{k_o A}{\Delta x}\left[(T_1 - T_2) + \frac{\beta_k}{2}(T_1^2 - T_2^2)\right] \tag{4.9}$$

or

$$q_k = \frac{k_{av} A}{\Delta x}(T_1 - T_2) \tag{4.10}$$

where k_{av} is the value of k at the average temperature $(T_1 + T_2)/2$.

The temperature distributions for a constant value of thermal conductivity $(\beta_k = 0)$, as well as for thermal conductivity increasing $(\beta_k > 0)$ and decreasing $(\beta_k < 0)$ with temperature, are shown in Fig. 4.3.

If heat is conducted through several layers in series, as for example through a multilayer wall

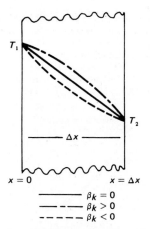

Fig. 4.3 Temperature distribution in conduction through a plane wall with constant and variable thermal conductivity. (*From F. Kreith and J. F. Kreider, Principles of Solar Engineering, Hemisphere, Washington, DC, 1978.*)

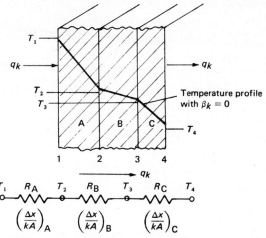

Fig. 4.4 One-dimensional heat conduction through a composite wall and corresponding thermal resistance network. (*From F. Kreith and J. F. Kreider, Principles of Solar Engineering, Hemisphere, Washington, DC, 1978.*)

as used in the construction of most houses, the analysis is only slightly more difficult. In the steady state, the rate of heat flow through all the sections must be the same. However, as shown in Fig. 4.4 for a three-layer system, the gradients are different. The heat-transfer rates can be written for each section and set equal to one another, or

$$q_k = \left(\frac{kA}{\Delta x}\right)_A (T_1 - T_2) = \left(\frac{kA}{\Delta x}\right)_B (T_2 - T_3) = \left(\frac{kA}{\Delta x}\right)_C (T_3 - T_4) \qquad (4.11)$$

If the intermediate temperatures T_2 and T_3 in Eq. (4.11) are eliminated, the rate of heat flow can be written in the form

$$q_k = \frac{T_1 - T_4}{(\Delta x/kA)_A + (\Delta x/kA)_B + (x/kA)_C} \qquad (4.12)$$

Equation (4.12) is a convenient starting point to introduce a different viewpoint for the analysis of heat transfer, a viewpoint which is used in later chapters. This viewpoint makes use of concepts developed previously in the analysis of electric-circuit theory and is called the electric analogy to the flow of heat. If the heat-transfer rate q_k is considered to be analogous to the flow of electricity, the combination $(\Delta x/kA)$ to be analogous to a resistance R, and the temperature difference to be analogous to a potential difference ΔT, Eq. (4.7) can be written in a form similar to Ohm's law in electric circuit theory, or

$$q_k = \frac{\Delta T}{R} \qquad (4.13)$$

where* $\Delta T = (T_1 - T_2)$
$\qquad\quad R = (\Delta x/kA)$

Similarly, for heat flow through several sections in series, as shown in Fig. 4.4, Eq. (4.12) can be expressed in the form

$$q_k = \frac{\Delta T}{R_A + R_B + R_C} \qquad (4.14)$$

*Some authors define the resistance as $\Delta x/k$.

where $\Delta T = T_1 - T_4$
$R_A = (\Delta x/kA)_A$
$R_B = (\Delta x/kA)_B$
$R_C = (\Delta x/kA)_C$

In Eq. (4.14) the rate of heat flow q is expressed only in terms of an overall temperature potential and the resistances of the individual sections in the heat-flow path. These values can be combined into what is generally called an overall transmittance or overall heat-transfer coefficient U, as shown below:

$$UA = \frac{1}{R_A + R_B + R_C}$$
(4.15)

Table 4.4 contains thermal conductivities of some common building materials.

The electric analog approach can also be used to solve more complex problems. One application is shown in Fig. 4.5 where heat is transferred through a composite structure involving thermal resistances both in series and in parallel. For this system, the resistance of the middle layer, R_2 in Fig. 4.5, becomes

$$R_2 = \frac{R_B R_C}{R_B + R_C}$$
(4.16)

and the rate of heat flow is

$$q_k = \frac{\Delta T_{overall}}{\sum_{n=1}^{n=N} R_n}$$
(4.17)

TABLE 4.4   Thermal Conductivity and Density of Some Building Materials

	ρ, kg/m^3	k, W/m · K
Asbestos cement sheet	1520	0.29–0.43
Asbestos felt	144	0.078
Asbestos insulating board	720–900	0.11–0.21
Asphalt, roofing	1920	0.58
Brick, common, dry	1760	0.70–0.81
Brick, wet	2034	1.67
Chipboard	350–1360	0.07–0.21
Concrete, gravel 1:2:4	2240–2480	1.4
Vermiculite aggregate	400–880	0.11–0.26
Cellular	320–1600	0.08–0.65
Cork, granulated, raw	115	0.046
Slab, raw	160	0.05
Fiberboard	280–420	0.05–0.08
Glass, window	2500	1.05
Glass fiber, mat	50	0.033
Hardboard	560	0.08
Plasterboard, gypsum	1120	0.48–0.50
Polystyrene, expanded board	15	0.037
Polyurethane foam	30	0.026
Polyvinyl chloride, rigid foam	25–80	0.035–0.041
Roofing felt	960–1120	0.19–0.20
Tiles, clay	1900	0.85
Tiles, concrete	2100	1.10
Tiles, PVC asbestos	2000	0.85
Urea formaldehyde foam	8–30	0.032–0.038
Vermiculite granules	100	0.065
Wilton carpet		0.058

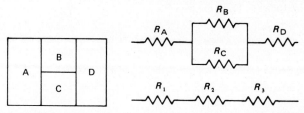

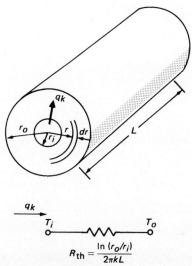

Fig. 4.5   Series and parallel one-dimensional heat conduction through a composite wall with corresponding thermal resistance network. (*From F. Kreith and J. F. Kreider, Principles of Solar Engineering, Hemisphere, Washington, DC, 1978.*)

Fig. 4.6   Radial heat conduction through a hollow cylinder. (*From F. Kreith and J. F. Kreider, Principles of Solar Engineering, Hemisphere, Washington, DC, 1978.*)

where N = number of layers in series
 R_n = thermal resistance of the nth layer
 $\Delta T_{\text{overall}}$ = temperature difference between the inner and outer surfaces

The heat flowrate from a long rectangular duct of width W, height h, and thickness t is

$$q_{k,\,\text{rect}} = kL \left(\frac{2W + 2h}{t} + 2.16 \right) (T_i - T_o)$$

Cylinders and Spheres

Heat conduction through tubes, pipes, and spherical containers is of importance in many solar engineering systems. Consider first a long hollow cylinder of inside radius r_i, outside radius r_o, and length L (see Fig. 4.6). The temperature at the inside surface is T_i and at the outside surface T_o. To determine the rate of heat conduction Fourier's law can be used with appropriate coordinates. If the cylinder is sufficiently long for end effects to be negligible, heat flows only in the radial direction. Then, at a radial distance r the area through which heat is conducted in the cylindrical coordinate system is $A_r = 2\pi r L$ and Fourier's law can be written in the form

$$q_k = -k A_r \frac{dT}{dr} = -2\pi k L r \frac{dT}{dr} \qquad (4.18)$$

Separating the variables in Eq (4.18) gives

$$q_k \frac{dr}{r} = 2\pi kL \, dT \tag{4.19}$$

Equation (4.19) can be integrated subject to the boundary conditions

$$T(r_i) = T_i$$
$$T(r_o) = T_o$$

This yields the following relation for the rate of heat conduction through the cylinder:

$$q_k = \frac{2\pi kL \, (T_i - T_o)}{\ln (r_o/r_i)} \tag{4.20}$$

In terms of the electric analog, Eq. (4.20) can be written

$$q_k = \frac{T_i - T_o}{R} \tag{4.21}$$

where

$$R = \frac{\ln (r_o/r_i)}{2\pi k \, L}$$

The thermal resistance concept can be extended easily to multiple-layer cylindrical walls. For example, in a two-layer system (see Fig. 4.7), such as a pipe covered with insulation, the rate of heat conduction can be written in the form

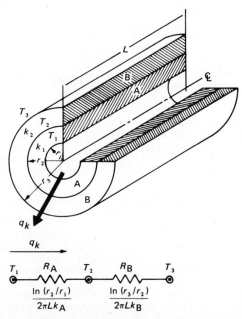

Fig. 4.7 Heat conduction through a tube covered with insulation and corresponding resistance network. (*From F. Kreith and J. F. Kreider, Principles of Solar Engineering, Hemisphere, Washington, DC, 1978.*)

$$q_k = \frac{T_1 - T_3}{R_A + R_B} \tag{4.22}$$

where $\quad R_A = \ln (r_2/r_1)/2\pi L k_A$
$R_B = \ln (r_3/r_2)/2\pi L k_B$.

Spherical systems can be treated in a similar manner. When the heat flow is radial, i.e., the temperature is only a function of the radial distance, the rate of heat conduction through a shallow spherical container of inner radius r_i and the outer radius r_o with a uniform temperature over the interior surface T_i and over the exterior surface T_o is

$$\frac{4\pi k (T_i - T_o)}{(r_o - r_i)/r_o r_i} \tag{4.23}$$

For the thermal network, the denominator can be treated as the thermal resistance in a multilayer spherical system.

Thermal Conductivity

The thermal conductivity is a material property defined by Eq. (4.6). Except for gases at low temperatures, it is not possible to predict this property analytically. Available information about thermal conductivity of materials is therefore largely based on experimental measurements. In general, the thermal conductivity of a material varies with temperature, but in many practical situations a constant value based on the average temperature of the system will give satisfactory results. Table 4.5 lists typical values of the thermal conductivities for some metals, nonmetallic solids, liquids, and gases to illustrate the orders of magnitude to be expected in practice.

The mechanism of thermal conduction in gases is explained qualitatively by the kinetic theory. All molecules in a gas are in random motion and exchange energy and momentum when they collide with one another. However, since higher temperatures are associated with molecules possessing more kinetic energy, when a molecule from a high-temperature region moves into a region of lower temperature, it transports kinetic energy on a molecular scale to the lower-temperature region. Upon impact with a molecule of lower kinetic energy, an energy transfer occurs which is seen as a transfer of heat from a macroscopic viewpoint. The mechanics of conduction in liquids are qualitatively similar but the picture is even more complex than in gases.

Figure 4.8 shows how the thermal conductivity of some gases varies with temperature. The thermal conductivity of gases is almost independent of pressure, except near the critical point. According to a simplified analysis based on a kinetic exchange model, the thermal conductivity of gases will increase as the square root of the absolute temperature.

Figure 4.9 also shows the thermal conductivity of some liquids as a function of temperature. It can be seen that except for water, the thermal conductivity of liquids decreases with increasing temperature, but the change is so small that in most practical situations the thermal conductivity

TABLE 4.5 Thermal Conductivities of Some Metals, Nonmetallic Solids, Liquids, and Gases

Material	Thermal conductivity at 300 K, W/m · K
Copper	386
Aluminum	204
Carbon steel	54
Glass	0.75
Plastics	0.2–0.3
Water	0.6
Ethylene glycol	0.25
Engine oil	0.15
Freon (liquid)	0.07
Hydrogen	0.18
Air	0.026

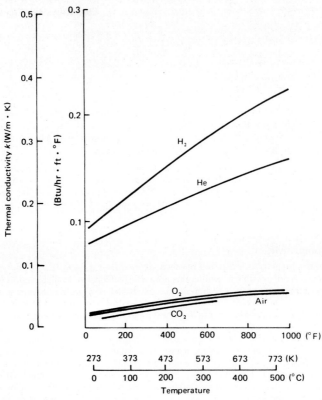

Fig. 4.8 Thermal conductivity of common gases. (*From F. Kreith and J. F. Kreider, Principles of Solar Engineering, Hemisphere, Washington, DC, 1978.*)

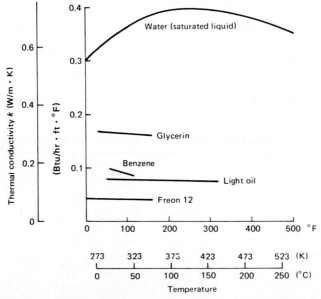

Fig. 4.9 Thermal conductivity of common liquids. (*From F. Kreith and J. F. Kreider, Principles of Solar Engineering, Hemisphere, Washington, DC, 1978.*)

may be assumed constant at some average temperature. There is no appreciable dependence on pressure.

Figure 4.10 shows the thermal conductivities of some metals. In solids, thermal energy is transported by free electrons and by vibrations in the lattice structure. In general, the movement of free electrons is the more important transport mode, and since in good electrical conductors a large number of free electrons move within the lattice structure, good electrical conductors are also good heat conductors, e.g., copper, silver, and aluminum. On the other hand, good electrical insulators are also good thermal insulators, e.g., glass and plastics. The best types of thermal insulators, however, rely for their insulating effectiveness on trapping a gas within a porous structure. In those materials the transfer of heat may occur by several modes: conduction through a fibrous or porous solid structure, conduction and/or convection through air or other gas trapped in the void spaces, and radiation between portions of the solid structure. The last-mentioned mechanism is especially important at high temperatures or in evacuated enclosures. Special types of superinsulation materials have been developed for cryogenic applications at very low temperatures, down to about 25 K. These superinsulators consist of several layers of

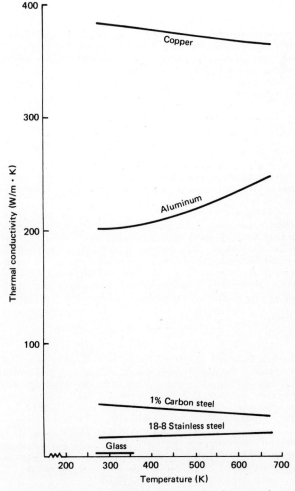

Fig. 4.10 Thermal conductivity of common solids. (*From F. Kreith and J. F. Kreider, Principles of Solar Engineering, Hemisphere, Washington, DC, 1978.*)

highly reflective materials, separated by evacuated spaces to minimize conduction and convection, and can achieve effective conductivities as low as 3×10^{-4} W/m · K. They may some day become very important if solar energy can be used to produce hydrogen which would have to be stored as a cryogenic liquid for end use in a "hydrogen economy."

CONVECTION HEAT TRANSFER

When a fluid comes in contact with a solid surface at a different temperature, the resulting thermal energy exchange process is called "convection heat transfer." This process is a common experience, but a detailed description and analysis of the mechanism is complicated. In this chapter we will not attempt to treat analytical procedures for predicting convection heat transfer, but we will rather concentrate first on developing an intuitive grasp of the mechanism and then present correlations of experimental data which can be used to calculate the rate of heat transfer in those subsystems of complete solar heating and cooling systems in which convection occurs.

There are two kinds of convection processes: natural or "free" convection and "forced" convection. In the first type, the motive force is the density difference in the fluid which results from its contact with a surface at a different temperature and gives rise to buoyant forces. Typical examples of such free convection are the heat transfer between the wall or the roof of a house on a calm day, the convection of the fluid in a tank in which a heating coil is immersed, and the heat loss from the cover surface of a solar collector when there is no wind blowing.

Forced convection occurs when an external force moves a fluid past a surface at a higher or lower temperature than the fluid. Since the fluid velocity in forced convection is higher than in free convection, more heat can be transferred at a given temperature difference. The price to be paid for this increase in the rate of heat transfer is the work required to move the fluid past the surface. But irrespective of whether the convection is free or forced, the rate of heat transfer q_c can be obtained from Newton's law of cooling:

$$q_c = \bar{h}_c \, A \, (T_s - T_f) \tag{4.24}$$

where $\bar{h}_c$ = average convection heat-transfer coefficient over the surface A in W/m² · K
  (Btu/h · ft² · °F)*
 A = surface area in contact with the fluid in m² (ft²)
 T_s = surface temperature in K (°F)
 T_f = undisturbed fluid temperature in K (°F)

Table 4.6 lists some approximate magnitudes of convection heat-transfer coefficients.

Before attempting to calculate a heat-transfer coefficient, we shall examine the energy transport process in some detail and relate the convection of heat to the flow of the fluid. Figure 4.11 shows a heated flat plate cooled by a stream of air flowing over it, with the velocity and the temperature distributions in the air stream. The velocity decreases in the direction toward the surface as a result of viscous forces, and the fluid does not move relative to the fluid-solid interface. Since the velocity of the fluid layer adjacent to the wall is zero, the heat transfer at the interface between the surface and the adjacent fluid layer is by conduction, or

$$\frac{q_c}{A} = - k_f \frac{\partial T}{\partial y}\bigg|_{y=0} = \bar{h}_c \, (T_s - T_f) \tag{4.25}$$

Although this viewpoint suggests that the convection process can be viewed as conduction at the interface, the temperature gradient at the surface, $(\partial T/\partial y)\big|_{y=0}$, is determined by the rate at which the fluid removed from the wall can transport the energy into the mainstream. Thus the temperature gradient at the wall depends on the flow field, with higher velocities being able to produce larger temperature gradients and higher rates of heat transfer. At the same time, however, the thermal conductivity of the fluid plays a role. For example, the value of k_f for the water is an order of magnitude larger than that of air; thus, as shown in Table 4.6 the convection heat-transfer coefficients for water are also larger than for air.

The situation is quite similar in free convection, as shown in Fig. 4.12. The principal difference is that in forced convection the velocity far from the surface approaches the free-stream value

---

*5.677 W/ m² · K = 1 Btu/h · ft² · °F.

TABLE 4.6 Approximate Ranges of Convective Heat-Transfer Coefficients $\bar{h}_c$

	w/m² · K	Btu/h · ft² · °F
Air, free convection	6–30	1–5
Superheated steam or air, forced convection	30–300	5–50
Oil, forced convection	60–1800	10–300
Water, forced convection	300–6000	50–1000
Water, boiling	3000–60,000	500–10,000
Steam, condensing	6000–120,000	1000–20,000

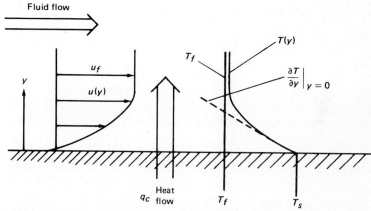

Fig. 4.11 Temperature and velocity distribution in forced convection air flow over a flat plate. (*From F. Kreith and J. F. Kreider, Principles of Solar Engineering, Hemisphere, Washington, DC, 1978.*)

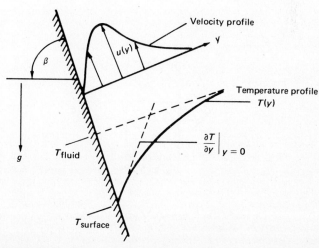

Fig. 4.12 Temperature and velocity distribution for free convection over a heated flat plate inclined at angle β from the horizontal. (*From F. Kreith and J. F. Kreider, Principles of Solar Engineering, Hemisphere, Washington, DC, 1978.*)

imposed by an external force, whereas in free convection the velocity at first increases with increasing distance from the heat-transfer surface and then decreases. The reason for this behavior is that the action of viscosity diminishes rather rapidly with distance from the surface while the density difference decreases more slowly. Eventually, however, the buoyant force also decreases as the fluid density approaches the value of the unheated surrounding fluid. This interaction of forces will cause the velocity to reach a maximum and then approach zero far from the heated surface. The temperature fields in free and forced convection have similar shapes, and in both cases the heat-transfer mechanism at the fluid-solid interface is conduction.

The preceding discussion indicates that the convection heat-transfer coefficient will depend on the density, viscosity, and velocity of the fluid as well as on its thermal properties (thermal conductivity and specific heat). Whereas in forced convection the velocity is usually imposed on the system by a pump or a fan and can be directly specified, in free convection the velocity will depend on the temperature difference between the surface and the fluid, the coefficient of thermal expansion of the fluid (which determines the density change per unit temperature difference), and the body-force field, which in solar systems located on earth is simply the gravitational force.

Also, convection heat transfer can be treated within the framework of a thermal resistance network once the heat-transfer coefficient is known. The thermal resistance to convection is

$$R_c = \frac{1}{\bar{h}_c A} \tag{4.26}$$

and this resistance of a surface-to-fluid interface can easily be incorporated into a network. For example, the heat transfer from the interior of a room at T_i through a wall to atmospheric air outside at T_o is shown in Fig. 4.13.

Heat is first transferred by free convection to the interior surface of the wall, then by conduction through the wall to the exterior surface, and finally from the exterior surface to the air outside. Thus, there are three resistances in series, and the rate of heat transfer is

$$q = \frac{T_i - T_o}{\sum\limits_{i=1}^{i=3} R_i} = \frac{T_i - T_o}{R_1 + R_2 + R_3} \tag{4.27}$$

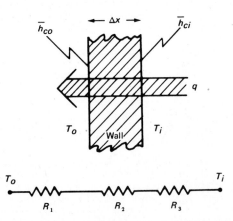

Fig. 4.13 Thermal circuit for heat transfer from the interior of a building at T_i, through a wall, to the exterior environment at T_o. (*From F. Kreith and J. F. Kreider, Principles of Solar Engineering, Hemisphere, Washington, DC, 1978.*)

where

$$R_1 = \frac{1}{\bar{h}_{co}\,A}$$

$$R_2 = \frac{\Delta x}{k_w\,A}$$

$$R_3 = \frac{1}{\bar{h}_{ci}\,A}$$

Experimental data for convection heat transfer are usually correlated in terms of dimensionless quantities. The heat-transfer coefficient, nondimensionalized in terms of the Nusselt number Nu, is defined as

$$\overline{\mathrm{Nu}} = \frac{\bar{h}_c\,L}{k_f} \tag{4.28}$$

where L is a length dimension characteristic of a system in m (ft) (e.g., the diameter for flow in a tube or the distance between the collector surface and the cover in free convection between these surfaces), and k_f is the thermal conductivity of the fluid in W/m·K (Btu/ft·h·°F), usually evaluated at the arithmetic mean between the surface and the bulk fluid temperature.

A third dimensionless parameter of importance is the Prandtl number Pr, defined as

$$\mathrm{Pr} = \frac{c_p \mu_f}{k_f} \tag{4.29}$$

where c_p is the specific heat of the fluid in J/kg·K (Btu/lb$_m$·F) and μ_f is the viscosity in N·s/m² (lb/ft·s). The Prandtl number represents the ratio of momentum diffusivity to thermal diffusivity in a substance.

In general, the Nusselt number $\overline{\mathrm{Nu}}_L$ for forced convection can be related to the Reynolds number Re_L and the Prandtl number Pr by a relation of the form

$$\overline{\mathrm{Nu}}_L = C\,\mathrm{Re}_L^{\,n}\,\mathrm{Pr}^{\,m} \tag{4.30}$$

where C, n, and m are empirically determined constants and subscript L indicates a value based on a characteristic length dimension. Sometimes a functional relation to account for variations in geometry and physical properties is also incorporated into this correlation.

In free convection, the velocity is not given explicitly, and the Reynolds number, which can be calculated only when the velocity is known, cannot be used. The dimensionless parameter which relates the buoyant inertial and viscous forces and thus determines the velocity indirectly is the Grashof number Gr_L, defined by

$$\mathrm{Gr}_L = \frac{\rho^2\,g\beta_T\,(T_1 - T_2)L^3}{\mu^2} \tag{4.31}$$

where $\rho =$ fluid density in kg/m³ (lb/ft³)

$g =$ gravitational constant, equal to 9.8 m/s² (32.2 ft/s²)

$L =$ characteristic length dimension in m (ft)

$\beta_T =$ coefficient of expansion of the fluid in 1/K (1/°R); for ideal gases, β_T equals the reciprocal of the absolute temperature, i.e., $\beta_T = 1/T$

The temperature difference $(T_1 - T_2)$ depends on the system. For a vertical wall, T_1 is the surface temperature and T_2 is the temperature of the undisturbed fluid. For free convection between two parallel surfaces, T_1 and T_2 are taken as the two surface temperatures.

The gravitational constant must be modified when a flow does not occur in the direction of the force field. Thus, for a plate inclined an angle β relative to the horizontal, $g \sin \beta$, is the gravitational force along the surface. Also, in free convection the flow can become turbulent when the Grashof number is large. Normally, for convection from a vertical surface of height L, the flow will be laminar as long as $\mathrm{Gr}_L < 10^9$.

In general, the Nusselt number in free convection can be related to the Grashof number and the Prandtl number by a relation of the form

$$\overline{Nu}_L = C(Gr_L\, Pr)^n \tag{4.32}$$

where C and n are constants determined empirically for given system geometries. In solar energy systems, convection heat transfer plays an important role in the transfer of heat from the absorber plate of a flat-plate collector to the working fluid, the heat loss from the outer cover of a collector, the transfer of heat between the collector surface and the transparent cover, the heat transfer in heat exchangers, and the heat losses from buildings, to mention just a few typical cases. A key problem in constructing the thermal circuit and predicting the rate of heat transfer is to evaluate the heat-transfer coefficients or unit-surface conductances. The evaluation of the heat-transfer coefficient in engineering practice is based on the correlation of experimental data by relations such as Eqs. (4.30) and (4.32). Whenever one uses a correlation based on experiments, it is important to make sure that the conditions of the real system are geometrically and dynamically similar to and within the range of the conditions for which the empirical or semiempirical correlation applies. The correlation equations presented below are mainly for geometries and flow conditions encountered in solar energy conversion systems and associated equipment. A more general treatment is presented in Refs. 5 and 6.

Laminar Flow inside Tubes and Ducts

Because of the small heat flux in flat-plate collectors, the fluid flow is usually laminar. For laminar flow in tubes or ducts, the following equation can be used to determine $\overline{h}_c$ as long as $(Re\, Pr\, D_H/L) > 10$:

$$\overline{Nu}_{D_H} = \frac{\overline{h}_c D_H}{k_b} = 1.86 \left(Re_{D_H}\, Pr\, \frac{D_H}{L} \right)^{1/3} \left(\frac{\mu_b}{\mu_w} \right)^{0.14} \tag{4.33}$$

where D_H = hydraulic diameter of the duct
L = length of the tube or duct
k_b = fluid thermal conductivity at the bulk temperature
μ_b = viscosity of the fluid at the bulk temperature, halfway between inlet and outlet
μ_w = viscosity of the fluid at the wall temperature; the other parameters have previously been defined

When the duct is short and $(Re_{D_H}\, Pr\, D_H/L) > 100$, the following equation applies:

$$\overline{Nu}_{D_H} = \frac{\overline{h}_c D_H}{k_f} = \frac{Re_{D_H}\, Pr\, D_H}{4L} \ln \left[\frac{1}{1 - \dfrac{2.654}{Pr^{0.167}\, (Re_{D_H}\, Pr\, D_H/L)^{1/2}}} \right] \tag{4.34}$$

In the preceding relation and in Eq. (4.33), all physical properties should be evaluated at the mean temperature of the fluid in the duct, defined by

$$T_{mean} = \frac{T_{bi} + T_{wi} + T_{bo} + T_{wo}}{4} \tag{4.35}$$

where T_{bi} = inlet bulk temperature
T_{wi} = wall temperature at the inlet
T_{bo} = outlet bulk temperature
T_{wo} = outlet wall temperature

In view of the low velocities in solar convection apparatus, the velocity produced by buoyant forces, i.e., free convection, can affect the convection heat exchange. In this so-called mixed-flow region, when Gr/Re^2 is between 1 and 10 and $L/D_H > 50$, the heat-transfer coefficient for flow in horizontal ducts and ducts inclined not too far from the horizontal can be evaluated from the relation

$$\overline{Nu}_{D_H} = \frac{\overline{h}_c D_H}{k} = 1.75 \left(\frac{\mu_s}{\mu_w}\right)^{0.14} [Re_{D_H} Pr\, D_H/L + 0.012(Gr_{D_H}^{1/3} Re_{D_H} Pr\, D_H/L)^{4/3}]^{1/3} \quad (4.36)$$

with properties and symbols the same as for Eq. (4.33), except for Gr_{D_H}, which is defined as

$$Gr_{D_H} = \rho^2 g \beta_T (T_w - T_b) D_H^3/\mu^2 \quad (4.37)$$

Turbulent Flow inside Tubes and Ducts

When the Reynolds number is above 6000 in a conduit, the flow is turbulent and the rate of heat transfer will be appreciably greater than in the laminar regime. A dimensionless correlation applicable to fluids with Prandtl numbers between 0.7 and 700 and $(L/D_H) > 60$ is

$$\overline{Nu}_{D_H} = \frac{\overline{h}_c D_H}{k} = 0.023\, Re_{D_H}^{0.8} Pr^{1/3} \quad (4.38)$$

where all properties are to be evaluated at the mean temperature.[6] When the duct is so short that $L/D_H < 60$, the heat-transfer coefficient calculated from Eq. (4.38) should be multiplied by $[1 + (D_H/L)]^{0.7}$.

For turbulent flow between two parallel flat plates with only one surface heated, the relationship

$$\overline{Nu}_{D_H} = 0.0196\, Re_{D_H}^{0.8} Pr^{1/3} \quad (4.39)$$

is recommended on the basis of experimental data for air.[7]

In turbulent-flow heat transfer, the physical properties of the fluid do not undergo as large a variation as in laminar flow. It is therefore sometimes possible to simplify the evaluation of $\overline{h}_c$ for a given fluid by evaluating all the physical properties in Eq. (4.38) for a specific fluid at a given temperature and incorporating them into the coefficient. Note, however, that these simplified equations are applicable only over a limited temperature range, are approximations useful primarily for preliminary design, and are dimensional.

For air at atmospheric pressure flowing through a long duct in turbulent flow at temperatures between 300 K (80°F) and 380 K (223°F) use

$$\overline{h}_c = 3.5 \frac{V^{0.8}}{D_H^{0.2}} \qquad\qquad \overline{h}_c = 0.50 \frac{V^{0.8}}{D_H^{0.2}} \quad (4.40)$$

if D_H is in m	if D_H is in inches
V is in m/s	V is in ft/s
$\overline{h}_c$ is in W/m² · K	$\overline{h}_c$ is in Btu/h · ft² · °F

For atmospheric air temperatures between 300 K (80°F) and 380 K (223°F) flowing between two flat plates spaced a distance D apart, with one side heated, use

$$\overline{h}_c = 2.6 \frac{V^{0.8}}{D^{0.2}} \qquad\qquad \overline{h}_c = 0.372 \frac{V^{0.8}}{D^{0.2}} \quad (4.41)$$

if D is in m	if D is in inches
V is in m/s	V is in ft/s
$\overline{h}_c$ is in W/m² · K	$\overline{h}_c$ is in Btu/h · ft² · °F

For water flowing turbulently through a long duct of inside diameter D in the temperature range between 278 K (40°F) and 378 K (220°F), use

$$\bar{h}_c = 1056(0.02T - 4.06)V^{0.8}/D^{0.2} \qquad \bar{h}_c = 150(1 + 0.011T)V^{0.8}/D^{0.2} \qquad (4.42)$$

if T is in K	if T is in °F
V is in m/s	V is in ft/s
D is in m	D is in inches
$\bar{h}_c$ is in W/m² · K	$\bar{h}_c$ is in Btu/h · ft² · °F

For flow in a helical coil, the value of the heat-transfer coefficient calculated for flow in a straight tube should be multiplied by $[1 + 3.5 (D_{tube}/D_{coil})]$ (Ref. 6).

Single Cylinder in Cross Flow

The average Nusselt number for air flowing over a single cylinder can be obtained for $\mathrm{Re}_D > 500$ from the relation

$$\overline{\mathrm{Nu}}_D = 0.46\,\mathrm{Re}_D^{1/2} + 0.00128\,\mathrm{Re}_D \qquad (4.43)$$

where the Nusselt and Reynolds numbers are based on the cylinder diameter D and the free-stream velocity, and the fluid properties are to be evaluated at the mean temperature between the surface and the air. This geometry is encountered in cross-flow heat exchangers in a solar system.

Flow across Tube Bundles

Flow across tube bundles is encountered in shell and tube heat exchangers which are used industrially to transfer heat from one fluid to another. Also, in solar space heating this type of exchanger is widely used. The flow and the details of the convection process are quite complex, but for bundles with 10 or more tubes at Reynolds numbers between 100 and 40,000, the average Nusselt number can be estimated from the relation

$$\overline{\mathrm{Nu}}_D = 0.33\,\mathrm{Re}_{max}^{0.6}\,\mathrm{Pr}^{0.33} \qquad (4.44)$$

In Eq. (4.44) the Nusselt and Reynolds numbers are based on the tube diameter and the physical properties are based on the mean temperature. However, the velocity is taken at the maximum value the fluid achieves in the tube bank. This maximum occurs at the smallest free-flow area between adjacent tubes. Equation (4.44) gives only an approximation, and for further information the reader is referred to Refs. 6, 8, and 9.

Airflow over a Flat Surface

An interesting case of flow over a surface occurs when air blows over and parallel to the roof or the wall of a building or when the outermost cover of a flat-plate collector is exposed to wind. For the heat loss from a flat plate exposed to wind, the heat-transfer coefficient in W/m² · K is related to the wind speed V in m/s by the approximate dimensional relation

$$\bar{h}_c = 5.7 + 3.8V \qquad (4.45)$$

For a more general discussion of convection in flow over surfaces, see Refs. 5 and 6.

For flow of air over inclined and yawed square plates of side S, Sparrow and Tien (*J. Heat Transfer*, vol. 99, November 77, pp. 507–712) found experimentally that the relation

$$\bar{h}_c = 1.16\,\rho\,C_p\,U_\infty/\mathrm{Re}^{1/2}$$

where $\bar{h}_c$ is the average convection coefficient for the upper surface and $\mathrm{Re} = U_\infty \rho S/\mu$, correlates with experimental data for inclination angles β between 0 and 90° and yaw angles between 0 and 45°. For rectangles, no experimental data are available, but a reasonable approach would be to take S equal to the mean between the longer and the shorter side. The above correlation indicates that convection heat losses from solar collectors are only about one-half the value predicted by standard equations for flow over an infinitely wide flat plate.

Free Convection from a Flat Surface

Free convection correlations are used to calculate the heat flow to and from buildings on calm days, and the heat loss from solar collectors. For heat transfer to air at atmospheric pressure from a heated plate of length L inclined an angle β from the horizontal,

$$\bar{h}_c = 1.42 \, (\Delta T \sin \beta / L)^{1/4} \qquad (4.46)$$

gives the heat-transfer coefficient in laminar flow with $10^4 < \text{Gr}_L < 10^9$ if L is in m and ΔT in K, while the relation

$$\bar{h}_c = 0.95 \, (\Delta T \sin \beta)^{1/3} \qquad (4.47)$$

applies when the flow is turbulent and $\text{Gr}_L > 10^9$.

Free Convection between Two Parallel Surfaces

Free convection between two parallel flat plates has been investigated experimentally by Hollands et al.[10,11,12] and by Buchberg et al.[13] Their results are correlated for inclination angles β from the horizontal between 0 and 75° by the relation

$$\overline{\text{Nu}}_L = 1 + 1.44 \left[1 - \frac{1708}{\text{Ra}_L \cos \beta} \right]^{+} \left[1 - \frac{1708 \, (\sin 1.8 \beta)^{1.6}}{\text{Ra}_L \cos \beta} \right] + \left[\left(\frac{\text{Ra}_L \cos \beta}{5830} \right)^{1/3} - 1 \right]^{+} \qquad (4.48)$$

where L is the distance between the plates at temperatures T_1 and T_2, respectively, and

$$\text{Ra}_L = \frac{2g \, (T_1 - T_2) \, L^3}{\nu^2 \, (T_1 + T_2)} \, \text{Pr}$$

(Ra_L is called the Rayleigh number.) It should be noted that when $\text{Ra}_L < (1708/\cos \beta)$, the Nusselt number in Eq. (4.48) is exactly equal to unity. Since by definition

$$q_c = A \, \bar{h}_c \, (T_1 - T_2) = A \, \overline{\text{Nu}} \, \frac{k}{L} \, (T_1 - T_2)$$

the condition $\overline{\text{Nu}} = 1$ implies that the heat transfer is by conduction only.

For inclinations between $75° < \beta < 90°$ the recommended relation for air is

$$\overline{\text{Nu}}_L = \left[1, 0.288 \, [\sin \beta \, A \text{Ra}]^{1/4}, \, 0.039 \, (\sin \beta \, \text{Ra})^{1/3} \right]_{\text{max}} \qquad (4.49)$$

where the subscript max indicates that, at any given value of Ra, the largest of the three quantities separated by commas should be used. The quantity A in Eq. (4.49) is the aspect ratio of the air layer, defined as the ratio of the thickness L to the length along the layer H, measured along either the heated or cooled bounding surface in the upslope direction.

Figure 4.14 is a dimensional plot of Eq. (4.48) for air illustrating the effect of gap width L and inclination angle β on the free-convection coefficient h. The quantities ϕ_1, ϕ_2, ϕ_3 are functions of the surface temperatures T_1 and T_2 in K. For air these functions are

$$\phi_1 = \frac{137}{(T_m + 200)^{1/3} \, T_m^{1/2}} \qquad (4.50)$$

$$\phi_2 = \left(\frac{T_1 - T_2}{50} \right)^{1/3} \qquad (4.51)$$

$$\phi_3 = \frac{1428 \, (T_m + 200)^{2/3}}{T_m^2} \qquad (4.52)$$

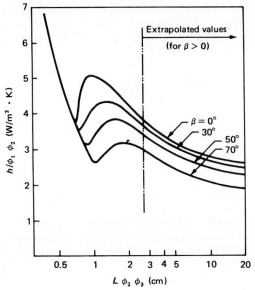

Fig. 4.14 Plot of the free convection heat transfer coefficient across an air gap versus gap spacing L. (*From K. G. T. Hollands, G. D. Raithby, and T. E. Unny, Studies on Methods of Reducing Heat Losses from Flat Plate Collectors, Final Rept., ERDA Contract E (11-1)-2597, University of Waterloo, Waterloo, Ontario, Canada, June 1976.*)

where $T_m = (T_1 + T_2)/2$. To maximize the thermal performance of a flat-plate collector, h_c should be as small as possible. A spacing of 5 cm (2 in) appears to be a reasonable compromise, but for more precise design the curves in Fig. 4.14 show that the smallest convective loss corresponds to $L_{min} = 0.71/(\cos \beta \, \phi_2 \, \phi_3)$.

Free Convection across Inclined Air Layers
Constrained by Honeycombs

A review of the state of the art in the application of honeycombs for flat-plate solar collectors is given in Refs. 14 and 15, and the theory of the honeycomb collector is presented in Ref. 16. Briefly, the honeycomb is used to suppress free-convection heat transfer across the air gap between a collector plate and its glass cover and to reduce radiation losses from the collector. For an inclined square honeycomb, such as shown in Fig. 4.15, the Nusselt number depends on the Rayleigh number, the inclination, and the aspect ratio of the honycomb, $A = L/D$. For the range $0 < \mathrm{Ra} < 6000 \, A^4$, $30° < \beta < 90°$, and $A = 3, 4$, and 5, the Nusselt number for air is given

$$\overline{\mathrm{Nu}} = \frac{\bar{h}_c L}{k} = 1 + 0.89 \cos (\beta - 90°) \left(\frac{\mathrm{Ra}}{2420A^4} \right)^{(2.88 \,-\, 1.64 \sin \beta\,)} \tag{4.53}$$

The above relation may also be used for hexagonal honeycombs if D is replaced by the hydraulic diameter. For engineering design the honeycomb should be chosen to give a Nusselt number of 1.2 according to Hollands. For air at atmospheric pressure and moderate temperature (370 K $> T_m >$ 280 K), the optimum geometry is found from

$$A = C(\beta) \left(1 + \frac{200}{T_m} \right)^{1/2} \left(\frac{100}{T_m} \right) (T_1 - T_2)^{1/4} \, L^{3/4} \tag{4.54}$$

if L is in m and T in K. The function $C(\beta)$ is plotted in Fig. 4.16.

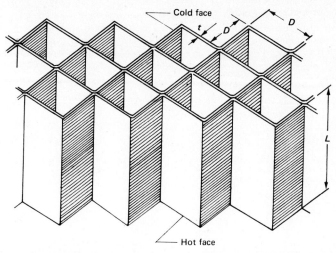

Fig. 4.15 Schematic diagram of a honeycomb. (*From K. G. T. Hollands, G. D. Raithby, and T. E. Unny, Studies on Methods of Reducing Heat Losses from Flat Plate Collectors, Final Rept., ERDA Contract E (11-1)-2597, University of Waterloo, Ontario, Canada, June 1976.*)

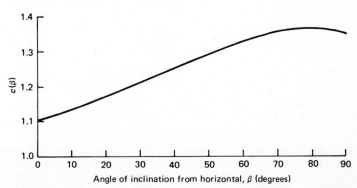

Fig. 4.16 Plot of C (β) versus β for use in Eq. (4.54). (*From K. G. T. Hollands, G. D. Raithby, and T. E. Unny, Studies on Methods of Reducing Heat Losses from Flat Plate Collectors, Final Rept., ERDA Contract E (11-1)-2597, University of Waterloo, Waterloo, Ontario, Canada, June 1976.*)

Free Convection from Cylinders and Spheres

The average heat-transfer coefficient by free convection from a horizontal cylinder of diameter D can be calculated from the relation

$$\overline{\mathrm{Nu}}_D = \frac{\overline{h}_c D}{k} = 0.53 \, (\mathrm{Gr}_D \, \mathrm{Pr})^{1/4} \qquad (4.55)$$

over a range $10^3 < \mathrm{Gr}_D < 10^9$.

The heat-transfer coefficient in free convection to or from a sphere can be obtained from the equation

$$\overline{\mathrm{Nu}}_D = 2 + 0.45 \, (\mathrm{Gr}_D \, \mathrm{Pr})^{1/4} \qquad (4.56)$$

For vertical cylinders, Eqs. (4.46) and (4.47) apply.

Natural Convection in Annuli

Very little experimental or theoretical work has been reported applicable to long cylindrical solar collectors other than for annuli formed by concentric circular cylinders. On the basis of flow patterns, temperature profiles, and overall heat-transfer correlations obtained with air, water, and silicone oil contained in annuli of radial gap $L = (R_o - R_i)$ formed by long isothermal concentric cylinders, the following correlations are recommended:[13]

$$\mathrm{Nu} = \frac{k_{\mathrm{eff}}}{k} = 0.135 \left(\frac{\mathrm{Ra\ Pr}}{1.36 + \mathrm{Pr}} \right)^{0.278} \tag{4.57}$$

for

$$3.5 \leq \log \left(\frac{\mathrm{Ra\ Pr}}{1.36 + \mathrm{Pr}} \right) < 8.0 \quad \text{and} \quad 0.25 \leq \frac{(R_o - R_i)}{R_i} \leq 3.25$$

and $\mathrm{Nu} = 1$ when

$$\log \left(\frac{\mathrm{Ra\ Pr}}{1.36 + \mathrm{Pr}} \right) < 3.0$$

Note that

$$\frac{k_{\mathrm{eff}}}{k} = \frac{(dq/dl)\ \ln (R_o/R_i)}{2\pi k\ (T_i - T_o)}$$

where dq/dl is the heat loss per unit length.

To properly analyze cylindrical collectors that have been proposed, additional data must be obtained for nonisothermal concentric circular cylinders and nonisothermal noncircular cylinders. Until such data are available, it is recommended that the correlations given by Eq. (4.57) be used. If the inner cylinder is noncircular, one might use the hydraulic radius in the Rayleigh number instead of annulus width $(R_o - R_i)$. An extensive correlation of data for free convection between concentric cylinders is presented in Ref. 17.

Flow and Convection Heat Transfer in Fixed Beds

In addition to conventional heat exchangers in which the hot and cold fluids are separated by a solid wall, some solar energy systems also make use of another type of heat-exchange device. This device consists of a bed of solid particles which can be heated by the working fluid when it is hot. However, the particles can transfer the energy stored in them to the same working fluid at a later time when it is cold and must be heated. The most common example of this type of heat-exchanger–storage collector system in solar engineering is the "pebble-bed" of a solar collector system using air as the working fluid.

In general, two types of particle heat-exchange systems are encountered in engineering: the fixed bed and the fluidized bed. In the former, the particle position is not changed by the fluid flowing through the voids. However, if the fluid is flowing upward and its velocity is increased sufficiently, the net lift force will eventually counteract the gravitational force, and the particles will become suspended. This condition is called a fluidized bed, characterized by thorough mixing of particles which eliminates any temperature gradient through the bed. This type of operation is often sought in the chemical industry, but would be undesirable in a rock storage bed for a solar system.

To analyze the flow and convection heat transfer in a fixed bed of pebbles, certain simplifications are usually made. The particles are considered to be approximately spherical. For such a pebble bed Löf and Hawley[27] investigated the heat transfer and recommended the following empirical relation to evaluate the heat-transfer coefficient:

$$h_v = 650\ (\dot{m}/A_b D_s)^{0.7} \tag{4.58}$$

where h_v = *volumetric* heat transfer coefficient in $W/m^3 \cdot K$
  $\dot{m}_b$ = mass rate of flow in kg/s
  A_b = cross-sectional area of the pebble bed in m^2
  D_s = equivalent spherical diameter of the particles in m

D_s is given by

$$D_s = \left(\frac{6}{\pi} \times \frac{\text{net volume of particles}}{\text{number of particles}} \right)^{1/3}$$

Experimental verification of the range of variables over which this relation is applicable extends specifically over entering air temperatures between 100 and 200°F (38 and 93°C), gravel sizes between 0.19- to 1.5-in diameter, airflow rates between 12 and 66 ft^3/min per square foot (60 and 330 L/s m^2) of empty bed cross section, and voids obtained by normal filling.

For a bed of approximately spherical particles the total surface area of all particles A_p is $6(1 - \epsilon_v)A_bL/D_s$ where L is the bed length. Handley and Heggs[28] measured the heat transfer and frictional pressure drop in flow through packed beds and correlated their results with those of previous investigators. They found that the heat-transfer coefficients for a bed of approximately spherical particles can be correlated by the dimensionless equation

$$j_h = 0.255/\epsilon_r \, Re^{0.33}$$

where ϵ_r = void fraction of the bed
  Re_{D_s} = Reynolds number (GD_s/μ)
  D_s = sphere diameter
  j_h = Stanton number $(Nu_{D_s}/Re_{D_s} \, Pr^{1/3})$
  G = superficial mass velocity of the gas $(\dot{m}/A_b)$

For air (Pr = 0.71), this relation becomes

$$Nu_{D_s} = \frac{0.23 \, Re_{D_s}^{2/3}}{\epsilon_r} \tag{4.59}$$

The pressure drop in a bed of length L can be determined from the relation

$$\frac{\Delta p}{\rho V_s^2} = \frac{L}{D_s} \frac{(1 - \epsilon_r)^2}{Re_{D_s}\epsilon_r^3} \left[1.24 \frac{Re_{D_s}}{1 - \epsilon_r} + 368 \right] \tag{4.59a}$$

where ϵ_r = void fraction (or porosity) of the bed and $V_s = (\dot{m}/A_b\rho_f)$, the superficial velocity. For pebble beds used in solar systems, ϵ_r is between 0.35 and 0.45.

More detailed analysis of the transient heat transfer and data on the effect of particle shape on the heat-transfer coefficients are presented in Ref. 29.

RADIATION HEAT TRANSFER

The material presented in this section has been selected from textbooks on heat transfer or radiation (e.g., Refs. 5, 18, 19) to provide sufficient background in radiation for the thermal design and engineering analysis of solar energy systems. Some material from quantum mechanics and electromagnetic theory is used, but no prior familiarity with these subjects is assumed. However, the treatment of these topics is principally oriented towards the utilization of results to calculate heat transfer by radiation.

All radiation travels at the speed of light, but the actual wavelength depends on the frequency and refractive index. The speed of light is equal to the product of the wavelength and the frequency of radiation, and this product in turn equals the speed of light in a vacuum divided by the refractive index of the medium through which it travels:

$$c = \lambda \nu = c_o/n$$

where λ = wavelength, m (or micrometers, $1\ \mu\text{m} = 10^{-6}\ \text{m}$)
ν = frequency, s^{-1}
c_o = speed of light in a vacuum
n = index of refraction of the medium

Thermal radiation is a form of electromagnetic energy, and all bodies emit thermal radiation by virtue of their temperature. When a body is heated, its atoms, molecules, or electrons are raised to higher energy levels called excited states. However, they tend to return spontaneously to lower energy states, and in this process energy is emitted in the form of electromagnetic waves. Changes in energy result from rearrangements in the electronic, rotational, and vibrational states of atoms and molecules. Since these rearrangements involve different amounts of energy changes and these energy changes are related to the frequency, the radiation emitted by a body is distributed over a range of wavelengths. A portion of the electromagnetic spectrum is shown in Fig. 4.17. The wavelengths associated with the various mechanisms are not sharply defined, but thermal radiation is usually considered to fall within the band from about 0.1 to 100 μm, whereas solar radiation is concentrated in the wavelength range between 0.1 and 3 μm.

For some problems in solar engineering the classical electromagnetic wave theory is not suitable. In such cases, e.g., in photovoltaic or photochemical processes, it is necessary to treat the energy transport from the point of view of quantum mechanics. In this view energy is transported by particles or photons, which are treated as energy units or quanta rather than waves. The energy of a photon E_p of frequency ν_p is

$$E_p = h\nu_p$$

where h is Planck's constant ($6.625 \times 10^{-34}\ \text{J}\cdot\text{s}$).

A heuristic picture of quantum radiation propagation may be obtained by viewing each quantum as a particle, similar to a molecule in a gas, and the radiation as the transport of a "photon gas" flowing from one place to another. By treating radiation as a gas, it can be shown from quantum statistical thermodynamics (see, for example, Ref. 20) that the energy density of radiation per unit volume and per unit wavelength $u_{b\lambda}$ is given by the relation

$$u_{b\lambda} = \frac{8\pi\ h\ c_o\ n^3}{(e^{hc_o/\lambda kT}-1)\lambda^5} \tag{4.60}$$

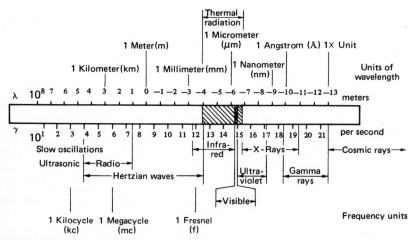

Fig. 4.17 The electromagnetic radiation spectrum. (*From F. Kreith and J. F. Kreider, Principles of Solar Engineering, Hemisphere, Washington, DC, 1978.*)

where k is Boltzmann's constant ($k = 1.38 \times 10^{-23}$ J/molecule·K) and T is the absolute temperature in K. It can also be shown that the energy density at a given wavelength is related to the monochromatic radiation emitted by a perfect radiator, usually called a blackbody, according to the relation

$$E_{b\lambda} = \frac{u_{b\lambda}c_0}{4} = \frac{C_1}{(e^{C_2/\lambda T} - 1)\lambda^5 n^2} \tag{4.61}$$

where $C_1 = 3.74 \times 10^8$ W$(\mu m)^4/m^2$ [1.19×10^8 Btu$(\mu m)^4/h \cdot ft^2$] and $C_2 = 1.44 \times 10^4 \mu m \cdot$ K [$2.59 \times 10^4 \mu m \cdot$ R]

The quantity $E_{b\lambda}$ has the units of W/m²·μm (Btu/h·ft²·μm), and it is called the monochromatic emissive power of a blackbody, defined as the energy emitted by a perfect radiator per unit wavelength at the specified wavelength, per unit wavelength, per unit area, and per unit time at temperature T.

The total energy emitted by a blackbody can be obtained by integration over all wavelengths. This yields the Stefan-Boltzman law

$$E_b = \int_0^\infty E_{b\lambda} d\lambda = \sigma T^4 \tag{4.62}$$

where $\sigma =$ Stefan-Boltzman constant $= 5.67 \times 10^{-8}$ W/m²·K⁴ (0.1714×10^{-8} Btu/h·ft²·°R⁴) and $T =$ absolute temperature in K (or °R).

When radiation strikes a body, a part of it is reflected, a part is absorbed and, if the material is transparent, a part is transmitted as shown in Fig. 4.18. The fraction of the incident radiation reflected is defined as the reflectance ρ, the fraction absorbed as the absorptance α, and the fraction transmitted as the transmittance τ. According to the first law of thermodynamics, these three components must add up to unity, or

$$\alpha + \tau + \rho = 1 \tag{4.63}$$

Opaque bodies do not transmit any radiation, and $\tau \equiv 0$.

The reflection of radiation can be "specular" or "diffuse." When the angle of incidence is equal to the angle of reflection, the reflection is called specular, whereas when the reflected radiation is uniformly distributed into all directions, it is called diffuse (see Fig. 4.19). No real surface is either specular or diffuse, but a highly polished surface approaches specular reflection whereas a rough surface reflects approximately diffusely.

Using these radiation properties, one can relate the emissive power of a real surface to that of an ideal "black" surface. Suppose an enclosure with perfectly black walls at temperature T contains a body. The enclosure would emit radiation in accordance with Eq. (4.62). After equilibrium is reached, the body inside the enclosure also must be at temperature T, and the

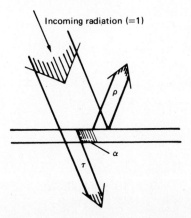

Fig. 4.18 Sketch showing reflection, absorption, and transmission of incident radiation. (*From F. Kreith and J. F. Kreider, Principles of Solar Engineering, Hemisphere, Washington, DC, 1978.*)

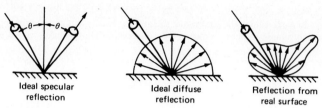

| Ideal specular reflection | Ideal diffuse reflection | Reflection from real surface |

Fig. 4.19 Sketch showing specular, diffuse, and real reflection. (*From F. Kreith and J. F. Kreider, Principles of Solar Engineering, Hemisphere, Washington, DC, 1978.*)

energy emitted by the body must equal the energy it receives. If we let the radiation flux received at some point in the enclosure be q_i, equilibrium is established when the radiation emitted is equal to the radiation received, or

$$EA = q_i A \alpha \tag{4.64}$$

If the real body inside the enclosure is now replaced by a blackbody of the same shape, it too will achieve equilibrium at the temperature of the surrounding walls. Since $\alpha_b = 1$, we get

$$E_b A = q_i A (1.0) \tag{4.65}$$

It can now be shown that at a given temperature the absorptance of a blackbody is equal to its emittance. Dividing Eq. (4.64) by Eq. (4.65), one obtains

$$\frac{E(T)}{E_b(T)} \equiv \epsilon(T) = \alpha(T) \tag{4.66}$$

since $E/E_b \equiv \epsilon$ by definition. It should be noted that the absorptance and emittance discussed above are total values for the material. This means that they are representative of the behavior over the entire radiation spectrum.

Real substances not only emit less than blackbodies, but their emittance varies with wavelength. It is therefore necessary to define monochromatic properties. The monochromatic emittance is defined as the ratio of the monochromatic emissive power of the real body to the monochromatic emissive power of the blackbody at the same wavelength and temperature, or

$$\epsilon_\lambda = \frac{E_\lambda}{E_{b\lambda}} \tag{4.67}$$

Since the emittance defined by Eq. (4.67) applies to all the radiation emitted into the hemispheric half space above the emitting area, it is strictly speaking the hemispherical emittance, but generally the word hemispherical is omitted. The total emittance on a surface ϵ is related to the monochromatic emittance by the relation

$$\bar{\epsilon} = \frac{E}{E_b} = \frac{\int_0^\infty \epsilon_\lambda E_{b\lambda} d\lambda}{\int_0^\infty E_{b\lambda} d\lambda} = \frac{\int_0^\infty \epsilon_\lambda E_{b\lambda} d\lambda}{\sigma T^4} \tag{4.68}$$

A special type of surface, called "gray" is defined as having a monochromatic emittance which does not change with wavelength. For a gray body, $\bar{\epsilon} = \epsilon_\lambda = \alpha_\lambda = \bar{\alpha}$. For real surfaces, however, both the emittance and the absorptance vary with wavelength, and to obtain the total values an integration is necessary. The integration is usually performed numerically with the aid of radiation tables discussed below.

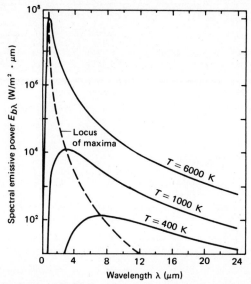

Fig. 4.20 Spectral distribution of blackbody radiation. (*From F. Kreith and J. F. Kreider, Principles of Solar Engineering, Hemisphere, Washington, DC, 1978.*)

Radiation Function Tables

Engineering calculation of radiative transfer are facilitated by the use of radiation function tables which present the results of Planck's law in a more convenient form than Eq. (4.61). A plot of the monochromatic emissive power of a blackbody as a function of wavelength is shown in Fig. 4.20 for temperature of 400, 1000, and 10,000 K. Notice that the peak of $E_{b\lambda}$ occurs at progressively shorter wavelength as the temperature is increased. These peaks, or maximum points, are uniquely related to the body surface temperature. By differentiating Planck's distribution law, Eq. (4.61). and equating to zero, the wavelength corresponding to the maximum value of $E_{b\lambda}$ can be shown to occur when

$$\lambda_{max}T = 2897.8 \ \mu m \cdot K \ (5215.6 \ \mu m \cdot °R) \tag{4.69}$$

Frequently one needs to know the amount of energy emitted by a blackbody within a specified range of wavelengths. This type of calculation can be performed easily with the aid of the radiation functions mentioned previously. To construct the appropriate radiation functions in dimensionless form, note that the ratio of the blackbody radiation emitted at wavelengths between 0 and λ and between 0 and ∞ can be made a function of the single variable (λT) by using Eq. (4.61) as shown below (where the refractive index is unity):

$$\frac{E_{b,\,0-\lambda}}{E_{b,\,0-\infty}} = \frac{\int_0^{\lambda} E_{b\lambda}\, d\lambda}{\sigma T^4} = \int_0^{\lambda T} \frac{C_1 d\,(\lambda T)}{\sigma\,(\lambda T)^5\,(e^{\,C_2/\,\lambda T}-1)} \tag{4.70}$$

The above relation is plotted in Fig. 4.21 and the results are also shown in tabular form in Table 4.7. In this table the first column is the ratio of λ to λ_{max} from Eq. (4.69), the second column is the ratio of $E_{b\lambda}$ to $E_{b\lambda,\ max}$ from Eq. (4.61), and the third column is the ratio of $E_{b,\,0-\lambda}$ to σT^4 from Eq. (4.70). For use on a digital computer, Eq. (4.70) can be approximated by the following polynomials:[18]
$\nu' \geqslant 2$:

$$\frac{E_{b,0-\lambda}}{\sigma T^4} = \frac{15}{\pi^4} \sum_{m=1,2,...} \frac{e^{-m\nu'}}{m^4} \{[(m\nu' + 3)\,m\nu' + 6]m\nu' + 6\} \tag{4.71}$$

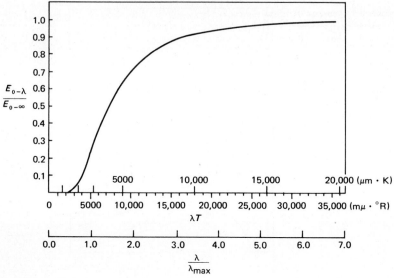

Fig. 4.21 Plot of radiation function $(E_{b,0-\lambda}/E_{b,0-\infty})$ versus λT and λ/λ_{max}. (*From F. Kreith and J. F. Kreider, Principles of Solar Engineering, Hemisphere, Washington, DC, 1978.*)

$\nu' < 2$:

$$\frac{E_{b,0-\lambda}}{\sigma T^4} = 1 - \frac{15}{\pi^4} \nu'^3 \left(\frac{1}{3} - \frac{\nu'}{8} + \frac{\nu'^2}{60} - \frac{\nu'^4}{5040} + \frac{\nu'^6}{272,160} - \frac{\nu'^8}{13,305,600} \right)$$

where $\nu' \equiv C_2/\lambda T$.

Intensity of Radiation and Shape Factor

The emissive power of a surface gives the total radiation emitted in all directions. To determine the radiation emitted in a given direction, we must define another quantity, the radiation intensity I. It is defined as the radiant energy passing through an imaginary plane in space per unit area, per unit time, and per unit solid angle perpendicular to the plane as shown in Fig. 4.22. I is defined by the relation

$$I = \lim_{\substack{dA' \to 0 \\ d\omega \to 0}} \frac{dE}{dA' \, d\omega} \tag{4.72}$$

Radiation intensity is a vector and as such has both magnitude and direction. It can be related to the radiation flux, defined as the radiant energy passing through an imaginary plane per unit area per unit time in all directions. Note that the area dA' is perpendicular to the direction of the radiation and the area dA is located in the center of a hemisphere through which all of the radiation passes. The solid angle subtended by dA' is given by $d\omega = dA'/r^2$. The radiation flux q_r emanating from dA can be obtained by integrating the intensity over the hemisphere. As shown in Fig. 4.22, the unit projected area for I is $dA \cos\theta$, and the differential area dA' on the hemisphere is $r^2 \sin\theta \, d\theta \, d\phi$; thus

$$q_r = \int_0^2 \int_0^{\pi/2} I \cos\theta \sin\theta \, d\theta \, d\phi \tag{4.73}$$

For the special case of a diffuse surface, for which I is isotropic, Eq. (4.73) gives

$$q_r = \pi I \tag{4.74}$$

TABLE 4.7   Thermal Radiation Functions*

λ/λ_{max}	$\dfrac{E_{b\lambda}}{E_{b\lambda, max}}$	$\dfrac{E_{b\lambda,0-\lambda}}{\sigma T^4}$	λ/λ_{max}	$\dfrac{E_{b\lambda}}{E_{b\lambda, max}}$	$\dfrac{E_{b\lambda,0-\lambda}}{\sigma T^4}$	λ/λ_{max}	$\dfrac{E_{b\lambda}}{E_{b\lambda, max}}$	$\dfrac{E_{b\lambda,0-\lambda}}{\sigma T^4}$
0.00	0.0000	0.0000	1.50	0.7103	0.5403	2.85	0.1607	0.8661
0.20	0.0000	0.0000	1.55	0.6737	0.5630	2.90	0.1528	0.8713
0.25	0.0003	0.0000	1.60	0.6382	0.5846	2.95	0.1454	0.8762
0.30	0.0038	0.0001	1.65	0.6039	0.6050	3.00	0.1384	0.8809
0.35	0.0187	0.0004	1.70	0.5710	0.6243	3.10	0.1255	0.8895
0.40	0.0565	0.0015	1.75	0.5397	0.6426	3.20	0.1141	0.8974
0.45	0.1246	0.0044	1.80	0.5098	0.6598	3.30	0.1038	0.9045
0.50	0.2217	0.0101	1.85	0.4815	0.6761	3.40	0.0947	0.9111
0.55	0.3396	0.0192	1.90	0.4546	0.6915	3.50	0.0865	0.9170
0.60	0.4664	0.0325	1.95	0.4293	0.7060	3.60	0.0792	0.9225
0.65	0.5909	0.0499	2.00	0.4054	0.7197	3.70	0.0726	0.9275
0.70	0.7042	0.0712	2.05	0.3828	0.7327	3.80	0.0667	0.9320
0.75	0.8007	0.0960	2.10	0.3616	0.7449	3.90	0.0613	0.9362
0.80	0.8776	0.1236	2.15	0.3416	0.7565	4.00	0.0565	0.9401
0.85	0.9345	0.1535	2.20	0.3229	0.7674	4.20	0.0482	0.9470
0.90	0.9725	0.1849	2.25	0.3052	0.7777	4.40	0.0413	0.9528
0.95	0.9936	0.2172	2.30	0.2887	0.7875	4.60	0.0356	0.9579
1.00	1.0000	0.2501	2.35	0.2731	0.7967	4.80	0.0308	0.9622
1.05	0.9944	0.2829	2.40	0.2585	0.8054	5.00	0.0268	0.9660
1.10	0.9791	0.3153	2.45	0.2447	0.8137	6.00	0.0142	0.9790
1.15	0.9562	0.3472	2.50	0.2318	0.8215	7.00	0.0082	0.9861
1.20	0.9277	0.3782	2.55	0.2197	0.8290	8.00	0.0050	0.9904
1.25	0.8952	0.4081	2.60	0.2083	0.8360	9.00	0.0033	0.9930
1.30	0.8600	0.4370	2.65	0.1976	0.8427	10.00	0.0022	0.9948
1.35	0.8231	0.4647	2.70	0.1875	0.8490	20.00	0.0002	0.9993
1.40	0.7854	0.4911	2.75	0.1780	0.8550	40.00	0.0000	0.9999
1.45	0.7477	0.5163	2.80	0.1691	0.8607	50.00	0.0000	1.0000

*
$$\lambda = \text{wavelength in } \mu m$$
$$\lambda_{max} = \text{wavelength at } E_{b\lambda, max} \text{ in } \mu m = 2898/T$$
$$E_{b\lambda} = \text{monochromatic emissive power in W/m}^2 \cdot \mu m$$
$$= 374.15 \times 10^6/\lambda^5 \left[\exp\left(14{,}387.9/\lambda T\right) - 1\right]$$
$$E_{b\lambda, max} = \text{maximum monochromatic emissive power in W/m}^2 \cdot \mu m$$
$$= 12.865 \times 10^{-12} T^5$$
$$E_{b\lambda, 0-\lambda} = \int_0^\lambda E_{b\lambda}\, d\lambda$$
$$\sigma T^4 = E_{b\lambda, 0-\infty} = 5.670 \times 10^{-8} T^4 \text{ W/m}^2$$
$$T = \text{absolute temperature in K}$$

Since all black surfaces are diffuse,

$$E_b = \pi I_b \tag{4.75}$$

Equation (4.75) can, of course, also be written for monochromatic radiation as well as for total radiation, or

$$E_{b\lambda} = \pi I_{b\lambda} \tag{4.76}$$

The rate of radiation heat transfer between two surfaces also depends upon their geometric configurations and relationships. The influence of geometry in radiation heat transfer can be

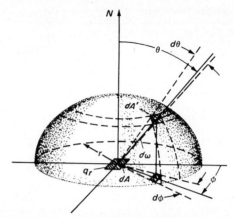

Fig. 4.22 Schematic diagram illustrating radiation intensity and flux. (*From F. Kreith and J. F. Kreider, Principles of Solar Engineering, Hemisphere, Washington, DC, 1978.*)

expressed in terms of the radiation shape factor between any two surfaces 1 and 2, defined as follows:

F_{1-2} = fraction of radiation surface 1 which reaches surface 2

F_{2-1} = fraction of radiation leaving surface 2 which reaches surface 1

In general, F_{m-n} = fraction of radiation leaving surface m which reaches surface n. If both surfaces are black, the energy leaving surface m and arriving at surface n is $E_{bm} A_m F_{m-n}$, and the energy leaving surface n and arriving at m is $E_{bn} A_n F_{n-m}$. If both surfaces absorb all the incident energy, the net rate of exchange, $q_{m \rightleftarrows n}$ will be

$$q_{m \rightleftarrows n} = E_{bm} A_m F_{m-n} - E_{bn} A_n F_{n-m} \tag{4.77}$$

If both surfaces are at the same temperature $E_{bm} = E_{bn}$ and the net exchange is zero, $q_{m \rightleftarrows n} \equiv 0$. This shows that the geometric radiation shape factor must obey the reciprocity relation

$$A_m F_{m-n} = A_n F_{n-m} \tag{4.78}$$

The net rate of heat transfer can therefore be written in two equivalent forms:

$$q_{m \rightleftarrows n} = A_m F_{m-n} (E_{bm} - E_{bn}) = A_n F_{n-m} (E_{bm} - E_{bn}) \tag{4.79}$$

The evaluation of geometrical shape factors is generally quite involved. For a majority of solar energy applications, however, only a few cases are of interest. One of these is a small convex or flat object of area A_1 surrounded by a large enclosure A_2. Since all radiation leaving A_1 is intercepted by A_2, $F_{1-2} = 1$ and $F_{2-1} = A_1/A_2$.

Another case is the exchange of radiation between two large parallel surfaces. If the two surfaces are close to each other, almost all of the radiation leaving A_1 reaches A_2, and vice versa. Thus, $F_{1-2} = F_{2-1} = 1.0$ according to the definition of the shape factor.

A third case of importance is the exchange between a small surface ΔA_1 and a portion of a large space A_2, e.g., a flat-plate collector tilted at an angle β from the horizontal viewing the sky dome. For this situation we refer to the definition of radiation flux (see Fig. 4.22). The portion of the radiation emitted by ΔA_1 which is intercepted by the surrounding hemisphere depends on the angle of tilt. When the surface is horizontal, $F_{1-2} = 1$; when it is vertical, $F_{1-2} = 1/2$ ($\beta = 90°$). For intermediate values, it can be shown[19] that

$$F_{1-2} = 1/2 (1 + \cos \beta) = \cos^2 \frac{\beta}{2} \tag{4.80}$$

If the diffuse sky radiation is uniformly distributed and assumed to be black, then a small black area A_1 receives radiation at the rate

$$A_1 F_{1-\text{sky}} E_{\text{sky}} = \cos^2 \frac{\beta}{2} \sigma T^4_{\text{sky}} \qquad (4.81)$$

and the net radiation heat transfer is

$$q_{\text{sky} \rightleftarrows 1} = A_1 F_{1-\text{sky}} \sigma (T^4_{\text{sky}} - T^4_1) \qquad (4.82)$$

If the receiving area is gray, with an absorptance α equal to the emittance ϵ, the net exchange is

$$q_{\text{sky} \rightleftarrows 1} = A_1 F_{1-\text{sky}} \alpha \, \sigma (T^4_{\text{sky}} - T^4_1) \qquad (4.83)$$

A simple method to evaluate shape factors for two-dimensional surfaces that are infinitely long in one direction and have a uniform cross section normal to the infinite direction has been developed by Hottel and Sarofim. The method, called the *crossed-string method*, has applications to trough-type concentrators, which will be discussed in Chapter 9. According to this method, the shape factor F_{1-2} is equal to the sum of the lengths of the cross strings stretched between the ends of the two surfaces minus the sum of the uncrossed strings divided by twice the length of A_1. The shape factor F_{1-2} is therefore

$$F_{1-2} = \frac{1}{2L_1} [(\overline{ac} + \overline{bd}) - (\overline{ab} + \overline{cd})]$$

Radiation Exchange between Two Parallel Gray Flat Plates

The net rate of radiation heat transfer between two *gray* parallel flat plates, A_1 and A_2, with emittances ϵ_1 and ϵ_2, respectively, can be obtained by tracing the radiation emitted by each of the two surfaces. Surface A_1 emits radiation at the rate $\epsilon_1 A_1 \sigma T^4_1$, and since $F_{1-2} = 1$, all of it reaches A_2 (see Fig. 4.23), where $\alpha_2 \epsilon_1 A_1 \sigma T^4_1$ of the radiation is absorbed and $(1-\alpha_2) \epsilon_1 A_1 \sigma T^4_1$ is reflected back to A_1. Noting that $\epsilon_2 = \alpha_2$ for gray surfaces, a continued tracing of the radiation to obtain the total amount absorbed by A_2 gives

$$q_{1 \rightarrow 2} = A_1 \sigma T^4_1 [\epsilon_1 \epsilon_2 + \epsilon_1 \epsilon_2 (1-\epsilon_1)(1-\epsilon_2) + \epsilon_1 \epsilon_2 (1-\epsilon_1)^2 (1-\epsilon_2)^2 + \cdots] \qquad (4.84)$$

which is a geometric series whose sum is

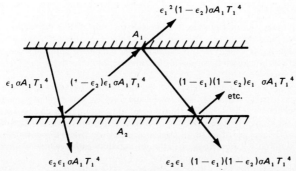

Fig. 4.23 Ray-trace of radiation between two parallel gray surfaces. (*From F. Kreith and J. F. Kreider, Principles of Solar Engineering, Hemisphere, Washington, DC, 1978.*)

$$q_{1 \to 2} = A_1 \sigma T_1^4 \frac{\epsilon_1 \epsilon_2}{1 - (1-\epsilon_1)(1-\epsilon_2)} = A_1 \sigma T_1^4 \frac{1}{(1/\epsilon_1) + (1/\epsilon_2) - 1}$$

Repeating the radiation trace for the radiation emanating from A_2 and absorbed by A_1 gives

$$q_{2 \to 1} = A_2 \sigma T_2^4 \frac{1}{(1/\epsilon_1) + (1/\epsilon_2) - 1} \tag{4.85}$$

so that the net rate of heat transfer between the two surfaces becomes

$$q_{1 \rightleftarrows 2} = \frac{A \sigma (T_1^4 - T_2^4)}{(1/\epsilon_1) + (1/\epsilon_2) - 1} \tag{4.86}$$

It is apparent from the preceding discussion that the rate of heat transfer by radiation is not a linear function of the temperature difference, but rather of the difference between the fourth powers of the surfaces exchanging radiation. However, if the temperatures of the surfaces exchanging radiation are not high and the temperature difference between them is relatively small, say less than 20 K, Eq. (4.86) can be linearized and radiation heat transfer can be incorporated into the thermal circuit concept. To linearize Eq. (4.86), note that

$$T_1^4 - T_2^4 = (T_1^2 - T_2^2)(T_1^2 + T_2^2)(T_1 - T_2)(T_1 + T_2)(T_1^2 + T_2^2) \approx 4\overline{T}^3 (T_1 - T_2)$$

Using this equality, Eq. (4.86) can be expressed in the form

$$q_{1 \rightleftarrows 2} = \frac{T_1 - T_2}{R_r} \tag{4.87}$$

where

$$R_r = \left(\frac{\dfrac{1}{\epsilon_1} + \dfrac{1}{\epsilon_2} - 1}{4 A_1 \sigma \overline{T}^3} \right) \tag{4.88}$$

$$\overline{T} = \frac{(T_1 + T_2)}{2}$$

Chapter 5 contains lengthy tabulations and descriptions of absorptances α and emittances ϵ of opaque materials. Optical properties of transparent materials are also introduced in Chap. 5. The next section provides additional details on the optical properties of multiple layers of transparent media.

Transparent Materials

As mentioned previously, for transparent materials the sum of the absorptance, reflectance, and transmittance must equal unity [see Eq. (4.63)]. However, the mechanism of transmission of radiation through a transparent material such as glass is a rather complex process and depends on the wavelength of the radiation, the angle of incidence, the refractive index, and the extinction coefficient.

Transparent covers in various shapes are used to reduce heat losses from the radiation-absorbing surfaces of most solar collectors (see Chaps. 7 to 10). Consequently, an understanding of the process and laws which govern the transmission of radiation through a transparent medium is important for the design of solar collectors. Ideal materials for this type of application should be able to transmit solar radiation at all angles of incidence, have dimensional stability, and have long-term durability under exposure to the sun and the weather. Other desirable properties include low weight, low infrared transmittance, and low cost.

Index of refraction and extinction coefficient   The calculation of transmission of light through a transparent medium is governed by principles laid down long ago by Fresnel, Snell, and Stokes; these concepts are outlined below.

The optical behavior of a substance can be characterized by two wavelength-dependent physical properties—the index of refraction n and the extinction coefficient K. The index of refraction determines the amount of light reflected from a single surface, while the extinction coefficient determines the amount of light absorbed in a substance in a single pass of radiation.

Figure 4.24 defines the angles used in analyzing reflection and transmission of light. The angle i is the angle of incidence and angle θ_r is the angle of refraction. The incidence and refraction angles are related by Snell's law:

$$\frac{\sin i}{\sin \theta_r} = \frac{n'_r}{n_i} = n_r \tag{4.89}$$

where n'_i and n'_r are the two refractive indices and n_r is the index ratio for the two substances forming the interface. Typical values of refractive indices for various materials are given in Chap. 5. For most materials of interest in solar applications, the values range from 1.3 to 1.6, a fairly narrow range.

The reflectance ρ from a surface of a transparent substance is related to the refractive index indirectly by the values of i and θ_r according to Snell's law, Eq. (4.89). The reflectance has two components corresponding to the two components of polarization resolved parallel and perpendicular to the plane of incidence. The perpendicular ($\perp$) and parallel ($\|$) components are given by the relations

$$\rho'_\perp = \frac{\sin^2 (i - \theta_r)}{\sin^2 (i + \theta_r)} \tag{4.90}$$

$$\rho'_\| = \frac{\tan^2 (i - \theta_r)}{\tan^2 (i + \theta_r)} \tag{4.91}$$

For normal incidence the two components are equal, or

$$\rho'_\perp = \rho'_\| = \frac{(n_r - 1)^2}{(n_r + 1)^2} \tag{4.92}$$

where n_r is the index of refraction ratio. For grazing incidence ($i = 90°$), both components are also equal:

$$\rho'_\perp = \rho'_\| = 1 \tag{4.93}$$

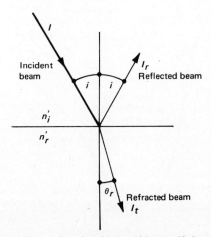

Fig. 4.24 Diagram showing incident, reflected, and refracted beams of light and incidence and refraction angles for a transparent medium. (*From F. Kreith and J. F. Kreider, Principles of Solar Engineering, Hemisphere, Washington, DC, 1978.*)

The reflectance of a glass-air interface, common in solar collectors, may be reduced by a factor of 4 by an etching process. If glass is immersed in a silica-supersaturated fluosilicic acid solution, the acid attacks the glass and leaves a porous silica surface layer. This layer has an index of refraction intermediate between that of glass and that of air. Consequently, reflectance losses can be reduced significantly by this gradual change in refractive index.

The efficacy of the process depends upon solution temperature and composition, immersion time, and surface pretreatment. Mar et al.[26] have studied these effects in detail and have devised a repeatable process for producing glass with reflectance of 1 percent per interface (2 percent per pane of glass). They also found that heat treatment at 100°C enhances the durability of the coating significantly. Figure 4.25 shows the spectral reflectance of a pane of glass before and after etching. Attempts to etch plastics to reduce reflection have been unsuccessful to date.

When radiation passes through a transparent medium such as glass or the atmosphere, the decrease in intensity can be described by assuming that the attenuation is proportional to the intensity in the medium and the distance traveled through the medium. If $I_\lambda(x)$ is the monochromatic intensity after radiation has traveled a distance x, Bouger's absorption law is expressed by

$$-dI_\lambda(x) = I_\lambda(x) \, K_\lambda \, dx \tag{4.94}$$

where K_λ is the monochromatic extinction coefficient. If the transparent medium is a slab of thickness L and the intensity at $x = 0$ is designated by the symbol $I_{\lambda,o}$, the monochromatic transmittance τ'_λ for the absorption process alone is equal to the ratio of the intensity at $x = L$ to $i_{\lambda,o}$. An expression for $I_\lambda(L)$ can be obtained by integrating Eq. (4.94) between 0 and L which gives

$$\ln \frac{I_\lambda(L)}{I_{\lambda,o}} = -K_\lambda L \tag{4.95}$$

or

$$I_\lambda(L) = I_{\lambda,o} \, e^{-K_\lambda L}$$

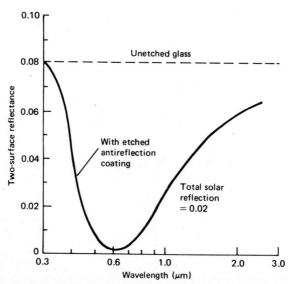

Fig. 4.25 Reflection spectra for a sample of glass before and after etching. (*From B. Mar, Optical Coatings for Flat Plate Solar Collectors, NTIS, PB-252-383, September 1975.*)

Then

$$\tau_\lambda' \equiv \frac{I_\lambda(L)}{I_{\lambda,o}} = e^{-K_\lambda L} \tag{4.96}$$

If the transmittance is sensitive to the direction of the incoming radiation, $I_{\lambda,o}(\theta, \phi)$ must be used in Eq. (4.96) and the transmittance will also be a function of θ and ϕ. Fortunately, the absorption for glass and plastic films is not a strong function of direction. Moreover, the spectral transmittance of low-iron glass is almost a constant over the solar spectrum, i.e., between 0.3 and 3 μm. Consequently, it can be treated as a transparent gray medium with $K_\lambda = K$, a wavelength-independent property.

Bouger's law (Eq. 4.96) can be used to calculate the absorption α' of homogeneous substances as follows:

$$\alpha' = 1 - e^{-KL} \tag{4.97}$$

The optical path length L is the thickness of the material t divided by the cosine of the angle of refraction (see Fig. 4.24), i.e.,

$$L = \frac{t}{\cos\theta_r} \tag{4.98}$$

It is very difficult to evaluate the extinction coefficient for materials used in solar collectors since they are expressly selected to have very small values of K. In order to measure small values of K, a large material thickness L is required. However, thick sections made for measurement may have properties different from those for thin layers used in solar collectors. Alternatively, if many layers of a thin material are used, the surface reflection from each dominates the optical attenuation of the stack, and no increase in accuracy in the measurement of K is achieved. This problem is particularly acute with plastic films. Consequently, few data exist for transparent materials used in solar collectors. A few values of K are given in Table 4.8.

Stokes' equations The quantities ρ' and α' above apply to single surfaces, single passes, and single reflections only. In practice, a layer of glass or other transparent material will have multiple interreflections of radiation which must all be accounted for in calculating the total reflectance ρ, transmittance τ, and absorptance α as shown in Fig. 4.26.

The total reflectance ρ, the flux leaving the top surface divided by the incident flux, in Fig. 4.26 can be calculated by summing the infinite series of components making up the total flux. For simplicity of notation $\tau' = e^{-KL}$ and ρ' denotes both $\rho'_\perp$ and $\rho'_\parallel$. The two series to be summed can be written from the figure; they are

$$\rho = \rho(\rho', \tau') = \rho' + \tau'^2(1 - \rho')^2\rho'[1 + \rho'^2\tau'^2 + \rho'^4\tau'^4 + \cdots] \tag{4.99}$$

$$\tau = \tau(\rho', \tau') = (1 - \rho')^2\tau'[1 + \rho'^2\tau'^2 + \rho'^4\tau'^4 + \cdots] \tag{4.100}$$

The terms in brackets are geometric series with the first term unity and ratio $\rho'^2\tau'^2$. Using the formula for the sum of an infinite geometric series from elementary algebra, we have

TABLE 4.8 Extinction Coefficients for Transparent Materials in the Solar Spectrum

Polyvinyl fluoride (Tedlar)*	1.4 cm^{-1}
Fluorinated ethylene propylene terephthalate (Teflon)*	0.59 cm^{-1}
Polyethylene (Mylar)*	2.05 cm^{-1}
Polyethylene	1.65 cm^{-1}
Ordinary window glass	~0.3 cm^{-1}
White glass (< 0.01 % Fe$_2$O$_3$)	~0.04 cm^{-1}
Heat-absorbing glass	1.3–2.7 cm^{-1}

*Trademarks of the duPont Company, Wilmington, DE; data from Edlin, notes for S. U. N. Seminar, Arizona State University, Tempe, 1975.

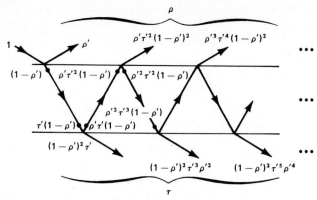

Fig. 4.26 Ray trace diagram for a transparent medium showing the reflected and transmitted fractions accounting for multiple interreflections. Total fraction reflected and transmitted denoted by ρ and τ, respectively; $\tau' \equiv e^{-KL}$. *(From F. Kreith and J. F. Kreider, Principles of Solar Engineering, Hemisphere, Washington, DC, 1978.)*

$$\rho(\rho', \tau') = \rho'\left[1 + \frac{\tau'^2(1-\rho')^2}{1-\rho'^2\tau'^2}\right] \tag{4.101}$$

$$\tau(\rho', \tau') = \tau'\left[\frac{(1-\rho')^2}{1-\rho'^2\tau'^2}\right] \tag{4.102}$$

The total absorptance is simply unity decreased by $(\rho + \tau)$, or

$$\alpha(\rho', \tau') = 1 - \rho(\rho', \tau') - \tau(\rho', \tau') \tag{4.103}$$

Equations (4.101–103) were first developed by G. G. Stokes[21] and are called the Stokes' equations.

Most materials used for optical transmission of solar energy have very small values of KL as described above. Therefore $\tau' \approx 1$ and Eq. (4.102) can be simplified by separating the absorptance and reflectance effects,

$$\tau(\rho', \tau') \approx \tau'\left(\frac{1-\rho'}{1+\rho'}\right) \tag{4.104}$$

Eq. (4.104) is accurate to a few percent for most materials and incidence angles used in solar applications. This simplification can also be used when more than one absorbing-reflecting layer is involved [see Eq. (4.107)].

Care must be taken in the use of the Stokes' equations to account for polarization properly. The average reflectance $\bar{\rho}$ and the average transmittance $\bar{\tau}$ for both components of polarization, if they are of equal magnitude, are

$$\rho = \frac{1}{2}[\rho(\rho'_\perp, \tau') + \rho(\rho'_{||}, \tau')] \tag{4.105}$$

$$\tau = \frac{1}{2}[\tau(\rho'_\perp, \tau') + \tau(\rho'_{||}, \tau')] \tag{4.106}$$

It is incorrect to calculate an average ρ' for both polarization components and use this value in Eqs. (4.101) and (4.102) directly to calculate the average properties. Errors of up to 18 percent in the calculated transmittance of a pane of glass can result from this improper averaging. Figure 4.27 shows the calculated transmittance $\bar{\tau}$ of a layer of glass vs. incidence angle i if absorption losses are negligible.

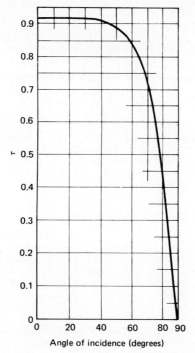

Fig. 4.27 Transmission versus angle of incidence, for glass neglecting absorption ($KL = 0$) with refractive index of 1.53. (*From F. Kreith and J. F. Kreider, Principles of Solar Engineering, Hemisphere, Washington, DC, 1978.*)

Multiple transparent layers Stokes[21] also derived expressions for reflection and transmission of multiple layers of transparent media using the results for one layer. The analysis involves multiple reflectance ray tracing like that used in the case of a single layer shown above and will not be repeated here. The effective transmittance and reflectance equations are used to generate the properties of multiple layers and are derived by simply increasing the number of layers one by one or by doubling repeatedly. The general equations for $N + M$ layers are

$$\tau_{N+M}(\rho', \tau') = \frac{\tau_N \tau_M}{1 - \rho_N \rho_M} \tag{4.107}$$

$$\rho_{N+M}(\rho', \tau') = \rho_N + \frac{\rho_M \tau_N^2}{1 - \rho_N \rho_M} \tag{4.108}$$

in which ρ_M and ρ_N are calculated from repeated application of Eq. (4.101) and τ_M and τ_N from Eq. (4.102).

For example, for two transparent sheets

$$\tau_2(\rho', \tau') = \frac{\tau_1 \tau_1}{1 - \rho_1 \rho_1} \tag{4.109}$$

and

$$\rho_2(\rho', \tau') = \rho_1 + \frac{\rho_1 \tau_1^2}{1 - \rho_1 \rho_1} \tag{4.110}$$

where ρ_1 and τ_1 are from Eqs. (4.101) and (4.102).

Then for three layers ($M = 2$, $N = 1$) we would have

$$\tau_3(\rho', \tau') = \frac{\tau_2 \tau_1}{1 - \rho_2 \rho_1} \tag{4.111}$$

and

$$\rho_3(\rho', \tau') = \rho_1 + \frac{\rho_2 \tau_1^2}{1 - \rho_1 \rho_2} \tag{4.112}$$

Values of ρ_1 and τ_1 are as above; ρ_2 and τ_2 are from Eqs. (4.109) and (4.110). This process can be carried out to calculate the effective transmittance and reflectance of any number of layers from the two basic properties K and n_r. Note that each component of polarization must be treated separately in this analysis as was done for a single layer. For example, the average transmittance of a three-layer system is

$$\bar{\tau}_3 = \frac{1}{2} [\tau_3(\rho'_\perp, \tau') + \tau_3(\rho'_\parallel, \tau')] \tag{4.113}$$

Equations (4.107) and (4.108) can also be used if one surface is opaque as in a solar collector absorber. In that case $\tau_M = 0$ and $\rho_M = \rho_S$, where ρ_S is 1 minus the absorber surface solar absorptance. The quantity $(1 - \rho_{N+S})$ represents the radiation absorbed by a collector and its covers. The ray trace approach can also be used to calculate the effective optical transmittance-absorptance transfer function for only the absorber plate of a collector in closed form.

If the layers of material of interest are very thin or have a very small extinction coefficient, the amount absorbed in each layer will be negligible, i.e., $\tau' \approx 1$. In that case the Stokes equations for single and multiple layers can be simplified. For example, for single layers we have analogs to Eqs. (4.101) and (4.102),

$$\rho(\rho') = \frac{2\rho'}{1 + \rho'} \tag{4.114}$$

$$\tau(\rho') = \frac{1 - \rho'}{1 + \rho'} \tag{4.115}$$

$$\alpha(\rho') = 0 \tag{4.116}$$

For multiple layers, Eqs. (4.107) and (4.108) reduce to

$$\tau_{N+M}(\rho') = \frac{1 - \rho'}{1 + (2N + 2M - 1)\rho'} \tag{4.117}$$

$$\rho_{N+M}(\rho') = 1 - \tau_{N+M}(\rho') \tag{4.118}$$

$$\alpha_{N+M}(\rho') = 0 \tag{4.119}$$

These simpler equations can generally be used in solar designs where good, low-iron glass or thin plastic sheets are the material of choice.

Note that in the foregoing analysis, the magnitude of the parallel and perpendicular polarization components have been assumed equal. Although not strictly true in all cases, this assumption is sufficiently accurate for engineering calculations. Wehner,[22] in an elegant study, has prepared charts and diagrams depicting the degree of polarization of the daytime sky. His measurements indicate that, although the degree of polarization is small by engineering standards, it is large enough to be used by some insects for navigation. He also notes that it is possible to use the small polarization differential, by viewing the sky through the mineral corderite as a polarizing filter, for navigational purposes. The Vikings may have used this navigational aid in the Middle Ages.

HEAT EXCHANGERS

Conventional heat exchangers are devices in which two fluid streams, separated from each other by a solid wall, exchange thermal energy: one stream is heated while the other is cooled. There are a number of arrangements used to transfer heat from one fluid to another. The simplest arrangement is the double-pipe heat exchanger shown in Fig. 4.28. It will be discussed first.

In this system fluid A flows inside a tube of inner radius r_i and outer radius r_o. Fluid B flows in the annulus formed between the outer surface of the inner tube, A_o, and the inner surface of the outer tube. Individual convection heat-transfer coefficients can be calculated by methods outlined previously, and an overall heat-transfer coefficient U can be calculated from the thermal circuit shown in Fig. 4.28. The overall heat-transfer coefficient may be based on any convenient area of the exchanger, but usually the outside area $A_o = 2\pi r_o L$ of the inner tube is most convenient. Then,

$$U_o A_o = \frac{1}{\dfrac{1}{h_i A_i} + \dfrac{\ln r_o/r_i}{2\pi k L} + \dfrac{1}{h_o A_o}} \qquad (4.120)$$

and at any cross section the local rate of heat transfer across the tube is

$$dq = U_o \, dA_o \, (T_A - T_B) = U_o \, 2\pi r_o dx \, (T_A - T_B) \qquad (4.121)$$

Simplifications in the evaluation of U_o are possible when one or two of the thermal resistances dominate. For example, the thermal resistance of the tube wall, R_2 in Fig. 4.28, is often small compared with the convective resistances, and sometimes one of the convective resistances is negligible compared with the other.

After a period of operation the surfaces of a heat exchanger may become coated with a deposit or become corroded. Such a coat or scale can offer a substantial thermal resistance over and above those considered in Eq. (4.120) and thereby cause a decrease in the thermal performance. This condition is called *fouling*, and a fouling resistance must be added to calculate the overall heat-transfer coefficient under these conditions.

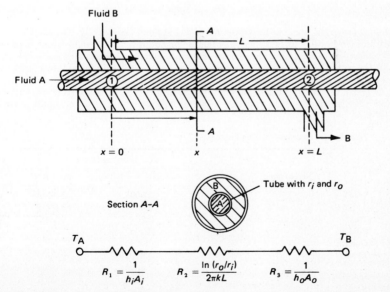

Fig. 4.28 Schematic diagram of double-pipe heat exchanger. (*From F. Kreith and J. F. Kreider, Principles of Solar Engineering, Hemisphere, Washington, DC, 1978.*)

Fouling resistances, often called *fouling factors*, can be measured experimentally by comparing the values of U for clean and dirty or corroded exchangers. The fouling factor R_f can then be calculated from

$$R_f = \frac{1}{U_{\text{dirty}}} - \frac{1}{U_{\text{clean}}} \tag{4.122}$$

A list of fouling factors for several fluids is given in Table 4.9.

The flow arrangement shown in Fig. 4.28, where both fluids enter from the same end, is called "parallel flow," and Fig. 4.29a shows the temperature distribution for both fluid streams. If one of the two fluids were to enter at the other end and flow in the opposite direction, the flow arrangement would be "counterflow." The temperature distribution for the latter case is shown in Fig. 4.29b. If a counterflow arrangement is made very long, it approaches the thermodynamically most efficient possible heat-transfer condition.

There are two basic methods for calculating the rate of heat transfer in a heat exchanger. One method employs a mean temperature difference between the two fluids in the exchanger, ΔT_{mean}, and then determines the rate of heat transfer from the relation

$$q = U A \, \Delta T_{\text{mean}} \tag{4.123}$$

This mean temperature difference is called the logarithmic mean temperature difference (LMTD for short). It can only be evaluated directly when the inlet and outlet temperatures of both fluid streams are specified. The use of the LMTD approach is explained in Ref. 5. The other method is called the effectiveness-NTU (ENTU) method. The ENTU method offers many advantages over the LMTD approach and will be discussed below.

First, we define the exchanger effectiveness ϵ:

$$\epsilon = \frac{\text{actual rate of heat transfer}}{\text{maximum possible rate of heat transfer}} \tag{4.124}$$

The actual rate of heat transfer can be determined by calculating either the rate of internal energy loss of the hot fluid or the rate of internal energy gain of the cold fluid, i.e.,

$$q = \dot{m}_h \, c_h \, (T_{h.\text{in}} - T_{h.\text{out}}) = \dot{m}_c \, c_c \, (T_{c.\text{out}} - T_{c.\text{in}}) \tag{4.125}$$

where the subscripts h and c denote the hot and cold fluid, respectively. The maximum rate of heat transfer possible for specified inlet fluid temperatures is attained when one of the two fluids undergoes the maximum temperature difference in the exchanger. This maximum equals the difference in the entering temperatures for the hot and the cold fluid. Which of the two fluids can undergo this maximum temperature change depends on the relative value of the product $(\dot{m}c_p)$, the mass flow rate times the specific heat of the fluid at constant pressure. This product is called the heat rate $\dot{C}$. Since a thermodynamic energy balance requires that the

TABLE 4.9   Average Fouling Factors*

Type of fluid	Fouling factor	
	hr · ft² · °F/Btu	m² · K/W
Seawater, below 125°F	0.0005	0.00009
above 125°F	0.001	0.0002
Treated boiler feedwater above 125°F	0.001	0.0002
Fuel oil	0.005	0.0009
Quenching oil	0.004	0.0007
Alcohol vapors	0.0005	0.00009
Steam, non-oil-bearing	0.0005	0.00009
Industrial air	0.002	0.0004
Refrigerating liquid	0.001	0.0002

*From Ref. 25.

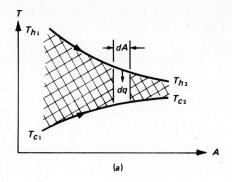

(a)

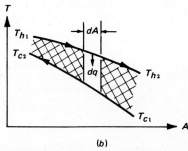

(b)

Fig. 4.29 (a) Temperature distribution in a double-pipe heat exchanger in parallel flow; (b) temperature distribution in a double-pipe heat exchanger in counter flow. (*From F. Kreith and J. F. Kreider, Principles of Solar Engineering, Hemisphere, Washington, DC, 1978.*)

energy given up by the one fluid must be received by the other if there are no external heat losses, only the fluid with the smaller value of C can undergo the maximum temperature change. Thus, the fluid which may undergo this maximum temperature change is the one which has the minimum value of the heat capacity rate, and the maximum rate of heat transfer is

$$q_{max} = \dot{C}_{min} (T_{h,\,in} - T_{c,\,in}) \qquad (4.126)$$

where $\dot{C}_{min} = (\dot{m}\,c_p)_{min}$. The fluid with the minimum $\dot{C}$ value can be either the hot or the cold fluid. Using the subscript h to designate the effectiveness when the hot fluid has the minimum $\dot{C}$ and the subscript c when the cold fluid has the minimum $\dot{C}$ value, we get for a parallel-flow arrangement

$$\epsilon_h = \frac{\dot{C}_h (T_{h1} - T_{h2})}{\dot{C}_h (T_{h1} - T_{c1})} = \frac{T_{h1} - T_{h2}}{T_{h1} - T_{c1}} \qquad (4.127)$$

$$\epsilon_c = \frac{\dot{C}_c (T_{c2} - T_{c1})}{\dot{C}_c (T_{h1} - T_{c1})} = \frac{T_{c2} - T_{c1}}{T_{h1} - T_{c1}} \qquad (4.128)$$

where the subscripts 1 and 2 refer to the left- and right-hand sides of the heat exchanger as shown in Fig. 4.28. Similarly, for a counterflow arrangment,

$$\epsilon_h = \frac{\dot{C}_h (T_{h1} - T_{h2})}{\dot{C}_h (T_{h1} - T_{c2})} = \frac{T_{h1} - T_{h2}}{T_{h1} - T_{c2}} \qquad (4.129)$$

$$\epsilon_c = \frac{\dot{C}_c (T_{c1} - T_{c2})}{\dot{C}_c (T_{h1} - T_{c2})} = \frac{T_{c1} - T_{c2}}{T_{h1} - T_{c2}} \qquad (4.130)$$

Kays and London[24] have calculated effectiveness ratios for various types of heat exchanger. The most important results of this analysis are presented in the charts of Figs. 4.30 to 4.35 and in Table 4.10, where the analytical expressions from which the charts were prepared are summarized. To use these charts it is necessary to know something about heat-exchanger geometry. The double-pipe parallel and counterflow arrangements (Fig. 4.29) have already been discussed. The gas heating and cooling cross-flow exchangers such as shown in Figs. 4.36 and 4.37 are widely used in solar heating and cooling systems. In the arrangements shown in Fig. 4.37, one fluid is contained within tubes whereas the other flows across the tube bundle. The fluid within the tubes is said to be unmixed whereas the gas flowing across the tubes is not constrained and can mix freely while exchanging heat. The effectiveness for this arrangement is given in the charts of Fig. 4.32. When a fluid is unmixed, there will be a temperature gradient both parallel and normal to the flow direction, whereas when the fluid in the exchanger is mixed, its temperature in the direction normal to the flow tends to equalize as a result of the mixing. This is illustrated in Fig. 4.36b, where the temperature profile for the gas flowing through the heat exchanger configuration of Fig. 4.36a is shown schematically for the gas being heated. The overall rate of heat transfer in an exchanger depends on whether a fluid is mixed or unmixed, because the average temperature difference in an exchanger is a function of the mixing arrangement as illustrated by the difference in effectiveness.

In the arrangement shown in Fig. 4.36, both fluids are restrained from mixing, one fluid by the tubes and the other by the fins attached perpendicular to the tubes. This arrangement is widely used in solar air conditioning applications. The effectiveness of heat exchangers with both fluids unmixed is shown in Fig. 4.33.

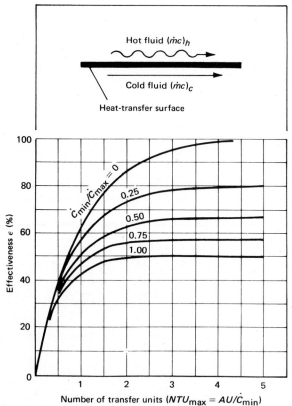

Fig. 4.30   Effectiveness versus NTU_{max} for a parallel-flow heat exchanger. (*From W. M. Kays ana A. L. London, Compact Heat Exchangers, McGraw-Hill, New York, 1964.*)

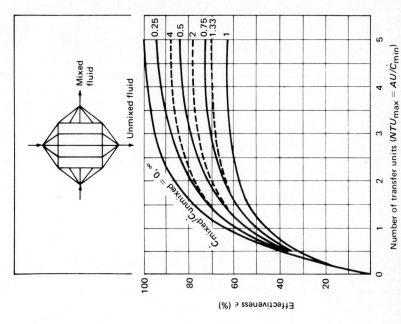

Fig. 4.31 Effectiveness versus NTU_{max} for a counterflow heat exchanger. (*From W. M. Kays and A. L. London, Compact Heat Exchangers, McGraw-Hill, New York, 1964.*)

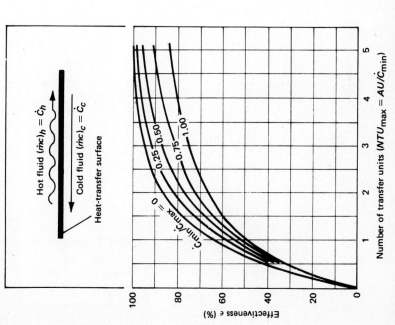

Fig. 4.32 Effectiveness versus NTU_{max} for a cross-flow heat exchanger with one fluid mixed, the other unmixed. (*From W. M. Kays and A. L. London, Compact Heat Exchangers, McGraw-Hill, New York, 1964.*)

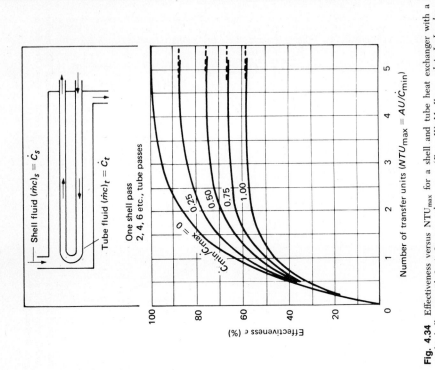

Fig. 4.33 Effectiveness versus NTU_{max} for a cross-flow heat exchanger with both fluids unmixed. (*From W. M. Kays and A. L. London, Compact Heat Exchangers, McGraw–Hill, New York, 1964.*)

Fig. 4.34 Effectiveness versus NTU_{max} for a shell and tube heat exchanger with a single-shell pass with 2, 4, 6, etc., tube passes. (*From W. M. Kays and A. L. London, Compact Heat Exchangers, McGraw–Hill, New York, 1964.*)

*From Ref. 24.
aFrom (24, 29).

TABLE 4.10 Heat-Exchanger Effectiveness Relations*

$$N = NTU = \frac{UA}{\dot{C}_{min}}, \quad C = \frac{\dot{C}_{min}}{\dot{C}_{max}}$$

Flow geometry	Relation
Double pipe	
Parallel flow	$\epsilon = \dfrac{1 - \exp[-N(1+C)]}{1+C}$
Counterflow	$\epsilon = \dfrac{1 - \exp[-N(1-C)]}{1 - C\exp[-N(1-C)]}$
Cross flow	
Both fluids unmixed	$\epsilon = 1 - \exp\left\{\dfrac{C}{n}\left[\exp(-NCn) - 1\right]\right\}$ where $n = N^{-0.22}$
Both fluids mixed	$\epsilon = \left[\dfrac{1}{1 - \exp(-N)} + \dfrac{C}{1 - \exp(-NC)} - \dfrac{1}{N}\right]^{-1}$
C_{max} mixed, C_{min} unmixed	$\epsilon = (1/C)\{1 - \exp[C(1 - e^{-N})]\}$
C_{max} unmixed, C_{min} mixed	$\epsilon = 1 - \exp\{(1/C)[1 - \exp(-NC)]\}$
Shell and tube	
One shell pass, 2, 4, 6 tube passes	$\epsilon = 2\left\{1 + C + (1 + C^2)^{1/2}\,\dfrac{1 + \exp[-N(1+C^2)^{1/2}]}{1 - \exp[-N(1+C^2)^{1/2}]}\right\}^{-1}$

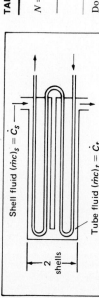

Shell fluid $(\dot{m}c)_s = \dot{C}_s$

Tube fluid $(\dot{m}c)_t = \dot{C}_t$

Two shell passes
4, 8, 12 etc., tube passes

2 shells

$C_{min}/C_{max} = 0$

0.25
0.50
0.75
1.00

Effectiveness ϵ (%)

Number of transfer units $(NTU_{max} = AU/\dot{C}_{min})$

Fig. 4.35 Effectiveness versus NTU_{max} for a shell and tube heat exchanger with two-shell passes and 4, 8, 12, etc., tube passes. (*From W. M. Kays and A. L. London, Compact Heat Exchangers, McGraw-Hill, New York, 1964.*)

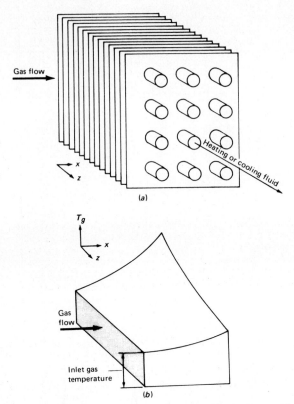

Fig. 4.36 Cross-flow heat exchanger with both fluids unmixed—schematic diagram (*a*) and temperature distribution (*b*). (*From F. Kreith and J. F. Kreider, Principles of Solar Engineering, Hemisphere, Washington, DC, 1978.*)

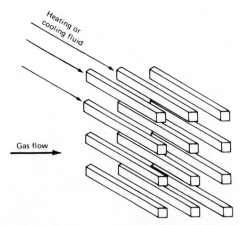

Fig. 4.37 Cross-flow heat exchanger with one fluid mixed, one unmixed. (*From F. Kreith and J. F. Kreider, Principles of Solar Engineering, Hemisphere, Washington, DC, 1978.*)

Fig. 4.38 Shell and tube heat exchanger with one-tube pass and one-shell pass. (*Courtesy of the Young Radiator Company.*)

The type of heat exchanger most widely used in commercial practice consists of a shell containing a tube bundle as shown in Fig. 4.38. One fluid flows inside the tubes while the other is forced across the tubes by baffles, as shown. The arrangement at the ends where the tubes emerge, called the header, determines the number of tube passes.* If one header is halved and the other is open, there will be two tube passes and one shell pass. The effectiveness of such a unit with a single shell pass and with two or a multiple of two tube passes is shown in Fig. 4.34.

REFERENCES

1. J. Kestin (ed.), "The Second Law of Thermodynamics," Halsted Press, New York, 1976.
2. J. W. Gibbs, "The Collected Works of J. W. Gibbs," vol. 1, p. 77, Yale University Press, New Haven, CT, 1948.
3. T. H. Keenan, "Thermodynamics," Wiley, New York, 1948.
4. J. E. Parrot, Theoretical Upper Limit to the Conversion Efficiency of Solar Energy, *Solar Energy*, **21**: 227 (1978).
5. F. Kreith, "Principles of Heat Transfer," 3d ed., Intext Publishers, New York, 1976.
6. W. M. Rohsenow and J. P. Hartnett (eds.), "Handbook of Heat Transfer," McGraw-Hill, New York, 1973.
7. W. M. Kay, "Convective Heat and Mass Transfer," McGraw-Hill, New York, 1966.
8. E. D. Grimson, Correlation and Utilization of New Data on Flow Resistance and Heat Transfer for Cross-flow of Gases over Tube Bands, *Trans.* ASME, **59**: 583–594 (1937).
9. A. A. Zhukauskas, V. Makarevicius, and A. Schlanciankas, "Heat Transfer in Banks of Tubes in Cross-flow of Fluids," *Mintis*, Vilnius, Lithuania, 1968.
10. K. G. T. Hollands, L. Konicek, T. E. Unny, and G. D. Raithby, Free Convection Heat Transfer Across Inclined Air Layers, *J. Heat Transf.*, **98**: 189–193 (1976).
11. K. G. T. Hollands, G. D. Raithby, and T. E. Unny, Studies on Methods of Reducing Heat Losses from Flat Plate Collectors, Final Rept., ERDA Contract E(11-1)-2597, University of Waterloo, Waterloo, Ontario, Canada, June 1976.
12. G. D. Raithby, K. G. T. Hollands, and T. E. Unny, Free Convection Heat Transfer Across Fluid Layers of Large Aspect Ratios, ASME Paper 76-HT-45, *16th National Heat Transfer Conference*, St. Louis, August 8–11, (1976).
13. H. Buchberg, I. Catton, and D. K. Edwards, Natural Convection in Enclosed Spaces: A Review of Application to Solar Energy Collection, ASME Paper 74-WA/HT/12, 1974.
14. K. G. T. Hollands, Honeycomb Devices in Flat Plate Solar Collectors, *Solar Energy*, **9**: 159 (1965).
15. H. Buchberg and D. K. Edwards, Design Considerations for Solar Collectors with Cylindrical Glass Honeycombs, *Solar Energy*, **18**:193–204 (1976).
16. H. Buchberg, O. A. Lalude, and D. K. Edwards, Performance Characteristics of Rectangular Honeycomb Solar-Thermal Converters, *Solar Energy*, **13**: 193–221 (1971).
17. G. D. Raithby and K. G. T. Hollands, A General Method of Obtaining Approximate Solutions to Laminar and Turbulent Free Convection Problems, in "Advances in Heat Transfer," vol. 11, Academic Press, New York, 1975.

*If both headers are open, there will be one tube pass.

18. J. R. Howell and R. Siegel, "Thermal Radiation Heat Transfer," McGraw-Hill, New York, 1972.
19. E. M. Sparrow and R. D. Cess "Radiation Heat Transfer," McGraw-Hill, New York, 1978.
20. J. P. Holman, "Thermodynamics," 2d ed., McGraw-Hill, New York, 1975.
21. G. G. Stokes, On the Intensity of the Light Reflected from a Transmitter through a Pile of Plates, *Proc. Roy. Soc. London*, 11:546–556, 1860–62.
22. R. Wehner, Polarized-light Navigation by Insects. *Scientific American*, **235**:106–115, 1976.
23. "Standards of Tubular Exchanger Manufacturers Association," 4th ed., TEMA, New York, 1959.
24. W. M. Kays and A. L. London, "Compact Heat Exchangers," 2d ed., McGraw-Hill, New York, 1964.
25. J. D. Parker, J. H. Boggs, and E. F. Blick, "Fluid Mechanics and Heat Transfer," Addison-Wesley, Reading, MA, 1969.
26. B. Mar, Optical Coatings for Flat Plate Solar Collectors, NTIS, PB-252-383, September 1975.
27. G. O. G. Löf and R. W. Hawley, Unsteady State Heat Transfer between Air and Loose Solids, *Ind. Eng. Chem*, **40**:1061–1066 (1948).
28. D. Handley and P. J. Heggs, Momentum and Heat Transfer Mechanisms in Regularly Shaped Packings, *Trans. Inst. Chem. Eng.*, **46**:251–264 (1968).
29. D. Handley and P. J. Heggs, The Effect of Thermal Conductivity of the Packing Material on the Transient Heat Transfer in a Fixed Bed, *Int. J. Heat-Mass Transf.*, **12**:549–570 (1969).

Chapter **5**

Optical Properties of Materials Used in Solar Energy Systems

FRANK EDLIN
Consulting Engineer, Sun City, AZ

INTRODUCTION

Refracting and reflecting materials used in the solar environment can be described by their optical and spectral qualities. There can be considerable change in optical qualities with change in latitude, elevation, the time of day and year, and atmospheric quality. The evaluation of the performance of a material in changing its use from one situation to another requires a spectral analysis based upon the spectral properties of local solar radiation. Solutions to problems involving diverse spectra can be made with computers. Alternatively, a series of graphs are developed in this chapter which avoid the use of computers and which can simplify the evaluation with results of the accuracy required for most solar energy work.

Some data to be used for the accurate development of the spectral properties of materials are scarce or not well developed and are to be applied with caution. For example, only two indexes of refraction for different wavelengths may be found for near-ambient temperatures. The spectra reported for materials are not usually measured with a spectrophotometer having a reflective cavity, and therefore dispersed rays are not measured. Measurements of properties in the solar spectrum do not indicate zenith angle and latitude or air mass. Selective surface properties are not measured at operating temperatures. If considerable expenditure is contemplated, the properties of materials should be evaluated at their conditions of use and adjustments be made spectrally.

The terminology used for optical properties is defined below. The suffix "-ance" (e.g., transmittance) is the specific quantitative measure at or near the conditions of use; "-ivity" (e.g., absorbtivity) is the physical property of a material; and "-ion" (e.g., reflection) is the general effect of the action of a material upon a beam of radiation.

The general method used for evaluating the properties of materials in the solar environment has evolved since 1960. The method has been applied to pigments, paints, coating, dyes, plastics and films, camouflage materials, flames and flame retardants, and the evaluation of outdoor weather test sites.

THE SOLAR ENERGY ENVIRONMENT

The solar spectral environment is described below for direct solar radiation with respect to altitude, latitude, declination, and air mass and is used graphically to calculate spectrally averaged properties. Also a general spectral graph is developed, as an example, for a portion of the spectrum having interest for solar design. Solar geometry is given only to the extent that is necessary for the evaluation of materials. Chapter 2 develops solar astronomy analysis in detail.

The Blackbody Equal Energy Graph

The general solution   Planck's equation expresses the monochromatic energy radiated from a blackbody $E_{B\lambda}$ at wavelength λ and absolute temperature T,

$$E_{B\lambda} = \frac{C_1}{(e^{c_2/\lambda T} - 1)\lambda^5} \tag{5.1}$$

C_1 is 3.705×10^{-5} erg·cm²/s, 0.889×10^{-12} cal·cm²/s, or 1.76×10^{-8} Btu/ft² · h·cm⁴; and C_2 is 1.439 cm·K, or 2.589 cm·°R. By rearrangement and dividing by T^5 the expression can be transformed to

$$\frac{E_{B\lambda}}{T^5} = \frac{C_1}{(\lambda T)^5(e^{c_2/\lambda T} - 1)} \tag{5.2}$$

The independent variable is seen to be λT. The Stephan-Boltzmann equation evaluates the total radiant energy from a blackbody E_B at absolute temperature T

$$E_B = \sigma T^4 \tag{5.3}$$

The Stephan-Boltzmann constant is 5.771×10^{-5} erg/cm²·h·K, or 0.173×10^{-8} Btu/ft²·h·°R⁴, or 5.67 W/m² · K⁴. The sum of the monochromatic energies of the Planck equation is equated to the energy of the Stephan-Boltzmann equation to give

$$E_B = \sum_{\lambda=0}^{\lambda=\infty} E_{B\lambda} \Delta\lambda = \sigma T^4 \tag{5.4}$$

and a ratio summed to one is given by

$$\sum_{\lambda T=0}^{\lambda T=\infty} \frac{E_{B\lambda}}{\sigma T^5} \Delta\lambda T = 1 \tag{5.5}$$

The monochromatic energies $E_{B\lambda}/T^5$ for any temperature T are evaluated with Eq. (5.2) for values of λT using small intervals of the wavelength λ which begin and end at low energy levels for the spectrum of the source temperature T. Values of $E_{B\lambda}/T^5$ are shown graphically on rectangular coordinates in Fig. 5.1. The characteristic blackbody radiation curve represents monochromatic energies at any temperature. Siegel and Howell[1] evaluate monochromatic energies for values of λT from 10^3 to 10^5. This relationship between $E_{B\lambda}/T^5$ and λT is used to develop the blackbody equal energy graph.

Development of the graph   The blackbody equal energy graph depicts equal amounts of energy represented by equal areas. These areas are equal proportions of the area of the whole. The development of the equal energy graph for any radiation wave band is shown by the simplified example in Fig. 5.2. Select a small increment of wavelength representing the maximum energy $E_{B-\max}$, at λ_2 in Fig. 5.2. For blackbody radiation shown in Fig. 5.1, Wien's displacement law, which is derived from Eq. (5.1), identifies the blackbody radiation wavelength of maximum energy as $\lambda T_{\max} = 2885$ μm·K. Other wavelengths in the wave band such as λ_1, λ_3, λ_4 are decreased in width or length along the abscissa by their ratios $E_{B1,2,4}/E_{B-\max}$, graphically producing a rectangle of height $E_{B-\max}$ and length $\Sigma \, \Delta\lambda$ in Fig. 5.2. The incremental values of $\Delta\lambda$ are summed and converted to cumulative percent of the total, as shown by Fig. 5.2. The ordinate is then made equal in length to the abscissa and decimally divided (Fig. 5.2). Either the ordinate or the abscissa then represents the percent of energy as a ratio. In the development of a graph small and equal wavelengths are chosen across the wave band with no area omitted or included twice. It is not practical to continue evaluation from zero to infinity. To ensure that the abscissa and ordinate represent the total energy, it is evaluated from Eq. (5.3), and the small difference at spectral extremes is proportioned and added at each end of the graph.

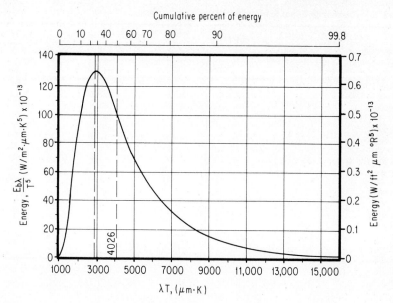

Fig. 5.1   Spectral energy at wavelengths for all black-body temperatures with cumulative percent energy.

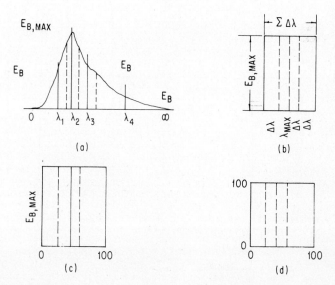

Fig. 5.2   Development of an equal-energy graph with four intervals. (a) a radiation plot; (b) convert to rectangle of equal area; (c) convert to percent; and (d) make ordinate equal abscissa.

TABLE 5.1   Blackbody Equal Energy Plot Values

λT, μm·°K	$E_{B\lambda}/T^5 \times 10^{+13}$ W/ m²·μm·K	$E_{B\lambda}/T^4\sigma$, $(\mu$K$)^{-1}$	λT, μm·°K	$E_{B\lambda}/T^5 \times 10^{+13}$ W/ m²·μm·°K	$E_{B\lambda}/T^4\sigma$, $(\mu$K$)^{-1}$
1,600	44.38	0.028	4,800	77.13	0.615
1,800	66.90	0.052	5,000	71.37	0.642
2,000	87.90	0.083	5,200	66.00	0.666
2,200	105.03	0.120	5,400	60.98	0.685
2,400	117.36	0.161	5,600	56.34	0.702
2,600	124.91	0.202	5,800	52.05	0.723
2,800	128.28	0.249	6,000	48.10	0.740
3,000	128.28	0.294	6,500	39.60	0.774
3,200	125.74	0.338	7,000	32.68	0.808
3,400	121.38	0.381	7,500	27.15	0.832
3,600	115.83	0.421	8,000	22.65	0.856
3,800	109.57	0.459	9,000	16.05	0.887
4,000	102.95	0.495	10,000	11.63	0.910
4,200	96.24	0.528	12,000	6.48	0.941
4,400	89.62	0.559	14,000	3.87	0.956
4,600	83.22	0.589	16,000	2.44	0.966
			∞	—	1.000

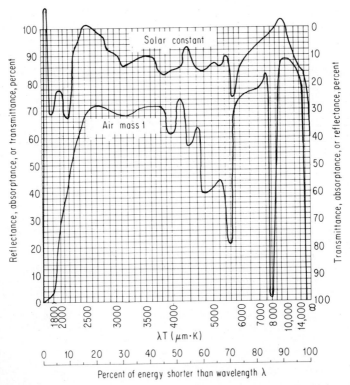

Fig. 5.3   Values of the solar constant and direct radiation, air mass 1, on the blackbody equal energy graph.

Values of monochromatic energies $E_{B\lambda}/T^5$, for the blackbody curve Fig. 5.1, and cumulative decimal fractions of energy, both with respect to λT, are given in Table 5.1. The blackbody equal energy graph is produced from Table 5.1 and is shown in Fig. 5.3.[2]

Application The blackbody equal energy graph can be used for thermal radiation transfer to and from solar energy systems. It is used conveniently for thermal radiation transfer by absorption, emission, reflection, and transmission and used also for recording thermal radiation spectra at test conditions.

The irradiance spectra for the extraterrestrial solar spectrum and for direct solar radiation at air mass 1 (for turbidity coefficients $\alpha = 0.66$, $\beta = 0.085$) are shown in Fig. 5.3.[3] The abscissa for these plots is displaced 0.12 below the abscissa of the coordinate system so that the plots fall within the boundary. The spectral properties of a material can be plotted at any temperature on the blackbody equal energy plot as decimal fractions. The area either above or below the plot is then calculated relative to the area of the whole. A planimeter may be used for this purpose, or a transparent overlay divided decimally to the same scale as the ordinate of the graph and the small squares and fractions are then counted by hand.

Thermal radiation is considered to approach blackbody radiation closely only under careful laboratory conditions. Most bodies emit radiation which is lower in energy content and is specularly dissimilar from that absorbed. The emitted radiation is called "gray-body" radiation. The blackbody equal energy graph also evaluates the response of materials to gray-body radiation even though the radiation wave band and the response of the material are substantially different in spectral character and temperature. The spectral energies of the gray body are graphed as a fraction of those of a blackbody. The spectral property of the material is graphed as a ratio of the lengths of the gray-body wavelength ordinates. The reaction of the optical property to the gray-body radiation is then proportional to its relative area. Evaluations of any optical properties can be made using this technique. It is preferable in dealing repetitively with other spectra to prepare special equal energy graphs for each.

The spectral transmittance of Tedlar plastic film is known at 25°C for values from 0.4 to 30 μm. It is graphed for thermal radiation from a source at 100°C in Fig. 5.4 to evaluate a mean transmission of 0.231. This compares favorably with blackbody radiometer measurements of 0.22 for thermal radiation at 65°C measured by M. Telkes.[4] The emittance of a (Ni-Zn)S "nickel black" selective surface[5] is shown also in Fig. 5.4 at 25°C.

The Solar Equal Energy Graph

Thekaekara[3] has made an evaluation of data on incident solar energy which has generally replaced earlier work. His evaluation has been adopted by the American Society for Testing and Materials, Standard E 490-73a (Ref. 6), and by the National Aeronautics and Space Administration as design values for its space vehicles design criteria.[7]

This work provides the basis for the solar equal energy graph. An equal energy graph is made for direct solar radiation by selecting an abscissa of desired length representing the total amount of energy at the selected atmospheric conditions. The line is proportioned by the cumulative sums of the wavelengths from $\lambda = 0$ to $\lambda = \infty$ as reduced in width by the ratios of $E_{S\lambda}/E_{S\lambda(max)}$. An ordinate of equal length is divided decimally (Fig. 5.5).

Cumulative energy ratios by wavelength are given in Table 5.2, for air mass $m = 1$, from Thekaekara's evaluation using turbidity coefficients $\alpha = 0.66$ and $\beta = 0.085$. This atmospheric model is selected for the solar equal energy graph because it has a total atmospheric transmittance of direct solar radiation of 0.657 at 1 air mass, a value commonly found with a cloudless and somewhat hazy sky. Atmospheric transmittance for Thekaekara's three other evaluations of air quality at air mass 1 are 0.707, 0.684, and 0.591. (See p. 2–18 for the definition of air mass.) Absorption of direct solar radiation by air mass is given in Fig. 5.6 for the four types of atmosphere used by Thekaekara. The cumulative sums of the wavelengths as reduced in width for the ratios of monochromatic energy to the wavelength of maximum energy are given in Table 5.2 for air masses 4, 7, and 10.

The solar equal energy graph, Fig. 5.5, is used later for listing optical properties. For example, the reflectances of an apple leaf, clean galvanized iron, and a translucent window shade are represented by the areas *above* the plots. It is obvious that substantial changes in the spectra for atmospheric conditions make a considerable change in the transport of solar radiation.

Solar Intensity Parameters

The declination of the sun, the eccentricity of the earth's orbit, and the equation of time with respect to the time of day and date are needed often to evaluate data and equations dealing

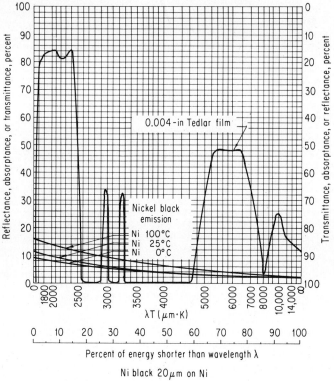

Fig. 5.4   Transmission and emission of thermal radiation.

with the qualities of the atmosphere and its effects on the optical properties of materials. These parameters may be obtained from equations in Chap. 2. The following section describes altitude effects and the calculation of air mass.

Air mass   It is customary to define "air mass 1" as being the air mass with the sun at the zenith, at sea level, and at standard barometric pressure. It is common practice to define the air mass m at other zenith angles of the sun as the secant of the zenith angle, i.e. (90° − altitude angle; see Chap. 2) $m = \sec z = 1/\cos z$. This expression is valid for z from 0 to 60°, but it is less accurate for $z > 60°$, particularly as $\sec z$ becomes infinite at 90°.

Robinson[9] develops air mass values from the air density–pressure ratio. Robinson's air masses, interpolated, are given in Table 5.3 for 1° increments of the zenith angle from 0 to 89°. The 90th degree, representing the last 4 min before sunset with an air mass of 40, is omitted as it is not of practical interest in solar energy. Its use adversely increases the mean air mass summation for the day. The air mass for a single observation at any zenith angle is obtained from Table 5.3. The air masses are summed by degrees from 89 to 0°, also in the table. The average air mass is found in the third column for any day at any latitude. The zenith angle at solar noon, z_n, is estimated from equations in Chap. 2 and is used to define the average air mass for the day. Daily mean air mass is shown in Fig. 5.7 for latitude and time of year.

The density of air and the corresponding optical air mass decrease with altitude of the site. The decrease in all air mass values for the elevation of the site is[10]

$$\frac{m}{m_0} \propto \frac{p}{p_0} = (1 - kH)^{4.256} \tag{5.6}$$

The constant k is 2.2539×10^{-5} when the altitude H is in meters, and 6.87×10^{-6} when in feet. The constants assume a sea-level temperature of 15°C and a decrease in temperature

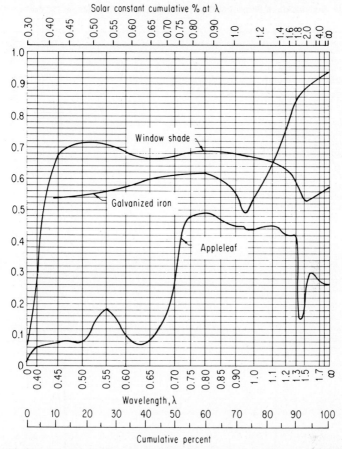

Fig. 5.5 Equal energy graph for solar radiation at air mass 1; extinction coefficients $\alpha = 0.86$, $\beta = 0.085$.

of 1.98°C per 305 m. Air densities and pressures vs. altitude are given in Fig. 5.8. The altitude correction can then be made to air masses given in Table 5.3.

Atmospheric Absorption and Spectral Effects

The absorption and dispersion of direct solar radiation for a given atmospheric quality and air mass defines the spectral nature of a solar environment.

Bouger's exponential law of absorption[11] states that the intensity of a beam I_x traversing thickness l of a medium is expressed for an initial beam intensity I_0 as

$$I_x = I_0 e^{-al} \qquad (l \leqslant 1) \tag{5.7}$$

The exponent a is the absorption coefficient at the value $l = 1$. A similar equation for l greater than 1 is

$$I_x = I_0 e^{-al} \qquad (l \geqslant 1) \tag{5.8}$$

Equation (5.8) stated for absorption α_m of direct radiation by the air mass, where transmittance of the air mass is τ_m and absorption coefficient is a for 1 air mass, is

$$\alpha_m = 1 - \tau_m = 1 - \frac{I}{I_0} = e^{-a/m} \tag{5.9}$$

TABLE 5.2 Summation in Percent of Direct Solar Energy by Wavelength for Various Air Masses

λ, μm	M = 0	M = 1	M = 4	M = 7	M = 10
0.35	4.52	1.4	—	—	—
0.40	8.73	4.7	1.14	0.30	0.08
0.45	15.14	10.7	5.6	—	—
0.50	22.60	18.7	9.6	5.0	2.2
0.55	29.38	26.6	16.0	9.4	3.9
0.60	35.68	34.1	23.2	16.0	10.8
0.65	41.54	41.4	31.3	25.3	20.3
0.70	46.88	48.2	—	—	—
0.72	—	51.1	43.5	36.6	30.8
0.75	51.69	54.7	46.8	40.3	34.8
0.80	56.02	59.7	52.1	45.4	40.0
0.85	—	64.7	57.7	51.7	45.2
0.90	63.37	69.7	63.8	56.3	50.0
1.0	69.49	75.7	74.2	64.0	57.5
1.1	—	81.8	77.1	71.6	65.6
1.2	78.40	85.8	84.6	79.9	75.3
1.3	—	90.1	—	—	—
1.4	84.33	—	—	—	—
1.5	—	92.6	89.3	85.2	81.1
1.6	88.61	—	—	—	—
1.7	—	97.4	—	—	—
1.8	91.59	—	—	—	—
2.0	93.49	98.7	92.7	88.2	86.8
3.0	97.83	—	96.4	93.1	90.0
∞	100.0	100.0	100.0	100.0	100.0

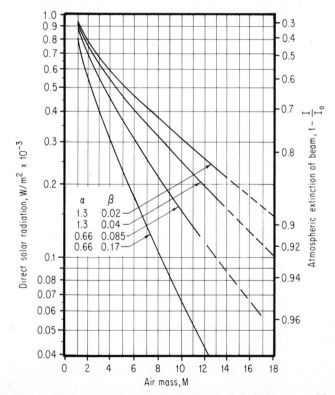

Fig. 5.6 Absorption of direct solar radiation by the air mass. (*By Thekaekara, courtesy of the publisher.*)

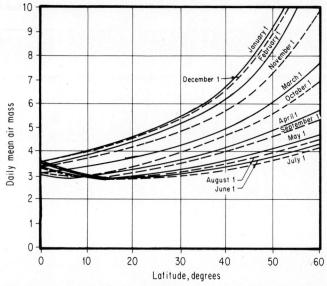

Fig. 5.7 Daily mean air mass for latitude and declination; standard atmosphere at sea level.

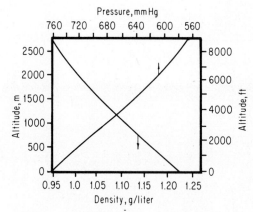

Fig. 5.8 Atmospheric pressure and density for altitude.

Thekaekara evaluates the atmospheric transmittance for the four qualities of the atmosphere shown in Fig. 5.6 and warns that values do not conform logarithmically. The values estimated from Eq. (5.9) are shown as a dotted line in Fig. 5.9 for air masses to 40. It is evident that Eq. (5.9) does not adequately represent absorptance and dispersion in the atmosphere. The abscissa of Fig. 5.9 is linear and represents varying values of the exponent X evaluated from Thekaekara's work. The ordinate is logarithmic. Values for Thekaekara's curve, $\alpha = 0.66$, and $\beta = 0.085$, and developed for Fig. 5.6 are extended graphically to 40 air masses. If the periods for air mass are shortened from the logarithmic period of Fig. 5.9 to produce a straight-line function and this revised scale is used in Fig. 5.10, Thekaekara's evaluations of the spectra for air mass provides the solar equal energy chart for direct radiation as follows.

Diffuse Radiation Spectral Distribution

In outer space the sun is a fireball and the sky is black, while on earth the sky has the light of diffuse radiation from the atmosphere. This diffuse radiation comes from two general conditions of the atmosphere—atomic and molecular scattering and larger particulate scattering. Each of

TABLE 5.3 Direct and Mean Air Mass for Zenith Angle at Local Solar Noon

Zenith at solar noon, degrees	Air mass at zenith angle	Summation*	Mean air mass	Zenith at solar noon, degrees	Air mass at zenith angle	Summation*	Mean air mass
0	1.000	262.401	2.916	46	1.439	210.700	4.7886
1	1.000	261.401	2.937	47	1.466	209.261	4.8665
2	1.000	260.401	2.9591	48	1.494	207.795	4.9475
3	1.001	259.401	2.9816	49	1.524	206.301	5.0317
4	1.002	258.400	3.0046	50	1.556	204.777	5.1194
5	1.003	257.398	3.0282	51	1.589	203.221	5.2107
6	1.004	256.395	3.0523	52	1.624	201.632	5.3061
7	1.006	255.391	3.0770	53	1.662	200.008	5.4056
8	1.007	254.385	3.1022	54	1.701	198.346	5.5096
9	1.010	253.378	3.1281	55	1.743	196.645	5.6184
10	1.015	252.368	3.1546	56	1.788	194.902	5.7324
11	1.019	251.353	3.1817	57	1.836	193.114	5.8519
12	1.022	250.334	3.2094	58	1.887	191.278	5.9774
13	1.026	249.312	3.2378	59	1.942	189.391	6.1094
14	1.031	248.286	3.2669	60	2.000	187.449	6.2483
15	1.035	247.255	3.2963	61	2.063	185.449	6.3948
16	1.040	246.220	3.3273	62	2.130	183.386	6.5495
17	1.046	245.188	3.3587	63	2.203	181.256	6.7132
18	1.051	244.134	3.3907	64	2.281	179.053	6.8866
19	1.058	243.083	3.4237	65	2.366	176.772	7.0709
20	1.064	242.025	3.4575	66	2.458	174.406	7.2669
21	1.071	240.956	3.4921	67	2.559	171.948	7.4760
22	1.079	239.885	3.5277	68	2.669	169.389	7.6995
23	1.086	238.806	3.5643	69	2.79	166.72	7.9390
24	1.095	237.720	3.6018	70	2.92	163.93	8.1965
25	1.103	236.625	3.6404	71	3.22	161.01	8.474
26	1.113	235.522	3.6800	72	3.41	157.79	8.766
27	1.122	234.409	3.7208	73	3.65	154.38	9.081
28	1.133	233.287	3.7627	74	3.65	150.73	9.420
29	1.143	232.154	3.8058	75	3.29	146.94	9.796
30	1.155	231.011	3.8502	76	4.00	142.94	10.210
31	1.167	229.856	3.8958	77	4.51	138.72	10.670
32	1.179	228.689	3.9429	78	4.85	134.21	11.184
33	1.192	227.510	3.9914	79	5.25	129.36	11.760
34	1.206	226.318	4.0414	80	5.63	124.11	12.411

35	1.221	225.112	4.0929
36	1.236	223.891	4.1461
37	1.252	222.655	4.2010
38	1.269	221.403	4.2577
39	1.287	220.134	4.3163
40	1.305	218.847	4.3769
41	1.325	217.542	4.4355
42	1.346	216.217	4.5045
43	1.367	214.871	4.5717
44	1.390	213.504	4.6414
45	1.414	212.114	4.7136

81	6.01	118.48	13.16
82	6.88	122.47	14.06
83	7.67	105.59	15.08
84	8.82	97.92	16.32
85	10.51	89.10	17.82
86	12.73	78.59	19.64
87	16.20	65.86	21.95
88	21.31	49.56	24.78
89	28.35	28.35	28.35
90	—	—	—

*Air mass summation for 89° to 0° zenith angle used to calculate the mean air mass.

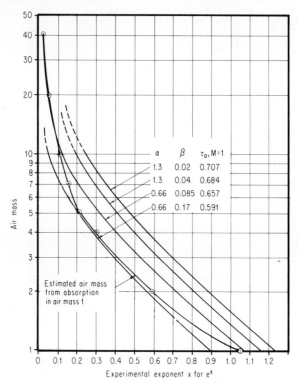

Fig. 5.9 Experimental absorption of direct solar radiation with air mass.

these two general classes are divided into parts. Atomic and diatomic molecules scatter radiation, known as "Rayleigh scatter," in accordance with an expression developed by Rayleigh. Larger molecules such as ozone, water vapor, and carbon dioxide absorb radiation at certain wavelengths, and both reemit and scatter radiation. Man-made pollution may contribute hydrocarbons, nitrogen oxides, and other molecules in sufficient amounts to obscure the scatter of all but the diatomic molecules, nitrogen and oxygen. Particulate matter ranges in size from the smallest carbon particles of smoke to very large aerosols in some clouds.

Rayleigh[12] developed the expression for the atomic and molecular scattering coefficient τ_R as

$$\tau_R \propto e^{-Ns^2/\lambda^4} \tag{5.10}$$

where N is the total number of atoms in the optical path and s is a scattering function of the atom. He continued by evaluating the scattering function s from the study of the refraction of liquids and obtained the first good value for Avogadro's number of molecules in a g mole of gas.

The area of the hemisphere of the sky is $2\pi R^2$ and of a smaller, central area is $2\pi R^2 - R(R \cos \theta)$, where θ is the central angle. On a clear day with the sun at the zenith, neglecting particulate dispersion, the Rayleigh attenuation of radiation by one-half and one-fourth is, respectively, within circles of the radii of 20.36° and 5.07°. The number of molecules N in the ray path increases proportionally with air mass. Liu and Jordan[8] developed an expression for the relationship between the atmospheric transmittances of diffuse τ_d and direct τ_D radiation, from measurements at Minneapolis, MN; Blue Hill, MA; and Hump Mountain, NC,

$$\tau_d = 0.2710 - 0.2939 \tau_D . \tag{5.11}$$

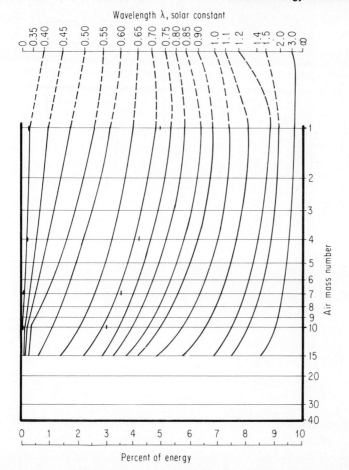

Fig. 5.10 Spectral distribution of direct solar energy for air mass number; extinction $\alpha = 0.86$, $\beta = 0.085$, Thekaekara's visible spectrum shown with dashes at M 1, 4, 7, and 10.

This diffuse radiation transmittance includes Rayleigh scatter and particulate scatter (Mie scatter) effects. Dave[13] suggests that Rayleigh scatter at air mass 1 may be 36 percent of the total, increasing to about 50 percent at 76° or air mass 4. This value is approximate because the scattering function s changes a little with wavelength, the ratios of diffuse and Rayleigh scatter are approximate, and there is some scatter of data by Liu and Jordan related to varying air mass.

In reviewing the literature for spectral distribution of diffuse solar radiation in respect to air mass and atmospheric quality it has not been possible to select a model to construct a solar equal energy graph for diffuse and then for total radiation. Until such time as this information is available, it is suggested that the equal energy graphs for direct radiation be used for total radiation. It appears that the contribution of diffuse radiation at air mass 1 will change the midline of the graphs by not more than 2 or 3 percent toward the infrared, and less at higher air masses.

Equal energy graphs are developed easily for other parts of the solar spectrum having a particular interest. The ultraviolet portion of the spectrum for direct radiation is shown in Fig. 5.11. This is of interest in the study of photodegredation of organic materials used outdoors in the solar environment. Photons in the ultraviolet have sufficient energies to cause chemical reactions both within the composition and between it and the atmosphere. Upon exposure,

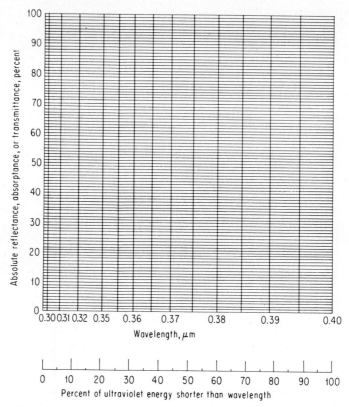

Fig. 5.11   Solar equal energy graph of the ultraviolet spectrum at air mass 1.

organic base materials undergo loss in transmission and increase in absorbtion, sometimes at wavelengths that can be associated with chemical bond energies, and have been correlated in degredation in a general way to the amount of local ultraviolet exposure. Equal energy graphs of other parts of the solar spectra have interest in the band gap of photovoltaic solar cells, for photographic processes, for correlating accelerated weathering machines, and for thermal energy transfer.

OPTICAL PROPERTIES OF MATERIALS

The spectral properties of reflectance, transmittance, absorptance, and emittance tabulated in the literature are usually plotted on rectangular charts to obtain intermediate and extended values. This approach can lead to error. A similar problem is found for most organic materials which have strong and weak transmission and absorbtion bands with slow and rapid rates of change, difficult to tabulate or to plot conventional coordinates. An average value of transmission or absorption can be obtained only by converting the data in proportion to the energy contained in equal units of wavelength—the equal energy graphs.

An energy balance of the optical properties of materials is stated for a monochromatic wavelength or for a wave band of constant monochromatic character for transmittance τ, reflectance ρ, and absorbance α as

$$\tau + \rho + \alpha = 1 \tag{5.12}$$

The evaluation of any two properties defines the third by difference. When one property, such as transmittance, is zero, then the measurement of one property defines the other. Nearly every solid material has imperfections and heterogeneity on its surface and within it. The outside surfaces of an extruded plastic film tend to be a little more dense, and "fish eyes" of polymer within it may represent localized changes in molecular weight in sizes down to a small fraction of a micrometer. Such imperfections cause scatter of the resultant beam, and the angles of scatter are not predictable. If the energy balance is to be maintained, both the beam and its scatter in the hemisphere of the resultant beam must be measured. Such properties are referred to as "hemispherical" properties. In the laboratory the resultant beam from a spectrophotometer-type instrument must be collected and measured in a reflective cavity. Usually these cavities are coated at frequent intervals with magnesium oxide from burning magnesium ribbon. A commercial block of spectrographic grade magnesium carbonate having known spectral reflectance is used as a sample to calibrate the instrument. For average values of hemispherical properties a proper type of radiometer is used with a sufficiently large sample, and a blackbody cavity at a corresponding temperature compares resultant radiation to blackbody energy.

In practice, the sun is often used to obtain an average value for an optical property. The sample must be large compared with the radiation sensing element and placed as close as possible to the element, so that nearly all of the scatter is measured. The sample must be at least 4 ft² for plastic films or glass sheets. The quality of solar radiation, as defined by air mass and atmospheric transmittance, is noted for the tests. The sample is placed in a removable frame which may be adjusted in angle with an alt-azimuth support, and has a shadow gauge marking angles and tilt. A black cloth is fastened to the frame or its support and hangs down to exclude ground reflection to the underside of the sample. A recording voltmeter with a wide strip chart and relatively fast speed is used with the pyrheliometer. Readings are taken with the frame (a) removed, (b) in place, and (c) then removed, all in the shortest time of the instrument response. When as few as four sets of data are taken for air mass and atmospheric transmittance, the measurements can be extended to other sites and weather conditions.

The amount of extant data for the optical properties of materials is exceedingly large and diverse. One important compilation has been made by Touloukian and Witte.[14] Three of these volumes dealing with the thermal radiation properties of materials have over 4700 pages of graphs and tabulated data. The data which are reported in the following sections provide results which may be of special interest in solar energy work.

Index of Refraction

For diathermous materials that transmit a substantial part of the incident ray, Snell's law[15] states that a refracted ray of refractive angle i' lies in the plane of the ray incident at i, and the sine of the angle of refraction bears a constant ratio to the sine of the angle of incidence; thus $\sin i/\sin i'$ is constant. When the incident ray approaches the boundary in a vacuum or in air, this constant becomes the index of refraction n. In solar energy work the index of refraction is also used to evaluate reflectance. When the index of refraction is known only for a single wavelength of monochromatic radiation, then reflectance can be evaluated accurately only for that wavelength. Quite small changes in the index of refraction (1 part in 100) result in substantial changes in reflection. It is conventional to report indexes of refraction for the strong sodium D line of the spectrum, $\lambda = 0.58932$ μm, at 25°C and at an incident angle of 20°.

When three indexes of refraction are known for the same conditions of measurement and they are well spaced across the wave band, the average value for that wave band is evaluated. Several rather complex expressions have been developed from dispersion and quantum theory for the evaluation of the index of refraction with respect to wavelength. A simple and usually quite valid expression was developed by Cauchy[16] in 1836,

$$n = A + \frac{B}{\lambda^2} + \frac{C}{\lambda^4} \tag{5.13}$$

where A, B, and C are constants that are characteristic of the material. The three constants are evaluated from measured indexes of refraction at several wavelengths by the use of regression methods. The index of refraction can then be evaluated for other wavelengths. An index of refraction can undergo an unpredictable change in value usually for unknown reasons. At times absorption will eliminate reflection of part of the wave band, and the corresponding values of the index of refraction will be affected.

A scale of appropriate interval is placed on the ordinate of an equal energy graph, and the graph is used to plot and measure the average index of refraction for that wave band. A number of indexes of refraction are shown in Fig. 5.12 and are identified in the legend.

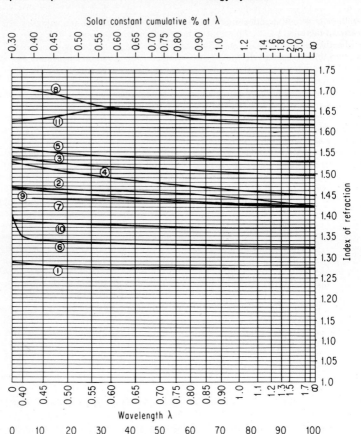

Fig. 5.12   Index of refraction on a solar equal energy graph, air mass 1. Index of refraction, Fig. 5.12: 1. air, dry, 760 mm, 15°C, $(n-1)\,10^7$ with the value one added for plotting [17]; 2, fused quartz [17]; 3, common window glass; 4, methyl methacrylate sheet; 5, sodium chloride crystal, rock salt [17]; 6, distilled water [17]; 7, fluorite, CaF_2 [17]; 8, calspar, calcium carbonate crystal, ordinary ray [17]; 9, Tedlar PVF plastic film; 10, Teflon FEP plastic film; 11, Mylar plastic film.

Reflection*

The reflectance ρ from a diathermous material is evaluated by the Fresnel expression for a wave band using the average value of the index of refraction. At normal incidence,

$$\rho = \frac{(n-1)^2}{(n+1)^2} \tag{5.14}$$

Reflectance for angles of incidence i and of refraction i' other than zero are

$$\rho = \frac{\sin^2(i-i')}{2\sin^2(i+i')} + \frac{\tan^2(i-i')}{2\tan^2(i+i')} \tag{5.15}$$

where $\sin i$ equals $n \sin i'$. A small error is introduced by using total solar radiation as the average of beam and diffuse radiation, having an angle of incidence which differs from that for the beam radiation.

*For a detailed analysis of reflection in and between several layers of transparent media, see Chap. 4.

From the energy balance of Eq. (5.12), if the reflectance from a single surface of a pane is ρ, the transmittance is $1 - \rho$.

The reflection of radiation for metals at normal incidence may be estimated using two constants representing the minimum values of the vectors of polarization and an index of refraction for the metal. The values for each are wavelength-dependent, involving many calculations based on experimentally determined coefficients. The need for this computation and its inaccuracies becomes reduced as carefully measured values are published.

Diffuse reflection from earth materials is measured with a pyrheliometer directed at a substantial expanse of the material. A large mirror is placed first over the material at a time and place where the angle of incidence is about 30 to 40°. The pyrheliometer is provided with an axis shadow gauge and is aligned normal to the reflected beam, and as close to the surface as its heterogeneous nature will permit. The mirror is removed and a reading is made of the diffuse reflection of the material. The reflectance of the mirror is evaluated by directing the pyrheliometer at the sun. Values vary somewhat with different samples of the same class of earth materials and with high angles of incidence. Values of reflectance from earth materials measured in this manner are given in Table 5.4. The spectral reflectances of several materials are given in Fig. 5.13.

Transmission

Spectral transmission is measured in the laboratory with a spectrophotometer and reflective cavity and with a collimated light source. Plots of the data on solar equal energy graphs compare closely with solar calorimeter measurements.[2] Measurements of solar transmittance with respect to angles of incidence up to 75° are given for several materials in Table 5.5 at an air mass of 1.30 to 1.35 and total atmospheric transmittance of 0.782. In addition, the hemispherical thermal transmittances of 0.003-in-thick Tedlar PVF plastic film measured at 400, 500, 600, and 700°R are 0.17, 0.20, 0.23, and 0.26, respectively. The thermal transmittance of quite clear, thin plastic films is usually proportional to their masses per unit of area.

When the solar transmittance is known for normal incidence, a pseudo–index of refraction can be evaluated with Eq. (5.14) and used in Eq. (5.15) to obtain approximate values at other angles of incidence. Then absorption is estimated for a ray path corresponding to the angle of incidence and a computation of reflectance and transmittance from the true and pseudo–indexes of refraction. The values of transmittance for multiple panes of materials in Table 5.5 do not equate to simple relationships of values for a single pane. It is observed that transmittance is slightly higher at angles of incidence between 20 and 40° when compared to normal incidence, and it should be lower according to theory. This is believed to result from the angular component of diffuse radiation having a more favorable angle of incidence to the pane than direct radiation, since the sun is not at the zenith. Additional transmittance data are tabulated in the Appendix of this chapter.

Absorption and Emission

The relation between the emission ϵ and absorption α of various bodies was first stated by Kirchhoff in 1859.[18] Light emitted at any wavelength can be absorbed by a body at a lower

TABLE 5.4   Hemispherical Reflectance of Solar Energy from Earth Materials, Incident Angle about 30°

Sheet steel, no rust	0.72–0.76	Plowed earth, dry	0.12–0.20
Copper, new sheet	0.75–0.81	Limestone wall, gray	0.18
White lead paint, new	0.85–0.88	Limestone wall, white	0.40
Zinc oxide paint, new	0.81–0.84	Apple leaf	0.27
Light cream paint, new	0.75	Apple tree, large	0.18
Light yellow paint, new	0.65	Asphalt pavement, old	0.13–0.16
Light green paint, new	0.60	Brick wall, light buff	0.40
Aluminum paint, new	0.65–0.68	Brick wall, red	0.30
Gray paint	0.35	Brick wall, dark red	0.20
Black flat paint	0.06–0.08	Snow, fresh	0.65–0.72
Wood, pine, weathered	0.26–0.30	Grass turf, mowed	0.28–0.35
Sandy loam, packed	0.24–0.26	Concrete walk	0.18–0.25
Plowed earth, moist	0.08–0.14		

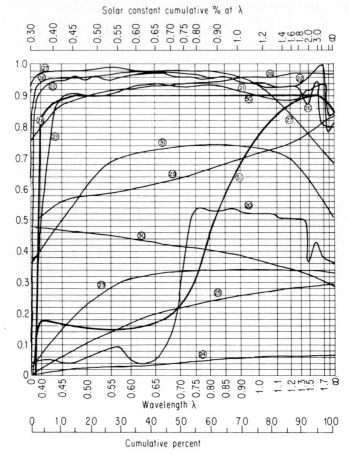

Solar constant cumulative % at λ

Wavelength λ

Cumulative percent

Fig. 5.13 Solar reflection of materials. Solar reflectance, Fig. 5.13: 20, holly leaf; 21, outside white titanium dioxide base paint; 22, PKT, potassium tetratitanate insulating pigment; 23, lampblack flat paint; 24, acetylene lampblack; 25, black velvet cloth; 26, spectral grade magnesium carbonate black; 27, magnesium oxide smoked surface; 28, polished silver [17]; 29, polished monel metal [17]; 30, Indiana Limestone [17]; 31, white paper [17].

temperature. The emission and absorption energies of a body are equal to a constant, $\epsilon/\alpha = k$. When absorption is complete, α becomes unity and there is no reflection. Kirchhoff called such a body a perfect "blackbody." Truly blackbody conditions are approached but never attained by experiment. Equal absorption and emission occur only in a wave band of the same spectral energies, and thus at the same radiative conditions. The equations and references for absorptance are given in the paragraph above on air mass.

Absorption and emission generally increase with temperature. Often the rate of change follows the periodicity of temperature to the $1/5$ power, $T^{1/5}$. When this periodicity occurs, spectral data are plotted directly on the blackbody equal energy graph (Fig. 5.3) vs. λT for any temperature covered by the data and for reasonable extensions of it. The periodicity P, is evaluated

$$\frac{dP/dT}{\Delta T} = \left[\frac{T^{0.20}}{(T-1)^{0.20}} - 1 \right]^{-1} \tag{5.16}$$

The periodicity of $T^{1/5}$ is given in Table 5.6 for ΔT intervals of 25°C. Any part of this periodicity may be used to match the data when two requirements are met. First, the intervals

TABLE 5.5   Total Solar Transmission through Diathermous Panes

Material	Thickness, in	0°	15°	30°	45°	60°	75°
1 Polyethylene, clear[a]	0.002	0.936	0.936	0.936	0.925	0.872	0.704
2 Polyethylenes, clear[a]	0.002	0.894	0.894	0.894	0.889	0.795	0.581
3 Polyethylenes, clear[a]	0.002	0.838	0.838	0.834	0.824	0.721	0.485
1 Tedlar PVF	0.003	0.947	0.947	0.947	0.934	0.868	0.710
2 Tedlar PVFs	0.003	0.902	0.902	0.902	0.890	0.805	0.585
3 Tedlar PVFs	0.003	0.857	0.857	0.857	0.845	0.755	0.481
1 Mylar[b]	0.003	0.868	0.868	0.873	0.857	0.823	0.689
2 Mylars[b]	0.003	0.806	0.806	0.814	0.782	0.745	0.602
3 Mylars[b]	0.003	0.761	0.761	0.765	0.747	0.711	0.538
1 PVC [c]	0.004	0.825	0.828	0.828	0.800	0.740	0.595
1 PVC [c]	0.008	0.780	0.781	0.783	0.778	0.719	0.585
1 Teflon FEP	0.001	0.965	0.977	0.977	0.971	0.941	0.748
1 Teflon FEP	0.002	0.969	0.977	0.977	0.971	0.926	0.668
1 Teflon FEP	0.005	0.966	0.972	0.972	0.967	0.922	0.665
1 Window glass[d]		0.921	0.921	0.921	0.902	0.887	0.751
1 Window glass[e]		0.891	0.897	0.890	0.881	0.865	0.720
1 Tedlar, dry[f]	0.003	0.916	0.916	0.894	0.845	0.723	0.525
1 Tedlar, wet[g]	0.003	0.948	0.944	0.933	0.923	0.831	0.569

[a] "Crystal Clear" polyethylene.
[b] Weatherable Mylar, treated one side.
[c] PVC Weatherable, Monsanto.
[d] Single strength glass, air mass 1.2.
[e] Same glass, same day, air mass 3.8.
[f] Tedlar, dry, interference grating.
[g] Tedlar, wet, interference grating.

TABLE 5.6   Periodicity of Temperature* for 25° Intervals

1.667	2.163	3.079	5.341
1.703	2.223	3.203	5.724
1.740	2.287	3.336	6.165
1.778	2.354	3.481	6.680
1.819	2.426	3.640	7.288
1.861	2.501	3.813	8.019
1.905	2.582	4.004	8.912
1.952	2.668	4.215	10.103
2.010	2.760	4.450	11.467
2.052	2.859	4.712	13.385
2.106	2.965	5.007	16.077

* In Kelvins.

of comparison are the same for the data and the periodicity, and second, the axis of the periodicity is the same as the axis of the data—the zero values of both are parallel. The absorptance of silver and the emittances of aluminum, 17-7 stainless steel, and copper, selected for their lowest consistent values, data numbers, 2, 24, 58, and 13 (Ref. 14), are plotted in Fig. 5.14 coinciding with the periodicity of $T^{1/5}$ and are extended.

To measure absorptance of transparent media in the laboratory, the beam of a spectrophotometer is passed through one, two, and then progressively more panes cut from the same material, with the panes being cemented together in progression. The thinnest possible amount of liquid cement is used, a hydraulic press is helpful, and the cement is chosen to have the same index of refraction as the sample, as nearly as possible. The monomers of plastics samples can often be used. A microscope grade of Canada balsam oil, diluted with appropriate solvents, is used with glasses to reduce interface reflections to a minimum. The spectral quality of the cement is not exactly that of the sample, and it can usually be differentiated in the plotting of measurements of one and two panes.

Single films of varying thickness are measured both in the laboratory and in panels with the sun. The results are not always dependable, usually due to differences among batches of material

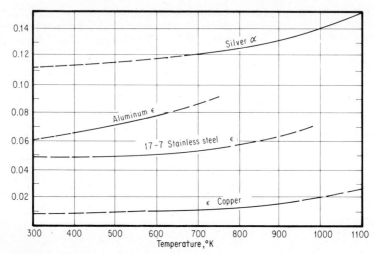

Fig. 5.14   Absorptivity and emissivity of metals are related to the periodicity of $T^{1/5}$.

and processing. Another measurement technique uses plastic film rolled very tightly on a mandrel. Values are obtained thereby for hundreds of thicknesses of film with essentially very little air at the interface. The results compare favorably with the first method above. Absorptance and emittance values for many materials are tabulated in the Appendix of this chapter.

Special Materials

Several novel types of materials have been proposed or used successfully for solar energy applications.

Photochromic glass   (The Corning Glass Works)   Silver chloride with other ions is dispersed in a glass pane in the form of very small particles. When the glass is not irradiated the silver chloride and the glass are substantially clear. Solar radiation converts the silver chloride into silver, and the glass becomes more opaque to the visible spectrum. A plot of spectral transmittances by Smith and Justice[19] is shown as curves 50 in Fig. 5.13. The transmittance decreases with higher temperatures and increases with higher radiant energy.

V-corrugated glazings   Gibbett and Hollands[20] find that the transmittance of a flat glazing is increased if it is V-corrugated. A sheet of Teflon FEP film is corrugated with an opening angle of 30° and placed between the outer glazing and the absorber. Theoretical predictions based upon a Monte Carlo—type analysis compare closely with experimental values. Transmittance divided by the transmittance of the outer cover plate is compared to the zenith angle of incidence β and the azimuth angle of corrugations, shown in Fig. 5.15.

Plastic glazings subtending glass   Ribbans[21] places a thin Teflon FEP plastic film between a glass glazing and the absorber and obtains 5 to 30 percent increase in collected energy over that obtained with two panes of glass. The cost of the Teflon FEP is estimated to be about one-sixth that of a glass pane. Teflon FEP has temperature stability up to 400°F (204°C) and an index of refraction of 1.376 at 25°C resulting in low reflectance.

Whillier[22] evaluates various assemblies of panes of glass and Tedlar PVF film and concludes that Tedlar PVF film is a fully acceptable substitute for glass in flat-plate solar collectors from the efficiency of collection point of view. It is probable that the addition of an interval plastic glazing reduces convection and radiation heat transfer to effect increased energy collection.

Interference glazings   The author ruled 1400 lines per inch on Tedlar film and observed little reduction in the transmission at solar wavelengths from 0.4 to 2.0 μm, but did obtain reduction in transmission in the infrared wave band from 8 to 35 μm. The most economical methods for approximating this roughness are by "graining" with a water slurry of fritted alumina No. 125 screen size and a rubbing block, and by dry "sand-blasting" with this grit. Talley-surf measurements of roughness showed a roughness similar to that of the ruled sample. The abrasion made the plastic film quite hydrophilic, or wettable. Transmission of solar radiation for sand-

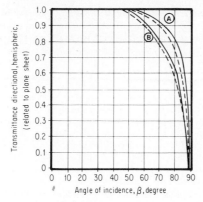

Fig. 5.15   Transmittance ratio for V-corrugated plastic glazings; Grooves at 90°A and 0°B azimuth angles.

blasted Tedlar, both wet and dry, are given in Table 5.5. Production of water from "development" solar stills is unaccountably higher than as indicated by the effects of the grating and the water film, opaque to infrared radiation. Some 6×10^5 ft² of solar stills have been built, all of them using this treatment.

Wettable surfaces   In some solar energy processes such as solar distillation it is desirable to have a highly wetted glazing surface, either for low-angle run off of condensate or to eliminate dropwise condensation and reflectance. Plastic films are made wettable with a chemical treatment. The film is passed through concentrated sulfuric acid at a temperature and a rate to produce a slight darkening of the film. Clearness is restored by subsequent treatment with diluted hydrogen peroxide. The drop contact angle for these treatments is less than that for clean glass, and the films are more easily wetted at lower inclination angles.

The wettability of glass is improved either by treating one surface with dilute hydrogen fluoride or by coating with sodium metasilicate and firing.[23] The transmission of solar radiation is relatively unaffected by these treatments.

Fresnel lenses   A Fresnel lens is the transposition of segments of equal depth and decreasing width from a spherical, transparent lens to a transparent sheet (see Chaps. 8 and 9). Each segment has a spherical curvature to provide the same focal length. The Fresnel lens presents a vertical wall at each segment which is observed to interfere and disperse rays passing through it. The loss of incident energy results from optical reflection and absorption plus the scatter from walls, a function of the focal length-to-diameter ratio or the f-number. The transmission of good acrylic lenses are f 1.6, 0.80; f 1.1, 0.70; f 0.82, 0.60; and f 0.6, 0.45. Any displacement from normal incidence increases dispersion and lowers transmission.

Total internal reflecting prisms   Rabl[25] developed the parameters for total internal reflection by prisms and found promising applications for the system, which has long been used in binoculars, for heliostats, line-focus concentrators, and in some instances compound, parabolic concentrators (see Chap. 8). Aluminum reflectors may have values of 80 to 90 percent, and silvered mirrors up to 95 percent. The prisms lose only the radiation that is absorbed by the material of the prism. A problem is that reflection occurs for a restricted range of incident angles, limited by a low value of the index of refraction.

Honeycomb structures   Several experimenters have worked with thin-walled plastic or aluminum hexagon-shaped structures of honeycomb nature. Hollands[26] presents a theoretical study of a flat-plate collector and shows that natural convection losses are suppressed and thermal losses are reduced substantially, depending upon the nature of the honeycomb and the length-to-diameter ratio. He develops a form factor for radiant energy exchange with respect to the honeycomb aspect ratio, and develops the solar energy efficiency factors for four models.

MACROECONOMIC AND SOCIAL IMPACTS OF SOLAR MATERIALS SELECTION

Sociological impacts will relate to the availability of materials in adequate amounts, an acceptable ratio of the energy needed to manufacture systems to the energy obtained from them, and the social acceptance of solar energy as a dependent energy resource. The failure to meet any of these demands will certainly lead to a reduced and even minimal solar energy use as an energy resource. It must be presumed that such economic requirements can be met.

Materials in Critical Supply

In considering solar energy systems which involved the direct use of solar energy, including flat-plate-type collectors, reflective collectors up to large, central power plants, and silicon-base photovoltaic systems, the largest demands are likely to be steel, glass, and aluminum. There is an ample supply of these raw materials in the perspective of percentage demand on present production. Fibrous mineral insulation and closed-cell foam insulation could be in short supply, temporarily because of manufacturing capacity. Likewise, the small amounts of specialty materials such as chromium, nickel, and cobalt which might be used for selective surfaces are very minor additions to current usage. Silicon is amply available. The requirements for energy storage are not known. If the hydrous salts are used for low-temperature storage systems there will be ample supplies of most of the candidates. As the energy potential of storage is increased, the availability of ample supplies becomes more doubtful. If most solar collectors are made with copper, as well as the piping systems, copper may become a critical material.

The United States produces about 3,000,000 tons of milled product copper materials annually, of which about 600,000 tons went into the building industry. About 55 percent of this copper is newly mined, and 90 percent of that is obtained in the United States. If solar heating were to be added to a million homes annually, with an average collector area of 1000 ft^2 and using current designs, the new copper requirement would be 400 lb per dwelling including piping or 200,000 tons annually. If new supplies of recycled copper are not available, copper consumption would be increased by 12 percent. Unmined copper reserves are about 80,000,000 tons. If 15,000,000 dwellings (equivalent total copper collectors) are to use ultimately 3,000,000 tons of copper in solar collectors, the present U.S. copper resource of about 48 years is reduced by 2 years. Considering the unidentified total copper resource, copper is not a critical material for solar collectors.

Energy Content of Materials

It must become a requirement that solar energy systems are designed and built with a useful life expectancy that is comparable to dwellings and their major auxiliaries such as heating and cooling systems, kitchen appliances, e.g., stoves, and large electric appliances. This suggests a life with reasonable maintenance of 20 to 35 years. When this condition is met there is no problem concerning the energy content of solar devices. Present commercial systems range from 2 to 5 percent material energy content to the energy produced. Therefore, most systems require about 1 year's use to recover the energy used to produce them.

REFERENCES

1. R. Siegel and J. R. Howell, *Thermal Radiation Heat Transfer*, McGraw-Hill, New York, 1975.
2. F. E. Edlin, Selective Radiation Reflection Properties of the Fiberous Alkali Titanates, Amer. Soc. Mech. Eng., 64-WA/Sol 8, September 1964.
3. M. P. Thekaekara, *Data on Incident Solar Energy*, Inst. Environ. Sci., Mount Prospect, IL, 1974.
4. M. Telkes, Infrared Transmission of Tedlar and Mylar Films (letter to author), Curtiss-Wright, Inc., Dec. 29, 1959.
5. J. Jurison, R. E. Peterson, and H. Y. B. Mar, Principles and Applications of Selective Solar Coatings, *J. Vac. Sci. Technol.*, vol. 12, no. 5, Sept./Oct. 1975, p. 1011.
6. Anon., Standard Specification for Solar Constant and Air Mass Zero, Solar Spectral Irradiance, Amer. Soc. Test. Mater. Std. E490-73a, 1974, *Annual Book of ASTM Standards*, Part 4, ASTM, Philadelphia, PA, 1974.
7. Anon., Solar Electromagnetic Radiation, National Aeronautics and Space Administration, SP 8005, Washington, DC, May 1971.
8. B. Y. H. Liu and R. C. Jordan, The Interrelationship and Characteristic Distribution of Direct, Diffuse and Total Solar Radiation, *J. Sol. Energy Sci.*, vol. 4, no. 3, July 1960.
9. N. Robinson, *Solar Radiation*, Elsevier, Amsterdam, 1965.
10. J. H. Perry, *Chemical Engineering Handbook*, 2d ed., McGraw-Hill, New York, 1941, p. 773.
11. P. Bouger (1698–1758), *Fundamentals of Optics*, 2d ed., F. A. Jenkins and H. E. White (ed.), McGraw-Hill, New York, 1950, p. 197.
12. J. W. S. Rayleigh (1842–1919), *Scientific Papers*, 6 vol., Royal Society, London, England.
13. J. V. Dave, Validity of the Isotropic Distribution Approximation in Solar Energy Estimations, *J. Sol. Energy Sci.*, vol. 19, no. 4, p. 331, 1977.

14. V. S. Touloukian and D. P. Witte, *Thermal Radiation Properties of Materials*, vol. 7, "Radiative Properties, Metallic Elements and Alloys," Plenum Press, New York, 1970.
15. W. Snell (1591–1626), published by Rene Descartes, *La Monde*, vol. 1, "Light," Paris, 1633.
16. A. L. Cauchy, *Oeuvres Completes d'Augustin Cauchy*, 1836; also F. A. Jenkins and H. E. White, *Fundamentals of Optics*, 2d ed. McGraw-Hill, New York, 1950, p. 465.
17. R. C. Weast and S. M. Selby, *Handbook of Physics and Chemistry*, 49th ed., Cleveland, Ohio, 1968.
18. G. R. Kirchhoff, in W. Ostwald's *Klassiker No. 100, V. Leipsig, Leipsig, Germany, 1898*.
19. G. Smith and B. Justice, Photochromic Glass; Light and Heat Control with Variable-Transmittance Glazing, Amer. Soc. Mech. Eng., 64-WA/Sol 6, October 1965.
20. B. E. Gibbett and K. G. T. Hollands, Radiant Transmittance of V-Corrugated Transparent Sheets with Application to Solar Collectors, Amer. Soc. Mech. Eng., 76-WA/Sol-1, November 1976.
21. R. C. Ribbans, Letter June 3, 1977, Plastic Prod. and Resins Dept. DuPont Co., Chestnut Run, DE.
22. A. Whillier, Plastic Covers for Solar Collectors, *J. Sol. Energy*, vol. 7, no. 3, July 1963.
23. M. H. Bahadori and F. E. Edlin, Improvement of Solar Stills by The Surface Treatment of the Glass, *J. Sol. Energy*, vol. 15, no. 1, 1973.
24. F. E. Edlin, Reflector for Solar Heaters, U.S. Patent 3,058,394, Oct. 16, 1962.
25. A. Rabl, Prisms With Total Internal Reflection as Solar Reflectors, SOL 76-04, Argonne Nat. Lab., Argonne, IL, May 1976.
26. K. G. T. Hollands, Honeycomb Devices in Flat Plate Solar Collectors, *J. Sol. Energy*, vol. 9, no. 3 1965.

APPENDIX

TABLE 5.A1 Emittances and Absorptances of Materials[a]

Substance	Short-wave absorptance	Long-wave emittance	$\dfrac{\alpha}{\epsilon}$
Class I substances: Absorptance to emittance ratios less than 0.5			
Magnesium carbonate, $MgCO_3$	0.025–0.04	0.79	0.03–0.05
White plaster	0.07	0.91	0.08
Snow, fine particles, fresh	0.13	0.82	0.16
White paint, 0.017 in, on aluminum	0.20	0.91	0.22
Whitewash on galvanized iron	0.22	0.90	0.24
White paper	0.25–0.28	0.95	0.26–0.29
White enamel on iron	0.25–0.45	0.9	0.28–0.5
Ice, with sparse snow cover	0.31	0.96–0.97	0.32
Snow, ice granules	0.33	0.89	0.37
Aluminum oil base paint	0.45	0.90	0.50
White powdered sand	0.45	0.84	0.54

Note: See page 5.25 for footnote.

TABLE 5.A1   Emittances and Absorptances of Materials[a] (Continued)

Substance	Short-wave absorptance	Long-wave emittance	$\frac{\alpha}{\epsilon}$
Class II substances: Absorptance to emittance ratios between 0.5 and 0.9			
Asbestos felt	0.25	0.50	0.50
Green oil base paint	0.5	0.9	0.56
Bricks, red	0.55	0.92	0.60
Asbestos cement board, white	0.59	0.96	0.61
Marble, polished	0.5–0.6	0.9	0.61
Wood, planed oak	—	0.9	—
Rough concrete	0.60	0.97	0.62
Concrete	0.60	0.88	0.68
Grass, green, after rain	0.67	0.98	0.68
Grass, high and dry	0.67–0.69	0.9	0.76
Vegetable fields and shrubs, wilted	0.70	0.9	0.78
Oak leaves	0.71–0.78	0.91–0.95	0.78–0.82
Frozen soil	—	0.93–0.94	—
Desert surface	0.75	0.9	0.83
Common vegetable fields and shrubs	0.72–0.76	0.9	0.82
Ground, dry plowed	0.75–0.80	0.9	0.83–0.89
Oak woodland	0.82	0.9	0.91
Pine forest	0.86	0.9	0.96
Earth surface as a whole (land and sea, no clouds)	0.83	—	—
Class III substances: Absorptance to emittance ratios between 0.8 and 1.0			
Grey paint	0.75	0.95	0.79
Red oil base paint	0.74	0.90	0.82
Asbestos, slate	0.81	0.96	0.84
Asbestos, paper		0.93–0.96	—
Linoleum, red-brown	0.84	0.92	0.91
Dry sand	0.82	0.90	0.91
Green roll roofing	0.88	0.91–0.97	0.93
Slate, dark grey	0.89	—	—
Old grey rubber	—	0.86	—
Hard black rubber	—	0.90–0.95	—
Asphalt pavement	0.93	—	—
Black cupric oxide on copper	0.91	0.96	0.95
Bare moist ground	0.9	0.95	0.95
Wet sand	0.91	0.95	0.96
Water	0.94	0.95–0.96	0.98
Black tar paper	0.93	0.93	1.0
Black gloss paint	0.90	0.90	1.0
Small hole in large box, furnace, or enclosure	0.99	0.99	1.0
"Hohlraum," theoretically perfect black body	1.0	1.0	1.0

TABLE 5.A1   Emittances and Absorptances of Materials[a] (Continued)

Substance	Short-wave absorptance	Long-wave emittance	$\dfrac{\alpha}{\epsilon}$
Class IV substances: Absorptance to emittance ratios greater than 1.0			
Black silk velvet	0.99	0.97	1.02
Alfalfa, dark green	0.97	0.95	1.02
Lampblack	0.98	0.95	1.03
Black paint, 0.017 in, on aluminum	0.94–0.98	0.88	1.07–1.11
Granite	0.55	0.44	1.25
Graphite	0.78	0.41	1.90
High ratios, but absorptances less than 0.80			
Dull brass, copper, lead	0.2–0.4	0.4–0.65	1.63–2.0
Galvanized sheet iron, oxidized	0.8	0.28	2.86
Galvanized iron, clean, new	0.65	0.13	5.0
Aluminum foil	0.15	0.05	3.00
Magnesium	0.3	0.07	4.3
Chromium	0.49	0.08	6.13
Polished zinc	0.46	0.02	23.0
Deposited silver (optical reflector) untarnished	0.07	0.01	
Class V substances: Selective surfaces[b]			
Plated metals:[c]			
Black sulfide on metal	0.92	0.10	9.2
Black cupric oxide on sheet aluminum	0.08–0.93	0.09–0.21	
Copper (5×10^{-5} cm thick) on nickel or silver-plated metal			
Cobalt oxide on platinum			
Cobalt oxide on polished nickel	0.93–0.94	0.24–0.40	3.9
Black nickel oxide on aluminum	0.85–0.93	0.06–0.1	14.5–15.5
Black chrome	0.87	0.09	9.8
Particulate coatings:			
Lampblack on metal			
Black iron oxide, 47 µm grain size, on aluminum			
Geometrically enhanced surfaces:[d]			
Optimally corrugated greys	0.89	0.77	1.2
Optimally corrugated selectives	0.95	0.16	5.9
Stainless-steel wire mesh	0.63–0.86	0.23–0.28	2.7–3.0
Copper, treated with $NaClO_2$ and NaOH	0.87	0.13	6.69

[a]From Anderson, B., "Solar Energy," McGraw-Hill Book Company, 1977, with permission.

[b]Selective surfaces absorb most of the solar radiation between 0.3 and 1.9 µm, and emit very little in the 5–15 µm range—the infrared.

[c]For a discussion of plated selective surfaces, see Daniels, "Direct Use of the Sun's Energy," especially chapter 12.

[d]For a discussion of how surface selectivity can be enhanced through surface geometry, see K. G. T. Hollands, Directional Selectivity Emittance and Absorptance Properties of Vee Corrugated Specular Surfaces, *J. Sol. Energy Sci. Eng.*, vol. 3, July 1963.

TABLE 5.A2 Thermal and Radiative Properties of Collector Cover Materials

Material name	Index of refraction (n)	τ (solar)[g] (%)	τ (solar)[a] (%)	τ (infrared)[b] (%)	Expansion coefficient (in/in · °F)	Temperature limits (°F)	Weatherability (comment)	Chemical resistance (comment)
Lexan (polycarbonate)	1.586 (D 542)[c]	125 mil 64.1 (± 0.8)	125 mil 72.6 (± 0.1)	125 mil 2.0 (est)[d]	3.75 (10^{-5}) (H 696)	250–270 service temperature	Good: 2 yr exposure in Florida caused yellowing; 5 yr caused 5% loss in τ	Good: comparable to acrylic
Plexiglas (acrylic)	1.49 (D 542)	125 mil 89.6 (± 0.3)	125 mil 79.6 (± 0.8)	125 mil 2.0 (est)[e]	3.9 (10^{-9}) at 60° F; 4.6 (10^{-6}) at 100° F	180–200 service temperature	Average to good: based on 20 yr testing in Arizona, Florida, and Pennsylvania	Good to excellent: resists most acids and alkalis
Teflon F.E.P. (fluorocarbon)	1.343 (D 542)	5 mil 92.3 (± 0.2)	5 mil 89.8 (± 0.4)	5 mil 25.6 (± 0.5)	5.9 (10^{-5}) at 160° F; 9.0 (10^{-5}) at 212° F	400 continuous use; 475 short-term use	Good to excellent: based on 15 yr exposure in Florida environment	Excellent: chemically inert
Tedlar P.V.F. (fluorocarbon)	1.46 (D 542)	4 mil 92.2 (± 0.1)	4 mil 88.3 (± 0.9)	4 mil 20.7 (± 0.2)	2.8 (10^{-5}) (D 696)	225 continuous use; 350 short-term use	Good to excellent: 10 yr exposure in Florida with slight yellowing	Excellent: chemically inert
Mylar (polyester)	1.64-1.67 (D 542)	5 mil 86.9 (± 0.3)	5 mil 80.1 (± 0.1)	5 mil 17.8 (± 0.5)	0.94 (10^{-5}) (D 696-44)	300 continuous use; 400 short-term use	Poor: ultraviolet degradation great	Good to excellent: comparable to Tedlar
Sunlite[f] (fiberglass)	1.54 (D 542)	25 mil (P) 86.5 (± 0.2) 25 mil (R) 87.5 (± 0.2)	25 mil (P) 75.4 (± 0.1) 25 mil (R) 77.1 (± 0.7)	25 mil (P) 7.6 (± 0.1) 25 mil (R) 3.3 (± 0.3)	1.4 (10^{-5}) (D 696)	200 continuous use causes 5% loss in τ	Fair to good: regular, 7 yr solar life; premium, 20 yr solar life	Good: inert to chemical atmospheres
Float glass (glass)	1.518 (D 542)	125 mil 84.3 (± 0.1)	125 mil 78.6 (± 0.2)	125 mil 2.0 (est)[d]	4.8 (10^{-5}) (D 696)	1350 softening point; 100 thermal shock	Excellent: time proved	Good to excellent: time proved

Note: See page 5.27 for footnotes.

TABLE 5.A2 Thermal and Radiative Properties of Collector Cover Materials (Continued)

Material name	Index of refraction (n)	τ (solar)[g] (%)	τ (solar)[a] (%)	τ (infrared)[b] (%)	Expansion coefficient (in/in · °F)	Temperature limits (°F)	Weatherability (comment)	Chemical resistance (comment)
Temper glass (glass)	1.518 (D 542)	125 mil 84.6 (± 0.1)	125 mil 78.6 (± 0.2)	125 mil 2.0 (est)[d]	4.8 (10^{-6}) (D 696)	450–500 continuous use; 500–550 short-term use	Excellent: time proved	Good to excellent: time proved
Clear lime sheet glass (low iron oxide glass)	1.51 (D 542)	Insufficient data provided by ASG	125 mil 87.5 (± 0.5)	125 mil 2.0 (est)	5.0 (10^{-6}) (D 696)	400 for continuous operation	Excellent: time proved	Good to excellent: time proved
Clear lime temper glass (low iron oxide glass)	1.51 (D 542)	Insufficient data provided by ASG	125 mil 87.5 (± 0.5)	125 mil 2.0 (est)	5.0 (10^{-6}) (D 696)	400 for continuous operation	Excellent: time proved	Good to excellent: time proved
Sunadex white crystal glass (0.01% iron oxide glass)	1.50 (D 542)	Insufficient data provided by ASG	125 mil 91.5 (± 0.2)	125 mil 2.0 (est)	4.7 (10^{-6}) (D 696)	400 for continuous operation	Excellent: time proved	Good to excellent: time proved

[a]Numerical integration ($\Sigma \tau_{avg} F_{\lambda_1 T - \lambda_2 T}$) for $\lambda = 0.2$–4.0 μM.

[b]Numerical integration ($\Sigma \tau_{avg} F_{\lambda_1 T - \lambda_2 T}$) for $\lambda = 3.0$–50.0 μM.

[c]All parenthesized numbers refer to ASTM test codes.

[d]Data not provided; estimate of 2% to be used for 125 mil samples.

[e]Degrees differential to rupture $2 \times 2 \times \frac{1}{4}$ in samples. Glass specimens heated and then quenched in water bath at 70° F.

[f]Sunlite premium data denoted by (P); Sunlite regular data denoted by (R).

[g]Compiled data based on ASTM Code E 424 Method B.

[h]Abstracted from Ratzel, A. C., and R. B. Bannerot, Optimal Material Selection for Flat-Plate Solar Energy Collectors Utilizing Commercially Available Materials, presented at ASME-AIChE Natl. Heat Transfer Conf., 1976.

SOURCE: F. Kreith and J. F. Kreider, Principles of Solar Engineering, Hemisphere Publishing Corporation, Washington, D.C., 1978. Used with permission.

Chapter 6

Energy Storage for Solar Applications

CHARLES J. SWET
Consulting Engineer
Mt. Airy, MD

GENERAL ATTRIBUTES

Needs and Roles

Storage is essential to any system that depends largely on solar-derived energy or that needs such energy at specific times. It adjusts temporal mismatches between the load and an intermittent or variable energy source, thereby improving system operability and utility and permitting increased solar contribution to the total energy demand.

System operability can be improved by a small amount of buffering storage for transient smoothing. An example is the addition of thermal mass between the collector and turbine of a hybrid solar thermal power plant to bridge brief periods of inadequate insolation caused by cloud passage (see Fig. 6.1a), thereby reducing the cycling frequency of the fossil backup.

System utility can be improved by providing enough storage to shift the period of energy delivery from the usual period of source availability to one of greater demand. With more thermal mass, the example power plant of Fig. 6.1 could delay the generation of solar-derived power to follow peak loads later in the day (see Fig. 6.1b).

In the examples of Fig. 6.1a and b, storage does not significantly increase f_s beyond no-storage values (f_s is the solar load fraction). The same system could alternatively use storage to lengthen the period of solar-derived power generation, thereby increasing f_s but requiring a larger solar collector (see Fig. 6.1c). It would then no longer be called a hybrid. Larger amounts of storage could extend the period of solar power production to 24 h or longer, with further increases in both f_s and collector size. Still larger amounts would permit yearly averaging of solar input and load, allowing f_s to approach unity (no backup requirement) without greatly increased collector size (see Fig. 6.1d).

Modes and Applications

The four basic modes of technological (nonbiological) energy storage are thermal, electrical, mechanical, and chemical, for example, molecular hydrogen. Thermal energy can be stored as sensible heat, as latent heat of phase change, or in reversible chemical reactions. Electrical energy is usually stored electrochemically in batteries. Mechanical energy can be stored inertially as kinetic energy in flywheels or as potential energy in compressed air or elevated liquids.

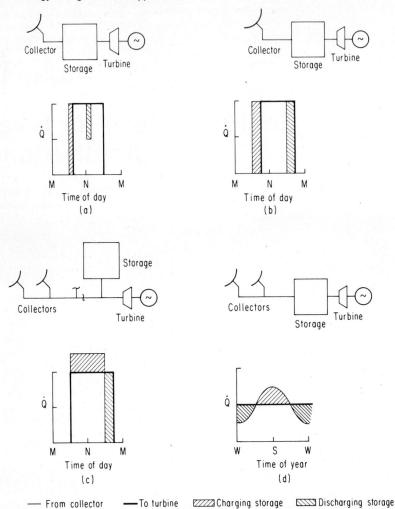

Fig. 6.1 Examples of storage strategies applied to solar thermal power. (*a*) Buffering—smoothes transients with little effect on f_s or collector size; (*b*) output time shift—delays period of power production with little effect on f_s or collector size; (*c*) output time extension—permits power generation during and beyond sunlight period with increased f_s and collector size; (*d*) yearly averaging—large amount of storage permits continuous power generation with unity f_s (no backup required) and collector sized for mean yearly insolation.

Hydrogen may be stored as a compressed gas, a cryogenic liquid, a hydride, or with fixed nitrogen as liquid ammonia or urea. These four modes of storage can be integrated into solar energy systems in various ways (see Fig. 6.2) to produce heat, cold, electrical and mechanical power, and hydrogen fuel or feedstock.

Generic Characterizations

Viewed as an element in the larger solar energy conversion system, any storage system can be broadly characterized by the following parameters:

Form of energy in and out (thermal, electrical, mechanical, H_2)

Quantity of energy in and out (first-law efficiency)

Quality of energy in and out (second-law efficiency)

Input and output power

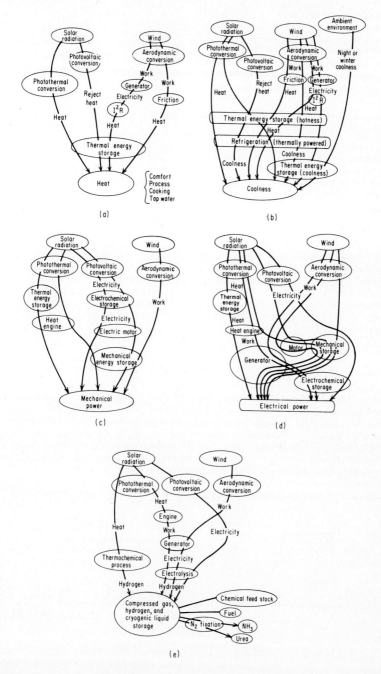

Fig. 6.2 Energy paths through storage: (*a*) heating applications; (*b*) cooling applications; (*c*) shaft power applications; (*d*) electrical power applications; and (*e*) hydrogen applications.

Storage cycle duration
Energy density per unit dedicated volume
Investment cost per unit energy out
Investment cost per unit input and output power
Operating and maintenance costs

Specific and derived parameters peculiar to each storage mode and application are applied where appropriate in subsequent sections.

THERMAL ENERGY STORAGE AS SENSIBLE AND LATENT HEAT

Thermal energy in the form of "hotness" or "coldness" can be stored in various media as sensible heat (temperature change), as latent heat (isothermal phase change), or by a combination of the two. It is often stored in identifiable subsystems that typically include a storage medium, container, insulation, heat exchangers, heat-transfer fluid, pumps or blowers, and controls. For such systems the first-law efficiency over a complete storage cycle is

$$\eta_1 = \frac{Q_{out}}{Q_{in}} = \frac{Q_{in} - \text{losses}}{Q_{in}}$$

Efficiency based on second-law considerations of energy quality or availability is variously expressed in terms of absolute input temperature T_{in}, output temperature T_{out}, and a sink temperature T_{sink}. For systems that convert heat to mechanical or electrical power, it is expressed most rigorously as the ratio of mechanical or electrical energy produced by heat from storage to that which could have been produced during the same period by bypassing storage. It can also be expressed as the ratio of ideal Carnot efficiencies integrated over the discharge period, or simply as

$$\eta_2 = \frac{T_{in}(T_{out} - T_{sink})}{T_{out}(T_{in} - T_{sink})}$$

if T_{in} and T_{out} do not change (see Chap. 4 for other second-law efficiency formulas).

Isothermal discharge from storage is desirable for reasons of solar collection efficiency as well as second-law considerations (see Fig. 6.3). This condition is approached naturally in systems that store thermal energy primarily by the latent heat of phase change, but the decay of outlet temperature can also be minimized in sensible-heat systems by stratification or by the use of multiple storage vessels, as shown later. Spatially uniform temperature within a storage system is not always desirable, particularly in solar thermal steam-power-plant applications (see Fig. 6.4) where substantial temperature gradients are beneficial.

Heat storage can also be dispersed throughout the overall system, as in the thermal mass of a building, with relatively undefined subsystem boundaries. In such cases, storage subsystem

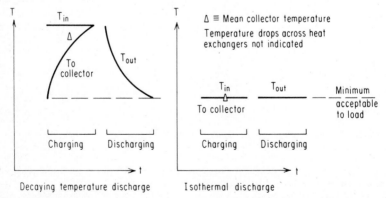

Fig. 6.3 Benefit of isothermal discharge from storage. Collector performance is improved by low mean collection temperature.

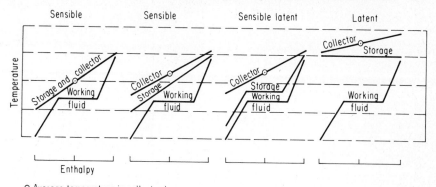

○ Average temperature in collector loop

Fig. 6.4 Thermal energy storage alternatives for subcritical Rankine cycle solar power plants. Mean collection temperature is minimized by (1) large maintained ΔT in storage medium and (2) using the same medium for heat transport and storage.

performance ordinarily is not separately evaluated, and storage media parameters such as ρ, k, c_p, and heat of fusion ΔH_s are integrated into the overall system model.

Sensible-Heat Storage

The sensible heat ΔQ gained or lost by a material in changing temperature from T_1 to T_2 is

$$\Delta Q = m \int_{T_1}^{T_2} c_p \, dT = V \int_{T_1}^{T_2} \rho c_p \, dT$$

Although ρ and c_p vary with temperature, average values are usually assigned so that $\Delta Q = c_p \, \Delta T$ per unit mass or $\rho c_p \, \Delta T$ per unit volume. Commonly used figures of merit are $\$/c_p$ (for relative media cost), ρc_p (volumetric heat capacity using the lowest value of ρ at the highest expected operating temperature, for relative container size), and k (for relative heat-exchanger size).

Solid media Table 6.1 characterizes some plentiful and economically competitive solid materials for sensible-heat storage. Common applications are rock-bed air heating (see Chap. 12) and the "Trombe wall" (see Chap. 16). Long-duration underground storage in undisturbed rock and dry earth may also be feasible, but validated system design data are presently unavailable.

Liquid media Table 6.2 characterizes some plentiful and economically competitive liquids for sensible-heat storage. Of these, water is clearly superior for temperatures below 100°C (212°F) for solar space and water heating, for example. Water sometimes remains economically competitive at higher temperatures despite the need for pressure containment, especially when stored in aquifers (treated separately in a later section). The cost of antifreeze in large water tanks is seldom justified, since only the collector loops (which can be isolated) commonly require protection. Oils and molten salts are used mainly in solar thermal power applications (see Fig. 6.5).

Dual media Solid and liquid sensible-heat storage materials can be combined in various ways. Rock beds and water tanks have been used jointly in hybrid storage systems for solar space heating. Rocks and oil have been used in a single vessel for solar thermal power applications (see Fig. 6.6), to improve stratification and reduce the required amount of relatively expensive liquid. This is an alternative to the multiple tanks shown in Fig. 6.5.

Effects of stratified storage on system performance In practice, the extent of stratification is poorly predictable and never complete, but its effects can be examined in the limit by comparing hypothesized fully stratified, fully mixed, and infinite-capacity conditions. Two distinct types of stratification are considered here: time ordered ("first in-first out," as in a rockbed) and temperature ordered ("hottest at top, coldest at bottom," as in an unagitated water tank). Fully mixed storage has spatially uniform temperature, as in a well-stirred water tank, instantaneously reflecting heat addition to and removal from storage. With infinite storage capacity, the storage temperature always equals that of the return load. The three distinct ways

TABLE 6.1 Solid Media for Sensible-Heat Storage

Storage medium	Material cost, $/kg ($/lb_m)	Density ρ, kg/m³ (lb_m/ft³)	Heat capacity c_p, J/kg·°C (Btu/lb_m·°F)	Thermal conductivity k, W/m·°C (Btu/h·ft·°F)	ρc_p, MJ/m³·°C (Btu/ft³·°F)	Cost/c_p, $·°C/kJ ($·°F/Btu)	Remarks (see footnotes)
Concrete (sand and gravel)	0.02 (0.01)	2240 (140)	1130 (0.27)	0.9–1.3 (0.5–0.75)	2.53 (37.8)	0.02 (0.04)	4
Rocks (granite)	0.01 (0.005)	2640 (165)	880 (0.21)	1.7–4.0 (1–2.3)	2.32 (34.7)	0.015 (0.03)	1, 4
Cast-iron brick	0.66 (0.30)	7900 (494)	837 (0.20)	29.3 (16.9)	6.62 (98.6)	0.79 (1.5)	2, 3, 4
Magnesia brick	0.32 (0.15)	3000 (187)	1130 (0.27)	5.07 (2.93)	3.39 (50.5)	0.284 (0.56)	3, 4, 5, 6

1. Typical values for 0 to 100°C, various sources.
2. Requires dry, nonoxidizing containment.
3. Mean values for 550 to 816°C range, from Ref. 2.
4. No allowance for voids; 1979 $.
5. Dusting problem with gas turbines.
6. Carbon steel OK, but requires internal insulation for temperatures > 300°C.

TABLE 6.2 Liquid Media for Sensible-Heat Storage

Media	Temperature Melting point °C (°F)	Maximum @ 1 atm °C (°F)	Material cost, $/kg ($/lb$_m$)	ρ @ maximum temperature, kg/m³ (lb$_m$/ft³)	c_p, J/kg·°C (Btu/lb$_m$·°F)	k, W/m·°C (Btu/h·ft·°F)	ρc_p, MJ/m³·°C (Btu/ft³·°F)	Cost/c_p, $·°C/kJ ($·°F/ Btu)	Remarks (see Notes below where indicated by letters)
Water	0 (32)	100 (212)	Nil	960 (60.0)	4200 (1.0)	38°C 0.63 (100°F) (0.363)	4 (60)	Nil	(a) 120°C(249°F) at 2 atm 134°C(273°F) at 3 atm
Therminol 66 ®	−27* (−18)*	343 (650)	2.0 (0.93)	750 (46.8)	343°C 2740 (650°F) (0.655)	343°C 0.106 (650°F) (0.0613)	2.05 (30.7)	0.75 (1.42)	Modified terphenyl ®Monsanto. *Pour point (a), (d)
Dowtherm A®	12 (54)	260 (500)	0.22 (0.10)	867 (54.1)	2200 (0.526)	260°C 0.112 (500°F) (0.0645)	1.91 (28.6)	0.1 (0.19)	73.5 diphenyl oxide, 26.5 diphenyl ®Dow. (a), (d) 292°C(557°F) at 2 atm 317°C(603°F) at 3 atm
Hitec®	142 (288)	540 (1000)	0.44 (0.20)	1680 (105)	540°C 1560 (100°F) (0.37)	0.61 (0.35)	2.62 (39.0)	0.28 (0.54)	40 NaNO$_2$, 7 NaNO$_3$, 53 KNO$_3$ ®Dupont. Also called HTS (b), (d); Ref. 3
Draw salt	220 (428)	540 (1000)	0.25 (0.11)	1733 (108)	1560* (0.37)*	0.57 (0.33)	2.70 (40.3)	0.16 (0.30)	46 NaNO$_3$, 54 KNO$_3$ *Assumed same as Hitec (b), (d); Ref. 3
Sodium	98 (208)	760 (1400)	0.57 (0.26)	960 (60)	1260 (0.30)	67.5 (39)	1.21 (18)	0.45 (0.87)	(c), (d)

Notes: (a) Carbon steel OK.
(b) Carbon steel OK to about 315°C(600°F). Special alloys above.
(c) Special alloy containment; nuclear technology.
(d) Requires inert atmosphere blanket.
Costs are in 1979 $.

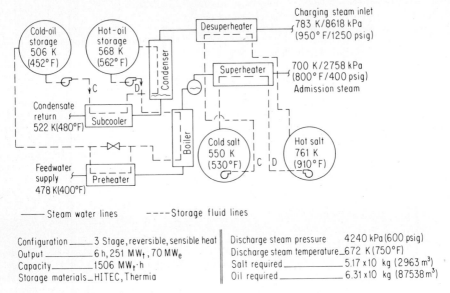

——— Steam water lines   - - - -Storage fluid lines

Configuration ____ 3 Stage, reversible, sensible heat
Output ____ 6 h, 251 MW$_t$, 70 MW$_e$
Capacity ____ 1506 MW$_t$·h
Storage materials _ HITEC, Thermia

Discharge steam pressure   4240 kPa (600 psig)
Discharge steam temperature _ 672 K (750°F)
Salt required ____ 5.17 x 10 kg (2963 m³)
Oil required ____ 6.31 x 10 kg (87538 m³)

Fig. 6.5 Liquid sensible heat storage system for solar thermal power plant. A subscale prototype of this system concept was built and tested at the Georgia Institute of Technology Engineering Experiment Station.

in which stratification improves system performance can be described in terms of heat delivery per unit collection area (see Fig. 6.6).

Collection efficiency is equally improved by both time-ordered and temperature-ordered stratification when there is no *load draw during collection and storage charging.* For constant load draw during collection, the improvements are also equal, but smaller. In both cases, stratification delays warming of the fluid returning from storage to the collector inlet, thereby reducing collector losses. The improvement is greater for air-cooled collectors because of their relatively large coolant temperature rise and the correspondingly large effect of inlet temperature reduction.

Both types of stratification also produce equal increases in *nighttime* delivery to load and in the corresponding daytime collector output to storage alone. These effects are the same for intermittent (shown) and continuous (not shown) nighttime load draw, and are maximal for domestic hot-water or warm-air space-heating systems with effective heat transfer to load. The stratified system delivers its stored heat at high temperature, hence with minimal load flow, leaving the collector inlet cooler in the morning so that more heat can be collected for delivery at night.

Temperature-ordered stratification is uniquely effective in systems with intermittent daytime load draw, such as water heating for a school or laundry. After an hour or so of solar collection with little or no load draw, even the relatively cool bottom portion of a temperature-ordered tank may be substantially warmer than the cold-water supply. With the advent of a large load draw, the incoming cold water goes directly to the collector, quickly increasing its efficiency. At the same time, the hottest water available is delivered to load, so that both collection and delivery of heat are simultaneously maximized. (A similar increase in collector efficiency would occur with time-ordered stratification, but heat delivered to load would be less because of the decreasing collector outlet temperture.) Moreover, if the load draw is large enough and/or the prior water temperature in the tank bottom is high enough, the collector outlet coolant may be directed to the tank bottom to cool the pool of available collector coolant. After the load draw ends, the collector can then continue to operate at enhanced efficiency until the water in the tank bottom has again reached the prior high temperature. Thus the benefits of stratification can be enhanced by appropriate flow control strategies.

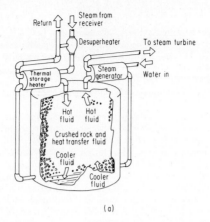

(a)

1 Day 337 H.M.S. 16.54.39	2 Day 337 H.M.S. 17.19.58	3 Day 337 H.M.S. 17.51.36
4 Day 337 H.M.S. 18.25.00	5 Day 337 H.M.S. 18.56.55	6 Day 337 H.M.S. 19.30.00
7 Day 337 H.M.S 20.00.00	8 Day 337 H.M.S. 20.31.55	9 Day 337 H.M.S. 21.09.42

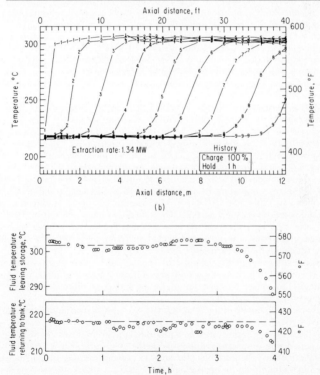

Fig. 6.6 Dual media storage system for solar thermal power plant. (*a*) Dual media concept. Poor diffusion through rocks maintains good stratification (sharp thermocline). (*b*) Typical extraction thermoclines. From tests of subscale prototype unit. (*c*) Typical performance data during extraction. (*Courtesy Rocketdyne, Division of Rockwell International.*)

Latent-Heat Storage

A material can gain or lose heat by isothermal phase change between solid and liquid (heat of fusion Δh_f), liquid and vapor (heat of vaporization Δh_v), or solid crystalline phases (heat of solid-solid transition Δh_s). This method of storing heat is particularly advantageous in applications that call for small storage volume with small temperature excursions.

Heat of fusion   Table 6.3 lists the properties of some heat-of-fusion materials that have been shown to be suitable for building heating and cooling, either alone or in combination with other substances.

Ice is an obvious choice for seasonal storage of coolness from the winter ambient environment, and many ways of improving on traditional methods of gathering and storing ice have been explored. These include water sprays, cooling towers, immiscible fluids, heat pipes, vibrators, and scrapers, but the economic feasibility of such refinements remains to be demonstrated. Ice is not a good choice for storing the coolness produced by absorption chillers because its low freezing temperature would excessively degrade the chiller performance.

Calcium chloride hexahydrate ($CaCl_2 \cdot 6H_2O$) melts congruently, i.e., the anhydrous salt and lower hydrates are soluble in free water, therefore settling of solids does not occur and the initial heat of fusion is retained after many thermal cycles. It can be packaged in plastic bottles, in steel cans, and as "chub sausages" in plastic-laminate films, with barium hydroxide added as a nucleating agent to minimize supercooling.[5] It can also be infused into sealed porous cement blocks and tiles for use as building components with barium chloride as the nucleating agent. Pumice cement blocks consisting of 60 percent by weight nucleated $CaCl_2 \cdot 6 H_2O$, with urethane coal-tar sealant, have been used in Trombe-wall construction for passive solar heating systems. Foamed cement tiles containing 80 percent by weight phase-change material, with polyester sealant, can be ceiling-mounted for passive maintenance of comfort-zone temperatures, with the melting point lowered and spread to 19 to 25°C (66 to 77°F) by adding 6 percent by weight saturated solution of sodium chloride. Such blocks and tiles have experienced more than 1000 melt-freeze cycles without observable degradation of thermal or mechanical properties[6] but problems of long-term degradation caused by salt-cement chemical incompatibilities remain to be solved.

Sodium sulfate decahydrate ($Na_2SO_4 \cdot 10H_2O$), also called Glauber's salt, has been widely investigated because of its exceptionally low cost and high energy density. It melts incongruently, and in its pure form loses much of its latent heat after a few cycles because of settling of undissolved solids. It also tends to supercool, but the supercooling can be reduced by adding 3 percent by weight borax and settling can be prevented for at least 1000 cycles by adding 7 percent by weight attapulgite clay thickener in a specific sequence.[7] The nominal melting temperature is too low for most active heating systems, but can be lowered to the comfort zone or to coolness storage temperature by adding inexpensive salts such as sodium and ammonium chloride as shown in Table 6.3.[1,8] Glauber's salt mixtures have been packaged in plastic trays, tubes, and laminated films (see Fig. 6.7). They have also been encased in polymer concrete ceiling tiles, but with only about 30 percent retained heat of fusion after repeated cycling. Clathrates and waxes have also been investigated for coolness storage, but none has yet been shown to be economically competitive.

Paraffin waxes melt congruently without supercooling, can be packageable in metal containers, and can be formulated to melt over a wide range of heating as well as cooling temperatures but these advantages are offset by their low densities and poor thermal conductivities and by considerations of resource depletion.

Sodium thiosulfate pentahydrate ($Na_2S_2O_3 \cdot 5H_2O$), also known as STP or photographer's hypo, is used in the University of Delaware Solar 1 house solar heating system, packaged in plastic trays. Supercooling is prevented by a proprietary nucleating device. Melting is nominally incongruent, since the dihydrate is insoluble in the free water, but in typical solar heating applications the melting rate is sufficiently rapid to prevent settling. Candidate phase-change materials for higher temperatures up to about 100°C (212°F) include other inorganic salt hydrates and their eutectics, waxes, urea eutectics, fatty acids, and naphthalene, but the economic attractiveness of phase change storage at those temperatures seems small.

Many organic polymers with high heats of fusion melt at temperatures in the range of 110 to 130°C (230 to 266°F), which is suitable for hot-side storage in solar absorption or Rankine cooling systems. When high-density polyethylene (HDPE) is partially cross-linked it retains nearly all of its original heat of fusion, but passes through the phase transition without changing form.[11]. Form-stable pellets of this and similar materials "melt" without flowing or agglomerating, and can be used in packed beds with direct heat transfer to a gas or liquid. For more

Material	Melt temperature, °C (°F)	Cost, $/kg ($/lb$_m$)	Heat of fusion Δh_f, kJ/kg (Btu/lb$_m$)	Density ρ, kg/m³ (lb$_m$/ft³) Solid	Liquid	Heat capacity c_p, J/(kg·°C) (Btu/(lb$_m$·°F)) Solid	Liquid	Thermal conductivity k, W/m·°C (Btu/h·ft·°F) Solid	Liquid	$\rho\Delta h_f$, MJ/m³ (Btu/ft³)	Cost/Δh_f, $/MJ ($/10² Btu)	Remarks	Reference
Ice	0 (32)	Nil	334 (144)	920 (57.5)	1000 (62.4)	5270 (1.26)	4220 (1.01)	0.62 (0.36)	2.26 (1.30)	308 (8250)	Nil		
Glauber's salt eutectic	13 (55)	0.07 (0.03)	146 (63)	1470 (92)	N/A	1420 (0.34)	2680 (0.64)	N/A	N/A	215 (5800)	0.5 (0.5)	By wt: 74.5 Na$_2$SO$_4$·10 H$_2$O/6.7 NaCl/6.2 NH$_4$Cl/2.6 Borax nucleator/7.9 thickener/1.8 Boric acid pH stabilizer/0.3 tetrasodium pyrophosphate dispersing agent	1, 8
CaCl$_2$ · 6H$_2$O	27 (81)	0.07 (0.03)	190 (82)	1800 (112)	1560 (97)	1460 (0.35)	2130 (0.51)	1.09 (0.63)	0.54 (0.39)	296 (7950)	0.4 (0.4)		5
Na$_2$SO$_4$ · 10H$_2$O (Glaubers salt)	32 (89)	0.04 (0.02)	225 (97)	1460 (91)	1330 (83)	1760 (0.42)	3300 (0.79)	2.25 (1.30)	N/A	300 (8050)	0.2 (0.2)	With nucleator and thickener	7
Paraffin wax (Sunoco P-116)	47 (117)	0.15 (0.07)	209 (90)	820 (51)	770 (48)	2890 (0.69)	2510 (0.60)	0.14 (0.08)	N/A	161 (4300)	0.8 (0.8)		12
Na$_2$S$_2$O$_3$ · 5H$_2$O	48 (118)	0.18 (0.08)	209 (90)	1650 (103)	N/A	1460 (0.35)	2380 (0.57)	0.57 (0.33)	N/A	345 (7200)	0.9 (0.9)		8, 10
MgCl$_2$ · 6H$_2$O	120 (247)	0.15 (0.07)	169 (73)	1560 (97)	N/A	1590 (0.38)	2240 (0.68)	N/A	N/A	250 (7000)	0.9 (1.0)		1
Cross-linked, high-density polyethylene	132 (270)	0.57 (0.26)	230 (99)	960 (60)	900 (56)	2500 (0.6)	2500 (0.6)	0.36 (0.21)	0.36 (0.21)	207 (5550)	2.5 (2.6)	Electron beam cross-linking. Form-stable pellets. Heat released at 120°C (248°F)	11

*N/A means not available.

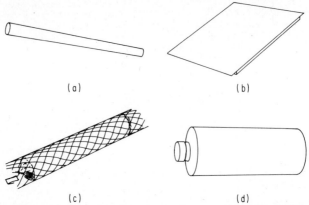

Fig. 6.7   Packaging concepts for low-temperature phase change materials. (*a*) Plastic tube for Glauber's salt mixture, used in Solar 1 house; (*b*) plastic tray for sodium thiosulfate pentahydrate, used in Solar 1 house; (*c*) "Chub" sausages of plastic laminate film; (*d*) plastic bottles.

conventional packaging, $MgO_2 \cdot 6H_2O$ is one of the few promising storage materials for solar cooling.

Many interrelated external considerations enter into the selection of a specific heat-of-fusion material and packaging concept, and into the choice between phase-change and sensible-heat storage. These include parallel questions of space heating vs. cooling, tap-water heating, air- vs. liquid-cooled collectors, warm-air vs. hydronic heating, Rankine vs. absorption cooling, the need for coolness storage, the importance assigned to volume, new construction vs. retrofit, active vs. passive space conditioning, storage duration, and system size. Table 6.4 summarizes three comparisons of solar-heating volumes using various phase-change and sensible-heat materials illustrating the wide disparity between presumably comparable estimates.

The heat-transfer salt HTS, or Hitec® (40 percent $NaNO_2$, 7 percent $NaNO_3$, and 53 percent KNO_3 by weight) is listed in Table 6.2 as a high-temperature liquid for sensible-heat storage, but its properties near the freezing point—81 kJ/kg (35 Btu/lb$_m$) at 142°C (288°F)—also qualify it as a candidate heat-of-fusion material for absorption cooling.

Table 6.5 lists the properties of some heat-of-fusion storage materials that melt at higher temperatures suitable for power generation or industrial processes.

Mixtures of sodium nitrate (Chile saltpeter) and sodium hydroxide (caustic soda) have been extensively investigated as storage media for Rankine-cycle power generation with evaporation temperatures in the 250 to 300°C (482 to 572°F) range. Sodium nitrate with 1 percent by weight NaOH freezes as a slush that can be scraped off the steam-generation heat exchanger. This

TABLE 6.4   Relative Storage System Volumes for Solar Space Heating

Source		Ref. no. 8	Ref. no. 10	Ref. no. 12†	
Storage medium \	Heat-transfer medium	Not specified	Liquid*	Liquid	Air
Rocks		17			4
Water		8	6	2	
P-116 wax			2.6	2+	2
$Na_2S_2O_3 \cdot 5H_2O$			1		
$Na_2SO_4 \cdot 10H_2O$		1		1	1

*50–50 water/glycol, 11.1°C (20°F) ΔT.
†ΔT as needed to satisfy equal thermal loads; very slight differences in f_s resulting from choice of medium.

TABLE 6.5 Heat-of-Fusion Storage Materials for High-Temperature Applications

Material	Melt temperature, °C (°F)	Cost, $/kg ($/lbm)	Heat of fusion Δhf, kJ/kg (Btu/lbm)	Density ρ, kg/m³ (lbm/ft³) Solid	Liquid	Heat capacity cp, J/(kg·°C) (Btu/(lbm·°F)) Solid	Liquid	Thermal conductivity k, W/(m·°C) (Btu/(h·ft·°F)) Solid	Liquid	ρΔhf, MJ/m³ (Btu/ft³)	Cost/Δhf, $/MJ ($/10² Btu)	Remarks	Reference
NaNO₃	307 (585)	0.17 (0.08)	181 (78)	2260 (141)	1900 (119)	1880 (0.45)	1840 (0.44)	0.57 (0.33)	0.61 (0.35)	344 (9280)	1.0 (1.0)	M5 OK. Freezes as slush with 1% added NaOH.	1, 13
NaOH	318 (605)	0.33 (0.15)	315 (136)	2030 (127)	1760 (108)	2000 (0.48)	2090 (0.50)	0.92 (0.53)	0.92 (0.53)	555 (14,700)	1.0 (1.1)	M5 OK with corrosion inhibitor. Freezes as slush with 1% added NaNO₃.	1, 13, 14
Mixed chlorides	385 (725)	0.20 (0.09)	292 (126)	2250 (140)	1630 (102)	960 (0.23)	1040 (0.25)	1.5–1.6 (0.87–0.92)	0.81–1.0 (0.47–0.59)	476 (12,900)	0.7 (0.7)	14.5 KCl/22.3 NaCl/ 63.2 MgCl₂ by wt. OK in sealed M5 cans.*	1, 15
Mixed carbonates	505 (941)	0.88 (0.40)	343 (148)	2130 (133)	2010 (125)	1340 (0.32)	1760 (0.42)	N/A	1.89 (1.09)	690 (18,500)	2.6 (2.7)	35 Li₂CO₃/65 K₂CO₃ by wt. 304 S5 OK.†	25
Aluminum	660 (1222)	1.17 (0.53)	400 (172)	2565 (160)	2387 (148)	1290 (0.308)	N/A	211 (122)	91 (53)	955 (25,500)	2.9 (3.1)	1978 material price. Limited experience with Inconel 718 and 600 in Omnium-G collector.‡	26
Mixed fluorides	832 (1530)	0.18 (0.08)	615 (265)	2570 (161)	2090 (131)	1420 (0.34)	1380 (0.33)	4.2–8.4 (2.4–4.8)	4.65 (2.69)	1290 (34,700)	0.3 (0.3)	67 NaF/33 MgF₂ by wt. From "free" fluosilicic acid. 316 S5 OK with Al getter.*	25, 27

*Handling difficult and hazardous. Must have near-zero O₂ and H₂O, with good quality control on container fabrication and sealing.
†High material cost offset by high energy density and relative ease of containment when used for buffering storage.
‡High thermal conductivity and consequently low HX cost offsets high material cost.

approach was used in a 31.9-MWh (thermal) subscale research experiment, without superheat storage, for a 10-MW (electric) central receiver pilot plant.[13] Tank and tube materials were carbon steel. A blend of 91 percent by weight NaOH with 8 percent $NaNO_3$ and 1 percent anticorrosion constituents is called Thermkeep®. Its latent heat is distributed over a temperature range (see Fig. 6.10) that closely matches the T-H (temperature–enthalpy) diagram of a toluene Rankine cycle. A 0.89-MWh (thermal) unit is being developed for the Albuquerque solar total-energy test facility (see Fig. 6.8).[14]

A ternary chloride eutectic melts at a temperature suitable for steam generation at a higher pressure. A 2-MWh (thermal) unit has been built that contains the mixed chlorides in multiple steel cans, with vaporizing m-terphenyl as the heat exchange and transport fluid.[15] All of the storage capacity is in the heat of fusion.

For still higher temperatures various carbonates, fusible metals, and metal fluorides show promise. Those with which most experience has been accumulated are characterized in Table 6.5.

Heat of vaporization   Storage as latent heat of evaporation of a liquid offers attractively high energy per unit mass of storage medium, but containment of the necessarily large volumes of vapor tends to be excessively expensive. An exception is the use of "steam accumulators" (see Fig. 6.9), which heat pressurized water by direct contact with steam, then deliver flashed steam when needed by slightly reducing the pressure.[16] For steam solar thermal power systems, an important advantage is that the thermal-storage medium can also be the power-cycle working fluid and the collector coolant. Underground accumulators in excavated caverns have been proposed for extended-duration thermal storage.

Heat of solid-solid transition   Some materials accept and reject heat through isothermal changes in their solid crystalline structure. Examples are anhydrous sodium hydroxide and

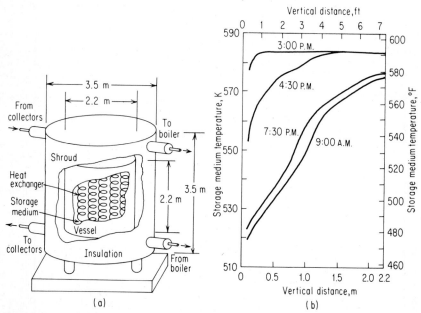

Fig. 6.8 Preliminary design for an experimental Thermkeep® unit for solar total energy system. (*a*) Configuration. Helical heat exchanger tubes contain Therminol for heat transport and transfer between collectors and toluene boiler for Rankine cycle. Unit contains 15,000 kg of storage medium. (*b*) Calculated variation in storage temperature over a daily cycle for Thermkeep® thermal storage unit. Daily cycle is as follows: 6:00 A.M.—power system on, operates from thermal storage; 7:30 A.M.—solar collectors on, power system operates from storage and collectors; 9:00 A.M.—solar collectors begin to provide excess output recharging of storage begins; 3:00 P.M.—solar collectors become inadequate, power system operates from storage and collectors; 4:30 P.M.—solar collectors off, power system operates from storage; 6:00 P.M.—power system off, begin overnight standby period. (*Courtesy Comstock & Wescott, Inc.*)

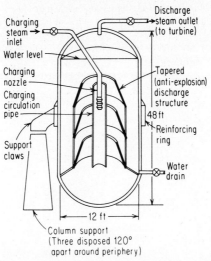

Fig. 6.9 Steam accumulator thermal-storage vessel. Systems using such devices may be designed to deliver constant or varying pressure steam from storage.

sodium sulfate, but these materials have greatest use in applications where most of the storage is in heat of fusion and/or sensible heat (see next section). Form-stable cross-linked polymers are more precisely described as heat-of-fusion materials.

Combined Sensible and Latent Heat Storage

Certain thermal energy storage media that experience large temperature excursions also undergo solid-liquid and/or solid-solid transitions, so that the total amount of heat gained or lost is the sum of the changes in sensible and latent heat. Figure 6.10 compares the cost and volumetric

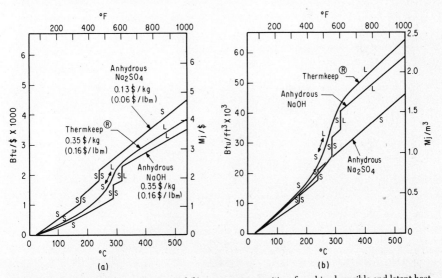

Fig. 6.10 (a) Volumetric heat capacities and (b) storage cost capacities of combined sensible and latent heat storage materials. (All 1978 prices. NaOH and Thermkeep® components are FOB destination freight equalized, in 55-gal drums. Na_2SO_4 price is bulk at farthest U.S. destination. Price at origin is about 0.07 $/kg (0.032 $/lb$_m$) bulk or 0.09 $/kg (0.04 $/lb$_m$) in 100-lb$_m$ bags.)

heat capacities of sensible/latent heat storage materials that have been used or proposed for solar thermal power applications. The relative merit of these media also depends largely on how well they match the temperature-enthalpy diagrams of specific heat-engine cycles (see Fig. 6.4).

Seasonal Storage in Aquifers

Aquifers are geologic formations which contain and conduct water. They are abundant in many parts of the world at depths ranging from a few meters to several kilometers, and may be fresh or saline depending on depth and geographic region (see Figs. 6.11 and 6.12). Confined aquifers

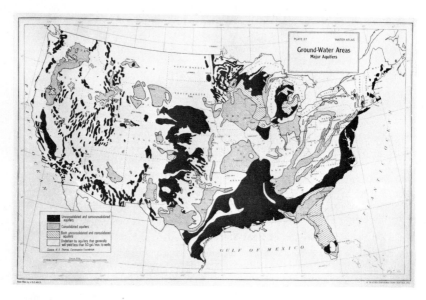

Fig. 6.11   Ground-water areas—major aquifers. (*Courtesy Water Information Center, Inc.*)

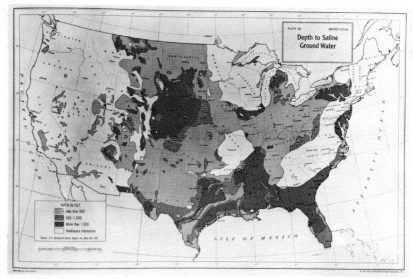

Fig. 6.12   Depth to saline ground water. (*Courtesy Water Information Center, Inc.*)

are those which are bounded above and below by impermeable layers and are saturated by water under pressure.

The natural flow is usually very small—on the order of a meter per year for aquifers deeper than those generally used as sources of potable water. These factors favor the use of confined aquifers for long-duration storage of large amounts of heated or chilled water (see Fig. 6.13). Thus water can be heated by the sun and placed in storage throughout the year, and withdrawn as needed for comfort cooling in the summer. A combined heating and cooling concept has been proposed that uses four thermally isolated sections of an aquifer (see Fig. 6.14).

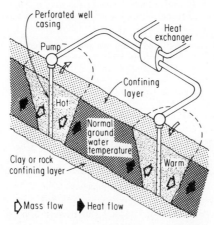

Fig. 6.13   Doublet heat storage in an aquifer (heat withdrawal phase shown).

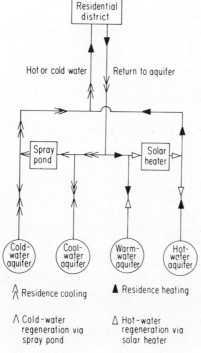

Fig. 6.14   Schematic of the Solaterre system.

Analytical models of aquifer storage systems have been developed and are being validated by full-scale experiments using natural aquifers. Table 6.6 lists the representative property parameters used in an analytical model of hot-water storage, the results of which are shown in Fig. 6.15. For systems of about this capacity, the estimated investment cost including wells, pumps, and associated piping is less than 1 ¢/MJ.

Insulation

Desirable characteristics for thermal insulation include low thermal conductivity, relatively unchanging with temperature; high thermal capacity, so that additional thermal storage can be obtained; no deterioration at working temperature; and ease of handling. The insulation must also be available and reasonably inexpensive. Most industrial insulating materials are made from

TABLE 6.6   Property Parameters Used in Hot-Water Storage Model

Property	Reservoir (sandstone)	Caprock-bedrock (mudstone)
Porosity	0.20	1×10^{-20}
Density, kg/m^3	2.6×10^3	2.7×10^3
Heat capacity, J/kg · °C	9.70×10^2	9.30×10^2
Thermal conductivity, W/m · °C	2.894	1.157
Permeability, m^2	1×10^{-13}	1×10^{-40}
Specific storage, m^2/N	1×10^{-3}	1×10^{-12}

Fluid parameters

Viscosity, cP	T, °C	Heat capacity, J/kg · °C	T, °C
1.005	20	4.127×10^3	25
5.45×10^{-1}	50	3.894×10^3	75
2.80×10^{-1}	100	3.652×10^3	125
1.82×10^{-1}	150	3.341×10^3	200
1.35×10^{-1}	200		

Expansivity, °C^{-1} 3.17×10^{-4}

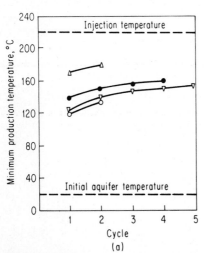

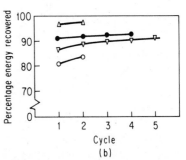

Δ Daily cycle (24 h)
● Semiannual cycle, partial penetration
▽ Semiannual cycle, full penetration
o Yearly cycle, full penetration

Fig. 6.15 (a) Temperature at the end of each production period (minimum production temperature) versus cycle number. (b) Percentage of energy recovered over energy injected versus cycle number.

the following basic materials: asbestos, magnesium carbonate, diatomaceous silica, vermiculite, rock wool, glass wool, cork, cattle hair, and wool.

Eighty-five percent magnesia (a mixture of approximately 85 percent magnesium carbonate and 15 percent asbestos fiber) is the most commonly used material for insulation applications up to 315°C (600°F). In addition to having a low thermal conductivity, 85 percent magnesia is light in weight, easily cut and fitted, unaffected by steam or water leakage, and strong enough to withstand ordinary use. Properties of this and other commonly used insulating materials are given in Table 6.8.

Heat loss through insulation   The rate of heat loss through insulation applied to cylindrical surfaces can be calculated by

$$Q = \frac{k2\pi L(T_s - T_a)}{\ln(R_2/R_1) + (k/h_t R_2)} \tag{6.1}$$

where Q = rate of heat loss, kJ/h (Btu/h)
k = thermal conductivity of the insulation at average temperature, W/m·°C (Btu·ft^{-1}·h^{-1}·°F^{-1})
L = length of cylindrical surface, m (ft)
T_s = temperature of surface covered with insulation, °C (°F)
T_a = temperature of the ambient air, °C (°F)
R_1 = inside radius of insulation, m (ft)
R_2 = outside radius, m (ft)
h_t = combined coefficient for convection and radiation from insulation surface, W/m^2·°C (Btu/h·ft^2·°F)

Some values of h_t for vertical cylindrical surfaces in a room at 21°C (70°F) are given in Table 6.7 (see Chap. 4 for other geometries and their h_t values).

Economical thickness   As the thickness of insulation is increased, the rate of heat loss from the surface is decreased, but the cost of the insulation is increased accordingly. The most economical thickness of insulation is that for which the sum of the life-cycle cost of the heat loss plus the life-cycle cost of the insulation is a minimum (see Chap. 28 for economic analysis methods).

The heat loss per hour per unit length for a vertical cylindrical tank can be calculated for several assumed thicknesses of insulation by using Eq. (6.1). The cost of the heat loss equals the product of this heat loss per hour multiplied by the hours of operation per year multiplied by the cost of the heat per kilojoule (Btu). The cost of the insulation equals the cost of the insulation (applied) multiplied by the fraction of this cost to be amortized each year (see Chap. 28). Results of these calculations can be plotted and the most economical thickness of insulation determined.

STORAGE IN REVERSIBLE CHEMICAL REACTIONS

Thermal energy can be stored in chemical bonds by means of reversible thermochemical reactions. Advantages of thermochemical storage systems include high energy density (much

TABLE 6.7   Combined Coefficient for Convection and Radiation from Insulation Surface*

Insulation skin temperature		Combined coefficient h_t	
°C	°F	kJ/m^2 · h · °C	Btu/ft^2 · h · °F
38	100	9.53	1.68
65	150	11.7	2.07
93	200	13.5	2.38
121	250	15.1	2.67
149	300	16.7	2.95

*For vertical cylindrical surface in ambient still air at 21°C (70°F). From Ref. 9.

TABLE 6.8   Thermal Conductivities of Insulating Materials*

Material	Apparent density, kg/m³ (lb/ft³)	Temperature, °C (°F)	Thermal conductivity k, W/m · °C (Btu/ft · h · °F)
85% magnesia	272 (17)	38 (100)	0.068 (0.039)
	272 (17)	260 (500)	0.081 (0.047)
Felted rock or glass wool	256 (16)	38 (100)	0.052 (0.030)
	256 (16)	315 (600)	0.099 (0.057)
Aluminum foil (⅜-in air spacing)	3.2 (0.2)	38 (100)	0.043 (0.025)
	3.2 (0.2)	182 (350)	0.066 (0.038)
Wool felt	136 (8.5)	0 (32)	0.038 (0.022)
	136 (8.5)	93 (200)	0.057 (0.033)
Cork (molded)	120 (7.5)	0 (32)	0.036 (0.021)
	120 (7.5)	50 (122)	0.042 (0.024)

*From Ref. 9.

greater than is possible with sensible- or latent-heat storage), low storage-related investment cost, ambient temperature storage (for indefinitely long periods without thermal loss), and potential for heat pumping and long-distance energy transport. Their principal disadvantages are high power-related investment cost (in most instances) and their present technological immaturity. They are conveniently categorized in terms of application: thermochemical energy storage (primarily long-duration storage of thermal decomposition products), thermochemical energy transport (closed-loop "chemical heat pipes"), and "chemical heat pump" storage.

Thermochemical Energy Storage

This type of storage is illustrated by a hypothetical reversible reaction $AB \rightleftharpoons A + B$ (see Fig. 6.16) with a "turning temperature" $T^* = \Delta h/\Delta S$ above which the equilibrium shifts to the right (endothermic decomposition) and below which the shift is to the left (exothermic recombination). Such reactions are most suitable for high-temperature processes such as photothermal power generation, since ΔS must be fairly low for easy reactant separation and the desired high value of Δh is achievable only with a correspondingly high value of T^*. This storage can be less expensive than sensible or latent heat storage if near-unity f_s is required, in which case the capacity-related economies (high-energy density, inexpensive reactants, and ambient temperature storage) dominate the costly power-related items (reactors, heat exchangers, separators, etc.). Some of the candidate reactions are listed in Table 6.9. Current developmental emphasis is on the SO_3/SO_2 system (see Fig. 6.17), largely because it can be used either with a paired collector/converter as shown or in a distributed collector system where it can also provide an energy transport function (see next paragraph). For very large storage capacities the released oxygen can be discharged to and released from the atmosphere as noted in the table, requiring only the storage of two liquids.

Thermochemical Energy Transport

In some geographical regions, areas of high insolation are only a few tens of kilometers from areas of high population and industrial density but low insolation. Transport of solar-derived thermal energy at ambient temperature, by means of closed-loop reversible chemical reactions, is a potential solution to these inequities. Such "chemical heat pipes" are illustrated by a hypothetical reaction $A + B \rightleftharpoons C + D$ (see Fig. 6.18), in which the system elements are essentially those shown in Fig. 6.16 but with the storage vessels replaced by pipelines through which

(Reaction: AB = A + B)

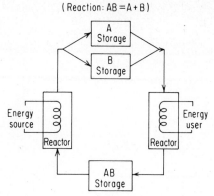

Fig. 6.16 Thermochemical energy storage concept.

TABLE 6.9   Candidate Reactions for Long-Duration Thermal Energy Storage

Reaction Exothermic → ← Endothermic	Δh at 20°C		Temperature,* °C (°F)
	kJ/kg (Btu/lb$_m$)	kJ/m^3 (Btu/ft^3)	
CO + 3H$_2$⇌CH$_4$ + H$_2$O	7345	3710	832
SO$_2$ + Air ⇌ SO$_3$	1544	16,920	765
SO$_2$ + ½O$_2$⇌SO$_3$	1517	9760	746
NH$_3$ + H$_2$O + SO$_3$⇌NH$_4$HSO$_4$			
CaO + H$_2$O ⇌ Ca(OH)$_2$	880	18,610	547
SO$_3$ + H$_2$O⇌H$_2$SO$_4$	885	1442	356
H$_2$O + H$_2$SO$_4$⇌H$_2$SO$_4$ · H$_2$O	230	327	

*Turning temperature where $\Delta h = \Delta G$.

the reactants and products are transmitted. Candidate reactions for transmission of high-grade thermal energy for industrial processes or power generation are listed in Table 6.10. The first-listed reaction (steam reforming of methane) is in an advanced stage of development in Germany for the transmission of nuclear-generated process heat, and may be suitable for solar applications if the detrimental effects of frequent temperature cycling on the endothermic reactor can be mitigated by "banking" or other means. Energy transmission costs via chemical heat pipes may compare favorably with all-electrical transmission when allowance is made for the availability at the destination of heat that otherwise would be wasted.

Chemical Heat-Pump Storage

These systems combine the functions of thermal energy storage and heat pumping for solar heating and/or cooling. In all versions a high-temperature reaction A is simultaneously coupled by a carrier gas to a lower temperature reversible process B that has a lower heat of reaction, with the coupled reactions delivering heat at an intermediate temperature (see Fig. 6.19a). The ratio of the two heats of reaction largely determines the COP and the fraction of total input energy that is stored for sunless operation. Reactions A and B can both be solid-gas equilibria, such as

Reaction A: MnCl$_2$·6NH$_3$ ⇌ MnCl$_2$·2NH$_3$ + 4NH$_{3,\text{ gas}}$
Reaction B: CaCl$_2$·8NH$_3$ ⇌ CaCl$_2$·4NH$_3$ + 4NH$_{3,\text{ gas}}$

where $\Delta h_B/\Delta h_A = 0.89$. Reaction A can be solid-gas and reaction B the condensation-evaporation of the carrier gas, such as

Reaction A: CaCl$_2$·2CH$_3$OH ⇌ CaCl$_2$ + 2CH$_3$OH$_{\text{gas}}$
Reaction B: 2CH$_3$OH$_{\text{liq}}$ ⇌ 2CH$_3$OH$_{\text{gas}}$

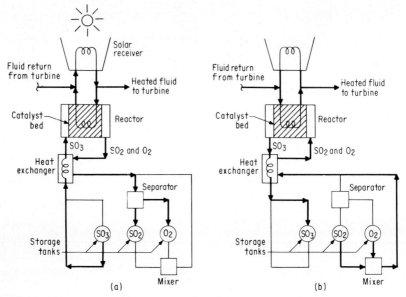

Fig. 6.17 Simplified schematic of SO_3/SO_2 thermochemical storage system for a central receiver solar power plant. Heavy lines and arrows represent primary flows. In a distributed collector system there would be many endothermic reactors and one exothermic reactor. (*a*) Storage charging—endothermic decomposition and (*b*) storage discharging—exothermic recombination.

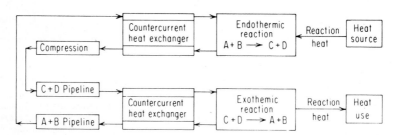

Fig. 6.18 Closed-loop chemical energy transport. The basic features of a chemical heat pipe are illustrated, with the hypothetical reaction A + B $\rightleftarrows$ C = D used as an example. The components of the system are essentially identical to those described in Fig. 6.16, with the storage tanks replaced by pipelines through which the reactants and products are transmitted.

TABLE 6.10 Candidate Reactions for Chemical Heat-Pipe Systems

| Reaction | Δh at 20°C | | Temperature, |
| Exothermic → | | | °C |
← Endothermic	kJ/kg (Btu/lb$_m$)	kJ/m³ (Btu/ft³)	(°F)
$CO + 3H_2 \rightleftarrows CH_4 + H_2O$	7345	3710	832
$N_2 + 3H_2 \rightleftarrows 2NH_3$	2695	1320	
$SO_2 + \frac{1}{2}O_2 \rightleftarrows SO_3$	1517	9760	
$C_6H_6 + 3H_2 \rightleftarrows C_6H_{12}$			295

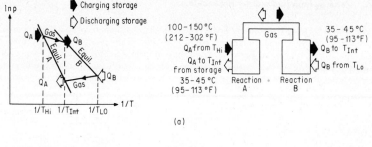

(a)

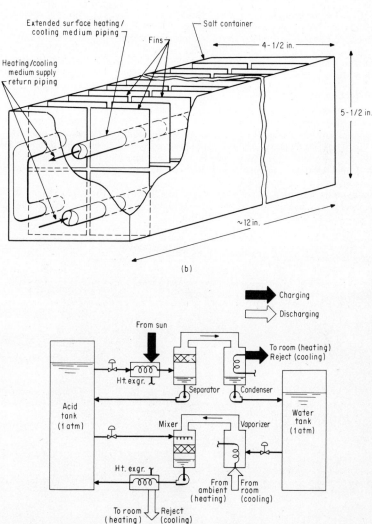

(b)

(c)

Fig. 6.19 Thermochemical heat pump storage: (*a*) Heat pumping and storage concept: T_{HI} is temperature delivered by solar collector, T_{INT} is temperature delivered to room for heating, rejected to ambient for cooling, and T_{LO} is temperature extracted from ambient for heating, extracted from room for cooling; (*b*) subscale prototype salt bed/heat exchanger for calcium chloride-methanol system (*Courtesy EIC Corp.*); and (*c*) liquid-gas system based on heat of dilution of sulfuric acid.

where $\Delta h_B / \Delta h_A = 0.54$. Systems involving solid-gas reactions have few if any moving parts, but require expensive heat exchangers (see Fig. 6.19b). Both reactions can be liquid-gas equilibria (see Fig. 6.19c) to facilitate heat and mass transfer. Such systems are especially suited for yearly averaging storage because the power-related and energy-related elements can be sized independently, but their use is probably limited to large buildings with maintenance and operating personnel. The sulfuric acid-water system (see Fig. 6.20) is closest to commercial availability.

MECHANICAL ENERGY STORAGE

Energy can be stored mechanically as kinetic energy in rotating flywheels or as potential energy in compressed air or elevated liquids (pumped hydro). Although the energy to and from flywheels, compressor/expanders, and pump/turbines is always shaft power (work), such devices are commonly used to store electrical energy (see Fig. 6.2d); therefore the defined storage subsystems typically include associated motor/generators, power conditioners, and controls. Much of the material in this section is adapted from Ref. 24.

Flywheels

Flywheel energy-storage systems are attractive because of their fast response, high input/output power, high efficiency, and environmental acceptability. Their main disadvantages are high cost and limited storage duration. In most solar applications they have electrical inputs and outputs (see Fig. 6.21), but they can be linked mechanically to heat engines or wind turbines to increase inertia and provide short-duration buffering storage. Capacities from a few kilowatthours (for single-family residences) to hundreds of megawatthours (for utilities) are feasible, but only the residential sizes are technologically current.

Wheel characteristics   The energy stored in a flywheel is $E = I\omega^2/2$, where I is the polar moment of inertia and ω is the angular velocity. In terms of flywheel parameters independent

Fig. 6.20   Engineering test model of sulfuric acid-water system. The glass tanks are for inspectional convenience rather than corrosion resistance. In this early model the acid tanks are also used for separation and mixing; two were installed for operational flexibility. (*Courtesy Rocket Research Co.*)

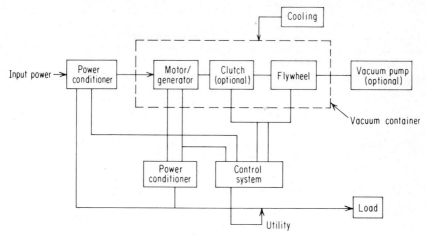

Fig. 6.21   Flywheel storage system.

of rotational speed, the energy per unit mass is $E/m = f(k_s \sigma / \rho)$, where k_s is a rotor shape factor, σ is the allowable rotor material stress, and ρ is the rotor material density. A cone-shaped disk (see Fig. 6.22a) is generally preferred geometry for rotors made of isotropic materials such as steel or aluminum. Hollow or solid cylindrical disks appear best for higher-performance experimental composite rotors (see Fig. 6.22b). Table 6.11 presents rotor cost and performance estimates for current and near-term technology.

Electrical elements   The motor charges the flywheel by converting input electrical power into mechanical power; the generator reverses this process. A single inductor-type synchronous ac machine can perform both functions in a vacuum environment. Since the output voltage and frequency vary with flywheel speed, a power conditioner is needed to maintain the constant voltage and frequency required by the load. Similarly, a variable-frequency constant-current input is required to accelerate the motor and wheel from standstill or low-speed operation, which calls for equipment to convert dc power (from a photovoltaic array) or constant-frequency ac power (from a wind generator) to a variable frequency.

Mechanical elements   In nearly all applications, the flywheel must operate in a vacuum to minimize windage loss. Generally the vacuum need not exceed the capability of a mechanical pump. Preferably the motor/generator should also be within the vacuum housing to avoid rotary seals and continuous pump operation.

If the rotor material is isotropic, provision must be made for safe containment in case of failure; with composite materials the consequences of failure are less severe. An electromagnetic clutch can decouple the motor/generator and flywheel to further reduce bearing, windage, and core losses. It would be energized to engage only during charge and discharge periods. For safety and ease of assembly, the flywheel shaft should be vertical, but this requires a thrust bearing that must support the entire rotating system. Magnetic bearings capable of supporting several hundred pounds are suitable for small residential systems; heavier assemblies will require hydrostatic or antifriction bearings. External cooling must remove the heat produced by windage, frictional, and electrical losses, and adequate solid conductive paths must be provided from the motor/generator and clutch to the vacuum-housing walls.

Compressed Air

Compressed-air energy-storage systems merit consideration if the required storage duration is very long, since they have essentially no "shelf losses," or if the capacity is large enough to justify storing the air in underground caverns or aquifers. Their main disadvantages are low efficiency, high cost (if constructed tanks are required), compressor noise, and the inherent hazards of high-pressure pneumatics. In all currently visualized solar applications, they have electrical inputs and outputs. Capacities from a few kilowatthours (storage in steel tanks, for single-family residences) to many hundreds of megawatthours (storage in underground caverns or aquifers, for utilities) are feasible with current technology, but solar applications have received little attention.

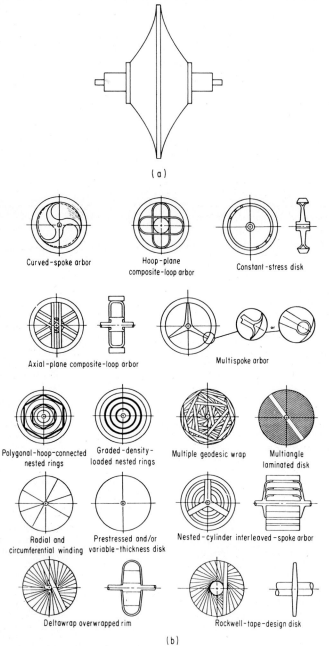

(a)

Curved-spoke arbor

Hoop-plane
composite-loop arbor

Constant-stress disk

Axial-plane composite-loop arbor

Multispoke arbor

Polygonal-hoop-connected
nested rings

Graded-density-
loaded nested rings

Multiple geodesic wrap

Multiangle
laminated disk

Radial and
circumferential winding

Prestressed and/or
variable-thickness disk

Nested-cylinder interleaved-spoke arbor

Deltawrap overwrapped rim

Rockwell-tape-design disk

(b)

Fig. 6.22 (a) Isotropic flywheel; (b) various composite flywheel designs.

TABLE 6.11   **Flywheel Rotor Materials**

Material	σ, N/m² (lb/in²)	ρ, kg/m³ (lb/ft³)	E/m, kW/kg (kWh/lb$_m$)	Material cost $/kg ($/lb$_m$)	$/kWh
Composite E-glass/epoxy	172×10^6 (250,000)	2160 (135)	0.117 (0.053)	2.20 (1.00)	19
Composite S-2/S-glass	241×10^6 (350,000)	2130 (133)	0.154 (0.070)	4.07 (1.85)	26
Composite Kevlar	152×10^6 (220,000)	1410 (88)	0.145 (0.066)	14.50 (6.60)	100
Aluminum Alloy	55×10^6 (80,000)	2770 (173)	0.057 (0.026)	5.50 (2.50)	80
Maraging steel	275×10^6 (400,000)	8000 (500)	0.097 (0.044)	11.00 (5.00)	110

1. Includes 0.60 $/lb fabrication cost (mass-produced simple shape with epoxy resin).
2. Rough cost estimate for high-volume production.
3. Fabricated cost.
SOURCE: Friedericy and Raynard, *Proc. 1975 Flywheel Technology Symp.*, Lawrence Livermore Laboratory.

Tank air-storage systems   Only one installation has been reported: an experimental wind-powered system by the Hydro-Quebec Institute of Research (IREQ). An illustrative system concept suitable for wind and/or photovoltaic power (see Fig. 6.23) is based on this example. The unoptimized IREQ system has an estimated "wire-to-wire" efficiency of 25 to 50 percent, which could be improved by using the compressor coolant for space heating or other purposes. Power-related costs of such systems are about 225 $/kW. Energy-related costs, based on a spherical steel tank, are about 240 $/kWh. Improvements in both efficiency and power-related cost appear possible by using a motor/generator and compressor/expander to replace the motor, compressor, expander, and alternator.

Underground air-storage systems   Large underground compressed-air storage systems are used by electric utilities to meet daily peak loads with stored off-peak energy supplemented by fossil fuel combustion (see Fig. 6.24). The air can be stored in a confined aquifer, or an excavated or solution-mined cavern. The first such plant, at Huntorf, Germany, became operational in 1977 using a solution-mined (in salt) cavern. About two-thirds of the delivered peak power comes from the stored compressed air, whch contains roughly one-half the energy

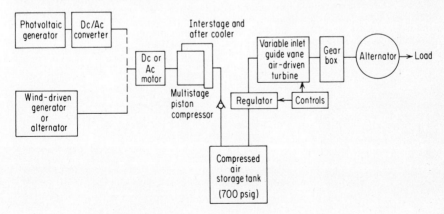

Fig. 6.23   Simplified block diagram of pneumatic storage system.

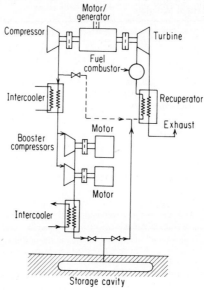

Fig. 6.24   Underground compressed-air storage system with fuel assist.

drawn from the grid to compress and store it. This scheme could also be used by utilities that include solar-derived power in their generating mix, but the overall efficiency of solar energy storage is probably unacceptably low, although the storage investment cost is attractively low for large systems. An acceptably efficient solar storage system must recover and store the presently wasted heat of compression. Such near-adiabatic systems are still in the feasibility-study stage.

Pumped Hydro

This is the traditional method of wind energy storage, in which the product (water) is pumped by a windmill into an elevated storage tank where its potential energy makes it available without pumping when there is no wind. The principle can be extended to wind or solar electric power systems of any size, but appears best suited for utilities applications because of the large storage volume required per unit of stored energy. Above-ground pumped hydro storage systems are commonly used by electric utilities to meet daily peak loads with stored off-peak energy (see Fig. 6.25a), terrain and esthetics permitting. Consideration is also being given to underground systems (see Fig. 6.25b) that circumvent these problems. Power-related costs, including plant substations, range from 95 to 150 $/kW for above-ground and from 120 to 165 $/kW for underground. Storage-related costs range from 2 to 10 $/kWh for above-ground and from 10 to 15 $/kWh for underground. Efficiencies of both types range from 70 to 75 percent.

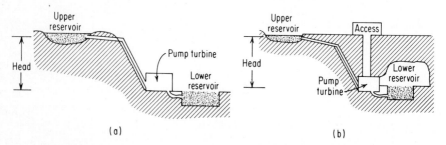

Fig. 6.25   (a) Aboveground pumped hydro storage; (b) underground pumped hydro storage.

ELECTROCHEMICAL STORAGE

Electrical energy derived from the sun can be stored electrochemically by currently available secondary (rechargeable) batteries of the lead-acid type. Advantages of lead-acid batteries, in addition to current availability, are proven reliability, long storage duration (shelf life), moderately high efficiency, and high customer acceptance. Disadvantages include high cost, limited lifetime with daily cycling, and the inherent hazard of high-voltage electrical power. Advanced batteries of potentially higher efficiency and lower cost are under development. Since batteries require a dc input they are particularly compatible with photovoltaic arrays (see Fig. 6.26), but require considerable power conditioning when used with ac wind generators or in utility grids using power from any source (see Fig. 6.27). Much of the material in this section is adapted from Ref. 24.

Lead-Acid Batteries

Lead-acid batteries have lead peroxide positive plates and spongy lead negative plates, with a sulfuric acid electrolyte. Both plates typically have a lead-antimony alloy grid for current collection, but cells for standby service now use a lead-calcium alloy grid to extend battery life beyond 20 years. A recent "hybrid" type containing a lead-antimony positive grid and a lead-calcium negative grid is expected to survive daily cycling for 10 years. Energy storage is accomplished by an electrochemical process in which lead sulfate and water are converted to lead peroxide, lead, and sulfuric acid by an electrical current. The reaction is reversed when an external electrical circuit is connected across the cell so that the lead peroxide and lead react with the sulfuric acid to release electrical energy.

A voltaic efficiency of about 85 percent is associated with the difference between the cell voltage required for charging and that available at the terminals for discharging. There also is a coulombic efficiency of about 95 percent associated with irreversible side reactions such as the decomposition of water, making the total efficiency for a charge-discharge cycle about 81 percent. Inclusion of power-conditioning losses typically reduces the system round-trip, first-law efficiency to about 70 percent (see Fig. 6.27).

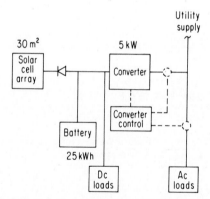

Fig. 6.26   Single-family battery storage system with photovoltaic power source.

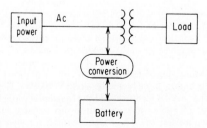

Fig. 6.27   Battery storage system with ac power source.

Lead-acid batteries operate at room temperature, therefore there is no need for insulation, warm-up or cool-down time, or energy expenditure to maintain elevated temperatures. Their weight tends to prohibit vertical stacking or use in upper floors, and provision must be made for venting the hydrogen generated by water decomposition. In large installations, water-replenishment systems are desirable. Representative power-related capital costs are 70 $/kW, while storage-related costs range from 73 $/kWh for residential sizes to 66 $/kWh for utility applications. Operating and maintenance (O&M) costs of 2 mils/kWh are generally used. Power-related and O&M costs are expected to remain fixed, but storage-related capital costs may be reduced by 1985 to 60 $/kWh for residential systems and 50 $/kWh for larger installations.

Advanced Batteries

Advanced aqueous (ambient-temperature) and nonaqueous (high-temperature) batteries are under intensive development for utility energy storage and electric vehicle propulsion. Development goals are superiority over lead-acid batteries in terms of lifetime (>2500 cycles), efficiency (>70 percent), cost (30 $/kWh), and energy density. These goals apply also to solar applications.

Nonaqueous systems include the sodium-sulfur, lithium–iron sulfide, and sodium–antimony chloride batteries. They operate similarly to the lead-acid cell, but the active electrode materials are liquid and must be kept hot (200 to 450°C, 400–850°F) to remain so and maintain good electrical conductivity. The volume of such batteries is at least as great as that of lead-acid batteries, but their much lower weight permits stacking and consequently a smaller "footprint." Since no hydrogen is involved, and since the battery must have its own temperature-control system, neither ventilation nor ambient temperature is critical.

Aqueous systems include the zinc-chloride battery and redox batteries. The zinc-chloride cell operates like a lead-acid cell except that the active materials are not stored within the cell but in separate reservoirs or storage tanks. It has a high theoretical energy density (829 Wh/kg) and operates at room temperature, but the chlorine is stored as a hydrate at 10°C and a rather complex piping system is required to circulate the chlorine and the zinc-chloride electrolyte. Although the cell weight and volume are significantly less than for an equivalent lead-acid cell, the extensive auxiliary equipment might nullify these advantages. Redox batteries, in which the positive and/or negative active materials are dissolved in the electrolyte, use various inorganic couples in aqueous solutions. Their principal potential advantage is that external reactant storage in tanks minimizes the storage-related capital costs.

HYDROGEN STORAGE

Hydrogen differs from other solar energy storage media in that it does not constitute a fixed reusable inventory, but is repeatedly produced from other substances and ultimately consumed as a fuel or a chemical feedstock. From a solar viewpoint it is seen primarily as a clean, storable, and transportable fuel that can be made from sunlight and water. As a storage medium, it has the advantage of indefinitely long shelf life at ambient temperature, but compares poorly with other storable fuels in energy density. At standard temperature and pressure, the energy density of gaseous hydrogen is only $12,100$ kJ/m³ (325 Btu/ft³) compared with $37,200$ (1000) for methane. Justifications for the use of hydrogen storage in solar applications must therefore invoke the more compelling advantages of clean combustion, transportability, and producibility from renewable resources.

Hydrogen Production

Hydrogen can be produced from sunlight and water by electrolysis (via solar electrical power generation), by closed-cycle thermochemical water splitting (via high-temperature solar heat), or by photochemical and biological processes. Of these processes, only water electrolysis has been shown to be both technologically and economically feasible.

The electrolytic process of decomposing water into hydrogen and oxygen is the reverse of hydrogen combustion, and the theoretical amount of energy required is the same as the heat of combustion. Ideally, the energy input would be 39.3 kWh/kgH$_2$ (17.86 kWh/lbH$_2$), of which only 33.3 kWh (15.11 kWh) need be electrical and the remainder could be heat. The theoretical cell decomposition voltage of water corresponding to this minimal electrical energy input is 1.23 V at atmospheric pressure and 25°C (77°F), and varies inversely with temperature. Excess voltage above 1.23 V, usually referred to as "overvoltage," consists of three principal components:

the ohmic resistance of the electrolyte and overvoltages at the hydrogen and oxygen electrodes. The latter two depend on the electrode construction and materials. For most currently available industrial electrolyzers, operating voltages in the range of 1.7 to 2.2 V are required.

Since the current efficiency of most electrolyzers is nearly 100 percent, the overall energy efficiency of hydrogen production is largely determined by the operating voltage. The "thermal efficiency" of electrolysis is the cell potential corresponding to thermoneutral operation (1.4 V at 25°C and 1 atm with no heat input) divided by the actual operating voltage. This corresponds to the ratio of the higher heating value of hydrogen produced to the electrical energy input. Both the efficiency and unit cost can be changed by simply altering the cell operating conditions: voltage (efficiency), rate of hydrogen production (current), and capital cost, all of which depend on the current-voltage characteristic of each type of electrolyzer. Much of the technology advancement is aimed at increasing current densities at lower operating voltages by means of improved and lower-cost electrode structures and by operating at higher temperature (larger fraction of energy input as heat) and higher pressure (to reduce the compressor power requirements).

Electrolyzers in existing large commercial installations are commonly of filter-press construction with bipolar electrodes. Figure 6.28 compares the cell operating performance of existing commercial installations (upper curves) and advanced technology (lower curves). A representative advanced system schematic is shown in Fig. 6.29. Capital investment cost for the electrolyzer plus auxiliaries is typically about 250 $/kW of equivalent hydrogen energy out for commercially available units, with a projected 1980 cost of 84 $/kW (out) for the advanced General Electric solid polymer electrode (SPE) system. If the electrical source is ac at line voltage, the additional cost for transformers and rectifiers is 45 $/kW (in).

Hydrogen Storage Methods

Hydrogen is routinely stored industrially as a cryogenic liquid and as a compressed gas in steel vessels. Methods of storing it as a hydride are under development, and the storage of very large volumes of gaseous hydrogen in underground caverns has been suggested.

Hydrogen can be stored most compactly and inexpensively as a liquid, in double-walled pressure vessels with vacuum insulation at −250°C (−423°F) and about 10 atm. The cost of storage tanks with stainless steel or aluminum alloy inner liners ranges from about 2.3 $/kWh for 10 MWh to about 0.3 $/kWh for 1000 MWh. However, a large fraction of the cost and energy loss associated with liquid hydrogen storage is in the liquefaction process. Theoretically, the work required to liquefy gaseous hydrogen is 3.94 kWh/kg (1.79 kWh/lb), all of which is in principle recoverable upon regasification. With current technology, however, the practically

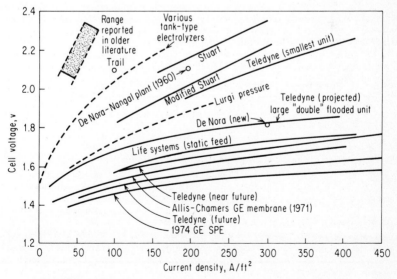

Fig. 6.28 Comparative performances of electrolyzer systems. (From Ref. 24.)

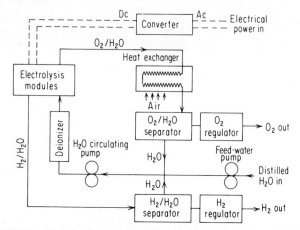

Fig. 6.29 Simplified SPE electrolytic H_2 generation plant schematic. (From Ref. 24.)

achievable minimum is 10.8 kWh/kg (4.92 kWh/lb), most of the irreversibility being attributable to the inefficiencies of reciprocating compressors. Thus the unrecoverable work of liquefaction is two-thirds of the net energy in the stored liquid. Equipment cost for liquefaction plants is approximately 300 $/kWh for capacities of 2.71 t/day (30 tons/day) and larger, to which must be added 50 to 75 percent for installation. Yearly operating and maintenance cost is estimated at 6 to 10 percent of the total liquefaction plant capital cost.

The storage of hydrogen as a compressed gas typically requires two major components: pressure vessels and compressors. If the electrolyzer operates at the intended storage pressure, the compressors can be replaced by less-expensive water pumps with much lower energy requirements. Steel vessels are available for gaseous storage at three pressures (Table 6.12). Annual operating and maintenance costs for cylinder storage ranges from about 1 to 3 percent of the total installed cost. Hydrogen compressors are usually of the nonlubricated reciprocating type, with one to five stages depending on the compression ratio. The compression cycle efficiency (net heating value of the compressed hydrogen minus the compressor work divided by the initial heating value) ranges from 93 to 98 percent, depending on gas flow rate, inlet pressure, and final storage pressure. For electrically driven compressors the overall efficiency will be about 95 percent of these values. Typical capital equipment costs for reciprocating compressors driven by low-speed synchronous motors are plotted in Fig. 6.30.

Hydrogen storage as metal hydrides is still in the developmental stage, with iron-titanium being the currently favored hydriding metal. During the hydriding process heat is released;

TABLE 6.12 Properties of Cylindrical Steel Hydrogen Storage Vessels

Properties			
Storage pressure, N/m²	3.45×10^6	8.92×10^6	16.9×10^6
(psi)	(500)	(1295)	(2450)
OD, m	0.61	0.61	0.61
(in)	(24)	(24)	(24)
Length, m	12.22	12.22	6.25
(ft)	40	40	20.5
Vessel cost, $	3300	3500	3900
Mounting cost, $	540	540	540
Manifolding cost, $	250	275	300
Number of vessels required for 10^6 SCF	265	116	136
Total cost for 10^6 SCF, $ $\times 10^6$	1.08	0.50	0.65

SOURCE: Courtesy U.S. Steel Corp.

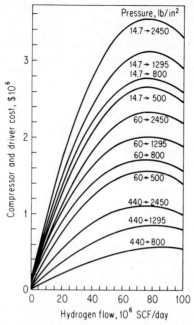

Fig. 6.30 Compressor and driver equipment costs as a function of daily throughput for varying compression pressure levels. (From Ref. 24.) (*Courtesy Worthington Corp.*)

TABLE 6.13 **Provisional Matchings**

Application	Storage method
Space heating only	
▪ Diurnal	Water tanks or rockbeds
▪ Seasonal (large)	Aquifers
Space cooling only	
▪ Diurnal (hot side)	$MgCl_2 \cdot 2H_2O$
	Form stable polymers
	Pressurized hot water
▪ Seasonal (large)	Aquifers
Space heating and cooling	
▪ Active	Chemical heat pumps
▪ Passive	Phase change in building materials
Photothermal power	
▪ Short duration	Sensible or latent heat
▪ Long duration	Thermochemical
Wind Power	
▪ Grid feeding	No dedicated storage (grid-mix)
▪ Autonomous	Batteries
Photovoltaics	
▪ Grid feeding	No dedicated storage (grid-mix)
▪ Autonomous	Batteries

in the reverse reaction the hydride is dissociated endothermically to release hydrogen. Hydrogen storage densities approaching that of liquid hydrogen can be achieved in this manner. The heat required for dissociation is approximately 9.3 to 13.9 MJ/kg (4000 to 6000 Btu/lb) of discharged hydrogen, which may be obtainable from otherwise wasted reject heat in subsequent power generation processes. Power requirements for associated compressors and heat exchangers are expected to be about 2.2 kWh/kg (1 kWh/lb) of hydrogen processed, so cycle efficiencies greater than 90 percent appear possible if waste heat is available. Hydride storage system capital costs are still speculative, but have been estimated at 5 $/kWh of output hydrogen HHV (higher heating value) for a fixed-bed system operating at 32 atm with Fe-Ti alloy at 1.10 $/kg (0.50 $/lb).

APPLICATION—TECHNOLOGY MATCHING

In Fig. 6.2 and in the foregoing discussions, the various combinations of application and storage technology have generally been presented without comparative evaluation or recommendation. To do so entails considerable speculation, since many of the more promising storage technologies are relatively immature and their costs cannot yet be confidently estimated, but the provisionally superior matchings listed in Table 6.13 can serve as an aid to final ranking and selection by the practitioner that will be possible when harder data become available and special considerations are known.

REFERENCES

1. *Proceedings of Third Annual Thermal Energy Storage Contractors' Information Exchange Meeting*, DOE report, 1979.
2. Advanced Thermal Energy Storage Concept Definition Study for Solar Brayton Power Plants, vol. 1, Boeing Engineering and Construction, DOE Rept. SAN/1300-1, November 1977.
3. M. D. Silverman and J. R. Engel, Survey of Technology for Storage of Thermal Energy in Heat Transfer Salt, Oak Ridge National Laboratory Rept. ORNL/TM-5682, January 1977.
4. R. C. Mitchell and G. R. Morgan, Gravel and Liquid Storage System for Solar Thermal Power Plants, *Proc. of Sharing the Sun Conf.*, American Section of ISES, 1976.
5. G. A. Lane et al., Macro-Encapsulation of PCM, DOE Rept. ORO/5217-8, November 1978.
6. D. W. Conner and R. O. Mueller, Simulation of Stratified Heat Storage in Solar Heating Systems, Argonne National Laboratory Rept. ANL/SPG-7, 1979.
7. P. F. Kando and M. Telkes, Characterization of Sodium Sulfate Dekahydrate (Glauber's Salt) as a Thermal Energy Storage Material, DOE Rept. COO-4042-16, January 1978.
8. M. Telkes, Storage of Solar Heating/Cooling, *ASHRAE Transactions*, vol. 80, pt. II.
9. C. L. Segaser, MIUS Technology Evaluation-Thermal Energy Storage Materials and Devices, Oak Ridge National Laboratory Rept. ORNL-HUD-MIUS-23, November 1975.
10. H. G. Lorsch, Thermal Energy Storage Devices Suitable for Solar Heating, *Proc. of 9th IECEC*, 1974.
11. R. A. Botham et al., Form-Stable Crystalline Polymer Pellets for Thermal Energy Storage, ERDA Rept. ORO/5159-10, July 1977.
12. D. J. Morrison and S. I. Abdel-Khalik, Effects of Phase-Change Energy Storage on the Performance of Air-Based and Liquid-Based Solar Heating Systems, *Solar Energy*, **20**:57-67.
13. R. T. LeFrois and H. V. Venkatassetty, Inorganic Phase Change Materials for Energy Storage in Solar Thermal Program, *Proc. of Sharing the Sun Conf.*, American Section of ISES, 1976.
14. R. E. Rice and B. M. Cohen, Phase Change Thermal Storage for a Solar Total Energy System, *Proc. of ISEC 77, International Solar Energy Congress*.
15. J. J. Nemecek et al., Demand Sensitive Energy Storage in Molten Salts, *Proc. of Sharing the Sun Conference*, American Section of ISES, 1976.
16. W. Goldstern, "Steam Storage Installations," Pergamon Press, 1970.
17. P. V. Gilli et al., Thermal Energy Storage Using Prestressed Cast Iron Vessels (PCIV), DOE Rept. COO/2886-2, November 1977.
18. Geraghty et al., "Water Atlas of the United States," Water Information Center, Port Washington, NY, 1973.
19. R. R. Davison et al., Storing Sunlight Underground—The Solaterre System, *Chemical Technology*, **5**:736-741, December 1975.
20. Chin Fu Tsang et al., Underground Aquifer Storage of Hot Water from Solar Energy Collectors, *Proc. of ISEC 77, International Solar Energy Congress*.
21. Chemical Energy Storage for Solar-Thermal Conversion, Rocket Research Co., Interim Rept. Number 2 77-R-558, April 13, 1978.
22. H. B. Vakil and J. W. Flock, Closed Loop Chemical Systems for Energy Storage and Transmission (Chemical Heat Pipe) DOE Rept. COO-2676-1, February 1978.
23. P. OD. Offenhartz, Chemically Driven Heat Pumps for Solar Thermal Storage, *Proc. of ISEC 77, International Solar Energy Congress*.

24. Applied Research on Energy Storage and Conversion for Photovoltaic and Wind Energy Systems, General Electric Space Division, NSF Rept. HCP/T-22221-01/3, January 1978.
25. H. C. Maru et al., Molten Salt Thermal Energy Storage Systems, DOE Rept. COO-2888-3, March 1978.
26. C. E. Birchenal, Heat Storage Materials, DOE Rept. COO-4042-16, December 1977.
27. J. L. Eichelberger, Investigation of Metal Fluoride Thermal Energy Storage: Availability, Cost and Chemistry, ERDA Rept. COO2990-6, December 1976.

Chapter 7

Nonconcentrating Solar Thermal Collectors*

FRANK KREITH
SERI, Golden, CO

and

JAN F. KREIDER
Jan F. Kreider & Associates, Boulder, CO

ENERGY BALANCE FOR A FLAT-PLATE COLLECTOR

The thermal performance of any type of solar thermal collector can be evaluated by an energy balance that determines the portion of the incoming radiation delivered as useful energy to the working fluid. For a flat-plate collector of area A_c this energy balance is

$$I_c A_c \bar{\tau}_s \alpha_s = q_u + q_{loss} + \frac{de_c}{dt} \tag{7.1}$$

where I_c = solar irradiation on a collector surface
$\bar{\tau}_s$ = effective solar transmittance of the collector cover(s)
α_s = solar absorptance of the collector-absorber plate surface
q_u = rate of heat transfer from the collector-absorber plate to the working fluid
q_{loss} = rate of heat transfer (or heat loss) from the collector-absorber plate to the surroundings
de_c/dt = rate of internal energy storage in the collector

The instantaneous efficiency of a collector η_c is simply the ratio of the useful energy delivered to the total incoming solar energy, or

$$\eta_c = \frac{q_u}{A_c I_c} \tag{7.2}$$

---

*Adapted from F. Kreith and J. F. Kreider, "Principles of Solar Engineering," Hemisphere Publishing Corp., Washington, DC, 1978, with permission.

In practice, the efficiency must be measured over a finite time period. In a standard performance test this period is on the order of 15 or 20 min, whereas for design, the performance over a day or over some longer period t is important. Then we have for the average efficiency

$$
\bar{\eta}_c = \frac{\displaystyle\int_0^t q_u \, dt}{\displaystyle\int_0^t A_c I_c \, dt}
\tag{7.3}
$$

where t is the time period over which the performance is averaged.

A detailed and precise analysis of the efficiency of a solar collector is complicated by the nonlinear behavior of radiation heat transfer. However, a simple linearized analysis is usually sufficiently accurate in practice. In addition, the simplified analytical procedure is very important, because it illustrates the parameters of significance for a solar collector and how these parameters interact. For a proper analysis and interpretation of these test results an understanding of the thermal analysis is imperative, although for design and economic evaluation the results of standardized performance tests are generally used.

Reference 3 surveys the history of the flat-plate collector.

Collector Heat-Loss Conductance

In order to obtain an understanding of the parameters determining the thermal efficiency of a solar collector, it is important to develop the concept of *collector heat-loss conductance*. Once the collector heat-loss conductance U_c is known, and when the collector plate is at an average temperature T_c, the second right-hand term in Eq. (7.1) can be written in the simple form

$$
q_{\text{loss}} = U_c A_c (T_c - T_a)
\tag{7.4}
$$

The simplicity of this relation is somewhat misleading, because the collector heat-loss conductance cannot be specified without a detailed analysis of all the heat losses. Figure 7.1 shows a schematic diagram of a double-glazed collector, while Fig. 7.2a shows the thermal circuit with all the elements that must be analyzed before they can be combined into a single conductance element shown in Fig. 7.2b. The analysis below shows how this combination is accomplished.

Figure 7.3 shows the qualitative temperature distributions in a flat-plate collector. Radiation impinges on the top of the plate connecting any two adjacent flow ducts. It is absorbed uniformly by the plate and conducted toward the flow duct, where it is then transferred by convection to the working fluid flowing through the ducts. It is apparent that at any cross section perpendicular to the direction of flow, the temperature is a maximum at the midpoint between two adjacent flow ducts and decreases along the sheet toward the tube. Since heat is transferred to the working fluid, the temperature of the fluid as well as that of the entire collector system will increase in the direction of flow. For example, the increase in temperature at the midpoint between the two tubes is shown qualitatively in Fig. 7.3c. The temperature distribution in both the x and y directions is shown in a three-dimensional view in Fig. 7.3d.

In order to construct a model suitable for a thermal analysis of a flat-plate collector, the following simplifying assumptions will be made:

1. The collector is thermally in steady state.
2. The temperature drop between the top and bottom of the absorber plate is negligible.
3. Heat flow is one-dimensional through the covers as well as through the back insulation.
4. The headers connecting the tubes cover only a small area of the collector and provide uniform flow to the tubes.
5. The sky can be treated as though it were a blackbody source for infrared radiation at an equivalent sky temperature.
6. The irradiation on the collector plate is uniform.

For a quantitative analysis consider a location at x,y on a typical flat-plate collector as shown in Fig. 7.3. Let the plate temperature at this point be $T_c(x,y)$ and assume solar energy is absorbed at the rate $I_s \alpha_s$. Part of this energy is then transferred as heat to the working fluid, and if the collector is in the steady state, the other part is lost as heat to the ambient air if $T_c > T_a$. Some of the heat loss occurs through the bottom of the collector. It passes first through the back insulation by conduction ($R_1 = l/k$) and then by convection ($R_2 = 1/h_{c,\,\text{bottom}}$) to the

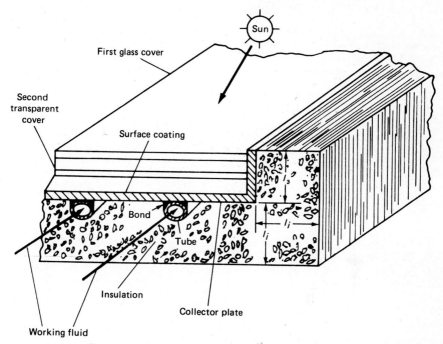

Fig. 7.1 Schematic diagram of solar collector with two covers.

environment. If the collector is part of a sloping roof, the environment for the back losses is the air inside the building; if the collector is installed on a flat roof, the environment is the same air to which the heat is lost from the upper surface of the plate. In a well-insulated collector the external convective resistance is much smaller than the insulation resistance and the back-loss conductance then becomes simply

$$U_b = \frac{1}{R_1 A_c} = \frac{k_i}{l_i A_c} \tag{7.5}$$

where k_i and l_i are, respectively, the thermal conductivity and thickness of the insulation.

Evaluation of edge losses is quite difficult for most collectors. However, in a well-designed collector, edge losses are small and need not be predicted with precision. If the insulation around the edges is of the same thickness as the back, the edge losses can be estimated by assuming one-dimensional conduction around the perimeter, but adding a constant term for the edge. Thus, for a collector of area $l_1 l_2$ and thickness l_3, and with a layer of insulation l_i thick at the bottom and along the sides (see Fig. 7.1), the effective back loss would be approximately

$$q_{\text{back, edge}} = \frac{A_c k_i}{l_i}(T_c - T_a)\left[1 + \frac{(2 l_3 + l_i)(l_1 + l_2)}{l_1 l_2}\right] \tag{7.6}$$

where k_i is the thermal conductivity of the insulation.

The conductance for the upper surface of the collector can be evaluated by determining the thermal resistance R_3, R_4, and R_5 in Fig. 7.2a. Heat is transferred between the cover and the second glass plate and between the two glass plates by convection and radiation in parallel. Except for absorptance of solar energy by the second glass plate, the relations for the rate of heat transfer between T_c and T_{g2} and between T_{g2} and T_{g1} are the same. Thus, the rate of heat transfer per unit surface area of collector between the absorber plate and the second glass cover is

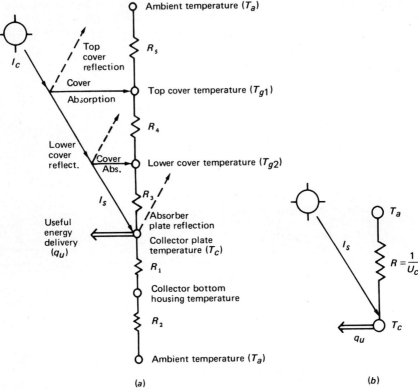

Fig. 7.2 Thermal circuits for flat-plate collectors shown in Fig. 7.1: (a) detailed circuit; (b) approximate, equivalent circuit to (a). In both circuits, the absorber plate absorbs incident energy equal to $a_s I_s$, where $I_s = \bar{\tau}_s I_c$. Collector assumed to be at uniform temperature T_c.

$$q_{\text{top loss}} = A_c \bar{h}_{c2}(T_c - T_{g2}) + \frac{\sigma A_c (T_c^4 - T_{g2}^4)}{1/\epsilon_{p,i} + 1/\epsilon_{g2,i} - 1} \qquad (7.7)$$

where h_{c2} = convection heat-transfer coefficient between the plate and the second glass cover
$\epsilon_{p,i}$ = infrared emittance of the plate
$\epsilon_{g2,i}$ = infrared emittance of the second cover

As shown in Chap. 4, the radiation term can be linearized; then Eq. (7.7) becomes

$$q_{\text{top loss}} = (\bar{h}_{c2} + h_{r2})A_c(T_c - T_{g2}) = \frac{T_c - T_{g2}}{R_3} \qquad (7.8)$$

where

$$h_{r2} = \frac{\sigma(T_c + T_{g2})(T_c^2 + T_{g2}^2)}{1/\epsilon_{p,i} + 1/\epsilon_{g2,i} - 1}$$

A similar derivation for the rate of heat transfer between the two cover plates gives

$$q_{\text{top loss}} = (\bar{h}_{c1} + h_{r1})A_c(T_{g2} - T_{g1}) = \frac{T_{g2} - T_{g1}}{R_4} \qquad (7.9a)$$

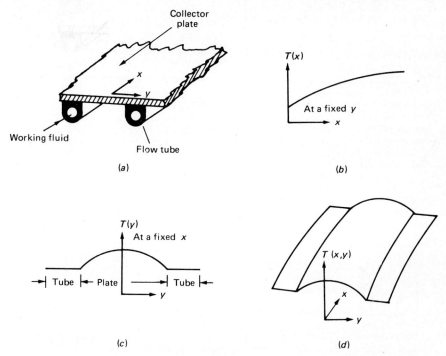

Fig. 7.3 Qualitative temperature distribution in the absorber plate of a flat-plate collector. (*a*) Schematic diagram of absorber; (*b*) temperature profile in the direction of the flow of the working fluid; (*c*) temperature profile at given *x*; (*d*) temperature distribution in the absorber plate.

where

$$h_{r1} = \frac{\sigma(T_{g1} + T_{g2})(T_{g1}^2 + T_{g2}^2)}{1/\epsilon_{g1,i} + 1/\epsilon_{g2,i} - 1} \qquad (7.9b)$$

and $\bar{h}_{c1}$ = convection heat-transfer coefficient between two transparent covers.

The emittances of the cover will, of course, be the same if they are made of the same material. However, economic advantages can sometimes be achieved by using a plastic cover between an outer cover of glass and the plate, and in such a sandwich construction, the radiative properties of the two covers may not be the same.

The equation for the thermal resistance between the upper surface of the outer collector cover and the ambient air has a form similar to the two preceding relations, but the convection heat-transfer coefficient must be evaluated differently. If the air is still, free-convection relations should be used, but when wind is blowing over the collector, forced-convection correlations apply as shown in Chap. 4. Radiation exchange occurs between the top cover and the sky at T_{sky}, whereas convection heat exchange occurs between T_{g1} and the ambient air at T_{air}. For convenience we shall refer both conductances to the air temperature. This gives

$$q_{\text{top loss}} = (h_{c,x} + h_{r,x})(T_{g1} - T_{air})A_c = \frac{T_{g1} - T_{air}}{R_5} \qquad (7.10)$$

where

$$h_{r,x} = \epsilon_{g1,i}\sigma(T_{g1} + T_{sky})(T_{g1}^2 + T_{sky}^2)\left[\frac{(T_{g1} - T_{sky})}{(T_{g1} - T_{air})}\right]$$

The total heat-loss conductance U_c can then be expressed in the form

$$U_c = \frac{1}{R_1 + R_2} + \frac{1}{R_3 + R_4 + R_5} \tag{7.11}$$

for a double-glazed flat-plate collector.

Evaluation of the collector heat-loss conductance defined by Eq. (7.11) requires iterative solution of Eqs. (7.9a) and (7.10), because the unit radiation conductances are functions of the cover and plate temperatures, which are not known *a priori*. A simplified procedure for calculating U_c for collectors with all covers of the same material, which is often sufficiently accurate and more convenient to use, has been suggested by Hottel and Woertz (10) and Klein (13). It is also suitable for application to collectors with selective surfaces. For this approach the collector top loss in watts is written in the form

$$q_{\text{top loss}} = \frac{(T_c - T_a)A_c}{N/\{(C/T_c)[(T_c - T_a)/(N + f)]^{0.33}\} + 1/h_{c,\infty}}$$

$$+ \frac{\sigma(T_c^4 - T_a^4)A_c}{1/[\epsilon_{p,i} + 0.05N(1 - \epsilon_{p,i})] + (2N + f - 1)/\epsilon_{g,i} - N} \tag{7.12}$$

where $f = (1 - 0.04h_{c,\infty} + 0.0005h_{c,\infty}^2)(1 + 0.091N)$
$C = 365.9(1 - 0.00883\beta + 0.00013\beta^2)$
$N = $ number of covers
$h_{c,\infty} = 5.7 + 3.8V$
$\epsilon_{g,i} = $ infrared emittance of the covers

The values of $q_{\text{top loss}}$ calculated from Eq. (7.12) agreed closely with the values obtained from Eq. (7.11) for 972 different observations encompassing the following conditions:
$320 < T_c < 420 \text{ K}$
$260 < T_a < 310 \text{ K}$
$0.1 < \epsilon_{p,i} < 0.95$
$0 \leq V \leq 10 \text{ m/s}$
$1 \leq N \leq 3$
$0 \leq \beta \leq 90°$

The standard deviation of the differences in $U_c = q_{\text{top loss}}/A_c(T_c - T_a)$ was $0.14 \text{ W/m}^2 \cdot \text{K}$ for these comparisons.

Thermal Analysis of Flat-Plate Collector-Absorber Plate

In order to determine the efficiency of a solar collector the rate of heat transfer to the working fluid must be calculated. If transient effects are neglected (11, 13, 21), the rate of heat transfer to the fluid flowing through a collector depends on only the temperature of the collector surface from which heat is transferred by convection to the fluid, the temperature of the fluid, and the heat-transfer coefficient between the collector and the fluid. To analyze the rate of heat transfer consider first the condition at a cross section of the collector with flow ducts of rectangular cross sections as shown in Fig. 7.4. Solar radiant energy impinges on the upper face of the collector plate. A part of the total solar radiation falls on the upper surface of the flow channels, while another part is incident on the plates connecting any two adjacent flow channels. The latter is conducted in a transverse direction toward the flow channels. The temperature is a maximum at any midpoint between adjacent channels and the collector plate acts as a fin attached to the walls of the flow channel. The thermal performance of a fin can be expressed in terms of its efficiency. The fin efficiency η_f is defined as the ratio of the rate of heat flow through the real fin to the rate of heat flow through a fin of infinite thermal conductivity, that is, a fin at a uniform temperature. We shall now derive a relation to evaluate this efficiency for a flat-plate solar collector.

If U_c is the overall unit conductance from the collector-plate surface to the ambient air, the rate of heat loss from a given segment of the collector plate at x,y in Fig. 7.4 is

$$q(x,y) = U_c[T_c(x,y) - T_a] \, dx \, dy \tag{7.13}$$

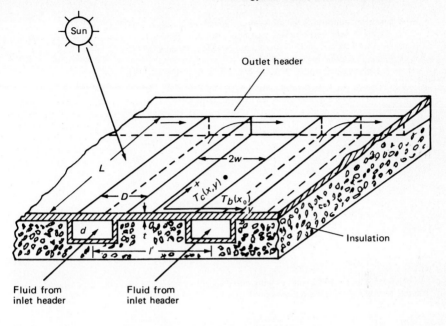

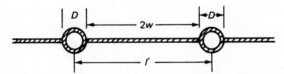

Fig. 7.4 Sketch showing coordinates and dimensions for collect plate and fluid ducts.

where T_c = local collector-plate temperature ($T_c > T_a$)
$\quad\quad T_a$ = ambient air temperature
$\quad\quad U_c$ = overall unit conductance between the plate and the ambient air

U_c includes the effects of radiation and free convection between the plates, the radiative and convective transfer between the top of the cover and the environment, and conduction through the insulation. Its quantitative evaluation has been previously considered.

If conduction in the x direction is negligible, a heat balance at a given distance x_0 for a cross section of the flat-plate collector per unit length in the x direction can be written in the form

$$\alpha_s I_s \, dy - U_c(T_c - T_a) \, dy + \left(-kt \frac{dT_c}{dy} \bigg|_{y,x_0} \right) - \left(-kt \frac{dT_c}{dy} \bigg|_{y+dy,x_0} \right) = 0 \quad\quad (7.14)$$

If the plate thickness t is uniform and the thermal conductivity of the plate is independent of temperature, the last term in Eq. (7.14) is

$$\frac{dT_c}{dy} \bigg|_{y+dy,x_0} = \frac{dT_c}{dy} \bigg|_{y,x_0} + \left(\frac{d^2 T_c}{dy^2} \right)_{y,x_0} dy$$

and Eq. (7.14) can be cast into the form of a second-order differential equation:

$$\frac{d^2 T_c}{dy^2} = \frac{U_c}{kt} \left[T_c - \left(T_a + \frac{\alpha_s I_s}{U_c} \right) \right] \quad\quad (7.15)$$

The boundary conditions for the system described above at a fixed x_0 are
1. At the center between any two ducts the heat flow is 0, or at $y = 0$, $dT_c/dy = 0$.
2. At the duct the plate temperature is $T_b(x_0)$, or at $y = w = (l' - D)/2$, $T_c = T_b(x_0)$, where $T_b(x_0)$ is the fin-base temperature.

If we let $m^2 = U_c/kt$ and $\phi = T_c - (T_a + \alpha_s I_s/U_c)$, Eq. (7.15) becomes

$$\frac{d^2\phi}{dy^2} = m^2\phi \tag{7.16}$$

subject to the boundary conditions

$$\frac{d\phi}{dy} = 0 \quad \text{at} \quad y = 0$$

$$\phi = T_b(x_0) - \left(T_a + \frac{\alpha_s I_s}{U_c}\right) \quad \text{at} \quad y = w$$

The general solution of Eq. (7.16) is

$$\phi = C_1 \sinh my + C_2 \cosh my \tag{7.17}$$

The constants C_1 and C_2 can be determined by substituting the two boundary conditions and solving the two resulting equations for C_1 and C_2. This gives

$$\frac{T_c - (T_a + \alpha_s I_s/U_c)}{T_b(x_0) - (T_a + \alpha_s I_s/U_c)} = \frac{\cosh my}{\cosh mw} \tag{7.18}$$

From the preceding equation the rate of heat transfer to the conduit from the portion of the plate between two conduits can be determined by evaluating the temperature gradient at the base of the fin, or

$$q_{\text{fin}} = -kt \left.\frac{dT_c}{dy}\right|_{y=w} = \frac{1}{m}\{\alpha_s I_s - U_c[T_b(x_0) - T_a]\}\tanh mw \tag{7.19}$$

Since the conduit is connected to fins on both sides, the total rate of heat transfer is

$$q_{\text{total}}(x_0) = 2w\{\alpha_s I_s - U_c[T_b(x_0) - T_a]\}\frac{\tanh mw}{mw} \tag{7.20}$$

If the entire fin were at the temperature $T_b(x_0)$, a situation corresponding physically to a plate of infinitely large thermal conductivity, the rate of heat transfer would be a maximum, $q_{\text{total, max}}$. As mentioned previously, the ratio of the rate of heat transfer with a real fin to the maximum rate obtainable is the fin efficiency η_f. With this definition, Eq. (7.20) can be written in the form

$$q_{\text{total}}(x_0) = 2w\,\eta_f\,\{\alpha_s I_s - U_c[T_b(x_0) - T_a]\} \tag{7.21}$$

where $\eta_f \equiv \tanh mw/mw$.

The fin efficiency η_f is plotted as a function of the dimensionless parameter $w(U_c/kt)^{1/2}$ in Fig. 7.5. When the fin efficiency approaches unity, the maximum portion of the radiant energy impinging on the fin becomes available for heating the fluid.

In addition to the heat transferred through the fin, the energy impinging on the portion of the plate above the flow passage provides useful energy. The rate of useful energy from this region available to heat the working fluid is

$$q_{\text{duct}}(x_0) = D\{\alpha_s I_s - U_c[T_b(x_0) - T_a]\} \tag{7.22}$$

Thus, the useful energy per unit length in the flow direction becomes

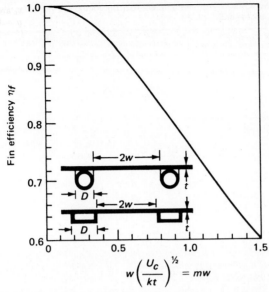

Fig. 7.5   Fin efficiency for tube and sheet flat-plate solar collectors.

$$q_u(x_0) = (D + 2w\,\eta)\{\alpha_s I_s - U_c[T_b(x_0) - T_a]\} \tag{7.23}$$

The energy $q_u(x_0)$ must be transferred as heat to the working fluid. If the thermal resistance of the metal wall of the flow duct is negligibly small and there is no contact resistance between the duct and the plate, the rate of heat transfer to the fluid is

$$q_u(x_0) = 2(D + d)\overline{h}_{c,i}[T_b(x_0) - T_f(x_0)] \tag{7.24}$$

Contact resistance may become important in poorly manufactured collectors in which the flow duct is clamped or glued to the collector plate. Collectors manufactured by such methods are usually not satisfactory.

Collector efficiency factor   To obtain a relation for the useful energy delivered by a collector in terms of known physical parameters, the fluid temperature, and the ambient temperature, the collector temperature must be eliminated from Eqs. (7.23) and (7.24). Solving for $T_b(x_0)$ in Eq. (7.24) and substituting this relation in Eq. (7.23) gives

$$q_u(x_0) = l'F'\{\alpha_s I_s - U_c[T_f(x_0) - T_a]\} \tag{7.25}$$

where F' is called the collector efficiency factor. It is given by

$$F' = \frac{1/U_c}{l'\,[1/U_c(D + 2w\,\eta_f) + 1/\overline{h}_{c,i}(2D + 2d\,)]} \tag{7.26}$$

Physically, the denominator in Eq. (7.26) is the thermal resistance between the fluid and the environment, whereas the numerator is the thermal resistance between the collector surface and the ambient air. The collector-plate efficiency factor F' depends on U_c, $\overline{h}_{c,i}$, and η_f. It is only slightly dependent on temperature and can for all practical purposes be treated as a design parameter. Typical values for the factors determining the value of F' are given in Table 7.1.

The collector efficiency factor increases with increasing plate thickness and plate thermal conductivity, but decreases with increasing distance between flow channels. Also, increasing

TABLE 7.1   Typical Values for the Parameters that Determine the Collector Efficiency Factor F' for a Flat-Plate Collector in Eq. (7.26)

	U_c	
Two glass covers	4 W/m² · K	0.685 Btu/h · ft² · °F
One glass cover	8 W/m² · K	1.37 Btu/h · ft² · °F
	kt	
Copper plate, 1 mm thick	0.38 W/K	0.72 Btu/h · °F
Steel plate, 1 mm thick	0.046 W/K	0.087 Btu/h · °F
	$\bar{h}_{c,i}$	
Water in laminar-flow forced convection	300 W/m² · K	52 Btu/h · ft² · °F
Water in turbulent-flow forced convection	1500 W/m² · K	254 Btu/h · ft² · °F
Air in turbulent forced convection	100 W/m² · K	17.6 Btu/h · ft² · °F

the heat-transfer coefficient between the walls of the flow channel and the working fluid increases F', but an increase in the overall conductance U_c will cause F' to decrease.

Collector heat-removal factor   Equation (7.25) yields the rate of heat transfer to the working fluid at a given point x along the plate for specified collector and fluid temperatures. However, in a real collector the fluid temperature increases in the direction of flow as heat is transferred to it. An energy balance for a section of flow duct dx can be written in the form

$$\dot{m}c_p(T_f\big|_{x+dx} - T_f\big|_x) = q_u(x)\,dx \tag{7.27}$$

Substituting Eq. (7.25) for $q_u(x)$ and $T_f(x) + dT_f(x)/dx\,dx$ for $T_f\big|_{x+dx}$ in Eq. (7.27) gives the differential equation

$$\dot{m}c_p\frac{dT_f(x)}{dx} = l'F'\{\alpha_s I_s - U_c[T_f(x) - T_a]\} \tag{7.28}$$

Separating the variables gives, after some rearranging,

$$\frac{dT_f(x)}{T_f(x) - T_a - \alpha_s I_s/U_c} = \frac{-l'F'U_c}{\dot{m}c_p}\,dx \tag{7.29}$$

Equation (7.29) can be integrated and solved for the outlet temperature of the fluid $T_{f,\text{out}}$ for a duct length L and for the fluid inlet temperature $T_{f,\text{in}}$ if we assume that F' and U_c are constant, or

$$\frac{T_{f,\text{out}} - T_a - \alpha_s I_s/U_c}{T_{f,\text{in}} - T_a - \alpha_s I_s/U_c} = \exp\left(-\frac{U_c l'F'L}{\dot{m}c_p}\right) \tag{7.30}$$

To compare the performance of a real collector with the thermodynamic optimum, it is convenient to define the heat-removal factor F_R as the ratio of the actual rate of heat transfer to the working fluid to the rate of heat transfer at the minimum temperature difference between the absorber and the environment. The thermodynamic limit corresponds to the condition of the working fluid remaining at the inlet temperature throughout the collector. This can be approached when the fluid velocity is very high. From its definition, F_R can be expressed as

$$F_R = \frac{Gc_p(T_{f,\text{out}} - T_{f,in})}{\alpha_s I_s - U_c(T_{f,in} - T_a)} \tag{7.31}$$

where G is the flowrate per unit surface area of collector $\dot{m}/A_c$. By regrouping the right-hand side of Eq. (7.31) and combining with Eq. (7.30), it can easily be verified that

$$F_R = \frac{Gc_p}{U_c} \left[1 - \frac{\alpha_s I_s / U_c - (T_{f,\,out} - T_a)}{\alpha_s I_s / U_c - (T_{f,\,in} - T_a)} \right]$$

$$F_R = \frac{Gc_p}{U_c} \left[1 - \exp\left(- \frac{U_c F'}{Gc_p} \right) \right]$$

(7.32)

Inspection of the above relation shows that F_R increases with increasing flowrate and approaches as an upper limit F', the collector efficiency factor. Since the numerator of the right-hand side of Eq. (7.31) is q_u, the rate of useful heat transfer can now be expressed in terms of the fluid inlet temperature, or

$$q_u = A_c F_R [\alpha_s I_s - U_c (T_{f,\,in} - T_a)]$$

(7.33)

This is a convenient form for design, because the fluid inlet temperature to the collector is usually known or can be specified.*

Example 1

Calculate the averaged hourly and daily efficiency of a water solar collector on January 15, in Boulder, Colorado. The collector is tilted at an angle of 60° and has an overall conductance of 8.0 W/m² · K on the upper surface. It is made of copper tubes, with a 1/cm inside diameter, 0.05 cm thick, which are connected by a 0.05-cm-thick plate at a center-to-center distance of 15 cm. The heat-transfer coefficient for the water in the tubes is 1500 W/m² · K, the cover transmittance is 0.9, and the solar absorptance of the copper surface is 0.9. The collector is 1 m wide and 2 m long, the water inlet temperature is 330 K, and the water flowrate is 0.02 kg/s. The horizontal insolation (total) I_h and the environmental temperature are tabulated below. Assume that diffuse radiation accounts for 25 percent of the total insolation.

Solution

The total radiation received by the collector is calculated from equations in Chap. 2:

$$I_c = I_{c,\,diffuse} + I_{c,\,beam} = I_h \times 0.25 \left(\cos^2 \frac{60}{2} \right) + I_h (1 - 0.25) R_b$$

Time, h	I_h, W/m²	T_{amb}, K
7–8	12	270
8–9	80	280
9–10	192	283
10–11	320	286
11–12	460	290
12–13	474	290
13–14	395	288
14–15	287	288
15–16	141	284
16–17	32	280

The tilt factor R_b is obtained from its definition in Chap. 2

$$R_b = \frac{\cos i}{\sin \alpha} = \frac{\sin (L - \beta) \sin \delta_s + \cos (L - \beta) \cos \delta_s \cos h_s}{\sin L \sin \delta_s + \cos L \cos \delta_s \cos h_s}$$

where $L = 40°$, $\delta_s = -21.1$ on January 15 (from Chap. 2), and $\beta = 60°$. The hour angle h_s equals 15° for each hour away from noon.

The fin efficiency is obtained from Eq. (7.21):

---

*Useful heat collection can be expressed as a function of heat exchanger fluid inlet temperature if a collector delivers heat to storage or load via a heat exchanger. The heat exchanger factor F_{hx} (sometimes called $[F'_R/F_R]$) is discussed in this chapter's Appendix.

$$\eta_f = \frac{\tanh m(l' - D)/2}{m(l' - D)/2}$$

where

$$m = \left(\frac{U_c}{kt}\right)^{1/2} = \left(\frac{8}{390 \times 5 \times 10^{-4}}\right)^{1/2} = 6.4$$

$$\eta_f = \frac{\tanh 6.4(0.15 - 0.01)/2}{6.4(0.15 - 0.01)/2} = 0.938$$

The collector efficiency factor F' is, from Eq. (7.26),

$$F' = \frac{1/U_c}{l' [1/U_c(D + 2w\eta_f) + 1/h_{c,i}\pi D]}$$

$$= \frac{1/8.0}{0.15[1/8.0(0.01 + 0.14 \times 0.938) + 1/1500\pi \times 0.01]} = 0.92$$

Then we obtain the heat-removal factor from Eq. (7.32):

$$F_R = \frac{Gc_p}{U_c}\left[1 - \exp\left(-\frac{U_c F'}{Gc_p}\right)\right]$$

$$F_R = \frac{0.01 \times 4184}{8.0}\left[1 - \exp\left(-\frac{8.0 \times 0.922}{0.01 \times 4184}\right)\right] = 0.845$$

From Eq. (7.33), the useful heat delivery rate is

$$q_u = A_c F_R[\alpha_s I_s - U_c(T_{f,\text{in}} - T_{\text{amb}})]$$

In the relation above, I_s is the radiation incident on the collector. If the transmittance of the glass is 0.9,

$$I_s = \tau I_c = 0.9 I_c$$

and

$$q_u = 2 \times 0.845 [I_c \times 0.81 - 8.0 (T_{f,\text{in}} - T_{\text{amb}})]$$

Time, h	I_h, W/m²	R_b	I_d, W/m²	$I_{b,c}$, W/m²	I_c, W/m²	q_u, W	T_{amb}, K	η_c
7–8	12	10.9	0.4	98	98	0	270	0
8–9	80	3.22	2.5	193	196	0	280	0
9–10	192	2.44	6.0	351	357	0	283	0
10–11	320	2.18	10.0	522	532	133	286	0.126
11–12	460	2.08	14.4	717	732	460	290	0.314
12–13	474	2.08	14.8	739	754	491	290	0.326
13–14	395	2.18	12.3	645	657	331	288	0.252
14–15	287	2.49	9.0	525	534	163	288	0.153
15–16	141	3.22	4.4	341	345	0	284	0
16–17	32	10.9	1.0	261	262	0	280	0

The efficiency of the collector is $\eta_c = q_c/A_c I_c$ and the hourly averages are calculated in the table above.

Thus, $\Sigma I_{\text{tot}} = 4467$ W/m² and $\Sigma q_u = 1578$ W. The daily average is obtained by summing the useful energy for those hours during which the collector delivers heat and dividing by the total insolation between sunrise and sunset. This yields

$$\overline{\eta}_{c, \text{day}} = \frac{\Sigma q_u}{\Sigma A_c I_c} = 100 \frac{1578}{2 \times 4467} = 17.7 \text{ percent}$$

Transient Effects

The preceding analysis assumed that steady-state conditions exist during the operation of the collector. Under actual operating conditions the rate of insolation will vary and the ambient temperature and the external wind conditions may change. To determine the effect of changes in these parameters on the performance of a collector it is necessary to make a transient analysis that takes the thermal capacity of the collector into account.

As shown in Ref. 14, the effect of collector thermal capacitance is the sum of two contributions: the *collector storage* effect, resulting from the heat required to bring the collector up to its final operating temperature, and the *transient* effect, resulting from fluctuations in the meteorological conditions. Both effects result in a net loss of energy delivered compared with the predictions from the zero capacity analysis. This loss is particularly important on a cold morning when all of the solar energy absorbed by the collector is used to heat the hardware and the working fluid, thus delaying the delivery of useful energy for some time after the sun has come up.

Transient thermal analyses can be made with a high degree of precision, but the analytical predictions are no more accurate than the weather data and the overall collector conductance. For most engineering applications a simpler approach is therefore satisfactory. For this approach it will be assumed that the absorber plate, the ducts, the back insulation, and the working fluid are at the same temperature. If back losses are neglected, an energy balance on the collector plate and the working fluid for a single-glazed collector delivering no useful energy can be written in the form

$$(\overline{mc})_p \frac{d\overline{T}_p(t)}{dt} = A_c I_s \alpha_s + A_c U_p [\overline{T}_g(t) - \overline{T}_p(t)] \tag{7.34}$$

where $(\overline{mc})_p$ is the sum of the thermal capacities of the plate, the fluid, and the insulation, and U_p is the conductance between the absorber plate at $\overline{T}_p$ and its cover at $\overline{T}_g$. Similarly, a heat balance on the collector cover gives

$$(mc)_g \frac{d\overline{T}_g(t)}{dt} = A_c U_p [\overline{T}_p(t) - \overline{T}_g(t)] - A_c U_\infty [\overline{T}_g(t) - T_a] \tag{7.35}$$

where $U_\infty = (h_{c,\infty} + h_{r,\infty})$ [see Eq. (7.9b)] and $(mc)_g$ = thermal capacity of the cover plate.

Equations (7.34) and (7.35) can be solved simultaneously and the transient heat loss can then be determined by integrating the instantaneous loss over the time during which transient effects are pronounced. A considerable simplification in the solution is possible if one assumes that at any time the collector heat loss and the cover heat loss are proportional, as in a quasi-steady state, so that

$$U_\infty A_c [\overline{T}_g(t) - T_a] = U_c A_c [\overline{T}_p(t) - T_a] \tag{7.36}$$

Then, for a given air temperature, differentiation of Eq. (7.36) gives

$$\frac{d\overline{T}_g(t)}{dt} = \frac{U_c}{U_\infty} \frac{d\overline{T}_p(t)}{dt} \tag{7.37}$$

Adding Eqs. (7.34), (7.35), and (7.37) gives a single differential equation for the plate temperature

$$\left[(\overline{mc})_p + \frac{U_c}{U_\infty} (mc)_g \right] \frac{d\overline{T}_p(t)}{dt} = \{\alpha_s I_s - U_c [\overline{T}_p(t) - T_a]\} A_c \tag{7.38}$$

Equation (7.38) can be solved directly for given values of I_s and T_a. Since meteorological conditions are usually known only as hourly averages, this restriction is no more limiting than the available input data. The solution to Eq. (7.38) then gives the plate temperature as a function of time, for an initial plate temperature $T_{p,0}$, in the form

$$\bar{T}_p(t) - T_a = \frac{\alpha_s I_s}{U_c} - \left[\frac{\alpha_s I_s}{U_c} - (T_{p,0} - T_a)\right] \exp\left[-\frac{U_c A_c t}{(\overline{mc})_p + (U_c/U_\infty)_c (mc)_g}\right] \qquad (7.39)$$

Collectors with more than one cover can be treated similarly, as shown in Ref. 4.

For a transient analysis the plate temperature $\bar{T}_p$ can be evaluated at the end of a specified time period if the initial value of $\bar{T}_p$ and the values of α_s, I_s, U_c, and T_a during the specified time are shown. Repeated application of Eq. (7.39) provides an approximate method of evaluating the transient effects. An estimate of the net decrease in useful energy delivered can be obtained by multiplying the effective heat capacity of the collector, given by $(\overline{mc})_p + (U_c/U_\infty)(mc)_g$, by the temperature rise necessary to bring the collector to its operating temperature. Note that the parameter $[(\overline{mc})_p + (U_c/U_\infty)(mc)_g]/U_c A_c$ is the *time constant* of the collector[4] and small values of this parameter will reduce losses resulting from transient effects.

Example 2

Calculate the temperature rise between 8 and 10 A.M. of a 1- × 2-m single-glazed water collector with a 0.3-cm-thick glass cover if the heat capacities of the plate, water, and back insulation are 5, 3, and 2 kJ/K, respectively. Assume that the unit surface conductance from the cover to ambient air is 18 W/m² · K and the unit surface conductance between the collector and the ambient air is $U_c = 6$ W/M² · K. Assume that the collector is initially at the ambient temperature. The absorbed insolation $\alpha_s I_s$ during the first hour averages 90 W/m² and between 9 and 10 A.M., 180 W/m². The air temperature between 8 and 9 A.M. is 273 K and that between 9 and 10 A.M., 278 K.

Solution

The thermal capacitance of the glass cover is $(mc)_g = (\rho V c_p)_g = (2500 \text{ kg/m}^3)(1 \times 2 \times 0.03 \text{ m})$ $(1 \text{ kJ/kg} \cdot \text{K}) = 15 \text{ kJ/K}$. The combined collector, water, and insulation thermal capacity is equal to

$$(\overline{mc})_p + \frac{U_c}{U_\infty}(mc)_g = 5 + 3 + 2 + 0.333 \times 15 = 15 \text{ kJ/K}$$

From Eq. (7.39) the temperature rise of the collector between 8 and 9 A.M. is

$$\bar{T}_p - T_a = \frac{\alpha_s I_s}{U_c}\left\{1 - \exp\left[-\frac{U_c A_c t}{(\overline{mc})_p + (U_c/U_\infty)_c (mc)_g}\right]\right\}$$

$$= 15\left[1 - \exp\left(-\frac{2 \times 6 \times 3600}{15,000}\right)\right] = 15 \times 0.944 = 14.2 \text{ K}$$

Thus, at 9 A.M. the collector temperature will be 287.2 K. Between 9 and 10 A.M. the collector temperature will rise as shown below:

$$T_p = T_a + \frac{\alpha_s I_s}{U_c} - \left[\frac{\alpha_s I_s}{U_c} - (T_{p,0} - T_a)\right] \exp\left[-\frac{U_c A_c t}{(mc)_p + (U_c/U_\infty)_c (mc)_g}\right]$$

$$= 278 + \frac{180}{6} - (30 - 9.2)\, 0.056 = 306.8 \text{ K } (91°\text{F})$$

Thus, at 10 A.M. the collector temperature has achieved a value sufficient to deliver useful energy at a temperature level of 306 K.

Experimental Testing of Collectors

The performance of solar systems for space heating and cooling depends largely on the performance of the solar collectors. Thus, the measurement of the performance of solar collectors is an important and necessary step for an understanding of the total system function. Although analytical design procedures, particularly for flat-plate solar collectors, are in reasonably good agreement with test results, it is preferable to base a system design on actual collector test data whenever available.

There are two basic methods for testing collectors: the *instantaneous* and the *calorimetric* procedures. Each allows determination of fundamental collector characteristics, and each has advantages and disadvantages.

For the instantaneous method it is only necessary to measure simultaneously the mass flowrate of the heat-transfer fluid flowing through the collector, its temperature difference, the collector inlet and outlet temperatures, and the insolation incident on the plane of the collector. The instantaneous efficiency can then be calculated from the relation[7.19]

$$\eta_c = \frac{q_u/A_c}{I_c} = \frac{Gc_p(T_{f.\,out} - T_{f.\,in})}{I_c}$$ (7.40)

where η_c = solar collector efficiency
q_u = useful heat output, W (Btu/h)
A_c = cross-sectional area, m^2 (ft^2)
I_c = total solar energy incident upon the plane of the collector per unit time per unit area, W/m^2 (Btu/h · ft^2)
G = mass flowrate of the heat-transfer fluid per unit of collector cross-sectional area, kg/s · m^2 (lb/h · ft^2)
c_p = specific heat of the heat-transfer fluid, J/kg · K (Btu/lb$_m$ · °F)
$T_{f,out}$ = temperature of the heat-transfer fluid leaving the collector, K (°F)
$T_{f,in}$ = temperature of the heat-transfer fluid entering the collector, K (°F)

The instantaneous method is particularly useful in testing a sample of a collector field on site prior to installation of the entire array, because the measurements are relatively simple and can be carried out with commercially available instrumentation to an accuracy of approximately 10 percent. Moreover, with the instantaneous procedure, one only needs to take measurements on and around the collector, which is an advantage for on-site procedures.

The calorimetric procedure employs a closed system in which the time rate of change of temperature of a constant thermal mass is measured and related to the incident solar energy by the relation

$$\eta_c = \frac{q_u/A_c}{I_c} = \frac{(m/A_c)\,c_p'\,dT/dt}{I_c}$$ (7.41)

where m = the mass of the medium in the calorimeter, kg (lb$_m$)
c_p' = the average specific heat of the medium in the calorimeter, J/kg · K (Btu/lb$_m$ · °F)
T = the average temperature of the medium, K (°F)

For the calorimetric procedure, one has to measure the incident solar radiation and the time rate of change of temperature of the mass in the system. However, if the calorimetric procedure is used, a very careful analysis of the calorimeter must be conducted beforehand in order to eliminate errors resulting from temperature gradients inside and to minimize or at least accurately determine the stray thermal losses from the calorimeter. The calorimetric procedure is limited primarily to collectors using a liquid working fluid, because the heat capacity of a gas such as air is too small to be easily measurable and the collector energy cannot readily be stored and determined without transfer to another medium. A complete review of testing procedures was prepared in 1976 by the National Bureau of Standards (NBS).[8]

From Eq. (7.33), the efficiency of a flat-plate collector operating under steady-state conditions can be described by the relationship

$$\eta_c = F_R \bar{\tau}_s \alpha_s - F_R U_c \left(\frac{T_{f,\,in} - T_a}{I_c} \right)$$ (7.42)

Alternatively, by means of the collector efficiency factor defined by Eq. (7.25) and an average fluid temperature between inlet and outlet, an approximate relation, satisfactory for liquid collectors only, is

$$\eta_c = F' \bar{\tau}_s \alpha_s - F' U_c \frac{[(T_{f,\,in} + T_{f,\,out})/2] - T_a}{I_c}$$ (7.43)

Regardless of the relation used, it is apparent that for a constant value of U_c, if the efficiency is plotted against the appropriate $\Delta T / I_c$, a straight line will result. In reality, of course, U_c will vary with the operating temperature of the collector and the ambient weather condition, causing some deviations from the straight-line relation. For most flat-plate collectors, however, this deviation is not a serious problem over the normal operating conditions, as shown in Fig. 7.6, which presents a typical correlation of test results taken from Ref. 19 by means of Eq. (7.43) for a flat-plate water collector.

NBS standard test method   In order to compare the performances of different collectors, the NBS devised a standard test method, which was officially adopted in modified form by the American Society of Heating, Refrigerating , and Air Conditioning Engineers (ASHRAE). The tests that comprised this procedure can be conducted outside under real sun conditions as well as indoors with a solar simulator described in Ref. 19. The test procedure is written in a format consistent with other standards adopted within the United States for testing building heating and cooling equipment to facilitate transfer to solar energy systems. Figure 7.7 shows the recommended testing configuration for the solar collector when the heat-transfer fluid is a liquid; Fig. 7.8 shows a comparable configuration to be used when the heat-transfer fluid is air. Reference 15 presents recommended procedures for concentrating collectors in which the fluid may undergo a phase change. Detailed requirements of the apparatus with specifications for instrumentation to be used in standard measurements of incident radiation, temperature, temperature difference, liquid flowrate, air flowrate, pressure, pressure drop, time, and weight are given in Ref. 7 and in ASHRAE standard 93–77.

A key requirement for reliable data is the time duration of the test. The NBS recommends that a series of tests should be conducted, each of which determines the average efficiency for 15 min over a range of temperature differences between the average fluid temperature and the ambient air. The efficiency should then be calculated from the relationship

$$\overline{\eta}_c = \frac{\displaystyle\int_0^t \dot{m} c_p (T_{f,\text{out}} - T_{f,\text{in}})\, dt}{\displaystyle\int_0^t I_c A_c\, dt} \tag{7.44}$$

The flowrate for the duration of each test must be constant to within ± 1 percent and the heat-transfer fluid must have a known specific heat, which varies by less than 0.5 percent over the temperature range of the fluid during a particular test period. At least 16 data points are required for a complete test series and they must be taken symmetrically with respect to solar noon to prevent biased results from possible transient effects.

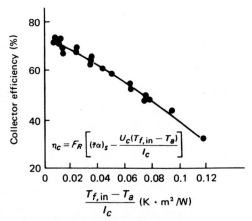

Fig. 7.6 Efficiency curve for a double-glazed flat-plate liquid-heating solar collector with a selective coating on the absorber $\dot{m} = 0.136$ kg/s m²; $T_a = 29°C$; $T_{f,\text{in}} = 38-101°C$; $I_c = 599-977$ W/cm²; wind = 3.1 m/s. (The tests were run indoors using a solar simulator.)

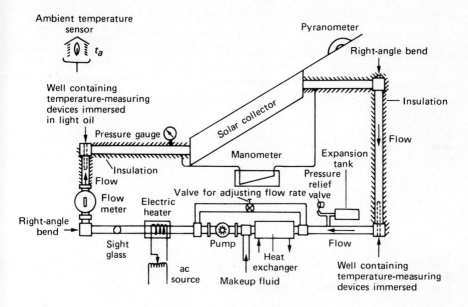

Fig. 7.7 Recommended testing configuration for the solar collector when the heat-transfer fluid is a liquid. (*From Ref. 8.*)

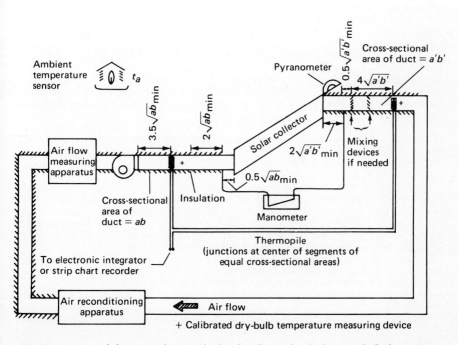

Fig. 7.8 Recommended testing configuration for the solar collector when the heat-transfer fluid is air. (*From Ref. 8.*)

During each test period, the incident solar radiation must be *quasi-steady*, as shown in Fig. 7.9. Conditions during a day in which cloud cover causes a time distribution, such as shown in Fig. 7.10, will not be acceptable. Each data point must be taken with an average insolation larger than 630 W/m² (over the 15-min average), and the incident angle between the sun and the normal from the collector surface must be less than 45°. Also, the range of ambient temperatures for the test series must be less than 30°C (54°F).

The procedure also requires that each of the measurements made and the calculated efficiency for each data point be reported in tabular as well as graphic form. This procedure may be adapted for use with concentrating collectors as well as flat-plate collectors. Some modifications required for application to concentrating collectors[15] are the following:

1. Measure the direct component as well as the total incident radiation in the plane of the collector and include a curve like Fig. 7.6 for each: one in which the insolation value in the abscissa is the total radiation and one in which it is only the direct component of insolation.

2. Measure the quality as well as the temperature and pressure of the working fluid so as to allow for a change of phase of the working fluid as it passes through the collector.

3. Measure separately the reflectance of the reflector surface and the absorptance and emittance of the selective surface on the absorber if used.

Comparison of analysis and test results   The results of thermal performance tests for solar energy collectors are generally presented by plotting the efficiency η_c as a function of the difference in temperature between the inlet to the collector and the ambient air divided by the solar flux incident on the collector I_c. Figure 7.11a shows the effect of temperature difference on efficiency at given insolation values; Fig. 7.11b shows the effect of insolation on the efficiency for specified temperature differences between the collector fluid at the inlet and the ambient air; and Fig. 7.11c shows the correlation plot of η_c vs. $\Delta t/I_c$. Note that the $\Delta T/I_c$ plot collapses the several curves of Figs. 7.11a and 7.11b onto a single curve within a narrow band.

A quantitative comparison of the heat losses from a flat-plate collector for different designs is presented in Fig. 7.12. The three cases illustrated are (1) single glass cover, black absorber (Fig. 7.12a), (2) single glass cover, selective absorber (Fig. 7.12b), and (3) double glass cover, selective absorber (Fig. 7.12c).

The collector heat-loss conductance can be evaluated from Eq. (7.11) or (7.12). For the three configurations that represent a spectrum of commercial practice, the following general conclusions may be drawn:

1. If both surfaces have high emittance ($\epsilon \approx 1$), the radiative heat-transfer loss is the most important part.

2. If the absorber surface has a low emittance ($\epsilon \approx 0.1$), the convective heat loss is more important than the radiative heat loss. The first step toward substantially reducing the heat transfer between two plates is therefore to cover the heat-transfer surface with a low-emittance layer.

3. Convective and conductive heat transfer in a flat-plate design can be reduced by introducing an additional cover plate, which normally also reduces the radiative heat transfer. However, this measure will reduce the $\alpha\bar{\tau}$ intercept and decrease the collector performance at small differences in temperature between collector fluid and ambient, that is, at low collector temperatures.

Figure 7.13 shows the performance curve for a flat-plate collector with a selectively coated absorber surface and two glass covers at 0° incident flux angle. On this graph $F_R\alpha_s\bar{\tau}_s$ corresponds to the intercept of the best fit through the experimental data points on the ordinate, while $F_R U_c$ is the absolute value of the slope of the line. The values for absorptance and transmittance used in an analysis may differ somewhat from the actual values and one cannot expect perfect correlation between the experimental and analytical values.

The efficiency of a collector under real operating conditions may be less than the values obtained in a performance test such as that shown in Fig. 7.7 because of the following:[19]

1. The optical efficiency decreases with increasing incidence angles (measured from the normal to the collector surface); that is, the collector curve is displaced toward the abscissa (see Fig. 7.13).

2. The optical efficiency decreases with an increasing ratio of diffuse-to-beam radiation.

3. The optical efficiency decreases when dust collects on the outer cover or when the outer cover becomes dirty.

4. Transient effects may decrease the amount of useful energy delivered.

5. Selective surface coatings may deteriorate with time.

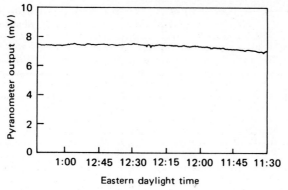

Fig. 7.9   Incident solar radiation on a horizontal surface at the National Bureau of Standards site in Gaithersburg, MD. Steady-state conditions. (*From Ref. 8.*)

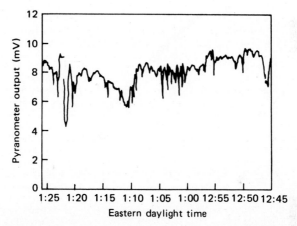

Fig. 7.10   Incident solar radiation on a horizontal surface at the National Bureau of Standards site in Gaithersburg, MD. Variable conditions. (*From Ref. 8.*)

Air-Cooled Flat-Plate Collector Thermal Analysis

The basic air-cooled flat-plate collector shown in Fig. 7.14 differs fundamentally from the liquid-based collectors described in preceding sections because of the relatively poor heat-transfer properties of air. For example, in turbulent flow in a given conduit for a fixed value of Reynolds number, the convection heat-transfer coefficient for water is about 50 times greater than that for air. As a result it is essential to provide the largest heat-transfer area possible to remove heat from the absorber surface of an air-heating collector.

The most common way to achieve adequate heat transfer in air collectors is to flow air over the entire rear surface of the absorber as shown. The heat-transfer analysis of such a collector does not involve the fin effect or the tube-to-plate bond conductance problem, which arises in liquid collectors. The heat-transfer process is essentially that of an unsymmetrically heated duct of large aspect ratio (typically 20 to 40).

Malik and Buelow[17, 18] surveyed the fluid mechanics and heat-transfer phenomena in air collectors. They concluded that a suitable expression for the Nusselt number for a smooth air-heating collector is

$$Nu_{sm} = \frac{0.0192 \, Re^{3/4} \, Pr}{1 + 1.22 \, Re^{-1/8} \, (Pr - 2)} \tag{7.45}$$

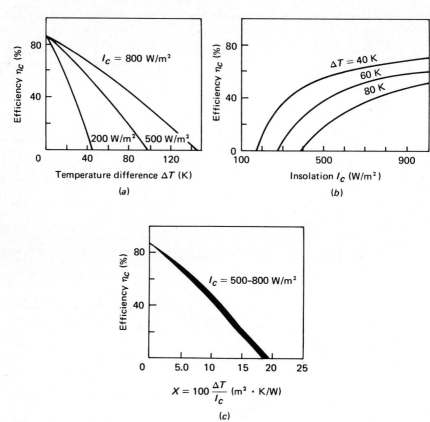

Fig. 7.11   Collector efficiency η_c versus (a) temperature difference ΔT; (b) total incident radiation I_c; (c) $\Delta T/I_c$.

where Re is the Reynolds number and Pr is the Prandtl number. If the surface is hydrodynamically rough, they recommended multiplying the smooth-surface Nusselt number by the ratio of the rough-surface friction factor f to the smooth-surface friction factor f_{sm} from the Blasius equation

$$f_{sm} = 0.079 \, Re^{-1/4} \tag{7.46}$$

The convective coefficient h_c in the Nusselt number is based on a unit absorber area but the hydraulic diameter D_H in the Nusselt and Reynolds numbers is based on the *entire duct perimeter*.

Air-collector efficiency factor   The collector efficiency factor F' is defined as the ratio of energy collection to the collection rate if the absorber plate were at the local fluid temperature. F' is particularly simple to calculate, since no fin analysis or bond-conductance term is present for air collectors. For the collector shown in Fig. 7.14, F' is

$$F' = \frac{\overline{h}_c}{\overline{h}_c + U_c} \tag{7.47}$$

where U_c is calculated from Klein's equation, Eq. (7.12), and the duct convection coefficient $\overline{h}_c$ is calculated from Eq. (7.45).

Air-collector heat-removal factor   The collector heat-removal factor is a convenient parameter, since it permits useful energy gain to be calculated by knowledge of only the easily

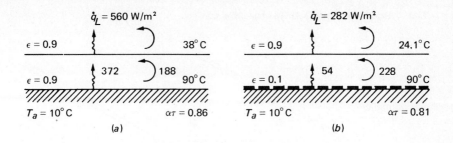

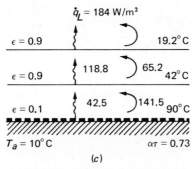

Fig. 7.12 Heat losses and temperature profiles for three different flat-plate collectors (ambient temperature $T_a = 10°C$, wind velocity = 3.75 m/s): (a) single glass, black absorber; (b) single glass, selective absorber; (c) double glass, selective absorber. (*From Ref. 12.*)

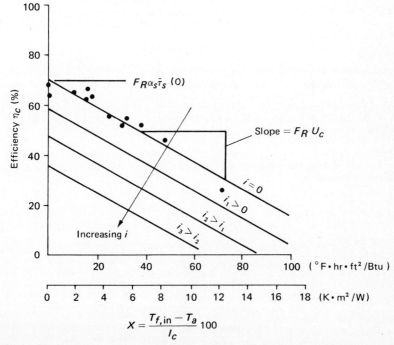

$$X = \frac{T_{f,\text{in}} - T_a}{I_c} 100$$

Fig. 7.13 Typical collector performance with 0° incident flux angle. Also shown qualitatively is the effect of incidence angle I, which may be quantified by $\bar{\tau}_s \alpha_s(i)/\bar{\tau}_s \alpha_s(0) = 1.0 + b_0(1/\cos I - 1.0)$, where b_0 is the incidence angle modifier determined experimentally (ASHRAE 93-77) or from the Stokes and Fresnel equations. (See Chap. 4.)

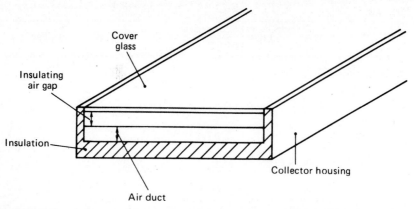

Fig. 7.14 Schematic diagram of a basic air-heating flat-plate collector with a single glass (or plastic) cover.

determined fluid inlet temperature as shown in Eq. (7.32). The heat-removal factor F_R for a typical air collector such as the one shown in Fig. 7.14 is

$$F_R = \frac{\dot{m}_a c_{p,a}}{U_c A_c} \left[1 - \exp \left(-\frac{F'U_c A_c}{\dot{m}_a c_{p,a}} \right) \right] \tag{7.48}$$

where F' is given by Eq. (7.47), m_a is the air flowrate, kg/s, and $c_{p,a}$ is the specific heat of air. The heat-removal factor for air collectors is usually significantly less than that for liquid collectors because h_c for air is much smaller than for a liquid such as water.

Example 3

Calculate the collector-plate efficiency factor F' and heat-removal factor F_R for a smooth, 1-m-wide, 5-m-long air collector with the following design. The flowrate is 0.7 m³/min · m²_c (2.1 ft³/min · ft²_c). The air duct height is 1.5 cm (0.6 in), the air density is 1.1 kg/m³ (0.07 lb/ft³), the specific heat is 1 kJ/kg · K (0.24 Btu/lb · °F), and the viscosity is 1.79×10^{-5} kg/m · s (1.2×10^{-5} lb/ft · s). The collector heat-loss coefficient U_c is 18 kJ/h · m² · K (5 W/m² · K; 0.88 Btu/h · ft² · °F).

Solution

The first step is to determine the duct heat-transfer coefficient h_c from Eq. (7.45). The Reynolds number is defined as

$$\text{Re} = \frac{\rho \overline{V} D_H}{\mu}$$

in which the average velocity $\overline{V}$ is the volume flowrate divided by the flow area:

$$\overline{V} = \frac{0.7 \times 1 \times 5}{1 \times 0.015} = 233 \text{ m/min} = 3.89 \text{ m/s}$$

and the hydraulic diameter D_H is

$$D_H = \frac{4(0.015 \times 1)}{1 + 1 + 0.015 + 0.015} = 0.0296 \text{ m}$$

$$\text{Re} = \frac{(1.1)(3.89)(0.0296)}{1.79 \times 10^{-5}} = 7066$$

From Eq. (7.45) the Nusselt number is

$$Nu = \frac{0.0192(7066)^{3/4}(0.72)}{1 + 1.22(7066)^{-1/8}(0.72 - 2.0)} = 22.0$$

The heat-transfer coefficient is

$$h_c = Nu \frac{k}{D_H} = \frac{Nu\, c_p \mu}{Pr\, D_H}$$

The Prandtl number for air is ~ 0.72. Therefore,

$$h_c = \frac{(22.0)(1.0)(1.79 \times 10^{-5})}{(0.72)(0.0296)} = 0.0185 \text{ kJ/m}^2 \cdot \text{s} \cdot \text{K}$$

$$= 66.5 \text{ kJ/m}^2 \cdot \text{h} \cdot {}^\circ\text{C} \ (3.26 \text{ Btu/h} \cdot \text{ft}^2 \cdot {}^\circ\text{F})$$

The plate efficiency is then, from Eq. (7.47),

$$F' = \frac{66.5}{66.5 + 18} = 0.787$$

and the heat-removal factor F_R can be calculated from Eq. (7.48). The mass flowrate per unit area is

$$\frac{\dot{m}}{A_c} = \frac{\rho q}{A_c} = \frac{1.1 \times 0.7}{60} = 0.0128 \text{ kg/s} \cdot \text{m}_c^2$$

Then

$$F_R = \frac{(0.0128)(1)}{(18/3600)} \left\{ 1 - \exp\left[-\frac{(0.787)(18/3600)}{(0.0128)(1)} \right] \right\} = 0.677$$

That is, the particular collector in question can collect 67.7 percent of the heat it could collect if its surface were at the air-inlet temperature. F_R varies weakly with the fluid temperature through the temperature effect upon U_c [see Eq. (7.12)].

Other air collectors are described in Chap. 12 along with complete heating systems using air as the working fluid. The optical efficiency of liquid and air flat plates is unaffected by the special heat-removal features required in air collectors.

TUBULAR SOLAR ENERGY COLLECTORS

Two general methods exist for significantly improving the performance of solar collectors above the minimum flat-plate collector level. The first method increases solar flux incident on the receiver; it will be described in the next chapter on concentrators. The second method involves the reduction of parasitic heat loss from the receiver surface. Tubular collectors, with their inherently high compressive strength and resistance to implosion, afford the only practical means for completely eliminating convection losses by surrounding the receiver with a vacuum on the order of 10^{-4} mmHg. The analysis of evacuated tubular collectors is the principal topic of this section.

Tubular collectors have a second application. They may be used to achieve a small level of concentration—1.5 to 2.0—by forming a mirror from part of the internal concave surface of a glass tube. This reflector can focus radiation on a receiver inside the tube. Since such a receiver is fully illuminated, it has no parasitic "back" losses. This low-level concentrator can also be warranted for performance improvement. Performance may be improved to levels between those for full evacuation and those for no evacuation, without vacuum and its sealing difficulties, by filling the envelope with high-molecular-weight noble gases. See Chap. 4 for heat-transfer correlations usable with heavy gases. External concentrators of radiation may also be coupled to an evacuated receiver for improvement of performance over the simple evacuated tube. Collectors of this type are described briefly below.

Evacuated-Tube Collectors

Evacuated-tube devices have been proposed as efficient solar energy collectors since the early 20th Century. In 1909, Emmett[6] proposed several evacuated-tube concepts for solar energy collection, two of which are being sold commercially today. Speyer[20] also proposed a tubular evacuated flat-plate design for high-temperature operation. With the recent advances in vacuum technology, evacuated-tube collectors can be reliably mass produced. Their high-temperature effectiveness is essential for the efficient operation of solar air-conditioning systems and process heat systems.

Figure 7.15 shows schematic cross sections of several glass evacuated-tube collector concepts. The simplest design is basically a small flat-plate collector housed in an evacuated cylinder (Fig. 7.15a). If the receiver is metal, a glass-to-metal seal is required to maintain a vacuum. In addition, a thermal short may occur from inlet to outlet tube unless special precautions are taken. Alternatively, an all-glass collector can be made from concentric glass tubes as shown in Fig. 7.15b. This collector avoids a glass-to-metal seal but has very limited working fluid pressurization capability. Some investigators have proposed the use of a square absorber circumscribed within the circular region shown. An increased concentration effect would result, but the pressurability of the absorber is reduced.

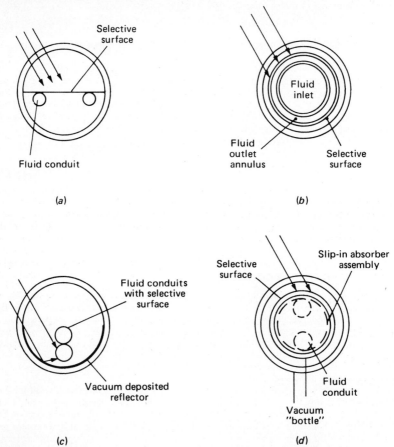

Fig. 7.15 Evacuated-tube solar energy collectors: (a) flat plate; (b) concentric tubular; (c) concentrating; (d) vacuum bottle with slip-in heat exchanger contacting rear surface of receiver.

Mildly concentrating, tubular collectors can be made using the design of Fig. 7.15c. Either a single flow-through receiver with fins or a double U-tube as shown can be used. Concentration ratios of from $2/\pi$ to 2.0 can be achieved ideally with this design, but a glass-to-metal seal is required.

One of Emmett's designs is shown by Fig. 7.15d. It consists of an evacuated vacuum bottle much like an unsilvered, wide-mouth Dewar flask into which a metal heat exchanger is inserted. The outer surface of the inner glass tube is the absorber. The heat generated is transferred through the inner glass tube to the metal slip-in heat exchanger. Since this heat transfer is through a glass-to-metal interface that has only intermittent point contacts, significant axial temperature gradients can develop, thereby stressing the glass tube. In addition, a large temperature difference can exist between the inner and outer glass tubes. At the collector ends where the two tubes are joined, a large temperature gradient and consequent thermal stress can exist.

The level of evacuation required for suppression of convection and conduction can be calculated from basic heat-transfer theory. As the tubular collector is evacuated, reduction of heat loss first occurs because of the reduction of the Rayleigh number. The effect is proportional to the square root of density. When the Rayleigh number is further reduced below the lower threshold for convection, the heat-transfer mechanism is by conduction only. For most gases the thermal conductivity is independent of pressure if the mean free path is less than the heat-transfer path length.

For very low pressure, the conduction heat transfer in a narrow gap[5] is

$$q_k = \frac{k\,\Delta T}{g + 2p} \tag{7.49}$$

where g is the gap width and p is the mean free path. For air the mean free path at atmospheric pressure is about 70 nm. If 99 percent of the air is removed from a tubular collector, the mean free path increases to 7 μm and conduction heat transfer is affected very little. However, if the pressure is reduced to 10^{-3} torr the mean free path is 7 cm, which is substantially greater than the heat-transfer path length, and conduction heat transfer is effectively suppressed. The relative reduction in heat transfer as a function of mean free path can be derived from Eq. (7.49):

$$\frac{q_{\text{vac}}}{q_k} = \frac{1}{1 + 2p/g} \tag{7.50}$$

where q_k is the conduction heat transfer if convection is suppressed and q_{vac} is the conduction heat transfer under a vacuum. Achieving a vacuum level of 10^{-3} to 10^{-4} torr for a reasonably long period of time is within the grasp of modern vacuum technology.

Thermal and optical analyses of tubular collectors are based on the principles used to analyze other collectors in previous sections. The evacuated-tube flat-plate design (Fig. 7.15a) and the reflecting design (Fig. 7.15c) can be analyzed using exactly the same procedure as that used for flat-plate collectors. The convection term in the receiver and envelope energy balances is simply set to 0. Analysis of the slip-in heat-exchange type is problematic since the glass-to-metal point contact is difficult to quantify so that production tolerances and surface precision and smoothness can be included.

The concentric-tube collector (CTC) illustrates several important principles in collector analysis and will be analyzed in detail in the next section. This collector type is manufactured in the United States by Owens-Illinois, Inc. (OII). Test data on the OII collector will be presented in order to verify the analysis that follows.

Optical Analysis of the CTC

Since close packing of CTC tubes in an array can result in shading losses at any angle other than normal incidence, it is cost-effective to space the tubes apart and to use a back reflector in order to capture any radiation passing between the tubes. Figure 7.16a shows the geometry of part of a CTC array with tube spacing d; Fig. 7.16b is a cutaway drawing of one tubular assembly. Evacuated-tube flat-plate designs and concentrating designs should be closely packed to optimize solar energy collection. Beekley and Mather[1] have analyzed the CTC in detail and their analysis is the basis of this section.

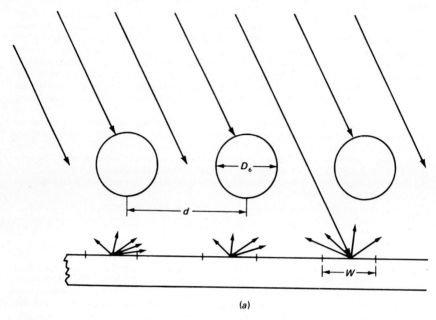

D_6

d

w

(a)

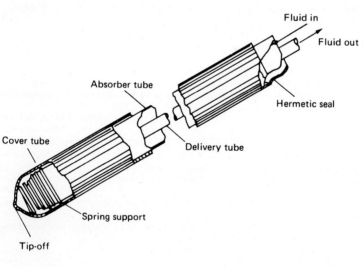

Fluid in

Fluid out

Absorber tube

Hermetic seal

Cover tube

Delivery tube

Spring support

Tip-off

(b)

Fig. 7.16 (a) Schematic diagram of concentric-tube collector optics; (b) cutaway view of evacuated-tube solar collector manufactured by Owens-Illinois, Inc. For important dimensions see Table 7.3. (*From Ref. 1.*)

CTC arrays can collect both direct and diffuse radiation. Each radiation component must be analyzed in turn. The optical efficiency η_o may be expressed as

$$\eta_o = \frac{\tau_e \alpha_r I_{\text{eff}}}{I_{b,c} + I_{d,c}} \tag{7.51}$$

where I_{eff} is the effective solar radiation both directly intercepted and intercepted after reflection from the back reflector and $I_{b,c}$ and $I_{d,c}$ are respectively, the beam and diffuse radiation components intercepted per unit collector aperture area. τ_e and α_r are the envelope transmittance and receiver absorptance, respectively.

The beam component of radiation intercepted by the tubes (with no shading loss) is

$$I_{b,t} = I_b \cos i_t \tag{7.52}$$

where i_t is the tube incidence angle.[1]

The reflection of beam radiation from a diffusely reflecting back surface can be calculated by assuming that an illuminated strip W reflects reflectance ρ onto the rear surfaces of the nearby tubes. The amount of reflected radiation intercepted can be evaluated using radiation shape factors (see Chap. 4). Although the analysis of this radiation problem is quite lengthy, it can be shown that the amount of reflected radiation intercepted by all tubes $I_{b,r}$ is

$$I_{b,r} = I_{b,c} \rho \Delta \frac{W}{D_6} \tag{7.53}$$

where Δ is a sum of shape factors equal to about 0.6 to 0.7 for tubes spaced one diameter apart, that is, $d = 2D_6$.

Diffuse radiation intercepted directly by the tubes $I_{d,t}$ (per unit collector plane area) is

$$I_{d,t} = \pi I_{d,c} F_{TS} \tag{7.54}$$

where F_{TS} is the radiation shape factor from a tube to the sky dome. F_{TS} depends on tube spacing and lies in the range for closely packed to infinitely spaced tubes (0.27 to 0.50, respectively). For tubes spaced one diameter apart, $F_{TS} \approx 0.43$.

Diffuse radiation reflected from the array back surface $I_{d,r}$ can be expressed as

$$I_{d,r} = \pi F_{TS} I_{d,c} \rho \overline{F} \tag{7.55}$$

The total effective insolation I_{eff} can be calculated by summing directly intercepted and reflected radiation, that is,

$$I_{\text{eff}} = I_{b,c} \left(\cos i_t + \cos i_c \, \rho \Delta \frac{W}{D_6} \right) + I_{d,c} \left[\pi F_{TS} \left(1 + \rho \overline{F} \right) \right] \tag{7.56}$$

The optical efficiency η_o [Eq. (7.51)] is not, therefore, a simple collector property independent of operating conditions, but rather a function of time through the incidence angles i_t and i_c, with values increasing away from solar noon for a given solar radiation level.

Thermal Analysis of the CTC

The heat loss from a CTC occurs primarily through the mechanism of radiation from the absorber surface. The rate of heat loss per unit absorber area q_L then can be expressed as

$$q_L = U_c (T_r - T_a) \tag{7.57}$$

Total thermal resistance $1/U_c$ is the sum of three resistances:

R_1—radiative exchange from absorber tube to cover tube
R_2—conduction through glass tube
R_3—convection and radiation to environment

The overall resistance is then

$$\frac{1}{U_c A_c} = R_1 + R_2 + R_3 \tag{7.58}$$

The conductances R_i^{-1} are

$$R_1^{-1} = \frac{1}{1/\epsilon_r + 1/\epsilon_e - 1} \sigma(T_r + T_{ei})(T_r^2 + T_{ei}^2) A_c \tag{7.59}$$

$$R_2^{-1} = \frac{2k}{D_r \ln (D_6/D_{ei})} A_c \tag{7.60}$$

$$R_3^{-1} = [h_c + \sigma\epsilon_e(T_{eo} + T_a)(T_{eo}^2 + T_a^2)] \frac{D_6}{D_r} A_c \tag{7.61}$$

where the subscript e denotes envelope properties, T_r is the receiver (absorber) temperature, and h_c is the external convection coefficient for the envelope. Test data have shown that the loss coefficient U_c is between 0.5 and 1.0 W/m² · °C, thus confirming the analysis.

The CTC energy delivery rate q_u on an aperture area basis can be written as

$$q_u = \tau_e \alpha_r I_{\text{eff}} \frac{A_t}{A_c} - U_c (T_r - T_a) \frac{A_r}{A_c} \tag{7.62}$$

where A_t is the projected area of a tube (its diameter) and A_r is the receiver or absorber area. The receiver-to-collector aperture area ratio is $\pi D_r/d$. Therefore

$$q_u = \frac{D_r}{d} [\tau_e \alpha_r I_{\text{eff}} - \pi U_c (T_r - T_a)] \tag{7.63}$$

TABLE 7.2 Physical and Thermal Characteristics of the Owens-Illinois Evacuated-Tube Solar Collector*

Component	Size or value
Glass cover tube	
OD	2.09 in
ID	1.93 in
Glass absorber tube	
OD	1.69 in
ID	1.54 in
Delivery tube	
OD	0.47 in
ID	0.35 in
Tube active length	42.0 in
Overall collector module size	24 tubes
	4 × 8 ft
	27.4 ft² net area
Tube spacing in module	4.18 in on centers
Weight	3.7 lb/ft² dry
	6.7 lb/ft² water-filled
Vacuum level	10^{-3} to 10^{-4} mmHg
Absorber solar absorptance	0.9 (approximate)
Absorber solar emittance	0.1 or less
Cover transmittance	0.92
Specific heat	2.5 Btu/ft² · °F (approximately 5 times typical flat-plate value)

*From Ref. 1.

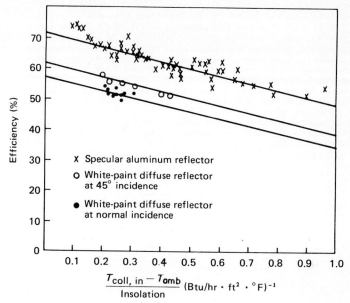

Fig. 7.17   Owens-Illinois, Inc., evacuated-tube solar collector efficiency. (*From Owens-Illinois brochure, 1977.*)

Beekley and Mather[1] have shown that a tube spacing one envelope diameter D_6 apart maximizes daily energy gain. Test data on such a configuration with properties shown in Table 7.2 are shown in Fig. 7.17 for both a diffuse reflector and a specularly reflecting, cylindrical back surface. The specular reflector improves performance by 10 percent or more.

REFERENCES

1. D. C. Beekley and G. R. Mather, Analysis and Experimental Tests of a High Performance, Evacuated Tube Collector, Owens-Illinois, Toledo, OH, 1975.
2. R. W. Bliss, The Derivations of Several "Plate Efficiency Factors" Useful in the Design of Flat-Plate Solar-Heat Collectors, *Sol. Energy*, vol. 3, p. 55, 1959.
3. F. deWinter, Solar Energy and the Flat Plate Collector, *ASHRAE Rept.* S-101, 1975.
4. J. A. Duffie, and W. A. Beckman, "Solar Energy Thermal Processes," Wiley, New York, 1974.
5. S. Dushman, in "Scientific Foundations of Vacuum Technology," 2d ed., J. M. Lafferty (ed.), Wiley, New York, 1962.
6. W. L. R. Emmett, Apparatus for Utilizing Solar Heat, U.S. Patent 980,505,1911.
7. J. E. Hill and E. R. Streed, A Method of Testing for Rating Solar Collectors Based on Thermal Performance, *Sol. Energy*, vol. 18, pp. 421–431, 1976.
8. J. E. Hill, E. R. Streed, G. E. Kelly, J. C. Geist, and T. Kusuda, Development of Proposed Standards for Testing Solar Collectors and Thermal Storage Devices, *NBS Tech. Note* 899, 1976.
9. H. C. Hottel and A. F. Sarofim, "Radiative Transfer," McGraw-Hill Book Company, New York, 1967.
10. H. C. Hottel and B. B. Woertz, Performance of Flat-Plate Solar-Heat Collectors, *Trans. Am. Soc. Mech. Eng.*, vol. 64, p. 91, 1942.
11. H. C. Hottel and A. Whillier, Evaluation of Flat-Plate Collector Performance, *Trans. Conf. Use Sol. Energy*, vol. 2, part 1, p. 74, 1958.
12. E. Kauer, R. Kersten, and F. Madrdjuri, Photothermal Conversion, *Acta Electron.*, vol. 18, pp. 297–304, 1975.
13. S. A. Klein, Calculation of Flat-Plate Collector Loss Coefficients, *Sol. Energy*, vol. 17, pp. 79–80, 1975.
14. S. A. Klein, J. A. Duffie, and W. A. Beckman, Transient Considerations of Flat-Plate Solar Collectors, *J. Eng. Power*, vol. 96A, pp. 109–114, 1974.
15. F. Kreith, Evaluation of Focusing Solar Energy Collectors, *ASTM Stand. News*, vol. 3, pp. 30–36, 1975.

16. F. Kreith and J. F. Kreider, *Principles of Solar Engineering,* McGraw-Hill, New York, 1978.
17. M. A. S. Malik and F. H. Buelow, Hydrodynamic and Heat Transfer Characteristics of a Heated Air Duct, in "Heliotechnique and Development." (COMPLES 1975), vol. 2, pp. 3–30, Development Analysis Associates, Cambridge, MA, 1976.
18. M. A. S. Malik and F. H. Buelow, Heat Transfer in a Solar Heated Air Duct—A Simplified Analysis, in "Heliotechnique and Development" (COMPLES 1975), vol. 2, pp. 31–37, Development Analysis Associates, Cambridge, MA, 1976.
19. F. F. Simon, Flat Plate Solar Collector Performance Evaluation with a Solar Simulator as a Basis for Collector Selection and Performance Prediction, *NASA Rept.* TM X-71793, 1975; see also *Sol. Energy,* vol. 18, pp. 451–466, 1976.
20. F. Speyer, Solar Energy Collection with Evacuated Tubes, *J. Eng. Power,* vol. 87, p. 270, 1965.
21. H. Tabor, Radiation, Convection, and Conduction Coefficients in Solar Collectors, *Bull. Res. Coun. Isr.,* vol. 6C, p. 155, 1958.

APPENDIX: HEAT EXCHANGER PENALTY FACTOR
USED IN LINEAR SOLAR COLLECTOR MODEL

Although a heat exchanger is not strictly a part of a solar collector, it is present as a part of many solar collector loops. In this appendix a simple method of incorporating heat-exchanger effects into the linear model of a flat-plate collector is given. Heat exchangers can ideally execute a no-loss energy exchange between fluid streams, but the exchange is always accompanied by a temperature decrement and resultant loss in available energy. This temperature penalty will require a solar collector to operate at a higher temperature than a system lacking a heat exchanger, in order to deliver the same fluid temperature to the load. Since higher collector temperatures result in lower collection efficiency, the use of a heat exchanger results in lower solar system energy delivery. This phenomenon is illustrated in the following example.

EXAMPLE 7.A1

Calculate the efficiency at which the solar collector in Fig. 7.A1 operates in order to deliver energy to a working fluid at 65°C for several values of the approach temperature difference. The approach temperature is the difference between the incoming cool fluid and the exiting warm fluid at the heat exchanger. The temperature rise through the heat exchanger is 10°C and is equal to the fluid temperature rise through the solar collector.

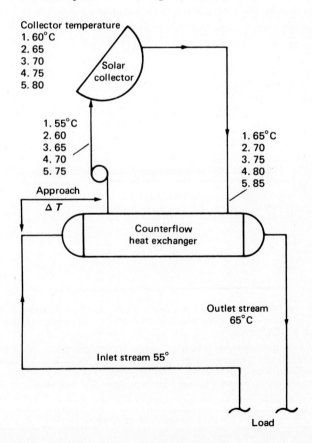

FIGURE 7.A1.   Fluid stream and solar collector temperatures for example. *(From F. Kreith and J. F. Kreider, Principles of Solar Engineering, Hemisphere, Washington, DC, 1978.)*

The collector has a $\tau\alpha$ product of 0.80 and loss coefficient U_c of 5 W/m^2 · K. If the solar radiation normal to the collector surface is 500 W/m^2 and the ambient temperature T_a is 20°C, calculate the collector efficiency for the following five values of approach ΔT:

Case	Approach ΔT, °C
1	0 (thermodynamic limit)
2	5
3	10
4	15
5	20

SOLUTION

Fluid stream temperatures are shown at several points for each case in Fig. 7.A1. The collector temperature T_c can be taken as the average of inlet and outlet fluid temperatures for purposes of the example as shown in the figure.

Collector efficiency can be calculated from the following equation

$$\eta = \tau\alpha - U_c\left(\frac{\Delta T}{I_c}\right)$$

where $\Delta T = \overline{T}_c - T_a$

For the collector specified here

$$\eta = \left(0.8 - 5\frac{\Delta T}{I_c}\right) \times 100\ \%$$

The collector efficiency can be calculated in tabular form as shown in Table 7.A1.

TABLE 7.A1   Collector Efficiency Calculation for Example

Case	$\overline{T}_c$, °C	ΔT, °C	$\Delta T/I_c$, K · m^2/W	η, %
1	60	40	0.08	40
2	65	45	0.09	35
3	70	50	0.10	30
4	75	55	0.11	25
5	80	60	0.12	20

Note that the heat-exchanger design has a very strong effect on collector efficiency; for relatively small changes of approach temperature difference, efficiency and, therefore, energy delivery change significantly.

A general analytical discussion of heat exchangers is contained in Chapter 4. In this section the integration of heat exchangers into solar-thermal systems is discussed. DeWinter* has developed a closed-form expression for the energy delivery penalty resultant from the use of a collector-to-storage or collector-to-load heat exchanger. The heat-exchanger penalty F_{hx} is defined as

$$F_{hx} = \frac{q_u\ (\text{with exchanger})}{q_u\ (\text{with no exchanger and collector fluid inlet equal to the heat-exchanger inlet temperature from storage})}$$

*Solar Energy, 17, pp. 335–337, 1975.

The analytical expression for F_{hx} is

$$F_{hx} = \frac{1}{1 + [F_R U_c A_c/(\dot{m}c_p)_c][(\dot{m}c_p)_c/(\dot{m}c_p)_{min}\epsilon_{hx} - 1]} \qquad (7.A1)$$

where F_R = the collector heat-removal factor
U_c = the collector loss coefficient
A_c = the collector area
$(\dot{m}c_p)_c$ = the fluid capacitance rate through the collector array
$(\dot{m}c_p)_s$ = the fluid capacitance rate from load or storage
$(\dot{m}c_p)_{min}$ = the minimum $[(\dot{m}c_p)_c, (\dot{m}c_p)_s$
ϵ_{hx} = the heat-exchanger effectiveness (see Chapter 4)

The collector thermal properties are contained in the group $F_R U_c A_c$, while the information for operating conditions (flowrate) is contained in the two groups $(\dot{m}c_p)_c$ and $(\dot{m}c_p)_s$. The heat-exchanger effectiveness is defined as the ratio of heat exchanged to the maximum possible heat exchange amount limited by the second law of thermodynamics. Figure 7.A2 is a plot of the heat-exchanger penalty factor F_{hx}. Equation (7.A1) shows that the heat-exchanger penalty F_{hx} decreases as heat-exchanger effectiveness increases. It is therefore desirable to use a heat exchanger with largest possible effectiveness. The heat-exchanger factor F_{hx} is sometimes denoted by (F'_R/F_R).

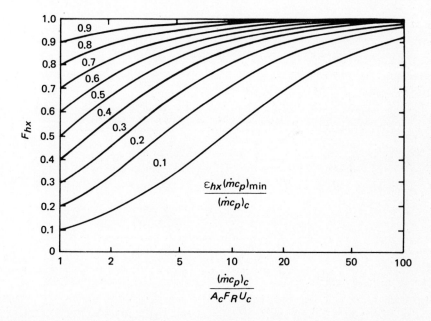

FIGURE 7.A2   Heat exchanger penalty factor F_{hx}. In most solar-heating systems $F_{hx} > 0.90$. (*From F. Kreith and J. F. Kreider, Principles of Solar Engineering, Hemisphere, Washington, DC, 1978.*)

Chapter **8**

Intermediate Concentration Solar Thermal Collectors

ARI RABL
SERI, Golden, CO

INTRODUCTION

Concentration of solar radiation becomes necessary when higher temperatures are desired than can be obtained with a flat-plate collector or when, as in the case of expensive receivers with evacuated envelopes, the cost of the receiver, per unit area, is higher than the cost of mirrors, per unit aperture area.

Concentration Ratio

Two definitions of concentration ratio (CR) are natural and have been in use; to avoid confusion a subscript should be added whenever the context does not clearly specify which definition is meant. The first definition is strictly geometrical, the ratio of aperture area A_a to receiver area A_r:

$$CR = CR_{area} = \frac{A_a}{A_r} \tag{8.1}$$

and the names "geometric concentration," or "area concentration" for short, are recommended. The second definition, in terms of the ratio of intensity at aperture to that at receiver

$$CR_{flux} = \frac{I_a}{I_r} \tag{8.2}$$

involves absorption effects in addition to geometry and should be referred to as "flux concentration." While flux concentration is a useful concept in photovoltaic work, the geometrical definition is more appropriate for solar thermal collectors; therefore, throughout this chapter, "concentration" shall mean "area concentration" even if not explicitly stated.

Closely related to the concentration is the "acceptance angle," i.e., the angular range over which all or almost all rays are accepted without moving all or part of the collector. The second law of thermodynamics requires that the maximum possible concentration for a given collector acceptance half angle θ_c is[1,2]

$$CR_{ideal, 2D} = 1/\sin \theta_c \qquad (8.3)$$

for two-dimensional (troughlike) concentrators, and

$$CR_{ideal, 3D} = 1/\sin^2 \theta_c \qquad (8.4)$$

for three-dimensional ones (cones, dishes, pyramids).*

Since the angular radius of the sun is $\Delta_s \approx \frac{1}{4}°$, the thermodynamic limit of a tracking solar concentrator is about 200 in two-dimensional (line focus) geometry, and about 40,000 in three-dimensional (point focus) geometry. The concentration achievable in practical systems is reduced by a number of factors:

1. Most conventional concentrators, in particular focusing parabolas or lenses, are based on optical designs which fall short of the thermodynamic limit by a factor of 2 to 4.

2. Tracking errors and errors in mirror contour and receiver alignment necessitate design acceptance angles considerably larger than the angular width of the sun.

3. No lens or mirror material is a perfectly specular reflector and, therefore, the acceptance angle must be enlarged further; the nonspecular effect is aggravated by dirt and dust.

4. Due to atmospheric scattering, a significant portion of the solar radiation may come from directions other than the solar disk itself.

The concentration ratio in a solar concentrator involves a compromise between optical and thermal performance. The absorber should be chosen as small as possible to reduce heat loss, yet large enough to intercept all, or almost all, incident radiation. One therefore has to consider the rays with the largest expected deviation θ_c from the design direction, i.e., the direction from collector aperture to center of sun. This angular deviation θ_c is due to the finite size of the sun and to mirror and tracking errors. The example of the parabolic-trough reflector with cylindrical absorber tube, in Fig. 8.1, illustrates the definition of θ_c. The absorber tube is

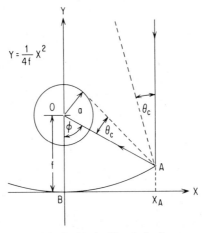

Fig. 8.1 Focusing parabola showing focal length f and acceptance half-angle θ_c.

placed concentrically around the focal line. If the ray with the largest deviation is to just reach the absorber, as shown by the dashed line in Fig. 8.1, then the concentration must be

$$CR_{2 D, parab., cyl. abs.} = \frac{2x_A}{2\pi a} = \frac{\sin \phi}{\pi \sin \theta_c} = \frac{\sin \phi}{\pi} CR_{ideal} \qquad (8.5)$$

where ϕ is the rim angle $\angle AOB$. The maximum occurs at $\phi = 90°$ and falls a factor π short of the ideal limit. This is typical of all single-stage focusing concentrators; i.e., they achieve only one-fourth to one-half of the thermodynamic concentration limit. For practical installations, geometric concentration is not the only design criterion, and slightly different rim angles will

*The optics of a concentrator can be described in terms of the f-number. See Chap. 3 of this handbook.

be used. For example, rim angles beyond 90° and undersized adsorbers can be used if the reflector cost is low.

There is, however, a class of nonimaging concentrators, the CPC described in the following section, which actually reaches the thermodynamic limit of concentration. Furthermore, a conventional focusing system with a matching CPC as second-stage concentrator can closely approach the thermodynamic limit. The choice of optimal concentration for a given application involves evaluation of many other factors—optical, climatic, thermal, economic, etc.—and it is unlikely that any single concentrator type will be the best for all applications. It is therefore appropriate to analyze many solar concentrator types. In the present chapter, the emphasis will lie on systems with low to intermediate concentration. The distinction between low and intermediate concentration on the one hand and high concentration on the other is made most naturally in terms of tracking and accuracy requirements. In other words, the present chapter will be concerned with concentrators which have a large acceptance half angle, on the order of 3° and larger, and thus great tolerance to mirror and tracking errors. Their concentration ratios will be in the range of 1 to about 10 in two-dimensional, and to about 100 in three-dimensional geometry. The low concentration range is of interest because the magnitude of the daily and yearly solar motion permits concentration ratios up to about 2 with fixed concentrators, and up to 10 with nontracking concentrators whose tilt is adjusted from day to day.

Acceptance of Diffuse Radiation

A very important property of solar collectors with low concentration is their ability to accept a significant portion of the diffuse sky radiation. A precise calculation of this effect would require detailed information about the angular distribution of diffuse sky radiation. In view of the limited data available on this distribution, one is justified in assuming that the total insolation I_T is the sum of the direct component I_B ("beam") and an isotropic background I_D (diffuse)

$$I_T = I_B + I_D \tag{8.6}$$

From a simple argument involving radiation-shape factors, it follows that the fraction of the isotropic component I_D which is accepted by a concentrator of concentration CR can be taken to be 1/CR, regardless of any detail of the optics.[2-5]

Due to the predominance of near-forward scattering in the atmosphere, the sky radiation tends to be centered around the sun, and therefore the isotropic model will slightly underestimate the actual acceptance for diffuse sky radiation. Let us designate by γ the fraction of the total radiation I_T which falls within the acceptance angle of a solar concentrator. In terms of γ the above discussion can be summarized by the lower bound

$$\gamma \geq \frac{I_B}{I_T} + \frac{1}{CR}\left(1 - \frac{I_B}{I_T}\right) \tag{8.7}$$

Data taken at Argonne National Laboratory indicate that γ is not much larger than this lower bound. Preliminary values of γ for different acceptance angles and weather conditions are listed in Table 8.1.

TABLE 8.1   Estimates of Energy Collection for Different Collector Designs*

		Energy available to collector as percent of insolation			
				CPC collector¶	
Conditions	Total insolation† (W/m²)	Direct beam‡	CR = 10	CR = 5	CR = 3
Light haze-blue sky	≈1000	88	89	91	92
Heavy haze-white sky	≈ 920	79	82	85	87

*Courtesy of Dr. K. A. Reed of Argonne National Laboratory.
†Pyranometer at normal incidence.
‡Normal incidence pyrheliometer.
¶The values of CR heading each column refer to concentration ratio.

TYPES OF CONCENTRATORS

Compound Parabolic Concentrators

Concentrators which reach the thermodynamic limit of concentration

$$CR \quad 1/\sin \theta_c \qquad \text{in two dimensions} \tag{8.8}$$

$$CR = 1/\sin^2 \theta_c \qquad \text{in three dimensions} \tag{8.9}$$

with an acceptance half angle θ_c have been called ideal concentrators due to their optical properties. In the solar energy literature, names such as "compound parabolic concentrators" (abbreviated as "CPC") and "nonimaging concentrators" have also been used. We shall refer to all concentrators of this class as CPC, even though some of them are not exactly parabolic in shape.

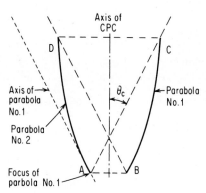

Fig. 8.2 Compound parabolic concentrator (CPC) showing foci, axes, and parabola branches.

The first example of a CPC, shown in Fig. 8.2, was found[7] independently in the United States, Germany, and the U.S.S.R. in the mid-1960s. It consists of parabolic reflectors which funnel the radiation from aperture to absorber. The right and left half belong to different parabolas, as expressed by the name CPC. The axis of the right branch, for instance, makes an angle θ_c with the collector midplane, and its focus is at A. At the end points C and D, the slope is parallel to the collector midplane.

Tracing a few sample rays reveals that this device has the following angular acceptance characteristic: all rays incident* on the aperture within the acceptance angle, i.e., with $|\theta_{in}| < \theta_c$, will reach the absorber, while all rays with $|\theta_{in}| > \theta_c$ will bounce back and forth between the reflector sides and eventually re-emerge through the aperture. This property, plotted schematically by the solid line in Fig. 8.3, implies that the concentration is equal to the thermodynamic limit, Eqs. (8.8) and (8.9).

Subsequent to the discovery of the basic CPC, Fig. 8.2, several generalizations of the ideal concentrator have been described which are relevant to special applications. These generalizations concern:

1. The use of arbitrary receiver shapes,[8,9] for example, fins and tubes (the latter being important because of their ability to carry a heat transfer fluid); see Fig. 8.4.

2. The restriction of exit angles θ_{out} at the receiver[10] to values $|\theta_{out}| \leq \theta_2 < \pi/2$ (important because some receivers have poor absorptance at large angles of incidence); see Fig. 8.5.

3. The asymmetric orientation of source and aperture[2] (for the design of collectors with seasonally varying outputs); see Fig. 8.6.

4. The matching of a CPC to a finite source of radiation[10] (because second-stage concentrators, discussed below, must collect radiation from a source, viz., the first stage, which is a finite distance away); see Figs. 8.7 and 8.21.

---

*For rays not in a cross-sectional plane, θ_{in} is the projection of the incidence angle i onto the cross-sectional plane of Fig. 8.2.

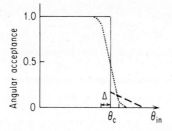

Fig. 8.3 Fraction of the radiation incident on aperture at angle θ_{in} which reaches absorber, for ideal concentrators in two dimensions, with acceptance half-angle θ_c, assuming reflectance $\rho = 1$; full line: untruncated ideal concentrator with perfect reflectors; dashed line: truncated ideal concentrator with perfect reflectors; dotted line: untruncated ideal concentrator with reflector surface erors Δ.

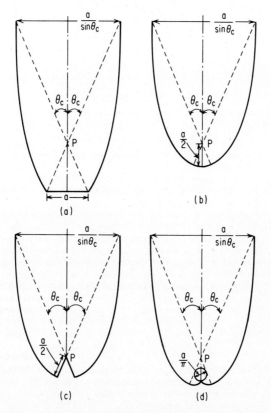

Fig. 8.4 CPC's with different absorber shapes: flat one-sided (*a*), fin (*b*), wedge (*c*), tube (*d*), all drawn to scale for the same aperture and concentration ratio.

The design goal is to maximally concentrate radiation, subject to any of the subsidiary conditions 1 to 4 that may have been specified. The concentrator should consist of reflecting surfaces only, and the number of reflections should be as small as possible. The design of two-dimensional concentrators is determined uniquely by the extreme or edge rays (see Chap. 3). Extreme rays are defined as rays coming from the edge of the source. In three dimensions (geometry of cones and pyramids), the design is in general overdetermined, but at least for flat

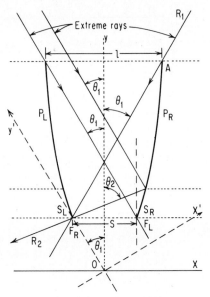

Fig. 8.5 $\theta_1 - \theta_2$ transformer, consisting of parabolic section P_R and P_L and of straight sections S_R and S_L. The slope γ of the straight section is $\gamma = (\theta_2-\theta_1)/2$. $\theta_1 = \theta_c$ is the acceptance half-angle.

receivers a good compromise is achieved by choosing a surface of revolution whose cross section has been determined by the two-dimensional solution[1].

The general solution is to maximize the slope of the profile curve of the mirror, subject to the condition that extreme rays illuminate the absorber within the prescribed angular limits $\pm\theta_2$. This implies that rays originating at angle θ_2 from a convex absorber (flat absorbers are treated as limiting cases of convex absorbers, and in nonconcave absorbers cavities are replaced by flat chords stretched across cavity openings) emerge from the concentrator as extreme rays after undergoing at most one reflection. The procedure is illustrated in Fig. 8.7. Radiant energy emanating from diffuse source S and entering the aperture BB' is concentrated onto the absorber R. The mirror curve must have a slope such that a ray emitted from any point P of the absorber at an angle $+\theta_2$ $(-\theta_2)$ may be reflected toward the edge $A'(A)$ of the source. The mirror starts at the edge $C(C')$ of the absorber, and it terminates at the intersection $B(B')$ with the limiting rays $AC'(A'C)$ to avoid shadowing of the absorber. Then the concentration achieved is given by

$$\text{CR} = \begin{cases} n \sin \theta_2/F_{A-S} & \text{in two dimensions} \\ (n \sin \theta_2)^2/F_{A-S} & \text{in three dimensions} \end{cases} \tag{8.10}$$

where n = index of refraction of medium surrounding absorber and F_{A-S}-the radiation shape factor (see Chap. 4) for diffuse radiation from aperture to source. The concentrator slope is uniquely determined by a first-order differential equation with one boundary condition (reflector curve must pass through edge of absorber).

As to the choice among various CPC types, the configurations with fin or tube absorbers are preferable for most solar applications. Not only is the absorber material used more efficiently than in other designs, but heat losses through the back are low. This may be quite an important advantage, because it may not be cost-effective to reduce the effective U-value of the back of a collector much below 0.5 W/m² · K. Compared to frontal U-values for CPC-type solar collectors[11]—approximately 3 W/m² · K for threefold and 1.4 W/m² · K for tenfold concentration— the losses through the back are indeed significant. Thus the reduction in backlosses possible with configurations in Figs. 8.4 (b) and (d) will more than compensate[12] for the slightly higher optical losses [the average number of reflections for all configurations of Fig. 8.4 (b), (c), and (d) can be obtained within an acceptable approximation by adding 0.5 to $<n>$ for the CPC of Fig. 8.4 (a)].

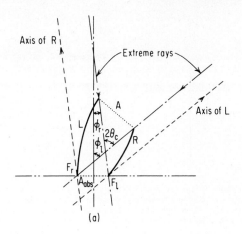

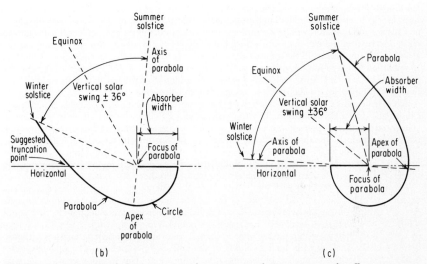

Fig. 8.6 (a) Asymmetric ideal concentrator with acceptance angle $2\theta_c = \phi_l + \phi_r$; the effective concentration varies with angle of incidence, A = aperture, A_{abs} = absorber, R = right parabola, L = left parabola, F_r = focus of R, F_1 = focus of L; (b) "sea shell" concentrator with variable effective concentration for maximal output in summer; and (c) in winter.

Examples of asymmetric CPC's are shown in Fig. 8.6. Due to variations of the incidence angle cosine factor from summer to winter, the "sea shell" collector of Fig. 8.6 has maximal output in summer while design (c) peaks in winter. Optimal match to the hot water, air conditioning, or heating load can be provided by the general asymmetric CPC of Fig. 8.6 (a), which is intermediate between Fig. 8.6 (b) and Fig. 8.6 (c). Its acceptance angle is $2\theta_c = \phi_l + \phi_r$ and its geometric concentration is

$$CR = [\sin \tfrac{1}{2} (\phi_l + \phi_r) \cos \tfrac{1}{2} (\phi_l - \phi_r)]^{-1} \tag{8.11}$$

At central incidence the effective concentration, including cosine factor, is $1/\sin \theta_c$.

All CPC types are almost the same in their optical properties. The same relation, as expressed by Eqs. (8.8) and (8.9), exists between their concentration and angular acceptance, with the sharp cutoff implied by Fig. 8.3. The flux distribution at the absorber depends on the angle of incidence at the aperture and on absorber shape, and must be determined by detailed ray

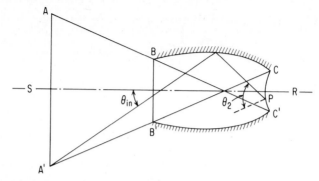

Fig. 8.7 Example design of an ideal concentrator for a finite source. Radiant energy from diffuse source (S) enters aperture BB' and is maximally concentrated onto the absorber (R). The angle of incidence on R is restricted to $|\theta_{out}| \leqslant \theta_2$. A typical extreme ray originates from A', impinges on the aperture with angle of incidence θ_{in}, and, after one reflection, is directed to point P on the absorber with angle of incidence θ_2.

tracing. However, the following important statement can be made about all CPC's without any need for ray tracing: If the radiation incident on the aperture is uniformly spread over the entire acceptance angle, then it will be isotropic when it reaches the absorber—unless the design was chosen to restrict the exit angles to values below $\theta_2 < \pi/2$, in which case the radiation at the absorber will uniformly fill the angular range from $-\theta_2$ to $+\theta_2$. This consideration of uniform illumination[12] is very important because it gives a simple and reliable estimate of the average performance of a CPC solar collector. For certain angles of incidence, hot spots of high local flux concentration (up to about 40) may appear on the absorber.

CPC's have a large reflector area relative to other concentrators. Fortunately this disadvantage can be alleviated by truncation: The top portion of a CPC can be cut off with little loss in concentration since it does not intercept much radiation. Figure 8.8 shows the effect of truncation on concentration and on the ratio of reflector area to aperture area for several design acceptance half angles θ_c. Similarly, Fig. 8.9 shows the effect of truncation on the depth (expressed as height to aperture width ratio) of a CPC. Both figures were calculated for the CPC configuration of Fig. 8.4 (a), but they are approximately correct for all other CPC types.[12]

The number of reflections varies both with angle of incidence θ_{in} and with point of incidence on the aperture. A very good estimate of the transmittance is given by the simple formula $\tau = \rho^{<n>}$ where $<n>$ is the average number of reflections. For incidence at angle θ_{in} one needs $<n(\theta_{in})>$, the average over all aperture incidence points at angle θ_{in}, while $<n>$, the average of $<n(\theta_{in})>$ over all θ_{in} within the acceptance angle, is relevant for average performance. Figure 8.10 shows $<n>$, along with the high and low values of $<n(\theta_{in})>$ for several acceptance angles (reflector profiles), as a function of concentration (truncation). The variation of $<n(\theta_{in})>$ with θ_{in} decreases with truncation. This feature is important because small variation is desirable for the sake of uniform collector output. In practice one can often neglect the variation of $<n(\theta_{in})>$ and simply work with the average $<n>$. Figure 8.10(a) is based on the standard CPC configuration of Fig. 8.4(a). If fin or tube absorbers are employed, $<n>$ will be higher by about 0.5 for concentration and truncation values of practical interest, i.e., CR between 1.5 and 10; this is shown in Fig. 8.10 (b).

If a CPC is truncated, some rays outside the acceptance angle ($|\theta_{in}| > \theta_c$) can reach the absorber, while of course no rays with $|\theta_{in}| < \theta_c$ are rejected. The resulting increase in angular acceptance is, however, insignificant in most practical applications, as shown in Fig. 8.3. For example, if direct sunlight enters a CPC of Fig. 8.4 (a), moderately truncated to a concentration CR', with $\theta_{in} > \theta_c$, the fraction of radiation reaching the absorber is less than 1/CR', and under these conditions the collector is useless for thermal and marginal for photovoltaic applications. The fraction of isotropic radiation which is accepted is, of course, 1/CR' independent of any details of the concentrator.

Regarding sensitivity to mirror-surface errors, the analysis is equally simple for all CPC's, because their geometry implies that all rays incident near the cutoff angle, i.e., with $|\theta_{in}| \approx \theta_c$, undergo exactly one reflection on their way to the absorber. In almost all practical applications, the acceptance half angle θ_c will be larger than 5°, and it is reasonable to assume

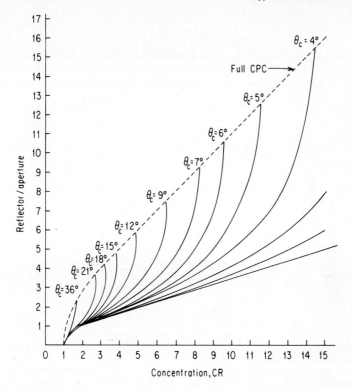

Fig. 8.8   Reflector/aperture ratio as function of concentration for full and for truncated CPC's.

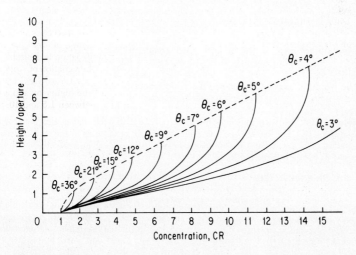

Fig. 8.9   Height/aperture ratio as function of concentration and acceptance half-angle θ.

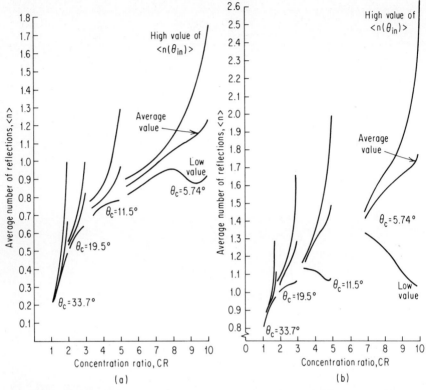

Fig. 8.10 (*a*) Number of reflections for full and for truncated CPC with one-sided flat receiver. The average over all point of impact was taken at each angle of incidence θ_{in} in order to find $\langle n(\theta_{in})\rangle$. For each acceptance half-angle θ_c in this graph, the high and low values of $\langle n(\theta_{in})\rangle$ are shown in addition to the average $\langle n \rangle$ over all $|\theta_{in}| < \theta_c$. For example, if a CPC with $\theta = 11.5°$ is truncated to a concentration of 4, the average number of reflections ranges from a low of 0.76 to a high of 0.86 with a mean of 0.82; (*b*) average number of reflections for CPC with fin receiver [Fig. 8.4(*b*)]. The center curve shows the average value; upper and lower curves show extreme values. Average number of reflections is approximately the same for fin configurations Fig. 8.4 (*b*) and for tube configurations Fig. 8.4 (*d*).

that the mirror surface errors Δ will be fairly small compared to θ_c. Therefore, all of the rays with $|\theta_{in}| < (\theta_c - 2\Delta)$ and none of the rays with $\theta_{in} > (\theta_c + 2\Delta)$ will reach the absorber, while in the transition region, $(\theta_c - 2\Delta) < |\theta_{in}| < (\theta_c + 2\Delta)$, some rays are accepted and some are rejected. The resulting angular acceptance is shown schematically by the dotted line in Fig. 8.3. Further details on optical and geometrical properties of CPC's can be found in Refs. 8 to 13.

The "involute" (Ref. 14) with unit concentration (CR = 1) is included in this chapter as a special case of the CPC. The acceptance half-angle is 90°. The examples in Fig. 8.11 distribute incident radiation over the surface of a tubular (*a*) or fin (*b*) absorber, and the average number of reflections is $\pi/4$ for isotropic radiation.[13]

V-Trough and Side Reflectors

V-trough concentrators and side reflectors[15-17] can be considered straight-line approximations to the symmetric or asymmetric CPC, and they are interesting for some applications because of their simple manufacture. In analyzing their multiple reflections, the method of images is convenient. The angular acceptance of the V-trough of Fig. 8.12(*a*) is shown schematically by the solid line in Fig. 8.12(*b*). Compared to a CPC of the same concentration CR = 1/sin ($\phi + \delta$), the useful acceptance angle δ of a V-trough is significantly smaller. Only when the trough angle ϕ approaches zero, does the acceptance angle approach that of the CPC; however, in this limit the trough becomes too deep and the reflection losses become excessive.

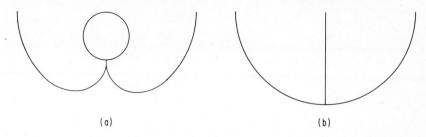

Fig. 8.11 Involute reflector with unit concentration, for tube absorber (*a*), and fin absorber (*b*).

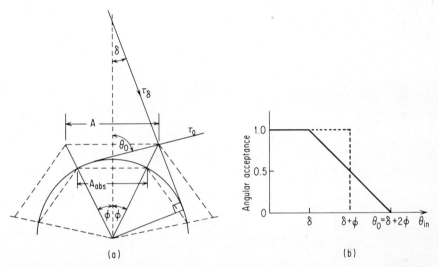

Fig. 8.12 V-trough concentrator, with mirror images and reference circle (*a*). The rays τ_o and τ_δ have angle of incidence θ_o and δ, respectively; they pass through the edge of the absorber and are tangential to the reference circle; angular acceptance of V-trough (*b*) [schematic, neglecting difference between polygon and circle in Fig. 8.12(*a*)].

The higher the desired concentration, the greater the relative advantage of the CPC over the V-trough. The upper limit of concentration for a practical V-trough is about 3 (as a nontracking collector with daily tilt adjustments). With summer/winter adjustments only, the V-trough is limited to concentration values below 2, and for a completely fixed collector a V-trough gives almost no concentration. As for absorber shapes, the V-trough is limited to flat one-sided absorbers.

Straight-side reflectors have been used to boost flat-plate collector performance in summer or winter. Figure 8.13 shows the performance typical of a flat-plate collector with straight-side reflector during the course of the year.

Parabolic Reflectors

One of the best known solar concentrators is the parabolic reflector[18]; it can be built either as trough (two-dimensional geometry, line focus, one-axis tracking) or as dish (three-dimensional geometry, point focus, two-axis tracking). The absorber can take a variety of shapes, the most common being flat or round.

The sizing of the receiver for a given geometry and acceptance half angle has been discussed in the introduction, based on the example of a line-focus parabola with tubular absorber. The same design principle applies for other absorber shapes, but the resulting concentration will be different. Table 8.2 lists the appropriate formulas for the most important geometries. The

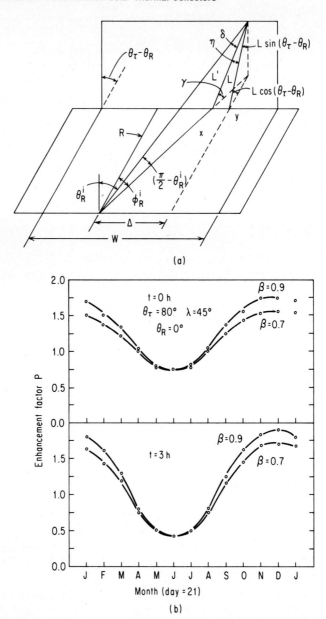

Fig. 8.13 (a) Reflector-collector geometry for calculation of the effective reflector length ratio $R(t)/L$ and enhancement factor P described by Eq. (14) of Ref. 17. $\theta_R{}^i$ and $\theta_R{}^i$ are the zenith and azimuth angles, respectively, of a reflected ray measured as shown on the figure; (b) enhancement factor P plotted as a function of time throughout the year. The abscissa labels time in terms of the twenty-first day of each month. A collector angle of $\beta = 80°$ and a reflector angle of $0°$ were assumed, with latitude = $45°$. The maximum value of R/L was 3.94 for 21 December at $t = 3$ h. Most of the winter months required R/L to be less than 2.0. [*From Ref. 17 D.K. McDaniels et al., "Enhanced solar energy collection," Solar Energy 17:277 (1975).*]

TABLE 8.2 Concentration CR and Reflector/Aperture Area Ratio (A_r/A_a) as a Function of Rim Angle ϕ and Acceptance Half-Angle θ_c for Parabolic Reflectors. The Flat Absorber is One-Sided; Concentration for Two-Sided Flat Absorber Can Be Obtained from ($CR_{2-\text{sided}} = \frac{1}{2}\,(CR_{1-\text{sided}} + 1)$.

Type / Absorber shape	CR	$\frac{1}{4}°$	$\frac{1}{2}°$	$1°$	CR/CR_{ideal}	A_r/A_A
2D (trough)						
Tube	$\dfrac{\sin\phi}{\pi\sin\theta_c}$				$\dfrac{\sin\phi}{\pi}$	$\left\{\left[\dfrac{1}{\cos(\phi/2)} + \left(\cos\dfrac{\phi}{2}\right)\right] \times \log\cos\left(\dfrac{\pi}{4}-\dfrac{\phi}{4}\right)\right\}/2$
Flat	$\dfrac{\sin\phi\cos(\phi+\theta_c)}{\sin\theta_c} - 1$				$\sin\phi\cos(\phi+\theta_c) - \sin\theta_c$	
With optimal ϕ:						
Round tube $\phi_{\text{opt}} = \dfrac{\pi}{2}$	$\dfrac{1}{\pi\sin\theta_c}$	73	37	18	$\dfrac{1}{\pi}$	1.15
Flat $\phi_{\text{opt}} = \dfrac{1}{2}\left(\dfrac{\pi}{2} - \theta\right)$	$\dfrac{1}{\dfrac{1}{2\sin\theta_c} - \dfrac{3}{2}}$	113	56	27	$\dfrac{1}{2} - \dfrac{3}{2}\sin\theta_c$	1.03
3D (dish)						
Sphere	$\dfrac{\sin^2\phi}{4\sin^2\theta_c}$				$\dfrac{\sin^2\phi}{4}$	
Flat	$\dfrac{\sin^2\phi\cos^2(\phi+\theta_c)}{\sin^2\theta_c} - 1$				$\sin^2\phi\cos^2(\phi+\theta_c) - \sin^2\theta_c$	$\dfrac{2\,[1/\cos(\phi/2) - \cos^2(\phi/2)]}{3\sin^2(\phi/2)}$
With optimal ϕ:						
Sphere $\phi_{\text{opt}} = \dfrac{\pi}{2}$	$\dfrac{1}{4\sin^2\theta_c}$	13000	3300	820	$\dfrac{1}{4}$	1.22
Flat $\phi_{\text{opt}} = \dfrac{1}{2}\left(\dfrac{\pi}{2} - \theta_c\right)$	$\dfrac{1}{\dfrac{1}{4\sin^2\theta_c} - \dfrac{1}{2\sin\theta_c} - \dfrac{3}{4}}$	13000	3200	790	$\dfrac{1}{4} - \dfrac{1}{2}\sin\theta_c - \dfrac{3}{4}\sin^2\theta_c$	1.04

concentration is also expressed in terms of CR/CR_{ideal} for $CR_{ideal} = 1/\sin\theta_c$ or $1/\sin^2\theta_c$, and numerical values are given for several values of θ_c. For practical installations, geometric concentration is not the only design criterion. For example, rim angles beyond 90° and undersized absorbers can be used if the incremental reflector cost is low. Parabolic collector design is discussed in detail in Chap. 9.

Fresnel Reflectors and Central Receiver Collectors

The smooth optical surface of a reflector or lens can be broken into segments, a trick invented by Fresnel. Optically a Fresnel mirror, shown schematically in Fig. 8.14, can approximate a focusing parabola. The mirror segments can be flat if high concentration is not needed and/or if the number of segments is sufficiently large. Even with flat segments, the number of mirrors need not be larger than about 100 for line-focus systems or 10,000 for point-focus systems, because at that point the aberrations become comparable to the angular width of the sun. An example of a point-focus Fresnel mirror system is the power tower or central receiver, sketched in Fig. 8.14 and described in Ref. 19.

A detailed analysis of Fresnel mirrors is complicated by shading and blocking effects.[20,21] "Shading" occurs if direct sunlight fails to reach a mirror because the mirror is in the shadow of another mirror; "blocking" occurs if light reflected by a mirror fails to reach the absorber because it is intercepted by the backside of another mirror. In the interest of efficient mirror utilization, the reflector segments should be spaced far enough apart to minimize shading and blocking, within the constraints of relative cost of total area and reflector area. The effective aperture is obtained by multiplying the total area of the Fresnel mirror by a "ground cover factor" ψ. Values of ψ between 0.3 and 0.6 are practical for Fresnel reflectors which track around the north-south axis or about two axes. For linear Fresnel reflectors with east-west tracking axes, the shading and blocking effects are less severe and ψ can be on the order of $0.9/\cos(\beta - L)$ where β is the tilt of collector array and L is the latitude.

For solar power generation, steady output throughout the day and year is desirable; this can be achieved by optimizing the reflector field for early afternoon operation at winter solstice. The resulting diurnal and seasonal power variation is shown in Fig. 8.15 for the power tower.

The use of Fresnel reflectors is advantageous for very large installations because of reduced wind loading and simplified manufacture. For linear Fresnel reflectors, the tracking motion can be accomplished by a mechanical linkage, because all reflector segments undergo the same angular excursion at the same time. In point-focus Fresnel systems the synchronization of the mirror motion can also be accomplished by mechanical linkage, but this is unlikely to be practical in large installations.

Fresnel Lenses

Fresnel lenses[22,23], shown schematically in Fig. 8.16, can be used both in line-focus and in point-focus systems. To minimize accumulation of dirt, the surface exposed to the environment should be smooth. The prism facets need not be curved if their number is sufficiently large (on the order of 100), or if high concentration is not needed.

If the incident radiation is not parallel to the optical axis of a Fresnel lens, the effective focal length shortens due to aberrations and the size of the focal spot increases. Figure 8.17 shows the variation in focal length of a line-focus Fresnel lens as a function of the angle $\theta_{\parallel}$ of incident radiation from the cross-sectional plane in Fig. 8.17. For a two-dimensional concentration system (line focus or cylindrical geometry) with east-west axis, $\theta_{\parallel}$ changes from $-60°$ (at 8 A.M.) to $+60°$ (at 4 P.M), and the corresponding variation in focal length makes the use of Fresnel lenses impractical for this configuration.

On the other hand, in a system with *polar* tracking axis, the seasonal variation in $\theta_{\parallel}$ is less than $\pm40°$ for the same cutoff times, and a linear Fresnel lens can be used with low concentration ($CR \lesssim 6$, or up to approximately 10 if a second-stage concentrator is used). If higher concentration values (20 to 40) are to be reached with a line-focus Fresnel lens, the tilt of the tracking axis must be adjusted seasonally.[24]

The optics of a refractive element (lens) differ in two important aspects from those of a reflective element (mirror). First the off-axis aberrations are different. In particular, the focal length of a linear Fresnel lens changes with nonplanar incidence, as mentioned above, whereas the focal length of a line-focus parabolic mirror remains constant.

Secondly, the effect of surface and tracking errors is different. If the slope of a reflector element differs from the nominal value by angle δ, the reflected ray will deviate from the design direction by 2δ. In a Fresnel lens, on the other hand, the corresponding deviation from the

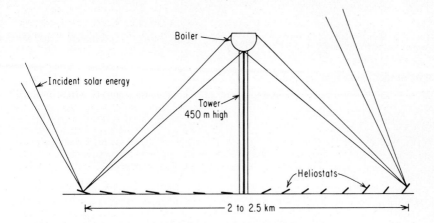

Fig. 8.14 Central receiver collector; an example of a Fresnel mirror system. (The heliostats and boiler are not to scale.)

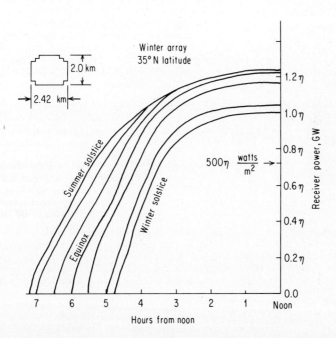

Fig. 8.15 Diurnal variation of power redirected to the receiver by the field of Fig. 8.14 for the twenty-first day of each month. Typical values of η are 0.65 to 0.8. At noon this array could produce 250 to 300 MW of electricity from a ground area of 4.3 km² supporting 1.46 km² of reflecting surface. The receiver radius is 10 m, the optical errors are 3 mrad, and tower height 450 m. (From Ref. 19.)

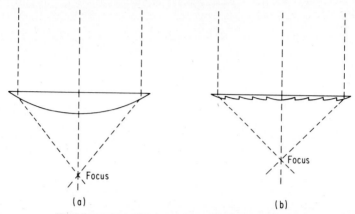

Fig. 8.16   Cross sections of a convex lens and a Fresnel lens.

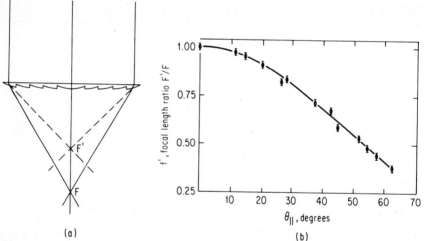

Fig. 8.17   (a) Linear Fresnel lens showing an incident beam in the cross-sectional plane (solid line, focus F), and an out-of-plane ray (dashed line, focus F'); (b) focal length f' of linear Fresnel lens for nonplanar incidence ($\theta_{\parallel}$= elevation from cross-sectional plane), from Ref. 24.

design direction will be $(n-1)\delta$ for rays passing through the center of the lens where n is the index of refraction. Near the edge of the lens the deviations will be somewhat larger, but in general a Fresnel lens is a factor of 3 to 4 less sensitive to surface and tracking errors than a reflector. In practice, this effect will more than compensate for the chromatic aberrations (see Chap. 3) of simple Fresnel lenses as compared to reflectors (which are inherently free from chromatic aberrations).

The most important factor in transmission losses of a Fresnel lens is the reflection loss (unless antireflection coatings are used). The reflection losses at the second surface are larger because of greater incidence angles at the prism faces, the exact value depending on the f-number of the lens. Some rays will be lost if the prism edges are not perfectly sharp, or if rays reach the wrong prism faces, as illustrated in Fig. 8.18 (a). With some production processes, the latter defect can be avoided by the so-called undercut design of Fig. 8.18 (b). Additional losses arise from scattering or absorption of light in the lens material. Figure 8.18 (c) indicates typical loss factors[25] achievable in cast acrylic lenses; in practice overall transmission factors $\tau \sim 0.85$ are possible.

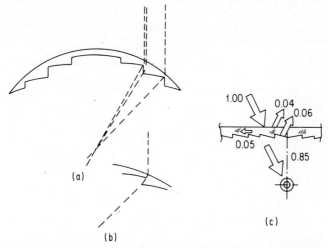

Fig. 8.18 Possible source of optical loss between facets in a Fresnel lens (a); avoidance of this loss by "undercut" design (b); (c) typical losses if the sun's energy incident on Fresnel lens = 1.00; reflection losses are about 0.04 at first surface and, due to facet angles, 0.06 at the second surface, absorption and scattering losses are about 0.05, and energy at the collector plane is 0.85. (From Ref. 25.)

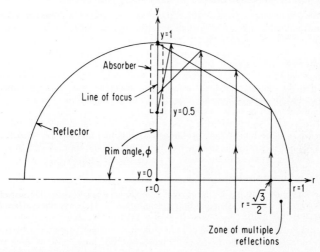

Fig. 8.19 Ray pattern for spherical mirror with incident rays perpendicular to plane of aperture.

Systems with Fixed Reflector and Moving Receiver

As an alternative to moving a (large) reflector or refractor, one can design systems where only the (small) receiver moves. Two such systems are known; both have linear receivers. The first, shown in Fig. 8.19, uses a hemispherical reflector[26,27] and a receiver which tracks around the center of the sphere, always pointing in the direction of the sun. The other fixed mirror system was invented by J. Russell.[28] In this system up to approximately 100 parallel flat mirror slats are mounted with fixed slope along a cylindrical trough, as shown in Fig. 8.20. If the mirror slats slope correctly, the system will always produce a perfect line focus (in the limit of infinitely many infinitesimally narrow slats), regardless of direction of incidence. The tracking motion is simple since the receiver moves on a circular arc.

Both of these fixed mirror systems can reach fairly high concentration values and are described in greater detail in Chap. 9.

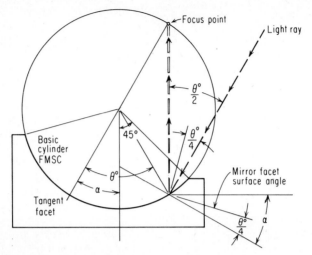

Fig. 8.20   Diagram of fixed-mirror-slat concentrator with tracking receiver showing ray trace geometry. (*Adapted from G. H. Eggers et al., Ref. 28.*)

Approximation by Reflector Shapes of Easy Manufacture

Some manufacturing processes naturally lend themselves to the reproduction or approximation of certain reflector geometries. Flat float glass mirrors, for example, can be used to approximate the profile of a CPC reflector. Reflector surfaces of near-spherical shape[29] can be obtained by stretching a metallized plastic film over a circular frame and applying a small pressure to one side. The profile is never exactly spherical because the boundary conditions imposed by the fixed frame prevent the film from stretching uniformly in all directions. A very slight pressure difference, e.g., less than 1 percent of atmospheric pressure, is adequate for maintaining the proper shape, and the surface is close enough to a true paraboloid to yield concentration values of several hundred with two-axis tracking. Air-inflated reflectors may also be suited for some of the fixed-reflector systems described in the previous subsection (see Ref. 30).

The catenary or hyperbolic cosine profile can be obtained by suspending metallized plastic film from a frame. A parabolic trough reflector can also be approximated by applying an appropriate bending moment to the edges of an elastic reflector sheet.

Second-Stage Concentrators

With increasing concentration CR, the ratio of reflector area to aperture area of a CPC increases as $(1 + a\,CR)$, with $a = 0.5$ to 1.0 depending on truncation, and the average number of reflections $\langle n \rangle$ increases as $\log \sqrt{CR}$. Therefore, a single CPC trough (cone) is likely to be impractical for concentrations above ten (one hundred). For higher concentrations, a two-stage system becomes advantageous because it reaches almost the ideal limit, without excessive reflector area and high transmission loss.

To illustrate the design of second-stage concentrators, we consider the line-focus parabola in Fig. 8.21 (*a*). Without a second stage, the receiver would be the flat surface from B to B'. If the acceptance half angle is Δ and no radiation is to miss the receiver, the concentration $\overline{AA'}/\overline{BB'}$ is

$$CR_1 = \frac{\sin \phi \cos (\phi + \Delta)}{\sin \Delta} - 1 \qquad (8.12)$$

where ϕ is the rim angle of the principal concentrator. The last term of Eq. (8.12) accounts for the shading of the aperture by the receiver. Since the present discussion is most relevant for systems with moderately high concentration, we can assume $\Delta \ll 1$ and neglect the second term. A second-stage CPC with acceptance half-angle ϕ boosts the concentration for the system as a whole. For small rim angles the concentration of the two-stage system approaches the thermodynamic limit, and even for ϕ as large as $30°$, it is still within 15 percent of the limit.

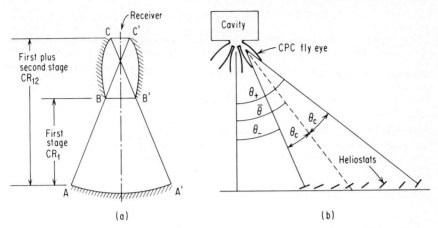

Fig 8.21 (a) Ideal second stage concentrator for focusing parabola showing both stages of concentration ratio CR_1 and CR_{12}; and (b) CPC "fly eye" second stage concentrator for the power tower. Drawings are schematic, and second stage is shown much too large in proportion to first stage.

Two further points should be noted about the optical design of this collector. First, the extreme angles of incidence of radiation from the first stage onto the second stage are not all equal to ϕ because the first stage is only a finite distance away from the second stage (the angle of incidence is larger for the ray AB' than for the ray AB). Therefore, the CPC of Fig. 8.2 does not have the optimal shape for use as second stage. Rather, the second stage should be designed such that rays from the edge A of the first stage become extreme rays at the receiver. For example, if the receiver is flat, extending from C to C', and if θ_2, the maximum angular spread of radiation at the receiver, is chosen to be $\pi/2$, then the second stage must reflect all rays from A to C. The reflector must therefore be elliptic from C' to B' with focal points at A and C, and it may be called a compound elliptical concentrator (CEC).[10]

The second concern is the rim angle ϕ. Equation (8.12) implies that a second stage is not very effective if ϕ is too large. On the other hand, in some systems, values of ϕ up to 65° may be desirable. For example, the cost of the tower for the central receiver can be justified only if radiation is received from a sufficiently wide mirror field. This requirement can easily be met without sacrificing concentration, if several intermediate CPC's are combined to form a "fly eye" second-stage concentrator. Such a design is well suited for use with a cavity absorber, and Fig. 8.21 (b) includes this feature. For the best aiming strategy the mirror field is divided into zones, one for each CPC, and in each zone the mirrors are aimed at their respective CPC secondary concentrator.

The fact that the concentration of the CPC is about four times as high for a given acceptance angle 2Δ as for a system without a second-stage concentrator is an obvious advantage for the design of ultrahigh-temperature power plants (using power cycles such as high-temperature Brayton cycle gas turbines, magnetohydrodynamics or thermionic conversion). This effect can also be important for solar collectors of low or intermediate temperature, because the acceptance half angle Δ can be doubled for a specified concentration, thus allowing a very significant relaxation of mirror surface accuracy. For example, a power tower with effective mirror and tracking error $\Delta_m = \Delta_s = 4.7$ mrad in its heliostats and with CPC second stage achieves as high a concentration as a power tower without CPC but with perfect mirrors. Contour errors of the second-stage CPC are insignificant as long as they are small compared to the acceptance half angle θ_c; this is a crucial difference between an imaging Cassegrain system and the nonimaging CPC system. Since the heliostats constitute a large fraction of the total cost of a central receiver solar power plant and since this cost depends strongly on the accuracy requirements, significant savings may be possible.

Since the CPC is relatively deep and narrow it can act as convection suppressor if the absorber is above the aperture. In addition, the overall concentration of the two stage system is nearly independent of rim angle, thereby increasing design flexibility. For instance, in some locations a mountain slope or a building wall may provide a natural support for a Fresnel mirror field, and a matching CPC second stage would guarantee high concentration for almost any geometry.

The tradeoffs between advantages and disadvantages of a second stage have been discussed elsewhere.[12] A CPC with an acceptance half-angle of 15° requires one reflection on the average, causing optical losses which may necessitate cooling of the mirrors. For most line-focus systems, especially if operation temperatures above 200°C are desired, a second stage will be cost-effective. For point-focus systems a second stage will probably be advantageous for very-high-temperature operation above 500°C.

COLLECTOR EFFICIENCY CALCULATION

The instantaneous efficiency of a solar concentrator is a convenient figure of merit useful for design. This section describes efficiency calculation methods.

Measurement of Insolation

It is natural to define the collector efficiency from a first law viewpoint as

$$\eta_c = \frac{q_u}{I_{\text{acc}}} \tag{8.14}$$

where I_{acc} is the insolation incident on the collector aperture within the collector acceptance angle and q_u, the useful heat collected, is the difference

$$q_u = q_{\text{abs}} - q_{\text{loss}} \tag{8.15}$$

between absorbed flux q_{abs} and heat loss q_{loss}. A troublesome ambiguity with this definition arises from the mismatch between the angular acceptance characteristics of the collector on one hand and of the insolation measuring instrument of the other. With respect to their angular acceptance, all insolation meters that are or have been generally available fall into one of two classes

1. The "pyranometer" with an approximately hemispherical field of view (i.e., acceptance half-angle $\theta_c = \pi/2$) measures the total insolation I_T.

2. The "pyrheliometer" whose field of view is limited by a blackened cylinder of aspect ratio (ratio of length/diameter) of 10 (acceptance half-angle $\theta_c \approx 2.8°$, approximately because cutoff is not sharp) measures the so-called "direct" or "beam" insolation I_B. These properties are sketched in Fig. 8.22 (a) and (b). Typical acceptance angles of solar concentrators are shown in Fig. 8.22 (c) and 8.22 (d). The CPC in Fig. 8.22 (c) has an acceptance half-angle much larger than 5° and the proper value of I_{acc} would lie between I_B and I_T. The high-concentration system in Fig. 8.22 (d) on the other hand has a much smaller acceptance angle than the pyrheliometer and I_{acc} is smaller than I_B.

The difference between I_B as measured by a pyrheliometer and the radiation from the solar disk itself is called "circumsolar radiation" (Ref. 31). To date little is known about the radial distribution of circumsolar radiation, or about the distribution of diffuse sky radiation in general. The circumsolar radiation has been measured by the solar energy group at the Lawrence Berkeley Laboratory, and their preliminary data indicate that on clear days the circumsolar component is small (a few percent), but that it may be quite significant (more than 10 percent on days with some atmospheric haze, even though the reading for I_B may be identical in both cases).

The variability of clouds and haze can cause significant scatter in the ratios I_{acc}/I_T and I_{acc}/I_B for a particular collector which will be reflected in the efficiency data if they are based on I_B or I_T. One can reduce or eliminate this scatter by designing solar radiometers with the same angular acceptance properties as the collector to be tested. Nonetheless it is desirable, in the interest of facilitating comparison between different collector types, to report all efficiencies with respect to either I_T or I_B. In principle the choice between I_T and I_B is somewhat arbitrary. But in order to minimize scatter of data points due to variable atmospheric conditions, the following convention is recommended: The efficiency of flat-plate collectors should be based on I_T (pyranometer) while that of tracking concentrators should be based on I_B (pyrheliometer), as has been done traditionally. Nontracking concentrators such as the CPC should be based on I_T if their tilt is fixed (*concentration less than 2*) and on I_B if their tilt requires adjustments (*concentration larger than 2*). To avoid any misunderstanding it is advisable to indicate clearly, for example by subscripts, the type of insolation on which the efficiency is based. The conversion from

$$\eta_B = \frac{q_{\text{out}}}{I_B} \qquad (8.16)$$

to

$$\eta_T = \frac{q_{\text{out}}}{I_T} \qquad (8.17)$$

is

$$\eta_T = \frac{I_B}{I_T}\,\eta_B \qquad (8.18)$$

Even on clear days I_B and η_B will be at least 10 percent less than I_T and η_T. For the purpose of stating instantaneous efficiency it is best to use data at normal incidence, supplemented by a separate curve showing the effect of incidence angle θ on optical efficiency. If nonnormal incidence data are to be used in Eqs. (8.14) to (8.21), I_B must be multiplied by $\cos i$.

Optical Efficiency

The *optical efficiency* η_o is defined as that fraction of the solar radiation I_{acc} which reaches the receiver and is absorbed there

$$\eta_o = \frac{q_{\text{abs}}}{I_{\text{acc}}} \qquad (8.19)$$

In the flat-plate collector literature[32] η_o has also been called the $(\tau\alpha)$ product. In most collectors η_o is the collector efficiency $\eta(0)$ at $\Delta T = 0$, $\Delta T = T_r - T_a$ being the temperature difference between receiver surface and ambient. In some collectors, however, the solar radiation absorbed in the reflector near the receiver may raise the reflector temperature above the receiver temperature if it is near ambient. In that case the receiver can gain thermal energy from the reflector and $\eta(0)$ can be larger than η_o. In certain collectors $\eta(0)$ can exceed the optical efficiency η_o by several percent.

Combining Eqs. (8.15) through (8.19) one obtains the efficiency with respect to "beam" radiation (pyrheliometer) as

$$\eta_B = \gamma_B\,\eta_o - \frac{q_{\text{loss}}}{I_B} \qquad (8.20)$$

$$\text{where} \qquad \gamma_B = \frac{I_{\text{acc}}}{I_B} \qquad (8.21)$$

is a weather-dependent factor which accounts for the mismatch between the optical properties of collector and pyrheliometer.

Collector Heat Loss Conductance (*U*-Value)

The collector heat loss conductance (*U*-value) is defined as

$$U_c = \frac{q_{\text{loss}}}{\Delta T} \qquad \text{with} \qquad \Delta T = T_r - T_a \qquad (8.22)$$

and it is based on net collector aperture area.

The *U*-value increases slightly with temperature and with temperature difference, but approximation by a single constant will usually be adequate to describe collector performance under a small range of operating conditions. This is a result of the requirement that the heat losses be small enough to assure operating efficiencies of at least 40 to 50 percent under full sunshine, in order for the collector to be economical. Even in evacuated collectors, which display a relatively large temperature variation of *U*-value, the error introduced by working with a constant effective *U*-value at average operating temperature is likely to be smaller than the variation in performance due to variable weather conditions. On the other hand, the stagnation temperature (also called the no net energy delivery temperature) of a collector, corresponding to zero efficiency, may be seriously overestimated if a low-temperature U-value is used for extrapolation to high temperature where radiation heat exchanges become important.

Since the U-value is defined with respect to the receiver plate temperature, the expression

$$\eta_B = \gamma_B \eta_o - U_c (T_r - T_a)/I_B \tag{8.23}$$

gives the efficiency with respect to average receiver temperature T_r. On the other hand, for the purpose of calculating the heat extracted by the heat transfer fluid, an expression based on fluid temperature is more useful. In terms of the average bulk fluid temperature $\overline{T}_f$ in the collector the efficiency is (see Chap. 7)

$$\eta_B = F'[\gamma_B \eta_o - U_c (\overline{T}_f - T_a)/I_B] \tag{8.24}$$

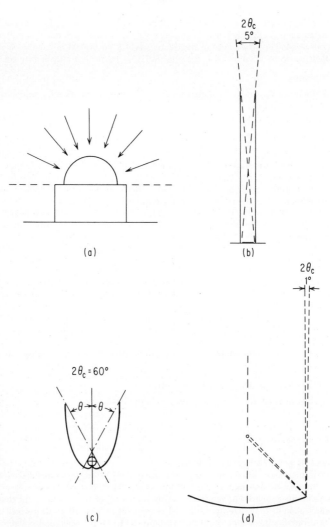

Fig. 8.22 Typical acceptance angles: (a) pyranometer, $2\,\theta_c = 180°$; (b) pyrheliometer, $2\,\theta_c = 5°$; (c) nontracking CPC, $2\,\theta_c = 60°$; (d) tracking parabola, $2\,\theta_c = 1°$.

where F', the so-called collector efficiency factor, is

$$F' = \frac{R_{ra}}{R_{fr} + R_{ra}} \tag{8.25}$$

where $R_{ra} = 1/U_c$ and R_{fr} is the thermal resistance between the working fluid and the receiver surface conduits. In this formula all resistances are defined relative to collector aperture area.

For some systems studies the inlet fluid temperature $T_{f,\text{in}}$ is needed. In terms of $T_{f,\text{in}}$ the efficiency is given by the equation (see Chap. 7)

$$\eta_B = F_R[\gamma_B \eta_o - U_c(T_{f,\text{in}} - T_a)/I_B] \tag{8.26}$$

with the heat removal factor

$$F_R = \frac{\dot{m}c_p}{U_c}\left[1 - \exp\left(-\frac{U_c F'}{\dot{m}c_p}\right)\right] \tag{8.27}$$

F_R depends on flowrate $\dot{m}$ (per unit aperture area) and fluid heat capacity c_p in addition to the collector efficiency factor F' and the U-value.

HEAT TRANSFER IN SOLAR CONCENTRATORS

Methods of Reducing Heat Loss

In general, all three heat transfer modes[3-5]—radiation, convection, and conduction—must be considered in calculating the thermal performance of a solar collector. From a nonselective absorber at high temperature, radiation is the dominant heat loss; it can be reduced by an order of magnitude by means of a selective coating (see Table 8.3 and Chap. 4). Black chrome, for example, seems very stable and is available with $\alpha \approx 0.95$ and $\epsilon \lesssim 0.1$ (at operating temperatures below 300°C). Since only a relatively small absorber surface need be coated, the use of selective coatings is likely to be cost-effective in all concentrators of low and intermediate concentration.

Convection and conduction losses to the air surrounding the absorber are significant, making it desirable to use evacuated receivers. The design of evacuated receivers is made difficult by the thermal cycling of the absorber. If a large gap between the absorber surface and its glass envelope can be established, then a moderate vacuum (about 10^{-2} to 10^{-3} atmospheric pressure) will suppress convection but not conduction. Such a vacuum level could be maintained by continuous or intermittent pumping. On the other hand, if the evacuated glass envelope is to be made as small as possible, the vacuum must be very good, of the order of 10^{-7} atm, to eliminate conduction through the residual air.

Several evacuated collectors have been developed by Owens-Illinois,[33] General Electric,[34] and by Philips GmbH Research Laboratory[35] [Fig. 8.23 (a) through (c)]. They are essentially of Dewar flask design (i.e., one end of the outer glass tube is closed) and are hermetically sealed. They also contain getters and are expected to maintain a vacuum of 10^{-7} atmospheric pressure or better for a period of at least 20 yr. Corning[36] has developed an alternative design with copper absorber plate and a single glass-to-metal seal, as shown in Fig. 8.23 (d).

The relative magnitude of radiation and convection and the relatively low cost of selective coating compared to the cost of evacuated tubes suggest that evacuation should be used only in conjunction with a selective coating. As an alternative convection suppression approach, honeycombs[37,38] will reduce convection and radiation losses (see Chap. 7).

Convective and radiative losses can also be reduced by use of so-called transpired absorbers. These absorbers consist of a porous material (the matrix). Cold incoming heat transfer fluid is pumped inward through the porous structure (away from ambient). The outer surfaces of these absorbers remain cool and heat losses are low. Transpired collectors are appropriate only for applications in which the fluid inlet temperature is relatively low. Individual collector modules must be interconnected for parallel flow.

It is impossible, in the brief space of this chapter, to analyze heat transfer in all concentrator types that have been built. However, most concentrators can be classified in terms of the three basic absorber configurations shown in Figs. 8.24 through 8.26:

TABLE 8.3 Chart of Selective Surfaces Based on Informal Discussions of a Working Group at the American Electroplating Society Solar Symposium, Atlanta, GA, November 9–10, 1976

Type of coating	α	ϵ 100°C	Maximum temp. air/vac, °C	Life-yrs air/vac	Possible problems
Interference	0.94	0.10	350/400	5/30	Cost
Black chrome	0.96	0.12	350/400	15/30	Quality control
Black nickel	0.90	0.07		?	Humidity sensitivity stability
CuO	0.88	0.15	200/	?	High temperature stability
Enamels	0.95 +	0.8–0.9	500/500?		Weight, high emittance
PbO$_2$	0.98	0.3	200/300	20/	Temperature limit
SS (INCO)	0.90	0.4	150/	30/30	Cost of substrate (SS) and high emittance
Anodic aluminum	0.95	0.80	500/	50/	High emittance
Aluminum conversion	0.93	0.35	200/200	10–20/	Temperature, lifetime, high emittance
Paints	0.97	0.91	150–175/	5–20/	High emittance
Tungsten dendrites	0.99	0.3 at 500°C	250/500	?	Cost and angular acceptance

SOURCE: Courtesy of John Allen, Argonne National Laboratory.

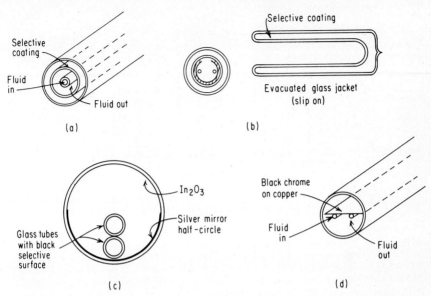

Fig. 8.23 Commercial evacuated tubular collectors: (*a*) Owens-Illinois, all glass, (*b*) General Electric using copper tubes and fin contacting mirror wall of glass jacket, (*c*) Philips (Aachen) GmbH with IR reflector In$_2$O$_3$, (*d*) Corning using glass to metal seal.

Fig. 8.24 Planar receiver configuration for a solar concentrator showing (*a*) receiver relative to reflector and (*b*) receiver detail.

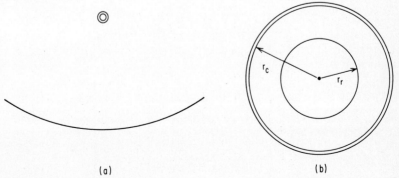

Fig. 8.25 Tubular receiver with concentric glass cover for solar concentrator showing (*a*) receiver relative to reflector and (*b*) receiver dimensions.

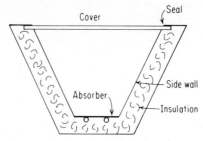

Fig. 8.26 Concentrating enclosure configuration schematic diagram. Cover may be flat glass or a Fresnel lens, side walls may be straight or curved.

1. Planar configuration: flat absorber, covered by parallel glass plate of approximately the same size. Figure 8.24 (*b*) shows receiver, Fig. 8.24 (*a*) typical orientation relative to focusing parabolic dish or trough reflector.

2. Tubular configuration: tubular absorber, surrounded by a glass tube, as shown in Fig. 8.25 (*b*); Fig. 8.25 (*a*) shows typical orientation relative to focusing parabolic trough reflector.

3. Concentrating enclosure configuration: flat or tubular absorber, surrounded by insulated sides and a flat or curved cover which is much larger than the absorber, as indicated schematically in Fig. 8.26. Practical realizations include the V-trough or CPC with flat absorber, the CPC with tubular absorber, and the Fresnel lens with flat absorber.

In the following subsections we give correlations which can be used for calculating the heat loss from the absorber in each of the three basic configurations. In some collectors, more than one configuration is relevant. For example, the heat loss from the nonevacuated CPC in Fig. 8.27 is found by calculating the thermal resistance R_{ii} between absorber and glass tube based on configuration (ii), then adding it to the resistance R_{iii} between glass tube and cover, based on configuration (iii).

Configurations (i) and (ii) have been thoroughly covered in the heat transfer literature, and reliable calculations can be made. Configuration (iii), on the other hand, has not been well explored, and the lack of sufficient data on natural convection may introduce significant errors in the calculation of the heat loss.

Heat Loss in Planar Configurations (i)

Radiative heat transfer between receiver and cover in Fig. 8.24 (*b*) is determined by

$$Q_{rc,\text{rad}} = A_c F_{cr}\, \sigma\, (T_r^4 - T_c^4) \tag{8.28}$$

Subscripts c and r refer to cover and receiver, respectively. F_{cr}, if edge effects can be neglected, is determined by

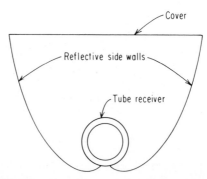

Fig. 8.27   CPC reflector coupled to tubular receiver.

$$F_{cr} = \frac{1}{1/\epsilon_c + 1/\epsilon_r - 1} \tag{8.29}$$

but, if edge effects must be considered, a more complicated analysis is required. For the value of the cover emittance ϵ_c of glass, Tabor[39] has recommended a value of 0.88.

The heat transfer coefficient h, for natural convection between large parallel plates separated by a distance L, is obtained from the relation between Nusselt number $\text{Nu} = hL/k$ and Grashof and Prandtl numbers

$$\text{Nu} = c \, (\text{Pr}\,\text{Gr}_L)^a \tag{8.30}$$

c and a are constants given in Table 8.4. $\text{Gr}_L\,\text{Pr} < 1700$ indicates conduction as the heat transfer mode.

Theoretically the exponent a is ¼ for laminar flow and ⅓ for turbulent-free convection. In the temperature range of interest (250 to 600 K), the temperature variation of conductivity k, kinematic viscosity ν, and expansion coefficient β, which are needed to compute the heat transfer, can be parameterized by

$$\beta = \frac{1}{T} \tag{8.31a}$$

$$\nu = \nu_o T^{1.7} \tag{8.31b}$$

with $\nu_o = 9.76 \times 10^{-10} \ \text{m}^2 \cdot \text{s}^{-1} \cdot \text{K}^{-1.7}$ \tag{8.31c}

$$k = k_o T^{0.7}$$

with $k_o = 4.86 \times 10^{-4} \ \text{W} \cdot \text{m}^{-1} \cdot \text{K}^{-1.7}$

Tabor[39] has critically examined the available convection data and recommends the correlations in Table 8.4 for large absorber widths. In the case of downward heat flow, instabilities may develop and considerable uncertainty exists regarding the proper heat transfer correlation. For small absorber widths, as often used in concentrating collectors, predicted h values can be in error owing to edge effects. Reference 37 gives new calculated and measured values for Nu for convection in rectangular enclosures for various values of the aspect ratio (height/width). Figure 8.28, taken from Ref. 37, shows the variation of Nu with aspect ratio A assuming constant Rayleigh number. For small A the large plate correlations of Table 8.4 are applicable, but for $A = O\,(1)$ the convection increases by about 20 percent; with large values of A, i.e., narrow absorber/large gap, convection is suppressed with a pure conduction regime for $A \gtrsim 5$.

Heat Loss in Tubular Configurations (ii)

Radiative heat transfer between concentric tubes is given by

$$Q_{rc,\text{rad}} = \frac{A_p\sigma\,(T_r^4 - T_c^4)}{\dfrac{1}{\epsilon_r} + \left(\dfrac{r_p}{r_c}\right)^k \left(\dfrac{1}{\epsilon_c} - 1\right)} \tag{8.32}$$

TABLE 8.4   Nusselt Number Correlations for Natural Convection Between Parallel Plates

Horizontal planes, heat flow upwards
$\text{Nu} = 0.168 \, (\text{Gr Pr})^{0.281} = 0.152 \, \text{Gr}^{0.281}$ for Gr from 10^4–10^7

45° planes, heat flow upwards
$\text{Nu} = 0.102 \, (\text{Gr Pr})^{0.310} = 0.0925 \, \text{Gr}^{0.310}$ for Gr from 10^4–10^7

Vertical planes
$\text{Nu} = 0.0685 \, (\text{Gr Pr})^{0.327} = 0.0616 \, \text{Gr}^{0.327}$ for Gr 1.5×10^5–10^7
$\text{Nu} = 0.0369 \, (\text{Gr Pr})^{0.381} = 0.0326 \, \text{Gr}^{0.381}$ for Gr 1.5×10^4–1.5×10^5

SOURCE Tabor, Ref. 39.

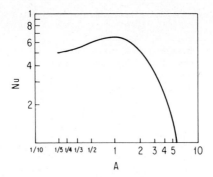

Fig. 8.28 Variation of Nusselt number Nu with aspect ratio A for a rectangular enclosure with conducting walls (Rayleigh number, GrPr $= 5 \times 10^5$).

where $k = 0$ if the outer tube is a specular reflector and $k = 1$, if diffuse. A value of $\epsilon_c = 0.88$ is recommended for bare glass. End effects will be negligible in all practical situations.

If the gap between inner and outer tubes is not evacuated, the gap width should be chosen as large as possible without causing convection; in other words, the Rayleigh number—$\text{Ra}_L = \text{PrGr}_L = \text{Pr}(g\ \beta L^3\ \Delta T/v^2)$, with $L = r_c - r_r$—should be on the order of 1700. The convective heat transfer is not affected greatly if the gap is increased, but, for economic reasons, it is impractical to use larger glass tubes than necessary.

Heat Loss in Concentrating Enclosure Configuration (iii)

Analytical and experimental determination of heat transfer coefficients for the concentrator configuration of Fig. 8.26 is difficult, and little information is available. The problem is complicated by the interaction between three transfer modes: radiation, convection, and conduction (through the side walls). However, satisfactory results have been obtained by the simplified approach described in Ref. 11. Radiation and convection are treated separately, and conduction effects from absorber plate to ambient, are included only as an effective conductive loss coefficient $U_{r-a,\text{cond}}$.

The radiative heat loss calculation requires the effective emittance ϵ_{eff}

$$Q_{rc,\text{rad}} = A_r \epsilon_{\text{eff}} \sigma (T_r^4 - T_c^4) \tag{8.33}$$

It involves the infrared emittance ϵ_m of the side walls. For diffusely reflecting side walls, standard shape factor analysis is appropriate. For specular straight side walls, the method of multiple images in V-troughs serves satisfactorily. For curved specular side walls of the CPC type, a simplified approximate formula can be used (Ref. 11), with results summarized in Table 8.5. The two cases of greatest practical interest, namely first surface reflectors ($\epsilon_{ir} \lesssim 0.1$) and second surface reflectors ($\epsilon_{ir} \approx 0.9$) with a cover of untreated glass ($\epsilon_c = 0.9$) show almost no variation of ϵ_{eff} with side wall reflectance or concentration. This suggests the reflector shape is immaterial for these cases and that the approximation

$$\epsilon_{\text{eff}} = \frac{1}{1/\epsilon_c + 1/\epsilon_r - 1} \tag{8.34}$$

should be adequate. (In this configuration ϵ_c is slightly larger than the value 0.88 recommended by Tabor for the flat-plate case, because the angles of incidence are restricted.)

To estimate the convective heat transfer for the concentrating enclosure configuration greatly simplified models must be used, since no experimental convection heat transfer coefficients have been obtained. If all interactions with the side walls can be neglected, correlations for heat transfer coefficients of flat plates in an infinite environment can be used to calculate the convection both from the absorber plate to the enclosed air and from the enclosed air to the cover c. The use of parameterization [Eqs. (8.31)] for the thermophysical properties of air yields the expression

TABLE 8.5   Variation of Effective Emittance ϵ_{eff} with ϵ_c, ϵ_m, and ϵ_r, and with Concentration*

ϵ_c	ϵ_r	ϵ_m		
		0.1	0.5	0.9
		0.06	0.06	0.06
0.1	0.1	0.07	0.08	0.08
		0.07	0.09	0.09
		0.11	0.13	0.13
0.1	0.5	0.15	0.21	0.22
		0.18	0.29	0.31
		0.16	0.14	0.14
0.1	0.9	0.17	0.26	0.28
		0.22	0.37	0.42
		0.09	0.09	0.09
0.5	0.1	0.09	0.10	0.10
		0.09	0.10	0 10
		0.34	0.36	0.36
0.5	0.5	0.36	0.40	0.42
		0.37	0.43	0.45
		0.50	0.53	0.52
0.5	0.9	0.52	0.63	0.66
		0.55	0.71	0.75
		0.10	0.10	0.10
0.9	0.1	0.10	0.10	0.10
		0.10	0.10	0.10
		0.47	0.46	0.43
0.9	0.5	0.48	0.48	0.46
		0.48	0.48	0.47
		0.82	0.79	0.69
0.9	0.9	0.82	0.83	0.77
		0.83	0.85	0.82

*The entries in the tables are numerical values of ϵ_{eff}, calculated from Eq. (II-9) of Ref. 11 for $CR = 1.6$, $\theta = 36°$ (top row of each entry), for $CR = 4.0$, $\theta = 11.5°$ (middle rows), and for $CR = 8.0$, $\theta = 5.74°$ (bottom rows).

$$Q_{\text{conv, }r} = A_r \frac{k_0}{\sqrt{\nu_0}} c (\text{Pr } g)^{1/4} \frac{(\Delta T_r)^{5/4}}{s^{1/4}} \left(T_r - \frac{\Delta T_r}{2} \right)^{-0.4}$$

with $\Delta T_r = \dfrac{T_r - T_c}{1 + CR^{-0.6}}$ and s, the receiver width.

(8.35)

for laminar flow. The constant c is found in Table 8.6. With $c = 0.56$ (for intermediate collector tilt), the grouping of constants in Eq. (8.35) has the value

$$\frac{k_0}{\sqrt{\nu_0}} c (\text{Pr } g)^{1/4} = 14.2 \text{ W} \cdot \text{m}^{-7/4} \cdot \text{K}^{-0.85}$$

(8.36)

This formula has been found to give agreement (within 20 percent) with measurement of convective heat transfer in a CPC configuration; however, the data available are too limited for any reliable conclusion, and Eq. (8.35) is recommended for order of magnitude estimates only.

Analysis of heat transfer in the concentrating enclosure configuration is further complicated by conduction of heat from the absorber into the side walls, and by absorption of solar radiation by the side walls. The latter mechanism raises the wall temperature and thus lowers the heat loss from absorber plate to cover; in collectors with aluminized reflectors ($\rho \approx 0.85$) and insulated

TABLE 8.6   Nusselt Number* Equation Coefficients for Natural Convection from Flat Surfaces Positioned at Various Angles

Tilted plates $(0° < \beta < 60°)$	Range of (Gr Pr)	c	a
Laminar flow	10^4–10^9	0.59	$\frac{1}{4}$
Turbulent flow	10^9–10^{12}	0.13	$\frac{1}{3}$

Horizontal plate	Range of (Gr Pr)	c	a
Laminar flow	10^5–10^7	0.54	$\frac{1}{4}$
Turbulent flow	10^7–10^{11}	0.15	$\frac{1}{3}$

*$Nu = c(Pr\ Gr)^a$. From Ref. 40.

side walls, the heat loss coefficient measured under full sunshine has been calculated and measured to be about 10 percent lower than the heat loss coefficient measured at night. However, conduction of heat from the absorber into the reflector wall can significantly increase the heat loss. This can pose a serious problem, since aluminum, the preferred reflector material by virtue of its durability, also has a very high thermal conductivity. If aluminum sheet is to be used in this configuration, it should therefore be as thin as possible, and a slight gap between absorber and reflector is recommended.

Heat Loss From Cover to Ambient

For heat transfer from cover c to the ambient environment a, an effective average heat transfer coefficient,[41] which includes both radiation and convection, is recommended:

$$h_{ca} = \begin{cases} 20\ \text{W/m}^2 \cdot \text{K} & \text{for wind}\approx 3\ \text{m/s} \\ 30\ \text{W/m}^2 \cdot \text{K} & \text{for wind}\approx 6\ \text{m/s} \end{cases} \tag{8.37}$$

Calculating radiation and convection terms separately will produce greater accuracy. The radiation heat loss calculation

$$Q_{ca,\ \text{rad}} = A_c \epsilon_c \sigma \left(T_c^4 - T_a^4\right) \tag{8.38}$$

is fairly straightforward, except for the possible replacement of T_a by the sky temperature T_{sky}, which may be as much as 10 percent lower (in absolute temperature) than T_a.[45]

Convection heat transfer to ambient can be estimated from[3]

$$Q_{ca,\ \text{conv}} = A_c \frac{k}{L} Nu_L \left(T_c - T_a\right) \tag{8.39}$$

with

$$Nu_L \begin{cases} 0.56\ (Gr_L\ Pr)^{1/4} & \text{laminar, natural convection, } Gr_L\ Pr < 10^9 \\ 0.12\ (Gr_L\ Pr)^{1/3} & \text{turbulent, natural convection, } Gr_L\ Pr > 10^9 \\ 0.664\ Re_L^{1/2}\ Pr^{1/3} & \text{for laminar forced convection, } Re_L < 5 \times 10^5 \\ 0.036\ Pr^{1/3}\ (Re_L^{0.8} - 23{,}200) & \text{for } Re > 5 \times 10^5 \end{cases}$$

PRACTICAL CONSIDERATIONS

Materials

The most critical properties of materials for a concentrating collector are
1. Reflectance of reflectors
2. Transmittance of covers
3. α and ϵ of absorber coating

Aluminum with a total reflectance of 85 to 90 percent and silver with a total reflectance of 95 percent are the best reflective surfaces for solar energy applications, except for special cases where total internal reflection can be employed.[43,44] A precise value of reflectance can be obtained only by integrating the spectral reflectance over an appropriate model solar spectrum, for example a standardized air mass 1.5 spectrum (see Chap. 5). Reflectance measurements reported by different investigators may differ because of the assumed solar spectrum. In most cases an adequate engineering approximation is obtained by using a single effective reflectance value.

First surface mirrors require protection from the environment. Aluminum sheet can be protected very well by anodization which produces an oxide coating which can survive both outdoor exposure and washing. If an aluminum surface is not exposed to the environment, a thin coating of MgF_2 or SiO will be adequate protection. Acrylic and teflon coatings have also been used. Teflon is impractical because its electrostatic properties cause it to attract dust. Acrylic does not suffer from this problem and may be a good coating even for outdoor exposure of aluminum mirrors if discoloration problems can be solved.

Silver is difficult to protect except as a second surface mirror behind glass. An acrylic coating would be likely to have pinholes through which pollutants could penetrate and attack the silver. In such a second-surface mirror the glass must be traversed twice by each light ray and absorption losses are severe unless either low-iron glass or microsheet glass are used. Microsheet silvered glass appears to be the most promising reflector material if suitable techniques for handling it in the field can be developed (e.g., lamination to plastic film substrate).

A special comment needs to be added about reflectors in the immediate vicinity of a non-evacuated receiver in the concentrating enclosure configuration. In these cases one would like to use anodized aluminum sheet as reflector material because of its durability, especially in view of possible elevated service temperature. However, the high thermal conductivity of aluminum will cause excessive conductive heat losses unless great care is taken to minimize this heat leak by interposing a small air gap between reflector and absorber, and by using either a thin aluminum sheet or aluminized stainless steel. The optical losses incurred by using a gap between reflector and absorber will be analyzed shortly.

Transmittances of the most important glazing materials[45] are listed in Table 8.7. Glass is the most durable material; it should have low iron content to avoid excessive absorption losses. Of particular interest is etched glass, a low-cost (cost estimates for etching in mass production range from $0.10 to $0.50 per square meter) antireflection material. As a result of etching, reflection losses can be reduced from 4 percent per surface to about 1 percent at normal incidence. A double etching process can yield even better results, although at higher cost.[46] Etching of the cover for a concentrator receiver will always be cost-effective in mass production because of the relatively small area involved and because the cover is sufficiently well protected from the environment so as not to degrade by accumulation of dirt. If the cover is to be exposed to high temperatures, borosilicate glass (pyrex) is recommended; however, it is more expensive than soda lime glass. An etching process for borosilicate glass is also available.[47]

Acrylics are suitable as cover materials, particularly for those used as Fresnel lenses. Little degradation was found in acrylic exposed to desert conditions in Albuquerque, NM, for 17 yrs.[48] As the cover of the receiver, acrylic is less practical because its low service temperature limit (about 100°C), also UV damage may be excessive in the high flux area near the receiver. Slight improvement in solar transmission (about 1 percent) of acrylic has been obtained by applying a thin teflon coating which also enhances scratch resistance. While glass, acrylic and polycarbonate are highly specular transmitters, this is not necessarily true of other plastic such as Tedlar ™. It is however possible to treat their surfaces in such a way as to make them specular.[31]

Selective coatings are listed in Table 8.3. At the present time, black chrome (CrO_x) appears to be the most attractive candidate. It can be electroplated on steel, copper, aluminum and other metals, and values of $\alpha \simeq 0.95$ and $\epsilon \lesssim 0.1$ (at operating temperature of 300°C) can be achieved with proper quality control.[49] Its use is cost-effective in concentrating collectors of low or moderate concentration and operating temperatures above 100°C. It is stable both in vacuum and in air. A glass substrate will require special processes and multiple coatings.

Whether a high α_r or a low ϵ_r are more important will depend on collector type and on application. In nonevacuated collectors, conduction and convection will begin to dominate radiative losses when ϵ_r is reduced below about 0.2. On the other hand in evacuated collectors there is great incentive to achieve the lowest possible values of ϵ_r, provided α_r is at least 0.85. Better selective coatings are likely to be produced as a result of intensive research.

When interpreting quoted values of α_r and ϵ_r one should keep in mind the following:

TABLE 8.7 Thermal and Radiative Properties of Collector Cover Materials

Material name	Index of refraction $(n)^a$	τ (solar),^{a} %	τ (solar),^{b} %	τ (infrared),^{c} %	Expansion coefficient, in/in · °F	Temperature limits, °F	Weatherability (comment)	Chemical resistance (comment)
Lexan (polycarbonate)	1.586 (D 542)^d	125 mil 64.1 (± 0.8)	125 mil 72.6 (± 0.1)	125 mil 2.0 (est)^e	$3.75\ (10^{-5})$ (H 696)	250–270 service temperature	Good: 2-yr exposure in Florida caused yellowing; 5 yr caused 5% loss in τ	Good: comparable to acrylic
Plexiglas (acrylic)	1.49 (D 542)	125 mil 89.6 (± 0.3)	125 mil 79.6 (± 0.8)	125 mil 2.0 (est)^f	$3.9\ (10^{-9})$ at 60°F; $4.6\ (10^{-6})$ at 100°F	180–200 service temperature	Average to good: based on 20-yr testing in Arizona, Florida, and Pennsylvania	Good to excellent: resists most acids and alkalis
Teflon F.E.P. (fluorocarbon)	1.343 (D 542)	5 mil 92.3 (± 0.2)	5 mil 89.8 (± 0.4)	5 mil 25.6 (± 0.5)	$5.9\ (10^{-5})$ at 160°F; $9.0\ (10^{-5})$ at 212°F	400 continuous use; 475 short-term use	Good to excellent: based on 15-yr exposure in Florida environment	Excellent: chemically inert
Tedlar P.V.F. (fluorocarbon)	1.46 (D 542)	4 mil 92.2 (± 0.1)	4 mil 88.3 (± 0.9)	4 mil 20.7 (± 0.2)	$2.8\ (10^{-5})$ (D 696)	225 continuous use; 350 short-term use	Good to excellent: 10-yr exposure in Florida with slight yellowing	Excellent: chemically inert
Mylar (polyester)	1.64-1.67 (D 542)	5 mil 86.9 (± 0.3)	5 mil 80.1 (± 0.1)	5 mil 17.8 (± 0.5)	$0.94\ (10^{-5})$ (D 696-44)	300 continuous use; 400 short-term use	Poor: ultraviolet degradation great	Good to excellent: comparable to Tedlar
Sunlite^g (fiberglass)	1.54 (D 542)	25 mil (P) 86.5 (± 0.2) 25 mil (R) 87.5 (± 0.2)	25 mil (P) 75.4 (± 0.1) 25 mil (R) 77.1 (± 0.7)	25 mil (P) 7.6 (± 0.1) 25 mil (R) 3.3 (± 0.3)	$1.4\ (10^{-5})$ (D 696)	200 continuous use causes 5% loss in τ	Fair to good: regular, 7-yr solar life; premium, 20-yr solar life	Good: inert to chemical atmospheres

Material							
Float glass (glass)	1.518 (D 542)	125 mil 84.3 (± 0.1)	125 mil 78.6 (± 0.2)	125 mil 2.0 (est)[e]	4.8 (10⁻⁵) (D 696)	1350 softening point; 100 thermal shock	Excellent: time proved	Good to excellent: time proved
Temper glass (glass)	1.518 (D 542)	125 mil 84.3 (± 0.1)	125 mil 78.6 (± 0.2)	125 mil 2.0 (est)[e]	4.8 (10⁻⁶) (D 696)	450–500 continuous use; 500–550 short-term use	Excellent: time proved	Good to excellent: time proved
Clear lime sheet glass (low iron oxide glass)	1.51 (D 542)	Insufficient data provided by ASG	125 mil 87.5 (± 0.5)	125 mil 2.0 (est)	5.0 (10⁻⁶) (D 696)	400 for continuous operation	Excellent: time proved	Good to excellent: time proved
Clear lime temper glass (low iron oxide glass)	1.51 (D 542)	Insufficient data provided by ASG	125 mil 87.5 (± 0.5)	125 mil 2.0 (est)	5.0 (10⁻⁶) (D 696)	400 for continuous operation	Excellent: time proved	Good to excellent: time proved
Sunadex white crystal glass (0.01 % iron oxide glass)	1.50 (D 542)	Insufficient data provided by ASG	125 mil 91.5 (± 0.2)	125 mil 2.0 (est)	4.7 (10⁻⁶) (D 696)	400 for continuous operation	Excellent: time proved	Good to excellent: time proved

[a]Compiled data based on ASTM Code E 424 Method B.
[b]Numerical integration $(\Sigma \tau_{avg} F_{\lambda_1 T - \lambda_2 T})$ for $\lambda = 0.2$–4.0 μM.
[c]Numerical integration $(\Sigma \tau_{avg} F_{\lambda_1 T - \lambda_2 T})$ for $\lambda = 3.0$–50.0 μM.
[d]All parenthesized numbers to ASTM test codes.
[e]Data not provided; estimate of 2% to be used for 125-mil samples.
[f]Degrees differential to rupture $2 \times 2 \times \frac{1}{4}$-in samples. Glass specimens heated and then quenched in water bath at 70°F.
[g]Sunlite premium data denoted by (P); Sunlite regular data denoted by (R).

SOURCE: Ref. 45. Table data abstracted from Ratzel, A. C., and R. B. Bannerot, Optical Material Selection for Flat-Plate Solar Energy Collectors Utilizing Commercially Available Materials, presented at ASME-AIChE Natl. Heat Transfer Conf., 1976.

1. α_r depends on the assumed solar spectrum.

2. α_r will decrease with angle of incidence, but usually only normal incidence values are quoted. In concentrating collectors the angles of incidence of some rays at the absorber will deviate strongly from the normal direction (by 30° to 90° depending on concentrator optics). Black chrome maintains a high absorptance up to 60° incidence angle or better, but multilayer interference films tend to drop off significantly beyond 45°.

3. ϵ_r increases with temperature, because of spectral shift of the blackbody radiation toward the solar spectrum and because of physical changes of the surface.

4. A single number such as α_r/ϵ_r does *not* adequately characterize a selective surface (it only determines the stagnation temperature in a vacuum).

Gaps between Reflector and Absorber in CPC Collectors

The design of CPC collectors or of receivers with second-stage concentrators frequently requires a gap between reflector and absorber in order to incorporate an evacuated glass envelope or to reduce conductive heat losses to the reflector. Ideal optical design on the other hand demands placement of the reflector as close to the absorber as possible to avoid loss of solar radiation in the gap. For the CPC, simple formulas have been derived which are useful for finding the best compromise between optical and thermal losses, and a detailed discussion is given in Ref. 12. In most cases the best solution is to leave the basic reflector/absorber geometry unchanged but truncate the edge of the reflector adjacent to the absorber as in Fig. 8.29. The resulting optical loss corresponding to a particular angle of incidence would have to be calculated by ray tracing, but for most practical applications this is not necessary because only the average over all incident directions are relevant; this is certainly the case for second-stage concentrators and for the average performance of CPC collectors. The average optical loss OL, defined as

$$OL = \frac{\text{radiation lost in gap}}{\text{radiation reaching both absorber and gap}} \qquad (8.40)$$

is given by the following approximate formulas for the two-dimensional (troughlike) CPC's shown in Fig. 8.29:

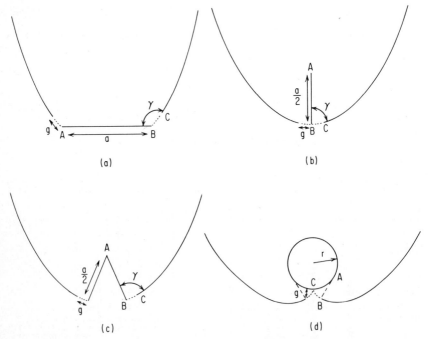

Fig. 8.29 Various CPC collectors showing gap g through which light can be lost.

$$OL = (1 + \cos\gamma)\frac{g}{a} \quad \text{for Fig. 8.29 }(a) \tag{8.41}$$

$$OL = \frac{g}{a} \quad \text{for Fig. 8.29 }(b) \text{ and }(c) \tag{8.42}$$

and

$$OL = \frac{1}{\pi}\left[\sqrt{\frac{2g}{r} + \left(\frac{g}{r}\right)^2} - \arccos\left(\frac{r}{g+r}\right)\right] \quad \text{for Fig. 8.29 }(d) \tag{8.43}$$

where the angle γ, the absorber perimeter a, the absorber tube radius r and the gap width g are defined in the figure. For practical configurations the gap optical loss is quite small, typically a few percent. The tubular configuration shown in Fig. 8.29 (d) is particularly tolerant to large gaps: even for g/r as large as 0.5, OL is less than 10 percent. By comparison, gap losses in a nonconcentrating, flat array of tubular collectors are of the order of g/r.

Heat Transfer Fluids

In view of the high temperature capability of most concentrating collectors there is a need to consider heat transfer fluids other than unpressurized water. Table 8.8 lists some of the most common heat transfer fluids and their temperature limitations. For heating and cooling applications air is attractive because it is noncombustible and noncorrosive. However, air has a significantly lower heat transfer coefficient than water or oils, resulting in a lowered collector efficiency factor F'. However, if evacuated receivers are used this effect is almost negligible and F' will be greater than 0.95.

TESTING OF CONCENTRATING COLLECTORS

Measurement of Efficiency

The operating efficiency is determined by measuring insolation I_T or I_B and thermal output[51-53]

$$q_u = \dot{m}c_p(T_{f,\text{out}} - T_{f,\text{in}}) \tag{8.44}$$

For collectors with *concentration below 2* the insolation is to be measured by a *pyranometer* and the efficiency is to be stated as

$$\eta_T = \frac{q_u}{I_T} \tag{8.45}$$

with respect to total insolation I_T. The pyranometer is to be mounted in the plane of the collector aperture and it should be calibrated for changes of sensitivity with tilt. For collectors with *concentration above 2* (requiring tilt adjustments or tracking) the insolation is to be measured by a *pyrheliometer* and the efficiency is to be stated as

TABLE 8.8   **Properties of Heat Transfer Fluids for Concentrators**

Fluid	Temperature range	Comments
Air (He)	Unlimited	Poor heat transfer coefficient
Water	100°C (atm), ~250°C (pressurized)	Excellent heat transfer, especially on boiling
Steam	To 500°C	
Water + ethylene glycol	~150°C	Also protects against freezing
Heat Transfer Oils (Dowtherm, Therminol)	150 to 300°C	Fairly good heat transfer, inflammable, very viscous at low temperatures

$$\eta_B = \frac{q_u}{I_B} \qquad (8.46)$$

with respect to beam radiation I_B. For higher accuracy an active radiation cavity is recommended instead of the pyrheliometer. The efficiency should be plotted as a function of $(\Delta T / I)$ using only test data for conditions with insolation above 700 W/m².

The most direct method of measuring q_u involves a determination of flowrate $\dot{m}$ and heat capacity c_p in addition to temperatures $T_{f,\text{out}}$ and $T_{f,\text{in}}$. Care must be taken to insure good mixing of the fluid in the vicinity of the temperature probe, otherwise the difference between wall and bulk fluid temperature may cause erroneous measurements. Special mixing tubes are commercially available and can be inserted just upstream of the temperature probe. Adequate mixing can usually be achieved by inserting spoilers such as copper wool or spirals in the flow channel if the flow is turbulent.

The most reliable flow measurements are obtained by using water as the heat transfer fluid and by measuring flowrates with bucket and stop watch. But since many concentrating collectors are expected to operate above 100°C, other heat transfer fluids have to be considered for tests at atmospheric pressure:

- Water plus ethylene glycol (good to about 120°C)
- Ethylene glycol (good to about 150°C)
- Heat transfer oils such as Dowtherm™, Therminol™, silicone oil (good for the 200 to 350°C range)
- Air (temperature range unlimited, but care must be taken to avoid large pressure drops, otherwise pumping power can obscure measurement of thermal output)

Relying on flow meters and on published values of heat capacity is not recommended. Flow meters must be calibrated carefully; also the heat capacity must be measured if commercial antifreeze mixtures are used in place of pure ethylene glycol.

A simpler and more reliable method of measuring q_u is by the use of a *calibrated heat source*[53] as shown in Fig. 8.30. An electric-resistance heater is used to preheat the heat transfer fluid before it enters the solar collector. (Placement downstream from the collector may result in

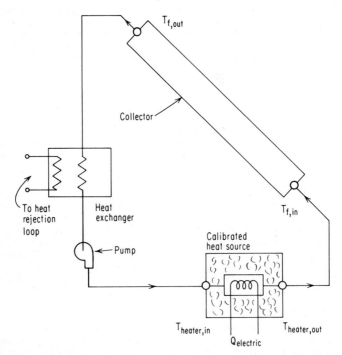

Fig. 8.30 Diagram showing the use of a calibrated heat source for measuring collector useful energy delivery.

excessive fluid temperatures.) If the electric heat source is well insulated against thermal loss to the environment, all the electric energy input

$$Q_{\text{electric}} = \text{voltage} \times \text{current} \tag{8.47}$$

raises the fluid temperature by an amount

$$Q_{\text{electric}} = Gc_p \left(T_{\text{heater, out}} - T_{\text{heater, in}}\right) \tag{8.48}$$

The useful collector output measurement

$$Q_u = \left(\frac{T_{f,\text{ out}} - T_{f,\text{ in}}}{T_{\text{heater, out}} - T_{\text{heater, in}}} \right) Q_{\text{electric}} \tag{8.49}$$

is thus reduced to a measurement of four temperatures and of electric power, both of which can be accomplished with ease and very good accuracy. (If alternating current is used in locations with significant phase angle between current and voltage, an ac power meter should be used.) The method above eliminates the need to know either flowrate or heat capacity, both of which are difficult to determine accurately.

Accurate measurements of efficiency under actual flow conditions at high temperature is a rather difficult task. The cost and effort involved in building a high-temperature collector test station can be justified only if many collectors are to be tested, otherwise the services of established test facilities should be called upon. However, the approximate performance of a concentrating collector can be determined indirectly by a number of tests which are relatively simple. These tests which are described in the following subsections can be carried out on small collector modules; thus they are valuable tools which permit evaluation of a preliminary design without the need to construct a complete (and expensive) collector prototype.

Indirect Measurement of *U*-Value

The heat loss coefficient U_c and the optical efficiency η_o are the two most important collector performance parameters. The U-value can be measured most easily by heating the absorber electrically either in the laboratory or outdoors at night. For this purpose, a heating wire (for example, nichrome) is inserted into the absorber flow channel. Electric power q_{el} and the resulting equilibrium absorber temperature T_r are measured. The U-value, relative to collector aperture (or cover) area A_c, is then given by

$$U_c = \frac{Q_{\text{el}}}{A_c \left(T_r - T_a\right)} \tag{8.50}$$

In nonevacuated collectors of the enclosed type with reflectors adjacent to the absorber, the "super-efficiency effect" due to hot reflectors discussed above may lead to an erroneous value of U_c. This effect is sufficiently small, however, not to detract from the value of the information gained by this test.

Determination of Collector Efficiency Factor *F'*

Experimental determination of F' requires measurement of absorber plate temperature T_r in addition to average fluid temperature T_f. In nonevacuated collectors this can usually be accomplished by attaching thermocouples to the back of the absorber plate. In some tubular configurations and in evacuated collectors this will not be practical because the addition of thermocouples would interfere with the properties of the selective coating or with the maintenance of the vacuum. In these cases F' can be calculated analytically. In well-designed collectors of this type F' will be so close to 1.0 (larger than 0.95 even with air as heat transfer fluid) that precise determination of F' will not be necessary.

Measurement of Optical Efficiency

Once the U-value is known, the optical efficiency can be determined by a "masked stagnation test." The solar flux reaching the receiver is reduced by placing a mask (e.g., a perforated sheet) of known transmittance τ_m over the collector aperture. The perforations must be small relative to receiver size. Under stagnation the efficiency is zero: therefore,

$$\eta_B = \gamma_B \eta_o - U_c(T_{r,s} - T_a)/(I_B \tau_m) = 0 \qquad (8.51)$$

Hence the optical efficiency is

$$\eta_o = \frac{U_c(T_{r,s} - T_a)}{\gamma_B I_B \tau_m} \qquad (8.52)$$

where $T_{r,s}$ is the receiver temperature at stagnation. An analogous formula with γ_T and I_T holds if a pyranometer is used for measuring insolation.

The following points should be remembered in the interest of accuracy:

1. The back side of the mask must be flat black to avoid multiple reflections between the mask and collector.

2. The mask should not interfere with the heat transfer from the collector to the environment.

3. The U-value to be used in Eq. (8.52) must have been measured at the same condition of T_r and T_a or corrections must be applied.

4. In collectors with heated reflectors τ_m should be as small as practical to insure that the U-value in Eq. (8.52) is as close as possible to the value measured by electric heating.

Measurement of stagnation temperature alone determines only the ratio η_o/U_c (or α_r/ϵ_r in evacuated collectors); in order to completely characterize collector performance, U_c (or ϵ_r) must have been measured independently.

In collectors with F' close to unity, the optical efficiency can also be measured directly by operating the collector as close to ambient temperature as possible. This can be accomplished by precooling the inlet fluid temperature to slightly below ambient to keep the average absorber temperature exactly at ambient.

Measurement of Optical Accuracy/Angular Acceptance

In addition to the (peak) efficiency measured at normal incidence, one must measure the angular response of a concentrator, that is, the performance at off-axis incidence. For concentrators with axial symmetry only, one angular variable need be considered. For all others, two angular variables are necessary: $\theta_\parallel$, the angular elevation of the incident direction from a cross-sectional plane; and $\theta_\perp$ = the angular deviation perpendicular to the collector trough axis in a cross-sectional plane.

The angular response is plotted as $\eta(i)/\eta(i=0)$ versus i where i is the incidence angle. As an example Figs. 8.31 and 8.32 show the measured angular response[54] of a troughlike CPC collector with concentration CR = 6.5 and design acceptance half-angle θ_c = 6.5°.

The angular efficiency scan can be done by any of the tests described above. These measurements are time-consuming and in some cases can be replaced by simple optical inspection.

Basically one looks at the collector from different angles and marks down those angles from which the whole collector aperture appears black. Despite its subjective element, this visual inspection provides essentially all the information needed for practical collector operation; furthermore, defective parts of the collector can be spotted at a glance. The distance from the eye to the collector aperture must be large compared to the width of the collector to ensure that the angle between rays from the eye to the collector is small compared to the acceptance angle which is to be measured. In principle, the optical accuracy and acceptance angle of low concentration collectors could be determined by laser ray tracing, but this method can be complex and costly.

Collector Time Constant

Knowledge of the collector time constant is needed to assess performance under intermittent sunshine; a small time constant is desirable. The thermal response to a sudden change in solar radiation is assumed to be exponential in character, and the time constant is defined as the time in which the collector output changes by 63 percent (= ln 2) of the total change which occurs when the collector is suddenly and completely shaded. The time constant depends on collector construction and materials, on heat transfer fluid, flowrate, and operating temperature, and therefore must be measured under realistic operating conditions. Short time constants have only a secondary effect on long-term, average collector performance, therefore, its precise determination is not essential. However, time constants longer than about 15 min do reduce long-term performance and should be avoided by appropriate collector design (long time constants also make the measurement of collector efficiency difficult).

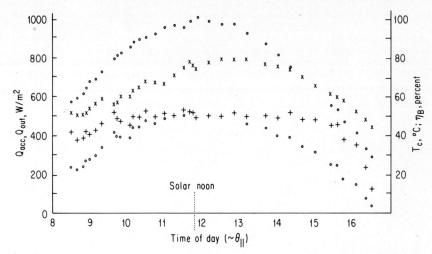

Fig. 8.31 Day long performance of a CR = 6.5 CPC with a nonevacuated tubular receiver during a clear day (1 March 1977): q_{acc}, top curve; q_u, bottom curve; efficiency η (+); collector temperature T_c (x).

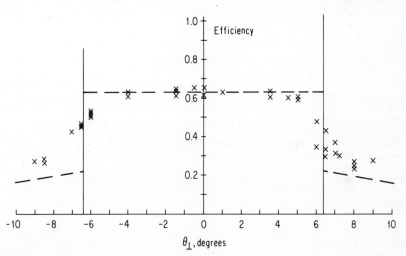

Fig. 8.32 Angular scan of a CR = 6.5 CPC collector showing optical cutoff at acceptance half-angle $\theta_c \approx$ 6.5°.

PERFORMANCE OF CONCENTRATING COLLECTORS

In this concluding section the performance of concentrating collectors is presented in graphical and tabular form.

Figure 8.33 shows the performance of several nontracking concentrators with concentration below 2, normalized to total insolation I_T. In a similar manner, but normalized to beam radiation I_B as measured by a pyrheliometer, Fig. 8.34 shows the performance of nontracking and of tracking collectors with concentration above 2.

Table 8.9 presents concentration and operating temperature in order of increasing tracking requirements. The operating efficiencies in these tables are based on the assumption that the collector be at least 40 percent efficient under full sunshine. In some collectors significant improvements are possible by means of a vacuum, selective coatings, improved reflectors, and antireflection coatings. For example, use of etched glass[46] in place of untreated glass can raise the entire efficiency curve of single glazed collectors by 3 to 6 percentage points; this can amount to a relative increase in operating efficiency of about 5 to 15 percent. Etched glass is particularly important for the covers of the receivers in concentrating collectors. Comparable improvements in efficiency could be achieved if aluminum reflector surfaces are replaced by silvered surfaces. The operating temperature can be raised if better selective coatings become available. The need for improvements is greatest for selective coatings deposited on glass. Presently available absorber coatings in evacuated all-glass collectors[33,34] have good emittance properties ($\epsilon \simeq 0.05$), but their absorptance is too low.

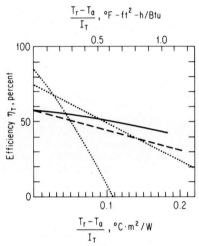

Fig. 8.33 Efficiency η_T (with respect to total insolation) for several fixed collectors. Dotted lines show upper limits of flat plate collectors, from single glazed nonselective to double glazed antireflection coated selective. The dashed line represents an evacuated nonconcentrating tubular collector (Owens-Illinois). The solid line represents an evacuated concentrating CR = 1.5 CPC.

TABLE 8.9   Summary of Applications of Different Concentrator Types

Tracking mode	Examples	Approx. range of CR	Approx. maximum operating temperature
None (tilt fixed)	CPC	1.5 to 2.0	Up to 100°C no vacuum, Up to 150°C vacuum
None (Two tilt adjustments per year)	V-trough CPC	1.5 to 2.0 3.0	Up to 150°C vacuum Up to 180°C vacuum
None (Tilt adjusted, seasonal to daily)	V-trough CPC	2 to 3 3 to 10	Up to 180°C vacuum 100 to 150°C no vacuum, 150 to 250°C vacuum
One-axis tracking	Fresnel lens Parabolic trough Fresnel mirror Fixed mirror moving receiver	6 to 30 15 to 50	100 to 200°C 200 to 300°C (with second-stage concentrator, vacuum, and selective absorber up to 400, perhaps 500°C)
Two-axis tracking	Fixed spherical reflector plus tracking receiver Fresnel lens Parabolic dish, Power tower	50 to 150 100 to 1000 500 to 3000	300 to 500°C 300 to 1000°C 500 to 2000°C

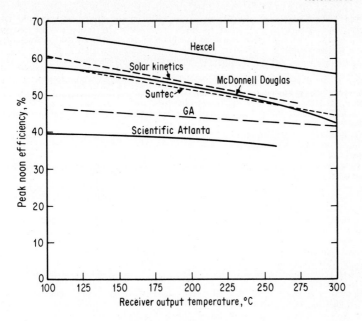

Collector Characteristics

Collector	Aperture area, m²	Secondary aperture, cm	Receiver length, m	Focal length, cm	Reflector surface
General Atomic	16.26	13.3	7.16	302.0	Silvered Glass
Hexcel	15.91	—	6.40	91.4	FEK-163 Acrylic
McDonnell Douglas	15.54	7.80	17.34	92.7	Cast Acrylic Fresnel Lens
Solar Kinetics	12.7	—	12.20	26.7	FEK-244 Acrylic
Scientific Atlanta	18.75	13.3	9.10	248.9	Silvered Glass
Suntec	35.97	—	12.20	305.0	Silvered Glass

Fig. 8.34 Comparison of peak efficiency vs. temperature for several concentrating collectors.

REFERENCES

1. R. Winston, "Light Collection within the Framework of Geometrical Optics," *J. Optic. Soc. Am.* **60:** 245 (1970).
2. A. Rabl, "Comparison of Solar Concentrators," *Solar Energy* **18:** 93 (1976).
3. F. Kreith, *Principles of Heat Transfer,* Intext Educational Publishers, New York, 1973.
4. E. M. Sparrow and R. D. Cess, *Radiation Heat Transfer,* Brooks Cole Publ. Co., Belmont, CA, 1970.
5. R. Siegel and J. R. Howell, *Thermal Radiation Heat Transfer,* McGraw-Hill, New York, 1972.
6. K. Reed, "Instrumentation for Measuring Direct and Diffuse Insolation in Testing Thermal Collectors," *Optics in Solar Energy Utilization II,* vol. 85, SPIE, 1976.
7. V. K. Baranov and G. K. Melnikov, *Soviet Journal of Optical Technology* **33:** 408 (1966); H. Hinterberger and R. Winston, *Rev. Sci. Instr.* **37:** 1094 (1966); M. Ploke, "Lichtfuehrungseinrichtungen mit starker Konzentrationswirkung," *Optik* **25:** 31 (1967).
8. R. Winston and H. Hinterberger, "Principles of Cylindrical Concentrators for Solar Energy," *Solar Energy* **17:** 255 (1975).

9. A. Rabl, "Solar Concentrators with Maximal Concentration for Cylindrical Absorbers," *Applied Opitcs* **15:** 1871 (1976).
10. A. Rabl and R. Winston, "Ideal Concentrators for Finite Sources and Restricted Exit Angles," *Applied Optics* **15:** 2880 (1976).
11. A. Rabl, "Optical and Thermal Properties of Compound Parabolic Concentrators," *Solar Energy* **18:** 497 (1976).
12. A. Rabl, N. B. Goodman, and R. Winston, "Practical Design Considerations for CPC Solar Collectors" (Jan. 1977), to be published in *Solar Energy.*
13. A. Rabl, "Radiation Transfer Through Specular Passages," *Int. J. Heat Mass Transfer* **20:** 323 (1977).
14. Patent by F. Trombe; also A. Meinel et al., presented at National Science Foundation Solar Thermal Review, March 1974.
15. K. G. T. Holland, "A Concentrator for Thin-film Solar Cells," *Solar Energy* **13:** 149 (1971).
16. H. Tabor, "Stationary Mirror Systems for Solar Collectors," *Solar Energy* **2:** 27 (1958); H. Tabor, "Mirror Boosters for Solar Collectors," *Solar Energy* **10:** 111 (1966).
17. D. K. McDaniels et al., "Enhanced Solar Energy Collection Using Reflector Solar-Thermal Collector Combinations," *Solar Energy* **17:** 277 (1975).
18. See, for example, D. L. Evans, "In the Performance of Cylindrical Parabolic Solar Concentrators with Flat Absorbers," *Solar Energy* **19:** 379 (1977); R. P. Stromberg, "A Status Report on the Sandia **17:** Laboratories Solar Total Energy Program," *Solar Energy,* 1975.
19. L. L. Vant-Hull and A. F. Hildebrandt, "Solar Thermal Power System Based on Optical Transmission," *Solar Energy* **18:** 31 (1976).
20. L. L. Vant-Hull, "An Educated Ray Trace Approach to Solar Tower Optics," *Proceedings of the Society of Photo-Optical Instrumentation Engineers* **85:** 111 (1976).
21. R. Riaz, "A Theory of Concentrators of Solar Energy on a Central Receiver for Electric Power Generation," *J. of Engineering for Power* **98:** 375 (1976). This reference derives closed analytical formulas for shading and blocking effects.
22. O. E. Miller, J. H. McLeod, and W. T. Sherwood, "Thin Sheet Plastic Fresnel Lenses of High Aperture," *J. Opt. Soc. Am.* **41:** 11 (1951).
23. D. T. Nelson, D. L. Evans and R. R. Bansal, "Linear Fresnel Lens Concentrators," *Solar Energy* **17:** 285 (1975).
24. M. Collares-Pereira, A. Rabl, and R. Winston, "Lens-Mirror Combinations with Maximal Concentration," *Applied Optics* **16:** 2677 (1977).
25. G. Nixon, "Cast Acrylic Fresnel Lens Solar Concentrator," *Report of Swedlow, Inc.*, Garden Grove, CA, 1977.
26. J. F. Kreider, "Thermal Performance Analysis of the Stationary Reflector/Tracking Absorber (SRTA) Solar Concentrator," *Journal of Heat Transfer* **97:** 451 (1975).
27. A. M. Clausing, "The Performance of a Stationary Reflector/Tracking Absorber Solar Concentrator," *ISES Solar Energy Conference, Winnipeg, Canada,* vol. 2, p. 304, Aug. 1976.
28. J. L. Russell, "Central Station Solar Power," *Power Engng.,* Nov. 1974. G. H. Eggers et al., General Atomic Report GA-A14209 (Rev.), 1977.
29. "A Pressure Stabilized Solar Collector," presented at ISES Solar Energy Conference, Florida, June 1977.
30. H. Tabor and H. Zeimer, "Low Cost Focusing Collector for Solar Power Units," *Solar Energy* **6:** 55 (1962).
31. D. F. Grether, A. Hunt, and M. Wahlig, "Results from Circumsolar Radiation Measurements," *Proceedings of the Joint Canada-American Conference, ISES,* vol. 1, p. 363, 1976.
32. J. A. Duffie and W. A. Beckman, *Solar Energy Thermal Processes,* John Wiley & Sons, New York, 1974.
33. D. C. Beekley and G. R. Mather, Jr., "Analysis and Experimental Tests of High Performance Tubular Solar Collectors," presented at ISES Conference in Los Angeles, CA, July 1975.
34. R. Bingman, Final Report for Contract 31-109-38-3805, General Electric Space Division, 1977.
35. E. Kauer, R. Kersten and F. Mahdjuri, *Photothermal Conversion,* Publication No. 32/75, Philips Forschungslaboratorium, GmbH, Aachen, October 1975.
36. U. Ortabasi and W. M. Buehl, "Analysis and Performance of an Evacuated Tubular Collector," presented at ISES Conference in Los Angeles, CA, July 1975.
37. K. G. T. Hollands, et al., "Studies on Methods of Reducing Heat Losses From Flat Plate Solar Collectors," Report No. EY-76-C-02-2597, University of Waterloo, Ontario, Canada, June 1976.
38. H. Buchberg and D. K. Edwards, "Design Considerations for Solar Collectors with Cylindrical Glass Honeycombs," *Solar Energy* **18:** 193 (1976).
39. H. Tabor, "Radiation, Convection and Conduction Coefficients in Solar Collectors," *Bulletin of the Research Council of Israel,* vol. 6C, no. 3, p. 155, August 1958.
40. T. Fujii and H. Imura, *Int. of Heat Mass Transfer,* **15:** 755 (1972).
41. *ASHRAE Guide and Data Book,* TH 7011 A5, chap. 24, ASHRAE, New York, 1965–1966.
42. Richard B. Pettit, "Characterization of the Reflected Beam Profile of Solar Mirror Materials," *Solar Energy* **19:** 733 (1977).
43. R. Winston, "Dielectric Compound Parabolic Concentrators," *Applied Optics* **15:** 291 (1976).
44. A. Rabl, "Prisms With Total Internal Reflection As Solar Reflectors," *Solar Energy* **19:** 555 (1977).
45. J. F. Kreider and F. Kreith, *Principles of Solar Engineering,* Hemisphere, Washington, DC, 1978.

46. R. E. Peterson and J. W. Ramsey, "Thin Film Coatings in Solar-Thermal Power Systems," *J. Vac. Sci. Technol.* **12:** 174 (1975).

47. An etching process for borosilicate glass (pyrex) has been developed by Corning Glass, Corning, N.Y.

48. L. G. Rainhart and W. P. Schimmel, "Effect of Outdoor Aging on Acrylic Sheet," *Solar Energy* **17:** 259 (1975).

49. R. B. Pettit and R. R. Sowell, "Solar Absorptance and Emittance Properties of Several Solar Coatings," *J. Vac. Sci. Technol.* **13:** 596 (1976).

50. Solar Total Energy Test Facility Project Semiannual Report for U.S. Dept. of Energy, Albuquerque, Oct. 1976–March 1977, Report No. SAND77-0738, August 1977.

51. J. E. Hill and E. R. Streed, "A Method of Testing for Rating Solar Collectors Based on Thermal Performance," *Solar Energy* **18:** 421 (1976).

52. F. F. Simon, "Flat Plate Solar Collector Performance Evaluation with a Solar Simulator as a Basis for Collector Selection and Performance Prediction," *Solar Energy* **18:** 451 (1976).

53. K. Reed, "Test Plan for Low Concentration Thermal Collectors," Argonne National Laboratory, 1977, to be published.

54. M. Collares-Pereira, et al., "Nonevacuated Solar Collectors With Compound Parabolic Concentrators," presented at U.S. Section ISES 1977 meeting, Florida, June 1977.

Chapter 9

High-Concentration Solar Thermal Collectors

D. L. EVANS
Arizona State University, Tempe, AZ

INTRODUCTION

It is clear from Chap. 1 that most early attempts at using solar energy for anything other than the passive heating of dwellings involved the use of optical devices to concentrate the solar rays. However, although some solar concentrators used as furnaces became interesting research tools, most other focusing devices were never much more than fascinating toys for their inventors. When interest in solar energy was renewed in 1973 with the advent of the world oil crisis, it was the flat-plate collector that received the most emphasis. Its capability of accepting diffuse radiation and the advantage of not requiring tracking of the sun were to its credit. The development of technology over the years permitted it to reach efficiencies adequate for many low-temperature purposes. The comment made in 1974 that the largest difference between flat-plate and concentrating collectors "lies in the fact that all of the thousands of practical solar energy devices that have been manufactured, sold and used in the world are based on flat-plate concepts and none on focusing systems (other than for short-time or laboratory uses)"[1] was indeed true.

Slowly but surely, however, interest in concentrating collectors began to develop when it was realized that flat-plate devices could not meet many thermal needs. This interest was stimulated, for example, by the need for process steam, the need for shaft work (including that necessary for the production of electricity), the attractiveness of total energy systems, and the potential for cost reductions in photovoltaic converters. All of these, except the last, require energy at temperatures well above those at which flat-plate collectors work efficiently.

It has been shown in Chap. 7 that the efficiency of a given thermal absorber increases as the intensity of the incident radiation increases. This is the premise on which concentrating thermal collectors are built. Thermal losses associated with large absorber areas are reduced, and increased optical losses associated with concentrating elements are accepted to achieve a net increase in overall efficiency. In photovoltaic systems, expensive absorber area (i.e., solar cells) is traded for cheaper concentrating elements in an effort to reduce system costs.

The subject of this chapter is the medium to high overall (geometrical) "concentration ratio" (CR) collector. This ratio is defined here, as it was in Chap. 8, as the aperture area of the collector divided by the intended absorbing area of the receiver. Medium to high is meant to imply a CR range of roughly 20 and above.* Although there is no clearly defined separation, the present CR range is intended to differ from the range discussed in Chap. 8 in that it requires more accurate tracking and more accurate optical components in order to use the incident solar radiation to good advantage. Some overlap with Chap. 8 is unavoidable, however, and the reader should review it as part of the study of this chapter.

Since the overall concentration ratio involves only area ratios, it is a common but not a completely adequate term used to describe collectors. For example, if the reflectance or transmittance of a concentrator is very low, the advantages of concentration may be negated. Likewise, if the accuracy of the optics is such that a sizable fraction of the radiation is directed to places other than the absorber, then the potential advantages may again not be realized. In view of the first of these examples, perhaps a more important parameter than aperture area is the product of aperture area and effective reflectance or transmittance† sometimes referred to as "effective aperture."[2] The term "intercept" or "shape factor," often used to quantify the problem elucidated in the second example, is defined as the ratio of energy intercepted (not necessarily absorbed) by the absorber to the energy reflected or transmitted by the concentrator.

Since concentrators can concentrate only the beam or direct radiation and not the diffuse, it is convenient to neglect the latter and define the collector thermal "efficiency" as the ratio of the useful thermal energy gain per unit time acquired by a fluid passing through the absorber to the total radiant energy (as beam radiation) entering the aperture per unit time, i.e., the product of the direct normal intensity, the aperture area, and the cosine of the incidence angle. Since the distribution of energy reaching the absorber is seldom uniform, it is also convenient to define the "local concentration ratio" as the ratio of the actual intensity reaching a point on the absorber to the intensity of the beam radiation normal to the aperture.

The material presented in this chapter is organized in the following way. The first section deals with concentrator shape and tracking mode, while the second discusses construction techniques of concentrators. The next two major sections deal with specific performance parameters of concentrating collectors and cover optical and thermal performance. The last major section deals with certain system related topics.

Space does not permit the listing of every possible type of concentrating collector that has been or could be devised to operate in the CR range above 20. For this reason the chapter will attempt to present basic information for the more common and practical types.

CONCENTRATORS: SHAPE AND TRACKING

Diurnal tracking is required in the subject CR range; however, there are several ways that this may be accomplished. The most common method, historically, has been to move the concentrating optics. This may entail moving of the receiver. In recent years concentrator designs which move only the receiver have been developed. Therefore, concentrator shape is discussed below in terms of the number of axes of tracking freedom and whether the concentrator optics are moved (and perhaps the receiver) or whether only the receiver is moved. Solar incidence angle equations for the various tracking modes can be found in Chap. 2 and will not be repeated here.

One-Axis Tracking-Moving Optics

One-axis tracking concentrators are generally linear or cylindrical in shape and are often referred to as "line focus" devices due to their ability to concentrate solar radiation more or less along a line parallel to the cylindrical axis. They usually involve single-curvature surfaces and can consist of reflective optics, refractive (or transmitting) optics, or combinations of reflective and refractive optics (in multielement systems). Such concentrators, depending on their design and accuracy, can operate in a wide CR range from very low (as discussed in Chap. 8) to moderate.

Line focus: reflective   Perhaps the most common linear focus device is the parabolic cylinder concentrator sometimes referred to as the "parabolic trough."[3-13] It is a reflective-type concentrator whose transverse cross-sectional shape is that of a parabola. The properties of the parabola in collecting incoming parallel light and focusing it along a line have been described in Chap. 8 and will not be repeated here. This ability to focus incoming parallel light rays on a line might tend to imply that such solar concentrators could lead to an infinite CR.

---

*The central receiver concept using multiple heliostats is not discussed here but is covered in Chap. 20.
†The word "effective" is meant to imply normal reflectance or transmittance losses plus others such as edge losses in Fresnel optics.

However, the finite size of the sun, even for perfect parabolic shaped concentrators, limits the maximum local CR to about 200. A photograph of a recent field of such troughs is shown in Fig. 9.1 (efficiencies for these collectors are in the neighborhood of 60 to 70 percent when operated about 150°C above ambient at near normal incidence).

Fig. 9.1 A field of parabolic cylindrical collectors. Collector construction is aluminum honeycomb between aluminum skins. Reflective surface is second surface aluminized acrylic film. (*Photo courtesy Hexcel Corporation.*)

Approximations to the parabolic shape are, of course, possible designs. One example is the use of segmented plane mirrors arranged to form the parabolic trough; such a concentrator has been described in the literature.[14] Two other possible examples are the circular-cylindrical shape and the Fresnel reflector. The former is limited to the lower CR range, since it suffers from spherical aberrations. In addition, its potential for manufacturing costs lower than those for the parabolic trough is questionable.[10] The Fresnel reflector, on the other hand, may be thought of as a compressed parabolic trough reflector in much the same way as the Fresnel lens may be thought of as a compressed lens (see Chap. 8). Like Fresnel lenses, Fresnel reflectors have an inherent reflection loss and/or edge blockage or scattering loss that decreases the effective aperture and optical efficiency. One advantage that they do have is that the various steps that make up the reflecting surface can lie essentially in a single plane which may offer some monetary savings in structure cost.

A novel variation on the linear Fresnel reflector is one designed for installation into the Sandia Laboratories Solar Total Energy System Test Facility (SLSTESTF).[9] A sketch is shown in Fig. 9.2. Each curved-slat reflector rotates about a longitudinal axis to concentrate the impinging solar rays on the fixed receiver. Although this concentrator has a nominal CR of about 40, the effective aperture will vary dramatically depending on the time of day and year.

Line focus: refractive   Refractive linear concentrators may also be considered. Cylindrical lenses, owing to their inherent massiveness, have received and probably deserve little attention. However, Fresnel lenses have been more favorably viewed (low-concentration lenses were discussed in Chap. 8). Analysis and measurements on such a device have been reported.[15] The measurements on a 56-cm-wide, $f/1.0*$ lens showed a peak local CR of about 45 for near normal incidence. Overall CR would depend upon absorber size but would typically be about half of

---

*The term "f-number" is defined in Chap. 3.

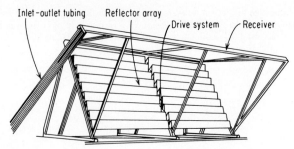

Fig. 9.2 Movable slat linear solar concentrating collector designed by Sheldahl, Inc. (*Figure from Ref. 9.*)

this. Another Fresnel lens of width 1.8 m (Ref. 16) has shown a peak local CR of about 60 with 90 percent of the transmitted energy focused in a 5-cm width (overall CR = 26). Measured thermal efficiencies on a system using this lens were in the 40 to 30 percent range for average fluid temperatures between 100 and 300°C. Tests on yet another linear Fresnel lens concentrating collector system[9] has shown efficiencies in the 60 to 50 percent range for fluid temperatures from 60 to about 300°C. A sketch of this collector is shown in Fig. 9.3.

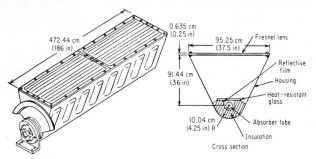

Fig. 9.3 Linear Fresnel lens solar concentrating collector designed by McDonnell Douglas Astronautics Company. (*Figure from Ref. 9.*)

Orientation of axis The axis about which the above concentrators are tracked could have any of several orientations. Most common, historically, have been axes which are parallel with the cylindrical axis of the concentrator. They are usually oriented east-west horizontal, north-south horizontal, or north-south and parallel to the earth's axis of rotation (polar axis). Of course, general north-south orientation with various tilts up from the horizontal are possible; peak energy interception results from seasonally adjusting these tilts (see Chap. 2). In thermal systems, losses (both thermal and pumping losses) in the interconnecting piping between the collectors (often called the "distributed" losses) can become sizable. In the horizontal orientations many collectors can be arranged in tandem so that the absorbers can be connected to one another, thus minimizing distributed losses. Such an arrangement is usually not economical or practical in collectors which have some degree of tilt from the horizontal due to structural problems. However, since tilted collectors can collect more energy than horizontal ones, an economic and system analysis must be made prior to the choice of the final system.

Vertical axis orientation is of interest, especially in concentrating photovoltaic systems.[17-20] The collector array, inclined at an angle to the axis, tracks the solar azimuth angle around the axis. This arrangement offers fairly good energy collection potential and may offer some structural advantages. However, for thermal systems, the problem of distributed losses may be severe.

Figure 9.4 shows the energy available to an aperture for the three tracking modes as a function of time of day. (March 21 or September 21 clear day assumed.) The peaked output of the east-west–oriented axis is due to the fact that incidence angles vary from −90° (−1.57 rad) to +90° (1.57 rad) over the day. For the other geometries the variation is much less.

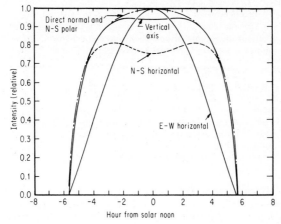

Fig. 9.4 Daily solar intensity variation at the apertures of several different tracking collectors. March 21 or September 21 clear day assumed.

End losses All one-axis tracked devices suffer end losses. That is, when the aperture normal does not point directly at the sun, the radiation reflected from the end of the concentrator most distant from the sun will not strike any absorber unless the absorber is longer than the concentrator. At the other end there will be a length of absorber which will be unilluminated and will have thermal losses but not thermal gains. The extent of this loss will depend upon the concentrator focal length-to-longitudinal length ratio and upon the orientation of the tracking axis. The longer the collector length for a given focal length the less significant are the end losses. For fixed collector lengths and focal lengths, east-west horizontal tracking axes suffer greater end losses than north-south axis orientation, since a wider range of incidence angles is encountered.

One-Axis Tracking-Moving Absorbers

A novel perturbation of the linear segmented concentrator has been devised by Russell.[21] It consists of placing narrow reflector strips parallel to the axis of a circular cylinder in contact with the surface of that cylinder. Each strip has a fixed tilt such that incoming parallel light from any direction focuses along a straight line that is always on the same cylinder. Figure 9.5 shows a cross section of the system. An absorber parallel to the axis of the cylinder can be tracked on the surface of the cylinder (extended) so as to intercept the reflected solar radiation. The CR is controlled by the rim angle and the mirror strip widths. The main advantage is that

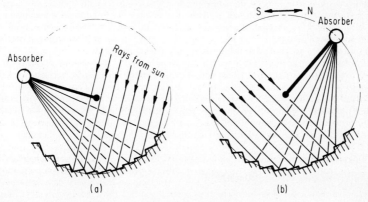

Fig. 9.5 East-west view of cross section of Russell linear fixed mirror solar concentrator. Configurations shown for (*a*) sun north of zenith and (*b*) sun south of zenith. Note individual reflector slats of various fixed tilts and also potential for edge losses.

the tracking mechanism need move only the absorber (on the arc of a circle) and not the mirror. The flexibility in mirror design is thereby greatly increased. Collectors of this type have been built that have mirror strips supported on a cast concrete structure.[9] Intensity distributions in the focal plane of an early prototype have been reported in Ref. 22. This concentrator is a member of a class of concentrators that have become known as "fixed mirror solar concentrators" (FMSC). Another member is discussed later.

Since the mirror remains fixed in this design, it suffers from the fact that energy collection potential is greatly reduced due to large incident angle cosine losses. Besides relying on the direct normal radiation falling on the fixed (i.e., nontracking) aperture, the collector has edge losses typical of Fresnel-type optics that can be sizable [23] (see Fig. 9.5). However, the ratio of cost to performance is the deciding criterion for any particular use. Even if energy collection is reduced, for sufficiently low costs, the system may be economically viable. This collector suffers more than most from dirt and snow buildup owing to its fixed reflector which cannot be stored in the inverted mode.

Two-Axis Tracking-Moving Optics

Any of the devices mentioned in the preceding section could be fastened to a two-degree-of-freedom mount and tracked so that the aperture normal always remains pointed at the sun. However, although such modes of operation would yield higher energy collection, higher efficiency (due to increased radiation), and more useful energy production, they would not produce higher overall concentration ratios. This parameter is controlled by concentrator optical design, not by the tracking mode.

Point-focus concentrators: reflective If higher efficiencies are required at high temperatures than can reasonably be reached with line-focus concentrators, a designer can use "point-focus" concentrators. These involve surfaces of double curvature which reflect or refract incoming rays into the general vicinity of a point. Surfaces of revolution are most commonly used for this purpose. It is interesting to note that most, if not all, very early solar energy collectors made use of such designs. This was undoubtedly due to their lack of a technology base which would permit development of thermally efficient lower CR systems. On the other hand, early inventors were plagued with the problem of constructing large curved surfaces. In most cases they approximated the desired surfaces with a large number of small plane mirrors. Examples are Pifre's solar-powered printing press exhibited at the 1878 Paris Exhibition[24] and Eneas' solar-powered irrigation pump[25] (see Chap. 1).

Paraboloidal mirrors (being parabolas of revolution) are perhaps the ultimate in their concentrating ability. Although their maximum CR is not unlimited, it is rather high with peak local concentrations near 4×10^4 for perfect optics and alignment. Like the parabolic trough, this concentrator has received much attention in the literature.[10,26–36].

Various approximations to paraboloidal shapes have been suggested or used. Circular Fresnel reflectors, for example, fall into this category. The segmented mirror approach used by early workers has continued to be used over the years, particularly by the Russians.[37–40] They have also experimented with inflated film concentrators that are essentially spherical in shape. Half of the spherical film is clear and half is silvered or aluminized to form the concentrator.[41]

A sketch of another approximate scheme is shown in Fig. 9.6. This collector, a part of the SLSTESTF[9] installation and described in Ref. 42, consists of about 200 small spherical mirrors that approximate a paraboloidal surface of 6.7-m-diameter aperture.

Point-focus: refractive For refractor optics, only the circular Fresnel lenses seem worthwhile to consider. Their potential for low cost is certainly attractive. Some work has been reported on this type of concentrator.[43,44]

Orientation of the axis Two-axis tracking works best, of course, if the two axes about which rotations are permitted are mutually perpendicular. Two commonly used configurations are (a) one axis vertical and one horizontal and (b) one axis parallel to the earth's axis (polar axis) and one horizontal. If shading by adjacent collectors is important, then rectangular apertures tracked by schemes (a) and (b) show significant differences in the optimum spacing and optimum aspect ratios of the apertures. This phenomenon is discussed in more detail later.

Two-Axis Tracking-Moving Absorbers

A FMSC that allows the absorber to be two-axis tracked has also been explored.[45–49] It makes use of a stationary concave-spherical mirror and a small movable, linear-cylindrical absorber. Most of the solar rays reflected from the sphere pass through or near a "focal" line that always coincides with the radius vector of the sphere that is colinear with a line to the sun. Obviously,

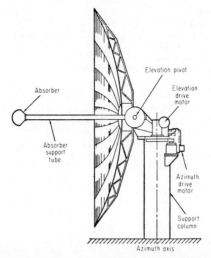

Absorber

Absorber
support
tube

Elevation pivot

Elevation
drive
motor

Azimuth
drive
motor

Support
column

Azimuth axis

Fig. 9.6 Segmented mirror approximation to paraboloidal dish designed by Raytheon, Inc. Paraboloid is approximated by about 200 spherical mirrors. Average CR is 118 while maximum local CR is 350. (*From Ref. 9.*)

as the sun moves through its diurnal motion, this focal line moves. Thus, with one end of an absorber fixed at the geometrical center of the sphere the other end could be tracked so that the absorber always coincides with the moving focal line. Ultimate CR is between 200 and 300.

As with the previous FMSC discussed, energy intercepted by the fixed aperture is considerably reduced from that available to a tracking aperture. Although performance computations have been made which show efficiencies in the 40 to 60 percent range at temperatures of 160°C[47] experimental performance and economical viability in a system context have not been explored.

CONSTRUCTION

Compared to most flat-plate collectors the structure of concentrating optical components, both reflective and refractive, is usually quite simple. There are no fluid flow passages, no insulation, no multiple cover glasses (that may require antireflection coating), no seals, no (perhaps selectively) absorbing surfaces, and no fluid connection ports. Some of these features may be incorporated on the receiver, but this is, physically, a much smaller item than the concentrator and can be manufactured separately. For example, the structure of reflecting concentrators often consists simply of a reflective surface on a substrate that provides rigidity. The advantages must be weighed versus the disadvantages of requiring tracking capability and of not using diffuse radiation.

Ideally, a concentrator should have good accuracy, be light in weight, be rigid (both in torsion and in lateral flexure), and be long lived. The effects of accuracy and rigidity will be discussed in the following sections. Small weight leads to three advantages: (1) smaller and less expensive support structure, (2) in general, smaller material costs, and (3) fewer tracking problems. Lifetime is of extreme importance, but few data exist. Of course, cost is important, but, again, cost-to-benefit ratios must be evaluated.

Single-Curvature Concentrators

For linear reflective concentrators some success has been experienced with formed thin sheets of aluminum, for example, Alcoa's Alzak,* simply clamped to a perimeter frame. More sophisticated concentrators, and perhaps sturdier ones, often have the reflector's rigid backing made as a sandwich consisting of a central core (solid or cellular) held between thin front and rear skins.

*Alzak is a trademark of the Aluminum Co. of America. It is electrically deposited pure aluminum (reflective) on an alloy substrate, protected by a clear anodized finish.

Table 9.1 lists some candidate reflectors which have been evaluated in a systematic way.[9] All were formed in a pair of precision metal parabolic-shaped molds of 0.6 × 1.2-m (2 × 4-ft) size and then evaluated for deviations in focal length and in surface-slope errors. The results are shown in the table. (Surface-slope errors will be discussed in the next section.)

Plywood was used as the core material for the original parabolic trough concentrators at the SLSTESTF.[9] Figure 9.7 shows the plys being inserted into a mold where heat and pressure

Fig. 9.7 Making plywood parabolic cylindrical concentrators. Workers are placing preglued plys in press for setting the glue. (*Photo courtesy Sandia Laboratories.*)

will be applied to set the glue. This structure was found to lack stiffness and to require backing with a rigid series of ribs when mounted in place. Large rim angle troughs are a particular problem. Also, if plywood serves as the front skin, it is generally too grainy to use with glued-on flexible reflective films. A smooth, preferably hard, layer should be laminated as the front skin in this case. In the original SLSTESTF trough collectors a thin aluminum sheet was edge-clamped over the plywood to reduce surface errors. Then aluminized tetrafluoroethylene (TFE) over polyethylene terephthalate was attached to the aluminum to serve as the reflective surface.

The collectors in Fig. 9.1 are made of an aluminum honeycomb core glued to aluminum front and back skins. The reflector is adhesively bonded, second-surface, aluminized acrylic. Each module of the collector is about 2.7 × 6 m in size and is lightweight and rigid. If installed in climatic regions that receive frequent hail, precautions would be required to protect these concentrators.[9] (The collectors shown in Fig. 9.1 can be rotated to face the ground, for example.)

Compound-Curvature Concentrators

Paraboloids in principle could be built in the same fashion as troughs, but the double curvature presents practical problems. Umarov[41] reviews some of the techniques that have been used. These include inflated film concentrators (discussed earlier), replication, electroforming, and the centrifugal method of casting. "Replication" is a common term used in the precision mirror industry to describe the casting or forming of replicas over a male master surface (for concave mirrors). A reflective surface is applied to the casting or replica after it is removed from the master. (This in essence is the process described above for constructing trough concentrators.)

Electroforming consists of plating a thin metal film on a master surface, attaching the film to a rigid frame, and then removing it from the master. A reflective surface is added later. Solar Fresnel reflectors have been made by this method.[50,51]

Centrifugal casting of paraboloidal surfaces can be done by placing an unhardened epoxy in a container that is rotating at a constant angular velocity about a vertical axis. The top surface will naturally form a paraboloidal shape owing to the combined effects of centrifugal and gravity body forces; when the epoxy has hardened, the container is removed and a reflecting surface is applied. Common problems in this technique are shrinkage and bubble formation and dirt accumulation on the surface.

Other work[10,36] has considered paraboloids manufactured from many materials by several processes. In the range of 10- to 30-cm aperture diameter, pressed glass with vapor-deposited aluminum appeared to be the best material-manufacturing combination. Larger sizes were found to be best made from aluminum sheets which would be diaphragm hydro-drawn in medium diameters (0.5–3 m) and spin-formed in large diameters (3.5–7.5 m). Building of larger sizes was found to be impractical because of the sheet size required and the clearance and size restrictions in transporting from factory to use site. The alternative of field erection was not appealing because of high costs.

Refractive Concentrators

Refractor optics must also be designed to meet the criteria given at the beginning of this section. In addition, it is desirable to have as high a refractive index as possible in order to promote refraction, so that short focal lengths can be used. However, reflection losses increase as index of refraction increases (see Chaps. 4 and 5).

Glass is a material proven to be particularly durable when exposed to the elements. It is a candidate for Fresnel refractors made by pressing, rolling, or extruding processes. Glass has an acceptable index of refraction (~1.4 to 1.7, depending on the type), good transmittance (see Chap. 5), and low cost. However, glass is not particularly light in weight (~2.2 kg/m^3) and tends to shatter easily. Also, in processing of the type mentioned above, it is difficult to maintain sharp corners (e.g., at the edge of Fresnel strips).

Clear plastics, on the other hand, tend to be light in weight (~1.2 kg/m^3) and more shatter resistant. Indexes of refraction tend to be very close to those for glass (see Chap. 5). For example, polymethyl methacrylate, an acrylic resin, has an index of refraction of about 1.49. Tests on sheets of this material exposed for nearly 18 years in the southwest United States have been reported.[52] Detrimental surface abrasion was found, but transmission loss of repolished samples was surprisingly small. Twenty-year lifetimes in the southwest were predicted, but harder surface coatings (such as glass resin) were recommended to reduce surface pitting and resultant transmission loss due to scatter. For second-surface aluminized acrylic reflectors the recommended surface coating should also prevent moisture penetration to the aluminized surface.

Acrylics can be processed in diverse ways, such as casting, extruding, molding, and thermoforming. Indeed, fairly large, circular Fresnel lenses are commercially available (e.g., up to 1-m diameter for operating at a CR of over 1000) but tend to be expensive in small purchases.

PERFORMANCE

Overall performance of concentrating collectors depends on a number of parameters. Changing these parameters in an effort to improve performance almost always involves changes in the cost. Economic analyses are therefore necessary in order to assess the value of proposed designs. Detailed energy audits are often beneficial in assessing where the most worthwhile improvements can be made. Example results of such an analysis are shown in Fig. 9.8.[12] It should be emphasized that the best system is the one that involves the lowest cost for delivered energy over a period of time for a particular application. Therefore, data of the type shown in Fig. 9.8 are but a partial representative of long-term performance, and simulation of the system over the ranges of its insolation values and fluid inlet temperatures should be considered. However, they do allow one to better understand a particular concentrator and thus anticipate possible improvements.

The desirable feature is, of course, the useful energy gain, which can be maximized if all loss

TABLE 9.1 Parabolic Test Section Fabrication and Laser Inspection Data

Reflector	Parabola construction			Reflector attachment	Remarks	Deviation in focal length,* mm	Slope error one standard† deviation, mrad
	Front skin	Core	Back skin				
Alzak†	—	7-ply 1/10Luan	—	During lamination	Plywood high forming pressure	+ 8.9	2.9
Alzak	—	7-ply 1/10 Luan	—	Vacuum held	Plywood high forming pressure	−30.9	7.9
Alzak	3-ply 1/10 Luan	2-in paper honeycomb	3-ply 1/10 Luan	During lamination	3/8-in cell honeycomb	+ 2.5	4.5
Alzak	Mel	7-ply 1/10 Luan	Mel	During lamination	High-pressure forming	−37.9	3.8
Alzak	3-ply 1/10 Luan	2-in paper honeycomb	3-ply 1/10 Luan	During lamination	3/8-in cell honeycomb	− 2.5	2.2
Alzak	—	2-in paper honeycomb	0.025 Al	During lamination	Al skinned paper	− 7.9	2.5
Alzak	—	1-in Al honeycomb	0.025 Al	During lamination	May be paper honeycomb	+11.7	2.3
—	Melamine	2-in paper honeycomb	Melamine	Vacuum held		+ 4.9	1.9
Alzak	Melamine	2-in paper honeycomb	Melamine	During lamination		−19.4	2.4
Alzak	Micarta††	2-in fiber glass honeycomb	Micarta††	After lamination	Low pressure	+ 5.8	2.0
Alzak	Fiber glass	2-in fiber glass honeycomb	Fiber glass	After lamination	Fiber glass–polyester skins	+17.3	2.5
Alzak	Micarta††	2-in paper honeycomb	Micarta††	After lamination		+31.8	2.0
Alzak	Melamine	2-in paper honeycomb	Melamine	After lamination		+24.1	1.9

Alzak	Fiber glass	{2-in fib... glass honeycomb}	After lamination	{This is one 4 × 4 ft reflector**}	+ 5.1	1.6
Alzak	Fiber glass		After lamination		+ 6.3	2.1
Alzak	Fiber glass	{2-in fiber glass honeycomb}	After lamination	{This is one 4 × 4 ft reflector**}	+ 2.8	2.3
Alzak	Fiber glass		After lamination		+ 2.8	2.7
Alzak	Fiber glass	{2-in fiber glass honeycomb}	After lamination	{This is one 4 × 4 ft reflector**}	+ 7.9	2.1
Alzak	Fiber glass		After lamination		+ 3.4	2.4
Ag-glass	—	6-in foam glass	During lamination	Back surface Ag-glass	+ 3.1	1.3
Ag-glass	—	6-in foam glass	During lamination	Back surface Ag-glass	+ 3.1	1.5
FEK-163§	0.014 Al	1.5 in Al honeycomb	After lamination	3/8 ox ACG honeycomb	+ 9.8	1.8
FEK-163	0.014 Al	1.5-in Al honeycomb	After lamination	Double sheet adhesives	+11.4	1.3
FEK-163	0.016 Al	1.5-in Al honeycomb	After lamination	Thicker skins and double adhesives	+ 2.1	1.0
FEK-163	0.016 Al	1.5-in Al honeycomb	After lamination	Thicker skins and double adhesives	+ 9.7	1.3
Alzak	Fiber glass	{2-in fiber glass honeycomb}	After lamination	{This is one 4 × 4 ft reflector**}	+12.6	1.5
Alzak	Fiber glass		After lamination		+15.3	1.5

*Mold focal length and desired trough focal length is 762 mm (30 in).

†Slope error is measured only in the trough transverse direction. Longitudinal slope error is ignored.

‡Alzak is a trade name for Alcoa's anodized aluminum.

§FEK-163 is a trade name for Minnesota Mining and Manufacturing's adhesively backed, second surface aluminized acrylic.

††Micarta is ® by Westinghouse Electric Corp.

**Separate measurements were made on each 2- × 4-ft section.

SOURCE: Ref. 9.

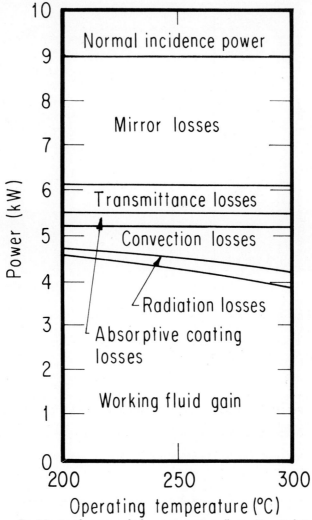

Fig. 9.8 Sample power audit for a concentrating collector. (*From Ref. 12.*)

mechanisms are minimized. These loss mechanisms separate conveniently into two categories, optical and thermal.

Optical Performance

Several factors affect the ability of an optical system to direct the incoming beam radiation to the absorber or target and add its energy to the heat-transfer fluid. These include absorption of radiation in the refracting or reflecting elements, surface irregularities of these elements, and size and positioning of the absorber, all of which make up the "mirror losses" shown in Fig. 9.8. The losses in the receiver envelope lead to the "transmittance losses," while incomplete absorption by the receiver leads to the "absorptive coating losses." Each phenomenon will be discussed in the sections that follow.

For reflecting optics, Fig. 9.9 shows two rays from diametrically opposed edges of the solar disk which meet at common points on reflecting surfaces. The maximum total included angle

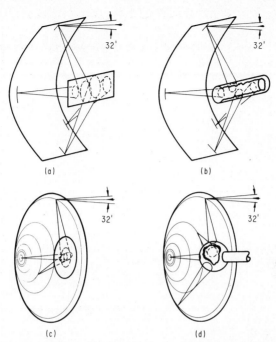

Fig. 9.9 (*a*) Parabolic cylindrical collector with flat absorber, (*b*) parabolic cylindrical collector with circular cylindrical absorber, (*c*) paraboloidal collector with flat absorber, and (*d*) paraboloidal collector with spherical absorber.

of approximately 32 minutes (9.3 mrad) is determined by the radius of the sun and the earth-to-sun distance. Perfect reflection from the reflector of all such rays from the edge of the sun yields cones of light of the same included angle and same intensity distribution as before reflection. The assumed intensity distribution within this cone depends upon the solar intensity model adopted.

Two commonly used solar intensity models are (a) the constant intensity circular disk and (b) the nonuniform intensity model suggested by Jose[35] based on data presented by Abetti[53] and given by

$$ I = I_o \frac{1 + 1.5641 \sqrt{1 - (R/R_o)^2}}{2.5641} \tag{9.1} $$

where R_o is the radius of the sun (or its image), R is the radial coordinate measured from the center, and I_o is the intensity at the center of the disk. The distribution of the intensity that impinges on the absorber depends upon the shape of the absorber and the intersection of elements of its surface with this cone. Figure 9.9 visualizes the intersections for four different reflector-absorber combinations.

An example of an analysis leading to the resulting intensity distribution produced by a cylindrical, parabolic reflector on a flat absorber is given in Ref. 8. For that geometry, peak local concentration ratios for perfect reflectors are given by (298 sin ϕ), where ϕ is the rim angle. Figure 9.10 shows the results of local concentration ratio for a perfect parabolic cylinder concentrator with flat receiver for various rim angles. Note that the rim angle ϕ is half the included trough angle.

Reflectivity and small-scale surface imperfections: dispersion Real surfaces are, of course, not perfect specular reflectors. That is, a parallel bundle of rays impinging on some small surface area of a concentrator becomes a cone of rays represented by some nonzero solid

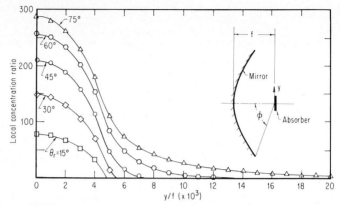

Fig. 9.10 Local concentration ratio distribution for focused on-axis performance of perfect parabolic cylindrical solar concentrator with flat absorber using solar model given by Eq. (9.1). (*From Ref.* 8.)

angle upon reflection or refraction at that surface.* This solid angle depends upon the incidence angle of the incoming rays and upon the physical condition and makeup of the actual surface.

Figures 9.11 to 9.14 show example data for surfaces of different "specularity."[54,55]† Figure 9.11 is for aluminized Dupont Mylar-S film which is only a fair specular reflector compared to an aluminized Dupont experimental Mylar film (Fig. 9.12). Both materials show some dependence on incidence angle i. Figure 9.13 is for a first-surface ground-glass mirror, while Fig. 9.14 shows several different surfaces compared at near normal incidence.

What this dispersion or lack of specularity does is to effectively increase the apparent size of the solar disk, after reflection, from its value of 32 minutes (9.3 mrad), the included angle before reflection. The data in the form presented in Figs. 9.11 to 9.14 are sometimes inconvenient to use. It can often be more easily used if it is assumed that the surface will scatter a beam in a normal distribution about the central outgoing ray. Then the reflectance could be written as

$$R(\Delta\theta) = \frac{R_1}{\sigma_r \sqrt{2\pi}} \exp \frac{-\Delta\theta^2}{2\sigma_r^2}$$

where $\Delta\theta$ is the angular departure from the central (specular) ray and σ_r is the standard deviation. For some types of reflectors the addition of a second normal distribution is necessary.[55] It would be given by

$$R(\Delta\theta) = \frac{R_1}{\sigma_{r1} \sqrt{2\pi}} \exp \frac{-\Delta\theta^2}{2\sigma_{r1}^2} + \frac{R_2}{\sigma_{r2} \sqrt{2\pi}} \exp \frac{-\Delta\theta^2}{2\sigma_{r2}^2}$$

Pettit[54] has analyzed his data to obtain representative monochromatic values for R_1, R_2, σ_{r1}, and σ_{r2}. They are shown in Table 9.2 for several wavelengths. Solar-spectrum weighted values

---

*A single ray reflected (or refracted) from a surface may leave in any direction depending on the orientation of the surface tangent at the point where the ray is reflected (or refracted). The important feature for concentrators is not what the distribution of outgoing rays is over all possible angles, but what the distribution of outgoing rays is near an outgoing central ray which lies in the plane of incidence and which obeys the law of reflection (or Snell's law for refraction) with respect to the "average" surface (the central ray would undergo true specular reflection). Thus a specular reflectance can be defined as the fraction of incoming energy with a particular incidence angle that leaves within a certain solid angle (or included angle) centered about a leaving ray that obeys the simple law of reflection (or Snell's law for refraction) relative to the "average" surface.

†References 54 and 55 differ in their measurement techniques; Ref. 54 used circular apertures to define the incoming and outgoing beams, while Ref. 55 used slits for defining these beams. The results cannot be compared directly unless the effects of the measurement instruments are taken into account. This was not done in the work of Ref. 54. Instrument effects are also included in the data shown in Fig. 9.14 from Ref. 55, but have been removed from the results shown in Table 9.2.

of σ_r (or σ_{r2} and σ_{r2}) can be obtained from these data. The intensity distribution of the sun can then be integrated with these distributions to obtain an effective model of the sun as seen by the absorber. If the sun is considered to have a standard deviation σ_s depicting its angular spread,* this convolution yields another normal angular spread distribution of standard de-

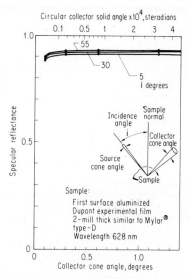

Fig. 9.11 Specular reflectance measurements on an aluminized Dupont Mylar® film. Circular apertures were used. (*From Ref. 54.*)

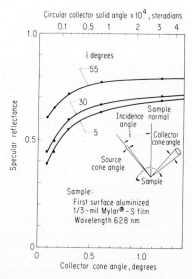

Fig. 9.12 Specular reflectance measurements on an aluminized Dupont experimental film. Circular apertures were used. (*From Ref. 54.*)

*σ_s is often taken as 8.6 minutes (2.5 mrad), or approximately one-quarter of the sun's total included angle.

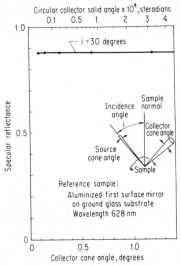

Fig. 9.13 Specular reflectance measurements on an aluminized glass mirror. Circular apertures were used. (*From Ref. 54.*)

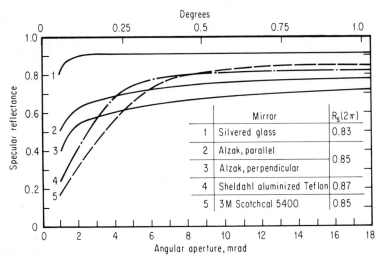

Fig. 9.14 Specular reflectance measurements on several mirrors at near normal incidence. Parallel slit apertures were used. (*From Ref. 55.*)

viation $\sigma = (\sigma_s^2 + \sigma_r^2)^{1/2}$ for the single-term reflectance model. This modified solar standard deviation is useful in many optical and performance calculations. R_1 is then the reflectance to be used in such calculations. The "effective" solar diameter could then be approximated by 4σ, which would contain 95 percent of the area under the normal curve. For the two-term model, essentially two calculations of performance need to be made, one for $\sigma = (\sigma_s^2 + \sigma_{r1}^2)^{1/2}$ and one for $\sigma = (\sigma_s^2 + \sigma_{r2}^2)^{1/2}$. The respective R_1 and R_2 need to be used for the reflectances in such performance calculations.

The hemispherical reflectance (or more accurately the angular hemispherical reflectance as defined in Ref. 56) is obtained by determining the asymptotes of data like that shown in Figs. 9.11 to 9.14 at large angles (2π sr). Solar averaged reflectances can be calculated by weighting

TABLE 9.2 Specular Reflectance Properties of Several Mirror Materials

Material	Measurement wavelength, nm	R_1	σ_{r1}, mrad	R_2	σ_{r2}, mrad	$R_s(2\pi)$
I. Second surface glass						
a. Laminated glass—Carolina Mirror Co.	500	0.92	0.15	—	—	0.83
b. Laminated glass—Gardner Mirror Co.						
Perpendicular to streaks	600	0.92	0.4*	—	—	0.90
	500	0.92	0.4*	—	—	
Parallel to streaks	800	0.88	< 0.05	—	—	
	500	0.92	< 0.05	—	—	
c. Corning microsheet (Vacuum Chuck)	550	0.77	1.1	0.18	6.2	0.95
II. Metallized plastic films						
a. 3M Scotchcal 5400	500	0.86	1.9	—	—	0.85
	600	0.86	2.0	—	—	
	700	0.82	2.1	—	—	
	900	0.84	1.9	—	—	
b. 3M FEK-163	500	0.86	0.90	—	—	0.85
	600	0.86	0.78	—	—	
	700	0.82	0.86	—	—	
	900	0.84	0.86	—	—	
c. Sheldahl Aluminized Teflon	400	0.73	1.4	0.15	12.1	0.87
	500	0.80	1.3	0.07	30.9	
	700	0.80	1.6	0.04	39.8	
	900	0.81	1.4	0.03	31.4	
III. Polished, bulk aluminum						
a. Alcoa Alzak†						
Perpendicular to rolling marks	670	0.66	0.39	0.21	9.7	0.85
	505	0.56	0.42	0.33	10.1	
	407.5	0.45	0.53	0.42	9.8	
Parallel to rolling marks	670	0.70	0.24	0.17	7.7	
	505	0.62	0.29	0.27	7.1	
	407.5	0.58	0.46	0.29	9.0	
b. Kingston Ind. Kinglux						
Perpendicular to rolling marks	498	0.65	0.37	0.23	16.1	0.85
Parallel to rolling marks	498	0.67	0.43	0.21	18.5	
c. Metal Fabrications Bright Aluminum	550	0.44	1.4	0.43	10.3	0.84

*σ_{r1} obtained from an area away from a major streak.
†Alzak is a trade name for Alcoa's anodized aluminum.
SOURCE: Ref. 54.

the monochromatic values over the solar spectrum (see Chap. 5). Figure 9.14 lists the solar hemispherical reflectances obtained by such a method.

Weathering. Exposure to the elements and cleaning operations can cause changes to any particular surface, which may dramatically affect its performance. Light accumulations of dust and dirt decrease the performance of concentrating collectors more than they would for flat-plate collectors, since such accumulations often seem to cause scattering more than absorption of incident radiation. Thus, although light dust and dirt often do not produce significant changes in the *hemispherical* reflectance, they can drastically reduce the *specular* reflectance. Washing can restore this latter property if the surface is hard enough to withstand the abrasion associated with washing. The aluminized TFE used in the original collector field of the SLSTESTF showed poor abrasion properties[9]; its specular reflectance changed as shown in Fig. 9.15. Aged samples

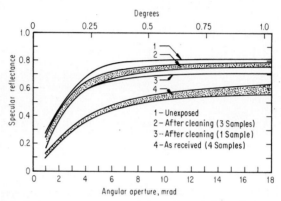

Fig. 9.15 Specular reflectance measurements on aluminized TFE before and after cleaning. Parallel slit apertures were used. (*From Ref. 55.*)

of this material were tested[57] both as received (curve band 4) and after cleaning by three methods which involved sprays and mists, but no wiping (curve band 2), and washing followed by wiping (curve 3). Curve 1 is for material that had not been exposed. Apparently the wiping scratched the surface and degraded specular performance. Wind-driven erosion may account for the difference between curves 1 and 2. It has also been noted[9] that electrostatic attraction and retention of dirt is much more of a problem with TFE-covered collectors than with glass or anodized-aluminum-covered collectors.

Of course, long-time exposure to the elements can cause degradation of the hemispherical reflectance owing to corrosion, discoloration, etc. Optical tests have been conducted and reported on the materials shown in Table 9.3.[6] Several samples reported were specially made and did not have the benefit of second- or third-generation development. This should be kept in mind, since the results depend somewhat on construction techniques as well as materials.

Samples 1, 2, and 3 (aluminized fiber glass with protective coating) either failed or demonstrated a significant loss in reflectance. The surfaces became pitted, and small green spots appeared which expanded in size to render the reflector useless. Sample 4 (aluminized acrylic) tested well in Arizona with the exception of some small scratches. Very slight "fogging" was apparent on the sample tested in Minnesota, while serious fogging occurred on the Florida-tested system. Samples 5, 6, and 7 (aluminized or silvered TFE) did not degrade in hemispherical reflectance but did show slight scratches in Arizona and fogging in Minnesota. Sample 8 (second-surface aluminized glass) survived extremely well as might be expected. Sample 9 (aluminized acrylic) showed hemispherical reflectance losses of from 4 to 10 percent during the approximate year of exposure testing. Tests on the hemispherical reflectance of Alcoa's Alzak exposed for 28 mo in Arizona showed only about 4 percent reduction from the original solar weighted value of 0.82. Figure 9.16 shows the wavelength dependence of the hemispherical reflectance of this material before and after test. These values must be weighted with the solar spectrum to obtain a total solar hemispherical reflectance. Again, however, it should be remembered that for high-concentration devices specular reflectance, not just hemispherical reflectance, is the important property.

TABLE 9.3 Weatherability Data on Reflective Surfaces

Sample type	Type*	Location	Material	Exposure, in weeks	Solar hemispherical reflectance original	Solar hemispherical reflectance dirty†	Solar hemispherical reflectance clean‡	% degradation
1	1st	Arizona	Aluminized fiber glass w/ protective coating	69	0.88	0.70	0.69	22
	1st	Florida	Same	<15	0.85	Failed	Failed	—
	1st	Minnesota	Same	<15	0.85	Failed	Failed	—
2	1st	Arizona	Aluminized fiber glass w/ protective coating	<45	0.83	Failed	Failed	—
	1st	Florida	Same	<15	0.83	Failed	Failed	—
	1st	Minnesota	Same	<15	0.81	Failed	Failed	—
3	1st	Arizona	Aluminized fiber glass w/o protective coating	74	0.92	0.75	0.76	17
	1st	Florida	Same	<45	0.93	Failed	Failed	—
	1st	Minnesota	Same	<15	0.91	Failed	Failed	—
4	2d	Arizona	Aluminized acrylic	66	0.86	0.79	0.86	0
	2d	Florida	Same	68	0.86	Failed	Failed	—
	2d	Minnesota	Same	74	0.85	0.74	0.84	1
5	2d	Arizona	Aluminized TFE	58	0.78	0.74	0.82	0
	2d	Florida	Same	60	0.79	0.78	0.80	0
	2d	Minnesota	Same	62	0.79	0.70	0.80	0
6	2d	Arizona	Silvered TFE	58	0.86	0.72	0.88	0
	2d	Florida	Same	60	0.86	0.85	0.89	0
	2d	Minnesota	Same	62	0.87	0.79	0.89	0
7	2d	Arizona	Aluminized TFE	58	0.78	0.61	0.78	3
	2d	Florida	Same	60	0.77	0.72	0.75	0
	2d	Minnesota	Same	62	0.77	0.67	0.77	0
8	2d	Arizona	Aluminized glass	63	0.76	0.69	0.76	0
	2d	Florida	Same	65	0.76	0.76	0.76	0
	2d	Minnesota	Same	81	0.76	0.73	0.76	0
9	2d	Arizona	Aluminized acrylic	58	0.80	0.69	0.72	10
	2d	Florida	Same	60	0.80	0.74	0.74	8
	2d	Minnesota	Same	62	0.80	0.69	0.77	4

*1st: first surface; 2d: second surface.
†Time since last cleaning was 21 weeks.
‡Cleaning was done with deionized water and a cotton swab followed by forced-air drying.
SOURCE: Ref. 6.

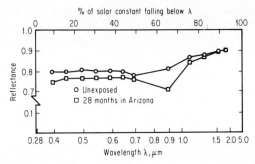

Fig. 9.16   Monochromatic hemispherical reflectance measurements on Alcoa's Alzak. (*From Ref. 6.*)

Large-scale surface imperfections   Large-scale surface imperfections can be extremely critical for collectors in the subject CR range. These are deviations of the "average" surface (neglecting the small-scale deviations discussed in the previous section that affect the specular reflectance) from its intended shape (e.g., parabolic). For the most part these are readily apparent to the unaided eye in the form of the distortions they cause in reflected images. The exception to this obvious deviation is the slow, smooth type of departure from ideal curvature.

Analytical Description.   Figure 9.17 demonstrates the problem caused by true surface tangents being different from the ideal. Instead of the central ray from the sun being directed to the center of the absorber, it is misdirected to another place. Note that an angle error δ in the slope causes an error of 2δ in the departing ray.

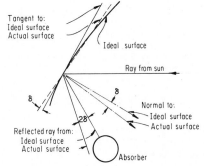

Fig. 9.17   Surface slope errors cause deviation of solar rays. A slope error of δ causes a reflected ray deviation of 2δ.

Surface-tangent errors are highly dependent on the method of manufacturing, handling, and mounting and on the stiffness of the concentrator. What seems like an excellent design from all of these standpoints can often be rendered nearly useless by some unplanned phenomenon such as a severe hail storm. According to Umarov,[41] the value of these large-scale deviations was not fully appreciated by workers in the solar furnace field until pointed out by Baum[58] at a 1954 New Delhi symposium.

Some of the departures of the actual surface from the intended are nearly random in occurrence, while others are systematic. The random ones are often dealt with by assuming that the differences between the true surface tangents or slopes and the ideal or intended surface tangents can be characterized by a normal distribution. Aparisi[59,60] was apparently the first to complete such an analysis on a paraboloidal concentrator focusing onto a flat disk. Umarov has given a review of such work (particularly the Russian work through the mid-1960s including that of Aparisi on surface inaccuracies as well as concentrators in general).[41] Aparisi found the local concentration ratio to be given by

$$CR\left(\frac{r}{f}\right) = \frac{R_1}{2}\left(\frac{\sin\phi}{2\sigma_c}\right)^2 \exp\left[-\frac{1}{8}\left(\frac{r(1+\cos\phi)}{f\,2\sigma_c}\right)^2\right] \tag{9.2}$$

where r is the radial coordinate measured from the center of the absorber, f is the focal length of the mirror, R_1 is the (specular) reflectance (e.g., from Table 9.2), ϕ is the rim angle, and σ_c is the standard deviation of the surface-tangent or -slope errors. Silvern[61] among others[4,62] who have also studied this problem, found a similar result differing only slightly in peak CR and spread.

Thus, the maximum concentration (i.e., at $r = 0$) varies as the inverse square of σ_c, while the width of the image grows rapidly with σ_c. This latter result means that for large surface errors (i.e., large σ_c), large receiver sizes would be required to intercept significant fractions of the radiation flux, thus potentially increasing the thermal losses. This result, as a function of σ_c, is not extendible to $\sigma_c \to 0$, since the sun was considered a point source. Therefore, $2\sigma_c$ in Eq. (9.2) should be replaced by* $(4\sigma_c^2 + \sigma_s^2 + \sigma_r^2)^{1/2}$ when σ_c becomes comparable to σ_s and σ_r. In three dimensions it takes two angles to define the orientation of a tangent plane, but deviations of one of these angles may be more significant than deviations in the other depending on how the concentrator was manufactured.[63]

Duff and Lameiro,[2] building on the work of Aparisi, have presented general expressions for several concentrator-absorber geometries which allow one to determine the location of the radiation in the region of the focal plane.† They also assume normally distributed mirror surface-slope errors. Their parameter, g, which is a measure of the spread of the radiation in the focal region, can be calculated from the expressions given in Table 9.4. This parameter is used in the expressions

$$\Gamma = 1 - \exp\left[- \pi a^2 / g\right] \tag{9.3}$$

for point-focus concentrators, and

$$\Gamma = \frac{1}{g\sqrt{2\pi}} \int_{-a}^{a} \exp\left[- \frac{1}{2}\left(\frac{y}{g}\right)^2\right] dy = \frac{1}{\sqrt{2\pi}} \int_{-a/g}^{a/g} e^{-Z^2/2} dZ \tag{9.4}$$

for line-focus concentrators to give the fraction Γ of the reflected radiation which falls within a receiver radius a, for point-focus concentrators, or halfwidth a, for linear concentrators, of the focal point or line. If a matches the appropriate dimension of an absorber, Γ is simply the intercept factor. Equation (9.4) can be easily solved by the use of probability tables.

Monte Carlo statistical techniques can also be used to investigate surface-slope errors. Figure 9.18 shows the results from Ref. 8 for a 30° (0.52 rad) rim-angle trough for different assumed normally distributed slope errors. Only errors in slopes defined in transverse planes were considered, since errors in longitudinal slope do not give rise to losses except near the ends of the troughs. Significant degradations can result even for modest slope errors.

The vulnerability to hail was mentioned earlier. This phenomenon can certainly change the distribution of surface-slope errors and needs to be considered in concentrator selection. Some preliminary qualitative data now exist for certain combinations of materials describing their resistance to simulated hail.[9] These are reproduced in Table 9.5. This table shows the poor resistance of honeycomb cores with aluminum skins and of all plywood collectors. Hard or resilient skins (such as fiber glass) tend to be better than aluminum.

Systematic surface errors are more difficult to deal with analytically than are the random errors. They may arise anywhere during the construction phase or may appear after installation (e.g., the sagging of plastic parts or the warping of wooden members). For optimum performance they need to be assessed and minimized. The systematic experimental study[9] on trough construction mentioned in the section on collector construction provides needed information in this area.

Studies to assess manufacturing techniques and projected costs of concentrators have concluded that surface accuracies of 0.1° (1.8 mrad) are readily achievable.[10] Most of the cost savings experienced in increasing this tolerance to 2° (35 mrad) would result from savings in the labor involved in manufacturing, handling, and installation. But since labor was minimized in the assumed, automated, high-volume production, large changes in labor result in only small changes

---

*A useful standard deviation for the angular spread of radiation from a point on the concentrator which includes sun size, lack of specularity, and random surface-slope errors is given by $(\sigma_s^2 + \sigma_r^2 + 4\sigma_c^2)^{1/2}$, where σ_s is the standard deviation of surface-slope errors; for nonzero tracking errors (to be discussed later) $4\sigma_t^2$ could be added to the terms in parentheses.

†Also, see Ref. 10.

TABLE 9.4 Beam Spread Parameter g

	Point–focus parabolas		Line–focus parabolic cylinders	
	Spherical target*	Flat target*	Round target*	Flat target†
g	$\dfrac{2A_p(4\sigma_c^2+\sigma_r^2+\sigma_s^2)(2+\cos\phi)}{3\phi_r\,\sin\phi}$	$\dfrac{2A_p(4\sigma_c^2+\sigma_r^2+\sigma_s^2)}{\sin^2\phi}\ddagger$	$\left(\dfrac{W_p^2(4\sigma_c^2+\sigma_r^2+\sigma_s^2)(2+\cos\phi)}{12\,\phi_r\,\sin\phi}\right)^{1/2}$	

$$\dagger\left\{\frac{W_p^2(4\sigma_c^2+\sigma_r^2+\sigma_s^2)}{4\phi_r\tan^2(\phi/2)}\left[\frac{-1}{3\sin^3\phi\cos\phi}+\frac{2}{3\sin^3\phi}+\frac{2}{\sin\phi}-\frac{\cos\phi}{3\sin^3\phi}-\frac{2\cos\phi}{3\sin^3\phi}+\frac{4\sin\phi}{3\cos\phi}-\ln\tan\left(\frac{\pi}{4}+\frac{\phi}{2}\right)+\ln\tan\left(\frac{\pi}{4}-\frac{\phi}{2}\right)\right]\right\}$$

*A_p, aperture area; W_p, aperture width; ϕ, rim half-angle.
‡Approximate expression.
†Approximate expression.
SOURCE: Refs. 2 and 10.

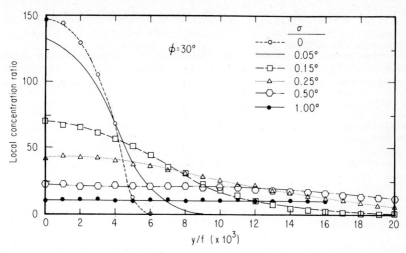

Fig. 9.18 Local concentration ratio distributions for surface slope error normal distributions of various standard deviations, σ. These are for parabolic cylindrical perfect reflectivity concentrators with flat absorbers having the geometry shown in Fig. 9.10. (*From Ref. 8.*)

TABLE 9.5 Typical Materials Tested with Simulated Hail

Reflector	Support skin	Structure skin	Hail size, mm	Approximate impact velocity, m/s	Comments
Alzak†	Micarta	Paper honeycomb	19	13.7 ± 2.1	Good hail resistance, conical dents
Alzak†	Melamine	Paper honeycomb	19	13.7 ± 2.1	Good hail resistance, conical dents
Alzak†	Fiber glass	Paper honeycomb	19	13.7 ± 2.1	Good hail resistance, conical dents
Alzak‡	Luan plywood	Luan plywood	19	13.7 ± 2.1	Poor hail resistance, spherical dents
Sheldahl§	Fiber glass	Fiber glass	19	13.7 ± 2.1	Moderate damage, conical dents
FEK-163§	Aluminum	Aluminum honeycomb	19	13.7 ± 2.1	Heavy damage, spherical dents
Alzak‡	0.025-in galvanized iron	Aluminum honeycomb	19	13.7 ± 2.1	Heavy damage, conical dents
FEK-163§	Fiber glass	Aluminum honeycomb	19	13.7 ± 2.1	Moderate damage, conical dents
FEK-163§	0.025-in galvanized iron	Aluminum honeycomb	25.4	19.2 ± 0.3	Good hail resistance, conical dents
Other materials tested with simulated hail glass receiver tube			38	16.0 ± 0.7	Fracture, poor resistance

Alzak is a trade name for Alcoa's anodized aluminum; Micarta is a trade name of Westinghouse Electric Corp. and FEK-163 is Minnesota Mining and Manufacturing's (3M) second-surface aluminized acrylic.
　†Bonded.
　‡Mechanical attachment (screws or tape) of reflector to support structure.
　§Adhesive backed.
　SOURCE: Ref. 9.

in total cost. For surface accuracies better than about 0.14° (2.4 mrad), costs were estimated to escalate rapidly. The surface-slope errors listed in Table 9.1 for collectors formed in molds can be seen to be in this range.

For given surface errors, image spread at the optical focus can be minimized by reducing the distance between the mirror and receiver.[12] This is demonstrated in Fig. 9.19 which shows various rim-angle concentrators, all having a common focus and fixed aperture. Since aperture size may be strongly related to cost, this may be a reasonable equal-cost method of comparison.

A 90° rim-angle parabola obviously has the smallest maximum distance from concentrator to focus. It can, however, be shown[12] that a rim angle of 120° (2.1 rad) has the shortest *average* distance between the concentrator surface and the focus. This does not necessarily mean that 120° (2.1 rad) is the optimum rim angle, since there are other considerations, such as cost. For example, the material added at large rim angles is inefficiently used and might be better added as an extension of the length of a trough or as an additional collector. Of course, rim angles larger than about 75° (1.22 rad) obviate the use of many types of receivers, such as flat and cavity types.

In refracting optics such as Fresnel lenses, an analogous line of reasoning would lead to shorter focal lengths. However, it can be shown that edge losses and/or reflection losses become significant at f-numbers below about 1 to 1.5.[17]

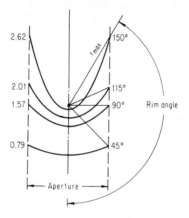

Fig. 9.19 Equal aperture area parabolas of various rim angles. (*From Ref. 12.*)

Experimental intensity distributions Several methods have been used to experimentally quantify the intensity distribution in the intended focal plane, or equivalently, measure surface-slope deviations. These methods show the effect of all anomalies, both large and small. Due to space limitations, only references to where additional information may be found will be given. The methods include photographic techniques using the sun as the source[1] and the moon as the source,[64] using a calorimeter detector with the sun as the source,[22] using grids of heat-sensitive paint[9] with the sun as a source, manual measurements using a small diameter laser beam as a source,[6] and, more exotically, methods using photodiode detection with a laser beam as a source.[9] This latter device, capable of in situ measurements on trough concentrators, can readily provide needed information on flux distributions at the receiver.

Receiver envelopes: transmittance losses Transparent tubes or plates are quite often used in concentrating collectors to enclose the absorber or receiver in order to decrease thermal losses. The reflection and absorption by such elements give rise to additional optical losses, not unlike those experienced in glazed flat-plate collectors (see Chap. 7). For flat-plate glazings, however, the incidence angle for beam radiation varies with time of day. The incidence angle for diffuse radiation covers all ranges at any time. In concentrators using only beam radiation, the incidence angle with respect to the aperture normal may vary with time of day, depending on the mode of tracking. However, the solar rays reaching the transparent tube or plate enclosing the absorber at any time may have a range of incidence angles with respect to the local tube or plate surface normals.* This is demonstrated in Fig. 9.9. High transmittance glass and antireflective coatings may be justified, since they may improve optical performance by several percent.

Receiver absorptance   Reflection of incoming photons at the receiver surface can also constitute an important loss mechanism. It can be reduced simply by using surfaces of high absorptance. It should be remembered that absorbers on concentrating collectors will operate at higher temperatures than absorbers in flat-plate use and that their lifetime could be affected. In addition, the higher the temperature, the greater the overlap of absorbed and emitted spectra, thereby reducing the potential for highly selected surfaces.

Absorber size and positioning-shape factor losses   Reduction of all the optical losses discussed above results in better thermal performance. In fact, when thermal losses are very low, as they usually are in concentrating collectors, improvements in these optical losses result in almost proportional gains in useful thermal energy collected in the heat-transfer fluid. This is not necessarily the case with the absorber size and placement. Therefore, these topics will be discussed in the section on system performance.

Thermal Performance

The convection and radiation losses shown on the example energy audit of Fig. 9.8 may be termed "thermal" losses. The analysis of these losses and the useful gain can be carried out by starting with a small control volume containing fluid in a length dx of absorber. For steady operation in many geometries, this results in an equation of the form

$$\dot{m}\frac{dh}{dx} = P_{a,i}\,h_{c,af}(T_a - T_f) \tag{9.5}$$

where x is assumed to be the axial coordinate along the absorber and changes in the kinetic and potential energies of the fluid stream and axial heat conduction in the fluid have been neglected.† Here $P_{a,i}$ is the inside perimeter of the flow area of the absorber and h is the fluid enthalpy. If the absorber temperature T_a is known, then this equation can be integrated to yield the useful working-fluid internal energy gain per unit time over the total absorber length L:

$$q_u = \dot{m}\,(h_{\text{out}} - h_{\text{in}}) = \int_0^L P_{a,i}\,h_{c,af}(T_a - T_f)\,dx \tag{9.6}$$

The heat-transfer coefficient $h_{c,af}$ can be obtained from existing correlations for the Nusselt number for internal flows (see Chap. 4). The immediate limitation to the use of this equation is the unknown absorber temperature T_a. This temperature floats at some level determined by the solar radiation at the aperture, the optical properties of the concentrator and absorber, the fluid inlet temperature, the geometry of the absorber and its surroundings (such as convection-suppressant envelopes), and q_u. Its determination relies on establishing separate energy balances for the absorber and any intermediate structure between it and ambient conditions. These energy balances are then solved simultaneously to yield the necessary intervening temperatures.

Example case   Depending on the geometry, a thermal analysis can become quite complex; it necessarily involves many assumptions and simplifications. The case of a horizontal, circular-cylindrical absorber surrounded by a cylindrical glass jacket, typically used as the receiver of trough concentrators, is discussed here as an example analysis. It is assumed that the collector is operating in steady state, that the properties of absorptance, reflectance, or transmittance are constant, and that all local temperature gradients within the absorber or envelope are negligible.

An energy balance on a section of the absorber tube of length dx and per unit absorber tube outside area then results in the equation

---

*Transmittance of nearly all materials is a function of incidence angle, as discussed in Chap. 5. It typically is fairly constant up to incidence angles of 50 to 60° (0.87–1.05 rad), whereupon it drops rapidly to zero at 90° (1.57 rad) incidence angle.

†A macroscale thermal-loss calculation method which does not consider absorber-axial temperature gradients as functions of x is contained in Chap. 8.

$$q_{s,a} = \frac{D_{a,i} h_{c,af}(T_a - T_f)}{D_{a,o}} + q_{c,ae} + q_{r,ae} \tag{9.7}$$

where

$$q_{s,a} = \frac{q_s \tau_{e,s} \alpha_{a,s}}{1 - \rho_{a,s}\rho_{e,s}} \tag{9.8}$$

is the solar energy absorbed by the absorber per unit area per unit time (including reflected radiation from the envelope) based on the energy q_s incident on the envelope per unit area of absorber. q_s, in harmony with the assumption of no circumferential temperature gradients in the absorber, is assumed uniform over the envelope and, thus, over the absorber. It is the product of the energy flux incident on the aperture, the effective reflectance (or transmittance), the intercept factor, and the overall concentration ratio. The symbols $\tau_{e,s}$, $\rho_{e,s}$, and $\alpha_{a,s}$ represent the transmittance and reflectance of the envelope and the absorber absorptance, respectively; $\rho_{a,s} = 1 - \alpha_{a,s}$. Also

$$q_{c,ae} = h_{c,ae}(T_a - T_e) \tag{9.9}$$

is the convective heat-transfer rate (per unit absorber area) between the absorber and the envelope at temperature T_e. Eckert,[64] for example, gives a correlation for $h_{c,ae}$ for natural convection in a cylindrical annulus applicable to the whole continuum regime (i.e., where the mean free path of the molecules is short compared to the radial separation between the cylinders). Also in Eq. (9.7), $q_{r,ae}$, the radiant exchange rate (per unit absorber area) between the absorber and the envelope, is given by

$$q_{r,ae} = \epsilon_{ae}\sigma(T_a^4 - T_e^4) \tag{9.10}$$

where the view factor ϵ_{ae} is found from

$$\epsilon_{ae} = \left[\frac{1}{\epsilon_{a,\,\text{ir}}} + \left(\frac{D_{a,o}}{D_{e,i}}\right)^k \left(\frac{1}{\epsilon_{e,\,\text{ir}}} - 1\right) \right]^{-1} \tag{9.11}$$

Here, $k = 0$ for a specularly reflecting envelope or $k = 1$ for a diffusely reflecting envelope. The ϵ's are the infrared (ir) emittances of the absorber and envelope.

The occurrence of the unknown envelope temperature T_e in Eq. (9.7) requires that an energy balance on the envelope be made. This gives, for the previous assumptions,

$$q_{c,ae} + q_{r,ae} + q_{s,e} = \frac{(q_{r,es} + q_{c,es})D_{e,o}}{D_{a,o}} \tag{9.12}$$

where

$$q_{s,e} = q_s \alpha_{e,s}\left(1 + \frac{\tau_{e,s}\rho_{a,s}}{1 - \rho_{a,s}\rho_{e,s}}\right) \tag{9.13}$$

is the solar radiation absorption rate for the envelope per unit absorber area (including reflected radiation from the absorber). Also

$$q_{r,es} = \epsilon_{e,\text{ir}}\,\sigma(T_e^4 - T_s^4) \tag{9.14}$$

is the energy radiated per unit envelope area to the surroundings at temperature T_s^*; $q_{c,es}$ is the energy convection rate from the envelope to the ambient air and is given by (per unit envelope area)

*The T_s that is used in Eq. (9.14) is called the temperature of the "surroundings" and is often taken to be the "effective" sky temperature. In this analysis it will be assumed to be the same as the ambient dry-bulb temperature. This will place it somewhere between the temperature of the reflector (which is slightly above ambient) and the effective sky temperature (which is below ambient).

$$q_{c,es} = h_{c,es}(T_e - T_s) \tag{9.15}$$

The convective film coefficient $h_{c,es}$ is a strong function of the external flow conditions in the vicinity of the envelope. These are usually taken to be identical to the ambient weather conditions but are often much more complicated due to secondary flows, etc., that may exist. Equations (9.7) and (9.12) can be rewritten as

$$q_{s,a} = \frac{D_{a,i} h_{c,af}(T_a - T_f)}{D_{a,o}} + K_{ae}(T_a - T_e) \tag{9.16}$$

and

$$q_{s,e} = \frac{D_{e,o} K_{es}(T_e - T_s)}{D_{a,o}} - K_{ae}(T_a - T_e) \tag{9.17}$$

where, using Eqs. (9.9), (9.10), (9.14), and (9.15),

$$K_{ae} = h_{c,ae} + \epsilon_{ae}\, \sigma(T_a^2 + T_e^2)(T_a - T_e) \tag{9.18}$$

and

$$K_{es} = h_{c,es} + \epsilon_{e,ir}\, \sigma(T_e^2 + T_s^2)(T_e - T_s) \tag{9.19}$$

The solution to the set of Eqs. (9.5), (9.16), and (9.17) can be accomplished numerically in the following way. Starting at the collector inlet, where $x = 0$ and $T_f = T_{in}$ (known), Eqs. (9.16) and (9.17) can be solved by iteration to yield T_a and T_e (the coefficients of the temperature difference terms are recalculated after each iteration using the latest T_a and and T_e). Once the correct temperatures have been converged upon, Eq. (9.5), written in finite difference form, can be used to calculate the enthalpy rise produced in a short length, Δx, of the absorber. Axial variations in q_s can also be handled in this manner.[47] Using thermodynamic equation-of-state data for the heated fluid, other properties (in particular, the temperature) can be calculated as the integration continues in x.[47,66].

The radiation losses shown in Fig. 9.8 are calculated from Eq. (9.14) and the convection losses from Eq. (9.15).* Since the film coefficients $h_{c,af}$ can often be rather large, the difference between the absorber temperature and the bulk fluid temperature is sometimes neglected in the study of convective and radiative control strategies. Equation (9.12), with T_s and T_a assumed known, then can be solved iteratively for T_e for use in Eqs. (9.14) and (9.15). Equation (9.7) then yields the useful thermal gain q_a by the fluid [q_a is the first term on the right-hand side of Eq. (9.7)].

Table 9.6 shows some example results for this cylindrical geometry under discussion. These values will certainly change with changes in the parameters, but several interesting features should be noted. First, it is obvious that thermal losses are greatly reduced by convection control devices. Also, the 6 to 11 percent (for this example) improvement by evacuating the gap has to be weighed in terms of the practical problems of maintaining the vacuum. Finally the large temperature difference between absorber and envelope (for this example) poses design problems due to differences in axial expansion of the two tubes.

Another approach to solving the energy equations is to eliminate T_e and T_a from Eqs. (9.5), (9.16), and (9.17) to yield

$$\frac{dq_u}{dx} = \dot{m}\, \frac{dh}{dx} = \pi D_{a,i} F'\, [CR q_s(\tau\alpha)_e - U_c(T_f - T_s)] \tag{9.20}$$

where, for this special problem, the collector geometrical efficiency factor is given by

$$F' = h_{c,af} \Big/ \left(h_{c,af}\frac{D_{a,i}}{D_{a,o}} + K_{ae} - \frac{K_{a,e}^2}{K_{ae} + K_{es}D_{e,o}/D_{a,o}} \right)^{-1} \tag{9.21}$$

---

*These give the losses from the envelope. Equations (9.9) and (9.10) can be used to calculate losses from the receiver.

TABLE 9.6 Example Envelope Temperatures and Heat Loss Results

T_a (°C)	Gap condition	$\epsilon_{a,ir} = .12$					$\epsilon_{a,ir} = .91$				
		T_e	$q_{r,es}$	$q_{c,es}$	q_{TOT}	q_u	T_e	$q_{r,es}$	$q_{c,es}$	q_{TOT}	q_u
300	evacuated gap	49	20	290	310	2430	66	42	543	585	2160
	gap at 1 atm air	58	32	424	456	2290	74	54	679	733	2010
	w/o envelope	—	54	1833	1887	828	—	406	1833	2239	476
450	evacuated gap	58	32	424	456	2290	103	109	1233	1342	1400
	gap at 1 atm air	69	47	602	649	2090	111	125	1367	1492	1250
	w/o envelope	—	143	2851	2994	(−278)	—	1083	2851	3934	(−1218)
		°C	W/m of length				°C	W/m of length			

SOURCE: $h_{c,ae}$ was taken from Ref 65; $h_{c,es}$ was taken as 85 W/m² (2.2 m/s wind). The following data were used: $\alpha_{a,s} = 0.96$, $\tau_{e,s} = 0.95$, $\alpha_{e,s} = 0.05$, $\epsilon_{e,ir} = 0.88$, $k = 0$, $D_{a,o} = 25.4$ mm, $D_{e,i} = 50.8$ mm, $D_{e,o} = 62.8$ mm, $q_s = 2830$ W/m of length, incident on envelope. (This corresponds to approximately a CR = 25 with an effective concentrator reflectance of 75%.)

$q_{r,es}$ and $q_{c,es}$ are the power radiated and convected, respectively, from the envelope. $q_{\text{TOT}} = q_{c,es} + q_{r,es}$.

q_u is useful power gain in fluid.

the loss coefficient by

$$U_c = \frac{K_{ae} K_{es} D_{e,o}/D_{a,o}}{K_{ae} + K_{es} D_{e,o}/D_{a,o}} \tag{9.22}$$

and the "effective" absorptance-transmittance product by

$$(\tau\alpha)_e = \eta_{opt} \left(\frac{\tau_{e,s} \alpha_{a,s}}{1 - \rho_{a,s} \rho_{e,s}} + K_{ae}\alpha_{e,s} \left[\frac{1 + \tau_{e,s} \rho_{e,s}/(1 - \rho_{a,s} \rho_{e,s})}{K_{ae} + K_{es} D_{e,o}/D_{a,o}} \right] \right) \tag{9.23}$$

Here η_{opt} expresses the efficiency of the concentrator in delivering the solar energy impinging on the aperture to the absorber envelope, (i.e., it is the product of the effective reflectance and the intercept factor Γ).

Equation (9.23) can be integrated for either fluid heating (increasing T_f) or boiler use (constant T_f). When it is possible to express the enthalpy in terms of the specific heat via $h = C_p T_f$, this equation becomes

$$\dot{m} C_p \frac{dT_f}{dx} = \pi D_{a,i} F' [q_s \mathrm{CR}(\tau\alpha)_e - U_c(T_f - T_s)] \tag{9.24}$$

which yields to integration in closed form when F', $(\tau\alpha)_e$, U_c, and C_p are constant. The final result for these assumptions is then the total collection Q_u,

$$Q_u = A_{ap} F_R [q_s(\tau\alpha)_e - \frac{U_c A_a}{A_{ap}} (T_{in} - T_s)] \tag{9.25}$$

which is identical in form to the Hottel-Whillier-Bliss (HWB) flat-plate equation developed in Chap. 7. Since U_c and, to a lesser extent, $(\tau\alpha)_e$ depend strongly on temperature, these values should be calculated at the arithmetic mean temperature between T_{in} and T_{out}. The heat-removed factor F_R is given by

$$F_R = \frac{\dot{m} C_p}{U_c A_a} \left[1 - \exp - \left(\frac{A_a U_c F'}{\dot{m} C_p} \right) \right] \tag{9.26}$$

Equation (9.25) is convenient to use for many purposes and can be adapted to other geometries. It makes the advantages of concentration readily apparent, since it shows that the coefficient of the thermal loss term $A_a U_c/A_{ap} = U_c/\mathrm{CR}$ reduces proportionately with $1/\mathrm{CR}$. It is also convenient for use in analyzing experimental collector performance data in order to ascertain $F_R U_c$ and $F_R(\tau\alpha)_e$.

The analyses above are useful for calculating the instantaneous performance of concentrators. Long-term performance can be calculated for any concentrator using methods described in the final section of Chap. 8.

SYSTEM PERFORMANCE

For a given concentrator there are many topics which a designer must consider in order to investigate system sensitivity. Discussed here are only four such things: receiver size, receiver position, collector tracking, and shadowing by adjacent collectors.

Receiver Size

A receiver placed in the focal plane will not intercept all of the radiation if it is too small, whereas one that is too large may have unnecessary thermal losses. Thus, one of the keys to optimizing the efficiency of the collector is to optimize the receiver. This is difficult to do analytically without resorting to computer calculations. It is also difficult to do experimentally without accurate test control and measurements and without the expense of building receivers of several sizes and shapes.

Cobble[11] studied the optimum receiver shapes that would intercept all of the reflected radiation when used with reflective parabolic concentrators. He found that parabolic-shaped receivers were best theoretically. The cross section of these consisted of the region formed by the intersection of two identical parabolas, one rotated 180° (3.14 rad) with respect to the other (they look much like the cross section of a football). Lumsdaine and Cherng[67] showed that elliptical-shaped cross-sectional receivers offered some advantages up to relative apertures (essentially f-numbers) of about 5. Figure 9.20 shows their results along with some earlier results of Cobble.[11]

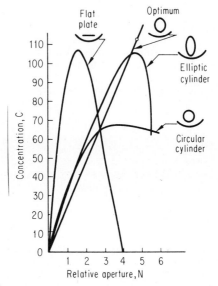

Fig. 9.20 Concentration ratio for receivers of various geometries. (*From Ref. 67.*)

Computer modeling on the original SLSTESTF[8] parabolic trough concentrators has shown that collector efficiency, fortunately, may not be a strong function of receiver size.[12] For example, a ± 25 percent change in receiver diameter produced roughly only a 2 to 3 percent change in efficiency. This study hints that the extra effort in fabricating noncircular receivers may not be justified. In essence, shape-factor losses (i.e., energy lost when radiation misses the absorber) will be partially balanced by improved thermal performance with good design.

Receiver Position

The positioning of the receiver within the focal area can be much more critical than the size of the receiver for concentrators in the CR range above 20. Figure 9.21 demonstrates the phenomemon of defocusing with a perfect parabolic–cross-sectional concentrator with a flat absorber. As with receiver sizing, the effect of absorber position on the thermal efficiency depends upon the magnitude of the thermal losses and upon the size of the absorber relative to image size produced by the concentrator.

Most important, perhaps, is the distance between the absorber and the collector centerline (at least for symmetrical concentrators). Computer studies on the original parabolic trough collectors of the SLSTESTF have shown an efficiency-vs.-off-center mounting distance curve that was bell-shaped with a (half-) halfwidth of about 3 percent of the focal length.[12] That is, if the receiver tube was moved ± 3 percent of the focal length (or 60 percent of the receiver diameter) out of the focal plane, the efficiency was reduced by one half. Thus, receiver sag between supports could be very important. These results, coupled with the results of receiver size studies, suggest that receivers should be oversized to compensate for focal-length variations.

For trough concentrators it is also important that the absorber be parallel to the longitudinal axis. If parallelism exists, any lateral displacement, assuming it is small, could possibly be compensated for by tracking adjustments. However, off-axis performance can degrade rapidly, especially at high concentration.

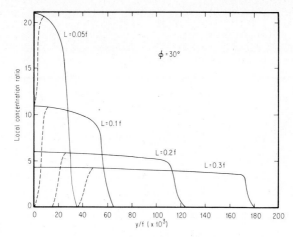

Fig. 9.21 Defocused performance of perfect 30° rim angle parabolic cylindrical concentrators with flat absorbers. L is the distance the absorber is displaced to the right of its position shown in Fig. 9.10. The dotted curves show the effects of self-shading of the absorber. (*From Ref. 8.*)

Tracking

The primary purpose of the tracking system is, of course, to keep the collector aperture pointed in the optimum direction so as to minimize the incidence angle and keep the solar image centered on the absorber as the sun moves across the sky. The accuracy with which the tracker does this task can be measured by the tracking error stated in terms of degrees (or radians) that the actual aperture normal is allowed to deviate from its optimum. It is usually thought of as either the maximum tracking error or the standard deviation of a random distribution of tracking errors.* In addition to this primary purpose, a tracking system should also have some secondary attributes which will be discussed shortly.

Tracking accuracy Theoretical studies done on high quality optics for parabolic troughs[8] and for linear Fresnel lenses[68] (both with flat absorbers) show that peak concentration drops off as tracking error increases and that the intensity distributions become skewed and move away from the optical axis. Figure 9.22 shows this result for the parabolic trough case with flat absorbers.

Tracking-error sensitivity tests performed on the original SLSTESTF parabolic collector (CR $\simeq$ 20) test field[9] showed that tracking errors of about ± 1° (17 mrad) decreased the thermal efficiency by about 20 percent.† In another trough prototype[6] (CR $\simeq$ 5) a ± 1.5° (26-mrad) error could be tolerated before similar decreases were noted. The need for much smaller errors and more accurate equipment was demonstrated in tests on a good quality 1.56-m-diameter paraboloid (CR $\sim$ 10,800) where it was found that thermal output was reduced by 20 percent of its optimum by a tracking error of only 20 minutes (5.8 mrad).[69] This latter experiment also showed that the thermal output had gone to zero by the time the tracking error had reached 1° (17 mrad).

Active tracking systems There are two commonly discussed methods of tracking control. These are the active, or closed-loop, control system and the passive, or open-loop, control systems. Systems of the active, or closed-loop, type "actively" make some measurement of where the collector is and then, if alignment is not proper, activate devices that can change the orientation of the collector. Unless the collectors are grouped together with some mechanical linkage, individual active control systems must be placed on each collector.

The sensing devices that make the system an active one are usually light-sensitive detectors such as photovoltaic (solar) cells or phototransistors. The former devices, if operated within

---

*A useful standard deviation for the angular spread of radiation off a point on the concentrator, which includes the standard deviation of the tracking error σ_t, in addition to the sun size, lack of specularity, and random slope errors, is given by $\sigma = (\sigma_s^2 + \sigma_r^2 + 4\sigma_c^2 + 4\sigma_t^2)^{1/2}$.

†This is a relative percentage change, not an incremental change.

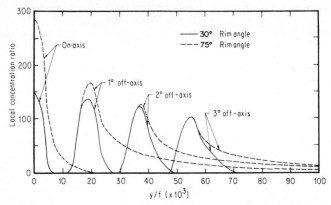

Fig. 9.22 Off-axis performance of perfect parabolic cylindrical concentrators with flat absorbers. (*From Ref. 8.*)

certain constraints, produce electrical currents and voltages that are related to the intensity of the sunlight falling on them. The latter devices change their impedance as the amount of sunlight falling on them changes.

Ideally these sensors should be placed on the receivers to insure that the receivers are always at the focal point. Due to receiver shape and radiation intensity problems, this is usually not possible in practice. The common alternative has been to use light-sensitive detectors in pairs mounted on the concentrator structure, so that a shadow cast by some mask or shield covers opposite halves of each detector. Motion of the sun causes the shadow to move off of one detector onto the other, creating an unbalanced condition that can be used to activate the tracking drive train through a differential amplifier or bridge circuit, for example. Adjustment of the detector package must be provided so that the receiver is aligned at the focal plane when the detector outputs are balanced. Such detector-electronic packages are commercially available.

Problems that are possible with this type of sensing include the unintentional tracking of bright clouds instead of the sun, mechanical misalignment of detectors, aging and drift of the detectors and electronics, and the possibility of constant hunting and rapid acceleration of the tracking system. The misalignment and aging problems may require adjustments that may become intolerable nuisances, especially in large collector fields. The problem of hunting and rapid acceleration depends on the response of the system to out-of-balance detection and on-system inertia. If serious, it may degrade the lifetime of the tracking drive train. Antibacklash and counterbalancing provisions are highly desirable.

Passive control systems   Systems of the passive, or open-loop, control type respond to preprogrammed control steps in order to track the sun. For example, a minicomputer working in conjunction with an accurate clock and programmed to calculate the necessary collector position as a function of time of day and year feeds information to the drive train. There is usually no feedback that would tend to make the system self-checking. One major advantage is the ability to command many collectors from just one computer, thus making the cost per collector potentially much lower than that required for individual sensors.

In such systems antibacklash and counterbalancing provisions are necessary, since there is no feedback signal. Also, wind or mechanically caused defocus may be more of a problem than with active systems, since the computer is never aware of the exact collector position at any instant and, therefore, cannot compensate for any inaccuracies.

Combination systems   A combined system using both active and passive controls may incorporate the good features of both systems. Active systems might provide the main tracking to keep the concentrators aligned, but computers, using information from position encoders, would constantly check on gross misalignment that might be due to a tendency to track clouds or to electronic malfunctions.

Other features   In addition to the primary task of keeping the collectors properly aligned, the tracking system should provide certain services and incorporate certain safety features. The services include the shutting down of the solar system on overcast days and the resetting of the collectors at night for use the next day. This latter task might include collector stowage at night in the lowest risk position (e.g., inverted with the reflector down). For safety reasons the tracking system should be capable of "detracking" or defocusing in the event that certain system problems arise. For example, the loss of power to any major component or loss of coolant or overheating of the energy collection fluid in the receiver must not be allowed to cause collector damage.

The work reported in Ref. 10 lead to the conclusion that there should be little difference in cost for tracking accuracies of from 2° (35 rad) to 0.14° (2.4 mrad). Costs for accuracies better than 0.14° (2.4 mrad) would escalate rapidly due to the increased costs of precision parts and assembly for sensors, motors, and gears. For the concentrating systems which were costed in that study, tracking accounted for typically less than 6 percent of the total costs.

Mention should be made here of the need for stiffness in the structural and concentrator system. Cited as an example is the need for either a good torque tube or a torsionally rigid system for long, linear concentrators that are to be tracked around some axis parallel to the longitudinal axis, especially if they are not counterbalanced about that axis. If the torsional rigidity is not great enough, one part of the system could be in focus (e.g., the region near the tracking sensor, if actively tracked) while other parts may be out of focus due to twisting of the concentrator. Wind loading can cause similar problems. Of course, adjustments should be provided to allow the removal of the set or torsional skewness that is often built into a linear concentrator in manufacturing. Such adjustors, as well as counterbalance weights, can be seen on the concentrators in Fig. 9.1.

Array Losses Due to Shading

If a solar energy collector is isolated from shadow-producing elements, then the particular shape and size of its aperture have no effect on the energy collected per unit area. However, when the collector is located in an array of collectors, the energy-collecting capacity will be reduced due to shadows cast on it by neighboring collectors. When the cost of land is taken into consideration, it can be expected that there will be optimum conditions for the shape and size of the collector and an associated optimum placement of the collectors in the array which will minimize the cost of energy produced by the system. The optimum configuration will experience some shading.

Thermal-energy cost calculations If it is assumed that all of the thermal energy produced is useful, a capital cost of that energy can be simply computed from

$$C_e = \frac{K(C_c + C_l/P_f)}{E_u}$$

where C_c is the capital cost* of the solar system per unit aperture, C_l is the cost of land per unit of land,† P_f is the ratio of aperture area to land area (sometimes referred to as the packing factor), and E_u is the useful energy produced per unit of aperture area (usually on a per year basis). K is known as the fixed charge rate but can be regarded as a constant for this discussion. C_e is often termed the "levelized" capital cost of the energy. If it is assumed that a constant efficiency η describes the ratio of useful energy (E_u) to the incident energy (E), with shading losses (E_i omits shading losses) and defining C_s, the spacing-related, relative cost, as

$$C_s = [1 + C_l/P_f/C_c]/E \tag{9.27}$$

C_e is then given by

$$C_e = \frac{KC_cC_s}{\eta} \tag{9.28}$$

Economic competitiveness requires the lowest C_e possible. It is perhaps obvious to suggest that C_e can be lowered by lowering the collector capital costs or raising the efficiency. Beyond these two parameters, lowering C_e requires finding the minimum C_s. Assuming a fixed ratio of land costs to collector costs (i.e., C_l/C_c), C_s varies in the following way. At one extreme as the collector field is packed progressively more tightly (P_f increasing), the denominator in Eq.

---

*This is assumed to include installation costs.

†This would include the cost of improvements such as access roads, drainage improvements, fences, etc.

(9.27) soon decreases faster than the numerator, and C_s begins to increase. At the other extreme, as the collectors become more and more widely separated (P_f decreasing), the numerator in Eq. (9.15) continues to decrease as the denominator levels out causing C_s to again increase.

Minimum-cost collector fields   Detailed calculations for a particular collector aperture geometry and tracking mode can be carried out over a period of time to determine when shadows are present and to establish losses due to shaded apertures. If this is done for different packing factors, the minimum C_s can be found.

Such calculations have been carried out for several collector geometries in Ref. 70, where concentrating photovoltaic collectors were investigated. Some of the results are unique to photovoltaics; however, those results which should mimic thermal systems are presented here. Figure 9.23 shows the grid for identical collectors that were assumed to surround the example collector of a height h and width w (or circular collector of diameter d). Table 9.7 gives a summary of the results for six different systems for a land cost that is 5 percent of the collector cost. Such a value might be low for some on-site solar systems where land might be quite expensive. On the other hand, it might be high for desert land for central-station types of plants.

Several things are apparent from Table 9.7 and are discussed here briefly. First, for all tracking geometries, at the minimum C_s, one could expect to lose 7 to 12 percent of the incident energy due to shading. Second, single-axis tracking with the axis oriented east-west yields the highest packing factor of any of the tracking modes considered. It can also be seen that optimum packing factors for two-axis tracked rectangular-shaped apertures do not depend on the orientation of the axis but do depend on the optimum shape of the rectangle and the north-south spacing. Table 9.6 also shows that circular apertures can be packed tighter than rectangular apertures for two-axis tracking. Staggering or offsetting every other east-west row improves the packing factor slightly for two-axis tracked devices. The opposite* is true for one-axis polar tracking.

The work presented in Ref. 70 also shows that placing the array of collectors on south-sloping ground greatly improves C_s and increases P_f. Latitude, as might be expected, is also a very sensitive parameter. Table 9.8 shows how much the results change with a change in latitude from 33.4°N to 45°N for two-axis tracking (one-axis vertical) of rectangular apertures. Although E/E_i does not change significantly between the two latitudes, E_i is less at the higher latitudes.

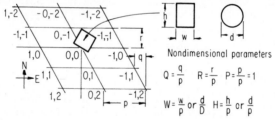

Fig. 9.23   The grid layout for shading studies. A collector identical to that located at 0,0 is assumed to exist at all grid points. East-west spacing of rows is specified by p; north-south spacing of rows is specified by r; staggering of adjacent east-west rows is specified by q.

SUMMARY

Many topics have been covered in this chapter, but many more have been left untouched. In particular, costs have not been explored, since they are not broadly known in the commercial sector, although many references have been made to the need for cost-benefit studies. Studies reported in Ref. 10 have projected automated, high-volume, installed collector costs in the range of $50 to about $70 per square meter of aperture area (in 1972 U.S. dollars), depending on type and size. No hardware was actually built, however.

It is encouraging to note that many different concentrating collector systems are now commercially available, and although they have not begun to approach the projections given above, the prices of several units are less than those for present high-performance flat-plate collectors (i.e., less than $150/m²). This will certainly extend their range of useful tasks to include low-temperature applications.

*It does not always lower the spacing-related, relative cost (C_s).

TABLE 9.7 Comparison of Collector Performance in an Array for Different Tracking Modes and Array Configurations*; Latitude = 33.4°N, Solar Altitude for Collection $\geq$ 5°, Cloudless Days, Diffuse Radiation Neglected, C_l/C_c = 0.05

System	Specified conditions Q	Calculated, R, W, H, E/E_i, for $(C_s)_{min}$					Notes†
		R	W	H	P_f	E/E_i %	
1. Horizontal EW axis tracked NS	0.0	1.0	1.0	0.53	0.530	98.3	1
2. Horizontal NS axis tracked EW	0.0	1.0	0.36	1.0	0.360	88.3	2
3A. Polar axis tracking	0.0	0.50	0.48	0.30	0.288	92.7	3
3B. Polar axis tracking	0.5	0.65	0.62	0.28	0.267	92.6	4
4A. Two-axis tracking with one polar	0.0	0.50	0.50	0.27	0.270	92.7	5
4B. Two-axis tracking with one polar	0.5	0.19	0.26	0.21	0.287	91.6	4
5A. Two-axis tracking with one vertical	0.0	0.90	0.72	0.33	0.264	93.5	
5B. Two-axis tracking with one vertical	0.5	0.31	0.43	0.21	0.291	93.5	
6A. Two-axis tracking with circular apertures	0.0	0.74	0.56	0.56	0.333	93.2	6
6B. Two-axis tracking with circular apertures	0.5	0.75	0.57	0.57	0.340	92.9	6

*Yearly hours for energy collection = 4069. Calculations based on projection from 26 days per year.
†*Notes:*
1. R was arbitrarily chosen as 1.0; H/R is the significant criterion.
2. $H/R = 1.0$ is the important criterion for R and H values.
3. This is a relative optimum for the given value of R which was arbitrarily chosen to be 0.5. There is a weak dependence of C_s on R such that C_s decreases with R; however, W/H becomes very large as R decreases. The ratio $H/R = 0.6$ is the most significant criteria and is virtually independent of C_l/C_c.
4. H/R is independent of C_l/C_c.
5. This case is similar to 3A in that the optimum is essentially dependent on H/R which in this case is equal to a constant of 0.543. The variation in C_s as R varies is less (R was selected as 0.5).
6. Optimum values of R, W, H, and performance are independent of the selection of the primary axis of the two axis system (i.e., polar or vertical). The effect of multiple shadows was not considered. Results are based on largest shadow.
SOURCE: Ref. 7.

TABLE 9.8 Comparison of Effects of Different Latitude on Optimal Parameters and Performance of a Two-Axis Tracking System with One-Axis Vertical for C_l/C_c = 0.05

Quantity	Latitude 33.4°N	Latitude 45°N
R	0.31	0.88
W	0.43	0.72
H	0.21	0.31
Pf	0.291	0.254
E, kWh/m²/year	2992	2780
E/E_i, %	93.5	93.6

SOURCE: Ref. 70.

REFERENCES

1. J. A. Duffie and W. A. Beckman, *Solar Energy Thermal Processes*, Wiley-Interscience, New York, 1974.
2. W. S. Duff, G. F. Lameiro, and G. O. G. Löf, Parametric Performance and Cost Models for Solar Concentrators, *Solar Energy* 17, 47 (1975).
3. G. O. G. Löf and J. A. Duffie, Optimization of Focusing Solar-Collector Design, *J. Eng. Power Trans. ASME* 85A, 221 (1963).
4. G. O. G. Löf, D. A. Fester, and J. A. Duffie, Energy Balances on a Parabolic Cylinder Solar Collector, *J. Eng. Power Trans. ASME* 84A, 24 (1962).
5. H. Tabor, Stationary Mirror Systems for Solar Collectors, *Solar Energy* 2, 27 (1958).
6. Research Applied to Solar-Thermal Power Systems, Series of Reports Prepared by University of Minnesota and Honeywell, Inc., under U.S. National Science Foundation Grant GI-34871 (1972–1974).
7. Aerospace Corporation Comparative Systems Analysis, Task 1, Report No. ATR-73(7283-01)-1 (November 1972).
8. D. L. Evans, On the Performance of Cylindrical Parabolic Solar Concentrators with Flat Absorbers, *Solar Energy* 19, 379 (1977).
9. Sandia Laboratories, Solar Total Energy Program Reports (1974–1976).
10. Solar Thermal Electric Power Systems, Prepared by Colorado State University and Westinghouse Electric Corporation, Final Report No. NSF/RANN/SE/GI-37815/FR/74/3, U.S. National Science Foundation, Washington, D.C. (1974).
11. M. H. Cobble, Theoretical Concentration for Solar Furnaces, *Solar Energy* 5 (2), 61 (1961).
12. G. W. Treadwell, Design Considerations for Parabolic Cylindrical Solar Collectors, *Sharing the Sun, Joint Solar Conference* 2, 235, Winnipeg (1976), or Sandia Laboratories, Report No. SAND76-0082, Albuquerque (1976).
13. R. W. Stineman, Optimum Reflector-Absorber Geometry for a Solar Generator, *Applications and Industry* 45, 332–337 (1959).
14. J. A. Sakr and N. H. Helwa, Experimental Measurements of Concentrated Solar Energy Pattern in Focus of a Plan Segments Concentrator, *COMPLES Bull.* 14 (1968).
15. J. L. Hastings, S. L. Allums, and R. M. Crosby, An Analytical and Experimental Evaluation of the Plano-Cylindrical Fresnel Lens Solar Concentrator, *Sharing the Sun, Joint Solar Conference* 2, 275, Winnipeg (1976).
16. L. J. Hastings, S. L. Allums, and W. S. Jensen, An Analytical and Experimental Investigation of a 1.8 by 3.7 Meter Fresnel Lens Solar Concentrator, *Proc. 1977 U.S. Section International Solar Energy Society* 3, 35-5, Orlando (1977).
17. Terrestrial Photovoltaic Power Systems with Sunlight Concentration, Prepared by Arizona State University and Spectrolab Division of Textron, Inc., Report No. NSF/RANN/SE/GI-41894/PR/74/4, U.S. National Science Foundation, Washington, D.C. (1978).
18. Photovoltaic Systems Concept Study, Prepared by Spectrolab Division of Textron, Inc., Report No. ALO-2748-12, U.S. Energy Research and Development Administration, Albuquerque (1977).
19. Conceptual Design and Systems Analysis of Photovoltaic Systems, Prepared by General Electric Company, Report No. ALO-3686-14, U.S. Energy Research and Development Administration, Albuquerque (1977).
20. Conceptual Design and Systems Analysis of Photovoltaic Power Systems, Prepared by Westinghouse Electric Corporation, Report No. ALO-2744-13, U.S. Energy Research and Development Administration, Albuquerque (1977).
21. J. Russell and R. Potthoff, Demonstration Model of Solar Power Concentrator, Final Report, General Atomics Corp., Report No. GA-A-13352 (1975).
22. S. Y. Harmon, C. E. Backus, R. Pinon, Characteristics of the Concentrated Solar Flux Produced by the FMSC, Prototype, *Sharing the Sun, Joint Solar Conference* 2, 291, Winnipeg (1976).
23. R. K. Bansal, *Theoretical Analysis of Fixed Mirror Solar Concentrators*, Master's Thesis, Arizona State University (1974). (Available from University Microfilms, Ann Arbor.)
24. J. F. Kreider and Frank Kreith, *Solar Heating and Cooling, Engineering, Practical Design, and Economics*, McGraw-Hill, New York, 1975.
25. C. F. Holder, Solar Motors, *Scientific American*, March 16, 1901 (p. 169).
26. V. B. Veinberg, The History of Soviet Solar Engineering, *Applied Solar Energy*, 3 (5) (1967).
27. J. I. Yellott, Power from Solar Energy, *Trans. ASME* 79 (1957).
28. D. L. Teplyakov, Effect of Longitudinal Target Defocusing on the Power Characteristics of Solar Reflector Systems, *Geliotekhnika* 3 (1), 10 (1967).
29. V. A. Baum, Prospects for the Application of Solar Energy and Some Specific Research Results in the USSR, *Proc. World Symp. Applied Solar Energy*, Phoenix (1955).
30. R. W. Bliss, Notes on the Performance Design of Parabolic Solar Furnaces, *J. Solar Energy Sci. Eng.* 1, 22 (1957).
31. R. C. Jordan, Mechanical Energy from Solar Energy, *Proc. World Symp. Applied Solar Energy*, Phoenix (1955).
32. B. Y. Liu and R. C. Jordan, Performance and Evaluation of Concentrating Solar Collectors for Power Generation, ASME Paper No. 63-WA-114 (1963), also *J. Eng. Power Trans. ASME* 87, 1 (1965).
33. Y. A. Dudko, Calculation of Concentrator-Receiver Systems with a Flat Receiver, *Geliotekhnika* 4 (2), 32 (1968).

34. R. E. DeLa Rue, Jr., E. Loh, J. L. Bremer, and N. K. Hiester, Flux Distribution near the Focal Plane, *Solar Energy* 1 (2, 3), 94 (1957).
35. P. D. Jose, The Flux Distribution through the Focal Spot of a Solar Furnace, *Solar Energy* 1, 19 (1957).
36. W. W. Shaner and H. S. Wilson, Cost of Paraboloidal Collectors for Solar to Thermal Electric Conversion, *Solar Energy* 17, 351 (1975).
37. V. V. Novikor and L. N. Skripkar, Adjustment Concentrator with Parabolic Facets, *Geliotekhnika* 5 (1), 20 (1969).
38. G. A. Umarov and A. S. Sharafi, Calculation of Faceted Solar Concentrators Based on a Paraboloid of Revolution, *Geliotekhnika* 3 (6), 5, 7 (1967).
39. D. N. Alavutdinov, A. K. Alimov, and G. Y. Umarov, Investigation of Solar Concentrators Composed of Doubly-Curved Facets, *Geliotekhnika*, 3 (3), 20 (1967).
40. G. Y. Umarov, A. K. Alimov, D. N. Alavutdinov, and N. F. Suleimanova, Study of Circular and Hexagonal Vacuum Film Facets for Solar Concentrators, *Geliotekhnika* 3 (1), 17 (1967).
41. G. Y. Umarov, Problems of Solar Energy Concentration, *Geliotekhnika* 3 (5), 32 (1967).
42. L. R. Paradis, A. L. Levine, and E. C. Vallee, Parabolic Collector for Total Energy System Application, *Proc. 1977 Meeting U.S. Section International Solar Energy Society* 3, 35-19, Orlando (1977).
43. I. Oshida, Step Lenses and Step Prisms for Utilization of Solar Energy, U.S. Conference on New Sources of Energy, No. 15/5/22, Rome (1961).
44. S. Harmon, Solar-Optical Analysis of a Mass-Produced Plastic Circular Fresnel Lens, *Solar Energy* 19, 105 (1977).
45. J. F. Kreider and W. G. Steward, The Stationary Reflector/Tracking Absorber Solar Concentrator, International Solar Energy Society Annual Mtg., Fort Collins, CO (1974).
46. F. Kreith and W. G. Steward, Optical Design Characteristics of a Stationary Concentrating Reflector/Tracking Absorber Solar Energy Collector, *Applied Optics* 14, 1509 (1975).
47. J. F. Kreider, Thermal Performance Analysis of the Stationary Reflector/Tracking Absorber (SRTA), Solar Concentrator, *J. Heat Transfer* 97, 451 (1975).
48. A. M. Clausing, The Performance of a Stationary Reflector/Tracking Absorber Solar Concentrator, *Sharing the Sun, Joint Solar Conference* 2, 304 Winnipeg (1976).
49. M. J. O'Neill, Optical Analysis of the Fixed Mirror/Distributed Focus (FMDF) Solar Energy Collector, *Proc. 1977 Meeting U.S. Section International Solar Energy Society* 3, 34-24, Orlando (1977).
50. F. K. Nabiulin, M. S. Sladkov, Z. M. Buzova, G. S. Zaslavskaya, G. B. Likhtsier, and B. Y. Rodichev, Manufacture of Parabolic Concentrators by the Electroformed Replica Method, 2 (3), 28 (1966).
51. R. E. Henderson and D. L. Dresser, Solar Concentration Associated with the Sterling Engine, in *Space Power System*, Vol. 4, *Progress in Astronautics and Rocketry*, Academic, New York, 1961.
52. L. G. Rainhart and W. P. Schimmel, Jr., Effect of Outdoor Aging on Acrylic Sheet, *Solar Energy* 17, 259 (1975).
53. G. Abetti, *The Sun*, Van Nostrand, Princeton, N.J., 1938.
54. R. C. Zentner, Performance of Low Cost Solar Reflectors for Transferring Sunlight to a Distant Collector, *Solar Energy* 19, 15 (1977).
55. R. B. Pettit, Characterization of the Reflected Beam Profile of Solar Mirror Materials, *Solar Energy* 19, 733 (1977).
56. E. M. Sparrow and R. D. Cess, *Radiation Heat Transfer*, Brooks/Cole Publishing Company, Belmont, CA, 1966.
57. R. B. Pettit and B. L. Butler, Semiannual Review ERDA Thermal Power Systems, Dispersed Power Systems, Distributed Collectors, and Research and Development, Mirror Materials and Selective Coatings, Sandia Laboratories, Report No. SAND77-0111, Albuquerque (1977).
58. V. Baum, Energie Solaire et Eolienne, *Aetes du Colloque de New Delhi, UNESCO, Discussion*, 172 (1952).
59. R. R. Aparisi, Experimental Device for Obtaining High Temperatures, in *Utilization of Solar Energy* (in Russian), ANSSSR; 1 Moscow (1957).
60. R. R. Aparisi, Candidate's Dissertation, *Concentration of Solar Energy in Solar Engineering Structures* (in Russian), Moscow (1955).
61. D. H. Silvern, An Analysis of Mirror Accuracy Requirements for Solar Power Plants, in *Space Power Systems*, Vol. 4, *Progress in Astronautics and Rocketry*, Academic, New York, 1961.
62. H. Oman and G. Street, Paper presented at IEE Meeting, San Diego (1960).
63. Y. A. Dudko and O. A. Dudko, Designing Concentrating Systems, *Geliotekhnika* 4 (6), 11 (1968).
64. T. Hisada, H. Mii, C. Noguchi, T. Noguchi, N. Nukuo, and M. Mizuno, Concentration of the Solar Radiation in a Solar Furnace, *J. Solar Energy Sci. Eng.* 1 (4), 14 (1957).
65. E. R. G. Eckert, *Analysis of Heat and Mass Transfer*, McGraw-Hill, New York, 1972.
66. M. W. Edenburn, Performance Analysis of a Cylindrical Parabolic Focusing Collector and Comparison with Experimental Results, *Solar Energy* 18, 437 (1976).
67. E. Lumsdaine and J. C. Cherng, On Heat Exchangers Used with Solar Concentrators, *Solar Energy* 18, 157 (1976).
68. R. M. Cosby, The Linear Fresnel Lens Solar Optical Analysis of Tracking Error Effects, *Proc. 1977 Meeting U.S. Section International Solar Energy Society* 3, 35-14, Orlando (1977).
69. I. M. Rubanovich, Effect of Tracking Accuracy on the Efficiency of Solar Devices with Paraboloidal Concentrators, *Geliotekhnika* 2 (4), 44 (1966).
70. Terrestrial Photovoltaic Power Systems with Sunlight Concentration, Prepared by Arizona State University on Contract No. 02-7850 to Sandia Laboratories, ASU Report No. ERC-C-76014 (1977).

Chapter 10

Nonconvecting Solar Ponds

HARRY TABOR
The Scientific Research Foundation, Jerusalem, Israel

and

ZVI WEINBERGER
Solmat Systems Ltd., Jerusalem, Israel

INTRODUCTION

The classical flat-plate solar collector, constructed of metal and other fabricated materials, is limited in size to a few square meters; large collecting areas are only possible by connecting an assembly of these units. Thus, to collect solar energy on a really large scale requires a radically different approach.

A mass of water, i.e., a pond, lake, or ocean, is potentially a horizontal collector of large area. Thus several attempts have been made in the past to use black-bottomed ponds as solar collectors. If the pond is not covered, evaporation—as well as convection and radiation to the sky—results in surface temperatures close to ambient. Attempts to reduce the evaporative loss by the use of an oil layer or monomolecular layer have not been successful due to the effect of wind removing the layer and, at best, have produced a temperature rise of only a few degrees. Covering such a pond with a transparent window of glass or plastic is a partial reversion to the classical flat-plate collector approach with its problems of limited size, removal of dirt, etc.

With this background, the concept of a nonconvecting pond, where elevated temperatures at the bottom are feasible because the water acts as its own insulator, has been particularly attractive. It is to be noted that this concept attacks three of the major technical problems (apart from cost) pertaining to a classical solar collector technology, namely, low energy density, dirt, and energy storage. The pond allows collection over a large area with negligible transport losses; there are no windows or mirror surfaces to be kept clean, and there is built-in storage adequate to smooth out diurnal and weekly fluctuations of output; even seasonal storage is not ruled out.

HISTORICAL BACKGROUND

There are numerous examples of natural and artificial lakes which possess density gradients due to salt concentration gradients which increase with increasing depth. In the limnological literature these lakes are called "meromictic lakes" and the salt concentration gradient, the "halocline."

If the halocline is sufficiently steep and if the surface of the pond is protected by surrounding geographical features from wind-effected mixing, then incident solar radiation can cause a considerable temperature rise above ambient in the body of the lake. The resulting temperature gradient, the "thermocline," parallels the halocline. The halocline assures the greater density of the lower depths even when heated by solar radiation.

The first natural solar lake described in the literature is probably also the most impressive. A.v. Kalecsinsky[1] described in 1901 the Medve Lake, which is situated near Szovata, Transylvania (42° 44′N, 28°45′E), in which temperatures of 70°C were recorded at a depth of 1.32 m at the end of the summer. The minimal temperature was 26°C during the early spring. In the winter the surface of the lake was washed by fresh water from surrounding springs. The bottom of the lake had a near saturation salinity of 26% NaCl.

More recently G. C. Anderson[2] has described a meromictic lake near Oroville in Washington. The lake, called appropriately enough "Hot Lake," lies in a wind-protected depression atop Kruger mountain (48°58′N, 119°29′W). The bed of this lake was mined for epsomite ($MgSO_4 \cdot 7H_2O$) during and immediately after the First World War. After mining operations ceased and the lake was reflooded, temperatures greater than 50°C during midsummer were recorded at a depth of 2 m. During the winter the surface of the lake is covered with ice.

Natural solar lakes are also to be found near Eilat, Israel,[3] in the Venezuelan Antilles,[4] and, under a permanent ice cover, in the Antarctic[5] (Lake Vanda 77°35′S, 161°39′E).

In 1954, R. Bloch, Research Director of the Dead Sea Works, suggested the study of solar lakes with a view toward practical utilization. Under controlled conditions, it was to be expected that higher temperatures and useful collection efficiencies could be achieved in artificial ponds. Subsequently, in experiments with small (1200 m²) ponds, temperatures greater than 103°C were measured and collection efficiencies greater than 15 percent for heat extraction at 70 to 90°C achieved.

The "solar pond," in spite of its simple description, is a complex physical system which interacts strongly with the local meteorology and geology. In particular, the solar pond must be protected from winds which will mix the pond from the top. The solar pond must be sealed at its bottom, for otherwise heat will be lost by seepage together with the saline solution of the pond bottom. Even worse, the seepage of salt solution might pollute a valuable underground aquifer.

Also, for thermal reasons, the pond must not lie too close to a flowing underground aquifer which will thereby convect heat from the pond. If the underground aquifer is rich in dissolved CO_2, then upon heating, copious quantities of gas may be released from the aquifer. The pressure buildup caused by the release of dissolved gas may be sufficiently great to modify the ground structure and destroy the pond bottom. (This seems to have occurred in an experimental pond constructed in Atlit, Israel.)

The saline and thermal gradients create environments which can nurture a flourishing halothermophilic floral community.[6] This flora can absorb solar radiation and possibly clog hydraulic machinery required for pond maintenance and energy extraction from the pond. We will turn our attention to these and other problems in the following subsections. It is difficult to claim that these problems have been satisfactorily solved. The purpose of this chapter is to define the potential of solar ponds and to describe the problems, which are generally geographically dependent and must be solved for local conditions.

Figure 10.1 illustrates, schematically, a solar pond and some of the physical parameters which we will describe in detail in the following subsections.

ESTABLISHING AND MAINTAINING THE NONCONVECTIVE POND

The Salt Concentration Gradients Required for Stability

The stability of a solar pond warmed by solar radiation is maintained by means of a sufficiently steep salt concentration gradient. It follows from a Rayleigh analysis of the small perturbations to which all natural systems are susceptible that the salt concentration gradient required for maintaining stability is given by

$$\frac{\partial s}{\partial z} \geq - \frac{(\nu + k_T)\,(\partial \rho / \partial T)\,(\partial T / \partial z)}{(\nu + k_s)\,\partial \rho / \partial s} \tag{10.1}$$

where ρ = density of saline solution, kg/m³
 z = depth in pond measured positive downward from the pond surface, m

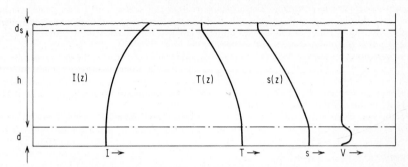

Fig. 10.1 Schematic diagram of the solar pond in which z = vertical coordinate, measured positive downwards; h = depth of nonconvective region of pond; d_s = depth of wind-effected mixed surface layer; d = depth of bottom mixed layer; $I(z)$ = insolation transmitted to a depth z; $T(z)$ = temperature profile; $s(z)$ = salinity profile; and V = flow velocity of bottom layer.

$$s = \text{salt concentration, kg/m}^3$$
$$T = \text{temperature, °C or K}$$
$$\nu = \text{viscosity, m}^2\text{/s}$$
$$k_T, k_s = \text{coefficients of temperature and salt diffusivities, respectively, m}^2\text{/s}$$

This criterion is more stringent than that expected from neutral buoyancy arguments. For dilute solutions at ambient temperatures,

$$\nu \simeq 7k_T \simeq 10^3 k_s$$

Equation (10.1) therefore implies a concentration gradient 14 percent greater than that required for neutral buoyancy.

Figure 10.2a and b shows the density dependence of $MgCl_2$ and NaCl solutions upon tem-

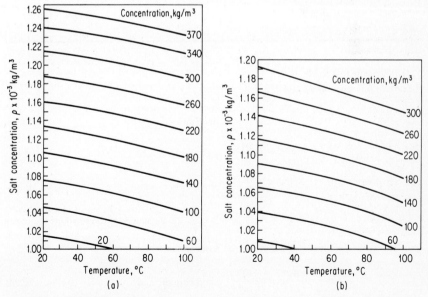

Fig. 10.2 (a) The density of $MgCl_2$-water solutions as a function of salt concentration and temperature. (b) The density of NaCl-water solutions as a function of salt concentration and temperature. [*Adapted from International Critical Tables, vol. III, p. 71 (1928).*]

perature and salt concentrations. For economic reasons it is probable that solar ponds will be constructed from either of these two salts. $MgCl_2$ is a major component of some terminal lakes and is the major residue salt in the end brines of common salt manufacturing processes which are based on the evaporation of sea water.

In evaluating the stability criterion for a solar pond we have to consider the local dependence of the parameters of Eq. (10.1). Of particular interest are the values of these parameters at the extremes of the pond. The temperature gradients are greatest at the surface, where the infrared portion of the spectrum is locally absorbed, and at the bottom of the pond, where up to 30 percent of the radiation may be absorbed.

From Fig. 10.2a and b, we find the representative values of $\partial\rho/\partial T$ and $\partial\rho/\partial s$ given in Table 10.1.

TABLE 10.1   Representative Values of $\partial\rho/\partial T$ and $\partial\rho/\partial s$ at the Pond Extremities

	Pond surface	Pond bottom, 20°C	Pond bottom, 90°C
	$MgCl_2$		
s	20 kg/m^3	300 kg/m^3	300 kg/m^3
$\dfrac{\partial\rho}{\partial s}$	0.75	0.65	0.68
$-\dfrac{\partial\rho}{\partial T}$	0.3 kg/m$^3 \cdot$ °C	0.25	0.45
	NaCl		
s	20 kg/m^3	260 kg/m^3	260 kg/m^3
$\dfrac{\partial\rho}{\partial s}$	0.8	0.62	0.52
$-\dfrac{\partial\rho}{\partial T}$	0.3 kg/m$^3 \cdot$ °C	0.5	0.51

In conjunction with Eq. (10-1), assuming that $\dfrac{\nu + k_T}{\nu + k_s} \simeq 1.14$ we obtain the following for stability in reference to an $MgCl_2$ pond:

$$\frac{\partial s}{\partial z} > 0.46\,\frac{\partial T}{\partial z} \qquad \text{pond surface} \qquad (10.2a)$$

$$\frac{\partial s}{\partial z} > 0.44\,\frac{\partial T}{\partial z} \qquad \text{pond bottom, cold} \qquad (10.2b)$$

$$\frac{\partial s}{\partial z} > 0.75\,\frac{\partial T}{\partial z} \qquad \text{pond bottom, warm} \qquad (10.2c)$$

and in reference to an NaCl pond:

$$\frac{\partial s}{\partial z} > 0.44\,\frac{\partial T}{\partial z} \qquad \text{pond surface} \qquad (10.2\,d)$$

$$\frac{\partial s}{\partial z} > 0.92\,\frac{\partial T}{\partial z} \qquad \text{pond bottom, cold} \qquad (10.2e)$$

$$\frac{\partial s}{\partial z} > 1.18\,\frac{\partial T}{\partial z} \qquad \text{pond bottom, warm} \qquad (10.2f)$$

The infrared part of the solar radiation can effect a remarkable temperature gradient near the surface.

From considerations which follow from Eqs. (10.30) and (10.42) to (10.44), we find that at the pond surface

$$\frac{\partial T}{\partial z} \sim 500°C/m \tag{10.3}$$

during a summer solstice noon at 32° latitude.

At the bottom of the pond the gradients can reach a maximum of 400°C/m for a pond without energy extraction and without allowing a convective zone at the ground bottom.

In general, limited convective zones of 0.2 to 0.5m are desirable for purposes of diurnal heat storage. If convective zones are allowed, then local temperature gradients less than 200°C/m are to be expected. Only if convective zones are allowed at the bottom of the pond will the temperature gradients be sufficiently small.

Equation (10.2) requires concentration gradients of approximately

$$\frac{\partial s}{\partial z} = 230 \text{ kg/m}^4 \tag{10.4a}$$

at the pond surface in both $MgCl_2$ and $NaCl$ ponds.

The salinity gradient required at the bottom of a warm $MgCl_2$ pond without a convective zone at its bottom is

$$\frac{\partial s}{\partial z} > 300 \text{ kg/m}^4 \tag{10.4b}$$

An $NaCl$ pond cannot be maintained without a convective zone at the bottom because greater-than-saturation solubilities would be required for ponds deeper than 0.75 m. If a 0.3-m convective zone is allowed, then $NaCl$ ponds will require a gradient of approximately

$$\frac{\partial s}{\partial z} > 230 \text{ kg/m}^4 \tag{10.5}$$

at the pond bottom.

The original density gradient is established at the time the pond is filled and is maintained against diffusive effects by controlling salt concentrations at the top and bottom surfaces.

Filling the Pond

Solar ponds are generally filled in layered sections with small saline differences between adjacent layers. Usually, solar ponds are built from the bottom upward, with the bottom layer filled first and successively lighter layers floated upon the lower denser layers. However, an experimental pond in Aspendale, Australia, was built by filling from the bottom with successively denser layers which lifted the lighter layers previously filled.

Soon after the stepwise filling process, the pond gradient smooths itself. N. Chepurniy and S. B. Savage have calculated the smoothing effected by diffusion.[8] According to these calculations, a 10-layer pond should have a smooth profile within a week after filling. The smoothing of the density profile is necessary because otherwise the individual layers themselves are unstable and convective cells can establish themselves in the individual layers.

Experimentally, it has been observed that a smooth gradient is established more rapidly than would be expected from diffusion considerations alone. This is explained by considering the partial mixing of adjacent layers caused by the kinetic energy of liquid flow injected into the pond during the filling process. Part of this kinetic energy is converted into gravitational potential energy by partially mixing the adjacent layers. (The rest of the kinetic energy is dissipated in viscous flow and at the walls of the pond.)

It is this partial mixing which hastens the desired smoothing of the concentration gradient.

Because the kinetic energy injected in the pond does cause mixing, there are limitations to the rate at which a solar pond may be filled. From energy considerations, it appears that the velocity of liquid flow into the pond is limited to approximately 0.12 m/s. We observe this from the following considerations.

If ρ_n is the density of the nth layer and d_n, the layer thickness, then the mean kinetic energy (KE) per unit pond surface area of the flowing layers is given by

$$\tfrac{1}{2} \rho_n d_n \overline{v^2} = \text{KE/unit surface area} \tag{10.6}$$

where $\overline{v^2}$ is the mean-square velocity of flow during the filling operation.

The potential energy (PE) of two adjacent unmixed layers of thicknesses d_{n-1} and d_n, respectively, is given by

$$\text{PE (unmixed)} = \tfrac{1}{2} g \left[\rho_n (2d_{n-1} + d_n) d_n + \rho_{n-1} d_{n-2}^2 \right] \tag{10.7a}$$

where the potential energy is referred to the bottom of the $(n - 1)$st layer. The first term in the right-hand member of Eq. (10.7) gives the contribution to PE of the nth layer, and the second term gives the contribution of the $(n - 1)$st layer. If, in the flow process, a quantity αd_{n-1} of the $(n - 1)$st layer is eroded and mixed uniformly with the nth layer, then the combined potential energy will be given by

$$\text{PE (mixed)} = \tfrac{1}{2} g \left\{ (\rho_n d_n + \alpha \rho_{n-1} d_{n-1}) \left[(2 - \alpha) d_{n-1} + d_n \right] + \rho_{n-1} (1 - \alpha)^2 d_{n-1}^2 \right\} \tag{10.7b}$$

The change in potential energy due to the fraction α of the $(n - 1)$st layer mixing with the nth layer is given by

$$\text{PE (mixed)} - \text{PE (unmixed)} = \tfrac{1}{2} g \alpha \left[\rho_{n-1} (3 d_n d_{n-1} - 2 d_{n-1}^2) - \rho_n d_n d_{n-1} \right] \tag{10.7c}$$

If the flow velocity v is sufficiently small, then we may neglect the viscous dissipation of the kinetic energy. For sufficiently large ponds, we may also neglect the energy dissipated at the walls.

Assuming then that all the kinetic energy is converted to potential energy, we obtain, by equating Eq. (10.6) and (10.7c) and allowing for simplicity that $d = d_n = d_{n-1}$,

$$\alpha = \frac{\rho_n \overline{v^2}}{gd(\rho_{n-1} - \rho_n)} \tag{10.8}$$

$1/\alpha$ is formally similar to the Richardson number of hydrodynamics for the stability of selective flow of adjacent layers.

Let us examine Eq. (10.8) for numerical parameters pertinent to a solar pond. If we assume that the pond is filled in 10-cm layers and that $\Delta\rho/\rho_n \simeq 0.03$, and if we allow $\alpha = \tfrac{1}{2}$, i.e., that half of the previous layer erodes into the freshly added layer, then

$$\overline{v} \sim 0.12 \text{ m/s}$$

Assuring a sufficiently low flow velocity is a hydraulic problem which must be considered in constructing a solar pond.

An important conclusion which we derive from Eq. (10.5) is that the greater the layer thickness d, the less will adjacent layers mix. Diffusion arguments, on the other hand, imply that the layer thickness should be reasonably small. The 0.1-m thickness chosen in our example appears to be a practical compromise.

Because we have neglected the dissipation due to viscous flow and the dissipation at the pond perimeter, the parameter given by Eq. (10.8) is conservative.

Maintaining the Salt Concentration Gradient

After filling a pond, the stabilizing salt concentration gradient is maintained by controlling the salt concentrations at the top and bottom extremities of the pond. In determining the salt concentrations required at the extremities, we must also consider the effect on the concentration gradient due to the vertical flow of pond solution. This flow may be either advertent or inadvertent. Inadvertent downward flow of the pond may be due to seepage of warm solution from the pond bottom. Upward flow can occur in systems which are fed by saline solutions near the bottom. Purposeful downward flow may be due to salt and water extraction processes proposed by S. Shachar (U.S. Patent 3,372,961) for the utilization of energy collected by the pond. An upward flow may be required in order to restrict the concentration gradient to the upper level of a deep pond. This has been considered by G. Assaf.[9]

The salt concentration gradient in the solar pond is governed by the diffusion equation:

$$\frac{\partial}{\partial z}\left(k_s \frac{\partial s}{\partial z}\right) = v\frac{\partial s}{\partial z} + \frac{\partial s}{\partial t} \qquad (10.9)$$

where v is the vertical velocity measured positive downward.

Restricting ourselves to steady-state systems, so that $\partial s/\partial t = 0$, a single integration of the diffusion equation then yields

$$q = vs - k_s \frac{\partial s}{\partial z} \qquad (10.10)$$

k_s is weakly dependent on salt concentrations, but strongly dependent upon temperature.

We will first consider the solution of Eq. (10.10) for a cool pond and subsequently for an operating pond.

The salinity gradient in the cool pond   At the start of operation we may consider k_s constant. The solution of Eq. (10.10) is, for $v = 0$,

$$s = s(o) - \frac{qz}{k_s}$$
$$\frac{\partial s}{\partial z} = \frac{-q}{k_s} = \frac{s(h) - s(o)}{h} \qquad (10.11)$$

where h is the total depth of the pond. For $v \neq 0$,

$$s = \left[s(o) - \frac{q}{v}\right]\exp\frac{vz}{k_s} + \frac{q}{v}$$
$$\frac{\partial s}{\partial z} = \frac{v}{k_s}\left[s(o) - \frac{q}{v}\right]\exp\frac{vz}{k_s} \qquad (10.12)$$

We observe from Eq. (10.12) that for v positive, the concentration gradient increases with increasing depth and that for v negative, the gradient decreases with increasing depth.

From Eq. (10.11) we have that for $v = 0$, $s(o) = 0$, and $\partial s/\partial z = 230$ kg/m^4 [as required by Eqs. (10.4a) and (10.5)],

$$h = \frac{s(h)}{230}$$

The maximum value of $s(h)$ is the saturation concentration s_{sat}. For MgCl$_2$ pond, $s_{sat} \approx 370$ kg/m^3, so that $h_{max} = \frac{370}{230} = 1.6$ m. For an NaCl pond, $s_{sat} = 300$ kg/m^3, or $h_{max} = \frac{300}{230} = 1.3$ m. The quantity of salt flux q required to maintain a cool pond is, from Eq. (10.11),

$$q = -k_s \frac{\partial s}{\partial z}$$

at ambient temperatures; for NaCl, $k_s \approx 1.39 \times 10^{-9}$ m^2/s and

$$q = 3.196 \times 10^{-7} \text{ kg/s} \cdot \text{m}^2$$

Then $q = 0.0276$ kg/m$^2 \cdot$ day upward. This is a large flux of salt.

We will find in the following subsection that for $v > 0$ (a downward flow of the pond body), the pond may be stable without a salt flux upward. During initial start-up operation, generally $v = 0$. The analysis for the maximum depths of ponds for which $v \neq 0$ and their accompanying salt fluxes will therefore be examined in the following section.

The salinity gradient in a warm pond   Table 10.2 shows k_s as a function of temperature for NaCl as compiled from various sources and for MgCl$_2$ as taken from W. Stiles.[10] The results for NaCl are in reasonable agreement with theoretical predictions that the diffusion coefficient is linearly dependent upon temperature. For NaCl, we may then write

Table 10.2   Salt Diffusivity Values

Temperature, °C	NaCl, $\times 10^{-9}$ m²/s	MgCl₂, $\times 10^{-9}$ m²/s
20	1.39	1.12
30	1.77	1.42
40	2.20	1.73
50	2.54	
75	3.60	
90	4.41	

$$k_s(T) = 1.39[1 + 0.029(T - 20°C)] \times 10^{-9} \text{ m}^2/\text{s} \tag{10.13a}$$

for better than 5 percent accuracy within the range 20 to 90°C. With less confidence, for MgCl₂, we write

$$k_s(T) = 1.12[1 + 0.027(T - 20°C)] \times 10^{-9} \text{ m}^2/\text{s} \tag{10.13b}$$

Independent studies of the dependence of k_s upon the concentration indicate that within a concentration range of 0.1 to 5 N, the value of k_s varies less than 10 percent. For the purposes of our analysis it will be permissible to consider k_s independent of concentration.

If, for the purpose of this analysis, we approximate the temperature profile of the pond by a profile linear with increasing depth, then k_s will also increase linearly with depth. We can therefore write

$$k_s(z) = k_s(20°C)\left(b + \frac{\mu\Delta Tz}{h}\right) \tag{10.14}$$

where $b = 1 + \mu(T_a - 20°C)$
T_a = mean daily ambient temperature
$T_a + \Delta T$ = temperature at $z = h$
$\mu = 0.029$ for NaCl
$\mu = 0.027$ for MgCl₂

Inserting Eq. (10.14) into Eq. (10.10) we obtain, for $v = 0$,

$$s(z) = \frac{-q}{bk_s(20°C)u} \ln(1 + uz) + s(o) \tag{10.15a}$$

where $u = \mu \Delta T/bh$

$$\text{and } \frac{\partial s}{\partial z} = \frac{-q}{bk_s(20°C)(1 + uz)} \tag{10.15b}$$

and for $v \neq 0$,

$$s(z) = \left[s(o) - \frac{q}{v}\right](1 + uz)^f + \frac{q}{v} \tag{10.16a}$$

where $f = \dfrac{vh}{k_s(20°C)\mu\Delta T}$

$$\text{and } \frac{\partial s}{\partial z} = \frac{v}{k_s(20°C)b}\left[s(o) - \frac{q}{v}\right][1 + uz]^{f-1} \tag{10.16b}$$

We observe from Eq. (10.16b) that for $f > 1$ the concentration gradient increases with increasing depth, whereas for $f < 1$ the gradient decreases with increasing depth.

The maximum depth of a pond is limited by (1) the concentration gradients desired at the extremities and (2) the saturation concentration of the salt. Let us examine Eq. (10.15b) for ΔT $\approx 70°C$, $T_a = 20°C$, and requiring a gradient of 230 kg/m^4 at the pond bottom as required by Eqs. (10.4a) and (10.5), then

$$230 \text{ kg/m}^4 = \frac{\partial s}{\partial z} = \frac{-q}{k(20)(1 + \mu \; \Delta T)} \tag{10.17}$$

for NaCl,

$$q = 9.69 \times 10^{-7} \text{kg/m}^2 \cdot \text{s}$$
$$= 0.0836 \text{ kg/m}^2 \cdot \text{day}$$

For a concentration of 300 kg/m^3 we find from Eq. (10.15a) upon substitution of numerical values for u and b that at $z = h$,

$$s(h) = 381\,h = 300 \text{ kg/m}^5$$

and therefore $h = 0.79$ m. For MgCl$_2$,

$$q = 7.44 \times 10^{-7} \text{ kg/m}^2 \cdot \text{s}$$
$$= 0.064 \text{kg/m}^2 \cdot \text{day}$$

and for a saturation concentration of 370 kg/m^3, we find upon substitution of appropriate numerical values into Eq. (10.16a) that

$$370 \text{ kg/m}^3 = 373h$$
$$h = 0.99 \text{ m}$$

Thus the maximum depths of warm ponds are limited unless we allow large convective zones at the bottom of the pond.

The salinity gradient in a pond with downward vertical flow For $v \neq 0$ there are several interesting cases. The first is for $f = 1$. For this case we observe from Eq. (10.16b) that the density gradient in the pond is linear and the concentration gradients, salt fluxes, and maximum depths of the pond are identical to those of a cool pond for which $v = 0$ and which is described by Eq. (10.11). For an NaCl pond we have found from Eq. (10.11) $h = 1.3$ m; therefore we have from Eq. (10.16b)

$$v = \frac{1.39 \times 0.029 \times 70}{1.3} \times 10^{-9}$$
$$= 2.17 \times 10^{-9} \text{ m/s} \tag{10.18a}$$
$$= 1.88 \times 10^{-4} \text{ m/day}$$

For an MgCl$_2$ pond,

$$v = \frac{1.12 \times 0.027 \times 70}{1.6} \times 10^{-9}$$
$$= 1.32 \times 10^{-9} \text{ m/s} \tag{10.18b}$$
$$= 1.14 \times 10^{-4} \text{ m/day}$$

With larger values of v the salinity gradient increases with increasing depth. For smaller values of v the salinity gradient decreases with increasing depth.

A rather interesting case is for $q = 0$. For this pond, the convection downward of salt solution at the top of the pond is just equal to the salt which diffuses upward because of the concentration gradient. At the bottom of the pond we must extract the salt and liquid added to the surface of the pond. How to do this is described by S. Shachar (U.S. Patent 3,372,691). In the literature this pond is referred to as "the falling pond."

We observe from Eq. (10.16b) that for $q = 0$, a concentration gradient may be maintained in the pond only if $s(o) > 0$. For $q = 0$, we have from Eq. (10.16a) that at $z = h$,

$$\frac{vh}{k_s(20)\,\mu\,\Delta T}\ln\left(1 + \frac{\mu\,\Delta T}{b}\right) = \ln\frac{s(h)}{s(o)} \tag{10.19}$$

and from Eq. (10.16b) at $z = 0$,

$$\left.\frac{\partial s}{\partial z}\right|_{z=0} = \frac{vs(o)}{bk_s(20°C)} \tag{10.20}$$

Substitution for v/k from Eq. (10.20) into Eq. (10.19) yields

$$h = \frac{-\mu\,\Delta Ts(o)\ln\,[s(h)/s(o)]}{b\left.\dfrac{\partial s}{\partial z}\right|_{z=0}\ln(1 + \mu\,\Delta T/b)} \tag{10.21}$$

The maximum value of h is obtained when

$$\frac{s(o)}{s(h)} = e^{-1} = 0.368$$

For $\left.\dfrac{\partial s}{\partial z}\right|_{z=0} = 230$ kg/m⁴, $\Delta T = 70°C$, then for NaCl,

$$h_{max} = 0.88 \text{ m}$$
$$v = 2.90 \times 10^{-9} \text{ m/s} = 2.5 \times 10^{-4} \text{ m/day}$$

and for $MgCl_2$,

$$h = 1.06 \text{ m}$$
$$v = 1.89 \times 10^{-9} \text{ m/s} = 1.6 \times 10^{-4} \text{m/day}$$

Such ponds will be of value where highly saline springs with $s(o) \simeq e^{-1}\,(s_{sat}) \simeq 110$ kg/m³ are plentiful and fresh water is in short supply. The salinity gradients of falling ponds increase with increasing depth.

The salinity gradient in warm ponds with upward vertical flow Ponds with large convective zones below the stabilizing gradients have been considered for seasonal heat storage. In these ponds, the temperature gradient at the interface with the convective zone is small (less than 70°C/m even during the summer). For these ponds it is necessary that v be negative in order to maintain a sufficient gradient in a pond where overall depth may be 4 or more meters.

From Eq. (10.10) we have for $s(o) = 0$ at $z = 0$ that

$$\left.\frac{\partial s}{\partial z}\right|_{z=0} = \frac{-q}{k_s(o)} \tag{10.22a}$$

At large depths the salinity gradient vanishes so that

$$s(\infty) = \frac{q}{v} \qquad (q, v \text{ negative}) \tag{10.22b}$$

We obtain from these relations that the vertical flow is given by

$$v = -\left.\frac{\partial s}{\partial z}\right|_{z=0}\frac{k_s(o)}{s(\infty)} \tag{10.23}$$

Assuming that the temperature gradient $\Delta T/h = 70°C/m$ is uniform from the surface to the top of the convective zone, we find, upon substituting for v, given by Eq. (10.23), in Eq. (10.16b), that for an NaCl pond,

$$\frac{\partial s}{\partial z} = \frac{\partial s}{\partial z}\bigg|_{z=0} (1 + 2.03z)^{-0.624} \tag{10.24a}$$

and for an MgCl$_2$ pond,

$$\frac{\partial s}{\partial z} = \frac{\partial s}{\partial z}\bigg|_{z=0} (1 + 1.89z)^{-0.67} \tag{10.24b}$$

For both NaCl and MgCl$_2$ solutions the salinity gradient at the bottom of the pond is half the salinity gradient at the surface.

Repair of a Disturbed Halocline

Solar ponds are exposed to disturbances which may be either environmental (for example, winds) or internal and related to the operation of the pond (for example, a fluctuation in rate of bottom layer flow required for the extraction of energy from the pond) or possibly due to negligence in the maintenance of the required concentrations at the extremities of the pond.

Disturbances can hold sufficient energy to a region of the pond so that the stability criteria discussed in the previous sections no longer apply and mixed layers can develop in the pond. The cause of mixed layers in the body of the pond, as opposed to mixed layers at the extremes of the pond, is not clear. Such layers have developed in operating ponds after severe winds or sudden cessation of bottom layer flow. The mixing has been attributed to resonant seiches in the pond.

Experience with operating ponds indicates that mixed layers in the top and bottom regions of the pond tend to grow if not attended to, whereas mixed layers in the body of the pond generally shrink in time. The remedial action for the top layer of the pond is to extract part of the mixed layer and replace part with a less saline solution. The effect is that of locally building the density gradient. Mixed layers at the bottom of the pond have been treated by introducing saturated saline solutions to the bottom of the pond.

THE AVAILABLE SOLAR IRRADIATION AND ITS ABSORPTION IN THE POND

In order to evaluate the performance of a pond at a particular site, it is necessary to estimate as accurately as possible the solar irradiation penetrating the pond surface and its absorption in the body of the pond.

The hourly irradiation incident on a horizontal surface is measured at many meteorological stations. If local measurements are not available, then reasonably accurate estimates may be obtained from tabulated meteorological data.

Reflection Losses

Part of the incident radiation will be reflected from the pond's surface. If the pond surface is smooth, or only slightly ruffled, then the fraction of direct radiation penetrating the surface will be given by Fresnel's formula:

$$P = 1 - \frac{1}{2}\left[\frac{\sin^2(z - \theta_r)}{\sin^2(z + \theta_r)} + \frac{\tan^2(z - \theta_r)}{\tan^2(z + \theta_r)}\right] \tag{10.25a}$$

where z = solar zenith distance ($90° -$ solar altitude α)
   θ_r = angle of refraction at the water surface
From Snell's law of refraction,

$$n \sin \theta_r = \sin z \tag{10.25b}$$

n being the index of refraction of the surface solution, which, for dilute solutions, is approximately 1.33. The solar zenith distance is tabulated in meteorological tables as a function of time and latitude. For purposes of analysis (see Chap. 2) we can approximate the solar zenith distance by

$$\cos z = \sin L \sin \delta_s + \cos L \cos \delta_s \cos h$$

where h_s = hour angle
L = latitude
δ_s = declination of the sun, in rad

From measured values of the direct radiation we calculate the irradiation penetrating the water surface by taking the product of this irradiation and P calculated from Eq. (10.25). The hourly values of z and θ_r are calculated by means of Eq. (10.26) and (10.25b), respectively. The total daily direct irradiation is found by numerical integration of the calculated hourly values.

Lacking local hourly measurements we can use measured (or tabulated) values for total daily direct irradiation and multiply these values by the daily transmission coefficient τ, where

$$\tau = \frac{\int Pa^{\sec z} \cos z \, dt}{\int a^{\sec z} \cos z \, dt} \tag{10.27}$$

The limits of integration are the hours of sunrise and sunset,

where a = atmospheric transmission coefficient
$\sec z$ = air mass

The numerator of Eq. (10.27) is proportional to the total daily direct irradiation just penetrating the water surface, and the denominator is proportional to total direct irradiation incident above the water surface. Table 10.3 gives τ as a function of latitude for equinox and solstice conditions.

TABLE 10.3   Transmission Coefficient τ for Various Latitudes

Latitude, degrees	Summer solstice	Equinox	Winter solstice
0	0.97	0.97	0.97
10	0.97	0.97	0.96
20	0.97	0.97	0.95
30	0.97	0.96	0.93
40	0.97	0.96	0.89
50	0.96	0.94	0.78

The diffuse sky radiation accounts for approximately 15 percent of the total irradiation when the sun is near the zenith and 40 percent of the total irradiation when the sun is near the horizon. If we assume that the sky is of uniform brightness, then the fraction of diffuse radiation just penetrating the water surface will be given by

$$\cdot \frac{\int_0^{\pi/2} P \cos z \sin z \, dz}{\int_0^{\pi/2} \cos z \sin z \, dz} = 0.93 \tag{10.28}$$

We observe, from Table 10.3 and Eq. (10.28) that the reflection losses are less than 10 percent for latitudes lower than 40°. Figure 10.3a shows the estimated irradiation just below the water surface at the Dead Sea for solstice conditions. These curves have been calculated from measured direct and diffuse irradiation and by applying Eqs. (10.25) and (10.26) to the direct radiation, multiplying the diffuse irradiation by 0.93, and summing the two values.

Table 10.4 shows the daily mean irradiance penetrating a smooth water surface by assuming direct and diffuse solar irradiation estimated from the "Smithsonian Meteorological Tables".[11] The direct irradiation was multiplied by the corresponding τ of Table 10.3 and the diffuse irradiation by 0.93.

We observe that the ratio of maximal to minimal daily mean irradiance is only 1.16 at the

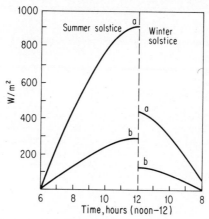

Fig. 10.3 (a) Solar radiation just penetrating a smooth water surface at 32° latitude. (b) Solar radiation penetrating to a 1-m depth of a pond at 32° latitude.

Table 10.4   Mean Daily Irradiance Penetrating Smooth Water

Latitude, degrees	Spring equinox, W/m²	Summer solstice, W/m²	Fall equinox, W/m²	Winter solstice, W/m²
0	307	264	302	281
10	301	296	297	235
20	283	319	279	183
30	252	331	248	129
40	213	333	211	74
50	166	321	163	30

equator but increases to 10.7 at high latitudes. This increased ratio severely reduces the usefulness of solar ponds at high latitudes.

The Absorption of Radiation in the Pond

The radiation penetrating the pond surface is attenuated by absorption in the pond so that only a fraction reaches the bottom. While the energy absorbed in the body of the pond also serves to warm the pond, the more transparent the pond solutions are, the greater will be the thermal efficiency of the pond.

The clarity of natural ocean water is remarkably similar to that of distilled water. Treated ocean waters, however, are quite turbid. The turbidity of treated ocean waters is probably due to the decomposition of ordinarily transparent microbial oceanic life.[12] The decomposition of microbial life has consequences for the clarity of solar ponds. Microbes such as diatomic algae present in the water solutions may not survive the temperature and saline conditions of the solar pond and may decompose. The products of decomposition may strongly absorb incident solar radiation.

Because of its density gradient, a solar pond will probably be dirtier and absorb incident solar radiation more strongly than homogeneous bodies of water located at the same geographical site. Dirt which falls onto the surface of a homogeneous lake will either float at the surface of the pond or, if denser than water, will settle to the bottom. Floating debris can be skimmed from the pond surface and that which settles does not effect the transparency of the pond. Dirt denser than water, does not necessarily settle to the bottom of a solar pond, but can sink to a level corresponding to its density and remain there. Also the temperatures and salt concentrations of solar ponds may afford a hospitable environment for thermophilic and halophilic

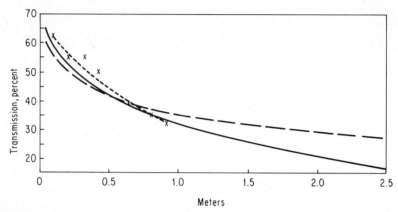

Fig. 10.4 Transmission of solar radiation through pond solutions. (————) Transmission assumed for the numerical examples and illustrations of the text; (————) Transmission assumed by Rabl and Nielsen (Ref. 14); (× × ×) Measured transmission in a solar pond at 32° latitude.

algae. Such algae abound in natural solar ponds and absorb incident radiation. They have not yet been found in artificial solar ponds.

Figure 10.4 shows theoretical transmission curves used by Weinberger[13] and Rabl and Nielsen[14] for the calculation of the solar pond temperature rise and also for the measured transmission in a solar pond located in an industrial complex near the shores of the Dead Sea. Figure 10.3b shows the hourly radiation reaching a 1-m depth in a solar pond for solstice conditions. These curves are based on the value of the solar irradiation penetrating the surface shown in Fig. 10.3a and by multiplying by Weinberger's theoretical transmission value shown in Fig. 10.4.

If z is the depth in the solar pond, then the optical path length x which is to be used for the ordinate of Fig. 10.4 is given by

$$x = z \sec \theta_r$$

$\sec \theta_r$ is calculated as a function of time by means of Eqs. (10.26) and (10.15b). Figure 10.5 shows the total daily irradiation reaching a 1-m depth throughout the year for a solar pond at 32° latitude and assuming Weinberger's transmission curve. Figure 10.4 has been calculated by means of numerical integration of average hourly values at a 1-m depth.

Figure 10.6 shows the mean daily radiation reaching a different depth in solar ponds at 32° latitude for summer and winter solstices. These curves are calculated from numerical integration of the hourly values of radiation penetrating to different depths in solar ponds.

The path lengths of the diffuse radiation has been assumed equal to that of the direct radiation in Figs. 10.3b and 10.4 to 10.6.

In order to evaluate the temperature rise in solar ponds, daily mean radiation values for the different depths in solar ponds as a function of date and latitude must be determined for use in heat transfer equations to be discussed in the following section. For analysis of the thermal performance of the pond, it is convenient to approximate the insolation by an analytic expression of the form

$$I = \sum_i A_i \exp - a_i z + \sum_{i,n} B_{i,n} (\exp -b_{i,n}z)\sin \frac{2\pi nD}{365} = \bar{I}(z) + \tilde{I}(z,t) \qquad (10.29)$$

The curves of Figs. 10.4 and 10.5 are approximated from 0.05- to 1.5-m depth in the pond by

$$
\begin{array}{ll}
A_1 = 110 \text{ W/m}^2 & a_1 = 0.48 \text{ m}^{-1} \\
A_2 = 50 & a_2 = 6.3 \\
B_{1,1} = 36 & b_1 = 0.094 \text{ m}^{-1} \\
B_{2,1} = 38.5 & b_2 = 3.1 \\
B_{i,n} = 0, i > 2, n > 1 &
\end{array}
$$

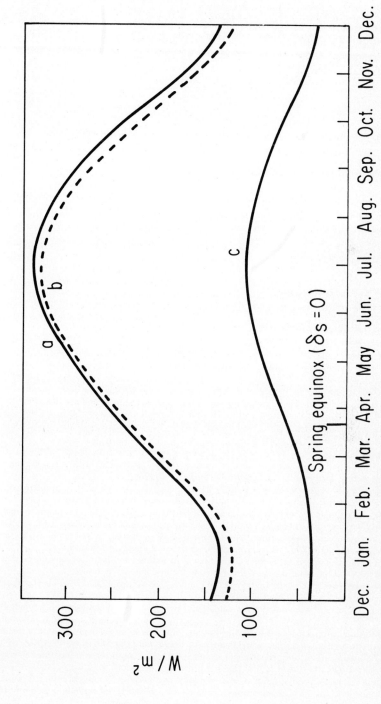

Fig. 10.5 (*a*) Daily mean solar irradiation incident upon a horizontal surface at 32° latitude (the Dead Sea); (*b*) daily mean solar irradiation just penetrating a water surface at the Dead Sea; (*c*) daily mean solar irradiation penetrating to a 1-m depth in a pond.

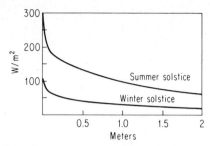

Fig. 10.6   Mean daily radiation penetrating to different depths in a pond at 32° latitude.

TEMPERATURES AND EFFICIENCIES OF SOLAR PONDS

If the lateral dimensions of a solar pond are sufficiently large with respect to its depth, then the temperature in the nonconvecting regions of the solar pond will be governed by the one-dimensional thermal diffusion equation:

$$\frac{\partial}{\partial z} \left(K \frac{\partial T}{\partial z} \right) = \rho c \frac{\partial T}{\partial t} - H(z,t) \tag{10.30}$$

where K = thermal conductivity
  c = heat capacity
  ρ = density
  $H(z,t)$ = heat absorbed in the body of the pond
    $= -\dfrac{dI}{dz}$
  $I(z,t)$ = irradiation penetrating to the depth z in the pond

In Eq. (10.30) we neglect the temperature effects of a possible vertical flow of the pond. The vertical flows in the pond are restricted by salt diffusion considerations and are described by Eqs. (10.18) and (10.23). Because the thermal diffusivity is approximately 50 times the salt diffusivity, the thermal effects of the maximal allowed vertical flows are negligible when compared to the other terms of Eq. (10.30).

Equation (10.30) is subject to boundary conditions at the surface of the pond and the pond bottom. At the surface of the pond we need to consider the heat exchanged with the atmosphere due to convection, evaporation, and long-wave surface radiation.

The accurate calculation of the heat transfer from a pond's surface is a difficult problem. We simplify our analysis by observing that the mean surface temperatures of convective lakes are usually several degrees warmer than the mean ambient temperatures. The mean surface temperatures of solar ponds built at 32°N latitude have also been found to be greater (by 2 to 4°C) than the mean ambient temperatures. The temperature of the surface water remains warmer than the mean ambient temperature even during periods of energy extraction.

For the purpose of analysis it is therefore sufficient to assume, as a boundary condition, that the temperature of the pond surface follows closely the ambient temperature.

The boundary condition at the bottom of the pond is that the irradiation $I(h,t)$ is absorbed there and that a quantity $U(h,t)$ of heat is extracted from the pond's bottom.

Parameters K, ρ, and c are functions of salinity and temperature and consequently a function of depth. The thermal properties of the ground, beneath the pond, can be considerably different from the thermal properties of the saline solution. The density of solar pond solutions as a function of temperature and salinity is given in Fig. 10.2a and b for $MgCl_2$ and $NaCl$.

Table 10.5 summarizes the properties of K and c for saline solutions as a function of temperature, and Table 10.6 shows typical thermal properties of various soils.

Because K, ρ, and c are not independent of depth, and because the insolation $I(z,t)$ is not usually available in the form of integrable functions, Eq. (10.30) is not directly integrable. It can, however, be solved numerically, and we will consider the numerical solution later [Eq. (10.41)]. If, however, we assume the thermal properties of the ground to be equal to the thermal

TABLE 10.5   Thermal Properties of Saline Solutions

The temperature and salinity dependence of the thermal conductivity is approximated closely by the empirical formulas*

$$K = K_w (1 - \alpha s)$$
$$K_w = 0.6 \, [1 + 2.81 \, (T - 20°) \times 10^{-3}] \, W/m°C$$
$$K_w = \text{thermal conductivity of pure water at temperature } T$$

For NaCl,

$$\alpha = 2.48 \times 10^{-4}$$

for $MgCl_2$

$$\alpha = 4.88 \times 10^{-4}$$

The heat capacity of saline solutions is given by the empirical formula†

$$c(s,T) = c(s,20°C) + \beta(s)(T - 20) + \gamma(s)(T - 20)^2$$

The values of $c(s,20)$, $\beta(s)$, and $\gamma(s)$ for NaCl and $MgCl_2$ are as follows:

	NaCl			$MgCl_2$	
s, kg/m³	$c(s,20)$, $\times 10^3$J/kg	β	γ, $\times 10^{-3}$	$c(s,20)$, $\times 10^3$J/kg	β
50	3.92	23	0	3.86	18.8
100	3.72	31	5.9	3.59	41
150	3.54	30	5.9	3.35	59
200	3.40	20	5	3.10	65
250	3.28	8.8	5	2.88	65
300				2.70	59

*Adapted from "International Critical Tables," vol. V, pp. 227–229, McGraw-Hill, New York, 1927.
†Adapted from *ibid.*, vol. II, p. 328.

TABLE 10.6   Thermal Properties of Various Soils at 20°

	K, W/m·°C	ρ, $\times 10^3$kg/m³	c, $\times 10^3$J/kg
Calcareous earth, 43% water	0.71	1.67	2.2
Quartz sand, 8% moisture	0.58	1.75	1.0
Sandy clay, 15% moisture	0.92	1.78	1.38
Soil, very dry	0.16–0.33		
Soil, wet	1.25–3.3		
Sandstone	2.6	2.6	0.5

SOURCE: Adapted from Ingersoll, Zobel, and Ingersoll, "Heat Conduction," p. 243, McGraw-Hill, New York, 1948.

properties of the saline solution, and that these values can be approximated for the depth of the pond by some mean value, and that $I(z,t)$ can be approximated by Eq. (10.29), then the steady-state time-dependent solution of Eq. (10.30) can be expressed in an analytic form. This analytic solution is extremely useful for estimating solar pond performance and efficiencies. We will consider the analytic solution for the mean daily insolation in the following section. We also require the analytic solution in order to estimate the peak temperature gradients which can occur at the surface and the bottom of the pond during the summer solstice, and which are necessary for analyzing the stability of the pond by means of Eq. (10.1).

The Steady-State Temperature Distribution in the Solar Pond

Expressing the daily mean irradiation reaching a depth z in the pond by Eq. (10.29) and the temperature of the surface of the pond by

$$T(o) = \overline{T(o)} + \Sigma \, \overline{T}_n(o) \sin \frac{2\pi n(D - \Psi_n)}{365} \tag{10.31}$$

and the mean power extracted from the bottom of the pond by

$$U = \overline{U}(h) + \Sigma \, U_n \sin \frac{2\pi n(D - \sigma_n)}{365} \tag{10.32}$$

then the steady solution of Eq. (10.30) is given by

$$T = \overline{T} + \hat{T} \tag{10.33}$$

where the time-independent part of T is given by

$$\overline{T}(z) = \frac{1}{K} \Sigma \frac{A_i}{a_i}[1 - \exp(-a_i z)] - \frac{\overline{U}z}{K} + \overline{T(o)} = \overline{T}_m(z) - \frac{Uz}{K} \tag{10.33a}$$

for $z < h_1$, $\overline{T}_m$ is the maximum mean temperature attainable at a depth z in a solar pond. For $z > h$

$$\overline{T}(z) = \overline{T}(h)$$

The steady-state time-dependent component is given by

$$\hat{T}(z,D) = \hat{T}_h + \hat{T}_b + \hat{T}_u + \hat{T}_a \tag{10.33b}$$

$\hat{T}_h$, the harmonic component of temperature rise in the pond due to solar radiation $\check{I}(h,D)$ absorbed at the bottom of the pond, is given by

$$\hat{T}_h(z,D) = \frac{\sqrt{2}}{4K} \Sigma_{i,n} \frac{B_{i,n}}{G} (\exp{-b_{i,n}h}) \left\{ \left[\exp{-(h,z)G} \right] \sin \left[\frac{2h\pi D}{365} - \frac{\pi}{4} - (h - z)G \right] \right.$$
$$\left. - \left[\exp{-(h + z)G} \right] \sin \left[\frac{2n\pi D}{365} - \frac{\pi}{4} - (h + z)G \right] \right\}$$

$$G = \left(\frac{\rho \, cn \, \pi}{3.16 \times 10^7 K} \right)^{1/2} \quad (3.16 \times 10^7 = \text{number of seconds in a year})$$

$\hat{T}_b$, the temperature rise due to the energy $-(\partial \check{I}/\partial z)$ absorbed in the body of the pond, is given by

$$\hat{T}_b(z,D) = \frac{\sqrt{2}}{4K} \Sigma_{i,n} \frac{B_{i,n} b_{i,n}}{G} \left[[G^2 + (b_{i,n} - G)^2]^{-1/2} \exp{-b_{i,n}z} \right.$$
$$\sin \left(\frac{2n\pi D}{365} - \frac{\pi}{4} + \beta_1 \right) - \left[\exp{-Gz} \right] \sin \left(\frac{2n\pi D}{365} - \frac{\pi}{4} - Gz + \beta_1 \right) \right]$$
$$+ [G^2 + (b_{i,n} + G)^2]^{-1/2} \left\{ \exp{[-(h - z)G - b_{i,n}h]} \right.$$
$$\sin \left[\frac{2n\pi D}{365} - \frac{\pi}{4} - (h - z)G + \beta_2 \right] - \exp{[-(h + z)G - b_{i,n}h]}$$
$$\sin \left[\frac{2n\pi D}{365} - \frac{\pi}{4} - (h + z)G + \beta_2 \right] - \left[\exp{-b_{i,n}z} \right] \sin \left(\frac{2n\pi D}{365} - \frac{\pi}{4} + \beta_2 \right)$$
$$+ \left[\exp{-Gz} \right] \sin \left(\frac{2n\pi D}{365} - \frac{\pi}{4} - Gz + \beta_2 \right) \right\}$$

$$\beta_1 = \tan^{-1} \frac{-G}{G - b_{i,n}} \qquad \beta_2 = \tan^{-1} \frac{G}{-(G + b_{i,n})}$$

$\hat{T}_u$ is the temperature rise due to the harmonic component of the rate of energy extraction U. Expressing $\hat{U}(D)$ in a series expansion,

$$\hat{U}(D) = \Sigma\, U_n \sin \frac{2n\pi(D - \sigma_n)}{365}$$

then

$$\hat{T}_u(zD) = \frac{\sqrt{2}}{4K} \Sigma_n \frac{U_n}{G} \left\{ \left[\exp -(h - z)\,G \right] \sin \left[\frac{2n\pi(D - \sigma_n)}{365} - \frac{\pi}{4} - (h - z)\,G \right] \right.$$
$$\left. - \exp -(h + z)\,G \sin \left[\frac{2n\pi(D - \sigma_n)}{365} - \frac{\pi}{4} - (h + z)\,G \right] \right\}$$

$\hat{T}_a(z,D)$ is the temperature rise effected by the harmonic component of the ambient temperature $\hat{T}(0,D)$. Expressing $\hat{T}(0,D)$ in a series expansion, we obtain

$$\hat{T}_a(z,D) = \Sigma\, T_n \left[\exp -Gz \right] \sin \left(\frac{2n(D - \psi_n)}{365} - Gz \right)$$

By a suitable choice of $\hat{U}_n$ and σ_n we observe from Eq. (10.33b) that we can achieve $\hat{T} = 0$. This means that we can extract heat from the pond at a constant temperature. $\hat{U}(h)$, the amplitude of the sinusoidal component of energy extracted, will be very nearly equal to $\Sigma_i\, B_{i,n} \exp (-b_{i,n}h)$, the amplitude of the sinusoidal component of the insolation reaching the bottom of the pond. The phase of the sinusoidal components of the heat extraction will lag behind the corresponding components of insolation.

The steady-state analytic solution for solar ponds possessing large convective zones at their bottom has been considered by Rabl and Nielsen.[14]

The Temperature of Heat Extraction and the Efficiency of the Pond

For a given temperature we can solve Eq. (10.33a) for U in order to determine the rate of heat extraction from the pond of depth h. Figure 10.7 shows the time-independent rate of heat extraction from the solar pond at a temperature of 89°C at 32° latitude as a function of the depth of the pond.

The time-independent part of the insolation is assumed to be (Eq. 10.29)

$$\bar{I}(z) = (110 \exp - 0.48z + 50 \exp - 6.3z)\,\text{W/m}^2$$
$$T_U = 26°C$$
$$K \simeq 0.6\,\text{W/m}\cdot°C$$

We observe that this curve has a maximum at 1.2 m. The insulation effect of the solar pond increases with increasing depth, but the insolation decreases. The optimum depth for energy

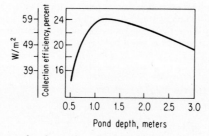

Fig. 10.7 Collection efficiency of a solar pond as a function of depth for an operating temperature of 89°C.

extraction at a temperature T_w is found by substituting $z = h$ into Eq. (10.33a) and differentiating with respect to h. For $\partial U/\partial h = 0$, we find

$$U = \Sigma A_i \exp -a_i h = \bar{I}(h) \tag{10.34}$$

i.e., that all the energy reaching the bottom of the pond is extracted and the pond is warmed only by the insolation absorbed in the body of the pond.

Substituting Eq.(10.34) into Eq.(10.33a), we obtain

$$\bar{T}_w = \bar{T}_m - I(h)\frac{h}{K} \tag{10.35}$$

Figure 10.8 relates the collection efficiency of the pond, its ideal depth, and temperature of operation. At 1-m depth we find from Eq. (10.34) and (10.35) that

$$T_w = 73°C \qquad \bar{U} = 66 \text{ W/m}^2$$
$$\text{For } \bar{T} = 10 \sin\frac{2\pi(D - 22)}{365}°C \tag{10.36}$$

and $\hat{I}(h) = (36 \exp -0.094z + 38.5 \exp -3.1z) \sin\frac{2\pi D}{365}$ then from Eq. (10.33b) for $T = 0$ we find that

$$\hat{U} = 34.4 \sin\frac{2\pi(D - 21)}{365} \text{ W/m}^2$$

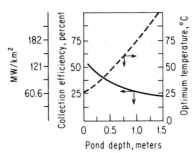

Fig. 10.8 Collection efficiency and ideal depth of the pond as a function of the temperature of energy withdrawal.

The mean irradiation $\bar{I}_s$ at the Dead Sea surface is 244.7 W/m^2. Thus the mean collection efficiency is greater than 27 percent, and the ratio of maximum to minimum utilization is 3.18:1.

Equations (10.34) and (10.35) give the optimum depth of a pond for utilizing energy at a particular temperature. We may, however, examine the question of optimum efficiency of a pond by examining the heat extraction from a pond of a particular depth where the value of the heat is weighted by its temperature of utilization.

If we attach a value $f(T)$ to the temperature of energy extraction, then the "worth" of the power U extracted from the pond is given by

$$\overline{J(T_w)} = \bar{U}f(T_w) \tag{10.37}$$

The maximum "worth" is given by

$$\frac{\partial J}{\partial \bar{U}} = f(T_w) + \bar{U}\frac{\partial f}{\partial T_w}\frac{\partial T_w}{\partial \bar{U}} = 0 \tag{10.38}$$

We substitute for $\partial T_w / \partial \overline{U}$ from Eq. (10.33a) into Eq. (10.37) to obtain

$$\frac{h}{K}\overline{U} = \frac{f(T_w)}{f'(T_w)} \tag{10.39}$$

For example, let $f(T_w) = (T_w - T_s)/(T_w + 273)$, i.e., the Carnot efficiency of operation between the temperature of energy extracted and a heat sink T_s. The "worth" of the energy is then the useful electrical or mechanical energy recoverable—in theory—from the pond.

The simultaneous solution of Eq. (10.33a) and (10.38) yields

$$J = Uf(T) = \frac{K}{h}[(\overline{T}_m + 273)^{1/2} - (\overline{T}_s + 273)^{1/2}]^2 \tag{10.40a}$$

$$\text{and } T_w + 273 = (\overline{T}_m + 273)^{1/2}(T_s + 273)^{1/2} \tag{10.40b}$$

i.e., the optimal Kelvin temperature is the geometric mean between the maximum Kelvin temperature which can be achieved at the depth h of the pond and the Kelvin temperature of the heat sink. Figure 10.9 shows the optimum recoverable power and the temperature T_w as a function of pond depth for Dead Sea conditions and $T_s + 273 = 300$K.

$$\text{For a 1-m-deep pond, } U(D) = 52.9 + 34.4 \sin \frac{2\pi(D - 21)}{365} \text{ W/m}^2 \tag{10.41a}$$

$$T_w = 98°\text{C} \tag{10.41b}$$

$$J(D) = 10.1 + 6.6 \sin \frac{2\pi(D - 21)}{365} \text{ W/m}^2$$

The mean efficiency of useful power extracted from the 1-m-deep pond is 4.2 percent. For the conditions described above the ratio of maximum to minimum energy exploitation is, for a pond without a convective storage zone, 4.8:1.

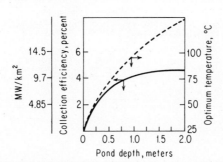

Fig. 10.9 Mean energy converted by means of a Carnot engine recoverable from ponds of various depths at the most efficient temperature of operation.

The Diurnal Temperature Variation and the Maximal Temperature Gradient in the Pond

In the previous sections we have considered the temperatures associated with the mean daily flux. Because of the diurnal dependence of the insolation, there will be a considerable temperature variation about the daily mean temperature. This temperature variation will decrease with an increase of the depth of a convective zone maintained at the bottom of the pond.

Associated with the diurnal temperature variation are the temperature gradients which are achieved at the extremes of the pond during a clear summer solstice. These temperature gradients determine the halocline required for stability. The temperature gradients derived in this section have been used in Eq. (10.1) for establishing the stability criteria.

The diurnal flux shown graphically in Fig. 10.3a and b may be approximated by

$$\Phi(z,t) = I(z,D) + \sum_i M_i \exp\left(-\eta_i z\right)\left(\frac{1}{2}\sin 2\pi\,\frac{(t-6)}{24}\right.$$

$$\left. - \sum_{n=2}^{\infty}\frac{2}{\pi(4n^2-1)}\,\frac{\cos 4n\pi\,(t-6)}{24}\right) \qquad (t \text{ in hours}) \qquad (10.42)$$

$\Phi(z,t)$ is formally similar to $I(z,D)$ except that we have taken a 24-h period instead of a 365-day period in the sinusoidal terms.

For summer solstice conditions,

$$M_1 = 110.4 \text{ W/m}^2 \qquad \eta_1 = 0.48 \text{ m}^{-1}$$

$$M_2 = 49 \text{ W/m}^2 \qquad \eta_2 = 6.3 \text{ m}^{-1}$$

$$M_3 = 36 \text{ W/m}^2 \qquad \eta_3 = 0.94 \text{ m}^{-1}$$

$$M_4 = 38.7 \text{ W/m}^2 \qquad \eta_4 = 3.14 \text{ m}^{-1}$$

The temperature variation about the daily mean The influence of the diurnal atmospheric temperature variation affects only slightly the diurnal temperature variation at a depth of 1 m in the body of the pond. The energy absorbed in the body of the pond also contributes only a small term to the diurnal variation.

We can, therefore, approximate the diurnal temperature variations at the bottom of the pond by considering the energy absorbed in the bottom of the pond alone. At the bottom of the pond we expect a convection zone of thickness d. This convective zone is necessitated by all envisaged energy extraction processes. A small convective zone does not affect the mean daily temperatures considered earlier, but it does affect the diurnal temperature variations. The equation of a well-mixed convective zone is given by

$$\rho cd\,\frac{\partial T}{\partial t} = K(z+d)\,\frac{\partial T}{\partial z}\bigg|_{z+d} - K(z)\,\frac{\partial T}{\partial z}\bigg|_{z} + I(z,t) - I(z+d,t) \qquad (10.43)$$

The nonconvective regions of the pond are connected analytically with the convective zone by the continuity of the temperature at the interfaces of the zones.

The harmonic part of the steady-state solution of Eq. (10.30) together with Eq. (10.43) for an insolation given by Eq. (10.42) is then

$$T_v = \sum_i M_i \exp-\eta_i\,h\left(\frac{[\exp-\gamma_1(h-z)]\sin[2\pi(t-6)/24-\gamma_1(h-z)-\delta_1]}{2\sqrt{\gamma_1 K' + [(2\pi/24)\rho\,cd + \gamma_1 K]^2}}\right.$$

$$\left. + \sum_n \frac{2\,[\exp-\gamma_n(h-z)]\cos[4n\pi(t-6)/24-\gamma_n(h-z)-\delta_n]}{\pi(4n^2-1)\sqrt{(\gamma_n K)^2 + [(2n\pi/24)\rho\,cd + \gamma_n K)]^2}}\right)$$

$$\gamma_n = \left(\frac{n\pi\,\rho c}{24\,K'}\right)^{1/2}$$

$$\delta_n = \cot^{-1}\frac{\gamma_n K'}{(2\pi n/24)\rho\,cd + \gamma_n K'} \qquad (K' = 3600 \times K)$$

Figure 10.10 shows the hourly temperature variation expected at the bottom of the pond without a convection zone, and for a pond with a 0.2-m convection zone.

The maximal temperature gradients in the pond By differentiating Eqs. (10.33a) and (10.33b) with respect to z and summing, we obtain the mean daily temperature gradient. Adding to this the maximal diurnal component obtained from Eq. (10.44) after differentiating with respect to z, we obtain, at the summer solstice, that

$$\frac{\partial T}{\partial z} \simeq 400°\text{C/m} \qquad \text{(for a pond without a convective zone)}$$

and

$$\frac{\partial T}{\partial z} \simeq 250°\text{C/m} \qquad \text{(for a pond with a 0.2-m convective zone)}$$

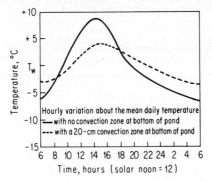

Fig. 10.10 The hourly variation about the mean temperature at the bottom of a 1-m-deep pond with and without a convective zone at the bottom.

Near the surface of the pond we must consider the steep gradients affected by the locally absorbed irradiation. If we solve Eq. (10.30) for the irradiation of Eq.(10.42) for $z = 0$, we obtain, for the summer solstice at 32° latitude,

$$\frac{\partial T}{\partial z}\bigg| \text{ surface} = 500°\text{C}/\text{m}$$

In the analysis following Eq. (10.2) we have assumed this value at the surface.

Heat Storage in the Solar Pond

In the previous section we examined the influence of a small convective zone at the bottom of the pond in smoothing the diurnal temperature variations. Larger convective zones are required to smooth out seasonal variations in temperature if the pond is to operate at more uniform rates of exploitation than those predicted by Eq. (10.41). Rabl and Nielsen have described analytically the storage effects of convective zones[14].

If energy is extracted from the storage zone at a uniform rate, then a surprisingly good approximation to the seasonal temperature variation in large storage zones is obtained by assuming the storage zone to be well mixed and perfectly insulated at its extremes.

With this approximation, we obtain for the temperature in the storage zone of depth d for D that

$$\hat{T} = \frac{3.16 \times 10^7 \, \hat{I}(z)}{2\pi \, dpc} \cos \frac{2\pi D}{365} \tag{10.45}$$

($3.16 \times 10^7 \simeq$ number of seconds in a year). For $\hat{I}(z) \simeq 34$ W/m² a storage zone of 10 m reduces the amplitude of temperature variation to 5.5°C.

Thermal Effects of a Convective Zone at the Surface of the Pond

In the preceding subsections we have assumed that there is no convective zone at the surface of the pond. A convective zone at the surface of the pond can have severe influences on both the temperatures and efficiencies of the pond.

The convective zone absorbs the irradiation $I(z,t)$ but does not thermally insulate the pond. For purposes of analysis we account for the thermal effects of the convective zone by measuring the depth in the pond from the bottom of the convective zone and replace the coefficients A and B in Eq. (10.29) by

$$A_i' = A_i \exp - (-a_i d_s)$$
$$B_i' = B_i \exp (-b_i d_s)$$

where d_s is the depth of the convective zone. The analytic procedure of the previous sections is then unchanged.

The numerical values of A and B used in the previous analysis as for approximating the irradiation of Figs. 10.4 to 10.6 have, in fact, assumed a 0.05-m convective zone near the surface of the pond.

Numerical analysis of the thermal diffusion equation   In order to account for variations of K, ρ, and c with respect to depth, to allow for convective regions of various thicknesses in the pond, and to experiment with different heat extraction possibilities, one requires a numerical method for solving Eq. (10.30).

A finite difference approximation to Eq. (10.30) is given by

$$
T_{l,m+1} = \frac{\Delta t}{\rho_{l,m} c_{l,m} \, \Delta z^2} \left[\frac{1}{4} (K_{l+1,m} - K_{l-1,m})(T_{l+1,m} - T_{l-1,m}) \right.
$$
$$
\left. + K_{l,m}(T_{l+1,m} - 2T_{l,m} + T_{l-1,m}) + (I_{l-1/2} - I_{l+1/2}) \, \Delta z \right] + T_{l,m} \tag{10.46}
$$

where $l \, \Delta z = z$
$\quad\quad m \, \Delta t = t.$

From the values of $T(z,t)$, $T(z + \Delta z, t)$, and $T(z - \Delta z, t)$ and from the values of ρ, c, and K appropriate for these temperatures, we calculate the temperature at $T(z, t + \Delta t)$. Temperatures as a function of time and depth are thus calculated on a step-by-step basis. It is useful to have $I(z,t)$, K, ρ, and c available in functional form. These parameters can, however, be stored as look-up tables in the computer memory.

A sufficient condition so that rounding-off errors will not grow and cause the step-by-step solution to become unstable is that

$$
\frac{K}{\rho c} \frac{\Delta t}{\Delta z^2} < 0.5
$$

For $\Delta z \simeq 0.1$ m, $K \simeq 0.6$ and $\rho c \simeq 4 \times 10^6$, a suitable time increment is $\Delta t \simeq 2.16 \times 10^4$ s ($= 6$h). The boundary condition at the top of the pond is the ambient temperature. At $z = h$, $I_{l+1} = 0$; i.e., that all the radiation reaching the bottom of the pond is absorbed there. There are numerical methods for considering an "infinite" slab of ground. It is probably sufficiently accurate to assume a constant temperature boundary at a depth of approximately 10h.

We treat energy extraction as follows: if, after an iteration, the temperature T_h at the bottom of the pond is greater than T_w, then we let the bottom temperature equal T_w and the power extracted is

$$
U(h,t) = \rho c \frac{\Delta z}{\Delta t} (T_w - T_h)
$$

Figure 10.11 shows a numerical experiment for which K, ρ, and c were allowed temperature and salinity dependence. The properties of the ground were assumed equal to those of the bottom solution and the insolation given by Eq. (10.29). Energy is extracted whenever the temperature at $h = 1.25$ m is greater than 100°C.

A 0.1-m convective zone was assumed at the top of the pond. Numerical evaluation of a convective zone is treated by approximating the analytic expression for the convective zone [Eq. (10.43)] by

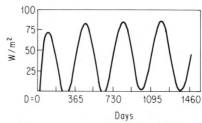

Fig. 10.11   Rate of energy extraction at 100°C as a function of time after exposure to insolation for a 1.35-m-deep pond with a 0.1-m convecting zone at the pond surface (at 32° latitude). The values given in the above illustration were calculated numerically by means of Eq. (10.46).

$$T_{l,m+1} = \frac{z\,\Delta t}{2\rho\,cd\,\Delta z}\,[(K_{l+2,m}+K_{l+1,m})(T_{l+2,m}-T_{l+1,n})$$
$$- (K_l + K_{l+1})(T_{l,m} - T_{l-1,n}) + (I_{l,m} - I_{l+1,m})\,\Delta z\,] + T_{l,m} \tag{10.47}$$

and $T_a = T_{l+1}$

$$\text{where } l\,\Delta z = z$$
$$l\,\Delta z + d = z + d$$
$$T(z+d) = T_{l+1}$$

For the nonconvecting regions of the pond we perform the step-by-step iterations using Eq. (10.46) and the analysis of the convective regions using Eq. (10.47). The regions are connected by the common values of T_l and T_{l+1} at the extremes of the convective region.

ENERGY EXTRACTION METHODS

Two methods have been proposed for the extraction of energy from the solar pond. The first is to place a heat-exchange system at the bottom of the pond. The second method is to selectively extract the bottom layer of heated brine through an exit port, remove the absorbed energy in an external heat exchanger, and return the cooled brines to the pond by means of an entrance port. This second method is intrinsic to the physical nature of hydrodynamic flow in the solar pond. In a liquid system possessing a stable density gradient, selective flow of the bottom layer can be maintained without requiring a mechanical separation between the flowing and stable regions of the system.

Extraction by Means of Heat Exchanger

The design of the heat exchanger placed at the bottom of the solar pond has been considered by M. S. Hipsher and R. F. Boehm.[15] The method suffers from two disadvantages. The first is the quantity of tubing required for efficient heat transfer, and the second is the difficulty in locating and repairing a damaged element of the heat exchanger at the pond bottom. We will therefore not consider this method of heat extraction in detail.

Stratified Flow in Solar Ponds

Experimental studies of stratified flow for the flow velocities and gradients expected in solar ponds indicate that a stable separated flow system is established when the Froude number Fr is unity, i.e.,

$$\text{Fr} = \frac{V}{d^2}\left(\frac{2\rho}{g\,\partial\rho/\partial z}\right)^{1/2} = 1 \tag{10.48}$$

and where V is the flow per unit width of the pond and D is the height of the mixed layer.[16] Equation (10.48) is valid for Reynolds number $\text{Re} = Vd/\nu < 1000$. V is related to the rate of heat extraction by

$$U = \frac{V\rho\,cd\,\Delta T}{L} \tag{10.49}$$

where ΔT is the temperature drop in the heat exchanger and L is length of the pond. Insertion of Eq. (10.49) into Eq. (10.48) yields

$$d^3 = \frac{Lu}{\rho c\,\Delta T}\left(\frac{2\rho}{g\,\partial\rho/\partial z}\right)^{1/2} \tag{10.50}$$

A numerical example is enlightening:

For $L = 1000$ m

   $\rho c = 7.4 \times 10^5\,\text{J/m}^3\,°\text{C}$

   $\rho = 1.2 \times 10^3\,\text{kg/m}^3$

   $U = 100\,\text{W/m}^2$

   $\Delta T = 10°\text{C}$

and $\dfrac{\partial \rho}{\partial z} = 200 \text{ kg/m}^4$

then $d = 0.12$ m

Equation (10.48) is based on experiments for which there was no horizontal density gradient superimposed on the vertical density gradients. A horizontal gradient may be presumed to exist due to the temperature difference ΔT across the pond effected by the heat extraction process.

In stratified flow extraction experiments in solar ponds, a difference between outflow and inflow temperatures of 10°C did not influence the height of the mixed flowing layer by more than a factor of two from that predicted by Eq. (10.50).

WIND-EFFECTED MIXING OF THE POND

Wind shear over a large body of water generates waves and surface drift. The kinetic energy of the waves and surface drift is partly converted into potential energy by mixing lighter upper-layer solution with denser lower-layer solutions and partly dissipated by viscosity.

A satisfactory theory of wind-effected mixing in a stratified liquid has not yet been developed. All that can be said with certainty is that the extent of the convective layer d_s does not grow more rapidly than[17]

$$d_s \leqslant 1.1 \times 10^{-2} \times \left(\frac{\rho \int v^3 \, dt}{g \; \partial \rho / \partial z} \right)^{1/3} \tag{10.51}$$

where v is the wind velocity and the numerical factor is formed from theoretical considerations and measurements of the dependence of the power transferred to the pond as a function of v^3. The units of the numerical factor are $\text{kg}^{1/3} / \text{m}^{1/3}$.

The arguments which are used in the derivation of the above equation do not include the effects of viscous dissipation. It may be that d_s is asymptotically limited, but we have neither theory nor sufficient experimental evidence to support this optimistic conjecture. If we examine the naturally occurring solar ponds, we observe that they are protected from prevailing winds by surrounding geographical features. Successful solar ponds in Israel have also been protected from the prevailing winds.

How to protect the pond from wind effects economically and effectively without simultaneously attenuating the incident insolation is still an unsolved problem. One way not recommended for the solution of this problem is to float a plastic sheet above or just below the water surface. In an experiment where this was tried, the wind-borne dirt and sand of a desert environment settled on the plastic sheet, which became opaque to light within 3 days.

GEOPHYSICAL AND ECOLOGICAL FACTORS

The construction of a solar pond on a land site requires the careful consideration of numerous interrelated economic, ecological, and geophysical factors. For economic reasons, we require that the solar pond be situated on relatively sterile land of no mineral importance and near salt and brackish water supplies which have otherwise little value. The land site should be relatively flat and easily worked so that a major earth-moving operation is not required for the construction of the solar pond. The site should not overlie too closely an underground aquifer. If the aquifer is of fresh water, then an accidental leakage of the brine from the pond bottom would seriously pollute a valuable water resource. Even if the underground aquifer contains brackish water, it can pose serious thermal and structural problems for the pond. If the underground flow is sufficiently rapid, then heat will be removed from the pond; i.e., the underground insulation is no longer "infinite" but limited in depth approximately to the aquifer surface.

The situation for a stagnant aquifer is even more serious. As the aquifer is heated to temperatures greater than 50°C, copious quantities of dissolved gases may be released from the aquifer solution. This appears to have occurred in an experimental pond constructed in Atlit, Israel, where the released CO_2 developed sufficient pressures to destroy the packed clay-bed structure of the pond bottom. Thus a stagnant underground aquifer could destroy an otherwise well-designed pond.

Other geophysical requirements are that the underlying earth structure should be homogeneous and free of stresses, strains, and fissures. If the underground structure is not homogeneous, then increases in temperature will cause differential thermal expansions which could result in earth movements. Similarly, the existing stresses and strains in the base rock have been in temperature and pressure equilibrium for several thousand years. The sudden warming of this structure can change the equilibrium conditions which could result in earth movements which might destroy the solar pond.

If an existing body of saline water is converted into a solar pond, then we must assure that the food chain of the surrounding flora and fauna are not disturbed by the disruption of the life cycle within the lake.

One problem which has not as yet been observed in artificial solar ponds is the growth of thermophilic and halophilic algae which prosper in naturally occurring solar ponds. These algae strongly absorb solar radiation.

INSTRUMENTATION OF A SOLAR POND

In order to monitor the stability, collection efficiency, and clarity of the solar pond, measurements of temperature, salinity, and insolation within the pond, and measurements of temperature, humidity, insolation, and wind velocity outside the pond are required. Temperatures are easily recorded with the help of the thermocouples or thermistors placed at 0.1-m intervals from the top to the bottom of the pond and at intervals of 0.3 m within the ground beneath the pond to a depth of about 2 m. In order to avoid collection of too much data, the recordings should be compatible with computer data processing techniques. Integrated hourly temperature measurements give an excellent indication of pond stability.

Salinity measurements are made by extracting liquid from chosen levels in the pond and measuring with a picnometer. After correction for the difference between the temperature of measurement and temperature at the level of extraction, accurate density and salinity profiles can be calculated with the help of density, salinity, and temperature graphs such as those shown in Fig. 10.2a and b.

It is more difficult to measure the insolation within the pond. Commercial spectrally flat radiometers are not designed to function in a hot saline liquid environment. One possibility is to use encapsulated silicon detectors. Silicon cells which are flat from the visible to 1 μ in the infrared are available commercially. These cells are insensitive to temperature if operated in the short-circuit mode. A knowledge (or assumption) of the sun's spectral composition allows a compensatory estimate to be made for that part of the radiation which reaches a particular depth but is outside the sensitivity of the silicon cell. The correction factor required for silicon cells as a function of hour angle and depth in the pond is the subject of a thesis by M. Kahn.[18] Measurements of temperature, humidity, and wind velocities outside the pond allow estimates of the heat losses from the surface of the pond and the correlation of the depth of the convective layer near the surface with the wind energy.

SPECIFIC APPLICATIONS

Heating

This is straightforward apart from the variations in output—from month to month—discussed earlier. If energy collected in summer is needed for winter use, a large storage capacity, i.e., a deep lower mixed zone is needed. Where no attempt is made to incorporate seasonal storage, the mixed extraction zone will generally provide enough storage to cover any reasonable succession of nonsunny days.

Cooling

Absorption machines are available which are capable of working at temperatures below 100°C, though the efficiency and, in particular, the capacity of a given machine fall rather rapidly as the input temperature is reduced. These characteristics depend upon the manufacturer.

Given the curves for the absorption machinery, the optimum operating temperature is that which results in the lowest total cost of pond plus cooling machine. For the case where the mean daily pond yield during the summer follows fairly closely the cooling load, the total cost is obtained as follows: Let

C_a = cost of the cooling machine per "nominal" rated ton of cooling
C_p = cost of pond per m^2
R = derating factor ($R < 1$) of cooling machine below nominal rating due to reduced operating temperature
COP = coefficient of performance at the chosen operating temperature
U = mean power output from pond, kW/m^2

Then the cost of the cooling machine per "actual" ton of cooling, due to derating is C_a/R. For a pond of area A (m^2) whose cost is AC_p, the mean yield is AU (kW). Then the mean cooling yield is

$$AU\,\text{COP(kW)} = AU\,\text{COP}/3.488 \text{ (tons)}$$
$$= L \qquad \text{the mean cooling load, tons (3.5 kW)}$$

Thus pond area required is

$$A = \frac{3.488L}{U\,\text{COP}}$$

and the total cost of pond plus cooling machine is

$$AC_p + \frac{LC_a}{R} = L\left(\frac{3.488C_p}{U\,\text{COP}} + \frac{C_a}{R}\right)$$

Remembering the U, COP, and R are all temperature-dependent, the operating point is chosen that yields a minimum value for the expression $3.488/U\text{COP} + r/R$, where $r \equiv C_a/C_p$. If the annual interest and amortization charges on the pond and the cooling machine are not equal, use r' for r, where $r' = C_aH_c/C_pH_p = rH_c/H_p$;

H_c = annual charges on cooling machine

H_p = annual charges on pond

both expressed as fractions (or percentages).

If the pond is used for both heating in winter and cooling in summer as, for example, in the heating and cooling of buildings, output in winter can be improved somewhat, as a lower operating temperature may be adequate.

Power Production

The theoretical efficiencies of solar pond operation for power production are given by Eq. (10.40). The effect of convective zones in damping seasonal temperature variations and allowing energy extraction from the pond at a fairly constant rate are described by Eq. (10.45).

The temperatures achieved in solar ponds are considerably greater than the ocean temperatures presently being studied for energy exploitation.

The results of the ocean thermal energy conversion studies indicate that there will not be serious engineering problems in the design and manufacture of a power production plant to operate between the temperatures achieved at the bottom of the solar pond and ambient temperatures.

REFERENCES

1. A. v. Kalecsinsky, Ungarische Warme und Heisse Kochsalzeen, *Ann. D. Physik*, **7** (4): 408–416 (1902).
2. C. G. Anderson, Limnology of a Shallow Saline Meromitic Lake, *Limnology and Oceanography*, 3:259–269 (1958).
3. F. D. Por, Solar Lake on the Shores of the Red Sea, *Nature*, **218**:860–861 (1970).
4. P. P. Huder and P. Sonnenfeld, Hot Brines on Los Roques, Venezuela, *Science*, **185**:440–442 (1974).
5. A. T. Wilson and H. W. Wellman, Lake Vanda, an Antarctic Lake, *Nature*, **196**:1171–1173 (1962).
6. Y. Cohen, W. Krumbein, and M. Shilo, Solar Lake (Sinai), *Limnology and Oceanography*, **22**:609–634 (1977).
7. G. Veronis, Finite Amplitude Instability in Thermohaline Convection, *J. Marine Res.*, **23**:1–17 (1965).
8. N. Chepurniy and S. B. Savage, Effect of Diffusion on Concentration Profiles in a Solar Pond, *Solar Energy*, **17**:203–205 (1975).
9. G. Assaf, The Dead Sea: A Scheme for a Solar Lake, *Solar Energy*, **18**:294–299 (1976).
10. W. Stiles, Indicator Method for the Determination of Coefficients of Diffusion, *Proc. Roy. Soc.*, **103**:260–275 (1923).
11. "Smithsonian Meteorological Tables," 6th rev. ed., pp. 420–421. U.S. Gov't Printing Office, Washington, D. C., 1948.

12. G. K. Clarke and H. R. Jones, Absorption of Light by Sea Water, *J. Opt. Soc. Amer.*, **29**:43–53 (1939).
13. H. Weinberger, The Physics of the Solar Pond, *Solar Energy*, **8**:45–56 (1964).
14. A. Rabl and C. E. Nielsen, Solar Ponds for Space Heating, *Solar Energy*, **17**:1–12 (1975).
15. M. S. Hipsher and R. F. Boehm, "Heat Transfer Considerations of a Non Convecting Solar Pond Heat Exchanger, ASME publ. 76-WA/Sol 4.
16. D. G. Daniels and M. F. Merriom, Fluid Dynamics of Selective Withdrawal in Solar Ponds, *1975 Intern. Solar Energy Soc. Cong.*, University of California, Los Angeles, Session 35.
17. H. Kato and O. M. Phillips, On the Penetration of a Turbulent Layer into Stratified Fluid, *J. Fluid Mech.*, **37**:643–655 (1969).
18. M. Kahn, "Broadband Underwater Radiometer," B. S. thesis, Jerusalem College of Technology, Jerusalem, 1977.

Chapter 11

Solar Water Heating

GARY SKARTVEDT
American Heliothermal Corporation,
Denver, CO

INTRODUCTION

Heating of water for bathing, washing, or commercial purposes is one of the oldest and most cost-effective uses of solar energy. The temperature levels required (40–60°C) can be produced efficiently by simple, relatively inexpensive collection devices. In addition, the demand for hot water tends to be uniform throughout the year, unlike space-heating demand which peaks during the winter months when solar energy is least available. The combination of moderate temperature requirements and uniform annual demand makes the production of hot water for domestic, commercial, and industrial uses a particularly attractive application for solar energy.

Historical Perspective

Solar heating of domestic water has occurred on a significant scale in the United States, Israel, Japan, and Australia. Solar domestic hot-water (DHW) systems appeared in Florida during the 1920s. The comparative costs of simple thermosiphon solar systems versus available alternate fuels in Florida created a viable solar hot-water heating industry. Thermosiphon systems were installed at an increasing rate until 1941 when it was estimated that approximately 60,000 installations had been made in the Miami area.[1] Sales of solar water heaters steadily decreased after 1941 until only a few new installations per year were made during the 1960s. The decay of the industry is attributed to three factors. First, the solar system lost its economic advantage as electric energy costs decreased and the cost of the solar equipment increased. Second, negative homeowner attitudes arose as steel storage tanks developed leaks. Third, large-scale builder-developers were reluctant to add the cost of a solar system to a house in a price-competitive market. The convergence of these three factors during the 1950s made the solar hot-water industry essentially dormant in Florida until renewed activity was spurred in the mid-1970s by rising energy costs and concerns about the long-term availability of fossil fuels.

Thermosiphon solar hot-water systems have been widely used in Israel since the early 1960s. These systems, consisting of two 1 × 2-m flat-plate collectors surmounted by a vertical cylindrical tank, are a prominent feature of the skyline in most Israeli cities. More than 100,000 small thermosiphon systems were installed in Israel between 1960 and 1975.

Energy Demand for Water Heating in the United States

A significant quantity of energy is consumed in the United States to heat water for domestic, commercial, and industrial purposes. The primary domestic uses of hot water include bathing, clotheswashing, and dishwashing. In addition, hot water is required for a variety of commercial and industrial applications. These include such diverse uses as electroplating, photographic processing, pulp manufacturing, dishwashing, degreasing, sterilization, etc.

Figure 11.1[2] shows the distribution of total energy consumption in the United States in 1973. These data indicate that approximately 4.1 percent of the total energy demand was used to heat water for residential and commercial use. This is roughly equivalent to 530 million barrels of oil per year or 20 percent of the oil imported by the United States in 1976. The supply of even 10 percent of the 1973 domestic and commercial hot-water requirement by solar systems would

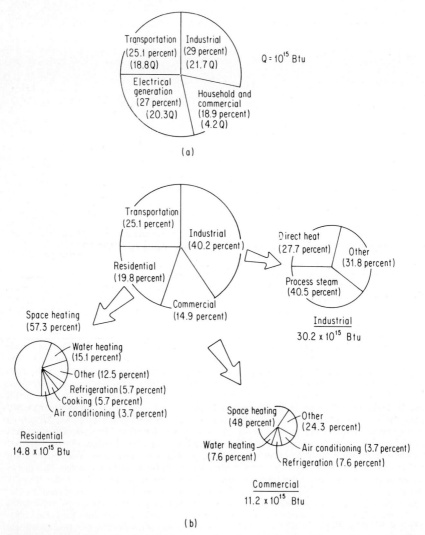

Fig. 11.1. Distribution of U.S. energy consumption, 1973 (1 Btu = 1.055 kJ). (a) Distribution of primary energy consumption; (b) distribution of end-use energy consumption (75 x 10^{15} Btu/year).

require approximately 90,000,000 ft² (9,000,000 m²) of flat-plate solar collector. These figures are stated simply to illustrate the enormous potential market for solar hot-water systems where such systems are economically competitive with available energy sources.

BASIC SYSTEM CONFIGURATIONS

A variety of techniques, ranging from very simple to complex, have been employed to produce hot water using solar energy. A blackened container filled with water and placed in sunlight represents possibly the simplest approach to a solar hot-water system.

Thermosiphon Systems

Almost as simple, and considerably more convenient and efficient, is the thermosiphon system shown in Fig. 11.2. This system consists of a storage tank placed above a flat-plate solar collector. As the sun heats the water in the collector, it becomes warmer (and less dense) than the water in the tank above. Buoyancy forces create a continuous convective circulation of water from the bottom of the tank to the bottom of the collector, up through the collector flow passages,

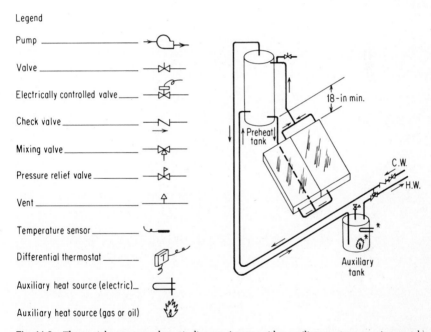

Fig. 11.2.   Thermosiphon system schematic diagram. * means either auxiliary energy source is acceptable.

and back into the top of the tank. Almost all solar hot-water systems in use prior to 1976 were of the thermosiphon variety. The systems were generally used in climates which do not have extensive freezing weather, and they were often not used in conjunction with an auxiliary system which would provide hot water on cloudy days. Although thermosiphon systems are simple and effective, the requirement for the storage tank to be above the collectors has proven to be a significant architectural and aesthetic deterrent to their use in many areas. This factor, plus the difficulty of preventing damage due to freezing in severe climates, has led to the development of several types of solar hot-water systems which use pumps rather than natural, convective circulation to force water (or air) through the solar collectors.

Forced-Circulation Systems

A forced-circulation, or pumped, system may be either "direct" or "indirect." A direct system is one in which the potable or service water is heated by circulation through the solar collectors.

In an indirect system, a fluid (such as antifreeze, air, distilled water, or an organic heat-transfer fluid) other than the potable or service water is circulated through the collectors; energy is transferred to the heated water through the collectors; and finally, energy is transferred to the water to be heated through some form of heat exchanger. Direct systems are generally used in climates which do not impose freezing problems, although methods such as draining collectors at night, trickle circulation, flexible collector waterways, etc., may be used to prevent expansion damage due to freezing in a direct system. An indirect system design permits use of a nonfreezing fluid in the collector circuit (or any of the methods available to the direct system) to circumvent the freezing problem, but the required heat-exchange process imposes a system thermal performance penalty–all other factors being equal.

Direct Systems

The simplest type of direct forced-circulation DHW system is the single-tank configuration shown in Fig. 11.3. In this system, tap water is circulated through the collectors and stored in a single tank which contains the auxiliary heating source (normally an electric element). As will be discussed in the next section, efficient operation of this system depends on maintaining

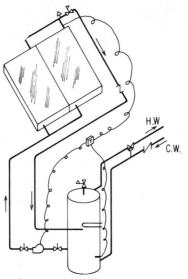

Fig. 11.3. Direct solar water heating with a single tank. See Fig. 11.2 for explanation of symbols.

thermal stratification in the tank. For this reason, the auxiliary heating element should be placed near the top of the tank, which precludes the use of oil or gas as the auxiliary energy source. A direct system with a preheat tank and separate auxiliary tank is shown in Fig. 11.4. Thermal efficiency of this system does not depend on stratification in the auxiliary tank; so gas, oil, or electricity may be used as the auxiliary energy source.

Since tap water circulates through the collectors in a direct system, methods of preventing damage due to freezing are limited. In mild climates, the pump may be turned on to cause circulation of warm water in the collectors when the ambient outside temperature drops below freezing. In a severe climate, this procedure would result in excessive energy loss and a significant risk of a freezing condition coincident with a power failure. Automatic draining of direct-system collectors and outside piping whenever the pump is not operating can be accomplished using solenoid valves, as shown in Fig. 11.5. This arrangement can be made fail-safe with respect to a power outage, but an undetected failure to close of either of the drain valves could cause a significant waste of water.

In addition to the freeze prevention problem, the presence of tap water in the collectors in a direct system imposes a constraint on the choice of materials which may be used for the collector waterways.

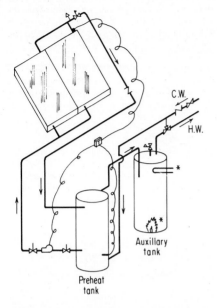

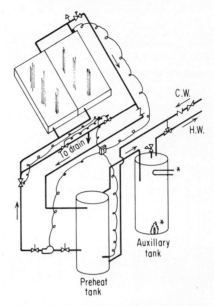

Fig. 11.4. Direct solar water heating with preheat tank and separate auxiliary tank. See Fig. 11.2 for explanation of symbols. * means either auxiliary energy source is acceptable.

Fig. 11.5. Direct solar water heating with preheat tank and automatic draindown. See Fig. 11.2 for explanation of symbols. * means either auxiliary energy source is acceptable.

Indirect Systems

A simple type of indirect DHW system is shown in Fig. 11.6. These systems are analogous to the single-tank direct system except that a heat exchanger is used to transfer solar energy into the lower portion of the tank. As will be shown in the section on thermal performance, proper design of the heat exchanger is essential to achieving high thermal efficiency. Since temperature stratification in the tank is also important to thermal efficiency, the auxiliary element should be placed in the upper part of the tank, precluding the use of oil or gas as a backup energy source.

A two-tank indirect system is shown in Fig. 11.7. Although a jacket heat exchanger is illustrated, the internal coil or external, pumped, shell-and-tube heat exchangers shown in Fig. 11.6 may also be used. The separate auxiliary tank permits use of electricity, oil, or gas as the backup energy source.

An indirect drain-down system can be configured as shown in Fig. 11.8. This concept requires an additional tank and pump compared to the direct drain-down system shown in Fig. 11.5 but the possibility of waste of tap water due to a valve failure is eliminated.

An indirect DHW system which employs air as the collector circulating medium is shown in Fig. 11.9. This concept may be used either with a single stratified tank or with separate preheat and auxiliary tanks as pictured. The use of air in the collectors and outside ducting eliminates freezing problems as well as the potential hazard of contamination of potable water with antifreeze or other nonfreezing heat-transfer liquids. The reduced efficiency of air as a heat-transfer fluid compared to liquids, however, will typically require a greater collector area to achieve a given level of thermal performance than would be needed in a well-designed liquid system.

Combined Collector-Storage Systems

A type of indirect DHW system which uses a heat pump to extract energy from a combination collector and storage tank is shown in Fig. 11.10. The water stored in the collector-tank serves as a thermal sink for the evaporator side of the heat pump. If the collector-tank is designed to permit freezing of the stored water without damage, the heat of fusion available from the ice-making process provides a large thermal-storage effect. This maintains a minimum heat-sink

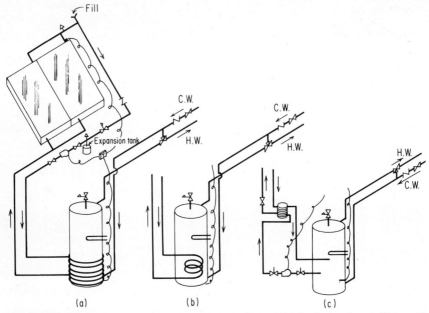

Fig. 11.6.   Indirect solar water heating with a single tank; (a) external jacket heat exchanger; (b) internal coil heat exchanger; (c) external shell and tube heat exchanger with pump. See Fig. 11.2 for explanation of symbols.

temperature of 0°C (32°F) for the heat pump even though the ambient temperature may drop well below freezing for considerable periods of time. Whenever the ambient temperature is above freezing, the collector draws heat from the surrounding air, rain, and moisture condensation as well as by absorbing the incident solar energy.

Heat-Pipe Systems

Figure 11.11 shows a solar water-heating system proposed by F. deWinter. The auxiliary tank is heated by a heat pipe activated by any auxiliary fuel. Because of the heat-diode characteristic of heat pipes, no heat is lost during periods of no-auxiliary use. In addition, the pipe connecting the solar preheat tank and the auxiliary tank acts as a heat diode very effectively separating these two tanks thermally, although they are part of the same assembly.

SYSTEM DESIGN CONSIDERATIONS

The objective of any solar hot-water heating system is to harvest free solar energy over an extended period of time with a minimum of maintenance and repair. As is the case with most electromechanical systems, achievement of an "optimum" system configuration involves a number of design compromises. For example, a pumped system may be more efficient thermally than a thermosiphon design, but over several years the simpler thermosiphon system may be more cost-effective because of lower maintenance costs. The following discussion attempts to define the significant solar hot-water system design issues and to provide guidance, both quantitative and qualitative, toward achieving effective design solutions for particular applications.

Thermal Performance

In order to determine the relative thermal performance of various solar hot-water system design concepts, detailed computer simulations were developed and exercised in three different U.S. climate areas (Boston, Denver, and Los Angeles). A brief description of the analytical model is presented so that the comparative system performance results can be viewed in proper perspective.

Simulation models Finite-element analytical models were constructed for the more commonly used liquid-collector forced-circulation systems. A typical nodal diagram of an indirect heating system is shown in Fig. 11.12. Annual performance of a particular configuration, i.e.,

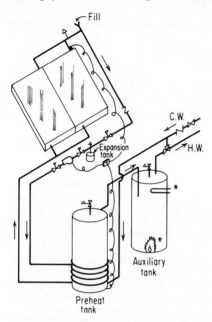

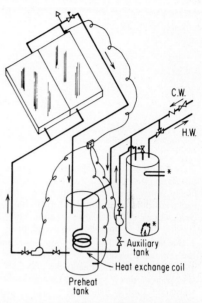

Fig. 11.7. Indirect solar water heating with a preheat tank and a separate auxiliary tank. See Fig. 11.2 for explanation of symbols. * means either auxiliary energy source is acceptable.

Fig. 11.8 Indirect solar water heating with vented (low pressure) automatic drain-down freeze protection. See Fig. 11.2 for explanation of symbols. *means either auxiliary energy source is acceptable.

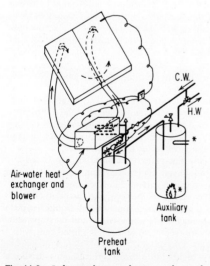

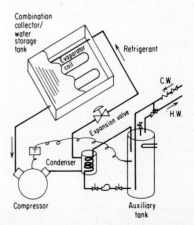

Fig. 11.9. Indirect solar water heating with air collector and air/water heat exchanger. See Fig. 11.2 for explanation of symbols. * means either auxiliary energy source is acceptable.

Fig. 11.10. Indirect solar water heating using electric heat pump and combined collector/storage unit which is permitted to freeze. See Fig. 11.2 for explanation of symbols.

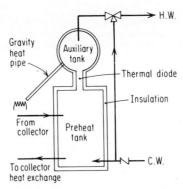

Fig. 11.11. deWinter water heating system using a heat pipe for backup interface and a heat diode to isolate the preheat and auxiliary tanks to prevent mixing.

the solar energy supplied to the hot water during a typical climatic year, was determined by calculating the temperatures of each node and the heat flows between nodes on an hourly basis for an entire year. The temperature and heat flow calculations were performed using the backward-differencing network relaxation option of the MITAS* thermal analyzer computer code. Details of the analytical treatment of the weather, hot-water loads, and system components are described in the following paragraphs.

Weather Data. The hourly insolation incident on the solar collectors was calculated using the approach presented in "Algorithms for Building Heat Transfer Subroutines," ASHRAE, 1975. This method estimates the effects of cloud cover on clear sky radiation using an hourly sky opacity index. The opacity data were obtained from magnetic tapes provided by the National Climatic Center at Asheville, NC. In order to check the accuracy of the procedure, the average daily insolation on a horizontal surface was computed for each month using the opacity factor. These values were compared to the *Climatic Atlas of the United States* (U.S. Gov't. Printing Office, 1968) monthly average values for the three selected cities (Boston, Denver, and Los Angeles). The initial computed results agreed well with the *Climatic Atlas* data for Denver and Los Angeles, but the computed values for Boston were significantly higher–as much as 20 percent in the winter months–than the *Climatic Atlas* data. An arbitrary adjustment factor was incorporated in the Boston calculations to bring the computed average daily values in closer agreement with the *Climatic Atlas* data. The resulting daily average insolation on a horizontal surface for each month computed using the method employed in the computer simulations and the corresponding values from the *Climatic Atlas* are listed in Table 11.1 for the three selected cities.

Hourly ambient temperature values were taken directly from the weather tapes. Wind was assumed constant at 7.5 mi/h.

Hot-Water Load Cycle. A hot-water load cycle which integrates to a daily total usage of 87 gal (330 L), typical of a single-family residence with four people was used for all of the comparative system performance analyses. The monthly average tap temperatures used for each city are listed in Table 11.2. A constant delivery temperature of 120°F (49°C) was assumed for all cases to determine the amount of energy supplied by the auxiliary or backup heater.

Solar Collector Model. The type of solar collector (flat-plate, concentrating, etc.) used in a solar hot-water system obviously may have a pronounced effect on system performance and cost-effectiveness. The objective of the performance investigations described herein, however, is to demonstrate typical system design sensitivities rather than to evaluate the relative merits of various collector configurations. To this end, a standardized conventional liquid flat-plate collector was assumed for all simulations. The collector efficiency curve and corresponding thermal characteristics shown in Fig. 11.13 are representative of a single-glazed unit with a selective absorber surface.

Hot-Water Storage Tank. The storage tank was divided into three vertical nodes, as shown in Fig. 11.12 to permit at least an approximate evaluation of the effects of temperature stratification. As will be shown in the following section on annual performance, stratification can

*Developed by the Martin-Marietta Corporation. Denver, CO.

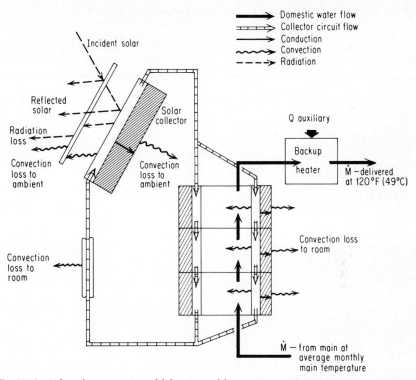

Fig. 11.12.   Indirect heating system nodal diagram used for computer simulation.

have a significant impact on thermal performance. The base-line 82-gal (310-L) tank configuration is shown in Fig. 11.14. Although a variety of heat-exchanger configurations may be employed in indirect heating systems, the external tank-jacket configuration was selected for the computer simulations because of its widespread use in commercially available solar domestic water systems.

Heat-transfer coefficients inside the jacket were computed by assuming the Nusselt number equal to 5.0 from Kays[4] for laminar flow between infinite parallel plates with one plate adiabatic.

TABLE 11.1   Monthly Average, Horizontal Insolation, Langleys/day

Month	Denver		Los Angeles		Boston	
	Climatic Atlas	Weather tape	*Climatic Atlas*	Weather tape	*Climatic Atlas*	Weather tape
Jan.	201	247	243	242	129	148
Feb.	268	296	327	272	194	207
Mar.	401	406	436	447	290	298
Apr.	460	513	483	506	350	359
May	460	554	555	559	445	501
June	525	579	584	542	483	515
July	520	542	651	621	486	463
Aug.	439	511	581	579	411	426
Sept.	412	420	500	463	334	351
Oct.	310	367	362	335	235	251
Nov.	222	226	281	306	136	159
Dec.	182	190	234	263	115	112

TABLE 11.2   Tap Water Temperature and Ambient Temperature Profiles, °F*

Month	Average Denver temperatures		Average Los Angeles temperatures		Average Boston temperatures	
	Ambient	Tap	Ambient	Tap	Ambient	Tap
Jan.	30.1	42	53.1	50	31.7	32
Feb.	27.3	45	57.9	50	29.1	36
Mar.	33.0	48	58.6	54	38.1	39
Apr.	46.9	50	58.8	63	45.3	52
May	59.6	52	62.3	68	60.3	58
June	65.2	55	68.2	73	66.5	71
July	75.9	56	70.6	74	70.5	74
Aug.	70.4	52	71.2	76	66.6	67
Sept.	62.4	48	67.5	75	61.6	60
Oct.	51.9	45	65.0	69	52.1	56
Nov.	39.3	43	59.1	61	44.2	48
Dec.	33.0	40	59.5	59	32.6	45

*°C = $\frac{5}{9}$(°F − 32)

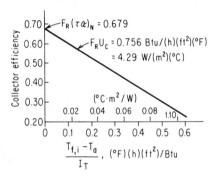

Fig. 11.13.   Characteristics of a liquid flat plate solar collector used in computer simulations; assumed parametric values include single glazing of low-iron sheet glass transmittance = 0.86, selective absorber surface: solar absorptance = 0.92, IR emittance = 0.19, three inches back surface insulation (R9).

The internal free-convection boundary-layer mass flow, mixed temperature, and heat-transfer coefficients on the inside tank wall can be estimated using the following equations from Ede.[5]

$$\dot{m} = \frac{CU\delta\rho}{12} \qquad (11.1)$$

where C is the circumference

$$U = 5.17\nu\left(\mathrm{Pr} + \frac{20}{21}\right)^{-1/2}\left(\frac{g\beta\theta_w}{\nu^2}\right)^{1/2}x^{1/2} \qquad (11.2)$$

$$\delta = 3.93\mathrm{Pr}^{-1/2}\left(\mathrm{Pr} + \frac{20}{21}\right)^{1/4}\left(\frac{g\beta\theta_w}{\nu^2}\right)^{1/4}x^{1/4} \qquad (11.3)$$

$$\mathrm{Nu}_x = 0.51\mathrm{Pr}^{1/2}\left(\mathrm{Pr} + \frac{20}{21}\right)^{-1/4}\mathrm{Gr}x^{1/4} \qquad (11.4)$$

Thermal tests were conducted on a jacketed 82-gal (310-L) tank similar in configuration to that shown in Fig. 11.14. The test results produced overall jacket-to-tank heat-transfer coefficients in the range of 14 to 19 Btu/h·ft²·°F (79–103W/m²·°C) tor flows both upward and

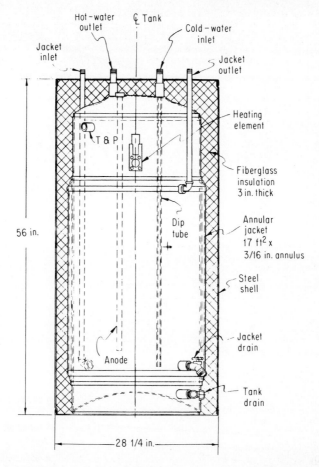

Fig. 11.14.   Cross section of a typical 82-gal solar hot water storage tank with annular heat exchanger.

downward through the jacket using 50% water and 50% propylene glycol in the jacket. As shown in Fig. 11.16, flow downward through the jacket resulted in somewhat higher heat-transfer coefficients, particularly at the lower flowrates.

Equations (11.1) to (11.4) predict a jacket-to-bulk heat-transfer coefficient of 15 Btu/h·ft²°F (85W/m²·°C) in the midrange of the test conditions, in good agreement with the test results.

Additional tests were conducted on the 82-gal jacketed tank to determine if the heated boundary-layer flow up the inside tank wall created significant mixing (destratification) of the bulk water inside the tank. A typical temperature-time history, shown in Fig. 11.15, indicates that the heated boundary layer apparently does not penetrate a layer of water warmer than itself and, therefore, does not cause mixing of the warmer layer with the colder layer below it.

Based on these results, the analytical model was constrained such that boundary-layer flow from the bottom or middle tank nodes could only enter a higher node when the boundary-layer mixed temperature exceeded the bulk temperature of the higher node.

Hot-Water Storage-Tank Losses. The energy loss to the surroundings from even a well-insulated hot-water storage tank can amount to a significant fraction of the energy required to heat the water over a year. For example, an 82-gal tank, insulated with 3 in of fiberglass, requires approximately 290,000 Btu/mo (306 MJ/mo) energy input simply to maintain the water temperature at 120°F (49°C) in a 70°F (21°C) room (no hot-water usage). This amounts to 16

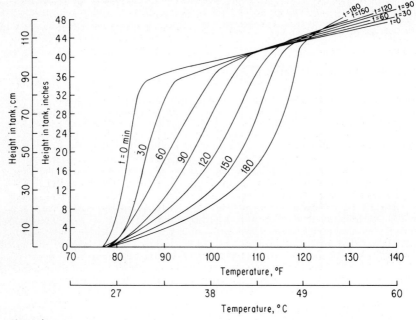

Fig. 11.15. Measured time-temperature history on tank centerline. Jacket flow from bottom to top; constant heat input to jacket of 4600 Btu/h (1.35 kW).

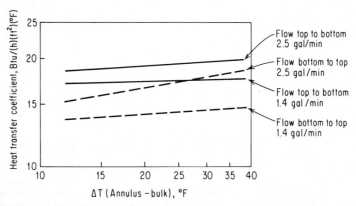

Fig. 11.16. Measured heat transfer coefficients for 82-gal solar water heater tank jacket. [1 Btu/(h) (ft²) (°F) = 5.67 (W/m²) (°C), 1 gal/min = 0.063 L/s]

percent of the total energy input of 1,800,000 Btu/mo (1.9 GJ/mo) required to deliver 87 gal (330 L) of heated water each day and maintain the storage tank at a constant 120°F(49°C). The energy lost to the room is normally a useful increment to space heating during cold months but is an undesirable energy expense during months with little or no space heating load. In order to achieve a realistic assessment of the effects of storage-tank losses in the analytical simulations, one-half of the lost energy is assumed to be a contribution to space heating and is not charged to the solar hot-water system auxiliary energy requirement.

Pump Energy. The small circulating pumps used in forced-circulation solar hot-water systems typically draw approximately 85 W ($<\frac{1}{10}$ hp). In a year, assuming an average of 5 h/day of operation, this amounts to an energy consumption of 155 kWh. Much of this energy goes either into the circulating water or into the surroundings where it can augment the space-heating requirement during cold months. The net nonuseful energy consumption of the circulating pump is, therefore, on the order of 50 kWh or less annually. This value is small compared to the 3000 to 4000 kWh of solar energy provided annually by a typical system; so the circulating pump energy consumption is neglected in the analytical performance simulations.

Annual Performance and Sensitivity Results

The overall thermal performance and cost-effectiveness of either direct or indirect solar hot-water systems are influenced by a number of basic design choices. Perhaps the most significant of these is whether (1) the water is to be "preheated" by the solar energy prior to being heated to the desired delivery temperature by the auxiliary energy source or (2) solar heat is directly added to water which is being maintained at the delivery temperature by the auxiliary.

Preheat system performance The potential advantage of the preheat system approach is that solar energy can be collected at a lower average collector operating temperature, with corresponding higher long-term collection efficiency. Thus, all other factors being equal, more solar energy can be collected by a two-tank preheat configuration (Figs. 11.4 and 11.7) than with a single-tank configuration in which the auxiliary maintains the complete storage tank at a set minimum delivery temperature. Since, in a new installation, the two-tank system costs more than a single-tank system (and has twice the storage-tank heat loss), the use of temperature stratification suggests itself as a means of obtaining at least partial preheat system performance with a single tank (Figs. 11.3 and 11.6). The single-tank preheat concept, which directs the solar energy into the lower, colder part of the tank and heats only the upper volume with an auxiliary *electric* element, is not as efficient from a solar collection standpoint as a two-tank system because significant downward conduction and convection of the auxiliary electric input occur. The single-tank preheat system also provides considerably less heated water reserve than the two-tank system. A typical 82-gal (310-L) tank, for example, might have only 20 gal (76 L) of water in the upper part of the tank maintained at the set delivery temperature by the electric element.

With the above considerations in mind, the obvious question is how do the various systems (nonpreheat, single tank preheat, two tank preheat) compare in terms of cost and energy savings? Annual thermal performance of each of the systems is based on the results of the hour-by-hour computer simulations described above. A cost figure-of-merit is developed for each system by dividing the net annual solar delivery by a representative installed system cost. This solar/cost ratio is not intended to provide economic justification of a particular system, which depends on fuel costs, interest rates, tax considerations, inflation rates, etc. (see Chap. 28). It does, however, provide a means of comparing *relative* cost-effectiveness of competing designs, and it can be used to judge the cost-effectiveness of incremental system additions such as two tanks versus one, or three collectors versus two. For purposes of this discussion, the installed initial cost of a base-line system (single tank, two collectors) is assumed to be $1500. Each additional tank or collector is assumed to add $350 to the total cost.

The calculated annual thermal performance and corresponding solar/cost ratio for the preheat and nonpreheat systems are given in Table 11.3.

As shown by the results in Table 11.3, the preheat configurations deliver almost twice the amount of solar energy as the nonpreheat system. It is of interest to note, however, that the two-tank preheat system is only 10 percent more efficient thermally than the single-tank preheat system. In a new installation where both tanks must be purchased, the two-tank system has a lower solar/cost ratio than the single-tank system and, therefore, cannot be justified for this example case on a cost-effectiveness basis, although it may be desirable because it provides a larger standby reservoir of hot water to meet high-peak-load situations. In the retrofit case the solar/cost ratios are equal, and the two-tank system would appear to be the preferred choice.

The reason for the small difference in thermal performance between the one- and two-tank preheat systems is that the second (auxiliary) tank requires significant energy to compensate for thermal losses. During the six summer months, the auxiliary tank energy losses are wasted and offset almost all of the additional solar energy collected by the two-tank configuration. In a warmer climate (Los Angeles, for example) the wasted auxiliary energy is larger and may cause the single-tank system to have a better annual net solar fraction than the theoretically ideal two-tank system.

As noted in the comparison table above, the preheat systems analyzed were indirect systems which employed the external jacket heat-exchanger configuration (Figs. 11.6 and 11.7). This heat-exchanger arrangement was selected because it provides the least amount of mixing of the

TABLE 11.3   Preheat versus Nonpreheat System Performance Comparison*

System type	Auxiliary energy,† kWh/year	Net solar energy,‡ kWh/year	Net solar fraction,§ %	Solar/cost ratio,¶ kWh/$/year
Single-tank non-preheat (maintained at 120°F by electric elements in top and bottom— potable water circulated through collectors)	4413	1681	28	1.12
Single-tank preheat (upper volume maintained at 120°F by electric element—jacket heat exchanger— Fig. 11.6)	2774	3320	54	2.21
Two-tank preheat (no auxiliary input to preheat tank—jacket heat exchanger—Fig. 11.7)	2449	3645	60	1.97
Two-tank preheat (same as above except retrofit installation in which auxiliary tank is in place; its cost not debited against solar system)	2449	3645	60	2.43

*Denver, CO (latitude 40°); 82-gal-storage tank; 40 ft² collector net area; 40° collector tilt; 87.2 gal/day hot-water load; auxiliary energy required with no solar system = 6094 kWh/year.

†Auxiliary energy = total energy input to auxiliary hot-water heater minus half of energy loss from storage tank (or tanks) to 70°F surroundings. (This assumes that half of the tank heat losses are credited to useful space heating.)

‡Net solar energy = auxiliary energy with no solar system minus auxiliary energy with solar system.

§Net solar fraction = net solar energy divided by auxiliary energy with no solar system.

¶Solar/cost ratio = net solar energy delivered divided by installed cost of system.

stratified water in the single-tank system. The direct single-tank preheat system shown in Fig. 11.3 must be designed with care to avoid excessive fluid mixing. If the circulating water in the lower part of the tank destroys the temperature stratification, performance will approach that of the single-tank nonpreheat system.

Effects of climate and collector area   The overall performance of any solar heating system is strongly influenced by the local weather conditions. In the case of solar hot-water systems, the temperature of the local ground or supply water also affects the amount of energy provided by a given system. For example, the available solar energy and ambient temperature are better (from a solar collection standpoint) in Los Angeles than in Denver. The average monthly tap temperatures in Los Angeles are higher than in Denver. Therefore, less energy is required to supply a given hot-water load. This means that for a given collector area, a system may have a higher solar/cost ratio in Denver than in Los Angeles even though the annual solar load fraction is higher in Los Angeles for the same system. Comparative results for the two-tank indirect preheat sytems (with jacket heat exchangers, Fig. 11.7) are given in Table 11.4.

As shown by Fig. 11.17, the increase in solar load fraction achieved by adding collector area diminishes rapidly as high solar load fractions are approached. As is the case with many solar systems, the best solar/cost ratios are achieved at solar fractions less than 70 percent.

The net solar energy delivery of these systems does not vary as significantly due to climate variations as one might expect. The average annual horizontal insolation in Los Angeles, for example, is approximately 1.35 times greater than in Boston. Yet in Los Angeles a system with 40 ft² (3.7 m²) of collector area delivers only 1.15 times the energy obtained in Boston. This is partly due to the fact that the 40 ft² (3.7 m²) is used more effectively in Boston at a lower solar fraction. Also, in this example all results are based on 40° collector tilt angle, which is more nearly optimum for Boston than Los Angeles, but, as will be shown later, the effect of tilt angle is small for DHW systems in the tilt range of latitude ± 10°.

TABLE 11.4   Effect of Solar Collector Area and Location on Water Heater Performance*

Location and size, ft²†	Auxiliary energy, kWh/year	Net solar energy, kWh/year	Net solar load fraction, %	Solar/cost‡ ratio, kWh/$/year
Los Angeles				
40	1517	3303	69	2.20
60	850	3970	82	2.15
80	563	4257	88	1.94
Denver				
40	2449	3645	60	2.43
60	1605	4489	74	2.43
80	1151	4943	81	2.25
Boston				
40	2813	2874	51	1.92
60	2256	3431	60	1.85
80	1890	3797	67	1.73

*Two-tank indirect system (Fig. 11.7); 82-gal preheat tank; 40° collector tilt; 87 gal/day usage.
†A retrofit installation is assumed; therefore only the cost of the preheat tank is debited against the solar system, since the auxiliary tank was already installed.
‡1 ft² = 0.093 m².

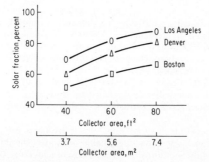

Fig. 11.17.   Solar fraction vs. collector area for example systems in Los Angeles, Denver, and Boston.

Direct versus indirect system performance   All indirect system heat-exchanger configurations which rely on free convention on one side of the exchanger have relatively low heat-exchanger effectiveness values (typically in the range of 0.3 to 0.5, see Chap. 4). It is of interest, therefore, to determine if the annual performance of indirect systems, such as those shown in Figs. 11.6 a and b, 11.7, and 11.8 is significantly less than that of direct, drain-down systems.

The ratio of net solar energy for an indirect system to that for a direct system is plotted as a function of net solar fraction for each of the three example cities in Fig. 11.18. These ratios are surprisingly high considering that the heat-exchanger jacket configuration used in the computer simulations has an effectiveness* of approximately 0.4. The heat-exchanger penalties shown in Fig. 11.18 are small enough that they do not significantly influence the design choice between an indirect and a direct drain-down configuration.

An internal coil heat exchanger (Fig. 11.8) should perform in a manner similar to the jacket configuration provided it has sufficient area to achieve a heat-exchanger effectiveness of 0.4 or greater. Since the shell-and-tube heat exchanger (Fig. 11.6) can only approach the direct system performance as a limit, it would not appear to be justified if it is more complex or costly than a jacket or internal coil configuration with a sufficiently large exchanger surface area.

*Effectiveness $= \dfrac{T_{\text{jacket in}} - T_{\text{jacket out}}}{T_{\text{jacket in}} - T_{\text{storage,average}}}$ See Chap. 4 for a discussion of heat-exchanger effectiveness.

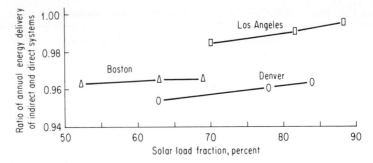

Fig. 11.18. Ratio of annual energy delivery of indirect system vs. direct system for three example cities.

Effects of storage-tank volume   The effects of preheat storage-tank volume on energy delivery for a two-tank system in Denver are plotted in Fig. 11.19. Increasing the preheat tank volume beyond a volume which roughly corresponds to the daily consumption causes little improvement in annual performance. In addition to the ratio of storage volume to daily usage, the ratio of storage volume to collector area influences performance. In general, it appears that this ratio should be maintained at a value of 1.5 gal of storage per square foot of collector or greater.

System Life and Reliability Considerations

Long life, low maintenance, and minimum operating costs are essential if a solar system is to be technically and economically justifiable. To this end, durability and simplicity must be primary design objectives for all elements of the system. If potable water is involved in a solar hot-water system, fail-safe means of preventing contamination and possible health hazards are also necessary. Design considerations which address the issues of system durability and reliability are discussed in the following paragraphs.

Materials compatibility   Corrosion processes can produce a failure in the storage tank or the collector circuit in a liquid-collector system if the system is not designed carefully. The design problem is to control rates of corrosion such that systems can operate for 20 years or longer before corrosion failures occur.

Proper selection of compatible materials and fluids is the key to low corrosion rates and long life. If the solar hot-water system is heating potable tap water, the fluid is not selectable and corrosion problems can only be addressed by selection of materials. This problem is common to the conventional nonsolar domestic hot-water system and has resulted in the widespread use of glass, porcelain enamels, and other nonmetallic coatings as liner materials for storage tanks. In addition, magnesium rods are frequently used in the tanks to provide sacrificial anodic protection against galvanic corrosion. These techniques are well developed and effective and should be used wherever possible in solar storage tanks. In many areas where the ground water is "nonaggressive," the material choice is not critical. In certain areas, however, the impurities in the tap water can cause rapid corrosion and/or buildup of scale in a short time with certain metals. Because of this, the consensus of technical opinion is that, as a general rule, copper should be used for the collector circuit waterways in direct solar heating systems in preference to steel or aluminum.

In an indirect solar heating system, a choice exists for both the circulating fluid and the materials it contacts. Selection of an inert nonelectrolytic fluid, such as a silicone heat-transfer oil, essentially eliminates internal corrosion concerns and permits use of almost any material with appropriate mechanical and thermal characteristics. The inert, heat-transfer fluids, however, are not as effective as aqueous fluids for heat-transport media. Treated aqueous liquids, such as inhibited glycol-water antifreeze mixtures, can be used in indirect systems to obtain a combination of freeze protection, efficient heat transfer, and low corrosion rates.

Effective treatment of aqueous solutions, however, is complex and requires detailed understanding of the inhibitor system and its interaction with the collector circuit materials. As an example, certain commercially available glycol antifreeze solutions use a phosphate inhibitor complex which is designed primarily to combat acids formed by the oxidation of glycol in high-temperature, oxygen-rich circuits (i.e., automotive cooling systems). Use of this particular inhibitor may not be advisable in a solar circuit depending on its design. If the solar circuit is

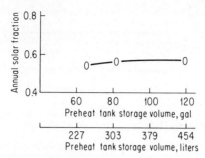

Fig. 11.19. Effect of storage size on two-tank preheat system performance for Denver, with 40 ft² collector at 40° tilt and 87-gal/day usage; jacket heat exchanger used.

closed to the atmosphere (non-drain-down), oxidation of the glycol is probably not the primary corrosion potential since an oxygen source does not exist. Galvanic corrosion is a greater problem. In this case, a different type of inhibitor, possibly a nonelectrolytic organic formulation, should be used in preference to the phosphate inhibitor. A nonelectrolytic inhibitor combined with glycol and deionized water should be used to control galvanic corrosion potential if dissimilar metals are present in the circuit.

Galvanized steel should be avoided if possible in both direct and indirect solar systems. The electromotive potential of the zinc and steel reverses above 130°F (55°C) causing the steel to corrode with respect to zinc. In addition, the zinc is incompatible with certain inhibitors, such as the phosphate system discussed above. Zinc combines with the phosphate inhibitor to form a solid precipitate, rapidly depleting the inhibitor and fouling the system. In addition, if the circulating liquid is electrolytic, local galvanic corrosion may occur in galvanized systems resulting in the formation of hydrogen and oxygen gas.

Tables 11.5 and 11.6 are included to provide general guidelines for the selection of materials for use in solar collectors, collector circuit piping, and heat exchangers in closed and open systems. A more detailed discussion of corrosion phenomena applicable to solar system design is presented in Ref. 6, pp. 5-26 through 5-30.

TABLE 11.5 Use Conditions for Metals in Direct Contact with Heat-Transfer Liquids in Open Systems*

Generally unacceptable use conditions	Generally acceptable use conditions†
Aluminum	
1. When in direct contact with untreated tap water with pH <5 or >9.	1. When in direct contact with distilled or deionized water which contains appropriate inhibitors and does not contact copper or iron.
2. When in direct contact with aqueous liquid containing less electropositive metal ions, such as copper or iron, or halide ions.	2. When in direct contact with distilled or deionized water which contains appropriate inhibitors and a means of removing heavy metal ions obtained from contact with copper or iron.
3. When specific data regarding the behavior of a particular alloy are not available, the velocity of aqueous liquid shall not exceed 4 ft/s.	3. When in direct contact with stable anhydrous organic liquids.
4. When in direct contact with a liquid which is in contact with corrosive fluxes.	

*Open systems are those in which air, in addition to that initially in the transfer liquid, can be absorbed into the liquid by contact with the atmosphere or air entrapped in the system. For more detailed information, see Ref. 6.

†The use of suitable antifreeze agents and buffers is acceptable provided they do not promote corrosion of the metallic liquid containment system. The use of suitable corrosion inhibitors for specific metals is acceptable provided they do not promote corrosion of other metals present in the system. If thermal or chemical degradation of these compounds occurs, the degradation products shall not promote corrosion.

TABLE 11.5 Use Conditions for Metals in Direct Contact with Heat-Transfer Liquids in Open Systems (Continued)

Generally unacceptable use conditions	Generally acceptable use conditions†

Copper

1. When in direct contact with aqueous liquid containing high concentrations of chlorides or sulfates or with liquid which contains hydrogen sulfide. 2. When in direct contact with chemicals that can form copper complexes such as ammonium compounds. 3. When in direct contact with an aqueous liquid having a velocity greater than 4 ft/s.‡ 4. When in direct contact with a liquid which is in contact with corrosive fluxes. 5. When in contact with an aqueous liquid with a pH lower than 5. 6. When the copper surface is initially locally covered with a copper oxide film or a carbonaceous film. 7. When operating under conditions conducive to water line corrosion.	1. When in direct contact with distilled, deionized, or low-chloride, low-sulfate, and low-sulfide tap water. 2. When in direct contact with stable anhydrous organic liquids.

Steel

1. When in direct contact with untreated tap, distilled, or deionized water with pH <5 or >12. 2. When in direct contact with a liquid which is in contact with corrosive fluxes. 3. When in direct contact with an aqueous liquid having a velocity greater than 6 ft/s.‡ 4. When operating under conditions conducive to water line corrosion.	1. When in direct contact with distilled, deionized, or low-salt-content water which contains appropriate corrosion inhibitors. 2. When in direct contact with stable anhydrous organic liquids. 3. When adequate cathodic protection of the steel is used (practical only for storage tanks).

Stainless Steel

1. When the grade of stainless steel selected is not corrosion resistant in the anticipated heat-transfer liquid. 2. When in direct contact with a liquid which is in contact with corrosive fluxes.	1. When the grade of stainless steel selected is resistant to pitting, crevice corrosion, intergranular attack, and stress corrosion cracking in the anticipated use conditions. 2. When in direct contact with stable anhydrous organic liquids.

Galvanized Steel

1. When in direct contact with aqueous liquid containing copper ions. 2. When in direct contact with aqueous liquid with pH <7 or >12. 3. When in direct contact with aqueous liquid with a temperature >55°C.	1. When adequate cathodic protection of the galvanized parts is used (practical only for storage tanks). 2. When in contact with stable anhydrous organic liquids.

‡The flowrates at which erosion/corrosion becomes significant will vary with the conditions of operation. Accordingly, the values listed are approximate.

TABLE 11.6 Use Conditions for Metals in Direct Contact with Heat-Transfer Liquids in Closed Systems*

Generally unacceptable use conditions	Generally acceptable use conditions†
Aluminum	
1. When in direct contact with untreated tap water with pH <5 or >9.	1. When in direct contact with distilled or deionized water which contains appropriate corrosion inhibitors.
2. When in direct contact with liquid containing copper, iron, or halide ions.	2. When in direct contact with stable anhydrous organic liquids.
3. When specified data regarding the behavior of a particular alloy are not available, the velocity of aqueous liquids shall not exceed 4 ft/s.	
Copper	
1. When in direct contact with an aqueous liquid having a velocity greater than 4 ft/s.‡	1. When in direct contact with untreated tap, distilled, or deionized water.
2. When in contact with chemicals that can form copper complexes such as ammonium compounds.	2. When in direct contact with stable anhydrous organic liquids.
	3. When in direct contact with aqueous liquids which do not form complexes with copper.
Steel	
1. When in direct contact with liquid having a velocity greater than 6 ft/s.‡	1. When in direct contact with untreated tap, distilled, or deionized water.
2. When in direct contact with untreated tap, distilled, or deionized water with pH <5 or >12.	2. When in direct contact with stable anhydrous organic liquids.
	3. When in direct contact with aqueous liquids of $5 < pH > 12$.
Stainless Steel	
1. When the grade of stainless steel selected is not corrosion resistant in the anticipated heat-transfer liquid.	1. When the grade of stainless steel selected is resistant to pitting, crevice corrosion, intergranular attack, and stress corrosion cracking in the anticipated use conditions.
2. When in direct contact with a liquid which is in contact with corrosive fluxes.	2. When in direct contact with stable anhydrous organic liquids.

*Closed systems are those in which the air initially absorbed in the transfer liquid is not replaced to a significant extent. In a closed system, there is no exposure of the liquid to the atmosphere except above the liquid in the expansion tank (in some cases a nitrogen blanket above the liquid in the expansion tank may be used); there is no entrapped air in the piping or storage systems and the expansion tank is isolated from the flow path between the collector and storage. In addition, liquid leakage requiring frequent make up is avoided. For more detailed information, see Ref. 6.

†The use of suitable antifreeze agents and buffers is acceptable provided they do not promote corrosion of the metallic liquid containment system. The use of suitable corrosion inhibitors for specific metals is acceptable provided they do not promote corrosion of other metals present in the system. If thermal or chemical degradation of these compounds occurs, the degradation products shall not promote corrosion.

‡The flowrates at which erosion/corrosion becomes significant will vary with the conditions of operation. Accordingly, the values listed are approximate.

TABLE 11.6   Use Conditions for Metals in Direct Contact with Heat-Transfer Liquids in Closed Systems (Continued)

Generally unacceptable use conditions	Generally acceptable use conditions†
Galvanized Steel	
1. When in direct contact with water with pH <7 or >12.	1. When in contact with water of pH >7 but <12.
2. When in direct contact with an aqueous liquid with a temperature >55°C.	

Brass and Other Copper Alloys
Binary copper-zinc brass alloys (CDA 2XXX series) exhibit generally the same behavior as copper when exposed to the same conditions. However, the brass selected shall resist dezincification in the operating conditions anticipated. At zinc contents of 15 percent and greater, these alloys become increasingly susceptible to stress corrosion. Selection of brass with a zinc content below 15 percent is advised. There are a variety of other copper alloys available, notably copper-nickel alloys, which have been developed to provide improved corrosion performance in aqueous environments.

Heat-Transfer Fluid Selection

As discussed in the preceding paragraphs, an indirect solar heating sytem offers a choice of fluid to be used in the solar collector circuit. Air is possibly the least expensive choice and certainly eliminates freezing and internal corrosion concerns. Air is also the least efficient of the candidate fluids with respect to heat transport, however, and may not be the most cost-effective choice depending on the particular application.

If an indirect liquid-collector system is selected, the basic fluid choice involves an aqueous versus a nonaqueous heat-transfer medium. Inert, nonelectrolytic, nonfreezing, low-vapor-pressure, nonaqueous heat-transfer liquids are commercially available. These fluids have the inherent advantage of essentially eliminating concerns relative to corrosion, freezing, and boiling. The disadvantages of the nonaqueous heat-transfer liquids are that they generally cost considerably more than the aqueous solution alternatives, and they are less effective heat-transport media and have lower surface tensions (i.e., lower specific heat, lower thermal conductivity, higher viscosity). Properties of some typical heat-transfer liquids are listed in Table 11.7.

The choice of aqueous versus nonaqueous liquids depends on whether the simplicity advantages of the nonaqueous candidates outweigh the lower cost and better heat-transfer properties of the aqueous liquids. This can only be determined by careful assessment of the total impact on system cost, performance, life, and reliability of the design features required to contend with the inherent freezing, boiling, and corrosion characteristics of aqueous solutions.

The relative thermal performance of aqueous versus nonaqueous fluids may be highly dependent on the type of heat exchanger employed. For example, the heat-transfer coefficient on the solar side of a jacket heat exchanger (Figs. 11.6 and 11.7) is directly proportional to thermal conductivity, k, of the fluid, whereas inside a tube coil (Figs. 11.6b and c, 11.8, and 11.10) the turbulent flow heat-transfer coefficient is proportional to $k^{0.67}\mu^{-0.47}c_p^{0.33}$. Estimating the effect of the relevant properties of 50 percent water, 50 percent ethylene glycol and those of a typical inert, organic heat-transfer liquid provides the comparison of respective heat-transfer coefficients shown in Table 11.8.

From the data in Table 11.8 it can be seen that the heat-transfer coefficient inside the jacket can be reduced by a factor of 3 by use of an organic heat-transfer fluid; but only a 30 percent reduction occurs in the tube coil configuration.

Safety Conditions

Prevention of contamination of heated potable water can pose a problem in an indirect solar hot-water system. This may occur from leaks in the solar collector fluid circuit or some other form of communication between the potable water and the fluid in the solar circuit. A number

TABLE 11.7 Typical Heat-Transfer Liquids

	Water	50% ethylene glycol/water	50% propylene glycol/water	Silicone fluid	Aromatics	Paraffinic oil
Freezing point, °F (°C)	32 (0)	−33 (−36)	−28 (−33)	−58 (−50)	−100 to −25 (−73 to −32)	
Boiling point, °F (°C) (at atm. pressure)	212 (100)	230 (110)	230 (110)	Very high	300–400 (149–204)	700 (371)
Fluid stability	Requires pH or inhibitor monitoring	Requires pH or inhibitor monitoring	Requires pH or inhibitor monitoring	Good	Good	Good
Flash point*, °F (°C)	None	None	600 (315)	600 (315)	145–300 (63–149)	455 (235)
Specific heat (73°F) [Btu/ (lb·°F)] (kJ/ kg·°C)	1.0 (4.18)	0.80 (3.35)	0.85 (3.55)	0.34–0.48 (1.42–2.00)	0.36–0.42 (1.51–1.76)	0.46 (1.92)
Viscosity (cstk at 77°F)	0.9	21	5	50–50,000	1–100	
Toxicity	Depends on inhibitor used	Depends on inhibitor used	Depends on inhibitor used	Low	Moderate	

SOURCE: Ref. 6. These data are extracted from manufacturers' literature to illustrate the properties of a few types of liquid that have been used as transfer fluids.
*It is important to identify the conditions of tests for measuring flash point. Since the manufacturers' literature does not always specify the test, these values may not be directly comparable.

TABLE 11.8 Relative Heat-Transfer Properties of Aqueous and Organic Fluids

	50% water/ 50% ethylene glycol	Organic heat-transfer liquid
Properties at 100°F k, Btu/h·ft·°F (W/m·°C)	0.23 (0.390)	0.076 (0.132)
μ, cP	2.5	0.7
c_p, Btu/ lb / ·°F (kJ/kg·°C)	0.84 (3.51)	0.45 (1.88)
Heat-transfer coefficient proportionality factor Tube flow: $h \propto k^{0.67}\mu^{-0.47}c_p^{0.33}$	0.23	0.16
Jacket flow: $h \propto k$	0.23	0.076

of preventive measures can be taken to minimize the possibility of potable water contamination and resulting health hazard:

1. Use a nontoxic fluid in the solar collector circuit.
2. Use appropriate pressure gradients to force flow *into* the toxic loops, not *from* it.
3. Use heat-exchanger designs which are fail-safe in event of collector fluid leakage. For example, a double-wall exchanger can achieve this purpose.

It should be noted that neither preventive measure above can be relied upon with a high degree of confidence under all conditions over an extended time period. An original fill of nontoxic fluid may become contaminated or be replaced with a toxic fluid over a long period of time. Repair activities or periods of disuse may reverse a protective pressure gradient long enough to permit a hazardous accumulation of contamination in the potable fluid zones.

Therefore, a fail-safe heat-exchanger design should be considered a mandatory requirement in any indirect system involving potable water. The jacket configuration (Fig. 11.14) is an example of a double-wall fail-safe design. The jacket has its own inner wall which is in mechanical contact with the storage tank outer wall, but there is no seal to prevent fluid from leaking through the jacket inner wall, flowing down between the jacket and tank wall, and appearing on the floor below the tank. Even though a performance penalty is incurred by the finite thermal contact resistance between the jacket and tank wall, the penalty for the entire jacket thermal resistance, of which the contact resistance is only a part, is small, usually less than 5 percent in annual solar heat delivery. A further discussion of heat-exchanger safety criteria and design requirements can be found in Ref. 6, p. 5-42.

Protection against scalding is a necessary concern in the design of a solar hot-water system, since temperatures approaching boiling can be achieved during periods of high solar input and low water use. Although numerous control schemes can be devised to prevent reaching excessive temperatures in solar storage, a simpler, more efficient approach is to provide a tempering valve at the hot-water outlet as shown in Figs. 11.2 through 11.10.

System Control

Any forced-circulation solar hot-water system requires some type of sensing and control to activate the collector pump when solar energy is available and to turn the pump off when solar energy cannot or should not be collected. The control system should be as simple as possible from a reliability and maintenance standpoint, yet should not be so primitive that thermal performance is seriously compromised.

A manual switch or simple timer may provide adequate system control in certain applications. A differential thermostat, however, can provide a more precise and efficient means of automatic control and is commonly used in most forced-circulation solar systems.

A typical differential thermostat controller consists of one sensor which measures collector temperature, a second which measures storage temperature, an electronic circuit which compares the two temperatures, and logic circuits which provide appropriate signals to pumps, blowers, valves, and dampers. The sequential daily operation of a typical liquid-system differential thermostat controller is shown in Fig. 11.20.

In this example, as the sun rises, the collector temperature increases rapidly in the absence of cooling fluid flow until the temperature reaches point 1. This corresponds to a preset difference above storage temperature called the "turn-on differential," designated ΔT_{ON} in the figure. When ΔT_{ON} is reached, the controller starts the collector circuit circulating pump. Since only the collector absorber plate is warm at this point, a transient dip in collector temperature occurs as the colder water in the supply piping reaches the collector sensor. If ΔT_{ON} is designed sufficiently large, the transient dip will not turn the pump off thus causing the start cycle to repeat.

As the solar intensity decreases later in the day, the collector temperature drops approaching the storage temperature. At point 3 the "turn-off differential," ΔT_{OFF}, is reached and the pump

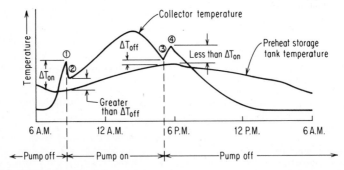

Fig. 11.20.   Time-temperature history of collector and storage fluids showing turn-on and turn-off temperature differences for controller design.

is turned off. Since some solar radiation is still incident on the collector, the collector temperature rises during a transient period until collector heat losses equal diminishing solar input. As can be seen from this example, the key to effective controller design is the proper selection of ΔT_{ON} and ΔT_{OFF}.

If ΔT_{ON} is made very large, the system is stable, since the starting transient will be less likely to cycle the system off. However, some otherwise collectable energy is lost while the collector is warming up to the starting condition if ΔT_{ON} is too large. Values of 15 to 20°F (8.3–11°C) for ΔT_{ON} and 3°F (1.7°C) for ΔT_{OFF} have been shown to provide stable operation without sacrificing thermal efficiency in typical small DHW systems. The "optimum" values, however, vary depending on system characteristics and may need to be determined by trial and error for particular installation. Therefore, the two differentials should be adjustable.

In any system, care must be taken to assure that the temperature sensors are properly placed. Installation of a collector sensor in the supply line only a few inches from the collector proper can cause a very long delay in achieving ΔT_{ON}, resulting in a significant reduction in solar energy collected. Installation of the tank sensor too close to the cold water inlet can cause the controller to be "fooled" by a small amount of entering cold water, and thus, to operate the system in a net energy loss condition.

The collector turn-on and turn-off temperature transients can be reduced (with corresponding improvement in control stability) by throttling the collector circuit flowrate. This can be accomplished either on a continuous basis or as required by proportional control of pump speed. Reduced flowrates decrease collector efficiency, but this effect may be at least partially offset by improved stratification in the storage tank. An estimate of the effect of reduced flowrate on annual net solar gain has been made using hour-by-hour computer simulations for a two-tank, direct, preheat system in Denver using a 40° collector tilt, 40-ft² collector area, and an 82-gal, direct-heated preheat tank. For a collector flowrate of 15 lb/h·ft², the annual net solar gain (kWh/yr) was 3821 and for 3 lb/h·ft², 3630 kWh were collected. These results indicate a 5 percent reduction in solar gain due to reducing collector flowrate by a factor of 5. This indicates that reduced flowrates may be used to improve control-system stability without seriously compromising thermal performance.

Rejection of Excess Solar Energy

The rejection of excess collected solar energy can present as many difficulties to the designer as the usual concern of maximizing useful gain. Simply turning a forced-circulation collector circuit off when the storage temperature reaches a set upper limit is a means of preventing boiling or overheating of the storage medium. This can cause problems in the collector circuit, however, as it results in stagnation of the collectors. Prolonged stagnation temperatures can be detrimental to certain types of collector absorber surfaces and working fluids. Stagnation can cause boiling of aqueous solutions if the liquid is not drained from the collectors when the pump is turned off. Repeated boiling of glycol can accelerate its oxidation and thermal decomposition resulting in formation of acids and depletion of inhibitors.

Based on these considerations, it would appear undesirable in most instances, particularly if aqueous transfer fluids are used, to shut the system down as a means of preventing excess energy collection.

Nocturnal heat rejection is one approach to maintaining storage at acceptable temperature levels during periods of low load without stagnating the collectors. This requires a control strategy which overrides the normal end-of-day turn-off signal if the storage temperature is above a predetermined limit. During periods of little or no load, this control strategy sets up an equilibrium 24-h cycle in which the storage tank may reach a maximum temperature at the end of each day's collection period. During late afternoon and evening, the collector circuit continues to operate rejecting the day's solar input until the storage reaches a preselected lower temperature at which point the system is turned off. This cycle can be achieved in most U.S. climate areas using conventional flat-plate collectors as long as sufficient storage is provided (1.5 lb of water per square foot of collector area or greater) to limit the temperature swing to manageable values. However, large amounts of electric power are used. A simpler heat rejection method simply dumps a volume of hot water, the entering cold water thereby placing a new load on the solar system.

System Startup after Power Outages

Even if a system is designed which prevents collector stagnation during normal operation, consideration must be given to nonnormal conditions such as a power outage. If collector circulation depends on electrical input to a pump, a power failure (or pump failure) during high insolation conditions will result in very high collector temperatures very rapidly. The system design, therefore, must accommodate stagnation conditions even if they do not occur during normal operation.

Apart from collector durability considerations, the primary system problem imposed by stagnation is how to accommodate, and recover from, boiling of aqueous heat-transfer fluids. Unless the entire collector circuit is designed for unusually high pressures, boiling of aqueous solutions cannot be suppressed by allowing the system pressure to build to saturation pressure levels. Stagnation absorber temperatures in excess of 400°F (~200°C) can be reached by conventional flat-plate collectors. The vapor pressure of 50% water–10% ethylene glycol is approximately 180 psia (1.24 MPa) at 400°F (~200°C), well in excess of the 75 psig (0.5 MPa) working pressure limit typical of commercially available residential hydronic system components.

An aqueous-fluid collector circuit must, therefore, either be drained during a power failure or be designed to accept boiling of some or all of the liquid in the collectors. The vapor generated by boiling can be vented through an automatic pressure relief valve or retained in an expansion tank of sufficient volume to limit the pressure increase to an acceptable value. The expansion tank would appear to be the preferred choice in most instances. It prevents nonrecoverable loss of heat-transfer fluid and improves the chances of automatic system recovery when power is restored. It should be noted that expansion tanks sized to accommodate collector boiling are significantly larger than those sized simply to accept thermal expansion of a nonboiling liquid. As an example, an expansion tank volume of as much as 6 to 8 gal (23–30 L) may be required to limit maximum pressure to 50 psig (345 kPa) in a small DHW system with a total aqueous-fluid heat-transfer circuit volume of 5 gal (19 L).

In any forced-circulation liquid-collector system, automatic recovery from a boiling or drain-down condition requires replacement of the vapor and noncondensable gases present in the collector circuit with the heat-transfer liquid. In the case of a drain-down system, the pump must have sufficient static head capability to "lift" the liquid through the vertical height that has been drained above the pump. The collector circuit must also be designed so that the return line completely fills if "siphon" operation is required for the pump to provide the design flowrate. If the return line does not have sufficiently high pressure drop to cause it to refill fully, the pump may not have enough flow capacity at the static lift head to refill this return line. This can result in a permanent "water fall" condition in the return line, which in turn causes a lower-than-design flowrate with consequent performance penalties.

Automatic recovery from a boiling condition should theoretically occur only after the sun has disappeared and the vapor generated by boiling has cooled and recondensed. If the expansion tank is sized and pressurized properly, sufficient pressure exists to expel liquid from the expansion tank and lift it back to the highest point in the system as the vapor condenses. In practice, however, the presence of noncondensable gases in the system can prevent a complete refill even when the system has cooled and can cause a small circulating pump to become "air locked." Air-lock problems can be minimized by selecting a pump with sufficient head capability to overcome the static lift requirement and by installing an air separator to divert entrained gases into the expansion tank.

One final design choice related to the presence of noncondensable gases in a liquid-collector circuit involves the choice of whether or not the expansion tank should be of the diaphragm type. In general, a diaphragm tank would appear to be suitable only if a nonboiling, low-vapor-pressure fluid is used. In the case of the aqueous fluids, in addition to possible boiling, it is almost impossible to prevent significant amounts of dissolved gases from coming out of solution and becoming entrained in the flow as the liquid temperature increases during collection periods. If any dissimilar metals are present, galvanic cell action can produce significant amounts of hydrogen and oxygen over extended periods of time. Because of these factors, a nondiaphragm expansion tank with a well-designed scoop to transfer entrained gases into the tank is recommended for closed, indirect, aqueous-fluid collector circuits.

Comparison of Hour-by-Hour Simulations with f-Chart Results

A detailed study has been made to determine how well the f-chart performance estimation method* agrees with hour-by-hour simulation results presented earlier. The basic f-chart

*See Chap. 14.

equations used, which are applicable to hot-water-only systems (not combined hot water–space heating), are

$$X = \frac{F'_R A_c U_c \, \Delta t \, (11.6 + 1.18 T_w + 3.86 T_m - 2.32 \overline{T}_a)}{L} \tag{11.5a}$$

In English units,

$$X = \frac{F'_R A_c U_c \, \Delta t \, (-66.2 + 1.18 T_w + 3.86 T_m - 2.32 \overline{T}_a)}{L} \tag{11.5b}$$

$$Y = \frac{F'_R A_c \overline{I}_c \, (\overline{\tau\alpha})}{L} \tag{11.6}$$

$$f = 1.029Y - 0.065X - 0.245Y^2 + 0.00182X^2 + 0.0215Y^3 \tag{11.7}$$

Where $F'_R = F_R \left[1 + \dfrac{F_R U_c A_c}{(\dot{m} C_p)_c} \left(\dfrac{(\dot{m} C_P)_c}{\epsilon_c (\dot{m} C_p)_{\min}} - 1 \right) \right]^{-1}$ (11.8)

A_c = collector area, m²

$F_R U_c$ = slope of collector efficiency curve, kJ/°C h · m²

$(\dot{m} C_p)_c$ = thermal capacity rate of collector circuit, kJ/n·m²· °C

$(\dot{m} C_p)_{\min}$ = thermal capacity rate of collector or storage heat-exchanger circuit, whichever is smaller.

ϵ_c = effectiveness of collector to storage heat exchanger

Δt = number of hours in month

L = monthly hot-water heating load, kJ/mo

$\overline{I}_c$ = monthly average insolation on collector plane, kJ/m²· mo

$(\overline{\tau\alpha})$ = monthly average collector transmittance-absorptance product, assumed

     = 0.95 times transmittance-absorptance product at normal incidence

f = solar load fraction, i.e., monthly ratio of net solar energy delivered to monthly hot-water heating load

T_w = minimum hot-water delivery temperature, °C

T_m = monthly average temperature of water supplied from mains, °C

$\overline{T}_a$ = monthly average ambient temperature, °C

The comparison of the f-chart results and the simulation results for two-tank direct and indirect preheat systems in each of the three cities is shown in Tables 11.9 and 11.10. An examination of the results indicates that the agreement between hourly calculations and monthly f-chart predictions is excellent at small solar fractions. However, the f-chart results become progressively higher compared to the hourly simulation results as the solar load fraction increases.

Examination of the month-by-month hourly simulation results indicated that even in the summer months, a solar load fraction of 85 percent was rarely exceeded. This is reasonable, since even during the summer in warm, clear areas, the auxiliary will be required at certain times. This is particularly true as the ratio of collector area to tank size increases, providing less relative available storage to ride through periods of high demand and/or low insolation.

Based on these observations, the f-chart correlation has been modified so that 85 percent was the maximum achievable solar fraction, rather than 100 percent, in any given month. This simple change produced excellent agreement between the f-chart results and the simulation results at the high solar fractions and did not affect the already good agreement at the lower fractions. An upper limit for f of 85 to 90 percent for a particular month is, therefore, rec-ommended as a modification to the hot-water-only f-chart method. It should be noted that agreement is good for both the direct and indirect systems, indicating that the F'_R factor adequately predicts heat-exchanger penalties, at least down to effectiveness values of 0.4.

TABLE 11.9 Comparison of f-Chart Results with Simulation Results for Three Cities

| Location | Collector area, ft²* | Solar load fraction, % | | % deviation $\dfrac{f\text{-chart}-\text{sim}}{\text{sim}} \times 100$ |
		f-chart	Simulation	
		Direct System, $F'_R/F_R = 1.0$		
Denver	40	60	59	+1.7
	60	79	74	+6.8
	80	92	83	+10.8
Los Angeles	40	67	66	+1.5
	60	85	81	+4.9
	80	94	87	+8.0
Boston	40	45	48	−6.3
	60	59	59	0
	80	70	66	+6.1
			Average deviation	+3.7
		Indirect System, $F'_R/F_R = 0.93$		
Denver	40	57	56	+1.8
	60	76	71	+7.0
	80	89	79	+12.7
Los Angeles	40	69	65	−1.5
	60	82	80	+2.5
	80	92	87	+5.7
Boston	40	42	46	−8.7
	60	57	56	+1.8
	80	67	64	+4.7
			Average deviation	+2.9

*1 ft² = 0.093 m².

TABLE 11.10 Comparison of f-Chart Results with Simulation Results for Three Cities—f-Chart Modified such that Maximum Monthly Solar Fraction = 85%

| Location | Collector area, ft²* | Solar fraction, % | | % deviation $\dfrac{f\text{-chart}-\text{sim}}{\text{sim}} \times 100$ |
		f-chart	Simulation	
		Direct System, $F'_R/F_R = 1.0$		
Denver	40	60	59	+1.7
	60	78	74	+5.4
	80	83	83	0
Los Angeles	40	67	66	+1.5
	60	81	81	0
	80	84	87	−3.4
Boston	40	45	48	−6.3
	60	59	59	0
	80	68	66	+3.0
			Average deviation	+0.2

TABLE 11.10   Comparison of f-Chart Results with Simulation Results for Three Cities—f-Chart Modified such that Maximum Monthly Solar Fraction = 85% (Continued)

Location	Collector area, ft²*	Solar fraction, %		% deviation $\dfrac{f\text{-chart} - \text{sim}}{\text{sim}} \times 100$
		f-chart	Simulation	
Indirect System, $F'_R/F_R = 0.93$				
Denver	40	57	56	+1.8
	60	76	71	+7.0
	80	82	79	+3.8
Los Angeles	40	64	65	−1.5
	60	80	80	0
	80	83	87	−4.6
Boston	40	42	46	−8.7
	60	57	56	+1.8
	80	66	64	+3.1
			Average deviation	+0.5

*1 ft² = 0.093 m².

CODES AND STANDARDS

Myriad federal, state, and local construction codes, as well as government and industry standards, exist which deal either directly or by implication with solar hot-water systems. A useful starting point toward obtaining familiarity with pertinent codes and standards is the 1977 edition of the HUD *Intermediate Minimum Property Standard Supplement for Solar Heating and Domestic Hot Water Systems*.[6] This document contains a detailed overview of applicable industry standards and federal specifications in its Appendix B.

Perhaps the greatest impact on solar hot-water systems is created by the local electrical, plumbing, and roofing building codes. These vary from place to place and cannot be fully cited here, but a simple example of a Denver, CO, roofing code illustrates the type of problem that can be encountered if these codes are ignored. The Denver code requires that if an object is installed on a shingled roof, it must be done in such a manner that the roof can be easily reshingled. The interpretation of this requirement vis-à-vis solar collectors is that they must either be installed at least 4 ft away from the roof or flashed into the roof to become an integral part of the weather seal. Simple ignorance of this type of code requirement can result in code violations which are costly to remedy.

SUMMARY: HARDWARE DESIGN CHECKLIST

The following section describes a set of general design guidelines for the selection and sizing of components for forced-circulation, liquid-collector, solar hot-water systems. It should be recognized that exceptions may exist to these rules, and that experience and/or the requirements of a particular application may dictate deviations from these guidelines. The intent is to provide a checklist to prevent oversights and unnecessary re-creation of past mistakes.

The functional arrangement of the component parts of a typical residential solar hot-water system is shown in Fig. 11.21. The tank shown can either be a preheat tank in a two-tank system or comprise a single-tank system with an electric auxiliary element installed in the upper part of the tank above the jacket. A typical packaging arrangement, which includes all of the ancillary components (except the collectors, fill vent, and collector temperature sensor) mounted on top of the storage tank, is shown in Fig. 11.22.

General design and selection guidelines for each of the system elements are listed in Table 11.11.

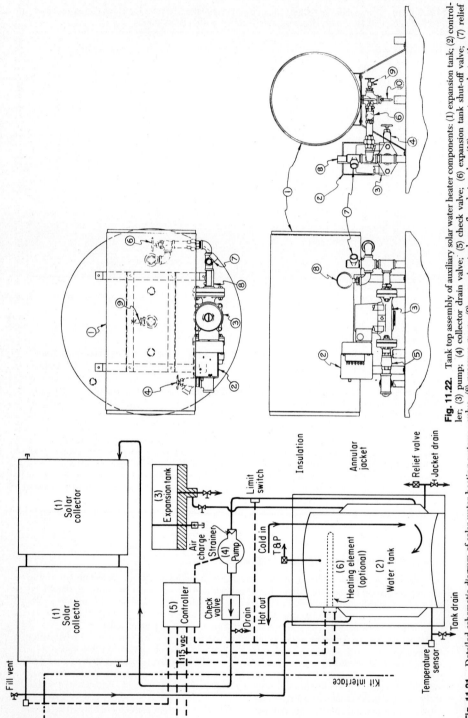

Fig. 11.22. Tank top assembly of auxiliary solar water heater components: (1) expansion tank; (2) controller; (3) pump; (4) collector drain valve; (5) check valve; (6) expansion tank shut-off valve; (7) relief valve; (8) pressure gauge; (9) expansion tank overflow drain valve; (10) expansion tank air valve.

Fig. 11.21. Detailed schematic diagram of solar water heating system.

Component	Recommended design guidelines
Steel storage tanks	1. Glass or stone liner.
	2. Minimum of 2.5 in of fiber glass insulation or equivalent.
	3. Magnesium anode.
	4. Internal piping, heat exchangers, and auxiliary heater (if any) arranged to preserve temperature stratification.
	5. Minimum preheat volume equal to 1.5 gal/ft² of collector area, or estimated daily hot-water usage, whichever is greater.
	6. Pressure-temperature relief valve in conformance with local codes.
	7. Tempering valve on outlet to hot-water distribution piping.
Collector circuit piping	1. Material compatible with heat-transfer fluid (copper if heat-transfer fluid is untreated tap water).
	2. Size compatible with collector design flowrate and pump head capacity.
	3. Provisions to vent air from high points—no air traps.
	4. Check valve (or equivalent) to prevent reverse thermosiphoning at night.
	5. Return line restricted to prevent "water fall" operation (if pump selected for siphon operation).
	6. Dielectric connections between dissimilar metals (if heat-transfer fluid is electrolytic).
	7. All exposed piping (inside and outside) insulated with 1/2-in fiber glass or equivalent.
	8. Manual drains at low points.
	9. Pressure relief valve.
Pumps	1. Head and flow capacity compatible with collector design flowrate, piping and heat-exchanger circuit head-loss, static lift requirement, fluid characteristics.
	2. Materials compatible with heat-transfer fluid (bronze or stainless steel if fluid is untreated tap water).
Expansion tanks	1. Sized to accommodate maximum possible heat-transfer fluid expansion under stagnation conditions (must accommodate boiling of aqueous fluids).
	2. Nondiaphragm configuration with air eliminator if heat-transfer fluid is aqueous solution.
	3. Should have liquid level indicator (overflow fitting or sight glass) and pressure gauge.
Differential thermostat controller	1. Turn-on and turn-off differentials set to minimize pump on-off cycling (may require experimental determination).
	2. Collector temperature sensor located to minimize response lag (either on absorber or inside absorber flow passage).
	3. Storage temperature located to measure preheat volume bulk temperature.
Heat exchangers	1. Fail-safe prevention of contamination of potable water (double wall, as example).
	2. Collector-to-storage heat-exchanger effectiveness = 0.4 or greater [effectiveness of jacket or submerged coil exchanger = $(T_{\text{EXCH IN}} - T_{\text{EXCH OUT}})/(T_{\text{EXCH IN}} - T_{\text{STORAGE}})$ at collector design flowrate and typical midday collector heat output].
	3. Physical arrangement should minimize disturbance of preheat volume temperature stratification (by either convection flow or pumped flow).
Solar collectors	1 Waterways compatible with heat-transfer fluid (copper if fluid is untreated tap water).
	2. Collector area (liquid flat plate) sized to provide 50 to 70 percent annual free-energy fraction—rough rule of thumb is 0.75 to 1.0 ft²/gal of daily hot-water load.
	3. Collector tilt = latitude ± 10° (tilt not less than 30°).

REFERENCES

1. J. E. Scott, R. W. Melicher, and D. Sciglimpaglia, "Demand Analysis Solar Heating and Cooling of Buildings, Phase I Report, Solar Water Heating in South Florida: 1923–1974," NSF-RA-N-74-190, 1974.
2. J. M. Hollander and M. K. Simmons (eds.), *Annual Review of Energy*, vol. 1, Annual Reviews, Inc., Palo Alto, Calif., 1976.
3. American Society of Heating, Refrigerating and Air-Conditioning Engineers, *ASHRAE 1976 Systems Handbook*, New York, 1976.
4. W. M. Kays and A. L. London, *Compact Heat Exchangers*, McGraw-Hill Book Company, New York, 1964.
5. A. J. Ede, *Advances in Free Convection*, 1967.
6. U.S. Department of Housing and Urban Development, *Intermediate Minimum Property Standards Supplement, Solar Heating and Domestic Hot Water Systems*, 1977 Edition, No. 4930.2, Government Printing Office, Washington, D.C., 1977.

Chapter **12**

Air-Based Solar Systems for Space Heating

GEORGE O. G. LÖF, P.E.
Solaron Corporation, Denver, CO

INTRODUCTION

"Solar air heating" is defined as a process for supplying heat from a solar collector through which air is circulated. In most space-heating applications, the heated air may be supplied directly to the living space as needed and may also be supplied to some type of heat storage device for later transfer to the living space.

The use of solar energy for residential hot-water supply in nonfreezing climates commenced nearly a century ago. Extension of this simple technology into the space-heating field required only an increase in the system size, the addition of pumps, and protection against freezing in the collector. Most of the space-heating developments since 1940 have accordingly involved liquid solar heaters, and the commercial manufacture and sale of solar equipment after 1975 has been based mainly on liquid systems.

The use of air as the medium for heat transfer between collector and storage (and also to the living space) commenced in 1944. A small dwelling in Boulder, Colorado, was retrofitted with a solar air-heating system developed by Löf.[1] A solar heater of the overlapped glass-plate type (see below) and a pebble-bed heat storage bin were used in the Boulder residence. In 1946, Telkes[2] installed a solar air-heating system on a new dwelling in Dover, Massachusetts. Vertical, glass-covered metal absorber plates served as solar collectors in the Dover house, with heat storage in stacked, 5-gal containers of Glaubers salt ($Na_2SO_4 \cdot 10H_2O$). A gas furnace provided auxiliary heat in the Boulder house, and electric heaters supplemented the solar system in the Dover house. Performance data on both systems were limited by lack of instrumentation and the short duration of the experiments.

In 1954, Donovan and Bliss[3] constructed a solar air heater of the matrix type (solar absorption in black cotton gauze) and a pebble-bed storage bin adjacent to a small house in Amado, Arizona. The system was operated several seasons, and performance data reported.

Not until 1957 was another full-size solar air-heating system constructed. In that year, Löf[4] installed a solar air-heating system comprising a 600-ft^2 overlapped glass-plate type of solar collector on a new house in Denver, Colorado. Pebble-bed heat storage (11 tons of 1.5-in-diameter gravel), an exchanger for hot-water supply, and a gas furnace auxiliary were used. This system is still functioning after 20 years of continuous operation. Extensive instrumentation, including solar radiation measurements, provided the data on which several reports of performance have been based.[5,6]

Solar-heated air has also been experimentally used in Australia for lumber drying and space heating.[7] A single-glazed, vee-corrugated absorber plate, with most of the air flow beneath the plate, supplied hot air directly to an industrial-type building or to a pebble bed for heat storage. The drying operation involves a similar system.

In 1975, a commercially designed solar air-heating system was installed in a new experimental house at Colorado State University in Fort Collins, Colorado. The house is one of three identical structures in the CSU Solar Village, each solar-heated by a different type of system. The collector is a double-glazed panel in which a black metal absorber plate serves to heat air passing beneath it. Pebble-bed heat storage, auxiliary gas heat, domestic hot water by heat exchange, and accessories complete the system. Extensive data[8,9] on the design and performance of this system have been published.

In 1974, commercialization of solar air heating commenced, and by late 1977, several hundred solar air-heating systems had been installed. In 1979, the largest solar-heated building in the world was constructed in Denver, Colorado, and used an air-heating system.

Some common misconceptions may have deterred researchers and practicing engineers from recognizing and utilizing several advantages of solar air-heating systems. One misunderstanding, even among solar specialists, has been that solar air-heating systems are less efficient than the liquid type; that is, less heat is deliverable from a unit collector area per year. However, as shown in detail below, air systems with pebble-bed storage usually have efficiencies equal to or higher than the liquid types. Another misconception has been that air systems require greater electricity use (for air circulation) than do liquid systems. Power requirements are usually comparable, but if the liquid system requires a heat exchanger between collector and storage (dual-liquid type), its power requirements are greater than those of the air type.

The larger space required in the air system for air ducts and the pebble bed than needed in the liquid system for pipes and water tank may impose design restrictions, but, as shown below, costs need not be substantially affected. With the absence of problems associated with freezing, corrosion, boiling, and damage from leakage in liquid systems, it is something of a paradox that air systems have not been more widely utilized.

The predominant method for heating dwellings in the United States is by forced-circulation warm air. Air is usually heated in a gas furnace, but oil furnaces, electric resistance heaters, and heat pumps are also used. Circulation of warm air through ducts leading to the rooms of the building and return of the cool air to the heater is an economical method providing a high level of comfort. Where air conditioning is also required, the same distribution system can be used to supply cool air from the air-conditioning unit to the living space.

Buildings are also heated by hot water or steam distributed to the living space by convective or pumped flow through pipes. Heat-exchange surfaces in the rooms may be of the baseboard-strip type, wall "radiators," window fancoil units, and occasionally coils of pipes in floors, walls, or ceilings. Baseboard-strip heaters are the most widely used.

Heat-supply temperatures are typically 140°F (60°C) in the air system, 180°F (82°C) in the baseboard hydronic system, and at levels usually between these extremes in the other types of hydronic systems. The compatibility of solar heat with these methods for heating the living space is strongly dependent on their temperature requirements. Solar-heated air can be efficiently provided at 140°F (60°C), and the lower temperature hydronic systems are compatible with flat-plate liquid solar collectors. The high-temperature operation of baseboard strips (required by the limited heat-exchange surface and relying on the relatively ineffectual mode of natural convection heat exchange), is not efficiently met by flat-plate solar collection systems unless additional heat-transfer surface is provided. It can be seen that a major advantage of a solar air system is its compatibility with conventional warm-air heating.

SOLAR AIR COLLECTORS

There are many ways in which air can be heated by solar energy. Air can be passed in contact with black solar absorbing surfaces such as flat metal plates, finned plates or ducts, corrugated or roughened plates of various materials, screens through which the air passes, and overlapped glass plates between which the air flows. Flow may be straight through, serpentine, above, below, or on both sides of the absorber plate, or through a porous absorber material. In its simplest form, the solar air collector differs from the liquid type only in the shape and size of the fluid passage in contact with the absorber. Figure 12.1 shows schematically several of the types which have been tested.

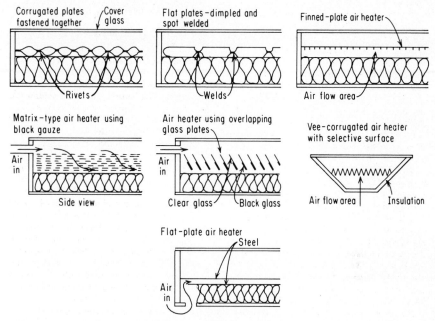

Fig. 12.1 Types of solar air heaters.

Physical Configurations

Because the heat-transfer coefficient between the absorber surface and the air is lower than that between the absorber and a liquid, the largest practical contact area, consistent with cost, is advantageous. Turbulent flow is also desirable for the same reason. The influence of flowrate on heat-transfer coefficient is much greater in air than in water collectors, so a good balance between heat-transfer coefficient, flowrate, temperature rise through the collector, and pressure loss must be established. Figure 12.2 shows the effect of changing the velocity in an air collector having a continuous smooth absorber plate under which air is circulating. As air rate is increased, efficiency rises, rapidly at low velocities but only slightly at rates above 4 $ft^3/min \cdot ft^2$ (~20L/s·m^2) of collector. At the higher flowrates, exit air temperature declines and, pressure loss increases. Although this graph is based on a specific collector design, other types would show the same general pattern.

The air collector is best employed in combination with a heat storage unit which provides a high degree of temperature stratification. Such a design permits the supply of air to the collector nearly always at room temperature, say 70°F (21°C). For acceptable temperatures in the air supply to the rooms, as well as for minimizing heat storage volume, collector flowrates should be such that a much higher temperature rise per pass is obtained than in liquid systems. At a flowrate of approximately 2 $ft^3/min \cdot ft^2$ of collector, a reasonably efficient collector will deliver air at a temperature of 140 to 150°F (60–66°C) in full sun. This rate of flow corresponds to 0.15 lb/ft^2 of collector per minute (0.01 kg/m^2·s), about the same mass flowrate as used in typical liquid collectors. Since the specific heat of air is one-fourth that of water, the temperature rise through the collector is approximately four times as great at the same collection efficiency.

Another difference between conditions in the two collectors is the approach of the fluid temperature to the plate temperature. A typical driving force in a liquid collector having reasonably closely spaced tubes (say 4 in, 10 cm apart) can be on the order of 10°F (5–6°C) between the fluid and absorber average temperatures. A smooth absorber plate being contacted by air on its lower side will operate at a temperature 40 to 50°F (22–28°C) higher than the adjacent air, for the same rate of heat transfer. In other words, the heat-transfer coefficient in the air collector is typically about one-fifth that in a good liquid solar heater. These relationships are illustrated in Fig. 12.3, where the temperature-distance relationships between the absorber surface and the fluid in these two types of collectors are illustrated.[10] It may also be observed

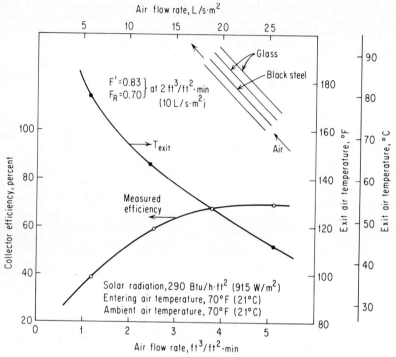

Fig. 12.2 Collector performance versus air flowrate.

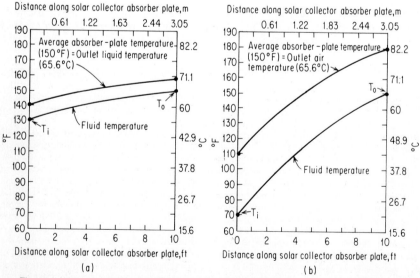

Fig. 12.3 Comparison of typical temperatures in (a) liquid- and (b) air-heating solar collectors.

that, under good solar conditions, the *average* absorber plate tempeɪature is about the same in both types when the storage conditions dictating supply temperatures to the collectors are such that fluid temperature at collector exits are comparable.

Performance Data

Test data on solar air collectors are not nearly as extensive as those on liquid types. Fewer models are commercially available, and the testing is somewhat more difficult. Figure 12.4 shows the performance of a leading commercial type involving a double-glazed porcelainized steel absorber plate below which air is circulated at 2 ft^3/min·ft^2 (10 L/s·m^2) in a passage 0.625 in (1.6-cm) high. The linear velocity through two 6.5-ft (2.0-m) panels in series is 8.33 ft/s (2.5 m/s).

Figure 12.5 shows the performance of three experimental air collectors in which absorber plates of irregular or corrugated contours were used.[11] The performance improvement achieved by greater turbulence and additional heat-transfer area is evident.

The efficiency of matrix-type collectors has not been extensively reported, but results of limited tests of heat-transfer rates are shown in Fig. 12.6.[12] The extended heat-transfer surface and the action of the air in cooling the exposed surface of the matrix both reduce heat losses

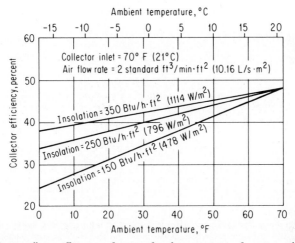

Fig. 12.4   Air collector efficiency as function of outdoor temperature for various solar inputs.

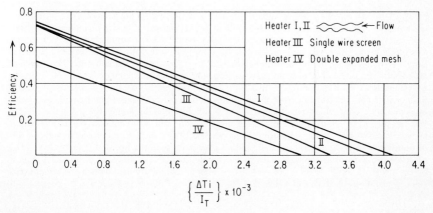

Fig. 12.5   Efficiency of solar air heaters, Heater I, steel; heater II, aluminum. $\Delta T_i = T_{f,i} - T_a$, $I_T =$ solar radiation, cal/cm^2·s.

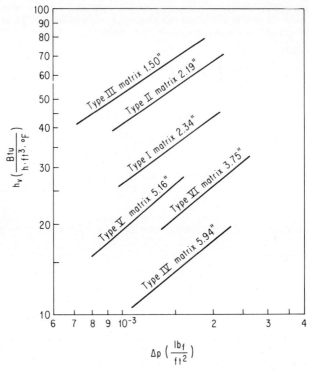

Fig. 12.6 Heat transfer and pressure drop in matrix-type solar air collectors. h_v = volumetric heat-transfer coefficient; p = pressure drop through matrix thick enough to absorb about 99.7 percent of incident radiation. Numbers on line show thickness of matrix; type of matrix refers primarily to size of openings in metallic foil. 1 Btu/h·ft³·°F = 18.63 W/m³·°C.

and increase efficiency. When operated at low ambient temperatures, however, losses are increased by contact between recirculated warm air and the cooler cover glazing. In a study of the overlapped glass-plate type of collector, efficiency and pressure drop were measured in a typical configuration.[13] The results are shown in Fig. 12.7.

As implied in the above discussion, the efficiency of a particular solar air heater can be substantially increased by use of higher volumetric air flowrates, but fan power requirements also increase and exit air temperatures decrease. Energy delivery can also be increased without a delivery temperature loss by increasing air velocity at constant volumetric flow per square foot of collector. The velocity increase can be achieved by flow through a series of collector panels or by reducing the cross-sectional area of the air passage in the collector. The question of air path length revolves around practical considerations in the application. If the collector panels are of a typical 6- to 8-ft (1.8- to 2.4-m) length, residential roof dimensions would usually limit the air path lengths to two or three such panels in series. Longer paths could, of course, be provided by suitable interconnections and manifolds. But the longer the path, the greater the pressure drop, so a reasonable balance must be accepted. Conventional residential warm-air heating systems involve pressure differences less than 1 in (2.5 cm) of water, and the use of reasonably sized motors for air circulation (normally below 1 hp, or 0.75 kW) limits the practical pressure drop through those collectors to about 0.5 in (1.2 cm) of water. A typical set of pressure requirements for various collector configurations involving a commercial smooth-plate type is shown in Table 12.1.[14]

AIR-HEATING SYSTEMS

As with all solar space-heating systems, the principal components are the collector and the heat storage facility. For reasons suggested above and outlined in more detail below, by far the most

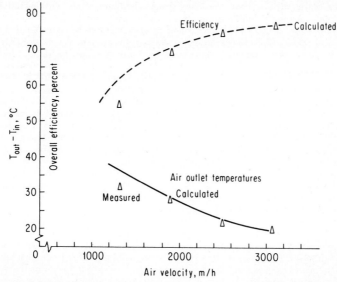

Fig. 12.7 Comparison of measured and calculated efficiencies and air outlet temperatures for various air velocities in overlapped plate collector.

practical and cost-effective storage medium, as well as being the one providing maximum collector efficiency, is a bed of loose, solid particles. Cost considerations invariably result in the choice of a bin of locally available gravel of uniform size.

The other components in a solar air-heating system are an air circulator, a control system with sensors and relays, and automatic dampers for properly directing the flow. If domestic hot water is to be solar-heated, a heat exchanger, pump, and hot-water storage tank are also required. Finally, as with all economically optimized solar heating systems, an auxiliary heater and a conventional heat distribution system are required.

Systems Configurations

A typical solar air-heating system containing the above components is shown in Fig. 12.8. An important distinction between air and liquid systems is the direct delivery of heat from the collector to the living space when air is used. The medium for heat transfer between collector

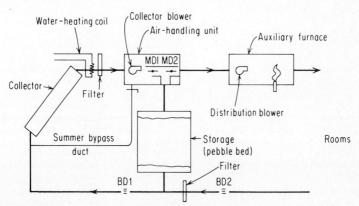

Fig. 12.8 Solar air-heating system schematic diagram showing collector, storage, and auxiliary components. MD1 and MD2 are motorized dampers. BD1 and BD2 are backdraft dampers.

and storage, warm air, also heats the living space. The temperature and flowrate from the collector usually match the building requirements reasonably closely.

Figure 12.9 shows diagrammatically the four primary methods of operation in a solar air-heating system. Figure 12.9 a shows the air circulation pattern when solar-heated air is supplied directly to the living space. Signals from the house thermostat and from the sensor which detects temperature difference between collector inlet and outlet operate the blowers and position the dampers to supply air directly from the collector to the rooms, as shown. Cool air returns from the living space to the collector for reheating. If necessary (but not usual), energy can be supplied to the auxiliary heater to increase the air temperature.

The control system diverts air from the room circuit to the warm end of the storage bed as in Fig. 12.9 b when the rooms no longer need heat and when solar energy is available. Heat is transferred to the pebbles, cool air returning from the opposite end of the pebble bed to the collector. This mode of operation usually predominates during sunny days. The very large heat-transfer surface in the pebble bed effectively removes heat from the air. Heat is stored in the rocks at progressively lower temperatures from the hot end to the cool end of the pebble bed. Except in spring and fall, when the solar system has surplus capacity, the temperature of return air to the collector is usually at approximately 70°F (21°C), that being room temperature and also the temperature of the cold end of the storage unit.

When the differential thermostat across the collector senses a rise insufficient for practical operation, the collector blower is shut off and a damper moves to prevent air flow through the collector. When heat is needed in the rooms, air is then circulated from the rooms through the pebble bed (in reverse direction, so that warm air leaves the heated end), through the auxiliary heater, to the rooms as shown in Fig. 12.9 c. If the heat furnished to the air by the pebble bed is sufficient to meet demand, the temperature of the rooms will rise past the thermostat set point, and the system will turn off. If the load is greater than can be met by the air temperature available from the pebble bed, the auxiliary heater is activated by a second, lower temperature point on the house thermostat so that additional heat is supplied to the air passing from the pebble bed to the rooms.

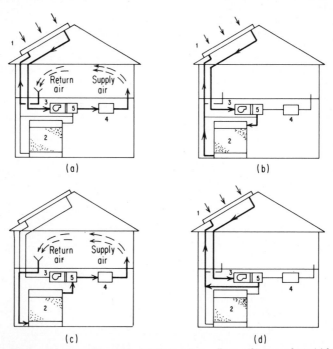

Fig. 12.9 Solar air system operation models. (a) Heating from collector; (b) storing heat; (c) heating from storage; (d) summer hot-water heating. Legend: 1, collector panel; 2, heat-storage container; 3, air-handling unit; 4, auxiliary heater; 5, water preheat heat exchanger.

It can be realized that these modes of operation permit (1) maximum use of solar energy, (2) use of auxiliary heat as a "booster" rather than as a substitute source, (3) maximum air exit temperature from storage even when a relatively small quantity of heat may be in storage, and (4) recirculation of cool air to collector with resulting high efficiency.

In residential space-heating systems, an air-to-water heat-transfer coil can be advantageously installed in the air passage leading from the collector, thus providing hot water for domestic use. Whenever air is being delivered from the collector *and* its temperature is above the temperature in the solar hot-water "preheat" tank, a small pump circulates water from the tank through the coil and back to the tank. Optionally, a gravity circulation loop can be used if the water tank is positioned above the heat-transfer coil. Depending on conditions in the tank and in the collector, the air temperature usually decreases 1 to 3°F (0.5 to 1.5°C) as it delivers heat in the hot-water coil. A typical design for a residence involves a two-row, finned coil having approximately 2 ft² (0.19 m²) of face area.

In order that solar-heated water may be supplied in summer when no space heating is required and that the pebble bed need not be heated during that season (with resulting heat loss into the house), a bypass duct with appropriate dampers is usually provided. Figure 12.9d shows this arrangement, with a manual damper blocking air flow to the storage bed and another manual damper opening a bypass to permit circulation directly back to the collector.

Use of Modular Air Handlers

A considerable cost saving, particularly in installation, can be made by use of a prefabricated "air handler" which contains the collector blower and motor, all automatic dampers, and the hot-water preheat coil. This one unit, completely factory-made and adjusted, eliminates the need for separate purchase and installation of the several components it contains. The requirements for tightly closed dampers are much more satisfactorily met by use of such an assembly. The installer then need only connect air ducts from the air handler to the collector, to the hot end of storage, and to the conventional heater. Another duct is installed between the conventional cold-air return and the collector inlet, with a branch to the cold end of the storage bed. As seen in the diagram, two backdraft dampers are needed in this cold-air return duct to prevent reverse flow.

Two types of air handlers are in common use. One is designed for systems which require only one blower, as illustrated in Fig. 12.10. In this arrangement, an internal bypass with appropriate dampers permits use of the single blower to move air in all of the necessary modes. The bypass makes it possible for the blower to supply solar-heated air *to* the pebble bed and also to withdraw hot air *from* the same end of the pebble bed.

In the two-blower design, there is an additional fan, usually in, or associated with, the

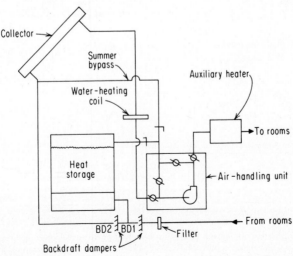

Fig. 12.10 Single blower system for solar air heating using an air-handling unit.

auxiliary heater. The design of the air handler is thereby considerably simplified, two automatic dampers are required rather than four, and control is simpler. As shown in Fig. 12.8, the blower in the air handler is used for circulating hot air from collector to storage during the storing mode, the second blower in the auxiliary unit is used for supplying heat to the living space from storage, and both blowers are operated when hot air is being supplied from the collector to the living space. In addition to simplifying the air handler and controls, this arrangement permits greater air flow to the living space than through the collector. The cost of the additional blower and its operation are more than offset by savings in the cost of the air handler.

Auxiliary Heat Systems

It is evident that any type of conventional air heater can be used as the auxiliary in this system. Gas furnaces are most common, but oil furnaces, electric resistance heaters, and conventional air-to-air heat pumps are also in use. The only modification in the conventional system needed for the solar combination is the substitution of a class B (rather than class A) furnace blower motor (if in the air stream) capable of operation in air at temperatures up to 150°F (66°C).

If a heat pump is used as the auxiliary, as shown in Fig. 12.11, the separate, two-component type with outdoor and indoor sections is most practical. This "parallel" system, involving *either* solar heat supply *or* heat-pump supply, has been found to require less electric energy than a dual source involving use of solar heat to supply the evaporator coil of the heat pump. Because of heat-pump compressor head pressure limitations, the temperature of solar-heated air is not "boosted" by the heat pump in the parallel mode. If solar heat is insufficient, the heat pump is used alone or in conjunction with resistance strip heaters.

Flow Distribution and Balancing

The extent to which series flow and parallel flow through the collector array are practiced depends on the design of the collector and its individual modules. The trend toward factory-made modules of convenient size, typically 3 × 6.5 ft(1 × 2 m), makes it necessary to consider various options for grouping the collector modules depending on mounting constraints. Flowrate per square foot of collector is the main determinant of air temperature rise (i.e., delivery temperature). If two or three modules are joined in series flow, the linear air velocity through the panel will be, respectively, two or three times as great as when only one panel is separately supplied. Collector efficiency is higher with series flow, because of higher air velocity, but pressure drop also increases. Table 12.1 shows that, with a typical air flowrate of 2 ft³/min ·ft² of collector, the pressure drop through two panels is about ¼ in water gauge (w.g.), and through

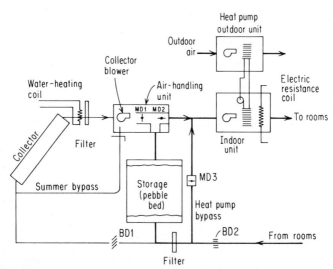

Fig. 12.11 Solar heating system with parallel air-to-air heat pump auxiliary heating building from storage with heat pump supplementary supply. MD1, MD2, and MD3 are motorized dampers. BD1 and BD2 are backdraft dampers. MD3 is partially open.

TABLE 12.1   Pressure Drop through Air-Collector Arrays*

Flowrate, ft³/min · ft²	Static Pressure Drop, in w.g.	
	Two panels in series	Three panels in series
2.0	0.23	0.77
2.2	0.27	0.91
2.4	0.32	1.06
2.6	0.36	1.22
2.8	0.41	1.39
3.0	0.47	1.57

*1 in w.g. = 2.54 cm w.g.; 1 ft³/min· ft² = 5.08 L/s· m².

three panels is about ¾ in. An array of 480 ft², for example, might therefore consist of 8 parallel channels of three 20-ft² (1.9-m²) series-connected modules, or 12 parallel passages of two panels in each series. If all 24 panels were connected in parallel, lower linear velocities would appreciably reduce heat-transfer coefficients and system efficiencies.

SYSTEM DESIGN

Many of the design principles involved with air systems are similar to, or modifications of, those associated with liquid systems. Collector tilt angles, orientation, glazing design, optimum sizing, auxiliary supply, and other factors are not significantly different. Coverage of these factors in Chaps. 7, 13, and 28 makes detailed discussion here unnecessary. Typical values of these important parameters for air systems are listed in Table 12.2.

Determination of Optimal Collector Area

As indicated in Chap. 28, the optimum size of a solar heating system depends on many factors, including the projected cost of conventional energy during the expected life of the solar heating system. Solar availability, heat requirements, equipment cost, interest rates, maintenance cost, and system durability are also important. Optimization of system size can be performed by use of one of the performance models described in Chap. 14. Use of these predictions of performance and an appropriate cost model permits determination of the collector size and the fraction of the total load carried by a solar system which provide the lowest lifetime cost of heating the structure.

TABLE 12.2   Typical Values of Solar Air-Heating System Parameters

Collector orientation	South facing
Collector tilt angle (from horizontal)	Latitude plus 15°
Storage volume	0.5 ft³/ft² (0.152 m³/m²) of collector
Storage mass	50 lb/ft² (244 kg/m²) of collector
Storage dimensions	Rectangular bin, 5 to 7 ft (1.52–2.13 m) depth
Storage material	Clean gravel or crushed rock uniformly sized
Specific heat of rock	0.2 Btu/lb °F (0.2 cal/g °C = 0.84 kJ/kg °C)
Storage particle size	0.75 in minimum to 1.5 in maximum (1.9 to 3.8 cm)
Air flowrate	2 ft³/min ft² (10.16 L/s m²) of collector
Static pressure loss, two panels in series	0.2 to 0.3 in w.g. (0.51 to 0.71 cm w.g.)
Static pressure loss, storage	0.15 to 0.25 inch w.g. (0.38 to 0.64 cm w.g.)
Superficial air velocity in storage	20 to 30 ft/min (6.1 to 9.1 m/min)
Air velocity in collector supply and return ducts	1000 ft/min (305 m/min)

Heat Storage with Air Systems

Although heat from solar air collectors can be stored in tanks of water or other liquids, in phase-change compounds, in numerous small containers of liquids, and in bins of loose solids, only a pebble bed is technically and economically advantageous. Figure 12.12 shows why pebble bed storage is so important in a solar air system.[9] The temperature-position history in a solar-heated pebble bed through a typical winter day shows storage of collected heat at temperatures of 100 to 140°F (38 to 60°C), while the cool end of the bed, from which air is returned to the collector, remains continuously at approximately 70°F (21°C). The temperature stratification which develops and persists in the pebble bed results in high collector efficiency even when air is being heated nearly to 150°F (66°C) and when heat is being stored at that high temperature.

Liquid storage requires a heat exchanger for transfer of collected heat to storage and another heat exchanger for heat delivery to the living space. These two exchangers, in addition to their cost, require pumps, pumping power, and temperature-driving forces which force a collector to operate at temperatures several degrees higher than storage. Temperature stratification is not practical in liquid storage systems, since the net result is an air-supply temperature to the collector usually well above 100°F (38°C) (determined by the average storage tank temperature) and a system efficiency substantially lower than that obtained with a pebble bed and its cooler collector inlet temperature.

Storage in phase-change materials has drawbacks similar to those of liquid storage. The isothermal characteristics of phase-change storage force the collector to operate at higher inlet temperatures and lower efficiencies. It is evident from Fig. 12.13 that if collector inlet temperature increases substantially, say from 70 to 110°F (21 to 44°C), with an ambient temperature of 40°F (4°C) and solar radiation of 300 Btu/h·ft² (946 W/m²), the decline in collector efficiency from 42 to 32 percent necessitates about one-fourth more collector area for the same net heat delivery. Even if liquid or phase-change storage were cheaper than pebbles, which is not the case, system costs would be greater because of increased collector area.

Figure 12.14 is a schematic drawing of a typical pebble-bed heat storage unit, approximately cubic in shape, of concrete, masonry, or wood, with plenums at top and bottom. Spaced concrete blocks on which wire screen or expanded metal lath is supported form the bottom plenum above which uniformly sized pebbles (normally used for concrete) are placed. Inlet and outlet air

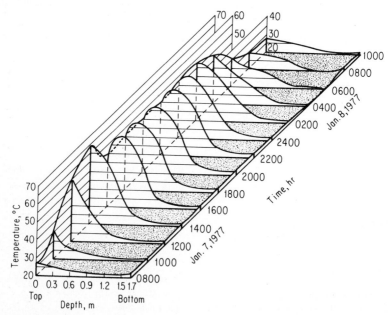

Fig. 12.12 Temperature profiles in the pebble bed during peak heating season.

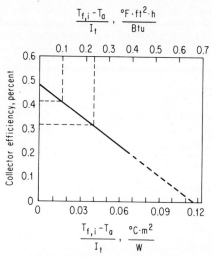

$$\frac{T_{f,i} - T_a}{I_t}, \quad \frac{°F \cdot ft^2 \cdot h}{Btu}$$

Fig. 12.13 Typical air-heating collector efficiency curve. $T_{f,i}$ = inlet air temperature; T_a = ambient temperature; I_t = solar radiation on tilted collector.

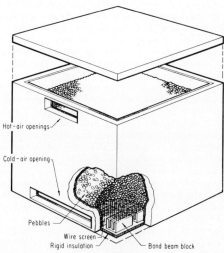

Fig. 12.14 Peeble-bed heat-storage unit, cutaway view.

connections and a tight-fitting cover are employed. Insulation equivalent to R-10 is advisable, but higher heat loss is not detrimental if the pebble bed is located in the heated space.

Typical pebble beds in single-family residences are cubical with volumes of 125 to 350 ft³ (3.5 to 9.9 m³), containing about 6 to 17 tons of rock. With 50 lb (23 kg) of gravel per square foot of collector, these pebble beds accompany collectors of 250 to 700 ft² (23 to 65 m²) and can store all of the solar heat collectible on a cloudless day.

Table 12.3 contains data on air pressure loss at typical flowrates through pebble beds of various size materials.[14] if space is not sufficient for the use of depths of 5 ft (1.5 m) or more, smaller rock in shallower beds should be used. The total pressure loss can thus be maintained at an acceptable level while obtaining satisfactory heat-transfer rates and temperature stratification.

TABLE 12.3   Static Pressure Loss through Pebble-Bed Storage Unit, in w.g.

Rock depth, ft	Rock size, in	Face velocity across rock box, ft/min		
		20	25	30
5	¾	0.14	0.23	0.34
	1½	...	0.12	0.17
5½	¾	0.16	0.24	0.37
	1½	...	0.13	0.19
6	¾	0.17	0.27	0.40
	1½	...	0.14	0.20

*1 in = 2.54 cm; 1 ft = 0.3048 m; 1 ft/min = 0.3048 m/min.

Since convection in a pebble bed is negligible when air is not being forced through it, no substantial difference has been observed in the performances of pebble beds heated from the bottom and of those heated from the top. The rate of temperature change at a given point in the bed is unaffected by the relative positions of the hot and cold zones. It has been observed, however, that undesirable channeling occurs in horizontal pebble beds, flow during the storing cycle tending to be in the upper part of the bed, whereas air tends to flow along the lower part of the bed in the heat removal cycle. Horizontal air flow through pebble beds is therefor inadvisable because of lower heat-exchange effectiveness than when vertical flow is used.

Auxiliary Heat

The use of a solar air-heating system permits auxiliary heat supply with maximum effectiveness. This fact is due to the capability of the system to use solar heat at low temperatures (70 to 90°F, 21 to 32°C), auxiliary then "boosting" this heat to usable temperature. The supply of auxiliary heat to air being delivered to the living space does not drive up the temperature of either the storage or the solar collector, so collector efficiency and storage capacity are not adversely affected. Solar collection even at temperatures only slightly above room temperature is useful; hence, the minimum solar intensity necessary for collector operation is lower and total solar collection is higher than in a system requiring the recirculation of warm fluids to the collectors.

Gas, oil, or electric heaters, usually with self-contained fans, having full capability for meeting the design heat load are supplied with warm air from the solar air handler. Cold air returning from the rooms, instead of entering the furnace, passes to the collector or to the cold end of the pebble bed. Furnace controls are wired directly to the solar controller, as explained below.

If a heat pump is used, it must be designed as an alternative heat source rather than as a booster, because excessive head pressure would be developed in the heat-pump compressor if warm air from the solar system were being further heated in the heat-pump condenser. If the solar system cannot meet the demand, the heat pump, with the aid of electric strip heaters if necessary, furnishes all the heat required in the building. Return air from the rooms is supplied through a bypass duct (Fig. 12.11) and automatic damper to the heat-pump condenser, while outdoor air is used as the low-temperature heat supply to the evaporator. Although the system does not have the capability of using low-temperature solar heat (below 90°F, 32°C), it nevertheless requires less electricity than a heat pump employing stored solar heat supply to the evaporator. Conventional (electric) cooling can also be provided in summer by reversing the function of evaporator and condenser, solar heat then being used only for water heating.

Controls

As in liquid solar systems, control of solar heat collection in an air system is performed by sensing the temperature difference between collector inlet and outlet. The outlet sensor must be located in the air passage within the collector panel, as shown in Fig. 12.15.[14] When the temperature at this point is approximately 30°F (17°C) higher than in the inlet duct, the collector blower motor is actuated (in the air handler), and dampers are positioned to direct the hot air either to the rooms or to storage, depending on whether the house temperature is above or below the room thermostat setting. Table 12.4, associated with Fig. 12.16, show the conditions corresponding to these two modes of operation.[10] The two-blower system is assumed in this table.

Adjustment of the deadband in the setting of the collector thermostat is such that the blower is not shut off until the increase in air temperature through the collector falls below about 10°F.

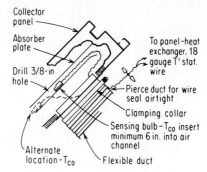

Fig. 12.15 Collector sensor placement.

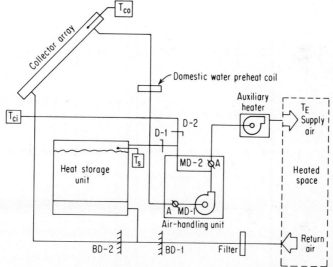

Fig. 12.16 Control dampers and sensors in solar air-heating system. T_{ci} = collector inlet temperature sensor; T_{co} = collector outlet temperature sensor; T_s = hot zone storage temperature sensor; T_E = room temperature sensor; D1, D2 = manual seasonal dampers; MD1, MD2 = motorized dampers; and BD1, BD2 = backdraft dampers.

At that point, the blower is turned off and the dampers are repositioned as shown in the table. If the house continues to require heat, the room thermostat actuates the blower in the auxiliary heater and positions the dampers to provide circulation of air from the rooms through the pebble bed, into the air handler, and on through the auxiliary heater to the rooms.

The auxiliary heat supply is called into use if room temperature continues to decrease, even though solar-heated air is being supplied. When a second contact in the room thermostat closes, at a temperature about 2°F (1°C) lower than the primary set point, fuel is supplied to the auxiliary furnace, or electricity is supplied to the auxiliary electric heater, thereby increasing the temperature of the air passing to the rooms. Since a full-capacity auxiliary heater is used, room temperature rapidly increases until the upper temperature setting is exceeded and the system is entirely shut off.

Unless an additional temperature sensor is provided in the pebble bed, the above sequence of events could occur even when there is little or no useful heat in the pebble bed. In such a situation, relatively cool air would be delivered to the rooms from the pebble bed until the room thermostat actuates the auxiliary heater. In particularly cold weather, room temperatures could fall considerably below the lower set point before auxiliary heat could reverse the direction

TABLE 12.4 Control Conditions for Solar Air-Heating System in Fig. 12.16*

$T_E < T_{R1}$†	$T_E < T_{R2}$	$T_{co} > T_{ci}$	$T_S > T_{SR}$	B_{main}	B_{aux}	MD_1	MD_2	BD_1	BD_2	D_1	D_2	GAS	HWP	Mode
1	0	1	1	1	1	0	1	1	0	0	0	0	0	Heating from storage
1	1	0	1	1	1	0	1	1	0	1	0	1	0	Heating from storage plus auxiliary
1	x	0	0	1	1	0	1	1	0	1	0	1	0	Heating from auxiliary
1	0	1	x	1	1	1	1	1	1	1	0	0	1	Heating from collector
1	1	1	x	1	1	1	1	1	1	1	0	1	1	Heating from collector plus auxiliary
0	1	1	x	1	0	1	0	0	1	1	0	0	1	Store heat
0	x	0	x	0	0	x	x	0	0	1	0	0	0	Do nothing
0	x	1	x	0	0	0	0	0	0	1	1	0	1	Domestic water heating, summer only
x	x	0	x	0	0	1	0	0	0	1	1	0	0	Do nothing, summer only

*In temperature table, (1) indicates condition is as stated, (0) indicates condition opposite to that stated, and (x) indicates either condition can exist. In component table, (1) indicates on or open, (0) indicates off or closed, and (x) indicates either condition can exist.

†Definitions of abbreviations:

T_E — Room temperature
T_{R1} — Room thermostat setting, first point
T_{R2} — Room thermostat setting, second point
T_{co} — Collector outlet temperature
T_{ci} — Collector inlet temperature
T_S — Storage temperature at hot end
T_{SR} — Storage thermostat setting, hot end

B_{main} — Blower in air handler
B_{aux} — Blower in auxiliary heater
MD_1 — Motorized damper in supply to main blower
MD_2 — Motorized damper in supply to auxiliary heater and rooms
BD_1 and BD_2 — Automatic (no motor) backdraft dampers
GAS — Auxiliary heat supply
HWP — Circulating pump for domestic water preheat coil

of temperature change in the living space. To avoid this excessive temperature decrease, another temperature sensor can be placed in the hot end (normally the upper end) of the pebble bed and set to require a minimum temperature, say 80°F (27°C), for first-stage (solar) heating from storage to be attempted. If the pebble-bed temperature does not exceed 80°F (27°C), a room temperature decline below the upper thermostat set point, say 70°F (21°C), causes immediate actuation of the auxiliary heater. A smaller temperature variation in the living space can therefore be achieved by use of this low-limit sensor in the storage unit.

Domestic hot-water supply can be controlled by sensing the difference in temperature of the air at the collector exit and the water in the solar hot-water tank. When this difference exceeds about 20°F (11°C), a small pump circulates water from the storage tank to the heat-exchanger coil in the hot air stream. If the temperature difference is below the preset value, the circulating pump does not operate. To avoid developing excessive temperatures in the solar preheat tank, near boiling for example, the tank temperature sensor can also be used as a limit control. Water circulation through the heat-exchanger coil is prevented if the storage tank temperature is above a certain level, say 180°F(82°C).

Figure 12.17 shows a wiring diagram of a typical controller for an air system.[14] Terminals for connection of all sensors, the air handler, solar hot-water heater, compression-type air conditioner, and the auxiliary heater are provided.

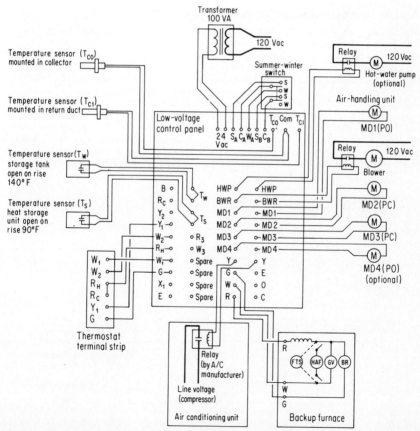

Fig. 12.17 Solar air system controller. Symbols: W_1, first-stage heating—solar; W_2, auxiliary heating; R, transformer power; S, summer; W, winter; Com, common ground; W1, first-stage heating; W2, second-stage heating; GV, gas valve; BR, blower relay; C, common—24 V ac; □, marked board terminal; ○, marked component terminal; MD, motorized damper; BWR, blower motor; HWP, hot-water pump; T_{co}, temperature—collector outlet; T_{ci}, temperature—collector inlet; T_w, temperature—water; T_s, temperature—storage; FTS, fan timer switch; HAF, heat-assisted fan switch; —, low-voltage wiring; —, line voltage.

If a heat pump is used as the auxiliary heat supply unit, an additional motorized damper is operated by the lower temperature set point in the room thermostat. When the solar system fails to provide adequate heat, the lower temperature contact in the room thermostat causes the collector blower to shut down, the motorized damper MD2 (Fig. 12.11) in the air handler to close, the air blower in the heat pump indoor unit (condenser) to operate, and damper MD3 to open. Cool air from the rooms is then supplied directly to the heat-pump condenser where it is heated, and warm air is returned to the rooms. The evaporator side of the heat pump, located outdoors, is operated in the usual manner. A third temperature set point in the room thermostat, a degree or two below the second point (or an outdoor temperature sensor), actuates the electric resistance heating (strip heaters) when neither the solar supply nor the heat-pump output is sufficient.

Factory-built controllers containing essentially all the functions outlined above are commercially available, with or without the necessary sensors.

RESULTS OF TESTING SOLAR AIR-HEATING SYSTEMS

Detailed performance results on two solar air-heating systems have been published. Although several hundred systems have been installed from 1975 to 1978, absence of monitoring instruments has limited data procurement from these systems.

Solar Air-Heating System in 1957 Denver House

One of the most detailed reports of air system performance is that based on the Denver installation operating since 1957. The results of a complete year of operation[4,5] show solar heat supply, fuel heat supply, hot-water use, and solar and weather data on daily, monthly, and annual bases. Figure 12.18 and Table 12.5 show, respectively, the principal design features of the house and heating system, as well as the monthly energy statistics for the 1959–1960 heating season. These results were based on continuous records of temperature, solar radiation, and operating mode and on daily readings of gas and electric meters.

The solar heating system was operated for 18 years without interruption or significant repair, supplying one-fourth to one-third of the annual heating requirements, depending on weather conditions. To reappraise the heating capability of the system after this service, the monitoring equipment was returned to use in the fall of 1975. System performance was again continuously monitored through the 1975–1976 heating season. Solar heat deliveries were then compared

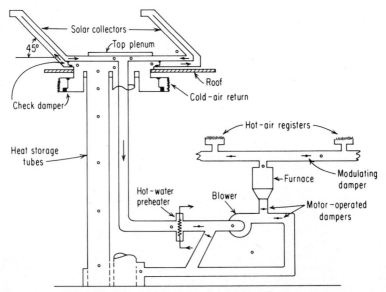

Fig. 12.18   Diagram of the Denver house solar air-heating system.

with those in corresponding months during the 1959–1960 season. Two columns so identified in Table 12.6 show that total solar heat delivery declined about 15 percent (45.53 to 38.92 × 10^6 kJ) during the 16-year period.[6] Most of the decline was attributed to dust deposition on the interior surfaces of the collector, some breakage of glass inside the collector, partial air blockage in about 10 percent of the collector panels, and air leakage.

In the fall of 1976, the system was overhauled by removing and cleaning all the collector glass and replacing broken plates, increasing duct sizes in the air-handling equipment to obtain additional air flow, rebalancing air flow throughout the system, and replacing modulating zone dampers with motorized dampers operated by a two-stage room thermostat. The collector thermostat sensor was moved from an exterior position to a location directly in the exit air stream from the collector.

Following completion of maintenance and remodeling, costs of which were closely monitored, system performance was again determined by continuous monitoring of pertinent operating data. The results are shown in the last columns of Table 12.6. It may be observed that delivery from the reconditioned solar system was about 15 percent higher (52.27 versus 45.53 × 10^6 kJ) than the output of the nearly new system 16 years earlier.

The conclusions that can be drawn from this important study are (1) a solar air-heating system can function for many years without appreciable repair requirements; (2) the annual decline in solar system heat supply, in this instance, appears to be about 1 percent per year; and (3) the normalized or adjusted cost of restoring this system to original performance was approximately $1350, equivalent to about 0.6 percent of estimated system cost per year.

Of importance is the fact that these figures apply to a system which involves a unique type of solar collector, not suitable for factory production and transport. The *system* is, however, very similar to commercial units now being installed. Full details of recent testing are available in Ref. 6.

Solar Air-Heating System in Solar House II, Colorado State University

A modern solar air-heating system on which extensive operating data are available is CSU Solar House II, pictured in Fig. 12.19. This installation in a new house completed in 1975 (shown schematically in Fig. 12.20) provides space heating, cooling, and hot water, supplemented by natural gas for heating and hot water.[15] Cooling is accomplished by use of the pebble bed as a regenerative heat exchanger in which heat is rejected to evaporatively cooled night air.

The principal design features of the house and heating system are shown in Table 12.7. The structure is nearly identical to two other adjacent buildings in which different types of solar heating systems are in operation. Digital logging of all essential data, including temperatures, flowrates, solar radiation, fuel supply, and other variables, provides data for hourly, daily, monthly, and seasonal energy supply from each source to each use.

A portion of the integrated monthly results of system operation is shown in Figs. 12.21 and 12.22. [9,16,19] Figure 12.21 shows, for each month, the total solar radiation intercepted by the collector, the amount of solar radiation intercepted when heat was being collected (collector blower in operation), and the solar heat recovery for space heating and water heating. The

Fig. 12.19   Colorado State University Solar House II heated by an all-air solar system.

TABLE 12.5 Energy Supply to Denver Solar House (1959–1960)

Winter 1959–60: all values in million Btu*

	September (18–30)	October	November	December	January	February	March	April	May	June (1–10)	Total*
1. Total solar incidence on 45°, 600 ft² collector area	9.93	26.84	25.81	21.98	25.48	22.10	30.43	26.56	29.61	8.13	226.86
2. Total solar incidence available on 45°, 600 ft² collector area when collection cycles operated	6.94	20.16	17.55	15.67	17.05	13.38	22.03	19.94	22.60	6.00	161.33
3. Gross collected solar heat	†	†	†	†	5.99	4.33	9.08	9.16	8.86	2.40	
4. Gross collector efficiency, %	†	†	†	†	34.7	32.4	41.1	45.9	39.8	40.0	
5. Useful collected heat	1.93	5.91	5.34	5.61	5.59	3.79	8.45	8.66	8.25	2.18	55.72
6. Net collector efficiency, %	27.9	29.4	30.3	35.8	32.7	28.3	38.3	43.5	36.5	36.4	34.6
7. Solar heat absorbed by storage tubes	1.12	3.04	2.77	2.79	2.64	1.92	3.99	3.68	3.01	0.50	25.46
8. Storage tube inventory	-0.03	0.07	-0.002	-0.008	0.017	-0.023	-0.008	0.07	0.17	-0.05	
9. Solar heat absorbed by water preheater	0.11	0.35	0.30	0.32	0.27	0.21	0.51	0.72	0.87	0.29	3.95
10. Heat delivered by natural gas for house heating	3.19	12.26	19.65	22.16	26.90	28.10	17.45	7.09	4.78	0.25	141.83

TABLE 12.5 Energy Supply to Denver Solar House (1959–1960) (Continued)

Winter 1959–60: all values in million Btu*

	September (18–30)	October	November	December	January	February	March	April	May	June (1–10)	Total*
11. Heat delivered by natural gas for water heating	0.67	1.49	1.79	2.01	1.84	2.68	2.60	3.23	3.24	0.88	20.43
12. Total heat load	5.79	19.66	26.78	29.78	34.33	34.57	28.50	18.98	16.27	3.31	217.98
13. Percent of useful collected heat absorbed by water preheater	5.7	5.91	5.63	5.7	4.84	5.55	6.04	8.2	10.56	13.4	7.09
14. Percent of total water heating load supplied by solar energy	14.1	19.0	14.4	13.75	12.80	7.26	16.4	18.25	21.20	28.40	16.25
15. Percent of house heat load supplied by solar energy (including water preheating but excluding water heating)	37.5	32.3	21.4	20.2	17.2	11.87	32.6	55.3	63.4	89.6	28.20
16. Percent of house heat load supplied by solar energy (including both water preheating and heating)	37.0	30.2	19.9	18.9	16.25	11.95	29.6	45.8	50.7	65.8	25.7

*1×10^6 Btu $= 1.055 \times 10^6$ kJ $= 293$ kWh. Collector area $= 55.74$ m^2 (based on outside dimensions) and 49.1 m^2 (glass area).
†Not determined.

TABLE 12.6 Energy Supply to the Denver Solar House (1975–1976), × 10⁶ kJ*

Month	1960	April 1975	April 1976	No.
Calendar year	1960	1975	1976	1
Total solar radiation on 45° collector area during month (49.1 m²)	24.75	30.43	26.00	2
When collection cycles operated	18.58	25.36	17.32	3
Useful collected solar heat	9.14	6.77	7.26	4
Solar heat absorbed by — Rock storage tubes	3.88	5.20	6.02	5
Solar heat absorbed by — Solar water preheater	0.76	0.77	0.35	6
Heat delivered by natural gas for — House heating	7.48	8.58	8.08	7
Heat delivered by natural gas for — Water heating	3.41	0.92	0.71	8
Total heat load — House heating — Solar plus natural gas	15.86	14.58	14.99	9
Total heat load — House heating — Calculated from monthly degree days	16.67	14.51	14.52	10
Total heat load — Water heating (solar + natural gas)	4.17	1.69	1.06	11
Total heat load — House heating + water heating (solar + natural gas)	20.02	16.27	16.05	12
Electrical energy used by air blower electric motor		1.60	1.79	13

TABLE 12.6 Energy Supply to the Denver Solar House (1975–1976), × 10⁶ kJ* (Continued)

Month		April		No.
Percent				
Solar collector efficiency, % (row 4 ÷ row 3)	49.2	26.7	41.9	14
Water heating furnished by solar energy, % (row 6 ÷ row 11)	18.2	45.6	33.0	15
House heating furnished by solar energy, % $\left(\dfrac{\text{row 9} - \text{row 7}}{\text{row 9}}\right)$	52.8	41.2	46.1	16
Portion of total heat load from solar energy, % (row 4 ÷ row 12)	45.8	41.6	45.2	17

No.	May			October			November			December			No.
	1960	1975	1976	1959	1974	1976	1959	1974	1976	1959	1974	1976	
1	27.59	26.55	26.26	25.01	26.07	28.98	24.05	22.89	23.77	20.47	22.16	24.54	1
2	21.06	22.77	16.82	18.79	21.55	25.29	16.35	19.22	17.11	14.60	19.20	19.34	2
3	8.70	6.56	7.90	6.24	7.25	9.45	5.63	5.33	7.46	5.92	4.01	6.20	3
4	3.18	4.88	7.28	3.21	6.05	8.45	2.92	4.30	6.34	2.94	2.84	5.29	4
5	0.92	0.85	0.27	0.37	1.19	0.59	0.32	0.74	0.31	0.34	0.64	0.35	5
6	5.04	3.30	2.90	12.93	6.75	5.04	20.73	18.67	15.33	23.37	25.43	17.73	6
7	3.42	1.13	1.14	1.57	1.41	1.98	1.89	0.57	1.62	2.12	0.14	1.35	7
8	12.83	9.01	10.53	18.84	12.81	13.49	26.05	23.26	22.48	28.96	28.80	23.59	8
9	12.04	4.52	9.50	19.01	9.31	15.45	26.09	21.54	21.26	27.50	29.04	24.70	9
10													10

TABLE 12.6 Energy Supply to the Denver Solar House (1975–1976), × 10⁶ kJ* (Continued)

No.	May			October			November			December			No.
11	4.34	1.98	1.41	1.94	2.61	2.57	2.20	1.31	1.93	2.46	0.78	1.70	11
12	17.16	10.99	11.94	20.78	15.41	16.07	28.25	24.57	24.41	31.42	29.58	25.29	12
13		1.36	1.22		1.29	1.76		1.76	2.20		2.09	2.37	13
14	41.3	28.8	47.0	33.2	33.6	37.4	34.5	27.7	43.6	40.5	20.9	32.1	14
15	21.2	43.1	19.4	19.0	45.7	23.0	14.4	56.5	15.9	13.7	82.0	20.5	15
16	60.7	63.3	72.4	31.1	47.3	62.6	20.4	19.7	31.8	19.3	11.7	24.8	16
17	50.7	59.7	66.2	30.0	47.0	58.8	19.9	21.7	30.6	18.8	13.6	24.5	17

No.	January			February			March			April through May and October through February Total			No.
	1960	1975	1977	1960	1975	1977	1960	1975	1977	1959–60	1974–75	1976–77	
1	23.75	21.35	24.74	20.59	22.38	29.21	28.36	26.24		166.21	171.83	183.50	1
2	15.89	17.71	20.66	12.47	18.92	18.54	20.53	20.84		117.74	144.73	135.08	2
3	5.90	4.09	5.96	4.00	4.91	8.04	8.91	5.50		45.53	38.92	52.27	3
4	2.79	2.91	3.70	2.03	3.85	6.73	4.21	4.85		20.95	30.03	43.81	4
5	0.28	0.49	0.08	0.22	0.54	0.15	0.54	0.61		3.21	5.22	2.10	5
6													6

TABLE 12.6 Energy Supply to the Denver Solar House (1975–1976), $\times 10^6$ kJ* (Continued)

No.	January				February			March		April through May and October through February Total			No.
7	28.38	24.92	24.54	29.65	22.14	12.55	18.41	17.23		127.58	109.79	86.17	7
8	1.94		0.84	2.83		2.19	2.74	0.63		17.18	4.17	9.83	8
9	33.99	28.53	30.42	33.42	26.51	20.45	26.79	22.11		169.95	143.50	135.95	9
10	34.19	28.41	29.51	34.41	26.12	21.01	26.26	22.45		169.91	133.45	135.95	10
11	2.23	0.49	0.93	3.05	0.54	2.34	3.28	1.24		20.39	9.40	11.94	11
12	36.22	29.01	31.34	36.47	27.05	22.79	30.07	23.36		190.32	152.88	147.89	12
13		1.96	2.57		1.87	2.14		1.84			11.93	14.05	13
14	37.0	23.1	28.9	32.1	25.9	43.37	43.4	26.4		38.7	26.9	38.7	14
15	12.8		10.0	7.3		6.31	16.4	49.2		15.7	55.5	17.6	15
16	16.5	12.6	19.3	11.3	16.5	38.6	31.3	22.1		24.9	23.5	36.6	16
17	16.3	14.1	19.0	11.0	18.1	35.3	29.6	23.5		23.9	25.5	35.3	17

*1×10^6 kJ $= 0.948 \times 10^6$ Btu $= 277.7$ kWh.

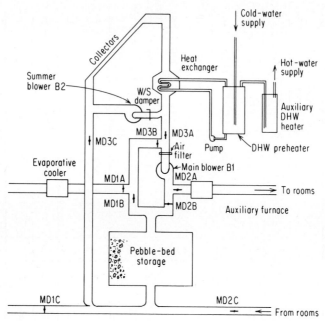

Fig. 12.20 Schematic arrangement of the solar system in Colorado State University Solar House II. MD = motorized damper.

TABLE 12.7 Design Characteristics of House and Solar Air-Heating System—Colorado State University Solar House II

Floor areas	
Lower level, half above grade	1382 ft² (128.5 m²)
Main floor	1382 ft² (128.5 m²)
Heating load	
UA, overall	419 W/°C; 19,060 Btu/°F·day (38.9 MJ/°C·day)
Design heat loss	60,000 Btu/h at −10°F (63.3 kJ/h at −23.3°C; 17.6 kW at −23.3°C)
Collectors:	
Site built, double-glazed, nonselective	722 ft² (67.1 m²) occupied area
South facing, 45° slope, directly on roof	690 ft² (64.1 m²) absorber area
Storage	
Insulated wood box	6 × 11 × 7.2 ft (1.82 × 3.35 × 2.2 m) high, inside
Rock volume	363 ft³ (10.2 m³)
Rock mass	36,300 lb (16,466 kg)
Pebble size	0.75 to 1.5 in (1.9 to 3.8 cm)
Furnace	
Gas-fired duct furnace	109,600 Btu/h (115.7 MJ/h) output,
79% efficiency	1000 to 2000 ft³/min air rate (28 to 56 m³/min)

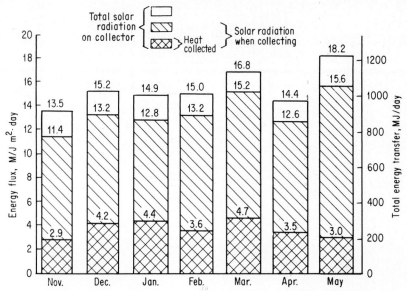

Fig. 12.21 Monthly average daily performance, Colorado State University Solar House II, 1976–1977.

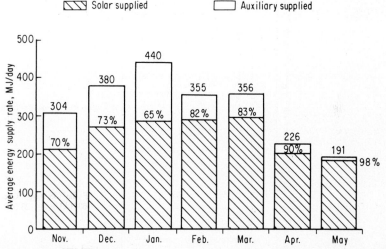

Fig. 12.22 Solar and auxiliary contribution to space and DHW heating load, Colorado State University Solar House II, 1976–1977.

number of days of usable data in each month is indicated. Malfunctions and adjustments of data logging equipment resulted in occasional losses of data.

Figure 12.22 shows, for each month, the total heat supply to the building from the solar system and from auxiliary fuel and the quantity supplied by solar. During the period 4 November 1976 to 16 May 1977, the solar system supplied 73 percent of the total space-heating and hot-water requirements. The numerical data depicted in Figs. 12.21 and 12.22, along with other results derived from these tests, are detailed in Table 12.8. It is evident that the solar system

TABLE 12.8 System Performance, CSU Solar House II All-Air System, 1976–1977*

| Month | † | Space and domestic hot-water heating | | | | | | | | | | | | Collector performance | | | | Electric energy | |
| | | Energy required | | | Solar energy use | | | Auxiliary energy used | | | Solar fraction | | | Total incident solar | Solar while collecting | Heat from collector | Avg. efficiency | Elect. energy for solar | Elect. energy for house |
		Space	DHW	Total	Space	DHW	Total	Space	DHW	Total	Space	DHW	Total						
Nov.	27	271	33	304	197	17	214	73	17	90	0.73	0.50	0.70	906	764	197	0.26	19.0	†
Dec.	27	331	49	380	238	39	277	93	10	103	0.72	0.80	0.73	1018	886	283	0.32	22.9	215.4
Jan.	31	378	62	440	233	53	286	146	8	154	0.61	0.87	0.65	1003	858	293	0.34	21.1	295.6
Feb.	28	290	66	356	231	60	291	58	6	64	0.80	0.91	0.82	1009	889	292	0.33	22.8	298.4
Mar.	31	298	58	356	245	52	297	52	7	59	0.82	0.89	0.83	1128	1017	315	0.31	27.5	327.9
Apr.	29	174	52	226	160	44	204	13	9	22	0.92	0.84	0.90	965	848	235	0.28	25.1	259.9
May	16	138	53	191	135	52	187	3	1	4	0.98	0.98	0.98	1220	1048	200	0.19	25.9	207.2

*Energy quantities in MJ/day.
†Number of days of usable data.
‡Not available.

supplied most of the heat needed for space heating and hot water and that the efficiency of collection averaged about 30 percent during midwinter months when heat requirements were high. It is also seen that the electric power requirements for air circulation were modest, averaging 5 to 10 percent of the solar energy delivered to use. A seasonal "coefficient of performance," that is, total heat delivery divided by electric energy use, was 13.5.

Figure 12.23 shows a comparison of performance of the system described above and of the solar heating system involving the use of flat-plate liquid solar heaters in CSU Solar House I.[17] The two collectors are approximately the same size, both are double-glazed, both have flat-black painted absorber plates, and both buildings are thermostated at approximately the same temperature. Although the heating requirements in the building heated by the air system are substantially larger, due to higher infiltration and ventilation, about the same fraction of the total load was supplied by the solar system in the two buildings. Total solar heat delivered by the air system exceeded solar heat delivery from the liquid system by about 30 percent. Collection efficiency, based on total solar energy intercepted by the collectors, averaged 25 percent in the air system and 17 percent in the liquid system. This difference is due largely to the fact, indicated in Fig. 12.21, that the air system can operate effectively, with useful heat delivery, at lower solar intensities and for longer daily time periods than can the liquid system.

Preliminary results on the use of nocturnal cooling with pebble-bed exchange are shown in Fig. 12.24.[9] Evaporatively cooled night air was supplied to the pebble bed whenever the ambient dry-bulb temperature was below 70°F. With a typical wet-bulb depression of 10 to 15°F (5 to 8°C), 55 to 60°F (~ 14°C) air could then usually be supplied to the pebble bed. House air was circulated through the pebble bed to cool the rooms during the daytime. This control

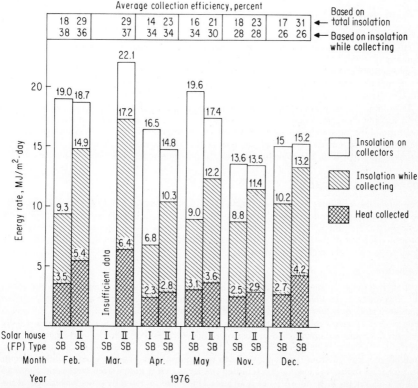

Fig. 12.23 Comparative performance of solar air and liquid systems (based on same days) for Colorado State University Houses I and II. Based on total insolation and insolation while collecting. Solar House I: site-built flat plate (SBFP) liquid collector, gross area 71.3 m²; Solar House II: site-built flat plate (SBFP) air collector, gross area 67.1 m². *Note:* Numbers on bars are based on gross area occupied by collectors.

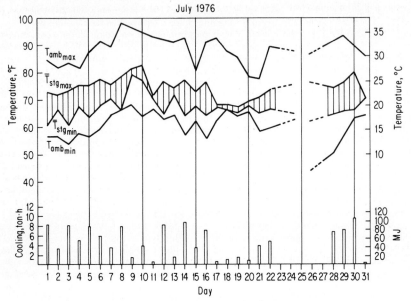

Fig. 12.24   Cooling performance for July 1976, Colorado State University House II.

strategy was not completely satisfactory, because greater cooling would have been possible if some type of wet-bulb temperature sensor had been used. It is expected that improved control instrumentation will augment the moderate amount of cooling provided by this system. It must be recognized that this type of cooling is not based on solar energy use, nor is it usable in humid climates where sufficiently low wet-bulb temperatures are not regularly available.

Use of the day-night exchange cooling system requires an additional air blower in the circuit, with suitable dampers, to permit circulation of hot air from the collector through the water-heating coil during summer periods. With such an arrangement, the main blower can be used for cooling by air circulation through the cool pebble bed, while the second blower is used for circulating hot air from the collector through the hot-water coil, back to the collector.

OPERATIONAL DETAILS

The advantages of solar air systems, particularly for residential buildings, may not be fully achieved in practice unless careful attention is given to details of design and construction. Air leakage in collectors and ducts and through dampers, pressure losses in poorly designed systems, excessive lengths of ductwork, dust pickup and deposition, and system noise, unless properly dealt with, can reduce air system benefits.

Leakage

Although leakage in air systems does not have the damaging consequences encountered in liquid systems, it can adversely affect operating efficiency. If the collectors are not well designed and fabricated, or if they are poorly installed, inflow of cold air or outflow of hot air will occur. For several reasons, collector operation at slight negative pressure is desirable and usually practiced. Cracks and openings in and between collectors result in inflow of cold air and dilution of the hot air stream delivered from collector to use. This infiltration is balanced by equal outflow of warm air from the living space to the atmosphere. The net result of this collector leakage is an infiltration load on the building similar to that resulting from cracks and openings in the building itself. In effect, there is an added heat load imposed by collector leakage. If the additional ventilation resulting from these leaks is not desired, good design and workmanship in manufacture and installation are essential.

Air leakage out of or into ducts which pass through unheated space, as in an attic, has the same detrimental effects as leakage in the collector. Careful installation and sealing of all duct seams and connections can eliminate this loss. Effective sealing of all joints in the pebble-bed container, with silicone or other durable caulking compounds, must be performed.

Leakage around imperfectly closed dampers can also reduce system performance. The most important requirement is a tightly closing collector damper (MD1 in Fig. 12.16). If this damper leaks when the building is being heated from storage, as at night, a portion of the air drawn from the rooms will pass through the collector and be cooled, possibly to subfreezing temperature. Heat is then being discarded from the house to the atmosphere through the cold collector. Cold air from the collector mixing with warm air from storage reduces the temperature of the room supply and increases the use of auxiliary heat. The tight closure of this damper is critical.

Insulation

Since the total surface area of the ducts between collector and storage, and the surface area of the storage bed itself, are substantial, insulation of these units is important. The hot-air duct, in particular, may have more than 100 ft^2 (9.3 m^2) of surface area, and if uninsulated and in unheated space, as much as 10,000 Btu/h (3 kW) could be lost. At least 1 in of effective insulation inside or outside these ducts or the use of ducts made of insulating material (molded fiber glass) is therefore essential. Even in the heated space, insulation is desirable for reduction of loss when heat is not desired. Pebble-bed insulation equivalent to R-10 is usually advisable.

Dust

With air filters normally used in warm-air heating systems, there is no accumulation of dust or lint either in the collector or in the pebble bed. Except in the matrix-type collector, or in designs involving flow of air across the upper surface of the absorber plate, there is no possibility of accumulation of household dust on the absorber plate or under the cover glass. With matrix collectors, very effective filters of the electrostatic type are recommended.

Odors

There have been suggestions that unpleasant odors might result from mold, mildew, and other growths in the pebble bed. If such growths occurred, their removal would be difficult. No such problems have been encountered, however, in the several hundred pebble beds installed inside buildings. In that location, the pebbles are never cooler than room temperature, so there is no possibility of moisture condensation in the bed. Dry pebbles will not support organic growths. If a pebble bed is used for cooling, however, as in CSU Solar House II, condensation is possible when room air is circulated through a cold pebble bed. Such a condition when accompanied by only limited temperature changes in the pebble bed during the cooling season might cause organic growths and odors. Experience with this cooling system is insufficient for evaluation of this problem.

Noise

Warm-air heating systems are inherently noisier than hydronic types. Provided that fan speeds are within proper ranges, the principal source of noise is air friction against ducts, grilles, and diffusers. If duct velocities are maintained below 1000 ft/min (5 m/s) and if the openings into the living space are designed so that velocities are below 500 f/min (2.5 m/s), noise is not noticeable. In well-designed solar air systems, the same generalizations apply. Noise levels are no higher than in conventional warm-air systems, particularly if a factory-built air handler is used. Sound transmission through ductwork can be further reduced by use of insulation *inside* the ducts rather than outside or by use of fibrous glass duct material with proper attention paid to fiber erosion into the airstream.

Maintenance

Maintenance requirements for solar air systems have been found comparable to those for conventional warm-air systems. Well-designed air collectors and pebble beds require no maintenance whatever and have long lives; only the air handler (and possibly the controls) has any parts subject to wear. Properly installed and adjusted motors, blowers, and motorized dampers deliver many years of maintenance-free service. Furnace gas valves and compressors in heat-pump systems, all of which are parts of *conventional* heating equipment, require more servicing than does a well-designed solar air-heating assembly.

Although a rare occurrence, if tempered glass is used, accidental breakage of the collector glazing must be repairable. In most air collectors, the modules are mounted so close together that access to an individual module requires the use of either a temporary working surface laid on top of adjacent collector panels (ladder, platform, scaffold) or a crane to suspend repair workers and materials from above.

Experience has shown that repair of solar air-heating equipment, once it has been put into operation and properly adjusted, is rarely required.

COSTS

Data on costs of solar air-heating systems are not as extensive as the information on liquid systems. Figures in 1978 dollars from one manufacturer show wholesale collector prices of about $14/ft^2 ($150/m^2), $700 for an air handler containing motor and blower, two motorized dampers, and a water heating coil, and $400 for the control system. A solar hot-water tank (80 gal, 300 L) and pump, two backdraft dampers, and two manual dampers for winter-summer switching add $350 to the wholesale price of components. The total wholesale price of a 500-ft^2 (47-m^2) solar collector and all other components, except for the pebble bed, is therefore about $8500.

Marketing of solar hardware through established heating and air-conditioning distributors and contractors adds 20 to 40 percent to these wholesale prices. By direct sale, small local fabricators could avoid some of the price markups, but additional material and labor costs, quality control problems, capital requirements, and other expenses tend to equalize final solar heat costs. It is not likely that solar heating systems will be marketed in a pattern substantially different from that by which conventional heating systems are distributed and sold.

To the retail, or customer, price of the above equipment must be added labor and materials for constructing the pebble bed, fabricating and installing the air ducts between collector, storage, and air handler, and for installing the solar collector on the roof or some other support. Materials and labor for the bin, the purchase and placement of the rock, and the installation and sealing of the cover involve an outlay of $500 to $1000 for a 250-ft^3 (7-m^3) bin, suitable for the 500-ft^2 (47-m^2) collector. The pebble bed thus involves a cost addition of $1 to $2 per square foot of collector.

With an internally manifolded air collector, installation on a roof involves a labor cost of about 50 cents to $1 per square foot of collector. The solar air ducts and the labor for their installation and insulation involve costs to the purchaser of $2000 to $4000 for a 500-ft^2 (47-m^2) collector system. The total installed cost of a typical solar air system including a 500-ft^2 (47-m^2) collector in a new residence therefore ranges from $12,000 to $17,000 in 1978 dollars. These costs are equivalent to $25 to $35 per square foot of collector.

A certain degree of confirmation of these prices may be found in the statistics of the U.S. Department of Housing and Urban Development Solar Heating and Cooling Demonstration Program. Table 12.9 contains a summary of fund allocations to the purchase and installation of solar air-heating systems in six residential buildings.[18] Although these figures do not represent actual project costs, they may be assumed to cover the anticipated costs of the solar heating systems.

It is highly unlikely there will be any significant cost reductions in solar heating equipment of the types described in this chapter. The figures cited above are based on carload prices of all materials, fabrication by use of large machine tools, and assembly on a production line.

TABLE 12.9   Total Contracted Costs of Residential Solar Air-Heating Systems Funded by U.S. Department of Housing and Urban Development, 1976 and 1977

Collector area, ft^2	Total funds allocated, $	System fund allocation, $/ft^2 collector
936	34,219	36.6
4,292	107,663	25.1
49,593	1,357,102	27.4
2,672	49,662	18.6
3,986	177,215	44.4
3,151	67,867	21.5
64,360 (5,979 m^2)	1,793,728	27.87 (average) [$300.00/m^2 (average)]

Virtually no reductions in these costs are possible. Conventional marketing methods are used, so savings in that sector are not likely. Increased experience by installing contractors can be expected to result in moderate reductions in the cost of installation labor until inflation obscures this trend. Total costs in the range of $25 to $35 per square foot ($270–375/m²) of collector should be applicable to solar air-heating systems during the early 1980s.

SUMMARY APPLICATIONS:

Solar air-heating systems are ideally suited to residential space heating and domestic hot-water supply. Their freedom from maintenance and repair requirements offers great advantage and convenience to the average homeowner. Corrosion and leakage problems are absent and durability is assured. Even extended vacancy in the building, during which time the system receives no attention or inspection, imposes no problems or hazards.

The size of ductwork and pebble bed in comparison with pipe and tank sizes in the liquid systems appears to be the only disadvantage. The pebble bed is approximately three times the volume of an equivalent water storage tank, and the ducts between collector and storage require a total cross-sectional area of about 4 ft² in a typical three-bedroom residence. The value of the occupied space, at present costs of unfinished (basement) residential area (at $10 to $15 per square foot of floor space) is in the range of $500 to $1000. The floor area required for a water storage tank costs about half these amounts. This cost difference is negligible in comparison with other savings.

In high-rise buildings and in very large commercial and industrial buildings on one or two levels, duct sizes and storage volumes may be prohibitively large. If air-distribution systems are used, however, ductwork between collector and storage would usually be no larger than the main distribution ducts, so solar heating of the air itself is a practical option. Space at ground level for one or more pebble beds would have to be valued in relation to competing uses for the space. In large buildings on one or two levels, multiple systems using collectors and pebble beds in various parts of the building provide economies through reduction in duct sizes and air-handling costs. Each zone then contains a separate solar air-heating system.

REFERENCES

1. G. O. G. Löf, "Solar Energy Utilization for House Heating," Office of the Publication Board, Washington, PB 25375, May 1946.
2. M. Telkes and E. Raymond, Storing Heat in Chemicals, A Report on the Dover House, *Heating and Ventilating*, November, 1949, p. 80.
3. R. W. Bliss, Design and Performance of the Nation's Only Fully Solar-Heated House, *Air Conditioning, Heating, and Ventilating*, vol. 92, October, 1955.
4. G. O. G. Löf, M. M. El-Wakil, and J. P. Chiou, Design and Performance of Domestic Heating System Employing Solar Heated Air — The Colorado House, *Proc. UN Conf. New Sources Energy*, vol. 5, 185, 1964.
5. *Ibid.*, Residential Heating with Solar Heated Air — The Colorado Solar House, *Trans. ASHRAE*, vol. 77, October, 1973.
6. J. C. Ward and G. O. G. Löf, Long-Term (18 Years) Performance of a Residential Solar Heating System, *Solar Energy*, vol. 18, no. 4, p. 185, 1976.
7. D. J. Close, R. V. Dunkle, and K. A. Robeson, Design and Performance of a Thermal Storage Air Conditioner System, *Mech. Chem. Eng. Trans., Inst. Eng. Australia*, vol. MC4, p. 45, 1968.
8. D. S. Ward and G. O. G. Löf, Design and Construction of a Residential Solar Heating and Cooling System, *Solar Energy*, vol. 17, p. 13, 1975.
9. S. Karaki, P. R. Armstrong, and T. N. Bechtel, "Evaluation of a Residential Solar Air Heating and Nocturnal Cooling System," Report prepared for U.S. Department of Energy, Committee on the Challenges of Modern Society, and the National Training Fund (COO-2858-3), Colorado State University, December, 1977.
10. Colorado State University, "Solar Heating and Cooling of Buildings, Design of Systems," S/N 003-011-00084-4, U.S. Government Printing Office, Washington, 1978.
11. C. L. Gupta and H. P. Garg, Performance Studies on Solar Air Heaters, *Solar Energy*, vol. 11, p. 25, 1966.
12. J. P. Chiou, M. M. El-Wakil, and J. A. Duffie, A Slit and Expanded Aluminum Foil Matrix Solar Collector, *Solar Energy*, vol. 9, p. 73, 1965.
13. K. Selcuk, Thermal and Economic Analysis of the Overlapped Glass Plate Solar Air Heaters, *Solar Energy*, vol. 13, p. 165, 1971.
14. Solaron Corporation, Solar Energy Systems, Application Engineering Manual, Denver, CO, April, 1977.
15. D. S. Ward, G. O. G. Löf, C. C. Smith, and L. L. Shaw, Design of a Solar Heating and Cooling System for CSU Solar House II, *Solar Energy*, vol. 19, no. 2, p. 79, 1977.

16. S. Karaki, G. O.G. Löf, and P. R. Armstrong, Space Heating with Solar All-Air Systems, *Proc. ISES Solar Energy Conf.*, *New Delhi*, Pergamon Press, New York, 1978, p. 1398.

17. S. Karaki, W. S. Duff, and G. O. G. Löf, "A Performance Comparison between Air and Liquid Residential Solar Heating Systems," Report prepared for U.S. Department of Energy and National Training Fund (COO-2858-4), Colorado State University, January, 1978.

18. Compiled from reports in *Solar Engineering Magazine*, December, 1977, p. 11.

19. S. Karaki, "Performance Evaluation of a Solar Air-Heating and Nocturnal Cooling System in CSU Solar House II," Department of Energy Report No. C00/2868/5, April, 1979.

Chapter **13**

Liquid-Based Solar Systems for Space Heating

FRED DUBIN, P.E.

and

SELWYN BLOOME, P.E.
Dubin-Bloome Associates, P.C., New York, NY

INTRODUCTION

Solar energy systems for space heating include active air-based systems, active liquid-based systems, and passive air or liquid systems. Energy conservation and some passive systems are frequently the most cost-effective measures to conserve nonrenewable fossil fuels and reduce operating costs for heating (see Chaps. 16 and 29). Conservation should be implemented first, to the greatest extent possible, before considering active solar energy systems, since it will reduce the cost and increase the efficiency of all active solar energy systems. Liquid-based solar energy systems are applicable to many existing[27] and new residential housing types and to most of the larger nonresidential building types where the space required for ductwork or a requirement for higher fluid temperatures for air conditioning preclude the use of an air-based system.

In a liquid-based system, water or an antifreeze liquid is circulated through the solar collector, but the building can be heated with either a hydronic system or an air distribution system. Liquid-based systems can readily supply combined space heating and cooling and domestic water heating. Chapter 11 deals with solar systems for domestic water heating. Chapter 12 describes air-based solar space-heating systems, and Chap. 15 analyzes solar space-cooling systems.

Economic, as well as performance, guidelines are summarized in this chapter. The initial capital, annual operating, and maintenance costs throughout the expected life of the buildings and facilities (life-cycle costing) are also considered. As the size and, hence, the cost of the solar energy heating system is roughly proportional to the building's thermal load, this load should be minimized wherever cost-effective. The solar energy system design must therefore be coordinated with energy-conservation design of the building envelope and an efficient heating, ventilating, and air-conditioning (HVAC) system.[16] Solar energy should be regarded simply as a further means to reduce consumption of fossil fuels. In many instances, active solar heating systems are not economically viable if natural gas is available. They are generally competitive with oil, propane, or electric resistance heating systems. However, natural gas and oil are not available in many areas and are becoming scarce in others, reducing their role as feasible alternative fuel sources. It is reasonable to assume that the cost of electricity and all fossil fuels will rise significantly. This, perhaps, is the most compelling argument in favor of solar heating.

This chapter is intended to be a summary of practical design guidelines for solar space-heating systems. Check-lists are used to ensure that no aspect of the system is overlooked during design and specification phases. Many analytical details are left for other chapters. Chapter 2 describes insolation calculations; Chap. 14, performance prediction; and Chap. 28, economic analysis and optimization. The material in these three chapters is not repeated herein.

COMPARISON OF AIR SYSTEMS AND LIQUID SYSTEMS

Liquid-based systems and air-based systems for solar heating have several advantages and disadvantages as summarized below.

Efficiency

A liquid collector has a higher efficiency than an air collector, at the same collector inlet temperature. However, the collector inlet temperature in an air system is generally lower than in a liquid- based system. The net effect is that under different conditions of temperature and insolation the collector efficiency of a liquid collector may be higher than, comparable to, or lower than that of an air collector.

A liquid-based solar heating system will require at least one additional heat exchanger. There may also be a second heat exchanger between collector and storage (e.g., with an antifreeze fluid in the collector). Each additional heat exchanger decreases the overall system efficiency and increases the system cost.

Simplicity

Air systems are simpler and easier to maintain than water systems. In water systems, precautions must be taken against leakage, corrosion, boiling, air binding, and freeze-up. These problems do not occur in air systems. However, air systems require more space for ducts.

Diversity

For domestic water heating, a liquid system is generally more efficient than an air system. Therefore, combined solar heating and hot-water systems often use liquid collectors and water-tank storage. Liquid systems are more adaptable for solar cooling; this is of particular significance for commercial buildings.

ENERGY REQUIREMENTS FOR HEATING

The size, type, configuration, and performance of a solar energy space-heating system depends on (1) the peak hourly heat load of the building and (2) the daily, monthly, and seasonal energy requirements to maintain the building at desired indoor temperatures during occupied and unoccupied hours. Energy conservation measures such as roof and wall insulation, reduced infiltration, double or triple glazing, and heat recovery systems reduce the hourly heat energy requirements, reduce the size of both the solar energy system and auxiliary backup system, and allow the use of the low temperature fluids to handle the load. A detailed discussion of energy conservation measures is given in Refs. 16 and 17.

The information required for calculating building energy requirements is as follows:
1. Climatic data
   - Wind velocity and direction
   - Winter design temperatures
   - Ambient temperatures
   - Heating degree-days (and degree-day base) or 5°F bin temperatures
2. Areas and heat-loss coefficients of walls, windows, doors, roof, and floors
3. Hot-water requirements (in gal/day) and cold-water supply temperature
4. Ventilation requirements
5. Details of mechanical systems in the building and their efficiency

Determination of Monthly and Annual Building Loads

Because accurate monthly and annual load determination procedures are not widely available, load models at two basic levels of sophistication and accuracy are recommended, depending on the design phase involved and the analytical tools available.

Steady-state models are generally quite accurate when used for residential and warehouse applications. Transient models, which account for hourly variations, are more accurate, but they require more detailed data input for their use. Hourly calculations account for the variation in solar availability, varying occupancy, equipment schedules, and the thermal capacity of the building and its contents—the effects of which are averaged in daily or monthly calculations. These transient effects can introduce time lags in the heating or cooling loads on the HVAC system, and must be accounted for if the energy consumption calculation is to be accurate in large buildings. The two basic model levels are described below.

Steady-state models   During the conceptual design phase, when only estimates of solar system performance, size, and fuel savings are needed, a steady-state load model may be used. The modified degree-day procedures[1] are restricted to small, single-story, envelope-dominated structures or large warehouses. For larger commercial or industrial buildings, where internal-cooling-only zones are prevalent, the bin method should be used.

Modified degree-day procedure   The degree-day procedure assumes that the steady-state heat loss or building heat load q_L is proportional to the equivalent heat loss coefficient (UA factor) of the building envelope; that is,

$$q_L = UA(T_B - T_A) - q_i \tag{13.1}$$

where T_B is the building interior temperature, T_A is the ambient temperature, q_i are internal heat sources, and UA is the building heat-loss conductance.

Traditionally for residential buildings an effective value of $T_B = 65°F$ (18.3°C) has been used to include heat-source effects in a 70°F (21°C) building. Recent research, however, indicates that monthly average internal and solar heat gains and improved insulation offset residential heat loss at a mean daily temperature below 65°F.[1] For commercial buildings, characterized by relatively large internal loads, T_B may be below 50°F (10°C). Therefore, the use of the modified degree-day procedure described in Ref. 1 is required for accuracy.

The building heat-loss parameter, UA in Eq. (13.1), is the space-heating loss (in Btu/h) at design conditions, divided by the design temperature difference. UA should include the ventilation and/or infiltration load based on the design temperature difference, calculated by traditional methods. (See Chap. 43 of Ref. 14.) Heating degree-days, referenced to a base temperature of 65°F, are given in the Appendix of this chapter.

Bin methods   Heating loads can be estimated using the "bin" method. The bin (or bin-hour) method consists of making instantaneous energy calculations at many different outdoor dry-bulb temperatures and multiplying the result by the number of hours of occurrence of each temperature bin. The bins are usually 5°F in size, and they are collected into three daily 8-h shifts. Because this method is based on hourly weather data rather than daily averages, it is considerably more accurate than the degree-day method. In addition, the bin method takes into account both occupied and unoccupied building conditions, and gives credit for internal loads by adjustment of the balance point. Weather data for bin method use are given in Chap. 6 of Air Force Manual 88-29.[24]

Transient load models   Massive buildings or buildings with small surface-to-volume ratios require more sophisticated load determination procedures. In such cases, the hourly differences in building thermal capacity (transients), solar loading on the building envelope, and heat generated internally by lights, equipment, and people significantly affect the building energy consumption. The diversity and sophistication of modern energy distribution and control systems, particularly solar energy systems, further contribute to energy consumption differences that can be found only by hourly calculation of the loads on the system and its response to them.

Two approaches to transient space-heating load determinations are recommended for the performance and sizing procedures that follow. Hourly building heating loads can be determined using computer dynamic analysis techniques. Computer programs that predict hourly, monthly, and annual loads on the basis of hourly weather data are available in both the public and private domain.

To date, the most widely used public domain programs have been NBS's NBSLD,[18] DOE-1 and -2, and NASA's NECAP.[19] NBSLD is strictly a load program, primarily a research tool, whereas NECAP is a full energy analysis program containing load, system, and equipment simulations. Both give good results, but the input is rather involved and computer run times are long. The DOE programs are computationally more efficient and include solar system subroutines.

Other programs to determine load or both load and energy analysis are privately available.

ASHRAE has published a bibliography of such codes.[21] Because some programs are designed to compute only annual heating and cooling energy requirements, they may have to be modified to give requirements on a monthly basis as required for solar design.

Energy Consumption Guidelines

For new design the Federal Energy Administration has suggested general guideline energy budget targets.[17] These targets, which the General Services Administration (GSA)[16] has adopted, indicate that new office buildings that consume less than 55,000 Btu/gross ft^2·year (624,000 kJ/m^2·year) are now being designed and constructed. A realistic energy budget goal seems to be 75,000 Btu/gross ft^2·year (850,000 kJ/m^2·year) for existing office buildings and 60,000 Btu/gross ft^2·yr (684,000 kJ/m^2·yr) for existing school buildings.

Buildings for which solar heating and/or cooling is being considered should be designed to meet these guidelines. For a small, single-story, well-insulated building, the annual thermal load should be in the range of 8 to 10 Btu/degree(°F)-day per square foot (160–200 kJ/°C-day·m^2) of floor area.

For an existing building, the exact load can be determined from past monthly and annual heating bills and degree-day values for the same period. Fuel consumption corrections should be made for furnace efficiency (typically 0.6) and for non-space-heating energy uses, which can be determined from summer fuel bills.[17]

SYSTEM COMPONENTS

The major components of a liquid-based solar energy space-heating system are:

- Solar collectors
- Insulated storage vessel(s)
- Heat-transport fluid
- Antifreeze system depending upon climate (fluid, heat exchanger, or drain-down mechanism)
- Controls and flow devices
- Circulation system (insulated pipes, pumps, ducts, fans)
- Conversion equipment and terminal devices (air-handling units, radiation, and air supply and return grilles)
- Backup and auxiliary heat source (electric resistance heaters, heat pumps, gas, oil, or wood furnaces, central steam, or hot water)

Solar Collectors

Solar collectors may be grouped into four classes—flat-plate collectors, concentrating collectors, hybrid collectors, and tubular collectors—as discussed in Chaps. 7 through 9.

Flat-plate collectors (Chap. 7) have the capability of collecting and using both the direct (or beam) and the diffuse radiation from the sun. Flat-plate collectors (see Fig. 13.1) with one or two cover plates, and flat black or selective surface absorber plates are most commonly used for space-heating applications due to their high efficiencies at low temperatures and low initial costs.

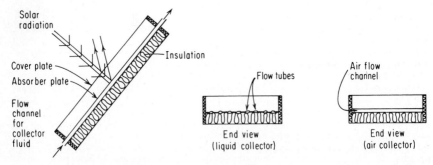

Fig. 13.1 Schematic drawings of flat-plate collectors.

If collectors are used for both heating and cooling, concentrating or evacuated tubular collectors should be considered because they can deliver higher temperature fluids required to operate absorption- or Rankine-cycle cooling machines. However, some flat-plate collectors with reflectors designed to increase concentration can be used effectively for heating and for solar cooling with those absorption refrigeration units which can develop full-load ratings at 195°F (90°C) generator temperatures.

Concentrating collectors are designed to concentrate the sun's radiation on a small absorbing surface and produce higher temperatures in the heat-transfer (transport) fluid than can be efficiently produced using a flat-plate collector. They are described in detail in Chaps. 8 and 9.

Hybrid collectors are combinations of flat-plate and concentrating collectors. An example is a flat-plate collector coupled to a Compound Parabolic Concentrator (CPC). In this system, the front of the flat-plate collector operates as a conventional flat-plate collector at higher insolation levels. The specially shaped curved reflector collects both direct and part of the diffuse radiation components within a certain range of directions and concentrates them onto the absorber plate.

Two kinds of tubular collectors have been designed:
1. Evacuated tube collectors
2. Nonevacuated tubular collectors

Evacuated tube collectors are designed to be efficient collectors at high temperature; the convection losses from the absorber are minimized and virtually eliminated by the vacuum between the absorber surface and the outer glass tube. Tubular collectors with a concentrating reflector and no vacuum are also available.

Collector performance standards   Various institutions have established standards of performance for solar heating and domestic hot-water systems. The Department of Housing and Urban Development (HUD) intermediate minimum property standards for solar heating and domestic hot-water systems have been adopted in several states as the minimum standards[64] a solar system should satisfy. It is also being widely required that collectors should have been tested according to the standards ASHRAE 93-77 (Ref. 4) and subjected to a 30-day exposure test. Examples of the requirements imposed by the HUD minimum property standards are given in Ref. 34.

Collector mounting   Solar collectors can be mounted on a sloping or flat roof surface (with appropriate supports to provide an optimum tilt angle), on the roof of a garage or porch or parking structure, or on the ground. Roof and support loads must accommodate the dead weight of collectors filled with fluid, the connecting piping, snow, and of utmost importance, wind loads. Wind loads are usually based on hurricane velocities. Supports can be fabricated of metal angles, tubes, or box struts or of treated wood. Particular care must be taken to properly flash all openings that penetrate roofs.

Solar collectors can be an integral part of the building wall or roof structure. In some cases the cost will be less than for separate collectors mounted on the structure. Examples of collector installations are shown in Figs. 13.2 and 13.3

A Checklist of Collector Installation Requirements

Support Structures and Design Loads. A framework to support the collector array is required unless the collectors are integrated with the roof. In either case, the support framework must be designed to withstand wind, snow, and hail, and roof-integral collectors must be able to support workers. All structural design should be based on generally accepted engineering practice and should be in compliance with ANSI A58.1 (Ref. 52). Design live and dead loads to be considered are discussed in Refs. 64 and 52.

Collector support structures may either be fixed or have an adjustable tilt angle. The struts and hinges of adjustable frames must be able to hold the collector in place at any angle of tilt without excessive vibration or deflection under design wind or snow conditions. The deflection magnitude must be related to the stresses and deflection that the collector glazing, absorber plate, and internal piping arrangement can withstand.

When collectors are mounted over a roofing membrane, the possible growth of fungus, mold, and mildew between the roofing membrane and the collector should be considered. Also, extreme temperature differentials could cause formation of ice dams which could back water under shingles or other roofing and cause rapid deterioration. Waterproofing membranes and proper flashing should be used. Flashing for collector panel supports that penetrate the primary roof membrane should be designed to prevent penetration by water or melting snow for the life of the roof system.

In very cold climates, water flowing from a warm collector may freeze on cold surfaces

Fig. 13.2 Solar-heated condominiums located in Boulder, CO, using liquid collectors and baseboard heating. The average utility bill for each 1100-ft² condominium was $11/month in 1978. Solar system designed by Dr. Jan F. Kreider, P.E.; architect, Joint Venture, Inc. (*Courtesy Dr. Jan F. Kreider.*)

Fig. 13.3 Aerial view of the largest solar-heating system in the world, located in Denver, CO; 40,000 ft² of collector provide 40 to 50 percent of the heat demand for the Denver Regional Transportation District garage located near the S. Platte River. Solar consultant was Dr. Jan F. Kreider, P.E.; architect, C.S. Sink. (*Courtesy Dr. Jan F. Kreider.*)

immediately below it (such as exposed eaves), thereby forming an ice dam that can cause water to back up under roofing or into the collector itself. This problem can be reduced by eliminating the cold surface or by using continuous flashing. Snow sliding from a collector may pile up below and cover part of the collector, thus increasing the possibility of thermal breakage of the glass cover plates. This may be prevented by providing space below the collector for snow pileup.

Ice Loads and Hail Loads. Ice loads may increase the collector support structure and roof loads. Means for computing ice loads are given in Ref. 50.

Collector cover plates should be designed to withstand the normal impact of hailstones falling at their terminal velocity. Reference 50 also gives probable hailstone masses and terminal velocities, as well as a map of the mean annual number of days with hail.

Wind Loads. Design wind loads should be calculated according to the procedures in ANSI A58.1 (Ref. 52). Flat-plate collectors mounted flush with the roof surface should use design loads that would have been imposed on the roof area they cover. Collectors mounted at an angle to or parallel to the roof surface on open racks can experience a lift force from wind striking their undersides. This wind load is in addition to the equivalent roof-area wind loads, and it should be determined according to accepted engineering procedures.

Suction loading can, and often does, exceed pressure loadings; so cover plate retainers must be adequately designed to keep them from being separated from the collector frame and to prevent cover-plate failure induced by suction loading.

Snow Loads. Collectors mounted flush with, parallel to, or at an angle to the roof surface or on racks in a saw-tooth arrangement should be designed to support the snow loads that would otherwise have been imposed on the roof area that they cover. If collector cover plates form steep slopes, snow shed from the collector or drifting snow may accumulate at the collector base, and this must be considered in the roof design.

Seismic Loads. Solar system components are subject to seismic forces generated by their mass. The design of all connections between components and the structure should allow for anticipated movements of the structure.

The seismic forces that collectors and other components should be designed to resist can be calculated by the procedures given in Ref. 64, which includes seismic zone maps.

Access and Circulation. Solar heating and/or cooling equipment must be designed for, and accessible to, maintenance, repair, or seasonal adjustment without disruption of major structural or mechanical elements. There should be enough clearance around solar equipment, based on potential maintenance tasks and equipment sizes, to permit examination, cleaning, adjustment, or servicing. Accessibility should reflect expected maintenance frequency. Roof-mounted collectors should have appropriate stairways, walkways, handrails, etc., to allow access by maintenance personnel. This is particularly important for multistory buildings. Solar system components should be located so that they do not block the primary means of occupant egress in case of fire, especially for roof-mounted solar equipment.

Shading. A solar collector should be so placed as to minimize shading by other on-site elements, particularly during those times of the day and year when the available energy is the greatest. Shade trees, chimneys, and other tall objects should not shade the collector array. Locating the solar buildings or collector in the northern part of a site decreases the probability of shading by future off-site development. Local codes should be reviewed for relevant height limits.

A careful study of seasonal shading characteristics is needed to determine the relative effects on ground-, roof-, or wall-mounted collectors. The choice of a sawtooth or planar array also may be dictated by shading considerations. Figure 13.4 shows the method of calculating spacing d for a sawtooth array

$$d \geq \frac{\sin (\gamma_{min} + \beta)}{\sin \gamma_{min}} L_c \qquad (13.2)$$

where γ_{min}, the profile angle, is given by

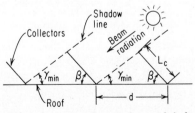

Fig. 13.4 Sawtooth collector arrangement spacing d required to avoid shading on the shortest day of the year when shadows are longest. The minimum profile angle γ_{min} is given by Eq. (13.3) and the minimum spacing d is given by Eq. (13.2).

$$\tan \gamma_{min} = \frac{\tan \alpha}{\cos a_s} \tag{13.3}$$

The solar altitude angle α and azimuth angle a_s are evaluated for 3 P.M. on the shortest day of the year. (See Chap. 2.)

Drifting Snow. Solar collectors, particularly those to be placed on the ground, should be located to avoid land depressions where frost or drifting snow can accumulate.

Prevailing Winds. As in determination of a building site, location of on-site solar collectors should take into account prevailing winds that can cause excessive collector heat loss. Protrusions such as parapets or fin walls windward of the collectors divert air flow away from the collector glazing. Also, fences, trees, and other on-site objects can act as windbreaks.

Collector Tilt and Orientation. Collector tilt and orientation are two important factors determining the solar energy collected per unit area of the collector. The best orientation of the collector depends on the particular application. For many cases, the best orientation has been found to be basically true south, or up to 10° east or west of south.

The optimum value of the collector tilt (from the horizontal) again depends on the application. When a roughly uniform, year-round usage of solar energy is planned (e.g., for hot-water heating), the best collector tilt is approximately equal to the latitude. When the primary application of the solar system is winter heating, the best collector tilt is roughly equal to the latitude plus 10 to 15°. For combination systems the optimum tilt angle must be determined from monthly performance predictions at varying tilt angles. (See Chap. 14.)

Collector Glass Transmittance. Improving glass transmittance by 6 percent by the use of "water white" glass should increase the annual solar heat collected by 2 to 5 percent, depending on the site. Improving transmittance by reducing glass reflectance should likewise have a similar effect. (See Chap. 4.)

Justification of the extra cost of double glazing will depend on climate. Collector area can be reduced only 5 percent by the use of double glazing in Phoenix, AZ, but by 25 percent in Bismarck, ND.

Collector Coolant Flowrate. Although not a critical design parameter, the flowrate should be designed to obtain a coolant temperature rise of about 20°F under peak conditions. Flows of 0.02 to 0.03 g/min·ft$_c^2$ (0.014– 0.02 L/s·m$_c^2$) are generally used.

Collector Back Insulation. U values of 0.1 BTU/h·°F·ft^2 (0.5 W/m^2·°C) or less should be assured by the collector design. This corresponds to 3 in (7.5 cm) of insulation having a thermal conductivity (k) of 0.025 Btu/h·ft·°F (0.043 W/m·°C).

Space planning for solar system components High-rise buildings often have insufficient roof area to permit use of solar collectors for anything other than water heating. Although it is possible to expand solar use with south-wall-mounted (curtain wall) collectors, incorporating collectors into the exterior walls of buildings presents very difficult engineering and architectural design problems.

Because flat-plate collectors may attain temperatures of 350 to 400°F under stagnation conditions, thermal expansion of the collector array must be accounted for in its location and mounting design. The collector array should not be constrained so that free expansion is prevented; otherwise, glass breakage and other damage may result.

Even with proper back insulation, the rear surface of a collector may attain temperatures significantly above ambient. Thus, collectors should be located so that radiation and convection from the rear surface will not damage adjacent materials and construction. If the collectors are an integral part of the roof, the space below will receive heat from the collector undersides; the impact of such heating must be accounted for in the adjacent spaces.

Services to Solar Equipment. Solar system components should be located so that they are accessible for cleaning, adjusting, servicing, inspection, replacement, or repair. In particular, storage containers may need periodic inspection or replacement; complete burial might seriously impede these procedures.

Reduction of Vandalism. Careful location of solar equipment with respect to circulation of people can reduce tampering and vandalism. Screening roof-mounted collector arrays with parapets is an effective means of discouraging vandalism. Such parapets may also be an aesthetic enhancement.

High-temperature and -pressure stability Temperatures above 350°F (175°C) probably will occur within flat-plate collectors that are not cooled by a heat-transfer fluid, as during normal shutdown, loss of fluid flow caused by pump or fan failure, system blockage, or loss of electric power. The designer should ensure that the collector materials can withstand the maximum temperatures and pressures or should provide temperature and pressure relief from thermal cycling or shock. Collectors that are not completely drained can generate steam and over-pressure. Heat that is collected but not needed because the storage capacity has been reached can be dumped to a standby cooling tower or to an air-cooled fan-coil unit.

Clearances in the collector should allow for thermal expansion of all components within the operating temperature range of the winter design temperature to the absorber-plate stagnation temperature. Performance should not drop more than 10 percent after collector exposure to an average daily solar flux of 1500 Btu/ft^2_c (17,000 k J/m^2_c) on a tilted collector under "no flow" conditions for 30 days. (See Ref. 4, Exposure Test.)

Safety hazards   Hazards that require special attention are the reflection of sunlight that creates glare or visual distraction (especially to passing motorists) and the projection of sharp edges that may cause injury from free-standing collectors.

Solar collectors often include smooth, slippery surfaces located in elevated positions at steep angles. These surfaces can heat up rapidly and loosen masses of snow or ice that may slide off. Methods for preventing harm to people or property, such as deflectors, restraints, or "safe fall" areas, should be considered.

Hot fluid from broken lines in liquid systems can be hazardous.

Reflective surfaces   Reflective surfaces may be placed adjacent to flat-plate collectors to enhance collector performance. Commonly used surfaces are aluminized Mylar, back-surface mirrors, Alzak aluminum sheet, and wood or metal covered with aluminum paint.

Such surfaces should not crack, peel, or otherwise deteriorate significantly during their service lives. The manufacturer should provide documentation indicating that the reflectance will not decrease by more than 10 percent from the design reflectance during the service life of the surface. Furthermore, hail impact should not significantly impair surface reflectance. Roof or ground surfaces in front of the collector covered with snow, light sand, or white gravel will also increase overall performance. Radiation gains due to reflectors as shown in Fig. 13.5 have contributed approximately 20 to 30 percent of the performance.

Building site selection

Site Surroundings. Because solar collectors can be large and visually dominant, they must be carefully located and designed to harmonize with the surrounding community. The building should be located so that chemical or particulate emissions from nearby facilities, such as incinerators and factories, do not significantly impair collector performance. The building site should be chosen to avoid the shading of collectors by nearby structures and/or vegetation. Solar rights and related legal aspects are the topic of considerable recent discussion, but the matter has not been resolved as of this date.

Prevailing Winds. Topography affects wind patterns by constricting the wind and increasing its velocity in certain areas while sheltering others. Of all climatic variables, wind is the most affected by local site conditions; general climatic data are probably insufficient for building design. Location of buildings that support solar collectors should take prevailing winds into account to avoid excessive heat losses because of wind and local low temperatures that impair collection of solar energy. Thus, natural wind paths should be avoided.

If the site is hilly, the midslopes are best, away from high winds at the crests and cold air

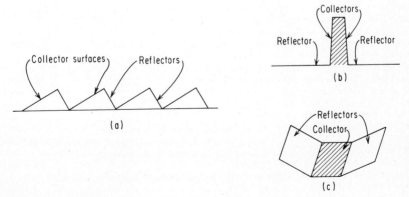

Fig. 13.5 Various reflector configurations for collector performance boost. (*a*) Sawtooth array for summer boost; (*b*) vertical collector-horizontal reflector for winter boost; (*c*) side wing using two reflectors.

that settles into low spots. Protection from prevailing winter winds can be provided by the shielding effects of the other buildings, walls, or windbreaks of evergreen trees. The building itself may be placed partly underground to reduce heat losses because of prevailing winds in extreme climates.

In summer, total daily insolation is approximately the same on northern, southern, eastern, or western grades of up to 10 percent. However, in winter, south slopes receive more insolation because of low sun angles, thereby raising local air temperatures. Thus, southern slopes are preferred as building sites in regions with cold winters.

Storage

The storage media that can be directly used with a liquid-based solar heating system are water and phase-change materials. Water is the most common storage medium. It is inexpensive, available and has a high specific heat-density product. It is easily transported to charge or discharge storage tanks. It has the disadvantage that it can be heated only to about 200°F (93°C) under normal pressures and that it gives rise to rust and corrosion.

Water can be contained in one or more storage tanks made of steel, concrete, wood, or fiber glass. The requirements for storage tanks are durability, resistance to corrosion, and insulation against heat losses. They must be able to withstand the maximum pressures that can arise in the system especially at high temperatures or high static heads.

Storage tanks normally have an interior protective coating or lining (e.g., of fiber glass, cement, or butyl) and exterior insulation. The protective coating must be resistant to the temperatures and pressures that the liquid attains and to any chemicals that may be present in the liquid.

If a metal storage tank is connected to pipes made of a different metal, neoprene gaskets or other dielectric material should be used for separating the two metals to reduce the rate of corrosion.

Stratification   System efficiency can be increased in some cases if the storage is stratified. The hot liquid from the collector can transfer most of its heat to the upper layers in the storage tank, while the fluid returning to the collector (from the lower end of the storage tank) is at a lower temperature. Stratification can be accomplished by using antiblending headers at the top and bottom or by baffles. Vertical storage tanks are better for stratification but require extra height for installation.

System efficiency can also be increased by using a form of stratification created by two storage tanks, one at a high temperature and the other at a relatively low temperature. Multiple storage tanks also allow the use of hot and cold storage, as shown in one of the case histories described later in this chapter. However, systems with multiple storage require a more complicated piping, valve, and control system and often cannot justify their higher costs by the performance increase.

Latent heat-of-fusion storage   Materials that undergo a change of phase within a suitable temperature range may be useful for energy storage if they can meet several criteria. The phase change must be accompanied by high latent heat and must be reversible over a very large number of cycles without serious degradation.

The most extensively tested material is "Glauber's salt" ($Na_2SO_4 \cdot 10H_2O$), which melts and freezes at 91.4°F (33°C) with a heat of fusion of 108 Btu/lb (250 kJ/kg). Other combinations of sodium salts and nucleating agents, as well as paraffins, have melting points that cover a wide range of temperatures, making some suitable for cold storage on the low-temperature side of an air-conditioning system, while others are useful for storing heat gathered by solar collectors.

Although heat-of-fusion storage systems show some promise for low-volume, low-cost storage, they are not sufficiently developed to be considered feasible for commercial applications. Advantages of the phase-change storage are that it requires less volume than water and operates over a small temperature range.

The main disadvantage of phase-change storage at the present time is that, after a number of phase-change cycles, the anhydrous salt tends to physically separate from the water, which slows down the reverse phase change and makes the storage much less effective. In order to prevent this separation, other substances are added to phase-change salts. A common additive is borax. Other thixotropic additives have been attempted, which prevent the anhydrous salt from settling during the phase-change cycle.

Design criteria for liquid storage systems   Standards for thermal storage units are given in Appendix B, Table B-4, of Ref. 64. Both above- and below-ground installations can be used. Liquid storage tanks should be leak-tested at 1.5 times the design pressure. Automatic relief valves should be incorporated to protect against overpressurization. Water inlets to storage tanks should be designed to minimize mixing in the tank.

Unsheltered, above-ground storage tanks should be designed to resist snow, wind, hail, and seismic loads. Completely enclosed tanks within buildings need only be designed to resist seismic loads. Underground tanks should be designed to resist soil and hydrostatic loads and foundation loads transmitted to them. Waterproof insulation is essential.

Storage units that are inaccessible for replacement should be designed to perform their function for the lifetime of the building in which they are installed. In general, however, storage units should be accessible for routine maintenance and repairs.

Tanks should be provided with means for emptying the liquid. Those above grade or floor level should have a valve at the lowest point, and buried tanks should have provisions for pump or siphon emptying. A vented storage tank can create pumping difficulties during operation near the boiling point of the storage water. Vapor lock can occur at the pump intake, especially if storage is below ground. The entire collector, pipe, heat exchanger, and tank system should include vacuum relief valves to protect against possible collapse caused by draining without venting.

Tanks should have level indicators. Tanks requiring makeup water from the potable water system but containing nonpotable water should be filled using an air gap or check valve to avoid contamination of the potable supply.

If flammable materials are used as storage media or in the storage container, they should conform to existing standards and fire codes. Thermal storage system materials should be chemically compatible enough to prevent corrosive wear and deterioration.

Tanks should be insulated so that the thermal losses do not exceed 2 to 3 percent of their maximum thermal energy capacity over a 24-h winter design day. This generally requires several inches of fiber-glass batts or rigid styrofoam insulation (at least R-20) for indoor installation. At least R-28 is recommended for outdoor installation. Heat losses from thermal storage containers, which are usually placed in basements or equipment rooms, must be accounted for in determining the thermal loading of these spaces. The surface temperature of any bare surfaces, exposed tanks, pipes, or ducts should be less than 100°F (38°C). Refer to Table 13.1.

Where the water table is near the surface, heat loss may be large if the tank is located below grade. Heat from storage will diffuse into the adjacent earth beneath a tank; the presence of water in the soil will increase the thermal losses from storage as water is evaporated.

Thermal storage for much more than 1 sunny day's collection does not improve the yearly system performance much, but fairly severe performance losses are to be expected if the storage provides less than 1.2 gal of water/ft$_c^2$ (50 L/m$_c^2$). For most buildings, storage sizes are 1.5 to 2 gal/ft$_c^2$ (60–80 L/m$_c^2$). The storage tank water is heated directly if water is the collector heat-transfer medium, and indirectly through a heat exchanger if other fluids are used. The heat exchanger is usually an external shell-and-tube type that requires circulating pumps.

TABLE 13.1 Upper Temperature Limits for Insulation Materials for Collectors, Tanks, Pipe, and Fittings

Material	Density, lb/ft³*	Thermal conductivity at 200°F, Btu/h·ft²·°F/in†	Temperature limits, °F‡
Fiber glass with organic binder	0.6	0.41	350
	1.0	0.35	350
	1.5	0.31	350
	3.0	0.30	350
Fiber glass with low binder	1.5	0.31	850
Ceramic fiber blanket	3.0	0.4 @ 400°F	2300
Mineral fiber blanket	10.0	0.31	1200
Calcium silicate	13.0	0.38	1200
Urea-formaldehyde foam	0.7	0.20 @ 75°F	210
Urethane foam	2–4	0.20	250–400

*1 lb/ft³ = 16.0 kg/m³.
†1 Btu/h·ft²·°F/in = 0.083 Btu/h·ft·°F = 0.144W/m·°C.
‡°C = 5/9 (°F − 32).

Heat-Transfer Fluids

The desirable properties of collector and distribution system fluids are:

- A low freezing point [below −50°F (−46°C)] and a high boiling point [above 400°F (205°C)]
- Chemical stability over the same temperature range
- High specific heat, high thermal conductivity, and low viscosity
- A flash point and an ignition point higher than the highest collector temperature
- Noncorrosive, in contact with metals such as copper, steel, and aluminum and in contact with building and roofing materials; nonscaling
- Nontoxic; nonodorous
- Low vapor pressure at the highest temperatures it would be exposed to
- Not energy-intensive
- Inexpensive

The types of liquids used are:

1. *Water.* Water has many desirable characteristics: it is inexpensive, easily available, and nontoxic and has a high specific heat. However, it freezes in cold weather, boils at a relatively low temperature, and can cause corrosion. Freeze protection methods such as draining down the collector in cold weather can prevent freeze-ups, or a collector can be selected with coolant passages that can tolerate the expansion caused by freezing. Protection against boiling and overheating can be provided by pressurizing the system, providing a heat-dump mechanism, or by draining the collectors.

When anti-freeze solutions are used, corrosion can be reduced by adding suitable inhibitors. Different kinds of inhibitors are available, such as Corr-Shield K-7, Dearborn 537, CWT 110, Nalco 39, and Corrosion Inhibitor CS. Both nitrite-based inhibitors and chromate-based inhibitors have been used; the former inhibitors are the ones more acceptable to health authorities. An example of an inhibitor is a mixture of borax, sodium nitrite, benzotriazole, and inorganic corrosion inhibitors, with a minimum sodium nitrite content of 70 percent. The pH value of water with the corrosion inhibitor must be monitored regularly; acceptable pH values are slightly basic, in the range 7.0 to 8.5.

2. *Glycol solutions.* A mixture of water and glycol is often used as the antifreeze solution. However, the most common glycol—ethylene glycol—is very toxic. Propylene glycol is less toxic. Glycols degrade chemically over a period of years, especially when used at high temperatures. Glycol has a lower specific heat and a higher viscosity than water, resulting in a higher energy requirement for pumping. The increase in pumping energy consumption resulting from using a 50 % ethylene glycol-water mixture instead of water is about 30 percent. Care must be taken in selecting the corrosion inhibitor for use with a glycol system; for example, chromate-based inhibitors should *not* be used. (See Chap. 11.)

The use of a fluid other than water requires a heat exchanger in the collector loop between the collector circuit and the storage tank; this reduces the efficiency of the system (see Chap. 7 for F_{hx} factor). In systems incorporating domestic hot-water subsystems, double separation between the collector fluid and the potable water supply is normally required by local building and health codes.

3. *Hydrocarbon oils.* These have been used widely in many heat-transfer applications in industry. However, some of the more common hydrocarbon oils have flash points and ignition points close to 300°F(150°C), oxidize at high temperatures, and under these conditions may cause corrosion.

4. *Silicone fluids.* These are in the course of development and are in early stages of application. Some manufacturers offer a silicone heat-transfer liquid that is nontoxic, has high dielectric strength, and does not promote galvanic corrosion. The main disadvantages are low specific heat, high viscosity, and very low surface tension.

Table 13.2 summarizes the properties of collector coolant liquids. Methods of freeze protection are outlined below.

Terminal Load Devices

Terminal load devices for use with a liquid-based solar heating system must be designed not to require high temperatures. A forced-air heating system is an efficient terminal system for use with a solar heating system. Baseboard heaters and radiant panels have also been successfully used with liquid-based solar heating systems. The effect of load device effectiveness on annual energy delivery is described in Chap. 14.

TABLE 13.2 Collector Coolant Characteristics

Characteristic	Desired range*	Air	Water	85% ethylene glycol and water	50% propylene glycol and water	Thermia 15 paraffinic oil	UCON (polyglycol) 50-HB-280-X	Dowtherm J	Therminol 60
Freezing point	L	—	32°F	-37°F	-26°F	—	—	-100°F	-90°F
Pour point		—	—	—	—	10°F	-35°F	—	—
Boiling point (at atmospheric pressure)	H	—	212°F	265°F	369°F	700°F	600°F	358°F	—
Corrosion		Noncorrosive	Corrosive to Fe or Al; requires inhibitors			Noncorrosive	Noncorrosive	Noncorrosive	Noncorrosive
Fluid stability	H	—	Requires pH or inhibitor monitoring			Good†	Good†	—	—
Flash point	L	—	None	None	215°F	455°F	500°F	145°F	310°F
Bulk cost ($/gal) (December 1975)	L	—	—	2.35	2.50	1.00	4.40	4.00	4.00
Thermal conductivity at (Btu/h·ft·°F at 100°F)	H	0.0154	0.359	0.17	0.22	0.76	0.119	—	—
Heat capacity at (Btu/lb·°F at 100°F)	H	0.24	1.0	0.66	0.87	0.46	0.45	0.47	0.42
Viscosity at (lb/ft·at 100°F)	L	0.4626	1.66	15.7	7.5	28.5	143.1	0.7	0.59

†Requires an isolated cold expansion tank or nitrogen-containing hot expansion tank to prevent sludge formation.
‡Contains a sludge formation inhibitor.
*L = low; H = high.

Heat Exchangers   Heat exchangers are used in a liquid-based solar heating system between the collector and storage for freeze protection, or to reduce the pressure on storage tanks.

Since solar heating systems with flat-plate collectors usually operate at low temperature, it is important to minimize the temperature drop across the heat exchangers in the collector loop. This requires use of heat exchangers with a large surface area and appropriate flowrates. The values of the flowrates should be based upon the manufacturers' recommendations, increased to take into account the lower specific heat and higher viscosity of antifreeze liquids with proper attention to tube rattle.

Heat exchangers should comply with Tubular Exchanger Manufacturers Association (TEMA)[53] and other standards applicable to exchangers in similar service. Collector-to-storage heat exchangers are generally of the shell-and-tube type, either single or multiple pass, although in some cases (see below) finned-tube exchangers can be used. The high flowrates required for efficiency in single-pass exchangers cause pumping power expense and reduced storage-tank stratification because of mixing. However, efficient heat exchange is more important to overall system performance than is the establishment of storage tank stratification.

The heat exchanger between the collector and storage should be sized so that its effectiveness is not less than 0.70 (see Chap. 4). A glycol-to-water heat exchanger should be sized for a 5 to 15°F(3–8°C) approach temperature difference (the difference between the incoming *heating* fluid and the outgoing *heated* fluid). Exchangers sized to less than a 5°F(3°C) approach will improve the collector performance only very slightly for a comparatively large increase in exchanger size and cost. Heat exchangers sized for greater than a 15°F(8°C) approach will reduce collector performance excessively relative to the decreased exchanger cost.

Note that approach ΔT is a rough measure of heat-exchanger size. A heat exchanger selected for a 40°F(22°C) approach will be about one-fourth the size of one selected for a 10°F(11°C) approach. However, the cost difference is generally not proportionate. Considering the relatively high installed cost of solar collectors it would be unwise to sacrifice collector performance by using a large approach ΔT. The solar system designer will have to specify conditions of approach ΔT, expected liquid temperature and flowrates, allowable head loss, and fouling factors.

Backup and Auxiliary Heating Systems and Piping

It is possible to design a solar space-heating system to supply 100 percent of the annual energy requirements of a building without a backup system, given sufficient solar collector area and thermal storage capacity; however, it is not cost-effective to do so. A solar system is normally designed to provide from 50 to 80 percent or more of the annual energy requirements for heating. Backup systems can be oil- or gas-fired furnaces or boilers, electric resistance heat, wood furnaces or stoves, central steam or hot water generated off-site, or a heat pump. The nature of the backup system will affect system configuration and control. A backup system supplies auxiliary energy when solar storage is depleted. The backup system must be designed to carry 100 percent of the design heat loss but need not be so rugged as a conventional system without solar, since it operates for far fewer hours.

Backup interface   Although there are several possible locations at which auxiliary heat can be provided, *maximum advantage is gained by supplying it to the load loop rather than to the collector loop or storage unit.* Such a design minimizes consumption of auxiliary energy by using it only when necessary and only to boost the temperature of the fluid being supplied for use. Any other design makes the collector operate at a higher temperature (and corresponding lower efficiency) than necessary and causes some of the heat storage capacity to be used for auxiliary heat rather than the solar heat for which it is designed.

Auxiliary heat may be supplied by a conventional hot-water boiler in one of two modes. Solar-heated water can be pumped through the auxiliary heater on its way to the load (auxiliary in *series*) with a temperature increase provided by auxiliary energy if needed. In this mode, the temperature of the water returning to storage from the load may be higher than the storage temperature, thereby adding part of the auxiliary energy to storage. Continued operation would gradually drive the solar storage temperature up, thereby using storage capacity for auxiliary, rather than solar, heat and reducing collector efficiency.

A preferable arrangement in which the auxiliary heat is in *parallel* with the load uses solar-heated water exclusively whenever the storage temperature exceeds the required heating coil, convector, or radiant panel temperature. When the storage temperature is too low, circulation of solar-heated water is discontinued and auxiliary heat is used exclusively. Piping and valving must be arranged so that the water bypasses the storage tank when auxiliary heat is in use. (See Fig. 13.6.)

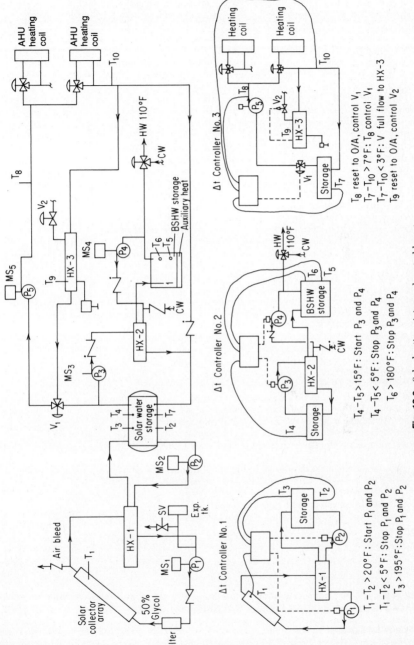

Fig. 13.6 Solar heating piping and control layout.

Δt Controller No. 1

$T_1 - T_2 > 20°F$: Start P_1 and P_2
$T_1 - T_2 < 5°F$: Stop P_1 and P_2
$T_3 > 195°F$: Stop P_1 and P_2

Δt Controller No. 2

$T_4 - T_5 > 15°F$: Start P_3 and P_4
$T_4 - T_5 < 5°F$: Stop P_3 and P_4
$T_6 > 180°F$: Stop P_3 and P_4

Δt Controller No. 3

T_8 reset to O/A, control V_1
$T_7 - T_{10} > 7°F$: T_8 control V_1
$T_7 - T_{10} < 3°F$: V full flow to HX-3
T_9 reset to O/A, control V_2

Where distribution is with air heated by a heating coil, another strategy is to use a two-coil arrangement. In this mode, solar-heated water supplies heat to the air in a preheater coil whenever the storage temperature is above a minimum, usually about 80°F(27°C). A second coil, immediately downstream of the solar preheater coil, is heated by auxiliary energy, increasing the air temperature to the required level. This method ensures maximum use of solar heat and minimum use of fuel because even low-temperature solar heat is applied usefully whenever it is available.

A variation on this last strategy uses a conventional warm-air furnace just downstream of the solar preheater coil. The furnace, designed to meet peak heating requirements, boosts the air temperature when necessary. A class B blower motor may be required.

Piping systems   For maximum efficiency, the collector should be supplied with the coolest available liquid, whereas the warmest should be supplied to the load. Therefore, water from the bottom of the storage tank is supplied to the collector (or collector heat exchanger), and heated water from the collector or exchanger is delivered to the top of the tank. To transfer heat from storage to the building air-transport system, heated water is pumped from the top of the storage tank through a second heat exchanger (water to air), a finned heating coil in the supply duct. The air thus heated reenters the occupied space through the air-handling system. The heat also can be taken from the top of the storage tank and passed through the water-to-air heat exchanger by natural convection, rather than being pumped, though at some loss in performance, or solar-heated water can be used in a baseboard heating system.

Pipe sizing and design should be in accordance with standard methods. Solar systems must tolerate wide temperature excursions, particularly in the collector loop. Liquid heat-transfer fluid systems should have means for removing air from high points in the piping system. To facilitate maintenance and repair, systems that use liquids should be capable of being filled, flushed, and drained conveniently. All piping should be pitched to drain completely. High operating temperatures may cause air binding of the transport fluid. System pressurization and use of air eliminators may resolve this problem.

Collector-to-storage distribution-pipe heat losses can be kept relatively low by insulating to a U value $\simeq 0.15$ Btu/h·°F·ft^2 (0.85 W/m^2·°C) of pipe or 2 in (5 cm) of insulation that has a conductivity of 0.025 Btu/h·ft·°F (0.04 W/m·°C).

A possible problem with collector circuits that include centrifugal pumps is heat loss resulting from reverse thermosiphon flow when the pumps are off. Thermosiphon flow will occur whenever the collector temperature is lower than the storage-tank temperature. This heat loss may be easily prevented by check valves. (See Fig. 13.6.)

Piping Arrangements. The optimum pipe arrangement is reverse return. To promote self-balancing, 10 percent of the array pressure drop should occur in the headers and 90 percent across the collector itself. Collectors are generally piped in parallel; series flow increases pumping head and reduces collector efficiency. Combination series and parallel flow arrangements, with balancing valves included, may, in certain circumstances, be more advantageous. Two-way, industrial-grade, automatic valves are recommended. Silicone fluids are especially difficult to contain. Organic coupling hoses should not be used.

Collectors available with internal headers will significantly reduce field labor and material, reducing field construction costs, and should be considered for all solar installations. Care in selecting collectors must be exercised to assure adequate internal header pipe size to handle the number of collectors to be piped in the array.

The pressure drop Δp_{TOT} across a parallel collector array (headers plus collectors) is given by

$$\Delta p_{\text{TOT}} = \frac{KW^2}{6}(2j + 1)(j + 1)j + \Delta p_{\text{coll}} \qquad (13.4)$$

where Δp_{coll} is the collector panel pressure drop, W is the mass flowrate per collector, K is a constant, and there are $j + 1$ collectors in parallel (see Ref. 1).

Expansion Tanks

Adequate provision for the thermal expansion of solar heat-transfer and storage liquids which may occur over the service temperature range should be incorporated into the heating system design. Each closed-loop piping circuit must be equipped with an expansion tank. Expansion tanks should be located on the inlet side of the pump. In a closed-loop, antifreeze system, two expansion tanks are required, i.e., one in the loop in the solar collector–heat exchanger circuit, and one for the solar water-storage-terminal subsystem circuit. In a drain-down system the only expansion tank required is in the solar storage-tank subsystem circuit.

For a maximum temperature of 250°F (120°C), allow for 8 percent expansion of total water inventory volume. Expansion tanks should be sized in accordance with the recommendation of Ref. 14. However, expansion tank size can be reduced if it is equipped with a valve for charging with compressed air. Expansion tanks in systems with glycol or other antifreeze fluids must be oversized to accommodate the increased expansion rate of these fluids.

SYSTEM CONFIGURATIONS

Freeze Protection in Closed and Open Systems

The collector-storage circuit may be a closed loop with (pressurized) antifreeze solution circulating in it, or it may be an "open-loop," drain-down system. An open loop is open to the atmosphere. In a closed-loop system, the antifreeze solution remains in the loop whether or not the collector circuit is in operation. On the other hand, in an open-loop, drain-down system, the water is drained as a freeze prevention measure from the collector circuit when the circuit is not in operation. Reference 3 (pp. 4-13 and 4-14) presents the detailed description of several drain-down systems and piping diagrams.

Some of the advantages of a closed-loop system are less corrosion and the capability of pressurization (so that the fluid can reach higher temperatures without boiling or flashing). A major disadvantage of the closed-loop system is the requirement for a heat exchanger between the collector and the storage. This decreases the efficiency of the system.

Design flowrates for water systems are generally 10 to 15 lb of water per hour per square foot of collector (ft_c^2) (0.02–0.03 g/m water · ft_c^2 (0.01–0.02 L/s · m_c^2). For various organic antifreeze solutions the flowrate may be increased by 15 to 150 percent depending upon specific heat, viscosity, specific volume, and other characteristics. Because of the lower specific heat and higher viscosity of the antifreeze solutions compared to water, a higher flowrate and larger pumping head are required for the same heat transfer.

Drain-down systems include a storage container and centrifugal pump located well below the collector, and piping arranged so that when the pump shuts off, the heat-transfer fluid flows into the storage container. Piping must be planned to permit positive air purging during pump startup, and the pump must be sized to provide a lift head to the top of the collector array. Moreover, the drain-down volume must be allowed for in the storage tank, and the pipes must be sized for drain-down (usually a friction head loss of less than ¼ ft per 100 ft is required).

Such systems should not be constructed of materials that corrode in air, or they should be suitably protected. Use of corrosion inhibitors and/or introduction of an inert gas, such as nitrogen, are means of preventing corrosion.

The drain-down system avoids the requirement for a heat exchanger in the collector-storage loop. Since no antifreeze is used in the collector circuit the system is therefore more efficient. However, the drain-down mechanism must be designed carefully, in order to avoid freeze-ups. Many drain-down systems have frozen, and costly damage to collectors is inevitable.

When a hot, dry collector is refilled with the heat-transfer liquid, there is a possibility of boiling or flashing the liquid; however, adequate pressurization (which is possible in a closed system) can prevent this. On the other hand, under stagnation conditions, the liquid in a closed collector loop can become overheated and may decompose chemically or cause corrosion.

In climatic regions where ambient temperatures below freezing occur infrequently, an intermittent circulation system, with warm water from storage periodically circulated through the collector when outdoor ambient temperature drops below 35°F (2°C), can be more efficient than a closed-loop system. However possible pump failure and night-sky radiation effects must be taken into account.

If nonpotable or flammable fluids are used, means should be provided for detecting leaks and warning occupants when they occur. These substances are to be treated like air-conditioning condensates and refrigerants [14] in providing for leak indicators. For instance, antifreeze agents should be treated with nontoxic dyes that distinguish them clearly. In antifreeze-type systems, makeup water should not be supplied automatically, as freezing might result because of dilution.

Control Systems and Modes of Operation

The modes of operation of a liquid-based solar heating system are the following:
1. Heat to storage from the collector
2. Heat to the load from storage

3. Heat to the load from the backup system
4. Excess heat rejection

Direct collector-to-load operation may be used, but is normally not done in a liquid-based solar heating system for controllability reasons.

The selection of the mode of operation is based on the temperature difference between collector and storage, between storage and load, and from solar to auxiliary heat. In a drain-down system, the controls also insure that the water will drain down from the collector into the storage when the outside temperature falls below about 35°F (2°C), and will be pumped back into the collector when the temperature is well above the freezing point.

A differential thermostat is used to control operation of the heat-transfer fluid pump. The collector should operate only when the useful collected energy exceeds a minimum value. In practice, this is accomplished by comparing the temperature of the fluid in the collector exit header to the temperature of the fluid in the exit part of storage. The pump or blower is operated only when this temperature difference exceeds a set value, usually 10 to 15°F (5–8°C) in liquid systems. A shutdown ΔT of about 3 to 5°F (1.5–3°C) is commonly used.

Thermostats designed to control introduction of heat to occupied spaces are often two-stage type. The first stage operates the system blower or pump to circulate fluid from storage to the space, whereas the second stage operates the auxiliary energy source.

SIZING THE SOLAR SPACE-HEATING SYSTEM COMPONENTS

The following steps may be used to arrive at an economical solar system size for a specific building.

- *Step* 1. Calculate the building heating load for each month, using the methods outlined above.
- *Step* 2. Select the collector tilt for latitude plus 10 to 15°. Select the collector orientation from due south to about 10° east or west of due south.
- *Step* 3. Use the Liu-Jordan method to calculate the monthly insolation on the tilted collector. (See Chap. 2.)
- *Step* 4. Select a collector which satisfies the specifications listed in the last section of this chapter. These specifications include minimum requirements for the instantaneous efficiency of the collector, which can depend on the details of the application, i.e., space heating alone or in combination with domestic hot water and/or cooling.
- *Step* 5. Use the f chart method (Chap. 14) to calculate energy delivery for a selected range of collector areas covering 30 to 90 percent annual solar load fraction.
- *Step* 6. Size the storage volume. A preliminary estimate for the storage-tank volume is about 1.5 to 2.0 gal of water for each square foot of collector. Long-term storage in steel or fiber-glass tanks is rarely cost-effective.
- *Step* 7. Select a value for the collector flowrate within the range recommended by the manufacturer. Typical values are 0.02 to 0.03 g/min·ft² (0.10–0.2 L/s·m²) of collector. If the collector fluid is an antifreeze solution, then the flowrate will be larger than that for water, because of the lower specific heat and higher viscosity. For a 50% ethylene glycol solution, the flowrate would be about 30 percent larger than for only water.

The remaining steps are to select the distribution system, backup heating system, pipe sizes, pumps, valves, expansion tank, heat exchanger, etc., using conventional methods. The controller design must be specified. Considerations involved in selecting these have been already discussed briefly.

The preliminary system design given above should be followed by an economic analysis, as follows.

- *Step* 8. Find the average life-cycle cost per year of the solar system, assuming a value of the discount rate and the financing period.
- *Step* 9. From the total annual heat demand, subtract the solar heat obtained per year. This gives the net auxiliary required per year. Multiply by the levelized life-cycle fuel cost.
- *Step* 10. Add the costs from *Steps* 8 and 9 to give the average total annual heating cost (solar plus fuel).
- *Step* 11. Repeat the complete procedure above for different values of the collector area. The optimum system is the one with the minimum value of the total heating cost. See Chap. 28 for details of the method and effects of taxes, insurance, etc.

SOLAR-ASSISTED HEAT-PUMP SYSTEMS

An auxiliary heat pump can improve the performance of either a liquid or an air system, but it adds complexity and expense. Its advantage over electric resistance heating as a solar supplement is the reduced electricity consumption; a coefficient of performance (COP) of about 3 or more is obtained on a seasonal basis. The heat pump is also available for power-operated summer cooling.

Low-temperature solar heat applied to one side of the heat-pump system (Fig. 13.7) evaporates the low-pressure refrigerant liquid. The compressor then raises the pressure and temperature of the vapor, which when next condensed, gives off heat at a higher temperature than that at which heat was provided. When this temperature difference is less than 30 to 40°F(20°C), a good heat pump can provide heat at 105 to 115°F with a COP of 3.5; that is, it provides 3.5 units of heat output for each equivalent heat unit of electrical energy input.

Solar-assisted heat-pump systems can be of several types. One type uses liquid in the solar collector loop, water storage, and a water-to-water heat pump, as shown in Fig. 13.7. Another type uses an air-to-air heat pump in conjunction with liquid-to-air heat exchangers and a liquid solar collector loop. Systems that use air in the solar collector loop with rock storage and an air-to-air heat pump also are possible.

Although solar-assisted heat pumps are normally used for heating only, the concept of a solar-assisted, double-bundle condenser, heat recovery chiller (see Fig. 13.8) can be included in the heat-pump category. In the summer, this system would operate as a conventional central station, chilled-water air-conditioning system; however, it could also operate at night under more favorable ambient conditions and possibly reduced off-peak power rates, and store the "coolness" in its storage tank. In the winter, solar heat, as well as heat from people and lights in the interior zone of a large building, would create the load on the chiller evaporator. This system, suitable for large buildings, has the advantages of a heat pump without the complications associated with the reversing cycle features, and it can simultaneously heat and cool different zones in the building.

References 26 and 32 contain considerable detail on solar-assisted heat-pump systems. See Ref. 13 for information on conventional heat-pump systems. While most of the small heat pumps on the market are air-air heat pumps, used for heating in the winter and as direct refrigeration units in the summer, there are available water-air and water-water unitary heat pumps in the size range of 0.5 to 5 hp (0.4–4 kW). The air-air unitary units can be used with liquid-based systems as follows:

1. When insufficient solar-heated water is available for direct heating, the air-air heat pump operates with or without auxiliary heat and the solar system is shut down. The heat pump is sized and controlled in accordance with standard heat-pump system procedures.

2. The solar energy system operates with a separate heating coil to supplement the solar

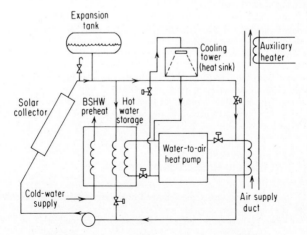

Fig. 13.7 Solar-assisted heat pump schematic diagram.

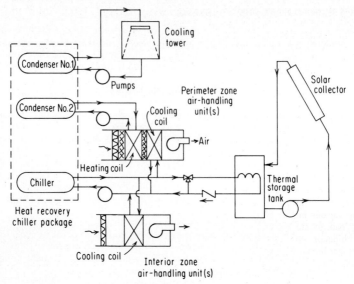

Fig. 13.8   Solar-assisted, double-bundle condenser chiller schematic diagram.

system. This system uses less energy in warm climates than the above system but is usually not cost effective in the smaller sizes due to the extra piping controls and coils required.

3. The solar-heated water is used directly to supply heat as long as its temperature is above the minimum system utilization temperature. It provides a heat source for the heat pump by a separate coil in the air-source air stream and increases the capacity and COP of the heat-pump. This system has been studied and is reported in Refs. 31 and 32. It was found to be excessive in initial cost and hence not cost effective. The unitary water-air heat source pumps operating in a closed loop are available from a number of manufacturers. Solar-heated water is employed as a heat source in combination with rejected heat to the loop from units which are operating in the cooling mode to improve the COP and capacity of the units operating in the heating mode. The loop temperature operates between 55 and 75°F (13 and 24°C). Auxiliary heat from boilers or electric resistance heaters are employed to boost the loop temperature when solar energy and other sources are inadequate to do so.

Heat pumps enhance the efficiency and lower the costs of the solar energy system by permitting the collectors to operate at low fluid temperatures with the heat pump boosting the air or water temperatures for delivery to the space at 105 to 115°F. The space-heating system must be designed for those low utilization temperatures.

In all cases, heat pumps are most cost effective when there are requirements for cooling as well as heating.

In the larger buildings the heat pump always utilizes the solar-heated water for a heat source to maintain high COP and capacity. A computer analysis is required to optimize the operating mode, collector and storage size, number of storage vessels, and control sequence. In some cases it may be more efficient to use all of the solar-heated hot water at low temperature as a source for the heat pump.

SUMMARY OF DESIGN SERVICES AND PROCEDURES FOR LIQUID-BASED SOLAR SPACE-HEATING SYSTEMS.

This section contains a list of steps to be followed in the design and economic analysis of solar heating systems.

■ *Step* 1. Obtain climatic and insolation data for the area, and select indoor and outdoor design temperature conditions.

■ *Step* 2. Calculate the peak and monthly building heating load. Calculate the average monthly domestic hot-water load for combined systems.

■ *Step* 3. Select the type of heating distribution system and auxiliary backup heating system for new buildings, and the necessary modifications to existing buildings, with low terminal temperatures for compatibility with a liquid-based solar energy space-heating system.

■ *Step* 4. Select a collector area to provide about 60 to 70 percent of the annual load.

■ *Step* 5. From the certified efficiency curve provided by the manufacturer for his collector, determine the amount of heat that can be delivered to load using the *f*-chart or other methods.

■ *Step* 6. Determine the thermal storage-tank(s) size in gallons or cubic feet based on storing 2 gal/ft$_c^2$ (80 L/m^2).

■ *Step* 7. Select a freeze protection system and the appropriate heat-transfer fluid. Use water if the system is an open drain-down system or an intermittent pumping system. Use an appropriate antifreeze mixture and heat exchanger if the system is closed-loop.

■ *Step* 8. Calculate the flowrates for each discrete piping circuit taking into account the type and characteristics of the heat-transfer fluid.

■ *Step* 9. Prepare an outline description of the proposed system including the control logic and various modes of operation, i.e., (a) solar collector heat to storage, (b) storage to building load, (c) storage to heat-pump evaporator (if used), (d) auxiliary or backup heat only or in combination with solar storage heat when storage heat is not capable of carrying the entire load, but can be used for preheating.

■ *Step* 10. Select differential controller, sensors, and other automatic control equipment.

■ *Step* 11. Prepare a schematic design showing solar collectors, collector supports, storage tanks, piping, pumps, and controls in a system flow diagram. Calculate preliminary expansion-tank size, piping lengths and sizes, and valve sizes and calculate the pumping head required under peak conditions. Select appropriate pumps from manufacturer's performance curves.

■ *Step* 12. Make a preliminary estimate of the initial cost of the system based upon the components selected.

■ *Step* 13. The collector area and storage volume should now be changed in increments of 10 percent; to determine the performance and cost sensitivity to these changes, the percent solar contribution to the heating load can be examined for values of 40 to 90 percent. Select the cost-optimal system.

■ *Step* 14 Prepare design development drawings for the system selected using the building floor plans, sections, and mechanical drawings. All points of interface with the building structure and the mechanical and electrical systems should be shown. Prepare a more detailed specification and cost estimate based on a material take-off in accordance with the accepted methods in HVAC and electrical system design.

■ *Step* 15. Upon approval of the design development drawings, prepare working drawings, detailed specifications, and final cost estimates for the system. Plans include all building system modifications or alterations, all system components final pipe and insulation sizes, and solar collector performance specifications and material.

■ *Step* 16. Follow bidding procedures, drawings, and construction phases in accordance with standard practices in the design and construction industry.

■ *Step* 17. Specify the water treatment and corrosion protection required. Prepare an operating, service, and maintenance manual to include complete instructions for service and maintenance personnel or the home owner or small business proprietor. Operating personnel should be trained for 2 to 5 days, depending upon system size, by the design engineer.

Specification for the Collector

The specifications should include, at a minimum,

  1. Total gross area of the collector array; maximum dimensions allowed
  2. Details of the collector array—number of rows, number of collectors in each row, and the maximum dimensions of each row
  3. Coverplates
     ■ Transmittance
     ■ Mechanical strength
     ■ Durability under different temperature conditions
     ■ Durability under exposure to radiation
     ■ Structural durability under wind and snow loading

- Number of coverplates
- Moisture removal method
- Requirements on sealing
4. Absorber plates
  - Mechanical and thermal performance criteria; absorptance and emittance
  - Design life requirements
  - Provision for draining all fluid passages
  - Provision for complete venting of air
  - Support of absorber within the container so that there is a thermal break between the absorber and the container
  - Maximum allowable temperature difference between any point on the absorber and the fluid
  - Test pressure that the absorber must withstand
  - Durability of absorber plate and coating in the specified temperature range and under stagnation conditions, in the presence of moisture and under extended radiation
5. Fluid flowrate; pressure drop across the collector
6. Methods of piping the collector array (e.g., reverse return) to ensure flow balance
7. Insulation:
   - Allowed materials
   - Location of insulation
   - Maximum allowed U value
   - Durability at high temperatures (with no outgassing); resistance to moisture
   - Method of attachment of the insulation to the container
8. Container
   - Allowed materials
   - Requirements on sealants
   - Treatment for corrosion prevention
   - Weather proofing, venting
   - Maximum allowed shading of the absorber plate
   - Durability
   - Provision for thermal expansion over the range operating temperatures
9. Specification of thermal performance
   - Must have been tested by a recognized testing facility, according to NBS-ASHRAE standards including stagnation temperature tests
   - Requirement of on-site test performance
   - Minimum requirements on instantaneous efficiency curve
   - Minimum requirements on overall day-long output for a specified orientation, specified weather conditions (temperature, wind speed), insolation conditions, and collector temperature and flowrate
10. Provision for freeze protection
11. Provision for protection against overheating
12. Installation of collectors
    - Location
    - Method of mounting the collectors
    - Provision for wind and snow loads
13. Provision for maintenance and repair

Specifications for Storage

1. Type of storage system, materials, dimensions, liner
2. Storage capacity
3. Temperature rise required in the storage
4. Insulation; maximum allowed heat loss from the storage tank
5. Location and size of piping connections
6. Antiblending devices

Specifications for the Distribution System

1. Type of distribution system
2. Sizing of pipes, ducts, and pumps
3. Insulation of pipes

4. Pump specifications, materials, and efficiency curve

5. Piping and appurtenance materials

Specifications for the Backup System

1. Type of backup system

2. Capacity required for the backup system

3. Automatic switching on of the backup system

4. Method of pumping

Other Specifications for the System

1. Provision for specified modes of operation including heat rejection

2. Provision for draining, venting, and filling the system

3. Specification of the type of controls and each operational mode; specification of what will be controlled manually and what automatically

4. Fail-safe requirements on valves

5. Requirements for relief valves, expansion tanks, etc., and their proper sizing and location in the system

6. Compliance with accepted HVAC requirements

7. Provision for makeup water (in a hydronic system)

8. Requirement for double separation (if the system uses a nonpotable fluid)

9. Compliance with all applicable building and other codes

10. Freeze protection requirements

Applicable Codes and Standards

The latest editions of the American Society of Heating, Refrigerating, and Air-Conditioning Engineers, Inc. (ASHRAE), Handbooks and the National Standard Plumbing Code at least, should be followed as basic design references. The following standards should be used in design:

1. "Interim Performance Criteria for Solar Heating and Cooling Systems in Commercial Buildings," U.S. Department of Commerce, National Bureau of Standards report NBSIR 76-1187 (November 1976).

2. "Thermal Data Requirements and Performance Evaluation, Procedures for the National Solar Heating and Cooling Demonstration Program," U.S. Department of Commerce, National Bureau of Standards report NBSIR 76-1137 (August 1976).

3. ASHRAE Standard 93-77, "Methods of Testing to Determine the Thermal Performance of Solar Collectors" (1977).

4. ASHRAE Standard 94-77, "Method of Testing Thermal Storage Devices Based on Thermal Performance" (1977).

5. "Intermediate Minimum Property Standards for Solar Heating and Domestic Hot Water Systems," U.S. Department of Commerce, National Bureau of Standards report NBSIR 76-1059 (April 1976). (Although this document was prepared for residential use, most of the standards also apply to commercial building use.)

6. ASHRAE 90-75, "Energy Conservation in New Building Design."

7. "Interim Performance Criteria for Commercial Solar Heating and Combined Heating/Cooling Systems and Facilities," published by NASA, George C. Marshall Space Flight Center, 1975.

ECONOMIC EVALUATION

Solar heating economics deals basically with optimizing the trade-off between solar system owning and operating costs and the future cost of the fuel saved by the solar system during its anticipated useful life. Life-cycle costing is the key to such an assessment. Methods of life-costing are discussed in detail in Chap. 28.

The life-cycle cost analysis must meet two basic design objectives. First, the optimal system size, which gives the lowest life-cycle heating cost, must be determined. Then, on the basis of the optimal system size and configuration, the economic feasibility of solar heating should be assessed. The designer should bear in mind the decreasing availability of conventional fuels and the savings of these fuels resulting from use of solar energy.

The major costs of a solar heating system, without auxiliary energy equipment, are the annual cost of owning the collector, storage unit, and associated controls, pumps, piping, etc.; the costs of operating the system, pumps, and blowers; and the costs of maintenance.

Operating costs are primarily for power needed to pump water and move air in the system, summed over the yearly operating time. Maintenance costs include repairs, replacement of glazing in collectors, and any other costs of keeping the system operable. Maintenance must be minimized if solar heating is to be economically feasible.

The real escalation of conventional fuel costs should be used in determining the cost of fuel saved by a solar system. Although estimating future fuel cost escalation is difficult, at best, a credible growth-rate range can be constructed to guide decision making. Because of the considerable uncertainty implicit in projecting future conventional fuel costs and solar system operating costs, one must resist inflexible break-even period requirements.

Solar System Capital Costs

In estimating the capital cost of solar system components, only the costs of items that are not normally part of a conventional HVAC system should be considered. Thus, the cost of the building's air-handling system would not be considered. Note that certain cost elements vary according to the size of a solar heating and/or cooling system, whereas others are relatively fixed, regardless of collector area or tank volume. Collector and storage tank costs are system-size-dependent items; others include heat-exchanger costs, and certain pump and piping costs. The additional control system cost associated with a solar energy system is an example of a cost difference that is largely independent of the collector area. The cost difference associated with the purchase and installation of an absorption chiller is also relatively independent of solar collector area, because for all but the smallest collector areas, selection of an appropriate cooling machine is dictated by the peak building-cooling load.

Because solar system costs depend on the purchaser's location and are also time-dependent, the designer should obtain actual price quotations from equipment manufacturers. However, for initial assessments, costs can be estimated from the data given below.

The subsystems to be considered are shown in Table 13.3, along with an estimate of the fraction of the total installed cost that each requires. If solar cooling is included, the incremental costs of a derated chiller and/or cooling tower are in addition to those listed.

TABLE 13.3   Solar System Cost Estimates for Commercial Buildings*

A.	Subsystem costs	Fraction of total solar cost, %
1.	Collectors and supports	35
2.	Storage and heat exchangers	20
3.	Piping, controls, electrical, and installation	45
B.	**Component costs†**	**Cost/ft$_c^2$**
1.	Collectors	
	a. Nonselective	$5.00–20.00
	b. Selective	$5.00–10.00
	c. Collector support structure	$10.00–20.00
2.	Heat exchangers	$3.00–10.00
3.	Collector fluid	$0.40– 0.80
4.	Storage tank and insulation	$0.15– 0.20
5.	Piping, insulation, expansion tanks, valves	$2.00– 5.00
6.	Pumps	$3.00– 6.00
7.	Controls and electrical ≈ $3,000–5,000	$0.40– 1.00
C.	**Systems costs by type**	**Installed cost/ft$_c^2$**
1.	BSHW only	$20–35
2.	Space and BSHW heating	$25–50
3.	Space heating and cooling	$35–65

*These costs do not include auxiliary energy equipment, engineering design and inspection, or contingency costs. Costs for retrofit projects are approximately 10 to 25 percent higher than the above. 1977 dollars.

†For approximate costs per square meter, multiply costs per square foot by 10.

A more detailed component-cost breakdown is provided in Ref. 62, which gives representative flat-plate collector and storage-tank costs. In lieu of actual manufacturer-quoted prices these may be used as first estimates.

In addition, one also must include solar-energy-related costs, such as those resulting from increased floor space required to accommodate solar equipment. Credit should be given for any roof area provided by roof integral collectors. Remember, moreover, that collectors must be mounted on some sort of structure. When all expenses are included, the installed system costs shown in Table 13.3, part C, result.

Recent experience indicates that the installed incremental system costs in Table 13.3, part C, would be expected for new construction; 10 to 25 percent more should be expected for retrofit construction.

Solar System Operating Costs

Solar heating and/or cooling system operating costs include pumping or blower costs for moving the heat transfer fluid from collector to storage, and maintenance costs. Estimated pumping (constant speed) costs for a large heating and cooling installation at Los Alamos are a few cents per ft_c^2 per year. By comparison, maintenance costs of from 1 to 3 percent of initial installed capital cost per year should be expected.

MAINTENANCE AND OPERATING PROCEDURES

Proper maintenance starts with proper design and operating performance of the system. System checkout tests for leaks and thermal output should be performed before the installation contractor leaves the job. The designer or the installation contractor should provide use and operating guidelines for the system and its equipment including control settings, flow diagrams, and valve tags and charts.

Routine Preventive Maintenance Check List

Filters. Check pressure gauge and replace filters when differential pressure matches design.

Lubrication. Follow manufacturer's recommendations on all motors and fans.

Fan Belts. Check for proper tension and replace when worn. Spare belts are recommended. Check pulleys for proper alignment and speed.

Air Heating Coils. Coils must be air-cleaned or flushed clean. Straighten bent and crushed fins by combing.

Automatic Dampers. Check operation and linkages. Make certain fresh air intake damper is not open beyond minimum position. Clean linkage with solvent and lubricate with graphite and penetrating oil.

Heat Exchangers. Remove heads and clean out scale. Pressure flush water tube passages.

Controls. Observe operation of entire control system in all modes. Check for abnormal conditions. Check room thermostat settings. Observe operating temperatures such as relationship of storage temperature to supply temperature and status of auxiliary energy system for excessive operation. Keep logs of operating temperatures and pressures to provide a base for abnormal detection.

Collector Cover Plates. Keep cover plates clean. Remove leaves, branches and wash off dust and soot as often as accumulated. Inspection and cleaning is more frequently required if the collector array is in the vicinity of new construction or unpaved roads due to the dust in the atmosphere. Inspect cover plates for leaks and seal with silicone rubber. Replace broken cover plates using same material as original manufacturer. If glass replacement is required, use iron-free glass.

Absorber Plates. Inspect absorber plate surface. Recoat any deteriorated areas with the same material as original manufacturer. Make sure that material used will withstand stagnation temperatures experienced.

Casing. Check that casing breathing filter is kept clear and is not iced over.

Water Treatment

All hydronic systems require chemical water treatment to protect the system against corrosion and buildup of solids.

A qualified water-treatment organization must be employed to recommend proper initial treatment and to follow up operational requirements for the life of the system. The scope of services shall include

- Sampling the water from each circulating system.
- Chemical analysis of samples.
- Establish chemical treatment rates by "dose" or by continuous injection.
- Treat systems for control of corrosion; scale; pH control; hardness, algae, and slime.
- Periodic analyses of water samples and adjustment of treatment. Report of analyses and changes in treatment.
- Instruction of owner's operating personnel in management of chemical treatment program. Periodic (monthly) blow-down of sediment from bottom of tanks and heat exchangers will maintain circulation and reduce corrosion.

Antifreeze Protection

Systems containing antifreeze require monthly sampling and checking to make sure that freeze protection has not been compromised by automatic water makeup into the system.

All systems using antifreeze should not have direct water makeup connections to prevent dilution by automatic makeup water and possible contamination of the water supply.

Samples are to be checked with a hygrometer measuring specific gravity and calibrated for various temperatures indicating freeze points. Sampling is to be done at operating temperatures. Addition of antifreeze solution should be by means of a solution tank.

The life of the antifreeze can be extended by seasonal drainage and storage in closed containers. Each reuse of the solution will require the addition of corrosion inhibitor. Check antifreeze for sludge formation, particularly after being subjected to stagnation temperatures.

Aluminum waterways used in a solar system requires extremely close control of chemical treatment, especially if antifreeze solution is required. Manufacturer's instructions must be followed. Aluminum should be avoided if possible.

Special antifreeze specified as suitable for use in automobiles with aluminum block engines can be used. The circulating solution should consist of a mixture of distilled water, the special antifreeze, and inhibitor. Antifreeze solution used in aluminum systems should not be maintained for more than one season. The selection of materials during initial design must be influenced by type of freeze protection system required and ability to maintain system.

The automatic drain-down method of protecting the system from freezing requires automatic temperature sensing controls, valves or drainage whenever the pump stops. The system should be checked for proper operation before the start of the winter season. The system utilizing automatic temperature controls should have a manufacturer's service contract which includes a prewinter operational check. The system using drainage through the pump can be observed for increase in tank water level. Each testing cycle should include a check to determine if water has actually drained; open drain valve at low point of frost-exposed piping to see if water remains.

CASE HISTORY: LIQUID-BASED SOLAR HEATING SYSTEMS

Given below is a detailed description of one solar heating project (the Cary Arboretum Project) which integrates the design principles given above.

This project, designed and engineered by Dubin-Bloome Associates, was to be for a building which is energy-efficient. Eighty-five percent of its load is supplied by solar energy. The architect was Malcolm Wells, Cherry Hill, N.J. The building was designed to be an integrated active-passive solar system with minimum energy use; it surpassed the requirements of ASHRAE 90-75 in all areas. Many energy-conserving features, including the use of earth berms, recycling of exhaust air, water, and wastes, and a passive heated greenhouse, thermal mass with insulation in the exterior surfaces of the outside walls, thermal barriers over the windows, task lighting, and other features were incorporated in the design.

The solar energy system is a solar-assisted heat-pump system and is designed to supply the heating load and the hot water. Well water is used for cooling in the summer (along with an electric chiller heat pump). Well water is also used as a heat source for the heat pump in the winter when solar hot-water temperature is too low to be used directly or as a heat-pump source. Technical data relating to the solar system are summarized below. A flow diagram of the system is shown in Fig. 13.9.

General System Data and Climatological Data

1. Type of building: Two-story office-laboratory building, 27,400 ft^2 (gross)

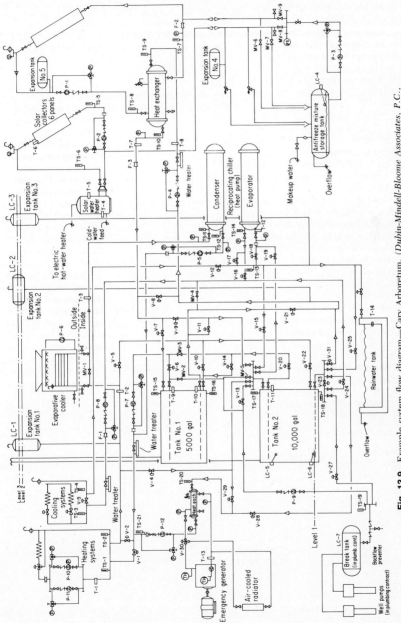

Fig. 13.9 Example system flow diagram—Cary Arboretum. *(Dubin-Mindell-Bloome Associates, P.C., Consulting Engineers, New York, NY.)*

2. Location: Millbrook, N.Y., 41°N latitude
3. Type of system: Solar-assisted heat-pump system for space heating
                                and domestic hot-water heating
4. Heating degree (°F) days: Yearly, 5800; January, 1100
5. Annual cooling hours: 2800 dry-bulb degree-hours (78° base)
6. Design (outside)   90°F; winter 0°F
   temperature:
7. Design heating load: 139,540 Btu/h
   Design cooling load: 60 tons (720,000 Btu/h)
8. Average wind speed: 10 mi/h
9. Peak daily insolation: 2650 Btu/ft^2 (June, horizontal surface);
                                2176 Btu/ft^2 (February, south-facing plane, 60° tilt)
10. Yearly sunshine: 55 percent

Collector Data

1. Type of collector
Two kinds of collectors were used:

a. A flat-plate, double-glazed collector with nonselective surface absorber plate. (Chamberlain Mfg. Co.)

b. Tubular, concentrating collectors (KTA), with a 4:1 concentration ratio. Total area was 5596 ft^2; net area was 4931 ft^2.

2. Type of heat-transfer fluid
Antifreeze solution: 50:50 mixture of water and propylene glycol; pH value: 7.2 to 7.3. Corrosion inhibitors added to the antifreeze solution.

Thermal conductivity: 0.25 Btu/h·ft^2·°F/ft.
Specific heat: 0.85; density 8.8 lb/gal.
Viscosity: 3.5 cP at 100°F.

3. Flowrate
0.031 to 0.012 g/min/ft^2 (or 15.5 to 6.0 lb/h · ft^2). Varied (using a two-speed pump) to maintain as closely as possible a 10°F temperature rise across the collector.

4. Collector temperature
Inlet: 90 to 200°F.
Outlet: 100 to 210°F.

5. Collector orientation
Azimuth: 4° west of due south.
Tilt: 57° above the horizontal.

6. Collector efficiency curve
See Fig. 13.10.

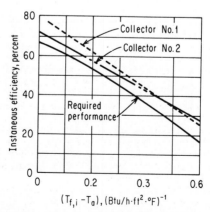

Fig. 13.10 Comparison of Cary Arboretum solar collector performance requirements with the performance of the collectors selected.

Storage

1. Tank dimensions:
   10 × 8 × 13 ft—capacity: 5,000 gal.
   19 × 10 × 13 ft—capacity: 10,000 gal
2. Tank materials:—12-in thick, reinforced concrete, liquid butyl lining, 4-in thick urea formaldehyde foam insulation on all surfaces, 7-ft buried depth.
3. Storage properties:
   Thermal storage medium—water
   Thermal conductivity—0.343 Btu/h·ft^2·°F/ft
   Viscosity—varies from 1.308 cP at 50°F to 0.432 cP at 150°F
   Specific heat—1.00 Btu/lb·°F
   Coefficient of cubical expansion—0.000115 at 70°F
   Density—62.4 lb/ft^3 at 50°F
4. Rated operating pressure: 13 psig at 250°F
5. Bursting pressure: 21 psig
6. Usable thermal capacity: 20,072,500 Btu
7. Average U value: 0.04 Btu/h·ft^2·°F

Heat loss will be less than 1°F in 24 h with storage temperature of 140°F.

Modes of Operation and Control System

The automatic control system is capable of providing the following modes of operation. (Note: Any individual mode is capable of independent operation; the solar collection and storage modes are compatible with any heating and/or cooling mode; heating mode 4 is capable of simultaneous use with cooling mode 9.)

- Mode 1. Solar energy collection and storage in either or both tanks
- Mode 2. Solar energy collection and storage in tank 1 only
- Mode 3. Solar energy collection for domestic hot water
- Mode 4. Heating with solar hot water from either or both tanks 1 and 2
- Mode 5. Heating with chiller acting as a heat pump with storage tank 2 as heat source (solar collection below 100°F)
- Mode 6. Heating with chiller acting as a heat pump with well water as the heat source
- Mode 7. Heating with waste heat from standby diesel generator
- Mode 8. Cooling with well water from tank 2
- Mode 9. Cooling with well water direct
- Mode 10. Cooling with chiller direct
- Mode 11. Storing chilled water in tank 2
- Mode 12. Cooling with chilled water stored in tank 2

The solar heating system will not operate until insolation measured by a pyranometer in the plane of the collectors exceeds 50 Btu/ft^2·h. On start-up, the primary and secondary pumps run at low speed. When the secondary supply temperature to the tanks exceeds the temperature at the top of either tank, flow is directed into the tanks. The tanks may be operated in series or each independently. This decision is accomplished automatically by comparison of top and bottom tank temperatures and the supply temperature. If the secondary supply temperature is less than the storage temperature in either tank, the flow bypasses the tanks. The primary and secondary pumps are two-speed and are controlled by the secondary system temperature differential across the heat exchanger. If the temperature differential rises above 15°F, the pumps run at high speed, and if the temperature differential falls below 7°F, the pumps run at low speed.

The domestic hot-water solar system is controlled by the pyranometer in the same manner as the main heating solar system but is piped and controlled separately from the main system. Temperatures are sensed in the panels themselves and in the hot-water storage tank. When the temperature of the solar panels exceeds the storage temperature by 10°F, the pump starts. The pump stops when the temperature difference drops to 3°F or less. The domestic hot-water solar system operates year-round. The heating solar system operates only during the heating season and will be drained during the summer.

Figure 13.11 shows some details of the solar system and the building; Figs. 13.12 and 13.13 illustrate monthly energy delivery and demand and other features of the system operation.

A computer operated control system is installed. Instrumentation has been installed to monitor the performance of the system.

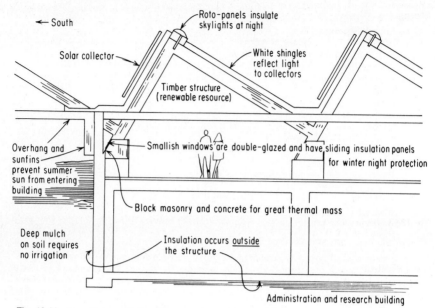

Fig. 13.11 Some features of the collector system and the building in the Cary Arboretum. Cross section at ¼-in scale.

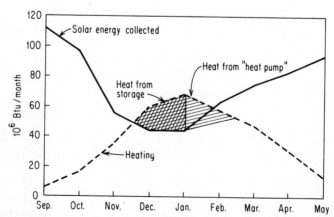

Fig. 13.12 Predicted solar collection and heat-load profiles in the Cary Arboretum building. (*Dubin-Mindell-Bloome Associates, P.C., Consulting Engineers, New York, NY.*)

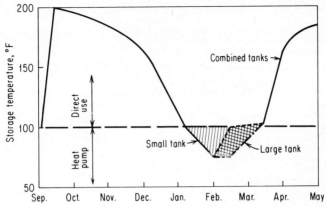

Fig. 13.13 Predicted variation of storage tank temperature with time of year in the Cary Arboretum project. (*Dubin-Mindell-Bloome Associates, P.C., Consulting Engineers, New York, NY.*)

REFERENCES

1. F. Kreith and J.F. Kreider, "Principles of Solar Engineering" (Hemisphere, Washington, DC, 1978).
2. J. Duffie and W. Beckman, "Solar Energy Thermal Processes" (Wiley-Interscience, New York, 1974).
3. "Solar Heating Systems Design Manual," ITT Training and Education Dept., Fluid Handling Division, Bulletin TESE-576 (1976).
4. "Methods of Testing to Determine the Thermal Performance of Solar Collectors," ASHRAE Standard 93-77 (1977).
5. R.C. Jordan and B.Y.U. Liu (eds.), "Applications of Solar Energy for Heating and Cooling of Buildings," ASHRAE GRP 170 (1977).
6. "ERDA's Pacific Regional Solar Heating Handbook," 2d ed. (ERDA's San Francisco Operations Office, November 1976) (Superintendent of Documents, Washington, DC, Stock No. 060-000-0024-7).
7. B.H. Hunn and D.O. Calafell, II, Determination of Average Ground Reflectivity for Solar Collectors, *Solar Energy*, vol. 19, no. 1, pp. 87–89 (1977).
8. K. Kimura and D.G. Stephenson, Solar Radiation on Cloudy Days, *ASHRAE Transactions*, part I (1969).
9. "Climatic Atlas of the United States," sponsored by the U.S. Dept. of Commerce, Environmental Science Services Administration, Environmental Data Service (U.S. Government Printing Office, Washington, DC, 1968).
10. I. Bennett, Correlation of Daily Insolation with Daily Total Sky Cover, Opaque Sky Cover, and Percentage of Possible Sunshine, *Solar Energy*, vol. 12, 391–393 (1969).
11. D.K. McDaniels, D.H. Lowndes, H. Mathew, J. Reynolds, and R. Gray, Enhanced Solar Energy Collection Using Reflector-Solar Thermal Collector Combinations, *Solar Energy*, vol. 17, no. 5, pp. 277–283 (1975).
12. H.E. Thomason and H.J.L. Thomason, Solar Houses/Heating and Cooling Progress Report, *Solar Energy*, vol. 15, p. 27 (1973).
13. Stanley F. Gilman (ed.), "Solar Energy Heat Pump Systems for Heating and Cooling Buildings," proceedings of a workshop conducted by The Pennsylvania State University, College of Engineering, University Park, PA, June 12–14, 1975. ERDA document No. C00-2560-1.
14. "ASHRAE Handbook and Product Directory" (1976).
15. "Solar Industry Index" (Solar Energy Industries Association, Inc., Washington, DC, 1977).
16. Dubin-Bloome Associates, "Energy Conservation Design Guidelines for Office Buildings" (General Services Administration, Public Building Service, Washington, DC, 1974); see also *Energy Conservation Standards*, McGraw-Hill, New York, 1978.
17. Dubin-Bloome Associates, Guidelines for Saving Energy in Existing Buildings, "Building Owners and Operators Manual," ECM 1, and "Engineers, Architects and Operators Manual," ECM 2, (Federal Energy Administration, Office of Energy Conservation and Environment, Washington, DC, June 1975).
18. T. Kusuda, "NBSLD, Computer Program for Heating and Cooling Loads in Buildings," National Bureau of Standards Report No. NBSIR 74-574 (November 1974).
19. R.H. Henninger (ed.), User's Manual, Part I, and Engineering Manual, Part II, "NECAP-NASA's Energy-Cost Analysis Program," NASA document No. CR-2590, parts I and II (U.S. Government Printing Office, Washington, DC, September 1975).
20. Z. Cumali, CAL-ERDA/1.0. A Public Domain Program for Energy Analysis in Buildings, paper presented at the Second ERDA-Wide Energy Conservation Symposium, San Francisco, CA, September 14, 1976.

21. "Bibliography on Available Computer Programs in the General Area of Heating, Refrigeration, Air Conditioning, and Ventilation," ASHRAE GRP 153 (October 1975).
22. D. Hittle, D. Holshouser, and G. Walton, Interim Feasibility Assessment Method for Solar Heating and Cooling of Army Buildings, "U.S. Pump Systems," paper 75-WA/Sol-3, presented at the American Society of Mechanical Engineers Winter Annual Meeting, Houston, TX, December 1975.
23. "Climatic Atlas of the United States" (see Ref. 9 above for full publication data).
24. "Engineering Weather Data," Air Force Manual (AFM) 88-29 (Superintendent of Documents, Washington, DC, Stock No. AFM 88-29) (supersedes AFM 88-8).
25. "Climatology of the United States—Decennial Census of United States Climate—Summary of Hourly Observations" (U.S. Dept. of Commerce, Washington, DC, 1970).
26. Fred S. Dubin, "Solar Assisted Heat Pump Installations in Cold Climatic Zones," paper delivered at The Pennsylvania State University, University Park, PA, June 1975.
27. Fred S. Dubin, Solar Energy Design for Existing Buildings, *ASHRAE Journal* (November 1975).
28. Dubin-Bloome Associates, The Atlanta School Project: Interfacing the Solar and Existing Mechanical Systems, *Specifying Engineer*, p. 59 (November 1976).
29. "Report on the Administration and Research Building for the Cary Arboretum of the New York Botanical Garden" (Dubin-Bloome Associates, New York, 1975).
30. "Design for Energy Conservation on a Life Cycle Basis—Two Projects for Cheney, Washington School District" (Dubin-Bloome Associates, New York, 1977).
31. "Solar and Heat Pump Systems Applications Handbook" (Dubin-Bloome Associates, New York, 1977) (prepared under contract to the U.S. Air Force).
32. "Solar Assisted Heat Pump Study for Heating and/or Cooling of Military Facilities" (Dubin-Bloome Associates, New York, 1977).
33. "Solar Energy Heat Pump Systems for Heating and Cooling Buildings," ERDA document No. C00-2560-2 (for full publication data see Ref. 13 above).
34. "HUD Intermediate Minimum Property Standards Supplement: Solar Heating and Domestic Hot Water Systems" (U.S. Government Printing Office, Washington, DC, 1977) (HUD document No. 4930.2).
35. J.F. Kreider and F. Kreith, "Solar Heating and Cooling: Engineering, Practical Design, and Economics," rev. 1st ed. (Hemisphere, Washington, DC, 1977).
36. "ASHRAE Handbook and Product Directory": Systems (1976); Applications (1974); Equipment (1975).
37. "Heating and Air Conditioning Systems Installation Standards for One and Two-Family Dwellings and Multifamily Housing, Including Solar" (Sheet Metal and Air-Conditioning Contractor's National Association, Inc., Vienna, VA, 1977).
38. "Solar Heating of Buildings and Domestic Hot Water," Civil Engineering Laboratory, U.S. Navy, Port Hueneme, CA, Technical Report No. R835 (January 1976).
39. J. Duffie and W. Beckman, "Solar Energy Thermal Processes" (John Wiley and Sons, New York, 1975).
40. W. Beckman, S. Klein, and J. Duffie, "Solar Heating Design" (John Wiley and Sons, New York, 1977).
41. "Solar Heating Handbook for Los Alamos," Los Alamos Scientific Laboratory Report LA-5967-MS (May 1975).
42. J. D. Balcomb, J. C. Hedstrom, S. W. Moore, and B. T. Rogers, "Solar Heating Handbook for Los Alamos," Los Alamos Scientific Laboratory Report No. LA-5967-MS (May 1976).
43. S. A. Klein, W. A. Beckman, and J. A. Duffie, A Design Procedure for Solar Heating Systems, *Solar Energy*, vol. 18, no. 2, pp. 113–127 (1976).
44. "Life Cycle Costing Emphasizing Energy Conservation," ERDA Division of Facilities and Construction Management Report No. ERDA-76/130 (September 1976).
45. V. C. Karman, T. L. Freeman, and J. W. Mitchell, Simulation Study of Solar Heat Pump Systems, "Proceedings of the Joint Conference of the American Section, International Solar Energy Society and Solar Energy Society of Canada, Winnipeg, Manitoba," vol. 3 (August 1976).
46. R. T. Reugg, "Solar Heating and Cooling on Buildings: Methods of Economic Evaluation," National Bureau of Standards Report No. NBSIR 75-712 (July 1975).
47. T. L. Freeman, J. W. Mitchell, W. A. Beckman, and J. A. Duffie, "Computer Modeling of Heat Pumps and the Simulation of Solar Heat."
48. D. Hittle and B. Sliwinski, CERL Thermal Loads Analysis and Systems Simulation Program, "User's Manual," vol. 1, U.S. Army Construction Engineering Research Laboratory Interim Report No. E-81 (December 1975).
49. H. C. Hottel and A. Whillier, Evaluation of Flat Plate Solar Collector Performance, "Trans. Conf. on Use of Solar Energy, University of Arizona, Tucson" (University of Arizona Press, Tucson, 1958).
50. "Interim Performance Criteria for Solar Heating and Cooling Systems in Commercial Buildings," National Bureau of Standards, Center for Buildings Technology Report No. NBSIR 76-1187 (November 1976).
51. "Method of Testing Thermal Storage Devices Based on Thermal Performance," ASHRAE Standard 94-77 (1977).
52. "Building Code Requirements for Minimum Design Loads in Buildings and Other Structures," ANSI A58.1-1972 (1972).
53. "Standards of the Tubular Exchanger Manufacturers Association" (Tubular Exchanger Manufacturers' Association, Inc., New York, 1968).
54. Solar Energy Utilization for Heating and Cooling, "ASHRAE Handbook and Product Directory, Applications," Chap. 59 (1974).

55. J. Duffie and W. Beckman, "Solar Energy Thermal Processes" (Wiley-Interscience, New York, 1974).
56. "Solar Heating Systems Design Manual" (see Ref. 3 for full publication data).
57. "Interim Feasibility Assessment Method for Solar Heating and Cooling of Army Buildings," U.S. Army Construction Engineering Research Laboratory, Champaign, IL, Technical Report No. E-91 (May 1976).
58. "ERDA's Pacific Regional Solar Heating Handbook" (see Ref. 6 for full publication data).
59. ASHRAE GRP 170 (see Ref. 5 for full publication data).
60. E. C. Boes, "Estimating the Direct Component of Solar Radiation," Sandia Laboratories, Albuquerque, NM, Report No. SAND75-0565 (November 1975).
61. J. E. Hill and T. Kusuda, Manchester's New Federal Building: An Energy Conservation Project, *ASHRAE Journal* (August 1975).
62. E. J. Beck and R. L. Field, "Solar Heating of Buildings and Domestic Hot Water," U.S. Navy Civil Engineering Laboratory, Port Hueneme, CA, Technical Report No. R835 (January 1975).
63. "Engineering Weather Data," Air Force Manual (AFM) 88-29, Chap. 6 (see Ref. 24 for full publication data).
64. "Intermediate Minimum Property Standards for Solar Heating and Domestic Hot Water Systems," National Bureau of Standards, Solar Energy Program Report No. NBSIR 76-1059 (April 1976).
65. S. V. Szokolay, "Solar Energy and Building" (John Wiley and Sons, New York, 1975).
66. "Proceedings of the Workshop on Solar Energy Storage Subsystems for the Heating and Cooling of Buildings," Charlottesville, VA, April 16–18, 1975, ASHRAE Report No. NSF-RA-N-75-041 (1975).
67. W. M. Kays and A. L. London, "Compact Heat Exchangers," 2d ed. (McGraw-Hill, New York, 1964.)
68. "ERDA Facilities Solar Design Handbook" (Energy Research and Development Administration—Division of Construction Planning and Support and Division of Solar Energy, Washington, DC, 1977).
69. "Handbook for the Design of Solar Energy Assisted Water-Source Heat Pumps for Non-Residential Buildings," prepared under grant from the Dept. of Energy at the Pennsylvania State University (1978).

APPENDIX: MONTHLY AND ANNUAL DEGREE-DAY DATA FOR CITIES IN THE UNITED STATES.

TABLE 13.A1 Normal Total Heating Degree-Days, Base, 65°F

	Location	Jul	Aug	Sep	Oct	Nov	Dec	Jan	Feb	Mar	Apr	May	Jun	Ann
AL	Birmingham	0	0	6	93	363	555	592	462	363	108	9	0	2,551
	Huntsville	0	0	12	127	426	663	694	557	434	138	19	0	3,070
	Mobile	0	0	0	22	213	357	415	300	211	42	0	0	1,560
	Montgomery	0	0	0	68	330	527	543	417	316	90	0	0	2,291
AK	Anchorage	245	291	516	930	1,284	1,572	1,631	1,316	1,293	879	592	315	10,864
	Annette	242	208	327	567	738	899	949	837	843	648	490	321	7,069
	Barrow	803	840	1,035	1,500	1,971	2,362	2,517	2,332	2,468	1,944	1,445	957	20,174
	Barter Is.	735	775	987	1,482	1,944	2,337	2,536	2,369	2,477	1,923	1,373	924	19,862
	Bethel	319	394	612	1,042	1,434	1,866	1,903	1,590	1,655	1,173	806	402	13,196
	Cold Bay	474	425	525	772	918	1,122	1,153	1,036	1,122	951	791	591	9,880
	Cordova	366	391	522	781	1,017	1,221	1,299	1,086	1,113	864	660	444	9,764
	Fairbanks	171	332	642	1,203	1,833	2,254	2,359	1,901	1,739	1,068	555	222	14,279
	Juneau	301	338	483	725	921	1,135	1,237	1,070	1,073	810	601	381	9,075
	King Salmon	313	322	513	908	1,290	1,606	1,600	1,333	1,411	966	673	408	11,343
	Kotzebue	381	446	723	1,249	1,728	2,127	2,192	1,932	2,080	1,554	1,057	636	16,105
	McGrath	208	338	633	1,184	1,791	2,232	2,294	1,817	1,758	1,122	648	258	14,283
	Nome	481	496	693	1,094	1,455	1,820	1,879	1,666	1,770	1,314	930	573	14,171
	St. Paul	605	539	612	862	963	1,197	1,228	1,168	1,265	1,098	936	726	11,199
	Shemya	577	475	501	784	876	1,042	1,045	958	1,011	885	837	696	9,687
	Yakutat	338	347	474	716	936	1,144	1,169	1,019	1,042	840	632	435	9,092
AZ	Flagstaff	46	68	201	558	867	1,073	1,169	991	911	651	437	180	7,152
	Phoenix	0	0	0	22	234	415	474	328	217	75	0	0	1,765
	Prescott	0	0	27	245	579	797	865	711	605	360	158	15	4,362
	Tucson	0	0	0	25	231	406	471	344	242	75	6	0	1,800
	Winslow	0	0	6	245	711	1,008	1,054	770	601	291	96	0	4,782
	Yuma	0	0	0	0	148	319	363	228	130	29	0	0	1,217
AR	Ft. Smith	0	0	12	127	450	704	781	596	456	144	22	0	3,292
	Little Rock	0	0	9	127	465	716	756	577	434	126	9	0	3,219
	Texarkana	0	0	0	78	345	561	626	468	350	105	0	0	2,533
CA	Bakersfield	0	0	0	37	282	502	546	364	267	105	19	0	2,122
	Bishop	0	0	42	248	576	797	874	666	539	306	143	36	4,227
	Blue Canyon	34	50	120	347	579	766	865	781	791	582	397	195	5,507
	Burbank	0	0	6	43	177	301	366	277	239	138	81	18	1,646
	Eureka	270	257	258	329	414	499	546	470	505	438	372	285	4,643
	Fresno	0	0	0	78	339	558	586	406	319	150	56	0	2,492
	Long Beach	0	0	12	40	156	288	375	297	267	168	90	18	1,711
	Los Angeles	28	22	42	78	180	291	372	302	288	219	158	81	2,061
	Mt. Shasta	25	34	123	406	696	902	983	784	738	525	347	159	5,722
	Oakland	53	50	45	127	309	481	527	400	353	255	180	90	2,870
	Pt. Arguello	202	186	162	205	291	400	474	392	403	339	298	243	3,595
	Red Bluff	0	0	0	53	318	555	605	428	341	168	47	0	2,515
	Sacramento	0	0	12	81	363	577	614	442	360	216	102	6	2,773
	Sandberg	0	0	30	202	480	691	778	661	620	426	264	57	4,209
	San Diego	6	0	15	37	123	251	313	249	202	123	84	36	1,439
	San Francisco	81	78	60	143	306	462	508	395	363	279	214	126	3,015
	Santa Catalina	16	0	9	50	165	279	353	308	326	249	192	105	2,052
	Santa Maria	99	93	96	146	270	391	459	370	363	282	233	165	2,967
CO	Alamosa	65	99	279	639	1,065	1,420	1,476	1,162	1,020	696	440	168	8,529
	Colorado Springs	9	25	132	456	825	1,032	1,128	938	893	582	319	84	6,423
	Denver	6	9	117	428	819	1,035	1,132	938	887	558	288	66	6,283
	Grand Junction	0	0	30	313	786	1,113	1,209	907	729	387	146	21	5,641
	Pueblo	0	0	54	326	750	986	1,085	871	772	429	174	15	5,462
CT	Bridgeport	0	0	66	307	615	986	1,079	966	853	510	208	27	5,617
	Hartford	0	6	99	372	711	1,119	1,209	1,061	899	495	177	24	6,172
	New Haven	0	12	87	347	648	1,011	1,097	991	871	543	245	45	5,897
DE	Wilmington	0	0	51	270	588	927	980	874	735	387	112	6	4,930
FL	Apalachicola	0	0	0	16	153	319	347	260	180	33	0	0	1,308
	Daytona Beach	0	0	0	0	75	211	248	190	140	15	0	0	879
	Ft. Myers	0	0	0	0	24	109	146	101	62	0	0	0	442
	Jacksonville	0	0	0	12	144	310	332	246	174	21	0	0	1,239
	Key West	0	0	0	0	0	28	40	31	9	0	0	0	108
	Lakeland	0	0	0	0	57	164	195	146	99	0	0	0	661
	Miami Beach	0	0	0	0	0	40	56	36	9	0	0	0	141
	Orlando	0	0	0	0	72	198	220	165	105	6	0	0	766
	Pensacola	0	0	0	19	195	353	400	277	183	36	0	0	1,463
	Tallahassee	0	0	0	28	198	360	375	286	202	36	0	0	1,485
	Tampa	0	0	0	0	60	171	202	148	102	0	0	0	683
	W. Palm Beach	0	0	0	0	6	65	87	64	31	0	0	0	253
GA	Athens	0	0	12	115	405	632	642	529	431	141	22	0	2,929
	Atlanta	0	0	18	127	414	626	639	529	437	168	25	0	2,983
	Augusta	0	0	0	78	333	552	549	445	350	90	0	0	2,397
	Columbus	0	0	0	87	333	543	552	434	338	96	0	0	2,383
	Macon	0	0	0	71	297	502	505	403	295	63	0	0	2,136

*"Climatic Atlas of the United States," U.S. Government Printing Office, 1968.

SOURCE: From Ref. 35. Original data from Ref. 9.

TABLE 13.A1 Normal Total Heating Degree-Days, Base 65°F (*continued*)

	Location	Jul	Aug	Sep	Oct	Nov	Dec	Jan	Feb	Mar	Apr	May	Jun	Ann
GA	Rome	0	0	24	161	474	701	710	577	468	177	34	0	3,326
	Savannah	0	0	0	47	246	437	437	353	254	45	0	0	1,819
	Thomasville	0	0	0	25	198	366	394	305	208	33	0	0	1,529
ID	Boise	0	0	132	415	792	1,017	1,113	854	722	438	245	81	5,809
	Idaho Falls 46W	16	34	270	623	1,056	1,370	1,538	1,249	1,085	651	391	192	8,475
	Idaho Falls 42NW	16	40	282	648	1,107	1,432	1,600	1,291	1,107	657	388	192	8,760
	Lewiston	0	0	123	403	756	933	1,063	815	694	426	239	90	5,542
	Pocatello	0	0	172	493	900	1,166	1,324	1,058	905	555	319	141	7,033
IL	Cairo	0	0	36	164	513	791	856	680	539	195	47	0	3,821
	Chicago	0	0	81	326	753	1,113	1,209	1,044	890	480	211	48	6,155
	Moline	0	9	99	335	774	1,181	1,314	1,100	918	450	189	39	6,408
	Peoria	0	6	87	326	759	1,113	1,218	1,025	849	426	183	33	6,025
	Rockford	6	9	114	400	837	1,221	1,333	1,137	961	516	236	60	6,830
	Springfield	0	0	72	291	696	1,023	1,135	935	769	354	136	18	5,429
IN	Evansville	0	0	66	220	606	896	955	767	620	237	68	0	4,435
	Ft. Wayne	0	9	105	378	783	1,135	1,178	1,028	890	471	189	39	6,205
	Indianapolis	0	0	90	316	723	1,051	1,113	949	809	432	177	39	5,699
	South Bend	0	6	111	372	777	1,125	1,221	1,070	933	525	239	60	6,439
IA	Burlington	0	0	93	322	768	1,135	1,259	1,042	859	426	177	33	6,114
	Des Moines	0	9	99	363	837	1,231	1,398	1,163	967	489	211	39	6,808
	Dubuque	12	31	156	450	906	1,287	1,420	1,204	1,026	546	260	78	7,376
	Sioux City	0	9	108	369	867	1,240	1,435	1,198	989	483	214	39	6,951
	Waterloo	12	19	138	428	909	1,296	1,460	1,221	1,023	531	229	54	7,320
KS	Concordia	0	0	57	276	705	1,023	1,163	935	781	372	149	18	5,479
	Dodge City	0	0	33	251	666	939	1,051	840	719	354	124	9	4,986
	Goodland	0	6	81	381	810	1,073	1,166	955	884	507	236	42	6,141
	Topeka	0	0	57	270	672	980	1,122	893	722	330	124	12	5,182
	Wichita	0	0	33	229	618	905	1,023	804	645	270	87	6	4,620
KY	Covington	0	0	75	291	669	983	1,035	893	756	390	149	24	5,265
	Lexington	0	0	54	239	609	902	946	818	685	325	105	0	4,683
	Louisville	0	0	54	248	609	890	930	818	682	315	105	9	4,660
LA	Alexandria	0	0	0	56	273	431	471	361	260	69	0	0	1,921
	Baton Rouge	0	0	0	31	216	369	409	294	208	33	0	0	1,560
	Burrwood	0	0	0	0	96	214	298	218	171	27	0	0	1,024
	Lake Charles	0	0	0	19	210	341	381	274	195	39	0	0	1,459
	New Orleans	0	0	0	19	192	322	363	258	192	39	0	0	1,385
	Shreveport	0	0	0	47	297	477	552	426	304	81	0	0	2,184
ME	Caribou	78	115	336	682	1,044	1,535	1,690	1,470	1,308	858	468	183	9,767
	Portland	12	53	195	508	807	1,215	1,339	1,182	1,042	675	372	111	7,511
MD	Baltimore	0	0	48	264	585	905	936	820	679	327	·90	0	4,654
	Frederick	0	0	66	307	624	955	995	876	741	384	127	12	5,087
MA	Blue Hill Obsy	0	22	108	381	690	1,085	1,178	1,053	936	579	267	69	6,368
	Boston	0	9	60	316	603	983	1,088	972	846	513	208	36	5,634
	Nantucket	12	22	93	332	573	896	992	941	896	621	384	129	5,891
	Pittsfield	25	59	219	524	831	1,231	1,339	1,196	1,063	660	326	105	7,578
	Worcester	6	34	147	450	774	1,172	1,271	1,123	998	612	304	78	6,969
MI	Alpena	68	105	273	580	912	1,268	1,404	1,299	1,218	777	446	156	8,506
	Detroit (City)	0	0	87	360	738	1,088	1,181	1,058	936	522	220	42	6,232
	Escanaba	59	87	243	539	924	1,293	1,445	1,296	1,203	777	456	159	8,481
	Flint	16	40	159	465	843	1,212	1,330	1,198	1,066	639	319	90	7,377
	Grand Rapids	9	28	135	434	804	1,147	1,259	1,134	1,011	579	279	75	6,894
	Lansing	6	22	138	431	813	1,163	1,262	1,142	1,011	579	273	69	6,909
	Marquette	59	81	240	527	936	1,268	1,411	1,268	1,187	771	468	177	8,393
	Muskegon	12	28	120	400	762	1,088	1,209	1,100	995	594	310	78	6,696
	Sault Ste. Marie	96	105	279	580	951	1,367	1,525	1,380	1,277	810	477	201	9,048
MN	Duluth	71	109	330	632	1,131	1,581	1,745	1,518	1,355	840	490	198	10,000
	International Falls	71	112	363	701	1,236	1,724	1,919	1,621	1,414	828	443	174	10,606
	Minneapolis	22	31	189	505	1,014	1,454	1,631	1,380	1,166	621	288	81	8,382
	Rochester	25	34	186	474	1,005	1,438	1,593	1,366	1,150	630	301	93	8,295
	St. Cloud	28	47	225	549	1,065	1,500	1,702	1,445	1,221	666	326	105	8,879
MS	Jackson	0	0	0	65	315	502	546	414	310	87	0	0	2,239
	Meridian	0	0	0	81	339	518	543	417	310	81	0	0	2,289
	Vicksburg	0	0	0	53	279	462	512	384	282	69	0	0	2,041
MO	Columbia	0	0	54	251	651	967	1,076	874	716	324	121	12	5,046
	Kansas	0	0	39	220	612	905	1,032	818	682	294	109	0	4,711
	St. Joseph	0	6	60	285	708	1,039	1,172	949	769	348	133	15	5,484
	St. Louis	0	0	60	251	627	936	1,026	848	704	312	121	15	4,900
	Springfield	0	0	45	223	600	877	973	781	660	291	105	6	4,561
MT	Billings	6	15	186	487	897	1,135	1,296	1,100	970	570	285	102	7,049
	Glasgow	31	47	270	608	1,104	1,466	1,711	1,439	1,187	648	335	150	8,996
	Great Falls	28	53	258	543	921	1,169	1,349	1,154	1,063	642	384	186	7,750
	Havre	28	53	306	595	1,065	1,367	1,584	1,364	1,181	657	338	162	8,700
	Helena	31	59	294	601	1,002	1,265	1,438	1,170	1,042	651	381	195	8,129
	Kalispell	50	99	321	654	1,020	1,240	1,401	1,134	1,029	639	397	207	8,191
	Miles City	6	6	174	502	972	1,296	1,504	1,252	1,057	579	276	99	7,723
	Missoula	34	74	303	651	1,035	1,287	1,420	1,120	970	621	391	219	8,125

TABLE 13.A1   Normal Total Heating Degree-Days, Base 65°F *(continued)*

	Location	Jul	Aug	Sep	Oct	Nov	Dec	Jan	Feb	Mar	Apr	May	Jun	Ann
NE	Grand Island	0	6	108	381	834	1,172	1,314	1,089	908	462	211	45	6,530
	Lincoln	0	6	75	301	726	1,066	1,237	1,016	834	402	171	30	5,864
	Norfolk	9	0	111	397	873	1,234	1,414	1,179	983	498	233	48	6,979
	North Platte	0	6	123	440	885	1,166	1,271	1,039	930	519	248	57	6,684
	Omaha	0	12	105	357	828	1,175	1,355	1,126	939	465	208	42	6,612
	Scottsbluff	0	0	138	459	876	1,128	1,231	1,008	921	552	285	75	6,673
	Valentine	9	12	165	493	942	1,237	1,395	1,176	1,045	579	288	84	7,425
NV	Elko	9	34	225	561	924	1,197	1,314	1,036	911	621	409	192	7,433
	Ely	28	43	234	592	939	1,184	1,075	977	672	456	225	7,733	
	Las Vegas	0	0	0	78	387	617	688	487	335	111	6	0	2,709
	Reno	43	87	204	490	801	1,026	1,073	823	729	510	357	189	6,332
	Winnemucca	0	34	210	536	876	1,091	1,172	916	837	573	363	153	6,761
NH	Concord	6	50	177	505	822	1,240	1,358	1,184	1,032	636	298 ·	75	7,383
	Mt. Wash. Osby.	493	536	720	1,057	1,341	1,742	1,820	1,663	1,652	1,260	930	603	13,817
NJ	Atlantic City	0	0	39	251	549	880	936	848	741	420	133	15	4,812
	Newark	0	0	30	248	573	921	983	876	729	381	118	0	4,859
	Trenton	0	0	57	264	576	924	989	885	753	399	121	12	4,980
NM	Albuquerque	0	0	12	229	642	868	930	703	595	288	81	0	4,348
	Clayton	0	6	66	310	699	899	986	812	747	429	183	21	5,158
	Raton	9	28	126	431	825	1,048	1,116	904	834	543	301	63	6,228
	Roswell	0	0	18	202	573	806	840	641	481	201	31	0	3,793
	Silver City	0	0	6	183	525	729	791	605	518	261	87	0	3,705
NY	Albany	0	19	138	440	777	1,194	1,311	1,156	992	564	239	45	6,875
	Binghamton (AP)	22	65	201	471	810	1,184	1,277	1,154	1,045	645	313	99	7,286
	Binghamton (PO)	0	28	141	406	732	1,107	1,190	1,081	949	543	229	45	6,451
	Buffalo	19	37	141	440	777	1,156	1,256	1,145	1,039	645	329	78	7,062
	Central Park	0	0	30	233	540	902	986	885	760	408	118	9	4,871
	J. F. Kennedy Intl.	0	0	36	248	564	933	1,029	935	815	480	167	12	5,219
	Laguardia	0	0	27	223	528	887	973	879	750	414	124	6	4,811
	Rochester	9	31	126	415	747	1,125	1,234	1,123	1,014	597	279	48	6,748
	Schenectady	0	22	123	422	756	1,159	1,283	1,131	970	543	211	30	6,650
	Syracuse	6	28	132	415	744	1,153	1,271	1,140	1,004	570	248	45	6,756
NC	Asheville	0	0	48	245	555	775	784	683	592	273	87	0	4,042
	Cape Hatteras	0	0	0	78	273	521	580	518	440	177	25	0	2,612
	Charlotte	0	0	6	124	438	691	691	582	481	156	22	0	3,191
	Greensboro	0	0	33	192	513	778	784	672	552	234	47	0	3,805
	Raleigh	0	0	21	164	450	716	725	616	487	180	34	0	3,393
	Wilmington	0	0	0	74	291	521	546	462	357	96	0	0	2,347
	Winston Salem	0	0	21	171	483	747	753	652	524	207	37	0	3,595
ND	Bismarck	34	28	222	577	1,083	1,463	1,708	1,442	1,203	645	329	117	8,851
	Devils Lake	40	53	273	642	1,191	1,634	1,872	1,579	1,345	753	381	138	9,901
	Fargo	28	37	219	574	1,107	1,569	1,789	1,520	1,262	690	332	99	9,226
	Williston	31	43	261	601	1,122	1,513	1,758	1,473	1,262	681	357	141	9,243
OH	Akron	0	9	96	381	726	1,070	1,138	1,016	871	489	202	39	6,037
	Cincinnati	0	0	54	248	612	921	970	837	701	336	118	9	4,806
	Cleveland	9	25	105	384	738	1,088	1,159	1,047	918	552	260	66	6,351
	Columbus	0	6	84	347	714	1,039	1,088	949	809	426	171	27	5,660
	Dayton	0	6	78	310	696	1,045	1,097	955	809	429	167	30	5,622
	Mansfield	9	22	114	397	768	1,110	1,169	1,042	924	543	245	60	6,403
	Sandusky	0	6	66	313	684	1,032	1,107	991	868	495	198	36	5,796
	Toledo	0	16	117	406	792	1,138	1,200	1,056	924	543	242	60	6,494
	Youngstown	6	19	120	412	771	1,104	1,169	1,047	921	540	248	60	6,417
OK	Oklahoma City	0	0	15	164	498	766	868	664	527	189	34	0	3,725
	Tulsa	0	0	18	158	522	787	893	683	539	213	47	0	3,860
OR	Astoria	146	130	210	375	561	679	753	622	636	480	363	231	5,186
	Burns	12	37	210	515	867	1,113	1,246	988	856	570	366	177	6,957
	Eugene·	34	34	129	366	585	719	803	627	589	426	279	135	4,726
	Meacham	84	124	288	580	918	1,091	1,209	1,005	983	726	527	339	7,874
	Medford	0	0	78	372	678	871	918	697	642	432	242	78	5,008
	Pendleton	0	0	111	350	711	884	1,017	773	617	396	205	63	5,127
	Portland	25	28	114	335	597	735	825	644	586	396	245	105	4,635
	Roseburg	22	16	105	329	567	713	766	608	570	405	267	123	4,491
	Salem	37	31	111	338	594	729	822	647	611	417	273	144	4,754
	Sexton Summit	81	81	171	443	666	874	958	809	818	609	465	279	6,254
PA	Allentown	0	0	90	353	693	1,045	1,116	1,002	849 ·	471	167	24	5,810
	Erie	0	25	102	391	714	1,063	1,169	1,081	973	585	288	60	6,451
	Harrisburg	0	0	63	298	648	992	1,045	907	766	396	124	12	5,251
	Philadelphia	0	0	60	291	621	964	1,014	890	744	390	115	12	5,101
	Pittsburgh	0	9	105	375	726	1,063	1,119	1,002	874	480	195	39	5,987
	Reading	0	0	54	257	597	939	1,001	885	735	372	105	0	4,945
	Scranton	0	19	132	434	762	1,104	1,156	1,028	893	498	195	33	6,254
	Williamsport	0	9	111	375	717	1,073	1,122	1,002	856	468	177	24	5,934
RI	Block Is.	0	16	78	307	594	902	1,020	955	877	612	344	99	5,804
	Providence	0	16	96	372	660	1,023	1,110	988	868	534	236	51	5,954
SC	Charleston	0	0	0	59	282	471	487	389	291	54	0	0	2,033
	Columbia	0	0	0	84	345	577	570	470	357	81	0	0	2,484

TABLE 13.A1 Normal Total Heating Degree-Days, Base 65°F *(continued)*

	Location	Jul	Aug	Sep	Oct	Nov	Dec	Jan	Feb	Mar	Apr	May	Jun	Ann
SC	Florence	0	0	0	78	315	552	552	459	347	84	0	0	2,387
	Greenville	0	0	0	112	387	636	648	535	434	120	12	0	2,884
	Spartanburg	0	0	15	130	417	667	663	560	453	144	25	0	3,074
SD	Huron	9	12	165	508	1,014	1,432	1,628	1,355	1,125	600	288	87	8,223
	Rapid City	22	12	165	481	897	1,172	1,333	1,145	1,051	615	326	126	7,345
	Sioux Falls	19	25	168	462	972	1,361	1,544	1,285	1,082	573	270	78	7,839
TN	Bristol	0	0	51	236	573	828	828	700	598	261	68	0	4,143
	Chattanooga	0	0	18	143	468	698	722	577	453	150	25	0	3,254
	Knoxville	0	0	30	171	489	725	732	613	493	198	43	0	3,494
	Memphis	0	0	18	130	447	698	729	585	456	147	22	0	3,232
	Nashville	0	0	30	158	495	732	778	644	512	189	40	0	3,578
	Oak Ridge (CO)	0	0	39	192	531	772	778	669	552	228	56	0	3,817
TX	Abilene	0	0	0	99	366	586	642	470	347	114	0	0	2,624
	Amarillo	0	0	18	205	570	797	877	664	546	252	56	0	3,985
	Austin	0	0	0	31	225	388	468	325	223	51	0	0	1,711
	Brownsville	0	0	0	0	66	149	205	106	74	0	0	0	600
	Corpus Christi	0	0	0	0	120	220	291	174	109	0	0	0	914
	Dallas	0	0	0	62	321	524	601	440	319	90	6	0	2,363
	El Paso	0	0	0	84	414	648	685	445	319	105	0	0	2,700
	Ft. Worth	0	0	0	65	324	536	614	448	319	99	6	0	2,405
	Galveston	0	0	0	0	138	270	350	258	189	30	0	0	1,235
	Houston	0	0	0	6	183	307	384	288	192	36	0	0	1,396
	Laredo	0	0	0	0	105	217	267	134	74	0	0	0	797
	Lubbock	0	0	18	174	513	744	800	613	484	201	31	0	3,578
	Midland	0	0	0	87	381	592	651	468	322	90	0	0	2,591
	Port Arthur	0	0	0	22	207	329	384	274	192	39	0	0	1,447
	San Angelo	0	0	0	68	318	536	567	412	288	66	0	0	2,255
	San Antonio	0	0	0	31	207	363	428	286	195	39	0	0	1,549
	Victoria	0	0	0	6	150	270	344	230	152	21	0	0	1,173
TX	Waco	0	0	0	43	270	456	536	389	270	66	0	0	2,030
	Wichita Falls	0	0	0	99	381	632	698	518	378	120	6	0	2,832
UT	Milford	0	0	99	443	867	1,141	1,252	988	822	519	279	87	6,497
	Salt Lake City	0	0	81	419	849	1,082	1,172	910	763	459	233	84	6,052
	Wendover	0	0	48	372	822	1,091	1,178	902	729	408	177	51	5,778
VT	Burlington	28	65	207	539	891	1,349	1,513	1,333	1,187	714	353	90	8,269
VA	Cape Henry	0	0	0	112	360	645	694	633	536	246	53	0	3,279
	Lynchburg	0	0	51	223	540	822	849	731	605	267	78	0	4,166
	Norfolk	0	0	0	136	408	698	738	655	533	216	37	0	3,421
	Richmond	0	0	36	214	495	784	815	703	546	219	53	0	3,865
	Roanoke	0	0	51	229	549	825	834	722	614	261	65	0	4,150
	Wash. Nat'l. Ap.	0	0	33	217	519	834	871	762	626	288	74	0	4,224
WA	Olympia	68	71	198	422	636	753	834	675	645	450	307	177	5,236
	Seattle	50	47	129	329	543	657	738	599	577	396	242	117	4,424
	Seattle Boeing	34	40	147	384	624	763	831	655	608	411	242	99	4,838
	Seattle Tacoma	56	62	162	391	633	750	828	678	657	474	295	159	5,145
	Spokane	9	25	168	493	879	1,082	1,231	980	834	531	288	135	6,655
	Stampede Pass	273	291	393	701	1,008	1,178	1,287	1,075	1,085	855	483	393	9,283
	Tatoosh Is.	295	279	306	406	534	639	713	613	645	525	431	333	5,719
	Walla Walla	0	0	87	310	681	843	986	745	589	342	177	45	4,805
	Yakima	0	12	144	450	828	1,039	1,163	868	713	435	220	69	5,941
WV	Charleston	0	0	63	254	591	865	880	770	648	300	96	9	4,476
	Elkins	9	25	135	400	729	992	1,008	896	791	444	198	48	5,675
	Huntington	0	0	63	257	585	856	880	764	636	294	99	12	4,446
	Parkersburg	0	0	60	264	606	905	942	826	691	339	115	6	4,754
WI	Green Bay	28	50	174	484	924	1,333	1,494	1,313	1,141	654	335	99	8,029
	La Crosse	12	19	153	437	924	1,339	1,504	1,277	1,070	540	245	69	7,589
	Madison	25	40	174	474	930	1,330	1,473	1,274	1,113	618	310	102	7,863
	Milwaukee	43	47	174	471	876	1,252	1,376	1,193	1,054	642	372	135	7,635
WY	Casper	6	16	192	524	942	1,169	1,290	1,084	1,020	657	381	129	7,410
	Cheyenne	19	31	210	543	924	1,101	1,228	1,056	1,011	672	381	102	7,278
	Lander	6	19	204	555	1,020	1,299	1,417	1,145	1,017	654	381	153	7,870
	Sheridan	25	31	219	539	948	1,200	1,355	1,154	1,054	642	366	150	7,683

One of the most practical of weather statistics is the "heating degree-day." First devised some 60 years ago, the degree-day system has been in quite general use by the heating industry for more than 40 years.

Heating degree-days are the number of degrees the daily average temperature is below 65°F. Normally heating is not required in a building when the outdoor average daily temperature is 65°F. Heating degree-days are determined by subtracting the average daily temperatures below 65°F from the base 65°F. A day with an average temperature of 50°F has 15 heating degree-days (65 − 50 = 15) while one with an average temperature of 65°F or higher has none.

Several characteristics make the degree-day figures especially useful. They are cumulative so that the degree-day sum for a period of days represents the total heating load for that period. The relationship between degree-days and fuel consumption is linear, i.e., doubling the degree-days usually doubles the fuel consumption. Comparing normal seasonal degree-days in different locations gives a rough estimate of seasonal fuel consumption. For example, it would require roughly 4½ times as much fuel to heat a building in Chicago, Ill., where the mean annual total heating degree-days are about 6,200 than to heat a similar building in New Orleans, La., where the annual total heating degree-days are around 1,400. Using degree-days has the advantage that the consumption ratios are fairly constant, i.e., the fuel consumed per 100 degree-days is about the same whether the 100 degree-days occur in only 3 or 4 days or are spread over 7 or 8 days.

Chapter 14

Performance Prediction for Solar Heating Systems

W. A. BECKMAN

S. A. KLEIN

and

J. A. DUFFIE
University of Wisconsin, Madison, WI

This chapter outlines the f-chart method for rapidly estimating the thermal performance of active solar space-heating systems (using either liquid or air as the working fluid) and solar water-heating systems. The system configurations which can be evaluated by the methods presented here are expected to be common in residential applications. The material in this chapter is presented and explained in detail in *Solar Heating Design by the f-Chart Method* by Beckman, Klein, and Duffie.[1]

INTRODUCTION

It is technically possible to build a solar heating system which could supply the total annual heating load of a building. A system sized to meet maximum winter loads would be greatly oversized during other times of the year, however. As solar heating systems are capital-intensive, oversizing can result in a severe economic penalty. Thus, most solar heating systems consist of a combination of solar-conventional (auxiliary) sources. The basic problem treated in this chapter is to determine the fraction of total heating load which will be supplied by solar energy for a particular solar heating system size and type. The primary design variable is collector area. Secondary variables—collector type, storage capacity, fluid flowrates, and heat exchanger sizes—can be determined by references to Chaps. 11 through 13 of this handbook. The information resulting from this analysis is generally presented in the form of the annual solar load fraction F, the fraction of the annual load supplied by solar energy, as a function of collector area, for a given system; this information is then used with the economic analysis methods of Chap. 28 to optimize system sizing and design.

The f-chart method is essentially a correlation of the results of many hundreds of computer simulations of solar heating systems. The conditions of the simulations were varied over the ranges anticipated in practical system designs. The resulting correlations give the monthly solar load fraction f, the fraction of the monthly heating load (for space heating and hot water) supplied by solar energy as a function of two dimensionless parameters involving collector characteristics, heating loads, and weather. The performance predictions of f-chart have been compared with predictions made by very detailed simulations in 14 locations in the United States; the standard error of the differences between the simulation and the f-chart results was about 2.5 percent. The correlations have also been compared with the few long-term system performance data available in 1977 (Ref. 2).

Large scale solar energy applications and systems which are not standard in configuration must still be designed using simulation methods (see Chap. 17). The necessary thermal performance of standard heating systems can be determined with the f-chart method by hand or more simply with the aid of an inexpensive calculator that can evaluate exponential functions. Alternatively, programs for small computers and hand-held programmable calculators are available to do these calculations.

Meteorological data required for the f-chart method are: (1) monthly-averaged, daily radiation on the collector which can be derived from the most commonly available solar radiation data—average solar energy on a horizontal surface—as shown in Chap. 2; and (2) monthly-averaged ambient temperatures. The collector data required are the slope and intercept resulting from standard collector tests, i.e., $F_R U_c$ and $F_R(\tau\alpha)_n$ (see Chap. 7).

Information on the f-chart method, its background, and its development can be obtained from Refs. 1–4.

THE STANDARD SYSTEMS

In order to develop the f-chart, standard system configurations were assumed. F-chart results apply only to these systems and no others.

Liquid Systems

A schematic diagram of the assumed liquid-based solar heating system is shown in Fig. 14.1. This system uses an antifreeze solution as the heat transfer fluid in the collector loop and water as the storage medium. Energy is stored in the form of sensible heat in the water tank. Alternatively, collectors may be drained at night or during periods of excessive cloudiness in which case water is used directly in the collectors and a collector heat exchanger is not needed.

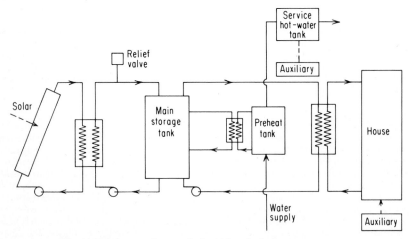

Fig. 14.1   Schematic diagram of a liquid-based solar heating system.

The water-to-air, load heat exchanger is used to transfer heat from the storage tank to the building. A liquid-to-liquid heat exchanger is used to transfer energy from the main storage tank to a domestic hot water preheat tank which, in turn, supplies solar heated water to a conventional water heater. A conventional furnace, or heat pump which acts as the auxiliary heater is provided to meet the space-heating load when the energy in the storage tank is depleted. Controllers, relief valves, pumps, and pipes make up the remaining equipment. For further information on liquid systems, see Chap. 13.

Air Systems

The assumed configuration for a solar air-heating system with a pebble-bed storage unit is shown in Fig. 14.2. Other arrangements of fans and dampers can result in an equivalent flow circuit. Energy required for domestic hot water is provided by heat exchange from the hot air leaving the collector to a domestic water preheat tank, as in the liquid system. The hot water is further heated, if required, by a conventional water heater. During summer operation, it is best not to store solar energy in the pebble bed, so a seasonal, manually operated, storage bypass is a usual part of this design (not shown in Fig. 14.2). For further information on air systems, see Chap. 12.

Domestic Water-Heating Systems

The standard configuration of a solar domestic water-heating system is shown in Fig. 14.3. The collector may heat either air or liquid. The collected energy is transferred via heat exchanger

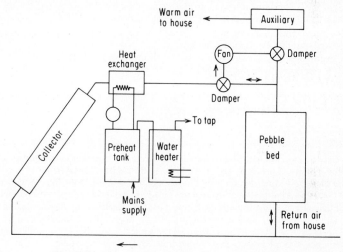

Fig. 14.2 Schematic diagram of a solar air heating system.

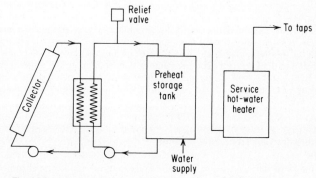

Fig. 14.3 Schematic diagram of a domestic solar water heating system.

to a domestic water-preheat tank, which supplies solar-heated water to a conventional water heater. The water is further heated to the desired temperature by conventional fuel if necessary. Water-heating systems are described in Chap. 11.

THE f-CHART METHOD

The f-chart design method for solar heating systems is based on detailed computer simulations used to develop correlations between dimensionless variables and the long-term performance of the systems. The correlations are in equation form or graphical form referred to as the "f-charts."

The fraction of the monthly space and water heating loads supplied by solar energy, f, is empirically related to two dimensionless groups:

$$X = A_c F'_R U_c \, (T_{ref} - \overline{T}_a) \, \Delta t / L \tag{14.1}$$

$$Y = A_c F'_R (\overline{\tau \alpha}) \, \overline{I}_T N / L \tag{14.2}$$

where A_c = the area of the solar collector, m² or ft²
  F'_R = the collector-heat exchanger efficiency factor (Chap. 7)
  U_c = the collector overall energy loss coefficient, (W/°C· m² or Btu/h· ft²· °F) (Chap. 7)
  Δt = the total number of seconds (or hours) in the month, depending on the units of U_c.
  $\overline{T}_a$ = the monthly average ambient temperature, °C or F
  L = the monthly total heating load for space heating and hot water, J or Btu
  $\overline{I}_T$ = the monthly-averaged, daily radiation incident on the collector surface per unit area, J/m² or Btu/ft²
  N = the number of days in a month
  $(\overline{\tau \alpha})$ = the monthly average transmittance-absorptance product
  T_{ref} = reference temperature (100°C or 212°F)

The equations for X and Y can be rewritten for convenience in calculations:

$$X = F_R U_c \cdot (F'_R / F_R) \cdot (T_{ref} - \overline{T}_a) \cdot \Delta t \cdot A_c / L \tag{14.3}$$

$$Y = F_R (\tau \alpha)_n \cdot (F'_R / F_R) \cdot (\overline{\tau \alpha}) / (\tau \alpha)_n \cdot \overline{I}_T N \cdot A_c / L \tag{14.4}$$

$F_R U_c$ and $F_R (\tau \alpha)_n$ are obtained from collector test results by the methods noted in Chap. 7, F'_R / F_R by the method of Chap. 7, and $(\overline{\tau \alpha}) / (\tau \alpha)_n$ by the method noted below. $\overline{T}_a$ is obtained from meteorological records for the month and location desired. (See the Appendix to this chapter.) $\overline{I}_T$ is found from the monthly-averaged, daily radiation on a horizontal surface by the methods of Chap. 2. The monthly loads, L, can be determined by the degree-day method or any other method the user prefers. (See Chap. 13 for degree-day tabulations on a monthly basis.) Values of A_c, the collector area, are selected for the calculations. Thus, all of the terms in these two equations are readily determined from available information.

The two dimensionless groups have physical significance. Y is related to the ratio of the total energy absorbed on the collector plate surface to the total heating load during the month. X is related to the ratio of a reference collector energy loss to the total heating load during the month.

To determine f, the fraction of the heating load supplied by solar energy for a month, values of X and Y are calculated for the collector and heating load in question. The value of f is determined at the intersection of X and Y on the f-chart, or from the equivalent equation. This is done for each month of the year. The solar energy contribution for the month is the product of f and the total heating load, L, for the month. Finally, the fraction of the annual heating load supplied by solar energy, F, is the sum of the monthly solar energy contributions divided by the annual load:

$$F = \Sigma fL / \Sigma L \tag{14.5}$$

Effect of Orientation on Transmittance and Absorptance

The transmittance of the transparent collector cover system, τ, and the absorptance of the collector plate, α, depend on the angle at which solar radiation is incident on the collector surface. Collector tests are usually carried out with the radiation incident on the collector in a nearly perpendicular direction. Thus, the product of F_R, τ, and α determined from collector tests ordinarily corresponds to the transmittance and absorptance values for radiation at normal incidence, $F_R(\tau\alpha)_n$. Depending on the collector orientation and the time of the year, the monthly-averaged values of the transmittance and absorptance can be significantly lower than the values for radiation at normal incidence. The f-chart method requires a knowledge of monthly-averaged values, or more specifically, $(\tau\alpha)/(\tau\alpha)_n$, the ratio of the monthly average to normal incidence transmittance-absorptance products.

A method of determining $(\tau\alpha)/(\tau\alpha)_n$, which is useful for most common situations, is as follows. For a collector oriented with a slope equal to the latitude plus or minus 15°, facing within 15° of due south, the ratio $(\tau\alpha)/(\tau\alpha)_n$ is about 0.96 for a single-cover collector and 0.94 for a two-cover collector for all months during the space heating season. For a discussion of a more detailed method of evaluating $(\tau\alpha)/(\tau\alpha)_n$ for situations not covered by this approximation, see Ref. 1.

Example 14.1

A solar heating system is to be designed for Madison, Wisconsin (lat., 43°N). The solar collectors considered for this system have two glass covers with $F_R(\tau\alpha) = 0.68$ and $F_R U_c = 3.75$ W/°C·m^2 (0.66 Btu/h · ft^2 · °F) as determined from standard collector tests. (See Chap. 7.) The collector is to face south, at a slope of 58° with respect to the horizontal. The average daily radiation on a 58° surface for January in Madison is 13.2 MJ/m^2 (1160 Btu/ft^2). The average ambient temperature is -7°C (19.4°F). The total (space and hot water) heating load is 36.0 GJ (34.1 MMBtu) and the collector-heat exchanger penalty factor, F'_R/F_R, is 0.97. The ratio of the monthly average to normal incidence transmittance-absorptance product $(\tau\alpha)/(\tau\alpha)_n$ is 0.94 for two-cover collectors at this orientation. Calculate X and Y for these conditions for collector areas of 25 and 50 m^2 (269.1 and 538.2 ft^2).

From Eq. (14.3) and (14.4) with $A_c = 25$ m^2,

$$X = 3.75 \text{ W/C}\cdot m^2 \times 0.97 \times [100 - (-7)]°C$$
$$\times 31 \text{ days} \times 86400 \text{ s/day} \times 25 \text{ m}^2/36.0 \times 10^9 \text{ J}$$
$$= 0.724$$

$$Y = 0.68 \times 0.97 \times 0.94 \times 13.2 \times 10^6 \text{ J/m}^2 \cdot \text{ day}$$
$$\times (31 \text{ days}) \times (25 \text{ m}^2/36.0 \times 10^9 \text{ J})$$
$$= 0.176$$

Note that X and Y are dimensionless variables and, as a result, their values are not dependent upon the system of units. In English units,

$$X = 0.66 \text{ Btu/h} \cdot °F \cdot ft^2 \times 0.97 \times (212 °F - 19.4 °F)$$
$$\times 31 \text{ days} \times 24 \text{ h/day} \times 269.1 \text{ ft}^2/34.1 \times 10^6 \text{ Btu}$$
$$= 0.724$$

$$Y = 0.68 \times 0.97 \times 0.94 \times 1160 \text{ Btu/ft}^2 \cdot \text{ day} \times 31 \text{ days}$$
$$\times 269.1 \text{ ft}^2/34.1 \times 10^6 \text{ Btu} = 0.176$$

For 50 m^2 (538.2 ft^2), the values of X and Y are proportionally higher.

$$X = 0.724 \times 50/25 = 1.45$$
$$Y = 0.176 \times 50/25 = 0.352$$

THE f-CHART FOR LIQUID SYSTEMS

The fraction, f, of the monthly total load supplied by the solar space- and water-heating system shown in Fig. 14.1, is given as a function of X and Y in Fig. 14.4, the f-chart for liquid systems. The relationship between X, Y, and f in Fig. 14.4 can be expressed in equation form as:

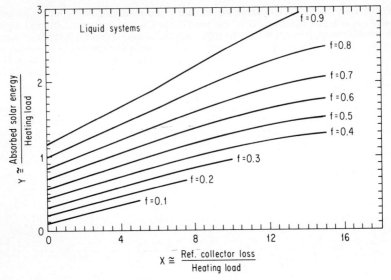

Fig. 14.4   F-chart for liquid systems.

$$f = 1.029\,Y - 0.065\,X - 0.245\,Y^2 + 0.0018\,X^2$$
$$+ 0.0215\,Y^3 \tag{14.6}$$
$$\text{for } 0 \leqslant Y \leqslant 3.5,\ 0 \leqslant X \leqslant 18,\ \text{and } 0 \leqslant f \leqslant 1$$

Example 14.2

The solar heating system described in Example 14.1 is to be a liquid system, as shown in Fig. 14.1. The storage and load heat exchanger sizes are to be the same as those used in the development of the f-chart. (See below.) What fraction of the annual heating load will be supplied by solar energy for a collector area of 50 m² (538.2 ft²)?

From Example 14.1, the values of X and Y are 1.45 and 0.352 respectively in January. From Fig. 14.4 or Eq. (14.6), $f = 0.24$. The total heating load for January is 36.0 GJ (34.1 MMBtu). Thus, the energy delivery from the solar heating system in January is

$$fL = 0.24 \times 36.0\ \text{GJ} = 8.6\ \text{GJ}$$
$$\text{or} \qquad fL = 0.24 \times 34.1\ \text{MMBtu} = 8.2\ \text{MMBtu}$$

The fraction of the annual heating load supplied by solar energy is determined by repeating the calculation of X, Y and f for each month, and summing the results as indicated in Eq. (14.5). Table 14.1 shows the results of these calculations for this example. The annual fraction of the load supplied by solar energy $F = 87.4/203.2 = 0.43$.

In order to determine the economically optimum collector area, the annual load fraction corresponding to several different collector areas is determined in the manner described here. The annual load fraction is then plotted as a function of collector area, as shown in Fig. 14.5. The information in this figure can then be used for economic optimization calculations, as shown in Chap. 28.

Storage Capacity

Annual system performance has been found to be relatively insensitive to storage capacity, as long as capacity is more than about 50 liters (L) of water per square meter of collector (1.23 gal/ft²). When the costs of storage are considered, it appears that there are broad optima in the range of 50 to 100 L of water per square meter of collector (1.23 to 2.46 gal/ft²).

The f-chart has been developed for a storage capacity of 75 L of stored water per square meter of collector area (1.85 gal/ft²). It can be used to estimate the performance of other systems

TABLE 14.1   Monthly and Annual Performance of an Example Liquid Heating System in Madison*

Month	Load		f	fL	
	GJ	MMBtu		GJ	MMBtu
Jan.	36.0	34.1	0.24	8.6	8.2
Feb.	30.4	28.8	0.30	9.1	8.6
Mar.	26.7	25.3	0.44	11.7	11.1
Apr.	15.7	14.9	0.56	8.8	8.3
May	9.2	8.7	0.81	7.5	7.1
Jun.	4.1	3.9	1.00	4.1	3.9
Jul.	2.9	2.7	1.00	2.9	2.7
Aug.	3.4	3.2	1.00	3.4	3.2
Sep.	6.3	6.0	1.00	6.3	6.0
Oct.	13.2	12.5	0.74	9.8	9.3
Nov.	22.8	21.6	0.31	7.1	6.7
Dec.	32.5	30.8	0.25	8.1	7.7
Total	203.2	192.5		87.4	82.8

*$A = 50 \text{ m}^2$ (468.2 ft²)

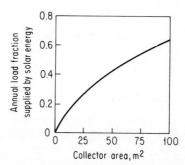

Fig. 14.5   Annual load fraction versus collector area.

with other short-term storage capacities in a narrow range by modifying the dimensionless group X by the storage-size correction factor given in Fig. 14.6 or Eq. (14.7).

$$\begin{array}{c}\text{Storage size} \\ \text{correction factor}\end{array} = (X'/X) = \left[\frac{\text{Actual storage capacity}}{\text{Standard storage capacity}}\right]^{-0.25} \qquad (14.7)$$

$$\text{for } 0.5 < \left[\frac{\text{Actual storage capacity}}{\text{Standard storage capacity}}\right] < 4$$

where the standard storage capacity is 75 L of water per square meter of collector area (1.85 gal/ft²).

Example 14.3

For the conditions of Example 14.2, what would be the annual solar contribution if the storage capacity of the tank is doubled to 150 L/m² (3.70 gal/ft²)?

To account for changes in storage capacity, the value of X calculated in the previous examples must be modified by Eq. (14.7) or Fig. 14.6. Using the equation, with the ratio of actual storage size to standard storage size equal to 2.0,

$$X'/X = 2^{-0.25} = 0.84$$

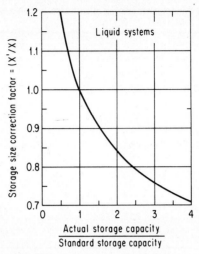

Fig. 14.6 Storage size correction factor.

The corrected value of X is then

$$X' = 0.84 \times 1.45 = 1.22$$

The value of Y is unaffected by a change in storage capacity, so $Y = 0.35$. From Fig. 14.4 or Eq. (14.6), $f = 0.26$. The solar contribution for January is

$$fL = 0.26 \times 36.0 \text{ GJ} = 9.4 \text{ GJ}$$

Repeating these calculations for the remaining 11 months results in an annual solar-load fraction of 0.45 (compared to 0.43 for the standard storage size).

Load Heat Exchanger Size

The effectiveness of the load heat exchanger can significantly affect the performance of the solar heating system. When the heat exchanger used to heat the building air is reduced in size, the storage tank temperature must increase to supply the same amount of heat. This results in higher collector fluid inlet temperatures and reduced collector efficiency. A measure of the size heat exchanger needed for a specific building is provided by the dimensionless parameter $(\epsilon_L C_{min}/UA)$. Here ϵ_L is the effectiveness of the water-air load heat exchanger; C_{min} is the minimum fluid capacitance rate (mass flowrate times the specific heat of the fluid) in the load heat exchanger, and is generally that of the air; UA is the building overall energy loss coefficient-area product used in the degree-day space heating load model.

The optimum value of $(\epsilon_L C_{min}/UA)$ from a technical standpoint is infinite. However, system performance is asymptotically dependent upon the value of this parameter, and for values of $(\epsilon_L C_{min}/UA)$ greater than 10, performance will be nearly the same as that for the infinitely large value. The reduction in performance due to an undersized load heat exchanger will be appreciable for values of $(\epsilon_L C_{min}/UA)$ less than about 1.0. Practical values of $(\epsilon_L C_{min}/UA)$ are generally between 1 and 3 when the cost of the heat exchanger is considered.

The f-chart for liquid systems has been developed using a value of $(\epsilon_L C_{min}/UA)$ equal to 2. The performance of systems having other values of $(\epsilon_L C_{min}/UA)$ can be estimated from the f-chart by modifying the dimensionless group Y as indicated in Fig. 14.7 or Eq. (14.8)

$$\begin{aligned}
\text{Load heat exchanger correction factor} &= (Y'/Y) \\
&= 0.39 + 0.65 \exp\left[-0.139/(\epsilon_L C_{min}/UA)\right] \quad (14.8) \\
&\text{for } 0.5 < (\epsilon_L C_{min}/UA) < 50
\end{aligned}$$

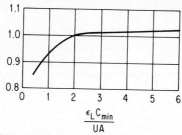

Fig. 14.7 Load heat exchanger size correction factor.

Example 14.4

For the conditions of Example 14.2, what will be the solar contribution if a crossflow heat exchanger (to transfer heat from the storage tank to air for the building) is used under the following circumstances: the air flowrate is 520 L/s (1102 ft³/min · ft²), water flowrate is 0.694 L/s (660 gal/h), and the heat exchanger effectiveness at these flowrates is 0.69. UA, the building overall energy loss coefficient-area product, is 470 W/°C (21400 Btu/°F · day).

First, the value of C_{min} must be determined. This is generally the capacitance rate of the air, which in this case is

$$C_{min} = 520 \text{ L/s} \times 0.69 \times (632 \text{ W/°C})/(470 \text{ W/°C})$$
$$= 632 \text{ W/°C}$$

Then

$$\epsilon_L C_{min}/UA = 0.69 \times (632 \text{ W/°C})/(470 \text{ W/°C})$$
$$= 0.93$$

The heat exchanger considered here is smaller than that used in developing Fig. 14.3. A correction factor is applied to the Y coordinate to account for the load heat exchanger size. From Fig. 14.7 [or Eq. (14.8)] with $\epsilon_L C_{min}/UA = 0.93$,

$$Y'/Y = 0.95$$
$$Y' = 0.95 \times 0.35 = 0.33$$

The value of X is unaffected by a change in load heat exchanger size. From Fig. 14.3, $f = 0.23$ for January. The solar energy contribution for January is

$$fL = 0.23 \times 36.0 \text{ GJ} = 8.3 \text{ GJ}$$

The annual solar-load fraction is found to be 0.41. (Note that if both the storage and load heat exchanger sizes differ from those used to develop the *f*-chart, the correction factor discussed in Example 14.3 and this example would both have to be applied.)

THE *f*-CHART FOR AIR SYSTEMS

As with liquid systems, the fraction of the monthly total heating load supplied by the solar air-heating system shown in Fig. 14.2 has been correlated with the dimensionless groups X and Y using computer simulations. The correlation is given in Fig. 14.8 and Eq. (14.9). It is used in the same manner as the *f*-chart for liquid-based systems. The definitions of X and Y given in Eqs. (14.1), (14.2), (14.3), and (14.4) apply to both air and liquid systems.

$$f = 1.040\,Y - 0.065\,X - 0.159\,Y^2 + 0.00187\,X^2 - 0.0095\,Y^3$$
$$\text{for } 0 \le Y \le 3.0,\ 0 \le X \le 18,\ \text{and } 0 \le f \le 1 \tag{14.9}$$

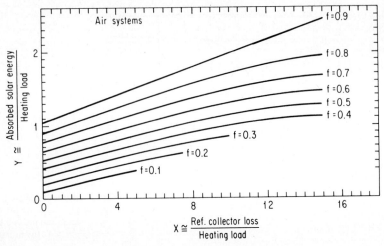

Fig. 14.8 F-chart for air systems.

Example 14.5

A solar heating system is to be designed for Madison, WI, with two cover collectors facing south at an inclination of 58° with respect to horizontal. A standard air system is considered, as shown in Fig. 14.2. The air-heating collectors have the following characteristics: $F_R U_c =$ 2.84 W/°C · m² (0.50 Btu/h · ft² · °F) and $F_R(\tau\alpha)_n = 0.49$. The total space and water-heating load for January is 36.0 GJ (34.1 MMBtu) (as in the previous examples in this chapter). What fraction of the load would be supplied by solar energy with a system having a collector area of 50 m² (538.2 ft²)?

For air systems, there is no heat exchanger penalty factor, and, in effect, $F'_R/F_R = 1$. As in the previous examples, $(\overline{\tau\alpha})/(\tau\alpha)_n = 0.94$ for a two-cover collector. From Eqs. (14.3) and (4.4),

$$X = 2.84 \text{ W/°C} \cdot \text{m}^2 \times 1 \times [100-(-7)]°\text{C} \times (31 \text{ days})$$
$$\times (86400 \text{ s/day}) \times 50 \text{ m}^2/36.0 \times 10^9 \text{ J}$$
$$= 1.13$$
$$Y = 0.49 \times 1 \times 0.94 (13.2 \times 10^6 \text{ J/day} \cdot \text{m}^2)$$
$$\times (31 \text{ days}) \times 50 \text{ m}^2/36.0 \times 10^9 \text{ J}$$
$$= 0.26$$

Then f for January, from Fig. 14.8 or Eq. (14.9), is 0.19. The solar energy supplied by this system in January is

$$fL = 0.19 \times 36.0 \text{ GJ} = 6.8 \text{ GJ}$$

As with the liquid systems, the annual system performance is obtained by summing the energy quantities for all months, as indicated in Eq. (14.5). The result is that 37 percent of the annual load is supplied by solar energy.

Collector Air Flowrate

The collector heat removal efficiency factor, F_R, which appears in X and Y, is a function of the collector air flowrate. Because of the higher cost of power for flowing fluid through air collectors than through liquid collectors, the capacitance rate used in air heaters is ordinarily much lower than that in liquid heaters. As a result, air heaters generally have a lower value of F_R. Values of F_R [and thus $F_R(\tau\alpha)_n$ and $F_R U_c$] corresponding to the expected air flowrate in the collector must be used in calculating X and Y.

An increase in air flowrate tends to improve collector performance by increasing F_R, but tends to decrease system performance by reducing the degree of thermal stratification in the pebble bed. The f-chart for air systems is based on a collector air flowrate of 10.1 L/s of air per square meter of collector area (2 standard ft^3/min · ft^2 of collector area). The performance of systems having different collector air flowrates can be estimated by using the appropriate values of F_R in both X and Y and by further modifying the value of X as indicated in Fig. 14.9 or Eq. (14.10) to account for the change of stratification in the pebble bed.

$$\text{Collector air flowrate correction factor} = X''/X$$

$$= \left[\frac{\text{Actual air flowrate}}{\text{Standard air flowrate}} \right]^{0.28} \tag{14.10}$$

for the bracketed ratio in the range [0.5, 2.0], where the standard air flowrate is 10.1 L/s per square meter of collector area (2 ft^3/min · ft^2).

Example 14.6

The system of example 14.5 is to be designed using a collector air flowrate of 15 L/s per square meter of collector (2.97 ft^3/min · ft^2) [rather than 10.1 L/s · m^2 (2 ft^3/min · ft^2) used in developing the f-chart for air systems]. Estimate the change in annual performance of the system resulting from the increased air flowrate.

The effects of increasing the air flowrate are twofold. First, it changes F_R, and second, it changes the amount of thermal stratification in the pebble bed. The effects on F_R and thus on $F_R U_c$ and $F_R(\tau\alpha)_n$ must be determined by collector tests at the correct air flowrate. In this case for the higher flowrate $F_R(\tau\alpha) = 0.52$ and $F_R U_c = 3.01$ W/°C · m^2 (0.53 Btu/h · ft^2 · °F). The correction factor to account for pebble bed stratification is found from Eq. (14.10) or Fig. 14.9. The factor is 1.12, to be applied to X.
Thus

$$X'' = 1.13 \times 3.01/2.84 \times 1.12 = 1.34$$

$$Y = 0.262 \times 0.52/0.49 = 0.28$$

From Fig. 14.8 or Eq. (14.9), $f = 0.19$ for January. Repeating these calculations for the remaining 11 months, it is found that 38 percent of the annual load is supplied by solar energy (compared with 0.37 at the standard air flowrate). This small performance increase must be evaluated in light of the increased fan power required at the higher air flowrate.

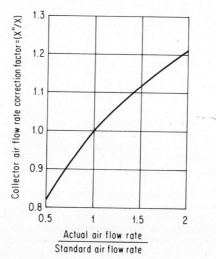

Fig. 14.9   Collector air flowrate correction factor.

Pebble Bed Storage Capacity

The performance of air systems is less sensitive to storage capacity than that of liquid systems for two reasons: (1) air systems can operate in the collector-load mode, in which the storage component is not used; and (2) pebble beds are highly stratified and additional capacity is effectively added to the cold end of the bed, which is seldom heated and cooled to the same extent as the hot end.

The f-chart for air systems is for a storage capacity of 0.25 m³ of pebbles per square meter of collector area (0.82 ft³/ft²). The performance of systems with other storage capacities can be determined by modifying the dimensionless group X as indicated in Fig. 14.10 or Eq. 14.11.

$$\text{Storage size correction factor} = (X'''/X) = \left[\frac{\text{Actual storage capacity}}{\text{standard storage capacity}} \right]^{-0.30}$$

$$\text{for } 0.5 < \left[\frac{\text{Actual storage capacity}}{\text{standard storage capacity}} \right] < 4.0$$

(14.11)

where the standard storage capacity is 0.25 m³/m² (0.82 ft³/ft²).

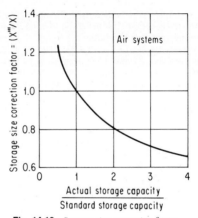

Fig. 14.10   Storage size correction factor.

Example 14.7

The system of Example 14.5 is to have a storage capacity which is 60 percent of the storage capacity used in the development of the f-chart for air systems. What fraction of the annual heating load will this system supply?

The storage size correction factor, from Fig. 14.10 or Eq. (14.11), is 1.17. Thus, for January

$$X'''/X = 1.17$$

$$X''' = 1.17 \times 1.13 = 1.32$$

Y is unaffected by a change in storage capacity and so $Y = 0.26$. From Fig. 14.8 or Eq. (14.9), $f = 0.18$ and $fL = 0.18 \times 36.0 \text{ GJ} = 6.5 \text{ GJ}$. The fraction of the annual load supplied by solar energy is 0.35 (compared to 0.37 for the standard storage size).

SOLAR WATER-HEATING SYSTEMS

With modification, Fig. 14.3 can be used to estimate the performance of solar water-heating systems having the configuration shown in Fig. 14.3. The mains water supply temperature, T_m, and the minimum acceptable hot water temperature (i.e., desired delivery temperature), T_w, both affect the performance of solar water-heating systems. Both T_m and T_w affect the average system operating temperature level and thus the collector energy losses. The dimensionless group X, which is related to collector energy losses, can be redefined so as to include these effects. If monthly values of X are multiplied by the correction factor in Eq. (14.12), the f-chart for liquid-based solar space- and water-heating systems [Eq. (14.6) or Fig. 14.4] can be used to estimate monthly value of f for water-heating systems.

In SI units (all temperatures in °C)

$$\text{Water-heating correction factor} = X_w/X$$
$$= \frac{11.6 + 1.18\,T_w + 3.86\,T_m - 2.32\,\overline{T}_a}{100 - \overline{T}_a} \tag{14.12a}$$

In English units (all temperatures in °F)

$$\text{Water heating correction factor} = X_w/X$$
$$= \frac{-66.2 + 1.18\,T_w + 3.86\,T_m - 2.32\,\overline{T}_a}{212 - \overline{T}_a} \tag{14.12b}$$

This method of estimating water heater performance is based on storage capacity of 75 L/m² (1.84 gal/ft²), and on a typical day's distribution of hot water needs. If other distributions of use occur, a system may not perform as well as indicated by the f-chart. If storage size is increased, system performance may improve more than is suggested by Fig. 14.6 and Eq. (14.7).

Example 14.8

A solar water-heating system is to be designed for a residence in Madison, Wisconsin (lat. 43°N). The collectors considered for this purpose have two covers with $F'_R\,(\tau\alpha)_n = 0.60$ and $F'_R U_c = 3.64$ W/°C · m² (0.64 Btu/h · m² · °F). The collectors are to be oriented facing south at a 45° angle with respect to horizontal. The estimated water-heating load is 325 L per day (86 gal/day) heated from 11°C to 60°C (52°F to 140°F) plus auxiliary storage tank losses of 0.0155 GJ/day. The storage capacity of the preheat tank is to be 75 L of water per square meter of collector area (1.84 gal/ft²). The problem is to estimate the fraction of the annual heating load supplied by solar energy for this system with a collector area of 10 m² (107.6 ft²).

The necessary meteorological data are monthly-averaged, daily radiation on the tilted collector surface and monthly-averaged ambient temperatures. The monthly water-heating load in this case is simply the product of the average daily hot water use, the specific heat of water, the difference between the design and mains supply water temperatures, and the number of days in each month. Values of X_w and Y are calculated for each month from Eqs. (14.12) and (14.2). For January,

$$L = (320 \text{ L/day}) \times (1.0 \text{ kg/L}) \times (4190 \text{ J/kg} \cdot {}^\circ\text{C})$$
$$\times (60{}^\circ\text{C} - 11{}^\circ\text{C}) \times (31 \text{ days}) + 0.0155 \text{ GJ/day} \times (31 \text{ days/mo})$$
$$= 2.55 \text{ GJ}$$

$$X_w = (10 \text{ m}^2) \times (3.64 \text{ W/}{}^\circ\text{C} \cdot \text{m}^2)/(2.55 \times 10^9 \text{ J})$$
$$\times [11.6{}^\circ\text{C} + 1.18 \times 60{}^\circ\text{C} + 3.86 \times 11{}^\circ\text{C}$$
$$- 2.32 \times (-7{}^\circ\text{C})]/[100{}^\circ\text{C} - (-7{}^\circ\text{C})]$$
$$\times (31 \text{ days}) \times (86, 400 \text{ s/day})$$
$$= 5.40$$

$$Y = (10 \text{ m}^2) \times (0.60) \times (13.2 \times 10^6 \text{ J/day})$$
$$\times (31 \text{ days})/(2.55 \times 10^9 \text{ J})$$
$$= 0.96$$

REFERENCES

1. W. A. Beckman, S. A. Klein, and J. A. Duffie, *Solar Heating Design by the f-Chart Method*, Wiley-Interscience, New York, 1977.
2. S. A. Klein, "A Design Procedure for Solar Heating Systems," Ph.D. Thesis, University of Wisconsin-Madison, 1976.
3. S. A. Klein, W. A. Beckman, and J. A. Duffie, "A Design Procedure for Solar Heating Systems," *Solar Energy* 18: 113 1976.
4. S. A. Klein, W. A. Beckman, and J. A. Duffie, "A Design Procedure for Solar Air Heating Systems," *Solar Energy*, 19 (1977).

APPENDIX: AVERAGE AMBIENT TEMPERATURES FOR U.S. CITIES

TABLE 14.A1 Normal Monthly Averge Temperature, °F: Selected Cities* (In Fahrenheit degrees. Airport data unless otherwise noted. Based on standard 30-year period, 1931 to 1960.)

	Location	Jan	Feb	Mar	Apr	May	Jun	Jul	Aug	Sep	Oct	Nov	Dec	Ann
AL	Mobile	53.0	55.2	60.3	67.6	75.6	81.5	82.6	82.1	77.9	69.9	58.9	54.1	68.2
AK	Juneau	25.1	26.8	30.4	38.0	45.6	52.3	55.3	54.1	48.9	41.6	34.3	28.4	40.1
AZ	Phoenix	49.7	53.5	59.0	67.2	75.0	83.6	89.8	87.5	82.8	70.7	58.1	51.6	69.0
AR	Little Rock	40.6	44.4	51.8	62.4	70.5	78.9	81.9	81.3	74.3	63.1	49.5	41.9	61.7
CA	Los Angeles	54.4	55.2	57.0	59.4	62.0	64.8	69.1	69.1	68.5	64.9	61.1	56.9	.61.9
	Sacramento	45.2	49.2	53.4	58.4	64.0	70.5	75.4	74.1	71.6	63.5	52.9	46.4	60.4
	San Francisco†	50.7	53.0	54.7	55.7	57.4	59.1	58.8	59.4	62.0	61.4	57.4	52.5	56.8
CO	Denver	28.5	31.5	36.4	46.4	56.2	66.5	72.9	71.5	63.0	51.4	37.7	31.6	49.5
CT	Hartford	26.0	27.1	36.0	48.5	59.9	68.7	73.4	71.2	63.3	53.0	41.3	28.9	49.8
DE	Wilmington	33.4	33.8	41.3	52.1	62.7	71.4	76.0	74.3	67.6	56.6	45.4	35.1	54.1
DC	Washington	36.9	37.8	44.8	55.7	65.8	74.2	78.2	76.5	69.7	59.0	47.7	38.1	57.0
FL	Jacksonville	55.9	57.5	62.2	68.7	75.8	80.8	82.6	82.3	79.4	71.0	61.7	56.1	69.5
	Miami	66.9	67.9	70.5	74.2	77.6	80.8	81.8	82.3	81.3	77.8	72.4	68.1	75.1
GA	Atlanta	44.7	46.1	51.4	60.2	69.1	76.6	78.9	78.2	73.1	62.4	51.2	44.8	61.4
HI	Honolulu	72.5	72.4	72.8	74.2	75.9	77.9	78.8	79.4	79.2	78.2	75.9	73.6	75.9
ID	Boise	29.1	34.5	41.7	50.4	58.2	65.8	75.2	72.1	62.7	51.6	38.6	32.2	51.0
IL	Chicago	26.0	27.7	36.3	49.0	60.0	70.5	75.6	74.2	66.1	55.1	39.9	29.1	50.8
	Peoria	25.7	28.4	37.6	50.8	61.5	71.7	76.0	74.3	66.4	55.3	39.7	29.1	51.4
IN	Indianapolis	29.1	31.1	38.9	50.8	61.4	71.1	75.2	73.7	66.5	55.4	40.9	31.1	52.1
IA	Des Moines	19.9	23.4	33.8	48.7	60.6	71.0	76.3	74.1	65.4	54.2	37.1	25.3	49.2
KS	Wichita	32.0	36.3	44.5	56.7	66.0	76.5	80.9	80.8	71.3	59.9	44.4	35.8	57.1
KY	Louisville	35.0	35.8	43.3	54.8	64.4	73.4	77.6	76.2	69.5	57.9	44.7	36.3	55.7
LA	New Orleans	54.6	57.1	61.4	67.9	74.4	80.1	81.6	81.9	78.3	70.4	60.0	55.4	68.6
ME	Portland	21.8	22.8	31.4	42.5	53.0	62.1	68.1	66.8	58.7	48.6	38.1	25.8	45.0
MD	Baltimore	34.8	35.7	43.1	54.2	64.4	72.5	76.8	75.0	68.1	57.0	45.5	35.8	55.2
MA	Boston	29.9	30.3	37.7	47.9	58.8	67.8	73.7	71.7	65.3	55.0	44.9	33.3	51.4
MI	Detroit	26.9	27.2	34.8	47.6	59.0	69.7	74.4	72.8	65.1	53.8	40.4	29.9	50.1
	Sault Ste. Marie	15.8	15.7	23.8	38.0	49.6	59.0	64.6	64.0	55.8	46.3	33.3	20.9	40.6
MN	Duluth	8.7	10.8	21.3	37.0	49.2	58.8	65.3	63.8	54.2	44.6	27.3	14.0	37.9
	Minneapolis-St. Paul	12.4	15.7	27.4	44.3	57.3	66.8	72.3	70.0	60.4	48.9	31.2	18.1	43.7
MS	Jackson	47.9	50.5	56.5	64.9	73.1	79.8	82.3	82.0	76.5	67.0	55.5	49.4	65.5
MO	Kansas City	31.7	35.8	43.3	55.7	65.6	75.9	81.5	79.8	71.3	60.2	44.6	35.8	56.8
	St. Louis	31.9	34.7	42.6	54.9	64.2	74.1	78.1	76.8	69.5	58.4	44.1	34.8	55.3
MT	Great Falls	22.1	23.8	30.7	43.6	53.0	59.9	69.4	66.8	57.4	47.5	34.3	27.3	44.7
NE	Omaha	22.3	26.5	36.9	51.7	63.0	73.1	78.5	76.2	66.9	55.7	38.9	28.2	51.5
NV	Reno	30.4	35.6	41.5	48.0	53.9	60.1	67.7	65.5	58.8	49.2	38.3	31.9	48.4
NH	Concord	21.2	22.7	31.7	43.8	55.5	64.5	69.6	67.4	59.3	48.7	37.6	25.0	45.6
NJ	Atlanta City	34.8	34.7	41.1	51.0	61.3	70.0	75.1	73.7	67.2	57.2	46.7	36.6	54.1
NM	Albuquerque	35.0	39.9	45.8	55.7	65.1	74.9	78.5	76.2	70.0	58.0	43.6	37.0	56.6
NY	Albany	22.7	23.7	33.0	46.2	57.9	67.3	72.1	70.0	61.6	50.8	39.1	26.5	47.6
	Buffalo	24.5	24.1	31.5	43.5	54.8	64.8	69.8	68.4	61.4	50.8	39.1	27.7	46.7
	New York†	32.2	33.4	40.5	51.4	62.4	71.4	76.8	75.1	68.5	58.3	47.0	35.9	54.5
NC	Charlotte	42.7	44.2	50.0	60.3	69.0	77.1	79.2	78.7	72.9	62.5	50.4	42.7	60.8
	Raleigh	41.6	43.0	49.5	59.3	67.6	75.1	77.9	76.9	71.2	60.5	50.0	41.9	59.5
ND	Bismarck	9.9	13.5	26.2	43.5	55.9	64.5	71.7	69.3	58.7	46.7	28.9	17.8	42.2
OH	Cincinnati†	33.7	35.1	42.7	54.2	64.2	73.4	76.9	75.7	69.0	57.9	44.6	35.3	55.2
	Cleveland	28.4	28.5	35.1	47.0	58.0	67.8	71.9	70.4	64.2	53.4	41.3	30.5	49.7
	Columbus	29.9	31.1	38.9	50.8	61.5	70.8	74.8	73.2	65.9	54.2	41.2	31.5	52.0
OK	Oklahoma City	37.0	41.3	48.5	59.9	68.4	78.0	82.5	82.8	73.8	62.9	48.4	40.3	60.3
OR	Portland	38.4	42.0	46.1	51.8	57.4	62.0	67.2	66.6	62.2	54.2	45.1	41.3	52.9
PA	Philadelphia	32.3	33.2	41.0	52.0	62.6	71.0	75.6	73.6	66.7	55.7	44.3	33.9	53.5
	Pittsburgh	28.9	29.2	36.8	49.0	59.8	68.4	72.1	70.8	64.2	53.1	40.8	30.7	50.3
RI	Providence	29.2	29.7	37.0	47.2	57.5	66.2	72.1	70.5	63.2	53.2	43.0	32.0	50.1
SC	Columbia	46.9	48.4	54.4	63.6	72.2	79.7	81.6	80.5	75.3	64.7	53.7	46.4	64.0
SD	Sioux Falls	15.2	19.1	30.1	45.9	58.3	68.1	74.3	71.8	61.8	50.3	32.6	21.1	45.7
TN	Memphis	41.5	44.1	51.1	61.4	70.3	78.5	81.3	80.5	73.9	63.1	50.1	42.5	61.5
	Nashville	39.9	42.0	49.1	59.6	68.6	77.4	80.2	79.2	72.8	61.5	48.5	41.4	60.0
TX	Dallas	45.9	49.5	56.1	65.0	72.9	81.3	84.9	85.0	77.9	67.8	54.9	48.1	65.8
	El Paso	42.9	49.1	54.9	63.4	71.9	81.0	81.9	80.4	74.5	64.4	51.2	44.1	63.3
	Houston	53.6	55.8	61.3	68.5	76.0	81.6	83.0	83.2	79.2	71.4	60.8	55.7	69.2
UT	Salt Lake City	27.2	32.5	40.4	49.9	58.9	67.4	76.9	74.5	64.4	51.7	36.7	30.1	50.9
VT	Burlington	16.2	17.4	26.7	41.2	53.8	64.2	69.0	66.7	58.4	47.6	35.3	21.5	43.2
VA	Norfolk	41.2	41.6	48.0	58.0	67.5	75.6	78.8	77.5	72.6	62.0	51.4	42.5	59.7
	Richmond	38.7	39.9	47.7	58.1	67.0	75.1	78.1	76.0	70.2	58.7	48.5	39.7	58.1
WA	Seattle-Tacoma	38.3	40.8	43.8	49.2	55.5	59.8	64.9	64.1	59.9	52.4	43.9	40.8	51.1
	Spokane	25.3	30.0	38.1	47.3	56.2	61.9	70.5	68.0	60.9	49.1	35.7	30.1	47.8
WV	Charleston	36.6	37.5	44.4	55.3	64.8	72.0	74.9	73.8	68.2	57.3	45.3	37.1	55.6
WI	Milwaukee	20.6	22.4	31.0	43.6	53.4	63.3	68.7	67.8	60.3	50.0	35.8	24.6	45.1
WY	Cheyenne	25.4	27.3	32.4	42.6	52.9	63.0	70.0	67.7	58.6	47.5	34.2	29.5	45.9
PR	San Juan	74.4	74.4	75.3	76.6	78.7	80.0	80.4	80.9	80.5	80.0	78.2	76.2	78.0

*From J.F. Kreider and F. Kreith, *Solar Heating and Cooling*, rev. 1st ed., Hemisphere, Washington, D.C., 1977. Data from U.S. Department of Commerce, NOAA, "Local Climatological Data." For more detailed data see U.S. Department of Commerce, NOAA, NCC, "Climatography of the United States," Publ. No. 84.

†City office data.

Chapter **15**

Solar Systems for Space Cooling

ALWIN NEWTON, P.E.
Borg-Warner, York Division (retired), and Consulting Engineer, York, PA

The most highly developed solar-powered cooling equipment is at present the lithium-bromide–water system which is readily available in commercial sizes. The performance characteristics of such equipment are discussed in detail. The possible gains from dual temperature storage are also shown. Limits of performance of other cooling equipment such as ammonia absorption, Rankine drives, open absorption, and adsorption are also described in engineering detail.

Another concern of this chapter is the need to treat the building to be cooled and the air-conditioning equipment energized by the solar system as an interacting combination. The load patterns of typical buildings are identified and the performance of auxiliary equipment such as cooling towers is shown over all normal operating conditions. The relationships between loads and varying ambient temperatures and the effect on the most effective energizing temperatures are shown in detail.

Conventional air-conditioning equipment utilizes either electrical or thermal energy to drive compressors or absorption equipment to produce a cooling effect. The thermal energy obtained from the sun by means of the various types of collectors described in Chaps. 7, 8, 9, and 10 can be used to directly energize the absorption-type equipment, and by means of heat engines to drive compressors. The variable nature of the solar energy requires a much more complete analysis of loads and equipment characteristics than is normally made for conventional applications. This chapter deals with performance characteristics of commercially available air-conditioning equipment and shows methods by which it can be used in solar-energized applications in the most cost-effective ways. It then closes with suggestions for specifying solar-cooling components and systems in such a manner as to assure that the desired equipment is obtained.

INTRODUCTION

A dynamic relationship exists between any air-conditioning system, the building it cools or heats, the users of that building, and the outdoor ambient conditions to which the building is exposed. The added variables of timing and the temperature level of energy availability are added when the energy to operate the equipment is supplied from solar collector and storage systems. The interactions are highly dynamic, and involve different phase relationships among influencing factors.

The cooling equipment itself operates over a wide range of efficiency as four conditions vary:

    1. The load which determines the capacity needed to maintain the building and its contents at the desired temperature and humidity conditions.

    2. The temperature level at which heat is received by the equipment from the conditioned space.

    3. The temperature level at which heat is supplied to energize the system.

    4. The temperature at which the spent heat and the heat taken from the conditioned spaces is rejected.

None of these four conditions are under the direct control of the system designer. The efficiency of collectors is also highly dependent on the temperature and insolation levels at which the energizing heat is collected. Maxima and minima for each of these factors occur at different times, and the proper use of these out-of-phase relationships can result in major improvements in the operating efficiency of the system and in the required sizes of the solar components.

Energy Requirements for Various Cooling Applications

A detailed knowledge of the expected load and ambient relationships is even more important for solar-energized systems than for conventional systems. This is particularly true for cooling. The standard design point established by the American Society of Heating, Refrigerating and Air-Conditioning Engineers and by the Air-Conditioning and Refrigeration Institute is 35°C ambient air temperature. In the United States and many other populated areas in similar latitudes the number of hours per year during which ambients reach 35°C is extremely small except in noncoastal desert areas. Figure 15.1 shows the hours per season within each 2 ½°C block of ambient temperature for typical latitudes of 35° to 42° in the United States. The solid lines show total hours for the season, the dotted lines the sunshine hours corresponding to a specific temperature range. Similar plots can be made for other locations using standard monthly weather summaries.*

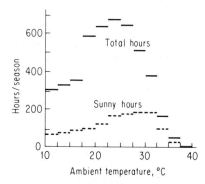

Fig. 15.1 Example of seasonal distribution of temperature—total hours and hours with sun (from Ref. 3).

It is important to determine how the load imposed by the building relates to ambient weather conditions. The sum of the internal and external loads of the building differs in sunny periods from the load in nighttime or cloudy periods. Particularly in commercial buildings, the internal load may be a major portion of the total. The total loads determine a changeover temperature below which no cooling is needed.

Figure 15.2 shows how the load patterns relate to ambient. Lines A and B in Fig. 15.2 show the load during periods of little or no internal load. Line A indicates the percentage of design load experienced at various outdoor ambients during periods of no sunshine, such as heavily clouded days or at night. Line B gives the same information for sunny days. The load relationships shown by the two lines, A and B, are typical of those exhibited by most residences. They are

*See for example, Air Force Manual 88/29, available from the Supt. of Documents, Washington, D.C.

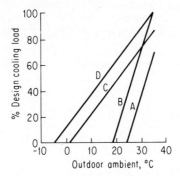

Fig. 15.2 Load-ambient temperature relationships: A, no sun, no internal sources; B, sun, no internal sources; C, no sun, internal sources; D, sun, internal sources.

also typical of many commercial buildings when unoccupied, unlighted, and with office machines and computers inoperative.

Lines C and D in Fig. 15.2 show the load relationships during periods of normal occupancy with maximum internal loads, as experienced in most office buildings. The line C applies when sun is not present, and the line D is typical of sunny periods.

Plots like those of Figs. 15.1 and 15.2 made for specific buildings permit development of the load patterns for the building in question. For example, Fig. 15.1 shows that for the location to which it applies a total of 630 h per season are expected to occur between 25°C and 27 ½°C, of which 175 h are expected to be sunny. This leaves 455 cloudy and nighttime hours. If one were to assume that 8 percent of the sunshine hours occur during building occupancy and that 30 percent of the hours without sunshine occur during occupancy, Figs. 15.1 and 15.2 taken together can be used to develop the load pattern for the ambient temperature range selected of 25°C to 27 ½°C. The results are shown in Table 15.1.

TABLE 15.1   Load Pattern, Typical Building, between 25° and 27½°C Ambient Temperature

Building condition	Hours per season	% Full load
Unoccupied, no sun	319	16
Unoccupied sunny	35	53
Occupied, no sun	136	64
Occupied sunny	140	81

Repeating this procedure for each applicable block of ambient temperature provides a load pattern for the entire season in terms of percentage of design load. The daily pattern of load and ambient is important in applying any solar-energized air-conditioning system. Individual days can be analyzed by determining their ambient pattern. For this purpose weather tapes which contain hourly cloud cover information are available from the National Climatic Center in Asheville, NC.

The daily ambient temperature variation is important, not alone because of its effect on load, but also for its effect in the optimization of the performance and selection of the best energizing temperature of some types of air-conditioning equipment. In the United States the range of daily ambient change is approximately 8°C for coastal regions, 12 to 15°C inland east of the Mississippi River, and 17 to 20°C in the plains area.

Heat rejection temperatures   The energy requirements for air conditioning of buildings are critically influenced by the temperature at which the spent heat and the heat taken from the conditioned spaces is rejected. In most installations this heat is rejected through a cooling tower to ambient air. Water sprayed into the ambient air stream in the cooling tower approaches the ambient wet-bulb temperature within 4°C at the usual design conditions. As the ambient wet-bulb temperature drops, the approach increases if the load remains constant, but at any wet-bulb temperature the approach is reduced by a reduction in the air-conditioning load. Thus the performance map of the cooling tower is needed for a complete energy analysis. Figure 15.3 shows the performance of a typical tower throughout the range of normal use.

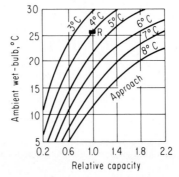

Fig. 15.3 Typical cooling tower performance curves showing capacity versus wet-bulb temperature.

The pattern of wet-bulb temperatures coincident with ambient temperature must be known to fully utilize the information given by Fig. 15.3. Where accurate weather data are available they should be used in determining the energy requirements. Such information may be available from local weather stations, but is not provided on most weather tapes. An indication of the probable relationships between dry and wet-bulb temperatures for the United States is given in lines 1 through 3 of Table 15.2. The expected approach of the cooling tower water to wet-bulb for the same conditions is shown in lines 4 and 5 of the table.

TABLE 15.2   Probable Dry-bulb, Wet-bulb, and Cooling Tower Water Temperature at Full Load

	Temperature, °C				
1. Ambient temperature	5	15	25	30	35
2. Maximum concurrent wet-bulb	5	14	24	25	24
3. Normal concurrent wet-bulb	4	10	18	20	18
4. CTW—wet-bulb, maximum	12	7	4.3	4.1	4.3
5. CTW—wet-bulb, normal	13	9.3	5.8	5.4	5.8

For other than full load, the appropriate wet-bulb temperature is taken from line 2 or 3 of Table 15.2 and used to determine the approach for the given load as a percentage of full load from Fig. 15.3. The value of approach so determined is then added to the wet-bulb temperature to obtain the temperature of the water leaving the cooling tower at the given load.

For example, for 15°C dry-bulb ambient, Table 15.2 shows 10°C as the normal wet-bulb. The approach for this condition is then determined for 60 percent load by the intersection of the 0.6 Relative Capacity line and the 10°C Ambient Wet-bulb line in Fig. 15.3, and read as 6.2°C. The leaving cooling tower water temperature is then 10°C + 6.2°C = 16.2°C.

Energy requirements The total energy required for the air-conditioning equipment to maintain the desired temperature and humidity conditions in a given building is usually reduced by modulation of the air-conditioning capacity. In modulation, the temperature at which the refrigerant evaporates is raised at reduced loads, and the temperature of heat rejection as in condensers and absorbers is lowered. Thermodynamically the cycle becomes more compact, and the efficiency is thus increased. Less energy is then needed to operate the equipment.

The hours of operation of auxiliary equipment such as pumps and fans are increased by modulation. For example, when operating in a modulated mode at half capacity the hours of operation for the same heat removal will be approximately doubled. To avoid the energy of the auxiliaries becoming great enough to negate significant savings in the operation of the main air-conditioning equipment, it is usually necessary to reduce the speed of the auxiliaries. Electrically driven pumps and fans can be controlled by solid state inverters which change the frequency and applied voltage, thus efficiently controlling the motor speed. Chilled water pumps, air-handling fans and blowers within the building, and cooling tower fans and circulating pumps in some heat exchangers can be reduced to 50 percent speed at 40 to 50 percent load without a major effect on the efficiency of components at reduced loads. Upon further load reduction there should be no further speed reduction because of a rapid decrease in efficiency. Figure 15.4, taken from Ref. 1, shows the performance of standard induction motors when driven at variable voltage and frequency. Except for changes in motor efficiency which are minor, the power to drive pumps and fans reduces as the cube of the speed. That is, at 50 percent speed the power is 12 ½ percent of the full-load power when serving the same piping or duct system. Thus at 50 percent of full-load speed such devices can operate for a period of time eight times greater with no net increase in energy consumption for the building as a whole.

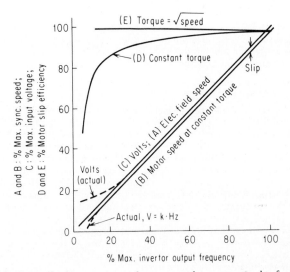

Fig. 15.4   Expected induction motor performance at voltage proportional to frequency.

Solar energy collection efficiency   It is a characteristic of all solar collectors that a greater amount of heat Q can be collected per day when the temperature difference between the fluid in the collector and the ambient is relatively lower than when higher. Thus the most efficient collector arrays are associated with cooling systems which can make use of thermal energy at the lowest temperatures. This fact is shown in Figure 15.5 from Ref. 2, which shows the relative area under the curve for 30° ΔT as 2.2 times the area under the curve for 70° ΔT. While Fig. 15.5 is drawn specifically for flat-plate collectors, the same relationships exist at higher ΔT's for concentrating collectors and for evacuated-tube collectors.

Only the direct-beam component of solar energy is collected by concentrating collectors. This fact must be considered when designing collector arrays in those geographical areas where clouds frequently cover major portions of the sky for a high percentage of the time. The dotted curve in Fig. 15.5 illustrates the amounts of heat which might be collected by a typical concentrating collector, and allows some comparisons to be drawn with the performance of the flat-plate collector.

The total seasonal energy needs of commercial or residential buildings are minimized by selection of equipment to operate efficiently over the entire range of conditions to which the

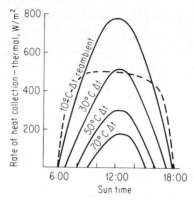

Fig. 15.5 Hypothetical collector performance at 35 percent diffuse radiation.

system will be exposed in its particular geographical location. The system designer should make the maximum use of the part-load capabilities of the air-conditioning equipment, and use the lowest possible energizing temperatures in order to enhance the amount of useful energy received from any given collector array.

Building Conditioning Loads

The cooling load is defined as the rate at which heat must be removed from the space to maintain the air in that space at desired wet- and dry-bulb temperatures. The heat gain is the rate at which heat enters into or is generated within a space, and may come from sensible heat additions which raise temperature, latent heat additions which add water vapor to the air, or from solar radiation. The radiative part of the heat gain does not appear immediately in the cooling load; it is absorbed on or transmitted through the surfaces which receive the radiation in the space or which form part of its enclosure. The radiation raises the temperatures of the walls, floor, and furniture on which it falls, which in turn warm the air in the space.

It is thus evident that the variables affecting cooling load calculations are numerous and intricately interrelated. Because any building operates many more hours at part load than at or near full load, and because it is most economical to operate the solar-energizing system to provide only the "base load" (see Chap. 29), it is important to determine the entire range of loads and the lengths of time they exist. Figures 15.1 and 15.2 have illustrated how typical building loads relate to ambient temperatures.

In office buildings a significant portion of the load may come from people during occupied periods at an average rate of about 75 W sensible heat and 60 W in latent heat represented by water vapor. A second large portion of load may come from lights and office machines, and a third large increment from ventilation air requirements. People and lights cause load only during working hours, and usually this is also true for the majority of ventilation.

Latent loads come primarily from people and from ventilation air in office buildings. In other types of buildings, processes may contribute to latent as well as sensible loads. Thus the major latent loads also occur only during working hours in most commercial applications.

Calculating Cooling Loads

Accurate heat gain and cooling load calculations are usually made by methods shown in the American Society of Heating, Refrigerating and Air-Conditioning Engineers' *Handbook of Fundamentals*, Ref. 3, which is updated every four years to reflect improved or new information. Methods presented give details for calculating loads either by summing the heat gains of the individual components of the buildings and its contents, or by the use of transfer functions which tend to take into account the time lags and heat capacities of the components. The same reference also contains all needed information regarding the physical properties of building materials, and of heat contributions from people, lights, office machinery, computers, and fan or pump motors of building machinery.

Determination of Daily Load Pattern

The daily cooling load changes over a much greater percentage of full load and at a faster rate than do heating loads. The cooling load drops to zero on a daily basis during night hours during much of the season in most localities. This situation is obvious from examination of Fig. 15.2 and from consideration of the daily cycles of ambient temperature.

The steps for constructing a load pattern diagram for any building under consideration are as follows:

1. Calculate the building load by components, using the heat gain method of the ASHRAE *Handbook of Fundamentals*.[3] The result of such a calculation for an example building is tabulated in column A of Table 15.3. Total load is 73.94 thermal kW.

2. Identify the fixed load components which are independent of the outdoor ambient, and determine their total. In Table 15.3 these loads are identified as lines 8, 9, 11, 12, and 15. Their total load contribution is 37.31 thermal kW (see column B).

3. Find the total of the variable load components, which may be assumed to decrease linearly as the outside temperature drops below the design temperature. In Table 15.3, these loads are identified as lines 1 through 7, 10, 13, 14, and 16. Their total is 36.63 thermal kW.

TABLE 15.3   Cooling Load Components of a Typical Building (kW$_t$)

Load source	A.*	B.†	C.
Envelope			
1. Roof	10.96		11.16
2. S. Wall	0.82		0.84
3. E. Wall	2.49		2.54
4. N. Wall (Exposed)	0.32		0.33
5. N & W Party Wall	0.47		0.48
6. Doors, NE	0.23		0.23
7. Doors, S	0.18		0.18
Solar transmission			
8. S. Glass	0.64	0.64	
9. N. Glass	0.26	0.26	
Internal load			
10. Infiltration	0.41		0.42
11. Lights	24.93	24.93	
12. People (85)	6.38	6.38	
13. Ventilation	7.80	———	7.94
Sensible Total	55.87	32.21	24.12
Latent load			
14. Latent Infiltration	0.64		0.65
15. People	5.10	5.10	
16. Ventilation	12.31	———	12.54
Latent total	18.05	5.10	13.19
Total load	73.94	+37.31	−37.31

*Full load—35°C.
†Zero load @ 12.79°C.

4. Find the ratio of the fixed load components to the variable load components: 37.31/36.63 = 1.019.

5. Note the difference between outdoor and indoor design temperature and multiply this difference by the ratio determined in step 4: (35 − 24) × 1.019 = 11.21.

6. Subtract the result of step 5 from the indoor design temperature to find the outdoor temperature at which the building cooling load equals zero: 24 − 11.21 = 12.79°C.

7. A load line for sunny days with full internal load, such as line D in Fig. 15.2, may now be drawn for the building under consideration. It will be drawn from 35°C and 100 percent load, to 12.79°C and zero load.

8. The full-load components which vary with ambient may be multiplied by the ratio of step 4 to determine their individual negative values at the zero load condition. The total should equal the total shown in column B for the fixed loads and is shown in column C of Table 15.3.

Construction of Load Pattern Diagram

To construct a fully occupied load line for cloudy days, such as line C of Fig. 15.2, the method of Ref. 3 is again used omitting all heat gains from the sun. The contributions for the individual components are again tabulated in a new table similar to Table 15.3. Step 1 through 8 above are then repeated to determine the zero-load temperature for the cloudy and night condition, and the appropriate load-pattern line added to the diagram for the building.

Using the same general procedures the load line for the unoccupied sunny condition and the unoccupied cloudy or night condition is determined. For the unoccupied conditions lines 11, 12, 13, 15, and 16 of Table 15.3 are either eliminated, or the appropriate ones reduced according to the operating pattern of the building being considered.

This load-pattern diagram will be used as shown later to determine the desirable operating temperatures and storage temperatures for cooling equipment for which complete operating characteristics are known. In doing so, the seasonal distribution of temperatures for the area in which the building is to be built will be used, such as shown in Fig. 15.1

Residential Conditioning Loads

Cooling loads for residences consist primarily of heat flow through structural elements and of air leakage and ventilation. Internal loads from occupants and lights are small compared to those in commercial buildings. Most residences are treated as single zones, and dehumidification is provided only when the air conditioner operates in response to thermostatic control. Design loads exist only a few days each year, and during night periods the load usually reaches zero, or becomes negative from the cooling viewpoint. A partial-load condition exists most hours of the season, and it is important not to oversize the air conditioner so that proper dehumidification and air circulation are provided.

Systems in most residences are operated as needed for comfort during the entire 24 h of the day, thus permitting full advantage to be taken of the flywheel effect of the building mass. Because of this, experience has shown that it is more economical to use a smaller unit set to operate at about 24°C than a larger unit set to operate at 26°C. In contrast, an oversized unit cycles frequently, and wide ranges of humidity result and cause a poor comfort condition.

Windows in residences facing toward the sun are a source of considerable load, but their effect can be reduced by use of overhangs large enough to shade the glass areas in the summer periods when the sun is high in the sky. The proper length of overhang will still allow the low winter sun to shine into the house to reduce heating load (see Chaps. 2 and 16).

The ASHRAE *Handbook of Fundamentals*[3] provides simplified methods for calculating the size of residential cooling equipment. The method does not actually calculate the cooling load of the residence, but instead determines the capacity of a unit which will prevent the space temperature from rising more than 2°C above the setting of the thermostat on a design day. Solar-energized systems are usually sized in the same manner for residential use.

Nevertheless residences have a definite zero load at ambients of about 12°C on sunny days and 18°C on cloudy days and at night. The equipment available in the 8 to 18 thermal kW capacity needed for residences in the early 1980s does not accept modulated energization as does commercial cooling equipment. Until residential cooling equipment is available with modulating capability, it is recommended that the ASHRAE method of sizing to residential needs be employed. When equipment with modulating ability does become available, the methods previously described for commercial buildings may be used.

Design Conditions

An indoor temperature of 24°C dry-bulb, and 18°C, wet-bulb, is considered a comfort optimum for both commercial and residential cooling applications. The trend to increase the design temperature to 26°C for energy conservation need not apply to solar-cooling applications when solar heat is available to provide the more comfortable temperature without drawing significantly on limited resources such as oil, gas, coal, or nuclear, which are considered nonrenewable.

In the continental United States outdoor design conditions range between 32 and 38°C except in desert areas and Alaska. In desert regions the design conditions may reach 43°C in the United States, and as much as 45°C in other parts of the world such as the mideast. Highest temperatures are normally experienced in the mid to late afternoon, typically at 15:00 to 16:00 sun time.

Results of many years of record keeping are tabulated in the ASHRAE *Handbook of Fundamentals.*[3] The temperatures which will not be exceeded 1, $2^1/_2$, and 5 percent of the time are given. In normal summers the 1 percent value is not exceeded more than 30 h, while the 5 percent value may be equalled or exceeded for up to 300 h. Since temperatures do not remain at the maximum levels for even one day, it is usually considered good practice to design at the $2^1/_2$ or 5 percent value.

Energy-Estimating Methods

For heating applications, a degree-day method of estimating energy requirements has been in use for many years. This method, or the refinement of it based on cooling degree-hours, has not produced good accuracy when used for cooling energy estimates. The ASHRAE *Handbook of Systems*[4] provides tables of equivalent full-load hours of operation for many locations, and as an alternate suggests a "bin" method.[4]

Computer models have the potential of still better accuracy in calculating loads and energy needs. No single model is known to exist which determines (1) loads, (2) the coincident performance characteristics of the components and the system in which they operate , and (3) the effects of varying heat rejection temperatures. Codes such as NECAP do provide energy analysis and a considerable amount of component simulation. A bibliography and description of existing computer codes is available from ASHRAE as GRP-153, and is periodically updated.[5]

SOLAR ABSORPTION COOLING

Absorption air-conditioning equipment has been in standard production for several decades. Residential water-ammonia and lithium-bromide–water units have been available for gas-fired applications and have been modified for solar energization by addition of circulating pumps for the absorbent fluid. Other modifications improve the range of application of this equipment when it is solar-energized.

Commercial units are available from at least three manufacturers in capacities from 250 to 4200 thermal kW, being highly refined for operation by heat at temperatures below 125°C. The most used commercial units are lithium-bromide–water units, and complete performance maps are known for this equipment at energizing temperatures from 40°C to 125°C, and for cooling-tower water temperatures from 5°C to 30°C.[6] Since these are the only production units available for solar operation in the early 1980s, their principles of design and operation will be discussed in detail.

How the Lithium-Bromide–Water System Functions

The main elements of a lithium-bromide–water system are shown schematically in Fig. 15.6. Water is the refrigerant, and the lithium-bromide-salt solution is the absorber. In operation the refrigerant is driven by heat from a strong solution in the generator at pressure p_1, flows as vapor to the condenser where it condenses at p_1, expands through a control valve into the evaporator at pressure p_2 from which it evaporates to cool the chilled-water circuit. The warm vapor is then absorbed at p_2 by the weak solution in the absorber which was left from step one and it is finally pumped back to the generator at p_1 where step one is repeated.

All presently made commercial units reject their heat to a water circuit which is usually cooled by a cooling tower. This circuit removes both the heat from the condenser and the heat produced in the absorber. Since the temperature of the absorber has a greater effect on the efficiency of the unit than does the temperature of condensing the heat-rejection fluid flow is

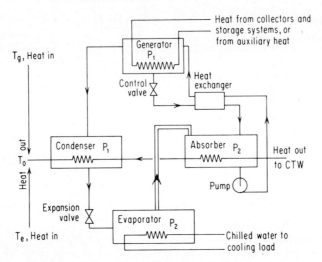

Fig. 15.6 Schematic diagram of lithium-bromide absorption equipment.

usually through the absorber first. The arrows at the left of Fig. 15.6 indicate that heat enters the unit at the energizing temperature and at the water-chilling temperature, and that all heat is removed at the intermediate temperature of the cooling-tower water.

For commercial building cooling applications absorption equipment provides chilled water temperatures from 5 to 12°C. Since the thermal energy to drive the cycle is injected directly into the unit it is thermodynamically equivalent to a power cycle. The heat rejected into the cooling-water circuit is equivalent to the heat rejected in the condenser of an electrically driven unit plus the heat removed at the condenser of the power plant to generate the amount of electricity needed to operate the unit. Thus the on-site heat rejection for a given cooling capacity is as much as twice that from the electrically driven equipment. This influences the cooling tower size, but does not double it as most manufacturers allow a greater temperature range through absorption equipment than through the corresponding electrical equipment.

Performance characteristics of Li-Br–water units Commercial practice for equipment energized by waste heat at the usual temperatures of 115 to 125°C has been to control the cooling tower water by means of by-pass valves to prevent it falling below 25°C. For solar systems it is much to be preferred to allow the cooling-tower water temperature to seek its natural minimum level which will be determined at any given time by the existing load and the ambient wet-bulb temperature. (See the cooling tower performance map of Fig. 15.3.) Reduced capacity is then obtained by lowering the temperature of the energizing fluid instead of its quantity. This permits operation at much lower temperature levels with accompanying improvements in efficiency of heat collection by solar arrays. Figure 15.7 provides a complete performance map for typical commercial absorption equipment of the lithium-bromide—water type. The meaning of the various lines shown is as follows:

1. The abscissa shows the capacity in terms of percent of peak performance at 30°C cooling tower water temperature and equivalent heat source temperature of 115°C which are the nominal rating points for most commercial equipment. Because of the heat transfer character-istics of the generator, the inlet temperature of the energizing water is between 2 and 4°C higher than the "equivalent heat source temperature."

2. The vertical lines labeled "Cooling tower water temperature" represent maximum limits of capacity at the designated temperatures. For example, during periods when the cooling tower water temperature is 15°C the unit must not be operated above 79 percent of the nominal rating to ensure proper system performance over a long period. The location of this line may

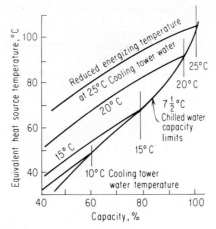

Fig. 15.7   Performance of typical absorption equipment.

vary slightly for units of different manufacturers since it relates to the maximum vapor velocity out of the generator which will not carry some of the solution into the condenser. Velocities are very high even though the generator and condenser may be located in the same shell as seen in Fig. 15.8 since the operating pressure is only 8000 N/m² (8 kPa).

3. The line labeled "7¹/₂° Chilled Water" intersects the vertical cooling tower water temperature lines at the equivalent heat source temperatures needed to produce the capacity limit for that specific cooling tower water temperature. Producing a 5°C chilled water temperature would require an increase in heat source temperature of 5°C, and producing a 10°C chilled water temperature would allow a decrease in heat source temperature of approximately 5°C. To illustrate, at 15°C cooling tower water temperature the required heat source temperature is read on the scale at the left as 68°C, and this temperature will produce 79 percent of the unit's nominal rating. In most buildings 79 percent of the rating needed at 25° or 30°C would be more than adequate for the 15°C cooling tower water temperature and a corresponding wet-bulb ambient of about 10°C needed to provide that water temperature.

4. The family of lines labeled "Reduced Energizing Temperature" provides a means to determine the equivalent energizing temperature needed when the building load is less than the capacity limit for a given cooling tower water temperature. The example above in paragraph 3 shows a capacity limit of 79 percent of nominal rating at the 15°C cooling tower water temperature. Had the building load been 50 percent, the reduced energizing temperature line running downward to the left from the intersection of the 15°C capacity limit line and the chilled water line shows that an energizing temperature of 45°C will provide the capacity to match the 50 percent load.

Coefficient of performance   Unless the energizing temperature of lithium-bromide–water absorption equipment is reduced as the building load reduces, it becomes necessary to cycle the unit on and off to prevent overcooling of the chilled water and possibly the building itself. *As the heat exchangers in absorption systems have long time constants, cycling reduces efficiency and, therefore, modulation of capacity is much to be preferred.* There is a second reason for modulation in solar-energized systems since it allows energization at much lower temperatures as was illustrated in the example in paragraph four above. We shall see that collectors are much more efficient at such lower temperatures, and that a collector array can provide much more cooling when the lower temperatures can be used effectively.

Figure 15.7 showed the relationships between available cooling tower water temperature, chilled water temperature, and desired capacity, and showed how they determine the energizing temperature needed to modulate the unit to follow different imposed loads. We now look at

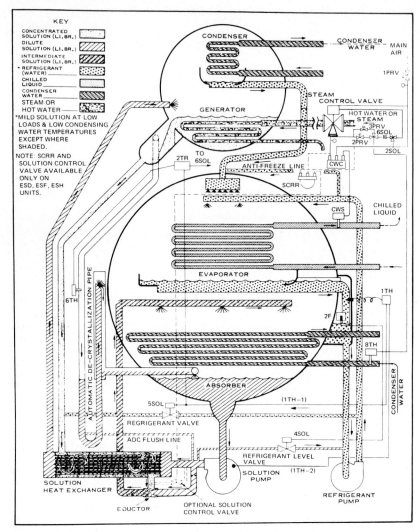

Fig. 15.8 Sectional diagram of typical commercial lithium-bromide–water absorption unit. (*Courtesy York Division, Borg-Warner Corporation.*)

the relationship of the coefficient of performance to these same variables. The coefficient of performance is defined as

$$\text{COP} = \frac{\text{Thermal kW cooling effect}}{\text{Thermal kW required to energize}} \tag{15.1}$$

Figure 15.9 provides a typical performance map relating COP to operating temperatures. The meaning of the various lines is as follows:

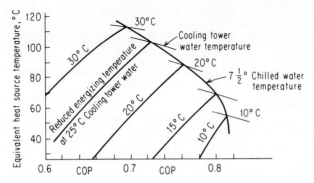

Fig. 15.9 Relationships among COP, cooling water temperature, and energizing temperature for a typical LiBr machine.

1. The lines labeled "Cooling tower water temperature" intersect the $7^1/_2°$ chilled water temperatures at the same equivalent heat source temperatures as shown in Fig. 15.7. The intersection also shows the resulting COP on the abscissa. For example, at the nominal rating at 30°C cooling tower water the COP is 0.69.

2. The "Reduced energizing temperature" lines running downward to the left from these intersections indicate the reductions in COP which will be experienced when the energizing temperature is reduced to match a load below the capacity limit set by the cooling tower water temperature. For example, if the load were 50 percent of nominal capacity when 25°C cooling tower water is used, Fig. 15.7 shows that the energizing temperature required is 72°C, and Fig. 15.9 shows a COP of 0.66 for this energizing temperature. This is the COP which would be experienced if the cooling tower water temperature is maintained above the 25° level as in some commercial practice.

Improved operating procedure   At 50 percent of nominal capacity used in the last example it would be likely that the cooling tower water temperature would reach 15°C if allowed to float with the load and ambient as indicated in Fig. 15.3. The required energizing temperature is then 45°C as obtained from Fig. 15.7, and the COP obtained from Fig. 15.9 is 0.76. Thus, allowing the cooling tower water temperature to float to a level dependent only upon the local wet-bulb temperature results in an improvement in COP from 0.66 to 0.76, while at the same time permitting the use of an energizing temperature of 45°C instead of 72°C. This reduction of energizing temperature is very beneficial to the operation of a solar-cooling system since both the COP and collector efficiency are higher.

The normal range of cooling tower temperature rise as it passes through the unit is 10°C at full load. The product $Q \cdot \Delta T$ is approximately 6×10^{-4} per kW of capacity at nominal full-load rating point whereas the COP is 0.69. Q is defined as flow in cubic meters/sec. This product increases 10 percent as the COP decreases to 0.60, and decreases by 8 percent as COP increases to 0.80. The normal range of chilled water drop as it passes through the evaporator is 5.5°C at full load. The product of $Q \cdot \Delta T$ for chilled water is 2.5×10^{-4} per kW of capacity.

These constants may be used to estimate the flow and temperature range for various capacities.

Other fluids for absorption equipment   Other combinations of salts and fluid refrigerants have been extensively examined over the years. Only a few have shown promise in workable and dependable equipment. Those which seem to be the most promising are discussed in the following paragraphs in which the absorbent is listed first followed by identification of the refrigerant

Water-ammonia.   The water-ammonia absorption system has been in use for nearly a century. It has found its greatest use in cold-storage applications wherein steam or waste heat was readily available. More recently it has been used in gas-fired residential air conditioners of the air-cooled type. The major components of the water-ammonia system are similar in function to those of the lithium-bromide system except for the addition of a reflux system. This is needed because ammonia and water are both liquids and do not separate sufficiently in the generator with a single stage of heating. A schematic diagram of a water-ammonia system is shown in Fig. 15.10.

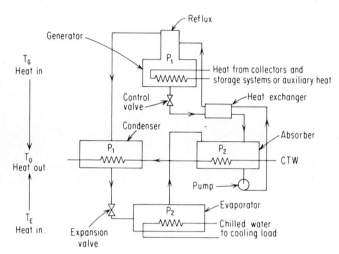

Fig. 15.10   Schematic of ammonia absorption system.

Both air-cooled and water-cooled water-ammonia equipment is under continuing research and development in residential sizes for use in areas where the equipment can be located outdoors. Gas-fired, air-cooled equipment has demonstrated COP of 0.28 to 0.32. The water-cooled equipment has indicated COP only slightly lower than those for lithium-bromide–water systems for the same energizing and heat rejection temperatures. The additional losses of the rectifier components must be coped with in these systems.

Before water-ammonia absorption systems can be considered for commercial-sized solar applications, the potential hazards of large quantities of ammonia in or near public buildings must be considered. New safety measures might be needed, and it is doubtful that present fire and safety codes would permit such installations.

Sodium-thiocyanate–ammonia.   The sodium-thiocyanate–ammonia systems have been researched by the Institute of Gas Technology for residential size equipment of the water-cooled type. Concerns exist for the longtime stability of the salt and as to methods of preventing corrosion in equipment. The system appears to have sufficient promise to justify further work. Performance approaches that of the lithium-bromide–water system.

Since the refrigerant in this system is ammonia the same concerns exist as to code acceptance for commercial size units as were mentioned for the water-ammonia system.

Lithium-bromide–methanol.   Methanol is substituted for water as the refrigerant in this combination. The higher vapor pressure of about 5000 Pa compared to 840 Pa for water at 5°C reduces flow velocity and allows a modest pressure drop without serious loss of performance. The system is nearly free of any problems of crystallization in all air-conditioning ranges, and the refrigerant is not subject to freezing.

The major problem with the lithium-bromide–methanol pair is the relatively high viscosity of the solutions which requires special heat transfer considerations. To reduce viscosity a combination of 2LiBr/1ZnBr has been used but with some increase in corrosion. The major continuing work is directed to solutions of these potential problems.

The thermodynamic properties of the lithium-bromide–methanol, and of the lithium-bromide–zinc-bromide–water combinations have been determined by Akers and Squires at Purdue University, and at the University of Osaka, Japan.

Dual Effect Absorption Equipment. The lithium-bromide–water commercial size units are also available for higher energizing temperatures in a dual cycle, in which the generator discharges into a condenser, the heat from which is used to energize the regular absorption cycle shown in Fig. 15.6. The energizing temperatures for this equipment are generally from 150–175°C, and they must therefore be used with concentrating or evacuated-type collectors. The COP can approach 1.00, and the advantages of the reduced energy requirement must be measured against the cost of more efficient collectors and the problems of handling the higher temperature in piping, pumps, and storage.

SOLAR-ENERGIZED ABSORPTION COOLING

It has been shown above that the cooling load in most buildings is highly variable, being close to maximum load during only a few hours per year. It has also been shown that absorption equipment operates best at the ambient temperatures which exist at part load with lowered energizing temperature. (See Figs. 15.1, 15.2, 15.7, and 15.9.)

Two basic types of systems exist to energize absorption equipment. In the most common system, heat is captured by a collector array and placed in storage for use as needed. In the second less common system little or no storage is used and the heat from the collector array is used directly to energize the absorption equipment. This second system is generally limited to desert-type areas where daily availability of adequate insolation is assured, and to buildings whose loads coincide closely with the insolation pattern. Only the first system will be discussed in detail since it offers the greatest opportunity to control heat collection and the operating sequence for maximum efficiency.

Collector Performance

The fact that buildings can be cooled by energizing equipment at temperatures well below those at design conditions has a major effect on the optimum design of solar cooling systems. This effect is much greater than it is for conventional systems since it allows major improvements in heat collection from solar arrays.

Figure 15.5 shows the rate of heat collection by a typical flat-plate collector at various times throughout a day for different temperatures of fluid within the collector. The collector inlet fluid temperature is determined by the temperature in the storage system from which the liquid is drawn. Thus the storage temperature determines the temperature at which the collector array operates. The lower the storage temperature the greater the total kWh of heat which can be collected during a day. Table 15.4 shows the effect of storage temperature on daylong heat collection for typical flat-plate collectors.

Table 15.4 shows that more than twice as much heat is collected when the liquid coming to the collectors from storage is at 70°C than when it is at 90°C, and that more than four times the amount of heat can be collected at 50°C than at 90°C. Systems thus benefit from collecting and storing heat at the lowest temperatures at which it can be effectively used.

TABLE 15.4   Example Daily Heat Capture at Various Collection Temperatures

Collection temp °C		90		70		50	
Time	Ambient	ΔT	kWh	ΔT	kWh	ΔT	kWh
8:00	15	75	0	55	0	35	0.160
10:00	21	69	0.050	49	0.200	29	0.395
12:00	32	58	0.210	38	0.390	18	0.615
14:00	35	55	0.210	35	0.400	15	0.610
16:00	32	58	0.030	38	0.120	18	0.260
18:00	30	60	0	40	0	20	0
	Daylong, kWh		0.900		1.900		3.660

For any given job the collectors under consideration should be analyzed for daylong performance in the location where they will be installed. The analysis must take into account the expected amounts of cloud cover and the time of day when direct sunshine may be obscured. Thus the degree to which a particular collector responds to diffuse irradiation becomes important. Chapters 7 and 8 describe methods of testing collectors to determine daylong performance.

The manner in which the basic types of collectors respond to varying insolation incident angles at conditions of high beam insolation and at 50 percent diffuse are shown broadly in Table 15.5.

TABLE 15.5   Effect of Collector Type on Heat Collection Rate*

| | | Heat collected, kW/m^2 | |
Type collector	Incident angle, °	Beam radiation	50% diffuse radiation
Flat plate	0	0.250	0.240
	45	0.217	0.220
Evacuated tube	0	0.525	0.510
	45	0.510	0.500
Concentrating collector	0	0.408	0.209
	45	0.030	0.020

*Based on aperture area.

Storage Requirements

Weather sequences are not predictable. Therefore, the number of hours each day during which a building load approaches the design load and the time of day that it does so are not known in advance. In most parts of the world the morning hours have relatively low ambient temperature. Thus cooling tower water will be low in temperature, and temperatures in the range of 50 to 70°C will operate the absorption unit at a higher COP than would result from maintaining a higher temperature cooling-tower water and using a higher energizing temperature.

Yet there will be a number of hours on some days during which higher energizing temperatures must be available. Efficient operation can be achieved by using two storage units. One storage should provide 70 to 75 percent of the total heat required at the lowest temperature which can be utilized effectively at the part-load conditions. Typical temperature may be from 50 to 70°C depending on the building load pattern and the expected pattern of ambient temperature.

The remaining 25 to 30 percent of the storage volume should be in a smaller tank with more insulation in order to store the heat collected in the 85 to 95°C range. Still higher temperatures may be used in this storage if it can be pressurized to prevent boiling, and if collectors are used which are capable of operating at higher temperature levels with good efficiency. Latent heat storage may be particularly worthwhile in the higher-temperature unit since it tends to reduce its physical size for a given amount of kWh stored, and provides more heat at the levels needed for full-load operation without significant temperature change.

Figure 15.11 shows schematically a dual storage system. The pump P circulates the liquid from either the low-or high-temperature storage. Valves 1 and 2 are opened when the system is to add heat to the low-temperature storage L, and valves 3 and 4 are opened for adding heat to the higher-temperature storage H. Control C determines when the pump operates and which valves are opened.

Figure 15.12 shows a function diagram of control wiring to accomplish the desired sequence. The pump is turned on whenever the difference in temperature between the sensors, S, and L or H is positive and large enough to indicate that sufficient insolation exists to begin to charge the storage. Choice is also made as to which storage is to receive the heat. The control sequence shown is typical and will be described in detail.

Several solid-state differential temperature controls are available for solar system use. Sensors may be resistance elements or thermocouples, but must have accuracy sufficient to track the difference between two sensors over a range of 100°C in order to respond *without crossover*. A suggested operating sequence is as follows:

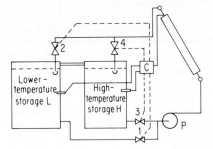

Fig. 15.11   Dual storage schematic diagram.

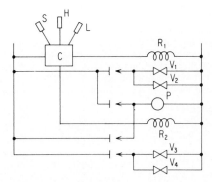

Fig. 15.12   Control of dual storage solar cooling system.

1. Sensor S measures a representative temperature of the collector array. It may, for example, be attached to the liquid outlet tube at the top of one of the collectors or to the collector absorber. Before sunrise it will be substantially at ambient temperature and therefore below the temperature of the low-temperature storage L.

2. As soon as the sun warms the collectors so that the sensor temperature exceeds that of the low-temperature storage sensor L, the solid state device energizes relay R_1 to start the pump and open valves V_1 and V_2. The stored liquid is then circulated to the low-temperature collector to add heat.

3. As insolation increases the temperature differential between S and L increases, until the control stops the pump at a predetermined difference indicative that heat can be collected in a significant amount at the higher temperature. At this time the controller closes all valves for a short time interval of some two to three minutes. The best time interval will vary from one installation to another and hence some adjustment of this interval is desirable.

4. If during the time interval of step 3 the sensor S reaches a temperature above that in the high-temperature storage as measured by sensor H, the solid state device restarts the pump and opens valves V_3 and V_4 to add heat to the higher temperature storage. If the required temperature level is not reached within the two- to three-minute time interval, the control restarts the pump and opens valves V_1 and V_2 to continue collection of heat into the lower temperature storage. The solid state device may include a time function which assures that the operation in the lower temperature mode will continue for some minimum time such as 15 or 30 min to prevent frequent cycling from high to low temperature storage.

5. The solid state device may also use the temperature sensed by either H or L as a high limit to discontinue heat into either storage when its maximum desired storage temperature

is reached. This action prevents boiling or other unsafe operation. If the system has a drain-back arrangement, circuitry can be added to Fig. 15.12 to hold either V_1 or V_2 open upon the termination of a limit operation. Likewise, V_1 can be kept open after each operation of the low-temperature storage sensor L if drain-back is used. Alternatively, an active heat rejector can be used.

Potential Benefits from Dual Storage

Separation of the storage into a high and low temperature system may increase the heat collected by a given collector array by a factor of 1.30 to 1.50, depending on location and type of collector. At the same time the COP on a seasonal basis may rise from approximately 0.65 to 0.75, a 15 percent improvement. Taken together these benefits may decrease the required area of the collector area to cool a given building by 30 to 40 percent—a very considerable savings.

Cold Side Storage

Storage of cold liquid, usually water, has had limited use. It appears to be applicable in areas of highly variable weather which permit excess cooling capacity for periods of two to three days followed by hot spells. However, it should be noted that only 5 to 7°C range can be allowed in chilled water temperature if system dehumidification ability is to be retained.

Before deciding on cold side storage it is suggested that a very careful analysis be made to compare the results of such operation to the results of dual hot storage for the more efficient operation of absorption equipment. The temperature range through which an intermediate temperature storage can operate at a COP in the absorption unit of 0.75 may be as great as 20 to 24°C—equivalent to a temperature range of 15 to 18°C in cold storage, which would be unattainable. On the positive side for cold storage is the fact that loss to ambient is usually less than with hot storage due to the lower temperature relative to ambient.

Supplementary Components

Most supplementary items of solar-energized absorption cooling systems are standard components of conventional building cooling systems. Their special requirements for solar system use do not make them "specials" as far as manufacture and availability.

Liquid-to-liquid heat exchangers   Liquid-to-liquid heat exchangers are used between storage and collector arrays when the circulating liquid is a nonfreezing or high boiling point liquid such as glycol solutions, silicone, or a heat transfer oil (see Chap. 4). It is usual to select heat exchangers for a 5 to 6°C leaving difference using the manufacturer's rating information for the particular liquid and temperature levels employed in the system.

Liquid-to-air heat exchangers   Standard finned coils and fan-coil equipment as manufactured for use throughout the conventional air-conditioning field should be used. In making coil selections, an average water temperature from 1 to 2°C higher than nominal for the installation may be assumed due to the variable nature of the energy source.

Flow control valves.   Valves for controlling chilled water flow to air handling units in the building are selected and applied as in any conventional system. Valves used to control flow between storage and collector arrays or between storage and the system *must* be of the "bubbletight" variety in order to prevent serious losses during off periods. Even a small leakage over a long period of time becomes a major loss.

DESIGN OF THE ABSORPTION COOLING SYSTEM

For optimum design of a solar cooling system the load pattern of the building must first be known. An analytical method of determining it was given earlier in this chapter under "Calculating Cooling Loads." The performance maps of the absorption equipment and of the cooling tower must also be determined from the manufacturers of each. The information should be as complete as that shown in Figs. 15.3, 15.7, and 15.9.

Numerical models to assist in design are available in several forms. TRNSYS provides digital simulation of solar cooling systems, and models each component in a subroutine so that input references can link each output to the corresponding input of another component. This program is available from the University of Wisconsin-Madison.

ASHRAE publication GRP-153[5] provides a bibliography of many other programs available for system design, and is updated as new programs become available. None of the programs can determine the minimum size of collector array needed to meet the building cooling needs unless it contains the complete load and capacity maps of the absorption equipment and the cooling tower.

Sizing Guidelines

In tropical and desert areas the solar cooling system is frequently designed to provide nearly all cooling needs without an auxiliary energy source. In such areas the heaviest loads occur during periods of maximum insolation. Certain classes of buildings such as hospitals may require auxiliary since they must have adequate capacity on a 24-h basis.

In temperate regions the year-long economics of combined heating and cooling must be considered. Typically, the contribution of solar to total cooling is as low as 60 percent during sequences of hot days, and 100 percent during cooler sunny periods. There is no clear-cut "best" percentage of solar contribution. Each building system needs individual analysis based on technical and economic criteria (see Chap. 28).

Collector Arrays

Proper choice of collectors depends on such factors as hours of beam radiation in the area in which the building is located, the percent of diffuse irradiation commonly experienced, and the period of the day in which it most likely occurs, and on whether heating or cooling is favored on an annual basis. The three basic types of collector—flat-plate, evacuated-tube, or concentrating—react differently in response to these variables.

Table 15.5 compared the rate of heat collection for representative collectors at liquid inlet temperature of 50°C above ambient, for incident angles of 0° and 45°, for total irradiation of 0.833 kW/m² at 100 percent beam, and at 50 percent diffuse. The values given for evacuated-tube and concentrating collectors are for single collectors, and both types require space between individual collectors to avoid shading and, in the case of concentrating collectors, to allow for the motion of tracking. The effective coverage of the area in the array may be as low as 50 percent of the collector plane area. Note that the performance of concentrating collectors falls off when diffuse irradiation is present. This fact should be considered for locations with a high incidence of diffuse irradiation.

Collectors rated for performance under the test methods of ASHRAE Standard 93-77 (Ref. 7) have a Hottel-Whillier efficiency curve to show their capacity over an extended range of insolation and fluid-to ambient ΔT. (See Chap. 7 and Ref. 6.)

Collectors are normally faced within 10° of true south. Tilt angles may be equal to the latitude as for water heating systems, but reduced 10 to 12° if it is desired to favor cooling performance over the heating cycle performance as is usually the case for southern locations.

Other Components

Storage systems   Storage of heat received by the solar array is needed to energize systems which must operate at night or during periods of little or no insolation. A few buildings such as schools which are occupied only during daytime in desert-type areas may dispense with storage. Nominal storage amounts for cooling purposes range from 0.08 m³/m² of collector area to 0.2 m³/m². Since storage is charged with heat only when it is available from the collector array in excess of the amount of heat needed to operate the cooling equipment a computer analysis is needed to accurately estimate instantaneous storage temperature. The TRNSYS program can provide such information.

Interface with back-up system   Each of the numerous methods of utilizing back-up systems provides different operating characteristics. For retrofitting, the existing cooling system may be used as the back-up. In new installations back-up can be achieved by use of a boiler to provide thermal energy to operate the absorption unit. The same boiler normally acts as back-up for the building heating system.

Common unit for solar and auxiliary input   Typical considerations for the more common back-up arrangements are as follows:

1. The auxiliary energy may be used to heat the storage to maintain it at a high enough temperature to operate the absorption unit. The larger the storage relative to system capacity, the slower the response in raising the temperature in storage and providing the level required to maintain the needed capacity. This is particularly true during a temporary increase in cooling tower water temperature or load which demands a higher energizing temperature than exists in storage. The advantage of introducing the auxiliary heat directly into storage is that the future collection of solar heat, albeit at lower efficiency, may be at a more useful temperature level than if storage remained depleted.

2. The auxiliary heat may be introduced into the output circuit between the storage and the generator of the absorption unit. One advantage is that the energizing temperature can then be controlled to provide the optimum operation at the existing cooling tower water temperature and load. If the storage temperature is below the needed energizing temperature but above the return temperature from the generator, only a portion of the energy need be supplied by the auxiliary to reach the energizing temperature. Thus this method of introducing auxiliary energy may be suitable in installations needing auxiliary energy only during short periods.

3. The auxiliary heat may be introduced in a separate loop isolated by valves from storage. The auxiliary then must furnish all the operating energy during the supplemental period, but storage is not further depleted and may be re-employed as cooling tower water temperature or load begins to reduce. Furthermore, whatever solar energy is available for increasing storage temperature may be collected while the auxiliary energy is operating the system, thus building temperature until storage can take over.

Use of separate cooling equipment for back-up   The use of separate equipment for back-up may be typical of some retrofit installations. If separate heat transfer coils are installed in the building equipment for solar-energized cooling, they should be positioned upstream of the coils used for the auxiliary system to enable the solar system to take as large a portion of the load as it can.

Steps in Selecting Absorption Equipment

After determining the building conditioning load patterns by such methods as those described on pages 15-6 through 15-9, the expected COP of the absorption unit should be determined by reference to performance information such as shown in Figs. 15.7 and 15.9. If a computer analysis is not available, the load patterns may be considered linear for each basic condition—sunny and occupied, unoccupied during night or cloudy, unoccupied and sunny, or occupied at night or during cloudy conditions. Assume optimum energizing temperatures for this analysis based on cooling tower performance similar to that of Fig. 15.3. Iterations are needed to perform such an analysis.

For each operating condition and COP, the amount of energy for operation of the unit is determined by dividing the corresponding load by the COP for that condition. For example, were the load 300 kW and 0.76 the COP for the condition, we have

$$\text{Input} = \frac{\text{Load}}{\text{COP}} = \frac{300}{0.76} = 394.75 \text{ kW}$$

This rate of energy input is thus required at whatever temperature level is shown by the performance requirements for the unit.

The size of collector array and volume of storage are determined by iteration based on known collector performance for key days, for example, for the 16th day of each month of the cooling season for the job under consideration. Daylong performance is determined for each day based on the Hottel-Whillier plot of the collector under consideration and on the cloud patterns obtained from N.W.S. records for the locality. The collector array is assumed to carry the building conditioning load and to add heat to storage whenever more heat is available than needed to operate the absorption unit.

Different size storage should be assumed to receive the stored heat, starting with a possible volume of 0.1 m³ per m² of collector area. Iterations should then determine the pattern of storage temperature throughout the season.

The absorption unit which will best meet the entire load pattern with the available energizing temperature pattern should then be chosen from available equipment. The entire selection procedure should be tried on large jobs for several areas of collector and percentages of full energy requirements from the collector array.

A control system must be selected to monitor the cooling tower temperature and determine the maximum allowable energizing temperature. Additional sensing should be provided to reduce the energizing temperature below this maximum level whenever the chilled water temperature drops below its design temperature.

Dual storage systems   When dual temperature storages are under consideration, separate analysis must be made for each combination of collector array and storage, assuming that the array is reconnected from one storage to the other as the relationship between storage temperature and collector temperature indicates.

An alternative to be considered for dual systems is to use a collector with higher temperature capabilities for charging the higher temperature storage, and a less costly collector for the low temperature storage.

Sizing the load device   Selection of air-handling units or finned coils for duct installation is made in the same manner as for conventional systems. If the location of the building permits, the chilled water temperature, as it leaves the chiller, should be chosen in the range of 8 to 9°C instead of the usual 6°C. This will permit some improvement in COP and energizing temperature level, and thus a more economical installation.

System example calculation   As of 1979 there were numerous solar-energized cooling installations using absorption equipment. While few if any are fully optimized for all aspects of equipment selection, control, operating sequence, and economics, the installation in the Department of Agriculture Building at New Mexico State University has many unique features and is described herein as a good example of solar cooling design.

Figure 15.13 shows the building and its collector array. The array includes 582 m² of American Heliothermal Miromit™ flat-plate collectors with black nickel selective surface on their absorber plates. Single-cover glasses of 4-mm thick-water white-tempered glass are used. In addition, the array has 50 m² of Northrup concentrating collectors in tracking mounts.

Two storage tanks, each of 53-m³ capacity are installed. Since the night wet bulb in this desert region is low enough to help chill the water in one tank, it is normally run as chilled storage with nighttime assist, and with cooling during daytime when available from the absorption units. The other tank is run as hot storage.

Fig. 15.13   New Mexico State University solar cooling system.

Two absorption units are installed. The first unit is a York Model ESF1A2A, rated at 176 kW at 86°C energizing temperature and 30°C cooling tower water temperature, or at 310-kW capacity at 85°C energizing temperature and 24°C cooling tower water temperature. The second unit is an Arkla-Servel WF-300, rated at 56 kW at 88°C energizing temperature and 29°C cooling tower water temperature. Figure 15.14 shows a portion of the absorption unit installation.

The back-up system is a Cleaver-Brooks gas-fired boiler of 30 thermal kW capacity. Its use for cooling was projected to be minimal, and it is reported that after two seasons of operation the gas lines were disconnected. In this installation it appears that the only future use of the back-up would be to provide cooling in the event of a serious failure of some component of the solar system itself.

SOLAR VAPOR-COMPRESSION COOLING METHODS

Figure 15.15 shows the elements of a system which uses the thermal energy of the sun to drive an engine which in turn drives a conventional refrigeration compressor to provide cooling of the building. In vapor-compression cycles a fluid is boiled by the heat input of the system, expands as a vapor through the engine to produce power, is then condensed at a lower pressure than in the boiler, and pumped back into the boiler to repeat the cycle. This cycle, called a Rankine cycle, is described in detail in Chap. 22. Smaller engines are likely to be of the piston-and-crank type, though they may be high-speed turbines or rotary engines. Large engines are almost all of the turbine type because of their relative simplicity. All such engines are characterized by having the heat input of the cycle external to the engine itself.

Other cycles have been in research and development for many years but have not yet reached commercial production for widespread application. One such cycle is the Stirling cycle in which

Fig. 15.14   Equipment room at New Mexico State University solar cooling installation. (*Courtesy Physical Science Laboratory, New Mexico State University.*)

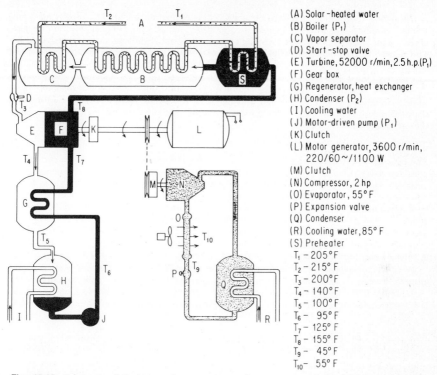

(A) Solar-heated water
(B) Boiler (P₁)
(C) Vapor separator
(D) Start-stop valve
(E) Turbine, 52000 r/min, 2.5 h.p.(P₁)
(F) Gear box
(G) Regenerator, heat exchanger
(H) Condenser (P₂)
(I) Cooling water
(J) Motor-driven pump (P₁)
(K) Clutch
(L) Motor generator, 3600 r/min,
       220/60~/1100 W
(M) Clutch
(N) Compressor, 2 hp
(O) Evaporator, 55°F
(P) Expansion valve
(Q) Condenser
(R) Cooling water, 85°F
(S) Preheater
T_1 – 205°F
T_2 – 215°F
T_3 – 200°F
T_4 – 140°F
T_5 – 100°F
T_6 –   95°F
T_7 – 125°F
T_8 – 155°F
T_9 –   45°F
T_{10}–   55°F

Fig. 15.15 Schematic of Rankine cycle-compression refrigeration system. (*Courtesy Barber Nichols Corporation.*)

the thermal energy is used to heat the engine itself instead of a boiler. Thus the Stirling engine is called an externally heated engine.

The efficiency of such engines is limited by the Carnot limit. If T_i is the absolute temperature of the fluid as it boils and as its vapor enters the engine, and T_0 is the absolute temperature at which the fluid condenses as it leaves the engine, the efficiency, η, which cannot be exceeded is, $\eta = (T_i - T_0) \cdot T_i^{-1}$. Because of the unavoidable losses in power cycles the actual efficiency is often less than half of the theoretical maximum. Thus the maximum possible efficiency for an engine operating between inlet temperature of 420 K and condensing at 320 K is the theoretical maximum efficiency which would be $(420 - 320) \cdot 420^{-1} = 24$ percent. The actual realized efficiency may be 12 percent. Refrigeration compressors have a similar theoretical maximum COP which depends on the temperature of evaporation and condensation of the refrigerant, $\text{COP} = T_e \cdot (T_c - T_e)^{-1}$. As an example, if the evaporator temperature is 278 K and the condenser temperature is 320 K, the maximum theoretical coefficient of performance is $\text{COP}_{max} = 278 \cdot (320 - 278)^{-1} = 6.619$. As in the engine, the maximum performance cannot be reached, and the best compression refrigeration cycles operate at no more than 60 percent of the theoretical, or in this case an operating COP of 3.91 could be expected.

System COP

Driving the compressor of the cooling system operating at a COP of 3.91 by the engine cited above at an efficiency of 0.12 yields a possible system COP of $3.91 \cdot 0.12 = 0.47$. The reciprocal of this COP, 2.12, is the number of thermal kilowatts at 420 K (147°C) which must be supplied by the solar system for each kilowatt of cooling capacity. The performance for other energizing temperatures and heat rejection temperatures can be determined as a first approximation by assigning the appropriate temperatures for energy input, heat rejection, and evaporator and completing the same sequence of calculations as shown in the examples.

The system COP will be higher as the energization temperature from the collector array and storage increases, and as the evaporator temperature increases. Lower heat rejection temperature, either when air cooling or a cooling tower is used, provides some improvement in COP. The degree of improvement will depend on the characteristics of the engine and of the compressor.

A trade-off exists between use of high temperatures to energize the engine and thus improve its efficiency, and lower collector efficiencies which result from collecting the solar energy at a higher temperature. More collector area might be needed for concentrating arrays than for flat-plate arrays in areas such as the northeastern United States which have a relatively large diffuse component.

Turbines for Solar Cooling Systems

The conventional steam turbine operating on steam evaporated in an unfired boiler can be considered as a drive for solar cooling systems. Steam turbines require several stages and are relatively large and costly. Use of denser fluids than steam permits the use of radial inflow turbines, which may usually be of the single-stage variety.

Numerous manufacturers produce centrifugal compressors for building air-conditioning systems, many of which are of single-stage design. Since refrigerants are more dense than water vapor, these compressors are physically small in the available sizes of 50 to 1000 kW. They are normally driven by two-pole 60-cycle motors, and hence have speed increasers built into them. A typical centrifugal compressor is shown in Fig. 15.16.

To convert conventional refrigerant compressors to radial-inflow turbines it is necessary to provide nozzles or vanes in the discharge passages and to reverse the flow direction of the working fluid. The shaft power thus produced is used as input to a similar compressor in the refrigeration circuit used to cool the building. Gears are not required if the rotating speed of

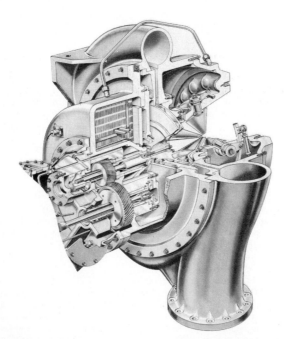

Fig. 15.16 Typical centrifugal refrigerant compressor section drawing. (*Courtesy York Division, Borg-Warner Corporation.*)

the radial inflow turbine and the compressor are equal. Turbine speeds in the 50-kW size using R-11 as a working fluid are about 18,000 r/min, and in the 1000-kW size about 8000 r/min.

For smaller cooling capacities from 10 to 150 kW of cooling effect, experimental systems have been built using the components originally designed for aircraft air conditioning.

Selection of Turbines and Compressors

Turbines and refrigeration compressors are usually manufactured from general designs for specific purposes. For example, wheel sizes are trimmed and come in various widths according to the working fluid and the speed and capacity requirements. For each design the manufacturer can be expected to furnish the detailed operating characteristics. No standard operating and performance data are available.

Part-load performance should also be obtained from the manufacturer since it is just as important to determine it for Rankine systems as for absorption systems. In general the turbine may be expected to become less efficient as the speed and capacity decrease, while the compressor may show a slight increase in efficiency, the net result being a decrease in COP of some five to ten percentage points.

Working fluids for compressors   Common refrigerants are R-11 for the 150- to 2000-kW cooling capacity compressors, R-114 for ranges from 1500- to 3000-kW cooling capacity, and R-12, or R-22 for the extremely large units. Other refrigerants are sometimes used, and there is no reason to change standard designs as far as requirements of the solar cooling system are concerned.

Working Fluids for Turbines   All available working fluids for driving turbines have combinations of properties which make it impossible to attain the theoretical maximum cycle efficiency. Any fluid can be used at high efficiency only within the portion of its pressure-enthalpy diagram which has sides nearly parallel below the saturation dome, unless rather massive heat exchangers are added to the cycle. Therefore, a high critical fluid temperature is desirable. Some of the desirable properties of working fluids are:

1. High critical temperature
2. High density
3. Stability at all temperatures to which it will be exposed in a system in contact with all metals and gasket materials, and with small quantities of oil and water
4. High latent heat
5. High heat transfer capability—evaporating, condensing, or in sensible heat exchange both as a liquid or a vapor.

Except for the high temperature tolerance, most of the properties listed are the same as those which are required for a good refrigerant. Thus it is not surprising that the list of the most promising working fluids comes from listings of available refrigerants. Table 15.6 lists some common working fluids and their critical temperatures. Good practice requires that the highest boiling temperature in the heat input portion of the cycle be at least 30°C below the critical temperature. Before choosing a working fluid it is recommended that information be obtained from those suppliers who maintain adequate research laboratories as to the stability of the fluid in both liquid and vapor form when in contact with the specific metals, plastics, or elastomeric materials at the expected temperature for each.

TABLE 15.6   Critical Temperature of Working Fluids for Rankine Cycles

Material	Critical temp., °C
R-22	96
R-12	112
NH_3	133
600a	135
R-114	146
Butane	152
R-11	198
R-113	214
H_2O	374

Isobutane (600a) and butane are included, but must be used with caution due to their flammability, and the possibility that fire codes would prohibit their use in some locations where solar cooling would be desirable.

Pressure-enthalpy diagrams for the common fluids are shown in the appendix of this chapter.

Collector Types for Solar Vapor-Compression Systems

As shown on page 15-24 engines or turbines which are used for driving refrigerant compressors operate more efficiently at high energizing temperature than at lower temperatures. Hence there is frequently an advantage in using concentrating- or evacuated-tube-type collectors which are capable of higher efficiency at elevated temperature than are flat-plate collectors. Some special considerations involving these collectors are (1) Some concentrating collectors must track the sun and hence need spacing between adjacent collectors to avoid interference, (2) evacuated-tube-type collectors with tubular absorbers do not occupy 100 percent of the space they need since closely spaced tubes would shade each other, and (3) the glass itself occupies some aperture area. For these reasons the space required for a given amount of energy collection can be greater for such collectors than for flat-plate designs.

As was shown in Fig. 15.15, the working fluid in the heat engine or turbine is boiled by a separate collector fluid. The collector fluid must be chosen to prevent an excessive pressure in the collector passages. Glycol solutions can achieve modestly high temperature for such use, but may need to be drained from the collector during periods when heat is not being collected. Silicone fluids and special heat transfer oils have low vapor pressures and have been used in solar cooling systems. The collector manufacturers' pressure ratings should be obtained and suitable fluids selected for each situation.

Storage Considerations for Vapor Compression Systems

Storage for engine driven systems may be either high or low temperature. Each installation needs separate analysis to determine the best choice using methods similar to those developed above for absorption systems.

High temperature storage   Pressurized water may be considered for storage by interposing a heat exchanger between the collector fluid circuit and the storage water (Fig. 15.17). The storage tank becomes an unfired pressure vessel from a code viewpoint and must therefore be designed and installed to meet all code and safety requirements. The high temperature water is used to boil the working fluid usually at a design temperature below the peak storage temperature to allow for the drop in storage temperature as storage becomes depleted. Because of the large difference in temperature between a high temperature storage tank and the surrounding ambient, heavy insulation is necessary. (See Chaps. 4 and 6.)

Low temperature storage   The vapor compression system may be operated when insolation is available even when no cooling of the building is required in order to remove heat from a cold storage. The lower differential between tank temperature and ambient reduces losses, but in general the smaller useful range of the cold storage temperature swing limits its utility.

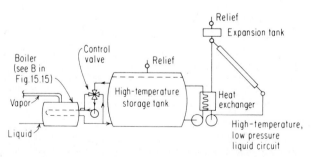

Fig. 15.17   Pressurized water thermal storage Rankine system.

Low Temperature Energization of Vapor Compression Systems

Systems using radial inflow turbines have been operated at temperatures available from flat-plate collectors, but over a more limited range of energizing temperatures than absorption systems, which do not have mechanical restrictions on pressure ratio and flow of the working fluid. Figures 15.18 and 15.19 show the capacity and COP of a typical combination of radial inflow turbine and centrifugal compressor rated at 275 thermal kW when energized by 100°C working-fluid vapor and supplied with 24°C cooling tower water.

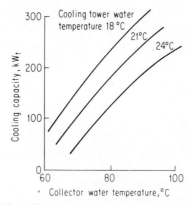

Fig. 15.18 Cooling capacity of 270-kW$_t$ Rankine-vapor compression.

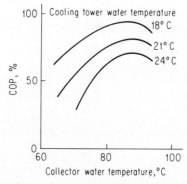

Fig. 15.19 Measured COP of 270-kW$_t$ Rankine-vapor compression system at 7.2°C chilled-water temperature.

The rapid reduction in capacity and COP as energizing temperature drops is quite characteristic of the Rankine-cycle systems, and the effect on required collector size must be considered in design of such systems. The peak COP which is at approximately 90°C for this unit can be shifted to higher temperatures by design of the turbine if it is intended to use high temperature energization, but the curve characteristically retains a sharp peak.

Design of Vapor Compression Systems

Performance of engine compressor combinations varies widely with the design of the turbine or engine and the characteristics of the compressor used. The variation between combinations is greatest at part-load conditions. Therefore, the performance maps such as shown in Figs. 15.18

and 15.19 must be obtained from the manufacturers of the individual units. Thereafter the same general design steps are followed as are shown on pages 15-18 through 15-21.

Manufacturers' information will show that positive displacement compressors such as reciprocating or rotary units can be capacity-modulated by varying the drive speed as the cooling tower water temperature varies and the load varies. Centrifugal compressor information will show that speed can be varied as the condenser temperature changes, but that some other device such as built-in pre-rotation vanes must be used to provide final capacity control.

Interface with back-up system   Vapor compression systems offer an additional type of back-up not available to heat-operated systems such as absorption or adsorption equipment. Figure 15.15 shows a motor generator connected to the drive shaft of the turbine-compressor. Electrical energy supplied to the motor can provide any deficiency of energy from the solar system. When no thermal energy is available at a suitable temperature either from storage or from the collector array, the electric motor takes over and operates the cooling system directly. This back-up method needs only a very simple controller.

Solar-powered heat pumps   The Rankine-vapor compression system can be operated advantageously as a heat pump in many cases. The overall COP must be greater than unity (1.00) for the cycle to be useful. Thermal energy to operate the turbine is converted to mechanical energy to drive the compressor. Thermal energy rejected at the condensing temperature, which can be chosen sufficiently high, can be used to heat the building. The mechanical energy drives the heat pump compressor in the usual fashion, at a high COP. Figure 15.20 shows a typical block diagram of such a system.

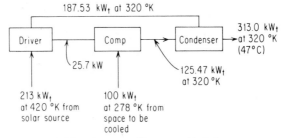

Fig. 15.20   Solar-powered heat pump block diagram.

The example given on pages 15-23 and 15-24 showed an engine efficiency of 12 percent and a refrigeration COP of 3.91. Referring to Fig. 15.20 it is assumed that 100 kW$_t$ is taken from ambient at 278 K, which at a COP of 3.91 requires 25.47 kW (100/3.91) of mechanical energy to drive the compressor. At 12 percent efficiency, the engine requires an input of 25.47/0.12 = 213 kW$_t$. The heat recovered in the condenser from the engine working fluid is 213 − 25.3 = 187.5 kW$_t$. The heat discharged into the condenser by the compressor at 320 K (47°C) is 100 + 25.5 = 125.5 kW$_t$, giving a total heat discharge into the condenser of 313 kW$_t$h for 213 kW$_t$h required to operate the engine. The heat pump COP = 313/213 = 1.47 for this example condition.

ADSORPTION-DESORPTION COOLING METHODS

Evaporative cooling has long been used in hot and dry climates with good results, by evaporation of water into the outdoor stream and thus cooling it to temperatures needed to cool the building. The same results can be accomplished with better control in areas where hot and dry air does not occur naturally by using solid or liquid sorbents to remove most of the water vapor from the air. The sorbent materials can be regenerated by solar heat. Water is then used to cool the air stream by partial saturation and air is then in condition to cool the building.

Solid materials like silica gel, activated alumina, or molecular sieves have the ability to adsorb water vapor by virtue of their structure, which consists of billions of tiny pores, and with very large total surface area. Some materials have as much as 1000 m^2 of area per gram of material.

Liquid sorbents function by presenting lower vapor pressure than water at a given temperature, thereby reducing the equilibrium dew point. Both liquid and solid sorbents require that heat be applied to reactivate them after they have taken on moisture before they can again be made effective. Some or all of the reactivation heat can be supplied by solar means.

Solid-Phase Systems

The water vapor equilibrium curves for the more common solid adsorbents are shown in Fig. 15.21 for temperatures between 0°C and 80°C. The curves show the percentage of water vapor contained in air which can be removed by the adsorbent for given relative humidity. For example, if air at 24°C and 55 percent relative humidity is passed through a bed of molecular sieve material, 19 percent of the water vapor contained in it will be removed. The air stream and the molecular sieve will be made warmer by the latent heat of the water removed and by a small additional amount of heat of wetting. To re-activate the adsorbent material it must be heated to a temperature of 120 to 200°C by an air stream. The hot air drives off the water vapor and, upon cooling, the adsorptive material is ready for the next adsorption cycle.

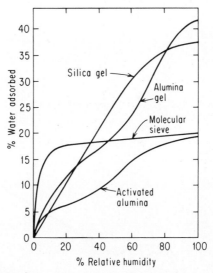

Fig. 15.21 Water vapor-desiccant equilibrium curves. (*Courtesy ASHRAE.*)

The Munters Cycle A demonstration unit of a system conceived by Munters of Stockholm and under development by the Institute of Gas Technology is shown in Fig. 15.22. In this particular installation evacuated-tube collectors were used, but other installations have used flat-plate collectors.

The components of the system include the molecular material in a rotating wheel somewhat over a meter in diameter, a heat-exchanger wheel, two humidifiers, a solar heating coil, and a fuel-fired, second-stage heater. These elements are located between fixed components for directing two air streams through them as shown in Fig. 15.23. The molecular sieve is a hydrate of sodium aluminum silicate from which the water is driven off. The remaining structure of "pores" is highly selective to the water molecule.

The process used is psychrometric in nature, and is best shown on a psychrometric chart as illustrated in Fig. 15.24. In the cycle shown two separate air streams are employed, the first being recirculated from the conditioned space through the unit and back to the room, while the second consists of outdoor air passing through the unit in the opposite direction and returning to outdoors.

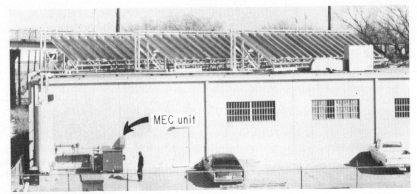

Fig. 15.22 Installation using "Solar MEC" cooling unit with evacuated-tube solar collectors. (*Courtesy Institute of Gas Technology.*)

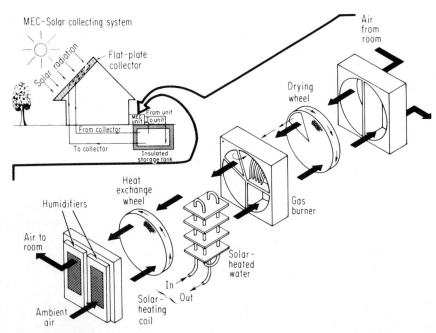

Fig. 15.23 "Solar MEC" schematic diagram. (*Courtesy Institute of Gas Technology.*)

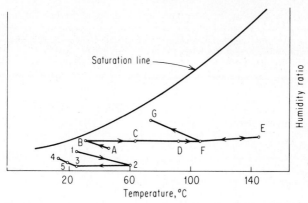

Fig. 15.24 Psychrometric cycle analysis of solar-energized Munters system.

The Room Air Path.   The cycle will be traced simultaneously through the diagram of the equipment in Fig. 15.23 and on the psychrometric chart of Fig. 15.24. The room air at condition 1 enters and passes through the portion of the drying wheel which has been activated. The air is simultaneously heated and dehumidified along the line 1–2. The air stream then passes through the cooled heat exchanger wheel as shown along the line 2–3, from which it emerges as both dry and cool at point 3, with a low enthalpy. At this point the air stream would be capable of providing useful dehumidification but not of doing sensible cooling of the room. To provide sensible cooling, the air stream is then passed through the humidifier which adds water adiabatically along the line 3–4. The air is now at a temperature of 13°C which is within the range supplied to air-conditioned rooms by conventional mechanical units.

A slope drawn between points 1 and 4 in Fig. 15.24 would show the same sensible heat ratio as would be obtained from conventional equipment. A unique capability of the adsorption cycle is that the sensible heat ratio can be controlled. While not shown in the figures, this is done by by-passing any desired portion of the dehumidified air at point 3 around instead of through the humidifier. For example, if 50 percent of the air is by-passed as indicated at point 5, the sensible heat ratio will be much lower, and a relatively large amount of dehumidification can be done without excessive sensible cooling. The slope of a line from point 5 to point 1 gives this ratio. Thus the comfort under cool but damp conditions can be better controlled with adsorption systems than by conventional systems which do not provide reheat.

The Ambient Air Path.   The ambient air passes first through the humidifier wherein it is adiabatically cooled along line A–B, Figs. 15.23 and 15.24. It then goes through the heat exchange wheel to remove the heat deposited thereon as the room air was cooled from 2–3. In so doing, the ambient air stream is heated from B–C, and thus begins to become hot enough to regenerate the drying wheel.

The solar heating coil then injects additional heat at a somewhat higher temperature between points C and D. The ambient air stream is then split by the baffles in the spacer next to the drying wheel, a portion of it passing through a section heated by a gas burner. This portion is heated to point E, after which it partially mixes with the solar-heated air, with an average temperature of F. The moisture deposited in the drying wheel is driven off by this hot air stream and discharged outdoors. Figure 15.23 shows that a small amount of the room air passing through the hot-test portion of the drying wheel is diverted to the burner compartment and used as combustion air. This airflow precools the rest of the drying wheel to make it more effective and provides a small amount of ventilation for the conditioned spaces. The products of combustion from the gas burner are discharged along with the ambient air stream.

COP of adsorption systems   In determining the COP of the adsorption-type systems the adiabatic portions of the cycle, such as 3–4 and A–B, can be neglected. Knowing the mass flow rate $\dot{m}$ and the enthalpy change between the points 1 and 3, $h_1 - h_3$, the cooling effect is immediately determined.

Likewise, the heat input can be determined by knowledge of the ambient air mass flow and enthalpy changes caused by solar input and fuel burned. The solar input is the enthalpy change from C –D, and the fuel input is the change from D–E. The fuel input must be increased to account for the efficiency of the burner. This is done by dividing the enthalpy difference between F and E by the efficiency, which may be in the 85 to 90 percent range since the flue gases remain in the heat exchange stream. The COP for the conditions shown in Fig. 15.23 should be between 0.6 and 0.75 depending on losses from air leakage between the two air streams.

It is common to consider two values of COP. The solar COP is the cooling effect divided by the solar heat input as determined by the mass flow rate and the enthalpy difference between points C and D. The auxiliary fuel COP is the cooling effect divided by the total fuel input which can be taken as the mass flow rate times the enthalpy change between D and E, then divided by the burner efficiency. These factors vary widely with load and ambient.

In considering the overall seasonal COP one must account for the energy used by the blowers which move the air through their circuits. Air pressure drop is usually several mm of mercury and the kW input needed to operate the blowers is significant.

Other adsorbents and cycles Other adsorption cycles exist with numerous other dessicants. The aim of each of the cycles is to produce an extremely dry air at a temperature approaching room temperature which may be sprayed with water to provide a full-load delivery temperature of air at about 10 to 12°C. Equilibrium data for the various dessicants can be found in the ASHRAE *Handbook of Fundamentals* and in the McGraw-Hill *Chemical Engineer's Handbook*. In practice, equilibrium is seldom approached more closely in economic designs.

Liquid-Phase Systems

Both organic and inorganic solutions have hygroscopic properties which permit some of them to be used for solar-energized cooling. The absorbent solutions are sprayed into contact with air in most such systems. Figure 15.25 shows a block diagram of a typical unit.

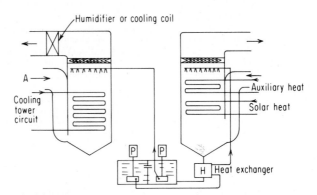

Fig. 15.25   Liquid-absorbent cooling-system block diagram.

In Fig. 15.25 air to be conditioned A is blown through a spray chamber and a coil over which concentrated solution is applied. The coil is cooled by a cooling tower or other means to keep the air and solution at a low enough temperature for dehumidification to take place. The weakened solution drops into one section of the sump from which it is pumped through a heat exchanger H to a second spray chamber through which scavenging air is passed to reconcentrate the solution.

The concentrated solution passes back through the heat exchanger to a separate section of the sump from which it is pumped back into the first chamber for dehumidifying more of the conditioned air stream. The sump may contain a partition to separate the strong and weak solutions, but is usually furnished with a suitably sized hole to equalize the levels within the sump.

Figure 15.26 shows a cycle diagram for one version of the system shown in Fig. 15.25. Return air at condition A is cooled and dehumidified to point B. Thermodynamically this is equivalent to drying to point C and then cooling by means of the cooling tower circuit to point B as shown by the dotted lines. The air is then spray-humidified, cooling it to point D for delivery to the conditioned spaces.

A second cycle is shown in Fig. 15.27. In this cycle point B is chosen to be at the same moisture level as point D. The air exiting at point B from the spray chamber is cooled by mechanical refrigeration to point D by the coil shown in the exit duct in Fig. 15.25.

In both cases the solar energy to concentrate the fluid plus any auxiliary energy needed, as by the coils in the scavenging chamber S, determines total energy input.

In both versions the net cooling effect produced by the adsorption portion of the system is the enthalpy difference between points A and B in the cycle diagrams. The heat input depends on such characteristics of the fluid as equilibrium dew points, and heats of solution. Chemical engineering handbooks can provide such information and enable calculation of the COP, usually between 0.3 and 0.5.

The overall COP for the cycle of Fig. 15.27 must include the energy input to the refrigeration system which performs the dry cooling between points B and D. The electric or other energy is minimized by use of a refrigeration system with as high a COP as possible.

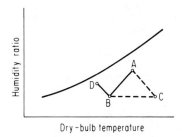

Fig. 15.26   Cycle of liquid-absorbent system with rehumidification.

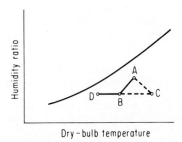

Fig. 15.27   Cycle of liquid-absorbent system with refrigerated second-stage cooling.

SPECIFYING SOLAR COOLING SYSTEMS

For maximum use of solar energy for any cooling system, attention must be given to part-load performance and to performance during periods of lower input temperature than would be the case for conventionally energized systems. A vendor of equipment should be required to provide capacity and COP throughout the whole range of intended operating temperature from thermal storage or direct from the collector array as the case may be. The specification should clearly define the relationship between load, expected heat rejection temperature for typical loads throughout the range from the changeover point to full load for both sunny and cloudy periods.

It is suggested that several items of importance to the performance of the solar cooling system be specified in separate sections of a building's mechanical specifications to avoid any possibility of "buck passing" or of modifying the relationships between components of a well-designed system. These components are covered separately below.

Specifying the Cooling Tower

A complete map of cooling tower performance similar to that shown in Fig. 15.3 should be obtained from possible cooling tower manufacturers for analysis. The important factors can be incorporated in the specification in the section on cooling towers. The required capacity obtainable from one or more known cooling towers and the corresponding approach for the wet-bulb temperatures at which capacity will be needed should be specified.

If modulation of the cooling tower fan or blower is to be included as a means of energy savings, the specified values of capacity and approach under the pertinent conditions should also be specified. The vendor should state the degree to which all values are the result of tests on the model bid, or on the line of which it is a part. Other specification items should be those normally employed in specifying cooling towers.

Specifying the Absorption Cooling Equipment

For residential systems there is little that can be done except to require that the vendor supply models of his line which are specifically designed for energization from solar systems. The unit should include a mechanical circulating pump for the solution flow between the absorber and the generator. It should be designed for useful operation at energizing temperatures at least as low as 80°C with a cooling tower water temperature of 25°C.

Commercial absorption equipment should be specified to be capable of operation throughout all areas of known performance (see Figs. 15.7 and 15.9). The limits of operation for each of several cooling water temperatures which cover the expected operating range should be provided by the vendor. The COP for each of these conditions should be given, as well as the relationship with capacity below the limit for the cooling temperatures of interest.

Commercial units should provide automatic de-crystallization, and have adequate solution control for the range of capacities and operating conditions expected. The unit should be capable of operation without installing a bypass or other control in the cooling tower water circuit.

The vendor should state the degree to which the performance information he provides for meeting the specifications is a result of actual test at the conditions reported. The ability to service the equipment without removing or breaking into the refrigerant charge of the unit is important, and this should extend to replacement of pump motors.

Since absorption units have special design, production, and application considerations, the specifier may consider requiring that any model submitted shall have been in regular production for some minimum period, such as 5 yr.

Specifying Controls

The proper control sequence for operating an absorption system is the key to economical performance and to minimizing the required area of the collector array. Hence it should be accurately specified in all modes. The control system should monitor the cooling tower water temperature entering the unit and program the maximum energizing temperature according to performance characteristics of the unit submitted in the manner shown in Fig. 15.7. It is desirable, but not imperative, that the energizing temperature be further modulated by the chilled water temperature at the maximum energizing temperature. Otherwise the limit for the highest chilled water temperature expected in normal use should be the determinant.

In addition the control system should sense the chilled water temperature, and as it falls within the throttling range of the controller, the energizing temperature should be further reduced just sufficiently to maintain control of chilled water to maximize efficiency.

Storage controls   Controls for storage at single temperature may be similar to those for any solar system. When dual storage is used, the controls should be specified to produce the sequence shown on pages 15-16–15-18.

Control of types of systems   The controls for vapor-compression and absorption systems depend on the type of equipment used and its particular operating characteristics. It is usually important to relate the energizing temperature to performance characteristics of each type of equipment as it is with absorption systems. These requirements should be included in the specifications.

Specifying the Solar Collectors

Uniform testing of collectors for rating is needed if the performance of several types or makes are to be compared in a meaningful way. During 1977 the ASHRAE-ANSI Standard 93–77 was issued and it is suggested that the specification section on collectors require the complete rating test as required in 93–77. Most types of collectors including flat-plate designs, evacuated tubes, or concentrating types can be tested by the methods of this standard.

The specifications should require the ratings to be provided on the basis of the *gross* area occupied by the collector. In the case of flat-plate collectors the gross area is the overall area, not just the aperture area. In the case of evacuated-tube collectors, several tubes are usually mounted as an assembly with spaces between them, and ratings should be on the basis of the gross area of the assembly. Concentrating collectors need clearance to track the sun, and the gross area should include the swept area, or the area of assembly of several collectors if shipped pre-assembled.

It may be desirable to specify the cover materials if flat-plate collectors are to be used. Outer-glass covers are usually of tempered glass with 5-mm nominal thickness. Other materials such as glass fiber-filled plastic, certain plastic sheets, or teflon are in use but may not be suitable for high collector temperatures encountered in solar cooling systems or during stagnation. The vendor should be required to state the reduction in transmissivity of the cover material over a period of 5 or 10 yrs. If the collectors are to be allowed to stagnate, the vendor should provide assurance that the cover material will not be damaged thereby and that it will not stretch or shrink significantly during stagnation, or thereafter.

In order to estimate daylong performance of collectors it is necessary to know their rating at at least one incident angle other than normal, say 45°. The vendor should be able to provide this rating since it is a part of the standard 93–77 test. To provide daylong performance on partially cloudy days, the performance of the collector under conditions of 45 to 55 percent diffuse irradiation should be requested.

If silicone or certain other organic liquids are to be used in the collector circuit, soldered or SAE flare connections should be specified, since screwed fittings and hose connections are very difficult to maintain leak-free for such fluids.

As of 1979 no industrywide certification standards have been initiated, although a program which will give a performance certification is being developed. When available, the vendor should be required to provide his ratings at the standard rating points specified in the standard.

Until a certification program is available a certifying test of the collector at any of the several qualified testing laboratories for solar collectors may be required for any collector before its acceptance.

Warranty Guarantee

Components and subsystems provided by a manufacturer are usually warranted against defects in materials and workmanship for period of one year. The warranty period may run from the date of delivery, or the date of delivery plus a fixed period to allow for installation. Warranties of solar equipment thus are similar to those of other cooling equipment.

Guarantees of worksmanship and conformance of the installation to the plans are the responsibility of the contractor for his specific portions of the work.

MAINTENANCE OF SOLAR COOLING SYSTEMS

Solar cooling equipment requires the same type of maintenance as does the same equipment in nonsolar installations. Some considerations for different components are worth noting.

Absorption Cooling Equipment

The lithium-bromide–water units operate under a vacuum and any leakage will be into the equipment. Large equipment is never hermetically tight, and hence provision is made for automatic purging of air from the unit. The purge unit should be maintained according to the manufacturer's instructions.

Lithium-bromide solutions are inherently corrosive and provision is made in all units for inhibitors to prevent damage to the equipment or deterioration of the solutions. This is a specialized area, and should probably be handled by the manufacturer's service representative. Larger systems may well consider a service contract which provides all needed inspection and actions.

Rankine Equipment

Most Rankine equipment is piped in the same manner as refrigerant piping and should need little maintenance. Lubrication in centrifugal equipment is provided by the internal sumps and pumps which are a part of the design. Any leaks should be noted and immediately repaired. Both the lubricant and the working fluid should be restored to normal amounts if leaks have occurred. Cooling systems for seals, if provided, must be maintained.

Cooling Towers

Cooling tower maintenance is important to the longtime performance of cooling equipment. An adequate blowdown is essential in any location, but the amount should be determined according to known local practice due to the wide variability in salt content of make up water in various locations. Dirt and debris should be removed whenever it accumulates. A definite schedule is recommended with special attention after any major storms.

The pH of the cooling tower water should be maintained at the levels specified by local water treatment specialists. Water treatment to minimize scale formation is usually applied. Keeping the cooling tower water in good condition helps assure the long operating life of the cooling equipment and prevents plugged tubes in condensers, or in the absorbers of absorption equipment.

REFERENCES

1. Newton, A. B., "Variable Speed Drives for Centrifugal Compressors," *International Institute of Refrigeration, Proceedings of XII International Congress of Refrigeration*, Washington, DC, 1971.
2. Newton, A. B., "Optimization of Cooling Systems Energized by Solar Heat," *International Institute of Refrigeration, Proceedings Joint Meeting Commissions* C_2, D_1, D_2, D_3 & E_1, Melbourne, Australia, September 1976.
3. ASHRAE, *Handbook of Fundamentals*, American Society of Heating, Refrigerating and Air-Conditioning Engineers, New York, 1976.
4. ASHRAE, *Handbook of Systems*, American Society of Heating, Refrigerating and Air-Conditioning Engineers, New York, 1976.
5. ASHRAE, *Computer Codes for Building Energy Analysis* (ASHRAE GRP153), American Society of Heating, Refrigerating and Air-Conditioning Engineers, New York, 1975.
6. Newton, A. B., "The Meaning and Use of Consensus Standards in Rating and Certification of Solar Collectors," *ASHRAE Journal*, November 1978.
7. ASHRAE, *Methods of Testing to Determine the Thermal Performance of Solar Collectors*, American Society of Heating, Refrigerating and Air-Conditioning Engineers, New York, 1977.

APPENDIX: PRESSURE-ENTHALPY DIAGRAMS FOR VARIOUS FLUIDS

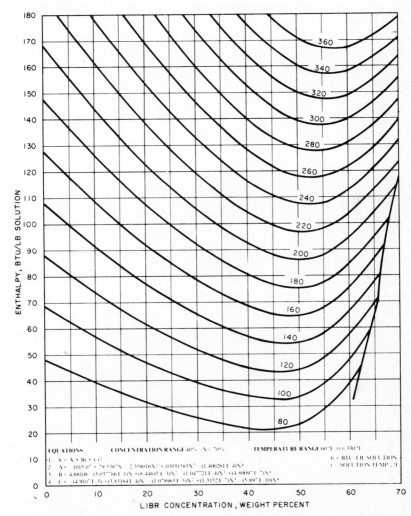

Fig. 15.A1 Enthalpy-concentration diagram for lithium-bromide–water combination. (*Courtesy ASHRAE.*)

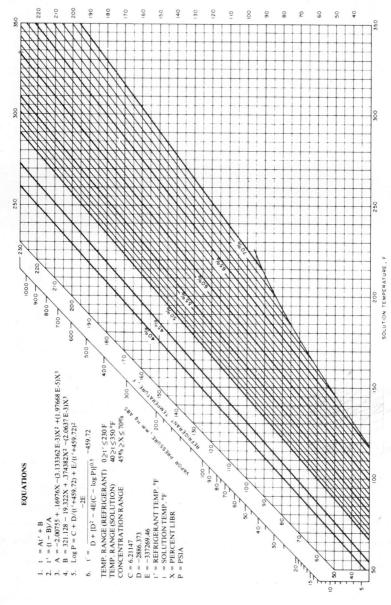

EQUATIONS

1. $t = At' + B$
2. $t' = (t - B)/A$
3. $A = -2.00755 + .16976X - (3.133362 E-3)X^2 + (1.97668 E-5)X^3$
4. $B = 321.128 - 19.322X + .374382X^2 - (2.0637 E-3)X^3$
5. $\text{Log } P = C + D/(t'+459.72) + E/(t'+459.72)^2$

6. $t' = \dfrac{-2E}{D + [D^2 - 4E(C - \log P)]^{0.5}} - 459.72$

TEMP. RANGE (REFRIGERANT)	$0 \geq t' \leq 230\text{ F}$
TEMP. RANGE (SOLUTION)	$40 \geq t \leq 350\,^\circ\text{F}$
CONCENTRATION RANGE	$45\% \geq X \leq 70\%$

$C = 6.21147$
$D = -2886.373$
$E = -337269.46$
$t' = $ REFRIGERANT TEMP. $^\circ$F
$t = $ SOLUTION TEMP. $^\circ$F
$X = $ PERCENT LIBR
$P = $ PSIA

Fig. 15.A2 Equilibrium chart for aqueous lithium-bromide solutions. (*Courtesy Carrier Corporation.*)

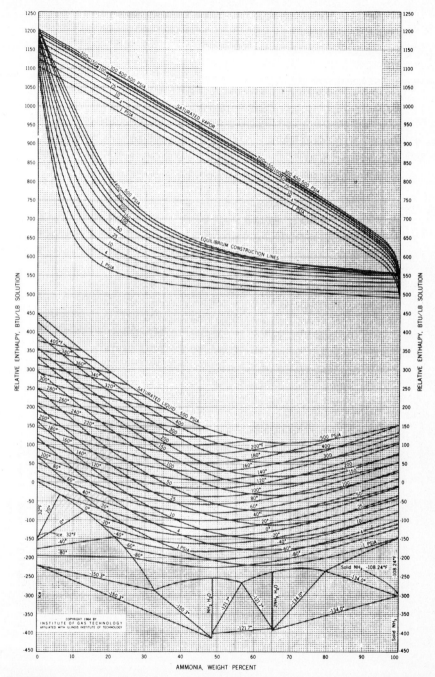

Fig. 15.A3 Enthalpy-concentration diagram for ammonia-water mixture. (Copyright 1964, Institute of Gas Technology and the American Gas Association. Used with permission.)

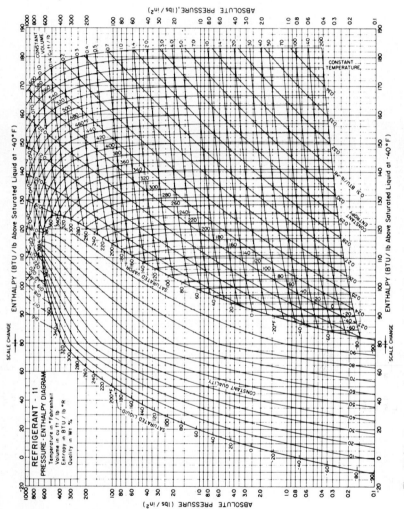

Fig. 15.A4 Pressure-enthalpy diagram for R-11. (Copyright 1965, E. I. du Pont de Nemours & Co., Inc. Used with permission.)

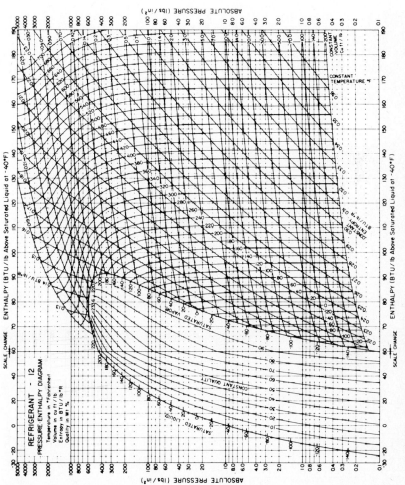

Fig. 15.A5 Pressure-enthalpy diagram for R-12. (Copyright 1955 and 1956, E.I. du Pont de Nemours & Co., Inc. Used with permission.)

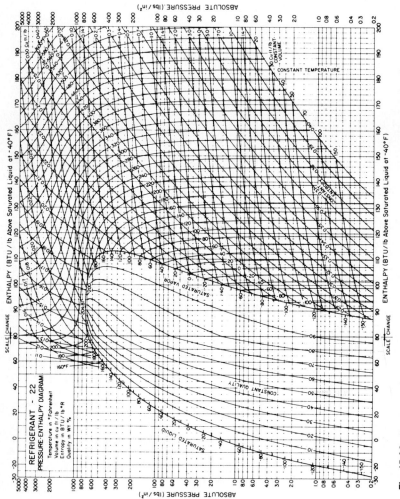

Fig. 15.A6 Pressure-enthalpy diagram for R-22. (Copyright 1964, E.I. du Pont de Nemours & Co., Inc. Used with permission.)

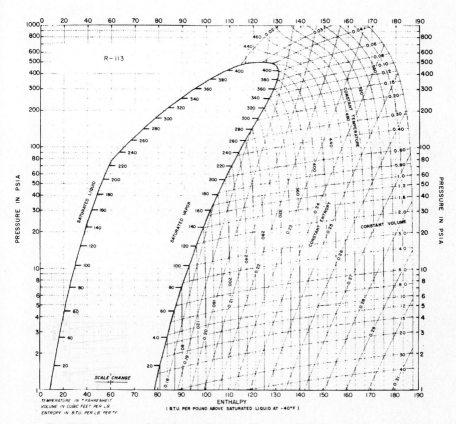

Fig. 15.A7 Pressure-enthalpy diagram for R-113. (*Courtesy Allied Chemical Corporation.*)

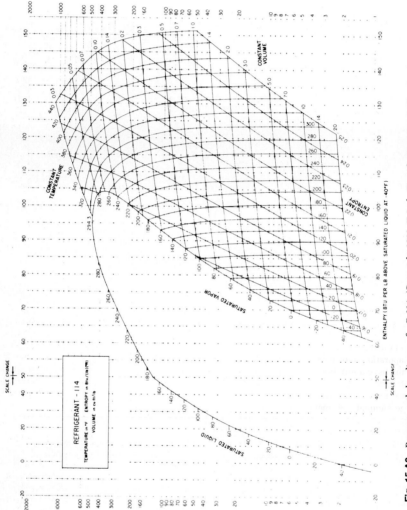

Fig. 15.A8 Pressure–enthalpy diagram for R-114. (Copyright 1966, E.I. du Pont de Nemours & Co., Inc. Used with permission.)

Chapter 16

Passive Solar Energy Systems for Buildings

J. DOUGLAS BALCOMB

Los Alamos Scientific Laboratory of the University of California, Los Alamos, NM

INTRODUCTION

Passive solar systems are ones which work with nature in a natural way to provide heat for the building in the winter and/or cooling for the building in the summer without the requirement for external energy to cause the system to operate. Most passive solar heating designs, for example, simply use south-facing glass in the building as the solar collection element and the mass of the building as the thermal storage element. Natural cooling systems use infrared radiation or evaporation in order to achieve heat rejection from the building, generally at night. Passive systems have a number of tremendous advantages: (1) Their operation is natural and normally maintenance-free, (2) the principles are simple and easily understood, (3) the cost of the system may be lower than that of active systems if proper soil conditions exist to support massive storage elements, (4) many passive designs are aesthetically more attractive than ordinary solar collectors and would be more acceptable to many buyers, and (5) the system will continue to operate even in the event of a utility system failure.

The principal disadvantage of passive designs is that they usually require integration of the solar collection and storage function into the architecture of the building. Thus the solar energy system is an integral part of the building. This integration of functions is not the contemporary approach to architectural design, and a major change in building design philosophy must take place before passive techniques will have a major influence on building energy consumption.

The performance of a passive system is generally higher than an engineer's intuition might normally lead him to believe, and passive approaches do represent a viable design alternative to active solar systems throughout the range of climates to be found within the 48 contiguous United States. Without movable insulation placed over the glazing to reduce heat losses at night, the performance of a passive system is generally less than that of a well-designed active system. However, with the use of movable insulation, a well-designed passive system has a net thermal performance which is competitive with that of the best active solar heating systems (for the same glazed aperture area).

Another current misconception is that the occupant of a passive solar building must accept a lower comfort standard than might be achieved in an active system. Although this has been the case in some designs, it is not an intrinsic property of passive systems, and, in fact, the addition of mass in the building could lead to an increased comfort standard compared to a low-mass building with a cycling active system for air distribution.

Passive Systems Defined

For the purposes of this handbook a passive system can be defined simply as one in which the thermal energy flow is by natural means. This definition precludes the use of either a pump to force a heat transfer liquid through a loop, as in the case of solar water-heating systems, or the use of a fan to blow air through a loop, as in the case of solar air-heating systems. The elements which makes these systems active rather than passive are the pump or the fan and the controller which require external energy to drive the unit. In a passive system the thermal energy flows from one location to another entirely by natural means. This is generally either by conduction through material, convection from a surface to air or liquid, or radiation heat flow between surfaces. Passive systems to be discussed fall into two broad categories: passive solar heating systems, and natural cooling systems.

Passive solar heating   Solar energy incident on a building is utilized to help make up heat losses from the building in order to maintain the building's internal temperature higher than the ambient. There is confusion on this issue since many people regard the mere use of insulation in a building as a passive system. Clearly this is so. However, it is not a passive solar system; the use of insulation to retain heat within a building comes under the category of energy conservation. Energy conservation constitutes a key element of any solar energy system design, be it an active or passive system; however, this is not the major subject of the present discussion of passive solar heating (see Chap. 29).

It is assumed in this chapter that one has designed a thermally tight envelope for the building. The objective of the passive solar heating system is to provide external energy from the sun to make up the thermal losses through that envelope into the environment. It is exactly analogous to the design objective of an active solar heating system. In many passive solar heating designs the passive solar collection element constitutes part of the thermal envelope of the building. The nighttime losses through the passive element may exceed those of a well-insulated opaque wall which might have been used in its place; therefore, the daytime solar gains must more than offset these losses in order for the system to be viable.

Natural cooling   A natural cooling system is not, strictly speaking, a solar energy system at all. The effect of the design is to encourage heat energy losses from the building so as to counteract, if anything, the effect of solar gains (and other internal heat sources) which are generally minimized in the design. The inclusion of these systems in this chapter comes about because many of the same design methodologies are applied as for passive solar heating, i.e., the use of mass in the building to store daytime heat. In addition, a well-designed building will achieve both objectives, providing passive solar heating in the winter and natural cooling in the summer. Thus the two concepts should not be considered separately in the design.

Figures of Merit for Passive Systems

Efficiency   The first law efficiency of any solar heating system, active or passive, is defined as the ratio of the total solar energy incident on the solar collection surface to the net useful heat energy supplied to the building. The efficiency of a natural cooling system is correspondingly defined as the ratio of the net energy removed from the building to the total energy lost from the radiative cooling element of the building. Over an extended time period, the efficiency of a natural cooling system must be unity.

Solar heating contribution   This is perhaps the most useful and least ambiguous figure of merit. It is simply the *net* energy contributed to the building by the passive solar element. This figure of merit is needed to compute the cost effectiveness of the system.

Solar heating fraction   There are at least three ways to define solar heating fraction. For the purpose of this chapter the most pessimistic definition is chosen as follows. The solar heating fraction is the ratio of net energy provided to the building by the passive solar elements to the total heat required. Thus the solar heating fraction is equal to one minus the ratio of the auxiliary heating required to the total load of the building, *excluding the passive solar element from the total load calculation.*

The justification of this definition is that the net energy flow through a "normal" south wall of a building is usually quite small. This is because solar gains through the normal fraction of glazing which would be expected to be found on a south wall are frequently enough to compensate for the total losses from the south wall. In order to avoid the problems of defining "normal" in this context, and also to avoid the complications of having to calculate what the net energy flow of such a "normal" wall might be, the issue is bypassed altogether with this rather more simple definition.

Historical Perspective

Passive designs have evolved in several early cultures which could not rely on plentiful fossil fuels, equipment, and automatic controls for their thermal comfort systems.

Early Anasazi dwellings   Examples of passive solar heating designs can be found in the structures of the Southwest American Indians constructed during the period 1100–1400 A.D.

Passive solar benefits were derived from cliff dwellers who built under overhanging cliffs along canyons in the Southwest, such as the structures in Grand Gulch, Canyon de Chelly, Batatican, and Mesa Verde. The low winter sun penetrated these overhanging caves and heated the mass of the building walls as well as the surrounding rock. This undoubtedly increased the temperature of the dwellings in the winter as well as affording protection from the north winds. In the same way, the high summer sun rays would be shaded by the overhang of the cliff, and thus the structure would not be heated during the day. Energy would be lost by natural radiation at night, providing a structure with a cooler-than-ambient average temperature in summer.

It is not clear that these Indians made a selection of these sites based solely on thermal considerations, since cliffs in the Southwest are generally eroded into overhanging caves only on southern faces because of the nature of the freeze-thaw erosion cycle.

A few, although by no means all, Pueblo Indian structures were also designed so as to benefit from passive solar gains in the winter. The most notable of these is the large structure at Pueblo Bonito in Chaco Canyon, NM. This semicircular apartment structure is four stories high on the north with a very thick wall and is cupped to face the south so as to provide some element of natural solar collection. Although it is evident that these Indians benefited thermally from these designs, it is not obvious that this was a motivating factor in the design, since, according to some theories, many structures were designed as astronomical observatories.

The performance of many of these early structures would be very low compared to what is possible now utilizing the main two ingredients which make passive solar heating very effective: glazing materials in large sizes, and thermal insulation to effectively retain the solar heat which has been collected. A key element, common between early structures and present day, is the use of thermal mass, frequently in adobe earth, brick, or rock, for thermal storage.

Eastern cooling systems   Several concepts utilizing natural cooling have evolved within the countries of the Middle East where the climate is hot and arid.

Before the advent of mechanical refrigeration, ice was made in Iran by utilizing night-sky radiation. A large wall was erected to provide permanent shade to a pond north of the wall. The pond would be flooded, a few inches at a time, and would freeze (even in summer) due to the fact that the energy losses by sky radiation and evaporation would exceed the gains due to air convection and conduction from below. Such a concept could only work in a very clear area and would work much better at high elevation than low.

A very sophisticated technique of natural building cooling using air towers has evolved and is still in use in Iran. Cool night air is drawn through the building by the effect of a tall chimney which terminates in a tower on top of the building. The walls of the chimney are massive and are warmer than the night air temperatures and thus heat the air, causing it to rise. During the day the natural air flow is reversed in direction—hot outside air falls through the chimney and, being cooled by the chimney walls, discharges into the rooms. The cooling effect is frequently augmented by evaporative cooling from ground water seeping through the chimney walls.*

TYPES OF PASSIVE SYSTEMS

For our purposes it is useful to separate the various types of passive systems according to five generic categories.

Direct Gain

In a direct-gain structure solar energy is transmitted through south-facing glazing which is frequently vertically placed. This is shown schematically in Fig. 16.1. This solar energy provides for the daytime heat requirements of the building, and the excess is stored in the form of

*For more details on these passive cooling ideas, see M. H. Bahadori, Passive Cooling Systems in Iranian Architecture, *Scientific American*, vol 238, p. 144 (1978).

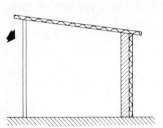

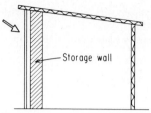

Fig. 16.1 Direct gain approach to passive solar heating. The sun shines in through south glazing and thermal storage is in massive floors and interior walls.

Fig. 16.2 Thermal storage wall approach to passive solar heating. The sun shining through south glazing strikes a massive wall which absorbs and stores the heat, transferring the heat by conduction through the wall to the building.

sensible heat in massive elements of the building, such as a concrete, brick, or tile floor and masonry walls, partitions, or roof. The building is a live-in solar-collector/thermal-storage unit. The advantage of this approach is its simplicity. Several disadvantages are deterioration of fabrics and other materials by the direct solar radiation, moderately large day-night temperature fluctuations, lack of privacy, and glare. Despite these drawbacks it is probably the most popular passive solar design approach.

Thermal-Storage Wall

In the thermal-storage wall approach, the material which serves as thermal storage in the building is placed immediately behind the south-facing glazing, as shown in Fig. 16.2. The sun shines directly on thermal storage, heating it during the day. Thermal energy is transported through the storage material and subsequently transferred to the building by convection and radiation.

Masonry walls The thermal-storage wall concept has been popularized by the Trombe-Michel houses at the village of Odeillo-Font-Romeu in the Pyrenees Mountains of Southern France. These small residences utilize 60-cm (2-ft) thermal storage walls of poured concrete. (The French concept includes a convective loop. Small vents are cut at the bottom and top of the wall to provide for thermocirculation of air from the floor of the building up through the space between the glazing and the wall, and out through vents at the top of the wall, returning the heated air to the ceiling space. These vents are not an intrinsic part of the thermal-storage wall concept and may or may not be beneficial in any given application.)

Water walls In several thermal-storage wall installations the wall material used for thermal storage is water contained in drums, cans, bottles, tubes, or tanks.

The advantage of the thermal-storage wall approach is in shielding the building environment from extremes. Temperature fluctuations are reduced and the rooms behind the wall are protected from the glare and material damage associated with direct sun. Mixtures of direct gain and thermal-storage wall are possible by blocking only a portion of the south glazing. For example, windows can be cut through the thermal-storage wall or alternating sections of thermal-storage wall and window can be used.

Solar Greenhouse

A combination of the thermal-storage wall and direct gain is the solar greenhouse shown schematically in Fig. 16.3. The greenhouse structure is the southernmost element of the building

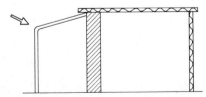

Fig. 16.3 Solar greenhouse approach to passive solar heating. The building consists of two distinct thermal zones, a south direct-gain room (which can serve as a greenhouse) and a north thermally buffered zone. The two zones are separated by a massive thermal storage wall.

and is a direct-gain room. The greenhouse is separated from the living part of the house by a thermal-storage wall. From the standpoint of the main living area of the house, the system performs much like a thermal-storage wall. Temperature fluctuations in the house are small, and this area is protected from extremes of solar radiation, glare, and temperature fluctuation. By contrast the greenhouse room experiences a large fluctuation in temperature and large radiation levels both of which are appropriate for a greenhouse space. Although the room need not necessarily be used as a growing greenhouse, it is an especially good environment for that application.

Convective Loop

This design approach, as illustrated in Fig. 16.4, is perhaps most akin to the normal solar energy system utilizing separate solar collectors and separate thermal storage. By placing the collector below the thermal storage element and providing sufficiently large areas for flow, a natural circulation loop is set up transferring the hot fluid from the collector to the thermal storage and return. Both water and air designs have been built. The classic thermosiphon water heater is of this type. (See Chap. 11 for details.) In addition, large thermosiphon air systems have been built for home heating utilizing rock as the thermal storage.

Thermal Storage Roof

In this concept the thermal storage and solar collection function are both located in the roof or ceiling of the building (see Fig. 16.5). This concept has been popularized by the Hay-Yellott test building in Phoenix and the Skytherm house in Atascadero, CA. In these examples the thermal storage is in plastic bags filled with water located on a flat steel roof deck of the building. In this concept some form of movable insulation is absolutely essential since the sun angles are altogether wrong for optimum solar gain in winter or solar rejection in summer. However, with the use of movable insulation, the thermal balance can be made positive in the winter in a mild climate and negative in the summer in a relatively dry climate. Thus the system is most suitable for a combined passive solar heating and natural cooling system in appropriate climate zones. The movable insulation is used to cover the building roof during the night in winter and rolled back to expose the water bags during a winter day. Thus heat gains from the day are retained through to the night. Correspondingly the insulation can be left opened at night in the summer and closed during the daytime. Heat from the building is stored in the water during the day and lost, principally by radiation, at night.

DESIGN CONSIDERATIONS

Basic Design Elements

Thermal insulation of the building As in any solar building design it is necessary to minimize the thermal load by proper insulation and air change control (see Chap. 29).

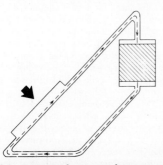

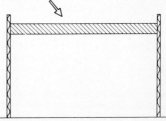

Fig. 16.4 Convective loop approach to passive solar heating. A heated fluid rises from the solar collector depositing its heat in a thermal storage volume located above the collector and returns by natural circulation to the collector inlet. Both air- and liquid-based systems have been built.

Fig. 16.5 Thermal storage roof approach to passive solar heating. The thermal storage is located on the building roof and is heated directly by the sun from above. When used in conjunction with movable insulation, this concept can provide for both winter heating and summer cooling, by use of night radiation and also convective and possibly evaporative cooling from the roof. When the thermal storage is in the form of water, this concept is referred to as a roof pond.

Solar energy collection   As contrasted with active solar heating systems, where it is generally possible to dump waste heat, in a passive design it is necessary to obtain a good relation between the amount of solar energy collected and the thermal requirement of the building. As will be shown, this can be achieved through vertical south-facing glazing.

Thermal storage   Since the amount of energy collected during the day is generally far in excess of the daytime thermal requirements of the building, it is necessary to provide for heat storage in order to carry this excess energy over for nighttime or cloudy-day use. (See Chap. 6.)

Design Methodology for Direct-Gain Buildings

The design approach to be developed here for solar heating is to base the design on a clear winter day. The design may provide for up to 100 percent solar heating on such a day. The principle advantage of such an approach is that it is relatively amenable to hand calculations, given a few assumptions. If the net clear-day solar gains exceed the net thermal losses over a 24-h period, then one of three situations will result: (1) the building will overheat, (2) extra remote storage must be provided, or (3) the extra energy must be dumped, for example, by opening a window.

Having designed for a clear winter day, then the question will arise: What is the performance to be expected for a normal winter pattern of variable weather? This is a rather more complex problem generally needing a complex analysis. Unfortunately this requires both more weather data than is generally available to most designers and a computer to make the required analysis tractable. A reasonable approach is to rely on the results of such a computer analysis for a nearby location and a similar design as a basis for estimating the annual performance. The results of many such calculations are given in subsequent sections of this chapter.

Solar Gains

Sun angle considerations.   The single most important design consideration to be factored into passive solar heating systems is planning to take best advantage of the location of the sun in the sky at different times of year and different times of day. Fortunately, glazing placed vertically in a due-south orientation (in the Northern Hemisphere) is tremendously effective for maximizing winter gains and minimizing summer gains. This general effect is illustrated in the diagram of Fig. 16.6, showing the daily sun path across the celestial dome on both the winter and the summer solstice. The diagram is drawn for a latitude of 40°N. As can be seen, the sun shines predominantly on the south vertical face of the building in the winter but scarcely shines at all on that same wall during the summer. The location of the sun in the sky can be calculated relatively easily from equations and charts contained in Chap. 2.

A sun chart for 36°N is shown in Fig. 16.7 and illustrates the point to be made. Similar charts for other latitudes can be found in the Appendix to this chapter.

Solar penetration through glazing.   Most glazing is highly transmissive for sun rays which strike the surface within a 60° angle to the perpendicular. Beyond this angle the transmission falls off markedly and most of the energy is reflected from the glazing surface depending on the type of glazing (see Chaps. 3, 4, and 5). This effect enhances the winter-summer selective characteristic of south-facing, vertical glazing. In the summer the solar incidence angle is very large, and thus most of the energy is reflected, whereas in the winter the angle is relatively small and most of the energy will be transmitted.

Clear winter day solar penetration values.   The foregoing considerations lead to curves such as that shown in Fig. 16.8 for the total amount of clear-day solar energy penetrating various double-glazed surfaces of a building. Note that the solar energy transmitted through south-facing, double glazing is generally in phase with the thermal loss of the window and far greater in magnitude. The loss curve is drawn for Los Alamos, NM. This is a cold, clear climate at 36°N latitude with a total heating load of approximately 6400°F·day (~3560°C·day). Thus, even under these severe conditions one can see that there is abundant excess energy through south-facing double glazing on a clear day for use by the rest of the building. Tabulated values of solar gains through single glazing are to be found in the ASHRAE *Handbook of Fundamentals*, 1977. Note that the solar energy incident on other surfaces of the building is totally out of phase with the requirements for heating. Glazing on the east or west face of the building contributes little to the winter heating of the building but creates a major problem in terms of overheating during the summer months.

A normal approach used to obtain monthly averages of the solar radiation, accounting for actual cloud cover conditions, is to multiply curves such as those shown in Fig. 16.8 by a "cloudiness factor." This is a dubious approach because the effect of some clouds may be to

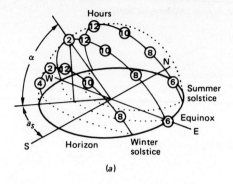

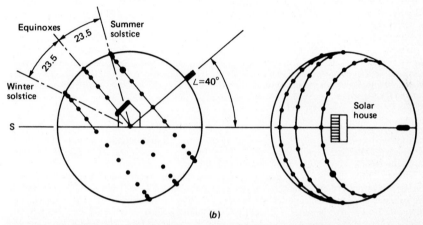

Fig. 16.6 Solar paths in winter and summer at 40° north latitude. The dotted lines show the sun path on the winter and summer solstice. (Reprinted from F. Kreith and J. F. Kreider, "Principles of Solar Engineering," Hemisphere, Washington, DC, 1978. Used with permission.)

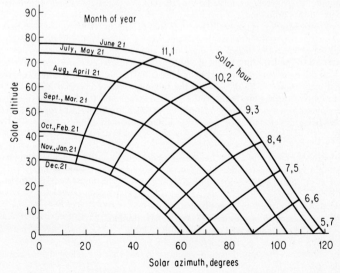

Fig. 16.7 Sun chart for 36° north latitude. Solar altitude and azimuth are shown for different values of suntime and for different times of the year.

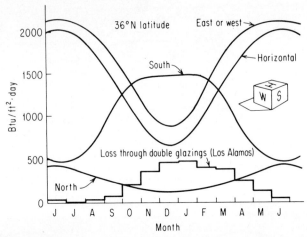

Fig. 16.8 Clear-day solar gains through double glazing for various orientations for 36° north latitude. Solar gains are shown for different months of the year. The loss curve shown applies for a 6400°F-day climate (Los Alamos, NM) for double glazing in any orientation. The net gain would be the difference between the total gain and the loss curve.

drastically decrease the direct component of the sunlight and actually increase the diffuse component. The cloud cover factor does, however, give some quasiquantitative information about the obscuration of the sun in different locations.

Shading. Note that the solar gains through south glazing illustrated in Fig. 16.8 are small during the summer months but clearly not negligible. A small overhang to shade the wall in the summer and early fall months can be very effective in reducing these gains which might otherwise lead to overheating. A difficulty with shading occurs since, even with south-facing glazing, there is some phase difference between the building load (which remains relatively high during the months of February, March, and April but is low during the months of August, September, and October) and the solar gains which are symmetric around the summer solstice (June 22). Many passive solar buildings have been observed to have their largest heating loads in the spring and their greatest overheating in the fall. An adjustable shade or overhang could be used to avoid this problem.

The use of some shading is generally recommended. The exact amount will depend on the detailed balance between the desire to maximize winter gains and minimize solar gains in the particular climate where the design is being implemented.

Thermal load of the building without solar gains   A detailed discussion of the intricacies of thermal load estimation in buildings is beyond the scope of this handbook. However, reasonable thermal load estimation can be done using the techniques developed by ASHRAE. An over-simplified example is used here to illustrate the basic concepts. If a better thermal load value can be determined by a more accurate technique, so much the better, but for the simple design methodology being developed here, a more accurate treatment is not warranted.

Example 16.1

An example using the clear-winter-day design methodology is summarized below. For simplicity, assume a simple, frame construction building with the following specifications:
- Floor area: 1320 ft^2; 22 ft NS × 60 ft EW
- Walls: 2 × 6 frame, R19 fiberglass batts; celotex; 8-ft high
- Shed roof over attic: 8-ft loose-fill insulation
- Windows: 5 percent of wall area E, W, and N: double
          50 percent of wall area on south: double
- Floor: Slab on grade, perimeter insulation
- After adding up the appropriate R-values, suppose that the following are obtained:
- Walls: R = 22 (U = 0.045)
- Ceiling: R = 31 (U = 0.032)

Then the overall products of $U \times A$ are as follows:
- Opaque walls 1030 ft^2 $\times$ 0.045 = 46
- Ceiling 1320 ft^2 $\times$ 0.032   = 42
- Windows 282 ft^2 $\times$ 0.55   = 155
- Perimeter 164 ft $\times$ 0.17   = 28
- Total Conductance (Btu/h· °F) = 271   (143 W/°C)

Infiltration of $^1/_2$ air change per hour adds an effective 91 Btu/h · °F (48 W/°C) for a total effective conductance of 362 Btu/h·°F. (191 W/°C). This means that the heat required to make up for conduction and infiltration losses is estimated to be 362 Btu/h for each degree F of temperature difference between the inside and outside. The actual requirement may be greater or less than this depending on quality of materials used, details of construction, care in construction, and the living habits of the occupants.

Thus the inside temperature could be maintained at 70°F for an outside temperature of 0°F by a heat source of 25,300 Btu/h or 7.4 kW. Note that this is quite a lossy building compared to current energy conservation practice, primarily because of the south glazing which makes up 43 percent of the total load.

Requirement for thermal storage. The solar energy transmitted through the south glazing of a passive solar-heated building will greatly exceed the thermal losses on a clear winter day. In our example problem the total energy transmitted through 240 ft^2 (22-m^2) of south glazing is approximately 1500 Btu/ft^2 (17,000 kJ/m^2) on a clear winter day for a total of 360,000 Btu (380 MJ). If we simply divide this total energy transmission by the total thermal load of the building, we can determine the average 24-h temperature difference which can be maintained between the inside and outside temperature. This yields:

$$360,000/(24 \times 362) = 41°F \ (23°C)$$

Thus the inside temperature could be maintained at 70°F if the outside temperature averaged 28°F over the 24-h period.

Let us assume that the inside temperature is somewhat higher during the daytime; for example, 75°F. Let us also assume that the outside temperature is lower; for example, 20°F. Therefore the average energy lost by the house over the eight daytime hours can be calculated as follows:

$$(75-20) \times 362 \times 8 = 160,000 \text{ Btu (169 MJ)}$$

Thus less than half of the total energy collected during the day is lost during the day. If a large and effective thermal storage mass is not included within the envelope of the building it will simply overheat to the point where the energy losses balance the net incoming energy. The occupants would take remedial action by opening the windows or, if this were not done, temperatures well over 100°F could be anticipated.

In a properly designed passive solar-heated building the 200,000 Btu of excess energy will be stored in the sensible heat of building materials. The mass of material required depends on a thermal storage capacity, or specific heat c_p, and on the temperature change of the material. Thus:

$$\text{Stored solar energy} = \Sigma M_s c_p \, \Delta T$$

where M_s = Mass of storage material
c_p = Specific heat of storage material
ΔT = Temperature swing of storage material

There are two difficult aspects to the use of this simple equation. The first problem is to determine what value to predict for ΔT, the temperature swing of storage material. The second problem is to determine how much of the material of the building to count in determining the mass, M_s. These problems are really related since the actual problem is to determine how heat penetrates the material and is then returned to the surface at a later time. Thus the depth of material used to calculate M in the above equation must not be greater than the equivalent dynamic thickness based on a 24-h period. A method for estimating this thickness and the associated effective ΔT will be presented in the following discussion.

Types of thermal storage. The most effective heat storage is in walls or floors which are directly or indirectly illuminated by the sun. Calculation of this storage is complicated because of the changing angles of incident sunlight and the determination of internal reflections and absorptances. It is good design practice to make the massive storage walls a dark color and other materials in a sunny room light in color so that the sunlight is reflected and eventually absorbed in the storage walls.

Interior walls which are not irradiated by direct or indirect sun can also store heat. Usually the air in the building is warmer during the day than at night, owing to solar heat transferred from lightweight materials. In order for this heat to be stored in the walls of the room, it must first be transferred to the surface of the wall and then conducted into the wall. The process at night is the reverse—conduction out of the wall to the surface and then transmission back into the room by convection and radiation.

Many questions arise on considering the design of interior heat storage walls. How much heat can be stored? How thick should the walls be? How should they be constructed? What materials are best? The purpose of the following discussion is to suggest some techniques for providing partial answers to these questions.

Qualitative assessment. To be an effective heat storage element a wall should have the following characteristics:

1. It should have a high thermal heat capacity. For a given thickness, this means the product of density ρ_s and specific heat c_p should be large.

2. It should have a high thermal conductivity, k. The deeper portions of the wall cannot participate in the charging and discharging cycle if they are isolated from the room by a layer of low-thermal-conductivity material.

Since materials which have a high density also usually have high thermal conductivity, it follows that wall materials which are good insulators are poor for thermal storage. Materials such as styrofoam and fiberboard are nearly worthless for heat storage walls, wood is moderately poor, and concrete, rock, brick, and adobe are relatively good. Thus the very properties which make a wall perform well as an (interior) thermal storage element make it perform poorly as an (exterior) insulating element. Unfortunately common construction practice is to make interior partitions of lightweight frame construction and place the more massive construction (if used at all) in exterior walls. Frequently the most massive elements are placed outside the thermal insulation; for example, the use of a brick exterior over an insulated, frame wall.

The concept of thermal admittance. The property of thermal admittance of a wall is a measure of the ability of the wall to absorb and store heat during one part of a cycle and then to release the heat back through the same surface during the second part of the cycle. Thus the property of thermal admittance is coupled to the cyclic nature of the give and take of heat at the surface of the wall. Technically, the thermal admittance is the ratio of the amplitude of a sinusoidal wave of heat flow to the amplitude of the corresponding sinusoidal wave of surface temperature.

In the illustration of Fig. 16.9, the wall surface temperature varies sinusoidally around a constant average value and the heat flow into the surface varies sinusoidally around an average value of zero. The magnitude of the temperature cycle is $\Delta T_s/2$ so that the wall surface temperature varies from $\overline{T}_s + \Delta T_s/2$ to $\overline{T}_s - \Delta T_s/2$ (where $\overline{T}_s$ is the average storage surface temperature) and the magnitude of the heat flow is Δq so that the heat flow varies from $+ \Delta q/2$ (into the wall) to $- \Delta q/2$ (out of the wall). Note that the two sinusoidal waves are slightly out of phase. Typically the temperature wave lags the heat flow wave by one-eighth of a cycle (45 angular degrees) for a thick wall, and by one-fourth of a cycle (90 angular degrees) for a very thin wall.

The thermal admittance, a, is simply the ratio $\Delta q/\Delta T_s$. For a thick wall it is given by the formula:

$$\frac{\Delta q}{\Delta T_s} = \sqrt{\frac{2\pi k \rho c}{p}} = a \tag{16.1}$$

where p is the period of the sine wave oscillation.

Generally, we are most concerned with cyclic oscillations which have a period of 1 day or 24 h. This implies a charging of the wall during the day and a discharging of the wall at night. The fact that the wave is not truly sinusoidal is not of great consequence. First-order effects can be obtained by considering a pure sine wave solution which leads to simple answers.

How much heat is stored? The total heat Q stored during the half cycle when q is positive is obtained by multiplying the peak heat flux by $p/2\pi$.

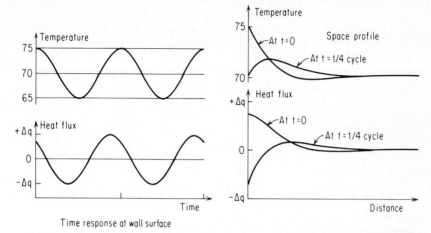

Time response at wall surface

Fig. 16.9 Thermal response characteristics of a thick wall. The upper curves show the time response at the wall surface. The temperature curve lags the heat flux curve by one-quarter cycle. The lower curves show the space profiles of both temperature and heat flux at two times; the curve marked T = 0 is a time at which the temperature peak is at a maximum, and the curve marked T = 1/4 cycle is at a time when the temperature difference is 0 and decreasing.

$$\frac{Q}{\Delta T_s} = \left(\frac{p}{\pi}\right)\left(\frac{\Delta q}{2\,\Delta T_s}\right) \tag{16.2}$$

For a thick wall the formula is:

$$\frac{Q}{\Delta T_s} = \sqrt{\frac{pk\,\rho c}{2\pi}} \tag{16.3}$$

How thick should the wall be? Consider what happens inside the wall. At each point the temperature variation is sinusoidal and the heat flow through any plane parallel to the surface is also sinusoidal. The magnitude of the sine wave decreases rapidly as the distance from the surface increases. The phase of the sine wave also changes with distance into the wall. At some point, well into the wall, the sine waves are completely out of phase with the sine waves at the surface. Thus the deeper portions of the wall can be counteracting the effect of storage in the outer portions of the wall.

The solution of the problem of a wall of finite thickness is more complicated than for a wall of infinite thickness. The answer is illustrated in the graph of Fig. 16.10, which shows the ratio of the thermal admittance of a wall of finite thickness to a wall of infinite thickness.* This graph

*The formulas for the curves are as follows:

$$\frac{\Delta q}{\Delta T_s} = \sqrt{\frac{\cosh 2x - \cos 2x}{\cosh 2x + \cos 2x}}\,a$$

where

$$x = \sqrt{\frac{\pi \rho c\,\ell}{pk}}$$

The formula for the phase of the admittance is as follows:

$$\phi = 45° + \arctan\left(\frac{\sin 2x}{\sinh 2x}\right)$$

At the optimum thickness, the phase lag is 52.6°.

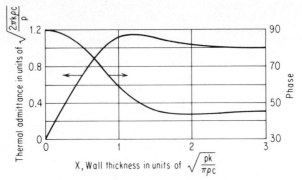

Fig. 16.10   Thermal admittance of a finite wall insulated on the back side.

shows that there is an optimum thickness for which the thermal admittance is the greatest. This optimum thickness is given by the following formula:

$$\ell_{\text{optimum}} = 1.18 \sqrt{\frac{pk}{\pi \rho c}} \tag{16.4}$$

Effect of the thermal resistance of the air film at the wall surface. So far the discussion has related the heat flowrate and the total heat storage to the wall *surface* temperature. However, the thermal resistance of the air film at the wall surface is large enough to significantly impede the flow of heat from the room to the wall surface. A wall surface resistance of $R = 0.67$ (Btu/ h ·ft^2·°F) is a good general average considering both convective and radiative heat transfer. This corresponds to a thermal conductance or U value of 1.5 Btu/ft^2·°F ·h (8.5 W/m^2·°C). Compare this to a thermal admittance of 3.13 Btu/ft^2·°F ·h (17.75 W/m^2·°C) for a concrete wall of optimum thickness. Since the air film resistance is larger, it will dominate the total heat transfer resistance of the combined pair in series.

To compute a net series conductance, we must account for the fact of the phase lag between the temperature and heat flow at the wall surface. This can be done by a vector addition of the two thermal impedances as shown in Fig. 16.11.

Thermal admittance of common materials. Table 16.1 can be used to estimate the heat stored in building materials based on two considerations:
1. Location of the material relative to the sun radiation
2. Properties of the wall material

If the material is located directly in the sun for several hours per day, then the surface temperature variation can easily exceed the room temperature variation—perhaps by as much as a factor of 2. In this case the appropriate parameter to use would be $Q/\Delta T_s$, the fifth row in

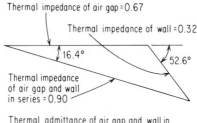

Fig. 16.11   Vector addition of wall impedances. Since the air gap and the wall are in series, the impedances are additive. Since the wall impedance has a phase lag the two must be added vectorially. Thus the net impedance is less than the sum of the magnitudes of the two parts. Note that the air gap impedance is the dominant effect.

TABLE 16.1 Thermal Admittance of Various Materials

Name	Symbol	Formula	Units	Material					
				Gravel and sand concrete	Limestone rock	Brick	Wood (pine)	Dry sand	Adobe
Density	ρ	—	$\dfrac{\text{lbs}}{\text{ft}^3}$	144	153	112	31	95	120
Specific heat	c	—	$\dfrac{\text{Btu}}{\text{lb}\cdot{}^\circ\text{F}}$	0.19	0.22	0.22	0.67	0.19	0.20
Thermal Conductivity	k	—	$\dfrac{\text{Btu}\cdot\text{ft}}{\text{ft}^2\cdot{}^\circ\text{F}\cdot\text{h}}$	1.05	0.54	0.40	0.097	0.19	0.332
Thermal admittance of infinite wall*	a	$\sqrt{\dfrac{2\pi k\rho c}{p}}$	$\dfrac{\text{Btu}}{\text{ft}^2\cdot{}^\circ\text{F}\cdot\text{h}}$	2.74	2.18	1.61	0.69	0.95	1.44
Energy stored daily $\dfrac{}{\Delta T_{\text{surface}}}$ (infinite wall)	$Q/\Delta T_s$	$\sqrt{\dfrac{pk\rho c}{2\pi}}$	$\dfrac{\text{Btu}}{\text{ft}^2\cdot{}^\circ\text{F}}$	10.4	8.3	6.1	2.6	3.6	5.5
Wall thickness for maximum heat storage	l_{optimum}	$1.18\sqrt{\dfrac{pk}{\pi\rho c}}$ See text	in	7.6	5.0	5.0	2.5	4.0	4.6
Admittance of air film and wall in series	$q/\Delta T_r$	See text	$\dfrac{\text{Btu}}{\text{ft}^2\cdot{}^\circ\text{F}\cdot\text{h}}$	1.11	1.04	0.92	0.57	0.70	0.87
Daily stored energy† $\dfrac{}{\Delta T_{\text{air}}}$	$Q/\Delta T_r$	$\dfrac{p}{\pi}\dfrac{\Delta q}{\Delta T_r}$	$\dfrac{\text{Btu}}{\text{ft}^2\cdot{}^\circ\text{F}}$	4.1	3.9	3.5	2.1	2.6	3.3

*In the table the value of sinusoidal period p is 24 h.

†For wall of thickness l_{optimum}, $U_{\text{air gap}} = 1.5$ Btu/h·°F·ft² (1 Btu = 1.055 kJ, 1 ft² = 0.093 m², 1 lb = 0.454 kg). For l_{optimum}, see Eq. (16.4).

Table 16.1, with a surface ΔT_s of perhaps twice the maximum allowable room temperature variation. If, however, the opposite is the case and the thermal storage material is located entirely out of the sunshine so as to be heated only by the room air then the appropriate parameter is $Q/\Delta T_r$, the last row in Table 16.1. This means that there can be as much as a factor-of-5 difference between the energy stored in these two cases.

An estimate of the storage can be obtained by assuming that the area directly heated by the sun is equal to the glazed area and by treating all other mass in the building as indirect storage. Thus:

$$E/\Delta T_r = 2 \times Q/\Delta T_s \times A_g + (A - A_g) \times Q/\Delta T_r \tag{16.5}$$

(assuming $\Delta T_s = 2\Delta T_r$) where A is the total surface area of all storage elements.

Carrying through the previous example will serve to illustrate this approach. The interior surface of the building is 2350 ft^2, not including the ceiling which has little mass and therefore little thermal storage value. Let us assume that the building walls are 8-in concrete, rather than frame, and are insulated on the outside to the same overall value of R-22. Of the interior surface, we consider that 240 ft^2 is directly illuminated by the sun. If we assume that this surface is concrete, then we can use the values in Table 16.1 as follows:

$$E/\Delta T_r = 2 \times 10.5 \times 240 + (2350 - 240) \times 4.1 = 13,700 \text{ Btu/°F}$$

Since the extra energy E to be stored is known, the room temperature ΔT_r can be calculated by dividing by $Q/\Delta T_r$:

$$\Delta T_r = \frac{\text{Stored energy}}{(Q/\Delta T_r)}$$

$$\Delta T_r = \frac{200,000}{13,700} = 14.6°F \ (8.1°C)$$

This value may be too large for comfort and can be reduced by additional mass walls in the building, such as interior partition walls, by a massive ceiling, or by adding other mass such as water in containers. To calculate the additional area of concrete required to reduce the variation to a desired value of room temperature variation, ΔT_r (desired), we can determine the heat stored by the exterior walls and floor and then provide a sufficient wall area to store the remaining energy. Assuming a ΔT_r (desired) of 12°F, we calculate:

$$E_{\text{remaining}} = E_{\text{excess}} - \Delta T_r \text{ (desired)} \times (E/\Delta T_r)$$

$$E_{\text{remaining}} = 200,000 - 12 \times 13,700$$

$$= 35,600 \text{ Btu}$$

$$A_{\text{required}} = E_{\text{remaining}}/(Q/\Delta T_r)/\Delta T_r \text{ desired}$$

$$= \frac{35,600}{4.1 \times 12} = 720 \text{ ft}^2 \ (67\text{m}^2)$$

Since both sides of a partition wall are effective for storage, the required area is one half this value or 360 ft^2 corresponding to 45 lineal ft of wall (two north-south partitions).

It would not have been a good idea to have used much more than 240 ft^2 of glazing for this building, without shading control, because to have done so would result in excessive temperature variations and gross overheating during warm spring or fall days.

Thermal admittance of layered walls. A comprehensive discussion of the thermal admittance of a wall made up of several layers of different materials is beyond the scope of this handbook chapter. Two general rules can be given, however:

1. A layer of insulating material in the wall has only a small phase lag. Its main effect is to isolate the mass located behind the layer from thermal interaction with the room.

2. A layer of heavy, thermally conductive material in the wall (for example, a layer of Sheetrock) will have a large phase lag. If the material is backed by an insulating layer, the temperature will lag the heat flux by as much as 90° (i.e., 6 h).

A technique for calculating the thermal admittance of a wall consisting of any number of layers has been developed by Davis.*

Horizontal thermal storage surfaces. Heat transfer between a horizontal surface, such as a ceiling or floor, and a room is quite different than for a wall and deserves special consideration.

Structurally, the floor is a convenient location for thermal storage mass. Poured concrete slab, dirt, brick, and other massive floors are inexpensive and common. However, the convective heat transfer from a warmer room to a cooler floor is essentially zero; the only mechanism for heating such a floor is by radiation—either by thermal radiation from the ceiling and walls or by direct solar radiation through windows. Conversely, the convective heat transfer from a warmer floor to a cooler room is high. Thus a warmer floor will tend to cool rapidly and, unless heated, will soon become cooler than the room. A floor is not very effective thermal storage unless it is uncovered, of a dark color, and in the direct sun, or unless some other means of heating is employed.†

A ceiling has the opposite characteristic. It is easily heated since the warm air in the room tends to rise and convective heat exchange is good. The return of heat to the room can normally occur only by radiation. This does not significantly reduce the effectiveness of the ceiling for thermal storage because the required heat exchange rates downward are much smaller. This is because the ceiling portion of the thermal load is not included and because the heat release period is generally much longer than the heat absorption period. Thus the ceiling is a good location for thermal storage. Of course, a massive ceiling is structurally awkward. Another problem is that it tends to increase temperature stratification in the room.

Internal energy sources. In the discussion and example used up to this point, no credit has been taken for internal sources in the building, such as lights, people, appliances, and so forth. Such sources can make an important contribution; they will raise the internal temperature by an amount equal to the source energy, Q_i, divided by the combined conduction and infiltration loss factor (UA product) of the building.

Example 16.2

Assume that a family of four lives in the example house and that the total generation of energy is 50,000 Btu/day. This will decrease the ambient temperature at which net heat loss to the environment occurs by

$$\frac{50,000}{362 \times 24} = 6°F\ (3.3°C)$$

Thus, for this building, one should use a reference temperature 6°F lower than the desired building inside temperature as a basis for calculating heating degree-day total. It has been conventional practice to use 65°F as a reference temperature for calculation of degree-day loads.

Annual solar performance   Estimation of annual energy performance would need to be based on simulations using actual weather data for the building location. This procedure is described in the final section of the chapter.

Thermal Storage Walls

The preceding example of thermal storage in a direct-gain passive solar structure has indicated the difficulty of reducing temperature variations. Even with extensive surface areas of thermal storage in the building, temperature variations of 12°F or greater over a day can be anticipated. This is an intrinsic character of direct-gain buildings. Filling the room with the normal accoutrements of living will tend to exacerbate the problem; lightweight materials such as carpets, furniture, pictures, wall hangings, etc., tend to convert the sun's energy into warming air and insulate the mass storage in the floor and walls from the desired thermal interaction in the building.

---

*M. G. Davis, "The Thermal Admittance of Layered Walls," *Building Science*, vol. 8, pp. 207–220. (Pergamon Press, London, 1973).

†An effective heating technique which has been used is to pass solar-heated air through an underfloor rock bed or through air passages in the floor. This usually is done using a fan (active system) although a passive air convective loop can be effective if the solar air-heating region is located at a lower level than the floor and if the air passages are large enough to allow air to flow with a minimum impedance.

Fortunately, a massive wall interposed between a solar-heated space and a living space can be used to tremendously decrease the magnitude of temperature fluctuations. This is the technique used in both the Trombe wall and solar greenhouse design approaches discussed previously. Its effectiveness has been proven in several passive solar-heated buildings.*

Results of simulation analysis   Thermal storage walls have been the subject of extensive computer simulation analysis. This method provides an accurate prediction but does require a computer due to the complexity of the analysis. Thermal network analysis techniques can be used. In the present analysis, the building temperature state is characterized by temperatures at various locations—air temperature, surface temperatures, glass temperatures, and the temperature of various thermal storage materials at various depths. Energy balance equations are written for each location accounting for thermal energy transport by radiation, conduction and convection, energy sources from the sun, lights, people, and auxiliary heaters, and sensible thermal energy storage in the material. The temperature history of each location is simulated by simultaneously solving these equations for given inputs of solar radiation, and ambient air temperature.

The simulation analysis emphasizes the thermal storage wall itself and the glazing. The building is represented as a simple conductance to the outside, here called the "load." For much of the analysis, this load is 0.5 Btu/°F · h per square foot (2.8 W/m² · °C) of thermal storage wall and associated double glazing, excluding the south wall.

A storage mass of 30 Btu/ft$_g^2$ · °F was chosen initially. This is equivalent to 30 lb of water or 150 lb of concrete per sq ft of glass.

The room temperature was allowed to vary by 5°F around a desired value of 70°F. Auxiliary energy was supplied, as necessary, to maintain the room temperature above 65°F and energy was dumped if necessary to prevent the room temperature from rising above 75°F.

It is instructive to observe the simulation results for a few days of cold weather. Figure 16.12 shows the 7-day interval between the 31st of December and the 6th of January for Los Alamos, NM. There was snow on New Year's Eve followed by two days of cloudy weather and then cold but sunny weather. Note that the storage temperature variations are much larger than the room temperature variations.

In order to study the basic performance characteristics of a Trombe wall, as compared to the case of a water wall shown in Fig. 16.12, the mathematical model was modified to describe the time and one-dimensional space-dependent thermal transport of heat through the wall. This was done by simulation of the masonry temperature at the wall surfaces and at several different distances into the wall.

The thermal properties used for the masonry were as follows (typical of dense concrete):

Heat capacity: 30 Btu/ft³ · °F
Thermal conductivity: 1 Btu/ft · °F · h

The calculated wall temperatures are shown on Fig. 16.13 a, b, and c for the same seven-day period shown in Fig. 16.12 for three different wall thicknesses—0.5, 1, and 2 ft. The daily fluctuations felt on the inside-wall surface are markedly different for the three cases, being very pronounced (~ 45°F) for the thin wall and almost nonexistent for the thick wall. The longer-term effect of the storm is observed on the inside of the thick wall as a 10°F variation in temperature.

Annual performance. A study of the effect of climate on annual solar heating performance is given in Table 16.2. These calculations are all for the following case:

- 18-in Trombe wall
- Wall thermal conductivity = 1 Btu/ft · h · °F
- Wall heat capacity = 30 Btu/ft³ · °F
- Thermocirculation vent size = 0.009 ft²/ft of wall (each row)
- No reverse thermocirculation
- Load = 0.71 Btu/ft$_g^2$ · °F · h (including 0.21 for the Trombe wall)
- Temperature band = 65 to 75°F

A 1-h time step is used to march through a one-year time period and energy flows are accumulated on an hourly basis.

Effect of different configurations. Annual simulation analysis results for five different configurations of thermal-storage walls are given in Table 16.3. The cases studied are as follows:

---

*Note that masonry walls may require up to three years to "cure," that is, to achieve a completed chemical hydration reaction so that nominal physical properties exist. Wall performance will lie below that predicted further on in this chapter until curing is complete.

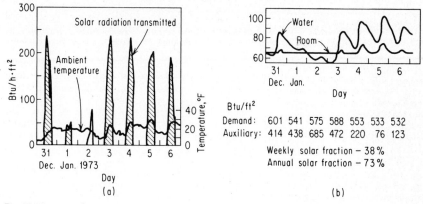

Btu/ft²

Demand: 601 541 575 588 553 533 532
Auxiliary: 414 438 685 472 220   76 123

Weekly solar fraction – 38%
Annual solar fraction – 73%

Fig. 16.12 Seven days in winter. Data are from Los Alamos, NM. The solar radiation is for a vertical south-facing double glazing. The lower curve shows the calculated response of a water wall with 30 lb of mass per sq ft (150 kg/m²) of glazing. Auxiliary energy is furnished as shown to maintain the room temperature at or above 65°F.

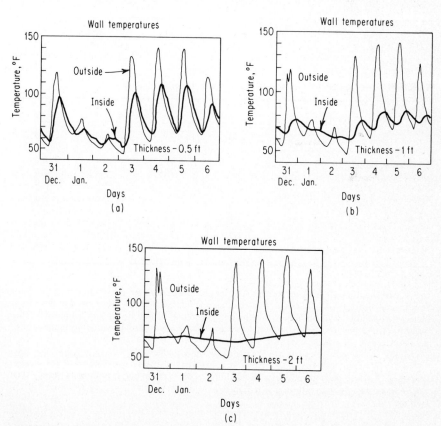

Fig. 16.13 Seven days in winter. The three curves show the calculated response of a solid masonry thermal storage wall of three thicknesses. Note that the damping affect of the masonry wall varies as the square of the wall thickness.

TABLE 16.2   Performance of a Reference Trombe-Wall Building

City	Year starting	Heating, °F · days	Latitude	Solar heating,* Btu/ft$_g^2$	Solar-heating fraction, %
Ft. Worth, TX	7/1/60	2467	32.8	38,200	80.8
Madison, WI	7/1/61	7838	43.0	44,900	41.6
Albuquerque, NM	7/1/62	4253	35.0	63,600	84.1
Fresno, CA	7/1/57	2622	36.8	43,200	83.3
Medford, OR	7/1/61	5275	42.3	47,400	56.1
Bismarck, ND	7/1/54	8238	46.8	53,900	46.4
New York, NY	6/1/58	5254	40.6	48,000	60.2
Dodge City, KS	7/1/55	5199	37.8	58,900	71.8
Nashville, TN	7/1/55	3805	36.1	39,500	65.2
Boston, MA	7/1/57	5535	42.3	47,100	56.8
Charleston, SC	7/1/63	2279	32.8	47,900	89.3
Seattle, WA	7/1/63	5204	47.5	42,400	52.2
Lincoln, NE	7/1/58	5995	40.8	53,500	59.1
Boulder, CO	1/1/56	5671	40.0	62,500	70.0
Vancouver, BC	1/1/70	5904	49.1	46,000	52.7
Edmonton, ALB	1/1/70	11679	53.5	37,700	24.7
Winnipeg, Man	1/1/70	11490	49.8	33,700	22.6
Ottawa, Ont.	1/1/70	8838	45.3	37,900	31.9
Fredericton, NB	1/1/70	8834	45.8	40,100	33.9

*The values in the solar-heating column are the net energy flow through the inner face of the wall into the building (1 Btu/ft^2 = 10.2 kJ/m^2).

TABLE 16.3   Performance of Reference Building with Different Variations

City	Annual % solar heating				
	WW	SW	TW	TW(A)	TW(B)
Santa Maria, CA	99.0	98.0	97.9	97.3	98.0
Dodge City, KS	77.6	69.1	71.8	62.8	73.6
Bismarck, ND	49.8	41.3	46.4	31.1	47.6
Boston, MA	60.0	49.8	56.8	44.9	56.7
Albuquerque, NM	90.8	84.4	84.1	81.8	87.5
Fresno, CA	85.5	82.4	83.3	78.0	83.4
Madison, WI	43.1	35.2	41.6	24.7	42.0
Nashville, TN	68.2	60.7	65.2	54.1	65.4
Medford, OR	59.0	53.3	56.1	42.2	56.8

- WW: Water wall. This might consist of cans or drums of water stacked to form a thermal-storage wall. Alternatively, vertical, free-standing cylindrical tubes could be used or any other means of containing water in a thermal-storage wall. When heated by the sun on one side, the water will freely convect to transport the heat across the wall horizontally, and thus temperature gradients across the wall will be very small.
- SW: Solid wall (18-in concrete) without thermocirculation.
- TW: Trombe wall: Thermocirculation is allowed only in the normal direction as previously described. Reverse thermocirculation, as would normally occur at night, is not permitted. (This can be implemented with a thin, plastic-film passive damper draped over the inside of the top opening.)
- TW(A): Trombe wall with vents open at all times: Note that if reverse thermocirculation is not prohibited then the vents are a net thermal *disadvantage* to the building.
- TW(B): Trombe wall with control: The result of the normal thermocirculation frequently is to overheat the building during the day. In this option, the vents are closed whenever the

building temperature is 75°F or greater. This greatly reduces the required venting. This strategy would require some passive or active control mechanism.

Although some configurations are clearly better than others, all seem to be viable approaches to solar heating in all the climates studied. The effectiveness of the thermocirculation vents is pronounced in the colder climates.

Parameter effects, water wall. In the following discussion the effect of variations in various parameters will be shown. In each case the simulation model was run repeatedly varying one parameter at a time while holding the others constant at the nominal values given above. The heavy dot on each graph represents the nominal case.

For this analysis, the solar and weather data used were for the Los Alamos, NM, year September 1972 to August 1973. For the year, the total radiation on a horizontal surface was 518,000 Btu/ft^2 and the space heating load (base: 65°F) was 7,350 degree-days (18 percent higher than normal for Los Alamos). This weather is more severe than average.

The effect of varying storage heat capacity and glass insulation is shown in Fig. 16.14. Most of the benefits of storage are obtained at a value of 30 Btu/°F · ft$_G^2$. The improvement obtained with double glazing is very dramatic. In fact, a single-glazed wall without night insulation can hardly be considered a viable passive solar-heating element since only 30 percent solar heating can be achieved even with very large storage, and the glass is a net loser at low storage. The increased effectiveness of insulating the glass at night is impressive. The cost-effectiveness of this approach needs further study. Night insulation can be seen to be far more important with single glazing than with double glazing. Single glazing becomes viable only with night insulation. A strategy of placing night insulation based on observed conditions rather than a time clock would result in only a small increase in performance ($\sim$ 2 percent).

The effect of varying the glass area is inverse to the effect of varying the building thermal load. This is shown in Fig. 16.15.

Parameter effects, Trombe wall. The net annual results of several solid Trombe-wall calculations are summarized in Fig. 16.16. The net annual thermal contribution of the three different thicknesses of walls are not markedly different. In fact the 1-ft thick wall is the best of the three—giving an annual solar-heating contribution of 68 percent. This compares with a value of 73 percent for a water wall with the same heat capacity. In each case, auxiliary cooling or heating was assumed to maintain the room temperature within the bounds given previously. Although the net thermal contribution of the thin-wall and thick-wall cases are nearly the same, the amount of control required for the thick wall is much less and the variation in room temperature within the set bounds is much lower.

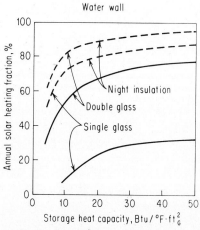

Fig. 16.14 Parameter variations for a water wall. This curve shows the effect of varying storage heat capacity for the four different cases of single and double glazing, with and without night insulation. For the night insulation cases, a value of R-10 was used between the hours of 5:00 P.M. and 8:00 A.M. every day.

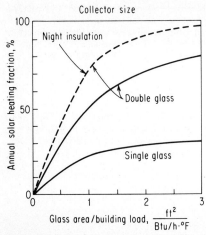

Fig. 16.15 Effect of glass area for a water wall case. The glass area is plotted as a ratio to the building load. In this case the building heat load includes all load except for the glass.

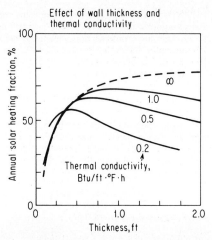

Fig. 16.16 Yearly performance of a thermal storage wall as a function of thickness for various thermal conductivities. The heat capacity was held constant at a value of 30 Btu/ft³ · °F.

Example 16.3

The example used previously to illustrate a direct-gain structure is now used for a Trombe-wall example. Suppose that the entire 60-ft long south wall is made into a Trombe wall. The "load" can be calculated as follows:

- Opaque walls: 790 ft²   × 0.045   35
- Trombe wall: 480 ft²   × 0.21   100
- Ceiling: 1320 ft²   × 0.032   42
- Windows: 42 ft²   × 0.55   23
- Perimeter: 164 ft   × 0.17   28
- Infiltration: 1/2 ACH   91
- Total load (Btu/h · °F)   319 (168 W/°C)

Suppose that only 94 percent or 451 ft² of the south wall is net useful collection area. Then the load, per square foot of Trombe wall, is:

$$\frac{319}{451} = 0.71 \text{ Btu/h} \cdot \text{°F} \cdot \text{ft}^2$$

This building is essentially the case used for computing Table 16.2. If the building, except for the Trombe wall, is of frame construction, then the assumption that the load is a simple conduction to the outside is reasonable. If the building were to be concrete, as in the direct-gain example, then this assumption is not very good because the lag of the building itself would be an important aspect of the dynamics.

From Table 16.2, we can predict that this building would be roughly 84 percent solar-heated if located in Albuquerque, 60 percent solar-heated if located in New York, or 41 percent solar-heated in Madison.

Other values of load. The results given in Table 16.2 are for a fixed load of 0.71 Btu/°F · ft² of Trombe wall. If the load is different than this, a reasonable procedure is to assume that $(1-f)$ is exponential (where f is the solar heating fraction). This leads to the expression:

$$f_2 = 1 - e^{(L_1/L_2)\ln(1-f_1)}$$

where   (16.6)

$$f_2 = \text{solar-heating fraction for load } L_2$$

$$f_1 = \text{solar-heating fraction for load } L_1$$

Example 16.4

The example used previously resulted in a total load of 362 Btu/°F · h for 240 ft² of south-glazed area for a ratio of 1.51 Btu/h · °F · ft² of glazing. Suppose this building is located in Albuquerque. Table 16.3 predicts a 90.8 percent solar fraction for a water wall with a load of 0.71 Btu/hr · °F · ft². A water wall with this larger load would be predicted to have a solar heating fraction of:

$$1 - e^{(0.71/1.51) \ln (1 - 0.908)} = 0.67 \quad \text{or} \quad 67 \text{ percent}$$

Note that the Trombe-wall example used 48 tons of concrete located behind 451 ft² of glazing to achieve an estimated 84 percent solar-heating fraction, whereas the direct-gain example used 153 tons of concrete located behind 240 ft² of glazing to achieve an estimated 67 percent solar-heating fraction. The comfort characteristics of both buildings would be comparable, having temperature swings of 10 or 12°F.

PERFORMANCE PREDICTION

The monthly Solar Load Ratio (SLR) provides an empirical means of estimating monthly solar and auxiliary energy requirements. The monthly Solar Load Ratio is a dimensionless correlation parameter defined as follows:

$$\text{SLR} \equiv \frac{\text{Monthly solar energy absorbed on the thermal storage wall surface}}{\text{Monthly building load (including the wall steady-state losses in the absence of solar gains)}} \tag{16.7}$$

The numerator is equal to the product of the total solar collection wall area times the monthly solar energy transmitted through 1 ft² of south glazing times the wall absorptance. (For direct gain the absorptance is taken as 1.0.) The denominator is equal to the building loss coefficient (including the steady-state conduction through the south solar collection wall) times the monthly heating degree days.

The SLR can be expressed as follows:

$$\text{SLR} = \frac{\text{Collector wall area} \times \text{absorptance} \times \text{monthly solar energy transmitted through the glazing}}{\text{Modified building loss coefficient} \times \text{monthly degree days}} \tag{16.8}$$

Step 1

Determine the building heat-loss coefficient (Btu/°F · day) and get a modified building loss coefficient by adding to it the term [24 × (solar wall area) × U_{tr}] where U_{tr} is taken from the following table:

Wall type	Plain double glazed	With R9 insulation added from 5:00 p.m. to 8:00 a.m.
Water wall	0.33	0.18 Btu/hr°F · ft²
18-in Trombe wall	0.22	0.12 Btu/hr°F · ft²

The value of U_{tr} is the steady-state conduction coefficient of the combined wall, glazing, and insulation, averaged over the day. For direct gain, no U_{tr} modification is needed.

Step 2

Determine the SLR for each month of the year. Solar radiation values generally available are measured on a horizontal surface, whereas the values required in order to determine the SLR are the actual solar radiation transmitted through the vertical south-facing surface. It is given by (for angles, in degrees)

$$\bar{I}_{\text{VERT}} = [0.2260 - 0.002512\,(L - \delta_s)$$
$$+ 0.0003075\,(L - \delta_s)^2] \times \bar{H}_h$$

for vertical walls with double glazing. (See Chap. 2 for symbol definitions.)

If the building does not face due south, then this equation cannot be used. It will be necessary to make another correction for building orientation. It is felt that a correction factor based on the ASHRAE clear-day tables would probably be a reasonable estimate. Those tables provide values for the clear-day conditions for southwest and southeast orientations as well as due south, as a function of latitude. For the time being, a straight proportional correction factor based on these tables is recommended. Note that a separate correction factor will be required for each month.

Step 3

Determine the monthly solar heating fraction f_s for each month of the year based on the values of SLR computed in Step 2 using the equations below:

$$f_s = a_1\,(\text{SLR}) \qquad \text{SLR} < R \tag{16.9}$$

$$f_s = a_2 - a_3\,e^{-a_4(\text{SLR})} \qquad \text{SLR} > R \tag{16.10}$$

such that the values are equal at SLR $= R$. The values of the parameters in the function give a minimum least-square error in the annual solar heating fraction.

The values of the least-squares coefficients are:

Wall type	R	a_1	a_2	a_3	a_4
Water wall	0.8	0.5995	1.0149	1.2600	1.0701
Night-insulated water wall	0.7	0.7642	1.0102	1.4027	1.5461
Masonry wall	0.1	0.4520	1.0137	1.0392	0.7047
Night-insulated masonry wall	0.5	0.7197	1.0074	1.1195	1.0948
Direct gain	0.1	0.6182	1.0097	1.0710	1.2208
Night-insulated direct gain	0.6	0.8865	1.0028	1.2646	1.6467

Effect of Internal Generation in the Building

Heat generated in the building, by people, lights, and equipment is effective in reducing the monthly load. This reduces both the auxiliary energy requirements and the monthly solar contribution.

The original basis for defining the degree-day base at 65°F was on the assumption that these internal energy sources would raise the building temperature from 65°F up to the accepted comfort standard of 72°F. This assumption can still be made in using the results from this section, namely, that the actual building temperature would be several degrees greater than the 65°F to 75°F band assumed in the analysis.

However, experience has been that most people now set their thermostat at lower levels. This is especially true of people who live in passive solar homes because the effect of the warm surrounding surfaces of these buildings increases the mean radiant temperature within the space so that one can be comfortable at a reduced air temperature. In any case, a 65°F thermostat setting seems more consistent with actual practice in the winter than the ASHRAE standard value of 72°F.

The user of the method can correct for this by subtracting the estimated internal energy generation from the monthly loads prior to computing the monthly Solar Load Ratio. The effect of this would be to increase the Solar Load Ratio, increase the monthly solar heating fraction, and decrease the auxiliary energy requirements.

Variations from the Assumed Reference Systems

The monthly SLR equations above are for very specific reference systems as defined below

- Thermal Storage = 45 Btu/°F · ft² of glazing
- Trombe wall has vents with backdraft dampers
- Double glazing (normal transmittance = 0.747)
- Temperature range in building: 65 to 75°F
- Building mass is negligible except for direct gain which uses mass on floor and non-south walls
- Night insulation (when used) is R9; 5:00 p.m. to 8:00 a.m.
- Wall to room conductance = 1.0 Btu/hr °F · ft²
- Trombe wall properties k = 1.0 Btu/ft · h · °F
  $$\rho c = 30 \text{ Btu/ft}^3 \cdot \text{°F}$$
- Foreground reflectance = 0.3

If it is desired to estimate the performance of the system which is different than one of these reference systems, then it is necessary to make a correction. The most reliable way of doing this is to refer to results of hour-by-hour calculations which are made for a specific system varying only the parameter of interest.

Effect of a Reflector

A tremendous performance advantage can be achieved through the use of a reflector to increase the total amount of solar radiation on the solar collection wall. The ratio of the total monthly solar energy transmitted with the reflector to that without the reflector is given by (for a specular reflector with $\rho = 0.8$)

$$R_{\text{refl}} = 1.0083 - 1.787\Delta + 19.16\Delta^2 - 40.31\Delta^3$$
$$+ 24.66\Delta^4 \tag{16.11}$$

where $\Delta \equiv (L - \delta_s)/100$.

If a reflector is used with a reflectance other than 0.8, the ratio R_{refl} can be computed from the above equation by assuming that the difference between unity and the calculated enhancement is proportional to the reflectance.

Discussion of Loads

Monthly heating degree-day values are used in the correlation procedure because they are the only indicators of heating load that are readily available in most localities.

It is possible to distinguish between two solar heat contributions from the solar wall: (1) the energy *saved* and (2) the energy *supplied*. In this method the energy saved is used to define the solar heating fraction even though it gives a lower value of f_s. The actual solar energy supplied by the solar collection wall will be greater than that estimated by taking the difference between the annual degree-day load and the auxiliary energy. The extra solar heat is the amount used to maintain the building above 65°F during a significant portion of the year. Since it is the actual auxiliary energy required which is the most important number to be estimated, this approach seems appropriate. In reality, the solar-heated building will generally be warmer than the nonsolar-heated building, assuming that the thermostat is set at 65°F in both cases.*

---

*Additional details of this method are given in Los Alamos Reports LA-UR-78-1159 (1978) and LA-UR-78-2570 (1978).

The altitude and azimuth of the sun are given by

$$\sin a = \sin \phi \sin \delta + \cos \phi \cos \delta \cos h \tag{1}$$

and

$$\sin \alpha = -\cos \delta \sin h / \cos a \tag{2}$$

where a = altitude of the sun (angular elevation above the horizon)
ϕ = latitude of the observer
δ = declination of the sun
h = hour angle of sun (angular distance from the meridian of the observer)
α = azimuth of the sun (measured eastward from north)

From Eqs. (1) and (2) it can be seen that the altitude and azimuth of the sun are functions of the latitude of the observer, the time of day (hour angle), and the date (declination).

Figure 16.A1 (b-g) provides a series of charts, one for each 5° of latitude (except 5°, 15°, 75°, and 85°) giving the altitude and azimuth of the sun as a function of the true solar time and the declination of the sun in a form originally suggested by Hand. Linear interpolation for intermediate latitudes will give results within the accuracy to which the charts can be read.

On these charts, a point corresponding to the projected position of the sun is determined from the heavy lines corresponding to declination and solar time.

To find the solar altitude and azimuth:

1. Select the chart or charts appropriate to the latitude.
2. Find the solar declination δ corresponding to the date.
3. Determine the *true solar time* as follows:
   (a) To the *local standard time* (zone time) add 4′ for each degree of longitude the station is east of the standard meridian or subtract 4′ for each degree west of the standard meridian to get the *local mean solar time*.
   (b) To the *local mean solar time* add algebraically the equation of time; the sum is the required *true solar time*.
4. Read the required altitude and azimuth at the point determined by the declination and the true solar time. Interpolate linearly between two charts for intermediate latitudes.

It should be emphasized that the solar altitude determined from these charts is the true geometric position of the center of the sun. At low solar elevations terrestrial refraction may considerably alter the apparent position of sun. Under average atmospheric refraction the sun will appear on the horizon when it actually is about 34′ below the horizon; the effect of refraction decreases rapidly with increasing solar elevation. Since sunset or sunrise is defined as the time when the upper limb of the sun appears on the horizon, and the semidiameter of the sun is 16′, sunset or sunrise occurs under average atmospheric refraction when the sun is 50′ below the horizon. In polar regions especially, unusual atmospheric refraction can make considerable variation in the time of sunset or sunrise.

Altitude and azimuth in southern latitudes. To compute solar altitude and azimuth for southern latitudes, change the sign of the solar declination and proceed as above. The resulting azimuths will indicate angular distance from *south* (measured eastward) rather than from north.

(a)

Fig. 16.A1 Description of method for calculating true solar time, together with accompanying meteorological charts, for computing solar-altitude and azimuth angles. (a) Description of method; (b) chart, 25°N latitude; (c) chart, 30°N latitude; (d) chart, 35°N latitude; (e) chart, 40°N latitude; (f) chart, 45°N latitude (g) chart, 50°N latitude. Description and charts reproduced from the "Smithsonian Meteorological Tables" with permission from the Smithsonian Institute, Washington, DC.

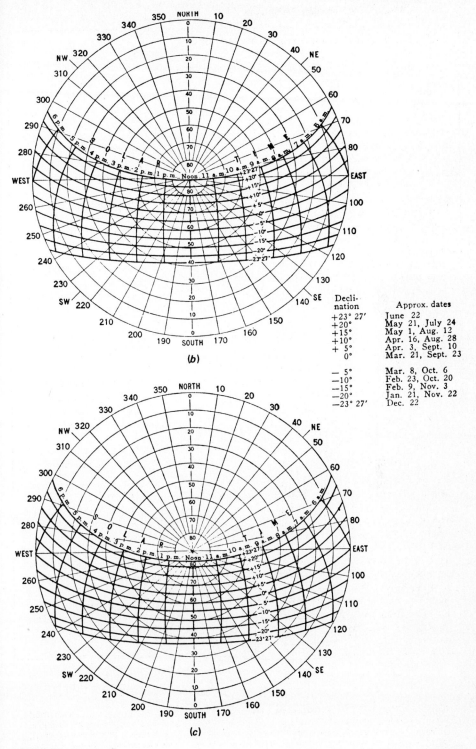

Declination	Approx. dates
+23° 27′	June 22
+20°	May 21, July 24
+15°	May 1, Aug. 12
+10°	Apr. 16, Aug. 28
+ 5°	Apr. 3, Sept. 10
0°	Mar. 21, Sept. 23
— 5°	Mar. 8, Oct. 6
—10°	Feb. 23, Oct. 20
—15°	Feb. 9, Nov. 3
—20°	Jan. 21, Nov. 22
—23° 27′	Dec. 22

(b)

(c)

Fig. 16.A1   *(Continued)*

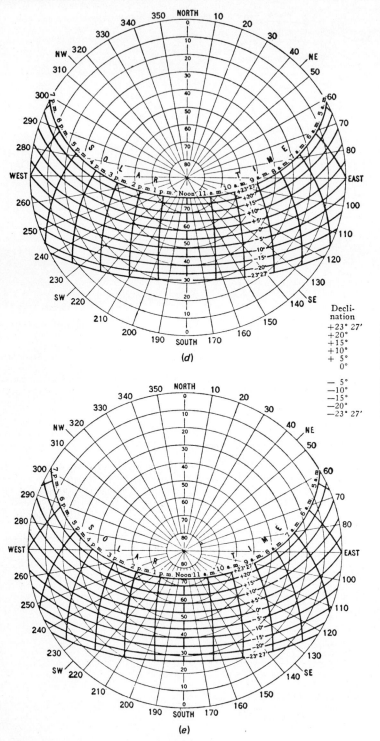

Decli- nation	Approx. dates
+23° 27′	June 22
+20°	May 21, July 24
+15°	May 1, Aug. 12
+10°	Apr. 16, Aug. 28
+ 5°	Apr. 3, Sept. 10
0°	Mar. 21, Sept. 2.
— 5°	Mar. 8, Oct. 6
—10°	Feb. 23, Oct. 20
—15°	Feb. 9, Nov. 3
—20°	Jan. 21, Nov. 22
—23° 27′	Dec. 22

(d)

(e)

Fig. 16.A1 (Continued)

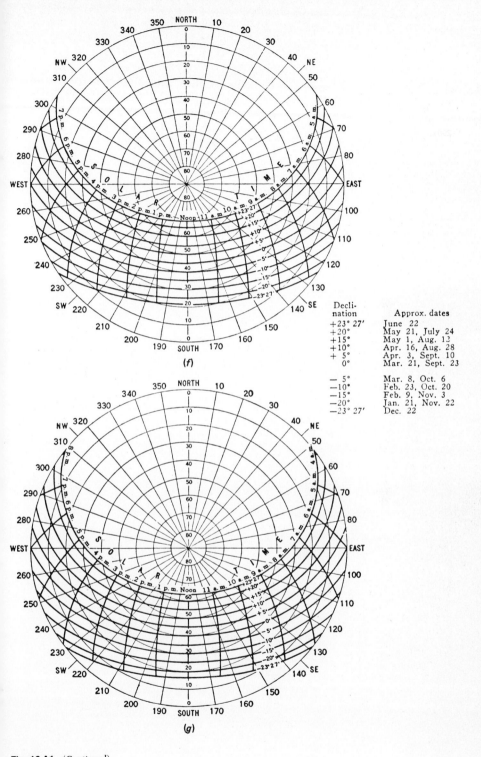

Decli-nation	Approx. dates
+23° 27'	June 22
+20°	May 21, July 24
+15°	May 1, Aug. 12
+10°	Apr. 16, Aug. 28
+ 5°	Apr. 3, Sept. 10
0°	Mar. 21, Sept. 23
— 5°	Mar. 8, Oct. 6
—10°	Feb. 23, Oct. 20
—15°	Feb. 9, Nov. 3
—20°	Jan. 21, Nov. 22
—23° 27'	Dec. 22

(f)

(g)

Fig. 16.A1 *(Continued)*

16-27

Chapter 17

Modeling of Solar-Thermal Systems

WILLIAM DUFF, P.E.
Colorado State University, Fort Collins, CO

and

BYRON WINN, P.E.
Solar Environmental Engineering Company, Fort Collins, CO

The intent of this chapter is to examine the reasons for using a modeling approach for solar thermal system design and to present the procedures involved in choosing a modeling approach and following it through to a useful application for a specific project. The topics covered in this chapter should aid the reader in selecting a modeling approach that is suited to providing needed results while at the same time not exceeding available economic and time resources.

Various models are used throughout this handbook in an exemplary fashion to best communicate the basic features of the systems discussed in each chapter. In contrast, the specific models and modeling approaches used in this chapter are not chosen as a structure to describe a specific system, but rather to illustrate the main points to be made in the procedural aspects of the modeling effort and to bring to the reader's attention some approaches that are widely used or particularly promising.

MODELING SYSTEMS

When confronted with a situation where modeling appears to be a useful approach, there is an inclination to jump immediately into the modeling effort, selecting a technique from the relatively few modeling approaches that have been used before. This tendency should be resisted since it may lead to an overcommitment of time and other resources and a model that is not appropriate to the objectives. It may not be possible to backtrack and take a more appropriate course of action if the available resources have been expended or it may not be possible to do so because of a public commitment to the approach. These problems can be avoided by taking time at the start to thoroughly determine the reasons for using a model and by being disciplined enough to consider modeling approaches that are unfamiliar.

Before looking at the reasons for using a model or examining the variety of modeling approaches that may be used, the question of how the system is to be represented will be addressed.

What Is a Systems Model?

The initial step in modeling a system is the derivation of a structure to be used to represent the system. It will become apparent that there is no unique way of representing a given system. Since the way the system is represented often strongly suggests specific modeling approaches, the possibility of using alternative system structures should be left open while the modeling approach selection is being made.

The structure that represents the system should not be confused with the real system. The structure will always be an imperfect copy of reality. However, the act of developing a system structure and the structure itself will foster an understanding of the real system.

In developing a structure to represent a system, system boundaries consistent with the problem being analyzed are first established. This is accomplished by specifying what items, processes, and effects are internal to the system and what items, processes, and effects are external. For example, a solar service hot-water system may be determined to consist of the collector, heat exchanger, preheat tank, pumps, associated piping, and controls. Alternatively, it could also include the domestic hot-water tank. In the first case the flow from the preheat tank—the system output—is an external effect if it is externally prescribed and, in the second case, it is an internal effect. The domestic hot-water demand is a system output and external effect for the second case and is only relevant to the second case. Solar radiation and the flow from the water supply are system inputs and external effects in both cases. The two cases are illustrated in Fig. 17.1.

The items, processes, and effects that are part of the system are next disaggregated into subsystems and interactions between subsystems. Subsystems are defined by their inputs, outputs, design specifications, and controls. For example, the preheat tank may be considered as a subsystem defined by its shape, dimensions, and the flowrates for the two inputs and two outputs shown in Fig. 17.1.

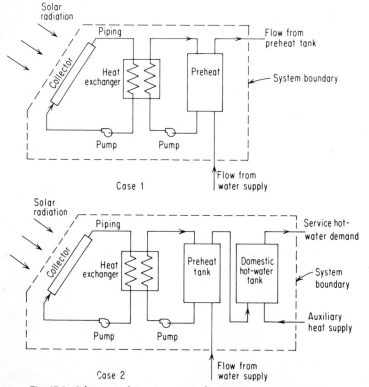

Fig. 17.1   Solar service hot-water systems showing component organization.

Obviously there is some flexibility and degree of arbitrariness in dividing the system into subsystems and interactions between subsystems. The pump and piping between the collector and heat exchanger in Fig. 17.1 could be considered as separate subsystems, or a single subsystem, or lumped into the specifications of the interactions between the subsystems. Usually the more subsystems that are used, the simpler the interactions become.

The electricity consumed by the pump would be attributed to one of the subsystems as a subsystem output and system output. Notice that electricity consumed would not be an external effect since it is an internally generated requirement. The electricity consumed could be translated into dollars and the dollar costs of the subsystems could be considered system outputs as well.

In addition to the flexibility available in dividing the system into subsystems, there are alternative approaches for organizing subsystems. The approach used in Fig. 17.1 was to organize by component. Another possibility is to organize by function. The case 1 solar service hot-water system of Fig. 17.1, organized by function, is shown in Fig. 17.2.

It may be argued that there is no difference between the two approaches. However, the functional organization does make it easier to think about the problem from a more general or abstract point of view. For instance, the Fig. 17.2 organization more clearly indicates that the choice of components may be considered as flexible design specification decisions. Also, in lumping the pipes and pump together into one subsystem, functionally labeled heat transport, it is made evident that there is no net mass transport through this subsystem. Thus the inputs and outputs of this subsystem could alternatively be represented in terms of temperatures and energy rates rather than mass flows and temperature differences.

A functional point of view is useful in converting the system representation into the system model. Such a point of view will enable one to more easily explore the range of possibilities in converting the system representation to the system model, since it more naturally brings the topic of different ways of specifying subsystems and interactions into the analysis process.

The functional approach also permits one to easily reorganize the system representation into other configurations. Figure 17.3 shows the solar service hot-water system of Fig. 17.2 reorganized into a two subsystems system. Such a reorganization could be brought about simply by considering the two heat transport and the heat exchange subsystems as the storage subsystem and having a more complicated model description of storage.

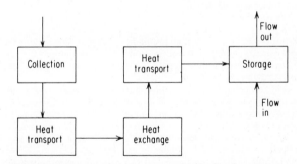

Fig. 17.2 Solar service hot-water system showing function orientation.

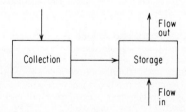

Fig. 17.3 Solar service hot-water system with a two subsystem organization.

Reorganizations of the system representation can also occur when models for describing subsystem operation are developed. If the system is modeled by mathematical equations, the equations for several subsystems may be combinable in such a way that the variables that represent the interactions between the subsystems are eliminated. The several subsystems can then be considered one system that performs according to the reduced set of equations which no longer explicitly includes the interactions. This variable elimination process can also be performed by using the additional structure imposed on the system when optimization is required, as discussed later in this chapter.

Another way of disaggregating a system is by dividing the system into elements. Elements are usually chosen so that the temporal and spatial parameters of a process, such as heat or mass transfer, can be treated using steady-state analyses within each element. This disaggregation of the system is used primarily in the finite element modeling approach where each element is considered a point or nodal source and sink for energy flows. Elements could, in fact, be considered subsystems under a functional organization and thus the term element will be avoided.

The system boundaries, subsystem interactions, and subsystem models, along with the system external effects, constitute the systems model. The actual development of a systems model need not be as sequential as is suggested by the organization of this section. The traditional approach to systems modeling is indeed fairly sequential where the subsystem organization is first established and then the models of each subsystem are developed.

The *analysis* philosophy of the traditional approach focuses most of the attention on modeling of an already organized set of subsystems. In contrast, a more recent approach having a *synthesis* philosophy emphasizes the process of determining how a system is assembled from its parts with potential modeling approaches for potential subsystems being an input to the system synthesis procedure. This latter philosophical approach would seem to provide a broader scope to the process of system model selection by permitting specific inclusion as part of the selection criteria of such considerations as the desired end uses of the model.

Why Should Systems Models Be Used?

When should systems models not be used or when should very simple models be used? Frequently there may not be enough time before a design must be finalized to develop a detailed systems model or learn to use one already available. Lack of appropriate expertise or access to computer facilities can be other reasons for not modeling. One of the simpler models should usually be used under these circumstances.

A more subtle reason for not undertaking a detailed modeling effort is when the cost of modeling does not justify the benefits gained by using the model. For example, a set of full-scale simulation runs would normally not be justified to design a single residential solar-heating system since the dollar consequences of inaccuracy due to using a simple sizing method would not be enough to pay for the simulation analysis. However, if a number of copies of the house are to be built, simulation may be justified.

Some effort should be made to determine the value of the information that will be generated by using a model. Often, as when simulation of a solar-heating system is being considered, experience will provide the estimate. However, sometimes only by developing the model itself can a good assessment of the value of the information be determined. Since it would obviously be an unsatisfactory way of answering the question, the next best possibility, an estimate of the maximum benefit that can be gained by modeling, can be made.

We will seldom be able to find a systems model that suits our purposes completely and many times we will be faced with choosing among several models that satisfy our objectives in different ways. Thus it is important that the question of "Why should systems models be used?" be asked. The way the possible answers to this question correlate with objectives will guide us in structuring the system and selecting the modeling approach. Many of these answers will be enumerated in this section.

Models can eliminate guesswork and reduce the expense of building prototypes

There is no question that, in many cases, prototype solar energy systems should be built since there are many problems that cannot be resolved or even anticipated by a systems model. For example, even the performance and reliability of off-the-shelf components, such as pumps, valves, fittings, and so forth, cannot be predicted when they are operated in a manner where there is no comparable experience. However, since prototypes and the data collection effort required to obtain useful operating information is expensive, efforts should be made to reduce the number of prototypes required. By substituting investigations using systems models for some of the prototypes, a considerable cost savings can often be realized.

A prototype that is a considerable departure from the final design may produce information that is of little value other than to indicate that the design is poor. A systems model can predict the performance of a system for various sizes and types of components and for various operating and control schemes. These predictions can narrow the component size ranges that need be considered, eliminate some component types from consideration, and permit selection of promising operating and control schemes. The narrowing of options can, in turn, reduce the number of prototypes built and insure that the data collected from those that are built will be worthwhile.

Systems modeling can organize complex systems into a more understandable format   Anyone who has tried to fully comprehend the detailed relationships involved in a large-scale complicated production process by looking at the blueprints of the production operations involved can appreciate the need for a systematic way of organizing the essentials of the process and interaction between operations into a simpler format. Industrial engineering techniques for accomplishing this simplification have been available for many years in the form of production and process flow work sheets. These techniques are related to the system structure organizing techniques discussed in the previous section.

Consider the added problem of comprehending the details of a process that is equally complicated but for which the blueprints have not been drawn and for which no comparable process exists. This difficulty exists with many solar thermal systems. The formality of defining boundaries, characterizing external and internal effects, determining subsystems, and establishing interactions between the subsystems, as was discussed in the previous section, will provide a discipline for arriving at a well-structured view of reality. Even if the subsequent system modeling is never performed, the system structure evolved will provide a logical framework for subsequent policy and operational decisions.

The use of systems models can generate insight into system operation and component interactions   Once a systems model has been developed it is then used to characterize the operation of the system for various designs. Some systems models may show only gross outcomes such as fraction of load carried by solar on a monthly or yearly basis, but in many models the performance of subsystems and performance at shorter intervals is also computed. By following the behavior of subsystem inputs, outputs, or other variables, an intuitive appreciation for the extent to which various interactions occur and under what conditions they occur can be gained.

For example, by following the hourly output from a residential solar-heating system simulation, one can gain an appreciation for the strong role that thermal storage tank temperature plays in the operation of the system. Collector efficiency falls off as tank temperature rises, providing a dampening action for high tank temperature tendencies. Tank temperature is also a direct measure of the capability of a system to sustain solar heating over periods of little sunshine. Relating this variable in the system model output to various input conditions can allow one to understand the system better and thus lead one to design changes that are advantageous.

Different systems can be tested under identical conditions by using systems models   When comparing the performance of prototype systems, the requirement of repeatable conditions necessary for a controlled experiment is usually not met. For example, to provide the same insolation and weather conditions, identical buildings with different solar-heating systems may be built side by side. If the buildings are occupied, the building loads will certainly be different. Even if the buildings are not occupied, there may be load differences due to variations in materials, construction, relative positions, accesses for maintenance, and so forth. Also, it may be argued that identical side-by-side systems are unnecessarily consumptive of research resources because other factors such as variations in location are not being tested. Thus the analyst invariably faces the problem of experiment differences, often temporal and geographic differences, when comparing the performance of prototype systems.

Under these circumstances the best that the analyst can do is to state comparative results in terms of confidence ranges, rather than single values, with a positive likelihood that the real results may not, in fact, be within the ranges. Alternatively, single values may be stated where it is understood that the conditions are not the same. In any event there can be a great deal of uncertainty when such comparisons are made.

Identical conditions can be specified when applying the same systems model to different systems or variations of the same system. The potential advantages in cost savings as compared to building prototype systems and in increased reliability and precision in such comparisons should be obvious. Furthermore, a systems model can frequently be very precise in reflecting performance differences between design variations of a system even though it may not be accurate in predicting absolute performance. However, when a systems model is used in this manner, care must be exercised to insure the model is internally consistent. This consideration, part of proper model building, is taken up in subsequent sections.

Systems models can be used to identify the most important design considerations   The more complex a system is, the more difficult it is to intuitively identify what aspects of the system should be given most attention in research and development. Moreover, intuitive judgments often turn out to be wrong. A systems model can be used to examine the quantitative consequences of a change in one or more design variables of a system. By comparing the consequences of changing each design variable, a judgment can be made as to their relative importance. This approach is called sensitivity analysis.

To perform a sensitivity analysis, a performance measure, the basis on which system performance is judged, must be specified. It is important that this measure reflect the objectives of the analysis, as the following example illustrates. In a residential solar-heating system, thermal storage tank size can be varied over a wide range without significantly affecting the fraction of the load carried by solar. However, large storage tanks are expensive. Thus, the system's performance is insensitive to storage tank size when the fraction of the load carried by solar is the performance measure and is sensitive to storage tank size when the system's economics are included in the performance measure. These points are illustrated in Fig. 17.4, where storage tank size is being varied while all other design variables are held constant.

Systems models permit systems optimization to be performed   As was the case in sensitivity analysis, the selection of a performance measure reflecting research goals makes it possible to ask some additional questions and use systems models to secure the answers. One question that might be asked is, "What is the best system?" Systems optimization refers to the steps, procedures, and techniques used to answer this question.

Systems optimization requires the implicit or explicit examination of the performance of a system for a multiplicity of designs under identical external conditions and the identification of the design that is the best according to the measure of performance. In most cases, only systems models can provide the optimization procedure with the capability for performing such an examination.

Systems models can be used to develop simplified design approaches   In using systems models to analyze a wide range of different systems and system design variations under difference external conditions, insight is generated not only into system operation and component interaction, but also into the role that the structure of the model itself plays in generating the results. Simplified solar design methods have been developed based on both kinds of insight. The *f*-chart procedure described in Chap. 14 is an example of the use of the latter kind of insight.

Modeling Approaches

A previous section described the process of organizing the system into a logical framework of subsystems, interactions, and boundaries. This process created the structure of the system representation and included the identification and interconnection of inputs and outputs. It did not, however, specify the manner in which the inputs and outputs varied. This is the function of modeling.

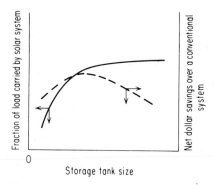

Fig. 17.4   Illustration of the effect that choice of performance measure has on a sensitivity analysis.

Limiting factors such as available time and resources, the degree of expertise of the modelers, and needs of users should be considered in choosing a modeling approach. The degree to which a model represents reality and manageability of the model will often be contradictory objectives. A balance should be struck that is consistent with modeling objectives. Many other factors will enter into the choice of a modeling approach, making it difficult to state general guidelines for the selection process. However, some recommendations will be made for various situations at the end of this chapter.

Mathematical models Realizing that a more general interpretation could be made for the word "mathematical," we will consider under the category of mathematical models only those representations of real systems, components, processes, and functions that can be directly expressed by algebraic equations, differential equations, integral equations, logical expressions, graphs, and tables.

The mathematical expressions that make up a mathematical model are usually derived from basic physical principles and, in fact, a mathematical model may be no more than the assembly of mathematical statements of well-known physical laws such as the first law of thermodynamics. At the other extreme is the regression or polynomial fit model. In this model an expression involving weighted sums of various linear, squared, cubed, trigonometric, exponential, and cross product terms of the design and input variables is equated to the outputs. The weighting coefficients are then determined from experimental results.

Many systems cannot be modeled mathematically, and other approaches such as simulation must be used. Even when a system can be modeled mathematically, the resultant expressions often cannot be solved directly.

Solving mathematical models in closed form. Solving a mathematical model in closed form refers to being able to derive a value for the dependent variables in an equation or set of equations simply by replacing the independent variables by their values and performing the indicated operations. In mathematical models of any degree of sophistication, including models of solar thermal systems, it is usually not possible to find a closed-form solution. Even when the model consists of a single equation in two variables, it may not be possible or perhaps desirable to find a closed-form solution. An example of the latter case would be when the value of y in the expression $ax^5y = bxy^2 = cx^2y^5 = 0$ must be determined for various values of x. Techniques exist for solving this problem, but it will be far less tedious to use an iterative procedure.

Sometimes a closed-form solution may be obtained for a fairly complex mathematical representation of a system by making simplifications in the mathematical form. It does not necessarily follow that the model becomes significantly less accurate. The accuracy may change very little or not at all. Discovering a closed-form solution is a creative process.

Another possibility for arriving at a model that can be solved in closed form is to make assumptions that allow one to simplify the derivation of the mathematical equations. An example is the derivation of the Hottel-Whillier-Bliss (HWB) equation (Chap. 7) by which solar collector performance can be calculated in closed form. Among the more crucial of the assumptions required to simplify a collector energy equation containing partial derivatives in time and two-dimensional space are that steady state conditions exist and that temperature gradients in the direction of flow and between the fluid tubes can be treated independently. The HWB equation is

$$q_u = A_c F_R \left[S - U_c \left(T_{f,\text{in}} - T_a \right) \right] \tag{17.1}$$

where F_R and U_c are functions of the fluid conditions, absorber geometry, and temperature. If one guesses an average fluid temperature, F_R and U_c may be found directly. With values for F_R and U_c, Eq. (17.1) may be solved in closed form for the dependent or output variable q_u, knowing only the dependent or input variables; absorbed solar energy per unit area S ($= I_c \tau\alpha$), collector area A_c, ambient temperature T_a, and inlet fluid temperature $T_{f,\text{in}}$.

Solving mathematical models iteratively. If a mathematical systems model cannot be solved in closed form, it often can be solved iteratively. For example, an inlet fluid temperature was guessed in the last section, permitting a closed-form solution for q_u. In some temperature ranges, however, fluid properties and U_c may change rapidly enough to warrant use of a more accurate value of average temperature. In the process of deriving Eq. (17.1), an expression for average fluid temperature is generated (see Chap. 7):

$$T_{f,\text{ave}} = \frac{S}{U_c} \left(1 - \frac{F_R}{F'} \right) + \left(T_{f,\text{in}} - T_a \right) \frac{F_R}{F'} + T_a \tag{17.2}$$

This expression is a function of U_c and various fluid properties, such as c_p. Since the fluid properties are being evaluated at the average fluid temperature, this temperature appears on the right-and left-hand side of Eq. (17.2). Since the fluid properties are complicated functions of $T_{f,\text{ave}}$, a closed-form solution to Eq. (17.2) cannot be obtained. However, the equation can be solved iteratively.

The iterative process used to solve for $T_{f,\text{ave}}$ is a cyclic one known as successive approximation and is illustrated in Fig. 17.5. When a value sufficiently close to the previous value of $T_{f,\text{ave}}$ is obtained from the iterative process we are satisfied and use that value to solve Eq. (17.2).

Different mathematical models will behave differently when this technique is applied. The convergence of the values can be quite rapid or quite slow and the values may diverge or convergence may be to an unrealistic value, depending on the value initially guessed. If these kinds of problems occur, other more sophisticated techniques may be required. The reader should be aware that these kinds of difficulties are possible and should be prepared to use the numerical analysis literature to find an appropriate solution technique.

Cyclic iteration is one means of solving a mathematical systems model that cannot be solved in closed form. Another means is sequential iteration. A mathematical systems model can sometimes be solved piecemeal and solutions to the separate parts aggregated into a solution to the total problem.

A mathematical model of a solar thermal system whose inputs are daily insolation and ambient conditions may yield to a sequential iterative approach. Insolation and ambient conditions may vary in degree from one day to the next, but in their basic patterns they display a periodic behavior from day to day. An example of such behavior is illustrated in Fig. 17.6. The solution of the solar thermal system mathematical model for a lengthy period, say a year, would be very difficult to accomplish in closed form or with cyclical iteration, since the mathematical expression for the totality of a year's information like that given in Fig. 17.6 would be quite complicated. For example, a Fourier series might be used but many higher order terms would be necessary for a reasonable solution. However, if individual days are considered, the mathematical representation of insolation and temperature can be much simpler. A sequential iteration scheme for this problem based on daily iteration is illustrated in Fig. 17.7. A technique developed by Bruno and Kersten,[1] described later in this chapter, is based on this approach.

Simulation models   In contrast to mathematical models that are directly expressible in mathematical equations systems simulation models, though they may use mathematical expressions, are primarily procedural in nature. In other words, the focus or superstructure of the

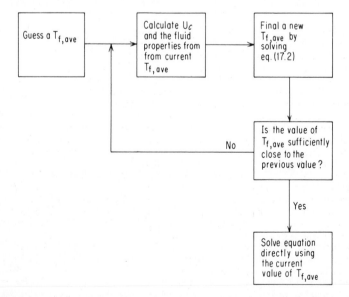

Fig. 17.5 Cyclic iterative approach to solving for the performance of a collector system.

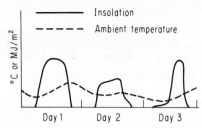

Fig. 17.6   Daily variation in insolation and ambient temperature.

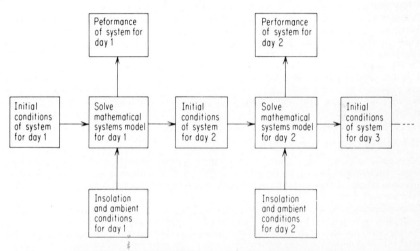

Fig. 17.7   Sequential iterative approach to the solar thermal problem.

simulation model is the procedure or logic whereby possible outcomes of the system either happen or do not happen.

Simulation models are invariably executed on a digital computer although it is also possible to use an analog computer or to perform simulations by hand. A simulation model of a complicated system will tend to be quite lengthy since, even though the languages are concise, a significant level of detail is required for a complete representation of the system. Because of the absolute rigor required in using the language and the amount of detail required, the process of debugging (removing all programming errors) is sometimes quite lengthy, often requiring months of effort. The level of detail and short time steps normally required when running the computer simulation model often consume extensive amounts of computer time. Therefore, simulation can be quite costly. This disadvantage is important when a large number of repetitive calculations with different designs is required as is the case when optimizing.

Simulation modeling is an important modeling approach for solar thermal systems, particularly for solar building heating and cooling systems, since these systems do not lend themselves to mathematical modeling approaches. However, there is some indication that mathematical models can be used in some solar thermal areas where simulation was previously required. Because of the importance and widespread use of simulation in modeling solar thermal systems, it will be treated in some detail later in this chapter.

Empirical models   Empirical models are those models derived from observation of results of actual systems or results of other systems models, where the derived relationships have no connection to, or perhaps a weak or incomplete connection to, the basic physical laws governing the system modeled. Regression or polynomial fit techniques are often used to establish these models. Empirical models are usually mathematical, but may take on virtually any form. A simplified example of such an equation is

$$Y = ax_1 + bx_2 + cx_1 x_2 + dx_1^2 + ex_2^2 + fx_1^2 x_2 + gx_1 x_2 + hx_1^3 + \ldots$$

In this example there are only two independent variables, x_1 and x_2, whereas in the actual case there are usually many more. For example, in building load analysis, Y could represent the load and variables, x_i would represent wind speed, building capacitance, ground capacitance, insolation, and so forth. Since existing models based on physics may not show good agreement with short-term behavior of a real building, the use of an empirical model may be advantageous.

Physical models   Sometimes a mathematical or simulation model will not be adequate to determine certain performance aspects of solar thermal systems. For example, forced convection heat transfer due to wind impinging on collectors at various angles and speeds has not been modeled very satisfactorily. If one needs to accurately determine heat transfer due to this mechanism, a wind tunnel test of a scale version of the collector may be required. This is an instance where a physical model rather than a mathematical or systems model has been chosen. Other examples of situations where a physical model might be preferred are where component maintenance or reliability requirements must be estimated. Also, reduced scale prototypes may be designed and built during mathematical or simulation modeling efforts to gain additional inputs on crucial aspects of the system so that the analytical models can be improved. Current capabilities of the mathematical or simulation models would be used to aid in the "best" design for the prototypes.

When using reduced-scale prototypes care must be exercised to insure that scaling effects are accounted for. For example, when a collector box is reduced in size, reduction of the space between the absorber plate and glazing will change convection losses. However, if the space is kept the same, then side losses will increase since the side areas will constitute a greater proportion of the area for losses.

The Role of Optimization in Modeling

A systems model is invariably used for performing systems optimization. If this is to be done sooner rather than later, it is important that this be one of the criteria used in the systems model selection process. Otherwise, the model selection effort may generate a systems model that is inappropriate to use for optimizing. This circumstance may seriously retard the achievement of overall objectives.

Implicit in optimization is the statement of a performance index. Measures of solar thermal system performance usually involve purely physical performance outputs of a system such as total energy generated and the fraction of energy supplied by solar. Alternatively an economic index such as dollars per unit energy generated may be used. A measure that recognizes the importance of system economics is usually preferred.

The optimization criterion is a statement, in terms of the measure of performance, of what the optimization technique is to accomplish. Usually a choice of design is sought that maximizes or minimizes the performance measure. Another alternative might be to minimize the cost of the worst possible outcome. This criterion might be appropriate in a study to determine federal standards for solar heating system components where the objective is to protect the consumer from catastrophic system failures. The primary purpose of an optimization technique is to efficiently progress to an optimum solution satisfying the selected criterion.

Two possible ways of integrating the fact that optimization is to be performed into the systems modeling procedure will be discussed in the remainder of this section. Each will tend to produce a choice of systems model and optimization technique that is quite different from the other.

Optimization as a parallel but separate effort from systems modeling   Optimization is usually a discipline in the operations research or systems optimization fields. Although there are a great number of different systems models used in the explanations of the various optimization techniques, these models take on a distinctly secondary role in the organizing process and give the distinct impression that the process of choosing an optimization model is distinct from the process of choosing a systems model.

The selection process and its consequences when the optimization and systems models are chosen in a separate but parallel process is illustrated in Fig. 17.8. As may be seen, the optimization model tends to provide a superstructure or means of guiding successive runs of the systems model. The optimization model selects the design and provides it to the systems model. The systems model then calculates the performance of the design using a fixed set of external conditions and provides it to the optimization model. The optimization model then uses this performance information, and possibly prior information, to select the next design which is input to the systems model for the next performance calculation. At some point in the

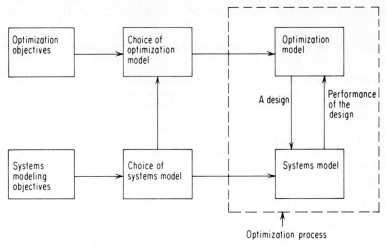

Fig. 17.8   Parallel optimization and systems model selection process.

cyclic iteration process between the two models, the optimization model will determine that the design that provides the best performance in terms of the performance measure has been obtained and the process is terminated.

The cyclic iterative process just described can be carried out only by a restricted subset of available optimization techniques. This subset consists of techniques classified as nonlinear programming and search. These techniques will be discussed later in the chapter. Implicitly rejected by the parallel approach to optimization and systems model selection are the remainder of the available optimization techniques.

Optimization as an integral part of systems modeling   Figure 17.9 illustrates an optimization and systems model selection process that avoids the problem of making a final choice of one model before considering the other. In the figure the single box for selection of both the optimization and systems model is meant to suggest that the process is carried out simultaneously. With every optimization model under consideration, compatible systems models are considered, and with every systems model under consideration potentially compatible optimization models are considered. Thus a more comprehensive range of possibilities is considered with the likelihood that overall objectives will be better satisfied than when the systems model is selected prior to the optimization model as in Fig. 17.8.

In performing the selection process this way, the performance measure and the way in which the optimization is performed become more naturally part of the system structure. The addition of constraints that occurs when an optimization structure is considered can reduce the scope

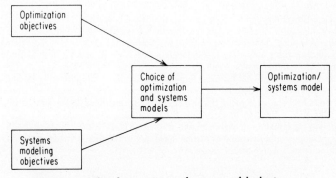

Fig. 17.9   Combined optimization and systems model selection process.

of alternatives, which in turn may point out that some aspects of the systems model will not be important. Thus, the systems modeling effort may be simplified. Also, the mathematical structure of the optimization technique, because it contains a process for choice, may permit preliminary analyses that generate some partial conclusions about optimum system design.

The optimization model that evolves from this integrating procedure may, in fact, be the same one that would have evolved from the parallel approach. However, a sequential rather than cyclic procedure is also possible and the mathematics of optimization will often be more firmly interwoven into the systems modeling. Under this approach one will often find the optimization technique conforming to subsystem interaction paths.

In addition to the nonlinear programming and search techniques from the optimization approach of Fig. 17.8, the broader perspective will lead to consideration of large-scale mathematical programming, dynamic programming, linear programming, and so forth. These techniques will be discussed in some detail later in this chapter.

MATHEMATICAL MODELS OF SOLAR THERMAL SYSTEMS

This section will discuss some mathematical models of solar thermal energy systems that are representative of the types described in the previous section. As will become obvious, much of the recent work to obtain solar thermal systems models that are easy to use, efficient to run, and still sufficiently accurate and detailed for most design purposes has been in the area of mathematical modeling.

Closed-Form Solutions

So far a closed-form solution for complex solar thermal systems, such as residential solar heating or cooling systems or solar thermal electric power systems, has not been consistent with the development of an accurate, flexible, and sufficiently detailed system model. The difficulties are due to complexities in interactions among subsystems; complexities due to serial interactions in time, that is, the influence of system histories; and complexities of the subsystems themselves. The additional mathematical structure imposed by optimization may ultimately produce a breakthrough, but the difficulties are formidable.

Iterative Solutions

The greater flexibility inherent in iterative mathematical approaches has resulted in the development of some systems models that are accurate and contain a useful level of detailed results and analyses. Two methods will be presented, both of which are relatively recent developments. One requires a cyclic iteration approach in its solution and the other uses sequential iteration.

Cyclic iterative—STOLAR   A STOchastic soLAR energy systems model STOLAR has been developed by Lameiro.[2] This approach consists of a statistical compression of hourly insolation and weather data and simplified system governing equations to form a mathematical systems model of residential solar heating and service hot-water systems. The model focuses on transitions in storage tank temperature as the driving force for system behavior. The resulting mathematical systems model is relatively simple, with numerical problems only associated with size and structure.

The statistical data compression of one or more heating seasons' hourly insolation and weather data was accomplished by using the data to estimate the coefficients of Markov chains for three external systems inputs: ambient temperature, insolation, and service hot-water demand. The transition period for the Markov chain was taken as one hour to match the hour-by-hour data available on many weather tapes. Figure 17.10 is an example of a Markov chain for ambient temperature transitions reduced to three temperature states to simplify its presentation. The temperature states have been defined as the temperature ranges shown. Each row is a conditional probability mass function expressing the probabilities of transitions to one of the three states, if the system is now in the state indicated by the row index.

System-governing equations have been kept simple in STOLAR to reduce required computer run time. Equation simplicity is not otherwise necessary to the model structure. The equations consist of the energy balance on the house,

$$q_u + q_{\text{aux}} - L = (mc_p)_S \, \Delta T_S + (mc_p)_H \, \Delta T_H \tag{17.3}$$

States	1	2	3
1	0.75	0.25	0
2	0.05	0.95	0.05
3	0	0.10	0.90

State 1: $T_a < 0°C$
State 2: $0°C \leq T_a < 20°C$
State 3: $20°C < T_a$

Fig. 17.10   Simplified Markov chain for ambient temperature transitions.

the energy balance on the storage tank,

$$q_u - q_{SH} = (mc_p)_S \, \Delta T_S \tag{17.4a}$$

the useful energy from the collector,

$$q_u = A_c I_c \eta_c (T_S, T_a, S) \tag{17.4b}$$

and the load energy requirements,

$$L = (UA)_H (T_H - T_a) \, \Delta t + \text{HW} \tag{17.5}$$

where the subscripts S and H refer to the thermal storage and house, respectively, ΔT_s is the temperature change in thermal storage, q_{SH} is the energy provided to the house from storage, Δt is 1 h, HW is the service hot-water load for 1 h, and η_c is the collector efficiency taken from a straight line plot of $(T_S - T_a)/I_c$ (see Chap. 7). A new storage tank temperature after an hour under conditions specified by $T_{S.\text{present}}$, T_a, S, and HW can then be calculated as:

$$T_{S,\text{new}} = T_{S,\text{present}} + \left[\frac{q_u - q_{SH}}{(mc_p)_S} \right] \tag{17.6}$$

by Eq. (17.3) through (17.5) where q_{SH} and q_{aux} are determined by using a thermal storage cut-off temperature.

Equation (17.6) is used to combine the Markov chains for each of the three external systems inputs, T_a, S, and HW, into a single Markov chain called the monitor matrix. The states of the monitor matrix are defined by a four-tuple consisting of the state definitions of the original three chains plus a thermal storage tank condition also expressed in terms of temperature ranges. A $T_{S,\text{new}}$ is calculated for each value of the four-tuple where $T_{S,\text{present}}$, T_a, S, and HW are taken as the range midpoint values. Equation (17.6) yields a unique value of $T_{S,\text{new}}$ for an individual value of the present four-tuple, and the original Markov chains provide a number of combinations of resulting values for the other three elements of the new four-tuple. The thermal storage temperature range that the value of $T_{S,\text{new}}$ falls into is taken as the new thermal storage condition in the new four-tuple. Thus each row of the monitor matrix will have only a few nonzero conditional probability entries.

When the monitor matrix is calculated, a vector of values of q_{SH} and a vector of values of L are calculated, where each element of the vector corresponds to a value of the four-tuple. In other words, there is a calculated hourly load and a calculated hourly energy provided to the house by solar heat corresponding to each monitor matrix row.

A stationary vector is then calculated from the monitor matrix. This vector expresses the long-term fraction of time that the system is in each of the states defined by different values of the four-tuples. By multiplying this vector by the vectors of values of q_{SH} and L associated with the monitor matrix rows, heating season load and fraction of load carried by solar can be calculated.

The monitor matrix is largely vacuous and can be efficiently specified by storing only nonzero values and beginning and ending matrix location indices. Even so, the computer storage required is quite large. The size of the matrix prevents a direct inversion to calculate the stationary vector, and an iterative cyclic successive approximations method is required. Furthermore, a straightforward application of this method was found to converge very slowly, and an exponential curve fit prediction routine was developed to substantially speed up convergence.

STOLAR results matched TRNSYS (see below) fraction of solar predictions within 6 percent absolute difference for three test cities. The load prediction agreement was even better, within 2.9 percent absolute difference.

STOLAR has a serious disadvantage in that it is very sensitive to selection of the ranges of T_S which characterize states in the monitor matrix. The sensitivity is due to threshold entry conditions for transitions from one state to another. A change in system design can trigger a jump to an unreasonable result necessitating the redefining of state ranges in the program. The STOLAR procedure includes a discussion of how to obtain the proper ranges, but the presence of this difficulty is an annoyance. The large memory requirements may also constitute a difficulty for some potential users of STOLAR.

Sequential iterative—Bruno's method   A mathematical systems model based on a sequential iterative solution scheme has been developed by Bruno and Kersten.[1] This approach requires only daily insolation and weather data. The daily behavior of insolation and other external inputs to the system are modeled using equations whose form yields an exact solution to the systems-governing equations. The systems-governing equations are similar to the systems-governing equations that are shown for the TRNSYS model description of the next section. The daily data values are substituted into the exact solution to the systems-governing equation yielding the day's performance as well as the final storage tank temperature and other day-end conditions. The analysis is then repeated for the next day using final conditions from the previous day as initial conditions. The process is sequentially repeated as shown in Fig. 17.7 day by day through the entire year.

The mathematical form of the equations used to model external inputs to the system were selected to give a good representation of the conditions being modeled while permitting an exact solution to the system governing equations. The insolation model is

$$
I_c(t) = a \left(\frac{\cos \dfrac{\pi}{12} t - \cos \dfrac{\pi}{12} \tau}{1 - \cos \dfrac{\pi}{12} \tau} \right)^+
\tag{17.7}
$$

where $\tau = b[\omega(o) - \omega(\beta)]/2$, t is the time from solar noon in hours, $\omega(\beta)$ is the length of time that beam radiation strikes a surface at slope β on a clear day, and $(x)^+$ is the positive part of x. The value of a and b are determined from a day's insolation. By adjusting b downward from 1, the effective duration of good insolation conditions for the day is reduced to conform to the actual data. After b has been determined, a is chosen so that the integral $\int I_c(t) dt$ over the day is equal to the total incident energy for the day. For example, in Fig. 17.6 the value of b would be close to one for days one and two and somewhat smaller than one for day three. The values of a would differ significantly for days one and two.

The mathematical model for ambient temperature as it affects the collector is taken as the average

$$
\overline{T}_a = \frac{1}{\tau} \int_{-\tau/2}^{\tau/2} T_a(t) dt
\tag{17.8}
$$

and for the wind speed as it affects the collector

$$
\overline{v}_w = \frac{1}{\tau} \int_{-\tau/2}^{\tau/2} v_w(t) dt
\tag{17.9}
$$

The heating load is modeled similarly to the insolation or as a step function that changes at only a few times during the day. Service hot-water load is calculated as instantaneous values at several times during the day.

The day-by-day data can be simplified to a few numbers including daily values of a, b, $\overline{T}_a$, and $\overline{v}_w$, given in Eq. (17.7) through (17.9), daily values of intensity and time of change for the heating load step function, and daily values of energy and time occurrence for the service hot-water function. This simplification only need be performed once for the year's data in a given location since these values are not affected by solar energy system design changes other than collector tilt. The simplified day-by-day data then operate the model sequentially and design calculations based on the solutions to the governing equations are performed. CDC 6400 computer run times for this latter operation average about two cpu seconds for a year of operation of a residential solar heating and cooling system.

The results of this model were compared to results of a finite element simulation for a wide range of different residential solar heating and hot-water systems in a number of different locations. The accuracy of annual thermal performance was always within 2.5 percent absolute error of the finite element model results and almost always within 1 percent. Results for yearly electrical consumption by the pumps were within 8 percent absolute error. This larger error is readily understandable in light of the fact that the pump on/off cycles are much more frequent than the daily cycle of the model. On a daily basis, absolute differences from the finite element results of up to 15 percent were observed. However, most differences tended to cluster to within a few percent of the finite element model results.

No major difficulties are perceived in the use of this systems model as long as the level of detail that the model provides is adequate. If more detail is required, such as a better estimate of pump electrical consumption, or if an hourly reporting of systems performance is needed, then other systems modeling approaches, such as simulations, should be used.

SIMULATION MODELS OF SOLAR-THERMAL SYSTEMS

The concepts behind the development of a simulation model, the inputs to a simulation model, methods for the use of simulation results and for the validation of simulation models, and finally a specific example of a simulation model TRNSYS are discussed in this section.

Overcoming Difficulties with Subsystem Interactions and Complexities

When faced with modeling a complex physical process, such as a solar thermal system, it is helpful to break the system into subsystems and then to represent the complete system by interactions between these subsystems. For example, the control system itself is a rather complex system but may be visualized quite simply as a collection of three subsystems: the sensor subsystem, the logic subsystem, and the output subsystem. The sensor subsystem consists of the sensors that typically measure temperatures and provide this information to the logic subsystem. The logic subsystem typically is comprised of comparators that compare the temperatures received from the sensor subsystem with various set point temperatures or with other temperatures and on the basis of these comparisons provides a logic signal to the output subsystem. The output subsystem simply provides a signal to turn on or off various subsystem components, such as pumps or blowers.

This relatively simple example may be expanded to increasingly complex systems. For example, a solar space-heating and domestic hot-water system may be represented pictorially as shown in Fig. 17.11. This figure represents schematically the various subsystems and their interactions.

When detailed performance information is to be provided by the systems model the system representation may become too complex and intricate to model in any simple way. The alternative is the procedurally oriented computer simulation approach that is well suited to representing subsystem interactions and complexities.

In the discussions of the TRNSYS simulation model that will follow, we shall illustrate how a system representation such as that shown in Fig. 17.11 may be used to construct an information flow diagram for a system simulation.

Simulation Structure

The procedural orientation of computer simulation languages allows the analyst to directly utilize the subsystem and interaction representation of the real system which was discussed at the beginning of this chapter. Computer subroutines are identified with subsystems and the logic and structure of the interactions between subsystems are written into computer statements that tell when and under what circumstances the subroutines are entered. While the transition

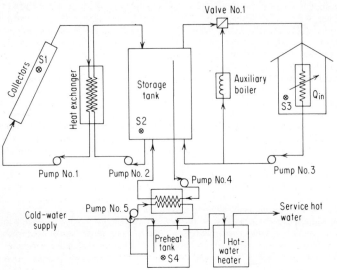

Fig. 17.11 Schematic representation of a typical liquid solar heating and cooling system.

from system representation to computer simulation model is a natural one, several fundamental structural issues must be resolved before the simulation modeling process is initiated.

Time steps Two different ways of incrementing a simulation can be used: incrementing by event and incrementing by fixed-time step. An event is an identifiable change in the status of a system, such as when a pump turns on. Calculations are performed only when a simulated time is incremented, and one means of shortening running times of simulation models is to insure that calculations are performed only when absolutely necessary. Since calculations must be performed when the systems status changes, event incremented simulations generally run the fastest. However, many of the processes in solar thermal systems are continuous, and identification of events under this circumstance is difficult. This is why most solar thermal systems simulations models are incremented by a fixed-time step.

Degree of disaggregation in simulation The disaggregation possibilities for general system representation discussed at the beginning of the chapter are applicable to the simulation modeling itself. The way that the system is disaggregated (into components, functions, or elements) may be different for different research objectives. Element breakdown usually yields the most accurate model, but because the breakdown is into more numerous parts, more calculations are required and therefore computer run times are longer. However, programming may actually be easier since there are few elements that are functionally different and thus only a few subroutines need be written.

Simulation Inputs

The inputs to a system simulation model are comprised of (1) parameters and (2) external inputs. The parameters are design specifications for the various subsystems. For example, storage size and insulation for the storage tank would constitute parameters. External inputs to a solar thermal system simulation include solar radiation, ambient temperature and wind speed. The external inputs drive the simulation model. That is, the purpose of the simulation model is to determine the response of the system to varying climatological inputs. These may be modeled as deterministic inputs, reduced data sets, or stochastic inputs. Each method of modeling the external inputs has its own degree of usefulness.

Deterministic inputs The most common way of treating the climatological data is using the actual measured climatological data. If records are available for the desired location, then these records may be used. The U.S. National Weather Service has established the SOLMET network which provides hourly measurements of solar radiation, wind speed, ambient temperature, and relative humidity. This information may be obtained from the National Climatic Center (see Chap. 2).

For locations other than the 38 included in the SOLMET network, one would need measurements from some other source or one could use a model for radiation data. Several models have been developed that estimate radiation based upon observations of other climatological variables as described in Chap. 2.

Reduced data sets   Even when several years of climatological data are available, one may not wish to make use of all of them in a systems simulation. Simulating the operation of a solar system for more than one year can be expensive. Consequently, one may wish to simulate the performance of a solar system over a shorter period of time by determining a reduced data set that would be representative of the complete data set. The reduced data set may be determined by selecting either a standard year from many years of observations or by simply selecting periods from the data that, when used in the simulation model, yield results that are representative. For example, periods for which both the insolation and temperature are low represent periods during which the solar energy system would not perform well. Whereas, periods during which the insolation and temperature are high represent periods during which the solar system would perform well. Studies have indicated that it is often possible to obtain results from simulations using reduced data sets that agree very closely with results obtained from the simulation using a complete year of data.

Stochastic inputs   It is often useful to drive a simulation model with stochastic inputs. The stochastic inputs may be derived by computing the statistical moments of the actual climatological data and then using these moments in connection with a random number generator to determine inputs to a simulation model. Since this approach can model the weather as it could actually vary from year to year, it could be used to determine the year-to-year variation in performance of a solar thermal system.

Analysis of Simulation Results and Validation of Simulation Models

A methodology for validating simulation models and analyzing simulation results will be discussed in this section. This methodology has been applied to the TRNSYS program discussed in the next section.

There are two aspects to checking a systems model. The first is model verification, that is, the process of determining if the model actually represents the system that it was intended to model. This means that performance data from the "real" system must be available to validate the model. To validate a model, one should consider all aspects related to the modeling process: (1) the quality of the input data used for model validation, (2) the validation of model assumption, (3) the validation of model logic, and (4) the validation of model behavior. The first area is related to data analysis and data selection techniques for model calibration and model validation. The second and third areas are related to techniques used in the modeling process, goals to be achieved by the model, and types and amount of input data available. The fourth area is needed to assure that the behavior of a model is in accordance with that of the real system.

Given a set of competing models to be used in the simulation of a particular process, one should next consider which of the models is the "best" in terms of fulfilling a specified purpose. This problem will also be discussed below.

A systematic approach to solar-heating system model validation based on clustering and statistical analysis of simulation residuals is detailed in the remainder of this section.

Solar building heating data analysis procedure   The need to have adequate and sufficient weather and solar component operating data is one of the most important requirements in any attempt to validate a systems design model. Following is a methodology for further screening and analyzing data collected in solar heated buildings.

1. A preliminary screening by simple thresholding is carried out to eliminate unreasonable values in the data set. This step will eliminate shot noise embedded in the collected data.

2. The data for each hour of each month are then fed to an outliers identification program, using a selected distance function, such as Bhattacharyya distance, for discriminant analysis. All outliers are identified, and their values considered as missing so that new values may be filled in.

A multivariate statistical approach can be used to fill in missing data points or to correct unreasonable data values. Other more sophisticated filtering techniques based on spectral analysis, estimation theory, or polynomial fitting techniques which are more suitable for generating large missing records can also be used.

Finally the whole data set is used as input to a statistical computer program package to compute the relative and cumulative frequency distribution functions for data of each month and for monthly averages of the data-year under study. The flowchart of the data analysis procedure is shown in Fig. 17.12.

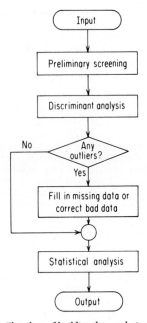

Fig. 17.12   Flowchart of building data analysis procedure.

Solar building design model validation procedure   The model validation process is a scientific approach evaluating the accuracy of a model. The necessary steps in the model validation process are presented in Fig. 17.13. Data used in calibrating the model should not be used in validating it. That is, given a large data set for use in the validation of a model, one needs to divide it into two separate sequences: a training set for model calibration purposes and a checking set for model validation purposes.

In the training sequence one must include points that represent system extremes to take into account all process characteristics in the model calibration phase. In contrast, the checking sequence must include points that represent the *average* process characteristics. Thus, in order to divide the data set into training and checking sequences, one needs to have a measure of separation or distance measured from the mean. The simplest normalized statistical distance from the mean is the variance, defined by:

$$D^2 = \left(\frac{X_1 - \overline{X}_2}{\overline{X}_1} \right)^2 + \left(\frac{X_2 - \overline{X}_2}{\overline{X}_2} \right)^2 + \cdots + \left(\frac{X_n - \overline{X}_n}{\overline{X}_n} \right)^2 \qquad (17.10)$$

Points with a large variance must then be included in the training sequence, and points with a small variance must be included in the checking sequence. The length of these data sequences is then determined from model characteristics, such as required time-period for the model to stabilize or for cyclic conditions to be achieved.

A flowchart of a computer program which constitutes one step in the validation process of solar house-design models is presented in Fig. 17.14. The outputs of each simulation model—the simulated values of the states and simulation residuals—are read into the validation program and the following steps are carried out: (1) plot (optional) the simulated values versus the

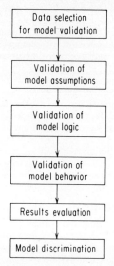

Fig. 17.13   Flowchart of necessary steps in the model validation process.

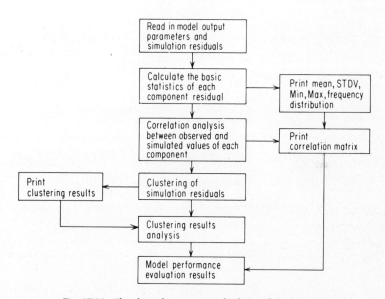

Fig. 17.14   Flowchart of a computerized aid to model validation.

observed values for each subsystem component; (2) compute the basic statistics, i.e., mean, standard deviation, third and fourth moments about the mean, minimum and maximum values, and correlation coefficients between simulated values and observed values for each subsystem component; (3) compute the same basic statistics for subsystem simulation residuals; (4) compute and plot the relative and cumulative frequency distributions for values of each subsystem component; (5) perform clustering analysis of simulated values and simulation residuals to examine for possible systematic errors in the model; and (6) output presentation for further analysis.

From this analysis one can identify what needs to be done to improve the model behavior, extend its domain of applicability, and put it into a more flexible form for other users.

Model discrimination   To assess the quality of a systems model, one needs to determine criteria for model comparison. Model selection criteria must be set up based on various model features which are presented in the validation steps discussed previously. Thus, the quality of a solar house design model is determined by capability to perform the desired tasks, input data requirements, calibration and production costs, feasibility of improvement, cost of improving the model, and payoff of the improved model.

Among those criteria for model comparison, the first is the most important, and, for design models, it should include the capability of automatic selection of subsystem components to optimize the system relative to some selected design and be able to conduct a life-cycle cost analysis for a given system.

Some approaches for systematically choosing among competing digital simulation models based on simulation performance include Bayesian discrimination based on the computation of the a posteriori probability of each candidate model to produce a set of observed values and a decision-making process based on the entropy concept.

Simulation Model Example—TRNSYS

TRNSYS, an acronym for "a transient simulation program," is a quasi-steady simulation model. This program was developed at the University of Wisconsin-Madison[3] by the members of the Solar Energy Laboratory. An early version of the program was released in 1974. The program consists of many subroutines that model subsystem components. The mathematical models for the subsystem components are given in terms of either ordinary differential equations or algebraic equations. For example, the mathematical model for a stratified fluid storage tank is represented by:

$$
\dot{m}_i c_{pf}\frac{dT_i}{dt} = \alpha_i \dot{m}_h c_{pf}(T_h - T_i) + \beta_i \dot{m}_i c_{pf}\ (T_L - T_i) + UA_i\,(T_{env} - T_{in})
$$

$$
+ \begin{cases} \gamma_i(T_{i-1} - T_i) & \text{if } \gamma_i > 0 \\ \gamma_{i+1}(T_i - T_{i+1}) & \text{if } \gamma_{i+1} < 0 \end{cases} q_i + \delta_i \Bigg| \quad \text{for } i = 1, N
$$

(17.11)

where $\dot{m}_i$ = fluid mass flowrate through the ith section
c_{pf} = specific heat of the fluid in the storage tank
T_i = temperature of the ith section
$\alpha_i, \beta_i, \gamma_i, \delta_i$ = control functions
$\dot{m}_h$ = fluid flowrate to and from the heat source
T_h = temperature of fluid entering the storage tank from the heat source
A_i = surface area of the ith tank segment
U = heat loss coefficient between the tank and its environment
T_{env} = temperature of the environment around the tank
q_i = rate of energy input by the internal heating element

The user of the TRNSYS program must first identify the system components that comprise that particular system to be simulated and then must construct an information flow diagram. The purpose of the information flow diagram is to facilitate identification of the components and the flow of information between them. An information flow diagram for the thermal storage unit is shown on Fig. 17.15.

Not all of the subsystems in the TRNSYS program are modeled by ordinary differential equations as was the case for the thermal storage subsystem. For example, the heat exchangers are modeled using algebraic equations. The equations for a counterflow heat exchanger are as follows:

$$
\epsilon = \frac{1 - \exp\left[-\dfrac{UA}{C_{min}}\left(1 - C_{min}/C_{max}\right)\right]}{1 - (C_{min}/C_{max})\exp\left[-\dfrac{UA}{C_{min}}\left(1 - C_{min}/C_{max}\right)\right]}
$$

$$
T_{ho} = T_{hi} - \epsilon\,\frac{C_{min}}{C_h}(T_{hi} - T_{ci})
$$

(17.12)

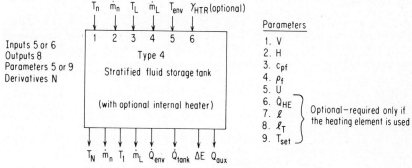

Inputs 5 or 6
Outputs 8
Parameters 5 or 9
Derivatives N

Fig. 17.15 Information flow diagram for a fluid storage tank.

$$T_{\text{co}} = \epsilon \frac{C_{\text{min}}}{C_c}(T_{\text{hi}} - T_{\text{ci}}) + T_{\text{ci}}$$

$$q_T = \epsilon C_{\text{min}}(T_{\text{hi}} - T_{\text{ci}})$$

An information flow diagram for the heat exchanger is shown on Fig. 17.16.

Additional subsystem components in the TRNSYS program include flat-plate solar collectors, differential controllers, pumps, auxiliary heaters, heating and cooling loads, thermostats, pebble-bed storage, tees, relief valves, heat pumps, and cooling devices. There are also subroutines for processing radiation data, performing integrations, and handling input and output chores.

Model validation studies have been conducted in order to determine the degree to which the TRNSYS program serves as a valid simulation program for a physical system. It has been shown by analyzing the results of these validation studies that the TRNSYS program provides an accurate simulation program for the analysis of selected solar heating and cooling systems. The mean error between the simulation results and measured results on actual operating systems is under 10 percent for all state variables considered. These state variables include storage temperature, collector outlet temperature, collector inlet temperature, collector mass flowrate, enclosure temperature, solar radiation on the tilted collector surface, and temperature in and out of various heat exchangers.

EMPIRICAL MODELS

In this section the use of empirical models applied to the design and analysis of solar thermal systems will be discussed. Included will be material relating to the development of models, material related to simplified rules of thumb that may be used in certain cases, and material related to models that are based upon simulation results obtained from other simulation models.

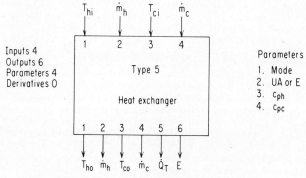

Inputs 4
Outputs 6
Parameters 4
Derivatives 0

Fig. 17.16 Information flow diagram for a heat exchanger.

Use of Experience

The use of experience and insight should not be discounted when considering the development of models for solar thermal systems. The first step in developing a mathematical model of a physical system is to formulate a conceptual physical model. That is, the essential physical characteristics should be considered in attempting to describe the physical process being modeled. Experience and insight play a significant role in this step.

As a specific example of this type of modeling, consider the Hottel-Whillier-Bliss model for a solar collector given in Eq. (17.1). This model represents a classic example of the application of experience and insight to the development of a model. The model, simply stated, represents the heat collected as the difference between the heat absorbed and the heat lost in the solar collector.

Equation (17.1) is a remarkable simplification of the actual system. A detailed description of the solar collector would involve a description of the transmission of radiation through the glazings to the absorber plate, a description of the absorption of the radiation by the absorber plate and radiation to the space outside the collector, reflections by and between the glazings, convective and radiative heat transfer between the absorber plate and the glazings and between the glazings themselves, conductive losses through the sides and back of the solar collector, and heat transfer to the fluid from the absorber plate. Each one of these processes taken by itself can represent a fairly complex process and the description of each of these processes is quite detailed.

The significant simplification was achieved by lumping all of the loss factors into the U_c term and all of the gain factors into the $\tau\alpha$ term. This enables one to characterize a collector by the two parameters, $F_R\tau\alpha$ and F_RU_c. These parameters may be determined quite simply by experiment, as suggested by the form of the model. That is, by dividing the useful energy per unit area by the incident radiation per unit area I_c one has a measure of the instantaneous efficiency as a linear function of the ratio $(T_{f,\ \text{in}} - T_a)/I_c$. The slope of the curve is represented by F_RU_c and the intercept is represented by $F_R\tau\alpha$. When one considers the physical processes involved in a solar collector, one can appreciate the beauty and simplicity of this particular model and the insight required to develop it.

It should be emphasized that the application to which a model is to be put must dictate the degree to which simplifying assumptions may be made. For example, while the Hottel-Whillier-Bliss model is quite adequate for describing the average performance of flat-plate solar heaters, it is not adequate for describing the detailed performance as a function of time of any given flat-plate solar heater because it ignores the transient effects. If one were interested in the dynamic performance of a concentrating collector, for example, then one would have to utilize a more detailed model.

Rules of Thumb

Rules of thumb represent useful modeling concepts. They are most valuable in terms of providing quick estimates of system performance. In general, rules of thumb should not be used for final system design but instead for obtaining initial estimates of factors relating to the final design of a system. Rules of thumb are generally developed from prior experience either from detailed simulation studies or from construction of actual systems.

Several rules of thumb have been developed for the design of space heating systems using either liquid or air as the transport medium in the collectors. These rules of thumb are listed on Table 17.1. These rules have been developed from observations of actual operating systems as well as from detailed simulation studies of solar-heating systems. It should be emphasized that they are approximations and should be considered as such.

Use of Models Based on Actual Data or Simulation Results

Models based on detailed simulation analyses have proven to be extremely useful for the rapid design of solar space-heating and domestic hot-water systems. The f-chart model developed by Klein et al.[4] is an example of a model that is based upon simulation results. The TRNSYS program was used to model two specific configurations of space-heating systems using both liquid- and air-heating collectors. The simulation studies were conducted using Madison, WI, climatological data. The f-chart model was then based upon correlations to results obtained from these simulation studies that were conducted by the use of the TRNSYS program. The correlation was developed between the monthly fraction of load provided by the solar system and two dimensionless parameters. These dimensionless parameters were defined as the ratio between the collector gains and the monthly load and the ratio between the collector losses and the monthly load. The f-chart is described in Chap. 14 in detail.

TABLE 17.1   Rules of Thumb for System Component Sizing

Solar air-heating systems

Collector slope	Latitude plus 15°
Collector air flowrate	1.2–2.5 ft^3/min·ft^2(6.0–12 l/s m^2) of collector
Pebble-bed storage size	0.5–1 ft^3 of rock per ft^2 (0.15–0.3 m^3/m^2) of collector
Rock depth	4–8 ft (1.2–2.5 m) in air flow direction
Pebble size	0.75–1.5 in (2–4 cm) washed and screened concrete aggregate
Duct insulation	1 in (2.5 cm) fiberglass minimum
Pressure drops:	
Pebble-bed	0.1–0.3 in (0.25–0.75 cm) W.G.
Collector (12 to 14 ft lengths)	0.2–0.3 in (0.5–0.75 cm) W.G.
Collector (18 to 20 ft lengths)	0.3–0.5 in (0.75–1.25 cm) W.G.
Ductwork	~ 0.08 in W.G./100 ft (~0.7 cm/100 m) duct length

Solar hydronic heating systems

Collector slope	Latitude plus 15°
Collector flowrate	~0.02 gal/min·ft^2 (0.015 L/s·m^2)
Water storage size	1–2 gal/ft^2 (40–60 L/m^2)
Pressure drop across collector	0.5–1.0 lb/in^2 (3–7 kPa)
Collector heat exchanger	F_{hx} greater than 0.9 (see Chap. 7)
Load heat exchanger	$\dfrac{\epsilon_L C_{\min}}{UA}$ greater than 1, less than 5 (see Chap. 14)

Solar domestic hot-water heating systems

Preheat tank size	1.5 to 2.0 times DHW auxiliary tank size
Air-water coil size	ϵ_{Hx}, greater than 0.2, less than 0.5
Water-water heat exchanger	ϵ_{Hx}, greater than 0.5, less than 0.8

The f-chart model is a very valuable tool in the design of solar space-heating and service hot-water systems since it provides a tool that may be used to very quickly estimate the fraction of the load that would be supplied by a solar system once one has specified the collector parameters and collector area. The results may be very quickly and inexpensively obtained, and many analyses have been conducted to demonstrate that the results obtained from the f-chart model compared rather closely with results obtained from the more detailed TRNSYS simulation model even for systems and locations different from the locations that were used to develop the correlation equations. Therefore, if one is interested in designing a solar-heating system of a type comparable to that which was assumed in the development of the f-chart model, then it is not necessary for one to use the more complicated and more expensive TRNSYS model in order to obtain good design results.

Numerical tests have demonstrated that on an annual basis, the average error in applying the f-chart model for space-heating systems of the type on which f-chart is based is in the order of 2 percent when compared with results obtained from the TRNSYS program.

The relative areas model, developed by Barley,[6] is similar to the f-chart model in that it is a correlation model. In fact, it is a correlation to a correlation. The motivation behind the development of the relative areas model was to provide the analyst directly with an optimal collector area or life-cycle cost. The relative areas model, which can be hand calculated, was developed to provide a tool for the analyst without access to a reasonably large-scale computational facility. It determines the optimal collector area for a given collector on a given system at a given location, the corresponding solar fraction, and the corresponding life cycle savings.

OPTIMIZATION MODELS

Consider a solar residential heating and service hot-water system. Collector area, storage tank size, collector type, control strategy, and plumbing configuration are all important design variables when system economics are included. To assure that an optimum is obtained, the design variables must be varied jointly and not separately. If four values of each of the five design variables are to be looked at, 4^5 or 1024 cases must be examined using the systems model. Four values of each of the design values are not sufficient to adequately characterize the solution space for optimization purposes. Assume that a still conservative 10 values would be adequate. Thus, by looking at four values of each design variable, only $1024/10^5$ or about one percent of the partial solution space has been examined. Optimization should, of course, not be conducted by this exhaustive enumeration approach that we have outlined here. It is the purpose of optimization models to move through the solution space toward an optimum in a manner that significantly reduces computational effort as compared to exhaustive enumeration.

Optimization Model Structure

As discussed previously, an optimization model includes the selection of a measure of performance and a specification or criterion that tells how the measure is to be used to find the best design. The measure and criterion together are called the objective function, which is usually expressed mathematically as a maximization or minimization operation on a weighted sum, product, or other aggregation of the system design variables. For example,

$$\text{Minimize } (c_1x_1 + c_2x_2 + \cdots + c_mx_m) \tag{17.13}$$

might represent a statement which says find the values of the design variables $x_1, x_2, \ldots, x_m$ that minimize the cost of the system where $c_1, c_2, \ldots, c_m$ are the unit costs of the design variables.

When designs are being varied, as is the case with optimization, certain combinations of values of the system variables do not make sense or are infeasible. These combinations are made "not allowable" in the optimization by means of constraints which are usually stated as mathematical inequalities. For example

$$a_1x_1 + a_2x_2 + \cdots a_n x_n \leq b \tag{17.14}$$

might represent a statement saying that we only have b people and design variables 1, 2, ..., n require $a_1, a_2, \ldots, a_n$ people per unit of $x_1, x_2, \ldots, x_n$. Both equations (17.13) and (17.14) are linear expressions. More complicated expressions may be required for an actual situation.

System Optimization

By system optimization we mean the process of finding a best design according to the objective function subject to the constraints for the *system*. Two of the most common mistakes made when performing systems optimization are optimizing over variables one at a time and not considering the total system when applying an optimization approach. This latter mistake is called sub-optimization and refers to when subsystems are independently optimized and the optima combined to form a "systems" optimum. There are ways of combining systems one at a time, such as through dynamic programming and large-scale programming, that assure a systems optimum is obtained. However, care must be exercised in setting up the systems optimization to assure strict observation of conditions specified in the optimization technique that guarantee a systems optimum.

An example of difficulty with a one variable at a time optimization is illustrated in Fig. 17.17. Level lines for two functions of x and y are shown in parts (a) and (b) of the figure. The functions are little different from each other, but the function in (b) will stymie one variable at a time optimization. The optimization is by search, first along x with the value of y fixed. The value of x that minimizes the value of the function is found for the fixed y value and then a search is made along y with x fixed at that minimizing value. The procedure is repeated alternating the roles of x and y until little or no further improvement is noted. As can be seen in (b), search in the y direction will not produce a decrease in the function. Thus it would be concluded that the optimum lay at point B whereas the optimum is in fact at point A. The difficulty is even more pronounced when there are more than two variables.

Subsystems can seldom be optimized independently to obtain a total systems optimum. For example, consider a concentrating collector. The system is separated into two subsystems, the concentrator and the receiver, and each is optimized separately on a minimum cost per unit

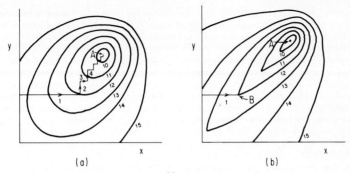

Fig. 17.17   One variable at a time optimization.

of performance basis. It might be concluded that a line focus concentrator is preferable to a point focus concentrator if a line focus concentrator were found that was cheaper than all point focus concentrators. However, this conclusion ignores the fact that the other subsystem, the receiver, tends to be much more expensive in the line focus case. This simple cost interaction is rather obvious and probably would not be missed by the analyst. The actual situation is less obvious because there are trade-offs between cheap, inaccurate concentrators coupled with expensive low-heat-loss receivers and expensive, accurate concentrators coupled with cheap, high-heat-loss receivers, as well as with intermediate options. In any event, one should never try to independently optimize subsystems without careful adherence to systems optimization principles in order to be certain that sub-optimization is avoided.

Another problem that often occurs in optimization of complex systems is that of multiple optima. An example is illustrated in Fig. 17.18. The optimization is again by search, this time along the path that provides the greatest local improvement in the objective function. As can be seen, different optima are found depending on where the search is begun, at point *A* or point *B*. The possibility of multiple optima must be investigated whenever a system is to be optimized. The difficulty can often be mitigated in a search approach by starting from a number of different points or by using an optimization technique that can be made free of the problem, such as dynamic programming.

Two categories of systems optimization techniques will be discussed. One is where the technique is largely independent of the systems model, as in Fig. 17.8, and the other is where the technique and systems model are integrated.

Superstructure optimization techniques   These techniques are generally referred to as nonlinear programming or search techniques and they operate as described in Fig. 17.8 by guiding the sequence of design choices in such a way that, in most cases, the optimum is found with a relatively small number of runs of the systems model. This type of optimization model

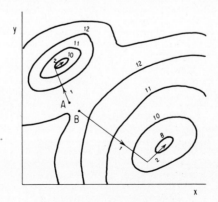

Fig. 17.18   Multiple optima example.

interacts with the systems model only after each model has been fully formulated and the actual optimization is being performed. Therefore, it works well with systems models, such as simulation, where optimization cannot easily be directly integrated into the model.

Nonlinear programming techniques usually operate in two distinct stages. First a direction is chosen that results in the greatest local improvement of the objective function. The direction is specified as a set of weighting coefficients of the design variables proportioned by a common scalar multiplier. These coefficients are selected by various methods such as estimation of directional derivatives or curve fits. Next, a one-dimensional search is carried out on the common scalar multiplier until a point is found where there is no further improvement in the objective function. A new direction is then determined at that point and the process repeated as many times as is necessary until no more improvement is obtained or when it is estimated that we are sufficiently close to the optimum. The process is illustrated in Fig. 17.18. The number of iterations required is usually greater than the two given in the figure, especially when there are more than two variables. The basic procedure illustrated in Fig. 17.18 is still valid when constraints are present. However, the implementation of the basic procedure can be a good bit more involved. The reader is referred to Wagner, Ref. 7.

Optimization techniques that are integral with the systems model   Linear programming, integer programming, dynamic programming, and large-scale programming are optimization techniques that are directly integrated into the systems model. Linear and integer programming are based on expression of the problem in the form of an algebraic linear objective function and algebraic linear equality or inequality constraints. Therefore, they may not be particularly suitable for solar thermal systems that tend to require nonalgebraic and nonlinear expressions. Dynamic programming and large-scale programming are not so tightly tied to mathematic structure and thus are better oriented to solar thermal systems optimization. These two approaches will be discussed below.

Large-scale programming. The basic idea of large-scale programming is illustrated in Fig. 17.19. The system/optimization is modeled as a master model and submodels. The master model provides sequentially changing structure specifications to the submodels, such as objective function coefficient or constraint parameter values. The submodels are solved for the specifications provided by the master model and then the optimum results are passed back to the master model. The master model uses these results to calculate new specifications that are again passed to the submodels. This iterative process continues until a systems optimum is reached.

The iterations between the master model and the submodels approach a systems optimum in two different ways. One way has the submodels return optimum, but infeasible, solutions to the master model. The master model, in changing the specifications of the submodels, pushes the solution toward feasibility. The second way causes the submodels to return feasible non-systems optimal solutions to the master model. The master model then pushes the solution toward a systems optimum by changing the submodel specifications.

The large-scale programming approach works well with complex systems having numerous design variables because the system is broken up into smaller problems with fewer variables.

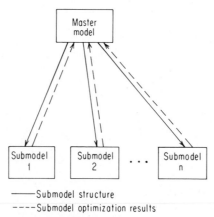

——————Submodel structure
- - - -Submodel optimization results

Fig. 17.19   Large-scale programming structure.

Though these problems have to be solved many times, the total effort is usually many orders of magnitude smaller than solving a single large problem. Moreover, a single large problem may not even be computationally feasible. The reader is referred to Geoffrion[8] for a detailed treatment of large-scale programming.

Dynamic programming. The dynamic programming approach, like the large-scale programming approach, breaks the system/optimization problem into smaller problems with fewer variables. However, instead of the cyclic iterative scheme depicted in Fig. 17.19, dynamic programming uses a sequential iterative process that successively collapses the submodels into each other, eliminating variables in the process.

Figure 17.20 shows a part of the system that has been organized into a serial set of submodels with linking variables. Variations on the serial structure, such as feedback, will still allow the application of dynamic programming, but a solution becomes more difficult to obtain. Assume that the costs of each submodel are summed to form the system objective function and we are seeking to minimize system costs. We can eliminate the design variables from the cost of submodel 1 by

$$R_1'\,(I_1,\,I_2) = \underset{D_1}{\text{MIN}}\; R_1'\,(I_1,\,I_2,\,D_1) \tag{17.15}$$

where the equation says to choose a design that minimizes the cost of submodel 1 when I_1 is the input and I_2 is the output. We have formed a parameters optimum and as long as I_1 and I_2 communicate all pertinent information about the performance of submodel 1 to submodel 2, we can collapse submodels 1 and 2 together and eliminate the design variables for submodel 2 by

$$R_1'\,(I_1,\,I_3) = \underset{D_2,I_2}{\text{MIN}}\; [R_2'\,(I_2,\,I_3,\,D_2) + R_1'\,(I_1,\,I_2)] \tag{17.16}$$

and for subsequent submodels by:

$$R_n'\,(I_1,\,I_{n+1}) = \underset{D_n,I_n}{\text{MIN}}\; [R_n(I_n,I_{n+1},D_n) + R_{n-1}'\,(I_1,I_n)] \tag{17.17}$$

This successive collapsing process finally yields the systems optimum $R_N'\,(I_1,I_{N+1})$ as a function of the system input I_1 and the system output I_{N+1}. As long as the linking variables communicate all pertinent systems information from one submodel to the next, we do not have suboptimization. Also, at some submodel if I_1 is not required for subsequent submodels, it can be treated like the design variables and minimized out, simplifying the remaining computations.

A distributed collector solar thermal electric power generation plant provides an example of the application of dynamic programming to a solar thermal system.[9] The system is disaggregated into submodels that consist of a concentrator ($i = 1$), an absorber ($i = 2$), the field ($i = 3$), and a turbine-generator-cooling tower ($i = 4$). The objective function for this system representation is

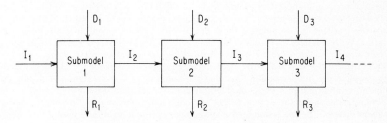

I_i — Linking variables between submodel $i-1$ and submodel i
I_1 — Input to system
D_i — Design variables for submodel i
R_i — Return or cost for submodel i

Fig. 17.20   Dynamic programming serial structure.

$$N(R_1 + R_2) + R_3 + R_4 \qquad (17.18)$$

where $I_i's$ and $D_i's$ have been suppressed. An additional design variable for the field, submodel 3, is the number of collectors, N. The fact that N appears where it does in Eq. (17.18) indicates that this system is not sequential, but includes convergent sequences of collectors in the field portion of the collapsing process.

This application is also of interest because the submodels correspond to functional subsystems in the solar thermal electric power system. To put it another way, the organizing structure was the optimization technique which was used to characterize the system interactions through the linking variables and the remainder of the systems modeling task was to mathematically model each subsystem individually. Subsystems models had to be expressed in terms of the specific linking variables as well as the design variables, making them a little more complicated or unusual than they otherwise might have been. In the aggregate though, the systems modeling process was greatly simplified over what it might have been had a systems modeling approach been used to model all aspects of the system.

One advantage to dynamic programming is that there is no requirement that the systems equations be continuous or differentiable as is the case for most other optimization techniques. A disadvantage for dynamic and large-scale programming is that it might be difficult to structure the system so that the techniques are applicable. Nonlinear programming approaches are somewhat more flexible in this regard.

USING THE MODELS

Trade-Offs Between Modeling Approaches

Model purpose   The purpose for which a model is intended plays a significant role in determining which model to use. For example, if the purpose in using a model is to size the collector array and other subsystem components for a solar heating or domestic hot-water system, then the f-chart or Bruno's method could be used to satisfy that purpose. The more detailed simulation models, such as TRNSYS, should not be used for that purpose because they are expensive and time consuming to operate. However, if the purpose is to obtain a detailed understanding of the performance of the various components in a system, then the more detailed simulation models should be used since the simplified models do not provide this level of detail in their outputs. Similarly, if the purpose is to optimize or to investigate the sensitivity in system performance to a parameter such as the collector loss coefficient, then again the simplified models should be used if possible due to the increased cost that would be incurred in operating the detailed simulation programs for that purpose.

Computational time   The computational time can become significant in some modeling approaches. For example, the detailed simulation programs such as TRNSYS typically require approximately 4 s to simulate one day of operation when using a 15-min integration time step. The computer time can increase significantly when modeling increasingly complex systems or when requiring a great deal of detailed output. Other approaches, such as the finite element approach, can also have large time requirements. One such simulation has been known to require approximately 18 h in order to simulate one year of operation of a solar-heating system. This extensive computation time could only be justified if it were necessary to have extremely detailed or very accurate outputs from the simulation model. On the other hand, the simplified approaches can lead to computation times on the order of one second to analyze one year of operation. With speeds of that order, it becomes feasible to use the simpler models for sensitivity analyses or other purposes that require a great many design variations.

Modeling effort—Is it worth it?   A primary purpose in developing and using a model is to gain insight and understanding regarding the behavior of a given system. It is not always feasible to build actual prototypes of given physical systems and use these prototypes to gain further understanding of the system. In fact, it is often too expensive to build even one prototype of a given system. The space and nuclear programs represent good examples of this situation. In such cases it is not only worth the effort to develop a system model, it is absolutely essential that the modeling methodologies be used in order to understand the performance of the actual system. In a classic example from the nuclear field, a complete system was designed with the aid of simulation models at a cost of only a few thousand dollars, whereas the cost of constructing the system was hundreds of millions of dollars. The value of using a systems model is obvious in such a case. On the other hand, it is not difficult to conceive of situations in which the cost of developing and operating a systems model would be so high as to negate the usefulness of a model. In such a case it is more realistic to consider actual construction of the system for the purposes of analysis and testing.

Accuracy of inputs   The quality of the inputs to models has an obvious effect on the results obtained from the models. It is not reasonable to expect models to have an accuracy better than that of the input data. This obvious statement has significance when considering the purpose for using a model. If the model is to be used for the purpose of design of solar systems, then it is unreasonable to expect the model to be more accurate than the weather data, for example.

Accuracy of outputs versus computational effort   The desired accuracy of the outputs from a systems model depends upon the purpose for which the model is intended. If the systems model is to be used for detailed system design, such as selection of a given control strategy from among various competing control strategies, then it is desirable that the model provide accurate outputs. However, if the purpose in using the program is to obtain long range performance predictions then it is not necessary that the program accurately model the detailed operation of the component but instead provide accurate results for average performance. The computational effort required to obtain accurate representations of detailed performance is usually quite high relative to the computational effort required to obtain accuracy in average performance. For example, when using the TRNSYS program, one usually makes a trade-off between computational accuracy and computer costs. In order that the numerical errors be reduced to less than 1 percent it is often necessary that considerable time and effort be expended in selecting an appropriate tolerance and integration step size.

Scenarios and Recommended Modeling Approaches

The recommended modeling approaches for various scenarios are presented in Table 17.2. This is a qualitative table and provides information regarding only the application of the different modeling approaches to different modeling problems.

Future Developments in Modeling

Efforts will continue toward developing more accurate mathematical and empirical models to provide more detailed results; (2) be directed toward developing models appropriate for use on four-function and programmable hand-held calculators; and (3) will also be directed toward developing simulation models that are both faster and more accurate than present simulation models. Reductions in computational time may be achieved through more efficient programming and possibly through changes in the mathematical models used to model subsystem components. Improvements in accuracy will be brought about through the efforts of model validation studies that will be conducted as more detailed and more accurate data become available regarding the performance of specific solar systems. Modeling progress is reviewed annually in the United States at a Systems Simulation Symposium sponsored by the Department of Energy.

REFERENCES

1. Bruno, R. and R. Kersten, "Models and Methods for the Analysis and Optimization of Solar Energy Systems," *Proceedings of the Energy Use Management Conference*, Tucson, Arizona, October 1977.
2. Lameiro, G. F., *Stochastic Models of Solar Energy Systems*, Ph.D. dissertation, Colorado State University, Solar Energy Applications Laboratory Research Report, Fort Collins, Colorado, March 1977.
3. *TRNSYS—A Transient Simulation Program*, Report 38, Solar Energy Laboratory, University of Wisconsin, Madison, Wisconsin, November 1976.
4. Klein, S. A., W. A. Beckman, and J. A. Duffie, "A Design Procedure for Solar Heating Systems," *Solar Energy*, vol. 18, pp. 113–127, 1976.
5. *SOLCOST—Solar Energy Computer Design Program for Non Thermal Specialists*, Martin Marietta Aerospace, Denver Division, Denver, Colorado, 1976.
6. Barley, C.D., *Relative Areas Analysis of Solar Heating System Performance*, M. S. thesis, Colorado State University, Fort Collins, Colorado, September 1977.
7. Wagner, H. M., *Principles of Operations Research*, 2d ed., Prentice-Hall, Englewood Cliffs, New Jersey, 1975.
8. Geoffriom, A. M., *Elements of Large Scale Programming*, Rand Report No. R-481-PR, Rand Corporation, Santa Monica, California, 1969.
9. Duff, W. S., "Minimum Cost Solar Thermal Electric Power Systems: A Dynamic Programming Based Approach," *Engineering Optimization*, vol. 2, pp. 83–95, 1976.
10. Lameiro, G. F. and P. Bendt, *The GFL Method for Sizing Solar Energy Space and Water Heating Systems*, SERI-30, Solar Energy Research Institute, Golden, CO, 1978.

TABLE 17.2 Recommended Modeling Approaches for Building

Scenario	Recommended modeling approach					
	Empirical			Detailed simulation	Mathematical	
	Rules of thumb	f-chart	G-chart*	TRNSYS	STOLAR	Bruno's Method
Initial estimation of collector area and storage size for a residential solar heating system	Good	Satisfactory but more time consuming than necessary	Satisfactory but more detailed than necessary	Too expensive	Satisfactory but large computer facility required	Satisfactory but computer facility required
Design of residential solar heating systems; monthly results desired	Not accurate enough	Good	Satisfactory for design but will not provide monthly results	Too expensive	Satisfactory for design but will not provide monthly results	Good
Determine optimal collector area for a residential solar heating system; also determine economics	Not applicable	Good	Good for those without access to a computer	Too expensive	Satisfactory but longer run times than f-chart	Good
Design of solar energy systems for commercial buildings	Poor	Good	Poor	Good	Not applicable	Good
Community solar energy projects	Not applicable	Not applicable	Not applicable	Good	Not applicable	Good
Industrial processes	Not applicable	Not applicable	Not applicable	Good	Not applicable	Not applicable at present
Design of residential solar cooling systems	Poor	SOLCOST—good f-chart—not applicable at present	Not applicable	Too expensive	Not applicable	Good

*G-chart is an annual correlation of month f-chart predictions developed by Lameiro.[10] It is available from SEDCOA, P.O. Box 1943, Ft. Collins, CO.

Chapter **18**

Agricultural and Other Low-Temperature Applications of Solar Energy

T. A. LAWAND
Brace Research Institute, Montreal, Quebec

Intermediate-technology solar systems use simple fabrication methods, can be built on site from nonfabricated components, usually have low unit cost, and operate at temperatures below 100°C. As such they find wide application in the developing world or by the do-it-yourself worker. Since the devices are of simple design, no detailed engineering analysis is required. The basic principles of heat transfer (Chap. 4) and solar radiation (Chap. 2) can be used to estimate performance.

This chapter discusses three intermediate-technology systems—*solar distillation, solar cooking,* and *solar crop drying.* Basic design guidelines and illustrations are provided but detailed theories are not developed. A lengthy bibliography is given for additional study.

SOLAR DISTILLATION

Solar distillation is a process whose principle was known to the ancients. The first mention of solar distillation known to the author was reported by Mouchot, who stated that the Arabs "se servaient de vases de verre pour opérer certaines distillations au soleil." He then elaborated on their method as follows: "Au dire des alchimistes, les Arabes pour opérer certaines distillations au soleil, se servaient de miroirs concaves, poli, fabriquées à Damas." In *l'Histoire naturelle,* published in 1551, Adam Loncier depicted by means of an illustration a similar procedure for distilling, among other items, the essential oils of flowers.

The next report on solar distillation comes in the excellent historical review of desalination by Nebbia and Menozzi. [46] They quote the work of Della Porta, published in 1589, whose apparatus was described in detail, for the distillation of herbs. Obviously, the distillation capabilities of solar energy were well understood although no specific reference to water desalination was made. It must be noted, however, that Della Porta published several other books on desalination experiments.

The first specific reference to the possibilities of solar distillation were made by an Italian, Nicolo Ghezzi, who wrote a short treatise in 1742 where he proposed the following, which has been freely translated from the original Italian script:

Perhaps placing a cast iron vase containing sea water in such a manner that the sun's rays will strike it (and during mild days and seasons, not an insignificant amount of vapor will be formed) and if the spout of the vase is shaded from the sun, it will result in a more copious and more extended flow of fresh water.

The next reference to solar distillation was given by Harding who reported on a 4800-m² still erected near Las Salinas, Chile. This unit was of the greenhouse- or roof-type solar still. No mention is made of what inspired the builder, a Mr. Wilson, regarding this design. It is known that the productivity of the still in summer was of the order of 5 kg of fresh water produced per square meter of evaporating surface per day. It is an interesting reflection on the simplicity of the process that productivities reported from solar stills recently built are of the same order of magnitude.

For a while after this, few reports of solar stills appear to have been published. In the decade following World War I, interest was renewed in solar distillation. Many publications have followed with reports on the process in general, often accompanied by descriptions of small stills of the roof type, v-covered, tilted-wick, inclined tray, suspended envelope, tubular or air-inflated design.

In order to increase the productivity, several workers have tried forced circulation systems to condense the water vapor externally from the still. Others tried to recapture the latent heat of evaporation through multiple-effect systems or humidification systems.

Several large solar distillation schemes have been proposed while others have considered combination plants which generated power as well as desalting saline water. Alternative uses besides desalination were also found for solar stills, such as regenerating solutions and obtaining fresh water from the ground. There are a number of plans and specifications for the building of solar stills which have been published. Several patents have also been issued in this field. With a few exceptions, they generally deal with small solar stills.

Small stills of under 50 m² in area, as have been described, are most useful for individual family units in isolated areas. Stills of this type have been extensively tested, particularly at the University of California, Berkeley, as reported in 1971 by McLeod et al. They give results of productivity of a small solar still for 7 yr, from 1952 to 1959. The Las Salinas still in Chile was reputed to have run for 40 yr, but records do not appear to have been published.

Work on larger solar stills was initiated through the efforts of the Office of Saline Water, U. S. Department of the Interior, reported mainly by Löf [38,54] and the Battelle Memorial Institute. This work was carried out chiefly at the Solar Research Station, Daytona Beach, FL. These stills were glass-covered units, and one was 250 m² in area. Although primarily designed as deep-basin evaporators with a depth of sea water up to 30 cm, there was continuous experimentation in solar still operation, which was excellently reported in the publications of this series.

Concurrently, several other designs of stills were tested, including the air-inflated plastic and tilted-wick stills. The former were tested for the Church World Service, who were instrumental, in 1964, in the installation of the first large plastic solar still on Symi, a small Greek island in the Dodecanese. Subsequently, several other stills were built on small Greek islands. This work was taken over by the Hellenic Industrial Development Bank which financed some of the largest and longest operating solar stills in the world.

Other significant activities have resulted in solar still installations in Spain and Australia. In both countries, glass-covered stills have been favored. The Australians have built a large still with 3800 m² of evaporator area at Coober Pedy, South Australia. In Pakistan, an even larger solar still has been built in Givadar.

The important work of Howe, [31,32] Tleimat, et al. in this field must be mentioned. In particular, their collaborative efforts with the South Pacific Commission in the testing and installation of solar stills on small islands in the Pacific Ocean have been most informative and useful in resolving problems affecting fresh water provision to small communities. These installations are not quite so large as the others, having been designed mainly for family use. One unit on Fiji was nearly 30 m² in area. The work of this group in stressing the importance of rainfall collection and storage in combination with solar stills clearly parallels the present study.

The Brace Research Institute has been associated with the construction, testing, and operation of several large solar stills in the West Indies at Petit St. Vincent (230 m²) and in a rural application in Haiti (300 m²). Both units are combined with rainfall collection.

One of the better overall assessments of this field is *Solar Distillation of Saline Water*, prepared by the Battelle Memorial Institute, June 1970, for the United States Department of the Interior. This manual reviews the whole subject and lists the different units, both experimental and practical, which have been built during the last few decades.

The short résumé given above is indicative of the very nature of solar distillation technology. It is by no means stagnant. Improvements are continuously being made which hopefully will reduce the costs and increase the productivity of solar stills. With this in mind, further advances in technology should make the future use of solar distillation even more feasible.

EXPERIENCES WITH OPERATING SOLAR STILLS

The manual, *Solar Distillation of Saline Water*, is the most complete document to date describing the different solar stills which have been built to date in different parts of the world. The situation is constantly changing due to the variable conditions under which these units are tested and installed.

Delyannis [15] at the 1973 Fresh Water from the Sea Conference in Heidelberg prepared a more up-to-date list given in Table 18.1, and illustrated in Figs. 18.1 through 18.7.

It is evident from this list that solar distillation technology has been tried on a fairly extensive basis in a number of different areas and climatic regions. Obviously not all these units will represent success stories. This is because by their very nature, they will be of a pilot plant nature as various organizations and research teams acquire increased operating knowledge. Nonetheless in order to ensure that this technology is adequately treated, it will be necessary to compile a comprehensive assessment of different systems and how they have performed in practice. It is equally important that the socio-cultural aspects of each of these installations also be monitored so that it will be possible to ascertain whether these have been given sufficient attention in the preparation of these projects. It is apparent that solar distillation is often an appropriate, intermediate technology. The question that must be raised, however, is whether this technology has been effectively and appropriately applied.

Many organizations have published details of operating experiences and productivity. The Hellenic Industrial Development Bank in Greece has monitored the performance of the solar still at Nissyros in the Aegean and these figures are given in Fig. 18.8. What this does show is a reduction of production of roughly 12 to 15 percent during the summer months over a period of 5 yr. What has caused this reduction? It is essential to determine these factors in order to illustrate the effectiveness of this type of technology. This, in addition, underlines the necessity for global assessment of the various components of these systems in order to ascertain their short- and long-term appropriateness.

Figures 18.9 and 18.10 illustrate single-sloped and inflated stills both under construction and completed. It is essential to realize that one of the prime advantages of solar distillation is its flexibility. It is exceedingly easy to increase the capacity of such solar stills by adding additional units. On the other hand, solar stills can be used in very small scale operations by providing fresh water for individual needs in units adjacent to residences or integrated directly into the roofs of buildings. A number of commercial firms now exist which provide hardware either in the form of installations on a turn-key basis or in the form of prefabricated units.

COMPONENTS OF SOLAR DISTILLATION SYSTEMS

There are many ways in which solar distillation systems can be built. It is generally agreed that it would be most advantageous for these systems to have a long life. At the same time full

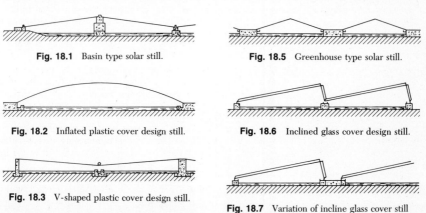

Fig. 18.1 Basin type solar still.

Fig. 18.5 Greenhouse type solar still.

Fig. 18.2 Inflated plastic cover design still.

Fig. 18.6 Inclined glass cover design still.

Fig. 18.3 V-shaped plastic cover design still.

Fig. 18.7 Variation of incline glass cover still design.

Fig. 18.4 Single sloped cover design still.

TABLE 18.1 A List of Some Important Solar Distillation Systems

Geographic location	Year built	Design (see Diagram 1)	Basin size, m²	Cover material	Feed	Specific productivity, L/day per m²
Australia:						
Muresk I	1963	1(e)	372	Glass	Brackish	2.24
Muresk II	1966	1(e)	372	Glass	Brackish	2.24
Coober Pedy	1966	1(e)	3160	Glass	Brackish	2.01
Caiguna	1966	1(e)	372	Glass	Brackish	2.10
Hamelin Pool	1966	1(e)	557	Glass	Brackish	2.17
Griffith	1967	1(e)	413	Glass	Brackish	2.20
Cape Verde Islands:						
Santa Maria	1965	1(c)	743	Plastic	Seawater	2.85
Chile						
Las Salinas	1872	1(e)	4460	Glass	Brackish	3.31
Quillagua	1968	1(e)	100	Glass	Seawater	4.01
Greece:						
Symi I	1964	1(b)	2686	Plastic	Seawater	2.82
Regina I	1965	1(c)	1490	Plastic	Seawater	2.85
Salamis	1965	1(c)	388	Plastic	Seawater	2.84
Patmos	1967	1(f)	8600	Glass	Seawater	3.04
Kimolos	1968	1(f)	2508	Glass	Seawater	3.02
Nisyros	1969	1(f)	2005	Glass	Seawater	3.02

Location	Year					
India: Bhavnagar	1965	1(e)	377	Glass	Seawater	2.21
Mexico: Natividad Islands	1969	1(d)	95	Glass	Seawater	3.98
Spain: Las Marinas	1966	1(a)	868	Glass	Seawater	2.96
Tunisia: Chakmou	1967	1(d)	440	Glass	Brackish	1.20
Mahdia	1968	1(d)	1300	Glass	Brackish	3.20
U.S.A.: Daytona Beach	1959	1(a)	228	Glass	Seawater	2.32
Daytona Beach	1961	1(a)	246	Glass	Seawater	2.31
Daytona Beach	1961	1(b)	216	Plastic	Seawater	1.75
Daytona Beach	1963	1(b)	148	Plastic	Seawater	4.10
U.S.S.R.: Bakharden	1969	1(e)	600	Glass	Brackish	2.72
West Indies: Petit St. Vincent	1967	1(b)	1710	Plastic	Seawater	2.88
Haiti	1969	1(d)	223	Glass	Seawater	3.40

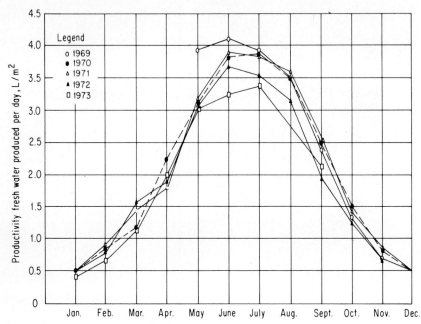

Fig. 18.8 Variation of productivity with time of solar distillation units, Nissyros, Greece.

appreciation and use must be made of locally available materials and technologies. In this manner, truly appropriate systems can be developed and maintained by the local population.

Specifications for Solar Still Components

Many designs of solar stills exist. One aim of this chapter is to evaluate and adapt various materials for use in still construction. Hence, it has been decided to split up the material requirements in relation to their function within the unit. These are listed below:

1. General specifications of solar stills
2. Transparent cover
3. Evaporator liner
4. Solar still frame
5. Sealants
6. Insulation
7. Auxiliaries—piping, pumping and reservoirs

General specifications of solar stills   There are certain basic requirements which must be met. In general, the unit:

a. Must be easily assembled in the field

b. Should be constructed of materials imported to the region that are packageable so that transportation costs will not prove excessive (this is particularly true not only for shipment from one country to another, but especially for internal movement within a given area, i.e., from the port to final destination)

c. Should be lightweight for ease of handling and shipping

d. Must have an effective life, with normal maintenance, of 10 to 20 yr

e. Must have access ports for ease of maintenance

f. Should not require or depend upon external power sources

g. Should serve as a rainfall catchment surface

h. Should be able to withstand the effects of severe storms

i. Must be manufactured of materials which will not contaminate the collected rain water or the distillate. (It must be continuously stressed that solar stills constitute the water supply system for the communities served and hence must be nontoxic in every respect to the fresh water produced)

(a)

(b)

Fig. 18.9 Single-sloped-cover solar still (a) under construction and (b) completed.

j. Must be fabricated so that the maximum size of solar still components can be directly related to economic shipping dimensions as specified by freight carriers

k. Must make use of as many local resources—whether material or labor—as possible

l. And must meet standard civil and structural engineering standards

Transparent cover   The cover serves to cover the distillation segment and transmits solar radiation to the interior.

Properties required are listed below:

a. The material must withstand the effects of weather—wind, sunshine, rain, dust, etc.

b. The material must have a transmittance for short-wave solar radiation (between the limits of 0.3 to 3.0 μm) of over 85 percent and preferably higher.

c. It must be nearly opaque to long-wave (over 3.0 μm) radiation.

(a)

(b)

Fig. 18.10   Plastic-covered, inflated, solar still (a) under construction and (b) completed.

d. It should not have a high water absorptance, both from its use as rainfall catchment surface on the outside and use as a condenser on the inside.

e. The solar reflectance at normal incidence should not exceed 10 percent where possible.

f. The solar absorptance of the material should be low, especially if the cover is also to be used as a condenser.

g. The thermal conductivity should be high in order to reduce the cover temperature.

h. The material properties should not alter with age.

i. The material should not possess electrostatic properties which would concentrate dust particles on the outside surface.

j. If the cover is to be used as a condenser, then the material must withstand temperatures of up to 80°C. In addition, one side will experience extremely high relative humidities (up to

100 percent), while the other surface must concurrently withstand the low humidities prevalent in arid regions.

k. The cover must be able to withstand the abuses of small animals.

l. The cover must be able to withstand a wind load corresponding to 45 m/s windspeed.

Evaporator liner The evaporator or basin liner serves as the absorbing surface for solar radiation as well as a container for the saline water. Materials used for this purpose should have the following properties and characteristics:

a. The liner must be impervious to water.

b. The liner must have a solar absorptance of the order of 0.95. Generally, black materials are used.

c. The surface should be fairly smooth so as to discourage the deposition of scale from the saline water. It also becomes easier to clean when this is necessary.

d. Because liners are often placed directly on the ground, the material should not deteriorate or decompose on contact with normal soils.

e. The material must withstand the effects of continuous immersion in hot, saline water. Temperatures should not exceed 80°C to 100°C upon being heated.

f. The basin liner should not emit any gases or vapors which could taint the taste of the fresh water distillate.

Solar still frame This section refers to materials which are used to form the frames of the evaporators. Any materials used in this fashion should possess the following characteristics:

a. They should be resistant to attack from the saline water or atmosphere.

b. In case they are exposed to the evaporator, they should be covered with a protective coating.

c. They should be sufficiently heavy so as to anchor the stills to the ground during periods of high winds.

d. Solar stills vary in width from 1 to 3 m and are generally up to 50 m in length. Frame components should be available in a series of sizes, which could be easily disassembled for shipping and erection on the site. In situ construction must also consider logical sizing.

e. The frames should be made of such materials as to permit ease of working or attachment.

f. These materials should not be affected by direct contact with the ground or exposure to normal weather conditions.

g. Because the frames separate different evaporator bays, they will more often than not be used as a walkway and must be able to withstand usage in this capacity.

h. Any sections of the frame exposed to the exterior will invariably serve as part of the rainfall catchment surface. In this regard, the material should neither absorb too much of the incident rainwater nor contaminate it.

Sealants This section includes materials used to seal transparent cover materials to one another as well as to the other components of the distillers. In addition, it includes any members used to support the superstructure of the distillation units as they will invariably come in contact with the transparent cover.

a. The materials should not be adversely affected by exposure to the weather on one face and by their possible exposure to the interior of the evaporator on the internal face.

b. If the transparent cover is to be used as a condenser, then the sealants and structural components should intercept a minimum amount of solar radiation in order to keep the efficiency high. In addition, all heat inputs to the cover area increase its temperature and reduce the evaporation potential of the system.

c. A minimum number of sealants should be utilized. Preferably the same sealant should be employed to bond the transparent cover materials as well as to seal other materials used in the solar still construction.

d. The sealants must be easily applicable under extreme field conditions, because it is likely that they will be utilized during erection phases on the site.

e. If structural cover supports are not used, the transparent cover material sealants must withstand the effects of winds of up to 45 m/s.

Insulation The insulation used in solar distillation is used beneath the sea water evaporator basins in order to reduce ground heat losses. The materials used for this purpose require the following properties:

a. They must be lightweight and structurally self-supporting.

b. They must be waterproof and basically water-impermeable.

c. They should insulate the edges as well as the base of the evaporator.

d. The insulation must withstand temperatures of up to 80°C and must not warp or change shape.

    e. The insulation must withstand the effects of the ground on which it is placed.

    f. Insulation materials could also serve as basin dividers in large solar still bays.

Auxiliaries—piping, pumping, and reservoirs   This section includes all fluid systems—gutters for rainfall and condensate collection, piping for feed and rain lines, and reservoirs for saline and fresh water. In addition, some form of pumping mechanism should be provided for water transfer.

    a. All auxiliaries in contact with either fresh or rain water should have a protective coating of inert material in order to avoid contamination or damage to the system.

    b. All internal gutters or piping systems must be of continuous, single-piece construction so as to avoid internal joints which are difficult to maintain.

    c. All joints which must be made to piping or gutters must be easily undertaken under field conditions.

    d. All auxiliaries and reservoirs must be so fabricated as to meet general shipping dimension regulations.

    e. Where conventional power sources do not exist, the pumps should be manually operated or wind powered units.

    f. The distillate reservoir should exceed at least the maximum daily production capacity by a factor of 3.

    g. The rainwater reservoirs should be rated to existing short-term rainfall intensities.

Freshwater reservoirs   In all cases, fresh water distillate and rainwater collection reservoirs should be provided. This will permit the measurement, collection, and storage of water from these sources. Generally, separate collection systems should be provided to avoid contamination from:

    a. dirt, dust, animal and bird droppings in case of the rainwater collection

    b. salt water overflow in case of the distillate production

The fresh water should be carefully handled and possibly sand-filtered if the need arises.

The above specifications may not cover all types of applications of solar stills. They should be viewed as criteria which should be adhered to if we wish to develop appropriate technology in solar distillation.

Costing Procedure for Solar Stills

In 1970, A. Delyannis[14] proposed the establishment of a standardized procedure for costing, applicable to all designs of solar stills and to all countries. This procedure was designed to allow the comparison of the cost of the various solar stills at a given locality, as well as the calculation of the cost of a given design in various localities. Therefore, such detailed information must be provided for each case to make this aim possible and any comparison reliable. A modified version of this procedure is given below.

A principal item of equipment for this comparison should be the distillation plant proper, as representing a particular design. Auxiliaries, more or less not directly dependent on the specific solar still design (e.g., site preparation, fencing, storage facilities, pumping, piping, etc.), must be quoted as well in detail, but as separate or independent items.

Units of necessary materials and total amounts needed must be reported separately for each item. Cost of materials and labor in the currency applicable to the proposed site must be given. A description of the proposed mode of operation, including continuous or batch feed, brine evacuation and renewal, necessity of cleaning cover or basin, etc. should be given. Special mention should be made if rainfall is to be collected. Only additional cost for this purpose, on top of the conventional solar distillation plant, should be included in this item.

The following standard form can be used as a design and economic checklist for solar still design.

Plant description

(a) Location:

    Place _____   Country _____

    Latitude _____   Longitude _____

    Elevation _____   m over sea level _____

(b) Feed:

    Type of water _____

    Salinity _____ mg/L CaO

    Total hardness _____ meq/L CaO

    Carbonate hardness _____ meq/L CaO

    Permanent hardness _____ meq/L CaO

(c) Plant Area:
Water-evaporating surface ——————————————————————— m²
Cover projected area ——————————————————————————— m²
Area inside boundary of the still ———————————————————— m²

(d) Available area for future enlargement ——————————————— m²

(e) Number of distilling units:
Water-evaporating surface ———————————————————— m² per unit
Cover projected area ————————————————————————— m² per unit
Depth of brine in basin ———————————————————————— mm
Cover material ———————————————————————————————————

(f) Mean daily output per year ——————————————————— L/m²·day or L/day
Maximum output ————————————————————————— L/m²·day or L/day

Construction of distillation units

(a) Basin structure per unit:
Gravel and sand————————————— m³ × ——————— = ——————
Cement ————————————————— kg ×——————— = ——————
Concrete mix ——————————————— m³ × ——————— = ——————
Precast concrete beams———————— pieces × ——————— = ——————
(No. of dimensions)
Precast concrete posts————————— pieces × ——————— = ——————
(No. of dimensions)
Steel for reinforcement———————— kg ×——————— = ——————
Other materials ———————————— ×——————— = ——————

(b) Lining:
Butyl-rubber sheet ————————— m² × ——————— = ——————
Asphalt mats ——————————————— m² × ——————— = ——————
Polyethylene, sheet ————————— m² × ——————— = ——————
Others—specify——————————— m² × ——————— = ——————

(c) Sealing materials:
Silicone rubber ——————————— tubes × ——————— = ——————
Asphalt cement ——————————— × ——————— = ——————

(d) Gutters and weirs:
Stainless-steel strip ————————— m × ——————— = ——————
Aluminum channel————————— m × ——————— = ——————
Plastic channel ——————————— m × ——————— = ——————
Asbestos—cement angles ————— m × ——————— = ——————
Asbestos—cement strips————— m × ——————— = ——————
Other specify ——————————— m × ——————— = ——————

(e) Insulation:
Polystyrene——————————— m² × ——————— = ——————
Other materials specify—————— × ——————— = ——————

(f) Labor for:
(1) Basin: concrete skilled ——————— mh* × ——————— = ——————
unskilled ————————— mh × ——————— = ——————
other work ————————— mh × ——————— = ——————
(2) Lining: skilled ————————— mh × ——————— = ——————
unskilled ————————— mh × ——————— = ——————
(3) Sealing: skilled ————————— mh × ——————— = ——————
unskilled ————————— mh × ——————— = ——————
(4) Gutters: skilled ————————— mh × ——————— = ——————
unskilled ————————— mh × ——————— = ——————
(5) Insulation: skilled ——————— mh × ——————— = ——————
unskilled ——————— mh × ——————— = ——————

*mh = man-hours.

(g) Any other (specify)

(h) Total cost of basin:

Cost _____ per m² evaporating surface

Cost _____ per m² of cover projected area

Cover construction

(a) Materials used:

Concrete curbs (dimensions) _____ pieces × _____ = _____

Aluminum angles (dimensions)_____ kg × _____ = _____

Aluminum T-ees (dimensions) _____ m × _____ = _____

Cover: Glass _____ m² × _____ = _____

Tedlar _____ m² × _____ = _____

Other plastics _____ m² × _____ = _____

Sealing materials:

Silicone rubber_____ tubes × _____ = _____

Silastic _____ tubes × _____ = _____

Other sealants _____ × _____ = _____

Primer _____ L × _____ = _____

(b) Labor for:

Cover structure— skilled _____ mh × _____ = _____

unskilled _____ mh × _____ = _____

Cover material— skilled _____ mh × _____ = _____

unskilled _____ mh × _____ = _____

(c) Total cost of cover:

Cost _____ per m² evaporating surface

Cost _____ per m² cover projected area

Cost of distillation units (total of 2 and 3)

(a) Basin _____

(b) Cover _____

Total of distillation units _____

Cost _____ per m² of evaporating surface

Cost _____ per m² of cover projected area

Site preparation

Minimum area required for projected output _____ m²

Cost_____ m² × _____ = _____

Removal and relocation of:

(a) Earthen materials _____ m² × _____ = _____

(b) Rocky materials _____ m² × _____ = _____

Type of mechanical means used:

Machine hours _____ h × _____ = _____

Labor skilled_____ mh × _____ = _____

Labor unskilled _____ mh × _____ = _____

Any other special _____

Total cost of site preparation _____

Cost _____ per m² of evaporating surface

Cost _____ per m² of cover projected area

Piping and pumps

(a) Salt water:

_____ m pipe_____ mm φ × _____ = _____

_____ m pipe_____ mm φ × _____ = _____

_____ m pipe_____ mm φ × _____ = _____

_____ valves _____ mm φ × _____ = _____

_____ valves _____ mm φ × _____ = _____

_____ valves _____ mm φ × _____ = _____

_____ fittings _____ mm φ × _____ = _____

_____ fittings _____ mm φ × _____ = _____

_____ fittings _____ mm φ × _____ = _____

(b) Distillate:

_____ m pipe_____ mm φ × _____ = _____
_____ m pipe_____ mm φ × _____ = _____
_____ m pipe_____ mm φ × _____ = _____
_____ valves _____ mm φ × _____ = _____
_____ valves _____ mm φ × _____ = _____
_____ valves _____ mm φ × _____ = _____
_____ fittings_____ mm φ × _____ = _____
_____ fittings_____ mm φ × _____ = _____
_____ fittings _____ mm φ × _____ = _____

(c) Pumping (specify per pump):

_____ salt water pumps_____ mm φ × _____ = _____
_____ distillate pumps_____ mm φ × _____ = _____
_____ windmill pumps _____ mm φ × _____ = _____

(d) Total cost of piping and pumping:

Cost _____ per m^2 of evaporating surface
Cost _____ per m^2 cover projected area

Storage

Capacity for salt water _____m^2
Capacity for distillate _____m^2

(a) Materials used (specify by item) _____
(b) Labor: skilled _____ mh × _____ = _____
        unskilled _____ mh × _____ = _____

(c) Total cost of storage:

Cost _____ per m^2 of storage capacity
Cost _____ per m^2 of evaporating surface
Cost _____ per m^2 of cover projected area

Fencing

Total area included inside fencing _____m^2
(a) Materials used (specify by item) _____

(b) Labor: skilled _____ mh × _____ = _____
        unskilled _____ mh × _____ = _____

(c) Total cost of fencing:

Cost _____ per m^2 of area included
Cost _____ per m^2 of evaporating surface
Cost _____ per m^2 of cover projected area

Other items of investment cost

(a) Facilities for pretreatment of salt water (specify by item) _____
(b) Facilities for post-treatment of distillate (specify by item) _____
(c) Transportation of materials to the site (specify by item) _____
(d) Engineering and design _____
(e) Supervision of construction _____
(f) Testing _____
(g) Brine disposal _____
(h) Power supply _____
                                        Total _____

Total cost of other items

Cost _____ per m^2 of evaporating surface
Cost _____ per m^2 of cover projected area

Summary   Cost of distillation units _____

Site preparation _____
Piping and pumps _____
Storage _____
Fencing _____

Other items _____

Total _____

Contingencies, 10 percent of total _____

Insurance _____

Interest during construction _____

Grand total _____

Cost _____ per m² of evaporating surface

Cost _____ per m² of cover projected area

When to Consider Solar Distillation

The need for water in arid areas of the world is taking on increasingly large proportions as the population of the world increases. There are 35,000 km of coastal deserts located in some of the most favored climatological regions of the world. The time will come when it will be necessary to make use of these areas not only for habitation but also for increased agriculture production. Given these conditions, technology which by large makes use of locally available resources will obviously be favored and decidedly more appropriate. The use of solar distillation could therefore be considered not only for the provision of water, but also for the production of agricultural produce when combined with structures like greenhouses which can control the often harsh natural climatological conditions experienced in these areas.

One would consider the use of solar distillation if the following conditions are met:

1. Adequate availability of saline water.

2. Small populations living in arid areas where inexpensive conventional sources of energy are not readily available.

3. No natural sources of fresh water, under the control of the local population, are easily exploitable.

4. For extended periods of time, there is adequate levels of solar radiation intensity and reasonably high ambient air temperatures.

5. Areas where the annual rainfall generally does not exceed 600 millimeters.

6. Land is available and has little opportunity cost.

It must be pointed out that generally solar distillation should only be provided in quantities less than 20 m³ per day. Even this constitutes an exceedingly large size of installation. It can be that the cost will be excessive and the size relatively unmanageable in attempting to build installations close to this upper level. This does not necessarily mean, however, that only small populations need be supported given that human beings can survive easily on 10 L of fresh water per day for his basic needs, approximately 2000 persons per community could be supported in a 20 m³ per day plant. In this method of operation however, all other forms of water would be derived from saline water sources.

What criteria should the decision makers in the developing arid areas utilize in considering solar distillation as an option for fresh water provision for rural communities. The prime advantages of solar distillation are that:

1. The units can be built to a large extent using locally available material or materials from manufacturers in the country or the general regions.

2. The local labor force can undertake all the principal jobs in the construction, installation, operation, and maintenance of the system.

3. Generally, apart from the amortization of the capital investment, the cost of the operation is not high if the unit has been appropriately constructed in the first instance.

No desalination process should even be envisaged until full exploitation has been made of natural fresh water sources—surface, ground, and rainwater. This applies as equally to solar desalination as other conventional processes. In the solar case, the energy cost is nil but the cost of collecting this "free" energy is not without value.

SOLAR COOKING

One of the primary demands for energy in all societies is that needed for the cooking of food. A wide variety of agricultural products are prepared by some form of transfer of heat to foods, including meat, fish, cereals, vegetables, dairy products, eggs, and the like.

There are a wide variety of cooking techniques used, which vary from location to location depending on cultural habits and traditions, as well as the availability of cooking fuels. The most common forms of cooking are described below.

Boiling   The ingredients are immersed in a fluid, generally water or juices, and heated. The source heat is transferred to the fluid through the walls of the container by conduction. Generally cooking occurs at temperatures approaching the boiling point which varies with the altitude of the location. Traditional cooking is in open or covered pots over a fire using a solid organic fuel. A modification to this process has been the use of pressure cookers, wherein the cooking temperature is raised by means of sealing the covers and slowly releasing accumulated steam generated within the cooking pot, in a controlled manner so as to maintain the pressure. Some domestic units operate at approximately 2 atm with a boiling point at mean sea level about 120°C.

Steaming   The food, held in a perforated dish, is cooked by means of steam, produced in a container located below the dish. It has been roughly estimated that over the world approximately 80 percent of the cooking is done either using the boiling or steam techniques.

Frying   Food is cooked in relatively small quantities on a hot pan by frying. Heat is transferred through the container by conduction to cook the food at temperatures running in the order of 180°C. Preparation of thin bread is also done by frying. This type of bread is widely used in the Middle East and the Asian subcontinent, and requires high cooking temperatures.

Baking   Heat is transferred to the food from heated air or hot surfaces generally in an oven. Temperatures run in the order of 125 to 250°C.

Roasting   Roasting is generally done in an oven, as is baking, with similar temperature ranges, depending on the type of food to be cooked.

Several fuels have been used in developing areas in providing the energy needs for cooking. These include:

Type	Caloric value
Dried animal dung	2,200 kcal/kg
Dry firewood	4,800 kcal/kg
Dry charcoal	8,000 kcal/kg
Coke	7,200 kcal/kg
Kerosene, oil	10,000 kcal/kg

It is important to recognize that each of the cooking sources has a varying degree of cooking efficiency. The caloric heat of combustion of dried animal dung for example is of the order of 2200 kcal/kg. It has been estimated in India that the effectiveness of this source as a cooking fuel is very low, in the order of 0.4 kWh or 344 kcal/kg. Not all fuel sources are as inefficiently used, however.

Regarding the classical method in which most civilizations prepare their food, Löf has stated the following:

Most foods contain a high proportion of water, and heating them to cooking temperatures requires nearly 1 cal per gram per °C (or 1 Btu per pound per °F). The higher the heat input rate to the food and container (and to any additional cooking liquid), the faster will the food be heated to cooking temperature. Then, except where water vaporization is a necessary part of the cooking process, as in bread baking, the speed of cooking is practically independent of heat rate as long as the temperature is maintained by a heat input rate equal to the thermal losses. It is therefore generally true that differences in the time required for cooking equal quantities of food on cookers having various heat supply capacities are due mainly to the different durations of the heating-up periods. Thus cookers of low and high heat supply rates may not show large differences in the time required for foods which must be cooked several hours.

The largest of the energy uses in cooking is usually the heat consumed in vaporizing water present in the food or added for cooking—nearly 600 kcal/kg or 1050 Btu/lb. Next in importance are convection losses from utensils and oven walls. If the energy source has limited capacity, control of these losses by use of covers on utensils, insulation on cooking chambers (ovens) and other means becomes important. If a food is cooked at boiling water temperature and with a surface area of 0.5 ft^2 per pound of container contents, the energy input for one hour of food boiling would be distributed roughly as follows if one-fourth of the water present is vaporized, and if the heat loss is about 600 Btu per square foot of utensil:

Heating materials to boiling temperature	20 percent
Convection losses from vessel	45 percent
Vaporization of water	35 percent

Although variation in the assumed conditions could alter this distribution, the figures show that most of the heat supplied in long duration cooking is dissipated.

The question of the transfer of energy from the heat source to the cooking pot needs to be examined in order to improve the efficiency of solar cookers. Until boiling occurs, the coefficient of heat transfer into the cooking pot is low due to the fact that the liquid is relatively stagnant. Burning often occurs, hence it is necessary to stir not only to increase the transfer of heat but to avoid localized build-up of heat and scorching, a phenomenon particularly true in the preparation of cereals.

A study undertaken by the Brace Research Institute for the United Nations Environment Program on meeting the water and energy needs of a rural village in Senegal indicated that cooking requirements formed about 80 percent of the basic energy needs of a village. It is essential therefore that this high energy demand be analyzed in a systematic way to achieve meaningful results. The mechanism of heat transfer into the cooking pot, the shape of the cooking pot, its size, and many other factors have a profound effect on the design of solar cooking devices. A combination of solar cooking with energy conservation is required.

Where electric or gas cooking is used, the normal burner has a capacity of approximately 1 kW, and is capable of bringing 2 L of water to boil in about 10 min. Automatic ovens for roasting or baking might have an installed capacity of approximately 2 kW. Therefore, in order to have performance comparable to existing systems, it would be necessary that a solar unit deliver roughly 1 kW to the cooking media. If this is not done, one must accept longer cooking times or the cooking of smaller amounts of materials at one time. Roughly 2 m² of solar collectors would be necessary at a 50 percent collection efficiency to give comparable normal cooking rates.

Types of Solar Cookers

There are a number of basic types of solar cookers. A brief review of the literature listed in the bibliography shows the degree to which researchers in developing areas have addressed this problem. In developed countries most attention has been paid to small production units for use by campers or weekend travelers. These units either of the collapsible umbrella reflector types or hot box types are designed to be portable and compact.

There are three general types of solar collectors:

Solar hot boxes   Insulated solar cooker with double glazing, generally in the form of a box set out in the sun and oriented manually. To increase the efficiency, reflectors are often added which permit the development of higher temperatures in the interior cooking chamber. One of the inherent disadvantages of this type of cooker is that it requires cooking to be done in the open, a feature which is often, but not always, socially unacceptable particularly during periods of warmer weather. (See Fig. 18.11.)

Reflector   Cooking rays are concentrated onto a focal point or area on which is placed a cooking pot or frying pan. This process requires also cooking under outside conditions. There is an even greater heat loss due to convection from the wind than with the hot box cookers. The reflectors can only concentrate direct radiation; therefore, these units are less efficient in areas which have high percentages of diffuse radiation. (See Fig. 18.12.)

Detached solar collection and cooking chamber units   The heat transfer fluid (water converted to steam or heat transfer oil) is heated in a separate collector. The heated fluid is transferred to a separate, insulated cooking chamber which can be located on the inside of the house where the cooking is done. In this way the social inconvenience of external cooking is avoided. Figure 18.13 shows a steam cooker which could be used in the detached mode. It should be noted that the limiting technical factor is the transfer of heat into the cooking pot. This applies to all systems mentioned above, though less severe in the case of the reflector cooking owing to the higher temperatures generated.

Finally as to economics, it was shown in the Brace Research Institute Senegal report for UNEP, cited above, that solar cooking with a smokeless wood stove as a backup (for the 10 to 15 percent of the time when climatic conditions did not permit their use) was by far the least expensive alternative compared to butane gas or charcoal.

Technical Considerations in the Design of Solar Cookers

Calculations should be made using the solar radiation incident on the collector area for the purposes of determining the amount of available energy. In the case of concentrating systems, the calculations will not differ very much from those dealt with under sections on parabolic or trough collectors (Chaps. 8 and 9). In the case of flat-plate collectors or hot-box collectors, calculations should be based on a thermodynamic balance of energy input and heat loss. The excess heat in each case must be transmitted to the cooking device, allowing for losses en route. An additional thermodynamic analysis must be undertaken around the cooking pot to determine the rates of heat transfer developed.

Fig. 18.11 Hot box solar cookers: (*a*) solar hot boxes with reflector; (*b*), (*c*) food warmer/dryer.

The overall efficiency of these units can then be determined by combining the heat balances on the collection system, the transmission system and the actual cooking system. In the case of the hot-box cooker, these systems are effectively combined in one, the limiting factor in all cases of course being the poor rates of heat transfer into the cooking recipient. In the case of concentrating cookers, the radiant heat is transferred directly to the cooking recipient, which is generally subject to high heat loss due to its external positioning.

SOLAR CROP DRYING

One of the oldest uses of solar energy since the dawn of civilization has been the drying and preservation of agricultural surpluses. The methods used are simple and often crude but reasonably effective. Basically crops are spread on the ground or platforms often with no pretreatment and are turned regularly until sufficiently dried so that they can be stored for later consumption. Little capital is required on the expenditure of equipment but the process is labor-intensive.

There is probably no accurate estimate of the vast amounts of material dried using these traditional techniques. Suffice it to say it is a widespread technology practiced in almost every country of the globe and at nearly every latitude. Diverse products such as fruit, vegetables, cereals and grains, skins, hides, meat and fish, and tobacco are dried using these simple techniques.

Fig. 18.12 Various concentrating solar ovens and cookers showing reflectors and cooking zones.

These technologies have originated in many of the developing countries so there is no major social problem in their acceptance, or in the use by the local populations of dehydrated foods for consumption. There are several technical problems, however, with this basic drying process; they include:

    a. Cloudiness and rain
    b. Insect infestation
    c. High levels of dust and atmospheric pollution
    d. Intrusion from animals and man.

In the more advanced segment of the society, whether in developing areas or in industrialized regions, artificial drying has in many cases supplanted traditional sun drying in order to achieve better quality control, reduce spoilage, and in general cut down on the losses and inefficiencies listed above.

The relatively high cost of labor in most industrialized areas and low costs of fossil fuels caused the development of artificial, large-scale drying processes. The cost of dehydration was added to the cost of selling the process materials. The advent of higher charges for fossil fuels as well as depletion and scarcity has stimulated renewed interest in solar agricultural dryers.

Fig. 18.13 Detached solar steam cookers.

It is estimated according to the *FAO World Book* that the amount of agricultural produce dehydrated in 1968 using solar energy amounted to 225 million tons. In that year alone Australia exported over 72,000 tons of sun-dried foods worth over 27 million dollars. Had this drying, or even part of it, been done using fossil fuels, it would have put an even greater strain on already limited reserves. Over the past three decades, increasing interest has been paid to the development of solar agricultural dryers which make use of known principles of heliotechnology in order to combat some of the principal disadvantages of classical sun drying.

In evaluating technologies which might be amenable to application in developing areas, one should distinguish between small- and large-scale operations. In general, small-scale systems would be used in those areas where land holdings are not large with the result that individual farmers, fishermen, and herdsmen only produce modest amounts of surplus products. The objective is to dehydrate these surpluses for use often only by the family of the producer or for sale in the local market in the immediate vicinity. At times, small-scale surpluses of certain products such as peanuts or rice are delivered to central facilities for processing, dehydration, and eventual marketing (as is indicated in the attached photos). These systems are generally well established and require a fair degree of organization in the industry. In many instances these amalgamated handling facilities do not exist. Therefore, in providing an overview of some of the technologies, one must differentiate between the existence of commercial and physical infrastructures within a given locality. A series of descriptions have been attached which describe some of both the small- and large-scale systems for use in developing areas.

Larger-scale systems invariably require the use of an external power source. Where conventional electric power supplies are available, reliable and not excessive in cost, it is logical to utilize these external sources for the operation of fans and blowers and vents and duct baffles in order to increase the efficiency and operating performance of a solar agricultural system. Some driers are of the portable, powered type wherein solar air heater collectors are fitted with electrically powered fans (this could be done using gasoline or diesel engines as well) and are taken directly to the areas of production for in-situ drying. Traditionally this process was used with fossil fuel—often butane or propane gas as the energy source. As the price of these systems increases, there has been a tendency to develop systems of this nature relying on solar energy to provide the bulk of the energy required for dehydration. In fact, in some instances fossil fuels are used to supplement these solar collectors in order to maintain optimum operating conditions in a system partially operated by solar energy.

The other major category applicable for dehydration in the industrialized sectors of developed and developing nations is to use the roof area of existing buildings as the solar collector, fitting the buildings with suitable blowers, ducts, collectors, and often storage mechanisms. In the United States of America a number of activities along these lines have been developed and interest has been generated in some of the prestigious industrial and academic institutions in the country.

Another system receiving increasing interest in this field, both in developed and the developing regions, is the use of greenhouses to dehydrate surplus produce. This combined effect of drying and greenhouse operations has much validity and has to be examined for each particular set of circumstances. A number of studies have been undertaken in this regard for specialized crops.

Technical Characteristics of Solar Agricultural Dryers

There are two principal aspects of the crop-drying process: (a) the solar heating of the working fluid (generally air) and (b) the drying chamber wherein the heated air extracts moisture from the material to be dried. The solar heating function consists of (a) separate solar air heater collectors using natural or forced convection to preheat the ambient air, and reduce its relative humidity; and (b) direct, in-situ, heating of air which in turn directly dehydrates the produce.

A discussion of drying theory is beyond the scope of this chapter, but a few principles will be outlined here. These are particularly applicable to direct-radiation drying. The principles involved in the indirect drying of materials in opaque enclosures by means of hot air, whether from a solar heater or some other type of heating unit, are well outlined in the drying literature. The first requirement is a transfer of heat to the surface of the moist material by conduction from heated surfaces in contact with the material, or by conduction and convection from adjacent air at temperatures substantially above that of the material being dried, or by radiation from surrounding hot surfaces or from the sun. Absorption of heat by the material is used to vaporize water from it (590 cal/g of water evaporated). Water starts to vaporize from the surface of the moist material when the absorbed energy has increased the temperature enough for the water vapor pressure to exceed the partial pressure of water in the surrounding air. Steady state is achieved when the heat required for vaporization becomes equal to the rate of heat absorption from the surroundings.

To replenish the moisture removed from the surface, diffusion of water from the center to the surface of the drying material must take place. The rate of this process depends upon the nature of the material being dried and upon its moisture content at any time. It may thus be rate-limiting in the drying operation, or if moisture diffusion is rapid, the rate of heat absorption on the surface or the rate of vaporization may be rate-limiting.

In the case of direct radiation drying, part of the radiation may penetrate the material and be absorbed within the solid itself. Under such conditions heat is generated inside the material as well as at the surface, and thermal transfer in the solid is accelerated.

For economic reasons, maximum drying rates are usually desired. Product quality must be considered, however, and excessive temperatures must be avoided in many materials. In addition, because drying occurs at the surface, those materials which have a tendency to form hard, dry surfaces relatively impervious to liquid and vapor transfer must be dried at a rate sufficiently low to avoid this crust formation. Close control of heat transfer and vaporization rates, either by limiting the heat supply or by control of the humidity of the surrounding air, must be provided.

The drying of a product simply by permitting relatively dry air to circulate around it, without the use of any direct or indirect heat source, is known as adiabatic drying. The heat required for vaporizing the moisture is supplied by the air to the solid material, thereby reducing the temperature of the air while increasing its humidity. Because of the low heat capacity of air, large volumes of air at reasonably low relative humidity must be used in this type of drying process. Air leaving the dryer is nearly saturated with water at the wet-bulb temperature. The air supply at its initial dry-bulb temperature and humidity is thus cooled and humidified toward its wet-bulb temperature, while the moist solids in contact with this air approach the wet-bulb temperature also.

The foregoing generalizations must be somewhat modified if the materials being dried are at all soluble in the water present. Fruits and other agricultural products contain salts and sugars which cause a lowering of the vapor pressure. The surface temperatures of these materials must therefore be higher than the wet-bulb temperature of the air in order for vaporization to take place. This means that the adiabatic drying of these solids requires air at lower relative humidities than do the materials having no solutes in the aqueous phase.

An important property of materials processed by direct radiation drying is their radiation absorptance. Most solids have relatively high absorptances which may change as drying proceeds, the surfaces of the materials becoming less or more "black" during the process. Also, there may be changes in opacity of the surface of the materials which are partially transparent to some of the wavelengths in the spectrum of the radiant source.

The thermal conductivity of the material is also an important property, particularly if the solids are dried in a layer of sufficient depth to require conduction of heat from particle to particle. If the thermal conductivity is poor, circulation of heated air through and between the particles of a moist solid would permit better heat transfer than direct radiation on the surface of a relatively deep bed of particles.

Solar dryers are classified according to their heating modes, or the manner in which the heat derived from the solar radiation is utilized. In this regard, several general categories have been set up which are defined below. In general, a dryer is classified according to its principal operating mode. Some of the direct and mixed-mode dryers also use circulating fans, and are strictly not totally passive systems.

Solar dryer classification

Passive systems. Dryers using only solar or wind energy for their operations.

Sun or natural dryers. These dryers make use of the action of solar radiation ambient air temperature and relative humidity and windspeed to achieve the drying process.

Solar dryers—direct. In these units, the material to be dried is placed in an enclosure, with a transparent cover or side panels. Heat is generated by absorption of solar radiation on the product itself as well as on the internal surfaces of the drying chamber. This heat evaporates the moisture from the drying product. In addition, it serves to heat and expand the air in the enclosure, causing the removal of this moisture by the circulation of air.

Solar dryers—indirect. In these dryers, solar radiation is not directly incident on the material to be dried. Air is heated in a solar collector and then ducted to the drying chamber, to dehydrate the product.

Solar lumber dryers. These dryers have been put in a special category as they constitute an important application of the technology. In most cases forced ventilation is used as proper circulation of air helps control the drying rate so as to avoid case hardening.

Chamber dryer. One in which the material to be dried is dried in an enclosure.

Rack or tray dryer. One in which the material to be dried is placed on wire mesh or similar holding racks.

Hybrid systems. Dryers in which another form of energy, such as fuel or electricity, is used to supplement solar energy for heating and/or ventilation.

Description of Several Types of Solar Dryers

Drying of grapes on racks In Australia, natural and sun drying of grapes on racks has been used for quite some time. Large-scale drying racks are widely used in the grape growing areas of Australia and in 1972, they dried about 100,000 metric tons of fresh grapes in 8 to 14 days. One 50-m drying unit is generally considered to provide enough rack space to dry the fruits from over 11,000 m^2 of vines during the drying season.

The drying rack consists of 8 to 12 galvanized wire netting tiers spaced vertically. At 3-m intervals along the racks, pairs of intermediate upright posts, imbedded in the ground, carry cross pieces that support the tiers. The wire netting is reinforced lengthwise along both edges with fencing wire. At each end of the rack, the load is taken by two heavy posts, sloped and stayed against the strain with part of their length below ground level.

It should be noted that the rack can be covered by a sheet metal roof. These roofs are often very practical as they protect the raisins against rain or excessive sun, thus leading to a better quality product. The roof is constructed of corrugated iron sheets fixed crosswise, with equal overhang on both sides of the rack. There should be no pitch to the roof so that when wind from any direction accompanies rain, it will blow the water on the roof away from the fruit. When there is no wind, the overhang ensures that water drips away from the drying rack. Certain raisin species obtain a superior quality when shade-dried, thus Hessian side curtains are often placed on the rack to provide this condition. These curtains are to be avoided in wet climates where excessive humidities will favor mold development.

Solar cabinet dryer This dryer is a widely available design of direct solar drying (Fig. 18.14). It is easy to build from almost any kind of available building materials and simple to operate, maintain, and control. It is a small-scale dryer versatile in operation, and it can be used to dry a wide variety of agricultural products. Tested and used in many countries and under many different climates, this dryer has proven to be a very effective and useful device for small-scale food preservation.

The dryer is essentially a solar hot box, in which fruit, vegetables or other matter can be dehydrated on a small scale. It consists of a rectangular container, insulated at its base and preferably at the sides, and covered with a double-layered transparent roof. Solar radiation is

Fig. 18.14   Solar cabinet dryer.

transmitted through the roof and absorbed on the blackened interior surfaces. Owing to the insulation, the internal temperature is raised. Holes are drilled through the base to permit fresh ventilating air entry into the cabinet. Outlet ports are located on the upper parts of the cabinet side and rear panels. As the temperature increases, warm air passes out of these upper apertures by natural convection, creating a partial vacuum and drawing fresh air up through the base. As a result there is a constant perceptible flow of air over the drying matter, which is placed on perforated trays on the interior cabinet base.

See-saw dryer   The see-saw dryer was originally developed on the Ivory Coast for the drying of coffee and cocoa beans. Its design was further refined under work sponsored by the Government of Ghana and the Food and Agriculture Organization (FAO) of the United Nations. This simple dryer is suitable for small-scale drying operations and can be easily operated. Its use is envisaged for tropical regions. The dryer can be operated in a simple manner in two positions along a central axis of rotation running north-south. This see-saw operation permits the drying material to face the sun more directly during both the morning and the afternoon. This increases the output of the unit and leads to a more evenly dried product.

This see-saw dryer consists of a rectangular wood frame divided lengthwise into parallel channels of equal width, and crosswise by means of retaining bars. The bottom of the dryer frame is made of bamboo matting painted black and receives the material to be dried. The cover of the frame is made of a film of transparent polyvinyl chloride (PVC), which provides a substantial screening effect against ultra violet light, thus reducing photodegradation of the drying product. All of the internal parts of the dryer are coated with a matte flat black paint. The drying tray is moved during operation in an east-west plane. It is mounted on a trestle whose height is equal to one-fourth of the frame length. The drying frame can then be tilted and fixed to face the sun in the east during the morning and to face west during afternoon. The effective area of the dryer is limited by two transverse retaining bars fixed at 200 mm from each end of the drying frame and two others set 300 mm apart, about the transverse central line of the drying frame. The three small black bands thus delimited by the retaining bars are left free from any drying product and their purpose is to convert the radiant energy from the sun into heat. The heated air is circulated by natural convection from the lower to the upper end of the dryer by means of gaps provided at each end of the frame. Additional air is drawn through the matting base of the dryer by the natural convective effect as well. Produce to be dried is loaded with the dryer in a horizontal position, up to the level of the crosswise retaining bars, allowing clearance under the cover in order to permit air flow.

Glass roof solar dryer   In general, the unit is similar to a regular greenhouse structure and has a special roof peak cap acting as a flue and protecting the inside of the dryer against rain. This cap, made of folded zinc sheet, allows the heated air charged with the moisture removed from the material to escape, thereby permitting the entry of fresh air through the side shutters provided in the structure. The dryer is aligned lengthwise along a north-south axis. Basically it consists of two parallel rows of drying platforms with a central passage for an operator. A fixed glass roof above the drying platform allows the radiation of the sun to penetrate inside the dryer and also prevents the ingress of rain or dew at night. All surfaces inside the dryer are painted black to facilitate the absorption of solar radiation. (See Fig. 18.15.)

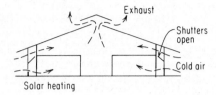

Fig. 18.15 Air circulation in the glass roof dryer for the solar heating mode.

The drying surface is made of galvanized iron wire mesh laid over wooden beams fixed across the platform. Strong metal wires stretched perpendicularly over the wooden beams and under the wire mesh provide additional support to the loaded wire mesh during the drying operation. This permits the solar or gas-heated air to pass easily through the wire mesh on which the drying product is spread.

In the prototype dryer, the gas heaters are situated underneath one of the two rows of drying platforms and are intended to serve during rainy or night periods only for shortening the drying time.

A free vertical space between the lower edge of the glass roof and the outer edge of the platform allows for the proper ventilation of the dryer. Six wooden shutters on hinges are located in this space along the length of the dryer on each side, and they can be opened or closed independently to regulate the air flow inside the dryer during the solar drying hours.

Solar fruit and vegetable dryer   This simple chamber type dryer operates in both the direct and indirect mode. It was designed to dry food for domestic needs or for a small-size restaurant, etc. It has been successfully used for drying a wide variety of food products, ranging from fruits and vegetables to herbs and meat.

Air, preheated in a solar air heater located at the base of the dryer, is admitted to the base of the drying enclosure. From there, it rises through the drying racks, dehydrating the product laid on them, and is then exhausted with its moisture content by natural convection through openings located at the top, rear wall of the chamber. The drying process is also carried out with the help of direct sun reaching the product through plexiglass sides, front, and top panels. The dryer faces south. (See Fig. 18.16.)

Solar wind ventilated dryer   The main feature of this dryer is its air circulation system. Air is drawn through the dryer by a wind-powered rotary vane located on the top of a chimney. Temperature and air flowrate are controlled by a damper.

The dryer can be described as a drying chamber through which warm air, heated in a solar air heater collector, is drawn by means of a rotary wind ventilator. The solar air heater collector

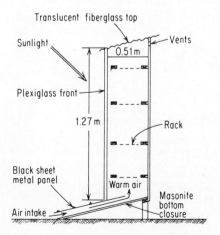

Fig. 18.16 Section view of the fruit & vegetable dryer.

used consists of a blackened hardboard sheet, insulated at the bottom and covered by a plastic (or glass) sheet. The collector is mounted facing due south, and tilted at an optimum angle for the area and particular season.

Air enters through the open bottom end of the collector. It passes up between the hardboard blackened bottom (absorber) and the cover. The effectiveness of the collector is increased by placing a perforated black mesh screen midway between the cover and the absorber: solar radiation which passes through the transparent cover is absorbed by both the mesh and hardboard. The mesh provides additional heat transfer surface area, and increased heat is supplied to the passing air. Collector efficiencies of over 75 percent have been achieved using this system.

The warm-air outlet of the collector is connected to the base of the drying chamber, which holds 12 trays placed in two adjacent six-tier stacks. Hot air circulates up through the drying produce, additional heating is obtained from solar radiation transmitted through transparent sheets which cover the east, south, and west sides of the drying cabinet. The rear vertical and bottom horizontal panels of the dryer are of blackened hardboard, which is insulated to reduce heat losses. A rotary wind ventilator is placed on top of a stack above the drying chamber. An adequate length of this stack is required both to achieve a chimney effect and to catch more wind.

The rotary wind ventilator is a moving corrugated vane rotor. As it spins in the wind, it expells air from the ventilator stack. The rotor is mounted on ball bearing suspension. The friction is low and momentum keeps the head spinning even in sporadic winds. Quantitative tests carried out using the ventilators indicate that the rotary ventilator keeps spinning between gusts, yielding a high, constant exhaust in spite of intermittent winds.

A stationary eductor placed on top of a chimney could be also used; however, it must be understood that it would rely solely on natural convection during periods of no wind.

Solar supplemental heat drying bin (Semiindustrial type)   The design presented here shows the possible transformation of a conventional bin dryer to an indirect solar bin dryer using the original structure of the drying bin. Considerable savings in fuel consumption can be made if this low-cost heating portion is added to the bin.

This drying bin is aligned longitudinally on an east-west axis, the south-facing side of the roof being used as the solar heat collector. This roof collector is sloped about 30° from the horizontal and is designed to produce an optimal temperature rise of 5 to 12°C over the outside air. The bin structure provides about 1 m² of collector area for each 2 m³ of grain. The author has found that this ratio provides an acceptable drying rate for shelled maize. The roof surface is painted black to absorb the solar energy. A transparent plastic film is supported about 8 cm above the roof by stretching it over the framing members set edgewise. (See Fig. 18.17.)

The air, drawn by a fan, enters the opening along the roof peak and moves through the collector roof down the south wall into the outside air duct. From there, the fan pushes the warmed air into the inside air duct and through the grain by way of a perforated floor. The bin is designed to dry half its depth of shelled maize at one time (1.22 m). The fan should be able to deliver about 2 m³ of air per minute for each cubic meter of corn to be dried, assuming the bin is full.

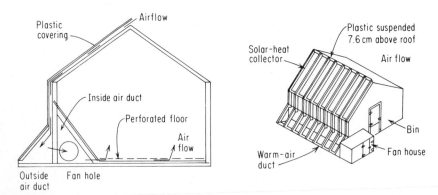

Fig. 18.17   Bin dryer cross section and isometric view.

Large-scale solar agricultural dryer (Barbados)   A large-scale agricultural dryer designed to provide livestock with locally produced feed operates by partly drying in a mobile cart. The material is then transferred to a room where drying is completed to the required moisture content.

Essentially the dryer is designed to operate in two stages. The first part consists of a mobile solar air heated drying cart, which should reduce the moisture content of the freshly harvested corn, for example, from 30 to 18 percent in the first day of operation. The corn is then transferred to a solar air heated storage bin, where the moisture content is gradually reduced to the equilibrium moisture content. Particular attention must be paid to the air flow design so as to cause minimum pressure drop and uniform flow through all parts of the air heaters and drying chambers.

A centrifugal fan blows air through a diffuser duct into a solar air collector. The collector can be fabricated from three plastic sheets, the top transparent, the center a black mesh with 50 percent openings, and the bottom a layer of insulation, sandwiched between two films, the upperside colored black, and the lower side aluminum. The sheets are sealed along the long edges, and supported by tension straps every meter. The center mesh rests on a rigid screen which is stretched between posts in the field. The collector is inflated on both sides of the tensioned layer and heats the air blown longitudinally through it. The end of the collector is connected to the mobile drying cart. The latter is insulated to reduce heat losses, and fitted with air flow dividers supporting a perforated drying floor. The fresh material is loaded into the cart, which is covered by a sloping double-layered plastic roof.

When the moisture content has been reduced to the required level, the partly dry matter is fed into a blower and transferred to the storage bin dryer, where its moisture is gradually reduced to the equilibrium level.

Solar timber seasoning kiln   This solar lumber dryer is designed to increase the drying rate of timber as compared to the traditional air-drying method. This particular design makes use of large quantities of low temperature heat (up to 60°C) and permits a rapid drying rate without undue degradation of the timber (cracks, warps).

The wood frame structure of the kiln is oriented lengthwise on an east-west axis, the higher wall facing north. Except for the north wall, the whole structure is covered with a double layer of transparent polyethylene sheeting separated by an air gap. The north wall is made of plywood sheeting. The roof is tilted towards south and above horizontal at an angle of 0.9 times the latitude. The drying space in a typical kiln can take about 3.5 m^3 of 25-mm thick planking (each time). Inside the kiln, a false ceiling covering the entire length of the kiln extends from the floor to the false ceiling and is provided with a hole in its center for housing the fan. The built-in interior surfaces within the kiln (the surfaces of wooden roof studs and pillars, the false ceiling and north wall partition, the north wall, the baffles and the concrete floor) are painted black for a maximum collection of heat. A fan is used for forced air circulation. The use of plywood baffles and movable partitions allows the dryer to be used either as a single-pass flow through dryer or as a recirculating dryer with partial venting. (See Fig. 18.18.)

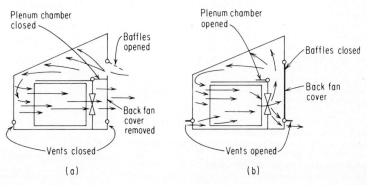

Fig. 18.18   Timber seasoning kiln working as (*a*) a single pass forced air dryer and (*b*) a recirculating air dryer with partial ventilating.

REFERENCES

1. Akyurt, M., and M. K. Selcuk, "A Solar Dryer Supplemented with Auxiliary Heating Systems for Continuous Operation," *Solar Energy*, vol. 6, no. 4, 1962.
2. Alward, R., and O. Goldstein, *Assembly Drawings for the Construction of SOLAR Steam Cookers*, 3 drawings, Brace Research Institute, February 1972.
3. Alward, R., Lawand, T. A., and P. Hopley, "Une Cuisiniere Solaire en Haiti," *Architecture Concept.*, vol. 28, no. 312, p. 2, March 1973.
4. Alward, R., Lawand, T. A., and P. Hopley, "Description of a Large Scale Solar Steam Cooker in Haiti," *Proceedings of the International Congress,"The Sun in Service of Mankind,"* UNESCO House, Paris, July 1973.
5. Anon., "Aprovechamiento de la Energia Solar en El Altiplano Peruano," *Agronomia*, vol. XXX, no. 4, 1963.
6. Baum, V. A., and R. Bairamov, "Prospects of Solar Stills in Turkmenia," *Solar Energy*, vol. 10, no. 1, pp. 38–40, 1966.
7. Beason, R. G., "Solar Cooking Turns Practical," *Mechanix Illustrated*, pp. 32–33, July 1976.
8. Bjorksten Research Laboratories Inc., *Development of Plastic Solar Stills*, U.S. Department of the Interior, Office of Saline Water, R&D Report No. 24, pp. 1–25, 1959.
9. Brace Research Institute Staff, *Study of the Feasibility of Establishing a Rural Energy Center for Demonstration Purposes in Senegal*, Brace Research Institute Report, August 1974.
10. Brace Research Institute, *Survey of Solar Agricultural Dryers*, Technical Report No. T-99, January 1976.
11. Cheng, K., Wong, H., and O. Tanaka, "Experimental Study of a Solar Steam Cooker," Project for Department of Mechanical Engineering, McGill University, Course No. 305-463A-464B, May 1973.
12. Cooper, P. I., "The Maximum Efficiency of Single-Effect Solar Still," *Solar Energy*, vol. 15, no. 3, pp. 205–219, 1973–74.
13. Della Porta, G. B., *Maeiae Naturalis Libri XX*, Libro X, p. 183 (Napoli 1589).
14. Delyannis, A., and E. Delyannis, "Solar Distillation in Greece," *Proceedings of First International Symposium on Water Desalination*, Session C, Paper No. SWD/87, Washington, DC, October 1965.
15. Delyannis, A., and E. Delyannis, "Solar Distillation Plants of High Capacity," *4th International Symposium on Fresh Water from the Sea*, vol. 4, pp. 487–491, 1973.
16. Duffie, J. A., *Reflective Solar Cooker Designs*, Engineering Experiment Station, University of Wisconsin, p. 9, undated.
17. Duffie, J. A., Lappala, R. P., and G. O. G. Löf, "Plastics in Solar Stoves," Reprinted from *Modern Plastics*, November 1957.
18. Duffie, J. A., Lappala, R. P., and G. O. G. Löf, *Plastics for Focusing Collectors*, Engineering Experiment Station, University of Wisconsin, Reprint No. 327, pp. 9–13.
19. Duffie, J. A., Löf, G., and B. Beck, "Laboratory & Field Studies of Plastic Reflector Solar Cookers," *Solar Energy, vol. VI, pp. 94–98, July 1962.*
20. Florida, N. *Solar Cookers and Water Heaters*, Appropriate Technology Series, Bardoli, India, 1974.
21. Ghai, M. L., "Solar Heat for Cooking," *Journal of Scientific & Industrial Research*, vol. 12A, no. 3, pp. 117–124, 1953.
22. Ghai, M. L., "Design of Reflector-Type Direct Solar Cookers," *Journal of Scientific & Industrial Research*, vol. 12A, no. 4, pp. 165–175, 1953.
23. Ghai, M. L., Khanna, M. L., Ahluwalia, J. S., and S. P. Suri, "Performance of Reflector-Type Direct Solar Cooker," *Journal of Scientific & Industrial Research*, vol. 12A, no. 12, pp. 540–551, 1953.
24. Ghai, M. L., and B. S. Phandher, "Manufacture of Reflector-Type Direct Solar Cooker," *Journal of Scientific & Industrial Research*, vol. 13A, no. 5, pp. 212–216, 1954.
25. Ghosh, M. K., "Sun Cookers for Villages," All India Solar Energy Working Group & Conference, Indian Institute of Technology, Madras, India, November 1973.
26. Gomella, C., "Practical Possibilities for the Use of Solar Distillation in Under-Developed Arid Countries," *Trans. of the Conference on the Use of Solar Energy*, vol. 3, part II, pp. 119–133, Tucson, AZ, 1955.
27. Gomella, C., "Possibilités d'extension des dimensions des distillateurs solaires," U.N. Conference on New Sources of Energy, Session III E, E/CONF.35/S/107, pp. 1–16, Rome, August 1961.
28. Gupta, Dr. J. P., "Studies on Solar Hot Box," All India Solar Energy Working Group and Conference, Indian Institute of Technology, Madras, India, November 1973.
29. Harding, J., "Apparatus for Solar Distillation," *Proceedings of Institute of Civil Engineers*, vol. 73, pp. 284–288, 1883.
30. Hirschmann, J. G., "Evaporateur et distillateur solaire au Chili," U.N. Conference on New Sources of Energy, Session III E, E/CONF.35/S/23, pp. 1–30, Rome, August 1961.
31. Howe, E. D., "Solar Distillation Research at the University of California," U.N. Conference on New Energy, Session III E, E/CONF. 35/S/29, pp. 1–22, Rome, August 1961.
32. Howe, E. D., "Pacific Island Water Systems Using a Combined Solar Still and Rainfall Collector," *Solar Energy*, vol. 10, no. 4, pp. 175–181, 1966.
33. Jenness, J. R., "Recommendations and Suggested Techniques for the Manufacture of Inexpensive Solar Cookers," *The Journal of Solar Energy Science and Engineering*, vol. 4, no. 3, July pp. 22–24, 1960.
34. Khan, E. U., "Practical Devices for the Utilization of Solar Energy," *Solar Energy*, vol. 8, no. 1, pp. 17–22, 1964.

35. Khanna, M. L., and K. N. Mathur, "Experiments on Demineralization of Water in North India," U.N. Conference on New Sources of Energy, Session III E, E/CONF.35/S/115, pp. 1–11, Rome, August 1961.
36. Khanna, M. L., "Solar Heating of Vegetable Oil," reprint from *Solar Energy*, vol. 6, no. 2, 1962.
37. Lawand, T. A., "Engineering and Economic Evaluation of Solar Distillation for Small Countries," M.SC. Thesis, Dept. of Agricultural Engineering, McGill University, August 1968.
38. Löf., G. O. G., "Design and Cost Factors of Large Basin Type Solar Stills," *Proceedings of Symposium on Saline Water Conversion*, National Academy of Sciences, National Research Council, Washington, DC, Publication No. 568, pp. 157–174, 1958.
39. Löf., G. O. G., "Use of Solar Energy for Solar Drying," United Nations Conference on New Sources of Energy, Rome, p. 17, June 1961.
40. Löf, G. O. G., "Fundamental Problems of Solar Distillation," *Proceedings of Symposium on Research Frontiers in Solar Energy Utilization*, National Academy of Sciences, Washington, DC, pp. 35–46, April 1961.
41. Löf., G. O. G., "Solar Energy for the Drying of Solids," *Solar Energy*, vol. 6, no. 4, pp. 122–128, 1962.
42. Löf., G. O. G., "Recent Investigations in the Use of Solar Energy for Cooking," *Solar Energy*, vol. 7, no. 3, pp., 125–133, 1963.
43. Lukes, T., *Research on the Application of Solar Energy to the Food Drying Industry*, Principal Investigator, Report No. NSF/RANN/SE/GI 42944/PR/74/3, California Polytechnic State University, San Luis Obispo, California, October 1974.
44. "Effect of Heat Input from Solar Cookers on the Ascorbic Acid Content of Peas," Final Year Project Course, Dept. of Food Science, McGill University, March 1974.
45. Mustacchi et al., "Solar Desalination, Status and Potential," *Proceedings A.S.I.S. Conference*, Rome, May 1976.
46. Nebbia, G., and G. Menozzi, "A Short History of Water Desalination," Aque Dolce Dal Mare, IIa Inchiesta Internazionale, Milano, pp. 129–172, April 1966.
47. Neubauer, L. Prof. Emeritus, and G. Williams, "*Solar Oven Economy for Farm Homes*," presented to 1976 annual meeting of American Society of Agricultural Engineers, University of Nebraska, Lincoln, paper No. 76-4021.
48. *References from United Nations Conference on New Sources of Energy, Solar Energy, Wind Power and Geothermal Energy. Agenda Item 111.C.4:*
    ■ "General Report:" G. O. G. Löf., E/CONF.35/GR/16(S)
    ■ "Use of Solar Energy for Cooking":
    ■ "Temperature Decay Curves in the Box-Type Solar Cooker:" M. Abou-Hussein, E/CONF.35/S/75
    ■ "Design and Performance of Folding Umbrella-Type Solar Cooker": G. O. G. Löf and D. A. Fester, E/CONF.35/S/100
    ■ "Laboratory and Field Studies of Plastic Reflector Solar Cookers": J. A. Duffie, G. O. G. Löf., and B. Beck, E/CONF.35/S/116
    ■ "A Cylindro-Parabolic Solar Cooker": A Salgado Prata, E/CONF.35/S/110
    ■ "Cheap but Practical Solar Kitchens": H. Stam, E/CONF.35/S/24
    ■ "Practical Solar Cooking Ovens:" M. Telkes and S. Andrassy, E/CONF.35/S/101
49. Sakr, I. A., "Elliptical Paraboloid Solar Cooker," *Proceedings of the International Congress, "The Sun in the Service of Mankind,"* UNESCO House, Paris, July 1973.
50. Sinson, D. A., *Design and Performance Evaluation of a 6 ft. × 4 ft. Parabolic Solar Steam Generator and Its Application to Pressure Cooking*, April 1964.
51. Stam, H., *Drawing of Solar Cookers*, A.W.S. Netherlands, 1957, 1959.
52. Swet, C. J., *A Universal Solar Kitchen*, Johns Hopkins University, Applied Physics Laboratory, Baltimore, July 1972.
53. Tabor, H., "Solar Cooker for Developing Countries," paper presented at Annual Meeting of Solar Energy Society, Boston, MA, March 1966.
54. Talbert, S. G., Eibling, J. A., and G. O. G. Löf, *Manual on Solar Distillation of Saline Water*, Sec. 3, "Examples of Large Basin Type Solar Stills," Battelle Memorial Institute, 1970.
55. Telkes, M., *The Solar Cooking Oven*, New York University, College of Engineering Research Divn., January 1958.
56. Telkes, Maria, *New and Improved Methods for Lower Cost Solar Distillation*, U.S. Department of the Interior, Office of Saline Water, R&D Report No. 31, pp. 1–38, August 1959.
57. Tschinkel, H., *Contribution a la Protection des Combustibles Ligneuz*, Min. de l'Agriculture, Institut National de Recherches Forestières, No. 4, Avril 1975.
58. University of Florida, "Solar Cooking Turns Practical," *Mechanix Illustrated*, July 1976. A concentrating, cotton seed oil cooker with storage.
59. Vickery, S., *The Effect of the Unbroiler on the Ascorbic Acid Content of Sweet Peas*, Department of Food Science, McGill University, March 1974.
60. VITA, *Solar Cooker Construction Manual*, VITA Union College Campus, Schenectady, NY, June 1967.
61. Ward, G. T., *La Cocina Solar Construide en Chucuito, Puno, Peru*, Por. el Ing. M. F. Pons de la F.A.O., diciembre 1963 (in Spanish).

62. Whillier, A., "Black Painted Solar Air Heaters of Conventional Design," *Solar Energy*, vol. 8, no. 1, pp. 31–37, 1963.
63. Whillier, A., *Preliminary Report on Solar Stove for Cooking by Boiling*, September 1963.
64. Whillier, A., *A Stove for Boiling of Foods Using Solar Energy*, April 1964.
65. Whillier, A., "A Stove for Boiling Foods Using Solar Energy," *Sun at Work*, vol. 10. no. 1, pp. 9–12, January 1965.
66. Whillier, A., *How to Make a Solar Steam Cooker*, January 1965. Revised February 1973. (Also in French.)

Chapter 19

Ocean Thermal Energy Conversion

GORDON L. DUGGER
Applied Physics Laboratory, The Johns Hopkins University, Laurel, MD.

FREDERICK E. NAEF
Lockheed Missiles and Space Company, Washington, DC

J. EDWARD SNYDER III
Ocean Energy Systems, TRW, Redondo Beach, CA

INTRODUCTION

An ocean thermal energy conversion (OTEC) power plant uses the temperature difference (ΔT_o) between warm surface water and cold deep water to drive a heat engine. In the closed-Rankine-cycle OTEC system conceived by d'Arsonval in 1881,[1] a working fluid (e.g., ammonia) is vaporized by heat exchange with warm water, drives a turbine generator, and is condensed by heat exchange with cold water drawn from a 700- to 1200-m depth (Fig. 19.1).

In 1930, Claude demonstrated an open Rankine cycle process at Mantanzas Bay, Cuba (Fig. 19.2).[2] Warm water was flash-vaporized at low pressure to drive a turbine and then was condensed by direct contact with cold water drawn from a 700-m depth by a 1.6-m-diameter, 1.75-km-long cold-water pipe (CWP). From a ΔT_o of 14°C, Claude's undersized turbine generated 22 kW$_e$. In 1956 a French team designed a 3.5-MW$_e$ (net) open-cycle plant for installation off Abidjan on the Ivory Coast of Africa and demonstrated the necessary CWP deployment. The project was marginally economic and was dropped when a hydroelectric plant was installed nearby.[3] Two problems with the open cycle are the size of the turbines required because of the low steam pressure of 0.03 atm (however, a lightweight rotor blade construction similar to that in sailplane wings may be possible[4]) and the development work needed on the direct-contact condensers.

In 1975 Beck[5, 6a] proposed an improvement on the open cycle: a "steam-lift water pump" analogous to an air-lift pump would raise the warm water to a substantial height from which it could fall to drive a water turbine. Zener and Fetkovich[6b,7] proposed an improvement whereby an additive to the warm seawater would cause it to foam. In a plant having an upper chamber evacuated to the vapor pressure of water at 5°C, foaming would begin at 25°C; as the foam rose

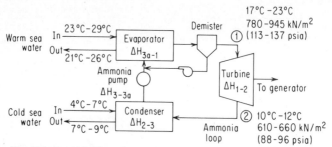

Fig. 19.1 Simplified loop diagram for the closed-Rankine cycle OTEC system.

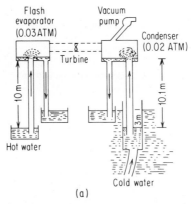

(a)

(b)

Fig. 19.2 (a) Claude's basic scheme for the open-Rankine cycle. Flash-evaporated seawater drives a turbine and is then recondensed by cold water falling like rain in the condenser (*from Ref. 2*) and (b) the Mini-OTEC first operated August, 1979 off Hawaii.

in an annulus between two cylindrical pipes and expanded, with gradual adiabatic cooling to 5°C, the water remaining as liquid would be carried in the corners of the foam cells. At the top of the annulus the foam would strike a barrier, causing the liquid to drain into the central pipe and fall to drive a turbine. The vapor would flow into an outer annular chamber between the second pipe and a third pipe to be condensed by contact with cold seawater at 5°C. In theory, the maximum power per unit horizontal area is developed when the foam breaker is at a 198-m height, with power generated at the rate of 0.15 kW/m². These approaches and a mist system proposed by Ridgway[6c] are in early stages of conceptual design and experiments. Such potential improvements to open-cycle OTEC systems are judged by most investigators to require more R&D than closed-cycle OTEC plants.

As the Andersons pointed out in 1966,[8] a closed-Rankine-cycle system (Fig. 19.1) producing electricity at competitive costs could be developed quickly using heat-engine technology drawn largely from the refrigeration/air-conditioning industry. In June 1973, at the first OTEC workshop, sponsored by the U.S. National Science Foundation (NSF),[9] early conceptual designs of OTEC plants by the University of Massachusetts (see also Ref. 10), the Andersons, Carnegie-Mellon University, and the Applied Physics Laboratory of The Johns Hopkins University (APL/JHU), were discussed along with the design of heat exchangers and other aspects. In 1974-75 NSF and the Energy Research and Development Administration (ERDA) supported competitive OTEC system studies by two industrial teams headed by Lockheed[11a] and TRW[11b] and some 20 other OTEC projects described at the second[12] and third[11] OTEC workshops. The initial system studies by Lockheed and TRW were constrained by NSF/ERDA to delivery of power to the bus bars on board. In 1975 the U.S. Maritime Administration (MarAd), Department of Commerce, supported APL/JHU to evaluate the maritime aspects of a tropical-ocean plant-ship that would "graze" slowly to seek out the highest ΔT_0 while producing energy-intensive products such as ammonia and aluminum on board.[13,14]

In 1979, *Mini-OTEC* achieved self-sustaining OTEC operations off Keahole Point, Hawaii, and generated 50 kW (gross) and 10 kW (net) from a temperature difference of 38.2 F. The system [Fig. 19.2(b)] was designed by Lockheed and outfitted with titanium plate type heat exchangers built by Alfa-Laval. The system contained all the essential components required for a full-scale OTEC plant, although it was designed as a test facility for developing and testing components and operations that could be used in the design of large-scale OTEC plants. The barge mounted plant was moored via a single anchor leg which incorporated the 24-in (OD) polyethylene pipe as an integral member of the moor. A flexible rubbber hose connected the top of the cold-water pipe to a moonpool in the center of the barge. *Mini-OTEC* was the first operational closed cycle system to operate *in situ*. The experiment provided valuable data on transient response, heat exchanger performance in various sea conditions, biofouling counter-measures, corrosion, plume dynamics, and deployment/operation problems.

The Ocean Thermal Resource

Oceans cover 71 percent of the earth's surface and receive the majority of the solar energy incident upon earth. In semitropical and tropical oceans, the available ΔT_0 is sufficient for OTEC operation and is available 24 h a day. The vast potential for OTEC implementation can be appreciated by noting that a quantity of electric power equal to the entire projected U.S. demand in the year 2000 (about 7×10^5 MW$_e$) could be obtained by extracting from the oceans in the $\pm$ 10° latitude band near the equator an amount of energy equal to only 0.004 percent of the incident solar energy.

Figure 19.3 shows typical temperature-vs.-depth profiles off Keahole Point, Hawaii; off Punta Tuna, Puerto Rico; and for a grazing pattern in an area in the South Atlantic Ocean 300 to 900 nmi east of Recife, Brazil (designated ATL-1).[15] Table 19.1 presents seasonal variations in ΔT_0 for the same three sites plus one 25 nmi east of Miami, Florida; for one in the Gulf of Mexico 150 nmi west-southwest of Tampa, Florida; and for a grazing pattern in an area in the North Pacific Ocean 200 to 400 nmi southwest of Acapulco, Mexico (designated PAC-2). The Puerto Rico and Hawaii sites are of interest in the U.S. OTEC program because of the needs of these islands for an alternative to imported fossil fuels (potential nuclear power plant sites are limited) and the possibility that one of them will be chosen by the Department of Energy (DOE), together with its local government and/or private interests, for an early demonstration of the "island industry" OTEC option discussed later. The grazing plants in ATL-1 and PAC-2 would seek out the highest ΔT_0 based on historical plots of monthly surface temperature contours plus daily information from satellites which would indicate the path to take.

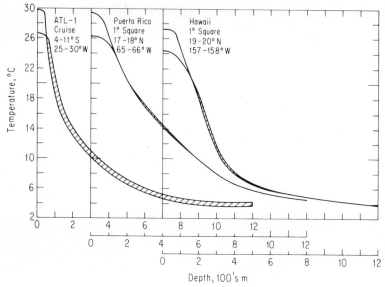

Fig. 19.3   Temperature-depth variations for three sites.

The variation of available ΔT_o with site and season is important because the gross power output from a given plant varies approximately with the square of ΔT_o (Ref. 6d):

$$P_{net} = P_{g_d} [\Delta T_{o_d} / \Delta T_o]^2 - P_{p_d} \qquad (19.1)$$

where P_{g_d} is the design or rated gross power output in MW_e at the annual average temperature difference ΔT_{o_d}, and P_{p_d} is the "parasitic power" needed for seawater pumping (the largest load), working-fluid pumping, and all other power requirements for operation of the plant-ship. The plant will be designed for some rated values of P_{g_d} and P_{p_d} to yield a desired rated annual average P_{net_d}, probably between 250 and 400 MW_e at ΔT_{o_d}. For a given plant designed for a site with a modest ($\pm$ 5 to 8 percent) seasonal variation in ΔT_o, P_p should be nearly constant and equal to P_{p_d}, and Eq. (19.1) should provide a good estimate for P_{net} at any time. This P_{p_d} will depend on details of the plant design, particularly the heat exchangers, and it will tend to be higher for a site with a lower ΔT_{o_d}. Typical values range from 25 to 40 percent of P_{net_d}. The larger P_p is, the greater will be the seasonal variation of P_{net} for a given site. However, its effect on P_{net} is second-order compared with the effect of ΔT_o. For example, if P_p is 30 percent of P_{net_d}, Eq. (19.1) predicts that a plant delivering P_{net_d} = 250 MW_e (P_{p_d} = 75 MW_e and P_{g_d} = 325 MW_e) at ΔT_{o_d} = 22°C will deliver P_{net} = 143 MW_e at 18°C. If P_{p_d} were 40 percent of P_{net_d} (P_{p_d} = 100 MW_e and P_{g_d} = 350 MW_e), the plant at this site would deliver 134 MW_e at 18°C. The first reduction due to ΔT_o is 107 MW_e or 43 percent; the additional reduction due to the higher P_p is 9 MW_e or 4 percent.

For simplicity, Table 19.1, which shows the seasonal variations in P_{net} for six selected sites, is based on P_{p_d} = 30 percent of P_{net_d} for all sites. For all sites in Table 19.1 except Tampa, the variations in P_{net_d} are in the $\pm$ 13 to 20 percent range. For Tampa and other sites in the Gulf of Mexico, the variations would be larger. The small variations in $P_{net\ ann\ avg}$ from 100 percent are a result of the character of the seasonal variations in ΔT_o. The bottom line in Table 19.1 shows $P_{net\ avg\ rel}$ for each site relative to that at ATL-1, where

$$P_{net\ avg\ rel}(\%) = \left[130 \left(\Delta T_{o_d} / \Delta T_{o_d\ ATL-1} \right)^2 - 30 \right] P_{net\ avg} / P_{net\ avg\ ATL-1} \qquad (19.2)$$

The $P_{net\ avg\ rel}$ values in Table 19.1 indicate that, other things being equal (except for the CWP depths indicated), a plant stationed off Miami would deliver 29 percent less power annually than if it were grazing in PAC-2 or ATL-1. Thus, its on-board bus bar electricity cost would

TABLE 19.1 Monthly Values of ΔT_o^*† and Power P_{net} (Percentage of Design Value) for Selected Sites

Month	Miami ΔT_o, °C	Miami P_{net}, %	Tampa ΔT_o, °C	Tampa P_{net}, %	ATL-1† ΔT_o, °C	ATL-1† P_{net}, %	PAC-2† ΔT_o, °C	PAC-2† P_{net}, %	Puerto Rico ΔT_o, °C	Puerto Rico P_{net}, %	Hawaii ΔT_o, °C	Hawaii P_{net}, %
January	19.1	78 ‡	19.7	71	23.5	96	22.6	86	21.3	88	20.6	90
February	19.5	82	20.6	80	24.3	104	22.9	89	20.9	84	20.1	85
March	19.7	84	20.6	80	25.2	115	23.6	97	20.9	84	20.0	84
April	19.8	86	20.9	83	25.2	115	25.1	113	21.2	87	20.6	90
May	20.4	93	23.0	107	24.2	103	25.1	113	22.2	99	21.3	99
June	21.3	104	24.3	123	24.4	105	24.5	108	22.6	104	21.3	99
July	22.8	123	24.6	127	23.6	97	24.1	102	23.0	108	22.4	112
August	22.8	123	25.0	132	24.3	104	23.9	100	23.0	108	22.7	116
September	23.0	126	24.2	122	22.4	84	23.8	99	23.6	116	22.7	116
October	22.3	117	23.0	107	23.3	94	23.6	97	23.9	119	22.5	114
November	20.8	98	21.8	93	22.6	86	24.1	102	23.3	112	21.5	101
December	20.0	88	21.1	85	23.3	94	23.2	92	22.3	100	20.7	92
Average	21.0	100.2‡	22.4	100.8	23.9	99.8	23.9	99.8	22.3	100.7	21.4	99.8
$P_{net\ avg\ rel}$		70.6		85.0		100		100		83.7		74.2

*ΔT between surface and 900-m depth for ATL-1 and PAC-2; 1000 m for Tampa, Hawaii, and Puerto Rico; 700 for Miami.[15]
†Data for ATL-1 (east of Recife, Brazil) and PAC-2 (southwest of Acapulco, Mexico) are for a grazing concept.
‡Average values for P_{net} vary slightly from 100% (design level); they are used in calculation of $P_{net\ avg\ rel}$ by Eq. (19.2).

be 42 percent higher. However, OTEC plant costs will vary with the site for various additional reasons. For example:

1. A plant designed for mooring off Miami or elsewhere in the Gulf Stream may be of a submerged spar design for hurricane survival and minimization of bending moments on the cold-water pipe. It will also require a high-current mooring system. Its pipe length (and ΔT_o) may be limited by local bathymetry (ocean floor depths); e.g., a depth of 700 m is available 40 to 75 km east of Miami in the Florida Strait.

2. A plant designed for use in the Gulf of Mexico might also be submerged for hurricane survival but would not be subject to high currents and could be bottom-mounted. Surface vessel designs are also being evaluated (see *The OTEC Platform*). Available depth and ΔT_o may be greater than they are off Miami, but will vary with distance from shore. Thus, the tradeoff between ΔT_o and electric cable costs will vary with site.

3. The Puerto Rico and Hawaii sites are special cases because good depths are available 3 to 10 km offshore. Available ΔT_o is slightly greater at Puerto Rico, but environmental conditions are less severe at Hawaii, so designs may differ.

4. A tropical ocean grazing (cruising) plant designed for operation in hurricane-free areas such as ATL-1 is expected to be a surface vessel to minimize cost. Its pipe length would not be limited by bathymetry, but would be chosen in a tradeoff between incremental gain in ΔT_o and the effects of length on (a) pipe bending moments during the worst-case, once-in-100-years storm, and consequent cost, and (b) the incremental drag and power requirement during normal cruising operation. The 900-m length used for Table 19.1 is judged to be near-optimal.

5. The early moored offshore plants are expected to deliver electricity directly to shore by cable, whereas the tropical grazing plant will produce an energy-intensive product on board for shipment to shore. The latter plant, being a lower-cost surface vessel, will also be able to accommodate the additional hull and deck requirements for on-board process equipment at lower cost. The use of thrusters for both grazing and station-keeping is more practical for these plants because a 0.5-kn speed (with 1-kn maximum capability) will usually suffice for grazing, and a once-in-100-year storm with a duration measured in hours would move it only a few tens of kilometers off course on high seas. In general, the plant's availability factor will be affected very little by the environment, whereas an offshore plant subjected to an average of six to seven hurricanes per year could suffer substantial reductions, depending on the nature and cost of its design and the survival tactics required.

6. The probability of earlier mass-production benefits for the tropical grazing plants is greater because (a) the sizes of suitable sites are far greater for the tropical grazing plants, (b) the variations among site areas such as ATL-1 and PAC-2 are smaller, (c) the possibility of adverse effects on the environment caused by a given number of plants is smaller, (d) the ability to use existing shipyards for construction may be greater, and (e) solutions to cable and mooring problems are not required.

From various of the foregoing considerations, one may expect the average capital cost of the basic OTEC plant to be between 10 and 30 to 40 percent lower for a tropical grazing plant, and the availability factor to be up to 10 percent higher. These factors together with the ΔT_o effect could combine to make the average on-board bus bar electricity cost (for a given type of financing) for, say, the first 50 tropical grazing plants to be lower by a factor near 2 compared with the first 50 offshore plants. Because so many factors are involved, including the institutional and financing considerations discussed briefly near the end of this chapter, such cost ratios can not yet be generalized.

Overall Thermal Efficiency

The overall thermal efficiency of an OTEC plant is low; therefore, large seawater pumps and large heat exchangers are required. This does not mean that power output will ever go to zero in a properly designed and maintained plant, as already indicated by Table 19.1. The ideal Carnot cycle efficiency is, typically,

$$\eta_C = \Delta T_o/T_{hw} = 22 \text{ K}/295 \text{ K} = 0.075 \tag{19.3}$$

where T_{hw} is the hot (surface) water temperature. This ideal efficiency allows for neither parasitic loss (P_{p_d}) nor temperature drops between seawater and working fluid for the evaporators and condensers (which, therefore, are assumed to be infinite in size). A rough rule of thumb is that about 25 percent of the ΔT_o should be allowed for the evaporators and 25 percent for the condensers, leaving 50 percent for the power turbine. For an 80 percent overall efficiency for the turbine-generator, the gross power output efficiency is approximately $0.075 \times 0.5 \times 0.8$ = 0.03, or 3.0 percent. Since the parasitic load P_{p_d} takes another 30 percent or so, the overall thermal efficiency for *net* power delivery may be $0.03/1.3 = 0.023$ or 2.3 percent.

Let us consider the ΔT_o effect. If Eq. (19.1), with P_{p_d} = 30 percent of $P_{net\,d}$, holds until ΔT_o falls to the point at which P_{net} reaches zero, ΔT_o would be 8.9°C for the Miami site. But the lowest ΔT_o anticipated is 19°C and yields approximately 78 percent of the annual average power output.

The heat-exchanger efficiency depends strongly on keeping the seawater side of the tubes relatively free of biofouling. Available data (noted later) indicate that it will be possible to keep the average fouling coefficient (see Chap. 4) near 0.000018 m² · °C/W (0.0001 ft² · °F · h/Btu) by regularly scheduled cleaning. The P_p used for Table 19.1 was based on this value and overall heat-transfer coefficients of the order of 2700 W/m² · °C (480 Btu/h · ft² · °F). If the tube-cleaning system malfunctioned and R_f were to increase by a factor of 10 before the tubes could be cleaned again, the U-value would be $(5700^{-1} + 5700^{-1} + 0.00018^{-1}) = 1880$ W/m² · °C. [The waterside coefficient for clean tubes and the ammonia (working-fluid) side coefficient are each of the order of 5700 W/m² · °C, therefore the U-value for very thin-walled tubes is $(5700^{-1} + 5700^{-1} = 0.000018^{-1}) = 2700$ W/m² · °C for the design case.] This 30 percent reduction is appreciable but not catastrophic and exceeds any reduction to be reasonably expected from biofouling.

Some Example OTEC Concepts

The plant concept initially developed by Lockheed Ocean Systems, Bechtel Corporation, and T. Y. Lin International[11a] and improved in 1976 is an anchored spar configuration (Fig. 19.4) of reinforced-concrete construction with a cold-water pipe reaching a 780-m depth for ΔT_{o_d} = 21.2°C. The main core vessel has a 57-m inside diameter (ID), and the maximum span across power modules is 126 m. Total displacement is approximately 300,000 t.* Four detachable power modules (see inset in Fig. 19.4) using titanium-tubed heat exchangers, with seawater inside and ammonia outside the tubes, generate a total P_{net_d} of 264 MW$_e$. The plant is conservatively designed to permit use in hurricane areas as well as at tropical sites (with mooring depth limited to 6000 m) with a 38-year average system life expectancy. The estimated average cost for the first 25 plants was $2000/kW$_e$ in 1978 dollars.

The team of TRW Ocean and Energy Systems, Global Marine Development, and United Engineers and Constructors briefly considered 15 alternative platform configurations before selecting a cylindrical surface vessel (Fig. 19.5) for their baseline 100-MW$_e$ (net) design.[11b] It is 103 m in diameter, weighs approximately 215,000 t, and is designed for a 40-year life. Four power modules, enclosed in the reinforced-concrete hull, employ heat exchangers using titanium tubes with ammonia outside the tubes. The 1200-m-long CWP made of fiber-reinforced plastic was selected to yield ΔT_{o_d} of 22°C. A dynamic positioning system (using the warm-water discharge and part of the cold-water discharge in shrouded jets) was chosen in preference to a mooring system for use in tropical oceans. The capital cost estimate for the first 100-MW$_e$ plant was $2100/kW$_e$ in late-1974 dollars. For 11.8 percent fixed charges for municipal-type costing, 40-year plant life, and 0.9 capacity factor, this gave an electricity cost at the bus bars of 35 mills/kWh. The designers proposed improvements—including use of aluminum heat exchangers (possibly with enhanced surfaces), placing the heat exchangers outside the hull, reduction of hull cost/kW$_e$ for larger plants, siting for a slightly higher ΔT_o, and learning-curve effects—that could reduce the cost of the fifth or sixth plant to approximately $1400/kW$_e$ in 1978 dollars.

A concept for a 100-MW$_e$ (net), tropical-ocean, OTEC/ammonia, demonstration-size plant-ship was developed by APL/JHU with assistance from the Sun Shipbuilding and Dry Dock Co., Hydronautics, Inc., Kaiser Aluminum and Chemical Corp., and others (Fig. 19.6).[13,14] The arrangement of the 60-m-beam by 150-m-long reinforced-concrete platform is symmetric about a central reinforced-concrete CWP, 18 m in diameter by 900 m long. Cold water is delivered by 19 pumps in the top of the CWP to head ponds located along the centerline fore and aft over the condensers; it flows down through the 20 condenser modules by gravity, and discharges below the platform at a depth of 20 m. Warm water is pumped from port and starboard intakes 15 m below the surface; 20 pumps supply 20 separate evaporator head ponds to provide gravity flow and discharge at 20-m depth.

*One metric ton (tonne) is 1000 kg.

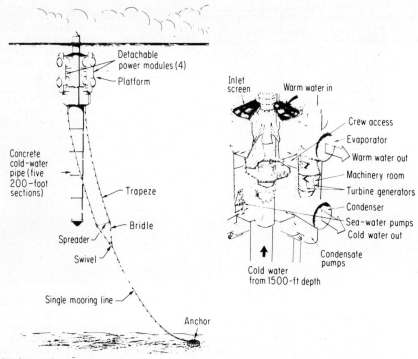

Fig. 19.4 The moored, spar-buoy type, 160-MW$_e$ plant design concept of the Lockheed/Bechtel Corporation/T. Y. Lin Team; inset in close view of one power module. (*From Ref. 11a.*)

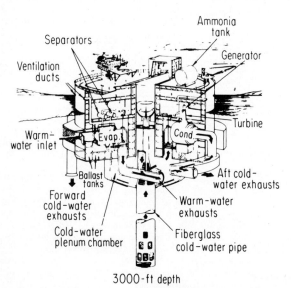

3000-ft depth

Fig. 19.5 The floating, cylindrical-surface-vessel, 100-MW$_e$ plant design concept of the TRW/Global Marine/United Engineers and Constructors Team. Shrouded-pipe water jets (condenser discharge and part of the evaporator discharge) control the plant position. (*From Ref. 11b.*)

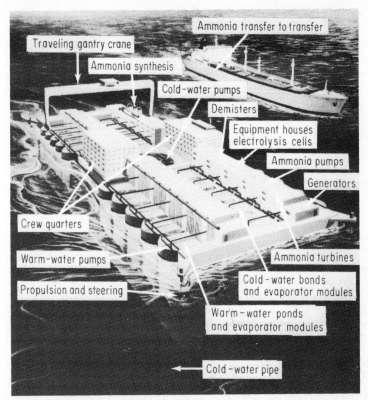

Fig. 19.6 The APL/JHU concept for a 100-MW$_e$ (net), 284-ton/day, OTEC/ammonia demonstration plant-ship for siting in tropical oceans. Seawater flows by gravity through the heat exchanger compartments. (*From Ref. 14.*)

Smooth concrete panels (an integral part of the ship structure) form the heat-exchanger compartment walls (the only heat-exchanger "shells" required in this design). There is clear vertical access to the heat exchangers for cleaning or for lift-out by a gantry crane for repair. The displacement is approximately 140,000 t including 10,000 t of product and 30,000 t of ballast water in hull tanks between the evaporator and condenser compartments. The heat exchangers, made of 7.6-cm outside diameter (OD), multipass aluminum tubes with ammonia working fluid inside the tubes, provide 15 percent of the buoyancy. A capability for grazing at 0.5 to 1 kn relative to the local current and for steering is provided by four electric-drive thrusters, one at each ship corner.

The estimated capital cost for the first-of-of-a-kind 100-MW$_e$ demonstration-size plant-ship to the power bus bars was \$1400/kW$_e$ (1978 dollars).[6e] On the basis of appropriate factors for economies of scale for a 325-MW$_e$ (net) commercial-size plant-ship, producing 1000 t of ammonia (NH$_3$) per day, and learning-curve factors (0.95 per doubling of units for the heat exchangers; 0.9 for the machinery, CWP, and ship outfits; 0.8 for the hull including reuse of the concrete forms for the modular construction), a cost of \$705/kW$_e$ (1978 dollars) was projected for the ninth commercial-size plant-ship.[14] The estimated capital cost with the NH$_3$ plant is \$299 million.

Working-Fluid Selection

Most investigators have selected ammonia as the working fluid for closed-cycle OTEC plants. Table 19.2 summarizes the relative merits of ammonia, propane, and a halogenated hydrocarbon, R-12/31.[11b] The superior thermodynamic characteristics of ammonia (described further under

TABLE 19.2   Working-Fluid Comparison (Ref. 11b)

	Ammonia*		Propane		R-12/31	
	Liquid	Vapor	Liquid	Vapor	Liquid	Vapor
Thermal conductivity, 10–21°C						
W/m · °C	1.65	0.085	0.40	0.06	0.28	0.034
(Btu/h · ft · °F)	(0.29)	(0.015)	(0.07)	(0.01)	(0.05)	(0.006)
Heat capacity, 10–21°C						
kJ/kg · °C	4.68	0.70	2.59	1.00	1.00	0.41
(Btu/lb · °F)	(1.11)	(0.29)	(0.62)	(0.24)	(0.24)	(0.097)
Heat of vaporization, 21°C						
kJ/kg	1185		325		163	
(Btu/lb)	(509)		(140)		(70)	
Materials compatibility	Wet NH₃ not compatible with copper		Excellent except some plastics		Excellent	
Toxicity	Severe but easily detected		Slight but difficult to detect		Slight but difficult to detect	
Flammability	Moderate		Explosion hazard		Not flammable	
Solubility in water	High		Low		Very low; hydrolyzes	
Effect on external environment	Slight		Undesirable local effects		Potentially severe problem	
Problem of contamination as working fluid	Moderate		Negligible		Negligible	
Availability	Good		Good		Potential problem	

*Ammonia properties revised to correspond to Table 19.6.

Heat Exchangers) result in advantages in component characteristics as illustrated by the propane/ammonia ratios in Table 19.3.[11a] Halogenated hydrocarbons would require even larger turbines and heat exchangers than propane and may pose the most severe environmental problems.[11b] Materials and techniques for the use of ammonia have been developed in the refrigeration industry (including use on fishing ships) and in the fertilizer industry. The DOE OTEC program is addressing various specific materials and environmental questions. (See also *Materials and Cost* in the following section.)

HEAT EXCHANGERS

The small available mean temperature difference ΔT_m between seawater and working fluid for OTEC exchangers (typically 4 to 5°C) results in a large surface area A requirement, since

$$A = \frac{q}{U \Delta T_m} \tag{19.4}$$

As a result, the heat exchangers typically represent one-third to one-half of OTEC plant capital costs, and their design is critical to plant economics. The value of ΔT_m is limited by site and seasonal characteristics (which determine ΔT_o) and by usable turbine operating ranges as discussed previously. Design of minimum-cost heat exchangers is concentrated on reducing either required area or cost per unit area. These factors depend upon the materials, exchanger thermal/hydraulic design, fabrication techniques employed, and the plan for integration with a minimum-cost platform. The emphasis in this section will be upon the thermal/hydraulic design considerations.

Classification

The primary geometric classes of OTEC heat exchangers are shell-and-tube, shell-less (e.g., plate-fin and open rack), and direct-contact types. Shell-and-tube exchangers have the working fluid on the shell (pressure vessel) side with seawater flow inside the tubes which run through tubesheets at each end. Tubes are typically arranged either horizontally (see, for example,

TABLE 19.3.   Propane/Ammonia Requirement Ratios for OTEC Turbines and Heat Exchangers 11ᵃ

	Propane/ammonia ratio
Turbine shaft power	1.0
Shaft rotational speed	0.37
Flowrate	3.25
Rotor diameter	1.5
Heat-exchanger volume	2.5

Fig. 19.7) or vertically, although at least one condenser design has tubes inclined at a 10° angle with the horizontal.[11a]

Alternative arrangements of tubes and baffles may be employed for structural reasons as discussed later, but most OTEC designs rely on a single pass on the tube side because of the small ΔT_m available and single working-fluid passes to avoid excessive pressure drop in large exchangers.

The aforementioned shell-less heat-exchanger types have a two-phase working fluid inside tubes or enclosed passages, with seawater on the outside.

A third alternative is the direct-contact exchanger. In the condenser of an open-cycle plant (Fig. 19.2), cold seawater is sprayed into the water vapor exiting from the turbine to cool and

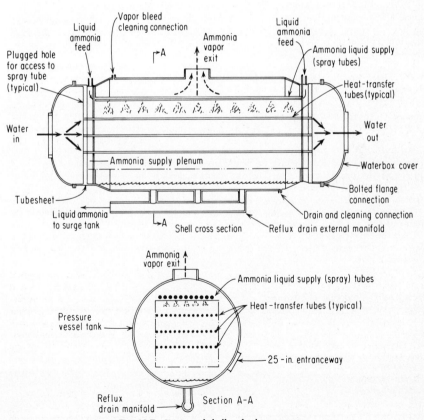

Fig. 19.7   Horizontal shell and tube evaporator.

condense it by direct contact. Because of the low pressures employed in open cycles, the noncondensable gases dissolved in the seawater will come out of solution and must be removed.

For binary-cycle OTEC plants using working fluids which are not highly soluble in seawater, direct-contact heat exchange is also possible. In an evaporator, liquid propane (for example) is introduced into a chamber which is filled with flowing warm seawater. The propane vaporizes and rises to the top of the chamber while seawater exits through the bottom. Most working fluids suitable for OTEC application are soluble to some degree in seawater. However, appropriate choice of a surfactant to reduce this solubility can keep solution rates within acceptable limits. The remainder of this section is devoted to closed-cycle exchangers.

Design Considerations

In the closed-cycle OTEC plant, exchangers operate between seawater and a two-phase working fluid. Most heat transfer takes place by means of a change of phase with water on the other side of a physical barrier. The overall heat-transfer coefficient U of Eq. (19.4) is determined from the sum of the individual thermal resistances due to seawater ($1/h_{sw}$), fouling or scale on the seawater side ($1/h_f$), the wall ($1/h_w$), fouling or scale on the working-fluid side ($1/h_s$), and the working fluid ($1/h_a$):

$$\frac{1}{U} = \frac{1}{h_{sw}} + \frac{1}{h_f} + \frac{1}{h_w} + \frac{1}{h_s} + \frac{1}{h_a} \tag{19.5}$$

where h is an effective coefficient for convective heat transfer based on the area base of U.

Depending on the relative magnitudes of the individual resistances, the heat-exchanger design (overall heat-transfer rate) may be fouling-limited ($1/h_f$ is the most significant or governing resistance), waterside-limited ($1/h_{sw}$ governs), or shell-side-limited ($1/h_a$ governs). For typical OTEC working fluids such as ammonia or propane, $1/h_s$ is usually negligible. The resistance $1/h_f \equiv R_f$ (usually called fouling factor) is largely caused by build-up of biological organisms and can probably be limited to the range $R_f \leq 0.00004$ m² · °C/W (0.0002 ft² · h · °F/ Btu) by suitable cleaning methods discussed later.

A summary of the physical properties of seawater is presented in Table 19.4. The heat-transfer coefficient for seawater can usually be determined from existing correlations and the properties can be determined from Table 19.4. For fully developed turbulent flow of water in round tubes with diameter d,

$$\text{Nu} = \frac{h_{sw} d}{k_w} = 0.058 \text{ Re}^{0.7} \text{ Pr}^{0.5} \tag{19.6}$$

Because of the Reynolds number Re dependence, h_{sw} is a strong function of water velocity through the heat-exchanger tubes. The coefficient h_{sw} can be increased by increasing water velocity, but the resulting increase in pressure drop will require greater seawater pumping power which reduces the overall system efficiency.

The waterside coefficient can also be increased or enhanced at a given water velocity by modifying the waterside flow geometry.[16] Such enhancement is usually accomplished by increasing the effective heat-transfer area (e.g., internal flutes or fins) or by introducing turbulence in the boundary layer to increase the degree of mixing near the tube wall (e.g., by surface roughness).

TABLE 19.4 Properties of Seawater[27]

Property	Evaporator average temperature, 25.6°C (78°F)	Condenser average temperature, 5.6°C (42°F)
ρ, kg/m³ (lb/ft³)	1021 (63.78)	1025 (64.01)
c_p, J/kg · °K (Btu/lb · °F)	4003 (0.9564)	3996 (0.9545)
μ, Pa · s (lb/ft · h)*	0.0571 (2.304)	0.0946 (3.815)
k, W/m · K (Btu/h · ft · °F)	36.45 (0.351)	34.47 (0.332)

*Viscosity μ values are from Ref. 28.

Both area enhancement and promotion of turbulence tend to increase pressure drop over that for a smooth tube with the same nominal internal diameter. The friction factor f for round tubes may be used as a measure of the frictional pressure drop due to enhancement:

$$f = \frac{\Delta p}{4(L/d)\,\rho V^2/2} \tag{19.7}$$

The ratio of enhanced-tube to smooth-tube friction factors (f/f_o) may be compared with the ratio of enhanced- to smooth-tube water-side heat-transfer coefficients (h/h_o) to form an efficiency ratio $(h/h_o)/(f/f_o)$.

Relative performance of several waterside enhancement techniques is presented in Table 19.5. The comparison is made at a constant Reynolds number of 60,000, typical of OTEC plant design conditions. Choice of waterside enhancement must consider the efficiency factor, increased cost over smooth tubes, and possible difficulties in cleaning the tubes.

The remaining coefficient, h_a, in Eq. (19.5) is attributable to working-fluid heat transfer and is a function of heat-exchanger configuration and operating conditions. A discussion of h_a therefore is included as a part of the following sections on types of OTEC heat exchangers.

Design of an OTEC exchanger is subject to constraints resulting from system requirements such as seawater mass flow and temperature for both evaporator and condenser, and working-fluid operating pressures and flowrates required by the turbine. Seawater pumping power depends on pressure drop through the heat exchangers, and cleaning techniques will impose other constraints such as minimum seawater velocity and flow areas.

Design of Shell-and-Tube Heat Exchangers

Design of an OTEC shell-and-tube exchanger to achieve a specified thermal duty can proceed according to established techniques[17,18] which include effects of:

- Shell type and shell-side flows
- Tube geometry (plain or enhanced), tube diameter and length
- Tube field layout and pitch
- Tube bundle and tubesheet geometry
- Baffle type and location

Evaporator design is more critical than condenser design; some means must be devised to assure liquid wetting of the evaporative heat-transfer surface so that vapor blocking or dry-out at the wall does not occur. Because vapor and liquid have different resistances to mass flow (vapor has a large specific volume), there is a tendency for two separate paths to develop—one for liquid and one for vapor. The resulting flow pattern makes the even distribution of liquid (i.e., uniform wetting) a difficult design problem.

In contrast, in the condenser, entering vapor is condensed as soon as it reaches saturation conditions. A dry condenser tube has little resistance to heat transfer, since most resistance is due to the condensate film thickness. Hence there is a natural driving force promoting condensation on all of the tubes in the bundle with a vapor velocity distribution being established which supplies vapor to regions of the condenser as required by the local condensation rate: areas with high available ΔT_m such as those near the seawater entrance region have higher condensation rates as required by Eq. (19.4).

In the discussion which follows, evaporator design is thus given major emphasis. Unless

TABLE 19.5 Performance of Internal Waterside Enhancements[16]

Type of enhancement	h/h_o	f/f_o	$(h/h_o)/(f/f_o)$
Spiral fins, radial and trapezoidal in shape, fin height of 0.89 mm	1.6	2.0	0.8
Straight fins, trapezoidal in shape, fin height of 0.91 mm	1.4	2.0	0.7
Corrugated tube having ratio of pitch to groove ID of 0.5 and ratio of height of corrugation to groove ID of 0.032	1.7	3.3	0.51
Rib-type rough surface on tube ID with ratio of pitch to rib height of 10 and ratio of rib height to tube ID of 0.02	2.4	7.2	0.33
Brazed spiral coil on tube ID with ratio of pitch to coil wire diameter of 10 and ratio of wire diameter to tube ID of 0.022	1.7	6.4	0.26

otherwise noted, ammonia (Table 19.6) is assumed to be the working fluid; available data for other fluids will be presented where appropriate.

Horizontal evaporators   The general operation of a horizontal OTEC shell-and-tube evaporator is shown in Fig. 19.7. Horizontal shell-and-tube evaporators may be divided into two general classes based on the liquid-distribution system. In the flooded-bundle design, the tubes are surrounded by liquid in a pool (hence the alternate designation of pool boiler). In the sprayed-bundle evaporator, liquid is supplied as a spray at various locations in the bundle.

An advantage of the flooded design is the relative ease of maintaining a uniform liquid distribution on all tubes in the bundle. Evaporation occurs as nucleate boiling. The major design conditions are:

1. Local liquid hydrostatic pressure increases with depth in the bundle, increasing the saturation or boiling temperature and decreasing the useful ΔT_m for the lower tubes. For ammonia, each 5 m of depth requires an additional 1°C of saturation temperature.

2. As the bundle depth increases, more vapor must bubble up through a given pool surface area. For bundle depths much beyond 5 m, the vapor tends to block liquid from the surfaces of the tubes in the upper rows, rendering them less effective. Vapor flows can also establish unstable two-phase flow regions where areas of the tube bundle are deprived of liquid by large vapor slugs. Forced liquid circulation may be used to alleviate these problems.

Modification of the tube surface to increase nucleation sites will increase h_a for pool boiling. Sandblasting, wire screening, or coating with a porous surface (e.g., a sintered metal matrix 0.25 to 0.50 mm thick) will increase h_a over that for a smooth tube.[6f] As shown in Fig. 19.8, a porous surface can increase h_a by a factor of 10 for a given temperature difference.[6g] When bubbles form on a smooth surface, the surface tension at the vapor-liquid interface creates an excess pressure on the vapor. Therefore, the liquid must be superheated in order for the bubbles to exist and grow. The porous matrix reduces this required superheat, since the cavities trap bubbles which are substantially larger than those found on smooth surfaces. These bubbles grow and rupture, leaving the nucleating bubbles intact within the matrix. This bubble growth is promoted by the highly conductive metallic wall surrounding each cavity.

In a sprayed-bundle evaporator, the liquid working fluid is sprayed onto the tubes and evaporation occurs at the surface of the liquid film as it flows down through the tube bundle by gravity. As film thickness increases, the initial laminar flow becomes turbulent. The theoretical value of h_a reaches a local maximum in the laminar region (Re ~ 10) where the film is thick enough to provide uniform tube wetting (Fig. 19.9). As the film thickness t increases, the thermal resistance through it (t/k) increases, and h_a decreases until the transition to turbulence in the film (Re ≈ 1000) occurs; thereafter h_a increases with increasing t. Because a complex system comprising many spray tubes and nozzles would be needed to maintain a uniform laminar liquid film at Re near 10 in a large heat-exchanger bundle, operation is contemplated with high individual tube liquid loadings in the thick-film regime. Design of a minimum-cost spray-bundle evaporator must trade off the cost of obtaining a high h_a using complex spray systems against the cost of recirculation pumps required for operation in the thicker-film regime.

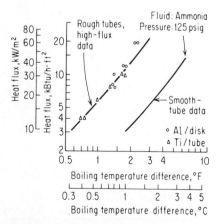

Fig. 19.8 Single tube evaporation data for submerged pool boiling in ammonia. (*From Ref. 6g.*)

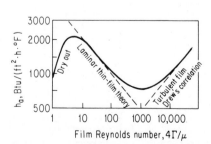

Fig. 19.9 Single tube theoretical h_a as a function of film Reynolds number.

TABLE 19.6 Properties of Saturated Ammonia

Temperature, °F (°C)	40	(4.44)	50	(10)	60	(15.56)	70	(21.11)	80	(26.67)
Liquid										
Vapor pressure, lb/in² abs. (kN/m²)	73.7	(508)	89.2	(615)	107.6	(742)	128.8	(888)	153.0	(1055)
Latent heat, Btu/lb (kJ/kg)	536.2	(1247)	527.3	(1226)	517.9	(1204)	508.6	(1182)	498.7	(1159)
Enthalpy, Btu/lb (kJ/kg)	86.8	(202)	97.9	(228)	109.2	(254)	120.5	(280)	132.0	(307)
μ, lb/h · ft (μN · s/m²)	0.431	(178)	0.406	(168)	0.381	(157)	0.359	(148)	0.337	(139)
k, Btu/h · ft · °F (W/m · °C)	0.306	(1.736)	0.298	(1.691)	0.291	(1.651)	0.284	(1.612)	0.276	(1.566)
c_p, Btu/lb · °F (J/kg · °C)	1.103	(4616)	1.110	(4645)	1.118	(4679)	1.127	(4716)	1.135	(4750)
ρ, lb/ft³ (kg/m³)	39.48	(632)	39.0	(625)	38.51	(617)	37.99	(609)	37.48	(600)
Vapor										
Enthalpy, Btu/lb (kJ/kg)	623.0	(1448)	625.2	(1454)	627.1	(1458)	629.1	(1463)	630.7	(1466)
μ, lb/h · ft (μN · s/m²)	0.0253	(10.5)	0.0258	(10.7)	0.0263	(10.9)	0.0268	(11.1)	0.0273	(11.3)
k, Btu/h · ft · °F (W/m² · °C)	0.013	(0.074)	0.014	(0.079)	0.015	(0.085)	0.016	(0.091)	0.016	(0.091)
c_p, Btu/lb · °F (J/kg · °C)	0.655	(2741)	0.676	(2824)	0.698	(2921)	0.725	(3034)	0.750	(3139)
ρ, lb/ft³ (kg/m³)	0.252	(4.03)	0.304	(4.86)	0.364	(5.82)	0.433	(6.92)	0.512	(8.19)

Data for spray-film evaporation of ammonia from a single smooth or porous tube are presented in Fig. 19.10. As with the pool boiler, a porous surface can increase h_a about tenfold. Photographs of both nucleate (pool) boiling and sprayed film evaporation of liquid ammonia are shown in Fig. 19.11.

Thermal/hydraulic considerations are important for sprayed-bundle evaporators. Recirculation or reflux can be provided to ensure full wetting of all tubes at all times. With high reflux ratios, a sprayed-bundle evaporator can become similar in operation to a pool-boiler without the problem of suppression of boiling by hydrostatic pressure within the pool.

The vapor velocity distribution is important in designing to minimize shell-side pressure drop and liquid maldistribution due to deflection of liquid streams and entrainment. Figure 19.12 illustrates vapor velocity distributions or profiles for typical evaporator-tube sections. Average velocity profiles can usually be calculated from considerations of continuity: local velocities between tubes can exceed the average value because of the channeling (venturi) effect of the tubes in the field. The distributions of liquid and vapor can be determined using a finite-element analysis and appropriate data for liquid deflection and entrainment.[19]

Data on liquid deflection for ammonia falling from rows of tubes subjected to a transverse ammonia vapor flow with an initially uniform velocity indicate that deflection is small for transverse velocities below 0.3 m/s. Liquid entrainment is a function of both droplet size and vapor velocity. Small ammonia jets ($\sim$ 1.0 mm in diameter or less) can produce a fog which would be entrained at vapor velocities greater than 0.3 m/s and should be avoided.

Shell-side baffles prevent excessive transverse vibrations in long tubes and provide multiple vapor paths. Because of the small ΔT_m in OTEC heat exchangers, use of a single pass on both water and shell sides yields the best compromise between minimum area and pressure drop. Thus, in evaporators, the major purpose of the baffles is to provide mechanical support for the tubes.

Horizontal condensers   Horizontal shell-and-tube condensers using ammonia have long been used in the refrigeration industry. The major shell-side design considerations are vapor distribution, condensate removal, pressure drop, and removal of noncondensibles. Other design parameters important to OTEC operation, such as tube-side water velocity and tube length and diameter, are similar to those in evaporator design as discussed previously.

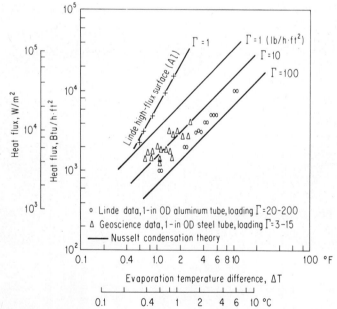

Fig. 19.10 Sprayed film evaporation data for single horizontal tubes in ammonia. (*From Ref. 29.*)

(a)

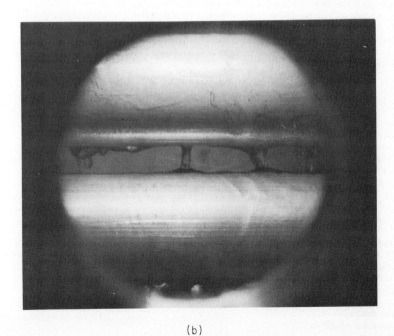

(b)

Fig. 19.11 Boiling of ammonia on smooth 1-in OD tubes: (a) pool boiling; (b) film boiling.

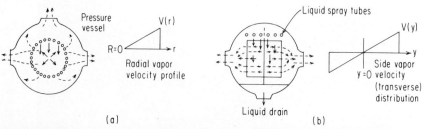

Fig. 19.12 Evaporator vapor velocity distributions: (*a*) radial removal—circular-tube bundle; (*b*) transverse (side) removal—square-tube bundle.

The Nusselt[20] analysis of film condensation on vertical banks of round tubes is applicable to the design of OTEC condensers. The film coefficient is

$$h_a = 0.728 \left[\frac{g\rho \, (\rho - \rho_r) \, k^3 h_{fg}}{\mu (\Delta T) \, d} \right]^{1/4} \tag{19.8}$$

Data for condensation of ammonia on horizontal smooth tubes result in a higher h_a than the Nusselt theory. Thus Eq. (19.8) yields a conservative design.

For tube rows below the first row in the bundle, the condensate loading will increase because of condensate flow from above, and h_a will decrease. The Nusselt analysis of this case indicates that the coefficient of the Nth tube in a vertical column $h_{a,N}$ will be related to that for the first (top) tube $h_{a,1}$ by

$$h_{a,N} = h_{a,1} N^{-1/4} \tag{19.9}$$

Experimental work[17] indicates that the exponent in Eq. (19.9) should be closer to $-\frac{1}{6}$ or smaller. Other data indicate smaller exponents for less viscous fluids such as ammonia.

Effects of bundle depth and variations in ΔT along the tube length must be considered. Pressure drop through the tube bundle is also a major design consideration for large OTEC condensers; it can be minimized with a single-pass downflow arrangement (Fig. 19.13), although the baffled cross-flow design might be preferred for units 1 MW$_e$ or smaller.

For large bundles, condensate removal is a potential problem. As the condensate falls through a large bundle, the liquid film becomes thicker on succeeding tubes and reduces performance. Providing condensate drains at various depths can alleviate this problem. Addition of radial fins, corrugations, or wire windings to the exterior of the tubes can also help to channel the condensate and increase the average value of h_a.

Vertical-tube, falling-film exchangers Heat-exchanger tubes can also be arranged vertically with the working fluid flowing down the outer surface by gravity. Considerable improvement of h_a in both evaporation and condensation results from use of axial flutes on the outside surface to promote a naturally occurring wave motion, as shown in Fig. 19.14. When the trough between waves passes by, surface tension thins the liquid film on the flute (and increases h_a there) by pulling the liquid into the rill. The waves contain disturbance eddies but the troughs are in stable laminar flow, a combination which appears to maximize the average heat-transfer coefficient.

Figure 19.15 compares single vertical-tube condensation data for smooth and fluted tubes. The flute geometry (see inset in Fig. 19.15) has a theoretical area ratio between fluted-to-smooth tubes of $\pi/2 = 1.57$ (the actual area ratio for the experiment shown was observed to be 1.5).

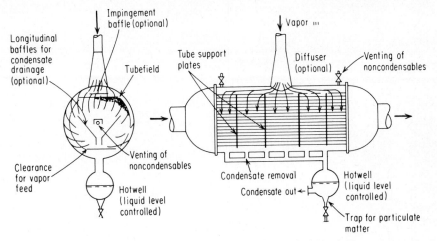

Fig. 19.13 Typical OTEC downflow horizontal shell and tube condenser.

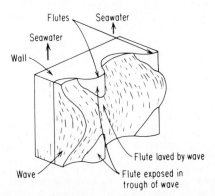

Fig. 19.14 Wave motion in vertical falling film on fluted surface.

The heat-transfer coefficient reported is a local average value representative of the test section. The smooth-tube results in the laminar range exhibit coefficients which are predictable from the classical Nusselt theory. The data for the two working fluids, R-11 and ammonia, are drawn together by normalizing the heat-transfer coefficients with $\phi = (k^3 \rho^2 g / \mu^2)^{1/3}$.

In the laminar region, fluted-tube coefficients are about 2.8 times greater than those for the smooth surface tube. Since the area ratio is only 1.5, this suggests a considerable fluid-mechanical contribution to the enhancement. (Heat-transfer coefficients reported here are based on the area of the smooth tube having the same nominal D_o.) As expected, a local maximum is reached at a Reynolds number of about 1500, increasing the coefficient by about 20 percent.

Evaporation heat-transfer data for smooth and fluted vertical tubes are presented in Fig. 19.16. The smooth-tube evaporator exhibits the expected dependence on Re, but the coefficients are approximately twice those of the smooth-tube condenser. This difference has not yet been explained in the literature but data for R-11 and ammonia show the same effect.

The fluted evaporator exhibits a greater dependence on Re; for Re ≤ 400, the fluted tube has lower coefficients than the comparable smooth tube, largely because, at low flowrates, the liquid does not wash the crests of the flutes. (Unlike the condenser, the evaporator has no way of wetting the crests except by liquid motion.) The coefficient rises steeply with increased Re, reaching a peak at Re ~ 3200. The R-11 and ammonia data are again correlated by the ϕ factor.

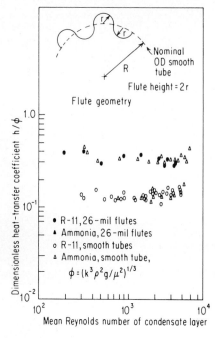

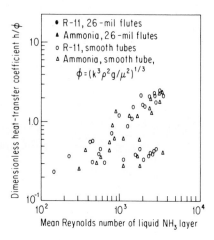

Fig. 19.15 Condensation of ammonia on single vertical tubes.

Fig. 19.16 Evaporation of ammonia on single vertical tubes.

Fluted vertical-tube, falling-film heat exchangers have been carried through conceptual design in the 6-MW$_e$ size range.[19] The evaporator (shown in Fig. 19.17) and condenser are similar except for the liquid ammonia metering orifice at the top of each evaporator tube. In addition to their compactness, these heat exchangers offer an advantage in that the upper water boxes are open, thus permitting access for mechanical cleaning of tubes without disturbing normal plant operation. Furthermore, they can be designed so that each tube performs independently of all others in the bundle; thus thermal design can be based on extrapolations from single-tube data. One disadvantage is the requirement for a complex liquid-feed arrangement for each tube in the evaporator which must be fabricated to tight tolerances. Vapor feed to the condenser is less critical, since the condensate will distribute itself on each tube. Some uncertainty exists regarding performance in sea-state-induced motions of the OTEC platform, since detachment of the liquid film in the evaporator could become a problem.

Shell-less Heat Exchangers

One method of eliminating the pressure vessel in heat exchangers is being developed by APL.[6e,21] In this concept, shown in Fig. 19.18, the working fluid (ammonia) is inside large-diameter (7.6-cm OD), horizontally folded multipass tubes. The seawater flows by gravity through the tube bundles (see also Fig. 19.6). Use of a large tube diameter reduces the number of tubes and joints required. In a typical design, seven multipass tubes are nested (following parallel paths) in a vertical plane to form one element. These elements or vertical tube rows are then assembled in a staggered array, up to the number required for a desired module size—e.g., 41 elements, each approximately 8 m long by 17 m deep, would provide 2.5-MW$_e$ net at $\Delta T_o = 23.9°C$ (43°F). A clear space approximately 3.2 cm wide is left between elements to permit cleaning of the outsides of the tubes by mechanical or possibly ultrasonic means.

In the evaporator, liquid ammonia is fed to the bottom of each tube via a manifold and a short riser. Following some liquid heat-up, evaporation begins in the second pass, and the two-phase mixture flows upward through the folded tube, with each pass accounting for a portion of the total increase in quality to approximately 70 percent leaving the last (e.g., twenty-fifth) pass. Part of the liquid is separated from the two-phase mixture in the exit manifold, and the

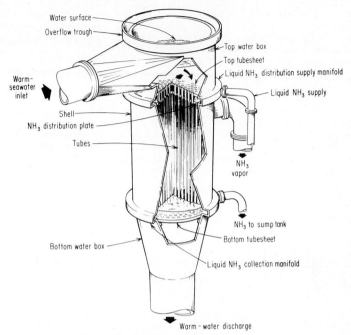

Fig. 19.17   Vertical tube falling film evaporator.

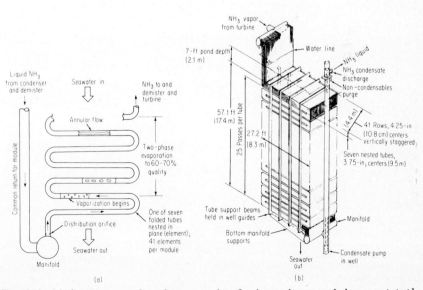

Fig. 19.18   (a) The APL concept for multipass, two-phase-flow heat exchangers with the ammonia inside the tubes. (b) Three-quarters view of 2.5-MW$_e$ (net) condenser module having 1.5×10^5 ft^2 (1.4×10^4 m^2) (nominal) surface area.

rest is taken out by a demister; this liquid is mixed with that coming from the condenser and is returned to the evaporator inlet. Full-scale tests of a single evaporator pass with appropriate variations of inlet conditions have shown that the initial vapor generation in the second pass may differ from the classical "onset of nucleate boiling" because of circumferential convective effects which carry warm liquid to the top of the tube, plus the increased turbulence induced by the U-bends. Some stratification may occur in the next few passes at low qualities, but the performance still matches the Chaddock-Brunemann correlation for two-phase flow (Fig. 19.19).[6e]

The condenser is geometrically similar to the evaporator, with ammonia vapor entering the top of each tube, condensing as it flows down the tube, and entering a sump.

The waterside heat-transfer coefficient h_{sw} is based on a correlation from Zukaukas which is conservative for the APL geometry as shown in Fig. 19.20.[21] The waterside pressure drop for the selected staggered tube array geometry is also consistent with data in the literature.[21] A water velocity near 0.8 m/s gives the best tradeoff between the waterside heat-transfer coefficient and pressure drop.

Another approach[11c] to shell-less heat-exchanger design is shown in Fig. 19.21(a). In this plate-fin design each element or panel comprises two flat plates bonded together by a fin system such as a thin corrugated sheet. The working fluid flows through the small (e.g., 3- to 9-mm) passages formed by the corrugations while the seawater flows in cross-flow between adjacent

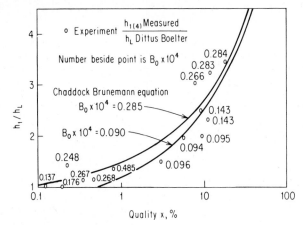

Fig. 19.19 Ammonia-side heat transfer data for APL evaporator.

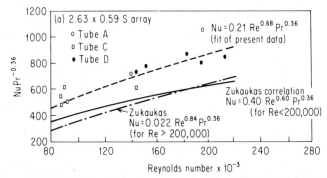

Fig. 19.20 Water side heat transfer data for APL bundle (external flow).

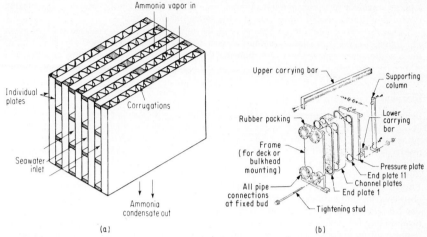

Fig. 19.21 (a) Section of a proposed type of OTEC plate-fin condenser and (b) Alfa-Laval heat exchanger used on Mini-OTEC.

panels.* In the condenser, ammonia vapor enters the corrugations from the top, and condensate drains by gravity to a collection manifold at the bottom. In an evaporator, liquid enters at the bottom of each panel through a distribution manifold. Evaporation occurs in two-phase flow and a high-quality mixture goes into a collection manifold at the top.

The main advantages of the plate-fin exchangers are their compactness (large surface area per unit exchanger volume) and ease of fabrication of multipass assemblies by brazing, which should lead to low cost. If mechanical cleaning is used, the spacing between elements (hence the compactness and efficiency) will have to be compromised to allow sufficient space for practical cleaning devices. A possible disadvantage of using the plate-fin units for OTEC is the complexity of both the ammonia manifolding and the seawater flow paths.[11c]

An exploded view of a plate heat exchanger built by Alfa-Laval and used on Mini-OTEC is shown on Fig. 19.21(b). The principle elements are a pack of thin metal plates, a frame, and means of keeping the plate elements together. The plates are suspended between horizontal carrying bars at top and bottom and compressed against the stationary frame plate by means of tightening bolts and a movable pressure plate. The frame plate is equipped with nozzles for inlet and outlet connections. Every plate is sealed around its perimeter with a gasket retained within a pressed track. Flow ports (or nozzles) at each of the plate corners are individually gasketed and thus divide the interplate spaces into two systems of alternating flow channels. Through these, the two media pass, the warmer medium giving up heat to the cooler by conduction through the thin plates. The plate, which is the basic element of the PHE concept, has a corrugated pattern stamped on it. These corrugations can be arranged to create an unlimited number of plate patterns. The specific pattern results from a careful tradeoff between pressure drop and convective heat transfer characteristics.

Phase Separators

As noted in the preceding section, the exit flows from two-phase-flow evaporators will contain substantial quantities of entrained liquid. In contrast, well-designed, full-scale pool-boiler or sprayed-bundle evaporators will have higher exit qualities† but still may require addition of

---

*Alternative designs might employ a gravity-flow system for the seawater (counterflow to the working fluid in the evaporator), depending on the manifolding and installation (system) concepts.

†Operation of small (30-kW$_e$ or 1-MW$_t$) units of pool-boiling and vertical-falling-film evaporators having large vapor disengagement spaces and low (< 0.5-m/s) exit vapor velocities has produced no measurable entrained liquid ammonia in the exit flow.[22]

a phase separator at the evaporator outlet to ensure essentially 100 percent quality entering the turbine to minimize turbine wear and maximize turbine output.

A phase separator forces impingement of liquid on a solid surface to coalesce small droplets which then flow back into the evaporator. Typical demisters include wire-mesh, hook-and-vane, and low-drag-vane types. The major demister design parameter is pressure loss as a function of the size distribution of droplets passed. Ammonia has very low surface tension and droplet sizes can be expected to be quite small (0.5 to 3 μm).

Biofouling and Cleaning

Biofouling on the seawater side of OTEC heat exchangers is expected to be more pronounced in the evaporators than in the condensers because the warmer tropical surface waters contain more marine organisms than the cold deep-ocean water. The increase in fouling resistance inside 25-mm-diameter tubes as a function of time for surface water near Keahole Point, Hawaii[23a] is shown in Fig. 19.22. Fouling resistance increased slowly during the first 4 to 6 weeks and at a growing rate thereafter, reaching 8.8×10^{-5} m^2 · (°C)/W (0.0005 ft^2 · °F · h/ Btu) in approximately 11 weeks. Although no effect of water velocity on rate of fouling was observed, the ability to clean with a brush after 16 weeks was a function of water velocity. For the 1.8-m/s (6-ft/s) case, clean-tube performance was restored, while for the 0.9-m/s (3-ft/s) case, fouling resistance was reduced only to 5.0×10^{-5} m^2 · °C/W with the same treatment because the tube contained larger organisms, including barnacles. For the 1.8-m/s case, the tube contained predominantly microbiological slime prior to cleaning.

Tests of biofouling on the exterior of 0.102-m OD alclad aluminum tubes arranged in an array simulating the APL heat-exchanger concept at 0.9-m/s velocity confirmed the slow build-up of slime in the first 4 to 6 weeks.[23b] This fouling was removed by a mechanical scrubbing device.

These test results indicate that with existing mechanical cleaning techniques, fouling resistance can be maintained at $\leq 3.5 \times 10^{-5}$ m^2 · °C/W (0.0002 ft^2 · °F · h/Btu) by cleaning at regular intervals of less than 6 weeks.

One mechanical cleaning technique for shell-and-tube heat exchangers, the AMERTAP system,[6i] relies on continuously circulating sponge balls which abrasively clean the inside of the tubes. The balls are injected into the inlet water boxes, collected at the outlet water box, and returned by a pump to the inlet injection point. This system requires a minimum tube water velocity of approximately 2 m/s, and the sponge balls must be replaced every 2 to 3 months. Another mechanical system, the M.A.N. brush system,[6j] employs shuttle brushes which are pushed through the tubes by the seawater, requiring a flow reversal to complete each cleaning cycle. This is difficult with the large quantities of water circulated through OTEC exchangers. The M.A.N. brushes are expected to be useful for several years.

In the vertical-tube exchanger design developed by TRW (see Fig. 19.17), rotating brushes

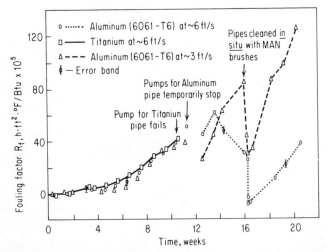

Fig. 19.22   Fouling factor versus time at Keahole Point, Hawaii, February 1977.

are placed on each tube at regular intervals with access through the open water boxes at the top of each exchanger. These brushes can be automatically indexed with multiple spindle (brush) heads providing rapid cleaning cycles.

Another method being explored by APL is ultrasonic cleaning. The heat exchangers in Fig. 19.18 can be cleaned by inserting vertical panels containing multiple transponders between the tube elements (vertical rows). Initial work indicates that the power requirement, ≤ 1 W/cm^2 for 15 s per weekly cleaning, will equal approximately 0.1 percent of the net power plant output.

Methods for alleviating or avoiding biofouling include chlorination, deaeration, ozonation, and use of biocide surface coatings. Because of the large volumes of seawater involved, the chemical water treatment methods may not be economically competitive with mechanical methods, but they are being investigated. Environmental effects also require analysis.

Chlorination is currently the most practical of these chemical control methods. Two chlorination processes have been used in seawater-cooled power plants: injection of chlorine-bearing chemicals and injection of gaseous chlorine. Gaseous injection is economically attractive but can leave toxic contaminants which are hazardous to the environment. For OTEC systems, chlorine can be generated *in situ* by electrolytic generation of sodium hypochlorite which is then injected into the seawater supply piping. The concentration of chlorine required is a function of the OTEC site. Existing coastal power plants require from 1 to 10 parts per million (ppm), but requirements in the open ocean are lower. Very low sublethal concentrations can be effective if injected continuously: hard-shelled organisms will be repelled and the growth rate of microorganisms will be inhibited. A concentration of 0.5 to 1.0 ppm has been used successfully in British power plants on a continuous basis.

Modularity

Selection of the size of OTEC heat-exchanger modules for a given power plant output is a major system design consideration. The low U-value for an unenhanced OTEC exchanger results in an enormous heat-transfer surface area. For a U of 2300 W/m$^2 \cdot$ °C (400 Btu/h $\cdot$ ft$^2 \cdot$ °F), a 25-MW$_e$ module requires about 1.4×10^5 m^2 (1.5×10^6 ft^2) of heat-transfer area for each exchanger. If 25-mm OD tubes, 9 m long, are used, about 180,000 tubes are required. For identical longitudinal and transverse pitches (a square array) of 32 mm, the bundle diameter, excluding any extra void, is 15 m. Heat exchangers of this size exceed the capabilities of existing heat-exchanger manufacturers and probably will require erection near waterways to avoid shipping difficulties. Such large bundles also will require special techniques to obtain uniform water feed to all the tubes and to avoid (by internal subdivisions) excessive liquid ammonia heads in pool boilers, or thick liquid films on the lower tubes in sprayed-bundle evaporators, or problems with ammonia liquid-feed distribution and vapor-quality degradation due to large vapor flows and consequent high vapor velocities toward the top or the periphery of the tube bundle.

Another consideration is ammonia-side pressure drop. The difference between the saturated vapor pressure of ammonia in an OTEC evaporator and condenser is 2.8 to 3.4 $\times 10^5$ N/m^2 (40 to 50 lb/in^2), and part of this is lost in intermediate plumbing. Hence, a loss of 7000 N/m^2 (lb/in^2) in the evaporator results in a loss of 2 or 3 percent of the turbine output. This loss will require an increase in the heat-transfer surface and related water and ammonia flows. The sources of this pressure loss are wall friction and momentum change due to the acceleration of vapor. Some but not all of the momentum loss in the evaporator may be recovered in the condenser.

All of the above considerations lead the designer to consider smaller heat exchangers which are interconnected in series or parallel to yield the total thermal duty required by the system, as well as use of enhanced surfaces, which can substantially reduce the number of tubes and hence the shell size per MW$_e$. For shell-less heat exchangers (Fig. 19.18), APL/JHU considers the optimal heat-exchanger module size to lie between 2.5 and 5 MW$_e$. The vertical-tube exchanger design (Fig. 19.17) developed by TRW[19] utilizes four 6.25/MW$_e$ heat-exchanger modules (two each of evaporators and condensers) to drive a single 12.5-MW$_e$ turbine-generator module. The learning-curve advantages of producing quantities of smaller modules on assembly lines can also compensate for what otherwise might be considered economies of scale for larger units. The integration of the heat exchangers with the other heat engine components and with a low-cost platform configuration must be considered from the outset and may, in itself, dictate an optimum module size.

Platform Interface

OTEC heat exchangers will be subject to platform motions. These motions in response to wind, wave, and current excitation depend strongly on both location and platform design. Designs should seek to avoid possible standing waves in liquid inventories especially in pool-boiling evaporators. For spray evaporators and condensers, it is recommended that the liquid be fully drained from the heat exchanger to a separate surge or holding tank where sloshing control can be maintained with baffles without adverse effect on heat-transfer performance.

Mechanical Considerations

Mechanical design of shell-and-tube exchangers for OTEC falls under the American Society of Mechanical Engineers (ASME) Unfired Pressure Vessel Code.[24] This code may be somewhat conservative in some areas and it is anticipated that some requirements may be modified, since OTEC is a relatively new application. Other codes and requirements will probably result from ocean operations (e.g., U.S. Coast Guard) and insurability [e.g., American Bureau of Shipping (ABS)].

In the conventional shell-and-tube units the tubes are usually expanded into grooves cut in the tubesheet. When tube materials such as titanium are chosen, a thin cladding is usually explosively bonded to the tubesheet to save the cost of a solid titanium tubesheet. When the tubes are welded to the tubesheet, this weld is made to the cladding. Future designs may involve materials such as fiber-reinforced plastics for the tubesheets with tube-to-tubesheet attachment by adhesive bonding. Such methods have not yet been certified in the ASME code.

Materials and Cost

Heat-exchanger materials must be compatible with seawater and working fluid, resistant to corrosion and erosion, and of adequate structural strength. High thermal conductivity is also required for heat-transfer surfaces. Materials need not satisfy all of the above requirements simultaneously. For example, bimetallic tubing having an ammonia-compatible material (e.g., carbon steel) on one side and a seawater-compatible material (e.g., a copper-nickel alloy) on the other could be used for heat-transfer surfaces.[25] Another example is the use of aluminum cladding on the seawater side of an aluminum alloy as a sacrificial layer to protect the primary alloy from corrosion pitting until essentially all of the cladding has been consumed.[26]

Two aspects of heat-exchanger cost are important. A low initial cost for a given thermal duty is important in itself to keep initial fund-raising requirements for the OTEC plant as reasonable as possible. The other aspect is total average operating cost including annualized initial cost, maintenance cost, and necessary replacement cost if heat-exchanger life is less than the required overall plant service life.

Candidate heat-transfer-surface materials include aluminum, titanium, stainless steel, and copper-nickel. Plastics with graphite or other fillers to improve thermal conductivity appear attractive because of their low initial cost,[11d] but low thermal conductivity and low structural strength are major drawbacks.

The selection of tube diameter is based on the thermal/hydraulic design considerations and materials cost. For an unclad material, a minimum tube-wall thickness t is determined for a given diameter based on the structural strength requirement (thickness t_A) plus allowances for corrosion and erosion and the requirement that the tube shall not be perforated by pitting at its end-of-life thickness t_A:

$$t = t_A + N\,(r_s + r_a + r_e) + (P - t_A) \qquad \text{when } P \geq t_A \qquad (19.10)$$

where N = number of years exposed (service life)
r_s = general corrosion rate in seawater
r_a = general corrosion rate in ammonia
r_e = erosion rate in seawater
P = maximum pitting corrosion depth, seawater side

Structural thickness t_A for shell-and-tube units can be determined by using the ASME pressure-vessel code[24] as a design basis. Erosion on the working-fluid side is assumed to be negligible, while erosion on the seawater side may occur because of both particles in the seawater and mechanical cleaning. Table 19.7 presents illustrative tubing material cost estimates for 38-mm OD metal alloy tubing used in typical shell-and-tube OTEC heat exchanger designs based on Eq. (19.10).[25] The values for r_s and r_e depend on seawater chemistry and thus may be site-dependent for a specific plant design. Aluminum is less expensive, has higher thermal conductivity, and is more plentiful than titanium, but the longer life for titanium has led to its selection for initial use in several OTEC conceptual plant designs.

TABLE 19.7   Relative Costs of 38-mm (1.5-in) OD Heat Exchanger Tubes Based on Eq. (19.10)[25]

Material	Wall thickness parameters, mm						Cost		N, year
	$A*$	Nr_s	Nr_a	Nr_e	P	t	$/m	$/m^2	
Ti-50A	0.71	Nil	Nil	Nil	Nil	0.71	6.07	50	30
5052-H32 Al	1.24	1.14	0.038	0.19	0.76	2.61	2.30	19	15
6063-T6 Al	1.24	0.76	0.038	0.19	1.52	2.51	2.21	18	15
Allegheny 6X CRES (stainless steel)	0.71	0.38	Nil	0.23	Nil	1.32	10.30	85	30
706 Cu (10/10 Cu-Ni)	1.02	0.38	0.38	0.38	Nil	2.16	8.00	66	30

*Based on Ref. 24 for a design pressure of 1.0×10^6 Pa.

The APL/JHU heat-exchanger concept is based on use of rolled-and-welded Alclad 3004 aluminum alloy tubing of 76-mm OD. Required wall thickness is approximately 2.0 mm including approximately 0.2 mm of cladding, at a tubing cost near $12/m^2 for 20-year service life. This lower tubing cost makes aluminum much more attractive than titanium for this concept; titanium tubing would cost about 7 times as much as aluminum for this type of heat exchanger.

AMMONIA TURBINES, GENERATORS, AND PUMPS

Emphasis in development of power turbines for the closed OTEC power cycle has been directed toward development of ammonia turbines for reasons discussed in connection with Tables 19.2 and 19.3. Experience with ammonia turbines and expanders is limited, but considerable operating experience exists for large ammonia compressors. Ammonia turbines can be designed based on experience with hydrocarbon and steam turbines using seals similar to those in ammonia compressors. No major problems are anticipated for the mild service condition, e.g., lower blade tip speeds, temperatures, pressures, and temperature variations than those in conventional steam or gas turbines.

Design options include axial flow or radial inflow, single or double flow, single stage or multistage, and various inlet and outlet nozzle orientations. Axial-flow turbines are preferred because large units of this type have been built and successfully operated in the 12- to 50-MW size range expected for OTEC turbines. Large radial-inflow turbines typically have small-diameter, high-speed wheels which require reducing gears to drive generators at their normal operating speeds. Since typical inlet/exit pressure ratios are small (near 1.4 for ammonia), a single stage can be used with little penalty.

Figure 19.23 presents a curve relating specific speed to adiabatic blade efficiency for state-of-the-art axial-flow turbines.[19] Specific speed N_s can be expressed as

$$N_s = NQ^{1/2}H^{-3/4} \tag{19.11}$$

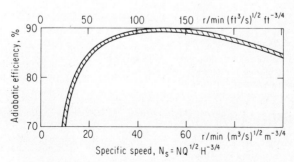

Fig. 19.23   Turbine specific speed versus adiabatic efficiency.

where N = operating speed, rpm
Q = volume flowrate at turbine wheel outlet, ft³/s
H = adiabatic head across the turbine wheel, ft

As can be seen from Fig. 19.23, maximum efficiency of 88 to 90 percent occurs in the range $90 < N_s < 150$ (or $36 < N_s < 62$ in SI units). A typical value for H is 15,000 ft (19.28 Btu/lb) or 4573 m for ammonia. For a 25-MW_e (net) power module, a typical flowrate is 6600 ft³/s (187 m³/s). For N = 1800 r/min, this combination yields N_s = 108 for a single/stage turbine. The same N_s would be obtained for a 6.25-MW_e (net) turbine (Q reduced by a factor of 4) at 3600 rpm.

Actual turbine efficiencies must include losses such as bypass flow around the blade tips, interstage losses, seal losses, and inlet nozzle and stator vane losses. Such losses will reduce the single-stage turbine efficiencies of Fig. 19.23 by a few percent. One four-stage, double-flow turbine design for 16-MW_e gross output, 1800 r/min, 0.79-m wheel diameter has an estimated efficiency of 89.6 percent.[19]

Off-design operation of a turbine optimized for a given specific speed will result in further reductions in efficiency. Two approaches can be used to prevent excessive drop in efficiency during off-design operation: reduced peak efficiency to produce a broader efficiency curve and use of multiple turbines with the number on-line based on available vapor mass flow.

Control of large steam turbines is usually through a throttling valve at the turbine inlet. The use of variable stator vanes could increase the efficiency for off-design OTEC operation over use of a throttling valve. The additional cost of variable stator vanes will probably be overcome by the advantage of higher average efficiencies over the operating lifetime of the plant.

Normal turbine-generator frequency control is achieved by lock-in to an alternating-current (ac) power grid. The possibility of direct-current (dc) transmission and direct (on-board) energy utilization for OTEC plants would require that turbine speed control be achieved on the basis of measured generator frequency or turbine speed. Present high-power application of generators favors ac over dc generators in sizes above 5 MW_e and voltages above 3000 V. AC generators are generally lower in cost, higher in efficiency, and more reliable than equivalent-power dc generators. Generators to match the turbines for sizes below 25 MW_e (net) are readily available for 3600-rpm speed. For 25-MW_e size or greater, 1800-rpm generators will have approximately 0.2 percent higher efficiency than the 3600-rpm units due to the reduced windage losses for the lower speeds.

Size (output) and cooling medium affect generator efficiency. The use of hydrogen gas for cooling as opposed to air will result in up to 1 percentage point of increased efficiency in 40-MW_e or larger units. The added initial cost of hydrogen cooling can be justified by increased efficiency over the life of the unit. Full-load generator efficiencies are expected to fall between 97 and 99 percent for power outputs between 5 and 50 MW_e. Costs for turbine-generator sets are expected to be in the range $100 to 200/$kW_e$ (net) (1978 dollars) depending on power module size and tradeoffs on the various features mentioned.

Ammonia-circulating pump-motor sets are available in appropriate sizes at a cost near $3/$kW_e$ (net). Multistage pumps can provide the required head to overcome the heat-exchanger, turbine, and piping pressure drops—55 to 60 m (180 to 200 ft)—at 80 percent pump efficiency; a suction head of 5 to 8 m (15 to 25 ft) is needed. Motor efficiency should be near 90 percent.

THE OTEC PLATFORM

Overall Design Considerations

The purpose of the platform is to support the power plant and to permit operations at a specified site in the most economical and reliable manner. Design differences can be expected as a function of application and site as discussed earlier in relation to Figs. 19.4 through 19.6 and Table 19.1. The design of the platform is an iterative process involving the power plant, the site characteristics, the hull, the cold-water pipe, and the energy utilization/transmission system. The dominant design goal of low cost must be achieved within the design and regulatory requirements in the following areas:

Economics—(*a*) minimum life-cycle costs including cost of materials and construction and maintenance costs including logistics; (*b*) ability to finance.

Operations—(*a*) provision of space for the power plant, energy transmission/utilization system, and crew accommodations; (*b*) position control and survival with minimum platform motions.

Compliance—(*a*) meeting applicable codes and regulations; (*b*) retaining flexibility to accommodate design improvements to reduce operating costs, increase productivity, and/or adapt to a new product or service mode during the design life.

Site consdierations—The environmental characteristics of the site are of great importance in the economic optimization of the hull and cold-water pipe. The key role of ΔT_o was emphasized in connection with Table 19.1. Additional typical design characteristics for five sites are shown in Table 19.8.[30]

System integration—Fundamental to meeting requirements in all the foregoing areas is the accomplishment of a systems integration based on thorough analysis and design of effective interfaces, without which the system remains merely a collection of major elements. The preferred configuration will combine adequate (not necessarily "best" or "ultimate") features of all systems and subsystems into a package which is most viable from a cost, risk, and schedule standpoint. In the design process it is essential to identify the several most reasonable alternatives for each system and to establish means of selection by performance/compliance/cost/risk/schedule tradeoffs.

Hull Structure

Numerous platform configurations and sizes have been evaluated, including those shown in Figs. 19.4 through 19.6. Studies have shown that the most cost-effective platforms are designed to accommodate heat-exchanger modules (in some cases, total power modules) that are immersed or submerged in seawater and are capable of being removed for replacement or repair. This conclusion results from consideration of the extensive structure that would be required to support the heat exchangers and associated water-filled ducting in a dry internal environment. The structural efficiency of mounting the heat exchangers external to the hull or in dedicated compartments within the hull (as shown in Fig. 19.6) has become a major design factor. The detachable heat-exchanger (or total power) module has the following favorable attributes: (1) elimination of large internal (dry) hull areas and volumes required for maintenance and replacement of power system components, (2) elimination of the need for large crane lifting capacity, (3) facilitation of module installation and replacement, (4) reduction of platform cost, (5) reduction of construction time because modules can be built concurrently with the platform, and (6) facilitation of financing of power modules by leveraged lease financing.

Some examples of size and displacement characteristics for the types illustrated in Figs. 19.4 through 19.6 plus four other generic types[30] are shown in Fig. 19.24.

Materials

Marine platforms have traditionally employed steel, but the large size and cost of OTEC platforms have caused designers to favor the use of reinforced concrete for the major underwater and/or load-carrying portions of the structure. The superstructure probably will be made of steel.

Large floating structures have been built in the size range of the proposed OTEC power plants, although not specifically of the hull configuration, internal arrangements, or weight distribution of the OTEC structures. Therefore, it is relevant to review the technology and required developments pertaining to the use of prestressed concrete for an OTEC platform. If the OTEC structure is built in the same manner as the North Sea platforms (CONDEEP, Ekofisk, Seatank) or existing barge-type structures (e.g., the ARCO barge[32]), there is a body of technical knowledge and experience. But if OTEC platforms incorporate new materials, configurations, and construction methods, or are used in conditions that are untried by the offshore industry, new knowledge and experience will be required.

The choice of materials is important because of the corrosive seawater environment which could cause serious damage to the concrete and steel. Seawater attacks concrete either through sulfates dissolved in the water acting on the tricalcium aluminate of cement, or by the removal of free lime by the hydration of calcium silicate.[32] The cement for an OTEC plant should therefore contain as little tricalcium aluminate as possible. High-standard quality control will be the key to success of a prestressed concrete OTEC platform. Concrete cover of 3 to 4 in (7.5–10 cm) over the reinforcement steel will be adequate for areas exposed to corrosive

TABLE 19.8 Example Wind, Wave, and Current Characteristics*

Criteria	Parameter	Keahole Point	New Orleans	ATL-1	Puerto Rico	Tampa
100-year storm	Maximum wind, m/s (kn)	34 (66)	52 (100)	31 (61)	48 (93)	59 (114)
	Gust, m/s (kn)	49 (95)	75 (145)		69 (135)	85 (165)
	Waves: $h_{1/3}$, m (ft)	11 (36)	18 (58)	9 (29)	13 (44)	14 (46)
	$t_{1/3}$, s	13	16	18	14	14
	Current, m/s (kn)	1.1 (2.2)	1.3 (2.5)	1.6 (3.2)	1.4 (2.8)	3.3 (6.4)
Typical	Annual average wind, m/s (kn)	2.6 (5)	2.6 (5)	2.6 (5)	2.6 (5)	2.6 (5)
	Waves: $h_{1/3}$, m (ft)	6 (20)	6 (20)	3.7 (12)†	20	20
	$t_{1/3}$, s	10.3	10.3	7.7†	10.3	10.3
	Current, m/s (kn)	0.6 (1.1)	0.6 (1.2)	0.3 (0.5)	0.6 (1.2)	2.4 (4.7)

*Data are from Ref. 30, except for the typical $h_{1/3}$ and $t_{1/3}$ for ATL–1 from APL.
†The significant wave height does not exceed this value 99% of the time (SSMO data 1963–1976).
‡Nominal wave period for 3.75 m significant wave height.
NOTE: $h_{1/3}$ = significant (1 in 3) wave height; $t_{1/3}$ = corresponding wave period.

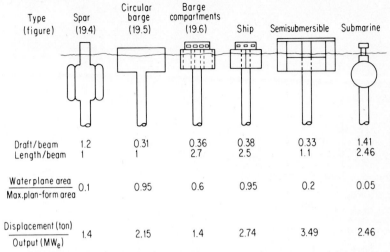

Type (figure)	Spar (19.4)	Circular barge (19.5)	Barge compartments (19.6)	Ship	Semisubmersible	Submarine
Draft/beam	1.2	0.31	0.36	0.38	0.33	1.41
Length/beam	1	1	2.7	2.5	1.1	2.46
$\dfrac{\text{Water plane area}}{\text{Max. plan-form area}}$	0.1	0.95	0.6	0.95	0.2	0.05
$\dfrac{\text{Displacement (ton)}}{\text{Output (MW}_e)}$	1.4	2.15	1.4	2.74	3.49	2.46

Fig. 19.24 Platform configurations for 100-MW$_e$ (net) OTEC plants. (*From Ref. 30.*)

environments.[32] However, in concrete near the surface, extra precautions must be taken due to the chloride ion buildup that occurs from wetting and drying.

An OTEC platform will be subjected to environmental conditions which will cause changes in the weight and displacement of the structure. Changes in hydrostatic pressure, salinity, temperature, marine growth, and seawater penetration must be considered in the design. Tests and experience have shown that seawater permeability does not pose a serious problem to good quality concrete when it is subjected to pressures near 14×10^6 N/m² (Ref. 32).

If a concrete platform is designed in a circular shape to achieve structural economy under uniform external pressure, the efficiency of such a design will depend on maintenance of circularity within tolerances. Deviation causes a reduction in pressure-resistance capacity and requires increased weight and consequently increased buoyancy. Hence, dimension controls must be established in the design stage and enforced during construction.

Increased weight of the vessel means the size must increase to provide buoyancy, and this in turn increases weight still further. Prestressing (and/or posttensioning) and reduced concrete density are both useful ways to reduce weight. Lightweight concrete may be particularly useful in the design and fabrication of the cold-water pipe to achieve near-neutral buoyancy and increased flexibility. However, lightweight concrete (or other alternatives including polymer-impregnated concrete and polymer-cement concrete) has not been used extensively in offshore applications, and development work is required.

Seawater Subsystems

Since a large portion of the platform is devoted to the seawater subsystem, it is necessary to find the optimal tradeoff among size, cost, and performance for the cold-water pipe, platform/CWP transition, interface structure, seawater pumps, water ducts or channels, and biofouling/corrosion control. The well-designed plant integrates these elements with each other and with the hull, power system, and transmission system, all within the constraints of the platform.

Cold-water pipe   The length and inside diameter of the CWP are determined by a complex tradeoff analysis which seeks the minimum life-cycle cost of the total OTEC plant for that site, since the required flowrate is a function of the heat-exchanger type, size, and cost (related to T_o and other factors), and the cost of delivering this flow is a function of inlet, internal-friction, depth (density head change) and exit losses of the CWP, seawater pump design including inlet and exit diffusers, and losses in water ducts or channels.

Candidate CWP materials under investigation include concrete, glass-reinforced plastic (GRP), steel, aluminum, and rubberized fabric. Steel-reinforced, precast concrete is often selected for its low maintenance and potentially low cost. Slip-cast, reinforced concrete may also be used. Research on low-density reinforced concrete (1100 to 1400 kg/m³ or 80 lb/ft³

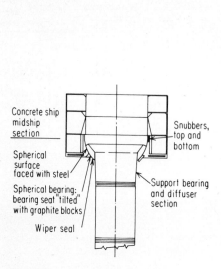

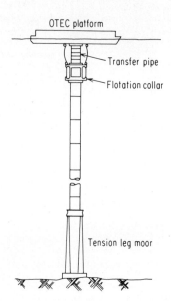

Fig. 19.25 Spherical bearing concept for the cold-water pipe/platform interface. (*From Ref. 30.*)

Fig. 19.26 Buoyant pipe in tension-leg moor with submerged transfer pipe.

value) is under way because such a material would permit construction of a near neutrally buoyant pipe (without the complexity of casting voids into the pipe) which would be lower in strength but also would be subjected to lower stresses because of its lower modulus of elasticity.

Other materials may permit reductions in the size and cost of the platform. For example, a rubberized fabric pipe comprising two walls connected radially by tension ribs (Fig. 19.27) may prove attractive.[30] The spaces between the walls are flooded and maintained at a slightly elevated hydrostatic pressure to enable the structure to resist the buckling forces due to platform motions, external currents, and the pressure differential across the pipe wall. The bottom of the pipe is weighted to keep the pipe nearly vertical when it is subjected to drag forces. The elasticity of the material accommodates bending loads and eliminates the need for flexible joints. However, much work is required to characterize materials, obtain design data, and determine life-cycle costs of rubberized fabric pipes.

Static and dynamic loads   A detailed description of the theoretical methods used to analyze sea-keeping response of OTEC platforms and consequent CWP loads is beyond the scope of this chapter, but the key features can be mentioned. For a given platform it is possible to prepare a weight summary, calculate the hydrostatic properties, generate the lines of the ship, calculate the hydrodynamic seaway response to head and beam seas, and develop the equations of motion.[33] These calculations and analyses follow the standard procedures of naval

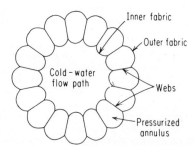

Fig. 19.27 Sectional (plan) view of a rubberized cold-water pipe.

architecture, except where the addition of the CWP introduces complications as discussed below. The design can then be analyzed for constructability, operability, and risks, and a cost estimate and construction schedule can be prepared. Some examples of significant motions are given in Table 19.9 for two plant sites.[6k] The platform motions and resulting moments on a rigidly attached, stiff CWP are shown for 500-MW_e (commercial-size) plants as well as for 100-MW_e (possible demonstration-size) plants.

To date, most published analyses have assumed the CWP to be an elastic circular pipe of constant diameter and sectional properties with a static tension which varies along the length of the CWP. Recent studies have begun to examine tapered and segmented pipes that employ a variety of materials, wall thicknesses, and flexible joints. Dynamic tension due to heaving of the CWP is usually neglected because it is so much smaller than static tension, although that assumption will have to be reviewed with the hybrid pipes. The CWP lateral motions are assumed to be excited by a combination of forces that include orbital ocean wave motions and the resulting surge and sway translations, as well as roll and pitch rotations of the CWP. The CWP hydrodynamic damping is linearized and therefore the dynamic lateral response can be described by a linear, partial differential equation of motion. For harmonic motion this equation reduces to a linear ordinary differential equation of motion which can be solved for the cases of unit wave excitation, translation, and rotation.[33] These solutions, which are transfer functions for the CWP attachment-point forces and moments, can be introduced into the equations of motion of the platform to determine the coupled motions of the platform and CWP. The resulting motions of the CWP attachment point can be used to calculate the complete response of the CWP. The methods of linear superposition can be used to determine the response of the coupled platform and CWP. The same analysis can be applied to a jointed CWP consisting of a number of segments connected by flexible joints by treating it as a continuous CWP with an equivalent modulus of elasticity that accounts for segment and joint flexibility.

The CWP is loaded statically by its own immersed weight, by the seawater moving inside, by the external current, by its attachment to the platform, and by wave action. The greatest loading is induced by the platform and depends on wave forces, platform motion response, heading, depth of pipe attachment, pipe material, size, stiffness, and platform/CWP attachment

TABLE 19.9 Significant Motions for Various Platform Types in Two Sizes, with a 610-m-deep CWP with Quasi-infinite Attachment Stiffness, for a Significant Wave Height $h_{1/3}$ of 18 m (Ref. 6k)

Response	500 MW$_e$ (net)					100 MW$_e$ (net)				
	Ship	Sub-marine	Semi-submers-ible	Disk	Spar	Ship	Sub-marine	Semi-submers-ible	Disk	Spar
Surge, m	2.6	1.3*	1.2	2.7	1.9*	2.5	2.2	1.9	3.5	2.7*
		1.2			1.7					2.4
Sway, m	2.9	2.3*	1.3	2.5	1.9*	3.7	3.0	2.1	3.5	2.7*
		2.0			1.7					2.4
Heave, m	6.4	2.6	1.7	6.4	1.9	6.9	3.3	2.3	7.1	2.6
Roll, degrees	0.4	0.2	0.1	0.4	0.2	0.5	0.4	0.2	0.5	0.3
Pitch, degrees	0.4	0.1	0.1	0.4	0.2	0.4	0.2	0.2	0.5	0.3
Yaw, degrees	1.4	1.0	0.5	0	0	1.8	1.5	0.7	0	0
Axial bending moments, GN · m:										
Attachment	15.6	21.2	9.8	3.8	9.3	4.9	5.1	2.8	13.0	2.0
Maximum	15.6	21.2	9.8	7.1	9.3	4.9	5.1	2.8	2.8	2.0
Lateral bending moments, GN · m:										
Attachment	10.2	7.0	1.4	3.8	9.3	2.3	1.4	3.9	13.0	2.0
Maximum	8.1	7.0	11.6	7.1	9.3	2.1	1.4	3.9	2.8	2.0

*Upper values at waterline, lower values at center of hull.

design. The maximum bending moment occurs at a depth which depends on the foregoing parameters. The amplitude of peak moment is approximately 2 to 3 times the significant value, which in turn is 4 times the rms moment. From each analysis, the maximum bending moment as a function of significant wave height can be obtained, and examples are shown in Fig. 19.28 for 1000-m-long, 16.7-m-diameter CWPs of four materials attached to a 100-MW$_e$ spar-type platform. These bending moments can then be converted to stresses.

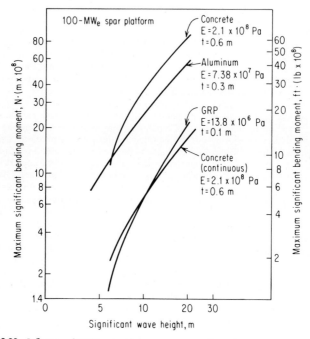

Fig. 19.28 Influence of CWP material on maximum significant dynamic bending moment.

A stress analysis for a segmented concrete CWP mounted on a 100-MW$_e$ ship hull operating off New Orleans (including both static and dynamic loads) is shown in Table 19.10. Loads include static weight, current shear and moment, dynamic shear, moment and axial forces due to platform and wave forces, and pressure differential. At the attachment depth of 13.7 m, the dominant stress component is axial tension due to pipe weight, while at 122-m depth the dynamic bending stress is 57 percent of the resultant stress. Based on the resultant stress, and depending on the acceptable safety factor, the designer can then estimate the CWP thickness. The advantages of multiple joints in reducing stresses are illustrated by Fig. 19.29. A typical

TABLE 19.10 Stresses in a Concrete CWP with 46-m Segments Separated by Flexible Joints, for 100-MW$_e$ Platform Operating South of New Orleans, 6-kn Current, $E = 2.1 \times 10^9$ Pa, $t = 0.6$ m, Stiffness of Attachment to Platform $= 1.4 \times 10^9$ N · m/rad in Roll and Pitch, $h_{1/3} = 18$ m (Ref. 30)

Depth, m	Static stress, 10^3 Pa				Dynamic stress, 10^3 Pa			Resultant stress, 10^3 Pa
	Axial	Shear	Bending	Hoop	Axial	Bending	Shear	
13.7	7619	231	1644	145	838	1322	483	11,466
122	6667	39	2	124	731	10,018	218	17,423
867	1000	1	0	19	110	10,342	141	11,383

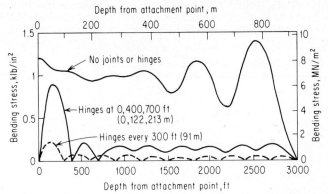

Fig. 19.29 Root-mean-square bending stresses for 30-ft (9.1-m) ID CWP on 10/20-MW$_e$ (net) pilot plant platform, $h_{1/3}$ = 29 ft (8.8 m), head sea, 1-ft (0.3-m)-thick CWP wall.

distribution of the rms bending stress for a nonjointed concrete CWP rigidly attached to the hull is compared to those for jointed CWPs with flexible attachments to the hull for the APL pilot plant concept.

The relationship between CWP wall thickness and significant wave height for four kinds of CWPs attached to a 100-MW$_e$, spar-type platform[30] is shown in Fig. 19.30. Buckling criteria determine the wall thickness for significant wave heights below 5 to 10 m, but wall thickness increases quickly with wave height when platform motions become larger. As illustrated for the concrete pipes, raising the attachment point of the CWP reduces wall thickness by 0.61 m (2 ft), and additional work is required to optimize the attachment depth.

The platform-CWP interface structure (see Figs. 19.4, 19.5, 19.25, and 19.26) transfers the supporting loads between the suspended CWP and the CWP foundations on the platform. Interface requirements include static and dynamic load distribution, axial, torsional, and rotational bearing stiffness, low wear rate, 40-year life in seawater environment, and compatibility with the CWP installation method. Concept options include gimbal, ball-joint, heave-compensated cable, CWP flange or elastomeric bearing, and an elastomeric section for the hydrostatic rubber pipe.

Seawater pumps   The seawater pumps are among the more critical elements in the OTEC plant since they consume most of the parasitic power. In general, the pumping power is directly proportional to pump head and flow rate. The pump head is required to overcome losses due

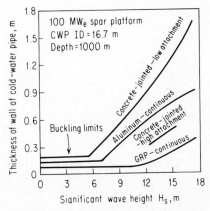

Fig. 19.30 Wall thickness of various CWPs for 100-MW$_e$ (net) spar platform as a function of significant wave height.

to water intake screens, contractions and expansions, turns, the heat-exchanger head loss, and the effluent kinetic energy loss. For the cold-water system, contributions also come from pipe-wall friction and the density differential effect for the cold water (which depends on pipe length), as shown in Table 19.11.

TABLE 19.11   Typical Distribution of Cold-Water Pump Head for 1200-m Depth[34]

	Head	
	m	%
Viscous losses due to intake screen, contractions, expansions, and turns	0.93	29
Heat-exchanger head	1.50	47
Cold-water-pipe wall friction	0.13	4
Seawater density change with depth	0.58	18
Effluent kinetic energy loss	0.06	2
	3.20	100

Although the cost per unit flowrate of pumps decreases as pump size increases, the merits of redundancy, serviceability, and operational convenience, as well as the cost of platform volume occupied by the pumps, must be considered. For these reasons, multiple pumps for both the warm- and cold-water systems are employed. As shown in Fig. 19.31, the value of the hull ($661/m³ or $18.75/ft³ in this example) tends to reduce the economy of scale. Assuming a 1.375 packing factor for the assembly of smaller pumps, the total cost is found to be essentially independent of the rated capacity of the pumps for this example. Since pumps of 5-MW$_e$ (net) size for OTEC plants already are in use for cooling systems of nuclear power plants, no appreciable size extrapolations from off-the-shelf designs will be needed for OTEC applications.

Position-Control Subsystem

Position control is required at all sites and in all applications. The three principal options are mooring, dynamic positioning, and a combination of the two. The systems requirements include: (1) ability to accommodate required depths (for mooring) and avoid bathymetric (bottom) obstructions, (2) orientation of the platform to minimize aero-hydro drag, platform seaway response, and recirculation of effluents, with ability to accommodate the extreme wind, current, and wave drag forces. The design of the position control subsystem requires an analysis of the

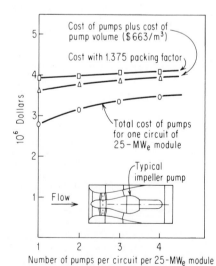

Fig. 19.31   Total cost of multiple seawater pumps for one circuit of a 25-MW$_e$ (net) power module.

horizontal loads on the platform to establish the orientation and/or holding power required in the extreme case, i.e., with the 100-year-storm wind, wave, and current acting in the same heading normal to the platform maximum area.

For a moored plant, position-holding will be vital unless tugs or other assistance are required to be available for call-out and use at first indication of a storm. The allowable mooring line load is 20 percent of breaking strength for normal operations and 50 percent for extreme conditions. The anchor must accommodate various sea floor conditions, and for an offshore power plant, the "watch circle" through which the plant swings around the mooring is limited to 5 percent of water depth to accommodate the electrical riser cable.[30]

For a tropical grazing plant, drift under proper orientation may be allowable until a plant approaches a shore or a hazardous bathythermal area, where tugs or other assistance might be required. A mooring system of modest capability may be included to cover periods when it is desirable to remain in a given favorable location without using thruster power.

Platform Support Subsystems

The platform support subsystems listed in Table 19.12 are similar to those employed on conventional vessels and petroleum production platforms, and constitute a well-developed body of knowledge within the field of naval architecture. In designing and selecting components for OTEC plants (as for any future power plants) it is important to maximize the energy efficiency. The overall economics of the plant are sensitive to the parasitic power requirements, which comprise the pumping power to move the warm and cold seawater and working fluid through the heat engine, propulsion and/or station-keeping, and the "hotel" load that is required to operate the platform support subsystems.

Construction, Deployment, and Operations

The main construction techniques planned for OTEC platforms are similar to those used in the construction of North Sea petroleum platforms and the ARCO barge.[31] The lower portion will be built in a shallow graving dock. After the dock is flooded, the platform will be floated to a mooring where construction can be completed. For spar-type platforms, slip-form casting will be used, and controlled ballasting of the structure will maintain stability and freeboard. However, this requires a protected deepwater construction site that is close to logistics terminals. In addition to slip-form casting, it is possible to prefabricate portions of the platform onshore and assemble the modules offshore.

Precast concrete modules speed construction time because they can be produced concurrently or ahead of time, permit greater quality and dimensional control, and may permit the use of hybrid materials and polymer-impregnated concrete. Problems include methods and procedures to prestress the modules as a unit, and to make a watertight joint between the precast modules. Field joints are possible and many designs and techniques are available. Waterstops are used,

---

TABLE 19.12 Platform Support Subsystems

Electrical:	Environmental control:
Auxiliary electrical power	Heating
Electrical distribution	Ventilation
	Air-conditioning
Mechanical:	Acoustics
Compressed air	
Ballast control	Lifesaving:
Fuel oil (for auxiliaries)	Lifeboats
Materials handling	Firefighting
Shops	Other emergency gear
Spare parts, storage	
Propulsion or mooring	Communication and navigation:
	Radio
Accommodations and life support:	Radar
Quarters and offices	Lights
Fresh water	Bridge
Galleys	
Medical	Transportation and logistics:
Recreation	Air to sea
Sanitation	Sea to sea

and it is possible to pressure-grout the joint with epoxy after the concrete is stressed. The number of field joints should be minimized and each should be properly prepared and painted with a bonding compound before the new concrete is poured.

If a large number of OTEC platforms are to be built, the North Sea method would not be economical and the available facilities and sites would be overstrained. One solution is to use a construction/launch barge as a base to fabricate the lower section of the platform. When the structure is large enough to float, the barge would be towed offshore and submerged by ballasting. When the concrete structure floats free, it is towed aside and the barge is refloated and returned to shore to reuse. Construction of the platform is completed while it is afloat as described previously.

CABLE TRANSMISSION OF ELECTRICITY TO SHORE

Cable Considerations and Estimated Costs

For OTEC plants connected by electric cables to a utility grid or to an industrial process on shore, the electrical demand may require certain voltage, current, and frequency characteristics which will impose design constraints on the generator, power-conditioning equipment, transmission cable, and land-based terminal. The cable lengths, power capacities, and water depths required for OTEC cable systems will exceed those of any existing installation. (An oil-impregnated-paper-insulated system of $\pm$ 250 kV cables, 122-km long with a maximum depth near 460 m, was installed between Denmark and Norway in 1976.[61])

The stationary cable from the shore to a fixed point on the ocean bottom is considered to be within the state of the art, but other components must be developed. The riser cable presents problems because the upper end of the riser must share some of the motion of the OTEC plant, and will experience considerable changes in stress and flexure. It is judged that a suitable flexible cable and attachment system can be developed.[61] The submarine connector between the riser and the stationary cable could be a swivel which would allow the plant to rotate about a single anchor. However, such a swivel would represent an element of unreliability and may require shutdown of the plant for maintenance or for replacement of a riser cable. An alternative solution is to employ a multipoint moor and/or dynamic positioning in order to maintain the plant relative to the anchor and within the capabilities of the riser.

For utility power grids on shore, the reliability of the transmission is rarely considered to be a variable parameter, because spare capacity is generally installed. However, in the case of OTEC, the high cost of spare cables, and the inaccessibility of the system, may require a different design philosophy. As the number of installed cables increases, the total capital and operating costs increase, thus increasing the cost of delivered electricity. At the same time, however, the availability of the system increases, thus decreasing the cost of delivered electricity. The optimum system design will, therefore, employ the number of cables that produces the minimum cost of electricity. The minimum cost will vary as a function of cable cost, OTEC plant cost, and the repairability of damaged cables.

An investigation[61] of probable system costs produced the following general conclusions about incremental transmission costs (with costs increased 15 percent for inflation, 1976 to 1978). DC transmission ($\pm$ 300 kV) is more economical for distances less than 32 km (20 miles); ac ($\pm$ 245 kV) is preferred for greater distances. The incremental transmission cost is between 15 and 20 mills/kWh when the plant is approximately 320 km (200 mi) offshore. This cost varies with the OTEC plant cost because of the variation in the cost of losses during transmission. Relative ac and dc transmission costs of typical cable installations are shown at the bottom of Fig. 19.32a for a 200-MW$_e$ OTEC plant and an oil-impregnated-paper-insulated cable (same type as used in the aforementioned Denmark-Norway installation) at 600-m (2000 ft) depth.[61] Larger OTEC plants would require one additional three-phase circuit for each 200-MW$_e$ increment (i.e., a 400-MW$_e$ plant would have two circuits, but delivered cost would not change) unless a cable with higher capacity is developed. Alternative cable types include a cross-linked-polyethylene-insulated cable, which may not be as reliable as the paper-insulated cable at the OTEC voltages, and a paper-insulated cable with a high-pressure, oil-filled (HPOF) core.[61]

Electricity Costs at Shore

Figure 19.32a also shows a band for estimated delivered costs of electricity at shore as a function of distance from shore for plants in the Gulf of Mexico. For this illustration the available ΔT_o is assumed to vary with distance from shore as indicated by the upper ΔT_o scale; corresponding

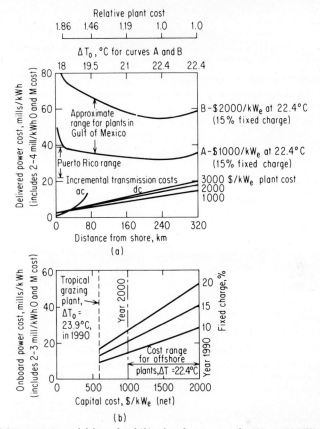

Fig. 19.32 (*a*) Transmission and delivered and (*b*) onboard power costs for 200- or 400-MW$_e$ (net) OTEC plants.

relative plant costs are shown by the top scale [e.g., a plant that would cost $1000/kW$_e$ at 22.4°C would cost $1860/kW$_e$ at 18°C according to Eq. (19.1)]. The lower cost ($1000/kW$_e$ for 22.4°C) is assumed to represent the year 2000 after many (> 50) plants have been built and there has been an appreciable cost reduction due to technological improvements and learning-curve benefits on mass production of modular components. The higher cost ($2000/kW$_e$) is assumed to be achievable by 1990. All delivered power costs in Fig. 19.32*a* are for a 15 percent fixed-charge rate and include an operation and maintenance (O&M) cost that varies from 2 mills/kWh at $1000/kW$_e$ plant cost to 4 mills/kWh for early, nearer-shore plants at ≥ $3000/kWh.

Note that a delivered power cost range of 22 to 39 mills/kWh (for OTEC plant costs of $1000 to 2000/kW$_e$) is estimated for Puerto Rico, where a ΔT_o of 22.3°C is available 10 to 50 km from shore. The corresponding range for Hawaii ($\Delta T_o = 21.4$°C) would be 25 to 43 mills/kWh. Such favorable ΔT_o values near shore give impetus to the concept for development of island energy complexes at Puerto Rico and Hawaii in semitropical waters and possibly at even more favorable island sites in tropical waters nearer the equator.

Figure 19.32*b* also indicates the lower capital cost estimated[14,35] for tropical grazing plants in ATL-1 or PAC-2, which yields on-board power costs in the 9- to 17-mill/kWh range, depending on the fixed charge rate (1978 dollars).

AMMONIA PRODUCTION ON BOARD

Ammonia can be produced as sketched in Fig. 19.33 by purifying seawater, electrolyzing it to obtain hydrogen, separating nitrogen from air, and reacting the hydrogen and nitrogen in the

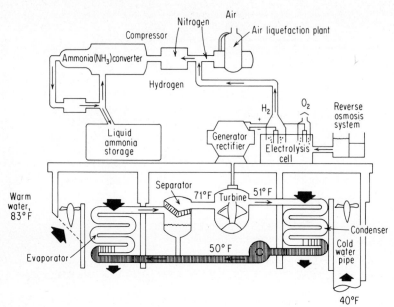

Fig. 19.33 Simplified schematic diagram of OTEC/ammonia plant ship. (*From Ref. 14.*)

presence of a catalyst (iron oxides) at approximately 370°C (700°F) and 1.4×10^7 N/m² (2000 lb/in² absolute). It is an attractive product because of the growing demand for fertilizers (75 percent of U.S. ammonia consumption) and chemicals and the fact that it is presently made from natural gas in the United States. It also can be employed as a hydrogen carrier as discussed later.

Ammonia Production and Transportation

Water electrolysis system The required purification of seawater for the feed to the electrolysis cells can be accomplished in several ways. Figure 19.35 indicates use of a reverse osmosis (RO) system which requires 18.6 kW$_e$h (0.2 percent of the OTEC net power) to produce the 1.64 m³ (432 gal) of fresh water needed per tonne of NH_3 output. The estimated capital cost is \$1.7 million for a 1000-t/day ammonia capacity, with an O&M cost of \$1.40/t ammonia, half of which is for membrane replacements in the RO cells (1978 dollars). In comparison, a conventional multistage distillation system would require a fuel energy input of 325 kJ/kg (140 Btu/lb) of water, or 0.53 GJ (0.50 MBtu) per tonne of ammonia, with an estimated capital cost of \$1.4 million and an O&M cost, including fuel at \$2.60 GJ (\$16/bbl), of \$1.70/t.[35] An electrically driven distillation system of the vapor-compression type made by the Aqua-Chem Co. operates as follows.[36] Feedwater is preheated by the outgoing distillate and waste brine in a plate-type heat exchanger. This warm feedwater plus recirculated brine is sprayed onto the tubes of a horizontal evaporator. The steam generated is drawn through a demister, compressed, driven back through the tubes of the evaporator wherein it is condensed, and the resulting pure water is cooled as it leaves through the feedwater preheater. The power requirement is approximately 70 kW$_e$h per tonne of ammonia, or 0.8 percent of OTEC net power; the capital cost is approximately \$2.0 million for a 1000-t/day plant. The O&M cost other than power should be less than that of the RO system.

Water electrolysis cells under development include improved filter-press type, 25 percent potassium hydroxide (KOH) electrolyte cells by the Teledyne Energy Systems Company,[37a] and solid polymer electrolyte (SPE) cells by the General Electric Company.[37b] Voltage-temperature curves achieved in bench-scale units (Fig. 19.34) indicate high voltage efficiencies at relatively high current densities. The voltage efficiency is the ratio of the thermoneutral voltage, which is 1.47 V at 25°C, to the required voltage. Thus, at 25°C and 100 percent voltage efficiency, the energy supplied would be equal to the heat of formation of liquid water, or the

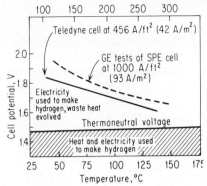

Fig. 19.34 Idealized operating conditions for electrolysis and curves from Teledyne and GE. (*From Ref. 35.*)

higher heating value of hydrogen [i.e., 1.419×10^8 J/kg (61,030 Btu/lb) or 39.5 kWh/kg (17.4 kWh/lb) of hydrogen]. At present most electrolyzers operate at nearly 2.0 V or 75 percent efficiency, but near-term advances promised by recent and planned developments appear to assure improvements to the 1.7- to 1.5-V range, corresponding to 46 to 40 kWh/kg or 86 to 98 percent voltage efficiency.

In the aforementioned Teledyne KOH system as proposed for near-term use (before 1985, with 2.5- to 3-year lead time), the installation for the first 100-MW_e or greater plant would be made up of 16-MW_e input unit plants, each containing five electrolysis modules run by one set of auxiliaries.[35] The H_2 would be generated at 8 kN/m^2 (100 lb/in^2 gage), separated from the gases in primary separators, cooled from 82°C to 74°C by a heat exchanger (used to preheat the reactant gases for the ammonia synthesizer), and filtered before reentering the electrolysis cells. With a cell voltage of 1.72 V and a power density of 74 W/m^2, the voltage and current efficiencies would be 86 percent and 98 percent, respectively, yielding an overall system efficiency of 84 percent. The space requirement and weight of a 16-MW_e unit (352 kg/h of H_2) would be $7.6 \times 7.6 \times 4.6$ m high and 365 t, respectively. Each electrolysis module could be run from 25 to 100 percent capacity by varying the voltage between 840 V and 860 V (1.68 to 1.72 V/cell) and the current between 950 and 3800 A. Control would be completely automatic with manual override.

The GE SPE system offers greater economic promise for succeeding plants beginning in 1985. The electrolyte is a solid plastic sheet of perfluorinated sulfonic acid polymer, 0.25 mm thick. On the basis of life testing of single cells up to 4 years with no maintenance and no degradation in performance, GE has proposed the 5-MW_t-output (in H_2 thermal equivalent, 127 kg/h) unit system depicted in Fig. 19.35.[35] The electrolysis module is 1 m in diameter by 2 m high, contains 580 cells of 0.27 m^2, operates at 105°C, and weighs 2.3 t. With adequate R&D and 3-year lead time, such units should yield 90 percent efficiency at 186 A/m^2 (2000 A/ft^2) by 1985. Projected cost (1978 dollars) is approximately $110/$kW_e$ input.

Ammonia synthesis system   Three parts of H_2 from the electrolysis cells will be combined with one part of N_2 (by volume) for the NH_3 synthesis plant. The N_2 will be obtained from an air separation (partial liquefaction) plant at an estimated installed cost of $8 million for a 1000-t/day ammonia plant. The ammonia synthesis plant will be essentially the back end of a natural-gas-reforming ammonia plant. However, because the gaseous hydrogen feed from the electrolyzers is of high purity, no purge is needed in the ammonia synthesis loop.[36]

Ammonia synthesis system requirements and costs were based on information provided by the staff of the TVA National Fertilizer Research Center, who have operated an experimental natural-gas-reforming, 225-t/day plant, and J. F. Pritchard, Inc.[35] The deck space and weight for a 1000-t/day plant are of the order of 1000 m^2 and 1000 t, respectively. With respect to cost, it is important to exceed the 600-t/day plant capacity at which the compressors are changed from the reciprocating to the centrifugal type; sizes $\geq$ 1000 t/day are most economic.[36,38] Since the basic OTEC plant is also believed to achieve substantial economies of scale by going to a few hundred MW_e,[11a,11b,38] a 325-MW_e plant producing 1000 t/day of anhydrous liquid ammonia was selected by APL/JHU for cost estimates.

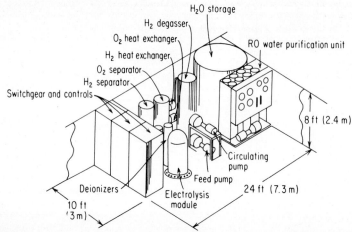

Fig. 19.35   Conceptual layout of General Electric's 5-MW$_t$-output SPE water electrolysis system. (*From Ref. 35.*)

Table 19.13 shows the APL/JHU estimates for capital cost and ammonia production costs for the first, ninth, and optimized twenty-fifth plants of this size. These costs are similar to the 1975-dollar costs given in Ref. 14 including 12 percent for escalation and interest during construction, escalated to 1978 costs by a 7 percent/year inflation rate, with an added estimate of $5 million for heat engines for waste-heat recovery from the electrolysis cells and the ammonia synthesizers. These costs also are based on operation in site ATL-1 with a 23.9°C (43°F) ΔT_o and use of the plant-ship and aluminum heat-exchanger concepts shown in Figs. 19.6 and 19.18 with 20-year plant life and a 0.000052 m^2 · K/W average fouling coefficient.

TABLE 19.13   Estimated Capital and Ammonia-Production Costs for 325-MW$_e$, 1000-mt/day, OTEC/Ammonia Plant-Ships (1978 dollars) for ΔT_{0_d} = 43°F (23.9°C)

	Ship		
	1st	9th	25th opt.
Capital costs, $M*			
Platform and CWP	120	75	65
Heat exchangers	121	98	74
OTEC machinery and miscellaneous	83	54	51
Deployment to site	2	2	2
Total basic ship, $M	326	229	192
($/kW$_e$)	(1003)	(705)	(591)
Ammonia production and transfer equipment, $M	81	70	67
Total plant-ship investment (PI), $M	407	299	259
Working capital (WC, 120 days), $M	10	7	6
Production costs, $/t, on 345,000 t/year			
Chemicals, labor, overhead	10	10	10
Fixed charge (12.8%)†	151	111	96
Interest on WC at 8%	3	2	2
Cost on board, $/t	164	123	108

*$M = millions of dollars.
†The 12.8% fixed charge on the total plant investment (PI) comprises: maintenance materials, 1%; insurance on hull, indemnity, and construction, 2.8%; depreciation (20 years), 5%; and interest, 8% of ½ PI. The PI includes 12% for interest and escalation during construction, which has been allotted to all components.

The ammonia plant costs in Table 19.13 include the GE electrolysis system ($123/kW or $40 million for the first ship), a reverse-osmosis fresh-water system ($2.5 million for the first ship), and the ammonia synthesis plant including waste-heat recovery ($39 million for the first ship). The cost reductions for all systems from the first to the ninth plant ships are based on learning-curve improvements as described in connection with Fig. 19.6. The overall improvements from the first to the "optimized twenty-fifth" plant-ship are based on a combination of an overall 93 percent learning curve for the total system and the assumption that a 15 percent reduction in the costs of the heat exchangers and the basic hull can be achieved between the ninth and twenty-fifth ships through ongoing, government-funded R&D and optimization of the integrated system.

The lower part of Table 19.13 shows the estimated ammonia production costs. These APL estimates are in reasonable agreement with those by IGT for "advanced long-range improved technology" (as APL assumed for the 1985-1990 time frame) at the APL on-board power cost of 12 mills/kWh (1978 dollars). All cost estimates are tentative until pilot/demonstration plants have been built and operated.

Ammonia transportation and storage costs   The Sun Shipbuilding and Dry Dock Co. made an estimate[35] of ammonia shipping costs for 70,000-t tankers making 12.5 round trips per year between ATL-1 (east of Brazil) and New Orleans (4000 nmi or 7410 km) at 14.5-km (26.9-km/h) speed with a 0.8 block factor (average speed reduced to 11.6 kn by allowances for loading and unloading, adverse weather, etc.). Each tanker was estimated to cost $60 million, with a resulting shipping cost of $12/t in 1975 dollars. This estimate becomes $15/t in 1978 dollars. For U.S. ships, an operating differential subsidy from the Department of Commerce could lower this cost to $12/t. The cost for delivery from PAC-2 (southwest of Acapulco, Mexico) to the U.S. west coast, or for other trips between plants and ports of approximately 1000 nmi distance, would be approximately $5/t.

Costs for terminal and storage facilities at U.S. ports are low—approximately $1/t in 1978 dollars.[36]

Estimated delivered ammonia costs and market   The foregoing OTEC-ammonia production and transportation cost estimates yield delivered costs at New Orleans ranging from $176 to $120/t ($160 to $109 per short ton) for the first to the twenty-fifth optimized tropical grazing plant-ship in the ATL-1 siting area. Delivered costs to San Diego from PAC-2 would be $7/t less. These costs compare to Gulf port FOB prices that have ranged from $243/t in 1975 (*Chemical Marketing Reporter*, converted to 1978 dollars using 7 percent/year for inflation) to spot prices considerably lower in 1978. Projections[38] have indicated that competitive entry into this market can be achieved in the mid-1980s if the government assumes half the cost of the first plant-ship and progressively lower shares of the next few. By the time the sixth to eighth 1000-t/day plant-ship is reached, a fully competitive position with all-private funding is foreseen. Cash-flow analyses indicate recovery of the full plant investment in 4 to 6 years.[38]

Figure 19.36 illustrates the anticipated competitive position of early tropical grazing OTEC plants vs. onshore plants which consume 38,500 standard ft^3 (1090 m^3) of natural gas per tonne of ammonia produced. In this comparison the general inflation rate is assumed to be 7 percent/year, and *additional* inflation rates of 2.8, 8, and 13 percent per year are assumed for natural gas, our scarcest fossil fuel. The circle on the figure indicates the projected production

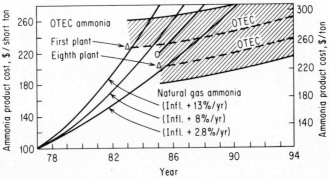

Fig. 19.36 Projected ammonia production costs-OTEC ammonia costs rise more slowly than onshore-natural-gas ammonia costs.

cost in January 1985 resulting from the anticipated U.S. government decision to decontrol natural gas prices at that time after a price rise averaging inflation plus approximately 4 percent/year up to that time. The triangles indicate initial costs for the first and eighth 1000-t/day OTEC plant-ships. Since 77 percent of the OTEC ammonia cost is a fixed charge related to capital cost and does not change once the plant is deployed, the OTEC ammonia curves rise much more gradually with time after deployment than the onshore plant curves.

The U.S. ammonia production capacity, demand, and the estimated OTEC penetration from 1984 to 2000 are shown in Fig. 19.37. Planned expansions will raise production capacity to 21.6×10^6 t/year in 1980.[38] Let us assume that no new natural-gas plants will be built after that time, and that the use of natural gas to make ammonia will be linearly phased out by the year 2000. Three levels of future demand are shown—a 5 percent/year increase ("high"), a 3.58 percent/year increase ("most probable," based on 3.1 percent/year for fertilizers, which represent 75 percent of demand, and 5 percent/year for ammonia for industrial chemical use[38]), and 2.8 percent/year ("low"). The high, most probable, and low OTEC shares are based on capturing 75, 60, and 50 percent, respectively, of the corresponding capacity *additions* and the same percentages of the 1.08 million t/year of replacements, on the rationale that the more economically attractive OTEC is, the greater the share it will capture. The resulting range of OTEC penetration is 13.5 to 33 million t/year or thirty-nine to ninety-six 1000-t/day OTEC plant-ships by the year 2000. The balance of the demand is made up by new onshore plants using coal as feedstock, by imports, or by natural gas plants retrofitted to use coal as feedstock. The dotted curve shows the most probable coal-share case.

The OTEC-ammonia plant-ships at site ATL-1 could also supply ammonia to coastal markets of South America, Africa, and Europe.[38] Plants at site PAC-2 could also supply markets in Central and South America and Canada. In the 1990s and beyond, additional OTEC–ammonia plants could be used to provide electricity on shore via the fuel-cell route as described in the following section.

Electricity generation from OTEC ammonia   Although the hydrogen gas produced on an OTEC plant-ship as previously described could be liquefied on board[35,36] and could compete in some limited existing markets, use of ammonia as a hydrogen carrier appears more attractive. This approach will avoid the need for development of new cryogenic tankers and cryogenic storage facilities near the point of use. Anhydrous liquid ammonia is readily shipped and stored at atmospheric pressure at $-33°C$ $(-28°F)$ in existing systems, and a pipeline distribution system already exists in the United States.[38] Ammonia also is readily decomposed to H_2 and N_2 gases, requiring approximately 14 percent of its energy content for the process.[14,38] The resulting hydrogen can be employed in H_2-O_2 fuel cells at efficiencies near 60 percent. For example, the GE SPE electrolysis cells previously described can be operated as fuel cells at this efficiency level.

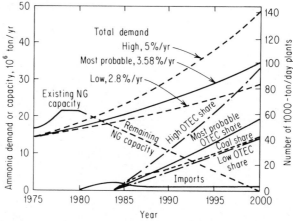

Fig. 19.37 U.S. ammonia production capacity from natural gas, forecast high, most probable, and low demand levels, and corresponding forecasts of OTEC market penetration. Most probable coal share and imports are also shown.

A more attractive possibility is use of the Deacon process, whereby the hydrogen would be reacted with chlorine rather than oxygen in a fuel cell. The resulting HCl from the fuel cell would be catalytically oxidized to regenerate the chlorine and water, which is removed by circulating sulfuric acid. Since the latter process can be conducted at 99.5 percent efficiency, and the fuel cell can operate at 80 to 85 percent efficiency, this system promises a power output per tonne of ammonia that is approximately one-third higher than that of the H_2-O_2 fuel cell concept. Although capital costs would be somewhat higher, the increased output would result in a 15 to 20 percent lower electricity cost. Estimates by APL indicate local bus bar electricity costs of 30 to 54 mills/kWh for ammonia costs in the range $110/t (when a utility owns the OTEC plant-ship) to $220/t. The concept is illustrated in Fig. 19.38. In comparison with oil-fired power plants operating at 33 percent efficiency, this process could be competitive in the 1990s. Since ammonia can be distributed over large distances at lower cost than electricity distribution, it will offer a great potential for the dispersed power approach.

ALUMINUM PRODUCTION, ON BOARD AND ASHORE

Onboard Aluminum Production Investigations

Several organizations including APL[35] and TRW[39] have considered conceptual designs for OTEC plant-ships which would reduce alumina (Al_2O_3) to aluminum metal on board. This process is attractive for OTEC application because (1) it is energy-intensive—approximately 15.4 kWh/kg (7kWh/lb) of aluminum is required for the electrolysis cells and ancillary electricity and (2) the alumina, made from bauxite ores by the Bayer process, is already imported over long distances, e.g., from Surinam, Jamaica, and Australia. OTEC alumina reduction plants could be located near alumina shipping routes, at Puerto Rico, Hawaii, tropical islands, South American coasts, or Australia.

The present Hall-Heroult process   In the commonly used Hall-Heroult process, 1.92 kg of alumina is required per kilogram of aluminum metal produced. The alumina is reduced by electrolysis of a molten (970°C) solution of alumina (2 to 5 percent) in cryolite (Na_3AlF_6). In a modern plant,[35] a 0.9-t/day (1-short ton per day) electrolysis cell consists of a rectangular steel tank, 4.3 × 9.1 × 1.5 m deep (14 × 30 × 5 ft) lined with refractory insulating bricks and an inner lining of baked carbon blocks. The active cathode is formed by the molten aluminum layer, 0.2 to 0.3 m (8 to 12 in) deep in the bottom. Floating on the liquid layer is 0.15 to 0.2 m of the cryolite bath. The anodes are rectangular carbon blocks suspended in this electrolyte

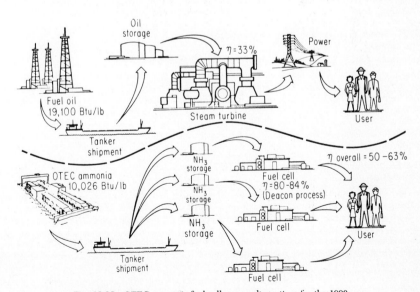

Fig. 19.38   OTEC-ammonia-fuel cell power alternatives for the 1990s.

from bus bars over the cell, which is operated at 4.3 V and 125,000 A. (Some plants using smaller cells run at $\leq$ 100,000 A.) Molten aluminum is tapped from the cells at 1- to 2-day intervals by suction into ladles for transfer to electrically heated holding furnaces located on the production deck below to feed the ingot casting machines.

Finely divided alumina is shipped to the OTEC plant and pneumatically pumped to tanks on an upper deck. The carbon anodes are consumed by oxidation over a 25- to 30-day period, and replacement anodes are made in a shore facility. Anodes are loaded on board the OTEC plant, and aluminum ingots are unloaded using a crane.

Relative motion of the molten aluminum cathode and carbon anode caused by sea-state-induced platform motion is a potential problem. Efficient reduction-cell operations require a constant anode-to-cathode spacing between 3.8 and 5.4 cm. Variations outside this normal operating range will result in inefficient cell operation and possible shorting of the cell, which may cause damage to the anodes and cell-insulating structure. The required spacing between the anode and the molten aluminum cathode might be maintained by hydraulically stabilizing the cell or by servo control of the anode position using the cell voltage as a feedback signal. Either solution will result in complex dynamics and should be verified experimentally.

Fluorides entrained in the cell exhaust gases are removed in land-based plants either by precipitators or wet scrubbers. For open ocean operations, some reduction in the stringency of pollution standards may be permissible. However, insulated fume hoods are provided which are integral with the top of each reduction cell, and the gases are scrubbed and exhausted through a stack.

The APL estimate[38] for the cost of aluminum produced on board a 369-MW_e 238,000-t/day plant-ship was $0.92/kg vs. a cost of $1.00/kg for equivalent T-ingots produced in a new onshore plant which paid 18.4 mills/kWh for electricity. The TRW estimate[39] was $1.41/kg for on-board production on a 100-MW_e, 57,000-t/day plant, with an OTEC power cost of 24.5 mills/kWh. (All 1975-dollar costs in Refs. 38 and 40 have been raised 7 percent/year for 1978-dollar costs.) The larger APL plant concept took advantage of economies of scale and waste-heat recovery not considered by TRW.

The Alcoa chloride process   An alternative electrolytic process being developed by the Aluminum Company of America (Alcoa, U.S. Patent 3,725,222) requires less power; 9.9 kWh/kg is cited at present. The process involves reaction of alumina, carbon, and chlorine to produce $AlCl_3$, CO, and CO_2. The $AlCl_3$ gas is fed to an electrolysis cell where metallic alumina and free chlorine are formed. Insufficient information was available for cost-estimating purposes. However, it is understood that this process would be affected less by OTEC ship motions, which could make the chloride process much more attractive for on board OTEC implementation. The lower electricity requirement would make it more attractive for either on-board or onshore use.

OTEC Island Energy Concept for Onshore Production

Aluminum appears to be the most attractive initial product for the OTEC island energy complex concept, i.e., use of OTEC plants off Puerto Rico, Hawaii, or tropical islands to supply electricity by cable to onshore industrial complexes. This approach would avoid the Hall-Heroult-cell stabilization problem but would lead to higher OTEC electricity costs because of transmission system costs and losses and to the lower ΔT_0 at Puerto Rico or Hawaii compared with ATL-1 or PAC-2 levels (possibly also available at some islands in Micronesia). Nevertheless, its use at Puerto Rico or Hawaii is judged most likely for initial demonstration of OTEC aluminum production. Puerto Rico also offers tax advantages and low labor rates at present.[40]

The Aluminum Market and Potential OTEC Penetration

As of 1978, U.S. aluminum production capacity exceeded demand. However, a demand growth rate between 3.3 and 5.1 percent/year is anticipated for the 1985 to 2000 period.[40] Based on a 15-year takeover time starting in 1990 and leading to a market capture of 50 percent of new additions by the year 2000, GE TEMPO projected that 1.7 to 4.5 GW_e of OTEC power would be used for aluminum smelting by the year 2000 as shown in Fig. 19.39. This OTEC power level would call for five to twelve 370-MW_e plant-ships.

Additional Possible OTEC Products

There could be sufficient market for liquid hydrogen (LH_2) by the year 2000 to warrant dedication of some plant-ships to its production and development of LH_2 tankers. For example, if LH_2

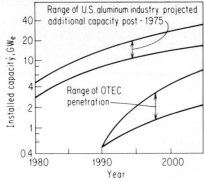

Fig. 19.39   Projected range of additions to U.S. aluminum production capacity and range of OTEC market penetration. (*From Ref. 40.*)

could be produced in the PAC-2 tropical grazing area for $6.80/GJ ($7.20/MBtu, from Ref. 35, corrected to 1978 dollars) and delivered over 1300 to 1800-km distances to Los Angeles at a transportation cost of approximately $0.60/GJ, it could be competitive then as a jet aircraft fuel.[40] It burns at very high efficiency in all kinds of engines with fewer pollution-control problems than hydrocarbon fuels. If hydrogen engines are operated fuel-lean to keep temperatures low and avoid formation of oxides of nitrogen, the only product of combustion is water. After entering a specialty market for, say, aircraft and space shuttle fuel, it could penetrate further for bus, truck, and eventually automobile fuel. The big problem would be expansion of cryogenic storage/service facilities. The literature on the "hydrogen economy" is substantial.

Many other products could be made using OTEC power. Some have been investigated, e.g., by DSS Engineers.[41] Table 19.14 lists some candidates, and Table 19.15 shows their assessment of a "sea chemicals" plant, with costs inflated from a 1975-dollar basis to 1978 dollars at 7 percent/year. They also considered an organic chemicals plant to produce vinyl chloride, ethylene oxide, polyethylene, and ammonia, for which they projected slightly less favorable economics (25 and 39 percent profit/sales ratios for OTEC power costs of 24.5 and 12 mills/kWh, respectively).

Several other possibilities have been evaluated, including on-board production of methanol, methane, gasoline, and specialty fuels, all of which appeared to be less competitive than ammonia or hydrogen.[42] Among systems considered for transporting thermal energy to shore for use as process heat, the HITEC molten salt [53 percent KNO_3, 40 percent $NaNO_3$, 7 percent $NaNO_2$, delivered at 517°C (900°F)] looked best but would cost approximately $12/GJ (1978 dollars) for 160-km transportation distance.[42]

Battery systems evaluated as "electrochemical bridges" to shore include the iron-titanium redox couple ($FeCl_3$ electrolyte transported to shore) and the lithium-water-air battery.[42] The former has a low electrolyte energy density (0.0133 kWh/kg) and costly battery materials which yield cost estimates in the 200 to 440 mill/kWh range for 70 percent electricity-to-electricity efficiency, an OTEC power cost of 22 mills/kWh, and transportation ranges of 320 to 800 km.[42]

TABLE 19.14   High-U.S.-Demand or High-Growth-Rate, Energy-Intensive Products[41]

Product	Demand, 1975, million t/year	Growth rate, %/year
Potash	4.5	5
Chlorine	10.7	6
Caustic	10.7	6.5
Magnesium	0.13	8
Vinyl chloride	1.8	4
Polyethylene (LD)	3.0	9
Ethylene oxide	1.7	6
Ammonia nitrate	7.9	7

TABLE 19.15   **Estimates for Outputs and Economics for OTEC Sea Chemicals Complex (Economics of Ref. 41 Revised to 1978 Dollars)**

Platform size, m	$140 \times 275 \times 45$ deep
OTEC power,* MW_e	274
Capacity factor	0.917
Products, 10^3 t/year	103
Na_2SO_4	
KCl	14
Mg	23
NaOH	420
Cl_2	440
Combined sales value, $M	211
Administrative sales cost (10%), $M	$- 21$
Net revenue, $M	190
Production costs, $M	$- 108*$ (78)†‡
Profit, $M	82 (112)
Chemical plant investment (CPI), § $M	326
Profit/CPI, %	25 (34)
Profit/sales, %	39 (53)

*OTEC power included at 24.5 mills/kWh.
†Numbers in parentheses are for the APL OTEC power cost of 12 mills/kWh.
‡Sea chemicals plant requires 2156 t/year of petroleum coke, 17,660 t/year chemicals, and 4.7×10^6 GJ/year of thermal energy.
§CPI includes working capital and interest during construction.

In the lithium system, $LiOH \cdot H_2O$ slurry (65 percent by weight) is transported to the OTEC plant for reduction to lithium metal. If the forecast specific energy consumption of 2.95 kWh/kg and electricity-to-electricity efficiency of 62 percent, with complete consumption of the lithium in the onshore discharge process, can be achieved, the process will be attractive for peaking-power application. Estimated costs of onshore electricity are in the 60 to 75 mill/kWh range for these assumptions and 22 mill/kWh OTEC power.[42] Because of the early R&D state of this process, it would come into contention later than ammonia or hydrogen systems.

Mechanical energy transport systems appear to be ruled out by transportation cost. For example, use of "superflywheels" installed in ships to transport energy to shore would lead to costs near $1/kWh.[43] However, use of superflywheels at onshore utility plants to store OTEC power received at night for peaking the next day could be very attractive as a nonpolluting means of fully utilizing available OTEC power. For a 5.5-h peaking duty per day, the capital cost computed as an addition to the OTEC plant cost would be approximately $300/kW_e of rated OTEC capacity.[43]

Environmental Impact

Like other solar energy systems, OTEC systems are expected to offer fewer environmental problems than nuclear-fission and fossil-fuel power systems. In contrast to other solar and nonsolar power plants, OTEC plants will use neither precious land area nor visible (from land) structures for power generation or for energy storage during noninsolation periods. An interesting possiblity of a favorable environmental impact is in the area of greater fish and shellfish yields in the oceans where OTEC plants are employed; indeed, mariculture operations might be associated with OTEC plants and might yield shellfish crops of values exceeding by several times the OTEC power value.[44] This possibility results from the fact that the cold water pumped by an OTEC plant is rich in nutrients.

Relative to the OTEC plant's ability to draw warm water from the mixed layer without drawing cold water from below the thermocline into the evaporators, Zener[45] has estimated that for a surface current of 0.03 m/s (0.06 kn), OTEC plants up to 500-MW_e size could operate without a problem. A larger current (or relative speed of a grazing plant) would permit an increase in plant size based on this criterion, but it seems unlikely that early plants will exceed 500 MW_e, because economies of scale probably will not be significant above that size.

With respect to ecological impact in the oceans, Bathen[46] used local weather data and surface water temperatures to calculate heat balances for the air-sea interface off Keahole Point, Hawaii, for average winter and summer conditions. Alteration in heat content of the mixed layer, rate of spreading of the discharged water, and potential biostimulation were analyzed for OTEC plants of 20-, 100-, and 240-MW_e (net) size. For the 240-MW_e plant, the cold-water discharge plume reaches initial stability 305 m from the point of discharge. The maximum heat load occurs in summer, with a northward advection of 0.89 kJ/m^2, equivalent to a 0.54°C surface temperature (T_s) decrease. The computed area of ΔT_s greater than the background diurnal fluctuations, at equilibrium, is 42 km^2. The maximum decrease in temperature of the mixed layer is 0.74°C, occurring during summer and slack water. Potential biostimulation is predicted over a 4.4-km^2 area, with nutrient concentrations increased by factors of 160 and 2.4 at distances ranging from the discharge point to 0.6 km downstream.

Many other investigations have been done (there are five more papers in this area in Ref. 6) or are under way. An important point is that the best way to assess environmental impacts of OTEC plants is to deploy several plants and establish an appropriate measurements program.

Financing and Institutional Considerations

The cost estimates made to date indicate that OTEC plants can be cost-competitive in several fields by the mid- to late-1980s if a priority program is established and government subsidies and/or other incentives are established for the first several plants. Like all solar technologies, OTEC power plants will be capital intensive. This means, among other things, that:

1. Great emphasis must be given to seeking overall system designs and implementations (e.g., site selections) which will minimize capital cost in $/$kW_e$ (net).

2. Since economies of scale will exist up to several hundred MW_e (net), the capital cost per plant will be large, $200 to 800 million. Thus, consortia of investors/owners will have to be established in most cases.

3. Obtaining classification as a vessel will be very important to take advantage of existing laws and incentives offered by the U.S. Maritime Administration (MarAd) for maritime vessels— e.g., mortgage guarantee and possible construction differential subsidy (perhaps requiring legislative change) under Titles XI and V, respectively, of the Merchant Marine Act of 1970.[35,38] Operating differential subsidies might be available to supporting product transporters (e.g., ammonia tankers) under Title VI.

4. The capital-intensive character is not a completely negative factor in competition with land-based systems using fossil fuel or fossil-fuel-generated power, as was discussed in connection with Fig. 19.36. Relative production costs for OTEC plants will become more favorable with each year of operation, and after the plant investment is recovered, the OTEC plant owner could undercut the onshore competition.

Various incentives have been proposed.[38,40,47] Since OTEC plant costs are expected to decline as numbers of plants with hundreds or thousands of identical components are built, government cost-sharing of several early plants could be undertaken on a progressively declining basis. Special federal income tax investment credits above the 1978 level (12 percent) could be very effective, as could more rapid depreciation allowances.[38,40] Take-or-pay contracts and leveraged-lease financing are possibilities.

A possible arrangement for outright ownership by a consortium of sponsors (corporations or partners) is illustrated in Fig. 19.40.

CONCLUDING REMARKS—OVERALL OTEC POTENTIAL

The engineering development of closed-Rankine-cycle OTEC plants is judged to be a straightforward task that can be accomplished as rapidly as funding permits. Component demonstrations, especially heat-exchanger and cold-water pipe tests, were conducted in 1978. These tests can be followed rapidly by pilot-plant construction beginning in 1980 and operation beginning in 1982 at the 10- to 40-MW_e (net) level in accordance with the present DOE plan.[47] A firm plan for follow-on commercialization steps is needed and should emerge from various investigations initiated in 1978.

The various potential market penetrations discussed earlier have been drawn together in Table 19.16 to illustrate that OTEC plants could provide approximately 2 to 8 $\times$ 10^9 GJ/year (or 2 to 8 Q_t/year, where 1 Q_t = 10^{15} Btu = 1.054 $\times$ 10^{18} J) by the year 2000, or 1.5 to 6 percent of total U.S. energy needs. The higher level might exceed the shipbuilding capacities of existing and new U.S. shipyards by the year 2000. It is emphasized, too, that all cost estimates are tentative until pilot plants of at least 10-MW_e size have been operated. But the potential is very great; the shipyard and merchant marine jobs that could be generated, and the favorable effects on the U.S. balance of payments that could result from reduced dependence on foreign oil and liquefied natural gas, could be of great national benefit.

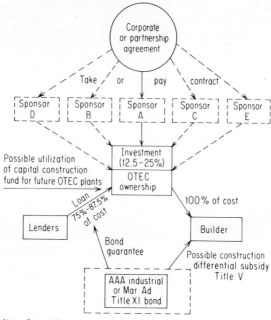

Note: Dotted lines to sponsors indicate that there may (but need not necessarily) be more than one sponsor entering into Take-or-pay contracts

Fig. 19.40 Outright ownership schematic: early OTEC/ammonia commercial development.

TABLE 19.16   Potential OTEC Capacity and Annual U.S. Fuel Energy Saving in the Year 2000

OTEC plant type/product	Product capacity, Mt/year	OTEC power, GW$_e$	Onshore plants avoided, GW$_e$	Fuel saved, 10^9 GJ/ year[a]
Tropical grazing plants				
Ammonia for fertilizer	14–33	13–30	· · · [b]	0.5–1.2[b]
$NH_3 \rightarrow H_2 \rightarrow$ electricity	10–30	9–35	5–22[c]	0.3–1.9
Tropical grazing or energy island plants				
Aluminum refining	1.1–2.9	1.7–4.5	2–5[c,d]	0.2–0.4
Smelting other metals (e.g., Ti, Cr, Ni, Mg)		1–4	1–4	0.1–0.3
Miscellaneous chemicals (e.g., Cl_2, NaOH, etc.)		1–5	1–5	0.1–0.4
Peaking power (e.g., Li/Li/OH)		1–8	1–8[e]	0.1–0.7
Electricity to utility grids		6–35	7–40[f]	0.6–3.3
Totals		32–121	16–84	1.9–8.2

[a] At 9490 kJ/kWh (9000Btu/kWh) heating rate for new fossil-fuel plants having 0.75 capacity factor compared with 0.95 for tropical grazing OTEC, 0.90 for offshore OTEC.

[b] Supplants 15–34 billion m³/year (540–1270 billion standard ft³/year at 900 Btu/standard ft³) of natural gas consumption; saving even greater for supplanting NH_3 made from coal.

[c] For these plants, base load capacity factor for onshore plants avoided is assumed to be the same as for OTEC plants.

[d] Supplanted onshore capacity exceeds the OTEC bus bar requirement because of improved plant integration and waste-heat recovery.[35]

[e] Assumes onshore peaking capacity replaced has same high *availability* factor as OTEC; however, OTEC ability to compete economically is a result of *utilization* factor (∼ 0.15) for peaking end, while OTEC end runs at 0.9–0.95.

[f] OTEC plants have 5% transmission loss; onshore avoidance/OTEC ratio = 0.95 (0.90/0.75) = 1.14 because OTEC seasonal output variation matches southeastern U.S. regional demand variation with season.[40]

REFERENCES

1. A. d'Arsonval, Utilization des forces naturelles. Avenir de l'electricité, *La Revue Scientifique*, Sept. 17, 1881, pp. 370–372.

2. Georges Claude, "Power from Tropical Seas," *Mechanical Engineering*, **52:** 1039–1044, December 1930.

3. Georges Louis Massart, The Tribulations of Trying to Harness Thermal Power, *MTS Journal*, **8,** no. 9, October-November 1974, pp. 18–21.

4. C. E. Brown and L. Wechsler, Engineering an Open Cycle Power Plant for Extracting Solar Energy from the Sea, Paper OTC 2254, *Offshore Technology Conf.*, Houston, May 5–8, 1975.

5. Earl J. Beck, Ocean Thermal Gradient Hydraulic Power Plant, *Science*, **189:** 293–294, July 25, 1975.

6. George E. Ioup (ed.), *Proc. Fourth Annual Conference on Ocean Thermal Energy Conversion*, OTEC, University of New Orleans, Mar. 22–24, 1977.

   (*a*) Earl J. Beck, Steam Lift Cycle at Very Low Mist Densities, pp. VIII-18–26.

   (*b*) Clarence Zener, The Foam OTEC System: A Proposed Alternative to the Closed Cycle OTEC System, pp. VIII-27–30.

   (*c*) S. L. Ridgway, The Mist Flow OTEC Plant, pp. VIII-37–41.

   (*d*) Gay Heit Lavi and Clarence Zener, A Comparison of Two Generic OTEC Systems and Missions, pp. II-11–25.

   (*e*) G. L. Dugger, H. L. Olsen, P. P. Pandolfini, and W. H. Avery, Experiments on and Design of Low-Cost Aluminum Heat Exchangers for OTEC Plant Ships, pp. VI-111–123.

   (*f*) C. M. Sabin and H. F. Poppendiek, Heat Transfer Enhancement for Evaporators, pp. VI-93–104.

   (*g*) A. M. Czikk, H. D. Fricke, and E. N. Ganic, Enhanced Performance of Heat Exchangers, pp. VI-71–92.

   (*h*) J. H. Anderson and J. H. Anderson, Jr., Compact Heat Exchangers for Sea Thermal Power Plants, pp. VI-3–14.

   (*i*) W. I. Kern, Increasing Heat Exchanger Efficiency Through Continuous Mechanical Tube Maintenance, pp. VII-64–65.

   (*j*) E. D. Nubel, Automatic Tube Cleaning System, pp. VII-61–63.

   (*k*) R. A. Barr and Pin Yu Chang, Some Factors Affecting the Selection of OTEC Plant Platform/Cold Water Pipe Designs, pp. V-11–22.

   (*l*) B. M. Winer and J. Nicol, Electrical Energy Transmission for OTEC Power Plants, pp. III-26–33.

7. Clarence Zener and John Fetkovich, Foam Solar Sea Power Plant, *Science*, **189:** 294–295, July 25, 1975.

8. J. Hilbert Anderson and James H. Anderson, Jr., Thermal Power from Sea Water, *Mechanical Engineering*, April 1966, pp. 41–46.

9. A. Lavi (ed), *Solar Sea Power Plant Conf. and Workshop*, Carnegie-Mellon University, Pittsburgh, June 27–18, 1973.

10. W. E. Heronemus, A Gulf Stream Based Ocean Thermal Differences Power Plant, *Conf. Proc.*, *Energy from the Oceans, Fact or Fantasy?*, Jerome Kohl (ed.), North Carolina State University Rep. 76-1, UNC-SG-76-04, Raleigh, NC, January 1976, pp. 90–105.

11. G. L. Dugger (ed.), *Proc., Third Workshop on Ocean Thermal Energy Conversion*, Houston, May 8–10, 1975, APL/JHU Rep. SR 75-2, 1975:

    (*a*) L. C. Trimble et al., Ocean Thermal Energy Conversion System Study Report. pp. 3–20. (See also Ocean Thermal Energy Conversion (OTEC) Power Plant Technical and Economic Feasibility, Rep. NSF/RANN/SE/GI-C937/FR/75/1, LMSC-DO56566, vol. I, Lockheed Missiles and Space Co., Inc., Sunnyvale, CA, Apr. 12, 1975.)

    (*b*) R. H. Douglass, Ocean Thermal Energy Conversion: An Engineering Evaluation, pp. 22–36. (See also Ocean Thermal Energy Conversion, Final Rep., Contract NSF-C 958, vol. 1, TRW Systems Group, Redondo Beach, CA, June 1975.)

    (*c*) W. P. Goss, W. E. Heronemus, P. A. Mangarella, and J. G. McGowan, Summary of University of Massachusetts Research on Gulf Stream Based Ocean Thermal Power Plants, pp. 51–63.

    (*d*) W. B. Suratt, G. K. Hart, and E. N. Sieder, Plastic Heat Exchangers for Ocean Thermal Energy Conversion, pp. 138–142.

12. H. P. Harrenstien (ed.), *Workshop Proc., Second Ocean Thermal Energy Conversion Workshop*, Sep. 26–28, 1974, Washington, DC, Clean Energy Research Institute, University of Miami, Coral Gables, FL.

13. G. L. Dugger, H. L. Olsen, W. B. Shippen, E. J. Francis, and W. H. Avery, Floating Ocean Thermal Power Plants and Potential Products, *Journal of Hydronautics*, **9,** no. 4, October 1975, pp. 129–141.

14. G. L. Dugger, E. J. Francis, and W. H. Avery, Technical and Economic Feasibility of Ocean Thermal Energy Conversion, *Solar Energy*, 20:259–274, 1978.

15. OTEC Thermal Resource Report for Western Gulf of Mexico, Area G29, Rept. TID 27949; for Hawaii, Area 5H, Rept. TID 27950; for Florida East Coast, Area 6A, Rept. TID 27951; for Puerto Rico, Area 4P, Rept. TID 27953; Ocean Data Systems, Inc., Monterey, CA, October 1977. Data for ATL-1 and PAC-2 sites deduced by APL/JHU from surface temperature contour plots provided by National Oceanographic and Atmospheric Administration, 1977.

16. A. E. Bergles, Survey and Evaluation of Techniques to Augment Convective Heat and Mass Transfer, "Progress in Heat and Mass Transfer," vol. I, pp. 331–424, Pergamon Press, 1969.

17. D. Q. Kern, "Process Heat Transfer," McGraw-Hill Book Company, New York, 1950.
18. "Standards of Heat Exchanger Manufacturer's Association," 5th ed., TEMA, Inc., New York, 1968.
19. "Ocean Thermal Energy Conversion (OTEC) Power System Development Utilizing Advanced, High Performance Heat Transfer Techniques," vol. I: Conceptual Design Report, prepared by TRW Systems and Energy for the U.S. Department of Energy under Contract EG-77-C-03-1570, May 12, 1978.
20. W. M. Rohsenow and H. Choi, "Heat, Mass, and Momentum Transfer," p. 58, Prentice-Hall, Englewood Cliffs, NJ, 1961.
21. P. P. Pandolfini, J. L. Keirsey, and J. L. Rice, Tests of the JHU/APL Heat Exchanger Concept, *Fifth Annual OTEC Conf.*, Miami Beach, Feb. 20–22, 1978.
22. N. F. Sather, L. G. Lewis, J. J. Lorenz, and D. Yung, "Performance Tests of 1 MWt OTEC Heat Exchangers," *Fifth Annual OTEC Conf.*, Miami Beach, Feb. 20–23, 1978.
23. Robert H. Gray (ed.), *OTEC Biofouling and Corrosion Symp.*, Oct. 9–13, 1977, Battelle Pacific Northwest Laboratories, Richland, WA, 1978:
    (a) J. G. Fetkovich, G. N. Grannemann, L. M. Mahalingam, and D. L. Meier, Degradation of Heat Transfer Rates Due to Biofouling and Corrosion at Keahole Point, Hawaii.
    (b) P. P. Pandolfini, W. H. Avery, J. Jones, and L. R. Berger, Effect of Biofouling and Cleaning on the External Heat Transfer on Large-Diameter Tubes.
24. "ASME Boiler and Pressure Vessel Code," ANSI/ASME BPV, VII, Division 1, The American Society of Mechanical Engineers, New York, 1977.
25. L. A. Rosales, T. C. Dvorak, M. M. Kwan, and M. P. Bianchi, Materials Selection for Ocean Thermal Energy Conversion Heat Exchangers, *Fifth Annual OTEC Conf.*, Miami Beach, Feb. 20–23, 1978.
26. F. L. LaQue, Qualification of Aluminum for OTEC Heat Exchanger Tubes, rep. to DOE Div. of Solar Technology, August 1978.
27. "Saline Water Conversion Engineering Data Book," 2d ed., prepared by M. W. Kellogg Co. for the Office of Saline Water, U.S. Dept. of the Interior, 1971.
28. Tables of Coefficients for A.T.T.C. Model-Ship Correlation and Kinematic Viscosity and Density of Fresh and Salt Water, Technical and Research Bulletin 1–25, Society of Naval Architects and Marine Engineers, New York, 1974.
29. J. Edward Snyder III, 1-MW$_e$ Heat Exchangers for OTEC, Status Report, February 1978, *Fifth Annual OTEC Conf.*, Miami Beach, Feb. 20–23, 1978.
30. OTEC Platform Configuration and Integration Study—Final Report, Lockheed Missile and Space Co. DOE Contract EG-77-C-01-4063, April 1978.
31. Arthur R. Anderson, A. 65000-ton Prestressed Concrete Floating Facility for Offshore Storage of LPG, *Marine Technology*, 15, no. 1, 1978, pp. 14–26.
32. Prestressed Concrete Hull Study for OTEC, Report by T. Y. Lin International to Lockheed Missile and Space Co., November 1977.
33. Calculated Seaway Response of Six Candidate OTEC Platforms, Tech. Rept. 7812-2, Hydronautics, Inc., November 1977.
34. Thomas E. Little, Cold Water Transport, Cold Water Pipe, and Deep Water Mooring Line Analysis—A Parametric Approach, *Proc. Fourth Annual Conf. on Ocean Thermal Energy Conversion*, George Ioup (ed.), University of New Orleans, New Orleans, July 1977, pp. V-40–48.
35. W. H. Avery, R. W. Blevins, G. L. Dugger, and E. J. Francis, Maritime and Construction Aspects of Ocean Thermal Energy Conversion (OTEC) Plant Ships, APL/JHU Rept. SR-76-1B, April 1976, available from National Technical Information Service, NTIS PB-257444/LL, Springfield, VA 22161.
36. A. J. Konopka, A. Talib, B. Yudow, and N. Biederman, An Optimization Study of OTEC Delivery Systems Based on Chemical-Energy Carriers, ERDA/NSF/00033-76/T1, Institute of Gas Technology, Chicago, December 1976.
37. T. Nejat Veziroglu (ed.), *Conf. Proc., 1st World Hydrogen Energy Conference*, Clean Energy Research Institute, University of Miami, Coral Gables, FL, March 1976:
    (a) J. B. Laskin and R. D. Feldwick, Recent Developments of Large Electrolytic Hydrogen Generators, vol. II, pp. 6B-3–17.
    (b) L. J. Nuttall, Conceptual Design of Large Scale Water Electrolysis Plant Using the Solid Polymer Electrolyte Technology, vol. II, p. 6B-21.
38. E. J. Francis, Investment in Commercial Development of Ocean Thermal Energy Conversion (OTEC) Plant-Ships, prepared by APL/JHU for Maritime Administration, U.S. Dept. of Commerce, December 1977.
39. J. E. Snyder III and R. H. Douglass, Jr., "Prospects for OTEC Energy Utilization," Paper T5-3, Spring Meeting/STAR Symp., San Francisco, May 25–27, 1977, Society of Naval Architects and Marine Engineers.
40. Ocean Thermal Energy Conversion Mission Analysis Study, Phase II, prepared by General Electric Co.–TEMPO for DOE Division of Solar Technology, Washington, DC, March 1978.
41. C. D. Hornburg et al., Preliminary Research on Ocean Energy Industrial Complexes, *Proc. of Joint ISES/SESC Conf., Sharing the Sun!*, 5:412–435, Winnipeg, 1976.
42. A. Konopka et al., Alternative Energy Transmission Systems from OTEC Plants, Rept. DSE/2426-20, Institute of Gas Technology, Chicago, September 1977.
43. D. W. Rabenhorst and G. L. Dugger, Superflywheel for Storing Energy from OTEC Plants, pp. 116–120 in Ref. 11.

44. S. Laurence and P. A. Roels, Potential Mariculture Yield of Sea Thermal Power Plants, pp. III-21–25 in Ref. 6.
45. C. Zener, Site Limitations on Solar Sea Power Plants, *J. Hydronautics*, **11**: no. 2, (1977).
46. K. H. Bathen, A Further Evaluation of the Oceanographic Conditions Found off Keahole Point, Hawaii, and the Environmental Impact of Nearshore OTEC Plants, pp. IV-79–99 of Ref. 6.
47. J. M. Niles and B. J. Washom, Economic Incentives for Commercialization of OTEC, pp. III-3–16 of Ref. 6.

Chapter **20**

Solar Thermal-Electric Power Systems

RICHARD CAPUTO
Jet Propulsion Laboratory
Pasadena, CA

INTRODUCTION

Electric Energy Use

The use of electricity in the United States has gradually increased over this century and now consumes nearly 29 percent of the total primary energy used. Current projections further increase this usage rate to about 44 percent by the year 2000.[1] The original use of electricity for lighting and mechanical drive has been enlarged to literally every sector of the economy, including resistance heating of water and buildings. These latter uses, which appear irrational on thermodynamic grounds, are increasing in spite of what appears to be unfavorable life-cycle economics. Thus U.S. society is increasingly attracted to the use of electricity for a host of reasons that are persuasive yet difficult to understand without a detailed knowledge of causality of energy use decision-making within each market sector.

Conventional Sources of Electricity

Most electricity (45 percent) is currently provided by using coal as an energy source in steam turbine-generator plants, and this source is estimated to continue to provide the bulk (53 percent) of use by the year 2000.[1] Until recently, light-water nuclear reactors (LWR) were considered by government and utilities to assume the supportive electric generation role from oil and gas over the rest of the century. Estimates of 1000 nuclear plants by the year 2000 have steadily been reduced, and current estimates are from 140 to 220.[1] Even these estimates may prove to be ambitious in light of what amounts to a marketplace moratorium on nuclear plants, which occurred after the 1973 oil-supply crisis. The reluctance to use nuclear energy is based on a number of factors related to lack of public acceptance because of perceived risk of accidents, long-term wastes, and fuel-cycle relationships to weapons proliferation. The utility industry is currently not choosing nuclear plants in spite of the preponderance of paper studies indicating that LWRs are the most cost-effective choice when compared with any other commercially available system.[2] Thus societal factors which value more than just commercial economics are for the first time playing a dominant role in utility decision-making. This is a major historical departure in an industry that is charged with supplying the least expensive form of energy to generate electricity.

Although coal plants or oil plants are being chosen in preference to nuclear, these choices have many adverse impacts. The domestic and global depletion of liquid fossil fuels has created a host of fiscal and political difficulties for importing countries such as the United States. These problems will grow as depletion continues. Coal reserves are large in the United States but coal is a rate-of-extraction-limited fuel. Recent estimates[1] place this rate at about 40 quads (1.6 billion tons of bituminous coal equivalent) per year, which is nearly 3 times the current extraction and use rate. This judgment is based on several factors relating to adverse social impacts of coal:

- Health and esthetic (visual) degradation in air basins because of air pollution caused by coal combustion.
- Occupational health impacts in mining coal and the difficulty in finding workers for deep-mining coal as the strippable coal is depleted toward the year 2000.
- Social resistance to Western land degradation because of extensive strip mining.
- Social resistance to coal transport, whether in the form of unit trains disrupting numerous small communities or the construction of electric transmission lines for mine-mouth coal plant or pipe slurry lines.
- Difficulty in generating sufficient railroad car and track capacity to move increasing amounts of coal.
- Limited Western water for uses such as coal mining.
- The perception that CO_2 and waste-heat generation will affect long-term global weather in adverse ways.

Thus, although the current situation is limiting nuclear plant use, oil and coal electric-generation plants will soon be faced with limitations to their use.

Background of Research Support

U.S. research on solar thermal systems for electricity generation started in 1972 with paper studies directed by the National Science Foundation (NSF). Strong support was maintained under the Energy Research and Development Administration (ERDA) and Department of Energy (DOE). Until fiscal 1978, the majority of research dollars went to a central-receiver (power-tower) approach which uses two-axis-tracking heliostats and conventional steam Rankine turbine-generator conversion. A small total-energy system was carried along in the research program using lower-temperature, one-axis-tracking trough collectors with organic Rankine heat-engine conversion.

In a parallel program, the Electric Power Research Institute (EPRI) conducted a solar thermal research program using gas (Brayton cycle) rather than steam turbines. However, the basic approach is similar to the DOE program—to design and develop solar thermal equipment that can be made compatible with existing or somewhat modified energy conversion equipment.

In fiscal 1978, a distributed-collector solar thermal-electric program was initiated which gave serious consideration to using two-axis-tracking parabolic dishes for electricity generation. Thus a research base exists for both distributed trough and dish systems as well as the central-receiver approach.

Overview of Chapter

The approach to this chapter will be similar to the comparative assessment study of several solar and conventional power systems contained in Ref. 3. This chapter will present several types of distributed- and central-receiver solar thermal-electric systems and perform a comparative evaluation among them and, to a limited extent, with conventional power plants. The primary factor which will be considered is a projection of *utility economics*. But consideration will be given several other factors such as technical development difficulty, utility interface, central vs. dispersed use, and utility vs. community or private ownership. A brief review will also be made of the major factors affecting social acceptance in comparison with advanced fossil, nuclear, and even solar orbital power systems. Due to the uncertainty of costs and, to some extent, the performance of these various solar systems, the results will be presented in two forms: the first will be based on a "best judgment" estimate of all economic and technical performance factors, and the second form will show parametric data for a range of values.

SYSTEM DESCRIPTION

Generic Systems

There are basically two types of solar thermal-electric power systems: (1) those that optically transmit radiant solar energy from a reflector (heliostat) to a centrally located receiver on a tall tower (central receiver) for conversion to heat and then to electricity; (2) those that convert radiant solar energy to heat at each local reflector site. Since the combination of reflector and receiver is called a collector, this latter generic system is called a distributed-collector system. The major variations on the distributed-collector approach are to have either a large, single, central energy conversion subsystem, or a smaller energy conversion subsystem at each collector. Thus a distributed collector can have a central or distributed engine with thermal energy transported via insulated pipes to the central engine for the former or can have electricity transported to a central point for the latter. These generic system types are summarized in Table 20.1.

Comparison of Generic Systems

Power plant site   Although the central-receiver approach is normally associated with a large plant (50 to 300 MW$_e$), it is technically possible to design a 100-kW$_e$ plant using an approach developed by Francia[4] if coupled to a small heat engine; large 1-GW$_e$ (1000-MW$_e$) plants using multiple (200-MW$_e$) receivers with outputs piped to a single large engine-generator have also been designed.

The distributed-collector approach is normally associated with a small plant (100 to 500 kW$_e$), but an extremely large plant could be built by using an extensive thermal transport system driving a large central engine.[5] Also a host of distributed engines can be ganged to form as large a plant as desired. By this modular approach, a distributed-collector system can be made very large. At the other extreme, a small plant size of 10 kW$_e$ is quite reasonable. Central-receiver systems should not necessarily be associated with only large plant sizes, nor should distributed-collector systems be associated with only small plants. The size is not limited for either the central or distributed systems by technical considerations alone.

Power plant application   The choice of the type and size of a solar thermal-electric system will depend on factors such as specific integration problems of a specific application (remote utility, on-site, community level, industrial, etc.) and economic competitiveness vs. type and size. This latter factor will be addressed in later sections. However, the question of integration into specific applications is discussed here in a preliminary fashion.

The central receiver has the unique characteristic of a tall tower, which could be several hundred meters high for the 100-MW$_e$ size. When used in large sizes, this system can be made compatible with conventional large steam- or gas-turbine technology developed in the utility industry. This is the strongest factor that directed the research by the federal government (steam turbine) and EPRI (gas turbine).

The distributed systems in general have a low physical profile, and the one-axis-tracking collector designs are easier to integrate into the built environment. The two-axis-tracking heliostat (central receiver) or dish collectors (distributed collector) are more difficult to integrate into structures, and this would be a negative factor in certain applications. The two-axis-tracking, distributed collectors which have a high temperature capability can be the most efficient system if an engine is used that takes advantage of this capability. Thus, this type of system would use the least amount of land—an important factor in certain applications. These considerations are summarized in Table 20.2.

Specific Systems

Introduction   Several different systems are being pursued for both central and distributed approaches, and many different collector designs are being explored. Since few of these systems have achieved even the pilot-plant stage of development, and commercial cost estimates are poor or nonexistent, it is impossible to predict which systems and components will reach

TABLE 20.1   Generic Systems

System type	Unique system characteristic
Central receiver	Optical transport from reflector to central-receiver/tower
Distributed collector	Receiver collocated with reflector
Central engine	Thermal transport to central energy conversion
Distributed engine	Engine-generator collocated at collector and electricity transport to central point for dispatch

TABLE 20.2   System Application Characteristics*

System type	Unique application characteristics*
Central receiver	Tall tower, or if short tower used with north-side field, there is a potential low-level glare and fire hazard
	Compatibility with utility conventional steam- and gas-turbine energy conversion system sizes
Distributed collector	Low profile
One-axis-tracking collectors	Easier to integrate into built structures
Two-axis-tracking collectors	Potentially more efficient if higher temperature capability is used in engine; would use less land area

*In specific applications, these system characteristics can have a direct influence on siting choices. Examples given in Table 20.3 identify esthetic, safety, land, and other factors.

TABLE 20.3   Application Influence on Type or Size of Plants

Type of application	Example of important factor*
Remote utility site	Size may be limited by environmental and land-use considerations. For example, above a certain size of contiguous land use, concerns for flora and fauna species impacts may dominate.
	Esthetic impact of tower height above a certain level or use of towers at all in a given recreational or scenic area may limit tower use, or plant size.
On-site residential, commercial, or industrial user	If a minimum of extra land is available, type and size may be dictated by collectors which can be physically integrated into a roof or other structure. This would tend to favor one-axis-tracking distributed collectors over two-axis-tracking collectors or heliostats.
	Choice of working fluid or engine type may be dictated by health and safety codes which may be quite different for different users in these market sectors.
	Central receiver fields may have additional problems of esthetics (glare) and safety hazards (heating of structures near the receiver) when located near the built environment.
Community-size plant owned by municipality, utility or neighborhood co-op	Different attitudes may cause strong preference for different esthetics of distributed vs. central systems.
	Different ability to perform maintenance may cause preference for one system vs. another.
	Limited or expensive land area will favor the most efficient system.

*Additional factors include cooling water limitations and the requirement to use dry cooling techniques; uses for waste heat in applications near the end user, favoring the system that can provide waste heat efficiently; and use in utility repowering, which dictates that the system produce steam at conditions compatible with existing power plants. These considerations along with economics vs. size and plant type will dominate the decision of which solar thermal-electric plant to use.

commercial maturity in the remainder of the 20th Century. Therefore, a representative selection of systems will be presented, and a best-judgment estimate of system performance and "eventual" commercial cost will be made. The uncertain parameters will be evaluated parametrically so that these data will be useful as future events remove some of the unknowns. The representative systems are shown in Table 20.4.

TABLE 20.4   Generic Types of Systems

Collector type	Energy transport	Storage	Energy conversion
Distributed systems Fixed V-trough	Organic fluid	*	Organic Rankine
One-axis Parabolic trough	Water or toluene	Sensible thermal	Central steam Rankine
Variable slats	Steam	Sensible thermal	Central steam Rankine
Two-axis Parabolic dish	Steam	Sensible thermal	Central steam Rankine
Parabolic dish	Electric	Advanced battery	Small heat engine mounted on dish
Central receiver Heliostat (two-axis)	Optical	Sensible thermal	Central steam Rankine

*Fixed system considered without storage.

The major factors which determine a system are grouped into two areas:
1. Type of subsystems
   (a) Collector
   (b) Storage
   (c) Energy conversion
2. Mode of operation
   (a) Stand-alone
   (b) Fuel saver
   (c) Repowering
   (d) Hybrid

Table 20.4 identifies the major subsystems used in the selected systems. Collectors range from fixed, to one-axis and two-axis tracking with distributed or central receivers (see Chaps. 7, 8, and 9 for a detailed treatment of these types of collectors). Energy transport schemes range from optical to pumped fluids (organic, steam, etc.) to electricity. Chemical transport (NH_3, SO_3, etc.) is also possible and eventually may be developed to accommodate long-term storage in chemical species.[5,6] The storage subsystem is either sensible-thermal, or electricity in advanced batteries, or pumped hydro for a large central application (see Chap. 6 for a more complete description). Energy conversion subsystems range from Rankine (organic or steam) to small heat engines such as a Brayton (gas turbine) or Stirling (reciprocating with external heat addition). (See Chap. 22 for a more thorough treatment of heat engines.) These combinations of subsystems are felt to be reasonable for achieving good system performance.[5,7-9]

Modes of operation

Stand-Alone. For the most part, a solar thermal plant will have storage which interacts with the solar energy collection subsystem so that the plant electrical output is at or near rated power. This is called the stand-alone solar plant and most closely represents the manner in which conventional plants operate. A schematic layout is shown in Fig. 20.1 for such a plant based on the central-receiver concept.

If the rated power of the energy conversion equipment is P_r, then the heliostat field-receiver combination is sized so that it can collect more heat than the energy conversion subsystem can use at P_r. The ratio of the energy collection capacity (in units of electric power equivalent) to the rated power is called the *solar multiplier*. The extra thermal energy collected is placed in storage and retrieved when needed. The power retrieved from storage may be either at the plant rated power or at some fraction of it. The central receiver plant shown in Fig. 20.1 takes 70 percent of rated power from storage since only part of the turbine is used.

The amount of storage capacity and the corresponding solar multiplier (size of heliostat field and receiver) can be increased while the same energy conversion rated power is maintained. This will increase the annual energy delivered and the capacity factor, which in turn will have a direct impact on plant capital cost, but will not affect the levelized energy cost over a wide range of capacity factor. The stand-alone plant is also considered to have some capacity displacement capability and a conservative numerical evaluation of this is developed shortly.

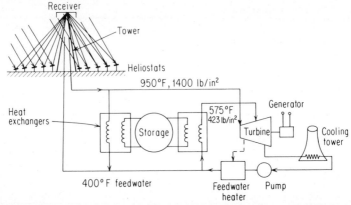

Fig. 20.1 Central receiver solar thermal-electric power plant, 10 to 200 MW$_e$.

Fuel Saver. If most or all of the storage system is removed from a stand-alone plant, it becomes a fuel saver. The assumption is that there will be no capacity displacement and this type of design can be compared economically only to the value of fuel displaced. If no storage exists in the utility grid, very limited capacity credit can be taken by a fuel saver. However, if there is storage elsewhere in the grid which is charged at off-peak periods and used later to reduce peaks, the synergistic interaction of a fuel-saver solar plant and this type of grid storage can give the fuel-saver type of solar plant full capacity displacement.

Repowering. The mode of operation is called repowering when both the storage and energy evolution subsystems are removed. The solar collection equipment (collectors, energy transport, heliostats, receiver, tower, etc.) generates steam or heated air which is used directly in existing fossil steam- or gas-turbine plants. This directly reduces the fuel consumption of the existing plant and the economic comparison would be with the levelized, inflating fuel cost over the life of solar equipment (see Chap. 28).

Recent studies[10,11] have estimated the number of possible sites for repowering gas and steam turbines in Southwestern United States. Table 20.5 shows the results from Ref. 10 where about

TABLE 20.5 Solar Retrofit to Steam and Gas Power Plants in the U.S. Southwest

	Installed capacity, GW$_e$	
	Gas	Steam
Available plants		
Catalog	7.2	41
Utility survey	1.9	18.4
Repowerable	1.1	
Land available	0.44	10.5
< 850 m distant from plant	0.37	7.0
at least 50% solar possible	0.14	4.6
Number of units*		
Texas		91
California	16	61
Arizona	7	21
Louisiana	1	25
New Mexico	1	22
Oklahoma	3	10
Colorado		9
Nevada		9
Utah		1
Other		2

*Based on totals of 1.1 GW$_e$ for gas and 18.4 GW$_e$ for steam.

one-third the capacity of the utilities that responded is suitable for repowering. Only one-fourth the capacity has at least the possibility of 50 percent minimum solar. However, utilities representing three-fifths the capacity in the Southwest did not respond to inquiries regarding interest in solar repowering opportunities.

Most of the repowering opportunities are with steam plants, and from a potential 41-GW$_e$ installed capacity in the Southwest, utilities with 18.4 GW$_e$ responded with some interest in repowering. Of this, 10.5 GW$_e$ of plants had suitable land available but only 7 GW$_e$ was within 850 m (0.5 mi) of the existing plant. Only 4.6 GW$_e$ could be powered by at least 50 percent solar. Most of these units were in Texas and California. Another study[6] raises the 4.6 GW$_e$ to 18 GW$_e$ by considering land availability within 2500 m (1.5 mi).

The maximum opportunity for repowering gas turbines is 7.2 GW$_e$ in the Southwest. Of this, utilities with 1.9 GW$_e$ of gas-turbine capacity responded with some interest in retrofitting. Only 1.1 GW$_e$ was repowerable and 0.44 GW$_e$ had land available within 850 m (0.5 mi). Only 0.14 GW$_e$ could be at least 50 percent solar.

Thus, of 10 percent of the national capacity potentially available for repowering in the Southwest, only 10 percent of this is currently of interest with at least a 50 percent solar fraction within 850 m.[10] This percentage more than triples when land within 2500 m is considered.[11] An evaluation of the entire national utility grid is required before the full potential of repowering is understood.

Hybrid. The fourth mode of operation is a hybrid combination of fossil and solar energy. The fossil plant can be based on a gas turbine, a steam turbine, or even a combined cycle with gas and steam turbines. Although similar to a repowering plant in many respects, the hybrid plant concept is usually limited to new plants.

With the hybrid solar-steam Rankine plant, there are several possibilities: (1) the solar equipment can act as a preheater at lower temperature in series with the fossil boiler which provides boiling and superheat, (2) the solar equipment can provide boiling as well as preheat with fossil fuel superheating, (3) the solar collectors can generate superheated steam and the fossil boiler be used in parallel or instead of the solar-heated steam if conditions warrant.

There is also the possibility of using solar feedwater heating at much lower temperatures where the solar equipment would operate at greater efficiency. The feedwater heaters could be either tracking or fixed collectors, or solar ponds for the first-stage heating. These four approaches are shown schematically in Fig. 20.2.

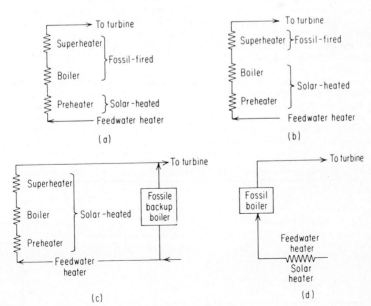

Fig. 20.2 Hybrid solar plants: (*a*) solar preheater, (*b*) solar boiler, (*c*) solar superheat, (*d*) solar feedwater heater.

Finally, a hybrid solar plant may combine the stand-alone solar plant (including storage) with a backup fossil boiler. This allows the solar equipment to provide a larger annual capacity factor yet have fossil backup to increase the plant reliability level to that of conventional plants.

The solar feedwater-heating approach shown in Fig. 20.2d is attractive since a significant amount of energy (25 percent) is used at low temperatures (30 to 250°C). Thus, simpler and lower-cost or more efficient solar collectors could be used. Table 20.6 shows the percent energy content of the various stages of energy input to the steam Rankine cycle.[9]

Steam turbines used for utilities have multiple bleed points where steam is removed for feedwater heating. Most utilities believe that feedwater heating is "free" and do not feel the solar feedwater heating adds anything to the cycle. However, by not bleeding steam off and sending it through the turbine, the turbine power can be increased. As can be seen in Table 20.6, about 25 percent more energy would be available for power production. However, most existing turbines are overdesigned by only 5 percent, while older turbines (before 1961) have a greater overdesign (~ 15 percent). Some manufacturers have retrofits available to provide for 10 percent overdesign. However, for a repowering application, there is a reluctance on the part of utility people to run older turbines at overdesign conditions. Thus, for preheating, the combination of utility industry reluctance to change turbine operating conditions, inability to use all of the 25 percent energy possible because of turbine overdesign limits, and supplying what is now "free" energy tends to cancel the opportunity of using simpler solar collectors to provide the energy used for feedwater heating in existing steam plants.[12]

Of the three other possibilities in Fig. 20.2, the solar superheater approach (c) is best with the fossil boiler in parallel.[12] However, ceramic-brick-lined boilers are used in many existing fossil steam plants, and they have limited ability to take the solar transients required in this mode of operation. To avoid cracking the ceramic boiler must be run in hot standby at from one-third to one-half full power rating. The obvious fuel inefficiencies of standing by under these conditions is a very serious drawback.[12] It is possible to install newer once-through auxiliary boilers which do not use ceramics and would not have this standby limitation.

When a gas-turbine hybrid is considered, the best arrangement is to use the solar equipment as a preheater in series with the fossil combustor.[12] The solar equipment will take compressed air (for an open-cycle turbine) and deliver energy at the highest temperature possible within the material limitation in the receiver. Metallic receivers under pressure can be used to 760°C (1400°F) or 815°C (1500°F). Early solar gas-turbine systems will be limited to these temperatures and have low efficiency as a result.

If a combustor is developed to add fossil heat to raise the temperature of 800°C solar-heated air to current turbine technology levels [1000°C (1800°F) to 1220°C (2200°F)], a more efficient gas-turbine cycle will result. If solar collectors heat air to 700°C (1400°F) and a fossil combustor further heats the air to 1000°C (1800°F) in a simple gas turbine, the solar equipment could provide 60 percent of cycle energy and save 6500 Btu/kWh.[12] A ceramic receiver for the solar equipment will be able to deliver the 1000°C temperature directly and a fossil combustor could then be used in parallel for hybrid operation.

Existing gas turbines are used for peaking power at low annual capacity factors (~ 0.10). The solar equipment could contribute an annual load factor of 0.15 to 0.35, depending on location and the temperature delivered by the solar collectors. A fossil combustor used in series could work with the solar equipment as discussed but still provide peaking power as required. The overall effect could be a gas turbine with a 0.20 to 0.35 annual capacity factor which is still a peaking plant but which also contributes additional solar energy similar to a fuel-saver plant.

TABLE 20.6   Energy Consumption Percent at Various Stages of Steam Rankine Cycles*

		Part of total energy, percent	
Stage	Temperature, °C	8.3 MPa (1200 lb/in²)	16.5MPa (2400 lb/in²)
Feedwater heating	30–250	25	25
Preheating	250–370	8	19
Boiling	370	36	18
Superheating	370–540	20	24
Reheat	540	12	12

*From Ref. 9.

Figure 20.3 illustrates the performance differences among three modes of operation other than feedwater preheat. The upper dotted line represents the thermal energy collected by the field of collectors (heliostat-receiver if the collector is a central-receiver design). If this energy is used to supplement the steam in a conventional plant it is in the repowering mode of operation. If this heat is used to drive energy conversion equipment built with the solar plant and little or no storage is used, this is the fuel-saver mode of operation, illustrated by the middle (solid) line in Fig. 20.3. Power is available only over the sunlight hours, and the equipment is rated at peak power P_p. However, if storage is used and the same collectors are used with conversion equipment rated at P_r, the excess energy collected is stored during the sunlight hours. The stored energy is used when needed as shown in the lowest line of Fig. 20.3. Hybrid operation could generate rated power very much like a conventional plant and be indifferent to the insolation, or be combined with a fuel-saver or even a stand-alone solar plant.

The early market in central utility applications will probably be repowering.[13] The economic tradeoff compares the cost of solar heliostats and receiver/tower (or collectors) with that of fossil fuel displaced. This approach will be encouraged if utilities are exempted from conversion to coal if solar repowering is used.

The next most attractive market is the fuel-saver plant introduced into the grid along with storage capacity such as pumped hydro. If the storage is about 4 h at the solar rated capacity, charged off-peak with baseload capacity, this synergism will give the solar plant approximately full-capacity displacement.[14] However, as fuel savers penetrate the grid to a point where baseload capacity is about to be displaced, then the stand-alone solar plant with larger amounts of on-site storage will be used increasingly. The use of on-site fossil backup (a hybrid plant) could be dictated as a result of detailed study considering: (1) distributed storage in the load center as part of dispersed solar equipment, (2) central storage, (3) storage at the solar central plants, (4) effects of weather throughout the utility service area, (5) effects of grid interties, etc.

Central-receiver system   The central-receiver system uses stream Rankine energy conversion with dry coooling-heat rejection and sensible-heat storage in a mixture of a heat-transfer salt (or oil) such as Caloria™-plus-rock as illustrated in Fig. 20.1.[7] The field of two-axis-tracking heliostats focuses insolation on a tower-mounted open receiver where steam is generated in a once-through boiler using forced convection. The steam at 510°C (950°F) and 10 mPa (1400 lb/in²) is piped down the tower to the high-pressure turbine inlet. After expansion through the turbine and heat-rejection subsystem, it is pumped through a series of feedwater heaters and back to the tower. The turbine for use in 10 to 100-MW$_e$ systems will probably be a two-port extraction industrial turbine when thermal storage is used. This design is suggested to accommodate the use of steam at lower temperature (300°C, 575°F) and pressure (29 MPa, 4231 lb/in²) from storage. Thermodynamic degradation always occurs when sensible heat is stored using two different fluids as suggested in this design. If the plant size is greater than 100 MW$_e$, separate turbines for high and intermediate pressure would probably be used, where the second turbine is chosen for compatibility with storage discharging conditions. For a plant size less

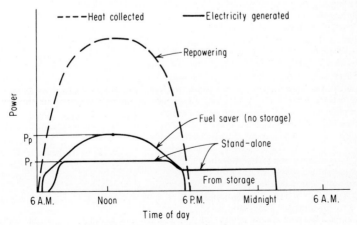

Fig. 20.3 Generic plant modes of operation.

than 10 MW_e, a turbine with one entry port will probably be used. In this case, the fluid used in the receiver may not be steam, but rather an organic fluid which may also be the sensible-heat storage medium with or without rocks. Thus, the fluid heated in the receiver will be stored directly in a one-tank (thermal-gradient) or two-tank (hot-and-cold) storage system. Steam would be generated only from storage at 15 to 40°C less than the receiver outlet temperature. This design would eliminate either the two-port turbine or multiple-turbine approach and is more appropriate for smaller systems.

For the 10- to 200-MW_e systems, the open (external-absorber) receiver has once-through flow with multiple-parallel-pass preheating/boiling with a second pass for superheating. Feed-water atemporators and feedwater flowrate are used to control exit steam quality over a wide range of demand and insolation (10:1). Receiver design is chosen as a compromise of the following factors:

- Lower receiver weight and cost
- Use of an all-around 360° field of heliostat
- Good but not the best receiver thermal performance
- Low tower height
- Illumination of structural elements between heliostats and receiver to be minimized
- Easy receiver maintenance

Other receiver approaches considered use north-facing cavity receivers with corresponding north-side heliostat field,[15] and a bottom-open cavity with 360° heliostat field.[16] Although each of these approaches is technically viable, the design chosen is considered to be the best compromise of all the design and cost factors in the 10- to 20-MW_e size range. For small systems ($<$10 MW_e) it may be more economical to use the north-face-open cavity receiver with north-field heliostats.

Storage subsystems for larger plants ($>$10 MW_e) are of the single-tank, Caloria®-rock sensible-heat approach. The stratified tank approach is described in Chap. 6.

The first-generation (and possibly the last-generation) heliostats were based on designs using relatively heavy steel structural members and a back-silvered glass reflector surface on a honeycomb substrate with a concrete foundation. The reflector area is approximately 40 m^2. Alternate designs use a lightweight approach with thin ($<$1 mil) aluminized Mylar® stretched over a metal frame inside a slightly pressurized Mylar® inflated bubble. Another possibility is using foam glass as the silvered-glass substrate. For a small system (100 to 500 kW_e), a field of many separate, faceted, kinematically driven heliostats has been developed on a test basis.[4] This type of collector is used as the basis for cost estimates at the lower bound of plant size for the central-receiver approach.

Dry cooling towers were not used on early systems, but future performance and costs are based on this type of heat-rejection system since cooling water availability will become a more severe problem over the next several decades. The performance degrades about 10 percent and energy conversion cost is increased by $50 to $100/$kW_e$ as dry cooling equipment is introduced, compared with wet cooling.

This description of the stand-alone solar plant can be modified to describe the hybrid, fuel-saver, or repowering type of plant by adding a fossil auxiliary boiler, removing storage, and removing energy conversion equipment, respectively.

Distributed systems   The distributed systems selected for analysis are shown in Table 20.4. Most of the approaches chosen use a central engine or energy conversion subsystem.

Central Engines.   The major factor differentiating the four distributed-collector, central engine systems identified in Table 20.4 is the type of collector. The optimum collector temperature (which corresponds closely to turbine inlet temperature), the transport fluid, storage medium, storage subsystem arrangement, and turbine working fluid are all derived relatively directly from the collector type.[5,8,9,17,18] The major exception to this is a temperature limit resulting from the choice of energy conversion subsystem.

The Paraboloidal Dish Collector System.   Figure 20.4 illustrates the use of a two-axis-tracking paraboloidal dish collector coupled to a steam Rankine conversion system. The upper limit for reasonable steam temperature (540°C, 1000°F) determines the receiver outlet temperature in this case. To minimize the loss of superheat and to minimize transport heat and pressure losses, the superheat function is separated from preheat and boiling. The collectors providing superheat are located at a minimum distance from the generator plant. Even then, a 45°C temperature drop is incurred in a 100-MW_e plant and a 345 kPa (50 lb/in²) pressure drop for a reasonably designed steam-transport subsystem.[19]

The energy transport system is optimized as a function of peak heat delivered, considering parasitic heat loss (operating and transient heat-up), pumping power, and direct capital cost.[19]

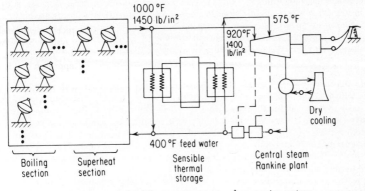

Fig. 20.4 Parabolic dish-steam transport and conversion system.

Steam transport is somewhat better than pressurized water transport, since higher temperatures are possible at the same pressure (an advantage for the paraboloidal dish collector) and there is a lower heat-up energy requirement. The somewhat larger pipe sizes needed to control pumping losses for steam transport are compensated by greater thermodynamic efficiency by operating at higher turbine inlet temperatures,[5] and thinner pipe walls can be used because of the lower pressure vs. that of pressurized water. The return line to the collector field is pressurized feedwater.

Variable-Slat Collector System. The use of variable-slat, one-axis-tracking collectors results in a plant schematic similar to that of Fig. 20.4. The only difference is the steam temperature exiting the field at 455°C (850°F). This is the optimum temperature to maximize the combined efficiency of the collector and engine as shown in Fig. 20.5.[20] The maximum efficiency of the primary equipment will essentially minimize system cost as a result of the direct relationship between efficiency, collector area, and cost.[17] (See Chaps. 9 and 22 for the basis of the efficiency-vs.-temperature characteristic of these subsystems.)

Parabolic-Trough Collector System. The use of a one-axis-tracking, parabolic-trough collector will cause the collector-engine temperature to optimize at the somewhat lower temperature

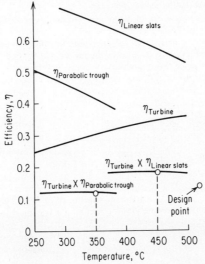

Fig. 20.5 Design point selection for linear concentrating systems-100 MW$_e$. (*From Ref. 17.*)

of 345°C (650°F) as shown in Fig. 20.5. A system layout more appropriate for this collector is shown in Fig. 20.6, where slightly superheated steam or pressurized water is used only to heat storage. Storage is used to generate steam for a one-port turbine as shown. Caloria™-rock storage would probably be the best thermal storage subsystem for this approach. If toluene heat-transport fluid is used in conjunction with an organic Rankine energy conversion subsystem, the collector temperature is limited to 318°C (600°F) because the fluid degrades above this temperature.

Fixed Collector System. The use of an advanced, fixed collector such as the CPC or V-trough reflector with a vacuum-tube receiver with selective coating (see Chap.8) dictates several changes in the basic system schematic shown in Fig. 20.6. The transport fluid would probably be pressurized water or toluene and the optimum fluid temperature would be 180°C (350°F) when a V-trough reflector is used[21] and 235°C (450°F) when the CPC reflector is used. An organic Rankine energy conversion subsystem would be used in place of steam Rankine if developed commercially. These fixed collectors would require a limited number of seasonal adjustments (2 to 10) to give adequate annual performance.

Thus a variety of approaches can be taken to the collector, transport, storage, and energy conversion subsystems for distributed systems with a central engine. The selected systems chosen are only a few of many, but do illustrate each generic type of system (two-axis, one-axis, and fixed collector design). The next section will summarize an economic comparison of these systems.

Distributed Engines. The traditional approach in the utility industry is to achieve minimized bus-bar costs by using larger, more efficient, more complicated and more capital-intensive energy conversion systems. This "bigger is better" approach has paid dividends for 60 years with decreasing electricity costs through 1970. After this time, a combination of factors have reversed the historic annual reduction in electricity costs, not the least of which is increasing direct cost and the cost associated with delays which afflict the bigger power plants.

Historically there has been a parallel path used in all industrialized states to achieve lower costs—mass production, or the "many small is better" approach. The use of many heliostats or collectors in a power plant described in the two previous sections will take advantage of the phenomenon, but a single engine is still used. The distributed engine system concept described here takes advantage of mass production techniques for both collector and engine.[5]

The distributed collector with distributed engine approach locates an engine at each collector. The system selected for consideration uses the two-axis tracking paraboloidal dish combined with a small heat engine such as a gas Brayton or Stirling engine. These small engines (10 to 25 kW$_e$) are located at or near the receiver of the collector. A generator is collocated with the engine, and electricity is generated at each collector. This electricity is collected and stored or dispatched to the load. Figure 20.7 illustrates this system. Electric transport and external, nonthermal storage are its distinguishing features.

A result of this concept is the use of a modular building unit that is identical for a power plant sized from 10 kW$_e$ to 1 GW$_e$ (1000 MW$_e$). This approach is a sharp departure from conventional utility industry practice, but has enormous flexibility in power plant size and takes full advantage of mass production gains. Special engineering and significant on-site construction will be reduced drastically since these modules can be factory-assembled. Society has the choice of a large- or small-sized power plant, and as will be shown in the next section, at almost the

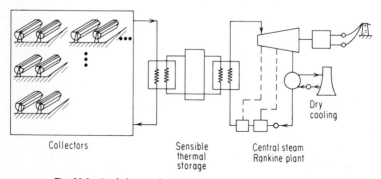

Collectors   Sensible thermal storage   Central steam Rankine plant   Dry cooling

Fig. 20.6   Parabolic trough-steam transport and conversion system.

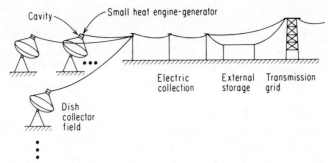

Fig. 20.7 Parabolic dish-electric transport plan.

same cost. This insensitivity of bus-bar (at the plant) cost to power plant size is contrary to utility industry experience and may cause a host of changes in the structure of the utility industry if these systems are widely adopted. The modular approach should prove to be very attractive to society because of its ability to meet large or small applications.

The paraboloidal dish has a higher temperature potential [815°C (1500°F) to 1400°C (2500°F)]. To take advantage of this potential, a higher-temperature-capability engine is necessary. The Brayton (gas-turbine) engine has this capability and has been developed in large sizes for peaking service in the utility industry and for propulsion in the aircraft industry. Smaller turbines have been developed for many industrial, aerospace, and automotive applications. References 22 and 23 indicate the potential of Brayton and Stirling engines for automotive use. Preliminary studies on integrating a Brayton engine with a dish collector indicate that location at the collector focal point (receiver) is somewhat preferable to other locations such as on the ground nearby, behind the reflector surface with a heat-transfer loop or optical transport (Cassegrainian method).[24] The optimum temperature is approximately 815°C (1500°F), unless limited to slightly lower temperature by materials problems. The optimization shown in Fig. 20.8 is relatively flat near the optimum. Thus, little system penalty is felt for a ±100°C variation in the operating receiver temperature. Major questions remain on the type of Brayton cycle (closed or open) and the receiver design (metal, ceramic, pressurized, or subatmospheric).

The Stirling engine, which is an old concept (Robert Stirling, 1816), has recently undergone development for automotive use. Its attractive features for this application are high efficiency (ideally the same as the Carnot engine's) and external combustion with low emissions. Upon development and application to a dish collector, the Stirling engine will probably provide for the most efficient and possibly the least expensive solar thermal-electric power system.

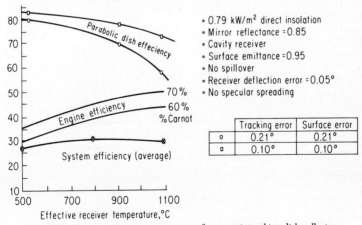

Fig. 20.8 Optimum receiver temperature for two-axis-tracking dish collector.

The electric transport subsystem is essentially the opposite of the distribution system used in a load center, and would use standard electrical components. The storage system would be standard batteries or advanced batteries (such as redox) for smaller systems, while pumped hydro or compressed air would be prime candidates for larger utility applications.[25] Apparently, there are large areas of the United States suitable for large-scale storage when pumped hydrostorage is sited with the upper reservoir on the surface and the lower reservoir in a cavern underground.[26] Other storage techniques are possible,[25] but are not considered here.

ECONOMIC COMPARISON

The basic steps in an economic comparison of these specific systems are development of subsystem performance data, system performance data, subsystem cost data, an optimizing technique, and system cost data.

Subsystem Performance and Cost Data

To aid in discussing these systems, the following subsystem definitions are used:

1. Collectors—concentrators (reflectors), receivers, tracking mechanisms, structure, foundation, and controls

2. Energy transport—pipelines and pumps or electrical collection network with associated control system

3. Energy conversion—heat engine (steam Rankine, organic Rankine, gas Brayton or Stirling) with generator and heat-rejection subsystem

4. Energy storage—sensible-heat storage or advanced battery storage

On the basis of data developed in a series of reports (5–9, 15–19, 21, 24–28) the performance and cost data for the systems described in the previous section were refined and collected in Ref. 29, which was technically directed by the author. The performance and cost data are summarized in Tables 20.7 and 20.8. It should be noted that the performance datum is the instantaneous efficiency (unless noted) at noon with clear-sky insolation. The collector efficiency is based on direct normal insolation, and there are significant off-axis effects for fixed, one-axis, and two-axis heliostats. Off-noon, off-peak load, and other effects such as transients

TABLE 20.7   Solar Subsystem Performance Data Plant Rating: 100 MW$_e$

						Efficiencies		
Type of plant		Fixed	One-axis			Two-axis		
							Dish electric	
Major subsystem	V-trough	Parabolic trough	Variable slots	Dish steam		Stirling	Brayton	Central receiver
Collectors[a]	0.34[b]	0.42	0.54	0.79		0.70	0.70	0.65[c]
Fluid/Temperature,°C	(177)	(350)	(450)	(537)		(810)	(810)	(510)
Energy transport[d]	0.95	0.93	0.92	0.87		0.94	0.94	0.95[e]
Storage throughput	None[f]	0.80	0.80	0.80		0.75[g]	0.75[g]	0.80
Energy conversion:[h]								
Turbine engine	0.20	0.27	0.30	0.35		0.42	0.35	0.36
Net subsystem[i]	0.19	0.24	0.27	0.31		0.36	0.30	0.32

[a] Combined effect of concentrator and receiver instantaneous efficiencies based on noon, normal insolation.
[b] Annual average efficiency based on concentration ratio of 3; peak efficiency is 43% twice a year.
[c] Bottom open-cavity receiver.
[d] Includes heat leak and pumping power.
[e] Includes small optical transport loss (1% absorption per 1000-ft line-of-sight) and thermal transport loss inside tower.
[f] Fixed system considered without storage.
[g] Includes inverters.
[h] Rankine system except for dish electric.
[i] Includes effect of dry cooling except for V-trough, auxiliary power, generator, etc.

TABLE 20.8   Solar Subsystem Direct Cost* Data Plant Rating: 100 MW$_e$, Year 2000 Plant Startup, Load Factor: 0.55

				Type of plant			
	Fixed	One-axis		Two-axis			
					Dish electric		
Major subsystems*	V-trough**	Parabolic trough	Variable slots	Dish steam	Stirling	Brayton	Central receiver
Collectors^a							
Concentrators, $/m²	28	103	130	182	182	182	145
Receivers, $/m²	35	26	41	7.6	11.5	11.5	
Energy transport, $/kW$_e$	100	185	185	305	77	77	252^b
Energy storage,^{c} $/kW$_e$·h		60	60	60	45	45	60
Energy conversion, $/kW$_e$	250	250	250	250	102^d	121^e	250
O&M cost^f 10^6 $/year	0.64	2.9	3.1	2.9	3.7^g	2.9	2.9

*Direct cost does not include spares and contingency, indirect cost, or interest during construction.
**Based on early estimates—considered optimistic compared to other systems.
^{a}Costs normalized to concentrator aperture area; diameter = 36 ft for dish systems.
^{b}Includes tower structure, receiver, and piping transport in tower, $/kW$_e$.
^{c}Storage cost normalized to rated storage output power of 70% plant rating.
^{d}Includes Stirling engine cost of $42/kW$_e$ plus generator, starter, switchgear, etc.
^{e}Includes Brayton engine cost of $61/kW$_e$ plus generator, starter, switchgear, etc.
^{f}First-year average cost without inflation and cleaning—levelized costs over 30-yr plant life are approximately 3 times higher due to inflation.
^{g}Includes cost of engine replacement every 5 y for Stirling/15 y for Brayton.

are not included in these nominal performance data. See Chaps. 9 and 22 for a better understanding of these effects. The data shown are for a nominal plant rating of 100 MW$_e$.

Collector efficiencies vary widely, but the dish collector has the greatest average efficiency because of its solar normal incidence angle, high concentration ratio (500 to 2000), and use of cavity receivers. The optical transport (central receiver), electric transport (dish electric with heat engine at collector), and low-temperature fluid all have high transport efficiency. The throughput efficiency of sensible-heat storage is somewhat better than the electric battery efficiency. Energy conversion efficiencies vary widely, but the Stirling engine has the possibility for the highest efficiency despite its small size.

The subsystem cost data shown in Table 20.8 are the direct costs to manufacture, ship, and install on a utility basis. They do not include factors such as spares and contingency allowance, indirect cost (engineering, construction supervision, and special construction equipment), or interest during construction. These costs were calculated on the basis of "eventual" commercial costs in mass production in 1975 dollars, but as if the subsystems were built today. Quantities are projected for the installed capacity equivalent of at least 1 GW$_e$ per year.

These cost estimates must be considered immature at this time, since all components except the steam Rankine subsystem and electric energy transport are not commercially produced components in the quantity indicated. Some subsystems such as the Stirling engine and small Brayton engine are available only in small quantity or on a technical demonstration basis. It should be noted that the manufacturing cost of a spark-ignition automobile engine is about $2/kW.[23] Development and commercial experience are greatest for the parabolic trough and least for the paraboloidal disk as a solar collector. These nominal costs will be varied parametrically so that system costs can be compared for a range of projected costs.

Collector Subsystem

Central-receiver heliostats   The performance estimate is based on Refs. 7, 15, 16, and 28. The nominal efficiency is stated as 65 percent based on a 360° field using a bottom-open cavity receiver. The open receiver has an efficiency reduced by several percentage points. The cost prediction is based primarily on the earlier projection for the one-axis-tracking systems.

The heliostat is a two-axis-tracking system that has a single structural support member with high cantilevered loads. The two-axis-tracking system is more complex than the one-axis type and the aiming requirements are much more stringent for a 1000:1 concentration ratio system vs. a 20:1 system. This combination of effects and the use of limited cost data for the central receiver lead to an estimate that the heliostat should be about 25 percent more costly than the average of the one-axis-tracking systems. The mass-production cost estimate is then $145/$m^2$ as compared with $116/$m^2$, which is the average estimated cost for the one-axis collectors.

Early prototypes for use in the 5-MW_e test facility at Sandia Laboratories provided by the Martin Company cost approximately $340/$m^2$ for several hundred heliostats. The DOE goal is in the $60/$m^2$ to $80/$m^2$ range; two studies [7,30] estimate that the DOE goal can be achieved.

Point-focusing dish   The performance and cost estimates for the two-axis-tracking, point-focusing dish have the greatest uncertainty associated with them. The device is similar in complexity to the two-axis-tracking heliostat except that the heliostat will probably have a flat surface, at least for larger plants ($\geq$ 10 MW_e). Further, the point-focusing dish introduces the additional factors of reflector surface curvature and receiver mounting on the collector. Smaller mirror facets may be necessary, and this would increase fabrication complexity.

Current microwave antenna dish costs are between $650 and $1100/$m^2$ in limited production ($\sim$ 100 per year) and for more stringent surface and aiming requirements than are necessary for heat collection. A near-term approach is to adapt these antennas to meet solar collection requirements.

On the basis primarily of the previous cost estimates and earlier studies,[5] the concentrator part of the dish collector is estimated to be 25 percent more costly than the heliostat and 60 percent more than the average cost of the one-axis linear concentrating collectors. Thus, the mass-produced concentrator part of the dish is estimated to cost $182/$m^2$ vs. $145/$m^2$ for the heliostat.

The receiver cost estimate depends on the temperature level, type of fluid, pressure level, and fabrication materials. The receiver cost varies from about $8/$m^2$ to $12/$m^2$ of aperture area. The lower cost is for a steam system at 550°C (1000°F), while the higher cost pertains to a gas at 815°C (1500°F). Thus, the total collector cost is about $200/$m^2$ for a mass-produced point-focusing dish.

The thermal performance estimate is based on earlier studies[5,24,6] and limited experimental data. These data include a room-temperature calorimeter test conducted with an 81 to 83 percent efficiency for a nine-facet (16 $\times$ 16 in each) mirror on a two-axis-tracking structure. This structure used back-silvered glass attached to a spherically curved, foamed-glass substrate. Therefore, the nominal collector efficiency is estimated to be 70 and 79 percent, based on 815°C (1500°F) and 550°C (1000°F) fluid exit temperatures, respectively. This was verified by another analysis.[31]

The estimates of the nominal performance and cost for the selected collectors have uncertainty associated with them. Upper- and lower-bound estimates are shown later in connection with an investigation of subsystem sensitivities. These bounds represent extreme best/worst limits in each area.

Parabolic trough   A parabolic trough that can achieve a noontime efficiency of 42 percent at an optimum operating temperature of 350°C (660°F) is the nominal design. Off-angle effects will reduce the efficiency at other solar times. The nominal mass production cost estimate is $103/$m^2$ for the concentrator part of the collector, which is made up of supporting structure (steel and concrete), reflecting surface and supports, tracking mechanism, shipping, and field assembly. The nominal receiver cost, as shown in Table 20.8, is projected to be $26/$m^2$ of concentrator aperture area and is based on a vacuum-tube receiver with selective coating. The total collector cost is then $129/$m^2$.

Variable slat   The basis for projecting the performance of the variable-slat linear concentrator is the work done by Francia at the University of Marseilles and later work at the Sheldahl (Suntec) and Itek companies. The noon instantaneous performance of 54 percent is estimated to occur at 450°C for optimum system operation, even though somewhat higher performance has been measured.[17]

It is believed that the slat-concentrator costs are higher by about 30 percent than those for the parabolic troughs because of increased mechanical complexity, increased number of reflector facets, and greater accuracy requirements. The receiver cost is also considered to be higher because of higher temperatures (450°C). The total collector cost is $171/$m^2$ vs. $129/$m^2$ for the parabolic trough—about one-third more. The same reservations expressed earlier exist here, in that no detailed mass production cost estimates have been performed on a specific design.

Advanced fixed-orientation collectors A tubular vacuum receiver containing a tube-and-fin absorber plate with a selective coating is used with V-shaped reflectors which are asymmetrical, so that by adjusting (reversing) the position of the reflector twice a year, the annual performance is enhanced while simplicity of design is preserved. The design achieves minimum system cost at about 350°F when coupled to a Rankine power plant.

Collector performance is based on detailed calculations.[21] At the optimum temperature (350°F) and concentration ratio CR=3, the annual average efficiency is estimated to be 34 percent, while the peak efficiency is 43 percent. The cost projection is heavily dependent on the cost of the vacuum absorber tube. One design with a copper tube-fin and a glass-to-metal seal was considered. Prototype costs were greater than $2.50/m² of absorber area, but the manufacturer predicted eventual costs of $1.00/m². The reflective surface is considered to cost $0.05/m² for aluminized plastic on a steel or aluminum sheet-metal substrate. The structural framing including concrete pads is estimated to cost $1.50/m² of frame area. Shipping and assembly are considered to be about $0.06/m².

For a concentration ratio of 3, these cost estimates are shown in Table 20.8 on the basis of the aperture area for the concentrator and receiver parts of the subsystem. The total collector direct cost is $63/m² of aperture area for the optimistic $1.00/m² absorber tube cost.

The CPC design is believed to be similar to the V-trough in that it is an advanced, fixed system. As the concentration ratio CR is increased, the annual number of concentrating surface tilt adjustments must increase; e.g., only two adjustments are needed for CR=3, while 12 are required for CR=5. Tilting increases performance but it also appears to require a more complicated tracking system than reflector reversal. A careful review of the CPC performance and cost characteristics needs to be conducted so that these issues can be explored in more detail and the present conclusion that the CPC is about as cost effective as the asymmetric V-trough approach can be checked.

Engines and Power Generation Subsystems[29]

Engine types for solar applications are identified[32] and divided into availability classes as shown below:

Near-term (1977–1985)
  Organic Rankine
  Steam Rankine
  Open Brayton
  Closed Brayton
Intermediate (1985–2000)
  Advanced Rankine
  Advanced Brayton
  Stirling
  Biphase
  Liquid-metal topping

Near-term engines are either in production or are proceeding successfully through development/demonstration phases. The intermediate class of engines encompasses advances to Brayton and Rankine systems, involving primarily higher-temperature designs having greater efficiency. This class also includes Stirling and biphase engines which are in the early stages of development. Far-term engines include those in the laboratory research stage and new concepts such as the sodium heat engine, thermionics, etc. The near-term engines are used in current baseline or alternative systems considered in this chapter. The intermediate-term engines are expected to reach a level of technological maturity such that they will be available for use in commercial power plants during the 1990–2000 period.

Far-term engines could reach the large-scale feasibility demonstration stage during 1990–2000, and it is expected that those which show promise will be in use after the year 2000. It is possible that some developments—for example, thermionic conversion systems—could be accelerated for other reasons such as use in spacecraft propulsion systems, but considerable progress toward achievement of higher efficiencies and reliable operation is required before these systems can be considered as viable candidates for terrestrial power applications.

The organic Rankine system is suitable for low-temperature sources including fixed-collector solar power systems, waste-heat recovery, and bottoming cycles for conventional power stations. Emphasis on energy conservation has recently stimulated developmental activities. The only system employing organic Rankine cycles is the fixed-collector V-trough power plant. Only a

100-MW$_e$ system is considered, and for this size, the same unit cost as for a conventional steam Rankine plant is estimated for a large-size organic Rankine power plant.

Steam Rankine power plants for ratings greater than 10 MW$_e$ are well-developed commercial systems. They can operate at temperatures of ~ 1000°F because of design optimizations involving cost and performance tradeoffs. The steam Rankine systems employed in this study are based on this proven technology.

For the advanced Brayton and Stirling engines, cost and efficiency projections are based on successful completion of development activities. The data of Table 20.7 correspond to cycle temperatures of 815°C (~ 1500°F). It is expected that solar receiver/engine systems could be developed without incurring major materials problems if temperatures are limited to this level. Therefore, these systems could reasonably be expected to be developed to a commercial status in the 1990–2000 time period. In the near term, existing open- and closed-cycle Brayton engines developed for nonsolar application could be adapted to the solar system with some decrease in performance as compared to Table 20.7. Near-term costs will also be higher than the Table 20.8 costs, which are predicated on high-volume production (10^5 to 10^6 units per year) in the 1990–2000 time frame. The use of high-temperature materials (e.g., ceramics) for both the receiver and heat-engine components would allow higher temperatures and greater efficiencies, but it is unlikely that these systems will reach a stage of development where they could be commercially implemented by 1990. Therefore, this high-temperature possibility is not considered further in this chapter.

Liquid-metal topping cycles[33] involving mercury and potassium have been pursued for space applications, and a substantial technology base has been developed. From this work, it appears that liquid-metal topping cycles could be implemented in the intermediate term. At present, costs for these systems are uncertain (e.g., mercury is expensive and has a historically unstable price structure), and there are problems such as toxicity, contamination, materials compatibility for potassium systems, turbine erosion, etc.

With regard to the engine data given in Tables 20.7 and 20.8, it is noted that these values are nominal estimates subject to a range of uncertainty. The effect of this uncertainty is examined as part of the subsystem sensitivity evaluation below.

Energy Storage Subsystem

As shown in Ref. 34, use of thermal storage systems is particularly advantageous for solar thermal power plants. If thermal storage is interposed between the collector field and energy conversion system, the storage can buffer insolation variations and thereby allow a more uniform level of energy input to the conversion system. In particular, the conversion system can now be sized to match this storage-buffered input energy level as opposed to being sized to accept peak insolation levels. This results in reduced conversion system capital costs which at least partially offset the cost of the storage system.

A survey of thermal storage systems[27] tends to confirm that sensible-heat systems are the most likely candidates for commercial implementation in the 1990–2000 time period. Latent-heat or phase-change storage systems offer higher energy density storage and therefore are potentially less costly. However, they require considerably more technological development regarding problems such as the long-term stability of eutectic salt mixtures, operation and maintenance considerations, and efficient means of heat transfer. The costs of thermal storage in Tables 20.7 and 20.8 based on Ref. 6 pertain to large storage systems capable of providing 6 h of power at a level of 70 MW$_e$ when coupled to a conventional steam Rankine power plant. Such systems are still in the early development stage and the cost performance values shown are based on projections and judgments. It is expected that unit costs of thermal storage systems will increase as the size of the system decreases, since these systems employ containment vessels which are more economical when sizes are large.

For the dish-electric system, a small heat-engine/generator is coupled directly to the receiver and is mounted at the focal point of the dish. This arrangement avoids the use of flexible lines to transport heat from the focal point to a ground-based conversion system. Location of a thermal storage system at the focal point will increase weight and size to the point where much of the advantage of the compact focal-point-mounted system will be lost. Therefore, inclusion of thermal storage for dish-electric systems is not considered in the present study. Instead, electrical energy from each of the dish-mounted engine/generators is collected and stored at a central location. Electrical energy can be stored by mechanical, chemical, and electromagnetic methods.[25] The mechanical approach includes pumped hydro storage, compressed air, and flywheels, while the electrical approach involves direct storage of electrical energy in super-conducting magnets.

Pumped hydro, compressed air, and lead-acid batteries are near-term candidates. Pumped-hydro and compressed-air systems require particular terrain and geology which tend to limit their application. Lead-acid batteries have lower energy densities than advanced battery concepts, and flywheels tend to be useful only for short-duration storage. Hydrogen systems require development of advanced electrolyzer–fuel-cell systems and appear to be advantageous when certain utility operating conditions exist. Electromagnetic systems require advanced technology development which offers high efficiency, but costs are highly uncertain.

The advanced battery system was selected along with pumped hydro as a baseline external energy storage system for this comparison, since pumped hydro is the only storage system presently in utility service and there is an extensive DOE/EPRI development program underway centered around the use of a large-scale battery energy storage test (BEST) facility, which is to be operational in the 1980s. Batteries are particularly attractive since they can be easily located at dispersed locations and have rapid response characteristics. Cost data in Tables 20.7 and 20.8 are from Refs. 6, 25, 27, and 33.

Even if advanced batteries do not attain projected performance and cost goals, other candidate storage systems are being pursued. Therefore, it appears likely that at least one system having performance and costs in the range of values projected for the battery will be available in the 1990–2000 time frame. The effect of performance and cost uncertainties associated with energy storage systems is discussed in a later section.

Energy Transport Subsystem

For distributed-collector concepts, energy is transported from the collector field either by pipelines or electric wires. For systems selected in this study, pipeline transport involves steam or organic fluids. For pipeline systems, the analysis procedure for distributed-dish (or point-focusing) systems is given in Ref. 19. This basic procedure was extended to encompass linear concentrator systems as part of the present study. The analysis was based on a square field arrangement, where 8480 linear collectors (100 m² each) were required for a 100-MW$_e$ plant.

For electrical collection involved in the paraboloid-dish-electric system, the basis for estimating costs and performance is given in Refs. 5 and 24. The collection network analysis includes both low- and high-voltage transformers, capacitors, circuit breakers, and cables.

Economic Methodology

The basic approach to an economic comparison is first to perform dynamic thermal simulation of plant performance; second, to develop an economic methodology to calculate plant costs; third, to establish a procedure for optimizing the plant design to minimize whatever is chosen as the cost parameter. The stand-alone plant is chosen as the mode of operation to evaluate. For the case where there is little or no storage, the fuel saver is simulated. In a later section, the grid reliability will be considered for hybrid operation. The only mode of operation that is not explicitly examined is the repowering solar plant (no storage and no energy conversion and integration with an existing fossil steam plant). Approximate calculations will be presented to aid in economic understanding of repowering. Thus thermal (internal) or external storage is used to absorb extra energy collected or generated so that the stand-alone plant does not produce more than the rated power.

Performance simultation The first step in the approach is to simulate the performance of the selected power plants. This involves the development of a computer simulation code based on an analysis of the collector field as a function of factors such as concentrator geometry and optics, solar-tracking characteristics, surface reflectances, and receiver heat losses. These factors are reflected in the nominal collector efficiency and efficiencies of the other system components as shown on Table 20.7. These are then combined to determine how much of the collected energy is converted to electricity. Off-angle effects on the collector performance, off-load effects on the engine efficiency, auxiliary power, and five modes of operation [normal, low insolation, intermittent clouds, night (using storage), and standby] are considered. During normal operation for a power plant with internal (heat) storage, there is sufficient solar energy collected to drive the heat engine as well as charge storage. Low-insolation operation modes do not charge storage but directly drive the heat engine at between 20 and 100 percent of rated power with additional energy taken from storage. The intermittent-cloud mode of operation uses storage to drive the heat engine while whatever heat is collected is placed in storage. At night or during an overcast, there is no energy collection, and storage is discharged to drive the heat engine. During standby, there is neither collection nor storage discharge. When necessary, energy from storage is used to maintain steam seals and other requirements. The systems using external storage (batteries, pumped hydro, etc.) can use the same five modes of operation with some minor differences in description.

The use of the system simulation concept[6] to determine system performance is illustrated in Fig. 20.9. Inputs to the code comprise weather conditions, insolation, and the electric demand of the utility grid. For this study, hourly direct insolation data for one year from Inyokern, California, are employed (3200 kW,h/m²year). The utility demand was set for baseload operation; i.e., the grid requested that the plant deliver rated power continuously. Under these conditions, the computer simulation control logic would allow the plant to operate and deliver rated power to the maximum extent possible within constraints of plant design characteristics. That is, the plant would deliver rated power during periods of insolation availability and store excess energy. The stored energy would then be delivered if insolation were insufficient. The total energy delivered during a year divided by the amount of energy that would have been delivered by continuous operation of the plant at the rated power is defined as the load or *capacity factor*.

Conventional nuclear and coal plants presently achieve capacity factors of 0.5 to 0.7 for baseload operation.[3] Investigation of utility interfacing as a function of demand characteristics involves complex considerations which are discussed to a limited extent. Some of the utility interfacing issues are identified and treated in a preliminary manner in Refs. 3, 6, 14, 36, and 37, but this basic work must be amplified and extended.

In operation of the simulation code, plant characteristics are first selected. These include collector type and area, plant rating, storage system size, and efficiencies of subsystems/components. The computer hour-by-hour simulation then determines the corresponding annual capacity factor. If collector area and storage size are varied, a performance map for a given type of power plant can be generated.

Utility economic methodology   The computer-generated performance map and the subsystem unit cost data (Table 20.8) are employed in determining power plant economics. The power plant costs depend on future cost escalation rates, discount rates, and the method of financing plant construction.

Therefore, an economic methodology is required so that comparisons can be conducted in a consistent manner. Such a methodology has been developed in Ref. 38 and is used here. It involves consideration of factors such as:

Capital costs
Direct
Contingencies and spares
Indirect costs
Interest during construction
Operation and maintenance
Other: insurance, profit, taxes, etc.
Differential inflation
Prior to start-up
During plant life

The methodology yields a levelized (discounted, average) energy cost over the plant life and allows the use of a constant-dollar base. Because of differential inflation effects, costs are a

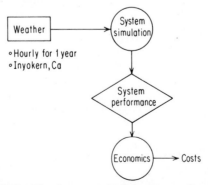

Fig. 20.9   Solar plant system simulation computer flowchart.

function of the year of plant start-up. All the plants have been analyzed on the basis of start-up in the year 2000 with a plant construction period of 6 years. As a consequence of the assumed differential inflation rates shown in Table 20.9 (discussed in detail in Ref. 3), costs for year 2000 start-up are 22 percent higher than costs for 1975 start-up. It is emphasized that all costs are in 1975 dollars. Also, the energy cost is calculated on the basis of 13.6 percent downtime for maintenance (scheduled and unscheduled) in addition to unavailability because of variations in insolation. Table 20.9 lists the economic factors used as input to the economic analysis described next.

The general economic methodology[38] considers differential escalations of not only initial capital cost but of the various cost streams over the life of the plant; however, constant 1975 dollars are used throughout. Thus, the predicted costs are not first-year costs of operation and maintenance (O&M) with levelized capital costs. It is necessary to predict the escalation rates of various components of the plant cost: operation, maintenance, and fuel.

For example, although the first-year, O&M cost of a typical solar plant is considered to be between 6 and 7 mills/kW_eh, the assumed escalation rates for O&M cause the levelized cost to be in the 18- to 20-mill/kW_eh range. If one does not wish to use this approach, the solar plant's O&M energy cost should be reduced to one-third of the value shown in Tables 20.10 and 20.12, and the energy costs should be similarly reduced in many of the figures. Escalation of operation cost is important for plants where fuel cost escalates sharply.

Example cost optimization The first step involves the use of the performance simulation code to generate a performance map. The map for the paraboloid dish-electric plant is shown in Fig. 20.10. The relationships among field size, storage capacity, and capacity factor are presented. The map shows that large field size results in excess energy collection during insolation availability periods. This excess energy can be stored in storage systems which are discharged during periods of insolation unavailability to result in higher capacity factors.

The storage capacity is arbitrarily sized at 100 percent of plant rated capacity, which is not possible for all 100-MW_e steam Rankine plants with two-port turbines. In this case, the storage capacity could be 70 percent rated capacity. Thus, the hours of storage required to achieve the plant annual capacity factors in Fig. 20.10 would simply be multiplied by 0.7. Thus the hours of storage in the figure would go from 1, 3, 6, 9, and 18 to 0.7, 2.1, 4.2, 6.3, and 12.6 h, respectively. The annual capacity factor without a maintenance allowance is 0.35 for a collective area of 0.40 km^2 as would be used in the fuel-saver type of plant.

TABLE 20.9   Economics Assumptions

Factor	Value
System operating lifetime, years	30
Annual "other taxes" as fraction of capital investment	0.02
Annual insurance premiums as fraction of capital investment	0.0025
Effective income tax rate	0.40
Ratio of debt to total capitalization	0.50
Ratio of common stock to total capitalization	0.40
Ratio of preferred stock to total capitalization	0.10
Annual rate of return on debt	0.08
Annual rate of return on common stock	0.12
Annual rate of return on preferred stock	0.08

	Annual inflation rates, %	
	1975–1985	After 1985
General price level	5.0	4.2
Labor (construction)	7.0	6.2
Manufactured goods	4.3	3.8
O&M (¾ labor, ¼ goods)	6.3	5.6
Other (insurance, taxes, profit, etc.)	5.0	4.2
Installed capital	6.2	4.8

TABLE 20.10   Projection of Baseload Power Plant Costs, 1975 Dollars

Cost	Coal[a]	Nuclear[b]	Photovoltaic[c]	Solar thermal[d]
Capital, $/kW_e	1200	2200	5500	3600
Energy, mills/kW_e · h				
Bus bar[e]	60	75	125	90
Delivered[f]	70[g]	90[h]	150[i]	110[j]

[a] Low-Btu gasification with combined-cycle gas and steam turbine, capacity factor = 0.74.
[b] Light water reactor or liquid metal breeder, capacity factor = 0.70.
[c] Silicon photovoltaics with 2:1 concentration using reversible V-trough at $0.50/W_p module, capacity factor using solar equipment = 0.70 with 12-h storage at 70% capacity. Low-Btu coal backup with gas turbine for reliability and to raise capacity factor to 0.86.
[d] Central receiver at 100 MW_e with $90/m² direct heliostat cost; capacity factor with solar equipment = 0.70 with 9-h storage at 70% capacity. Low-Btu coal used in energy conversion part of solar plant for reliability and to increase capacity factor to 0.86.
[e] Levelized over 30-year plant life including differential inflation for O&M.
[f] Includes average transmission distance and distribution within load center.
[g] 300-mi transmission distance.
[h] 1000-mi transmission distance.
[i] 2000-mi transmission distance from good insolation and inexpensive land in Southwest "sun bowl."

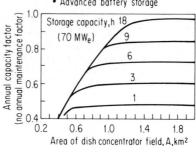

Fig. 20.10   Paraboloid dish-electric plant performance for various dish areas and storage sizes.

For utility power plant systems, it is necessary to achieve the lowest energy costs for any given operating load factor. Thus, an optimized design is one that minimizes levelized energy cost. The energy cost for solar plants is given below:

$$EC = \frac{CRF}{8760\ PL}\ (hI + f_1 O + f_2 M)$$

where CRF = capital recovery factor (see Chap. 28)
   h = factor that includes taxes and insurance
   I = total capital investment
   O = operation cost
   M = maintenance cost
   f = factor which creates present value of rising cost stream (subscript 1 is used for operation, while subscript 2 refers to maintenance)
   P = rated power
   L = annual capacity factor (energy generated/rated power × 8760) including 0.864 annual maintenance factor

From Table 20.9 the weighted interest rate is 9.6 percent the cost of capital after taxes is 8 percent, and the annualized fixed-charge rate is 14.8 percent with capital costs, taxes, insurance, etc. included. The annual maintenance factor of 0.864 is used in addition to plant unavailability because of insolation variations. This factor is appropriate for mature but large conventional plants. It may prove to be too low for a solar plant. Energy cost is the ratio of annualized costs to the annual energy delivered. When the collector field size and storage capacity are increased, capital costs increase, but the energy delivery denoted by the capacity factor also increases. In general, energy costs can be made to either decrease or increase, depending on the combination of field area, storage, and capacity factor. The combination yielding the lowest energy is sought. The method for determining this combination is depicted in Fig. 20.11. For a given concentrator or collector field area, storage capacity is increased until the lowest energy cost is achieved. Annual capacity factor (which includes a 0.864 annual maintenance factor) also increases as storage capacity is increased, as shown in Fig. 20.10. As storage size is increased beyond the value corresponding to minimum costs, energy costs rise rapidly while the capacity factor becomes essentially constant. This corresponds to the circumstance where the storage is oversized relative to the excess energy available for storage.

By considering a family of concentrator field sizes and determining the minimum energy cost for each size, one can construct an envelope curve (dashed line of Fig. 20.11) of minimum energy costs. This curve is relatively flat for annual load factors from 0.3 to 0.6 and thereafter rises rapidly. The knee in the curve occurs at a load factor of about 0.7, and the curve becomes asymptotic to a load factor of 0.864, which represents the maximum value possible in view of the assumed downtime for annual maintenance.

The rapidly rising portion of the minimum energy cost curve corresponds to a zone of diminishing returns. If the solar plant is forced to provide energy on an essentially continuous basis except for maintenance downtime, large field areas and large storage capacities are required and costs will be high. The reasonable operating range for solar plants would appear to be for load factors ≤ 0.7, which compares well to the 0.55 to 0.70 range of present (1974–1978) conventional baseload (coal and nuclear) plants.

For the dashed minimum curve in Fig. 20.11, the excess electric energy is assumed to be not stored. When storage is full, energy is rejected. If no energy is to be rejected, the storage system must be sized to handle the peak excess insolation occurring only once during the year, and for most of the year storage is in excess. Thus, under the ground rules of delivering a constant rated power level for baseload operation, minimum costs correspond to some energy rejection.

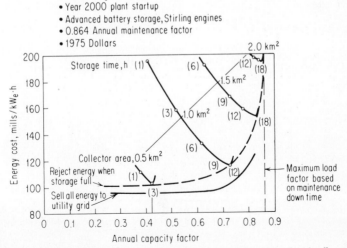

Fig. 20.11 Paraboloid dish-electric system characteristics showing the effect of total collector area and storage size on bus-bar energy cost for the Inyokern example.

For external battery or pumped-hydro storage systems, it appears likely that the excess electrical energy generated over rated power could be fed into the utility grid, and fossil plants feeding the grid could back down slightly and conserve fuel. If this strategy is used, the lower (solid) envelope curve of Fig. 20.11 would prevail and a 10 percent reduction in energy cost would result.

In addition to energy costs, capital costs are an important consideration in the implementation of power plants. Availability of capital can dictate the feasibility of constructing a power plant. Both capital and energy costs are shown in Fig. 20.12 for the minimum life-cycle energy cost. Capital costs increase rapidly with load factor, since both field size and storage capacity must be increased. As noted previously, energy costs increase only slightly up to a load factor of 0.6. Thus in this range, the price of achieving higher capacity factors is essentially manifest as higher capital costs.

Figure 20.12 also shows that the cheapest plant which would be considered initially by a utility would have little or no storage, since the energy and capital cost of such a plant are lowest. This is especially true if the capacity displacement were effective with use of either off-peak storage or inexpensive fossil backup. However, use of little or no storage results in low annual capacity factor (~ 0.3). To achieve a greater displacement of conventional energy sources, the storage and collector field size must be increased to achieve higher capacity factor. Thus, after the initial use of repowering and fuel-saver plants, stand-alone plants with increasing storage would be introduced with or without hybrid capability.

It should be noted that the external storage type of plant, such as the dish-Stirling plant, does not have the conversion capacity sized at the plant rated capacity. The energy converter must be able to instantaneously convert all the energy collected. Thus capacity is sized at the *peak* capability of the solar collectors. Only after conversion would any energy be stored externally (batteries, pumped hydro, etc.). Thus the energy conversion size and cost is scaled with collector capacity and collector area. The relationship between collector area and plant capacity is shown in Figs. 20.10, 20.11, and 20.12. These data are presented more clearly in Fig. 20.13, where the annual capacity factor, including the 0.864 assumed annual availability after allowance for maintenance, is shown vs. collector area. The plant design is optimized to minimize energy cost and the optimum amount of storage is shown at 100 percent plant rated capacity. Thus the no-storage (fuel-saver) plant has a collector area of 0.40 km² and an annual capacity factor of 0.30 with the 0.864 annual maintenance factor or 0.35 without.

The energy conversion subsystem rated power is 100 MW$_e$ at a collector area of 0.40 km² and a solar multiplier of unity. At an annual capacity factor of 0.55 the collector area is 0.70 km², and at a solar multiplier of 1.75 the peak engine capacity is 175 MW$_e$ for this 100-MW$_e$-

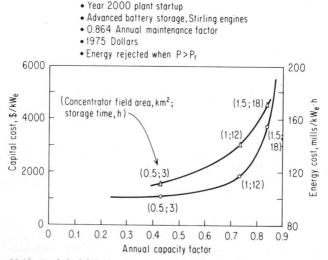

Fig. 20.12 Paraboloid dish-electric capital and energy costs for the Inyokern example.

- 100 MW$_e$ plant rating
- Advance battery storage
- Designed to minimize energy cost
- Hours – optimum storage at 100 % plant –rated capacity
- (assumed annual availability due to maintenance)

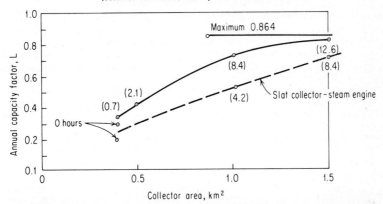

Fig. 20.13 Paraboloid dish-Stirling and slat-steam plant capacity factors versus collector area.

rated plant. About 0.90 km² of reflector surface is required for an annual capacity factor of 0.70, which is the expected value for baseload plants. The required ground area is about 3.3 times the reflector area, since a ground cover ratio 0.3 is typical in a Southwest location.

Also shown in Fig. 20.13 is area vs. capacity factor for the variable-slat, one-axis-tracking collector combined with a central steam Rankine plant. Since this is a less efficient system, greater area is required to achieve the same capacity factor. About 1.05 km² area will have a 0.55 annual capacity factor with the appropriate storage, rather than 0.70 km² for the dish-Stirling system.

System Comparative Results

Capacity factor sensitivity By the basic procedure just described, the energy and capital costs of the selected systems have been determined. At a fixed plant rating of 100 MW$_e$, the energy and capital costs of the systems are compared in Figs. 20.14 and 20.15. For distributed-

- Plant rating: 100 MW$_e$
- Year 2000 plant startup
- 1975 Dollars

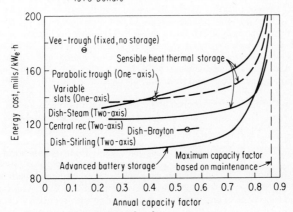

Fig. 20.14 Solar plant energy costs.

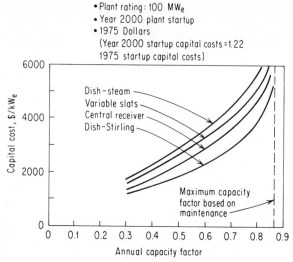

Fig. 20.15   Solar plant capital costs.

collector systems, the two-axis-tracking dish-Stirling arrangement potentially has the lowest energy costs. The dish-Brayton system is about 10 percent more expensive. The dish-steam and the one-axis-tracking variable slat and parabolic trough are approximately the same with an energy cost about 40 percent more than the dish-Stirling in the 0.3 to 0.6 capacity-factor range. The fixed orientation (nontracking) V-trough is 80 percent higher in cost even at $100/m^2 for the vacuum-tube collector.

At least part of this cost trend results from the use of higher-temperature (~ 1500°F, 815°C) technology for the Stirling and Brayton systems. The dish-steam concept employs conventional (~ 1000°F, 540°C) Rankine cycle technology. Use of advanced Rankine systems (e.g., topping cycles) could change the comparison. For one-axis-tracking systems, either variable-slat or parabolic-trough, achievable temperatures appear to be limited to the level of conventional Rankine technology. But even for these systems, modifications such as the addition of a bottoming cycle could improve plant economics.

The steam Rankine central receiver is between the dish-Stirling/Brayton systems and the dish-steam system(Fig. 20.14). Since the dish-steam system is based on comparable technology to the central receiver and is projected to have a higher cost, it appears that the central receiver (for the 100-MW$_e$ size) is the best candidate when conventional Rankine conversion systems are used. It is also noted that second- and third-generation central receiver systems may exhibit even higher performance than this system. Advanced central receiver designs using gas-turbine or liquid-metal technology could produce up to a 15 percent reduction in bus-bar energy cost.[39]

The capital cost comparison of Fig. 20.15 shows that the dish-Stirling system has the lowest costs. Although unit collector costs for two-axis-tracking dish-electric systems are high (Table 20.8), these systems collect a higher percentage of incident energy and are coupled to highly efficient cavity receivers and heat engines. High system efficiency allows the use of a smaller collector field for a given output, and these savings more than offset the higher unit collector costs.

The capital cost of the central receiver is between the dish-Stirling and the variable-slat/dish–steam systems. Thus, capital and energy costs exhibit the same relative rankings. The capital cost differences between systems is seen to be large; e.g., at a capacity factor of 0.5 the capital cost of the dish-steam system is more than 50 percent greater than that of the dish-Stirling. This indicates that substantial gains may result from technological advancement activities.

Plant size sensitivity   The comparative economics of Figs. 20.14 and 20.15 involved a 100-MW$_e$ plant, which is representative of central power applications. Solar plants are being considered for a wide spectrum of applications, ranging in plant size from 10 kW$_e$ to 1 GW$_e$. Thus, the variation of plant economics with size is an important consideration for evaluation.

Size trends for the selected systems are compared in Fig. 20.16. The dish-Stirling/Brayton systems are seen to be relatively insensitive to size because the arrangement is inherently modular with a single dish-receiver-engine/generator serving as the basic power unit which delivers only ~ 25 kW$_e$ at peak power rating. The slight increase in energy costs with decreasing size (Fig. 20.16) results from indirect costs which constitute a larger fraction of plant costs when sizes are small. Indirect costs combined with spares and contingency cost factors are shown in Fig. 20.17 as a function of plant size. The cost estimates are made on two bases: one-of-a-kind equipment such as a central engine and single-tank sensible-heat storage, and repetitive modular equipment with mass produced collectors and distributed engines.[20] Indirect costs include engineering, construction supervision, and special construction equipment. All other distributed-collector systems are based on conventional central Rankine cycle energy conversion plants. For these systems, costs increase with decreasing size, as shown in Fig. 20.18. For steam Rankine systems, most of the development emphasis over the last 60 years has been directed toward large-size units for central power applications. Therefore, the performance of present smaller-size Rankine systems is poor, but could be higher if developed and implemented for solar application in the 1990–2000 time frame. This technology development is accounted for in Fig. 20.16 to some extent; e.g., although present prototype 100-kW$_e$ Rankine engines have efficiency of ~ 20 percent, it was projected that engines of ~ 27 percent efficiency would

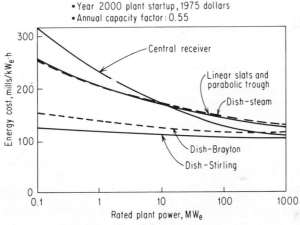

Fig. 20.16   Effect of plant size on power cost.

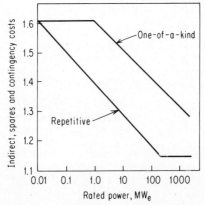

Fig. 20.17   Indirect costs, spares, and contingency costs versus plant size.

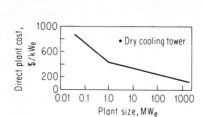

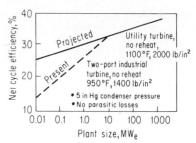

Fig. 20.18 Rankine energy conversion plant cost versus plant size.

Fig. 20.19 Rankine plant efficiency.

be available. The cost and performance assumptions used to evaluate solar plants with steam and organic Rankine energy conversion are shown in Figs. 20.18 and 20.19.[20]

The central-receiver system is shown to have an even steeper rise in energy costs as plant size decreases. In addition to higher Rankine engine costs, the heliostat field of the central receiver becomes more costly because of the need for smaller mirror facets and/or curved mirror surfaces. The cost assumption supporting this is shown in Fig. 20.20, where the heliostat cost is shown increasing as the plant size decreases. The cost estimate for the lower bound ($\sim$ 100 kW$_e$) is based on a preliminary projection of the Francia design. It may be possible in the smaller-size plant to use a different design than the field of 1.0-m^2 reflectors kinematically driven from a single control.

Breakdowns of capital and energy costs for four 100-MW$_e$ plants, including dish-Stirling, dish-steam, variable-slat, and the steam central-receiver, show that collectors (concentrators and receivers) dominate costs; e.g., for the 100-MW$_e$ dish-Stirling system, the collector subsystem cost of $139 million is 64 percent of the total capital cost of $217 million. For all the 100-MW$_e$ and 10-MW$_e$ systems, collectors constitute more than half of the total capital cost. Capital and O&M costs are comparable and second only to collector costs.

Subsystems for transport, conversion, and storage have relatively small contributions to capital and energy costs, but it is emphasized that their efficiencies have a direct influence on the size and cost of the collector field. For example, unit paraboloidal-dish concentrator costs (which comprise the bulk of the collector field cost) for the dish-Stirling and dish-steam systems are taken to be identical. However, the collectors for a 10-MW$_e$ dish-Stirling plant cost $14 million, compared with $21 million for the dish-steam. This difference of $7 million is primarily a result of the high conversion efficiency of the dish-Stirling system (Table 20.7) which results in a smaller collector field.

Parameter Sensitivity

Adjustments in the above costs can be made for nonnominal values of the first-order cost parameters.

Annual maintenance factor To consider an annual maintenance factor MF different from 0.864 for a specific annual capacity factor S, use Figs. 20.14 and 20.15 and enter the ordinate of the figures at an adjusted annual capacity factor AL:

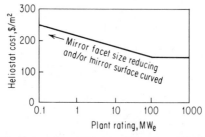

Fig. 20.20 Estimated central receiver heliostat cost characteristics.

$$\text{AL} = \left(\frac{0.864}{\text{MF}}\right) L$$

This first-order correction factor will give the capital and energy costs associated with the desired annual maintenance factor and annual capacity factor L.

Plant start-up time   To evaluate the effect of a start-up time different than 2000, evaluate the differential inflation of solar capital costs between 1975 and the desired plant start-up time T. With inflation factors from Table 20.9, the adjusted capital cost factor ACF is

$$\text{ACF} = \left(\frac{1.062}{1.05}\right)^{10} \left(\frac{1.048}{1.04}\right)^{T-1986} \qquad T > 1986$$

$$\text{ACF} = \left(\frac{1.062}{1.05}\right)^{T-1975} \qquad 1975 \leqslant T \leqslant 1986$$

The ACF used in this analysis is 1.22 for a projected year 2000 plant start-up. To correct capital cost, simply ratio the calculated ACF at a particular start-up time to 1.22, and multiply by the capital cost. For start-up times before 2000, the assumed annual generating capacity of 1 GW_e may not be achievable. Thus a correction may be necessary for certain subsystems that may not have achieved "eventual" commercial costs because full mass production has not been achieved.

Escalation factors   If escalation factors other than those assumed in Table 20.9 are desired, adjust the capital cost factor accordingly. The capital costs I may be adjusted with this new capital cost factor:

$$\text{AI} = I \left(\frac{\text{ACF}}{1.22}\right)$$

The energy cost should be corrected so that the part contributed by the capital costs is adjusted by (ACF/1.22), while the part due to O&M estimates is held constant.

Fuel-saver plant   To estimate the costs of a fuel-saver plant (no storage), refer to Figs. 20.13 through 20.15, and use the no-storage or left-most (lowest capacity factor) part of the curves shown for each type of plant. The data shown in Figs. 20.14 and 20.15 are for a 100-MW_e plant. The plant size effects are shown in Fig. 20.16, and a ratio of the cost at 100 MW_e to the desired size will be a first-order estimate at the proper plant size.

An example calculation may further clarify how the fuel-saver plant cost may be estimated. Consider a 10-MW_e plant of the paraboloidal dish-Stirling engine design. The no-storage $(L = 0.30)$ 100-MW_e plant size costs are \$1200/kW_e and 100 mills/kW_eh from Figs. 20.14 and 20.15 for the 2000 plant start-up in 1975 dollars. The ratio of plant cost from 100 MW_e to 10 MW_e is 1.05, from Fig. 20.16. Thus the no-storage (fuel-saver) 10-MW_e plant has an energy cost of 105 mills/kW_eh. If the O&M cost is almost entirely attributable to the collector-engine, the O&M energy cost is relatively constant and invariant with plant size and design capacity factor. The fuel-saver (no-storage) plant capital cost is then

$$\frac{105 - 20.5}{100 - 20.5} \times 1200 = \$1286/\text{kW}$$

where 20.5 is the levelized O&M cost.

Operation and maintenance   Although first-year O&M cost amounts to 6 to 7 mills/kW_eh, levelizing over the 30-year life of the plant with differential inflation causes the O&M cost to be about triple the first year values.

Subsystem uncertainty effects   Upper- and lower-bound performance and cost sensitivities have been determined for each subsystem. Additionally, upper and lower bounds are estimated for O&M costs. Each of the individual sensitivities is discussed below.

Collector Subsystem Uncertainty. The sensitivity of power cost to collector cost and efficiency is shown in Figs. 20.21 and 20.22 for 10-MW_e and 100-MW_e stand-alone plants at an annual capacity factor of 0.55. The greater uncertainty of the two-axis-tracking dish systems relative to the one-axis-tracking variable-slat system is clearly evident.

Energy Conversion Subsystem Uncertainty. Sensitivities to engine efficiency uncertainties are presented in Figs. 20.23 and 20.24 for both steam Rankine and Stirling engine systems.

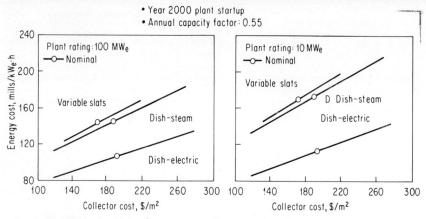

Fig. 20.21 Sensitivity of power cost to collector cost for 10- and 100-MW$_e$ plants.

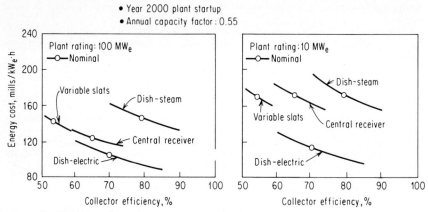

Fig. 20.22 Sensitivity of power cost to collector efficiency for 10- and 100-MW$_e$ plants.

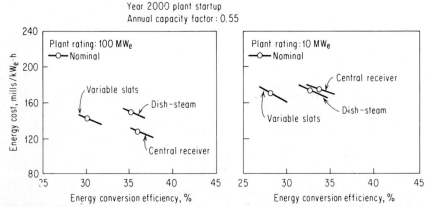

Fig. 20.23 Sensitivity of power cost to energy conversion efficiency for 10- and 100-MW$_e$ Rankine cycle systems.

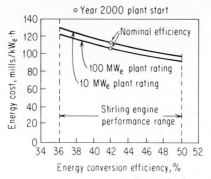

Fig. 20.24 Sensitivity of power cost to energy conversion efficiency for the paraboloid dish-electric system using a Stirling engine.

The nominal efficiency levels for the Rankine system were based on intermediate-term technology forecasts which were between the present and far-term potential. The Stirling engine efficiency range was taken to be from 36 to 50 percent according to Fig. 20.24, and the engine life (2 to 10 years) was based on automotive engine developments. Energy cost increases rapidly as engine life decreases below 2 years. Total replacement is assumed for initial estimating purposes; however, some cost savings may be possible by overhaul of engines as opposed to total replacement.

Energy Storage Subsystem Uncertainty. Cost sensitivities for both sensible-heat thermal storage and advanced batteries are presented in Fig. 20.25. The estimated cost ranges are from one-half to twice nominal for thermal storage, and two-thirds to twice nominal for advanced batteries. Since only a fraction of the collected energy is stored, the effect of storage efficiency is less important than for subsystems such as energy conversion.

Operation and Maintenance Cost Uncertainty. The sensitivity to O&M cost uncertainties is presented in Fig. 20.26. The estimates of O&M costs are extremely uncertain, and detailed studies are needed here. The nominal values used in this study correspond to the lower bound determined in Ref. 40, since it was interpreted to be most appropriate.

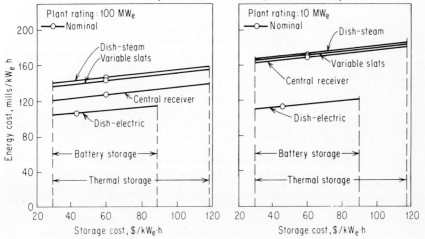

Fig. 20.25 Sensitivity of power cost to storage cost for 10- and 100-MW$_e$ plants.

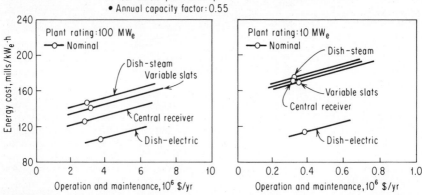

Fig. 20.26   Sensitivity of power cost to operation and maintenance cost for 10- and 100-MW$_e$ plants.

Summary. Subsystem sensitivity studies are useful in delineating the dominant parameters that affect system performance and economics. The cost and efficiency of the collector field are primary drivers for all systems (see Figs. 20.21 and 20.22). Conversion system efficiency is also a major driver since achievement of high efficiencies will enable a reduction in the size and cost of the collector field for a given plant rating. The storage and transport subsystems do not have a dominant effect, as shown by storage costs, for example, in Fig. 20.25.

The O&M cost (Fig. 20.26) has a sizable impact, and detailed studies should be conducted in parallel with subsystem/component development. This is particularly needed since there is a direct tradeoff between capital costs and O&M; i.e., a system designed for low maintenance will generally require higher capital costs.

It is possible to use these sensitivity analyses to predict plant energy costs. If more than one factor is involved (i.e., collector cost and efficiency, as well as energy conversion cost), a first-order approximation to correcting the system energy cost is to multiply the system cost ratios caused by each factor independently to arrive at an overall cost-correction factor. The greater the number of factors considered simultaneously, the greater the error in this simple, linear approach, however.

Grid Interaction

A limited number of studies are available which evaluate some aspect of the interaction of solar thermal power plants and a utility grid.[6,3,14,37,36] Apparently a good deal of additional work is required before there is a full understanding of the relationship.

Fuel-saver plants   A fuel-saver solar plant has no storage on site and thus has a solar multiplier of unity. The maximum thermal output of the field of collectors is sufficient to operate the energy conversion equipment at rated power. It is usually felt that this type of plant will have little or no capacity displacement when introduced into a utility grid. That is, the use of 1000 MW$_e$ of fuel-saver solar plants does not displace the introduction of 1000 MW$_e$ of conventional plants. Therefore, the cost of the solar plant can be compared only to conventional plant operation with fuel and maintenance costs, with little ability to displace capital costs. This tentative conclusion tends to favor the initial use of solar repowering plants (no storage or energy conversion) which generate steam for existing fossil plants because such plants minimize solar capital costs.

However, for fuel-saver plants used in conjunction with grid storage, which is possibly located elsewhere, a very different conclusion emerges. If about 4 h of storage is introduced at the rated power of the solar fuel-saver plant, the solar plant may achieve more than 100 percent capacity credit.[14] This is achieved by not operating the grid storage in any direct relationship to the fuel-saver solar plant, as with a stand-alone solar plant. The overall view is taken that the most effective way to use grid storage is to charge it with the least-expensive off-peak energy available, and to dispatch the stored energy against the most expensive peaking energy.

Reference 36 also indicates that a full capacity displacement is possible at low solar plant penetration into the grid ($\sim$ 5 percent). Thus the manner in which storage is dispatched, the total amount of solar penetration into a grid, and the total amount of storage have a direct bearing on: (1) the ability of a solar plant to displace both capacity and energy, (2) the type of conventional plants displaced by solar, and (3) the economic value of a solar plant to a utility. The analysis of Ref. 14 was limited to a 20 percent grid penetration, and it is not known at what penetration the combination of a fuel-saver solar plant and off-peak storage will no longer give full capacity displacement.

Stand-alone plant   Attempts to make a solar thermal-electric power plant become most like a conventional fossil or nuclear plant result in the stand-alone solar plant. Here storage is dedicated solely to the solar plant, and the collector field is oversized (solar multiplier greater than 1). The excess energy collected is stored for use when insolation is not sufficient to drive the energy conversion subsystem at rated power. The result is that the solar plant generates rated or near-rated power over most of the day.

As an initial attempt to develop a basis for understanding the relationship between a stand-alone solar plant and a utility grid, an approach was taken that required that the solar plant act like a conventional baseload plant.[3,6] A utility grid demand was created to simulate hour-by-hour demand in a future time frame.[36] The utility grid supplying power was simulated and included baseload, intermediate, and peaking plants. This combination of demand and generating capacity was simulated, and sufficient extra generating capacity (margin) was introduced into the grid to satisfy standard utility reliability requirements. The margin required for the conventional grid was 15 to 20 percent of peak demand. Solar plants were then introduced into the grid by substituting them directly for baseload capacity. The margin needed to keep the grid as reliable as it was with only conventional plants was introduced into the utility grid, along with required backup energy to run this extra margin. The total penetration of solar was varied, as was the designed annual capacity factor of the solar plants. The results of this analysis are shown in Fig. 20.27. (SCE refers to Southern California Edison.)

Several terms used in Fig. 20.27 are:

P_b, the extra margin or backup capacity needed so that the grid is as reliable with solar plants as it was with conventional plants.

P_r, the rated capacity of the solar plants introduced into the grid.

L, the solar plant annual capacity factor.

L_o, the conventional baseload plant annual capacity factor (usually 0.55 to 0.75).

E_b, the extra conventional energy required to operate the extra margin P_b so that the grid is as reliable with solar plants as it was with conventional plants.

E_r, the rated annual energy generated by the baseload plant operating at the assumed annual capacity factor L.

As can be seen from Fig. 20.27, the design solar load factor plays a dominant role in the performance of stand-alone solar "baseload" plants. If a solar plant has little or no storage

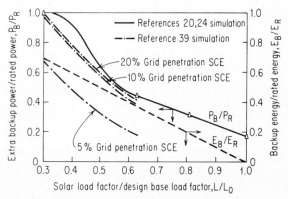

Fig. 20.27   Back-up power and energy for solar stand-alone plants (SCE refers to the Southern California Edison grid).

$(L/L_o < 0.3)$, the extra backup capacity (grid margin) is essentially equal to the rated solar plant capacity. That is, $P_b/P_r \approx 1.0$, and the solar plant provides little or no capacity displacement. The conventional energy required to run this backup capacity is about 0.7 of the annual energy used to run the baseload plant. This essentially is the fuel-saver mode of operation and is not effective at capacity displacement when substituted for baseload conventional plants, as expected.

At the right side of the curve in Fig. 20.27, the opposite situation occurs. If a solar plant is designed so that it has sufficient storage and a large solar multiplier, the annual capacity factor can be made to equal the traditional value for the conventional baseload plant capacity factor, $L/L_o \approx 1.0$. For $L/L_o = 1$ and L_p equal to 0.7, the corresponding point for the dish-Stirling solar plant would have 0.90 km^2 of collectors and about 7 h of storage at rated capacity (see Fig. 20.14). The extra backup power is between 15 and 20 percent of the rated capacity of the solar plant, and there is essentially no backup energy required. Thus a "baseload" solar plant designed at the same annual capacity factor as a conventional baseload plant required about the same extra margin for it to compensate for the variation of insolation as the conventional utility needs to compensate for traditional causes of unreliability. Therefore a "baseload" solar plant needs what might be called double margin for the solar and for the nonsolar sources of unreliability. As shown in Fig. 20.27, L/L_o values between 0.3 and 1.0 have varying requirements for extra backup capacity and energy for the solar plants. These results were achieved by use of a single site for all the solar capacity (Inyokern, California). If a variety of sites were used within the utility service area, the reliability requirements for the solar plants would be reduced.[36] Thus the results shown in Fig. 20.27 are conservative in that they are an upper bound for the backup (extra margin) requirements for solar "baseload" plants.

BEYOND ECONOMICS

The social acceptance of an energy system includes more than marketplace economics.[41] A host of factors are important, such as whether a system is based on a renewable or depletable fuel; foreign political dependency: environment impacts; occupational and public health effects; resource use, including material, land, water, and capital; employment intensity; weapon diversion potential; long-term toxic waste risk; energy ownership arrangements; the cost of R&D to develop a commercial system; the severity of boom-bust cycles; etc. An approach to evaluating competing energy systems is to consider the *total social cost* that is made up of both economic and social costs. Earlier sections discussed projections of economic costs of solar thermal-electric systems. An estimate of social costs would first involve identifying each characteristic of competing energy systems that has importance to some members of society. Then a method would have to be developed that actually sensed the cost of each of these energy system characteristics on a scale in some way convertible to economic cost. This would have to be done for a nationally averaged constituency, as well as for special-interest groups throughout society. The major elements that could constitute the ingredients of a total social cost accounting system are shown in Fig. 20.28. They are utility cost, RD&D cost, health costs, resource use, environmental cost, and "other." This last category contains items of considerable concern which are either nonquantitative, such as risk of sabotage, or are quantitatively known but whose effects are poorly understood, such as effects of CO_2 on global weather. These factors are evaluated for each competing system over the seven stages in the life of a system. This study matrix is shown in Fig. 20.29.

Estimates of marginal costing of bus-bar and delivered electrical energy for the year 2000 vary widely. Based on Ref. 3, the capital and energy costs for year 2000 start-ups of a new LWR, combined-cycle coal plant, silicon photovoltaic plant (at $0.50/$w_p$ module), and a solar thermal plant (at $90/m^2$ direct heliostat cost) are shown in Table 20.10 in 1975 dollars. (Differential inflation is included, and is assumed to be substantial for socially acceptable nuclear and coal plants.) The conventional plants appear to have a marketplace economic edge over the solar plants in baseload operation projected to the year 2000. This ranking may be reversed when social cost factors are considered.

A preliminary analysis of the maximum health impacts of each type of power plant is shown in Table 20.11.[3,42] This type of analysis is difficult at best, but these estimates are felt to roughly represent the maximum health impacts of these systems.

The solar plants are estimated to cause about five deaths over the plant life without fossil backup, while a nuclear plant is estimated about one order of magnitude more (50). The coal plant considered is one order of magnitude more than the nuclear plant (530). Using 10 percent

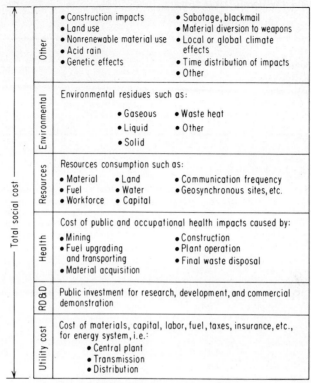

Fig. 20.28 Total social cost framework.

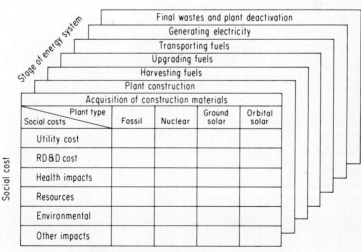

Fig. 20.29 Study matrix for total social cost assessment.

TABLE 20.11   Maximum Health Impacts

Stage of system	Coal[a]	Nuclear[b]	Photovoltaic[c]	Solar thermal[d]
Fuel cycle, PDL/MW_e year[e]	200[f]	15.6[g]	0[h]–4.4[i]	0[h]–4.4[i]
Construction and material acquisition, PDL/MW_e year	1	1.4	3–4.4	7–8
Total, PDL/MW_e year	201	17	3–9	7–13
Deaths/plant[j]	530	50	8–35	3–30

[a] See footnote a on Table 20.10.
[b] See footnote b on Table 20.10.
[c] See footnote c on Table 20.10.
[d] See footnote d on Table 20.10.
[e] PDL = person-days loss due to accident, illness, and death (6000 PDL = one death).
[f] Does not consider NO_x, CO, and other pollutants besides SO_x and particulates.
[g] Does not consider sabotage, blackmail, material diversion to a weapon, genetic effects, long-term-waste health effects.
[h] Stand-alone solar plant.
[i] Includes 10% backup energy from coal.
[j] Based on plant construction and 30-year life.

TABLE 20.12   Resource Use

Resource	Coal	Nuclear	Photovoltaic	Solar thermal
Land, m^2/MW_e year				
Plant and fuel cycle	2000–4700[a]	115[b]	1000[c]–2850[d]	2000
Total including transmission lines	2250–5000[e]	400[e]–1100[f]	1025[c,g]–4350[d,g]	2300[e]–4000[g]
Water, 10^6 liters/MW_e year	0.5–9.2[h]	1–24[h]	0.6[i]	0.9[j,k]
Construction materials, t/MW_e year				
Concrete	3	12.7	4.3	174
Steel	3	2.3	—	39
Other metal			55.7[l]	2.2
Glass			5[m]	6.3
Other			—	82[n]
Fuel, t/MW_e year	3500	100[b]	350[o]	350[o]
Worker power, worker-hours/MW_e year	2640	1120	2700	14,400

[a] Based on 30-year unavailability of land.
[b] Based on current uranium ore grades. Expected to increase as mining goes to lower-grade ores.
[c] 75 percent on buildings.
[d] 25 percent on buildings.
[e] 300-mi transmission.
[f] 1000-mi transmission.
[g] 2000-mi transmission.
[h] Range reflects going from dry to wet cooling towers.
[i] Washed every 10 weeks.
[j] Washed every 5 weeks.
[k] Dry cooling tower.
[l] Uses all-aluminum construction. Steel could be substituted.
[m] Half glass, half silicon.
[n] 90 percent granite rock, 10 percent heat-transfer oil.
[o] Assumes 10 percent backup energy using coal.

backup energy for baseload applications raises the solar plant estimates of deaths to the 30 to 35 range. Use of another type of backup fuel such as a biomass-derived fuel may reduce this range. The total person-days loss due to illness, accident, and death indicates a similar relationship.

The resources used are summarized in Table 20.12, and estimates are presented for land, water, construction material, fuel, and worker power. The required land depends on assumptions of transmission distance, longevity of land unavailability because of coal-mining activities, uranium-grade ore mined, and the fraction of photovoltaic system placed on buildings. Land required for the solar thermal plant is similar to coal and photovoltaic, but more than nuclear. Type of land is not reflected in these numbers.

Water consumption strongly depends on assumptions of the type of cooling towers and frequency of washing collectors. Once-through cooling, which is used extensively today, is an order of magnitude more consumptive than wet cooling-tower water requirements. Construction material use is also estimated for the solar thermal plant using collectors made of glass, steel, and concrete—the most material-intensive component. Some designs being considered are based on thin aluminized Mylar™ reflectors within plastic bubbles. This approach uses much less material and may prove to be the preferred commercial design. Fuel use is most intense in the fossil plant, while worker-power use is greatest for the solar thermal plant. This type of plant would provide more employment opportunities, which is a distinct benefit if unemployment is a persistent social problem (see Chap. 27).

Environmental residues and excess waste heat are shown in Table 20.13. Excess waste heat (MW_t year/MW_e year) is used to indicate the heat deposited in the biosphere. It is the amount of excess heat energy left in the biosphere by the plant divided by the electric energy generated. Excess heat is that which is deposited above that if there were no plant at the site. The solar thermal plant has least impact since it leaves the least amount of waste heat.

The magnitude of environmental residues is vastly lower for the solar plants compared with the coal and nuclear plants.

TABLE 20.13   Environmental Residues and Excess Waste Heat

	Coal	Nuclear	Photovoltaic	Solar thermal
Excess waste heat, MW_t year/MW_e year	1.7	2.1	0.5^a–1.2^b	0.25
Residues, t/MW_e year				
Water pollutants				
COD (chemical oxygen demand)	ND^c	ND	1.2	
Other dissolved solids	ND	ND	0.5	
Organic substances	ND	ND	0.2	
Acid	660–55,000			
Suspended coal	0–8			
Sludge	1.6–1.4			
Nonradioactive		260–4230		
Radioactive curies/MW_e year		0.1–4.5		
Air pollutants				
Particulates	4.8–44.9		11.2	5.7
NO_x	14.3–28.4	0.45		1.0
SO_x	12.1–41.9	1.2		
Hydrocarbons	0.8			
CO	0.6–2.4			0.2
Aldehydes			0.2	
Toxic metals	0.02			
Radioactive curies/MW_e year		4.7–600		
Solid pollutants				
Nonradioactive	1875–2316	105–000		
Radioactive				
High level, L/MW_e year		43–48		
Low level L/MW_e year		1530		
Intermediate level, L/MW_e year		30.7		
Buried solids m³/MW_e year		0.24		
Tailings curies/MW_e year		0.01–0.02		

^{a}75% photovoltaics on buildings.
^{b}25% photovoltaics on buildings.
^{c}ND: no data.
NOTE: No entry indicates residue less than 0.1 t/MW_e year.

Other important factors which would significantly affect the social cost of an energy system are listed in Table 20.14. These "other" social costs have characteristics that are nonquantitative or that are quantitatively known but for which the effects are poorly understood. An example of the first would be the degree of catastrophe associated with a health effect. There apparently is greater perceived social cost (impact) if an energy system's health effects occur all at once in time and location (i.e., nuclear core melt-down or an orbital launch vehicle falling on a population center), vs. a more even distribution of health effects (i.e., from coal plants). An example of a poorly understood but quantitatively known effect would be the amount of CO_2 and particulates which are released from a coal fuel cycle. The magnitude is known but the global climatic effects are not well known, nor are the ramifications of potential climate changes.

To deal to some extent with these types of characteristics of electric power systems, a rather simple subjective comparative evaluation is proposed. Social cost areas of this type were identified and a rating of low (L), medium (M), high (H), or very high (V) was given for each type of central electric power system based on the subjective judgment of the author. Such ratings are only an indication of the relative magnitude of the social impact of a particular impact area.

The first four areas, shown in Table 20.14, which have to do with sabotage, material diversion to weapons, catastrophic impacts, and duration of impacts, mainly affect nuclear power plants. The ratings are either high or very high. For nuclear plants, much speculation on these dangers is available publicly. The oil system is very susceptible to interruption militarily and will be increasingly so until the resource becomes fully depleted early next century.

The next category of other impacts is CO_2-particulate emissions and acid rain, which are residuals of fossil plants. Particulates and acid rain can be controlled to some extent and will be reduced in the reference coal plant, since about 99 percent of the sulfur is removed and a

TABLE 20.14 Summary of Relative Potential of Other Social Costs

	Fossil		Nuclear		Solar ground	
	Coal	Oil	LWR	LMFBR	Thermal	Photovoltaic
Sabotage, blackmail		L	H	V		
Material diversion to weapon			H	V		
Catastrophic impact of above or accident	L	L	H	V		
Duration of impact		L	V	V		
Military vulnerability	H	H				
CO_2 and particulate emissions	H	H				
Acid rain	H	H				
Net thermal emission	H	H	V	H	L	H
Long-term toxic waste			V	V		
Lack of flexibility in possible ownership arrangements	V	V	V	V	L	L
Life-cycle mass utilization	V	V	L	L	M	M
Nonrenewable resource use	V	V	V	L	L	L
Land use						
Area	H	M	L	L	H	H
Area and time	L	L	V	V	L	L
Local disruption						
Construction	M	M	M	M	H	H
Operation	H	M	L	L	M	M
Esthetic impact	H	L			M	H
Legal, liability			H	H		

KEY: L = low   H = high   No entry = nil or little
     M = medium   V = very high

clean gas is burned in the power plant. The effect of CO_2 and particulates on global climate is difficult to assess, as is the effect of acid rain on human health and vegetation.

Thermal emission effects are a characteristic of all energy systems, and the magnitude of excess waste heat was indicated earlier. Even the generated electrical energy should be included along with the excess waste heat, since it eventually becomes heat. Power plant heat islands, or increased moisture if wet cooling towers are used, will have some impact on local climate. The magnitude and nature of the impacts are very site-specific. In general, power plants contribute to the global heat burden. With continued growth, this heat burden could reach a significant fraction of global solar input in several centuries with profound global effects. Approximately 0.01 units of global solar input could be reached by 2070 at 5 percent growth of energy use. The LWR system produces the most net thermal emission since it is least efficient. Fossil, advanced reactors (LMFBR), and ground photovoltaics have less excess heat emission, while that for the ground solar thermal plant is an order of magnitude less than the LWR system. Although the relative magnitudes were shown earlier, the long-term climatic effects are unknown.

Long-term toxic waste disposal is a problem of nuclear systems; some waste products must be confined outside the biosphere for more than 100,000 years if they are not transmuted to substances with shorter half lives. The length of time of toxicity and the degree of toxicity of wastes in certain forms also contribute to social impacts.

Energy systems which have limited ownership arrangements (large, multinational corporations) lack flexibility to meet certain social needs which depend on more self-control over the circumstances of everyday life. Most solar systems are unique since they can be designed in many ways using basically the same equipment. That is, a solar system can be large or small, located centrally or in the load center, and owned/controlled by users, utilities, or other bodies. (The only exceptions are OTEC, satellites, and certain power-tower designs.) This technical flexibility contributes to a social-cultural flexibility in use, and is probably one of the strongest assets of solar energy. At this point it is a poorly understood characteristic, and one has the feeling that no matter what future social values are, solar energy will be adaptable to those values.

The use of nonrenewable resources is greatest for fossil and LWR systems. The breeder reactor would have as low impact at high breeding ratios as solar plants, since most of the materials can be recycled. Depriving future generations of nonrenewable fuels is a difficult impact to assess.

Another category of other social costs in Table 20.14 is local disruption during the construction phase; it is potentially large for ground solar systems because of the greater material and land requirements. The local disruption of the coal and nuclear plant construction is probably lower than that of the solar systems because of lower material and land requirements. Continued coal mining would sustain high impacts during operation.

The esthetic impact of coal mining is high, and large ground solar plants would change the appearance of large sections of the Southwest areas. Nuclear plants are compact and should have little adverse visual impact.

The liability area may become an increasing difficulty for nuclear systems because of the large potential damage from LWR core melt-down or LMFBR nuclear explosions which cannot be contained.

This preliminary compilation of other social costs is useful only in identifying some issues which could have a very strong bearing on the social acceptability of these power systems. A more careful development of these and other social costs is necessary and should be the subject of future work.

It is no longer possible to base energy system decisions just on a projection of marketplace economics. Additional factors must be considered in this social decision, and a broad input is required from a thoughtful society. If this is not done, there can be no assurance that the energy system coming on line 15 to 30 years from now will be socially acceptable.

REFERENCES

1. Impact Panel Report, *Domestic Council Review of Solar Energy*, October 1978.
2. A. D. Rossin and T. A. Riech, Economics of Nuclear Power, *Science*, vol. 201, no. 4356, Aug. 18, 1978.
3. R. Caputo, An Initial Comparative Assessment of Orbital and Terrestrial Central Power Systems, Final Report 77-44, Jet Propulsion Laboratory, Pasadena, CA, March 1977.
4. G. Francia, Pilot Plants of Solar Steam Generating Stations, *Solar Energy*, vol. 12, 1968, pp. 51–64.

5. R. Caputo, An Initial Study of Solar Power Plants Using a Distributed Network of Point Focusing Collectors, Internal Report 900-724, Jet Propulsion Laboratory, Pasadena, CA, July 1975.
6. R. Manvi, Terrestrial Solar Power Plant Performance, preliminary draft internal report, Jet Propulsion Laboratory, Pasadena, CA, March 1977.
7. "Central Receiver Solar Thermal Power Systems, Phase I," vols. I–VII, McDonnell Douglas Astronautics Co., May 1977.
8. J. C. Powell et al., Dynamic Conversion of Solar Generated Heat to Electricity, Honeywell, Inc., NASA CR-134724, August 1974.
9. "Solar Thermal Electric Power Systems," vols. I–III, NSF/RANN/SE/GI-37815/FR/74/3, Colorado State University–Westinghouse Electric, November 1974.
10. Solar Hybrid Repowering Project Review Meeting, Public Service Company of New Mexico, Albuquerque, NM, Feb. 16, 1978.
11. N. Lord et al., Feasibility of Solar Repowering Existing Electric Generating Stations in the Southwest, WP-12875, Mitre Corp., Metrek Div., March 1978.
12. H. Bloomfield and J. Calozeras, Technical and Economic Feasibility Study of Solar/Fossil Hybrid Power Systems, NASA TM-73820, December 1977.
13. P. A. Curto and Z. D. Nikodea, Solar Thermal Repowering, Mitre Corp., MTR-7861, May 1978.
14. C. R. Chowaniec et al., Energy Storage Operation of Combined Photovoltaic/Battery Plants in Utility Networks, Westinghouse Electric Corp.
15. "Central Receiver Solar Thermal Power System, Phase I," vols. I–VII, Martin Marietta Corp., June 1977.
16. "Central Receiver Solar Thermal Power System, Phase I," vols, I–VII, Honeywell Inc., June 1977.
17. M. K. Selcuk, A Preliminary Technical and Economic Assessment of Solar Power Plants Using Line Concentrators, Internal Report 900-704, Jet Propulsion Laboratory, Pasadena, CA, July 1975.
18. M. K. Selcuk, Preliminary Assessment of Flat Plate Collector Solar Thermal Power Plants, Internal Report 900-692, Jet Propulsion Laboratory, Pasadena, CA, March 1975.
19. R. H. Turner, Economic Optimization of the Energy Transport Component of a Large Distributed Collector Solar Power Plant, 11th IECEC, State Line, NV, September 1976.
20. R.S. Caputo et al., Projection of Distributed-Collector Solar-Thermal Electric Power Plants Economics to 1990–2000, Briefing Document to ERDA, 5102-39, June 21, 1977, Jet Propulsion Laboratory, Pasadena, CA.
21. R. S. Caputo, Solar Power Plants Using Vee-Trough Concentrators, preliminary draft EM 341-0060B, Jet Propulsion Laboratory, Pasadena, CA, December 1976.
22. Should We Have a New Engine? An Automobile Power Systems Evaluation, Jet Propulsion Laboratory JPLSP 43-17, August 1975.
23. D. G. Wilson, "Alternative Automobile Engines," *Scientific American*, vol. 239, 1, July 1978.
24. M. K. Selcuk, R. S. Caputo, R. French, and J. Finegold, Preliminary Evaluation of a Parabolic Dish-Small Heat Engine Central Solar Plant, Internal Report 900-749, Jet Propulsion Laboratory, Pasadena, CA.
25. T. Fujita, Preliminary Data on Electrical Energy Storage System for Utility Applications, Internal Report 900-738, Jet Propulsion Laboratory, Pasadena, CA, April 1976.
26. T. Fujita, Underground Energy Storage for Electric Utilities Employing Hydrogen Energy Systems, Internal Report 900-744, Jet Propulsion Laboratory, Pasadena, CA, June 1976.
27. R. Turner, Thermal Energy Storage for Solar Power Plants, Internal Report 900-754, Jet Propulsion Laboratory, Pasadena, CA, July 1976.
28. M. K. Selcuk, Survey of Several Central Receiver Solar Thermal Power Plant Design Concepts, Internal Report 900-417, Jet Propulsion Laboratory, Pasadena, CA, August 1975.
29. T. Fujita, N. El Gabalani, G. Herrera, and R. H. Turner, Projection of Distributed-Collector Solar-Thermal Electric Power Plants to Years 1990–2000, Jet Propulsion Laboratory, Pasadena, CA, December 1977, updated October 1978.
30. K. Kalheimer, Preliminary Cost Estimates for Heliostat Production Facilities, Battelle Northwest Laboratories, March 1978 (unpublished).
31. A. Wen and R. Caputo, Brief Review of Increasing Geometric Concentration Ratio (Decreasing Manufacturing Tolerances) versus Improving Receiver Surface Characteristics for Selected Solar Thermal Collectors, DOE Solar Thermal Program, Washington, DC, Jan. 24, 1978.
32. M. Baily, Near Term Engine Review, Briefing Document 5102-11, presentation to ERDA Thermal Power Systems, NASA LERC/JPL, Apr. 28, 1977.
33. M. Gutstein, E. R. Furman, and G. M. Kaplan, Liquid Metal Binary Cycles for Stationary Power, NASA TND-7955, Lewis Research Center, Cleveland, OH, August 1975.
34. R. Manvi and T. Fujita, Economics of Internal and External Energy Storage in Solar Power Plant Operation, 12th IECEC Proceedings, Jet Propulsion Laboratory, Pasadena, CA.
35. T. R. Schneider, An Assessment of Energy Storage Systems Suitable for Use by Electric Utilities, EM-264, EPRI Project 225/ERDA E(11-1)-2501, Public Service Gas and Electric Co., Newark, NJ, July 1976.
36. Solar Thermal Conversion Mission Analysis, Midterm Report, Report No. ATR-76 (7506-05)-1, The Aerospace Corporation, El Segundo, CA, Sept. 1, 1975.
37. G. W. Braun et al., Integration of Solar Thermal Power Plants into Electric Utility Systems, Southern California Edison, 76-RD, September 1976.

38. J. W. Doane et al., The Cost of Energy from Utility-Owned Solar Electric Systems, a Required Revenue Methodology for ERDA/EPRI Evaluations, Internal Report 5040-29, ERDA/JPL-1012-7613, Jet Propulsion Laboratory, Pasadena, CA. June 1976.
39. P. Mathur, Advanced Central Power Systems Analysis, *Semi-Annual Review of Solar Central Power Systems,* December 1976.
40. J. G. Herrera (ed.), Assessment of RD&D, Resources, Health and Environmental Effects, O&M Costs, and Other Social Costs for Conventional and Terrestrial Solar Electric Plants, Internal Report 900-782, Jet Propulsion Laboratory, Pasadena, CA, January 1977.
41. R. Caputo, Solar Power Plants: Dark Horse in the Energy Stable, *Bulletin of the Atomic Scientists,* May 1977, pp. 46–56.
42. K. R. Smith, J. Weyant, and J. Holdren, Evaluation of Conventional Power Systems, ERG75-5, University of California at Berkeley, July 1975.

Chapter **21**

Solar Process Heat Systems

M. D. FRASER
H. S. LIERS

InterTechnology/Solar Corporation, Warrenton, VA

The potential for the use of solar process heat in industry appears to be significant. Industrial process heat accounts for a significant fraction of the amount of energy consumed in the United States, and surveys have indicated that a certain portion of this process heat technologically could be provided by solar thermal energy.

The purpose of this chapter is to show how available solar technology can be utilized in an industrial environment. Other chapters of this handbook deal with the various elements of solar systems, such as collector types, storage, radiation, materials, and economics. Part of the goal in this chapter is to integrate this information so that a system designer can carry out a preliminary assessment of how solar energy could be used for a particular process heat application. More specifically, the material presented in this chapter should enable him to select a system conceptual design, including collector type, and storage type and size to estimate the amount of energy provided by a solar process heat system and to perform a rough cost estimate. A method is also presented for calculating the optimal collector field size.

Finally, conceptualized solar process heat system designs are shown to illustrate some design principles and to point out some important considerations. Examples of systems are described— one experimental system and another example based upon the information in this chapter— to illustrate the possible design and use of solar process heat systems. Possible process applications are also suggested.

USE OF PROCESS HEAT BY INDUSTRY

Among the various sectors of the U.S. economy, the largest energy user is the industrial sector. In 1968 it accounted for about 41 percent of the total national energy use, compared with about 19 percent in the residential sector, 15 percent in the commercial sector, and 25 percent in the transportation sector.

In the industrial sector, energy is used for a number of purposes, including the nonenergy use as a feedstock for chemical manufacture. In 1968 the total industrial use of energy was 2.64 $\times$ 10^{19}J (25.0 $\times$ 10^{15} Btu) for all purposes, including 2.3 $\times$ 10^{18}J (2.2 $\times$ 10^{15} Btu) used as feedstock.[1] The breakdown of this total industrial use of energy is as follows:

- Process steam   40.6 percent
- Electric drive   19.2 percent

- Electrolytic process   2.8 percent
- Direct process heat   27.8 percent
- Feedstock for chemicals   8.8 percent
- Other   0.8 percent

Process steam and direct process heat together accounted for 68.4 percent of the total industrial use of energy. It is thus clear that the greater portion of the energy used in the industrial sector is used in the form of thermal energy rather than in the form of electrical energy for power, and this indicates the significant potential for the use of solar thermal energy in industry.

However, one of the most important variables to consider in the application of solar process heat is the temperature that is required for the process.

Temperature Requirements for Industrial Process Heat

The temperature requirements for industrial process heat in the United States have been surveyed by two studies. In one study,[2] the survey was performed from the point of view of process requirements rather than from the point of view of current methods of using heat. Thus, the temperature of major interest was the required temperature of the process material rather than the temperature at which the heat is currently provided.

The data from this study were presented in the form of a cumulative process heat spectrum, which shows the percent of industrial process heat used as a function of terminal process temperature required. This cumulative process heat spectrum is shown in Fig. 21.1.

The data base for this curve consisted of process heat data from 78 different industries as defined by Standard Industrial Classification codes (SIC groups[3]) in both mining and manufacturing industries. The data base included process heat applications consuming 1.03×10^{19} J (9.8×10^{15} Btu) per year, about 59 percent of the estimated total use of process heat by U.S. industry of 1.75×10^{19} J (1.66×10^{16} Btu), based on 1974 data.

Of particular interest in Fig. 21.1 is the percent of process heat needed at terminal process temperatures below a temperature of 100°C—about 7½ percent—which could perhaps be provided by low-temperature solar thermal energy systems, and the percent of process heat

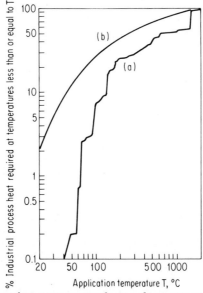

Fig. 21.1 Cumulative process heat requirements as function of process temperature: (a) all heat supplied at required process temperature, (b) heat supplied as preheat from 16°C up to T°C. (*From InterTechnology/Solar Corporation, "Analysis of the Economic Potential of Solar Thermal Energy to Provide Industrial Process Heat," p. 1243, Report No. COO/2829-76/1, U. S. Energy Research and Development Administration, February 7, 1977.*)

needed below a temperature of 200°C—about 26 percent—which could perhaps be provided by concentrating solar collectors.

Two observations should be noted about the terminal-temperature spectrum in Fig. 21.1. First, the temperature indicated is from the point of view of the process *requirement*, as opposed to the temperature of the present heat-delivering medium, which may be much higher. Second, in getting the temperature of the heat-delivering medium or the process material itself up to the terminal temperature, part of the heat could possibly be supplied by solar at a lower temperature as preheat. To illustrate the potential of using solar as preheat, the terminal-temperature curve was recalculated with the assumption that the amount of heat for each application could be supplied over a temperature range from a base of 16°C up to the terminal temperature required.

Another study[4] surveyed the process heat requirements of 20 particular industries. From this study, Table 21.1 shows the process heat requirements for these industries, grouped within SIC groups, based upon extrapolations of the detailed data obtained in the survey. This table shows a distribution of temperatures actually used rather than process temperatures and a breakdown of the type of process heat usage by industry.

Table 21.2 shows estimates of the future process heat requirements of these 20 industries by energy form and temperature. These requirements illustrate the relative potentials for solar process heat systems supplying either hot water, hot air, or steam.

Types of Fuels Used by Industry

The cost-effectiveness of solar energy used for process heat in competition with conventional fossil fuels depends upon the type of fossil fuel and its cost and availability. The types of fuels used by industry and the uses are shown in Table 21.3. This table shows that for fuel use,

TABLE 21.1   Summary of Process Heat Requirements for Selected Industries in the United States*

SIC Code	Description	Hot water <212°F	Steam 212–350°F	Steam >350°F	Direct heat/hot air <212°F	Direct heat/hot air 212–350°F	Direct heat/hot air >350°F	Rounded totals
1211	Bituminous coal and lignite						11	11
1477	Sulfur mining		44					44
20	Food and kindred products	60	275	100	10	95	15	555
22	Textile mill products	19	191	4		69	13	296
24	Lumber and wood products	5	21	4	106	4	70	210
26	Paper and allied products		465				94	559
28	Chemicals		1400			450	250	2100
2911	Petroleum refining		120	380			2600	3100
32	Stone, clay, and glass	8	20	37		24	1081	1170
3312	Blast furnaces and steel mills		65				1712	1777
3331	Primary copper	14.2			2.4		55.4	72
3334	Primary aluminum			38.4			63.1	102
3711 3712 3713	Automobile and truck manu- facturing	13	1.4		21.3	10	0.9	47
	Rounded totals	120	2600	563	140	652	5965	10,040
	Percent of total	1.2	25.9	5.6	1.4	6.5	59.4	

*Figures in 10^{12} Btu/year.

SOURCE: Battelle Columbus Laboratories et al., "Survey of the Application of Solar Thermal Energy Systems to Industrial Process Heat," p. 15, Report No. TID-27348/1, U.S. Energy Research and Development Administration, January 1977.

TABLE 21.2   Summary of Industrial Process Heat Requirements in the United States*

Energy form, temperature	Process-specific base data	20-industry extrapolated data		
		Present	Projected to 1985	Projected to 2000
Hot water < 212°F	73	120	148	201
Steam 212–350°F	1212	2600	3211	4351
Steam > 350°F	496	563	695	942
Direct heat/hot air < 212°F	98	140	173	234
Direct heat/hot air 212–350°F	140	652	805	1091
Direct heat/hot air > 350°F	5851	5965	7367	9981
Totals	7870	10,040	12,400	16,800
Totals below 350°F	1523	3512	4337	5877

*Figures in 10^{12} Btu/year.

SOURCE: Battelle Columbus Laboratories et al., "Survey of the Application of Solar Thermal Energy Systems to Industrial Process Heat," p. 16, Report No. TID-27348/1, U.S. Energy Research and Development Administration, January 1977.

TABLE 21.3   Industrial Consumption of Energy in the United States, 1974–2000*

Fuel	1974	1980	1985	2000
Coal	7831	9466	11,738	19,907
Fuel use	(4093)	(4600)	(4680)	(5460)
Nonfuel use	(115)	(200)	(250)	(45)
Electric power	(3623)	(4666)	(6531)	(10,288)
Synthetic gas†	—	—	(277)	(3230)
Synthetic liquids†	—	—	—	(479)
Petroleum	7485	9442	11,124	12,706‡
Fuel use	(3760)	(4380)	(4950)	(6200)
Nonfuel use	(2284)	(3120)	(3550)	(5400)
Electric power	(1441)	(1942)	(2579)	(2336)
Synthetic gas†			(75)	
Natural gas	12,520	10,762	10,124	9497§
Fuel use	(9495)	(9240)	(8920)	(10,360)
Nonfuel use	(1634)	(760)	(820)	(900)
Electric power	(1391)	(762)	(624)	(497)
Oil shale				
Synthetic liquids†	—	—	185	1281
Nuclear power				
Electric power	490	1733	4925	22,902
Hydro and geothermal power				
Electric power industrial	1296	1447	1602	3017
Total energy Consumption	29,600	32,900	39,700	69,300
Electric power as % of industrial total	27.8	32.1	41.0	56.4
Total industrial energy consumption as % of total U.S. energy consumption	40.5	37.8	38.2	42.4

*Figures in 10^{12} Btu/year.
†Gross energy requirements to manufacture synthetic fuels (industrial sector equivalent share).
‡Includes synthetic liquids, 1230×10^{12} Btu.
§Includes synthetic gas, 2260×10^{12} Btu.

SOURCE: Battelle Columbus Laboratories et al., "Survey of the Application of Solar Thermal Energy Systems to Industrial Process Heat," p. 29, Report No. TID-27348/1, U.S. Energy Research and Development Administration, January 1977.

natural gas is the predominant energy form. According to these estimates, more natural gas than any other fuel will continue to be consumed by industry as fuel, synthetic gas being manufactured and used to augment natural supplies.

With respect to regional variations in fuel availability, data indicate that industrial fuel mixes for contiguous states tend to be similar. Fuel use in any given industry tends to reflect a region's fuel distribution rather than an industry's national fuel mix and fuel distribution, although some processes may require a specific type of fuel for process reasons.

Costs of Fossil Fuels Used by Industry

The economic viability of solar thermal systems is in large part controlled by the cost of available competing fuels and their expected future costs. Future fuel costs are an important consideration to the solar designer because they are an integral part of the life-cycle cost analysis. Many different estimates of future fuel costs have been derived, although few are published in readily available literature.

Future fuel costs are generally estimated by applying an annual percent increase composed of two parts. One part is the annual increase due to inflation, and the other is a composite of social, political, economic, and technical factors which change with time. More detailed discussion of this model for estimating future fuel costs and the effect of inflation may be found in Chap. 28.

Projected future costs used in one study[5] for the three major fuels for the years 1985 and 2000 are:

$$\text{Cost of coal in } 1985 = (1976 \text{ unit cost}) (1 + r_i + 0.03)^9$$
$$\text{Cost of coal in } 2000 = (1985 \text{ unit cost}) (1 + r_i + 0.05)^{15}$$
$$\text{Cost of oil in } 1985 = (1976 \text{ unit cost}) (1 + r_i + 0.055)^9$$
$$\text{Cost of oil in } 2000 = (1985 \text{ unit cost}) (1 + r_i + 0.08)^{15}$$
$$\text{Cost of gas in } 1985 = (1976 \text{ unit cost}) (1 + r_i + 0.13)^9$$
$$\text{Cost of gas in } 2000 = (1985 \text{ unit cost}) (1 + r_i + 0.08)^{15}$$

where r_i = the general rate of inflation

Table 21.4 from another source shows estimates of ranges of future fuel costs in terms of constant dollars (1976 dollars).

TABLE 21.4   Estimated Price Ranges for Selected Fuels in the United States, 1976–2000*

Fuel	1976	1980	1985	2000
Petroleum				
Crude oil, composite @ refinery	1.70–1.80	2.25–2.50	2.50–3.50	3.50–4.50
No.2 distillate fuel @ terminal	2.00–2.30	2.45–2.75	2.70–3.75	3.70–4.75
Residual fuel, low sulfur	1.70–2.20	2.25–2.60	2.50–3.50	3.50–4.50
Natural Gas				
Industrial uses, average	0.85–0.95	1.70–2.25	2.75–3.25	3.50–4.50
Intrastate, new @ wellhead	1.90–2.10	2.25–2.75	2.75–3.50	3.50–5.00
LNG @ pipeline	1.90–2.00	2.25–2.50	2.50–3.25	3.50–4.50
Coal (steam)				
Utilities, average	0.80–0.90	1.00–1.50	1.25–1.75	1.50–2.50
East North Central Region	0.80–0.90	1.00–1.50	1.25–1.75	1.50–2.50
Mountain region	0.30–0.40	0.50–1.00	0.75–1.25	1.00–2.00
Synthetic Gas				
Pipeline quality @ pipeline	3.00–4.00†	3.50–4.50†	3.50–5.00	3.25–4.50
Low Btu, East North Central	3.00–3.50	3.00–3.25	3.00–3.25	3.00–4.25
Synthetic Liquid Fuel				
Oil shale	NA	NA	3.00–3.50	3.00–4.50
Coal	NA	NA	3.50–4.00	3.25–4.50

*Figures in constant 1976 dollars per million Btu.
†Produced from naphtha.
SOURCE: Battelle Columbus Laboratories et al., "Survey of the Application of Solar Thermal Energy Systems to Industrial Process Heat," p. 32, Report No. TID-27348/1, U.S. Energy Research and Development Administration, January 1977.

These future fuel costs indicate that solar process heat will be most cost-effective in competition with petroleum and natural gas, because they are, and will continue to be, the most expensive fossil fuels. One study concludes that it will be very difficult for solar to compete against coal strictly on the basis of fuel cost.[6] The advantages of solar as a clean fuel and a more detailed accounting of all of coal's costs such as for pollution control may permit solar to compete with coal in specific situations.

GENERAL PRINCIPLES OF SOLAR PROCESS HEAT SYSTEMS

Characterization of Solar Process Heat Systems

The basic building block in any solar system is the collector, and therefore the most significant part of the design of a solar process heat system is concerned with collector selection and sizing. Among the major variables affecting the choice of collector are: required amount of process energy, process energy demand pattern, required process temperatures, available solar energy, collector performance, and installed collector costs. A general procedure for evaluating various collectors is as follows:

  1. Conceptualize the solar system—e.g., devise a flow schematic for the system similar to those shown in the section on conceptualized designs in this chapter.

  2. From the information on the process, determine loads, load patterns, and temperatures.

  3. Determine available solar radiation and relevant ambient temperatures for the location (from Chap. 2, for example).

  4. From the collector manufacturer or other chapters of this handbook, find the collector efficiency relationship (efficiency is given as a function of the parameter $(T_{in} - T_{amb})/I$; see Chap. 7–9).

  5. Build a computer model which will allow system performance to be calculated.

  6. Exercise the model to determine performance (e.g., fraction of load supplied by solar) for different collector array sizes.

  7. Using an economic optimizing procedure, determine the best system configuration.

  8. Repeat the above for several alternative collectors, and determine which has the greatest cost benefits or gives the greatest rate of return.

The procedure described is both costly and time consuming but is normally carried out by design engineering firms for industrial projects. However, often only a preliminary assessment is needed, perhaps to determine whether further investigation is warranted or if an idea should be discarded. In such a case, a number of simplifying assumptions can be made, allowing most of the desired information to be presented in the form of charts and tables. This simplification has been done by InterTechnology/Solar Corporation in its solar process heat study and is presented here.

Base-line system for calculations   The most important simplifying assumption made is that for the purposes of comparison, a single simple system concept is adequate. Furthermore, both a fixed load and demand pattern are assumed, although performance is normalized to load so that other loads can be handled. A schematic of this base-line system is shown in Fig. 21.2. In this system, a constant demand for hot water is assumed; the temperature of the water is assumed to be raised from a city water temperature of 16°C to a desired temperature of T. The output temperature is held constant by either boosting it with an auxiliary heater or cooling it by mixing the hot water with unheated water in a mixing valve. The solar system is assumed to turn on whenever solar heat can be collected.

The storage tank is assumed to contain approximately 60 L per square meter of collector or a minimum of 3000 L, and is further assumed to be pressurized if necessary. During the day, city water is dumped into the tank as water is withdrawn; however, during the evening, water is only withdrawn so as not to lower the tank temperature.* In the morning, the tank is filled with city water. It is assumed that the load is continuous and uniformly distributed over the day. The effects of some of these assumptions upon system performance are noted in later sections of this chapter.

A computer program was written to calculate the performance of the base-line system. Calculations were made for a wide range of collector types and array sizes as well as for various delivery temperatures T. The program was written in modular fashion so that a minimum number of changes were required to insert different collectors into the system.

*This is the reason for a 3000-L minimum storage.

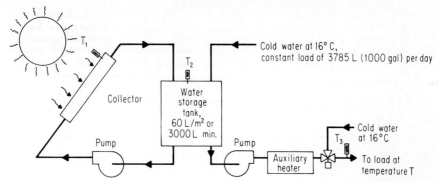

Fig. 21.2   Base-line solar system used for collector performance calculations. (*From InterTechnology/Solar Corporation, "Analysis of the Economic Potential of Solar Thermal Energy to Provide Industrial Process Heat," p. 73, Report No. COO/2829-76/1, U. S. Energy Research and Development Administration, February 7, 1977.*)

Collector performance model   The collection system performance model used for the calculations was a dynamic model in which calculations were done for the whole year at one-hour intervals (see Chap. 7–9, 17). Using the model the fraction of the hot-water load carried by solar was determined.

For systems involving different types of collectors, the basic model remained the same. Appropriate changes were made for different types of collectors. For example, for the concentrating collectors, it was necessary to use only the beam radiation.

Constant-Performance Solar Regions for Process Heat

Insolation data in the form of monthly average radiation falling on a horizontal surface are currently available for approximately 100 cities distributed throughout the United States (see Chap. 2). To carry out performance calculations for various types of collectors operating at several different temperatures, a regionalization procedure was adopted which allowed the United States to be divided into six "constant-performance" solar regions for process heat. The procedure followed was first to divide the country into six regions based upon insolation data alone. These six regions are shown in Table 21.5.

A specific set of calculations was then carried out for 90 different cities using the base-line system model in conjunction with a flat-plate, single-glazed collector which had a selective surface with an emittance of 0.4. The collector performance was calculated which provided 50 percent and 75 percent of the annual energy requirements to heat 3785 L/day from 16°C to 60°C. The results of these calculations were used to readjust the boundaries between regions so that there were more nearly "constant performance" regions. A final adjustment of boundaries was made so that state and county borders were followed for purposes of later data analysis. The resulting regional map is shown in Fig. 21.3.

TABLE 21.5   Definition of Six Solar Regions in the United States Based Upon Incident Insolation

Region	Insolation, $MJ/m^2 \cdot$ day
I	12.6 ± 0.8
II	14.2 ± 0.8
III	15.8 ± 0.8
IV	17.6 ± 0.8
V	19.2 ± 0.8
VI	>20

Fig. 21.3 Map of United States showing six "constant performance" solar regions. (*From InterTechnology/ Solar Corporation, "Analysis of the Economic Potential of Solar Thermal Energy to Provide Industrial Process Heat," p. 81, Report No. COO/2829-76/1, U. S. Energy Research and Development Administration, February 7, 1977.*)

The results of the calculations for the 90 cities are shown in Table 21.6. This table shows that there are isolated areas included within some regions according to the map but which by definition should be included in another region. The table may be used to make such corrections. Finally, a representative city was selected for each of the regions for which all further calculations for each region were carried out. This was done by finding the average system performance for all cities in the region and then selecting the city in which the system performance was closest to the average. The selected cities along with relevant insolation and weather data are shown in Table 21.7.

Competing fossil fuels in solar regions The economic viability of solar process heat systems is determined to a significant degree by the costs of available competing fuels. The definition of the constant-performance solar regions shows that performance and cost of solar systems can be characterized by geographical location. The mix of competing fossil fuels used by industry for thermal energy can also be characterized by geographical location.

Table 21.8 shows the estimated fuel mix used by industry for thermal energy as a function of constant-performance solar region. A more detailed breakdown of fuel mixes with respect to subregions is shown in the original reference. The table shows that in the better three solar regions—IV, V, and VI—solar will have to compete against oil and gas, which are the fuels rising most rapidly in cost and decreasing in availability. Solar should be able to compete best in these areas.

In addition, if individual industrial plants are unable to get sufficient oil or gas for existing process heat systems, such plants may turn to solar rather than incur the costs of installing new coal-fired systems and setting up new distribution and fuel-handling facilities. In comparing solar versus new coal-fired systems, the capital cost of coal-fired systems should be included in the economic analysis, a situation which would help offset the relatively low fuel cost of coal and enable solar to compete against coal.

COLLECTOR TYPES AND PERFORMANCE FOR PROCESS HEAT SYSTEMS

As discussed in Chaps. 7, 8, and 9, there are a variety of collectors available for use in solar process heat systems. Some guidelines for collector selection for any particular application are given here along with results of performance calculations for a few of the most frequently used collectors.

TABLE 21.6   Base-Line System Performance with Flat-Plate Collector for 90 Cities in the United States*

City	Collector output at solar fractions of 50% and 75%, $J \times 10^9/m^2 \cdot yr$		City	Collector output at solar fractions of 50% and 75%, $J \times 10^9/m^2 \cdot yr$	
	50%	75%		50%	75%
Albuquerque, NM (VI)	4.82	4.22	Laramie, WY (IV)	3.09	2.46
Allentown, PA (II)	2.56	1.84	Las Vegas, NV (VI)	5.13	4.48
Ames, IA (II)	2.50	1.79	Lemont, IL (II)	2.72	2.02
Apalachicola, FL (V)	4.22	3.57	Lincoln, NB (III)	3.11	2.43
Astoria, OR (I)	2.25	1.45	Madison, WI (II)	2.55	1.79
Atlanta, GA (III)	3.42	2.78	Manhattan, KS (III)	2.95	2.24
Bismarck, ND (III)	3.00	2.30	Medford, OR (III)	3.11	2.28
Blue Hill, MA (II)	2.36	1.66	Miami, FL (V)	4.46	3.81
Boise, ID (III)	3.26	2.54	Midland, TX (V)	4.36	3.72
Boston, MA (I)	2.32	1.56	Mt. Weather, VA (II)	2.70	2.02
Boulder, CO (III)	2.92	2.27	Nashville, TN (III)	3.14	2.39
Brownsville, TX (V)	4.20	3.54	New Orleans, LA (II)	3.03	2.27
Cambridge, MA (II)	2.32	1.64	New York, NY (II)	2.50	1.77
Caribou, ME (II)	2.23	1.49	Newport, RI (II)	2.67	1.91
Charleston, SC (IV)	3.70	3.04	North Omaha, NE (III)	2.98	2.31
Chicago, IL (I)	1.61	0.83	Oak Ridge, TN (III)	3.01	2.26
Cleveland, OH (II)	2.50	1.66	Oklahoma City, OK (V)	3.94	3.26
Columbia, MO (III)	3.14	3.27	Philadelphia, PA (II)	2.77	2.05
Columbus, OH (II)	2.31	1.60	Pittsburgh, PA (II)	1.82	0.95
Corvallis, OR (II)	2.50	1.64	Phoenix, AZ (VI)	5.30	4.65
Davis, CA (IV)	3.76	3.04	Portland, ME (II)	2.70	2.02
Dodge City, KS (V)	4.01	3.34	Prosser, WA (III)	3.18	2.42
East Lansing, MI (I)	2.19	1.36	Pullman, WA (III)	2.80	2.05
East Wareham, MA (II)	2.41	1.66	Put-in-Bay, OH (II)	2.35	1.61
El Paso, TX (VI)	5.11	4.41	Raleigh, NC (III)	3.18	2.53
Ely, NV (V)	4.04	3.43	Rapid City, SD (III)	3.39	2.78
Flaming Gorge, UT (IV)	3.48	2.85	Riverside, CA (V)	4.49	3.84
Forth Worth, TX (V)	4.18	3.48	San Antonio, TX (V)	4.14	3.47
Fresno, CA (V)	4.10	3.43	Santa Maria, CA (VI)	4.39	3.72
Gainesville, FL (IV)	3.97	3.28	Sault St. Marie, MI (II)	2.31	1.62
Glasgow, MT (III)	3.21	2.67	Sayville, NY (II)	2.87	2.16
Grand Lake, CO (IV)	2.92	2.26	Schenectady, NY (I)	1.91	1.16
Grand Junction, CO (V)	4.01	3.40	Seabrook, NJ (II)	2.76	2.04
Great Falls, MT (III)	3.05	2.38	Seattle, WA (I)	1.80	0.97
Greensboro, NC (III)	3.22	2.64	Shreveport, LA (III)	3.42	2.75
Griffin, GA (IV)	3.71	3.02	Spokane, WA (III)	2.89	2.16
Hatteras, NC (V)	4.03	3.40	St. Cloud, MN (II)	2.75	2.04
Indianapolis, IN (II)	2.70	1.90	State College, PA (II)	2.38	1.68
Inyokern, CA (VI)	5.82	5.16	Stillwater, OK (IV)	3.60	2.93
Ithaca, NY (II)	2.25	1.37	Summit, MT (I)	1.86	1.13
La Jolla, CA (III)	3.19	2.51	Tallahassee, FL (IV)	3.70	3.01
Little Rock, AR (III)	3.32	2.69	Tampa, FL (V)	4.43	3.77
Los Angeles, CA (V)	4.04	3.42	Tucson, AZ (VI)	5.02	4.36
Lake Charles, LA (IV)	3.86	3.14	Twin Falls, ID (III)	2.89	2.18
Lander, WY (IV)	3.70	3.04	Washington, DC (II)	2.92	2.20

*Calculations were performed using a delivery temperature of 60°C and a single-glazed collector with a selective surface of 0.4 emittance.

Equation (21.1) shows the most commonly used expression for the efficiency of collectors (see Chap. 7):

$$\eta = F_R \left[\alpha\tau - U_L \frac{(T_{in} - T_{amb})}{I} \right] \tag{21.1}$$

TABLE 21.7 Representative Cities Selected for Each of the Six "Constant-Performance" Solar Regions in the United States*

| | | Annual insolation on collector tilted at latitude, $J \times 10^9/m^2 \cdot yr$ | | | |
| | | | Single-axis tracking | | Annual average ambient temperature, °C |
Region	City	Fixed total	Total	Beam	
I	Schenectady, NY	4.46	6.60	4.80	10.2
II	Madison, WI	5.45	8.26	6.47	7.2
III	Lincoln, NE	6.00	8.41	6.67	13.2
IV	Stillwater, OK	6.41	9.62	7.75	17.9
V	Fort Worth, TX	7.00	10.34	5.47	18.8
VI	El Paso, TX	8.36	12.54	10.88	17.4

*Average annual insolation values and temperatures are shown.
SOURCE: InterTechnology/Solar Corporation, "Analysis of the Economic Potential of Solar Thermal Energy to Provide Industrial Process Heat," p. 86, Report No. COO/2829-76/1, U.S. Energy Research and Development Administration, February 7, 1977.

TABLE 21.8 Estimated Fuel Mixes Used by Industry in the United States for Thermal Energy as Function of Solar Region

| | Percent of thermal needs supplied by | | |
Constant-performance solar region	Coal	Oil	Gas
I	22.2	23.4	54.4
II	21.5	18.1	60.4
III	30.9	10.0	59.1
IV	6.1	7.8	86.1
V	1.2	3.3	95.5
VI	0.0	9.4	90.6
United States as a whole	14.5	11.8	73.7

SOURCE: InterTechnology/Solar Corporation, "Analysis of the Economic Potential of Solar Thermal Energy to Provide Industrial Process Heat," p. 31, Report No. COO/2829-76/1, U.S. Energy Research and Development Administration, February 7, 1977.

For very low temperature applications ($\leq$ 50°C) and in areas where the ambient temperatures and insolation are high, the loss-coefficient term $F_R U_L (T_{in} - T_{amb})/I$ becomes less important than the term $F_R \alpha \tau$. To maximize the term $F_R \alpha \tau$, it is important to choose a collector with good heat-transfer characteristics (usually a liquid collector), high cover-plate transmission [usually a single glazing of low-iron glass ($\tau = 0.88 - 0.92$) or clear plastic], and high absorber-plate absorptance (a good flat-black paint is best with $\alpha = 0.96 - 0.98$).

For high-temperature applications (> 100°C), the loss-coefficient term becomes important. To minimize the effect of losses, a concentrator system with a small-surface-area absorber is usually required. To be effective, such systems normally require a sun-tracking capability.

For applications requiring intermediate temperatures (50–100°C), a careful analysis needs to be done to determine the most economical system. In this temperature range, a concentrator system may be the most cost-effective.

For certain applications, specialized collectors may prove to be most effective. For example, a solar pond may be best where low tilt angles and temperatures are required and high insolation and ambient temperatures are experienced. An air collector may be best for an application which requires low-temperature air drying during daylight hours.

In the following discussion, a brief description of several collector types will be given along with results of performance calculations for the previously mentioned base-line system and computer model.

Flat-Plate Collectors

Although a flat-plate collector is a simple device, its performance is affected by many variables. Some of the major variables with ranges of values are shown below:

Collection fluid	Liquid, air
Number of cover plates	0, 1, 2
Transmittance of cover plate	0.8–0.92
Absorptance of absorber plate	0.85–0.98
Emittance of absorber plate	0.1–0.95

In addition, there are other items which affect collector performance. Among these are materials of construction, sensitivity to wind, size and location of manifolds and tubes, heat-transfer efficiency, and method of manifold interconnection. An additional factor of great significance is a collector's ability to withstand maximum dry-plate conditions. This occurs under no-flow conditions, at the highest coincident ambient temperatures and incident insolation. Although such conditions may not occur often in the lifetime of a system, one such occurrence (with temperatures as high as 225°C) can be catastrophic and as such cannot be tolerated. This fact alone can reduce to a small number those collectors which may be acceptable for any given application.

A collector which is useful for a large variety of applications is a single-glazed (high-transmittance glass) collector with a black chromium absorption surface. Such a collector combines good performance over a wide range of temperatures with good high-temperature stability. Performance calculations of the base-line system with this collector are presented in Fig. 21.4 (a–f) and in Table 21.9. The efficiency equation for this collector was assumed to be

$$\eta = 0.80 - 4.2 \frac{(T_{\text{in}} - T_{\text{amb}})}{I} \tag{21.2}$$

in SI units.

Recent measurements taken on this type of collector for several manufacturers have shown the loss coefficient to be somewhat higher (about 4.7 instead of 4.2 W/m² · °C). Such an increase will reduce the high-temperature and high solar-fraction performances by as much as 10 percent.

Table 21.10 gives typical cost ranges (1977 dollars) for collectors with 0, 1, and 2 cover plates and with selective and nonselective absorber coatings. These costs may be used as one input into the overall cost estimate of any particular solar system using flat-plate collectors.

Air Flat-Plate Collectors

Air flat-plate collectors have several advantages and disadvantages compared with liquid flat-plate collectors. These advantages and disadvantages are enumerated below.

The advantages of air collectors are:

1. Freedom from corrosion due to the coolant—corrosion can be a major problem with water-cooled collectors. Air and liquid collectors both must be protected against external corrosion due to rain leakage or other causes.

2. Freedom from freezing and boiling—freeze and overpressure protection is a major design consideration for liquid-cooled collectors.

3. The minor nature of leaks—leaks in air systems can be tolerated as long as they do not substantially decrease performance. Leaks in liquid systems can lead to a loss of coolant and pump failure as well as constitute an environmental and safety hazard.

Disadvantages of air collectors are:

1. Lower thermal performance—the thermal performance of air collectors is poorer than liquid collectors at a given temperature due to a reduced heat-removal factor F_R.

2. Higher installation costs for ductwork—installed, insulated air ducts are more costly than an equivalent amount of insulated pipe (with the same thermal transport capacity). They also take up more space.

3. Possibly higher operating costs for energy transport—the fan power requirement for

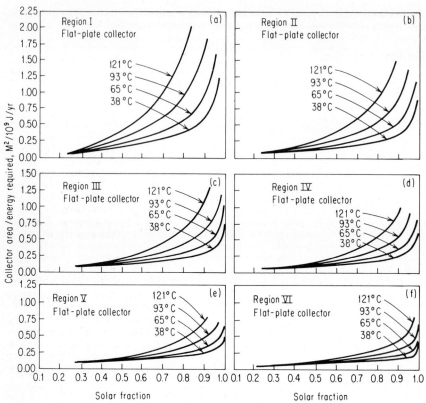

Fig. 21.4 Base-line system performance for a single-glazed flat-plate collector with a black chrome selective surface. (*a*) Solar Region I; (*b*) Solar Region II; (*c*) Solar Region III; (*d*) Solar Region IV; (*e*) Solar Region V; (*f*) Solar Region VI. (*From InterTechnology/Solar Corporation, "Analysis of the Economic Potential of Solar Thermal Energy to Provide Industrial Process Heat," pp. 114–124, Report No. COO/2829-76/1, U. S. Energy Research and Development Administration, February 7, 1977.*)

TABLE 21.9 Base-Line System Performance for Flat-Plate Collector in Six Solar Regions*

Region	Collector output, $J \times 10^9/m^2 \cdot yr$, for 50% solar fraction at desired water temperature, °C				Collector output, $J \times 10^9/m^2 \cdot yr$, for 75% solar fraction at desired water temperature, °C			
	38°	66°	93°	121°	38°	66°	93°	121°
I	3.06	2.33	1.78	1.32	2.49	1.63	1.08	—
II	3.82	3.09	2.47	1.89	3.44	2.38	1.60	1.01
III	4.58	3.73	3.05	2.39	4.12	2.99	2.12	1.38
IV	4.91	4.24	3.54	2.79	4.58	3.44	2.49	1.67
V	5.50	4.62	4.01	3.27	4.91	4.00	2.95	2.02
VI	6.19	5.63	5.12	4.21	5.89	4.99	4.03	2.88

*Calculations were performed for various delivery temperatures and for solar fractions of 50 and 75 percent for a single-glazed collector with a black chrome selective surface.

SOURCE: InterTechnology/Solar Corporation, "Analysis of the Economic Potential of Solar Thermal Energy to Provide Industrial Process Heat," p. 112, Report No. COO/2829-76/1, U.S. Energy Research and Development Administration, February 7, 1977.

TABLE 21.10   Costs for Several Types of Flat-Plate Collectors (1978 dollars)

	Collector costs, $/m²	
# Cover plates	Nonselective surface	Selective surface
0	15– 40	—
1	75–120	85–135
2	95–130	105–145

air systems can be from 0 to 100 percent higher than the pump power for an equivalent liquid system.

4. Thermal storage size—the volume of a rock-bed storage unit is approximately three times greater than that of a water tank of equivalent thermal capacity.

The base-line system previously described was not used to calculate the performance of an air collector, which normally uses rock-bed storage. To determine system performance for an air collector, the f-chart method, which is explained in Chap. 14, was used. Calculations were carried out using a single-glazed air collector with a black-chrome selective surface and an efficiency equation of

$$\eta = 0.60 - 3.4 \frac{T_{in} - T_{amb}}{I} \tag{21.3}$$

It was assumed that the solar system would provide either 50 or 75 percent of the annual energy required to heat a fixed volume of air per day from a temperature of 16 to 82°C. System performances for each region are shown in Table 21.11.

The costs of air flat-plate collectors are essentially the same as those shown in Table 21.10 for liquid flat-plate collectors.

Tubular Collectors

Tubular collectors are cylindrical in shape and normally have a relatively high vacuum ($\approx 10^{-4}$ torr) between an inner absorbing surface and an outer glass surface. The high vacuum eliminates a large fraction of the convective/conductive losses usually associated with flat-plate collectors. In addition, the absorbing surface can be coated with a selective absorber, thereby reducing the radiative loss considerably. Generally these collectors consist of a series of tubes mounted in a frame with a spacing of about one tube diameter between adjacent tubes. Either a white diffuse scattering surface or a curved reflective surface is placed behind the tubes to enhance the performance. Collectors such as these which incorporate insulated manifold piping have been selling at a cost of about $215 per square meter of collector aperture area (1978 dollars).

TABLE 21.11   Base-Line System Performance for Air Collector in Six Solar Regions*

Region	Collector output, J × 10⁹/m² · yr, for 50% solar	Collector output, J × 10⁹/m² · yr, for 75% solar
I	1.69	1.42
II	2.22	1.99
III	2.58	2.35
IV	2.78	2.52
V	3.06	2.78
VI	3.77	3.47

*Calculations were performed using the f-chart method for a delivery temperature of 82°C and for solar fractions of 50 and 75 percent for a single-glazed air collector with a black chrome selective surface.

SOURCE: InterTechnology/Solar Corporation, "Analysis of the Economic Potential of Solar Thermal Energy to Provide Industrial Process Heat," p. 134, Report No. COO/2829-76/1, U.S. Energy Research and Development Administration, February 7, 1977.

Calculations were performed for the base-line system using an Owens-Illinois collector with a diffuse back reflecting surface. The expression for collector efficiency is given in the literature.[7] Calculations were performed for Region III only at several delivery temperatures and for solar fractions of 50 and 75 percent so that comparisons with other collectors could be made. Results of these calculations are shown in Table 21.12.

Shallow Solar Ponds

A shallow *convecting* solar pond, which is basically a shallow body of water, is insulated on the bottom to prevent heat losses, blackened on the bottom to absorb sunlight, and covered by one or two plastic covers on top to prevent evaporative and convective/conductive losses. The primary advantages of such a solar collector are its simplicity and a low installed system cost of about $60/m² (1978 dollars). The main disadvantages are its tilt-angle limitations and relatively large loss coefficient. For these reasons, the shallow solar pond performs best in latitudes near the equator where the sun is overhead and where high ambient temperatures are experienced.

In one system studied in the literature,[8] the fluid in the pond was pumped to a storage tank two hours before sunset, and the pond was filled with water one hour after sunrise. The desired delivery temperature was obtained by either boosting the storage temperature with an auxiliary heater or by mixing with city water. The literature should be consulted for further details. The collection efficiency of the pond was assumed to be

$$\eta = 0.75 - 7.4 \frac{T_{pond} - T_{amb}}{I} \qquad (21.4)$$

if the pond temperature T_{pond} was greater than the ambient temperature T_{amb}. However, for pond temperatures less than ambient temperature, the efficiency equation was assumed to be

$$\eta = 0.75 + 2.8 \frac{T_{amb} - T_{pond}}{I} \qquad (21.5)$$

The results of the calculations are presented in Fig. 21.5 (*a–f*) and in Table 21.13.

Concentrating Collectors

Details on concentrating collectors are presented in Chaps. 8 and 9. A concentrating collector is required to produce temperatures in the range of 150–300°C. In most cases where the direct component of solar radiation is being concentrated, a sun-tracking capability is required, which reduces the incidence angle cosine loss of radiation striking the collector surface as compared to a flat-plate collector. However, a sun-tracking system has greater mechanical complexity which will lead to greater maintenance costs than in the case of a flat-plate collector.

In this section, the results of performance calculations are presented for three different types of concentrating collectors—a linear Fresnel-lens collector, a compound parabolic collector, and a tracking parabolic cylinder collector.

TABLE 21.12   Base-Line System Performance for Tubular Collector in Solar Region III*

Temperature, °C	Collector output, J × 10⁹/m² · yr, for solar fractions of 50% and 75% at various temperatures	
	50%	75%
66	3.44	2.91
93	3.01	2.68
121	2.84	2.49
149	2.64	2.21

*Calculations were performed for various delivery temperatures and for solar fractions of 50 and 75 percent for an evacuated tubular collector with a diffuse reflector.

SOURCE: InterTechnology/Solar Corporation, "Analysis of the Economic Potential of Solar Thermal Energy to Provide Industrial Process Heat," p. 143, Report No. COO/2829-76/1, U.S. Energy Research and Development Administration, February 7, 1977.

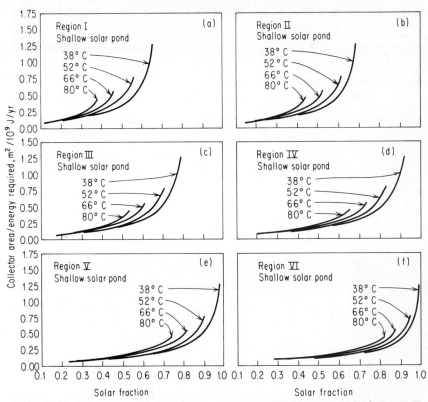

Fig. 21.5   Performance of a shallow solar pond. (*a*) Solar Region I; (*b*) Solar Region II; (*c*) Solar Region III; (*d*) Solar Region IV; (*e*) Solar Region V; (*f*) Solar Region VI. (*From InterTechnology/Solar Corporation, "Analysis of the Economic Potential of Solar Thermal Energy to Provide Industrial Process Heat," pp. 1449–1459, Report No COO/2829-76/1, U. S. Energy Research and Development Administration, February 7, 1977.*)

TABLE 21.13   **System Performance for Shallow Solar Pond in Six Solar Regions***

	Collector output, $J \times 10^9/m^2 \cdot yr$, for 50% solar fraction at desired water temperature, °C				Collector output, $J \times 10^9/m^2 \cdot yr$, for 75% solar fraction at desired water temperature, °C			
Region	38°	52°	66°	80°	38°	52°	66°	80°
I	1.45	1.09	—	—	—	—	—	—
II	1.87	1.65	1.21	—	—	—	—	—
III	2.46	2.16	1.90	1.50	0.92	—	—	—
IV	3.45	2.95	2.83	2.42	2.11	1.68	1.26	—
V	4.32	3.45	3.45	3.05	2.88	2.71	2.16	1.63
VI	5.53	4.15	4.14	3.78	3.77	3.37	3.11	2.66

*Calculations were performed for various delivery temperatures and for solar fractions of 50 and 75 percent.

SOURCE: InterTechnology/Solar Corporation, "Analysis of the Economic Potential of Solar Thermal Energy to Provide Industrial Process Heat," p. 167, Report No. COO/2829-76/1, U.S. Energy Research and Development Administration, February 7, 1977.

Linear Fresnel-Lens collector   A linear Fresnel-lens collector consists of a rectangular-shaped Fresnel lens; an absorber tube situated along the lens' focal line, an assembly which serves as both a structure for the lens and absorber tube and a means for insulating the absorber from the environment, and a structure for supporting the collector assembly. In addition, a tracking mechanism is needed since this device collects principally beam radiation. Normally this collector is mounted with its long axis in a north-south direction but tilted at some angle with respect to the horizontal to form a polar mount.

The cost of this type of collector including the supporting structure and tracking mechanism is approximately \$270/m² (1978 dollars).

Performance calculations were carried out in Solar Region III for the base-line system for various delivery temperatures and solar fractions of 50 and 75 percent. The efficiency equation used was

$$\eta = 0.680 - 11.3\,X^2 + 0.008\,X \tag{21.6}$$

where $X = [(T_{in} + T_{out})/2 - T_{amb}]/I$

This efficiency equation is representative of a collector with a Fresnel-lens transmittance of 0.86, absorber absorptance of 0.93, absorber emittance of 0.12, and concentration ratio of 6. Calculations were performed for a tilt angle equal to the latitude angle.

Results of the performance calculations are shown in Table 21.14.

Compound parabolic collector   The compound parabolic solar collector is a concentrating collector which does not produce an image. Instead, the collector funnels sunlight onto the absorber. It is expected that production-model collectors will cost in the range of \$200–300/m² (1978 dollars).

For the purpose of performance calculations for the base-line system, the following efficiency equation was assumed:

$$\eta = 0.56 - 6.13\,X^2 + 0.51\,X \tag{21.7}$$

where $X = [(T_{in} + T_{out})/2 - T_{amb}]/I$

This corresponds to a collector which has a concentration ratio of 3 and which uses an evacuated tube containing a black-chrome absorber to reduce conductive-convective and radiative losses. In addition, periodic tilt adjustments were assumed to enhance collection. The results of the performance calculations for Solar Region III at several temperatures and solar fractions are presented in Table 21.15.

Tracking parabolic cylinder collector   A parabolic cylinder collector consists of a specularly reflecting trough whose cross section is parabolic in shape. In this arrangement, the focus of the parabola forms a line which runs the length of the trough. The absorber is placed along this line and is usually a copper or stainless steel pipe. To reduce the radiative and convective/conductive losses, the absorber is sometimes enclosed in a glass tube, the annulus being either evacuated or nonevacuated.

TABLE 21.14   Base-Line System Performance for Linear Fresnel-Lens Collector in Solar Region III*

Collector output, J × 10⁹/m² · yr, for 50% solar fraction at desired hot-water temperature, °C				Collector output, J × 10⁹/m² · yr, for 75% solar fraction at desired hot-water temperature, °C			
66	93	121	149	66	93	121	149
5.15	4.91	4.66	4.12	4.64	3.97	3.43	2.81

*Calculations were performed for various delivery temperatures and for solar fractions of 50 and 75 percent.

SOURCE: InterTechnology/Solar Corporation, "Analysis of the Economic Potential of Solar Thermal Energy to Provide Industrial Process Heat," p. 191, Report No. COO/2829-76/1, U.S. Energy Research and Development Administration, February 7, 1977.

TABLE 21.15   Base-Line System Performance for Compound Parabolic Collector in Solar Region III*

Collector output, J × 10⁹/m² · yr, for 50% solar fraction at desired hot-water temperature, °C				Collector output, J × 10⁹/m² · yr, for 75% solar fraction at desired hot-water temperature, °C			
65°	93°	121°	149°	65°	93°	121°	149°
3.60	3.44	3.04	3.22	3.52	3.28	3.02	2.71

*Calculations were performed for various delivery temperatures and for solar fractions of 50 and 75 percent.

SOURCE: InterTechnology/Solar Corporation, "Analysis of the Economic Potential of Solar Thermal Energy to Provide Industrial Process Heat," p. 215, Report No. COO/2829-76/1, U.S. Energy Research and Development Administration, February 7, 1977.

These collectors track the sun. Tracking is accomplished by fixing the orientation of the absorber with respect to the earth and rotating the reflective cylinder about the absorber axis. In general, three different absorber axis directions have been considered: east-west, north-south, and north-south polar. Collectors mounted on the horizontal are simpler to install, requiring fewer support structures and less piping; however, the polar arrangement reduces "cosine losses" and therefore can collect significantly more energy. The costs for these types of collectors have been in the range of $150–300/m² (1978 dollars), including supports (for horizontal mount) and tracking mechanism.

For the purposes of calculations for the base-line system, a polar-axis mount was assumed in which the tilt angle is equal to the local latitude so that the absorber axis is parallel to the earth's axis. An efficiency equation of

$$\eta = 0.69 - 0.85 \frac{T_{in} - T_{amb}}{I} \tag{21.8}$$

was used.

This corresponds to a collector with the following parameters:
- Concentration ratio   = 15
- Parabolic surface reflectance   = 0.75
- Absorber cover tube transmittance   = 0.9
- Absorber absorptance   = 0.92 (black chrome)
- Absorber emittance   = 0.1 (black chrome)

Results of the performance calculations in all six solar regions are presented in Fig. 21.6 $(a-f)$ and in Table 21.16.

Comparative Analysis of Collector Performance

Results of the study[2] led to the following generalizations concerning collector performance. The calculations of collector performance for a simple base-line system indicated that for the same solar fraction or amount of thermal energy collected annually, greater than 2.5 times more collector area is required in Solar Region I than in Solar Region VI. Or, for the same collector area, a substantially greater amount of energy can be collected annually in Region VI than in Region I. This result illustrates the significant influence of geographical location upon collector and solar process heat system performance.

A most significant result of the study was that for all delivery temperatures above 50°C, the *polar-axis-mounted, single-axis-tracking concentrating collector gives a superior performance to all other collectors evaluated.* In particular, the parabolic cylinder collector with its low loss coefficient looks especially promising at the higher temperatures studied.

Below delivery temperatures of about 80°C, the best performing flat-plate collector deserves serious consideration. From these calculations, a single-glazed collector with a black-chrome selective surface appears to be the best performing and most cost-effective flat-plate collector. Although the polar-axis-mounted, single-axis-tracking concentrating collector gives a superior

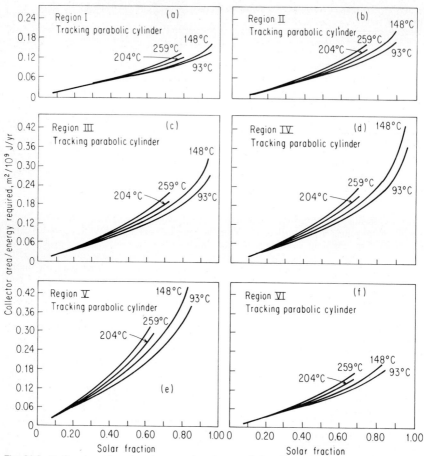

Fig. 21.6   Performance of a polar-axis-mounted tracking parabolic cylinder collector. (*a*) Solar Region I; (*b*) Solar Region II; (*c*) Solar Region III; (*d*) Solar Region IV; (*e*) Solar Region V; (*f*) Solar Region VI. (*From InterTechnology/Solar Corporation, "Analysis of the Economic Potential of Solar Thermal Energy to Provide Industrial Process Heat," pp. 239–244, 1501–1529, Report No. COO/2829-76/1, U. S. Energy Research and Development Administration, February 7, 1977.*)

performance at these lower temperatures and collects more energy than a flat-plate collector, the probable lower cost of the flat-plate collector may offset the performance difference and give delivered energy at a lower cost.

At temperatures less than 50°C, a shallow solar pond appears to be the most economical solar system in Solar Regions IV, V, and VI. In these regions, the higher ambient temperature and the lower latitude angle help increase the net amount of collected energy. In addition, the shallow solar pond has a much lower installed cost than any other solar system (by nearly a factor of 2).

SYSTEM COMPONENTS

In addition to the collectors, several other subsystems in a solar process heat system must be given careful consideration. Among these are storage, control, energy transport, and structural support subsystems. Each is discussed briefly to point out the primary factors which must be evaluated.

TABLE 21.16   Base-Line System Performance for Polar-Axis-Mounted Tracking Parabolic Cylinder Collector in Six Solar Regions*

Region	Collector output, J × 10⁹/m² · yr, for 50% solar fraction at desired hot-water temperature, °C						Collector output, J × 10⁹/m² · yr, for 75% solar fraction at desired hot-water temperature, °C					
	52°	66°	93°	149°	204°	260°	52°	66°	93°	149°	204°	260°
I	3.28	3.10	3.01	2.75	2.48	2.22	2.79	2.63	2.52	2.18	1.85	—
II	4.47	4.42	4.15	4.02	3.65	3.29	4.19	3.87	3.70	3.29	2.97	2.90
III	5.58	5.15	4.90	4.79	4.27	3.89	4.79	4.64	4.35	4.04	3.51	3.11
IV	5.70	5.53	5.34	5.15	4.67	4.31	5.25	5.05	4.81	4.36	3.73	3.34
V	6.22	6.15	6.01	5.89	5.31	4.84	5.58	5.53	5.15	4.95	4.38	3.85
VI	8.06	8.01	7.85	7.50	6.95	6.58	7.79	7.65	7.15	7.03	6.26	5.67

*Calculations were performed for various delivery temperatures and for solar fractions of 50 and 75 percent.
SOURCE: InterTechnology/Solar Corporation, "Analysis of the Economic Potential of Solar Thermal Energy to Provide Industrial Process Heat," p. 245, Report No. COO/2829-76/1, U.S. Energy Research and Development Administration, February 7, 1977.

Storage Subsystem

Several types of storage media can be used in conjunction with solar systems, including water, special high-temperature fluids, rock beds, and heat-of-fusion materials (see Chap. 6). It is also possible that for certain applications no storage may represent an optimal condition. Some of the important factors which influence the choice of storage system type are daily demand pattern of energy usage, the desired solar fraction, the temperature required for the process, and the form in which energy is carried to the process (e.g., air, water, or steam).

In all cases where storage is selected, it is necessary to insulate the storage container. Selection of the proper amount of insulation is made on the basis of a life-cycle cost-benefit analysis.

The choice of storage size should also be made on the basis of a life-cycle cost-benefit analysis. Experience has shown that too little storage for a given collector area forces the system to operate at high temperatures. In such a case, collector efficiency is always reduced, and heat is often rejected when maximum acceptable temperatures are exceeded. Too much storage for a given collector area causes the storage temperature to remain low and usually involves a large heat-loss surface which allows more heat to be conducted or radiated away. Figure 21.7 illustrates these effects.

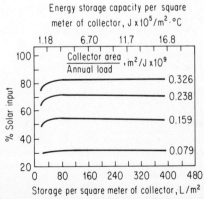

Fig. 21.7   Plot of solar fraction versus ratio of storage volume to collector area, for various ratios of collector area to annual load. (*From InterTechnology/Solar Corporation,"Analysis of the Economic Potential of Solar Thermal Energy to Provide Industrial Process Heat," p. 106, Report No. COO/2829-76/1, U. S. Energy Research and Development Administration, February 7, 1977.*)

These curves were obtained for the base-line system in conjunction with a flat-plate collector. The required delivery temperature was 66°C. These curves show that beyond about 80 L/m², little gain can be achieved and that for ratios of storage to collector area of less than 40 L/m², a sharp loss in energy supplied by solar is experienced. The curves also show that in the range of acceptable storage sizes, the higher solar fractions require greater amounts of storage. In addition, curves such as these represent the basis for selection of optimal storage. Basically the recommended method to follow is to find the point on a given curve at which the value of the additional energy supplied by solar over a prescribed time period is equal to the additional cost of storage capacity.

A rule of thumb which may be followed is that for water storage, about 60 L/m² should be used. Similar rules for other storage media may be deduced. However, such rules of thumb should not be followed blindly since analysis of a specific application may indicate a different amount of storage (for example, if a larger amount of storage is already in place).

Typical costs for storage subsystems range from 5 to 10 percent of the total system cost.

Control Subsystems

The control subsystem may be divided into two parts: that concerned with the collection of solar energy and that concerned with energy delivery. The controls concerned with energy delivery are normally application specific, and therefore it is not possible to make general definitive statements. However, as guiding principles, the lowest acceptable delivery temperature should be used and the simplest functional control system should be selected. The decision as to what is the simplest functional control system should be made on a cost-benefit basis; i.e., does an extra control pay for itself in energy savings in some prescribed time period?

The controls for the collection of solar energy can be very simple. Turn-on of the collector pump or fan should be done in such a way that the system will not cycle. Experience has shown that for water heating, the absorber plate temperature should be 8–10°C higher than the storage temperature before turning the collector pump/fan on. When the outlet temperature of the collector is about 2°C above the storage temperature, the collector pump/fan should be turned off.

Flowrates in a system may be fixed, or varied by a variable-speed or multi-speed pumping arrangement. Variable- or multi-speed pumping allows the system to collect 2 to 3 percent more energy per year. Control systems for accomplishing this should not cost more than the value of the additional energy collected over the lifetime of the system.

Additional controls which must be considered are those which protect the system from excessively high or low temperatures. This may be accomplished by gravity-draining the system whenever the collector pump/fan goes off and by turning the collector pump/fan off whenever the collector array outlet temperature exceeds a certain value. In such a case the collectors must be able to withstand the maximum temperatures generated (see Chap. 9).

Typical costs for controls range from 2 to 5 percent of the total system cost.

Energy Transport and Structural Support Subsystems

Experience with solar systems has indicated that energy transport and structural support subsystems can be major cost items. Typically, energy transport and structural subsystems cost in the range of 15 to 20 percent and 10 to 20 percent of the total system cost, respectively. Structural support requirements are normally dominated by wind loading consideration. Major considerations in energy transport subsystems are the pipe size required to meet flowrate and pressure-loss requirements, amount of insulation required to reduce energy losses, and piping configuration required to keep a balanced flow.

Other System Considerations

Many elements go into the design of a solar system. In addition to the subsystems already discussed, there are such considerations as collector array geometry, collector location, esthetics, possible shading, solar system process interface, and calculation of process energy requirements.

In designing a solar system, it is particularly important not to oversize the system. The reason for this is clear from the law of diminishing returns behavior of Fig. 21.8 where it is seen that beyond some point an increase in collector area gives very little marginal gain in solar fraction. Since solar systems are high-capital-cost items, the penalty for oversizing is great. It is clear from this discussion that it is important to determine accurately the process energy requirements so that the system is not oversized.

Expected cost ranges for various subsystems in a solar industrial process heat system are

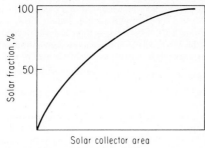

Fig. 21.8   Qualitative plot of solar fraction versus collector area.

shown in Table 21.17. From this table, it can be seen that installed collector costs are by far the most important single cost element in solar process heat systems.

The economics of the proposed solar system also play an important role in determining the proper size of the system.

ECONOMIC ANALYSIS OF SOLAR PROCESS HEAT SYSTEMS

To develop the optimum design for a solar process heat system for a specific application—or indeed, to decide whether to use solar process heat at all—an economic analysis of the potential system must be made. In this analysis, the basic difference in types of costs between a conventional system and a solar system must be considered. For a conventional system, initial costs are relatively low and operating costs relatively high, whereas for a solar system, the reverse is true. To compare the cost of a solar process heat system—primarily a capital cost—versus future expected savings in fossil fuel bills—an operating expense—life-cycle costing methods must be used (see Chap. 28 for a detailed exposition).

Life-Cycle Costs

The essence of a life-cycle cost analysis is to sum the costs involved in operating a system over its expected lifetime and to compare such total costs for all viable, alternative systems to determine the best investment. In calculating the total costs over the system's lifetime, future costs are discounted appropriately as a function of time to recognize the time value of money so that the present values of future cash flows may be added together on a uniform basis. The proper discount rate to use to calculate the present value of a future cash flow is the company's internal rate of return, which is the rate a company attempts to earn on its capital investments.

Two different types of life-cycle costs can be computed: the total cost of owning and operating the system in terms of present dollars—the "present equivalent" life-cycle cost or present value of the system—or the uniform annual equivalent cost. Both methods arrive at the same conclusion with respect to the relative attractiveness of two alternative investments. These methods are explained in detail in Chap. 28. For industry, the return on investment is an important quantity to consider, and items which influence the return on investment such as income taxes and depreciation must also be included in the economic analysis and the calculation of life-cycle costs.

TABLE 21.17   Summary of Costs for Solar Process Heat Systems (1978 dollars)

Item	Materials cost range	Installation cost range	Installed cost range as percent of total system cost
Collectors	\$75–\$300/m²	\$20–\$50/m²	40–60%
Support structures	\$10–\$30/m²	\$10–\$30/m²	10–20%
Storage	\$0.13–\$0.40/L	\$0.13–\$0.40/L	5–10%
Energy transport			15–20%
Controls			2–5%

Cost of process heat systems   The total cost of a combined conventional-solar process heat system includes the initial capital investment including the cost of any necessary process modifications; the total annual cost of the solar system including operating costs, maintenance, taxes, and insurance; and the total costs of operating the conventional system including the cost of the fossil fuel used. Certain credits may be applicable as well to reduce the total system cost, such as an investment tax credit and any expected salvage value which might be obtained at the end of the solar system's useful lifetime.

Under usually applicable conditions, the total costs of the conventional system are assumed to be the same with or without a solar system with the exception of the amount of fuel used. It is usually assumed that the conventional part of a combined process heat system must be a full capacity backup.

In calculating the total life-cycle cost or present value of a process heat system, it is important to recognize and to include in the analysis the effects of income taxes. Because annual operating expenses are deductible from gross income before income tax is paid, the net cash flow is affected by only a fraction of the annual operating expenses. Because depreciation is also a deduction, depreciation affects net cash flow as well. Thus, to calculate the net benefits and the after-tax rate of return on a proposed solar investment, the after-tax costs of alternative process heat systems must be calculated.

Life-cycle cost of fuel.   The major reason to use a solar process heat system with its high initial cost is to save on future fossil fuel costs, which are generally expected to escalate somewhat more rapidly than general inflation. The expected escalation of future fuel costs should be included in the calculation of the life-cycle cost and the annualized equivalent cost of the fossil fuel used in the conventional system. This calculation is discussed in detail in Chap. 28.

Incremental Cash Flow Analysis for Optimum Design

One very important objective of the economic analysis of a solar system, besides the calculation of the expected benefits and the total costs, is the determination of the optimum design for a given application. One way to do this optimization is to analyze the incremental cash flows for a proposed project as compared to a base condition.

Total incremental discounted cash flow   For this analysis, the cash flows are analyzed for an "incremental" project, which means that they are analyzed from a base condition, assumed to be a 100-percent fossil-fuel-fired system. An additional assumption is that the fossil fuel system used to backup the solar system must have 100 percent of the needed capacity. This means that the capital costs of the backup system are the same with or without solar. No credit is taken for a smaller conventional system; the cost of the solar system must be compared only with the savings in fossil fuel. For a specific case, this assumption may not be valid, but the effect of a reduced capital cost for the fossil fuel system may be included if it is appropriate for a specific application.

The general procedure is to calculate all of the cash flows occurring over the project lifetime. These future cash flows are discounted as a function of time so that their present values may be added together. The sum of these discounted cash flows (the present equivalent value) equals zero when the interest rate used to discount the cash flows is equal to the desired average internal rate of return on the entire investment. In the determination of the design for a solar system, this equation is one relationship between the initial investment I and the fraction solar f.

The other equation which is used to obtain a simultaneous solution is the curve of I/L versus f, which is obtained through a performance and cost analysis of the system. From the performance analysis, the curve of collector area per unit of load A/L versus f would first be derived. Cost data would then be used to calculate the investment per unit of load (I/L) from the collector area.

In the following discussion, individual cash flows are described, and analytical expressions for these flows, which are assumed to be paid at the end of the year except for the initial investment, and their present equivalent values are shown in Table 21.18. In this treatment, the cash flows are in dollars of constant purchasing power (i.e., net of inflation).

Initial Capital Investment.   The initial capital investment is assumed to be spent instantaneously at the beginning of the first year. Obviously, the investment is not made instantaneously, but the effect of a time lag can be included in a correction for less-than-capacity operation of the system during the first year. The present equivalent of the initial capital investment is simply the investment; no discounting is involved because the investment is made with present dollars at the present time.

TABLE 21.18 Cash Flows for Economic Analysis of Industrial Solar Process Heat Systems (In Constant Dollars)

Cash flow	Year of cash flow	Cash flow at end of year x	Net cash flow after federal income tax	Discounted cash flow in year x	Total discounted cash flow in n years
Initial investment	0	$-I$ (beginning of year)	$-I$	$-I$	$-I$
Investment tax credit (10% assumed)	1	$0.1I$	$0.1I$	$\dfrac{0.1I}{1+i}$	$\dfrac{0.1I}{1+i}$
Salvage	n	sI	$0.76\,sI$	$\dfrac{0.76\,sI}{(1+i)^n}$	$\dfrac{0.76\,sI}{(1+i)^n}$
Solar system annual cost (0.04 assumed)	$1-n$	$-0.04I$	$-0.52 \times 0.04I$	$\dfrac{-0.52 \times 0.04I}{(1+i)^x}$	$-0.52 \times 0.04I\left[\dfrac{(1+i)^n-1}{i(1+i)^n}\right]$
Depreciation† (sum-of-the-years-digits)	$1-n$	$\dfrac{I(n+1-x)}{1/2\,(n^2+n)}$	$\dfrac{0.48\,I(n+1-x)}{1/2\,(n^2+n)}$	$\dfrac{0.48I\,(n+1-x)}{1/2\,(n^2+n)\,(1+i)^x}$	$0.48I\left[\dfrac{n-\dfrac{1}{i}+\dfrac{(1+i)^n-1}{i(1+i)^n}}{1/2(n^2+n)}\right] \times \left[\dfrac{(1+i)^n-1}{i(1+i)^n}\right]$
Fossil fuel Cost savings (revenue)	$1-n$	$\dfrac{fLC(1+r)^x}{\eta}$	$0.52\,\dfrac{fLC(1+r)^x}{\eta}$	$\dfrac{0.52\,fLC(1+r)^x}{\eta\,(1+i)^x}$	$*$
Construction delay and start-up expenses as less-than-capacity operation of system	1	$-(1-\xi)\,\dfrac{fLC(1+r)^x}{\eta}$	$\dfrac{0.52\,(1-\xi)\,fLC(1+r)^x}{\eta}$	$\dfrac{-0.52\,(1-\xi)\,fLC(1+r)}{\eta\,(1+i)}$	$\dfrac{-0.52\,(1-\xi)\,fLC(1+r)}{\eta\,(1+i)}$

*When $r > i$ and $\dfrac{1+r}{1+i} = 1 + w$, then this sum is $\dfrac{0.52\,fLC\,(1+w)}{\eta} \cdot \left[\dfrac{(1+w)^n-1}{w}\right]$ and when $r < i$ and $\dfrac{1+r}{1+i} = \dfrac{1}{1+w}$ then this sum is $\dfrac{0.52\,fLC}{\eta}\left[\dfrac{(1+w)^n-1}{w(1+w)^n}\right]$.

When $r = i$, then this sum is equal to $\dfrac{0.52\,fLCn}{\eta}$.

†SOURCE: See source footnote above, pp. 210–217.
SOURCE: Gerald W. Smith, "Engineering Economy: Analysis of Capital Expenditures," pp. 53–55, The Iowa State University Press, Ames, Iowa, 1968.

Investment Tax Credit. An investment tax credit is assumed to be taken at the end of the first year. This credit may be in the range of 7 to 10 percent (in the United States) of the initial investment, depending upon the specific situation.

Effect of Income Taxes. Income taxes affect the net cost for the annual cost of the solar system and the annual expense for fossil fuel used for backup. Depreciation also enters into this calculation. The income tax rate for the federal income tax, 0.48, is assumed, but this figure can easily be adjusted to include the effect of state and local income taxes as well.

The effect of income taxes on cash flows is as follows:

$$\text{Taxable income} = \text{revenue} - \text{annual operating costs} - \text{depreciation}$$
$$\Delta\text{Taxable income} = \Delta\text{rev.} - \Delta\text{AOC} - \Delta\text{dep.}$$
$$\Delta\text{Net income} = 0.52\,\Delta\text{rev.} - 0.52\,\Delta\text{AOC} - 0.52\,\Delta\text{dep.}$$
$$\Delta\text{Cash flow} = \Delta\text{dep.} + \Delta\text{net income}$$
$$\Delta\text{Cash flow} = 0.48\,\Delta\text{dep.} + 0.52\,\Delta\text{rev.} - 0.52\,\Delta\text{AOC}$$

Cash flows for depreciation, revenue, and annual costs must be multiplied by factors of 0.48, 0.52, and 0.52 to reflect the net effect *after income taxes*.

Solar System Annual Costs. The annual costs of operating and maintaining the solar process heat system can be estimated as a fraction of its initial investment. These costs include operating costs, maintenance, replacements, insurance, and property tax. Little experience has been obtained with annual costs on solar systems, but preliminary estimates indicate that the total of these annual costs is in the range of 3 to 5 percent of the initial investment per year.

Depreciation and Salvage. Various depreciation methods may be used. In this analysis, the sum-of-the-years-digits method (SOYD) is assumed for depreciation over the n years of the project lifetime. To qualify for this method, a project must have a lifetime of at least seven years. An estimated salvage value of a fraction of the initial investment is included in the analysis. The salvage value is recovered at the end of the project lifetime and is discounted accordingly. Income tax must be paid on salvage (which is called a gain on disposal), although the applicable rate may be half the normal rate.

Fossil Fuel Cost Saving. Included in the expression for the annual cost saving for fossil fuel is an allowance for the escalation in its cost. Because the load L is in terms of the energy delivered to the process rather than the energy content of the fuel, the fossil fuel term includes an efficiency η for combustion and delivery to calculate the required energy content of the fuel. Because the cash flows in Table 21.18 are in terms of constant dollars, the rate of escalation r is in terms of constant dollars, i.e., the *real* escalation rate. This escalation rate is related to the rate of general inflation r_i and the escalation in cost above inflation r_f as follows:

$$\frac{1 + r_i + r_f}{1 + r_i} = 1 + r \tag{21.9}$$

Influence of Less-Than-Capacity Operation. During construction and the initial shakedown period, the solar system will not be operating at its design capacity. The solar load fraction during this period is $f\,(1 - \xi)$. Therefore, an additional amount of fossil fuel is used during this period and is an additional cost compared to the estimated saving. As shown in Table 21.18, this period is assumed to last less than one year.

Determination of the solar system design   The procedure for finding the optimum fraction of the load carried by solar f is as follows. The total discounted cash flow or present value for the project is set equal to zero, resulting in a linear relationship of the initial investment I to the fraction solar f. If this linear relationship is plotted on the performance-cost curve of I/L versus f, the fraction solar f satisfying the desired conditions is determined by the simultaneous solution of these two curves.

From Table 21.18, the present value for the project is

$$PV = -I + \frac{0.11I}{(1+i)} + \frac{0.76\,sI}{(1+i)^n} - 0.52 \times 0.04I\left[\frac{(1+i)^n - 1}{i(1+i)^n}\right]$$

$$\tag{21.10}$$

$$+\,0.48\,I\left[\frac{n - \dfrac{1}{i} + \dfrac{n}{(1+i)^{n-1}}}{1/2\,(n^2 + n)}\right]\left[\frac{(1+i)^n - 1}{i(1+i)^n}\right] + * - \frac{0.52(1 - \xi)fLC\,(1 + r)}{\eta(1+i)}$$

*See Table 21.18 for fossil fuel savings term.

If PV is set equal to zero, the equation can be arranged to the following form:

$$\frac{I}{L} = \frac{FC}{S}f \tag{21.11}$$

where S is a composite term of the costs related to the solar system—the initial investment, annual operating cost, depreciation, and salvage; F describes the effect of discounting and escalating future fuel costs.

In general, there may be two simultaneous solutions to the two equations. If the interest rate i is increased, the slope of the PV line should decrease until it is tangent to the I/L-versus-f curve. At this point, the investment in the solar process heat system should yield the maximum internal rate of return on the entire investment.

In a particular situation, rather than seek the point of maximum rate of return on the solar investment, management may wish to maximize the PV at a particular rate of return. For this condition, the following relationship results from the PV equation:

$$\frac{d\left(\dfrac{I}{L}\right)}{df} = \frac{FC}{S} \tag{21.12}$$

and the correct solution is at the point of tangency of this slope with the I/L-versus-f curve where the marginal rate of return is the desired rate. These relationships are illustrated in Fig. 21.9.

Annual equivalent cash flows   The annual equivalent method and the present equivalent method for calculating cash flows as shown above are the same in that each will arrive at the same design on the basis of the same information. Which method is used depends upon the form of the basic data and the output which is desired from the analysis. When basic data, such as the annual allocations of depreciation, are not uniform, such as in the case of the sum-of-the-years-digits method, the present equivalent of the irregular cash flows must be calculated, before the annual equivalent can be.

To calculate the annual equivalent cash flows, the terms for total discounted cash flow in Table 21.18 must all be multiplied by the capital recovery factor $i(1 + i)^n/[(1 + i)^n - 1]$, where n is the project lifetime.

Payback analysis   A *convenient but crude* method of evaluating a proposed investment is to calculate the time period required for the sum of the net cash flows resulting from the

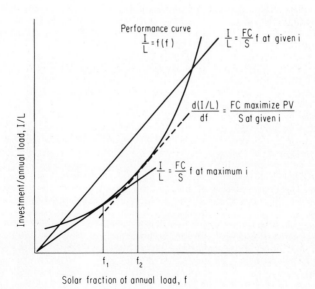

Fig. 21.9   Determination of optimum solar system design by means of economic analysis.

investment to just equal the investment without providing for return on investment or income taxes.[9] In terms of the cash flows in Table 21.18, the payback period is the time required for the sum of the investment tax credit and the fossil fuel savings less the solar system annual costs and the construction delay and startup expenses to equal the investment.

CONCEPTUALIZED DESIGNS OF SOLAR PROCESS HEAT SYSTEMS

The design of a solar process heat system and its integration with a conventional fuel backup system are influenced by many factors. Some of these factors are due to process operational requirements, such as required temperature, required temperature tolerance, process seasonality, or the continuous or intermittent nature of the process. The characteristics of the various types of collectors also influence system design—for example, flat-plate and concentrating collectors would use different types of storage, depending on the type of collecting fluid. Finally, because the performance of solar energy systems is influenced by climate, climate plays an important part in determining design.

All of these factors must be considered in developing a design for a specific application of a solar process heat system, and methods for performing the basic system design calculations are presented in other chapters. In spite of the numerous factors which influence the design of a solar system, it is possible to conceive basic generalized designs which would be suitable for many diverse applications. The same type of process operation—e.g., heating a liquid or a gas, drying a solid, providing steam—is common to many processes, although the hot liquid or gas, or steam may then be used in many ways. If designed as a plant utility system to provide simply the hot liquid, hot gas, hot air, or steam, a solar process heat system design can find numerous applications in many different processes and industries.

Little experience has been obtained with the design and operation of many solar systems in industrial plants. It is thus not possible to present specific guidelines for the design of solar systems in specific applications. However, because solar process heat systems designed as a utility system rather than as an integral part of the process will be economically competitive sooner, basic generalized designs conceived and presented in the literature[2] are described in this section to illustrate general principles involved in the design of solar process heat systems. In each case, a schematic block diagram has been drawn for the system, including the controls and the interface with the conventional fuel system. The operation of each system is described. Specific examples for applications are discussed, as well as potential operating problems.

System to Provide Process Hot Water

System description A schematic diagram of the generalized system is presented in Fig. 21.10. When insolation causes the absorber plate temperature $T2$ to rise above the storage tank temperature $T1$, the differential temperature controller closes the drain-down valve $V2$. The automatic air valve $V1$ vents air and vapor back to the storage tank as the collectors fill. When the plate temperature no longer exceeds tank temperature, or in the case of a power outage, the pump is shut off and the drain-down valve and the air vent return to their open position. As a result, the collector fluid drains back to the storage tank. This prevents both collector freeze-up and heat loss through reradiation. The drain-down system could operate by gravity if the collectors are physically at a higher elevation than the storage tank. If this is impossible, a sump pump must be used.

The heat storage is a large insulated water tank. The tank can be pressurized to allow heat storage with water at temperatures above 100°C. Maximum allowable pressure is set by the rating of the temperature/pressure relief valve $V5$. Utility water is charged to the tank by a normally closed automatic valve $V3$ controlled by a level indicator/controller in the tank. The tank is designed to have a certain amount of inert gas above the liquid level, even after collector drain-down. This acts in effect as an integral expansion chamber for the fluid system.

The hot-water delivery system includes a pump $P3$, auxiliary heating equipment for backup, and provision for tempering the heated water through addition of cold water by means of valve $V4$. The pump is controlled by some type of demand signal from the process. This could include sensing of flow, detection of fluid level in an accumulator tank, or manual startup by an operator. This signal should override all other signals to the heater; i.e., the heater should not function if the pump is not running.

The heater control loop is designed to accommodate a backup system with a finite startup time. Primary on/off control of the heater comes from the storage tank temperature sensor.

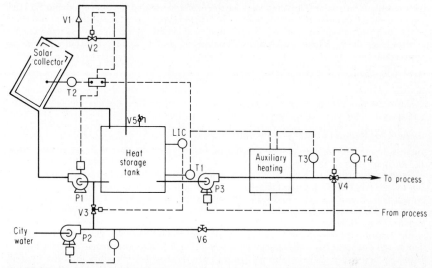

Fig. 21.10 Conceptual design of solar process heat system to provide process hot water. (*From Inter-Technology/Solar Corporation, "Analysis of the Economic Potential of Solar Thermal Energy to Provide Industrial Process Heat," p. 1289, Report No. COO/2829-76/1, U. S. Energy Research and Development Administration, February 7, 1977.*)

The warmup time lag can be compensated for by firing up the heater at some minimum heat rate when the storage tank is less than a few degrees above the required process temperature. This cut-in temperature would be a function of the lead time required, storage capacity, and the required delivery rate. Modulation of the firing rate would be controlled by a temperature sensor $T3$ on the downstream side of the heater.

During periods of high insolation or while the auxiliary is being brought on line, the temperature after the heater may exceed the desired process temperature. To compensate, cold utility water is passed through a pressure-reducing or balancing valve $V6$ and into the process line through the modulating (tempering) valve $V4$. This valve is controlled by a temperature sensor $T4$ immediately downstream.

System alternatives   The system has been designed with general applicability in mind. However, special process requirements may make various adaptations necessary. The use of a heat-exchange fluid other than water would require an additional heat exchanger. Similarly, the use of corrosion inhibitors in the collector fluid would necessitate a heat exchanger to prevent contamination of the water going to the process. This closed-loop collector system would also require an expansion tank and possibly a separate drain-down system. In addition to the added capital costs, this system is slightly less efficient because of the temperature drop necessary to transfer heat across the heat exchanger. Air collectors appear unsuitable because of the thermal inefficiencies inherent in the use of different collection and delivery fluids.

A variety of backup heating systems may be used. The simplest types could be characterized as instantaneous response, instantaneous on-line systems. Small oil- and gas-fired heaters would fall in this category. Indirect steam heat exchangers using line steam from a boiler are also capable of immediate response. In these situations, it would not be necessary to have the feed-forward control loop from the storage tank temperature sensor to the heater.

Large direct-fired heaters, however, might require a certain lead time to preheat tubes and furnace walls gradually. In this case, the time delay loop diagrammed would be useful in maintaining constant discharge during switchover. Boilers cannot be operated intermittently. Consequently, boiler systems used only for the same process as the solar system would have to be base-loaded and operated in parallel with solar. As storage temperature fell, load could be shifted to the boiler.

The generalized scheme in Fig. 21.10 is designed for an on/off batch-type of operation. Modulation of the discharge when needed could be controlled by installation of a valve downstream of the tempering valve. In situations where rapid on/off fluctuations might occur, a pump bypass or partial recirculation loop may be necessary around the delivery pump $P3$. This loop is common in industrial applications with centrifugal pumps. It avoids frequent restarting of the pump and maintains a required minimum cooling flow. While this necessitates another control loop, the technology is readily accessible and would already be installed in retrofit applications with this type of process flow.

System application   Once-through hot-water systems are useful in two classes of process operations. The hot water may be used by the unit operation and contaminated or dispersed beyond economic recovery. The other situation would involve the incorporation of the hot water in the product. Once-through systems are found throughout the food, leather, textiles, and specialty chemicals industries.

System for Indirect Heating of Process Fluids

Indirect heat exchange with process fluids is probably the single most common unit operation in industrial process operations.

System description   This system is similar to the solar system which provides once-through hot water except that a heat exchanger carrying the process fluid is contained within the heat storage tank. Figure 21.11 shows a schematic diagram of this system. The solar collector/storage system would be connected in series with the backup heating system for both retrofit and new installation applications.

The heat storage system in Fig. 21.11 uses the same fluid that is circulated to the collectors. This system offers several advantages over heterogeneous collector/storage systems. First, it is more efficient thermally because no temperature difference is necessary to transfer energy from the collector fluid to storage. Second, a separate tank is not needed for drain-down. Drain-down can be accommodated by maintaining a minimum inert gas space in the storage tank. Finally, the same space allows the storage tank to act as an expansion chamber for the collector. The space can be observed by use of a sight glass, with makeup to the closed system supplied when necessary by manual operation of the makeup valve $V3$.

The operation of the solar collector and the heat storage tank is similar to the operation of the once-through system. In the indirect system, the process liquid has two possible options with respect to the solar storage system. Normally, it would be pumped by means of $P2$ through a coil located in the storage tank to absorb the required heat. However, if the storage temperature is below the temperature of the process return line $T3$, a differential temperature controller bypasses the flow around the storage tank. Two valves—a control valve $V5$ and a check valve $V4$—are needed to divert the flow. Without the bypass, auxiliary heat could eventually end up heating the entire storage mass during extended periods of low insolation.

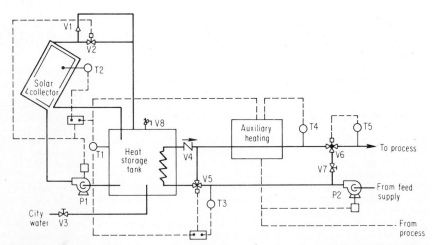

Fig. 21.11   Conceptual design of solar process heat system to heat process liquids by means of indirect heat exchange. (*From InterTechnology/Solar Corporation, "Analysis of the Economic Potential of Solar Thermal Energy to Provide Industrial Process Heat," p. 1297, Report No. COO/2829-76/1, U. S. Energy Research and Development Administration, February 7, 1977.*)

The same considerations regarding the use of the auxiliary backup system and its control discussed for the once-through system apply to this indirect system as well.

During days of peak insolation, the process liquid at the sensor $T4$ may exceed the desired process temperature. The same situation occurs during startup of the "finite response time" heater system. The fluid is tempered to the required delivery temperature by means of bypass valve $V6$. The pressure on the two inlets of the bypass valve is equalized by a manually operated balancing valve $V7$ in the bypass line. The bypass valve is controlled by temperature sensor $T5$ immediately downstream.

Proper design will preclude the necessity for much operator attention during routine operation. During maintenance or in emergencies, the system can be isolated from the process by simply opening the heat storage tank bypass valve. A power failure would result in the drain-down of the system, assuming the collectors can be located above the storage tank level. Stratification of the heat storage tank is not anticipated because of the agitation caused by pump intake and discharge. However, possible stratification can be avoided by placing at least the process liquid outlet near the top of the storage tank. With these precautions, no operational problems are anticipated.

System alternatives   Because of the high temperatures and pressures involved in systems with concentrator collectors, water may not be suitable for a liquid-phase collector fluid. Low-vapor-pressure heat-transfer fluids are available (Therminol^{TM}, Dowtherm^{TM}) for this high-temperature duty. These fluids can also have low pour points, eliminating the need for the automatic drain-down valve system to prevent freeze-up. If the cost per gallon allows, the collector fluid can also be used as the storage medium. This eliminates (1) the thermal inefficiency of driving a heat exchanger in a heterogeneous collector/storage fluid system, and (2) the need for a separate expansion chamber.

Three possible interfaces between the heat storage system and the process liquid could be used. The general case, outlined in Fig. 21.11, has the process liquid go through a coil actually located in the heat storage system. There are, however, many applications in which storage of the process fluid could act as the heat storage medium also. In this case, the collector loop would include a heat exchanger in the process storage tank and a small drain-down and expansion tank. Such a system would be useful, for example, in maintaining pumpability of oil in petroleum tank farms.

The third and most complex possibility involves a separate heat-exchange loop between the storage and process heat exchangers. This could be necessary if the process liquid were heat sensitive or if contamination were a major potential problem. System operation would be essentially the same diagrammatically as shown in Fig. 21.11. The process would just be "replaced" by a process-fluid heat exchanger.

System for Indirect Heating of Process Gases

Common uses for indirect heating of process gases can be divided into two general categories. In the first case, the gas is heated prior to use as a reactant. Petroleum refining and the petrochemicals industry provide numerous examples of this type of application. The second category involves heating the process gas prior to its use as a heat-transfer medium. Typical applications would be hot-air dryers or curing ovens. In these applications, the air may also have important secondary functions as a mass-transfer medium—e.g., for carrying away evaporated water or solvents.

System description   Figure 21.12 shows a schematic diagram of a solar system for heating process gases which has been designed to meet the heating requirements of a wide range of process applications. The system closely parallels the solar thermal system for indirect heating of process liquids discussed above.

To meet the widest range of temperature requirements, concentrating collectors can be used. Concentrators can develop fluid temperatures to over 260°C. Preliminary estimates indicate that they may be cost-competitive with flat-plate collectors at process delivery temperatures even below 100°C. This temperature range covers a large segment of process heating applications.

The system in Fig. 21.12 uses the collector fluid as the storage and final transfer medium. This eliminates the need for heat exchangers in the storage tank and raises the thermal efficiency of the system.

The collection and heat delivery loops operate independently of each other. Heat can be absorbed and transferred to the storage tank with or without concurrent removal of heat by the process control loop. There is no provision for bypassing the storage tank and piping directly

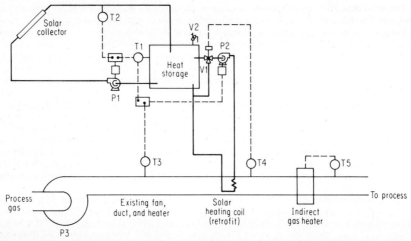

Fig. 21.12 Conceptual design of solar process heat system for indirect heating of process gases. (*From InterTechnology/Solar Corporation,* "*Analysis of the Economic Potential of Solar Thermal Energy to Provide Industrial Process Heat,*" *p. 1305, Report No. COO/2829-76/1, U. S. Energy Research and Development Administration, February 7, 1977.*)

from the collectors to the process heat exchanger. The operation of the collection loop and storage system is the same as the operation described above for the system for indirect heating of process fluids.

A differential temperature controller has been included in the generalized design to activate the heat delivery pump $P2$. When tank temperature $T1$ sufficiently exceeds the makeup gas temperature $T3$, the pump is activated and delivers hot fluid to the process heat exchanger. This control loop is designed to accommodate a waste-heat exchanger which may be located schematically between the process centrifugal blower $P3$ and the sensor $T3$. If the feed-gas temperature is essentially ambient, the pump can be controlled by the sensor $T4$ downstream of the solar heating coil in the process duct. When the process stream falls below the desired temperature, the pump would be activated. With the differential controller in the system, the downstream sensor $T4$ only functions to prevent temperature overshoot by opening the heat exchanger bypass valve $V1$ when the process stream exceeds its setpoint temperature.

The backup heater in Fig. 21.12 is an indirect heat exchanger in series with the solar coil. The only control necessary is a feedback loop from a downstream temperature sensor $T5$.

The system in Fig. 21.12 is based on a continuous-process heat load. Variation in process flow would not affect the operation of the system. In any case, a shutdown control would be necessary. This would involve a relay on the delivery pump $P2$ which would deactivate $P2$ should the process blower $P3$ shut down. Variation of temperature to the process would simply be a matter of either manually or automatically changing the setpoints on temperature sensors $T4$ and $T5$.

System application   The system in Fig. 21.12 is suitable for a wide variety of process applications. A partial list of potential operations includes the following: (1) aggregate drying in asphalt production, (2) coal and mineral drying, (3) food drying or baking, (4) paint drying ovens, (5) textile fiber drying, (6) lumber drying, (7) sulfuric acid concentrating, (8) leather drying, and (9) curing of sand cores and molds used in iron foundries.

Systems to Provide Steam

The importance and utility of process steam is evidenced by the fact that an estimated two-thirds of all industrial process heat is utilized in this form. If solar energy systems can be effectively interfaced with steam generating systems, then the potential impact on industrial consumption of conventional fossil fuels will be greatly increased.

There are numerous concepts of systems to provide process steam utilizing solar energy, each having its own advantages and disadvantages. The following paragraphs will describe three basic steam systems which were chosen because of their concept differences. System operation will be discussed together with the primary advantages and disadvantages of each.

Flash tank system   The system shown in Fig. 21.13 utilizes the flash tank concept, which requires that water be stored at high temperature and pressure. During operation, water from the high-temperature storage is flashed to a lower pressure and temperature to produce saturated steam which is delivered to the process.

If the flashing process yields an amount of water lower than the minimum rate required by the boiler, which must be fired at a nominal rate, flow controller $FC\,2$ will activate $V5$, allowing the process condensate return to enter the flash tank directly, thereby reducing the solar input. Since any condensate return introduced into the flash tank will not be at the temperature required by the process, temperature controller $TC\,1$ will activate valve $V3$, thereby injecting steam produced by the boiler into the flash tank to maintain equilibrium conditions. During normal operation, the level controller will activate valve $V2$, regulating the level of saturated water in the tank. Thus, as the temperature of the storage tank increases, the level of the water within the flash tank will decrease, causing the level controller to close down valve $V2$, reducing auxiliary-produced steam. As the temperature of the storage tank decreases, the level of the water in the flash tank will increase, causing the level controller to open valve $V2$, increasing the auxiliary-supplied steam.

If the temperature of the water in storage drops below the temperature of the process condensate, the differential temperature controller DTC will sense this, close valve $V4$ and valve $V6$, and open valve $V5$. Under this condition, the auxiliary boiler is supplying 100 percent of the process requirement. It should be pointed out that this particular design allows for auxiliary boiler water preheat. Thus, when no steam is produced during the "flashing" process, the level controller senses this and opens valve $V2$ fully, thereby allowing all flow through the auxiliary boiler. Process flow requirements are maintained through flow controller $FC\,1$ and valve $V1$.

The advantage of the system depicted in Fig. 21.13 is that the boiler water can be preheated and that the control system is relatively simple. The primary disadvantage of the flash tank concept is that relatively little steam can be produced efficiently through this process.

Simple boiler system   A solar steam-producing system utilizing a simple kettle evaporator or boiler is illustrated in Fig. 21.14. During operation, the level of the water in the boiler is maintained by a level controller which adjusts the amount of solar energy input by varying the flowrate through valve $V4$. This control is necessary to insure that for very high temperatures in solar storage and low process heat requirements, no more solar energy input than is required by the process is permitted; otherwise the boiler would boil dry, resulting in excessive scaling

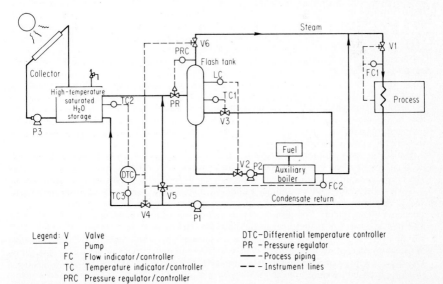

Legend: V   Valve   DTC–Differential temperature controller
——   P   Pump   PR –Pressure regulator
         FC   Flow indicator/controller   —— –Process piping
         TC   Temperature indicator/controller   —— –Instrument lines
         PRC Pressure regulator/controller

Fig. 21.13   Conceptual design of solar process steam system utilizing the flash tank principle. (*From InterTechnology/Solar Corporation, "Analysis of the Economic Potential of Solar Thermal Energy to Provide Industrial Process Heat," p. 1317, Report No. COO/2829-76/1, U. S. Energy Research and Development Administration, February 7, 1977.*)

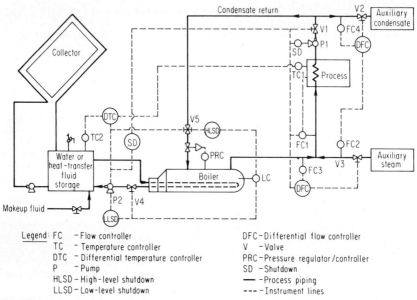

Fig. 21.14 Conceptual design of solar process steam system interfaced with a conventional auxiliary steam system for backup. (*From InterTechnology/Solar Corporation, "Analysis of the Economic Potential of Solar Thermal Energy to Provide Industrial Process Heat," p. 1320, Report No. COO/2829-76/1, U. S. Energy Research and Development Administration, February 7, 1977.*)

and unstable operation. This particular scheme illustrates the interface of a solar steam system with a particular process in which the auxiliary system is steam provided by the main plant utility system.

To interface the system properly with the auxiliary utility system, the auxiliary condensate return must be metered so that it remains equal to the amount of auxiliary steam supplied to the process. This requirement is met by flow controller *FC*2 and flow controller *FC*4, which are compared in the differential flow comparator, and valve *V*2 adjusted accordingly.

As the temperature of the storage system begins to drop, the level controller will open valve *V*4 to increase flow through the boiler, increasing solar input. If the amount of solar input cannot keep pace with the process steam requirements, the water level in the boiler rises, at which point the level controller will shut down pump *P*2 and close valve *V*5. Under this condition, all process requirements are supplied by the auxiliary steam system.

The other controls illustrated on the diagram are designed to accommodate various process flow requirements, shut down the solar system when required, turn the system on when the temperature in the storage tank reaches a critical level, and shut down the solar system should the level of water within the boiler fall below a critical point.

The advantages of this system are that a variety of fluids can be used as a solar storage medium, and that the system can be interfaced with typical processes without great difficulty. The main disadvantage of this system appears to be the complex control system requirements.

System interfaced with auxiliary boiler   A third solar steam system is illustrated in Fig. 21.15. This is another example of a solar steam system which is interfaced directly with the auxiliary boiler system. This design also provides for a minimum auxiliary-boiler firing rate. This particular system is designed so that flow from the solar storage through the boiler is at a maximum at all times except during one condition: When flow controller *FC*2 senses that the steam produced by the auxiliary boiler is dropping below the minimum required value, the rate of energy supplied by the solar storage is reduced by activating valve *V*4. With the resulting reduced flowrate from the storage tank through the boiler, the level of water within the boiler increases. The level controller senses the rising water level and accordingly opens valve *V*2, increasing the flowrate through the auxiliary system to an acceptable level.

During normal operation, solar input to the boiler can result in steam production rates greater than those possible from a flash tank. As the temperature of the storage tank falls, a point is

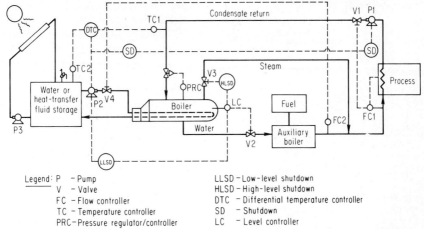

Fig. 21.15 Conceptual design of solar process steam system interfaced with a conventional auxiliary boiler. (*From InterTechnology/Solar Corporation, "Analysis of the Economic Potential of Solar Thermal Energy to Provide Industrial Process Heat," p. 1323, Report No. COO/2829-76/1, U. S. Energy Research and Development Administration, February 7, 1977.*)

reached at which boiling actually ceases. This condition is sensed by the level controller which closes valve $V3$ at the condition called high-level shutdown. At this condition, the auxiliary boiler is producing the steam required for the process; however, the solar system continues to preheat the process condensate return. Only when the temperature of the storage tank drops below the temperature of the condensate return does the differential temperature controller shut down the solar pump $P2$, at which time the auxiliary boiler supplies 100 percent of the process requirements. It should be noted that the process flow requirements are monitored and adjusted by flow controller $FC1$ and valve $V1$. System pressure is regulated by the pressure regulator/controller located at the boiler.

This particular design has several advantages. First, the control system is relatively simple and straightforward. Second, many different solar storage flowrates can be employed. Third, the system allows for maximum utilization of solar energy through the advantage of boiler auxiliary preheat. Another advantage of this particular design is its applicability to batch processing, provided the auxiliary boiler is so designed, since operation of the solar system is really unconcerned with what is taking place at the process end.

EXAMPLES OF SOLAR PROCESS HEAT SYSTEMS

Two examples of solar process heat systems are described in this section. The first example is an experimental shallow solar pond system being developed to provide hot water for a uranium-milling operation. This system has been studied and engineered extensively and is a good example of an industrial solar process heat system which can provide process heat at a cost competitive with fossil fuel in its location.

The second example described is a hypothetical system which illustrates the general design procedure described in this chapter and which uses one of the conceptualized generalized designs described above.

Experimental Shallow Solar Pond

A shallow solar pond system has been developed by the Lawrence Livermore Laboratory (LLL), Livermore, California, to provide process hot water for a uranium-milling plant in New Mexico.[10] The system has been extensively engineered and tested to develop a low-cost solar process heat system which can compete economically with fuel oil at $15 per barrel. The system is designed to heat water to a temperature of 60°C.

System description The design features of this shallow solar pond are shown in Fig. 21.16. To minimize the loss of heat through the bottom, it is insulated with a 3.8-cm slab of foamed glass, which was found to be preferable to polyurethane foam. The heat loss through the foamed glass is calculated to be about 6 percent of the heat collected daily.

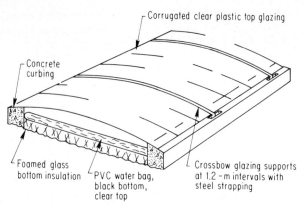

Fig. 21.16 Short length of a shallow solar pond module. A typical full-sized module measures 3.5 m × 60 m with a water depth of 10 cm. (*From W. C. Dickinson et al., "Shallow Solar Ponds for Industrial Process Heat: The ERDA-Sohio Project," p. 3, Lawrence Livermore Laboratory, Livermore, CA, Preprint UCRL-78288, Rev. 1, June 17, 1976.*)

The water bag is made of two layers of UV-stabilized polyvinylchloride (PVC), heat-sealed along the edges. The bottom layer is 0.5 mm thick and black in color to absorb solar radiation while the top layer is 0.3 mm thick and colorless. Containment of the water in a closed plastic bag serves to eliminate heat losses by evaporation. The water bag is shielded from most UV radiation by the top glazing and is expected to have an effective lifetime of 5 or more years. Natural convective mixing of the heated water causes the entire volume to be heated uniformly. The depth of water in the water bag is approximately 10 cm.

The top glazing suppresses upward convective and radiative heat loss as does a single glazing on a conventional flat-plate collector. A major difference between these two designs is the distance between the top of the water bag or plate and the middle of the top glazing. In the shallow solar pond, this distance is about 30 cm. Various greenhouse panel materials were tried for this top glazing, including clear fiberglass and clear acrylic paneling. A preferred material is a corrugated Filon® Supreme greenhouse panel manufactured by the Vistron Corporation, which is a 1.5-kg/m² weatherized fiberglass with a UV-absorbing Tedlar-bonded top surface. It has an 83 percent transmittance for normal sunlight, averaged over the corrugations, and a short-wave cutoff at 0.38 μm (10 percent transmission point).

Vertical elevation for the glazing panels as well as a solid attachment base is provided by concrete curbing. Crossbow supports are attached on both sides of the module at 1.2-m intervals. A steel banding tie-down system over the bows uses standard packaging and crating techniques and provides a tie-down (against aerodynamic lift) of the panels. The edge of the glazing is clamped down against a continuous neoprene rubber strip glued to the curbings and is sealed with galvanized sheet steel angle and hold-down clips.

Performance of the Sohio facility The prototype shallow solar pond facility built by Lawrence Livermore Laboratory at the Sohio uranium-milling plant near Grants, New Mexico, consists of three full-size modules of different designs, cold- and hot-water storage reservoirs, and instrumentation for measuring pond performance. Figure 21.17 is a picture of the facility. Each module heats 25 tons of water per day in a batch process, in which the pond is filled in early morning with cold water and is dumped in the afternoon when the water temperature reaches its maximum value, with the heated water going into an insulated storage reservoir.

In summer, the average water temperature at the end of the day is between 54 and 60°C, and in winter, between 29 and 32°C. Performance measurements indicate that on an annual average basis, the heat collected amounts to about 9.36 MJ/m² · day, with an average temperature of about 15°C for the water used to fill the ponds. Under these conditions, the annual collection efficiency is about 48 percent of the total insolation on a horizontal surface at the site.

Calculations indicate that if the average temperature of the cold water is 25°C, then the annual collection efficiency is about 39 percent.

Economics The total costs for the prototype facility consisting of three full-size modules, including site preparation, cold- and hot-water reservoirs, piping, pumps, and controls came to $75 per square meter of pond collector area.

Fig. 21.17 The prototype LLL shallow solar pond facility built and tested at the Sohio uranium mining and milling complex near Grants, NM. This facility consists of three shallow solar ponds of different designs. (*From W. C. Dickinson et al., "Shallow Solar Ponds for Industrial Process Heat: The ERDA-Sohio Project," p. 16, Lawrence Livermore Laboratory, Livermore, CA, Preprint UCRL-78288, Rev. 1, June 17, 1976.*)

However, for a 36-module system, half the projected full-size system with 72 modules, construction bids indicate that the cost can be reduced to about $60.20 per square meter. A breakdown of these costs is shown in Table 21.19.

Based on these costs, a discounted cash-flow cost analysis was made for the full-size 72-module system. Table 21.20 lists the input parameters that applied to the Sohio plant.

The results of this discounted cash-flow analysis are shown in Table 21.21. The results show a comparison of annual equivalent energy costs for a fuel-oil boiler and a combined solar/fuel-oil system as well as an analysis of the sensitivity of the results to variations in certain critical input parameters.

TABLE 21.19 Lowest Bid Contractor Prices for Construction of 36-Module Shallow Solar Pond System for Sohio*

Item	$ per m² of pond area ($1976)
Site preparation	8.81
Collector	
Materials	27.55
Installation	5.00
Hot-water reservoir	6.02
Pumping stations	2.28
Piping	4.62
Hot-water pipe to mill	2.67
Control and electrical	3.25
Total	60.20

*All prices quoted per unit area of pond modules.
SOURCE: W. C. Dickinson et al., "Shallow Solar Ponds for Industrial Process Heat: The ERDA-Sohio Project," p. 20, Lawrence Livermore Laboratory, Livermore, California, Reprint UCRL-78288, Rev. 1, June 17, 1976.

TABLE 21.20   Input Parameters for Discounted Cash Flow Analysis of Sohio Process Heat Facility

Annual requirement for process hot water = 1.05×10^5 GJ (10^5 MBtu)
Average annual contribution of solar heat = 50% of total
Delivered price of fuel oil = $15/bbl
Conversion efficiency of fuel oil to usable heat = 70%
Annual cost of oil (no solar) = $370,000 (~ 24,000 bbl)
Annual cost of oil (with solar) = $185,000 (~ 12,000 bbl)
Total solar pond collector area required = 15,390 m (~ 4 acres)
Cost of two oil boilers = $80,000 (same with or without solar facility)
Annual O&M (no solar) = $20,000
Annual O&M (with solar) = $30,000
Required after-tax real rate of return on investment = 8%
10% investment tax credit
Straight-line depreciation of solar system over 15 years
Industrial income tax rate = 50%
No salvage value for solar system

SOURCE: W.C. Dickinson et al., "Shallow Solar Ponds for Industrial Process Heat: The ERDA-Sohio Project," p. 21, Lawrence Livermore Laboratory, Livermore, California, Project UCRL-78288, Rev. 1, June 17, 1976.

TABLE 21.21   Overall Cost to Sohio of Process Heat under Various Conditions from Discounted Cash Flow Analyses*

		$/10^6 Btu†	
	Conditions	Constant oil price‡	5%/yr increase in oil price§
Plan A	100% fuel oil at $15/bbl	3.99	5.05
Plan B	50% solar/50% fuel oil with a 15-year lifetime on all parts of solar facility	3.78	4.53
Plan B	*Except* capital investment for solar pond system increases by 10% (from $60.20/m² to $66.22/m²)	3.84	4.59
Plan B	*Except* PVC water bags and top glazings must be replaced every 5 years	3.92	4.67
Plan B	*Except* solar contribution decreases from 50% to 45% (i.e., 9-month operation)	3.87	4.65
Plan B	*Except* 20-yr lifetime is assumed on all parts of solar facility	3.72	4.48
Plan B	*Except* 12% required real rate of return on investment	4.05	4.80

*All calculations made in constant 1976 dollars.
†Multiply by 0.95 to convert to $/GJ.
‡In 1976 dollars. Oil price would actually increase at general inflation rate.
§Above general inflation rate.
SOURCE: W. C. Dickinson et al., "Shallow Solar Ponds for Industrial Process Heat: The ERDA-Sohio Project," p. 22, Lawrence Livermore Laboratory, Livermore, California, Reprint UCRL-78288, Rev. 1, June 17, 1976.

The results indicate that the energy cost of the combined solar/fuel-oil system is competitive with the estimated cost of energy from fuel oil alone under a number of conditions.

Solar Process Heat System in Vegetable-Oil Refining Process

To illustrate the use of the information in this chapter for the design of solar process heat systems, a low-temperature solar system integrated into a vegetable-oil processing operation is described here as an example. The hypothetical plant is assumed to be located in Solar Region VI.

Raw oil from seed extraction operations is transported by tank car to the processing facility. It is stockpiled in small tank farms to provide assurance of constant availability of raw material to the processing plant. In winter—and to a lesser extent in the summer —oil is circulated to an external heat exchanger to help keep the viscosity in a pumpable range. The required temperature varies with the specific oil, but it is in the range of 38 to 71°C. The first processing operation is alkali refining. The oil must be at about 71 to 76°C for this operation.

The selected conceptual design is shown in Fig. 21.18. Except for the process piping arrangement shown, the application follows precisely the flow scheme of the generalized system shown in Fig. 21.11. Because the temperature required is relatively low, noncorroding flat-plate collectors would probably be used, with potable water as the collecting and storage medium. This design would be sufficient to provide heat at 71°C and would introduce no problem with contamination of the edible oil.

The backup system in this design consists of an indirect steam heat exchanger supplied by the central boiler. Heat rate is controlled by a normally closed automatic throttling valve. This valve is activated by an oil-temperature sensor immediately downstream. The valve is connected with the oil-storage pump so that the heater can only operate when the pump is running. The pump itself is controlled by both a temperature sensor and a demand signal from the process. In this way, oil is heated if storage temperature drops or if the alkali refining operation needs raw material.

The first step in sizing the collector array is to determine the total energy requirements. At the top of Fig. 21.18, it can be seen that the plant throughput is assumed to be 125,000 k per day. In the determination of process heat requirements, it was assumed that the oil needs to be heated from an ambient temperature of 16°C to a maximum temperature of 82°C. On this basis, the maximum annual energy requirements are 6.35×10^{12}J. More realistically, only about three-quarters of this amount will be required since ambient temperatures will be closer to 20–25°C and an oil temperature of 71°C will probably suffice. Therefore an annual load of 4.75×10^{12}J will be used for collector sizing.

For the purpose of finding the appropriate collector array size according to the economic analysis described above, the following assumptions have been made:

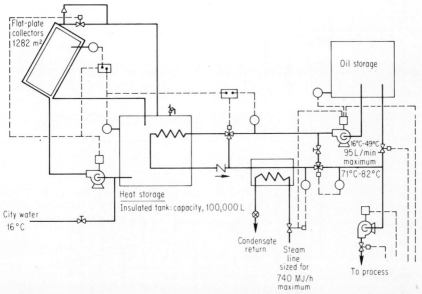

Fig. 21.18   Conceptual design of solar preheat system for vegetable oil processing operation. Typical plant for solar region IV: process throughput, 125,000 kg/day; process heat requirement, 1.74×10^{10} J/day; continuous operation; required temperature, 71°C. (*From InterTechnology/Solar Corporation, "Analysis of the Economic Potential of Solar Thermal Energy to Provide Industrial Process Heat," p. 1301, Report No. COO/2829-76/1, U. S. Energy Research and Development Administration, February 7, 1977.*)

TABLE 21.22 Potential Applications of Solar Thermal Energy Systems for Industrial Process Heat

Industry/process	Energy form*	Temperature, °F	Estimated time of technological readiness†	Shallow ponds or simple air heaters	Flat plates	Fixed compound surfaces	Single-tracking troughs	Central receivers
Aluminum								
Bayer process digestion	Steam	420	1985				X	
Automobile and truck manufacturing								
Heating solutions	Steam (water)	120–180	1980	X	X			
Heating makeup air in paint booths	Air	70–85	1980	X				
Drying and baking	Air	325–425	1985			X	X	
Concrete block and brick								
Curing product	Steam	165–350	1985			X	X	
Gypsum								
Calcining	Air	320	1985			X	X	
Curing plasterboard	Steam (air)	570	1990				X	X
Chemicals								
Borax, dissolving and thickening	Steam	180–210	1980	X		X		
Borax, drying	Air	140–170	1980		X	X		
Bromine, blowing brine/distillation	Steam	225	1985			X		
Chlorine, brine heating	Steam (water)	150–200	1980	X	X	X		
Chlorine, caustic evaporation	Steam	290–300	1985			X	X	
Phosphoric acid, drying	Air	250	1985			X		
Phosphoric acid, evaporation	Steam	320	1985			X		
Potassium chloride, leaching	Steam	200	1980		X		X	
Potassium chloride, drying	Air	250	1985			X	X	
Sodium metal, salt purification	Steam	275	1985			X	X	
Sodium metal, drying	Steam (air)	240	1985			X		
Food								
Washing	Water	120–160	1980	X	X			
Concentration	Steam (water)	100–200	1980	X	X			
Cooking	Steam	250–370	1985			X	X	
Drying	Steam (air)	250–450	1985			X	X	

Process	Form*	Temperature (°F)	Year†					
Glass								
Washing and rinsing	Water	160–200	1980	X	X		X	X
Laminating	Air	212–350	1985				X	X
Drying glass fiber	Air	275–285	1985		X		X	X
Decorating	Air	70–200	1980					X
Lumber								
Kiln drying	Air	150–210	1980	X	X		X	X
Glue preparation/plywood	Steam	210–350	1985				X	X
Hot pressing/fiberboard	Steam	390	1985		X			
Log conditioning	Water	180	1980					
Mining (Frasch Sulfur)								
Extraction	Pressurized Water	320–330	1985				X	X
Paper and pulp								
Kraft pulping	Steam	360–370	1985				X	X
Kraft liquor evaporation	Steam	280–290	1985				X	X
Kraft bleaching	Steam	280–290	1985				X	X
Papermaking (drying)	Steam	350	1985					X
Plastics								
Initiation	Steam	250–295	1985				X	X
Steam distillation	Steam	295	1985				X	X
Flash separation	Steam	420	1985					X
Extrusion	Steam	295	1985				X	X
Drying	Steam	370	1985					X
Blending	Steam	250	1985				X	X
Synthetic rubber								
Initiation	Steam (water)	250	1985				X	X
Monomer recovery	Steam	250	1985				X	X
Drying	Steam (air)	250	1985				X	X
Steel								
Pickling	Steam	150–220	1980	X	X		X	X
Cleaning	Steam	180–200	1980	X	X		X	X
Textiles								
Washing	Water	160–180	1980	X	X			
Preparation	Steam	120–235	1980	X	X		X	X
Mercerizing	Steam	70–210	1980	X	X		X	X
Drying	Steam	140–275	1980	X	X		X	X
Finishing	Steam	140–300	1980	X	X			X

*Preferred form (secondary form).

†Demonstrated thermal efficiency and reliability for 10 years.

SOURCE: Battelle Columbus Laboratories et al., "Survey of the Applications of Solar Thermal Energy Systems to Industrial Process Heat," pp. 22–24, Report No. TID-27348/1, U.S. Energy Research and Development Administration, January 1977.

1. General inflation rate = 7 percent; required real after-tax rate of return = 7 percent
2. Conventional fuel = oil; fuel inflation rate = 10 percent
3. Cost of fuel (1977) = $3.32/J × 10^9; boiler efficiency = 75 percent
4. Solar system lifetime = 20 years; first-year delay factor ξ = 0.5
5. Solar system cost = $30,000 + $250/m^2 × collector area; salvage factor = 5 percent

Based upon these assumptions, the following relationship has been developed for I/L and f according to Eq. (21.11):

$$\frac{I}{L} = \frac{17.2}{0.81} \times 3.32 \times f \tag{21.13}$$

With the solar system cost relationship given above in the general assumptions, Eq. (21.13) can be rewritten as

$$\frac{A}{L} = -0.025 + 0.282\,f \tag{21.14}$$

This curve may then be plotted on Fig. 21.4f for a flat-plate collector in Solar Region VI. At the points where this curve intersects the curve for a delivery temperature of 71°C, the values of the solar fraction are 0.93 and 0.40. A choice of either operating point will yield the same average rate of return. Suppose a solar fraction of 0.93 is chosen; then the corresponding value of A/L is equal to 0.27 m^2/J × 10^9. Since the annual load L is assumed to be 4750 × 10^9 J, the desired collector area is found to be 1282 m^2. Based upon the discussion of storage given above, a storage capacity ratio of 80 L/m^2 should be selected. The required storage capacity then amounts to approximately 100,000 L.

The actual solar fraction which would be achieved in this case would be somewhat less than 0.93. The reason for this is that the base-line system calculations were done without a heat exchanger between the solar storage tank and the process. A heat exchanger would reduce the solar fraction actually achieved to approximately 0.85.

APPLICATIONS OF SOLAR PROCESS HEAT SYSTEMS

Although studies indicate that the potential market for solar process heat systems is sizable, the actual impact and application of solar systems in industry will be affected by many variables. Such variables include energy recovery opportunities, particularly in industries which use and have available high-temperature heat; economics of solar systems; design considerations in applying such systems to industrial processes; and nontechnical issues. Some of these variables have already been discussed in this chapter. Nontechnical issues influencing the implementation of solar systems are discussed in Chap. 27. These issues are important to the industrial user of solar energy as well as to the residential and commercial user.

Expected Chronology of Potential Applications

Potential applications of solar process heat systems have been identified in surveys (Refs. 2 and 4). Table 21.22 shows the results from one such study and lists specific potential solar applications with suitable collector types and expected time of technological readiness.

In these surveys, many other specific potential applications were identified. Good potential applications for solar process heat can be found throughout industry as a whole.

REFERENCES

1. Stanford Research Institute, "Patterns of Energy Consumption in the United States," p. 6, Office of Science and Technology, Executive Office of the President, Washington, D.C., January 1972.
2. InterTechnology/Solar Corporation, "Analysis of the Economic Potential of Solar Thermal Energy to Provide Industrial Process Heat," Report No. COO/2829-76/1, U.S. Energy Research and Development Administration, Washington, D.C., February 7, 1977.
3. "Standard Industrial Classification Manual," Executive Office of the President, Office of Management and Budget, Washington, D.C., 1972.

4. Battelle Columbus Laboraties et al., "Survey of the Application of Solar Thermal Energy Systems to Industrial Process Heat," Report No. TID-27348/1, U. S. Energy Research and Development Administration, Washington, D.C., January 1977.
5. InterTechnology/Solar Corporation, op. cit., pp. 273–276.
6. InterTechnology/Solar Corporation, op., cit., p. 304.
7. InterTechnology/Solar Corporation, op. cit., pp. 139–140.
8. InterTechnology/Solar Corporation, op. cit., pp. 161–165.
9. Gerald W. Smith, *Engineering Economy: Analysis of Capital Expenditures*, p. 167, the Iowa State University Press, Ames, Iowa, 1968.
10. W. C. Dickinson et al., "Shallow Solar Ponds for Industrial Process Heat: The ERDA-Sohio Project," Lawrence Livermore Laboratory, Livermore, California, Reprint UCRL-78288, Rev. 1, June 17, 1976.

Chapter 22

Solar-Powered Heat Engines

ROBERT BARBER, P.E. and
DARYL PRIGMORE, P.E.
Barber-Nichols Engineering Company, Arvada, CO

HEAT ENGINES, WHAT ARE THEY?

Understanding the function of heat engines requires an understanding of their thermodynamics. However, many who may be interested in solar heat engines may not be interested in the details but only in general results. Therefore, the first two sections of this chapter are intended to provide general information and the later sections involve more detail for those with a thermodynamics background. If the reader wishes to study the thermodynamics of Rankine engines, several references are included in the bibliography.[1-4] References 5 to 12 deal with recent developments in the heat-engine field.

Heat engines are thermomechanical devices which, when operating between a high-temperature heat source and a lower-temperature heat sink, extract some of the thermodynamic heat energy from a working fluid passing from the high- to the low-temperature zones and convert it into shaft power. Only a portion of the heat energy added to the cycle can be converted to shaft work, and the maximum amount of energy which can be recovered is defined by what is termed Carnot cycle efficiency. The Carnot efficiency η_{Carnot} is the ratio of the temperature difference between the maximum cycle temperature and the minimum cycle temperature divided by the maximum absolute cycle temperature:

$$\eta_{\text{Carnot}} = \frac{T_{\text{hot}} - T_{\text{cold}}}{T_{\text{hot}}} \tag{22.1}$$

where T_{hot} = the maximum cycle temperature, K
T_{cold} = the minimum cycle temperature, K
K = °C + 273 = (°F + 460) (5/9)

Figure 22.1 shows the ideal Carnot efficiency for temperature ranges of interest for the solar heat engines. This figure shows several important trends for heat engines: (1) cycle efficiency increases as maximum cycle temperature increases, and the rate of increase is greater at low temperatures than at high temperatures; (2) decreasing minimum cycle temperature results in a larger increase in cycle efficiency than the same increase in maximum cycle temperature.

The Carnot efficiency is the maximum possible cycle efficiency for a given temperature range and is limited by the second law of thermodynamics. It cannot be met or exceeded by actual

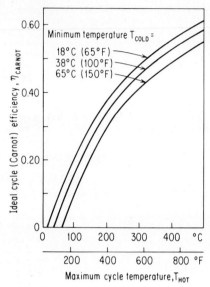

Fig. 22.1   Ideal (Carnot) heat engine efficiency versus temperature.

systems regardless of any promoter's contention that the means has been found to do so. This is very important since some enthusiasts for new engine approaches claim to exceed the Carnot efficiency for their particular devices in violation of the second law.

The actual efficiency obtained by real heat engines is substantially lower than the Carnot efficiency, since (1) real-fluid thermodynamics will vary from the ideal by 20 to 40 percent even with perfect (100 percent efficient) mechanical components, (2) the expansion device (e.g, turbine) will not be 100 percent efficient (50 to 80 percent is typical), (3) the feed pump or compressor will not be 100 percent efficient (20 to 60 percent is typical), and (4) fluid and mechanical friction exist in the engine components. When practical system component efficiencies and real fluids are considered, cycle efficiencies such as those shown in Fig. 22.2 are

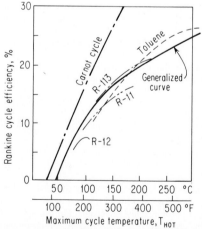

Fig. 22.2   Efficiency of Rankine cycle with real fluids and real components; assumptions: (1) expander efficiency, 80%; (2) pump efficiency, 50%; (3) mechanical efficiency, 95%; (4) minimum cycle temperature, 35°C (95°F); (5) regenerator efficiency, 80%; (6) pressure losses (a) 5% high side, (b) 8% low side.

obtained. Carefully designed engines are capable of reaching 60 to 70 percent of the Carnot efficiency for low temperature differences (80°C) and 50 to 60 percent for higher temperature differences. It is interesting to note that the actual efficiency for cycles operating with flat-plate solar collectors (approximately 100°C operation) is about 10 percent, while the efficiency for concentrating collectors at 150°C is 17 percent. Figure 22.2 shows that if a 10-kW engine is desired at an operating temperature of 100°C, then approximately 9 times that much thermal energy (90 kW) must be supplied to the heat engine; 80 kW of the supplied heat will be rejected to the heat sink.

Mechanical Description

The main components of a typical closed-cycle heat engine are shown in the schematic of Fig. 22.3. The basic cycle is very simple in principle and consists of a heater which transfers energy from a heat source, the solar collectors in this case, to a working fluid which is at relatively high pressure. The working fluid is expanded in a device where shaft power is extracted by use of a turbine or positive-displacement expander. The working fluid in turn flows to the cooler where heat is rejected to a heat sink (generally a cooling tower). After leaving the heat sink, the working fluid is pumped to high pressure before it returns to the heater, completing the loop.

The most common large heat engines are public utility electric-generating plants using steam as the working fluid (Rankine cycle), and aircraft and industrial gas-turbine engines in which the working fluid is a fuel-air mixture (Brayton cycle). Automobile engines use the internal-combustion Otto cycle. The Rankine steam engine is shown schematically in Fig. 22.3. The gas-turbine engine differs since it is an open cycle in which air from the environment is compressed, heated, expanded, and expelled to the atmosphere, eliminating the need for a cooler or cooling tower. If the working fluid is gaseous throughout the entire cycle, as it is in the gas-turbine cycle using air, it is termed a Brayton cycle. If the working fluid is condensed to a liquid, as in the steam cycle, it is termed a Rankine cycle.

In solar-powered systems the Rankine cycle is the most commonly used because it is more efficient than the Brayton cycle for small temperature differences, as will be described later. Other cycles such as the Stirling cycle are proposed for solar applications, but have created little interest to date since they require high operating temperatures and because of the complexity of their machinery and heat exchangers. All solar-powered heat engines of consequence which have been fabricated to date have been Rankine cycle engines, although some development work has been done on both solar-powered Brayton and Stirling engines.

Among the solar engines which have been developed, the primary difference is in the type

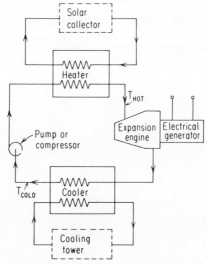

Fig. 22.3. Simplified schematic of typical heat engine to provide electric power.

of expander utilized. Expander types including the turbine, piston, vane, and gear have been used. The type of expander utilized in a heat engine affects the cycle only in the area of the expander efficiency, and all types should be in the 50 to 80 percent range. Therefore, large changes in cycle efficiency in the future because of very efficient expanders are not likely. However, for each application area an optimum expander should be selected as discussed in more detail in a later section.

Application Areas of Solar Power

Solar power systems are of interest in solar air conditioning, solar irrigation pumping, solar-powered heat pumps, and solar electrical generation systems. Solar air conditioning has the advantage that the cooling load is roughly in phase with high solar input periods in the summer, and the use of a solar collector system in the summer for cooling can result in year-round use of the collectors if heating is also required in winter. The Rankine solar air-conditioning system uses an electrical drive as supplemental power.

The geographic areas of application for solar power encompass the bulk of the populated tropical and midlatitude portions of the world. Solar air conditioning is applicable even in the northern climates. In these areas, the solar collectors (the most expensive component in the solar system) will also be used for solar heating in winter, thereby improving the system's load factor and economic viability.

Solar irrigation systems are of interest because the locale requiring irrigation is often an area which has very high solar availability. Some work has also been done in electrical power generation systems in the low-temperature range discussed herein. However, the bulk of the solar electrical power generation activity is being conducted with very high temperature concentrators using large heat engines as discussed in Chap. 20.

The cost of the first solar power systems can be quite high and may not be strictly competitive with other well-established power sources. However, if economic viability had been required for previous U.S. power source development, there would be very little hydropower (since the dams were paid for by flood-control dollars) and no nuclear power.

SOLAR-POWERED HEAT ENGINES

Several cycle types may be used in solar heat engines, including the Rankine, Brayton, Stirling, Vuilleumeir, and Ericsson cycles. Of these cycles, for the temperature range below 300°C only the Rankine cycle is presently of practical significance. Therefore, since an engine type must be selected in order to make a performance and cost optimization analysis of solar power systems, the Rankine cycle is considered in most detail in this chapter. A similar analysis can be made for other solar collectors and engine types when adequate practical information becomes available.

Solar Collectors and Rankine Engine Performance

Several types of solar collectors are presently available. The estimated efficiencies of these types are shown in Fig. 22.4 as a function of collector output temperature. In the design of heat

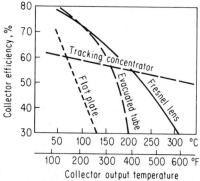

Fig. 22.4.   Solar collector performance of several types at an ambient temperature of 21°C (70°F) and direct solar intensity of 948 W/M² (300 Btu/hr·ft²).

engines, the output temperature of a collector is a more useful quantity than the more common input temperature used in Chaps. 7 to 9. One is easily calculated from the other. These efficiencies are applicable to high solar intensities and, in the case of the concentrating-type collectors, relate only to the direct components of the insolation. The collector efficiencies shown in Fig. 22.4 are felt to represent most commercially available collector types. In a subsequent section, a comparison is made of Rankine power systems using these several collectors.

In order to make a comparison of the various solar conversion approaches, it is helpful to develop a generalized curve for the Rankine cycle efficiency as a function of maximum cycle temperature such as the solid line presented in Fig. 22.2. The assumptions for the calculations are shown in the figure legend and are felt to be reasonable for practical Rankine systems. Several working fluids are compared to the generalized curve. R-113 most nearly follows the generalized curve at temperatures up to 200°C (400°F), which is near the critical point of R-113. Above 200°C, refrigerants decompose rapidly and are not suitable as working fluids. For applications above 200°C, possible working fluids are pure, fluorinated, or chlorinated toluenes or benzenes. The toluene curve of Fig. 22.2 is representative of the performance of this type of fluid. A detailed comparison of the fluids will be presented in a later section of this chapter. However, the data shown in Fig. 22.2 support the concept of a generalized curve which will be used in preliminary calculations.

When the efficiencies of a solar collector and the generalized Rankine cycle are multiplied, the curves shown in Fig. 22.5 result. It is assumed in this analysis that the Rankine cycle temperature is 5 percent less than the collector temperature; i.e., for a collector temperature of 93°C (200°F), the Rankine cycle temperature is 88°C (190°F). In order to obtain this small temperature difference, large heat exchangers are necessary. Another assumption in Fig. 22.5 is that concentrating collectors (the tracking concentrator and the Fresnel lens) have available to them only 853 W/m² (270 Btu/h·ft²) of beam radiation compared with 948 W/m² (300 Btu/h·ft²) for the flat-plate and evacuated-tube collectors, which accept both beam and diffuse sunlight.

Figure 22.5 exhibits several useful principles. The first is that the flat-plate collector has a maximum system efficiency of approximately 5 percent at slightly greater than 100°C (212°F). The evacuated tube, the Fresnel lens, and the tracking concentrator obtain approximately 10 percent system efficiency at the 150 to 200°C (300 to 400°F) range. The efficiency can be increased

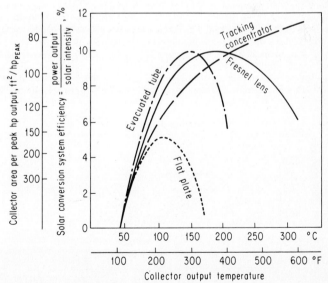

Fig. 22.5 Solar conversion system efficiency versus collector outlet temperature; assumptions: (1) solar intensity, 948 W/M² (300 Btu/hr·ft²) 90% direct; (2) maximum cycle temperature, 95% collector temperature; (3) collector efficiency from Fig. 22.4; (4) Rankine cycle with regeneration; (5) Fresnel and concentrator collectors use only beam radiation.

to approximately 11 percent at 315°C (600°F) with tracking concentrators. Consequently, the best matches of collector type and temperature are (1) low-temperature flat-plate collectors at approximately 93°C (200°F), (2) any of the higher-performance collectors in the 150 to 200°C (300 to 400°F) range, or (3) tracking concentrators at 315°C (600°F). Increasing the temperature above 200°C (400°F) results in a moderate increase in efficiency but will require more development, since common organic fluids and refrigeration components (heat exchangers and valves) cannot be used in higher temperature ranges. Of course, steam systems are being developed for use at 550°C (1000°F) for large solar generating plants, but the subject of this chapter is the design of lower-temperature systems.

Also shown in Fig. 22.5 is the collector area required to produce 1 peak-output horsepower. This peak horsepower would be obtained only when the solar intensity of 948 W/m² (300 Btu/h·ft²) (only a small percentage of the day). Consequently, areas larger than those shown would be necessary to produce average power at the levels shown for the peak power. It is interesting to note that 16 m²/hp (175 ft²/hp) is required for the 93°C (200°F) flat-plate collector while 8.4 m² (90 ft²) is required for the higher-temperature system. This 2:1 difference in the required collector area allows a higher-cost concentrator to compete with the flat-plate collector on a total system cost basis.

System Costs

Estimated costs in 1978 dollars of a large range of Rankine cycle power systems are shown in Fig. 22.6.[12] The system cost includes all the components necessary to produce shaft power and the generator with the associated controls required for electric power output. The costs of the collector system and the cooling heat sink are not included in the values shown in Fig. 22.6. In this figure, the effects of output power level and cycle temperature on Rankine system installed cost are shown. For comparison, the current cost of steam-turbine combined cycles, gas turbines, and diesel engines are also indicated. As shown in this figure, there is a large cost differential between a 93°C (200°F) cycle and a 200°C (400°F) cycle, while there is a smaller cost differential between a 200°C (400°F) cycle and a 315°C (600°F) cycle. This occurs since the heat-exchanger sizes and costs are decreasing rapidly between 93 and 200°C because of the large increase in cycle efficiency over the same range, whereas only a relatively small increase in cycle efficiency occurs between 200 and 315°C.

Table 22.1 shows component costs in a typical 200°C (400°F), 100-kW, Rankine cycle system. In this table it is shown that approximately 38 percent of the system cost is for heat exchangers (this would vary somewhat with cycle temperatures). The absolute costs of the turbine, gearbox, controls, pump, and miscellaneous costs are relatively insensitive to cycle temperature or output power. The generator, on the other hand, varies almost linearly with output power and costs approximately $40/kW at 100 kW and $30/kW at 1000 kW (1978 dollars). The installation cost shown in Table 22.1 is estimated to be 50 percent of the equipment cost, but could be as high as 150 percent in some locations. In Fig. 22.6 and all the following calculations, the 50 percent factor is used for the installation cost of both the collector and the Rankine subsystems.

Since the costs of solar collectors cover a wide range, the total solar power system installed cost was calculated for various collector costs up to $250/m² (approximately $25/ft²). These

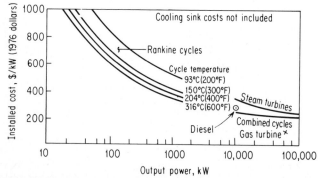

Fig. 22.6 Rankine cycle installed costs for production units as a function of capacity.

TABLE 22.1 Component Cost Distribution for 100-kW Rankine Electrical Power System Operating at 400°F

Component	% cost
Heat exchangers:	
Preheater	6
Boiler	14
Condenser	13
Regenerator	5
Pump	3
Turbine-gearbox	22
Controls	8
Miscellaneous—fluid, structure, assembly, test, etc.	22
Generator, including switchgear	7
	100
Installation cost for packaged unit (could be as high as 150% for some locations)	50

values are shown in Fig. 22.7 and Table 22.2 for flat-plate, Fresnel lens, and tracking concentrator systems. When the overall price is considered, the effect of power level is seen to be small at costs of $50 to $100/m² for flat-plate collectors. Hence, for other collector systems, only the 100-kW size is shown in Table 22.2.

One economic index to consider in the choice of flat-plate and higher-temperature collectors is the break-even cost. Figure 22.8 shows that the concentrating collector can cost 2 times the flat-plate collector for the same system cost if the temperature can be increased to 150°C (300°F). In order to have a cost advantage of 3:1, the collector temperature must be increased to 315°C (600°F). Since the cost of a 315°C collector may very well be 50 percent greater than that of a 150°C collector, the optimum system again appears to be in the area of 150 to 200°C, which is the limit of commercially available Rankine cycle equipment. Table 22.2 shows that the systems operating in the 150 to 315°C (300 to 600°F) range hold the greatest potential for reducing system installed cost to below the $1800/kW level, and all approaches could meet a $2000/kW cost for production units.

The break-even value for a solar power system is difficult to evaluate in the absence of a competitive market and varies widely, depending upon location and assumptions made in the

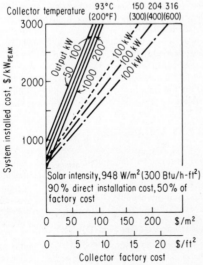

Fig. 22.7   Solar power system installed costs for various collector types.

TABLE 22.2   Installed Cost of Rankine Power System for Various Types of Collectors ($1978)

Collector type	Temperature, °F	Collector production cost, $/m²	100-kW installed cost, $/kW peak, from Fig. 22.7
Flat plate	200	40 to 60	1900 to 2400
Evacuated tube	300	100	2100
Fresnel lens	400	90	1800
Tracking concentrator	400	80 to 100	1600 to 1900
Tracking concentrator	600	100 to 150	1700 to 2300

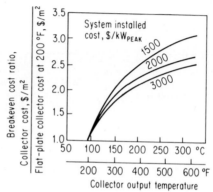

Fig. 22.8   Breakeven collector cost ratio versus collector output temperature for several installed costs.

analysis. However, a large industrial firm has evaluated the investment equivalent of utility for its U.S. plants and the resulting values were $1000 to $2100/kW, depending on the specific plant site. In other words, in energy conservation projects, if 1 kW of electrical power can be saved by installing $1000 to $2100 of equipment, the investment is a break-even proposition. These calculations were based on an estimate of the electrical cost in 1981, assuming an operating life of 12 years with zero salvage value and a 9-year depreciation period, and assuming 24-h usage of the equipment. When less than 24-h usage is contemplated, the break-even costs decrease. Since a solar power system has the potential of about a 25 percent use factor on the average, one would estimate that, perhaps, the break-even point would be roughly one-quarter to one-fifth of the value stated previously, or $250 to $600/kW. The analysis above does not apply to solar cooling systems if the collectors are charged to a heating system (i.e., if the collectors were purchased for the heating requirement), and cooling is to be considered as an extra benefit resulting from an extra investment only in cooling hardware.

RANKINE CYCLE ANALYSIS AND DESIGN

The following sections describe how to carry out a preliminary design analysis of a Rankine heat engine. The word *preliminary* must be emphasized, since this chapter is not intended to provide the information to complete a final design but rather to make a reasonable preliminary prediction of the performance and configuration of a Rankine engine. The detailed design techniques for each of the components (i.e., heat exchangers, pumps, expanders, etc.) are available in the literature, a guide to which is provided in the list of references.[13-21]

Thermodynamic Characteristics of Rankine Cycles

The general technique described here is applicable to all thermodynamic systems in terms of heat utilization to produce shaft power. As discussed earlier, the ideal heat engine does not exist, and at best only a fraction of the available energy existing between two temperature reservoirs can be realized as useful power. In general, heat can be most efficiently converted into useful power when the following guidelines are used:

1. Extract from the collection medium as great a fraction of heat gathered by the solar collector as possible for input heat to the cycle.

2. Add heat to the power cycle fluid at as high a solar collector temperature as possible.

3. Reject heat from the cycle at an average temperature which approaches the heat-sink temperature as closely as possible.

4. Maximize all component efficiencies and minimize all pressure drops in the system.

It is obvious that criteria 1 and 2 are conflicting. Maximum heat extraction from the collector heat-transfer medium requires a large temperature drop in the collector fluid as it passes through the vapor generator. But guideline 2 demands that latent-heat addition (vaporization) take place at as high a temperature as possible, requiring a minimal temperature drop in the collector heat-transfer medium. This enigma is called the pinch temperature difference ($\Delta \dot{T}_p$) problem and is discussed in detail in the section on fluid selection.

A typical Rankine cycle is shown on the pressure-enthalpy diagram of Fig. 22.9 for R-113.* The cycle points on this diagram relate to the cycle point numbers shown in the system schematic diagram of Fig. 22.10, and the tabulated data for the cycle shown in Table 22.3. Solar heat is added in the preheater to heat the working fluid from the pump exit from point 7 to point 8. At point 8 the working fluid is in the saturated liquid condition. It is vaporized between points 8 and 1 in the boiler also by the addition of heat from the solar collectors, to result in the saturated vapor condition at point 1. Pressure drop between the boiler exit and the expansion turbine inlet result in a slight temperature decrease between points 1 and 0 before the fluid is expanded through the turbine at an estimated 75 percent efficiency to the exit pressure at point 2. Point 2' shown in Fig. 22.9 and Table 22.3 is the isentropic (ideal) expansion point for a 100 percent expander efficiency. Because no expander is 100 percent efficient, the flow exits the turbine at point 2, not point 2'.

TABLE 22.3   Rankine Cycle State Points for 25-hp System of Figs. 22.9 and 22.10

Point	Temperature, °F	Pressure, lb/in² abs.	H, Btu/lb	ρ, lb/ft³
1	325	221.1	125.4	6.89
0	322.6	210.8	125.4	6.58
2	200	10	110.18	0.27
3	121	8.5	96.8	0.265
4	86	7.9	91.0	0.26
5	86	7.9	26.6	96.73 (6.33 gal/min)
6	91	231	27.7	
7	149	226	41.1	91.4
8	325	221	85.5	72.1
2		10	105.1	

$\dot{m}$ = 4918 lb/h η_P = 40%
q_B = 1.962 × 10⁵ Btu/h   Losses = 2.25 hp
q_{PH} = 2.184 × 10⁵ Btu/h q_R = 0.654 × 10⁵ Btu/h
q_c = 3.452 × 10⁵ Btu/h η_c = 15.3%
η_t = 75% ϵ_r = 0.70

Heat energy in the fluid leaving the turbine is recovered in the regenerator by the transfer of heat from the low-pressure vapor to the high-pressure liquid. In this process the vapor is cooled from points 2 to 3 while heating the liquid from points 6 to 7. The regenerator is a heat exchanger which can improve the cycle efficiency by 15 percent in the example given here and, as discussed later, may not be required or possible for all fluids. The low-pressure refrigerant flows to the condenser where it is condensed to a saturated liquid leaving the condenser at point 5. Finally the refrigerant is pumped through the boost and feed pumps, providing a high-pressure liquid at the regenerator inlet and thereby completing the cycle.

*The R-113 pressure-enthalpy diagram is for illustrative purposes only. More accurate R-113 thermo-dynamic data are included in Fig. 22.A3.

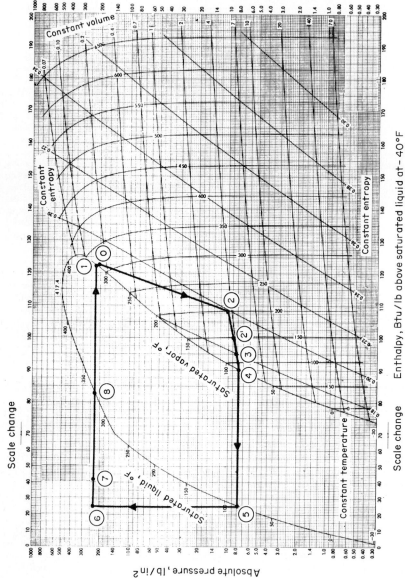

Fig. 22.9 Typical Rankine cycle diagram shown on a *p-h* chart for R-113.

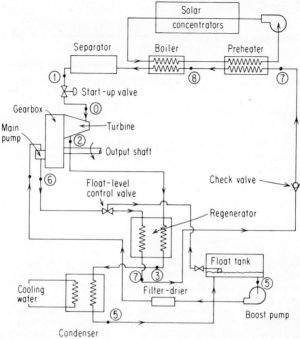

Fig. 22.10 Typical solar-powered Rankine cycle system schematic with state points from Fig. 22.9 shown.

In order to calculate the performance of a Rankine cycle, one must have:

1. Accurate thermodynamic fluid properties—calculations made from tabulated data are generally more accurate than those made from pressure-enthalpy or Mollier (enthalpy-entropy) diagrams.

2. Component efficiencies—these include expander, pump, and other mechanical components (i.e., gearbox, accessory belts, etc.) and regenerator effectiveness.

3. Component pressure drops—pressure losses in the vapor-phase lines (i.e., expander inlet, regenerator vapor side, condenser inlet, etc.) have a significant influence on cycle efficiency while losses in the liquid side result only in increased feed-pump work and a relatively small influence on cycle efficiency.

The result of a cycle analysis is the thermodynamic cycle efficiency. This value is often misunderstood because various definitions are used by various authors. Generally speaking, cycle efficiency is the ratio of the useful output to the heat added from the heat source. The variations among definitions occur since some authors define the useful work as that at the expander shaft after all mechanical losses and feed pump work is subtracted while others include only part or none of these losses. For purposes of this discussion, the cycle efficiency is defined in two manners (refer to Fig. 22.9 for state points):

1. Thermodynamic cycle efficiency

$$\eta_{TC} = \frac{\text{expander work} - \text{pump work}}{\text{heat added}} = \frac{(h_0 - h_2) - (h_6 - h_5)}{h_1 - h_7}$$

2. Cycle efficiency

$$\eta_C = \frac{\text{expander work} - \text{pump work} - \text{mechanical losses}}{\text{heat added}}$$

$$= \frac{\text{useful shaft output power}}{\text{heat added}}$$

For example, from Table 22.3,

$$\eta_{TC} = \frac{(125.4 - 110.18) - (27.7 - 26.6)}{125.4 - 41.1} = \frac{14.12 \text{ Btu/lb}}{84.3 \text{ Btu/lb}} = 0.167$$

$$\eta_C = \frac{(4918/2545) \, [(125.4 - 110.18) - (27.7 - 26.6)] - 2.25}{(4918/2545) \, (125.4 - 41.1)}$$

$$= \frac{27.25 \text{ hp} - 2.25 \text{ hp}}{162.9 \text{ hp}} = 0.153$$

Since the electrical draw for controls is impossible to predict at the preliminary design phase, it is generally not included in the efficiency definition. However, when the final evaluation of a solar system is made it *must* be included.

Working Fluid Selection

As discussed earlier, the maximum output of any energy-conversion system is limited by the second law of thermodynamics and is determined by the heat source and sink temperatures. The fraction of the Carnot value actually achieved by a practical system is governed in great part by the selection of a working fluid. References 22 through 24 deal with fluid selection criteria.

Ideally all Rankine cycle heat addition should take place at a constant temperature, preferably the highest temperature reached in the solar collector. In practice this is not possible because of the thermodynamic properties of real fluids and the required temperature across the vapor generator between the solar collector fluid and the Rankine cycle fluid. It is possible to vaporize the fluid in the collector itself, thus eliminating the need for a separate vapor generator. However, this scheme introduces problems of control instability, possible fluid overtemperature and decomposition, pressurized collectors, and high duct pressure losses associated with vapor flowing from the collector to the expander.

Figure 22.11 shows some of the thermodynamic implications of fluid choice. The latent heat of vaporization of fluid type A comprises a high percentage of the total heat addition, implying a high critical temperature relative to the peak cycle temperature. The latent heat of fluid B represents a much smaller fraction of the total heat addition. Fluid C is used in the supercritical phase (i.e., no phase change occurs). Both fluids B and C have critical temperatures of the same order of magnitude as the collector outlet temperature. Fluid A exhibits the highest cycle efficiency by virtue of the fact that its average heat addition temperature is highest. However, since its ΔT_p allows little temperature drop in the collector fluid, the average collector fluid return temperature is high, implying a relatively low efficiency for the collector. Also, for the same power level, a much higher collector flowrate is required by fluid A, producing higher parasitic power levels. Fluid C, on the other hand, exhibits low cycle efficiency because of low

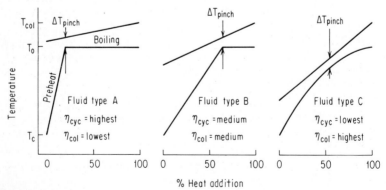

Fig. 22.11 Vapor generator temperature distribution for various fluid types where T_{col} = maximum collector temperature, T_o = maximum cycle temperature, T_c = cycle condensing temperature, ΔT_{pinch} = pinch temperature difference, η_{cyc} = cycle efficiency, η_{col} = collector efficiency.

average heat-addition temperature, but allows the highest temperature drop in the collector fluid. The resulting low average collector temperature results in higher collector efficiency.

It is apparent that in selection of the best Rankine cycle working fluid for a given application, it is important to examine the product of cycle efficiency and collector efficiency (η_{cyc} η_{col}), not either alone. The product η_{cyc} η_{col} is a good tool for fluid selection, but it does not provide a totally definitive comparison. Several factors are not included. For example, even though the comparison is made with a constant pinch temperature difference, the required vapor generator surface area may vary significantly. For similar cycle conditions, different fluids can exhibit very different expander and feed-pump performance characteristics. Pressure levels dictate pressure vessel and seal requirements and the direction of leakage in the event of its occurrence. Fluid compatability with seals, lubricants, and other system materials is important. Toxicity and flammability may exclude certain fluid choices for some applications. A rigorous fluid selection analysis must include all these factors.

Example Cycle Calculation

A solar collector system supplies 150°C (302°F) pressurized water to a Rankine heat engine, and 20°C (68°F) water is available for heat rejection from a cooling tower. Compare pyridine, R-114, and R-115 as working fluids for a Rankine cycle. These three fluids typify fluid types A, B, and C as discussed above.

In order to perform the calculations, the following assumptions are made:

Maximum cycle temperature, T_o = 135°C (275°F)
Condenser temperature, T_c = 26.7°C (80°F)
Expander efficiency, η_{ex} = 0.75
Feed-pump efficiency, η_p = 0.60
Regenerator effectiveness (if applicable), η_{reg} = 0.80
Pinch temperature difference, ΔT_p = 5°C (9°F)
Saturated expander inlet conditions, when possible
No system pressure losses

A complete system analysis would include optimization of each of the assumed parameters on an economic basis. Economic optimization requires knowledge of Rankine cycle and collector component costs and the value of produced power, all of which vary for each system size and configuration.

Beginning the cycle analysis at the expander inlet, one can obtain thermodynamic properties from the fluid diagrams provided in Figs. 22.A1–22.A6. Table 22.4 summarizes the calculations for this example. For the selected peak temperature T_o, the saturation pressure P_o, enthalpy h_o, and entropy s_o can be determined. If the cycle is supercritical, as is R-115 in this example, the peak cycle pressure P_o is selected independently of T_o on the basis of expander considerations to be discussed in a later section. The assumed condensing temperature T_c sets condensing pressure P_c, and thus expander exit pressure (also P_c). The isentropic enthalpy drop across the expander, $\Delta h'_{\text{ex}}$, is determined from the tables or charts. The real expander exit enthalpy h_2 is found by multiplying $\Delta h'_{\text{ex}}$ by η_{ex} and subtracting this real expander enthalpy change from h_o. The expander exit temperature T_2 and specific volume V_2 can then be determined. T_2 is necessary to size the regenerator (if applicable) and V_2 is used to compute expander specific speed N_s as discussed in the expander section.

The amount of energy available to the regenerator in the expander exhaust vapor is $h_2 - h_{\text{svc}}$, where h_{svc} is the condenser saturated vapor enthalpy. Regenerator efficiency η_{reg} multiplied by this maximum enthalpy difference ($h_2 - h_{\text{svc}}$) yields the regenerator enthalpy change Δh_{reg}. A regenerator is employed to improve cycle efficiency in cycles characterized by a substantial amount of superheat in the expander exhaust.

Heat recovered from the expander exhaust by the regenerator is supplied to liquid leaving the feed pump on its way to the boiler. The enthalpy level of fluid entering the vaporizer, $h_{\text{pre, in}}$, is found by taking the condenser saturated liquid enthalpy h_{slc} and adding the pump enthalpy rise Δh_p plus the regenerator enthalpy rise Δh_{reg}; Δh_p is calculated by multiplying the pump inlet specific volume by the pressure rise and dividing by the pump efficiency.

The value of vaporizer liquid enthalpy specifies its temperature at this point. The total heat added to the cycle is $\Delta h_{\text{add}} = h_o - h_{\text{pre, in}}$. Thermodynamic cycle efficiency is calculated by the equation

$$\eta_{TC} = \frac{\Delta h_{\text{ex}} - \Delta h_p}{\Delta h_{\text{add}}}$$ (22.3)

TABLE 22.4 Sample Rankine Cycle Analysis Results

	RC fluid	Pyridine	R-114	R-115
	Quantity	Fluid type A	B	C
T_o	Expander inlet temp., °F (°C)	275 (135)	275 (135)	275 (135)
P_o	Expander inlet press., lb/in² abs.	25.17	391	600
h_o	Expander inlet enth., Btu/lb	66.23	103.9	98.91
s_o	Expander inlet ent., Btu/lb.°R	0.02847	0.17077	0.16712
T_c	Cond. temp., °F (°C)	80 (26.7)	80 (26.7)	80 (26.7)
P_c	Cond. press., lb/in² abs.	0.451	32.66	138.1
h'_2	Ideal exp. exit enth. Btu/lb.	3.98	89.16	88.54
$\Delta h'_{ex}$	Ideal exp. enth. drop, Btu/lb.	62.25	14.77	10.37
Δh_{ex}	Real exp. enth. drop, Btu/lb.	46.69	11.07	7.78
h_2	Real exp. exit enth., Btu/lb.	19.54	92.85	91.13
T_2	Real exp. exit temp., °F	110	142	192
v_2	Real exp. exit sp. vol., ft.³/lb.	171	1.087	0.295
Δh_{reg}	Regen. enth. change, Btu/lb.	6.521	8.872	18.736
v_{slc}	Pump inlet sp. vol., ft.³/lb.	0.0172	0.01104	0.01255
Δh_p	Pump enth. rise, Btu/lb.	0.131	1.220	1.788
$h_{pre,\ in}$	Preheater inlet enth., Btu/lb.	196.1	36.957	49.059
$T_{pre,\ in}$	Preheater inlet temp., °F	98	120	150
Δh_{add}	Enth. added to cycle, Btu/lb.	262.3	66.97	49.85
η_{cyc}	Cycle efficiency	0.178	0.147	0.120
$T_{col,\ av}$	Av. coll. water temp., °F	289	277	233
η_{col}	Collector eff. { evac. tube	0.61	0.63	0.695
	tracking concen.	0.575	0.58	0.585
	Fresnel lens	0.66	0.67	0.705
$\eta_{col}\eta_{cyc}$	Overall eff. { evac. tube	0.109	0.093	0.084
	tracking concen.	0.102	0.085	0.070
	Fresnel lens	0.118	0.098	0.085

1 Btu = 1.055 kJ, 1 lb/in² = 6.9 kPa.
No pressure losses
η_{ex} = 0.75 (expander)
η_p = 0.60 (pump)
ΔT_p = 9°F (5°C)
η_{reg} = 0.80 (regenerator)

It is now possible to plot the temperature in the vaporizer as a function of percent heat transferred. This curve is necessary to determine the temperature profile of solar collector fluid in the vaporizer. The profile is fixed by the collector outlet temperature (150°C), the collector fluid flowrate, and the selected pinch temperature difference ΔT_p, in this case 5°C. Figures 22.12, 22.13, and 22.14 illustrate the vaporizer temperature profiles for pyridine, R-114, and R-115. The organic liquid temperature entering the vaporizer is different for each fluid because of the regenerator effect.

The average fluid temperature in the vaporizer is essentially the same as the average collector fluid temperature. Then, using the curves of Fig. 22.4, it is possible to estimate the collector efficiency η_{col}. Overall system efficiency is calculated by multiplying the collector efficiency with cycle efficiency $\eta_{col}\ \eta_{cyc}$.

Three collector types (evacuated tube, tracking concentrator, and Fresnel lens) are to be considered in this example and their resulting performance compared for each of the three fluids (see Table 22.4). The combination of collector type and fluid yielding the highest overall efficiency leads to the smallest collector area required for a given power output. Since the collector accounts for the major part of total system cost, it is important to maximize the long-term system overall efficiency for each set of heat source, heat sink, and power level conditions.

In this example the best fluid choice based on design point efficiency is pyridine regardless of whether an evacuated tube, tracking concentrator, or Fresnel lens is used for solar collection. This result indicates that, at least for this set of conditions, the cycle efficiency effect is stronger than the collector efficiency effect.

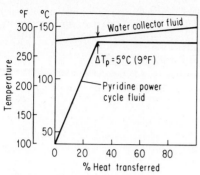

Fig. 22.12 Vaporizer temperature profile for fluid type A (pyridine) for $T_c = 26.7°C$ (80°F) and $\eta_{reg} = 0.80$.

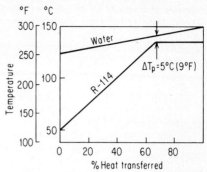

Fig. 22.13 Vaporizer temperature profile for fluid type B (R-114) for $T_c = 26.7°C$ (80°F) and $\eta_{reg} = 0.80$.

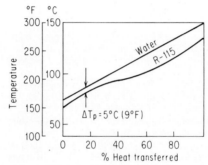

Fig. 22.14 Vaporizer temperature profile for fluid type C (R-115) for $T_c = 26.7°C$ (80°F) and $\eta_{reg} = 0.80$.

Design of Expanders, Pumps, and Speed Reducers

Expanders Expanders are devices which convert thermal energy to mechanical energy. For a specified set of cycle conditions, the performance of the expander component is one of the most important variables affecting overall system performance. There are two basic types of expanders. One uses the expansion energy directly in a positive-displacement (piston) process. The second converts internal energy to velocity and then to mechanical torque. In the positive-displacement expander, a fixed amount of working fluid is positively contained during its expansion. Expanders which utilize kinetic energy are turbines in which the fluid is not contained by a piston or sealed rotor, but continuously flows through the machine. Work is extracted by the dynamic effect of the working fluid.

To understand the selection and performance of expanders, it is convenient to introduce the concept of specific speed and specific diameter. The similarity parameters specific speed N_s, specific diameter D_s, Reynolds number Re, and either suction specific speed S or Mach number M serve as convenient parameters for presenting the performance criteria for expanders. These four parameters are sufficient to describe completely the performance of geometrically similar expanders. For a given volume flowrate and a given enthalpy change through an expander, the specific speed is a measure of the speed of rotation. Specific diameter is a measure of the size of the machine. Reynolds number expresses the ratio of inertial force to viscous force and reflects the physical properties of the Rankine cycle working fluid. The Mach number is the ratio of the velocity of the fluid to the acoustic velocity in the fluid.

It is difficult to present graphically the performance of an expander as a function of four parameters at one time. Fortunately, two of these variables, namely Reynolds number and Mach number have only a secondary effect on the performance of the expander. In addition, if the Reynolds number is above 1×10^6, the effect of the Reynolds number does not change. If the Mach number of the machine is less than 1, compressibility effects are small and the expander performance can be presented as a function of two parameters, specific speed N_s and specific diameter D_s. These similarity parameters are defined as follows:

$$N_s = \frac{N \, Q_3^{1/2}}{\Delta H_{ad}^{3/4}} \tag{22.4}$$

$$D_s = \frac{D \Delta H_{ad}^{1/4}}{Q_3^{1/2}} \tag{22.5}$$

where N = rotational speed, r/min
   Q_3 = rotor exit flowrate, ft³/s
   ΔH_{ad} = adiabatic enthalpy drop (no heat losses), ft-lb/lb
   D = diameter, ft

   The specific speed and specific diameter are not dimensionless in the form presented above hence English units are used since extant correlations are based on these units. However, the parameters can be made dimensionless when reduced to a form using angular velocity. An example of typical specific speed–specific diameter correlation for full-admission turbines is presented in Fig. 22.15. Speed N for turbines must be limited in some cases to keep wheel tip velocity below 365 m/s (1200 ft/s).

   An example of a general specific speed–specific diameter map for all expander types is presented in Fig. 22.16. This figure shows that for various ranges of specific speed, certain types of expanders offer better performance than others. For example, in the specific speed range between 30 and 100, radial turbines have performance equivalent to that of full-admission axial turbines. Figure 22.16 also shows that for a wide range of specific speed, expander efficiencies above 80 percent are obtainable with optimized machines. Examples of Rankine cycle expanders are shown in Fig. 22.17.

   Feed pump   A feed pump is required for every Rankine cycle, and its performance can significantly affect the overall performance of a solar heat engine. The performance of a feed pump, like that of the expander, can be described by four variables, three of which are the same as for the expander, namely specific speed, specific diameter, and Reynolds number. However, instead of Mach number, suction specific speed S is required to describe the performance of the feed pump. Suction specific speed is a parameter which indicates the possible

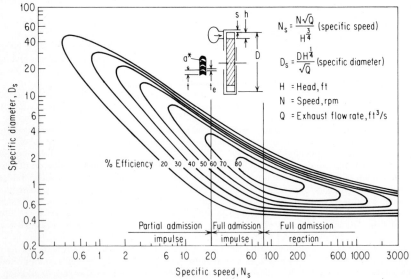

Fig. 22.15 Optimized performance chart for axial-flow turbines showing efficiency as a function of specific diameter D_s and specific speed N_s. (*Courtesy O. E. Balje.*)

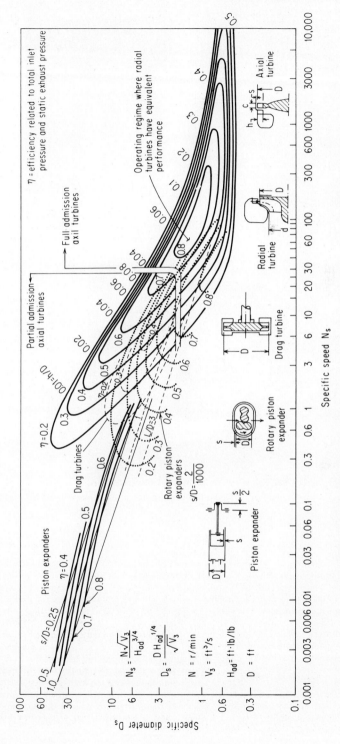

Fig. 22.16 $N_s D_s$ diagram for various expander types. Numbers on the curves are expander efficiency. *(Courtesy O. E. Balje.)*

η = efficiency related to total inlet pressure and static exhaust pressure

$$N_s = \frac{N\sqrt{V_3}}{H_{ad}^{3/4}}$$

$$D_s = \frac{D\,H_{ad}^{1/4}}{\sqrt{V_3}}$$

N = r/min
V_3 = ft³/s
H_{ad} = ft·lb/lb
D = ft

Specific speed N_s

Specific diameter D_s

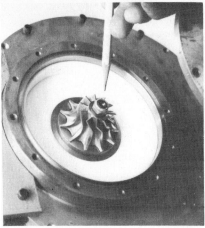

Fig. 22.17   Rankine cycle expanders: (a) 250-hp axial-flow turbine and nozzle block; (b) $2\frac{1}{2}$-hp radial inflow turbine.

occurrence of cavitation. Serious cavitation can significantly affect pump performance and its effect cannot be predicted only from similarity parameters. For pumps, the effect of Reynolds number is very nearly constant for Reynolds numbers above 10^7. As in the case of expanders, there are optimum configurations for use in various specific speed ranges. A general performance map for pumps and compressors is shown in Fig. 22.18. Components of a typical pump are shown in Fig. 22.19.

Speed reducers   Speed reducers are required for some Rankine cycle solar heat engines to match the speed and torque of the expanders and/or feed pump to the requirements of the load. For example, a speed reducer may be required between the turbine and the load, since high-performance turbines frequently operate at high rotational speeds. The load, for example, a generator, may operate at a much lower speed than the turbine. In high-power applications, the gearbox component efficiency is between 95 and 98 percent. However, in low power ranges, fixed parasitic losses in the speed reducer can cause the gearbox loss to be a significant portion of the expander output power. A typical gearbox is pictured in Fig. 22.20.

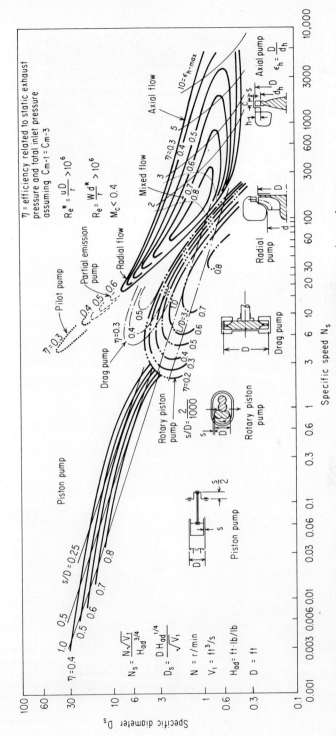

Fig. 22.18 $N_s D_s$ diagram for various pumps and compressors at low pressure ratios. Numbers on the curves are pump efficiencies. (*Courtesy O. E. Balje.*)

Fig. 22.19 Centrifugal pump with screw inducer.

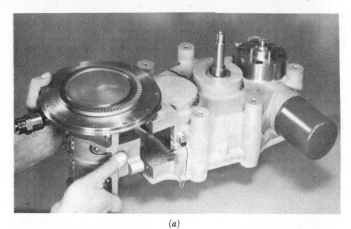

(a)

(b)

Fig. 22.20 Rankine cycle gear box showing gears, bearings, housings, and other components (a) assembled, and (b) disassembled.

Heat Exchangers*

Several heat exchangers are present in solar heat engine systems. The preheater and boiler are the interface between the Rankine cycle working fluid and the solar heat source. The preheater adds sensible heat to the pressurized Rankine cycle liquid, raising its temperature to saturation temperature for the corresponding peak cycle pressure. In the boiler additional heat from the solar collector fluid is added to vaporize the working fluid. The preheater and boiler are usually separate tube-and-shell-type heat exchangers. The flow passages, heat-transfer surfaces, and tube fitting (nozzle) sizes are optimized for the local fluid conditions.

The condenser is the Rankine cycle heat sink in which heat is rejected to air or cooling water. Air- and water-cooled condensers for solar Rankine cycles are similar in appearance and function to steam and refrigeration condensers.

The regenerator is used only in those Rankine cycles using "drying-type" fluids. With this type of fluid, vapor leaving the expander contains superheat, i.e., its temperature is higher than the condensing temperature. A significant improvement in cycle efficiency may be achieved by using a regenerator to transfer some of this sensible heat from the turbine exhaust to the pressurized, subcooled liquid before the liquid enters the preheater.

The performance specifications for Rankine cycle heat exchangers are determined from the cycle state-point diagram (i.e, Fig. 22.9), heat-sink fluid properties and temperatures, and collector-fluid properties and temperatures. Generating the final cycle state-point diagram is an iterative procedure. The state-point diagram is first made according to estimated Rankine cycle pressure drops. As the heat exchangers become better defined with successive iteration, and better estimates of pressure drops and fluid (heat sink, collector, and working fluid) conditions are generated, the state-point diagram is revised. The iterative procedure is completed when changes in all state points become negligible.

Heat exchangers are sized using the effectiveness-NTU method described by Kays and London in Ref. 18 (see also Chap. 4).

Preheater   After the preheater's specifications are estimated, its size is calculated. For systems using water as the collector fluid and a halocarbon-type refrigerant as the Rankine cycle working fluid, reasonable U-values are in the range of 70 to 200 Btu/h · ft² · °F.

Figure 22.21 shows the estimated preheater cost for commercially available units operating at temperatures less than 250°F and Rankine cycle pressures less than 100 lb/in² gage. One can see that as heat-exchanger sizes increase, the cost per square foot decreases. Figure 22.22, from Ref. 25, shows the approximate effect of Rankine cycle pressure on heat-exchanger cost. The curve assumes that the Rankine cycle fluid is in the tubes and that the collector fluid on the shell side is at a pressure below 150 lb/in² gage. It can be seen from the figure that increasing the Rankine cycle working-fluid pressure can have a significant effect on heat-exchanger and system cost. Since heat exchangers typically make up 25 to 40 percent of the total system cost, care must be taken not to select Rankine cycle fluid conditions which unnecessarily increase the cost of the heat exchangers.

Boiler   Typical values of overall heat-transfer coefficient for Rankine cycle boilers incorporating halocarbon refrigerants as the Rankine cycle working fluid and water as the collector fluid are in the range of 70 to 250 Btu/h · ft² · °F. When heat-transfer oils or water-glycol mixtures are used in the solar collector, the internal heat-transfer coefficient can be significantly reduced compared with that for water because of lower fluid thermal conductivity and higher viscosity. The effect of high viscosity can be particularly significant when high-temperature systems are started after a cold down-period. In these cases the oil viscosity can be up to 100 times greater than at the design point. Figure 22.23 shows estimated 1976 costs for refrigeration-type evaporators for use as Rankine cycle boilers. These costs are probably the minimum achievable costs for small quantities of Rankine cycle boilers, and if special construction features or a custom unit is required, the cost could easily be doubled. Figure 22.22 may again be used to estimate the effect of pressure on evaporator cost. In higher-temperature Rankine cycles, various metals or lubricants can act as catalysts to decompose some Rankine cycle fluids and heat-transfer oils. Materials of construction compatible with both the working fluid and the solar collector fluid must be carefully selected on the basis of temperature. Depending on the design conditions, a suitable vapor separator or demister may be required for some boiler configurations.

---

*This section was written by Douglas Werner, a project engineer at Barber-Nichols Engineering Co. and a Registered Professional Engineer.

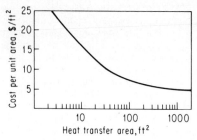

Fig. 22.21 Rankine cycle preheater cost versus area (1976 dollars).

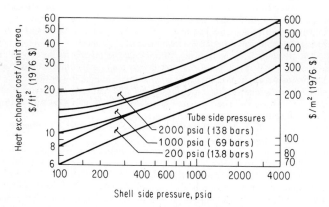

Fig. 22.22 Effects of pressure on heat-exchanger cost in 1976 dollars for unit size > 20,000 ft². Air-cooled exchanger costs based on bare tube area. (Taken from *Milora and Tester, Ref 25.*)

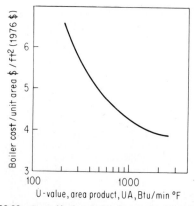

Fig. 22.23 Cost of boilers for Rankine cycles (1976 dollars).

Condenser Rankine cycle condensers may be classed into two major groups—water cooled and air cooled. Water-cooled condensers are recommended for use wherever feasible: they permit lower condensing temperatures and higher cycle efficiencies. For Rankine cycles incorporating refrigerants as the working fluid, the overall condensing heat-transfer coefficients are similar to those achieved in refrigeration cycles and are typically in the range of 70 to 250

Btu/h · ft² · °F. The usual practice is to place water in the tubes and the condensing vapor in the shell. The tube exterior surface may be either plain, wire wrapped, corrugated, or finned, depending on the condenser manufacturer. The number of water passes is a function of flowrate, heat-exchanger size, and allowable water-pressure drop. Design practice for Rankine cycle condensers for use with refrigerant-type working fluids is very similar to standard refrigeration practices. It is common, however, to use more than the customary 3 gal/min per 15,000 Btu/h at the condenser in order to reduce condensing temperature and improve cycle efficiency. Figure 22.24 shows typical water-cooled condenser costs for a range of UA values.

Air-cooled condensers are required when cooling water is not available in sufficient volume. They are not recommended for cycles incorporating peak cycle temperatures less than 250°F because the required higher condensing temperature results in low cycle efficiency and large collector areas, thereby significantly increasing overall system cost. Typical overall U-values for air condensers are in the range of 6 to 12 Btu/h · ft² · °F. Low fan power requires low air-side pressure drop and hence heat-transfer coefficients will usually be at the lower end of the range.

Regenerator   The regenerator transfers heat from the superheated vapor in the turbine exhaust to the subcooled, pressurized liquid leaving the feed pump before the liquid enters the preheater. Typical design U-values are in the 3 to 12 Btu/h · ft² · °F range based on the vapor-side area. Because the leaving temperature of the liquid is raised to a higher temperature than the leaving temperature of the vapor, a counterflow or cross-counterflow configuration is required. Usual practice is to flow the vapor on the shell side of a tube-and-shell-type heat exchanger. The vapor side is normally finned to provide additional heat-transfer surface. It is recommended that the regenerator be designed to prevent condensed liquid from collecting in the regenerator during cold start-up. The range of regenerator costs is shown in Fig. 22.25. The lower cost figures may be taken as representative of regenerators based on limited production while the higher range is applicable to one-of-a-kind units.

Performance Summary

Several factors cause Rankine cycles to have efficiencies less than the Carnot efficiency. Figure 22.26 summarizes these factors for a typical solar Rankine engine. One can see that at 215°F, the ideal (Carnot) efficiency is 19 percent. The fact that heat exchangers require a temperature difference to operate leads to an efficiency loss because maximum cycle temperature is decreased and minimum temperature is increased. For the example of Fig. 22.26, cycle efficiency drops to 16 percent. Thermodynamic properties of real fluids make it impossible to accomplish all of the heat addition at the peak cycle temperature and usually make it difficult to reject all of the heat at the lowest cycle temperature. This "real-fluid" loss drops the cycle efficiency to 13.5 percent. Turbine and feed-pump irreversibilities and flow pressure losses further decrease it to 9 percent, and mechanical losses external to the cycle (gearbox and drive train) cause the cycle efficiency measured at the gearbox output shaft to drop to 7 percent, for this example.

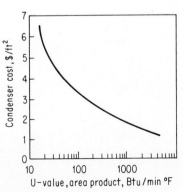

Fig. 22.24   Cost of Rankine cycle water-cooled condensers (1976 dollars).

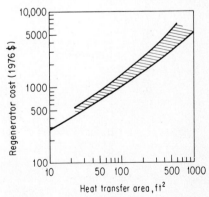

Fig. 22.25   Regenerator cost for P_{tube} = 150 psig and P_{shell} = 50 psig.

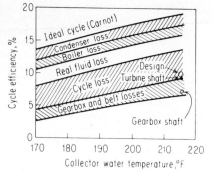

Fig. 22.26 Typical loss factors not including feed pump (ideal pump work 0.05 hp). Cooling water temperature, 85°F; condensing temperature, 95°F.

In another example, a sensitivity analysis for a theoretical solar Rankine engine revealed the relative importance of certain system parameters on cycle performance. Figure 22.27 shows how deviations from the design-point value affect the overall cycle efficiency. The analysis reveals that, by far, the most influential system parameters in terms of cycle efficiency are turbine inlet and condensing temperature and turbine efficiency. Much less important to performance are feed-pump efficiency, flow pressure losses, and regenerator efficiency.

Effect of Power Level

The overall performance and specific cost of a heat engine can be greatly affected by the level of power it produces. Generally speaking, as with most power systems, higher power production

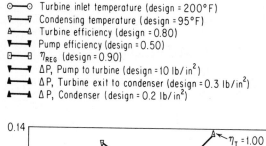

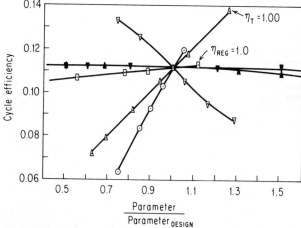

Fig. 22.27 Effects of various design parameters on cycle efficiency.

levels imply higher overall system efficiency and lower cost per unit power (specific cost). This is because of certain fixed costs and fixed system losses present in systems irrespective of size, and because mechanical components tend to become more efficient with increasing capacity. Smaller engines, however, can be located near their loads, so that power transmission penalties often associated with centralized plants are reduced.

Figure 22.6 shows the effect of output power level and cycle temperature on installed system cost. The results show that smaller, nonconcentrating (approximately 93°C or 200°F) solar heat engines producing about 60 kW cost about $1000 per kilowatt. If the power level is in the 10-kW neighborhood, the cost for nonconcentrating systems jumps to about $2500 per kilowatt. When concentrating collectors are used in the 10- to 60-kW power range, the cost drops into the $1200 to $700 per kilowatt cost range. In the 10,000- to 100,000-kW range, costs decrease to the $200 to $250 per kilowatt level.

Practical Concerns

Possible development problems are listed below in the order of importance and probable occurrence.

Leaks and loop cleanliness Fluids normally used in Rankine cycles require leaktight joints in order to prevent leakage of the working fluid from the loop or air into the loop. All connections should be considered as possible leak points. Threaded pipe joints should be avoided whenever possible, although many heat exchangers are supplied with threaded couplings. Threaded joints with dissimilar metals (i.e., steel/brass) are unsatisfactory if the temperature is above 150°F. Welded or sweat solder joints are preferred. Debris in the loops during the initial run (solder, steel filings, RTV) can easily plug and foul sensitive control valves or destroy accurately balanced rotating components.

Pumps Pumps can cause problems in two main areas. The first is inadequate feed-pump performance, either lower-than-necessary head or cavitation due to inadequate net positive suction head (NPSH) capability. Second, accessory pumps in the water loops often draw substantially more parasitic power than expected. Most commercial pumps are not suitable for operation on a nearly boiling fluid and consequently are not suitable for Rankine systems. In many cases the pressure drop tends to be somewhat higher than initially predicted. Consequently, if a centrifugal pump is utilized, provisions for additional pump stages should be made to account for this problem. When operating on the typical fluids used for Rankine systems which have low lubricity, piston pumps may have a short life and seal problems may occur. One solution is to select a working fluid with low pressure-rise requirements to reduce the demand on the pump. Centrifugal pumps with inducers that can be immersed in the hot well also work. Pumps without seals, such as those that utilize immersible motors as a pump drive, should be used if possible. Diaphragm pumps also show promise for this application, and the use of boost pumps to provide adequate NPSH for the main pump seems to be a reasonable solution. The pump efficiency should be in the 30 to 50 percent range.

Other standard components Specially designed components for the system receive considerable attention and often they have fewer problems than the "standard" items. These items include valves (including throttle, check, and relief valves), electronics, heat exchangers, and mechanical components (such as belts, idlers, shock mounts, etc.). Typical problems are leakage, lower than expected performance, and high pressure drop. The solution to these problems is to select the components carefully, check the vendor's claims analytically if possible, and conduct subcomponent tests. The vendor's information should not be relied on, since the solar-powered Rankine cycle application is novel and the vendor may be forced to extrapolate data.

Pressure drop It is not unusual to obtain loop or component pressure drops that are 2 to 3 times higher than the calculated values based on handbooks. The reason is presently unknown. However, it is prudent to maintain liquid velocity below 5 ft/s and vapor velocities below 30 ft/s. The only sure way to determine flow pressure drop is to set up a loop and measure it. If the feed pump is designed for a pressure rise of 200 lb/in² and the loop pressure drop is 10 lb/in² higher than expected, the system will not obtain maximum power and major loop modifications will be required to meet design point operation.

Start-up and fluid complement At start-up, the working fluid will be in the coldest part of the loop, not necessarily at the lowest point. Fluid distribution problems can be particularly severe since heat exchangers act as condensers until they reach normal operating temperatures. Therefore, components such as the regenerator and vapor plumbing must drain to the condenser during the critical start-up mode. The most common reason for unsuccessful start-up is the loss of working fluid in the boiler and subsequent overtemperature caused by liquid holdup at another point in the loop. Liberal use of sight gauges in development units will pay back in reduced test time. Another useful idea is to utilize an electric feed pump and bypass the main expander during start-up so that the loop can be brought up to operating conditions slowly. Then a switchover to the expander can be accomplished smoothly with the components at operating temperatures. When the Rankine loop has a feed pump driven by the expander, start-up problems are alleviated by the use of an electric-driven start pump which can be turned off at a later time.

On-site problems   No matter how much testing is done in the factory, problems occur when the unit is placed in the field. These problems generally result from limited training and marginal understanding by the operator. Typical problems are inadequate condenser water supply, condenser fouling by debris in the cooling water, backdriving of the expander from the load (solved by putting an overrunning clutch in the power train), and excessive boiler supply temperatures caused by the start-up problems in the system noted above. Additional system problems may involve interfaces between the Rankine engine and the external systems. For example, output shaft misalignment can cause bearing failure.

EXAMPLE DESIGN

A 25-hp Solar Engine Project—Summary

Low-temperature solar Rankine engines have been made in sizes from about 1 to 100 hp as of January 1979. These systems have been used with low-temperature flat-plate collectors (90 to 110°C) and high-temperature concentrating collectors (150 to 200°C). The applications have been air conditioning at 3 tons (Fig. 22.28) to 100 tons (Fig. 22.29) and irrigation pumping at 1 to 50 hp (Fig. 22.30).

The following project example summarizes a 25-hp Rankine engine which illustrates a typical solar application in detail.* The Rankine engine described here was developed by Barber-Nichols Engineering Co. for the U.S. Department of Energy Sandia Laboratories and was installed at Willard, New Mexico in April of 1977. A photograph of the solar installation is shown in Fig. 22.31. A Sandia publication explains that:

The experimental solar powered irrigation system, located south-east of Albuquerque, near the town of Willard, was a joint effort of the Energy Research and Development Administration (ERDA) and the State of New Mexico. The system engineering was under the direction of

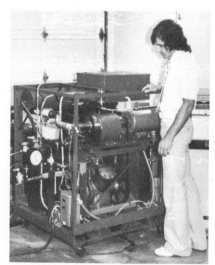

Fig. 22.28   Three-ton solar air-conditioning system.

---

*Since this project example was designed using English units, they are retained. To use SI units, see the handbook appendix for conversion factors.

Fig. 22.29   Seventy-seven-ton solar-powered water chiller.

Sandia Laboratories. When completed, the system pumped enough water to irrigate 100 acres of mixed crops. There was enough energy collected to also support compatible alternate season operations such as a greenhouse and fish farm. In 1977, there were over 160,000 irrigation wells in California, Arizona, New Mexico, and Texas powered by natural gas. With inevitable gas rate increases and potential shortages, many of these wells may become unprofitable to operate. When solar powered irrigation is proven to be technically and economically sound, it can provide a clean alternate source of energy for the nation's food production industry.

Generally, the Willard system consists of a solar collector field, energy storage (fluid tank), solar engine, irrigation pump, controls, and a water storage pond, all located on less than 4 acres of land. The solar engine, an organic Rankine cycle turbine, delivers 25 shaft horsepower to the irrigation pump. The pump will deliver 880 gpm from a well 75 feet deep during the 100-day irrigation season. A plastic-lined pond stores up to 4.5 acre-feet of water for demand as needed. The sun-tracking solar collectors raise the temperature of the heated fluid to 420°F in full sun.

When the solar-powered heat engine is not operating the irrigation pump, it can generate 10,000,000 Btu/day of heat or 20 kW (200 kWh/day) electrical power for other uses.

An electric motor serves as a backup drive for the pump, although rarely needed. The system is completely nonpolluting and environmentally safe since both heating and working fluids are sealed in their systems.

A unique feature of this experiment is that it was engineered and developed using available technology and hardware. Successful operation confirms the practicality of commercial production of similar systems without the need for any new materials and component technology. A 20-year lifetime is the design goal for the system.

The solar-powered irrigation system operates as follows: Preheated heat transfer fluid (an oil-like fluid which remains stable at high temperatures) is pumped into the collector field. Solar energy from the 6720 ft² of tracking parabolic trough collectors heats the fluid to 420°F. A thermal control valve then opens to allow the fluid to flow either to a thermal storage tank or to the boiler/heat exchanger.

In the Rankine cycle boiler the heat transfer oil boils the engine working fluid (R-113) at a temperature of 325°F and a pressure of 220 psi. This gas drives the turbine in turn operating the irrigation pump. In this closed-loop system, R-113 exhausted from the turbine is then circulated through a regenerator heat exchanger, a condenser, and through the cooling side of the regenerator. It is then returned to the boiler/heat exchanger as a liquid. The heat transfer fluid flows from the boiler to a thermal storage tank or to a mixing tank. The mixing tank limits the inlet temperature to the solar collectors and aids in controlling temperature of fluid leaving the collectors.

Selection of the Working Fluid

The vaporizer temperature profile for this project as a function of heat transfer is shown in Fig. 22.32. The percentage at the pinch point is different for each working fluid and, of course, is

Fig. 22.30 Fifty-horsepower irrigation system developed by Battelle Memorial Labs and Located at Gila Bend, Arizona: (*a*) aerial photo and (*b*) heat engine assembly.

different for each boiling temperature for a particular working fluid. These values are shown in Fig. 22.32 for refrigerants 11, 113, and 114. Also shown in this figure is the nominal Caloria (heat-source fluid) temperature change with heat transfer from 420°F entering the boiler to 180°F leaving the preheater. Refrigerant 114 would have a maximum cycle temperature of only 99°C (210°F) with a zero pinch point, while refrigerants 11 and 113 would have a maximum temperature of approximately 150°C (300°F). Consequently, refrigerant 114 is not suitable for this application. Also, refrigerants 113 and 114 require a regenerator, and consequently have a higher temperature entering the preheater than does refrigerant 11, which does not require a regenerator.

A supercritical fluid could be considered here since it could reach a higher temperature with a zero pinch point. However, studies have shown that unless substantially higher (75 to 100°C) temperatures can be reached with the supercritical fluid, the collector efficiency degradation in a supercritical cycle is greater than the cycle efficiency increases because of the higher temperature operation. Supercritical cycles, therefore, were not considered as viable alternatives in this application.

The cycle efficiencies of toluene and refrigerants 113, 11, and 12 are presented in   Fig. 22.2

Fig. 22.31   Twenty-five-horsepower solar-powered irrigation pump.

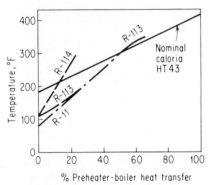

Fig. 22.32   Vaporizer temperature profile for example 25-hp system.

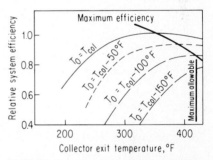

Fig. 22.33   Example power system thermal efficiency for various collector and sink temperatures; T_{col} is the collector exit temperature and T_o is the Rankine cycle maximum temperature.

as a function of the maximum cycle temperature. R-113 typically exhibits somewhat better performance than R-11, although a regenerator is required with R-113 to obtain good efficiency. In most cases the cost of the regenerator is offset by the saving in solar collector area because of the increase in system efficiency. Therefore, since the pinch points for R-11 and R-113 are the same, and the efficiency of R-113 is superior to that of R-11, R-113 was selected as the working fluid. Additionally, R-11 is known to be marginally stable chemically at 300°F and above, making it a poor second choice.

Selection of the Optimum Temperature of the Collector Working Fluid

When one combines the efficiencies of the Rankine cycle and the solar collector, the curves presented in Fig. 22.33 result. The upper curve of this figure corresponds to a situation when the Rankine cycle maximum temperature is equal to the collector exit temperature. This, of course, is the ideal situation and is not achievable when an intermediate heat-transfer fluid is required. However, it is interesting to note that the maximum efficiency occurs at a collector exit temperature of approximately 177°C (350°F). As the difference between the Rankine cycle maximum temperature and the collector exit temperature is increased, the relative system efficiency decreases, of course, and the maximum efficiency point shifts to the higher collector exit temperature.

At the nominal design condition, as shown in Fig. 22.33, it appears that a maximum Rankine cycle temperature of approximately 150°C (300°F) for a collector temperature of 216°C (420°F) is appropriate. The maximum practical temperature for utilization of the Caloria HT-43 was felt to be 216°C (420°F), based on thermal decomposition considerations. Since the optimum cycle temperature was also greater than 216°C, this temperature was selected for the remaining cycle analysis and parametric study. In general, optimum temperatures should be selected on the basis of a system cost analysis rather than maximum system efficiency. However, if the collector cost is approximately the same as the Rankine engine cost, as it is in this case, then the system efficiency optimum is equal to the cost optimum.

Selection of Optimum Rankine Cycle Maximum Temperature

To determine the optimum Rankine cycle temperature, the system cost and performance must be evaluated for various operating temperatures and various boiler pinch points. For this example project, maximum Rankine cycle temperatures varying between 121°C (250°F) and 178°C (350°F) were evaluated. Referring to Fig. 22.32, it can be seen that, for a given boiling temperature, as the pinch temperature difference is increased, the temperature drop of the heat storage fluid must be decreased. Therefore, a larger amount of storage fluid is required and, consequently, a cost penalty is incurred as pinch-point temperature differences are increased. As shown in Fig. 22.34, the heat storage cost is the dominant factor in this optimization, since the heat storage cost rises more rapidly than the heat-exchanger costs decrease. The sum of the heat storage cost and the cost of the preheater and boiler is presented in the upper part of Fig. 22.34, showing that a minimum occurs for this particular case at a pinch point of 10°F. For this project, the optimum pinch point varied between 11°C (20°F) for Rankine cycle peak temperatures below 138°C (280°F) and 5.6°C (10°F) at 149°C (300°F) and above.

Figure 22.35 shows that as the Rankine cycle maximum temperature is increased, the solar collector cost decreases because of an increase in cycle efficiency. The power system cost remains relatively constant, varying only slightly depending on the size of the heat exchanger, which is a small portion of the system cost. The cost of the heat storage system, on the other hand, increases as the temperature is increased, since the ΔT in the heat storage fluid decreases. The increase in cycle efficiency partially offsets this effect. However, the minimum system cost occurs at 160 to 170°C (320 to 340°F). Therefore, a 160°C (320°F) cycle was selected for the design point of the proposed irrigation pumping system.

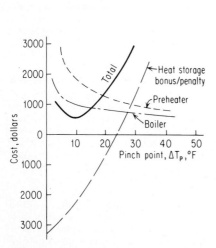

Fig. 22.34 Effect of pinch point ΔT on preheater, boiler, and heat storage cost total; $T_{col} = 420°F$ (215°C), $T_o = 300°F$ (149°C).

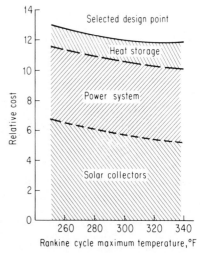

Fig. 22.35 Rankine cycle cost versus cycle peak temperature for $T_{col} = 216°C$ (420°F).

Description of the Selected System

The Rankine power system is shown in the sketch of Fig. 22.36 and is shown schematically in Fig. 22.37. The boiler is heated by circulation of hot Caloria HT-43 which transfers its heat to the R-113. The collector fluid is supplied at a constant 216°C (420°F).

Refrigerant 113 is supplied to the turbine at 160°C (320°F), where it is expanded to provide 25 shp at the gearbox output. The turbine shaft speed is 36,300 r/min and the output shaft speed is 1730 r/min. On the main output shaft a manual clutch is provided between the Rankine loop and the power takeoff.

Design-Point Operating Conditions   Design R-113 conditions throughout the Rankine power loop are shown in Fig. 22.9. As shown here, boiling takes place at 163°C (325°F) and 221 lb/in² abs. During expansion in the turbine, 29 hp is extracted from the fluid. The predicted turbine efficiency is 75 percent and the pump efficiency is 40 percent. There is a 2.25-hp loss in the bearings, seals, gears of the speed reducer, and the power takeoff. The R-113 flowrate through the loop is 4918 lb/h at the design point. Heat added in the preheater and boiler is 410,000 Btu/h, resulting in an overall cycle efficiency of 15.3 percent at a useful output of 25 hp. Figures 22.38 and 22.39 show assembled and disassembled views of the turbine-gearbox-pump for the 25-hp solar irrigation system.

Control   The control concept for this system is relatively simple since only design-point operation is required. The collector-fluid flowrate is controlled to maintain the exit temperature at 240°F. Refrigerant 113 flowrate is controlled by the float-tank arrangement feeding the boost pump. The loop has a balanced flowrate at all conditions since the amount of refrigerant boiled determines the flowrate through the loop.

It is not necessary to regulate the gearbox output speed since the irrigation-pump load characteristic provides a stable operating point. In other words, as the gearbox speed increases, the turbine torque decreases, and conversely, as the pump speed goes up, the torque increases. At 25 hp, therefore, speed control is not necessary since the operating load will match the power produced.

Test Results

Tests of the 25-hp system were run with heat supplied from a 10 million Btu/h steam boiler and cooling water supplied from a 300-ton cooling tower. The measured cycle conditions are

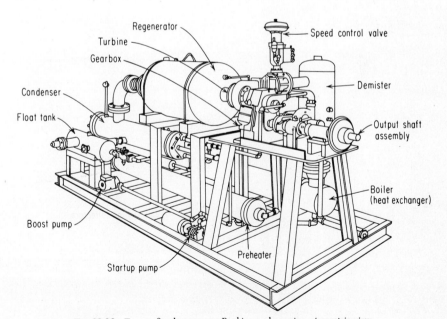

Fig. 22.36   Twenty-five-horsepower Rankine cycle engine—isometric view.

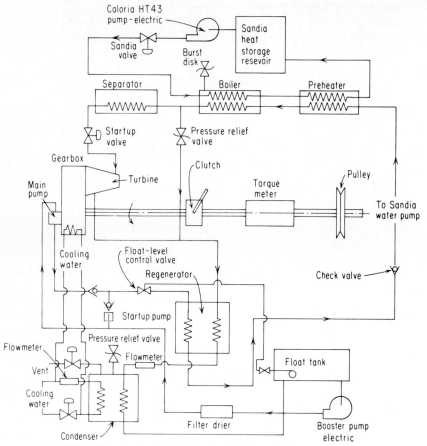

Fig. 22.37 Schematic diagram of 25-hp example solar-powered Rankine cycle system.

Fig. 22.38 Photograph of assembled turbine-gearbox-pump for 25-hp solar irrigation system.

Fig. 22.39   Photograph of dissassembled turbine gearbox.

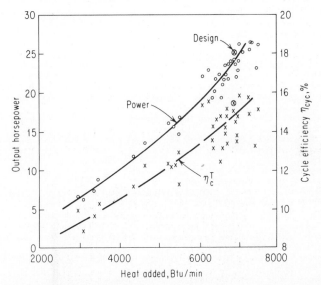

Fig. 22.40.   Measurements of output power and cycle efficiency versus heat addition rate for 25-hp example project; shaft speed 1650 < r/min < 1820 and cooling water temperature was 60°F (16°C).

presented in Table 22.5 along with corresponding design values for comparison. Figure 22.40 shows measured power and cycle efficiency data for this system as the amount of added heat was varied. For a range of about 85 to 115 percent of design output shaft speed, the power level for a given turbine inlet pressure remained relatively constant, indicating only small changes in turbine efficiency.

TABLE 22.5   Comparison of Design and Measured Cycle Conditions for 25-hp Engine

Location	Design value	Measured value
Caloria inlet	420°F	NA
Caloria outlet	240°F	NA
Demister outlet, R-113	325°F	324°F
Turbine inlet, R-113	323°F	323°F
Turbine outlet, R-113	200°F	196°F
Condenser inlet, R-113	121°F	120°F
Condenser outlet, R-113	86°F	82°F
Preheater inlet, R-113	149°F	153°F
Condenser inlet, H_2O	60°F	60°F
Condenser outlet, H_2O	70°F	71°F
Demister outlet, R-113	221 lb/in² abs.	212 lb/in² abs.
Turbine inlet, R-113	211 lb/in² abs.	211 lb/in² abs.
Turbine outlet, R-113	10 lb/in² abs.	8 lb/in² abs.
Condenser inlet, R-113	7.9 lb/in² abs.	8 lb/in² abs.
Boost pump outlet, R-113	8.4 lb/in² abs.	8 lb/in² abs.
Preheater inlet, R-113	226 lb/in² abs.	220 lb/in² abs.
Condenser inlet, H_2O	NA	14 lb/in² gage
Condenser outlet, H_2O	NA	11.2 lb/in² gage
Filter inlet, R-113	235 lb/in² abs.	235 lb/in² gage
Flow condenser water, no. 1	70 gal/min	70 gal/min
Refrigerant no. 2	7.6 gal/min	7.9/2.2 gal/min
Speed	1730 r/min	1730 r/min
Torque	910 in · lb or 76 ft · lb	874 in · lb

NA: Not applicable to test setup.

REFERENCES

1. Gordon Van Wylen and Richard Sonntag, "Fundamentals of Classical Thermodynamics," Wiley, New York, 1965.
2. John F. Lee and Francis W. Sears "Thermodynamics," 2d ed., Addison-Wesley, Reading, MA, 1963.
3. George Hatsopoulos and Joseph Keenan, "Principles of General Thermodynamics," Wiley, New York, 1965.
4. Ascher H. Shapiro, "The Dynamics and Thermodynamics of Compressible Fluid Flow," Ronald Press, New York, 1953.
5. Daryl Prigmore and Robert Barber, Cooling with the Sun's Heat—Design Considerations and Test Data for a Rankine Cycle Prototype, *Solar Energy,* vol. 17, pp. 185–192. Pergamon Press, Oxford, Great Britain, 1975.
6. J. N. Hodgson and F. N. Collamore, Turbine Rankine Cycle Automotive Engine Development, Society of Automotive Engineers Paper 740298, 1974.
7. W. P. Teagan and D. T. Morgan, 3 KW Closed Rankine Cycle Power Plant, U.S. Government Report TE 5092-99-72, Contract DA44-009-AMC-169(T), 1973.
8. J. P. Abbin, Jr., Rankine Cycle Energy Conversion System Design Considerations for Low and Intermediate Temperature Sensible Heat Sources, Sandia Laboratories Paper SAND76-0363 UC-93, 1976.
9. Ben Patterson, Jr. (Ed.), Solar Total Energy Test Facility Project Semiannual Report, Sandia Laboratories Paper SAND77-0738, 1977.
10. *Solar Cooling for Buildings–Workshop Proceedings,* National Science Foundation Publication NSF-RA-N-74-063, 1974:
    William Burriss, Solar Powered Rankine Cycle Cooling Systems.
    Henry Curran, Solar Rankine Cycle Powered Cooling Systems.
    Frank Biancardi, Needed Research on Solar Rankine Systems.
    Jerry Davis, Solar Rankine Powered Cooling Systems.
    Robert E. Barber, Solar Organic Rankine Cycle Powered Three Ton Cooling System.
11. *Second Workshop on the Use of Solar Energy for the Cooling of Buildings,* U. S. Energy Research and Development Administration Paper SAN/1122-76/2, 1976:
    J. Douglas Balcomb, Design and Performance Prediction of a 77 Ton Solar Air Conditioning System—System Predictions.
    Robert E. Barber, Design and Performance Prediction of a 77 Ton Solar Air Conditioning System—Rankine Cycle Predictions.
    R. Barber, Solar Air Conditioning Systems Using Rankine Power Cycles-Design and Test Results of Prototype Three Ton Unit.

S. E. Eckard, Performance Characteristics of a Three Ton Rankine Powered Vapor-Compression Air Conditioner.

Frank Biarcardi, Demonstration of a Three Ton Rankine Cycle Powered Air Conditioner.

12. R. E. Barber, Current Costs of Solar Powered Organic Rankine Cycle Engines, *Solar Energy*, vol. 20, pp. 1–6, Pergamon Press, Oxford, Great Britain, 1978.

13. O. E. Balje, A Study on Design Criteria and Matching of Turbomachines, Parts A and B, American Society of Mechanical Engineers Paper 60-WA-230, 1960.

14. D. G. Shepherd, "Principles of Turbomachinery," MacMillan, New York, 1956.

15. A. Stodola, "Steam and Gas Turbines," vols. I and II, McGraw-Hill, New York, 1927.

16. G. T. Csanady, "Theory of Turbomachines," McGraw-Hill, New York, 1964.

17. A. J. Stepanoff, "Pumps and Blowers—Two Phase Flow," Wiley, New York, 1965.

18. W. M. Kays and A. L. London, "Compact Heat Exchangers," 2d ed., McGraw-Hill, New York, 1964.

19. A. H. Church, "Centrifugal Pumps and Blowers," Robert E. Kreiger Publishing Co., Huntington, NY, 1972.

20. A. Kovats, "Design and Performance of Centrifugal and Axial Flow Pumps and Compressors," MacMillan, New York, 1964.

21. Frank Kreith, "Principles of Heat Transfer," 2d ed., International Textbook Co., Scranton, PA, 1965.

22. J. J. Jacknow, Analysis of Alternate Working Fluids for Rankine Cycle Engines, International Research and Technology Corp. Paper IRT-N-62, 1969.

23. D. R. Miller, Rankine Cycle Working Fluids for Solar-to-Electrical Energy Conversion, Monsanto Research Corp. Paper MRC-SL-399, 1974.

24. D. R. Miller, Optimum Working Fluids for Automotive Rankine Engines, vol. I–III, U.S. Environmental Protection Agency Paper APTD-1564, 1973.

25. S. L. Milora and J. W. Testor, "Geothermal Energy as a Source of Electric Power," The MIT Press, Cambridge, MA. 1976.

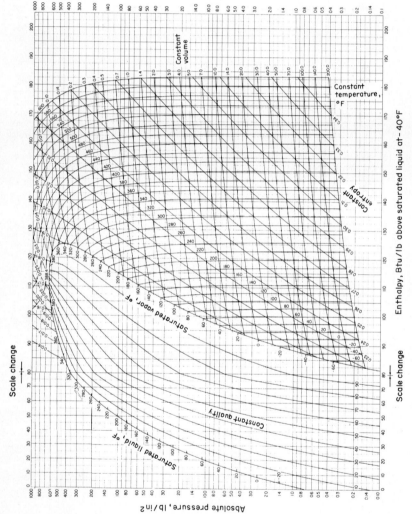

Fig. 22.A1 Thermodynamic properties for Freon® 113. (*Courtesy E.I. du Pont de Nemours & Co.*)

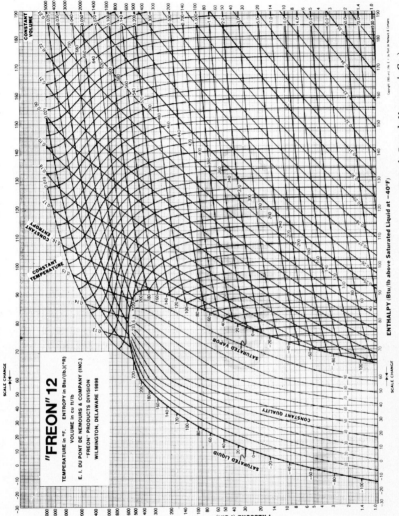

Fig. 22.A2 Thermodynamic properties for Freon® 12. (*Courtesy E. I. du Pont de Nemours & Co.*)

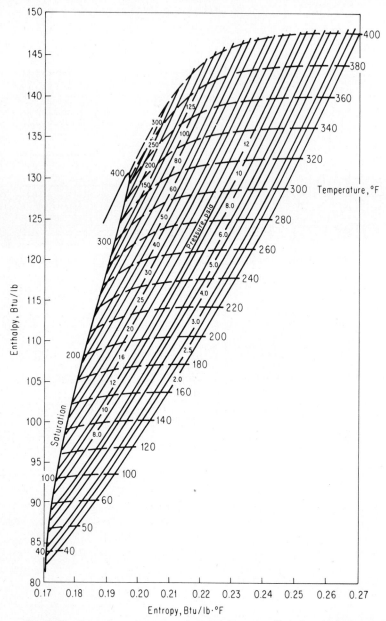

Fig. 22.A3 Thermodynamic properties for R-113. (*Courtesy Allied Chemical Corporation.*)

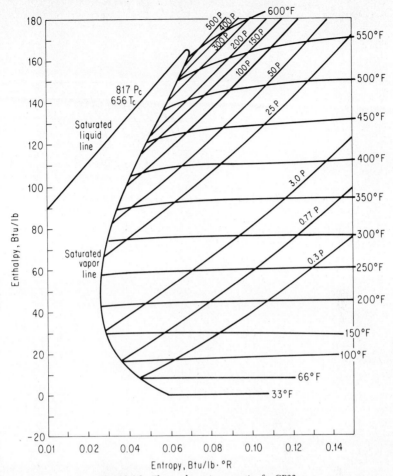

Fig. 22.A4 Thermodynamic properties for CP32.

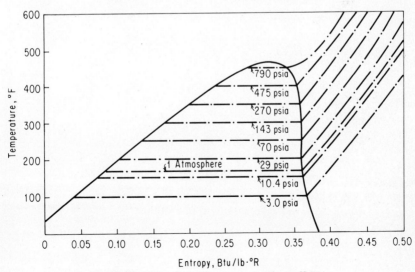

Fig. 22.A5 Thermodynamic properties for Fluorinol® 85.

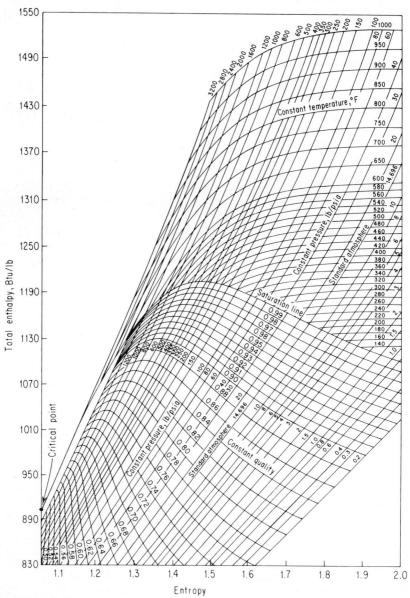

Fig. 22.A6 Thermodynamic properties for steam. (*From W. Severns and H. Degler, "Steam, Air and Gas Power," John Wiley and Sons, New York, 1948.*)

Chapter **23**

Wind Energy Conversion Systems

WILLIAM HUGHES
American Wind Turbine Co., Stillwater, OK

D. K. MCLAUGHLIN

and

R. RAMAKUMAR
Oklahoma State University, Stillwater, OK

APPLICATIONS OF WIND ENERGY SYSTEMS

This chapter describes wind energy converters and their integration into power systems. Many kinds of aerogenerators are described, the most practical being analyzed in detail. In addition, the history of wind energy use, site criteria, economics, and utility interfaces are described. The chapter is organized in three sections—introduction, aeroturbine design, and wind-electric systems.

To use wind energy effectively means to use it economically. That, in turn, means that wind energy must be able to do a particular job either more reliably, or less expensively, or with fewer undesirable side effects, than any other available source of energy capable of doing the same job.

Suppose there is a choice of (1) buying a windmill to generate electricity, (2) buying the electricity from an electric power company, or (3) generating electricity from a small, local Diesel- or gasoline-engine-driven generator. A windmill system will probably cost $2500 per kilowatt output when the wind is equal to 15 mi/h (6.7 m/s) or above. The cost of electricity from the electric power company averages 4.5¢/kWh and the capital cost of the engine-driven generator is $200.00/kW. Also, the engine-driven generator will last for 2000 h of successive running (a typical figure for a small gasoline-engine-driven plant). A windmill, with proper maintenance, will last for 20 yrs. Which of the three is to be selected?

First it must be recognized that the wind does not blow continuously. In a typical southwestern American city (for example, Oklahoma City), the wind is between 15 and 30 mi/h approximately one-third of the time. A distribution diagram of the Oklahoma City winds is shown in Fig. 23.1. Below 15 mi/h, little usable energy is available. With winds above 30 mi/h, significant energy may be available, but most practical windmill designs will not be able to take advantage of it.

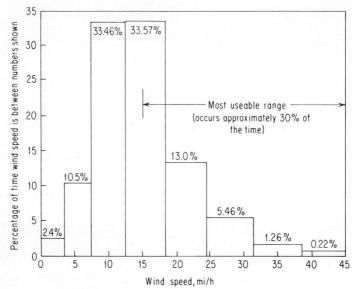

Fig. 23.1 Wind speed data for Oklahoma City.

Either they will be turned out for protection from damage or their maximum electrical output will be limited simply because it is not practical to build electrical generating systems to take advantage of winds which occur only 1 or 2 percent of the time. It is sufficient to say that in the Oklahoma City area, as a typical example, a 1-kW (at 15 mi/h) wind generation system will give about 4000 kWh of usable electricity per year.

If the wind system costs $2500/kW—borrowed at 8 percent and amortized over 20 yrs—the monthly payment to the bank will be $20.92 or $251.04/yr/kW. About 5 percent of the capital cost per year (or $125.00) is required as an operation and maintenance allowance. Thus the total annual cost is $376.04/yr, or about 9.4¢/kWh.

At first glance it appears that the windmill is not an economical source of electricity if utility electricity is available for 4.5¢/kWh. The problem is somewhat more complicated, however. Generally, electricity is available from the electric company whenever required, whereas wind-generated electricity is available when nature chooses to provide it, unless a storage system such as a lead-acid battery bank or some other suitable device is used. An investment in storage adds significantly to capital cost, however, and thus to the cost of electricity. As a first estimate it appears that a utility connection is cheapest.

Finally, the engine-driven generator costs $200.00/kW, but lasts only 2000 h. We conclude immediately that at least 10¢/kWh is required for capitalization. In addition, 15,000 to 20,000 Btu (16,000 to 21,000 kJ) of fuel will be used for each kilowatt-hour generated. One gallon (3.8 L) of gasoline will give about 7 kWh maximum, depending upon the engine. If the cost of fuel is 60¢/gal (16¢/L), it adds about 8.6¢ to the cost per kilowatt-hour, giving 18.7¢/kWh Maintenance will probably bring the total cost to 20 or 21¢/kWh. One would conclude that it would be worthwhile to have a windmill supplement for an engine-driven generator under these circumstances even if power on demand was an absolute necessity. If random intermittent power were satisfactory and if *power from the electric company were not available*, then the wind system alone might suffice.

The figures given in the preceding discussion were illustrative only. The actual costs of wind systems, engine-driven generation systems, and purchased electric power are changing and vary widely with location and time

It is important to evaluate the uses to which the wind energy may be put, especially if no storage system is contemplated. For example, if the wind energy is to be used to pump water into a storage tank for later use, a random energy source could be entirely satisfactory if the storage tank were adequate to handle peak demand. If the requirement were in a remote area where purchased electricity is simply unavailable (which is and will remain the case over a significant fraction of the earth) and where engine fuel is unavailable, wind energy or perhaps a solar system may well be the only viable alternatives.

Another application of wind energy is to provide a power source for unattended remote sensing stations, such as a weather station which periodically transmits meteorological data to a satellite, and is located in arctic regions. In this application, significant storage capability (probably batteries) is necessary to handle quiescent wind periods.

In these last applications the electricity cost is obviously secondary. Reliability becomes the most important criterion since a service trip might cost several times the initial purchase value of the power source.

Site Evaluation

A critical part of effective wind energy utilization is choosing an optimum site. It is clear that some parts of the world have significantly higher wind energy than others. Figure 23.2 is a map of the United States giving annual levels of wind energy based on decades of Weather Bureau and airport records. The energy figures are calculated from windspeed measurements, and one must therefore avoid giving a precise interpretation to such a map. Nevertheless, it is clear that the high-plains area east of the Rocky Mountains is a large and potentially significant region for wind energy exploitation. The Northeast and Northwest coasts appear excellent also. Certain parts of the Gulf Coast are promising, as are certain areas around the Great Lakes. Many parts of the rest of the nation, with isolated exceptions, appear to have at least some wind energy potential. For some jobs windmills are useful everywhere, but it is not coincidence that over the past 70 yrs most of the windmill sales in the United States have been in high-plains areas.

Within the general framework of a map such as that of Fig. 23.2, significantly more refinement is desirable. One generally wishes to know how the wind energy varies from month to month. Figure 23.3 gives the relative monthly variations of wind and solar energy for the Oklahoma City area as an example. It was noted from Fig. 23.1 that the wind was above 15 mi/h at this site for about one-third of the time. It now is observed from Fig. 23.3 that much of that time occurs in the spring, and that late summer has a very low value of total wind energy. This does not mean that significant winds do not occur in July and August, but rather that they do not occur as frequently as at other times of the year. Thus energy needs must be correlated with energy availability and appropriate supplementary sources must be arranged when required. In Fig. 23.3, it will be noted that wind energy is relatively low when solar energy is high and vice versa. Although the curves are for the Oklahoma City area, this phenomenon occurs in many parts of the world, and solar and wind energy systems could be complementary to each other.

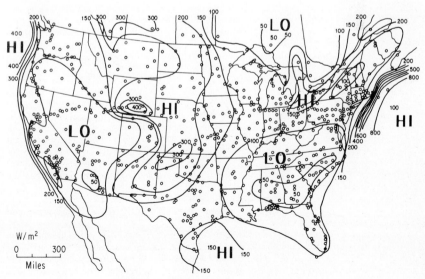

Fig. 23.2 Available wind power—annual average.

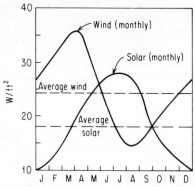

Fig. 23.3   Annual wind and solar energy distribution for Oklahoma City. (1W/ft² = 0.093 W/m²)

To fully evaluate the potential for wind energy, it is desirable to develop refined data. One nation that has done this is Egypt. Egypt has two coastal areas that show significant promise for wind energy exploitation—the north coast on the Mediterranean Sea and the east coast on the Red Sea. Meteorological records are relatively extensive for these areas, as they are in some inland areas, particularly along the Nile River.

In a joint effort with the National Science Foundation and the Oklahoma State University (in the early 1970s), the Egyptian Ministry of Electricity prepared detailed average monthly windspeed curves for a great many locations of interest. These locations were coastal, along the Nile River, and in certain interior locations in the Sahara Desert. Selected location curves are given as an example in Fig. 23.4. The best site, Hurghada, is on the Red Sea coast of Suez. Mersa Matruh, Sallum, and Alexandria are on the north (Mediterranean) coast, moving from west to east. Cairo is at the inland tip of the Nile Delta, and Minya is in southern Egypt, away

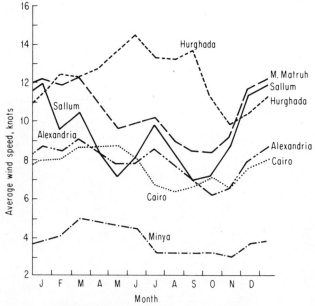

Fig 23.4   Average monthly wind speeds, knots.

from either coast. The general result of these studies obviously shows coastal areas to be most promising and inland areas, least promising.

Certain promising coastal locations were then selected for more detailed study. Curves were prepared of diurnal variations of windspeed throughout the year. Figures 23.5 to 23.8 give these curves for January, April, July, and October, respectively. It will be noted that with the exception of Hurghada the daily peaks come between 09:00 and 15:00 hours throughout the year. Hurghada invariably comes 3 to 6 h earlier. It is on the Red Sea whereas the other sites are on the Mediterranean coast in a different meteorological system.

In the selection of general areas for wind energy exploitation, data of the type presented in Fig. 23.1 to 23.8 are invaluable. Nevertheless, within any given general area, significant gains can be obtained by a detailed study of local orographic features. Any smooth-rising slope in the direction of the prevailing winds will generally give a significant advantage. For example, the coastal area between Alexandria and El Alemain gently rises for perhaps 150 m over a distance of 2 or 3 km. It then slopes downward more or less sharply, forming a smooth ridge. The wind on that ridge will be perhaps 30 percent higher than directly on the coast, giving a doubling of the energy density in the sea breeze. Grandpa's Knob, VT, the location of the Smith-Putnam windmill, was selected in part because of such considerations. It is important to carefully study the terrain to take maximum advantage of such rising conditions. If the rising terrain is covered with trees and verdure, the orographic advantage is significantly less than if it is smooth and barren. Surface roughness significantly decreases windspeeds close to the ground.

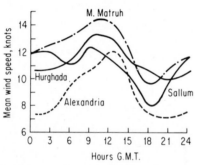

Fig. 23.5 Diurnal variation of wind speed in January.

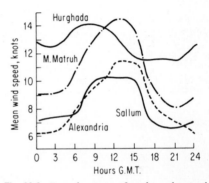

Fig. 23.6 Diurnal variation of wind speed in April.

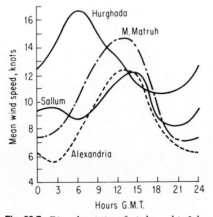

Fig. 23.7 Diurnal variation of wind speed in July.

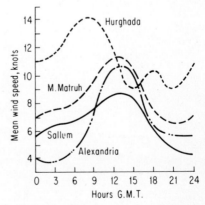

Fig. 23.8 Diurnal variation of wind speed in October.

Other Important Parameters

The most important considerations in wind energy use are cost, reliability, and minimization of undesirable environmental side effects. There are several different types of wind turbines, each with certain advantages and disadvantages. In some applications, a high starting torque is very desirable (such as in mechanical water pumping). In other applications, such as electric power generation, starting torque is much less important.

Generally speaking, significantly more wind energy is available if the wind turbine is raised 100 to 200 ft (30 to 60 m) above the surrounding terrain and obstructions. Tower costs (particularly when it is necessary to support significant weight and to withstand high turning moments) increase rapidly and nonlinearly with height. Therefore, it is often true that putting a windmill closer to the ground yields lower-cost energy, even though the windmill itself does not perform as effectively as possible. Usually the cost of wind energy in comparison with other sources will be the governing factor in choosing what kind of wind system to use or whether wind energy will be used at all.

AEROTURBINES

Wind-Axis Aeroturbines

The essential ingredient in a wind energy conversion system (WECS) is the aeroturbine, or wind turbine, traditionally called the windmill. Today, *wind-axis* and *crosswind-axis* are the predominant configurations in use throughout the world. The most modern wind-axis system today is the 100-kW Mod-0 wind turbine installed by the U.S. Department of Energy (DOE) and the National Aeronautics and Space Administration (NASA) Lewis Research Center at the Plumbrook site near Sandusky, OH.[1] Figure 23.9 shows a schematic of this aeroturbine, which has two blades, each of which is approximately 18 m (60 ft) long.

The blades have a high aerodynamic efficiency (low drag forces) and run at high rotational and peripheral speed. The blades also have an automatic pitch-change system which maintains the turbine speed constant during operation and feathers the blades at dangerously high wind-speeds. In addition, the entire turbine, gearbox, and generator system is rotated about the vertical axis to align the turbine with the wind direction.

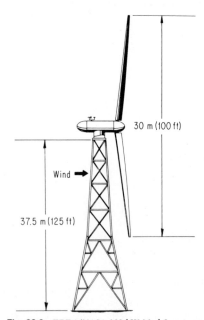

Fig. 23.9   ERDA/NASA 100-kW Mod-0 system.

The traditional American farm windmill, at the other extreme, has a large number of blades and a high solidity ratio σ (σ is the ratio of area of the blades to swept area of the turbine, πR^2). It typically operates at slower speeds and lower aerodynamic efficiencies than more advanced turbines. Modern multiblade turbines have been recently constructed[2] which have aerodynamically efficient airfoils. These turbines produce significantly more power than the old-style American farm windmill. The multiblade aeroturbines typically do not have pitch-change mechanisms and consequently operate variable-speed loads such as water pumps. They require systems (normally aerodynamic surfaces) which rotate the entire turbine out of the wind during excessive windspeeds.

Crosswind-Axis Turbines

There are three types of crosswind-axis turbines of major importance for WECS. The generally accepted advantage of crosswind-axis turbines is the elimination of the requirement to drive the axis of the turbine into the wind. The poorest performer of the three is the Savonious rotor composed of two semicylindrical offset cups rotating about a vertical axis. It is a slow-speed turbine with characteristically high starting torque accompanied by low overall aerodynamic efficiency.

The Darrieus rotor[3] looks somewhat like an eggbeater, as shown schematically in Fig. 23.10. The blades are typically high-performance symmetric airfoils formed into a gentle curve called a "troposkien". This shape is selected to minimize the bending stresses in the blades. There are usually two or three blades in a turbine and the turbines operate efficiently at high speed.

The third important crosswind-axis turbine is the cyclogiro aeroturbine.[4] The cyclogiro turbine is similar to the Darrieus rotor, with two important differences. First, the airfoils are straight, as shown in Fig. 23.11, rather than curved as in the Darrieus rotor. Second, the orientation (pitch) of the blades is continuously changed during rotation to maximize the torque induced by the wind. The peak power predicted for these turbines is greater than for any other turbine. A major advantage of the Darrieus rotor, namely complete insensitivity to wind direction, is not a property of the cyclogiro turbine. For maximum efficiency the cam mechanism which controls the local blade pitch must be continuously oriented into the wind.

The Performance of Wind-Axis Aeroturbines

In the performance analysis of wind turbines, the wind-axis devices were studied earliest, and their analysis set the present-day conventions for the evaluation of all turbines. Conventional analysis of wind-axis turbines begins with an axial momentum balance originated by Rankine,[5] using the control volume depicted in Fig. 23.12. In this figure, v is the wind speed which is

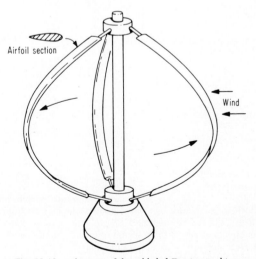

Fig. 23.10   Schematic of three-bladed Darrieus turbine.

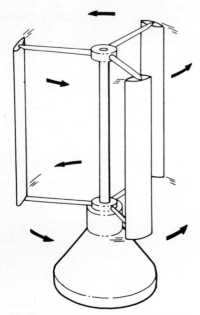

Fig. 23.11 Schematic of cyclogiro aeroturbine.

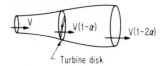

Turbine disk

Fig. 23.12 Control volume around aeroturbine.

decelerated to $v(1-2a)$ in the wake of the turbine; a is referred to as the interference factor. The momentum analysis predicts the axial thrust on the turbine of radius R to be

$$T = 2\pi R^2 \, \rho \, v^2 a(1-a) \tag{23.1}$$

where ρ is the air density (0.00237 $\mathrm{lb_f \cdot s^2/ft^4}$, 1.221 $\mathrm{kg/m^3}$) at sea-level standard atmosphere conditions.

Application of the mechanical energy equation to the control volume depicted in Fig. 23.12 yields the power P to the turbine:

$$P = 2\pi R^2 \, \rho \, v^3 a(1-a)^2 \tag{23.2}$$

This power can be nondimensionalized with the energy flux E in the upstream wind covering an area equal to the rotor disk, i.e.,

$$E = \frac{1}{2} \rho \, v^3 \, \pi R^2 \tag{23.3}$$

The resulting power coefficient is

$$C_p = \frac{P}{E} = 4a \, (1-a)^2 \tag{23.4}$$

This power coefficient has a theoretical maximum at $a = \mathrm{^1/_3}$ or $C_p = \mathrm{^{16}/_{27}} = 0.593$. This result was first predicted by Betz.[6]

The derivation above includes some important assumptions which limit its accuracy and applicability. First, the turbine must be a wind-axis configuration such that an average stream tube depicted in Fig. 23.12 can be identified. Second, the portion of kinetic energy which is in the swirl component of velocity in the wake is not included in the analysis. Third, the effect of the radial pressure gradient is excluded from the analysis. Partial accounting for the rotation in the wake has been included in the analysis of Glauert[7] with the resulting prediction of power coefficient as a function of turbine tip-speed ratio $\lambda = \Omega R/v$ (where Ω is the angular velocity of the turbine).

Blade Element Theory for Wind-Axis Turbines

Blade element theory[7,8] provides the mechanism for analyzing the relationship between the individual airfoil properties and the interference factor a, the power produced P, and the axial thrust T of the turbine. Instead of the stream tube of Fig. 23.12, the control volume consists of the annular ring bounded by streamlines depicted in Fig. 23.13. It is assumed that the flow in each annular ring is independent of the flow in all other rings.

A schematic of the velocity and force vector diagrams is given in Fig. 23.14. The elemental torque $d\tau$ which acts on all blade elements in an annular ring of chord c is

$$d\tau = \frac{B}{2} cr \, \rho \, W^2 \, (C_L \sin \phi + C_D \cos \phi) \, dr \tag{23.5}$$

The turbine is defined by the number B of its blades, by the variation in blade angle θ and by the shape of the blade section; a' is one-half of the angular velocity of the air just behind the turbine divided by the turbine angular velocity Ω, and W is the velocity of the wind relative to the airfoil. Equation (23.5) is derived for the case with no turbine coning, which can be accounted for if appropriate. The sectional lift and drag coefficients C_L and C_D are obtained from empirical airfoil data, and are unique functions of the local-flow angle of attack $\alpha = \theta - \phi$ and the local Reynolds number of the flow $\mathrm{Re} = Wc/\nu$.

Lift and drag coefficients are defined from:

$$dL = C_L \left(\tfrac{1}{2} \rho W^2\right) c \, dr \tag{23.6}$$

$$dD = C_D \left(\tfrac{1}{2} \rho W^2\right) c \, dr \tag{23.7}$$

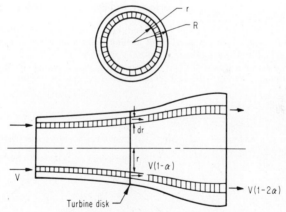

Fig. 23.13   Annular ring control volume.

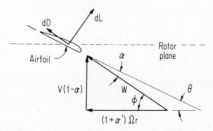

Fig. 23.14   Velocity and force diagram for blade element.

where dL is the lift force on the element of blade and dD is the drag force on the element of blade. In the Reynolds number, ν is the kinematic viscosity of air $= 160 \times 10^{-6}$ ft²/s (14.9×10^{-6} m²/s).

Power is computed by integrating Eq. (23.5) after multiplying it by the rotation rate Ω of the turbine. The result is

$$P = \rho \frac{B\Omega}{2} \int_0^R crW^2 (C_L \sin \phi - C_D \cos \phi)\,dr \qquad (23.8)$$

Similarly the total thrust force on the turbine is

$$T = \rho \frac{B}{2} \int_0^R cW^2 (C_L \cos \phi + C_D \sin \phi)\,dr \qquad (23.9)$$

The relative velocity of the wind with respect to the airfoil section W and the local angle of attack α are computed from the vector diagram of Fig. 23.13. To do this the axial interference factor a and the angular velocity fraction $a' = \omega/2\Omega$ must be calculated by relating the blade element forces to the momentum and energy equations applied to the annular control volume. The solution cannot be obtained in closed form and therefore a trial-and-error technique must be used. The idea is that the blade forces are responsible for blocking the wind and for instilling swirl in the wake. However, to compute the blade forces, the amount of blockage a and swirl a' must be known. Hence the trial-and-error requirement.

A typical solution for steady-state operation of a two- or three-bladed wind-axis turbine is shown in Fig. 23.15. When optimized, these turbines run at high tip-speed ratios and are thus referred to in this manner. The curve shown in Fig. 23.15 for the two- or three-bladed wind turbine is for constant blade pitch angle. These turbines typically have pitch-change mechanisms which are used to feather the blades in extreme wind conditions. In some instances, the blade pitch is continuously controlled to assist the turbine in maintaining constant speed. Turbines with continuous pitch control typically have flatter, and hence more desirable. operating curves than the one depicted in Fig. 23.15.

The performance curves of the other turbines discussed earlier are included in Fig. 23.15. As can be seen from this figure, several wind turbine designs have power coefficients in the range from $C_p = 0.35$ to $C_p = 0.60$.

Care must be taken not to put undue emphasis on the aerodynamic efficiency of the wind turbine configurations. Obviously, the most important criterion to be used in evaluating WECS

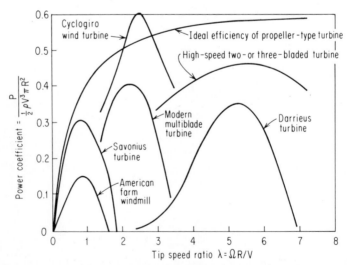

Fig. 23.15   Performance curves for various aeroturbines.

is the power produced on a per-unit-cost basis. The relationships between the turbine config-urations discussed here and their potential production costs are not presently well known. However, this type of information is no doubt forthcoming in the near future.

Wind augmentor systems are being studied in conjunction with wind turbines, also. Two of the most prominent systems are the diffuser augmentation and the vortex augmentation systems.

Airfoil Shapes for Aeroturbines

The aeroturbines discussed above typically use airfoils which are best suited to their specific configuration. The common feature of the airfoils is that they have a relatively high lift-to-drag ratio at low angles of attack (of the wind). Darrieus rotors have been successfully constructed[9] with the NACA 0012 airfoil. The shape of this airfoil is shown schematically in Fig. 23.16.

Wind-axis aeroturbines are typically constructed with a varying-shape airfoil. The airfoil at the tip of the blade is very similar to the NACA 0012 airfoil. However, at the root the blade is much thicker. The airfoil shapes at the tip and root of the ERDA/NASA Mod-0 aeroturbine are shown schematically in Figure 23.17.[10] Shown also in this figure is the blade twist, which is built into this particular turbine. Blade twist is required in all wind-axis turbines. The amount of twist is calculated from the computer model which optimizes the power using Eq. (23.8) as discussed earlier.

Structural Design Loads for Aeroturbines

The two most critical structural members of a WECS are the turbine and the support tower. Each of these members must be capable of withstanding the aerodynamic forces which they will encounter. There are normally two different design loads that the members must successfully support. The first is the load encountered during operation of the turbine at maximum operating windspeed. The turbine must be able to withstand the lift and drag forces on the blades as calculated by Eqs. (23.6) and (23.7). More important, however, the blades must be able to withstand dynamic forces associated with aerodynamic flutter and turbulence in the environ-mental winds. Such calculations are rather complicated. Reference 11 contains the methodology.

The tower must be able to withstand its major load, which is the weight of the turbine, gearbox, and generator system, and the thrust force of the turbine, which is calculated by Eq. (23.1) or (23.9). Also, the tower experiences environmental wind forces and a small component of oscillating turbine thrust force in most cases.

The second design load for both the turbine and the tower is encountered in the extreme wind situation. In most WECS the turbine is feathered or turned completely out of the wind. The wind loading on this configuration must be analyzed, and the structure must be designed to withstand the expected loading.

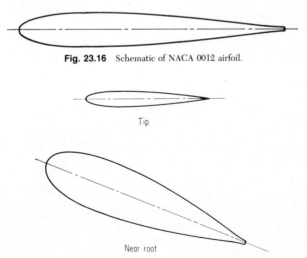

Fig. 23.16   Schematic of NACA 0012 airfoil.

Tip

Near root

Fig. 23.17   Airfoil sections at the tip and root of the ERDA/NASA Mod-0 aeroturbine.

WIND-ELECTRIC CONVERSION SYSTEMS

Introduction

Aeroturbines transform energy in moving air to rotary mechanical energy with ease and simplicity. Therefore, wind-to-electrical conversion is considered to be one of the primary means of wind energy utilization.[12]

Basic components of a wind-electric system are shown in block diagram form in Fig. 23.18. Energy extracted from moving air in rotary mechanical form is transmitted to an electrical generator via a mechanical interface consisting of a step-up gear and a suitable coupling. Electrical output of the generator is connected to the load or power grid as warranted by the particular application.

The controller senses wind speed and direction, shaft speeds and torques at one or more points, output power and generator temperature as necessary, and generates appropriate control signals for pitch control, yaw control (only in the case of horizontal-axis machines), and generator control to match the electrical output with the wind energy input and, in addition, protect the system from extreme conditions resulting from strong winds, electrical faults, generator overload, and the like.

For wind-electric systems, overall conversion efficiency, from the power available in wind to electrical power output, will be in the range of 20 to 35 percent. Figure 23.19 plots the swept area required to generate 1 kW of usable electrical power as a function of rated wind speed for different values of overall conversion efficiencies.[13] For an electrical output of P kW, the area obtained from this figure must be multiplied by P.

For practical systems, the ratio of optimal rated (full-load) windspeed to annual mean (hourly) windspeed can be anywhere from 1.25 to 2.5, depending on the size, location, wind regime, and the design philosophy employed. Annual energy outputs will be approximately 1.0 to 1.6 times the energy calculated using the mean annual wind speed.

As an example, consider a site where the annual mean wind speed is v. Let the wind-electric system be rated P kW at a windspeed of v_R where $v_R = (1.25 \text{ to } 2.5) \cdot v$. Then the annual energy production in kilowatthours is approximately $8760 \, P \, (v/v_R)^3 \cdot (1.0 \text{ to } 1.6)$. The annual average energy production factor (load factor) can be defined as (annual energy production)/(8760 P). Load-factor values range from as low as 0.1 to as high as 0.5.

Classification

Wind-electric conversion systems can be classified according to the following three basic factors:
1. Type of output
   - (*a*) Direct current
   - (*b*) Variable frequency, variable or constant voltage, alternating current
   - (*c*) Constant frequency, variable or constant voltage, alternating current
2. Aeroturbine rotational speed
   - (*a*) Constant speed with variable-pitch blades
   - (*b*) Nearly constant speed with simpler pitch-changing mechanisms
   - (*c*) Variable speed with fixed-pitch blades

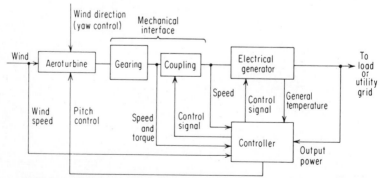

Fig. 23.18 Basic components of a wind-electric system.

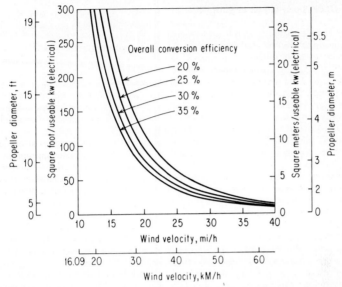

Fig. 23.19   Swept area required to generate 1 kW of usable electrical power from wind.

3. Utilization of the electrical energy output
   (a) Battery storage
   (b) Other forms of storage
   (c) Interconnection with conventional utility grids (fuel-supplement systems)

Generation Schemes

Selection of an appropriate generation scheme primarily depends on the type of output required and the mode of operation of the aeroturbine. Several combinations of speed and output have been proposed for wind-electric conversion, and some of the significant schemes are discussed below.

Permanent-magnet generator   One of the simplest generation schemes for wind-electric conversion is obtained by using a permanent-magnet alternator as illustrated in Fig. 23.20. The stator of the alternator is wound for polyphase (two- or three-phase) output, and the rotor consists of permanent magnets of alternating polarity mechanically embedded around the rotor periphery. The output frequency is equal to the rotational speed times one-half the number

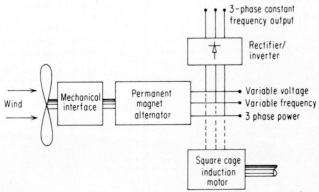

Fig. 23.20   Wind-electric conversion system employing a permanent magnet alternator.

of poles. Under open-circuit (no-load) conditions, output voltage is also proportional to the rotational speed. However, under load, armature reaction effects and internal impedance drops contribute to a departure from this proportionality.

If the aeroturbine is operated at a constant speed, the output under load will be at a constant frequency but variable voltage. If operated at variable speed, the ouput will be variable-voltage and variable-frequency. Either of these outputs can be converted to constant-voltage, constant-frequency ac by a solid-state rectifier-inverter combination (also known as an ac-dc-ac link) with appropriate controls.

It is well known that a conventional squirrel-cage induction motor can be operated in the variable-speed mode by supplying its stator with variable-voltage, variable-frequency power. To avoid saturating the magnetic circuit, the applied voltage should vary proportionally with the frequency. The air-gap flux will then be approximately constant, and satisfactory operation results. Hence, the output of a variable-speed wind-driven permanent magnet alternator is ideally suited for connection to an induction motor (as shown in dotted lines in Fig. 23.20).

The induction motor described in the previous paragraph can be made to drive a centrifugal water pump. The result will then be a wind-driven electrical water pump with electrical transmission between the aeroturbine and the pump. Such a system can be designed to extract maximum power from the wind by matching the power in the wind to the power required by the pump.[14]

Synchronous generators   The technology of synchronous machines for generating constant-frequency ac from a constant-speed prime mover is mature and well known. The generator must run at a constant speed, called the synchronous speed, which is related to the number of poles in the machine and the output frequency in hertz as given by 120/number of poles.

For synchronous machines operating in parallel with power grids, the requirement of constant speed is very strict and only minor fluctuations (about 1 to 2 percent) for short durations (fractions of a second) can be tolerated. Satisfying this requirement in the case of wind-driven synchronous machines is complicated by (1) constant fluctuations in the wind input, (2) sensitivity of the electrical output to speed changes, and (3) the capability of the machine to draw power from the grid and operate as a motor. These problems can be mitigated by using an appropriate coupling between the high-speed shaft of the step-up gear and the generator shaft.[15] Rate couplings (fluid-drive or eddy-current coupling) with a limit on maximum torque (usually less than 200 percent of rated) turn out to be the smoothest performance. Figure 23.21 shows a typical torque-slip characteristic of a torque-limited rate coupling.

Synchronizing a wind-driven generator with utility lines is a nontrivial problem in itself, especially with wind gusts present. Computer simulation studies have shown that random synchronization can be accomplished with limited transients by proper application of an effective speed-control system and an automatic synchronizer.[16]

Very high performance is needed in the blade-pitch-control loop to prevent overloads due to sharp wind gusts and to limit the output of the generator to rated value. In addition, excitation control can be used effectively to improve the electrical stability of a wind-driven synchronous generator operating in parallel with a power grid.

Induction generators   With the stator connected to a utility grid, if the rotor of an induction motor is driven at speeds above synchronous, the machine becomes a generator and delivers constant line-frequency power to the grid. The per-unit slip of an induction machine is defined as (synchronous speed − actual rotor speed)/(synchronous speed).

As a motor, the machine's full-load speed is slightly below synchronous speed and the per-unit slip usually lies between 0.0 and +0.05. As a generator, the machine's per-unit slip becomes negative with values between 0.0 and −0.05. Rated output conditions as a generator

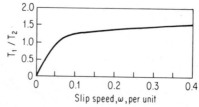

Fig. 23.21   Torque-slip characteristic of a typical torque-limited rate coupling.

are obtained at speeds slightly (less than 5 percent) above synchronous speed. Therefore, wind-electric systems employing conventional induction generators must operate "essentially" at constant speeds.

Induction generators are basically simpler than synchronous generators. They are easier to operate, control, and maintain, have no synchronization problems, and are economical. However, they draw their excitation from the grid and consequently impose a reactive-voltampere burden. Although this can be corrected by adding static capacitors or by rotor feeding, such a procedure adds to the cost and complexity of the generation scheme. Moreover, efficiencies of induction generators are slightly lower than the efficiencies of synchronous generators over the entire operating range.

Output of a wind-driven induction generator is uniquely determined by the operating speed. Once the output exceeds the rated value, blade pitch control should come into action and limit the output. Sudden disturbances such as wind gusts are automatically taken care of by load changes, and stable operation results as long as maximum torque (pullout) conditions are not reached. When that happens, the electrical output starts to decrease, speed continues to increase, and the system may "run away." Therefore, overspeed protection is vital for the safety of wind-driven induction generator systems.

If the aeroturbine is not self-starting (as in the case of Darrieus rotors), one can use one induction machine, operated as a motor for starting and as a generator after attaining the proper speed. Although such an arrangement is possible with smaller units, two separate machines are preferable in the case of large wind-electric systems because of the differences in the characteristics required for the two modes of operation.

To obtain constant-frequency output from an aeroturbine operated in the variable-speed mode, a variable-speed, constant-frequency (VSCF) generating scheme is necessary. Several such schemes have been developed by judiciously synthesizing rotating machines with mechanical (commutator) and/or electronic (thyristors and diodes) switching devices. The following schemes have been proposed for this application:

1. Field-modulated generator system
2. AC-DC-AC link
3. Double-output induction generator
4. AC commutator generator

These schemes are briefly described in the following paragraphs.

Field-modulated generator system   A field-modulated generator system employs a three-phase high-frequency alternator to generate single-phase power at the required low frequency. The rotating field coil of the alternator is supplied with alternating current at the desired low frequency instead of with conventional direct current. A parallel-bridge rectifier system connected across the machine stator terminals yields an output which consists mainly of a full-wave rectified sine wave at the frequency of modulation; in addition there is a small ripple at 6 times the basic machine frequency and a small dc component. The use of a parallel-bridge rectifier system eliminates line-to-line short circuits, and capacitors at bridge inputs reduce excitation requirements, and improve regulation and efficiency. If the ratio of the basic electrical rotational frequency to the modulation frequency is greater than 10, the ripple and the additional dc component become negligibly small. The required low-frequency output is obtained from the parallel-bridge rectifier output by using a four-thyristor switching circuit followed by a suitable filtering arrangement. The voltage at the input to the thyristor switching circuit is essentially a full-wave rectified sinusoid at the required low frequency. The thyristors switch at approximately zero voltage points of this waveform to produce the desired output. Figure 23.22 shows a block diagram of the field-modulated generator system with waveforms illustrated at different points.[17]

Power delivered by a wind-driven field-modulated generator system to a power grid can be controlled by controlling the magnitude and phase of excitation voltage applied to rotor terminals. This generation scheme has many unique features (such as the possibility of operation in single-, two-, or three-phase mode, self-excitation, etc.), providing for versatility in operation and control of wind-electric systems.[18]

AC-DC-AC link   With the advent of high-power solid-state devices and high-voltage dc transmission systems,[19] the technology of ac-dc-ac conversion is well established, and considerable literature is available on this topic. In these systems, naturally commutated (also known as line-commutated) inverters are used for converting dc into ac. They utilize an ac source (power lines) which periodically reverses polarity and causes commutation to occur naturally. Since frequency is automatically fixed by the power line, they are also known as "synchronous

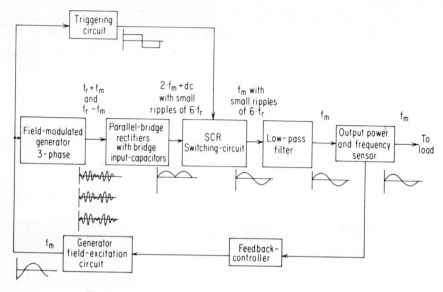

Fig. 23.22 Block diagram of the field modulated generator system.

inverters." Control of such converters is achieved by building controllable phase delay into the switching-control signal generator.

Figure 23.23 shows a schematic of a wind-electric system employing an ac-dc-ac link. The aeroturbine can be operated at constant, nearly constant, or variable speed, provided that proper controls are incorporated in the power-conditioning unit to obtain a constant-voltage, constant-frequency output.

Double-output induction generator As discussed earlier, wind-electric systems employing conventional induction generators must operate at essentially constant speeds. One approach to overcome this restriction and make the induction generator a variable-speed, constant-frequency device is to employ rotor power regeneration as illustrated in Fig. 23.24. The squirrel-cage rotor is replaced by a wound rotor with leads brought out through slip rings. Rotor power output at slip frequency is converted to line-frequency power by an ac-dc-ac link consisting of a rectifier and inverter as shown.

Output power is obtained from both stator and rotor, and hence this device is named double-output induction generator. Rotor output power has the electrical equivalence of an additional impedance in the rotor circuit. Therefore, increasing rotor outputs lead to increasing slips and higher speeds. This mode of operation broadens the operating speed range from synchronous to twice the synchronous speed. This corresponds to a per-unit slip range of 0 to -1.0.

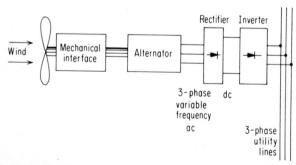

Fig. 23.23 Schematic of wind-electric system employing ac-dc-ac link.

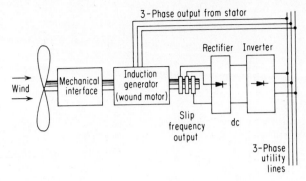

Fig. 23.24 Double output induction generator schematic.

In summation, a double-output induction generator is the result of combining two well-developed technologies (induction-machine technology and ac-dc-ac technology) to devise a variable-speed, constant-frequency generating system.[20]

AC commutator generator This device, also known as the Scherbius machine in the literature,[21] has two polyphase windings in the stator. One of the two, the exciting winding, provides the basic excitation to the magnetic circuit and sets up the rotating field. The other, known as the compensating winding, is connected in series with the brushes in such a way as to neutralize the armature reaction field set up by the armature (rotor) currents. The rotor winding is of the commutator type and is connected to the commutator as shown in Fig. 23.25.

For a given rotating field in the air gap, the frequency of the voltage and current collected by the polyphase brush gear resting on the commutator is the same as in the stator, irrespective of the rotational speed. In essence, the commutator acts as a mechanical frequency changer. Therefore, the brush gear can be connected in series with the stator (compensating) winding and the frequency of the output is independent of the shaft speed. It can be seen that the principle of operation is similar to that of a compensated dc generator except that instead of dc there is ac at a constant (line) frequency and there are three leads instead of two.

Additional windings such as interpole and damping windings may be provided to improve commutation. Basic problems in employing this device for wind-electric conversion are the cost and the additional maintenance and care required by the commutator and the brush gear.

Constant- vs. Variable-Speed Operation

Variable-speed wind-electric systems have the potential to extract a larger fraction of the energy in the wind. This can be seen in Fig. 23.26, which shows typical windspeed-duration and power-duration curves. The wind-electric system starts delivering power at a windspeed called the cut-in speed v_C, and the plant is shut down for windspeeds above a safe value called the furling speed v_F. For intermediate values of windspeeds, output power is determined by the

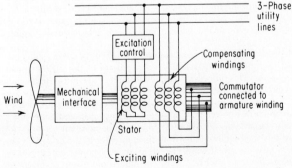

Fig. 23.25 Wind-electric conversion using an ac commutator generator.

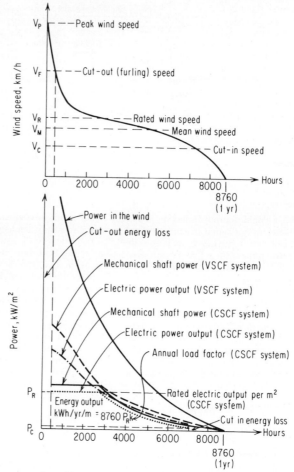

Fig. 23.26　Comparison of variable-speed and constant-speed modes of operation of wind-electric systems.

coefficient of performance of the aeroturbine and the efficiencies of the mechanical interface and the electrical generator.

At low windspeeds, constant-speed systems operate at a higher tip-speed ratio. Such an operation, in spite of appropriate changes in pitch angle of blades, results in C_P values that are less than optimum. Variable-speed systems operate at constant tip-speed ratio and consequently can maintain a high C_P even at low windspeeds. This accounts for the higher outputs at low windspeeds.

At high windspeeds, once the generator output reaches its rated value, it is maintained at that value for any additional increases in windspeeds up to the furling speed in the case of constant-speed systems. This process spills valuable energy in wind at high windspeeds. Spilling is tolerated to avoid overloading the electrical generator and to mitigate the electrical stability problems under constantly varying input conditions. The stability problems are particularly aggravated by bidirectional power flow between electrical grid and generator.

However, many variable-speed, constant-frequency generation schemes (discussed ealier) allow power flow only in one direction—from the generator to the grid. This characteristic may allow operation of variable-speed wind-electric systems at power outputs above rated value at high windspeeds, as illustrated in Fig. 23.26. The nature and extent of this increase will depend on the particular generation scheme employed and is not yet fully understood.

The discussion in the above two paragraphs leads one to conclude that variable-speed operation of wind-electric systems yields higher power outputs for both low windspeeds and high windspeeds. This results in higher annual energy yields per rated installed kilowatt. The percentage gain is highly dependent on the wind regime and the nature of the generation scheme employed. Both horizontal-axis and vertical-axis wind turbines will exhibit this gain under variable-speed operation.

The choice between constant-speed and variable-speed operation cannot be based solely on optimum energy collection. Depending on the size of the aeroturbine, certain other factors appear to dominate the design choice. For example, in the case of very large diameter (40 to 50 m and above) units, preserving mechanical stability and avoiding dangerous mechanical resonances may outweigh all other considerations. There may be more options available with small and medium-size systems than with large systems.

Utility Interfacing

The foremost application envisaged for large-scale utilization of wind power is in conjunction with electric utilities.[22] Wind-generated electrical energy in constant-frequency ac form is pumped into existing utility lines to save fuel. This approach earns "energy credit" for wind-electric systems. Several wind-driven generator units, distributed over a large geographical region, may yield a certain firm capacity depending on the diversity in wind regimes at different locations. Under these circumstances, wind-electric systems earn both energy credit and some "capacity credit".

Historical experience   Invariably, most of the significant past developments in wind power have been for generation of electrical energy. In 1929, at Bourget in France, a 20-m-diameter aerogenerator was built on a tower 20 m high to generate 15-kW dc at a windspeed of 6 m/s. In 1931, a 100-kW unit was designed and built on a tower 23 m high at Balaclava near Yalta on the Black Sea. The blades of this Soviet unit were 30.5 m in diameter and the ac output of the induction generator reached its rated value at a windspeed of 11.1 m/s. In Germany, the most recent significant work was conducted by Hütter and his group at the University of Stuttgart. They designed and constructed 100/kW units in the early 1960s with fiber-glass blades.

The largest wind-electric conversion system ever built prior to 1978 is the Smith-Putnam unit[23] constructed during World War II on Grandpa's Knob near Rutland, VT, in 1941. It had a two-bladed rotor 53 m in diameter, mounted on a 33-m tower and was rated at 1250 kW at a windspeed of 13.4 m/s or higher. In March 1945, this unit started pumping electricity routinely into the power grid of the Central Vermont Public Service Corporation. After a total of 1100 h of operation, a defective blade stopped further generation and the project was terminated. Contemporarily, Percy Thomas and his associates at the Federal Power Commission were developing numerous concepts in wind power generation for electric utilities.

During the decades of the 1920s, '30s, and '40s, thousands of small (under 5-kW) wind energy systems dotted the Midwestern countryside in the United States. They were used to pump water and generate electricity to be stored in batteries.

In Denmark, a 45-kW wind power installation was put into operation in 1952 on the island of Bogø. It was a three-bladed design, 13.5 m in diameter. Because of the success of this unit, a larger unit was installed near Gedser and put into operation in 1957. The uniform paddle-shaped three-bladed propeller had a swept area 24 m in diameter on a tower 26 m high. It operated at 30 r/min and turned a 200-kW, 750-r/min eight-pole induction generator with a slip of 1 percent at full load. The Gedser mill cut in at a windspeed of 5 m/s and reached its rated output at 15 m/s. It operated continually until it was shut down for economic reasons in 1968.

Wind power research in Great Britain was pioneered by Golding and his group.[24] The work lasted two decades—from 1940 to 1960. In the early 1950s, the English designed and built the 100-kW Enfield-Andreau wind generator unit at St. Albans in Dorset. It had two blades, 24 m in diameter, on a 30-m-high tower and generated its rated power at a windspeed of 13.4 m/s. The unit used a pneumatic transmission and an air turbine to drive an alternator.

Although wind power programs of varying sizes were in effect in many parts of the world, nothing of lasting significance resulted until the mid-1970s when interest was revived in the United States by NASA as part of the National Science Foundation's wind energy program. This has developed into a National Wind Energy Program of considerable significance. Since then, many other countries have followed suit with their own national wind energy research and development programs.

Control strategies   Wind-driven synchronous generators operating in parallel with power grids lock into synchronism and maintain a constant rotational speed irrespective of windspeeds. Suitable controllers must be used to sense the mode (generating or motoring) of operation and duration in that mode to initiate the necessary pitch control and other changes needed for proper operation. In general, input torque controls the power delivered by a synchronous machine and excitation current controls the operating power factor. Input torque changes with variations in windspeeds and it can also be changed by varying the pitch angle of blades. Once the output reaches the rated value, it is maintained constant at that value for any further increases in windspeeds by appropriate pitch control.

With induction generators, since the electrical output is uniquely determined by the operating speed, the primary purpose of pitch control is to ensure that the output does not exceed the rated value in high windspeeds.

The fundamental requirement of any control scheme for wind energy systems operating in the variable-speed mode is that it allow the system to deliver the right amount of electrical power needed to maintain a constant tip-speed ratio. The general nature of variation of aero-turbine angular speed, shaft torque, and mechanical power output for variable-speed systems is illustrated in Fig. 23.27. The control scheme should adjust the power delivered so that these ideal operating conditions are approached as closely as possible.

Generation Cost

For non-fuel-burning energy conversion such as wind and solar energy systems, the cost of electrical energy generated can be expressed as follows[25]:

$$c = \left[\frac{r(1 + r)^n}{(1 + r)^n - 1} + m \right] \frac{P}{8.76k} \tag{23.10}$$

in which

$$P = \left[\frac{(1 + r_1)^{t+1} - 1}{r_1} \right] \frac{(1 + e)^t}{(t + 1)} P_1 \tag{23.11}$$

where c = generation cost in mills/kWh
    e = average monthly inflation (escalation) rate, pu (per unit), during planning and construction periods
    k = annual load factor (also known as plant factor)
      = $\dfrac{\text{kWh generated per year by the unit}}{8760 \times \text{unit capacity}}$
    m = fraction of the capital cost needed per year for operation and maintenance of the unit
    n = number of years over which capital is amortized
    P = adjusted value of the capital investment in dollars per kilowatt of installed capacity at the year of commercial operation
    P_1 = capital cost in dollars per kilowatt at the start of planning (base-year cost)
    r = annual interest rate in pu
    r_1 = monthly interest rate in pu = $r/12$
    t = lead time (planning and construction) in months

Escalation and interest during planning and construction periods are considered by adjusting the capital cost figure P from its base-year value P_1. The ratio P/P_1 is plotted in Fig. 23.28 as a function of lead time for different interest rates and escalation rates. The total cash flow (excluding interest) amounting to $(1 + e)^t P_1$ is assumed to be in $t + 1$ equal monthly installments, spread over the planning and construction periods. Interest during this lead time is calculated by standard techniques.

For an amortization period of 20 yrs ($n = 20$), on the assumption that operation and maintenance cost is 5 percent ($m = 0.05$) of capital per year, generation cost c for wind energy systems calculated from Eq. (23.10) is plotted in Fig. 23.29 as a function of capital cost for different interest rates and plant (load) factors.

As an example of the use of these charts, let the base-year capital cost of the wind system be \$725/kW. For an interest rate of 7.5 percent and lead time of 12 months, P/P_1 is read as 1.1 from Fig. 23.28. This yields a value of \$797.5 for P. With this value of P, for a plant factor of 0.2, generation cost can be read as 68 mills (6.8¢) per kilowatthour from Fig. 23.29.

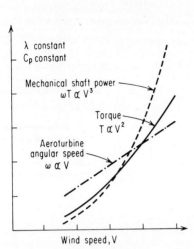

Fig. 23.27 Illustration of ideal operating conditions for variable-speed wind-electric systems.

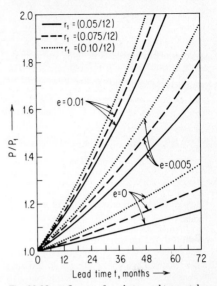

Fig. 23.28 Influence of escalation and interest during lead time.

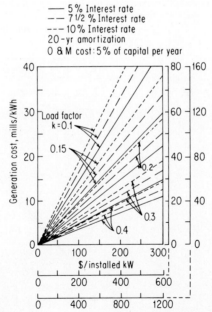

Fig. 23.29 Generation costs for wind-electric systems.

Wind systems operated in the fuel-saver mode earn energy credit. Cost of energy obtained from such "energy displacers" should be compared with only the fuel cost incurred in conventional power plants. Escalation in fuel prices over the amortization (study) period can be included by the use of an "escalation factor," defined as the ratio of average price over the study period to present price. This factor is related to the annual escalation rate as follows:

$$\text{Escalation factor} = \frac{100}{nx}\left[\left(1 + \frac{x}{100}\right)^{n} - 1\right] \qquad (23.12)$$

where n = study period in years and x = annual escalation rate, percent. Figure 23.30 shows a plot of this factor (for $n = 20$) as a function of annual escalation rate.

Fuel cost incurred in generating 1 kWh of electrical energy is given as

$$\text{Fuel cost in mills/kWh} = 3.413\,\frac{f}{\eta} \qquad (23.13)$$

where f = average fuel cost, \$/MBtu, and η = overall thermal efficiency, pu, or

$$\eta = \frac{3413}{\text{heat rate in Btu/kWh}}$$

Equation 23.13 is plotted in Fig. 23.31 for different efficiencies and fuel costs. By equating Eq. (23.10) and Eq. (23.13) break-even capital cost limits for fuel-saving wind energy systems can be obtained as a function of fuel cost for different interest rates and plant factors. This is presented in Fig. 23.32 for an amortization period of 20 years ($n = 20$) and operation and maintenance cost of 5 percent of capital ($m = 0.05$) per year.

To illustrate the use of Fig. 23.30 and 23.32, let the fuel cost be \$2/MBtu at present. If we assume an annual average escalation rate of 4 percent, then from Fig. 23.30, escalation factor is 1.5 and 20-year-average fuel price becomes \$3/MBtu. Corresponding to this 20-year-average fuel price, for a plant factor of 0.2 ($k = 0.2$) and annual interest rate of 10 percent ($r = 0.1$), break-even capital cost limit for fuel-saving wind energy systems can be read from Fig. 23.32 as \$305/kW.

The simplified economic analysis presented in this section does not consider taxes, insurance, and government subsidies such as investment tax credits. Government can make the price of energy (from any source) as cheap or as high as is necessary by appropriate tax policies.

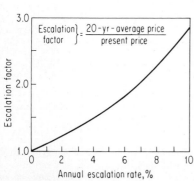

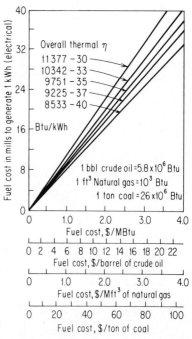

Fig. 23.30 Variation of escalation factor with escalation rate.

Fig. 23.31 Fuel cost to generate 1 kWh of electrical energy.

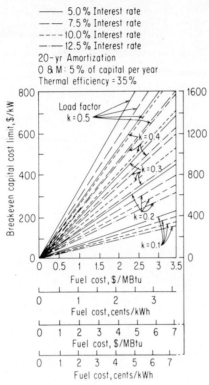

Fig. 23.32   Break-even capital cost limits for wind-electric systems.

REFERENCES

1. R. L. Puthoff, "Fabrication and Assembly of the ERDA/NASA 100-Kilowatt Experimental Wind Turbine," NASA Technical Memorandum, NASA TM X-3390, April 1976.
2. D. K. McLaughlin and D. G. Bogard, "Aerodynamic Engineering of Spoked Wheel Wind Turbines," *Workshop on Electrical Engineering Aspects of Wind Energy Systems*, Stillwater, OK, Oct. 26, 1976.
3. B. F. Blackwell et al., "Wind Tunnel Performance Data for the Darrieus Wind Turbine with NACA 0012 Blades," Report SAND 76-0130, Sandia Laboratories, Albuquerque, NM, May 1976.
4. R. V. Brulle, "Feasibility Investigation of the Giromill for Generation of Electrical Power," Report COO-2617-75/1 for ERDA, November 1975.
5. W. J. Rankine, *Transactions of the Institute of Naval Architects*, vol. 6, p. 13, 1865.
6. A. Betz, *Introduction to the Theory of Flow Machines*, Pergamon Press, New York, 1966.
7. H. Glauert, *Aerodynamic Theory*, W. F. Durand, ed., vol. 6, div. L, p. 324, Springer, Berlin, 1935.
8. R. E. Wilson and P. B. S. Lissaman, "Applied Aerodynamics of Wind Power Machines," Oregon State University Report, 1974.
9. L. V. Feltz and B. F. Blackwell, "An Investigation of Rotation-Induced Stresses of Straight and of Curved Vertical-Axis Wind Turbine Blades," Report SAND 74-0379, Sandia Laboratories, Albuquerque, NM, March 1975.
10. B. Kocivar, "World's Biggest Windmill," *Popular Science*, vol. 208, no. 3, March 1976.
11. W. D. Anderson, "100-kW Metal Wind Turbine Blade Dynamics Analysis, Weight/Balance, and Structural Test Results," NASA CR-134957, Lockheed LR 27230, June 1975.
12. F. R. Eldridge (ed.), "Wind Workshop²," *Proc. Second Workshop on Wind Energy Conversion Systems*, NSF-RA-N-75-050, MTR-6970, 1975.
13. R. Ramakumar et al., "Solar Energy Conversion and Storage Systems for the Future," *IEEE Trans. Power Apparatus and Systems*, vol. PAS-94, no. 6, pp. 1926–1934, 1975.
14. A. F. Veneruso, "An Electrical System for Extracting Maximum Power from the Wind," Sandia Laboratory Report SAND 74-0105, 1974.
15. C. C. Johnson and R. T. Smith, "Dynamics of Wind Generators on Electric Utility Networks," *IEEE Trans. Aerospace and Electronic Systems*, vol. AES-12, no. 4, pp. 483–493, 1976.

16. H. H. Hwang and L. J. Gilbert, "Synchronization of Wind Turbine Generators Against an Infinite Bus under Gusting Wind Conditions," IEEE Paper F77-675-2.
17. C. C. Tsung, "A Study of the Field Modulated Generator Systems," Ph.D. thesis, Oklahoma State University, Stillwater, 1976.
18. R. Ramakumar, "Wind-Electric Conversion Utilizing Field Modulated Generator Systems," to appear in *Solar Energy*.
19. E. W. Kimbark, *Direct Current Transmission*, vol. 1, Wiley-Interscience, New York, 1971.
20. T. S. Jayadev, "Windmills Stage a Comeback," *IEEE Spectrum*, vol. 13, no. 11, pp. 44–49, 1976.
21. B. Adkins and W. J. Gibbs, *Polyphase Commutator Machines*, Chap. 3, Cambridge University Press, Cambridge, England, 1948.
22. "Wind Energy Mission Analysis," Final Report COO/2578-1/1-3, General Electric for the U.S. Energy Research and Development Administration, 1977.
23. P. C. Putnam, *Power from the Wind*, Van Nostrand, New York, 1948; republished 1975.
24. E. W. Golding, *Generation of Electricity by Wind Power*, E. & N. Spoon, Ltd., London, 1955.
25. R. Ramakumar et al., "Prospects for Tapping Solar Energy on a Large Scale," *Solar Energy*, vol. 16, no. 2, pp. 107–115, 1974.

Chapter **24**

Photovoltaic Solar Energy Conversion Systems

MARTIN WOLF, P.E.
University of Pennsylvania, Philadelphia, PA

INTRODUCTION

The photovoltaic effect was discovered in 1839 and the first solid-state photovoltaic device was made in 1876. The utilization of this approach to solar energy conversion has started to gain significance only in the last 20 years, however. Since then, photovoltaic solar energy conversion, in which a voltage is generated when light falls on suitable materials, has provided the power for most of the spacecraft launched by all nations. In this application, solar energy has been found to be the most suitable energy source, and photovoltaic conversion devices have proven themselves as the key to a lightweight and highly reliable power supply system.

The principle technical feasibility of using photovoltaic solar energy conversion also for terrestrial applications has been well established, and the key remaining task is attaining economic feasibility for large-scale terrestrial utilization. This generally means reducing the initial costs of photovoltaic systems to the point at which they can generate electrical energy competitively with that obtained from other sources, such as fossil or nuclear fuels. The majority of the research and development work currently in progress is therefore centered on the task of cost reduction.

It is important to think of photovoltaic solar energy converters as systems rather than as "solar cells." What is required of a photovoltaic converter is the delivery of electrical power to satisfy the demand of a given load. Such loads may range from small, single-purpose devices, such as navigation lights, over single-family residences, commercial or public buildings, and industrial plants of any size, to a community or an entire utility network.

These loads generally require power delivery on demand, at a fixed voltage, and, in some cases, at precisely controlled frequency and phase. Consequently, the converter system may contain in addition to the solar collector, a voltage regulator, electrical energy storage, an inverter, and possibly other subsystems. Nevertheless, the collector forms the heart of the system. Except in a few specialty applications, the solar energy collector is a photovoltaic "array," in which numerous solar cells are connected in series to provide a desired voltage level, and in parallel to provide increased current. In lieu of parallel connections, the individual cells can be made larger, and series connections can be made within the semiconducting structure in the manner of integrated circuits widely used in electronics. A square meter of array area thus may deliver in full sunlight a current of approximately 1 A at 110 V.

History

The discovery of the basic effects behind the operation of solar cells has taken a span of approximately 100 years, starting with the discovery of selenium in 1817 by Berzelius, who was also the first to prepare elemental silicon. This was followed by the discovery of the photovoltaic effect in electrolytic cells by Becquerel in 1839,[1] and the discovery of photoconductivity in selenium in 1873 by Willoughby Smith.[2] This latter event spawned a flurry of activity which included the discovery of the spectral sensitivity of selenium photoconductors, the proposal of a light meter, and the observation of the photovoltaic effect in a solid-state selenium structure by Adams and Day in 1876.[3] Seven years later, the first selenium photovoltaic cell was described by Fritts,[4] who two years later also attempted the first simulation of the human eye response by a combination of selenium cells and color filters.[5] Then, in 1904, the photosensitivity of copper/cuprous-oxide structures was observed by Hallwachs,[6] and by 1917 the photovoltaic effect was connected with the existence of a barrier layer.[7]

With these discoveries the foundation was laid for the further development of a photovoltaic-device technology. However, more than a decade elapsed before a new period of concerted activity started. The development of the copper/copper-oxide rectifier led to new interest in the structure for photovoltaic devices.[8] Consequently, their characteristics were carefully explored, and first theories for their operation developed.[9] In the course of this work, the equivalent circuit was established which is still in common use.[10] Applications for the new device were developed, primarily in photometry and light control systems, and production lines were started.[11,12] On the heels of the development of the copper/copper-oxide photovoltaic device followed the perfection of the corresponding selenium device which quickly exceeded the performance of the former device by about an order of magnitude and, consequently, replaced the copper/copper-oxide photovoltaic devices. Solar conversion efficiencies of approximately 1 percent ultimately achieved with the selenium front-wall devices,[13] a value which was also reached with the thallium-sulfide photovoltaic device developed around 1941.

With the progressing development of the silicon technology, the "grown pn junction" technique led to the preparation of a single-crystal silicon photovoltaic device at Bell Telephone Laboratories in 1941.[14] However, it was not until the impurity diffusion method of pn junction formation had been developed 12 years later that the silicon solar cell became a practical device. A conversion efficiency of 6 percent achieved in the first year,[15] and within another two years, private industry started producing the devices with the hope for significant markets. Fabrication process improvements, improved understanding of the theory of device operation, and, accordingly, improved design of the device led to gradually increasing efficiencies which reached approximately 14 percent in terrestrial sunlight by 1958. From that point on, the prime engineering effort has shifted toward adapting the cells better to their use for space power systems, toward improving their reliability, and, in particular, their resistance to nuclear particle radiation, and, by no means least, toward reducing their fabrication cost.

When the silicon solar cells first appeared on the scene, they were thought to find a large market in terrestrial applications for light-signal detecting and metering applications as well as the much larger market of power generation. For the former applications, they were expected to quickly replace the selenium cells which exhibited lower performance, fatigue phenomena, and limited operating lifetime. It turned out, however, that the spectral response of the selenium cells made them ideally suited for applications in photographic light meters and in photometry in general. The higher output of the silicon cells was largely based on their broader spectral response. In order to reduce this to the "standard observer curve," optical filters are necessary, the addition of which made the silicon cell quite uncompetitive. Even in other applications, the positive attributes of the silicon cells did not outweigh their price disadvantage. For their use as power generators, the workers at Bell Telephone Laboratories originally foresaw a splendid future for the device as a low-cost terrestrial solar energy converter, with additional usefulness for the telephone system.[16] They intended to prove this capability by installing a solar-cell array on a telephone pole in Georgia to power a repeater amplifier. This array fulfilled its function satisfactorily for over a year, with bird droppings the only problem encountered. However, based on the original installation costs, the cost of the energy generated was not competitive with conventional power.

The two early commercial manufacturers had similar experiences with numerous expected applications until the fall of 1958 after the USSR had launched, in May of that year, a relatively large, successful spacecraft powered by solar cells. This launch had followed by only two months the launching of Vanguard I which was the first U.S. satellite to incorporate photovoltaic solar energy conversion. Its power system included, aside from the then standard dry-cell battery pack as the main power supply, a 5-W experimental system which included six small solar arrays assembled from commercially available 0.5-cm × 2-cm silicon solar cells. These events led to the development of the unexpected, but significant, market of photovoltaic systems for space which dominated solar cell production until 1974 (Fig. 24.1).

Principles of Photovoltaic Conversion

Photovoltaic conversion takes place in a thin, stationary layer of material, when light falls on it. Most of the materials used are solids, but do not have to be so. In this conversion process, electrical charges are freed and made to flow as a current through an outside circuit into an electrical load where they can perform work. The current will flow only as long as light falls on the device; no significant storage mechanism exists in the converter. If energy storage is required, it will have to be provided in a separate subsystem. Most photovoltaic devices are responsive to a broad range of light wavelengths (colors), and this wavelength range can be tailored to encompass the major part of the solar spectrum. As a generator, a solar cell can be represented by the equivalent lumped-constant circuit shown in Fig. 24.2. This circuit, although simplified by omitting components of lesser influence or of significance only for the transient response and by approximating certain distributed effects through lumped components, is adequate for most considerations of power generation from solar energy without or with relatively small optical concentration. The current-voltage characteristic corresponding to this equivalent circuit is described in rough approximation by

$$I = I_L - I_o \exp\left[\frac{q(V - IR_{sr})}{AkT}\right] \tag{24.1}$$

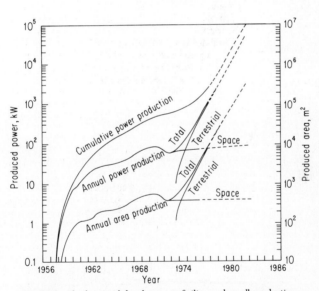

Fig. 24.1 The historical development of silicon solar cell production.

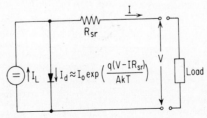

Fig. 24.2 Simplified equivalent circuit of the solar cell.

It has the general shape shown in Fig. 24.3. The light-generated current I_L is generally proportional to the instantaneous solar irradiance, assuming constant spectral distribution, and to the light-exposed area of the solar cell.

Solar cells are usually used in "arrays," which represent a circuit of N solar cells connected in parallel by M such parallel-connected groups connected in series. Such arrays may be composed of "subarrays" or "modules," which are series/parallel connected as just described to form an array, but where each in itself consists of series-/parallel-connected solar cells. Assuming all cells in an array to be identical and have a characteristic according to Eq. (24.1), the current-voltage characteristic of an array would be given by

$$I_{Ar} = NI = NI_o \exp\left[\frac{q\left(\dfrac{V_{Ar}}{M} - I_{Ar} R_{sr.Ar}\right)}{AkT}\right] \tag{24.2a}$$

$$V_{Ar} = MV \tag{24.2b}$$

The mechanism of photovoltaic device operation requires the presence of (1) a material in which mobile charge carriers can be generated by absorption of light, and (2) a built-in potential barrier by which these charge carriers can be separated from the region in which they were generated.

Semiconducting materials have properties fulfilling the first requirements, and pn junctions as used in rectifiers and transistors fulfill the second requirement. In order to obtain good efficiency in solar energy conversion, a host of other properties have to be in a suitable range. It is the task of the device designer to select these properties to the best compromise available with known materials, and to develop process methods which will permit close approximation to the desired or achievable properties.

A substantial number of semiconducting materials are known which have basic properties conducive to photovoltaic solar energy conversion. Important among these properties is a quantum-physical quantity called "energy gap." It has an influence both on the electrical properties and on the optical absorption of the material.[17,18] The energy gap is largely determined by the crystal structure of the materials, and its magnitude has to be in a certain range to yield a high efficiency of conversion of solar energy to electrical power.

The semiconducting materials are commonly grouped, according to their chemical nature, into elemental and compound semiconductors (Fig. 24.4). The former consists of only one chemical element, the latter of two (binary compounds) or more elements. Silicon and selenium are the only elemental semiconductors meeting the most basic conditions for solar energy conversion, while there are a number of suitable binary compounds, including gallium arsenide

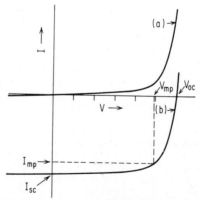

Fig. 24.3 Typical form of the current-voltage characteristic of solar cells (a) without and (b) with light incident on the active surface. Without supplying electrical energy to the cell from an external power supply, the maximum output current resulting from the incident light is the short circuit current I_{sc}, usually equal in magnitude to the light-generated current I_L; the maximum voltage is the open circuit voltage V_{oc}. The maximum output power is obtained at the point (I_{mp}, V_{mp}).

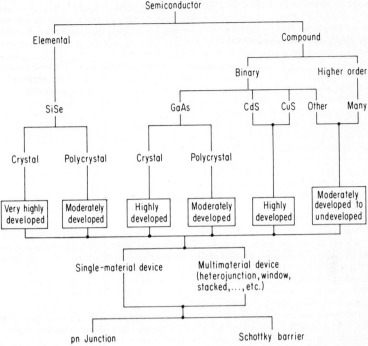

Fig. 24.4 Organization of the various semiconducting materials suitable for solar cell preparation—their crystal structure, their development status, and the device approaches possible. (Amorphous material structure not displayed; its status would be early development.)

(GaAs), cadmium telluride (CdTe), and copper sulfide (Cu_2S). Also, many higher-order compounds are appropriate for this purpose. Outside of inorganic materials, some organic materials have also shown semiconducting properties. They have, however, not yet provided any significant performance in energy conversion.

An important measure of a solar energy conversion device's performance is its conversion efficiency. The term "conversion efficiency" expresses the electrical power output available from the solar cell as a fraction of the total energy flux of the sunlight falling onto the cell. The maximum theoretical conversion efficiency potentially achievable with a single photovoltaic solar energy conversion device has been computed to be near 25 percent at room temperature in full, unconcentrated sunlight, and an efficiency of 22 percent has reportedly been achieved with GaAs cells.[19] If several photovoltaic devices of differing materials are suitably mounted behind each other, efficiencies up to about 40 percent may be obtainable.[17] Device operation at lower temperatures results in increased efficiency,[20] but operation below ambient temperature in the earth environment is energetically not advantageous.

A property having large influence on the conversion efficiency is the crystallinity of the material. The highest efficiencies obtained so far, 18 percent with silicon[21] and 22 percent with GaAs, were obtained in single-crystal solar cells. In very fine polycrystalline materials, including silicon and gallium arsenide, only small efficiencies, up to a few percent, have been obtained. However, in cells of materials which show ordering in the polycrystalline structure, such as cadmium (CdS) and cadmium telluride, an electrical performance up to approximately one-half the value obtained from the single-crystal devices has been achieved.[22] Also, solar cells from polycrystalline material consisting of relatively large crystallites—at least a few millimeters across in silicon seems to be the minimum requirement—have been prepared with efficiencies approximating those of comparable single-crystal devices.

A lack of crystalline order can influence the performance of the photovoltaic converter in two ways: by hindering the movement of charge carriers, and by affecting the properties of the necessary potential barrier. The latter determines the current-voltage characteristic of the device and thus the electrical power available from it. The most common form of potential barrier is the *pn* junction, which refers to a boundary layer within the crystal between two regions which have differing electrical properties. These electrical properties are generally obtained by the incorporation into the crystal of appropriate impurities in minute amounts. The use of a *pn* junction to obtain photovoltaic conversion is not an absolute requirement, since potential barriers can also be obtained by other means. A suitable metal layer on the surface of the semiconductor will generate a potential barrier at its interface with the semiconductor, called a "Schottky barrier." Similarly, a boundary layer between two different semiconductors can form a potential barrier, in this case called a "heterojunction." While the best performance in solar energy conversion has so far been obtained from devices containing *pn* junctions in highly ordered crystals, Schottky barriers seem less influenced by the crystal structure of the semiconducting materials and are therefore expected to yield a performance superior to that obtainable by *pn* junctions in polycrystalline semiconducting layers.

While amorphous semiconducting materials have, until recently, been considered even less likely to produce high-performance solar cells than fine-crystalline materials, recent findings have changed this situation somewhat. In especially prepared amorphous silicon which incorporates a significant percentage of hydrogen, some of the principally unfavorable properties of amorphous semiconductors have been reduced and, with the simultaneous existence of some favorable properties, experimental solar cells with Schottky barriers on this material have been prepared with efficiencies up to 5.5 percent. How much higher the performance of this cell type can be pushed, and which other amorphous semiconductors may give comparable or higher efficiency, is still uncertain.[23]

A beneficial attribute of the photovoltaic conversion process is that it requires no moving parts and consumes no material. Only electrons need to move; the atoms can stay in place. Thus, there is no wear due to friction as experienced in any machinery, and there is no change in the structure of the material due to the movement of atoms, as occurs in electrochemical batteries. Consequently, the basic operation of the device does not cause any limitation of its life. Nevertheless, secondary effects such as corrosion due to the influence of the environment will limit the operating life of the device, depending on the particular materials used in the device structure. To achieve an operating life of 20 or more years, the cells need to be encapsulated to protect them from the environment. Hermetic sealing, however, may not be necessary in most cases.

Photovoltaic conversion thus requires a device composed of one or more semiconducting layers with proper impurity content, optionally a metal layer as a Schottky barrier, and metallization of certain areas to provide contacts for extraction of the generated electric current. To reduce losses of incoming light, an antireflection coating is generally applied. Silicon solar cells are composed of a layer of silicon including a *pn* junction; the recent gallium arsenide solar cells also have a *pn* junction in addition to a window layer of gallium-aluminum arsenide. The copper-sulfide/cadmium-sulfide solar cells contain one layer each of these two materials, with a heterojunction between them. To avoid damage from wind forces, snow loads, etc., the array has to be equipped with an adequately strong backing and secure mounting, and, to reduce corrosion, must be equipped with an optically transparent cover.

Basic Photovoltaic Systems

The photovoltaic systems which have been proposed to supply large amounts of electrical power for terrestrial consumption are frequently grouped into three categories:

1. Small-to-medium-size photovoltaic systems dedicated to individual loads, often with the converters attached to or incorporated into a building structure. Their array size ranges from about 40 to 10,000 m², with average daily capacities ranging from 15 to 10,000 kWh.

2. Ground-based central systems,[24] with large photovoltaic collectors, serving either a distribution system or single, large consumers having a range of array sizes from about 1000 m² to 50 km², and average daily capacities ranging from 500 to 50,000 MWh.

3. Central systems in space, with power transmission to central ground stations and subsequent distribution. These systems, with a design capacity around 240,000 MWh daily, should not be expected to be operational in this century. They are now in the system study phase.

Systems of types 1 and 2, which are dedicated to a single user or a small community and are located near the load, are also called "on-site power systems." The application of photovoltaic arrays on or associated with buildings locates the generator at the place of the load, reducing the need for energy transmission and distribution with the associated losses and costs. The system thus matches the distributed nature of solar energy to the distributed pattern of energy consumption.

A possibility in this type of photovoltaic system is to design the photovoltaic arrays so that they are mounted directly on buildings or, even better, form part of their structure, the latter presenting economic and possibly esthetic advantages. The basic installation does not differ from that of solar thermal collectors. One attractive approach is to combine the photovoltaic array with a flat-plate thermal collector[25,26] since most buildings require both thermal energy (hot-water supply, space heating, absorption refrigeration, air conditioning, low-temperature process heat, etc.) and electrical energy (lighting, motive power, electronics, high-temperature process heat, etc.). In this case, which is illustrated in Fig. 24.5, the absorber surface of the thermal collector is formed by the solar array, which converts a portion of the incoming solar energy into electrical energy and permits collection of about 50 percent of the remaining energy in the form of heat. Although some of the efficiency of the solar cell array is traded off for thermal energy collection, this combination system provides the following advantages:

1. Up to about 60 percent of the available solar energy may be utilized.
2. The thermal collector uses the same land area as is occupied by the buildings.
3. The components fulfill several functions, yielding a more cost-effective system.

As in most terrestrial solar energy utilization systems, energy storage is generally required. In most cases, it is considered uneconomical to provide local storage capacity for more than an average day's requirement,[27] so that auxiliary energy will be needed at times. The overall system concept, shown in Fig. 24.6, includes several subsystems: solar array, electric power conditioning, energy storage, auxiliary energy, and possibly power line interfaces. However, it may be entirely impractical to expect the required auxiliary energy to be delivered in the form of electric power from utility systems, since this would pose new peaking problems and lead, through low utilization of the utility's generation and distribution capacity, to very high auxiliary energy costs. Storage can be provided by batteries, electrolysis cell fuel systems, flywheels, pressurized gases, hydraulic systems, etc.[28,29] (See Chap. 6.)

The maximum array area directly installable on buildings is approximately equal to the ground area covered by the buildings. Attention will have to be paid to the orientation, location, and shading of buildings in the same manner as for thermal collectors. This will make many existing buildings unsuitable for retrofitting, so that introduction may primarily occur on new construction. Further, of the new buildings, at most 90 percent may be suitable for solar energy collection. In addition, the highly fractionated nature of the construction industry can result in slow market growth and a long time period to saturation.

For ground-based central stations, including community-size stations and generating plants for large commercial or industrial installations, significant areas in regions of adequate insolation could be covered with photovoltaic arrays. Some may feed power into a distribution grid in the manner of a utility power plant.[24] Many variations have been proposed, including arrangements

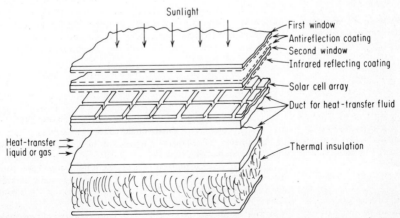

Fig. 24.5 Photovoltaic solar array mounted inside a thermal flat-plate collector for combination heating and photovoltaic power systems.

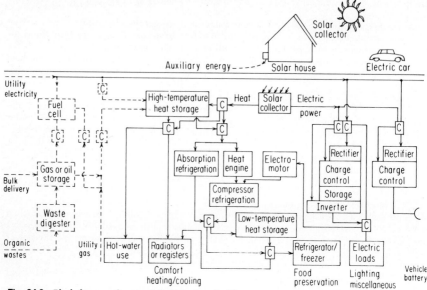

Fig. 24.6 Block diagram of a potential system for a building using solar energy both by thermal collection and direct conversion to electricity including ways of utilizing different forms of auxiliary energy. Also indicated is battery charging for a potential electrical automobile or other mobile equipment. In a given building, only certain parts of the conceptual system may be installed.

of the arrays for partial shading to permit agricultural activities beneath them, or distributed arrangements of the arrays interconnected to form a central station.[30]

The basic technology required for photovoltaic ground central stations does not differ significantly from that of the load-dedicated systems, except for scale and for site preparation and installation. There is also the potential need for precise voltage, frequency, and phase control, where direct connection with a distribution grid is planned.

THE PHOTOVOLTAIC SOLAR ENERGY CONVERSION DEVICE

The Mechanisms of Device Operation

A discussion of the operating mechanism of the solar cell is logically started from the viewpoint of the governing performance criterion, the efficiency for solar energy conversion. The overall conversion efficiency of a solar cell, like any efficiency, is determined by the relationship

$$\eta = \frac{P_{\text{out}}}{P_{\text{in}}} \tag{24.3}$$

$$\text{where } P_{\text{in}} = \int_0^\infty P_{\text{in}}(\lambda)\, d\lambda \tag{24.4}$$

integrated over all wavelengths comprising the intensity spectrum of the light incident upon the cell.[20] The maximum output power available is the product of a current density and a voltage:

$$P_{\text{out}} = j_{\text{mp}} V_{\text{mp}} \tag{24.5}$$

Here, all power values and the current density have been referred to unit area of light-exposed cell surface. The power output has been related to the energy gap E_G of the semiconductor used and to the light-generated current density j_L by means of two dimensionless factors:

$$P_{out} = j_L E_G (V.F.)(F.F.)/q \qquad (24.6)$$

$$\text{where V.F.} = \frac{V_{oc}}{E_G/q} \qquad (24.7)$$

$$\text{and F.F.} = \frac{j_{mp} V_{mp}}{j_L V_{oc}} \qquad (24.8)$$

The energy gap is a fundamental material constant of the semiconductor. It arises from quantum mechanical considerations and many physical material properties can be described in terms of the energy gap. V.F., called the "voltage factor,[17] and F.F., called the "fill factor," are largely determined by the characteristic of the rectifying potential barrier of the device (in the Si solar cell: a pn junction). However, shunt current and series resistance effects, among others, can degrade both factors below those which are based on the characteristics of an ideal barrier itself.

As a result of this observation, the voltage factor is best described as a product: V.F. = $(V.F.)_J (V.F.)_A$. The subscript J refers to the voltage factor as it would be obtained in an "ideal" solar cell, where only the characteristic of an ideal rectifying potential barrier within the device would determine the open-circuit voltage. However, artifacts such as process or design effects which result, for instance, in effective shunt currents, or nonideal barrier characteristics, can lower the open-circuit voltage. These effects are lumped in the second voltage factor designated by subscript A. In silicon solar cells, the shunt currents have generally been reduced to the point where they have negligible influence on the open-circuit voltage at room temperature or above. In this case, particularly in silicon solar cells of base resistivity $\geq 1\Omega\text{cm}$, $(V.F.)_A$ is frequently about 1, so that only the first factor needs consideration. Then, the subscript J is generally omitted, and the voltage factor is understood to be based on the junction characteristic only.

The fill factor can similarly be broken into two factors:

$$F.F. = (C.F.)(A.F.) \qquad (24.9)$$

where the "curve factor" (C.F.) has been used to designate the part based on the characteristics of the potential barrier,[17] and the "artifacts factor" A.F. has been used to account for additional "softening" of the IV characteristic, as caused by phenomena such as shunting, series resistance, or nonideal barrier characteristics.[31] Thus

$$\eta = \frac{j_L E_G F(J) G(A)}{P_{in}} \qquad (24.10)$$

where $F(J) = (V.F.)_J (C.F.)$ is a function dominated by the properties of the ideal potential barrier and less strongly dependent on the magnitude of j_L, while $G(A) = (V.F.)_A (A.F.)$ is a function primarily of the artifacts; j_L depends primarily on the material properties of the diffused and the base regions of the device and rather little on the parameters of the transition region associated with the potential barrier.

The study of the operating mechanisms of the solar cell serves to relate j_L and $F(J)$ to P_{in}, under the influence of the various physical and material parameters of the cell. This permits the designer to find an optimum combination of the selectable parameters, and to isolate the influence of the artifacts, which allows him to improve the production processes to obtain better performance.

The Light-Generated Currents

The light-generated current density is best described in terms of the incident photon flux, rather than the irradiance, since a free charge-carrier pair (electron-hole pair) is generated for each photon of suitable energy absorbed. Thus

$$j_L = q \int_0^\infty N_{ph}(\lambda) \gamma(\lambda) \, d\lambda \qquad (24.11)$$

or

$$j_L = \frac{q}{hc} \int_0^\infty P_{in}(\lambda) \gamma(\lambda) \lambda \, d\lambda \qquad (24.12)$$

where $\gamma(\lambda)$ is called the overall collection efficiency, since it relates the number of charge carriers collected, and made utilizable in current flow through an outside circuit, to the number of photons available for conversion. $N_{ph}(\lambda)$ is the number of photons incident on the unit area per unit time in the wavelength range $d\lambda$ around wavelength λ, h being Planck's constant (6.6×10^{-34} J · s), and c the velocity of light (3×10^8 m/s).

Not all of the incident energy is absorbed in the cell: some is reflected from the surface, another part is not absorbed, but effectively transmitted through the device. The reflected energy is

$$P_{refl}(\lambda) = P_{in}(\lambda)r(\lambda) \tag{24.13}$$

where $r(\lambda)$ is the reflection coefficient or monochromatic reflectivity at wavelength λ. The transmitted energy is given by Lambert's law of absorption

$$P_{tr}(\lambda) = P_o(\lambda)e^{-\alpha(\lambda)d} \tag{24.14}$$

where d is the thickness of the absorbing layer, or semiconductor wafer, P_o is the energy entering the layer, and $\alpha(\lambda)$ is the absorption coefficient at wavelength λ. The function $\alpha(\lambda)$ is specific for the semiconducting material used. It is strongly dependent on the energy gap E_G. (See Fig. 24.7.) For simplicity, it will be assumed that reflection at the back surface does not take place, although efforts have been made to design a back surface, particularly of very thin cells, so that a significant fraction of the photons are reflected and thus obtain a second chance for absorption in the wafer.

Then

$$P_{abs}(\lambda) = P_{in}(\lambda)[1 - r(\lambda)][1 - e^{-\alpha(\lambda)d}]d\lambda \tag{24.15}$$

is the energy actually absorbed in the layer of thickness d in the wavelength interval $d\lambda$ at λ. The relationships describing the effects of reflection and absorption are generally expressed in

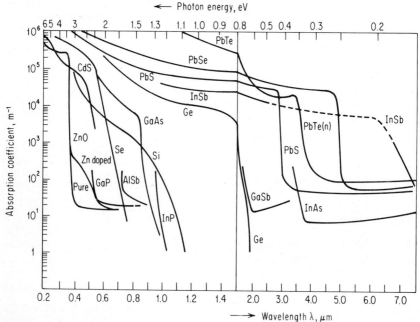

Fig. 24.7 Absorption characteristics of various semiconductors. Note the difference in curve shapes for Si and GaAs which are typical for "indirect" and "direct" bandgaps, respectively. (From Ref. 35 by courtesy of author.)

terms of light intensity, or irradiance, as in Eq. (24.15). It can readily be shown that these relationships can equally well be applied to N_{ph}, so that

$$j_L = q \int_0^\infty N_{ph}(\lambda) \, [1 - r(\lambda)] \, [1 - e^{-\alpha(\lambda)}] \, \eta_{coll}(\lambda) \, d\lambda \qquad (24.16)$$

where $\eta_{coll}(\lambda)$ is the collection efficiency of the device, related to the number of photons actually absorbed within it.

Since useful absorption does not take place beyond the absorption edge, the expression for the total light-generated current j_L can be written in the form

$$j_L = \int_0^\infty j_L(\lambda) \, d\lambda = \int_0^{\lambda_G} j_L(\lambda) \, d\lambda \qquad (24.17)$$

$$\text{where } \lambda_G = \frac{hc}{E_G} \qquad (24.18)$$

Figure 24.8 illustrates the effect of the cutoff of useful absorption relative to the spectral distribution of the solar radiation. The figure is idealized insofar as it would apply to a uniformly high absorption coefficient up to λ_G according to Eq. (24.18), where it would fall abruptly to zero, rather than to $\alpha(\lambda)$ curves as shown in Fig. 24.7. Thus, for every value of the energy gap, a cutoff line is obtained beyond which the photon energy is not sufficient to create electron-hole pairs which could contribute to a light-generated current. It is also observable that the smaller the energy gap of the material, the larger the portion of the energy spectrum of the sun that can be utilized.

This energy loss due to insufficient energy of the photons in part of the solar spectrum is simply

$$E_{ins} = \int_{\lambda_G}^\infty P_{in}(\lambda) \, d\lambda \qquad (24.19a)$$

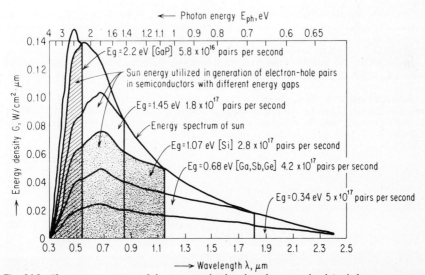

Fig. 24.8 The energy spectrum of the sun on a bright, clear day at sea level (excluding water vapor absorption) and the parts of this spectrum utilizable in the generation of electron-hole pairs in semiconductors with energy gaps of 2.25, 1.45, 1.07, 0.68 and 0.34 eV, respectively. Listed for each of these cases is the number of electron-hole pairs generated, obtained under the assumption of the existence of an abrupt absorption edge with complete absorption and zero reflection on its high energy side. (From Ref. 32.)

It may also be noted that a large number of the photons absorbed have more energy than necessary for the generation of an electron-hole pair. The energy needed for this photon-electron interaction is equal to the energy gap E_G of the semiconductor chosen. This excess energy of the photons contributes to lattice vibrations, meaning that it is dissipated in heat.

Figure 24.8 takes this energy loss into consideration by showing, in two cases as shaded areas, the actual part of the solar energy spectrum which can be utilized in the generation of electron-hole pairs for different energy gap materials. The smaller the energy gap, the more power is wasted near the peak of the sun spectrum. This energy loss is visualized by considering the factor $\lambda/(hc)$ which differentiates Eq. (24.12) from Eq. (24.11), and which expresses the inverse of the energy of a photon of wavelength λ. The total energy lost in the conversion process from photons to free charge carriers in the semiconductor because of this excess energy of the photons is

$$E_{\text{exc}} = \int_0^{\lambda_G} P_{\text{in}}(\lambda)\left(1 - \frac{\lambda}{hc}E_G\right)d\lambda \qquad (24.19b)$$

For a full evaluation of Eq. (24.16), the collection efficiency $\eta_{\text{coll}}(\lambda)$ needs to be known. This quantity describes how well a flux of photons which is absorbed within the solar cell is converted into a flux of charge carriers, or an electrical current, which can result in a current through an external circuit. The quantity $\eta_{\text{coll}}(\lambda)$ can be determined analytically, if several conditions are fulfilled. Otherwise, numerical methods have to be applied, which will not be discussed here.[32-36] Fortunately, in most practical cases, either the conditions are fulfilled, or assumption of their fulfillment provides an adequately close approximation to the real situation.

The first condition for validity of the analytical method is linearity of the integral of Eq. (24.16) in $N_{\text{ph}}(\lambda)$.[31,37] It is possible that the quantities under the integral, particularly $\eta_{\text{coll}}(\lambda)$, are not only functions of λ, but also of $N_{\text{ph}}(\lambda)$, including $N_{\text{ph}}(\lambda')$, where λ' can be any wavelength other than λ within or outside the range of integration. If this is the case, the integral is not linear in $N_{\text{ph}}(\lambda)$. This will occur, for instance, if trapping of charge carriers occurs to a significant degree, or if the total incident light intensity P_{in} is so large as to generate a minority carrier density in any region of the solar cell which is not small compared to the majority carrier density, so that minority carrier lifetime can no longer be considered independent of minority carrier density and, in general, the normal assumptions of small-signal theory cannot be made. Fortunately, in normal silicon solar cells, and probably in many other types of solar cells, the conditions for validity of the assumption are fulfilled, even for rather large light intensities as obtained by optical concentration of sunlight, up to concentration ratios above 100, depending on the design of the cell considered. Devices known to violate the assumption are the Cu_2S-CdS solar cell, or silicon solar cells after heavy proton radiation damage. It will therefore be necessary to test the validity of the assumption before applying the relationships given in the following. Although the general approach outlined here may be useful in these other cases also, modifications of some details will in general be necessary.

Averaged analysis   If the integral in Eq. (24.16) can be considered linear in $N_{\text{ph}}(\lambda)$, then weighted average values for the reflectance, the ratio of energy actually absorbed in the wafer to that entering it, and the collection efficiency η_{coll} can be introduced to describe the solar cell parameters in simpler terms, as is common practice:

$$j_L = q\,(1 - R_{\text{refl}})R_{\text{abs}}\,\eta_{\text{coll}}N_{\text{ph tot}} \qquad (24.20)$$

$$\text{where } N_{\text{ph, tot}} = \int_0^{\lambda_G} N_{\text{ph}}(\lambda)d\lambda \qquad (24.21)$$

$$1 - R_{\text{refl}} = \frac{1}{N_{\text{ph, tot}}}\int_0^{\lambda_G} N_{\text{ph}}(\lambda)[1 - r(\lambda)]\,d\lambda \qquad (24.22)$$

$$R_{\text{abs}} = \frac{1}{N_{\text{ph, tot}}(1 - R_{\text{refl}})} \qquad (24.23)$$

$$\cdot \int_0^{\lambda_G} N_{\text{ph}}(\lambda)\,[1 - r(\lambda)]\,\{1 - \exp[-\alpha(\lambda)\,d]\}\,d\lambda \qquad \text{(Fig. 24.9)}$$

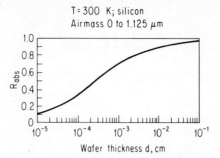

T = 300 K; silicon
Airmass 0 to 1.125 μm

Fig. 24.9 Ratio R_{abs} of the number of photons absorbed in a single pass through a silicon layer of thickness d to the number entering the layer, as contained in airmass zero sunlight up to 1.125-μm wavelength. (From Ref. 34.)

$$\text{and } \eta_{\text{coll}} = \frac{1}{N_{\text{ph, tot}}(1 - R_{\text{refl}})R_{\text{abs}}} \qquad (24.24)$$

$$\cdot \int_0^{\lambda_G} N_{\text{ph}} \lambda \, [1 - r(\lambda)] \, \{1 - \exp[-\alpha(\lambda)d]\} \, \eta_{\text{coll}}(\lambda) \, d\lambda$$

Thus, without fulfillment of the linearity condition, it is not meaningful to use a general collection efficiency η_{coll} for a solar cell, as it is to use a spectral response—in effect, an $\eta_{\text{coll}}(\lambda)$ curve. In that case, a spectral response is valid only for the conditions under which it was assumed, for instance, with bias light of a given intensity and spectral distribution.

Since the reflection can, in practice, be reduced to very small values (Fig. 24.10), R_{refl} is assumed to be zero in the following. However, actual reflection characteristics can readily be introduced into the subsequent equations by replacing $N_{\text{ph}}(\lambda)$ with $N_{\text{ph}}(\lambda)[1 - r(\lambda)]$. It should also be noted that in cases where $1 - r(\lambda)$ is significant and has a strong dependence on wavelength, the sample design curves given in the following will not be accurate and should be recomputed for the specific reflection characteristics.

Figure 24.9 shows a plot of R_{abs} as a function of wafer thickness d for silicon for the air-mass zero spectrum for $r(\lambda) = $ constant. For higher air-mass values, the shape of the curve would be slightly steeper.

Since contributions to the light-generated current j_L can be identified as originating in three different regions of the solar cell, it is practical to write

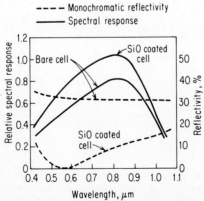

Fig. 24.10 The monochromatic reflectance of a bare and a silicon-monoxide-coated silicon solar cell, and the relative short-circuit current spectral response (constant energy basis) of a silicon solar cell taken before and after application of a silicon-monoxide coating at identical irradiance values. (From Ref. 41.)

$$j_L = \int_0^{\lambda_G} [j_{L,D}(\lambda) + j_{L,T}(\lambda) + j_{L,B}(\lambda)] \, d\lambda$$

$$= j_{L,D} + j_{L,T} + j_{L,B} \tag{24.25}$$

where the subscripts D, T, and B refer to the diffused region, transition region, and base region of the cell, respectively. In the following, it is tacitly assumed, without loss of generality, that the diffused region is the one adjacent to the light-exposed surface. If an "inverted" structure should be considered, it is merely necessary to enter the parameters of the diffused region into $\eta_{coll,B}$ and those of the nondiffused region into $\eta_{coll,D}$. Although many methods exist, outside of diffusion, to generate a layer of opposite impurity type as the base region, such as ion implantation, growth of an epitaxial layer, etc., the name "diffused layer" shall here be maintained as a general descriptor of the layer which is prepared in order to form a pn junction.

Three collection efficiencies can then be introduced, one each connected with one of the major regions and defined as in Eq. (24.17):

$$j_L = q \int_0^{\lambda_G} N_{ph}(\lambda) \{ [1 - e^{-\alpha(\lambda) x_{T,F}}] \eta_{coll,D}$$

$$+ [e^{-\alpha(\lambda) x_{T,F}} - e^{\alpha(\lambda) x_{T,R}}] \eta_{coll,T} \tag{24.26}$$

$$+ [e^{-\alpha(\lambda) x_{T,R}} - e^{-\alpha(\lambda) d}] \eta_{coll,B} \} \, d\lambda$$

Here, $x_{T,F}$ is the location of the front boundary of the transition region (depletion region) of the pn junction, that is, the one situated more closely to the light-exposed surface of the device, to which $x = 0$ has been assigned. $x_{T,R}$ is the location of the rear boundary surface of the transition region.

A second assumption has been made, that the entire structure is plane-parallel so that all surfaces of significance are planes and are parallel to the light-exposed surface, excluding the surfaces which form the edges of the water. Then

$$j_{L,D} = q \, R_{abs} R_D \, \eta_{coll,D} N_{ph,tot} \tag{24.27}$$

$$\text{where } R_D = \frac{1}{N_{ph,tot} R_{abs}} \int_0^{\lambda_G} N_{ph}(\lambda) [1 - e^{-\alpha(\lambda) x_{T,F}}] \, d\lambda \tag{24.28}$$

$$\text{and } \eta_{coll,D} = \frac{1}{N_{ph,tot} R_{abs} R_D} \int_0^{\lambda_G} N_{ph}(\lambda) [1 - e^{-\alpha(\lambda) x_{T,F}}]$$

$$\cdot \, \eta_{coll,D}(\lambda) \, d\lambda \tag{24.29}$$

$\eta_{coll,B}$ and $\eta_{coll,T}$ are defined similarly, with

$$R_B = \frac{1}{N_{ph,tot} R_{abs}} \int_0^{\lambda_G} N_{ph}(\lambda) [e^{-\alpha(\lambda) x_{T,R}} - e^{-\alpha(\lambda) d}] \, d(\lambda) \tag{24.30}$$

$$\text{and } R_T = 1 - R_D - R_B \tag{24.31}$$

Figure 24.11 shows R_D and R_B as functions of wafer thickness, and with $x_{T,F}$ or $x_{T,R}$ as parameters, for silicon under the air-mass zero spectrum.

Charge current density   The above analysis has laid the foundation for the presentation of collection efficiency data for the various aspects or parts of the device. R_{abs} permits evaluation of the direct effect of wafer thickness d on the overall collection efficiency. Introduction of R_D, R_B, and R_T permits separate optimization of the main regions of the solar cell, the design of which can be carried out relatively independently.

The relationships expressed by the various ratios R require the knowledge only of the absorption coefficient $\alpha(\lambda)$, the reflectance $r(\lambda)$ of the semiconductor material and the thickness of the wafer and the electrically effective layers within it. In addition, $P_{in}(\lambda)$, the spectral distribution of the light source, and $\eta_{coll}(\lambda)$ must still be determined. The latter quantity requires analyses of charge transport within the different regions of the solar cell. These analyses are carried out in an identical manner, except that the values for the various material and physical parameters differ according to the properties of the different regions. In each case,

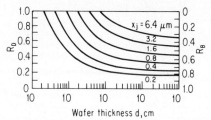

Wafer thickness d, cm

Fig. 24.11 Ratio R_D and R_B of the number of photons of airmass zero sunlight which are absorbed in the diffused region and the base region, respectively, as function of wafer thickness d. The ratio is taken relative to the number of photons absorbed in the wafer of thickness d, and is presented parametrically for various thicknesses x_j of the diffused region (depletion region width zero). Where the depletion region width is not zero, $x_{T,F}$ and $x_{T,P}$ will be used, in lieu of x_j, for the diffused region and the base region, respectively. (From Ref. 34.)

the method requires determination of the distribution $n(x)$ or $p(x)$ of the excess minority-carrier concentration within each region, because it results from the simultaneous effects of electron-hole pair generation as a result of the nonuniform photon absorption, depending on $\alpha(\lambda)$, of bulk recombination, and of transport to the region surfaces, where either drift across a boundary space-charge region or surface recombination may take place. Once the excess minority-carrier distribution is known, the current flow across the space-charge region, viz., a pn junction, can be determined as the second step. In general, the current is described by two terms, such as

$$j_p = q \left(-D_p \frac{dp}{dx} + \mu_p E_p p \right) \tag{24.32}$$

for a hole current, where the first term on the right-hand side describes charge transport by diffusion due to density gradients of the movable charges, and the second term transport by the drift due to the influence of electric fields.

The starting point for the derivation of the minority-carrier distribution is the continuity equation for holes as excess minority carriers,[38] i.e., the density of minority carriers above the equilibrium density:

$$\frac{\partial p}{\partial t} = \alpha(\lambda) N_{ph} e^{-\alpha(\lambda)x} - \frac{1}{q} \frac{\partial j_p}{\partial x} - \frac{p}{\tau_p} \tag{24.33}$$

where the first term on the right-hand side is the number of minority carriers generated by electron-hole pair generation due to photon absorption in a layer of unit area and thickness dx at the distance x below the surface of the solar cell. The second term covers the transport of minority carriers out of the same layer, described as a change in the magnitude of the electrical current j_p flowing through this layer, while the last term describes the recombination of excess minority carriers in this layer. The problem is generally reduced to the one-dimensional form shown in Eq. (24.33), since solar cells are usually thought of as being fabricated with rather constant parameters over their area as well as illuminated with rather uniform intensity. Since the solar cell is a large-area device with the length and width dimensions large compared with diffusion or drift lengths of minority carriers, the omission of surface effects on the edges in the derivation is a good approximation.

The assumption is made that the charge densities due to excess minority carriers are small compared with the charge densities normally present within the device, so that the modification of the built-in electric fields is negligibly small. This condition also assures nonsaturation of the recombination centers, resulting in independence of the minority-carrier lifetime from the minority-carrier density. The condition is fulfilled in normal solar cell operation, including that with concentrators, with the exception of very high concentration ratios, depending on the resistivity of the semiconductor layer. This can be readily verified by evaluation of the resulting excess minority-carrier distribution [see Eq. (24.25)]. An exception is found in a narrow portion of the pn junction space-charge region, but this is not sufficient to alter the results.

An analytical solution can be readily obtained only under the aforementioned assumptions and for the case of the electric field E zero or constant with respect to the independent variable x. This latter case corresponds to an exponential distribution of impurity density in the device, which represents a fair approximation to the impurity distributions usually obtained. It is to be further observed that both the minority carrier lifetime τ and the mobility μ are generally dependent on the impurity density and may therefore be functions of x. Again in the interest of ready solubility, these quantities are often assumed to be constant. The analytical derivations therefore yield calculations for models only, but nevertheless provide valuable information about the minority-carrier collection mechanism in solar cells and furnish a tool for device design.

With these assumptions, and for the steady-state condition as the one of main interest for solar cell operation, Eq. (24.33) becomes, after the insertion of Eq. (24.32),

$$\frac{d^2p}{dx^2} - \frac{qE}{kT}\frac{dp}{dx} - \frac{1}{L_p^2}p = -\frac{\alpha(\lambda)}{D_p}N_{ph}e^{-\alpha(\lambda)} \tag{24.34}$$

Use was made here of the Einstein relationship[39] between diffusion constant D and mobility μ:

$$D = \frac{\mu kT}{q} \tag{24.35}$$

and of the relationship between diffusion length L, diffusion constant D, and minority-carrier lifetime τ:

$$L = \sqrt{D\tau} \tag{24.36}$$

For the case of electrons as minority carriers in p-type material, Eq. (24.33) takes the form:

$$\frac{\partial n}{\partial t} = \alpha(\lambda)N_{ph}e^{-\alpha(\lambda)} + \frac{1}{q}\frac{\partial j_n}{\partial x} - \frac{n}{\tau_n} \tag{24.37}$$

$$\text{with } j_n = q\left(D_n\frac{dn}{dx} + \mu_n E_n n\right) \tag{24.38}$$

The sign inversions take place as a result of the flow of electrons as negative charge carriers in the direction opposite to the current flow j, in contrast to the flow of holes described in the earlier equations. This leads to

$$\frac{d^2n}{dx^2} + \frac{qE}{kT}\frac{dn}{dx} - \frac{1}{L_n^2}n = -\frac{\alpha(\lambda)}{D_n}N_{ph}e^{-\alpha(\lambda)} \tag{24.39}$$

The only significant difference between this differential equation and Eq. (24.34) for holes consists in a sign inversion of the term containing the drift field E.

The general solution of the nonhomogeneous differential equation (24.39) is

$$n(x) = A_1 e^{-(F_n + G_n)x} + A_2 e^{-(F_n - G_n)x} - \frac{\alpha(\lambda)N_{ph}e^{-\alpha(\lambda)}}{D_n G_n^2(b_n^2 - 1)} \tag{24.40}$$

where the following nomenclature has been used:

$$F_n = \frac{q}{2kT}E \tag{24.41a}$$

$$G_n = \sqrt{F_n^2 + \frac{1}{L_n^2}} \tag{24.41b}$$

$$b_n = \frac{-F_n}{G_n} \tag{24.41c}$$

The general solution of Eq. (24.34) is identical in form to Eq. (24.40), except that all quantities and subscripts n are replaced by p. The mentioned sign inversion affects only the quantity F:

$$F_p = - \frac{q}{2kT} E \qquad (24.41d)$$

A physical meaning is attached to the sign inversion of the F terms relative to the drift field E depending on the minority-carrier type. Positive values of F in the base correspond, for either minority-carrier type, to drift fields aiding the collection by excess minority-carrier diffusion, while negative values of F in the diffused region designate the same situation there. The sign inversion of F relative to the region of the device is related to the reversion of charge-carrier flow in the two regions for unidirectional current flow, arising from the fact that the two regions possess minority carriers of opposite polarity.

The constants A of the general solution are determined by the boundary conditions imposed by the device parameters.

1. At $x = 0$, *surface recombination takes place at velocity* s:

$$-D_n \left(\frac{dn}{dx} \right)_{x=0} - \mu_n E_n n (0) = - sn (0)$$

$$\text{or} \left(\frac{dn}{dx} \right)_{x=0} = \left(\frac{s}{D_n} - 2F_n \right) n (0) \qquad (24.42)$$

2. *A pn junction transition region boundary is located at* $x = x_\tau$ *and is maintained in the zero-bias condition (leading to short-circuit current) so that a perfect sink for excess minority carriers exists*:

$$n (x_\tau) = p (x_\tau) = 0 \qquad (24.43)$$

Operation under different bias conditions does not have to be considered here because of the assumed voltage-independence of the light-generated current and the linear superposition of the latter with the current flow across the pn junction in the opposite direction caused by various terminal voltages.

3. *Surface recombination also takes place at the back surface at* $x = d$:

$$-D_n \left(\frac{dn}{dx} \right)_{x=d} - \mu_n E_n n (d) = sn (d)$$

$$\text{or} \left(\frac{dn}{dx} \right)_{x=d} = - \left(\frac{s}{D_n} + 2F_n \right) n (d) \qquad (24.44)$$

Application of these boundary conditions results in two systems of two linear equations each, with the two systems applicable to the diffused layer and the base, respectively. The second boundary condition is used in both systems. The solution of these systems yields the constants of Eq. (24.40), which then takes the final form for the excess minority-carrier distribution for p-type material in the base:

$$n (x'_n, \lambda)_{\text{base}} = \frac{\alpha (\lambda) N_{\text{ph}}(\lambda)}{D_n G_n^2} e^{-(F_n G_n) x'_n}$$

$$\cdot \left\{ - \frac{e^{-x'_n}}{b_n^2 - 1} [e^{(1-b_n)} x'_n - e^{(1 - b_n) y_n}] \right.$$

$$+ \frac{e^{-b_n z_n} \sinh (x'_n - y_n)}{\cosh (z_n - y_n) + a_{n\,\text{base}} \sinh (z_n - y_n)}$$

$$\left. \cdot \left[\frac{1 - a_{n\,\text{base}}}{b_n^2 - 1} e^{(b_n-1)(z_n - y_n)} - 1 - \frac{1}{b_n + 1} \right] \right\} \qquad (24.45)$$

for $b_n \neq 1$ and with $y_n \leq x'_n \leq z_n$, with

$$x'_n = G_n x \tag{24.46a}$$

$$y_n = G_n x_\tau \tag{24.46b}$$

$$z_n = G_n d \tag{24.46c}$$

$$a_{n\,\text{base}} = \frac{1}{G_n}\left[F_n + \frac{s}{D_n}\right] \tag{24.46d}$$

$$a_{n\,\text{diff}} = \frac{1}{G_n}\left(F_n - \frac{s}{D_n}\right) \tag{24.46e}$$

and all other symbols as previously introduced.

The distributions for holes in n-type material are obtained by substituting a letter p for every n (including subscripts) in Eqs. (24.42), (24.44), (24.45), and (24.46a) through (24.46e) without any other changes necessary. The same Eq. (24.45) can be used for the excess minority-carrier distributions in diffused layers by setting $z = 0$ and by inserting a_{diff} for every a_{base} in Eq. (24.45). The distributions do not become singular for $b = 1$.

The current densities into the pn junction are obtained by application, at $x = x_\tau$, of Eq. (24.32) or (24.38), both for the diffused layer and for the base. Since the excess minority-carrier density falls to zero at $x = x$, according to boundary condition 2, only the diffusion term in Eq. (24.32) or (24.38) contributes to current flow. Thus, the current densities are proportional to the density gradients of the minority carriers at $x = x_j$:

$$j_{n_{\text{base}}}(\lambda) = \frac{q\alpha(\lambda)N_{\text{ph}}(\lambda)}{G_n}e^{-(F_n/G_n)y_n}\left[\frac{e^{-b_n y_n}}{b_n+1} + \frac{e^{-b_n z_n}}{b_n^2-1}\right.$$
$$\left.\cdot\frac{e^{-(1-b_n)(z_n-y_n)} - b_n - a_{n_{\text{base}}}(e^{-(1-b_n)(z_n-y_n)}-1)}{\cosh(z_n-y_n) + a_{n_{\text{base}}}\sinh(z_n-y_n)}\right] \tag{24.47}$$

for the current from the base.

As in Eq. (24.45), setting $z = 0$ and replacing every a_{base} with an a_{diff} provides use of the collected-current formula for diffused layers.

Inversion of the sign for the right-hand side of Eq. (24.47), corresponding to reversion of the direction of current flow, and exchange of all letters and subscripts n with p leads to the currents collected from n-type diffused layers or bases, respectively.

It is interesting to note that the final equations (24.45) and (24.47) for the minority-carrier distribution and for the collected-current density reduce to those for the field-free case,[17] as the factor $\exp[-(F/G)x']$ becomes 1, and G reduces to $1/L$. Beyond this, the constants a and b are simplified, as the drift field term $F = 0$.

For computer application, it has been found desirable to replace the hyperbolic sine and cosine functions of Eq. (24.47) by exponential functions, leading to the basic expression for the collected-current density:

$$j_{n_{\text{base}}}(\lambda) = \frac{q\alpha(\lambda)N_{\text{ph}}(\lambda)}{G_n}e^{-(F_n/G_n)Y_n}\left[\frac{e^{-b_n y_n}}{b_n+1}\right.$$
$$\left.+ \frac{2e^{-(z_n-y_n)}}{b_n^2-1}\frac{(1-a_n)e^{-z_n-(b_n-1)y_n} - (b_n-a_n)e^{-b_n z_n}}{1+a_n+(1-a_n)e^{-2(z_n-y_n)}}\right] \tag{24.48}$$

with variations for n-type material and for the diffused region as explained above.

Having thus obtained values for $j_{L,B}(\lambda)$ and $j_{L,D}(\lambda)$, we can readily deduce the desired collection efficiency data through

$$\eta_{\text{coll},B}(\lambda) = \frac{j_{M_{\text{base}}}(\lambda)}{N_{\text{ph}}(\lambda)\{\exp[-\alpha(\lambda)x_{T,R}] - \exp[-\alpha(\lambda)d]\}} \tag{24.49}$$

$$\text{and }\eta_{\text{coll},D}(\lambda) = \frac{j_{M_{\text{diff}}}(\lambda)}{N_{\text{ph}}(\lambda)\{1 - \exp[-\alpha(\lambda)x_{T,F}]\}} \tag{24.50}$$

Sample design curves resulting from such computations carried out with various material and dimensional parameters are shown in Figs. 24.12 to 24.17.

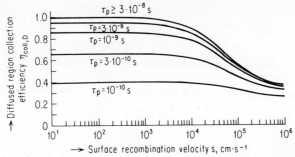

Fig. 24.12 Collection efficiency $\eta_{\text{coll},D}$ from the diffused region as function of surface recombination velocity, for several values of minority carrier lifetime τ_p. (Diffused region thickness $x_{T,F} = 0.2\ \mu\text{m}$.)

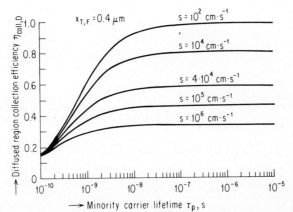

Fig. 24.13 Collection efficiency $\eta_{\text{coll},D}$ from the diffused region as function of minority carrier lifetime, for several values of surface recombination velocity s. (Diffused region thickness $x_{T,F} = 0.4\ \mu\text{m}$.)

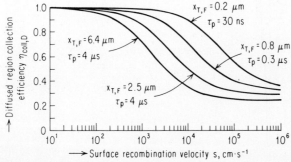

Fig. 24.14 Collection efficiency $\eta_{\text{coll},D}$ from diffused region as function of surface recombination velocity, for several values of diffused region thickness $x_{T,F}$ with minority carrier lifetime values τ_p which yield maximum light-generated current.

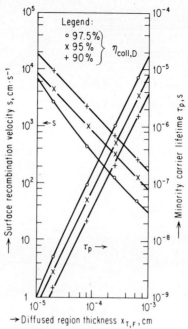

Fig. 24.15 Design curves for the material parameters as function of the diffused region thickness. The curves yield the values of minority carrier lifetime and surface recombination velocity s, for which a collection efficiency of 97.5, 95, and 90%, respectively, can be obtained, assuming the other one of these two parameters being adjusted so that this is possible.

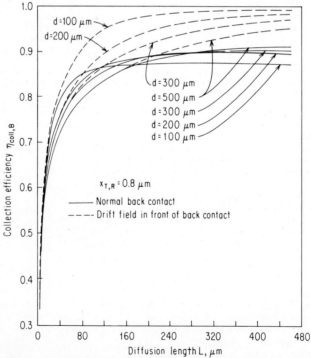

Fig. 24.16 Collection efficiency $\eta_{coll,B}$ from the base region as function of diffusion length for three values of thickness for silicon solar cells in airmass zero sunlight, $x_{T,R} = 0.8\ \mu m$.

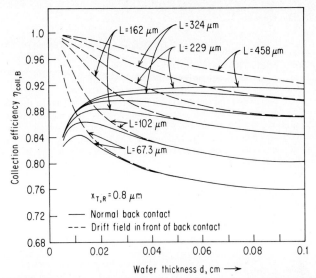

Fig. 24.17   Collection efficiency $\eta_{coll,B}$ from the base region as function of water thickness d, for three values of diffusion length in the base region, with and without a drift field in front of the back contact. Airmass zero sunlight, $T = 300$ K. $x_{T,R} = 0.8$ μm.

The Current-Voltage Characteristic   The light-generated current has been seen to be determined by the optical properties of the semiconducting material and the device structure, and by the minority-charge-carrier transport properties within approximately three diffusion lengths from the boundaries of the space-charge (transition) region. In contrast, the current-voltage characteristic is determined by the properties of the pn junction or other potential barrier, plus the light-generated current.

For the same linearity conditions discussed in the preceding section with respect to the light-generated current, the total current flowing from a solar cell into a load can be expressed, in accordance with Fig. 24.2, as

$$I = j_d A_j - j_L A_L \tag{24.51}$$

where j_d is the current density of the diode current which would flow without incident light as a result of a voltage V_d across the junction equal to that resulting from the combined effects of light and load resistance. A_j is the area of the space-charge region, measured normal to the direction of the potential gradient, and A_L is the effective surface area under which minority carriers are generated by absorption of light. Although the first term of Eq. (24.51) is usually thought of as expressing the properties of the pn junction, it is in reality determined by the same charge-carrier transport properties within approximately three diffusion lengths from the space-charge region as for the light-generated current.

Consequently, as the light-generated current can be expressed as a sum of several current contributions from different regions of the solar cell, so can the junction current j_d be expressed as a sum of several current contributions. In the most general case, the contributions do not only originate from different regions of the diode, but can also be based on various mechanisms. Thus

$$j_d = j_{d,D} + j_{d,T} + j_{d,B} \tag{24.52}$$

where the second subscripts refer to the diffused, transition, and base regions, respectively.

The most basic mechanism leading to dark diode current is the diffusion current, described by Shockley.[40] It results in the relationship

$$j_{d,D} + j_{d,B} = j_o \left[\exp\left(\frac{qV_d}{kT} \right) - 1 \right] \tag{24.53}$$

with the saturation current

$$j_o = j_{o,D} + j_{o,B}$$ (24.54)

Plotting $\ln j_o$ from Eq. (24.53) against V_d yields a straight line over many decades of current, and the plot deviates from this straight line only at very low current values, where the -1 term becomes significant (Fig. 24.18).

Equation (24.53) represents the classical diode equation which is frequently used exclusively for description of the solar cell current-voltage characteristics. As a minimum, however, series resistance effects also have to be accounted for. If the series resistance effects can be represented by a single lumped constant, as is generally the case in high-efficiency solar cells used in normal sunlight, without significant optical concentration, then the relationships stay relatively simple by expressing V_d of Eq. (24.53) as

$$V_d = V - j\,r_{sr}$$ (24.55)

where r_{sr} is the lumped series resistance for the unit cell area, expressed in Ω/cm^2 ($r_{sr} = R_{sr}A_j$), and

$$j = \frac{I}{A_j}$$ (24.56)

in accordance with Eqs. (24.1) and (24.51). V represents the terminal voltage as in Fig. 24.2. The series resistance effects cause a deviation from the straight-line plot of Fig. 24.18 at the high current levels, as indicated there.

The open-circuit voltage V_{oc} is determined by setting $I = 0$ in Eq. (24.51). It has been established that Eqs. (24.53) and (24.55) adequately describe the current-voltage characteristic of the silicon solar cell near the open-circuit-voltage point at normal light intensities and above,[31,41] with the exception of operation at temperatures below $-100°C$ or after serious damage from nuclear particle radiation. Thus

$$V_{oc} = \frac{kT}{q} \ln \left(\frac{j_L A_L}{j_o A_j} + 1 \right)$$ (24.57)

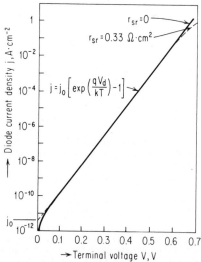

Fig. 24.18 The current-voltage characteristic of an ideal solar cell as described by Eqs. (24.53) and (24.59). Real solar cell characteristics are often roughly approximated by this presentation, but replacing kT with AkT as in Eq. (24.1), with A a constant chosen to provide a reasonable fit.

The ideal voltage factor defined by Eq. (24.7) can then be expressed as

$$(V. F.)_j = \frac{kT}{E_G} \ln\left(\frac{j_L A_L}{j_o A_j} + 1\right)$$ (24.58)

With Eq. (24.51) used in combination with Eq. (24.53) for determining the idealized limit of the voltage V_{mp} at which maximum power can be withdrawn from the solar cell, the structurally similar relationship is obtained:

$$V_{mp} = V_{oc} - \frac{kT}{q} \ln\left(\frac{qV_{mp}}{kT} + 1\right)$$ (24.59)

This equation lends itself to graphic or iterative solution. It is interesting to note that the quantity V_{mp} depends only on V_{oc} as an independent variable (at constant temperature), while V_{oc} is dominated by the light-generated current density j_L and the saturation current j_o. Thus, j_o merits special attention.

For an infinitely thick, diffused, n-type region adjacent to the depletion region, the saturation current contribution from this region is, according to Shockley,[40]

$$j_{o,D,\infty} = -q \frac{D_p}{L_p} \frac{n_i^2}{n_n}$$ (24.60)

$$\text{where } n_i^2 = 4\left(\frac{2\pi kT}{h^2}\right)^3 \left(m_n^* m_p^*\right)^{3/2} \exp\left(-\frac{E_G}{kT}\right)$$ (24.61)

contains the dominant part of the temperature dependence of the saturation current and consequently of the voltages in the current-voltage characteristic; m_n^* and n_p^* are the effective masses for electrons and holes, respectively. For silicon, $m_n^* = 1.1\, m_o$ and $m_p^* = 0.59\, m_o$, m_o being the electron rest mass. Exchange of all symbols and subscripts p and n in Eq. (24.60) yields the corresponding relationship for a p-type region.

In current solar cells, the cell surface is located at a distance from the space-charge region which is not large compared to the diffusion length, both in the diffused region and, particularly in thin cells (0.05 to 0.2 mm thick), in the base region. It is then necessary to take account of the surface recombination velocity. In many current cells, the surface recombination velocity is high everywhere, but in ultimate good cells, this should be the case only under the ohmic contact strip and the grid line, which by necessity have to exhibit a high surface recombination velocity.

A more detailed expression, applicable for the general case, can be written as

$$j_{o,D} = j_{o,D,\infty} F$$ (24.62)

where F is a factor based on the geometry and the surface properties of the region. It is expressed as

$$F = \frac{\sinh\left(\dfrac{x_T}{L_p}\right) + \dfrac{sL_p}{D_p}\cosh\left(\dfrac{x_T}{L_p}\right)}{\cosh\left(\dfrac{x_T}{L_p}\right) + \dfrac{sL_p}{D_p}\sinh\left(\dfrac{x_T}{L_p}\right)}$$ (24.63)

where x_T is, in general, the distance between the edge of the depletion region and the surface of the region from which the saturation current contribution is determined. The factor F becomes unity for $x_T/L_p \gg 1$. For $sL_p/D_p \gg 1$, F becomes $\coth(x_T/L_p)$, making Eq. (24.62) equal to the expression of Lindmeyer and Wrigley for the narrow-base diode.[42] On the other hand, for $sL_p/D_p \ll 1$, the factor F becomes $\tanh(x_T/L_p)$, which is less than 1 for small values of x_T/L_p (Fig. 24.19). Where the surface may be composed of different properties, such as the open surface and/or the contact and grid-line covered areas of the diffused region, Eq. (24.62) is to be expressed as

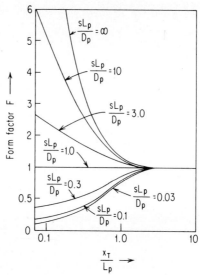

Fig. 24.19 Form factor F for the dark diode current as function of normalized diffused region thickness x_T/L_p with normalized surface recombination velocity sL_p/D_p as parameter.

$$j_{0,D} = j_{0,D,\infty} \frac{\sum\limits_{k} A_k F_k}{\sum\limits_{k} A_k} \qquad (24.64)$$

where A_k is the area of surface having a given factor F_k.

In most cases of real solar cells, additional effects exert an influence on the shape of the current-voltage characteristic, at least in some temperature and/or voltage ranges. These effects have their origin in or near the charge region, both within the bulk and at the surface of the device, in nonohmic behavior of the contacts, and in ohmic leakage paths past the junction.

It was noted early in semiconductor device development that many silicon-diode characteristics do not conform with the predictions of the diffusion theory, since they had $A > 1$ (Eq. 24.1) and j_0 values larger than predicted.[43] Sah et al. developed an analysis of junction characteristics which incorporated the effects of generation and recombination through a single trapping level in the depletion region.[44] This analysis yields A values as high as 2 for cases where recombination via deep traps inside the space-charge region dominates the forward current. Although it explains an effect which dominates the low forward-voltage portion of the I-V characteristic of many diode types as well as some silicon solar cells,[45] this analysis does not explain the characteristics observed on many silicon junction devices, including many solar cells, which exhibit A values greater than 2. Various phenomena have been suggested as the potential causes of these observed characteristics, including tunneling,[17] microplasma breakdown,[46] surface leakage, including metallic precipitates,[47] and surface channeling.[48] However, verification of these phenomena as the cause of the observations in individual solar cells has remained unsatisfactory.

In 1961, it was found[41] that the current-voltage characteristics of silicon solar cells, after elimination of the influences of series resistance and shunt resistance, are more accurately represented by a double exponential relationship of the form

$$j_d = j_{01} \left[\exp \left(\frac{qV_d}{A_1 kT} \right) - 1 \right] + j_{02} \left[\exp \left(\frac{qV_d}{A_2 kT} \right) - 1 \right] \qquad (24.65)$$

The first term on the right side represents Eq. (24.53), that is, diode current according to Shockley's diffusion theory, i.e., $A_1 = 1$, and j_{01} of proper magnitude and temperature dependence [Eqs. (24.50) to (24.63)]. The second term, with j_{02} generally 3 to 7 orders of magnitude larger than j_{01} and A_2 frequently greater than 2, has often been found difficult to explain by well-understood physical effects (Fig. 24.20). In a significant number of cases, however, a value

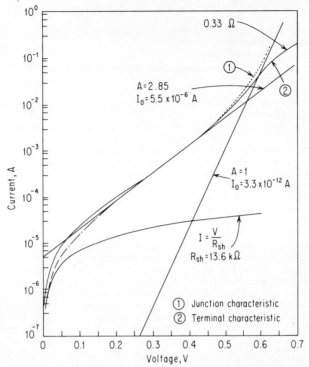

Fig. 24.20 The current-voltage characteristic of a real solar cell (heavy solid line), and the analytical component of which it is composed (light solid lines): the two exponential terms with different A and j_o values, the shunt resistance term, and the effect of series resistance. The dash-dot curve represents the characteristic without the influence of the shunt resistance, the dotted curve that without the influence of series resistance. (From Ref. 24.41.)

$A_2 \leq 2$ is found, and the second term on the right-hand side of Eq. (24.65) can be uniquely related, through all its characteristics, with recombination in the space-charge region of the pn junction.[49] In these cases, the center term on the right-hand side of Eq. (24.52) can be identified with this term of Eq. (24.65).

When other effects are simultaneously present, additional terms may have to be added to Eq. (24.65), once these effects can be analytically expressed. One of these is the voltage-proportional leakage currents which frequently alter the current-voltage relationship at the lower voltage levels. These leakage currents are usually expressed by addition of a term $(V - jr_{sr})/R_{sh}$ to either Eq. (24.53) or (24.65). This term represents the sources of these leakage currents as a lumped shunt resistance R_{sh} (Fig. 24.20).

Recently, interest in solar cells with Schottky barriers (metal-semiconductor junctions) has been renewed. The current-voltage characteristics of these cells follow Eq. (24.53), but their saturation current and its temperature dependence differ from Eq. (24.54) and Eqs. (24.60) and (24.61).

PERFORMANCE LIMITATIONS AND POSSIBLE IMPROVEMENTS

Although solar cells seem to be relatively simple semiconductor devices, containing only a single potential barrier, generally in form of a pn junction, a well-performing solar cell is a device which is finely tuned between optical and electronic material properties, and which is manufactured to much closer performance tolerances than most other semiconductor devices. This perfection could be obtained only through a good understanding of the device operation. On the basis of the analytical descriptions of the operating mechanisms in the solar cells, the dominant limitations on the conversion efficiency of photovoltaic solar energy converters can be listed, and their nature, their importance on the converter performance, and possibilities for their improvement can be discussed.[17]

The limitations on the efficiency of photovoltaic solar energy converters can be broken down into the following major factors:

1. Reflection losses on the surface.
2. Incomplete absorption of the photons available in the solar spectrum.
3. Utilization of only a part of the energy of the absorbed photons for the creation of electron-hole pairs.
4. Incomplete collection of the electron-hole pairs generated from photon absorption by the built-in potential barrier.
5. A voltage factor given by the ratio of open-circuit voltage to energy-gap potential difference.
6. A curve factor given by the ratio of maximum-power-point voltage times maximum-power-point current to open-circuit voltage times short-circuit current for an ideal pn junction.
7. Additional degradation of the current-voltage curve resulting from internal series resistance, shunt resistance, nonohmic contacts, depletion layer effects, surface effects, etc.

Factors 1, 2, and 4 combined form the "overall collection efficiency" defined in Eq. (24.12). All three of these factors are connected by the absorption characteristic of the material. Factors 1 through 4 together determine the amount of short-circuit current available from the device, while 5 to 7 describe losses based on the I-V characteristic.

Some of the seven factors governing the obtainable efficiency of photovoltaic solar energy converters are basic limiting factors, while others are mainly determined by manufacturing techniques, and improvement of these may make possible near elimination of the relevant factors' influence. Factors 1, 4, and 7 belong to this latter category, while factors 2, 3, 5, and 6 have absolute physical limitations beyond which improvement is not possible. It should be noted that the factors which have fundamental physical limitations are also partially influenced by fabrication technology, and in many instances better choices of parameters and improvements in techniques can be made in regard to these factors in order to approach the basic limits more closely. It has also been noted that, in order to make estimates of some of the basic limits, parameters have to be introduced into the calculations which are at least partially technique-influenced, such as the minority-carrier lifetime and mobility. Here, values have generally been used which appear reasonable by contemporary knowledge, but which may have to be revised as technology progresses.

Reflection Losses

The reflection losses given by the reflection coefficient $r(\lambda)$ as used in Eqs. (24.13), (24.16), and (24.22) are large on bare, clean semiconductor surfaces (see Fig. 24.10 for silicon). On practical solar cells, however, the reflection loss is reduced, by application of an antireflection coating, to a spectrally weighted average value near 7 percent, and to approximately 2 percent by additional "texturing" of the light-exposed surfaces of the cell. Texturing is forming a multicone structure on the surface by an etching process.[21]

Aside from its influence on reflection, optical refraction in the semiconducting material is not known to cause any deleterious effects in solar cell operation, at least not in devices prepared from a single semiconducting material. On the contrary, the high refractive index of silicon is utilized in the "nonreflective" cells having a textured light-exposed surface.

Incomplete Absorption

The discussions connected with Eqs. (24.17) to (24.19a) and Fig. 24.8 dealt with those photons in the solar spectrum which do not have enough energy ($E_{ph} < E_G$) to generate electron-hole pairs, and consequently are not absorbed in the semiconductor. Those discussions treated the effect in an idealized manner, assuming that all photons with $E_{ph} \geqslant E_G$, that is, right up to the absorption edge, are actually absorbed. This idealized view, which is closely approached in adequately thick wafers or layers, describes the fundamental part of the incomplete absorption. This fundamental part, however, is not only a function of the chosen material, through its energy gap, but is also determined by the detailed spectrum of the available sunlight. This spectrum is modified from that of sunlight in near-earth space by scattering and absorption in the earth's atmosphere, and thus is dependent on meteorological conditions. Consequently, the fraction of photons absorbed is normally referred to sunlight obtained under a standardized set of atmospheric conditions.

Outside of the fundamental part, there frequently is some additional incomplete absorption which depends on the thickness of the solar cell. This design-dependent part is more pronounced in semiconductors in which the absorption coefficient has relatively low values, e.g., below $10^4 cm^{-1}$, in a broader range of wavelengths below the "absorption edge" which corresponds to the energy gap [λ_G in Eq. (24.18)]. Semiconducting materials with such a broad range of low absorption coefficient values (see Si, Ge in Fig. 24.7) are said to have an "indirect band gap." Because of this property, solar cells from indirect-band-gap semiconductors must be made of thicker layers or wafers to achieve high performance than "direct-band-gap semiconductors." Silicon and aluminum arsenide belong to the former category, gallium arsenide and indium phosphide to the latter. Unity minus the "ratio absorbed" R_{abs}, defined by Eq. (24.23), expresses the nonideal losses due to incomplete absorption. As Fig. 24.9 for silicon indicates, these losses are a strong function of wafer thickness in a range of common thickness values in indirect semiconductors. The indirect semiconductors therefore are not appropriate for true "thin-film" solar cells.

Partial Utilization of the Photon Energy

A large number of the photons absorbed have more energy than necessary for the generation of an electron-hole pair. The energy needed for this photon-electron interaction is equal to the energy gap. This fact is illustrated in Fig. 24.8 and is expressed by Eq. (24.11), which includes the number of photons, rather than the photon energy, for all photons in the solar spectrum which have more energy than E_G. The excess energy of the photons contributes to lattice vibrations, meaning that it is *dissipated as heat*.

This situation persists for photons with energies up to $3 E_G$. Photons with higher energies lead to the generation of two electron-hole pairs per photon. The number of photons in the solar spectrum which have enough energy to cause this effect in the semiconductors of practical interest for solar cells is so small, however, that generation of two electron-hole pairs is generally not significant.

The considerations of the fundamental part of the incomplete absorption and of the partial utilization of the photon energy lead to a curve (Fig. 24.21) which gives, as a function of energy gap, that portion of the available solar energy which can be utilized in the generation of electron-hole pairs. One can see that the maximum is near 46 percent of total impinging energy at an energy gap of 0.9 eV. The same considerations also give the maximum possible number of electron-hole pairs generated by sunlight per square centimeter of exposed area per second, and the corresponding maximum light-generated-current density $j_{L_{max}}$, both as a function of energy gap of the semiconductor used. Figure 24.22 gives basic limits of these quantities which are completely independent of technique factors, and about which no assumptions about certain parameters have to be made for their evaluation.

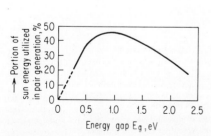

Fig. 24.21 The portion of the sun energy which can be utilized in electron-hole pair generation as a function of the width of the energy gap of the semiconductor.

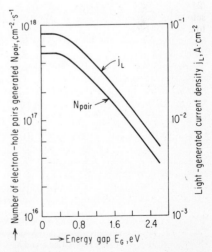

Fig. 24.22 The number N_{pair} of electron-hole pairs generated and the corresponding maximum theoretically possible light-generated current density jL versus energy gap.

Collection Losses

Most of the electron-hole pairs created by photon absorption are not generated within the space-charge region of the pn junction where the strong, built-in electric field effects the separation of the charges (generator action). Rather, the photons of high absorption coefficient (short wavelength) are absorbed in the layer between the surface and the space charge region (usually designated as the "diffused region"), and the photons of lower absorption coefficient are absorbed in the "base region" of the solar cell (Fig. 24.23).

The generation of electron-hole pairs alone does not change the electrical neutrality of a region, and they disappear by "recombination" at the end of a "minority-carrier lifetime" after their generation. If they are generated, however, within two to three "diffusion lengths" from the space-charge region (or "drift lengths" if an electric field dominates their movement), the minority carriers have a chance of reaching the space-charge region, being pulled across it by the built-in field, and being, as a majority carrier, no longer subject to recombination. Those minority carriers which recombine before reaching the space-charge region cause the collection efficiency [Eq. (24.16)] to fall below 100 percent.

The collection process is thus determined by material constants: the location of the pair generated by the absorption mechanism, and the diffusion and recombination by mobility and minority-carrier lifetime. The collection process is also determined by design: (1) the location of the space-charge region, (2) the resistivity, which influences mobility and lifetime, and (3) built-in electric fields. The processing methods applied in device fabrication can determine the material constants as well as the properties of the surfaces.

Because of the interplay of various effects, the collection efficiency should be expected to be strongly wavelength-dependent. This is illustrated in Fig. 24.24, which also shows the influence of a few important design and material parameters. It may be noted that the collection efficiency is well above 90 percent in the major part of the response spectrum in modern solar cells.

The Voltage Factor

The amount of energy utilized in the generation of electron-hole pairs is equal to the potential difference between the top of the valence band and the bottom of the conduction band, that

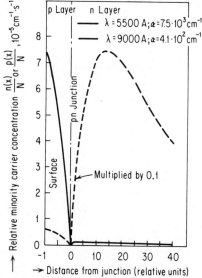

Fig. 24.23   Relative minority carrier distributions in the p- and np-layer at two different wavelengths of incident radiation: $\lambda = 5500$ Å, solid line; and $\lambda = 9000$ Å, dashed line. The distance from the surface to the pn junction was chosen as the unit for the abscissa. Note the change of scale on the abscissa at the p-n junction, as well as the compression by a factor of 10 of the hole-distribution in the n-layer for $\lambda = 5500$ Å. The distributions were calculated for cell #3-329, which had the following parameters: $x_1 = 2.8 \times 10^{-4}$ cm, $d = 5 \times 10^{-2}$ cm; $\tau_p = 4.2 \times 10^{-6}$ second; $\tau_n = 10^{-8}$ second; $\mu_p = 500$ cm²/V·s; $\mu_n = 80$ cm²/V·s; s $= 10^3$ cm s⁻¹.

np Solar cell
(computed)

No	x_j, μm	s, cm·s	τ_n, μs	t, mm	η_{coll}, AMO
1	0.4	10^5	3	0.3	0.71
2	0.1	10^5	3	0.3	0.79
3	0.2	10	3	0.3	0.83
4	0.2	10	12	1.0	0.88

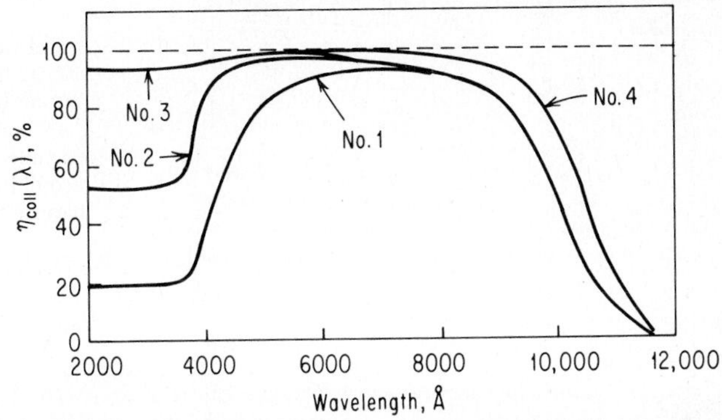

Fig. 24.24 Spectral collection efficiency η_{coll} (λ) of a silicon solar cell at various parameters of the diffused region.

is, the energy gap. The largest recoverable voltage, however, is *the open-circuit voltage which is always smaller than the energy gap.* The reason for this is twofold:

1. The barrier height, which is equal to the maximum applicable forward voltage on a *pn* junction, is determined by the difference in Fermi levels in the *n*- and *p*-type material on both sides of the junction. These Fermi levels are a function of impurity concentration and temperature in any given semiconductor, and are normally located inside the forbidden gap, so that the barrier height is less than the energy gap.

2. A voltage equal to the barrier height will only be obtained at extremely high injection levels which are orders of magnitude above those obtainable by photon absorption from direct sunlight.

Consequently, the open-circuit voltage falls generally between ½ and ¾ of the energy-gap potential difference.

The open-circuit voltage is primarily determined by the light-generated current I_L and the diode current I_d of the barrier [see Eqs. (24.51), (24.53), and (24.57)]. The series resistance of the cell is ineffective at open-circuit voltage ($j = 0$), so that the only other known effects to influence the open-circuit voltage are shunt resistance and other surface leakage, as well as punctures or other irregularities of the *pn* junction, such as may be caused by scratches, metal diffusion, etc. Since all of these effects are technologically determined, they can, in principle, be reduced to negligible levels. The diode current, in the region of the open-circuit voltage, is generally dominated by the diffusion current [first term in Eq. (24.65) with $A = 1$], since, in normal cell operation, the light-generated current density is high enough to cause the diode current contribution from recombination in the space-charge region to be negligible compared with the diffusion current in this voltage regime. Thus, the fundamental determinant of the open-circuit voltage is the diffusion current, whose magnitude, in turn, is determined by the saturation current [Eqs. (24.54), (24.60), and (24.61) to (24.64)]. Where the layer thickness is less than three diffusion lengths, the surface recombination velocity, which can also represent effects of an ohmic contact, can severely increase the saturation current and consequently lower the open-circuit voltage. This effect can be ameliorated by incorporation of properly designed drift fields which, when located adjacent to the back contact, result in the so-called BSF (back surface field) solar cell.

Equation (24.60) indicates that the saturation current resulting from minority-carrier diffusion should decrease proportionally to the inverse of the majority-carrier concentration (n_n or p_p) if $D_p/L_p eq \sqrt{D_p/\tau_p}$ were constant. Such a decrease of the saturation current and corresponding increase of the open-circuit voltage with increasing impurity concentration is actually observed down to resistivities of approximately 1 $\Omega \cdot$cm, where open-circuit voltages of approximately 0.6 to 0.62 V at the standard measuring temperature of 301 K are usually reached in silicon solar cells. At lower resistivities, an "apparent open-circuit voltage saturation" is observed whose origin is not yet fully clarified. Three effects are currently subjected to careful investigation so that this apparent open-circuit voltage saturation can be understood and perhaps overcome. These three effects are: (1) Auger recombination (involving one minority carrier and two majority carriers),[50] (2) general band-gap narrowing due to formation of an impurity band and smearing out of the otherwise sharp band edges,[51] and (3) band-gap narrowing in the depletion region only (Franz-Keldysh effect) because of the strong electric field existing there.[52] Without this saturation, open-circuit voltages well above 0.7 V could be expected, with corresponding efficiencies of 20 percent or higher in silicon solar cells at 301 K under normal, full sunlight.

As indicated by Eqs. (24.57), (24.60), and (24.61), the strong temperature dependence of the open-circuit voltage, which amounts to roughly -2.5 mV/K, results primarily from the saturation current which doubles approximately every 4.5 K.

Based on diffusion-current consideration only, and calculated for constant minority-carrier lifetime and diffusion constant, as well as impurity concentration, temperature, and irradiance, the open-circuit voltage exhibits an energy-gap dependence as shown in Fig. 24.25. Higher voltage factors can thus be expected, and generally are found, in semiconductors of larger energy gap.

The Curve Factor

The maximum power can be extracted from a photovoltaic device at that point where the largest rectangle can be inscribed into the current-voltage characteristic (Fig. 24.3). This point is given by the voltage and current values V_{mp} and I_{mp}. The ratio of the products $V_{mp} I_{mp}$ to $V_{oc} I_L$ is the fill factor [F.F., Eq. (24.8)]. The fill factor is influenced by the same effects which affect the voltage factor, plus the series resistance. In addition, recombination in the space-charge region in many cases also reduces the fill factor slightly. Abstracting from all these technology-determined influences which are expressed together in the artifacts factor [A.F., Eq. (24.9)] leaves the curve factor (C.F.), which again is determined by diode diffusion current [Eq. (24.53)]. Calculating the dependence of this idealized quantity (C.F.) on the energy gap under the same conditions as for the voltage factor leads to the curve in Fig. 24.25. Again, higher curve factors are obtained at larger energy gaps.

The product of the curve factor and the voltage factor has been called the "characteristic factor," which is also displayed as a function of energy gap in Fig. 24.25.

Series Resistance Losses

In actual silicon solar cells, the internal series resistance is large enough to cause a significant deviation of the current-voltage characteristic from its ideal curve. Contributions to this series resistance come from the contacts to the semiconductor n- and p-type regions, from the current flow through the bulk of the wafer (the base region), and, usually predominantly, from the lengthwise current flow in the thin diffused region whose thickness may be as low as $0.1\ \mu$m. To eliminate most of the series resistance of this layer, a metal grid contact structure (Fig. 24.26) is usually applied. However, it reduces the active area of the cell, since it is opaque to light. An optimization of such a grid contact structure for highest cell performance generally

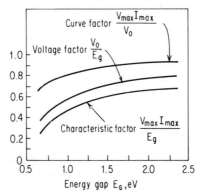

Fig. 24.25 Voltage factor, curve factor, and characteristic factor versus width of energy gap.

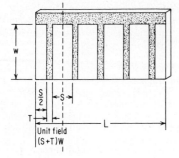

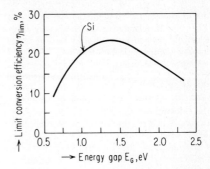

Fig. 24.26 Configuration of a contact grid structure.

Fig. 24.27 The limit conversion efficiency as a function of the width of the energy gap for single *pn* junction direct transition solar cells.

involves, after the minimum technically and economically desirable grid-line width has been chosen, the computation of grid-line spacing which provides the best compromise between the two performance losses due to series resistance and to active-area coverage, for a given set of solar cell operating conditions.

CONVERSION EFFICIENCY LIMITS

Combining the fundamental effects which determine solar cell performance, using Eq. (24.10) with $G(A) = 1$, j_L determined with $\eta_{coll} = 1$, and $F(J)$ obtained for the idealized, diffusion-current-dominated *pn* junction with optimized parameters, leads to the "limit conversion efficiency." This limit conversion efficiency, based on Fig. 24.25, is shown as a function of energy gap in Fig. 24.27. The limit conversion efficiency curve shows a fairly broad maximum at energy gaps between 1.1 and 1.6 eV. The band gap for silicon is at the lower edge of this range. The expected maximum-limit conversion efficiency for a *pn* junction device from a single semiconducting material is near 24 percent. It may be observed that the calculation of $F(J)$ includes some material parameters such as minority-carrier lifetime and mobility, which are technology-dependent and could possibly change with improvements in the state of the art. The limit conversion efficiency is also a function of device operating temperature (Fig. 24.28), with the

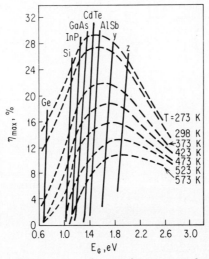

Fig. 24.28 Maximum efficiency η_{max} as a function of energy gap E_G for various temperatures.

maximum shifting to higher energy gaps with increasing temperature (Ref. 18, Loferski). In 1970, the performance of contemporary silicon solar cells was analyzed and compared with the idealized parameters as shown in Fig. 24.29.[53] Through this comparison, several potential improvements could be identified, which also are indicated in Fig. 24.29.

One of these improvements has been connected with the collection efficiency and is illustrated in Fig. 24.24, where curve 1 indicates the state of the art in 1970 and curves 3 and 4 combine the predicted possible improvement. This collection efficiency improvement was expected primarily from a large increase in the short-wavelength spectral response, obtainable by reducing the thickness of the diffused region and by reducing the influence of recombination at the surface of this region. These improvements have been made in the development of the "violet" cell,[54] which consequently required the development of a better antireflection coating and of a finer grid-line structure to reduce the series resistance originating from the increased sheet resistance of the diffused region. In the subsequent development of the "black" or nonreflective cell,[21] by Comsat Laboratories, the cell surface was shaped into a cone structure by an etching process, which reduced the optical reflectance to 3 percent, from approximately 7 percent in the violet cell. Simultaneously, because of refraction at the surface, the photons penetrate the wafer more obliquely, thus enhancing the long-wavelength absorption and collection. With these additional improvements, the originally set goals for collection efficiency and reflection losses have been met (Table 24.1). Further improvements in this area can thus only be of lesser significance.

The suggestion of improved open-circuit voltage, through use of lower-resistivity base material, has found considerable interest, without, however, having yielded significant results to date. It has not yet been clarified to what degree basic limitations in material properties, or process-induced effects are responsible for the observed difficulty in obtaining open-circuit voltages significantly above 0.6 V at 28°C. A considerable number of potentially effective mechanisms could cause the apparent open-circuit voltage limitation. Serious work toward understanding the potential influence of these mechanisms has been in progress for 3 to 4 years, and several of the effects have already been eliminated as likely causes of the observed voltage saturation limitations.

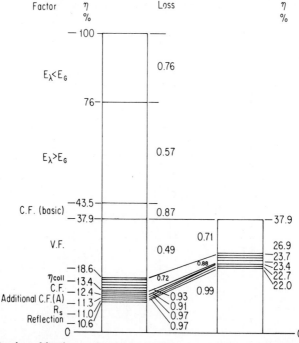

Fig. 24.29 Bar chart of distribution of energy losses of the present and the potential improved silicon solar cell.

TABLE 24.1 Performance Status of Current Silicon Solar Cells and Improvement Goals

Attribute	Goal	Efficiency contribution (1-Loss) 1970 comm'l cell	"Violet" cell	"Black" cell		Output values Goal	Black cell	Units
Basic losses	0.45	0.45	0.45	0.45	⎫ Light-Generated Current	63.0	63.0	mW/cm^2
Collect. eff.	0.88	0.71	0.79	0.88		49.0	49.0	mA/cm^2
Reflection	0.97	0.905	0.951	0.97		47.5	47.7	mA/cm^2
Grid-line cover	0.96	0.96	0.95	0.95	⎭	45.5	45.4	mA/cm^2
Voltage factor	0.61(?)	0.522	0.535	0.531	⎫ Power	0.675	0.591	V
Curve factor	0.86	0.82	0.825	0.822		26.3	22.1	mW/cm^2
Add'l curve fact.	1.0	0.91	1.0	0.99		26.3	21.8	mW/cm^2
Series resistance	0.97	0.96	0.985	0.984	⎭	25.6	21.4	mW/cm^2
Efficiency (AMO)	0.19	0.104	0.14	0.153				

The suspected effects can be located in either the base, the space-charge, or the diffused region of the cell. They can be caused by bulk minority-carrier recombination, by band-gap narrowing, surface recombination, or even the selected device structure. The principal basic phenomena which may be involved have been discussed in connection with the voltage factor. While these basic effects are being investigated, new design compromises are being developed which result in gradually increasing conversion efficiencies. With the achievement of 15.5 percent average-production-cell efficiency (air mass zero, space sunlight) recently announced, near-term improvements to 18 percent air-mass-zero conversion efficiency (20 percent for air mass 1, terrestrial sunlight) are being predicted. Simultaneous with these performance improvements, efforts are under way to reduce the mass of spacecraft solar cells by decreasing the cell thickness to as little as 50 μm (0.002 in.), and to reduce the cost of encapsulated modules for terrestrial applications from the current approximately \$12/W (peak) to \$2/W (peak) by 1982 and \$0.5 /W (peak) by 1986. Significantly advanced, automated production methods, based at least in part on newly developed technologies, are expected to provide these price reductions, while increasing the production rates by 3 orders of magnitude. These new solar cell technologies differ from those of the conventional semiconductor device industry primarily in the need to handle much larger volumes of material at high speed and low cost, while the requirements on material purity and perfection may not be significantly lower.

While the silicon solar cell will continue to be a strong contender in the flat-array market, the gallium arsenide cell with a wide band-gap window layer has recently achieved 25 percent conversion efficiency at high optical concentration ratios, and will be the likely candidate for applications with high-ratio optical concentrators.

REFERENCES

1. Becquerel, E., *Compt. Rend.*, **9**: 561 (1839).
2. Smith, Willoughby, *Nature*, Feb. 20, 1873, and *J. Soc. Tel. Eng.*, **2**: 31 (1873).
3. Adams, W. G. and Day, R. E., *Proc. Royal Soc.*, **A25**: 113 (1877).
4. Fritts, C. E., *Proc. Am. Assoc. Adv. Sci.*, **33**: 97 (1883), *Am. J. Science*, (3) **26**: 465 (1883).
5. Fritts, C. E., *La Lumière électrique*, **15**: 226 (1885).
6. Hallwachs, *Phys. Zeitschr.*, **5**: 489 (1904).
7. Kennard, E. H. and Dieterich, E. O., *Phys. Rev.*, **9**: 58 (1917).
8. Arondahl, L. O., and Geiger, P. H., *Trans. AIEE*, **46**: 357 (1927).
9. Schottky, W., and Deutschmann, W., *Phys. Zeitschr.*, **30**: 839 (1928).
10. Schottky, W., *Zeitschr. Techn. Phys.*, **11**, 468, and 913 (1930).
11. Lange, B., *Phys. Zeitschr.*, **31**: 964 (1930).
12. Lange, B., *Phys. Zeitschr.*, **32**: 850 (1931).
13. Billig, E., and Plesser, K. W., *Phil. Mag.*, **40**: 568 (1948).
14. Kingsbury, E. F., and Ohl, B. S., *BSTJ*, **31**: 802 (1952).
15. Pearson, G. L., *Bell Lab. Rec.*, **32**: 232 (1954). Chapin, D. M., Fuller, C. S., and Pearson, G. L., *J. Appl. Phys.*, **25**: 676 (1954).
16. Chapin, D. M., Fuller, C. S., and Pearson, G. L., *Bell Lab. Rec.*, **37**: 241 (1955).
17. Wolf M., "Limitations and Possibilities for the Improvement of Photovoltaic Solar Energy Converters," *Proc. IRE*, **48**: 1246–1263 (1960).
18. Wysocky, J. J., and Rappaport, P., *J. Appl. Phys*, **32**: 171 (1960). Loferski, J. J., "Principles of Photovoltaic Solar Energy," *Record of the 10th IEEE Photovoltaic Specialist Conference*, 1–4 (November 1973), IEEE Cat. No. 73 CH 0801-ED.
19. Stirn, R. J., private communication.
20. Wolf, M., and Prince, M. B., "New Development in Silicon Photovoltaic Devices and Their Application to Electronics," in *Solid State Physics*, vol. 2: *Semiconductors*, Academic Press, New York, 1960 part 2, pp. 1180–1196.
21. J. Haynos, et al., "The Comsat Non-reflective Silicon Solar Cell: A Second Generation Improved Cell," in *Proc. of the International Conference on Photovoltaic Power Generation*, Hamburg, Sept. 25–27, 1974, pp. 480–486; DLGR, Köln, Germany, 1974.
22. Palz, W., Bessen, J., and Duy, T. N., "Review of CdS Solar Cell Activities," *Record of the 10th IEEE Photovoltaic Specialist Conference*, pp. 69–76 (Nov. 13–15), IEEE Cat. no. 73 CH 0801-ED. Boer, K. W. et al., "Recent Results in the Development of CdS/Cu$_2$S Thin Film Solar Cells," ibid., pp. 77–84. Lebrun, J., "A New CdTe Thin Film Solar Cell,"; *Record of 8th IEEE Photovoltaic Specialist Conference*, pp. 30–39, Aug. 4–6, 1970, IEEE Cat. no. 70 C 32 ED.
23. Wronsky, C. R., Private communication.
24. Cherry, W. R., "The Generation of Pollution-free Electrical Power from Solar Energy," *Transactions of the ASME, Journal of Engineering for Power*, A94 (2): 78–82 (April 1972).
25. Altman, M., Telkes, M., and Wolf M., "The Energy Resources and Electrical Power Situation in the United States," *Energy Conversion*, **12**: 53–64 (1972).

26. Wolf, M., "Cost Goals for Silicon Solar Arrays for Large Scale Terrestrial Applications," *Record of the 9th IEEE Photovoltaic Specialist Conference*, pp. 342–350, May 2–4, 1972, IEEE Cat. no. 72 CHO 613-0-ED.

27. *Proceedings of the UN Conference on New Sources of Energy*, vol. 5, United Nations, New York, 1964.

28. Bruckner, A. et al., "Economic Optimizing of Energy Conversion with Storage," Paper no. 68CP134-PWR, presented at IEEE Winter Power Meeting, Jan. 28–Feb. 2, 1968, New York, N.Y.

29. Shinbrot, C. H., and Tonnelli, A. D., "Advanced Deployable Solar Cell Battery Power System Development," *Proceedings of 1972 Intersociety Energy Conversion Engineering Conference*, pp. 152–162, 1971.

30. Weber, N., and Krummer, J. T., "A Sodium-Sulfur Secondary Battery," *Proceedings of 1967 Intersociety Energy Conversion Engineering Conference*, pp. 913–916, 1967.

31. Wolf, M., "The Solar Cell Design Handbook," *Proc. 9th IEEE Photovoltaic Specialist Conference*, pp. 53–60, IEEE Cat. no. 72 CHO 613-U-ED.

32. Kleinman, D. A., "Considerations on the Solar Cell," *Bell Syst. Tech. J.*, **40:** 85–115 (January 1961).

33. Kaye, S., and Rolik, G. P., "Optimum Bulk Drift-Field Thickness in Solar Cells," *IEEE Trans. ED* **13:** 563–570 (July 1966).

34. Bullis, W. M., and Runyan W. R., "Influence of Mobility and Lifetime Variations on Drift-Field Effects in Silicon-Junction Devices," *IEEE Trans. ED* **14:** 75–81 (February 1967).

35. Van Overstraeten, R., and Nuyts, W., "Theoretical Investigation of the Efficiency of Drift-Field Solar Cells," *IEEE Trans. ED* **16:** 632–641 (July 1969).

36. Hauser, J. R., and Dunbar, P. M., "Performance Limitations of Silicon Solar Cells," *IEEE Trans. ED* **24:** 305–321 (April 1977).

37. Wolf, M., "Drift Fields in Photovoltaic Solar Energy Converter Cells," *Proc. IEEE* **51:** 674–693 (May 1963).

38. Shockley, W., *Electrons and Holes in Semiconductors*, Van Nostrand, New York, 1950.

39. Ibid., p. 300.

40. Shockley, W., *BSTJ*, **28:** 435–489 (1949).

41. Wolf, M., and Rauschenbach, H., "Series Resistance Effects on Solar Cell Measurements," *Adv. Energy Conversion* **3:** 455–479 (1963).

42. Lindmeyer, J., and Wrigley, C. Y., *Fundamentals of Semiconductor Devices*, Van Nostrand, Princeton, NJ, 1965, pp. 29–34.

43. Pfann, W. F., and Van Roosebroeck, W., *J. Appl. Phys.*, **25:** 1422 (1954).

44. Sah, C. T., Noyce, R. N., and Shockley, W., *Proc. IRE*, **45:** 1228 (1957).

45. Stirn, R. J., in *Proc. 9th IEEE Photovoltaic Specialist Conference*, pp. 72–82, IEEE Cat. no 72 CHO 613-0-ED, 1972.

46. Goetzberger, A., and Stephen, C., *Bull. Amer. Phys. Soc.*, **II, 6:** 106 (1961).

47. "A Study of Photovoltaic Solar Cell Parameters," Final Rep., Contract AF33(616)-7785, Rep. ASD-TDR-62-776, 1962.

48. Kirkpatrick, A. R., et al., NASA Rep. NAS 2-S516, vol. 1, 1970.

49. Wolf, M., Noel, A. T., and Stirn, R. J., "Investigation of the Double Exponential in the Current-Voltage Characteristic of Silicon Solar Cells," *IEEE Trans. Electron Dev.*, **24:** 419–428 (1977).

50. Fischer, H., and Pschuuder, W., "Impact of Material and Junction Properties on Silicon Solar Cell Efficiency," *Proc. 11th IEEE Photovoltaic Specialist Conference*, pp. 25–31, IEEE Cat. no. 75 CHO 948-0-ED, May 1975.

51. Lindholm, F. A., Li, S. S., and Sah, C. T., "Fundamental Limitations Imposed by High Doping on the Performance of PN Junction Silicon Solar Cells," ibid., pp. 3–12.

52. Rittner, E. S., "Improved Theory of the Silicon p-n Junction Solar Cell," *J. Energy*, **1:** 9 (January 1977).

53. Wolf, M., "A New Look at Silicon Solar Cell Performance," *Energy Conversion*, **II:** 63–73 (1971).

54. Lindmayer, J., and Allison, J., "An Improved Silicon Solar Cell—The Violet Cell," *Proc. 9th IEEE Photovoltaic Specialist Conference*, pp. 83–84, IEEE Cat. no. 72 CHO 613-0-ED, May 1972.

Chapter 25

Energy From Biomass

ANNE FEGE
Fege Forestry Consultants,
Silver Spring, MD

Since humans first existed as a distinct species, they have used biomass, or plant and animal materials, for various energy forms including food, shelter, clothing, and fuel. Today, about two-thirds of the world's population still uses biomass for their heating and cooking. Even now, when fossil fuels supply many of our chemical feedstock, shelter, clothing materials, and energy requirements, we are still dependent on plants—for fossil fuels are actually the remains of plant tissues which were converted to oil, coal, and natural gas over millions of years.

Obtaining energy from biomass has enjoyed renewed public interest and an increasingly widespread use, as the prices of fossil fuels rise. The use of biomass for energy offers significant advantages over the nuclear and fossil energy sources which Western countries have depended on during the past generation. Biomass is renewable, can be produced in most regions of the world, has negligible amounts of sulfur (resulting in much less air pollution than using coal), and has no major disposal problems.

Obtaining energy from biomass may be extremely simple, or very complex. A family may chop five cords of wood each fall from their woodlot, and use it in a Franklin stove during the winter for heating one room in their home. In the future, a commercial enterprise may grow biomass on a large energy farm, harvest the material, convert it into methanol, mix it with gasoline, and sell it as a transportation fuel. Regardless of the system, there are still three components of obtaining energy from biomass: (1) growing the biomass, (2) transporting it to the conversion site, and (3) converting it into a form of energy which man can use.

This handbook chapter will describe the basis for the conversion of solar energy into the chemical energy content of plants—the biomass resource base—the process for converting biomass to useful fuels, and one promising future system for increasing the amount of energy supplied by biomass. Most examples have been derived from United States experiences, but the concepts and potentials for obtaining energy from biomass may be applied in many parts of the world.

BIOMASS CONVERSION OF SUNLIGHT INTO CHEMICAL ENERGY

Green plants have a unique capability of capturing the energy of the sun's light and converting carbon dioxide and other inorganic molecules into chemical bond energy. Plants convert carbon dioxide and water into simple sugars by a series of chemical reactions called photosynthesis.

These sugars are used by the plant for (1) maintenance, (2) growth, and (3) reproduction. Both the rate at which the plant or biomass (or total living material) is increased and the allocation of the biomass to various parts of the plants are important for the growth of plants for food, fiber, chemicals, and energy.

Photosynthesis

In photosynthesis, green plants absorb the sun's energy and transform it into chemical energy. This is a two-step process: (1) in the light reaction, photo-oxidation of water releases oxygen and hydrogen and supplies energy which is stored in the form of adenosine triphosphate (ATP); and (2) in the dark reaction, the ATP energy is used to reduce carbon dioxide to form simple six-carbon sugars. The material and energy balance for photosynthesis is given by the equation:

$$6CO_2 + 12H_2O \xrightarrow{675 \; kcal} C_6H_{12}O + 6O_2 + 6H_2O$$

The photosynthetic mechanism in most plants is housed in the chloroplasts, which are cell organelles found primarily in the leaves of green plants. They consist of a series of folded membranes containing quantosome units and ribosomes. The quantosome units, which are responsible for the light reaction in photosynthesis, contain chlorophyll and are attached to these membrane surfaces. Ribosomes, which are also attached to membranes, contain the DNA or genetic material responsible for producing the enzymes which reduce carbon dioxide to form sugars in the dark reaction.

The sugars produced during photosynthesis and other molecules synthesized by the plant from these sugars are used by the plant cell either to maintain itself (respiration) or to increase its size and ultimately to divide in two. New cells increase the biomass and begin their own photosynthesis. The sugars and other molecules can also be secreted into the vascular system of higher plants and transported to other parts of the plant. In this way, the photosynthetic mechanism of the leaf provides chemical energy for the branches, stems, and roots to survive and grow.

Primary Productivity

The rate at which plants produce biomass in excess of their respiratory requirements is called net primary productivity. The amount of biomass which is present at a given time or is harvested is called the "standing crop" or net assimilation of organic matter. The standing crop or yield of agricultural crops or forests is usually measured at the end of a year or growing cycle, and thus is a measure of the net primary production over that time period. Only a small fraction of the sun's energy incident on a plant is actually converted into plant biomass (as shown in Table 25.1).

Man has increased net primary production substantially by supplying energy subsidies in the form of water, fertilizers, cultivation, disease and pest control, and improved seeds. These energy subsidies have primarily been provided in the form of fossil fuels, and are responsible for 10- to 40-fold increases in net primary production per hectare (Table 25.2).

Two interacting, complex factors determine total biomass and productivity: the genetic make-up of the plant, and the site environment. Within and among plant species, there may be

TABLE 25.1   Relationship between Solar Energy and Primary Productivity

	Theoretical maximum, %	Average favorable conditions, %	Average for biosphere, %
Total solar energy outside atmosphere	100	100	100
Energy reaching earth's surface	50	50	< 50
Gross primary production	5	1	0.2
Net primary production	4	0.5	0.1

SOURCE: Ref. 15.

TABLE 25.2   Edible Portion of Net Primary Production per Hectare

Level of agriculture	kg dry matter/hectare/yr
Food gathering agriculture	0.4–20
Agriculture without energy (fuel) subsidy	50–2000
Energy-subsidized grain agriculture	2000–20,000

SOURCE: Ref. 15.

considerable variation in the rate of photosynthesis, rate of growth, distribution of biomass among plant parts, and response to environmental stresses. A wide range of environmental factors affect total plant biomass: availability of nutrients and water, length of growing season, incident radiation reaching the plant, disease, and competition from other plants.

Allocation of Plant Biomass

The products of photosynthesis, or sugars, are translocated in the plant primarily to the actively growing areas, which are the leaf meristem, root meristem, vascular cambium, and reproductive tissues. The allocation of these photosynthetic products among the various growing parts at different times in the life cycle is determined by the genetic make-up of the plant and by hormones which are synthesized by various plant parts. Thus, during plant development from a seed to a seed-producing mature plant, growth and differentiation result in an orderly formation of leaves, roots, stems, and flowers.

Juvenile plants allocate most of their photosynthetic material to the leaves and roots, in order to increase the leaf area, height, and root system. As woody plants grow older, the vascular cambium activity increases, and the diameter of the stem is increased. At some point during the growing season or life cycle of the plant, the reproductive system is activated, and the plant produces either vegetative offspring or flowers and then seeds.

In using the total biomass of a plant for energy, the allocation of the biomass among various plant parts is relatively unimportant. In contrast, foresters have historically been most interested in the amount of biomass in the saleable bole of the tree [that part of the stem greater than 5 in (12.5 cm) in diameter], and agronomists have been interested in specialized plant parts, such as wheat seeds, corn kernels, potato tubers, or the fleshy fruit surrounding the seed.

Most plants exhibit their highest rates of growth during the first few months of the growing season, or during the first few years. Therefore, the highest net primary production or annual yield of a plant is obtained by harvesting it during the period of its rapid growth, before the rate of growth for that plant begins to decrease. The short-rotation concept of harvesting trees after 4 to 6 years, instead of after 20 to 60 years as is done in commercial forestry, takes advantage of this principle.

BIOMASS RESOURCE BASE

The current and future contribution of biomass to energy supplies will be based on the following systems: (1) growing promising terrestrial and aquatic plant species on energy farms, solely for their conversion to energy; (2) collecting forest, agricultural, and animal residues for energy; and (3) harvesting forest trees which are not suitable for lumber, paper, or other forest products. Table 25.3 summarizes the characteristics of these biomass resources.

The primary resources for growing plants on terrestrial energy farms are arable land and high-yielding plant species. Similarly, future aquatic energy farms would require fresh-water or marine water resources and high-yielding plant species. Other resource bases are forest, agricultural and animal residues, and existing noncommercial forests.

Land and Water Availability

The most important issue facing the production of terrestrial energy crops is the availability of land and water resources to be allocated to such enterprises. Factors which affect this availability are (1) the potential of land in various regions to grow energy; (2) competing demands for the land; and (3) future trends in the supply and demand for land.

The Soil Conservation Service (SCS) of the U.S. Department of Agriculture has developed

TABLE 25.3   Characteristics of Biomass Resources

Type of biomass	Theoretical availability, 10^6 dry tons†	Practical availability, 10^6 dry tons	Regional limitations
Terrestrial energy crops	206,000	100–300	Total United States
Aquatic energy crops	*	*	Coastal areas, open oceans, lakes, and ponds
Agricultural residues	322	100	Follows crop patterns
Forest residues	116	100	Follows crop patterns
Animal manures	36	26	Follows crop patterns
Standing vegetation	1000	20	Forested lands

*Long-term resource, cannot realistically project availability.
†1 ton = 0.91 metric ton.
SOURCE: Ref. 2.

a land classification system with eight broad groupings, or capability classes, as summarized in Table 25.4. These range from SCS Class I lands which have few limitations that restrict their use, to SCS Class VIII, which cannot be used for commercial agriculture or forestry. The limiting factors will be designated as subclasses by adding a small letter e, w, w, or c to the class number indicating, respectively, limitations due to erosion potential, wetness, poor soil quality, or climate. In general, land suitable for terrestrial energy farms would be that land in SCS Classes I to IV which is not currently used for food production.

Of the 2.3 billion acres (0.9 billion hectares) of land in the United States, 78 percent or 1.8 billion acres (0.7 billion hectares) are devoted to cropland, forestland or grassland. Figure 25.1 depicts the major uses of land in the United States in 1969. There were 268 million acres (107 million hectares) of SCS Classes I to IV in permanent pasture, forest, and range in 1969. It may be reasonable to expect that 10 percent of this land could be used for terrestrial energy farms, given changes in the economic, legal, institutional, and political factors which govern the use of land.[9]

The uses of land in the United States can change substantially over time. The most important current trends are the conversion of cropland to urban and suburban uses, drainage and irrigation of agricultural land, and the abandonment of cropland to pasture or forests. From 1967 to 1975, over 100 million acres (40 million hectares) changed use, as economic, legal, institutional and political factors operated.[26] Thus, it is clear that substantial acreages of land could be diverted over a short period of time to such a new use as the production of biomass for energy.

Water will be the limiting factor in many regions of the United States, and irrigation would be required during the entire growing season at many sites west of the 95° meridian. The

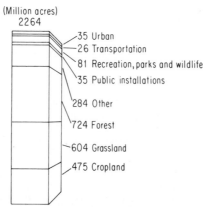

(Million acres)
2264

35 Urban
26 Transportation
81 Recreation, parks and wildlife
35 Public installations
284 Other
724 Forest
604 Grassland
475 Cropland

Fig. 25.1   Major uses of land, 1969, for all 50 states. Grassland includes some acreage not in farms—mostly federal land.

TABLE 25.4 Soil Conservation Service Land Classification System and Acreages

SCS class	Agriculture	Pasture	Watershed	Forestry	Urban, industrial	Range	Wildlife	Recreation	Total acreage in the United States,* 1000 acres†
I	x								46,868
II	x	x							287,167
III	x	x	x						296,735
IV	x	x	x						180,441
V			x	x	x	x			32,125
VI			x	x	x	x	x		277,585
VII			x	x	x	x	x	x	287,100
VIII			x		x		x	x	29,972
Noninventoried federal noncropland									759,602
Urban and built-up land									61,119
Water									71,117
Total acreage in the United States									2,268,215

*Includes Alaska, Hawaii, Puerto Rico, and Virgin Islands. Column does not add to total because of rounding.
†1 acre = 0.4 hectare.
SOURCE: Adapted from Ref. 7.

competition for water will continue to increase in the water-limited Western states, and may restrict the establishment of both terrestrial and fresh-water energy farms there. Many regions have an adequate annual rainfall for growth of plants for biomass, but it may be distributed unevenly and include drought periods during the growing season. Successful establishment of many hardwood species may require irrigation during those periods in the first few years. Careful examination of species which do not require irrigation for establishment must be made in those areas where irrigation is infeasible or uneconomic.

Aquatic energy farms could be established as (1) open-ocean systems (2) coastal systems, and (3) land-based fresh-water systems. There are no foreseeable limitations on the availability of the open oceans, except national and international laws and treaties which may be imposed in the future on the use of the oceans. Coastal systems would be subject to the extensive state and federal regulations on the development of wetlands and estuaries, and therefore this resource should be considered extremely limited. Fresh-water systems established in natural or man-made ponds and lakes would require both substantial land and water resources. Areas with low land prices and high incident solar radiation, such as the deserts of the Southwest, have been suggested for aquatic energy farms, but water is extremely scarce there and evaporation rates are high.

Promising Species for Biomass Production

The identification of species for biomass energy farms must include a search of plants which are currently cultivated for food or fiber, plants which only grow in natural systems and have not yet been exploited by man, and plants which are not grown in the United States, but which have high yields in other parts of the world. For comparison, the yields of various plant species and systems are summarized in Table 25.5.

Food and fiber crops   Many of the same species which have historically been cultivated for lumber or fiber production would be suitable for biomass energy farms, particularly if yields were increased by more intensive management. Table 25.6 presents the estimated annual production from species which could be considered for intensively managed short-rotation energy farms.

Similarly, some forage and crop species may have high enough yields per acre that it would be economical to cultivate and harvest them for energy. Table 25.7 summarizes the average yields which were obtained from some agricultural crops in 1976. For the most part, these plants have been selected and genetically improved to maximize the yield of the edible plant part, such as soybeans or corn kernels. Hybrids or varieties which maximize total biomass yield instead of the yield of the edible plant parts may be promising species for biomass energy farming.

Aquatic plant species   Marine and fresh-water plants have been grown for thousands of years for food, spices, fertilizers, medicines, and animal fodder. These plants include primarily the algae (blue-green, green, red, and brown) and the aquatic seed-bearing angiosperms. Productivity rates measured for these plants have been extremely varied, because growth rates depend on light, nutrient availability, temperature, and other factors. The optimal conditions provided during laboratory conditions cannot be achieved in large-scale commercial operations. Yields of promising marine and fresh-water plants are presented in Table 25.8.

New plant species   Plants which have not previously been valuable to man for food, fiber, chemical feedstocks, and other uses may offer promise for biomass energy farming. Kenaf, cattails, and giant reeds have received particular attention recently as species which should be considered. Because there are few reliable yield measurements for these plants in the literature, comprehensive species screening programs are required to identify species which have the highest net primary productivity rates when planted in various regions of the United States and the world. Subsequently, these plants could be cultivated and genetically improved, just as food and fiber crops have been.

Residues and Existing Biomass Resources

Forest and agricultural operations leave significant quantities of residues on the land after harvesting operations, residues which could be collected and converted into energy. In addition, there are underutilized forest stands in many regions of the United States that are currently not economical to harvest. With the development of an energy market for both residues and trees which are currently "noncommercial," significant quantities of these biomass resources could be economically collected and converted to usable energy forms.

TABLE 25.5 Aboveground Dry Biomass Yields and Energy Equivalents of Selected Plant Species or Systems

Species	Location	Tons/acre/year	Metric tons/hectare/year	10^7 Btu/acre/year*	10^7 kcal/hectare/year*
Exotic forage sorghum	Puerto Rico	30.6	68.7	46	29
Forage sorghum (irrigated)	Kansas	12	27	18	11
Sweet sorghum	Mississippi	9	20	14	8
Exotic corn (137-day season)	North Carolina	7.5	17	11	7
Hybrid corn	Mississippi	6	13	9	6
Kenaf	Florida	20	45	30	19
Water hyacinth	Florida	16	36	24	15
Sugarcane (state average)	Florida	17.5	39.3	26	16.3
Sugarcane (10-yr average)	Hawaii	26	58	39	24
Sugarcane (5-yr average)	Louisiana	12.5	28.1	19	11.7
Sudan grass	California	16	36	24	15
Alfalfa	New Mexico	8	18	12	7
Bamboo	Southeast Asia	5	11	7	5
Hybrid poplar (short rotation) 3 years old	Pennsylvania	8.7	20	13	8.1
American sycamore (coppice crop) 2 years old	Georgia	3.7	8.3	6	3.5
Black cottonwood 2 years old	Washington	4.5	10	7	4.2
Red alder (1–4 years old)	Washington	10	22	15	9.3
Eucalyptus spp	California	13.4	30.1	20	12.5
Algae (fresh-water and pond culture)	California	39	88	58	36
Tropical rain forest complex (average)		18.3	41.2	27	17.1
Subtropical deciduous forest complex (average)		10.9	24.5	16	10.2
World's oceans (primary productivity)		6	13	9	6

*Assuming 7500 Btu/lb.
SOURCE: Adopted from Ref. 1.

TABLE 25.6 Biomass Productivity of Short-Rotation Species

Species	Estimated annual biomass production, tons/acre/year*	Estimated rotation age, yrs
American sycamore	6	4
Eucalyptus	12	6
Loblolly pine	5	10
Populus spp		
Eastern cottonwood	5	9
Black cottonwood	6	4
Hybrid poplars	10	8–10
Red alder	9	4

*Oven-dry tons.
SOURCE: Ref. 9.

TABLE 25.7 Agricultural Crop Yields, 1975

	Oven-dry* residue factor bushels to tons	1000 acres harvested	Yield per acre grain bushels/acre	Ton/acre	Residue ton/acre
Wheat	0.0546	70,824	30.3	0.92	1.67
Rye	0.0504	814	22.0	0.62	1.11
Rice	0.0572/cwt	2,802	4,555 lb	2.28	2.61
Corn	0.031	66,905	86.2	2.41	2.67
Oats	0.0433	13,650	48.1	0.77	2.08
Barley	0.0546	8,711	44.0	1.06	2.40
Grain sorghum	0.0176	15,484	49.0	1.37	0.86
Soybeans	0.0257	53,606	28.4	0.852	0.73

SOURCE: Ref. 20.

TABLE 25.8 Yields of Selected Fresh-Water and Marine Species

	Total productivity	
	Dry metric tons/ hectare/year	Dry tons/ acre/year
Gracilaria foliifera (red algae)		
Florida (laboratory conditions)	127	56
Massachusetts (laboratory conditions)	73	32
Taiwan (commercial cultivation)	15	7
Porphyra spp (red algae)		
Japan (commercial cultivation)	1.8	0.8
Zostera marina (marine eel grass)		
North Carolina (laboratory conditions, less than one year)	3.5	1.5
Macrocystis spp (giant kelp)		
California (open ocean, natural conditions)	2.4	1.0
Eichhornia crassipes (water hyacinth)		
Florida (laboratory conditions)	42	18
Chlorella spp (fresh-water microalgae)	50	20

SOURCE: Ref. 23.

The cost of these biomass residues varies with the type of residue, collection system, and location with respect to the conversion facility. The prices for representative residues at several locations in the United States are presented in Table 25.9. The economics of these residues is also greatly dependent on the other uses for residues, which include animal feeds, fertilizer, fiberboard products, erosion control, and as soil conditioners.

Forest residues   There are four forms of residues generated in the forest products industry: mill residues, logging residues, intermediate cuttings, and noncommercial standing wood.

Mill residues include bark, sawdust, and wood scraps which are generated during the process of producing lumber and plywood. In 1970, 62 million dry tons of mill residues were used by the forest products industry, either for such products as particleboard and reconstituted wood, or for the energy requirements within the mill. Almost all mill residues are utilized within the forest products industry, and will increasingly be unavailable for other energy uses. Prices and availability of wood and bark mill residues fluctuate constantly, depending on the construction industry, export chip demand, domestic pulp and paper markets, and domestic production of other forest products.

Logging residues are the leftover material from logging operations, which do not have enough economic value to justify the expense of removing them from the forest. Most logging operations remove only that part of the stem or bole which is greater than 4 to 5 in. (10 to 12.5 cm) in diameter. Generally, the merchantable bole represents about 60 percent of the dry weight of the total tree; the stump and large roots from 15 to 25 percent; and the branches from 10 to 15 percent. Thus, logging residues include all those parts of the tree except the merchantable bole—tops, branches, foliage, stumps, and roots—as well as noncommercial tree species, trees of inferior quality, and understory brush. Table 25.10 summarizes the logging residues available in 1970.

Logging residues must be removed or reduced before a forest stand can be regenerated. Thus, collection of residues for energy would have an additional benefit of reducing the costs of preparing the site for regeneration. As the removal of tops and branches is increased the potential for soil nutrient depletion, erosion, and other environmental impacts also increases, particularly on sites with steep terrains and shallow soils.

Cuttings which are made at various times during the life of a commercial forest stand are called intermediate cuttings or thinnings. They usually involve removing trees of poorer quality or smaller size in order to leave more room for remaining trees to grow more rapidly into high-quality straight trees. This is currently a very small resource, as only 0.03 percent of all commercial forest land was treated annually with intermediate thinnings in the period 1968 to 1970.[21] More intensive management including fertilization and intermediate cuttings is increasing in many regions of the United States, and these residues may also increase in the future.

Large numbers of commercial trees die each year from disease, insect attack, fire, and storms. These are often individual trees or small acreages, thus are usually not salvaged because of high

TABLE 25.9   Selected Forestry and Agricultural Residue Purchase Prices and Component Costs

	Total delivered cost, $/ton*	Percent dry weight	Delivered dry cost	
Residue/location			$/dry ton	$/MM-Btu
Wheat straw/Colorado	11.50–24.50	91	12.50–26.50	0.80–1.80
Mill bark/California	4.00–5.50	50	8.00–11.00	0.40–0.60
Mill bark/Maine	2.00–4.00	50	4.00–8.00	0.20–0.40
Logging residue/California	12.00–22.50	50	24.00–44.50	1.30–2.50
Field corn residues/Illinois	19.00–28.00	53	35.50–52.50	2.40–3.50
Field corn residues/Delaware	14.00–22.00	53	20.50–39.50	1.30–2.70
Beef manure/Colorado	3.00–4.50	60	5.00–7.50	0.40–0.60
Poultry manure/Delaware	19.50–26.00	75	26.00–34.50	1.60–2.30

*Estimated cost for 15-mi transportation [except for 25-mi transportation in Maine], loading and unloading charges ($1978).
SOURCE: Adapted from Ref. 3.

TABLE 25.10   Total Logging Residues by Region, 1970

Region*	Total residues, 10^3 DTE	Energy equivalents, quads or 10^{15} Btu
New England	3,976	0.0068
Middle Atlantic	5,331	0.0091
Lake States	3,670	0.0062
Central States	4,530	0.0077
South Atlantic	11,813	0.0201
East Gulf	5,622	0.0096
Central Gulf	10,585	0.0180
West Gulf	9,694	0.0165
Pacific Northwest	18,361	0.0312
Pacific Southwest	4,937	0.0084
Northern Rocky Mountain	3,600	0.0061
Southern Rocky Mountain	1,079	0.0018
Total U.S.	83,198	1.415

*Regions defined as follows:
  New England—Maine, New Hampshire, Vermont, Massachusetts, Connecticut, Rhode Island
  Middle Atlantic—Delaware, Maryland, New Jersey, New York, Pennsylvania, West Virginia
  Lake States—Michigan, Minnesota, North Dakota, South Dakota (east), Wisconsin
  Central States—Illinois, Indiana, Iowa, Kansas, Kentucky, Missouri, Nebraska, Ohio
  South Atlantic—North Carolina, South Carolina, Virginia
  East Gulf—Florida, Georgia
  Central Gulf—Alabama, Mississippi, Tennessee
  West Gulf—Arkansas, Louisiana, Oklahoma, Texas
  Pacific Northwest—Alaska (coastal), Oregon, Washington
  Pacific Southwest—California, Hawaii
  Northern Rocky Mountain—Idaho, Montana, South Dakota (west), Wyoming
  Southern Rocky Mountain—Arizona, Colorado, New Mexico, Nevada, Utah
SOURCE: Ref. 8.

costs for removal and transportation of such a diverse resource, and because the quality may have been impaired. In areas with extensive storm damage or disease outbreaks, it may be economical to salvage these large acreages for fuel, since tree quality only marginally affects the value of such trees for fuel.

Agricultural residues   In the harvest of most agricultural crops, only a portion of the plant is removed from the field (note representative residue factors in Table 25.7). The residue remaining may either be (1) left on the ground to prevent wind and water erosion and improve soil productivity, (2) plowed into the soil as a fertilizer and soil conditioner, (3) removed for livestock feed, or (4) removed for conversion to energy. Residues can be burned directly, gasified, or anaerobically digested to produce methane. Table 25.11 summarizes the disposition of agricultural residues in various regions of the United States.

The maximum theoretical availability of residues in the United States for conversion to energy is far greater than the practical availability. Agricultural residues are extremely valuable in many regions of the United States to prevent wind and water erosion and for soil fertilizers— in many of these areas, significant quantities of residues will not be available for energy conversion. In some areas of the United States, however, the residues must be disposed of (usually by plowing) before planting the next crop, and significant quantities of residues can be removed annually from these croplands.

Animal manures   The manure from animals kept in confinement can be collected economically, then sold for fertilizer, added to the soil by the farmer, converted to methane by anaerobic digestion, or gasified to produce synthetic natural gas. An increasing number of animals are maintained in confinement systems such as cattle environmental feedlots or poultry houses. Table 25.11 includes the annual disposition of animal residues in various regions of the United States, and Table 25.12 presents the amount of manure produced daily by various kinds of livestock. Again, the availability of animal residues for conversion to energy depends on the residue's alternate uses and the costs of producing fuels from these resources.

TABLE 25.11   Annual Agricultural Residue Disposition, 1971–1973 Averages (1000 Dry Ton/ Year)

	Sold	Fed	Fuel	Returned	Wasted	Total
New England and Mid-Atlantic						
Crops	353	823	—	3,408	31	4,614
Manure	38	6	—	2,034	124	2,202
East North Central						
Crops	1,741	17,860	—	36,486	35	56,121
Manure	27	6	2	2,576	556	3,167
West North Central						
Crops	5,088	24,686	—	93,405	36	123,214
Manure	344	85	9	5,217	1,525	7,180
South Atlantic						
Crops	242	2,618	962	8,605	1,193	13,620
Manure	128	22	5	1,823	303	2,280
East South Central						
Crops	94	1,294	—	6,131	203	7,722
Manure	37	15	1	821	315	1,189
West South Central						
Crops	1,239	3,675	780	18,033	4,165	27,892
Manure	1,729	82	—	1,594	264	3,668
Mountain						
Crops	928	719	—	20,786	37	22,470
Manure	546	18	—	1,826	359	2,750
Pacific						
Crops	1,583	467	—	18,746	1,112	21,907
Manure	1,071	13	—	1,955	985	4,023

SOURCE: Adapted from Ref. 2.

TABLE 25.12   Manure Production by Livestock

Class of livestock	Pounds of manure produced per head per day— air-dry basis (ca 10% moisture)	VS,* %	Ash,† %	Btu per pound (dry)
Cattle				
Beef	7.5	82.8	17.2	6425
Dairy	10.9	80.0	20.0	7109
Poultry				
Broilers	0.05	70.0	30.0	5822
Layers	0.08	70.0	30.0	5822
Turkeys	0.20	72.4	17.6	5803
Swine				
Breeding	1.2	82.0	18.0	7308
Market hogs	1.04	83.5	16.5	7308
Sheep	1.5	84.7	15.3	7666

*VS—Volatile solids as a percentage of the air-dry pounds of manure.
†Ash—Ash as a percentage of the air-dry pounds of manure.
SOURCE: Ref. 25.

CONVERSION OF BIOMASS TO USEFUL ENERGY FORMS

Many processing options can be applied to the conversion of biomass to energy or chemicals. These processes range in state of development from laboratory scale to commercially proven processes. Figure 25.2 shows these alternative processing paths, as well as the range of products and markets. These products can be used directly, or can be upgraded further to serve markets

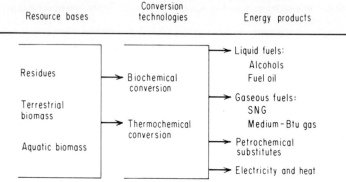

Fig. 25.2 Alternative processing paths and products for biomass resources.

with more critical needs or to produce more economically transportable products. For example, direct combustion of biomass can produce steam for direct use for electricity generation. Also, ethanol from fermentation of biomass can be used directly as a motor fuel or can be converted to ethylene, a major petrochemical feedstock for polymers.

Biomass can be converted to useful fuels by either biochemical or thermochemical processes. In the former, biological organisms are used to process the biomass, followed by initial chemical breakdown in some cases. High temperatures, catalysts, and sometimes high pressures are used to react the biomass chemically in thermochemical processes. In addition, chemical processes can be used to extract petrochemical substitutes from plant materials which have significant quantities of high molecular weight compounds, such as natural rubber or rosin. Each of the processing paths that produces a unique primary product has different conversion efficiencies, requires different levels of capital investment, and has distinct economies of scale.

Conversion Efficiency

Conversion efficiency varies considerably with the biomass feedstock which is used. The biomass will vary in heating value, ultimate analysis, proximate analysis, moisture content, and bulk density. Some of these characteristics differ only slightly among various forms of biomass, while others vary greatly.

The energy content or heating value of biomass varies with water content, chemical composition, and density. The fresh weights of plant tissue include range from 95 to 98 percent water (algae and water hyacinths) to 30 to 35 percent moisture for some woods. Because conifers (softwood tree species) usually have greater amounts of resins, oils, and tannins, they have a higher Btu value than deciduous woods (hardwoods) such as oak, maple, or ash. Oil-bearing tissues such as seeds or stems of oil plants (such as safflower, sunflower, and guyaule) have higher heating values than tissues without these oils or hydrocarbons. The greatest significant difference in heating values lies in the moisture content of the biomass which is to be combusted or converted. Heating values for some plant species are presented in Table 25.13.

The ultimate analysis of biomass describes its composition in terms of the amount of each constituent present. Analysis of several woods and barks are presented in Table 25.14, and it may be noted that there is little variation among most species. The ash content of wood is uniformly low, ranging from 0.2 to 2.2 percent,[8] which is an advantage in using wood as a fuel.

Bulk density varies greatly with the species of biomass, as well as the plant tissue and form of handling. Stems or trunks of woody plants are more dense than the stems of agricultural crops such as corn. Branches which have been piled together or chipped weigh less per unit volume, than whole tree trunks. The bulk density of biomass greatly affects the economics of using biomass for fuels, since those forms of biomass which are less dense are more costly to transport and require more storage space.

Biochemical Conversion

Those biomass resources with higher moisture contents, such as algae and animal manures, are more suitably converted into fuels by using the bioconversion processes of anaerobic digestion and fermentation.

TABLE 25:13   Energy Content of Plant Biomass

Plant species	Btu/lb*
Douglas fir	8,890
Western hemlock	8,410
White fir	8,210
Western red cedar	9,700
Southern pine	8,600
Ponderosa pine	9,110
Black cottonwood	8,800
Red alder	7,990
Beech	8,150
Elm	8,170
Hickory	8,050
Red maple	7,990
Red oak	8,050
White oak	8,150
Sycamore	7,990
Sugarcane (*Saccharum officinarum*) bagasse (12% moisture)	7,281
Sugarcane (*Saccharum officinarum*) bagasse (52 % moisture)	4,000
Bamboo (*Phyllostachys* spp) cane (10.5% moisture)	7,398

*Oven-dry wood.
SOURCE: Refs. 3 and 9.

TABLE 25.14   Ultimate Analysis of Oven-Dry Wood

	Percentage of fuel, dry weight					
	H	C	N	O	S	Ash
Western hemlock wood	5.8	50.4	0.1	41.4	0.1	2.2
Douglas fir wood	6.3	52.3	0.1	40.5	nil	0.8
Pine (sawdust only)	6.3	51.8	0.1	41.3	nil	0.5
Western hemlock bark	6.2	53.0	0	39.3	nil	1.5
Douglas fir bark	5.8	51.2	0.1	39.2	nil	3.7
Loblolly pine bark	5.6	56.3	*	37.7	nil	0.7
Longleaf pine bark	5.6	56.4	*	37.4	nil	0.7

*Included with oxygen measurement.
SOURCE: Ref. 8.

Anaerobic digestion   In the process of anaerobic digestion, microorganisms digest biomass directly to produce methane and carbon dioxide. The process is currently used for industrial and municipal waste treatment to reduce the volume of organic sludges prior to disposal. Anaerobic digestion requires either temperatures of 30 to 45°C for mesophillic bacteria, or temperatures of 50 to 65°C for thermophillic bacteria. Product gases are methane and carbon dioxide, which can be burned directly as low-Btu gas, or can be upgraded to high-Btu gas by removal of the carbon dioxide.

Methane fermentation is a sensitive microbial process because three major groups of bacteria are involved, many of which can only complete one stage of digestion: (1) conversion of complex insoluble biodegradable organic molecules to soluble organics; (2) conversion of soluble organics to acetate or propionate and CO_2 or acetate and H_2; and (3) production of CH_4 from H_2 and CO_2. The latter process is accomplished by methanogenic bacteria which are sensitive to various environmental conditions. Primary cost-controlling factors are the yields of methane gas from the biomass feedstock, and the retention time in the digestion vessel, as shown in Fig. 25.3.

Recently, the possibility of using anaerobic digestion to convert biomass other than animal manures or municipal wastes into usable energy has been considered, because almost all biomass has a relatively high moisture content before air drying. Research is in progress on such biomass

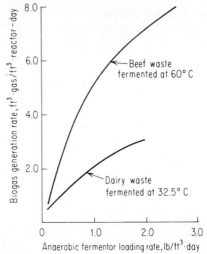

Fig. 25.3   Comparison of digestor retention time to methane generation ratio. 1 ft^3 = 0.0283 m^3. (*From Ref. 12.*)

feedstocks as marine algae fresh-water plants, corn stover, wheat straw, and other agricultural residues.[12,16,24]

Fermentation   The use of microorganisms to convert simple sugars into ethanol and a range of chemical feedstocks has evolved into a substantial chemical industry over hundreds of years. More recently, enzymatic and chemical hydrolysis processes have been developed to convert cellulose and other polysaccharides into fermentable sugars, primarily glucose. These sugars can then be fermented to ethanol, acetone, butanol, and a range of other chemicals. Ethanol can be blended with gasoline or used directly in modified internal combustion engines.

These processes thus enable the cellulose constituents of the vast wood and agricultural residues to be converted into ethanol and many other chemicals which are currently derived most economically from petroleum. The sugars extracted from sugar beets, sugarcane, and sweet sorghum, and the starch extracted from corn, wheat, and other grains can be converted easily to alcohols by fermentation, without the extensive need for hydrolysis. Research is currently focused on pretreatment of cellulosic materials to remove lignin (which is very resistant to chemical breakdown), identification of more effective cellulase enzymes and fermentation organisms, and decreasing commercial costs of producing ethanol so that it can compete economically with gasoline.[5,19,22]

Thermochemical Conversion

Thermochemical processing of biomass resources includes the processes of pyrolysis, gasification, liquefaction, and direct combustion. A wide range of energy products can be produced—synthetic natural gas, methanol, fuel oil, charcoal, heat, process steam, and electricity.

Pyrolysis   In pyrolysis, wood is heated in the absence of air to temperatures where the wood decomposes, producing combustible solids, liquids, and gases. In the past, pyrolysis of wood has been used to produce charcoal and methanol, and more recently, flash pyrolysis has been used to pyrolyze municipal refuse. The gases produced from pyrolysis have 100 to 300 Btu/scf (3700 to 11,200 kJ/m^3), and are considered low-Btu gases. Three pyrolysis processes are commercially available in an advanced stage of development: (1) the Nichols-Herreschoff furnace process, widely used as the basis for the production of barbecue charcoal; (2) the Tech-Air process, developed at the Georgia Institute of Technology; and (3) the Occidental Flash Pyrolysis process, developed for high-cellulose municipal solid waste.[4]

Gasification   Both medium Btu [approximately 300 Btu/scf (11,200 kJ/m^3)] and high-Btu [approximately 1000 Btu/scf (37,300 kJ/m^3)] gases can be produced from biomass, depending on the temperature and pressure of the reaction, and the subsequent upgrading of the initial product gases. Most biomass gasifiers currently under consideration operate at atmospheric pressure and can yield crude gas with up to 350 Btu/scf (13,000 kJ/m³) when partial oxidation is carried out only with oxygen. The Purox process, developed by Union Carbide Corporation for the partial oxidation of municipal solid waste, employs a moving-bed reactor-gasifier operating at approximately 3000°F (1650°C), through which oxygen is passed counter-currently to the biomass feedstock. The gas produced varies in heating value and composition with the biomass feedstocks which are used.

The medium-Btu synthesis gas produced by such gasification processes as the Purox process can be upgraded to substitute natural gas by converting the carbon monoxide and hydrogen present in the process gas to methane to the maximum extent possible. This technology has been developed for the synthesis gas produced from coal, and involves the reaction of the gas with a nickel catalyst at high temperatures, in a multistep process. The synthesis gas may also be upgraded to ammonia by removing any H_2O and CO_2 present and then reacting the nitrogen and hydrogen at high pressures to produce NH_4 (ammonia). In addition, the synthesis gas may be converted to methanol by reacting carbon monoxide and hydrogen to form CH_3OH. [4]

Liquefaction   Biomass can also be converted directly to oil by using sodium carbonate catalysts in a hydrogenation process, in the presence of carbon monoxide and steam. A process is currently under development for feeding a slurry of wood chips into the high-pressure reaction vessel. Product oil can be separated from unconverted material and water by centrifugation.

Heat, steam, and electricity   Direct combustion can be one of the most efficient methods of using the direct energy potential of biomass. Thermal efficiency is limited essentially only by the heat recovery equipment downstream of the combustion chamber. For direct-fired furnaces and steam boilers, thermal efficiencies as high as 85 percent can be achieved. Generally, biomass with moisture content below 30 percent can be burned directly, while those with higher moisture content may have to be predried or fired with supplementary fuel, either of which requires expenditure of additional energy and lowers the efficiency of the process.

Hot combustion gas can be used directly for drying or for heating, and indirectly to generate steam. The steam, in turn, can be used directly for power or heat, or indirectly to generate electricity. The overall direct combustion efficiency of biomass used to produce electricity is 20–30 percent, depending upon feedstock moisture content, ash content, and plant design. This efficiency is low compared to that of coal combustion plants, which are typically 35 to 40 percent efficient overall. [4]

Residential use of wood for heating   One hundred fifty years ago, essentially all homes used wood for heating and cooking. In 1970, wood accounted for about 1 percent of all energy used in the United States. One hundred fifty years ago, 30 cords (109 m³) of wood might be burned each year to heat the typical living space of a home. Today, the same size house with tighter construction and more insulation may be heated more uniformly using far less wood. As the costs of fossil fuels and electricity have been rising, the residential use of wood for heating has also been rising.

Three-quarters of the world's population still uses wood for all heating and cooking needs. In many countries there is a very serious firewood crisis due to the short supplies and intense pressures on forested areas. In many towns in India, central Africa, and Latin America, the task of collecting a family's daily wood once required several hours for one family member. As populations have grown, the cutting has been so intense that virtually no trees exist within 30 mi of some towns. Reforestation has not been widely attempted and, even where it has, new seedlings are often pulled up for firewood. Such serious side effects as soil erosion, loss of nutrients, increased runoff following precipitation, and greater flooding downstream follow. Whereas animal dung was previously spread on land as the only fertilizer, dung is often burned when wood becomes too expensive and difficult to obtain. Because of the more diverse natural resource and economic base, and less intensive population pressures, these situations do not occur as widely in Western countries.

Petrochemical Substitutes

Chemical feedstocks for a great range of products are derived from petroleum, yet many of these same chemical feedstocks can be derived from biomass. Historically, many of these chemicals were obtained from various plants, and it is expected that as the price of oil increases and its supply decreases, greater volumes of chemical feedstocks may be obtained from biomass. Some representative chemicals consumed in large volumes in industry are ammonia, ethanol, resins, and high-molecular-weight hydrocarbons, some of which are present in natural latex. Ethanol and other alcohols can be converted from biomass by fermentation or from petroleum by chemical extraction. Ammonia is produced by thermochemical processes from synthesis gas, which can be either natural gas or a medium-Btu gas derived from biomass.

Resins and turpentine   Oleoresins are natural products of most pine species and several other conifers, and may be separated into volatile turpentine and rosin during processing. Oleoresins have traditionally been collected in the gum naval stores industry by wounding trees, applying pine gum stimulants such as sulfuric acid paste, and collecting the resin flow from each tree. A substantial volume of resins and turpentine is also obtained during the process of making paper in the kraft pulp mills, and by collecting and processing large, virgin stumps which have high concentrations of resins stored in them.

In 1973, U.S. Forest Service research discovered that the herbicide paraquat would stimulate the production of oleoresins by pines. This concept has undergone widespread testing in the Southeast, and appears very promising. Nine to 12 months before pines are to be harvested for pulp, paraquat solutions are applied to a basal wound. Resin-soaking, or lightering, occurs as much as 10 to 15 ft above the wound as a response to the paraquat. As the trees must be harvested for pulp, the only increased cost is the initial treatment of trees with paraquat. One important impact to be considered is the increased insect infestation of the trees treated with paraquat.

Latex-derived hydrocarbons   Many flowering plants produce latex, which is a suspension or emulsion of varying chemical composition, including polyisoprene rubber, waxes, resins, proteins, essential oils, mucilages, organic acids, tannins, and alkaloids. The latex is usually found in specialized cells called latificers, and may be either excretory or secretory plant products.

Natural rubber is the most important economic product derived from latex. The Asian rubber tree, *Hevea brasiliensis* (grown primarily in Southeast Asia) and the guayule shrub *Parthenium argentatum Gray* (grown naturally in semiarid regions of North America) have been continuing sources of natural rubber. In 1910, more than 50 percent of U.S. rubber was extracted from guayule shrubs. During World War II, the United States initiated a massive "Emergency Rubber Project" when Japan invaded Southeast Asia. From 1942 to 1945, 32,000 acres (13,000 hectares) of guayule were planted in three states, and over 3 million lb (1.4 million kg) of resinous rubber was produced.[13] Interest in reviving the cultivating of guayule and improving techniques for extracting natural rubber has increased in the past few years.

Another semi-arid plant which could be cultivated as a source of petrochemical substitutes is the jojoba plant, *Simmondsia chinensis (link) Schneider*. The jojoba seeds contain up to 60 percent liquid wax that is almost identical to sperm whale oil, but is more uniform and contains fewer impurities.[14] In addition, several species of the Euphorbia genus (*E. lathyrus* and *E. tirucalli*) produce significant quantities of latex, from which valuable hydrocarbons may be extracted.

The synthesis of such complex molecules as polyisoprenes, waxes, and resins requires substantial energy input by the plant. Plants produce these high molecular weight compounds and expend less energy for other plant structures or biomass. Therefore, the yields of these plants per acre will substantially less than plants which produce less specialized compounds, and the price of petroleum-based hydrocarbons must increase substantially before it is economical to extract these compounds from plants.

FUTURE BIOMASS ENERGY FARMING

One of the most promising systems for increasing the production of biomass for energy is the establishment of terrestrial energy farms. Under this concept, plant species can be managed for their total biomass production, just as many are currently managed for their food or fiber content. The primary difference would be that the biomass from terrestrial energy farms would be converted to liquid fuels, gases, electricity, or other energy products.

While it is likely that much of the biomass which is converted into energy in the near future will be residues, it may be economical in the future to dedicate land acreage adjacent to biomass conversion facilities to the growth of the biomass feedstock. Woody species appear to have the most promise for the establishment of energy farms, primarily because they are more adapted to the marginal land areas which would not be occupied by agricultural crops. Nonwoody species, as well as mixtures of woody and nonwoody species, may also be grown on future terrestrial energy farms. Fresh-water and marine aquatic energy farms show future promise for energy production, but must be regarded as long-term alternatives.

Conceptual Design of Terrestrial Energy Farm

There is currently no terrestrial energy farm in existence, but considerable research has been supported by the U.S. Department of Energy and the U.S. Forest Service on the potential of intensively managed short-rotation silvicultural crops for energy production. A thorough evaluation of the potential of this concept and the costs of producing biomass for energy in this manner was made for the Department of Energy.[9] This conceptual design reflects the best available information on intensive silviculture, and is expected to be tested and modified over the next decade.

The proposed energy farm would consist of selected, rapidly growing tree species planted at close spacings. The crop would be harvested at appropriate intervals, or rotations, with the succeeding crops arising by coppicing (sprouting from stumps), precluding the need to replant after each harvest. Since most conifers do not coppice, selected hardwood species would be preferable in most cases. Intensive crop management would be practiced, including fertilization, irrigation, and weed and disease control. In this regard, energy crop production would be more similar to field crop production than to conventional forestry.

The size of the conceptual energy farm was chosen on the basis of the feedstock requirements of a conversion facility equivalent to a 50-MW power plant. A facility of this size requires 250,000 dry tons (227,000 metric tons) of biomass annually, or 500,000 green tons (450,000 metric tons), assuming 50 percent moisture content on a green weight basis. The critical factor limiting the size of a conversion facility is the distance which the biomass must be transported, since it has a low density and transportation costs are primarily based on volume rather than total weight. The biomass could be converted using a range of processes to steam, process heat, electricity, liquid fuels, synthetic natural gas, or chemical feedstocks.

Land Acquisition and Preparation

The amount of land required for an energy farm would depend on the productivity, or the number of dry tons produced per acre per year. For sites which can support 5 DTE (dry-ton equivalents) per acre per year, 50,000 acres (20,000 hectares) would be required for a 50-MW equivalent power plant, plus another 2000 to 3000 acres (800 to 1200 hectares) for irrigation lanes, work roads, and storage areas. Only 32,000 acres (12,800 hectares) would be required for sites supporting 8 DTE/acre/year.[10]

The costs of clearing land and preparing it for planting depend primarily on the vegetative cover present, since the cost of preparing wooded land is more costly than that for open land. Costs may range from $35/acre ($88/hectare) for cultivated land to $240/acre ($600/hectare) for mature forested land, not including the revenues obtained from selling timber on the forested land.[10] Land preparation is only required during the first year of establishing the plantation, and after 4–6 subsequent harvests when the coppicing vigor of the trees may have declined and it is desirable to replant. Land preparation includes the removal of vegetation, large stumps, and other debris, as well as making irrigation lanes, roads, and storage areas.

Stand establishment   The establishment of a healthy, well-stocked energy farm requires considerable labor and specialized operations. Generally, 12- to 16-in (30- to 40-cm) tall seedlings would be planted at 4 × 4 ft (1.2 × 1.2-m) spacings (planting density of 2725 plants/acre), and provided adequate fertilization, irrigation, and weed control to ensure their survival over the critical first two years.

Seedlings must be grown in nurseries from seed, at a cost which varies between $65 to $85 per 1000 seedlings for the hardwood species which are included in this conceptual energy farm. Many hardwoods can be regenerated by planting cuttings [10 to 12 in. (25 to 30 cm)] pieces of a young stem in the ground which will subsequently root, and grow into a new plant.

Irrigation is often critical for the establishment of hardwoods, either to insure the viability of seedlings or to stimulate and maintain the growth of cuttings. One of the most promising irrigation systems is the traveling sprinkler system, which is adaptable to irregular topography, provides uniform application to areas up to 200 feet (61 m) in radius, and would be able to supply water to the trees even after they had grown to 20 to 30 ft (6 to 9 m) high. Because rainfall is not distributed evenly over the growing season in many parts of the United States, irrigation would increase yields of many tree species if applied during the first 2–3 years of the growing cycle.

Weed control is also critical to the survival of the seedlings or cuttings, and may be accomplished either by application of herbicides during the land preparation activities, or mechanical cultivation for the first few years.

Crop management   The high yields of 8 to 10 DTE/acre/year (20 to 25 DTE/hectare/year) or more projected for the future are predicted on intensive management of these silvicultural crops, including fertilization, irrigation, weed control, and pest control. Irrigation and weed control are most critical in the first few years, before the trees are well established and have full root systems. As the trees grow larger and fully occupy the site, no sunlight will reach the ground between the trees, and weeds will survive poorly.

Optimum fertilization rates for various species will depend both on the species and the fertility of the soil where they are planted. Fertilization is critical both for obtaining the high yields which are projected, and to prevent the mining of nutrients from the soil by the succession of short-rotation harvests. Lime, urea (for nitrogen), and potassium chloride would be the most widely applied fertilizers applied by ground machinery for the first year of any rotation, and by air during the subsequent years of each rotation.

Disease and pest control may be required, if infestations break out in various tree species. Most crop diseases affect the value of trees for timber. As the entire above-ground plant would be harvested for energy, the appearance and quality of the plants would not be important, and there would be no need to control diseases which affect these characteristics. However, if large acreages of a few species of trees are planted, the risks of epidemic diseases and pest populations are indeed present, and must be treated.

Harvest and collection   Harvesting the biomass energy crop consists of cutting the stems at or near the ground following four to six growing seasons, chipping the cut stems and branches, and storing the chips in piles at field storage areas. Most of the harvesting would be done in the fall and winter, in order to take advantage of the full previous growing season, and to avoid affecting the ability of the trees to coppice (which is decreased if they are harvested in the middle of the growing season). A prime consideration in harvesting is to use light equipment and to reduce field traffic as much as possible, to reduce soil erosion and compaction.

Typical harvesting systems could consist of a self-propelled harvester to cut and chip the trees, pulling a wagon to collect the chips, and tractors to transport the wagons to field storage areas. Two primary problems are associated with wood chip storage: (1) the loss of total biomass over time, due to microbial degradation, and (2) possibilities of spontaneous combustion within the storage piles. It may be desirable to harvest almost the entire year to eliminate the chip storage problems, even if some biomass productivity were sacrificed, or to store whole trees or segments of trees. Transport of the biomass to the conversion facility could be accomplished by truck, rail, or a pipeline system.

Costs of Producing Biomass for Energy

Estimates of the costs of producing biomass by the intensive management of short-rotation hardwoods at 10 sites in the United States were made for the DOE study. The costs vary primarily with the productivity (DTE/acre/year) of the species chosen and the site, and range from $1.21 to $1.37 per MM Btu ($1.15 to $1.30 per GJ), as summarized in Table 25.15.

A cost breakdown for the sites in Wisconsin and Louisiana is presented in Table 25.16. The values for some of the inputs are listed in Table 25.17.

The energy yield for the silvicultural energy farm would be approximately 10 to 15 times as much as the energy consumed by the operations, as presented in Table 25.18. The energy

TABLE 25.15   Total Production Costs for Conceptual Silviculture Energy Farms

Site	Production costs, 1976 $/MMBtu
Wisconsin	1.78
Missouri	1.49
Louisiana	1.21
Georgia	1.37
New England	1.96
Washington	1.43
Illinois	2.47
California	2.00
Florida	1.24
Mississippi	1.25

SOURCE: Ref. 9.

TABLE 25.16 Costs of Producing Biomass in Conceptual Energy Farm at Wisconsin and Louisiana

Cost category	Production costs, 1976 $/DTE	
	Wisconsin	Louisiana
Planning and supervision	1.74	1.74
Land lease	2.09	1.06
Land preparation	1.26	0.60
Roads	0.16	0.07
Planting	0.60	0.53
Irrigation	6.04	2.52
Fertilization	5.00	5.45
Weed control	0.17	0.10
Harvesting	1.37	0.93
Tractors/wagons	0.68	0.51
Loading	0.45	0.39
Transportation	3.11	1.89
Interest	1.15	0.69
Taxes	3.44	2.10
Return on investment	3.23	1.97
Salvage value after 30 yrs operation	(0.15)	(0.08)
Total	30.33	20.47
Capital	3.27	1.93
O & M	19.40	13.86
Fixed costs	7.66	4.68
Total	30.33	20.47

SOURCE: Ref. 10.

TABLE 25.17 Typical Data Input for Site-Specific Production Costs

	Units	Wisconsin	Louisiana
Annual production	DTE	250,000	250,000
Annual yield	DTE/acre	5	12
Rotation period	Years	6	6
Level of debt financing	%	50	50
Level of equity financing	%	50	50
Interest rate on debt	%	9	9
Land lease rate	% of mkt value	5	5
Annual escalation rate			
Capital costs	%	5	5
O & M costs	%	5	5
Fertilizer/fuel costs	%	7	7
Value of land	%	5	5
Land required	Acres	70,900	23,700
Land value	$/acre	150	230
Clearing costs	$/acre	191	218
Tax rate	% value	2.58	0.55
Planting costs	$/acre	119	255
Harvesters	Field efficiency	0.47	0.52
Harvesting period	Days/year	148	100
Transportation distance	Miles (round trip)	48	27

SOURCE: Adapted from Ref. 10.

TABLE 25.18   Energy Balances for Conceptual Energy Farms

Operation	Material*	Energy consumed, 10^9 Btu/y Wisconsin	Louisiana
Supervision	Gasoline	1.67	1.67
Field supply	Diesel/gasoline	0.53	0.53
Harvesters	Diesel	9.43	7.87
Tractor haul	Diesel	5.45	5.49
Loading	Diesel	3.32	3.29
Transportation	Diesel	28.83	17.32
Irrigation move	Diesel	1.30	0.65
Irrigation pumping	Diesel	223.63	111.83
Manufacture	Urea	99.36	102.32
Manufacture	P_2O_5	8.03	8.17
Manufacture	K_2O	15.42	17.08
Ground operations	Diesel	1.16	0.06
Aircraft operations	Gasoline	0.15	0.16
Fertilizer transport	Diesel (rail)	0.98	0.64
Total energy consumption		399.26	277.62
Total energy yield		4250	4250
Net energy yield		3851	3972
Net energy efficiency		0.906	0.935
Energy in:Energy out		1:10.6	1:15.3

*Energy Content:
  Gasoline—124,000 Btu/gal
  Diesel fuel—138,690 Btu/gal
  Urea—27,730 Btu/lb · N
  P_2O_5—6,019 Btu/lb
  K_2O—4,158 Btu/lb
  Rail—800 ton · mi/gal
SOURCE: Ref. 10.

consumed by the supervisors and laborers and the energy required to manufacture the capital equipment used in various operations are not included, but would not add substantially to the energy requirements.

Aquatic Systems

The establishment of energy farms in fresh-water systems and open oceans have been considered extensively, based on similar systems which have been used to obtain food or other products. Energy farms are not likely to be established in coastal areas because of substantial Federal development regulations.

Microalgae can be cultivated in shallow fresh-water ponds, if adequate light and nutrients are provided. Generally, algae are harvested (1) by straining, (2) centrifugation, or (3) by chemically coagulating the algae, followed by filtration, sedimentation, or air flotation. The first two methods are currently too expensive for commercial applications, and the latter renders the algae unsuitable for anaerobic digestion. Following harvest, the microalgae can be anaerobically digested to produce methane.

The cost-controlling factors are the low yields of biomass when large commercial systems are installed, and the complex harvesting technologies which are necessary for the tiny algae plants. If the larger colinial or filamentous algae could be selectively cultivated, the harvesting cost could be reduced; this could be achieved by strict fertilization regimens or selective screening during harvesting to retain only the larger species.

Growing giant kelp plants for conversion to methane has been proposed for near-surface waters in the open oceans. A marine energy farm would consist of a support structure in the ocean, the establishment and growth of *Macrocystis spp.* (giant kelp) on the support structure, harvest of the giant kelp plants, transport of the plants to a processing facility, and conversion of the kelp into methane gas and by-products.

A major problem with this concept is that the surface layers in the open ocean are usually low in dissolved plant nutrients, thus sustaining only low growth rates of the kelp. As deeper ocean waters (off the California coast, where marine energy farms have been considered) have high nutrient concentrations, these nutrients could be upwelled to the surface, or pumped up to the ocean farm structure supporting the kelp plants. Many aspects of the concept remain to be demonstrated: the growth rates of the kelp with the heterogeneous upwelled water, technical feasibility of the support structure, and feasibility of constructing conversion facility on-shore and long-distance transport of the bulky harvested kelp.

CONCLUSION

As prices of fossil fuels continue to rise, the production of biomass for conversion to energy will become increasingly attractive. Currently, the forest products industries obtain about half of their own energy requirements from wood, and will increase this as their fossil fuel costs rise. Industries are considering establishing energy plantations to provide a guaranteed source of fuel. Homeowners are increasingly heating their homes with wood. Farmers are inquiring about the technologies for providing their own gasoline and grain-drying fuel needs from agricultural residues and grains. Plant biomass is a versatile, renewable energy resource—which will unquestionably supply increasing amounts of the U.S. and world energy requirements in the future.

REFERENCES

1. Alich, J. A. and R. E. Inman (Stanford Research Institute): *Effective Utilization of Solar Energy to Produce Clean Fuel*, Stanford Research Institute, Menlo Park, CA, 1974.
2. Alich, J. A. and R. E. Inman (Stanford Research Institute): *Crop, Forestry and Manure Residue Inventory*, vols. I–VIII, National Technical Information Service (Report Nos. TID 27162/1 through TID 27162/8), Springfield, VA, 1976.
3. Alich, J. A. and R. E. Inman (Stanford Research Institute): *An Evaluation of the Use of Agricultural Residues as an Energy Feedstock—A Ten Site Survey*, National Technical Information Service (Report No. PB-260-763), Springfield, VA, 1977.
4. Bliss, C. and D. Blake (Mitre Corporation): *Silvicultural Biomass Farms*, vol. V: "Conversion Processes and Costs," National Technical Information Service (Report No. MITRE-TR-7347-V5/LL), Springfield, VA. 1977.
5. Gaden, E. L., M. Mandels, E. T. Reese, and L. A. Spano: *Enzymatic Conversion of Cellulosic Materials: Technology and Applications*, John Wiley, New York, 1976.
6. Howlett, K. and A. Gamache (Georgia-Pacific): *Silvicultural Biomass Farms*, vol. II: "The Biomass Potential of Short-rotation Farms," National Technical Information Service (Report No. MITRE-TR-7347-V2-LL), Springfield, VA, 1977.
7. Howlett, K. and A. Gamache (Georgia-Pacific): *Silvicultural Biomass Farms*, vol. III: "Land Suitability and Availability," National Technical Information Service (Report No. MITRE-TR-7347-V3/LL), Springfield, VA, 1977.
8. Howlett, K. and A. Gamache (Georgia-Pacific): *Silvicultural Biomass Farms*, vol. VI: "Forest and Mill Residues as Potential Sources of Biomass," National Technical Information Service (Report No. MITRE-TR-7347-V6/LL), Springfield, VA, 1977.
9. Inman, R. E. (Mitre Corporation): *Silvicultural Biomass Farms*, vol. I: "Summary," National Technical Information Service (Report No. MITRE-TR-7347-V1/LL), Springfield, VA, 1977.
10. Inman, R. E., D. J. Salo and B. J. McGurk (Mitre Corporation): *Silvicultural Biomass Farms*, vol. IV: "Site-specific Production Studies and Cost Analyses," National Technical Information Service (Report No. MITRE-TR-7347-V4/LL), Springfield, VA, 1977.
11. Jewell, W. J. (Cornell University): *Bioconversion of Agricultural Wastes for Pollution Control and Energy Conservation*, National Technical Information Service (Report No. TID-27164), Springfield, VA, 1976.
12. National Research Council: *Guayule: An Alternate Source of Natural Rubber*, National Academy of Sciences, Washington, DC, 1977.
13. National Research Council: *Jojoba: Feasibility for Cultivation on Indian Reservations in the Sonoran Desert Region*, National Academy of Sciences, Washington, DC, 1977.
14. Odum, Eugene: *Fundamentals of Ecology*, W. B. Saunders Company, Philadelphia, 1971.
15. Pfeffer, J. T. (University of Illinois): *Biological Conversion of Biomass to Methane*, National Technical Information Service (Report No. COO-2917-5), Springfield, VA, 1977.
16. President's Advisory Panel on Timber and the Environment: *Report of the President's Advisory Panel on Timber and the Environment*, Government Printing Office, Washington, DC, 1973.
17. Shelton, J. and A. Shapiro: *The Woodburner's Encyclopedia*, Vermont Crossroads Press, Waitsfield, VT, 1976.
18. Underkofler, L. A. and R. J. Hickey (ed.), *Industrial Fermentations*, 2 vols., Chemical Publishing Company, New York, 1954.

19. U.S. Department of Agriculture: *Agricultural Statistics 1976*, Government Printing Office (Catalog No. A 1.47:976), Washington, DC, 1976.
20. U.S. Department of Agriculture (Forest Service): *The Outlook for Timber in the United States*, Government Printing Office (Forest Resource Report No. 20), Washington, DC, 1973.
21. Wilke, C. R. (ed.): *Cellulose as a Chemical and Energy Resource*, John Wiley, New York, 1976.
22. Personal communication with Dynatech Corporation, conducting algae systems study for the Department of Energy.
23. Ashare, E., D. L. Wise, and R. L. Wentworth (Dynatech Corporation): *Fuel Gas Production from Animal Residue*, National Technical Information Service (Report No. COO/2991-10), Springfield, VA, 1977.
24. Stanford Research Institute: *An Evaluation of the Use of Agricultural Residues as an Energy Feedstock*, National Technical Information Service (Report No. PB-260-764), Springfield, VA, 1976.
25. Didericksen, R. I., A. R. Hidelbaugh, and K. O. Schmude: *Potential Cropland Study*, Soil Conservation Service, U.S. Department of Agriculture (Statistical Bulletin No. 578), Washington, DC, 1977.

Chapter **26**

Solar Energy and The Law*

NORMAN L. DEAN
Bruce Terris, Attorney at Law
Washington, DC

GAIL BOYER HAYES
Environmental Law Institute
Washington, DC

FRANK O. MEEKER
Environmental Law Institute
Washington, DC

ALAN S. MILLER
American Bar Association
Washington, DC

and

GRANT P. THOMPSON
Conservation Foundation
Washington, DC

New laws are often born with bursts of enthusiasm, only to be quickly strangled by the tangled net of existing statutes, ordinances, regulations, and common law. This chapter looks at present laws and asks if they will thwart the future growth of solar usage in the United States.

It can be argued that most of the topics covered in this chapter—such as building codes, land-use control techniques, utility regulation, and home financing—present no legal obstacles that are unique to solar energy development. This argument has only limited value. While it is true, for example, that building codes require each structure to be examined and approved

*This chapter is a condensation of an Environmental Law Institute Report "Legal Barriers to Solar Heating and Cooling of Buildings," March, 1977; funded by ERDA, Contract No. E(49-18)–2528, EA–02–03, 57–60–91.

by a building inspector, it does not follow that solar energy systems will add no additional cost or uncertainty to the process.

Solar collectors need sunshine, of course, to function. There is presently no right to sunshine in the United States when the light passes over your neighbors' properties before reaching your own. The section on solar access and land use describes and criticizes suggested remedies to this problem.

Building codes also pose problems for solar users, because they do not yet deal specifically with solar equipment, and there is a lot of uncertainty as to how they will be applied to this new field. If rigidly interpreted, they could be used to bar solar systems.

Another major area of potential conflict is the appropriate role of utilities vis-à-vis solar energy. Areas of great sensitivity here include utility rate structures, service discrimination, and the scope of public utility commission jurisdiction.

Finally we examine the special difficulties involved with financing solar homes, problems with laws meant to exempt solar equipment from property taxation, and the possible financial hardship involved if solar systems were made mandatory on all new structures. Although this article was accurate when written in early 1977, the laws discussed are already the subject of much proposed legislation and are undergoing rapid change in the United States.

SOLAR ACCESS AND LAND USE

There is a consensus among experts on solar energy law on two broad points: (1) that property owners in the United States have no right to receive solar energy that would reach their land only after slanting across property owned by others; and (2) that some sort of access to solar energy should be legally guaranteed. Very little empirical evidence exists as to whether shadows on solar devices will be a major problem, but two limited studies of aerial photographs suggest that sun rights may be a smaller problem than sometimes feared (at least in suburban settings).[1]

The Scope of Protection Required

To date, discussions of legal issues relating to solar access have focused only on the requirements of active systems on single-family homes, but this chapter will also look at the solar access needs of multifamily residences and commercial and industrial structures. It is also vitally important to examine the legal needs of passive solar systems.[2]

Current active systems typically require a large southern exposure on a structure's roof, but some systems[3] rest against slopes with southern exposures. At least one engineering firm recommends using preexisting vertical wall panels as part of flat-plate collectors to heat industrial buildings or warehouses that do not have to be kept very warm.[4] Collectors may be attached to secondary structures, especially in retrofit situations, when a primary structure is poorly oriented. Since collector efficiency drops during the early morning and late afternoon, it may be less imperative to prevent shadows during these periods. Passive systems, on the other hand, have slightly different requirements. There are many varieties of passive systems, but the need for a southern exposure is common to most, and is the need most relevant to sun rights law.[5]

Doctrine of Ancient Lights and Prescriptive Easements

Very briefly, the doctrine of ancient lights grants property owners in England and parts of the Commonwealth a limited amount of indirect sunlight, if that light has been flowing through their windows, without interruption, for a given number of years (27 years in Great Britain).[6] In effect, the ancient lights doctrine gives a long-time recipient of sunshine a prescriptive easement over a neighbor's land.

As noted in *Fountainebleau Hotel Corp.* v. *Forty-Five Twenty-five, Inc.*,[7] the doctrine, and prescriptive easements in general, have been unanimously repudiated in the United States[8] on public policy grounds. For the doctrine to be useful in solar access cases, it would require such great modification that even a willing judiciary may refuse to make the leap without a legislative assist. For instance, the light guaranteed by the doctrine is not direct sunlight, but only enough indirect light to enable one to go about one's life indoors without grumbling. If the waiting period were substantially shortened to make it fair to owners of solar equipment, the result would probably be unfair to their neighbors, who would suffer great diminutions in value of their property. Furthermore, if someone built a fence or other structure just to keep a neighbor from acquiring a prescriptive easement for light, most U.S. courts would not enjoin the construction nor order the fence removed.[9] Even if an easement were established (in a court test), it might be lost if there were a substantial change in the size or location of the collector.

The right, if any, to receive radio and television signals is an interesting analogy to sun rights. Unfortunately, from the perspective of solar advocates, the leading cases have allowed broadcast signals to be blocked. In *People ex rel Hoogasian* v. *Sears, Roebuck & Co.*,[10] the Illinois Supreme Court refused to enjoin further construction of the Sears Tower, even though the completed structure would allegedly interfere with the reception of over a hundred thousand television sets in its future "shadow." The plaintiffs argued that the tall building would be a public nuisance, but the court held that a property owner (Sears) has a right to put land to any reasonable use, subject only to restrictive legislation to protect public health and welfare.

Nuisance Law

Ralph E. Becker, Jr. doubts that existing nuisance law would solve the solar access problem since courts seldom call a particular use of property a "nuisance" if the legislature, through zoning laws, specifically authorizes that use.[11] Although, because of the expense and future shortage of fossil fuels, there is an *indirect* impact on the public health, existing nuisance cases deal with more immediate dangers. Becker concludes, "unless exceptional circumstances existed, a court would probably be unwilling to grant a right to light based on grounds of nuisance."[12]

Other writers are equally pessimistic. Karin Hillhouse, for instance, concludes that "to succeed at [a private nuisance suit] a plaintiff must show irreparable damage and a greater hardship than would be caused by enjoining the defendant's activity, a standard a solar energy user probably could not satisfy."[13] Donald Zillman concurs, saying nuisance law "will not be of great help to either the party wishing to prevent solar use or the solar user wishing to secure his access to direct sunlight."[14]

Another limitation of nuisance suits is that only damages, and not injunctive relief, may be available in about half the jurisdictions (those using a "balance of conveniences" approach).[15] In the unlikely event that a court found shadows to be a nuisance, plaintiffs may have to prove they are not "hypersensitive" to injury, or relief may be denied.

Although it is doubtful that common-law nuisance approaches will be of any help to solar system owners, municipalities (or states) could simply *declare* shadows falling on solar collectors to be a *public* nuisance. Sandy F. Kraemer and James Felt say that "certainly the preservation of the community by providing alternate sources of energy and reducing the demand for fossil fuels would fall within the guidelines of the general police power."[16] A community considering such an approach may be wise to amend its zoning laws so that previously authorized uses are clearly prohibited by the new statutes.

At least one writer concludes that:
. . . the legislative power to expand the scope of nuisance beyond its common law configurations may prove an effective device for securing rights to sunlight for solar collectors. . . .[17]

One small town, Kiowa, Colorado, has passed a zoning ordinance allowing a property owner with a solar collector to have a structure declared a public nuisance if it interferes with the owner's collector.[18]

Even where legally possible, there are disadvantages to protecting solar access with a statutory public nuisance approach:

1. Lawsuits would be necessary to prove each instance of a nuisance.

2. In some cases, the owners of restricted property may truly deserve compensation, but no compensation may be available.

3. Injunctive relief would not be available in many jurisdictions.

4. If one tried to sue before going to the expense of installing a collector, the suit may be dismissed as not "ripe."

5. Since a public nuisance is a crime, a homeowner may have to wait for the state to sue.[19] Under some circumstances, however, private individuals may sue if they can show they were *particularly* damaged, in some way not shared by the public generally. The plaintiff's damage must be different in kind, rather than just degree, from the general public's.[20] It is uncertain whether shadows cast on a collector would meet this requirement. It is possible that a statute could get around this problem by stating that individual citizens may sue in the public interest.

6. A ticklish situation would arise if a bungalow owner living between two skyscrapers decided to put collectors on the roof. The majority view is that the person who in good faith

comes to an existing public nuisance has rights to have it abated—even though it was there long before he arrived. [21]

Just as nuisance laws will not be of great help to solar homeowners, they will not help those protesting solar homes. It is unlikely that a solar collector would be found to be either a public or private nuisance as "the mere unsightliness of defendant's premises" is usually insufficient to create a nuisance. [22]

Existing Legal Approach of Untested Value:
Transferable Development Rights

An innovative concept that is much discussed but that has received little actual application is transferable development rights (TDR). The development rights of any lot are governed by zoning laws that specify allowable heights, densities, setbacks, etc. To "transfer" such rights, the government must allow them to be sold separately from the land. What is sold is not air rights, but a governmental license to build. Everyone in an area could be given equal development rights, but may not be allowed to fully use them (by erecting a big or tall building) because of solar skyspace easements acquired by the municipality through condemnation or zoning. In other words, a specific governmental restriction would be imposed within a general, less limiting governmental restriction. Under a TDR approach, the restricted property owners would be allowed to sell any development rights that they could not use to owners of property zoned for more dense development. In effect, the government takes away with one hand what it gives back with the other.

TDR has been sparingly used to preserve unique historical sites and environmentally critical areas. It may have some limited applications to solar access planning. John Costonis, who helped develop the concept of TDR, has said:

It is conceivable that if we hold density down substantially in an area to prevent interference with the sun's rays, we may create a situation of basic inequity among landowners within that area. This TDR scheme may provide a basis for permitting a landowner not permitted to achieve certain densities to receive the cash equivalent for his loss. [23]

TDR may help avoid an unconstitutional taking [24] where the development on one lot is severely restricted for the benefit of another. Even if an unconstitutional taking is not involved, the public's sense of fair treatment may demand that TDR or some other form of compensation be applied. To conclude, TDR is a very complex approach with dramatic side effects. It probably should not be used just to secure solar access, but may have an appropriate role as part of a comprehensive land-use plan.

Express Easements

Easements are limited rights to use the land of another. A right to enjoy the unhindered flow of sunlight across a neighbor's property would be called an "express" easement because the two landowners involved would bargain intentionally and voluntarily for it.

Express easements for receiving light are recognized by the courts [e.g., West Annot. California Civil Code §801 (8)], and are available in most states. [25] If no termination provisions are written into the easement contract, the easement will bind all subsequent owners of either lot, unless one of a narrow set of conditions occurs. [26]

There are limits to the usefulness of express easements:

1. They are voluntary—courts cannot force their creation.
2. They may be prohibitively expensive.
3. Their enforcement may involve long, costly court proceedings.
4. They may give an unjustified windfall to an owner of "burdened" property who never had any intention of using the property in a manner that would block sunlight.
5. Express easements put the entire cost on the would-be solar homeowner. Public policy may suggest that the cost should be shared or that the builders of the interfering structures should pay solar equipment owners for their "resource" (solar energy) rights.

The major advantage of the express easement is that highly motivated individuals can usually obtain one through their own efforts without governmental action. Furthermore, it would provide security in many established neighborhoods as well as in new ones. Even if solar zoning laws are passed, property owners may want to negotiate express easements because (1) they want more protection than is afforded by the zoning law, or (2) they want a guarantee of permanence not found in easily changed zoning laws.

Restrictive Covenants

There are two legal terms that mean almost the same thing: "restrictive covenants" and "equitable servitudes." A covenant is simply a promise involving land. It is usually found in a deed and frequently controls esthetics—i.e., the appearance of property. Esthetics may also be controlled by laws and ordinances, but private controls are more common.

In practice, today, private promises are enforced as equitable servitudes, rather than as restrictive covenants. This is because money damages are usually less satisfactory than an injunction and because of the complicated ancient laws regarding covenants.

Covenants are of greatest potential use where new tracts are opened for development (they will be of almost no help in established neighborhoods). Subdivision developers realize that some homogeneity in a neighborhood will appeal to potential buyers. For instance, covenants may require shrub and tree plantings to conform to a general landscape plan. In large subdivisions, covenants could be incorporated that guarantee access to solar power for home heating and cooling. They could be worded like a solar easement, or could specify generous setback requirements and strict height limits on trees and structures. Large-scale developments could be *required* to provide such agreements. [27]

Restrictive covenants are included (or incorporated by reference) in every deed when individual lots are sold, and are also enforceable against future purchasers. The owner of a lot in the subdivision who would be harmed by a neighbor's breach of a covenant has standing to sue the neighbor, despite the lack of direct participation in the contract. [28]

Covenants to protect solar access can be routinely used in subdivision, mall, or industrial park situations. They cost nothing, and do not require unsophisticated individual property owners to draw up legal documents. The developer's lawyer has only to add a clause or two to the deeds.

Restrictive covenants could be used to hinder solar homes as well as encourage them, since they are used to create private architectural review committees with authority to reject changes in building appearance. This authority is often stated in extremely general terms, and could be exercised to prohibit use of a solar collector. The most likely course of relief against a recalcitrant board will be to seek declaratory relief in court, i.e., a statement from the court that the covenant is unenforceable, because it is not clearly and unequivocally expressed. Because the law on this subject is "highly technical, erratic, and in flux," [29] the outcome of such challenges will be difficult to predict.

There is little question that a city could prohibit *new* covenants among private parties that unduly interfere with the use of solar collectors. The enforcement of existing private covenants can also be affected by changes in zoning in some cases. [30] Generally, the more restrictive provision will govern. But zoning can influence a court in determining whether to invalidate a restrictive covenant because of changing conditions. [31] In a few states, courts have gone substantially further and created a rebuttable presumption that the zoning change reflects changed conditions. [32] Thus, a zoning ordinance defining areas in which use of solar energy is expressly encouraged might convince some courts that the review board must allow for use of solar collectors.

Zoning for Esthetic Purposes

Private, esthetic controls—in the form of prescriptive easements—were discussed in the above section. This section will briefly consider the state of public regulations for esthetic purposes, and possible remedies where such regulations could obstruct solar energy development. A right to sunlight falling on a roof is of little use unless the owner also has a right to put collectors on the roof to receive it.

Judicial consideration of esthetics regulation has a long, complicated history. [33] The majority of courts, until quite recently, were very reluctant to support regulation solely for esthetic purposes because of the inherently subjective nature of esthetic judgments. In many cases, esthetic values are joined with some other more acceptable public purpose, such as maintenance of property values. [34] This is particularly true where regulations seek to preserve areas attractive to tourists for their scenic or architectural beauty, such as in Santa Fe, New Mexico, and New Orleans, Louisiana. Courts usually uphold architectural controls under these narrow circumstances.

Another judicial approach to esthetics regulations is what Professor Williams terms the "least common denominator test." [35] Under this view, some land uses are so obviously discordant and disruptive that they clearly fall within the scope of permissible government regulation. Presumably, solar collectors would not come under this test.

A few states, including New York, Oregon, Massachusetts, and Wisconsin, have discarded all pretenses and upheld regulation for purely esthetic purposes.[36] The most recent major opinion in this area, *Donnelly* v. *Outdoor Advertising Board*,[37] characterized as "the modern trend in the law" that "esthetics alone may justify the exercise of the police power."[38]

In contrast to the trend in favor of architectural controls and restrictions on billboards, several courts have recently held ordinances invalid that required homeowners to keep grass and weeds below a specified height. These cases have been reported in the press, but to the best of our knowledge are not yet published.[39] The implication is that courts may apply a balancing test to restrictions on the appearance of property, rather than the traditional (more lenient) standard of reasonableness applied to zoning. If the property owner can assert some valid interest in the challenged use (appearance), courts may invalidate the ordinance even under the "modern view."[40]

Solar systems need not be ugly. One prominent solar engineer stated recently:

So long as the structure is still on the drawing board, I feel solar can be adapted to any architectural style. . . . We've designed applications for everything from salt boxes to contemporary designs to mountain cabins to a bank.[41]

Proposals for Federal Involvement

The federal government has both direct and indirect powers that could be used to protect solar access. For example, reduction of U.S. dependence on foreign fuels would justify an exercise of the right to provide for national defense.[42] A more likely source of authority, however, is the nearly unlimited power to regulate interstate commerce. This power has been held to extend even to activities such as farmers growing crops on their own farms for their own use. An analogy exists with homeowners collecting energy for their own use rather than buying it from a utility that may generate the power in another state or buy fuel from an out-of-state source. One writer believes that even "free-flowing sunlight, not yet reduced to usable forms of energy, may be subject to federal regulation even before its products are marketed interstate."[43]

The American Bar Foundation (ABF) reports that under the commerce, national defense, and other constitutional powers, Congress could claim the power to pass a federal act to guarantee unobstructed solar skyspace.[44] The same study says, however, that Congress is not likely to do this, because nationalizing airspace lower than is necessary for commercial aviation would conflict too greatly with private property rights.[45] The federal role in land-use regulation has historically been very narrow.

In spite of the lack of evidence of a need for federal intervention, federal legislation has been proposed that would flatly prohibit states from allowing any construction that would block sunlight needed by existing solar equipment being used for heating or cooling. From the viewpoint of the owner of existing solar equipment, this is an ideal law. It may not, however, be the best solution for society at large. The bills may result in leapfrog development, which would in turn, result in the use of more fossil fuel for transportation. Premature development may also be forced, since property owners would race to build while they could (the discussion on the disadvantages of prescriptive easements applies here). If a property owner failed to erect a building before a neighbor hooked up a solar system, the owner's land may be drastically reduced in value (particularly in densely built areas).

A more likely possibility is that Congress will use some of its *indirect* powers to aid solar collector owners. These are powers derived from other, specifically granted powers, such as the power to spend for the general welfare.[46] For instance, federal financial aid could be restricted to states that encouraged solar easements by providing for their inclusion in public records.

State and Local Approaches

Some of the approaches available at the state and local level have already been mentioned, such as solar easements and extensions of existing nuisance laws. Other methods might include legislation to ensure the effectiveness of private easements, various solar zoning schemes, land-use planning approaches, or even more esoteric schemes using analogous bodies of law, such as water rights.

The two taproots of local laws are the authority to zone and the power of eminent domain. Many municipalities have been delegated these powers by their state government,[47] but the state governments can also zone and condemn property.

Publicly negotiated skyspace easements The ABF study includes a suggested statute that would allow cities to negotiate or condemn skyspace easements, and either to borrow money to pay for them or to assess the costs against those benefited. Assessments would be made on the basis of the benefit (such as the energy supplied, or the surface area of the collector, or the increase in property values). Such municipal action would be a taking of property, of course, and both the U.S. and state constitutions would therefore require compensation. Such programs may be so expensive that they would be useful only in very limited circumstances. Municipalities may presently have the power to condemn such easements, but the ABF suggests state legislation that would make their authority clear.[48] A possible legal problem here is whether these skyspace easements would be a "public use," a required objective for any condemnation.[49] However, if a state legislature declares a use to be public it will seldom be challenged.[50]

Under the ABF approach, the skyspace easements would be transferred to the benefited private property owners. The new owners would then be assessed for the cost of the airspace. Violation of a skyspace easement would be considered a private nuisance. [51]

Privately negotiated skyspace easements A practical action which states can take might be to follow the example of Colorado and enact legislation guaranteeing express solar easements the status of regular easements. Although solar easements may be enforced in some states without the aid of special, facilitating legislation, property law contains many snares for the unwary. For instance, some writers are uncertain whether easements for light would be viewed as being "in gross" or "appurtenant." When property is sold, courts will usually apply only appurtenant negative easements against the new owner. A statute will guarantee that such agreements can be enforced against new owners of the benefited and burdened properties.

Analogy to water rights law A controversial state-level approach has been suggested by Mary D. White. [52] She suggests allocating sunlight as water is allocated in states that use the prior appropriation doctrine.

In prior appropriation jurisdictions (arid, Western states), water laws are more developed because there has been more litigation over a scarce resource. In general, property owners must give notice of their intent to appropriate, and of the actual appropriation. The water must be put to a beneficial use within a reasonable time.

Defining "beneficial use" is the subject of much water law. White suggests that domestic heating and cooling could be given priority over recreational or frivolous uses. She identifies the real question as whether a beneficial solar use would be preferred over a nonsolar use. For example, could the owner of an existing solar house enjoin construction of a conventionally heated but economically valuable structure that would shade the solar house's collector? Under the water-law analogy the newcomer would have to buy the solar homeowner's resource rights.[53]

The basic problem with this water-law approach is that it is grounded on the principle of "first-come, first-served." It seems unfair to allot sun rights on this basis when it may be possible to plan additions to structures so that they do not shade their neighbors. The application of a water-law analogy would drastically increase the value of some parcels, while slashing the worth of lots to their north. Such unequal treatment may be unconstitutional, since no compensation would be paid for lost development rights. It seems unreasonable to force property owners to install solar equipment prematurely just to protect their sun rights. Conversely, under a water-law analogy, property owners may have to build additions to their structures prematurely if their neighbors hint they may install solar equipment. Substantial conflict with land-use planning goals is possible.

Another problem with a water-law approach is that many more people would seek solar rights than presently file for a permit or go to court to secure water rights. Conceivably, nearly every property owner in the United States could try at some time to secure solar rights. In states like Colorado where court proceedings are necessary, courts could be overwhelmed.

To summarize, a simpler, more certain, and more equitable approach is necessary. Stretching water law to cover solar access issues may dampen enthusiasms for this new technology.

Solar zoning Although enabling legislation for zoning and some form of subdivision control exists in all 50 states, one expert estimates that only 5000 out of 60,000 jurisdictions with power over land use exercised zoning powers in 1974.[54] Existing authority to plan for solar energy may therefore be adequate but not in and of itself sufficient to solve the problem.

Zoning law can both facilitate and frustrate the collection of sunlight for heating and cooling structures. Relevant factors controlled by zoning include height, setback, and sideyard restrictions; percentage of lot area covered; use and accessory use; esthetics; structure orientation; etc. For instance, in residential areas only one accessory structure is often allowed, a regulation that rules out a detached collector if a garage or tool shed already exists. There are many other similar problems.

Most proposals for mandatory solar zoning are limited to areas zoned for single-family houses, provide for some sort of administrative appeal to relieve undue hardship, and would be enforced like regular zoning laws. Some give all homeowners sun rights; others use a first-in-time-wins approach. An example of the latter is Robbins' suggestion that prescriptive easements to perpetual use of skyspace arise "after seven full years of an actual collector's use, or notice of a proposal or seven years of official designation."[55]

Although zoning power is generally exercised on the local level, state legislation could be passed to *require* local governments to use their zoning power to facilitate solar energy utilization. The ABF has drafted a statute that would accomplish this.[56] Oregon has amended its city and county zoning enabling acts to *allow* local jurisdictions to adopt ordinances "protecting and assuring access to incident solar energy." [57]

ABF-suggested statutes would require municipalities to create three categories of solar overlay zones that would prevail over conventional zoning.[58] In Mandatory Use Districts, if a solar energy system was economically justified in a new structure, or if the energy system of an old structure was replaced, use of solar energy would be required. Skyspace would be protected by city action if private agreements could not be reached.[59] The same backup municipal protection would be offered in Affirmative Solar Use Districts. Building codes in both Mandatory and Affirmative districts would be revised to encourage solar use. The third category, Other Solar Use Districts, would find a city protecting solar skyspace for most uses, and also granting exemptions from other hindering regulations. "Where requirements for solar energy use are applied rationally without discrimination and in relation to a proven need to conserve energy, the statute should be upheld," the ABF concludes.[60]

Melvin Eisenstadt and Albert Utton feel that zoning is the most practical method of creating solar rights, particularly in established neighborhoods.[61] Where the zoning regulations may amount to a taking of property, the possibility of zoning with compensation is suggested. [62] This is a combination of police and eminent domain powers.

Disadvantages inherent to approaches based on zoning include:

1. Zoning boards are notoriously susceptible to local politics and special interest groups and often grant or refuse variances almost on whim.

2. It would be very expensive for a state or locality to intelligently redesign zoning plans.

3. If there are no restrictions in the enabling legislation, zoning laws can typically be changed by only three readings by the relevant local authority.

4. It can be expensive and difficult to appeal zoning decisions.

5. Blanket zoning for solar access may conflict with other energy-conserving techniques: compact, contiguous development, for example, cuts the amount of fossil fuels needed for heating and cooling structures and for transportation.

Land-use planning to provide for solar access   Many solar access conflicts may arise because the value of solar energy was not considered at the design stage. For example, the placement of a building on a lot may determine whether neighboring buildings are shaded.

Many existing controls on construction and land use might be used, with slight modification, to provide for the timely consideration of solar access. Several states—including Arizona, New Mexico, and Virginia—have information and promotional activities that could include educating builders about design criteria for solar energy.[63] A bill in Oregon suggests a more aggressive approach; the extension service program is directed to use county extension agents to disseminate information about solar energy.[64] A similar measure has been proposed at the federal level.[65] Land-use patterns appropriate to solar energy may not coincide with other methods to conserve energy. For instance, dispersed development is essential to some methods of utilizing solar energy. If lots are very large relative to the structures on them, neighbors will not shade one another. But such spread-out development means more fuel must be burned for transportation. Communities should be interested in minimizing total demand for nonrenewable fuels. A holistic approach is suggested by an ordinance adopted in Davis, California.[66] A more traditional approach is to require consideration of energy conservation objectives in comprehensive plans. Comprehensive plans are used in many states to guide long-range policy in local zoning.[67] A growing minority of states require localities to adopt comprehensive land-use plans that conform to standards set by the state.[68] Oregon, for example, requires that energy conservation be included as a goal in all local plans.[69]

Provision for solar energy in comprehensive plans was suggested by the American Bar Foundation,[70] and has so far been considered in at least two states.

Energy impact statements   Another approach that avoids direct regulation, but shifts more of the burden to the builder, is the use of an energy impact statement requirement. Since federal adoption of the National Environmental Policy Act in 1969, more than half the states and many localities have adopted requirements for environmental impact statements in some form.[71] Nine states explicitly require that impact statements discuss the effect of projects on the consumption of energy; two have specifically required discussion of energy conservation.[72]

Since large land developments will come under the impact statement requirement in most states, this procedure might be used to assure consideration of solar energy utilization. A bill in Colorado would require subdivision regulations to include standards and technical procedures for solar energy use. The builder would also have to demonstrate energy-efficient design, e.g., proper orientation of the structure to minimize energy consumption.[73]

Flexible zoning techniques   Although most communities have no direct provision for energy conservation in their land-use planning, many use flexible zoning procedures that could be modified to include solar access in the design process. Flexible land-use approaches encompass a variety of planning techniques characterized by a discretionary governmental review procedure.[74] For example, planned-units developments (PUDs) minimize zoning restrictions and allow developers to propose a layout, building design, and uses, all as one package. Often the local ordinance provides some criteria, but review of the site plan is performed through a flexible, case-by-case procedure. PUDs are flexible enough to incorporate any design objective, including solar access. Developers could be required to indicate the impact of shadows in their proposals and to justify any significant lack of solar access.

Other flexible zoning techniques that provide governmental rewards in return for the developer's attainment of specified objectives also may be applicable. This type of zoning, generally referred to as "incentive" or "bonus" zoning, is most appropriate when the public benefit could not be obtained directly by police power regulation. For example, a Milwaukee ordinance allows increased floor area in exchange for adding plazas, arcades, and other open space around office buildings.[75] This approach might be appropriate to encourage provisions for solar access in high-density areas.

Conclusion

The tentative conclusion is that a combination of approaches will probably work best. Protecting solar access is hardest to do in existing densely built communities. However, future innovations in solar technology may suggest presently undreamed-of solutions for the retrofit problem. Model statutes should be drafted before the prices of fossil fuels skyrocket and there is an immediate demand for solar energy retrofits. Such model statutes should be periodically reviewed in light of technological breakthroughs.

BUILDING CODES

The three most widely adopted model building codes may be significant barriers to widespread use of solar heating and cooling of buildings. These codes are: the *Basic Building Code* of the Building Officials and Code Administrators, International (BOCA), found mostly in the East and Midwest; the *Uniform Building Code* of the International Conference of Building Officials (ICBO), found mostly in the West; and the *Standard Building Code* of the Southern Building Codes Conference (SBCC), found mostly in the South. According to a 1970 survey by Field and Ventre of local building departments, 63 percent of the 919 cities reporting had adopted one of these three model codes if they had a building code at all.[76] Since then, a few states have adopted versions of one of these three codes as statewide mandatory building codes. Hence it is reasonable to study these codes as representative of building codes generally. Locally written codes would probably be less flexible and present equal or greater barriers to solar energy systems.

Nature of Building Codes

A building code is a set of regulations relating to building construction that defines terms, sets standards for materials and equipment, tells how materials and equipment may and may not be put together, and provides for enforcement through permits, inspections, etc. While primarily concerned with health and safety, codes incidentally protect the public from inferior building products.

The standards provisions are generally of two types. Specification standards provide for specific means: what kinds of materials and equipment may and may not be used, and how.

Performance standards, on the other hand, merely provide for ends: what the parts of a structure must be able to do. Specification standards are easier to administer, but are inflexible. Performance standards are flexible and allow for innovation, but require more trained personnel, time, and money to administer. For this reason it may be preferable to integrate solar systems into the present specification-oriented codes.

The principal hazards to health and safety from solar energy systems are derived from the risk of leakage, fire, or explosion (due to damage from high temperatures, pressures, corrosion, other component failure, or impact) and the risk of human contact with hot surfaces or broken glazing. There is also some risk of contamination of drinking water with toxic coolants if plumbing is not properly done, and there are potential structural problems from high winds or snow loads if collectors are not strong enough.

Possible Impediments to Solar Energy Systems

Impediments to innovation   A structural component that is unique to solar systems is the collector. Solar collectors are not dealt with by present building codes. Once heat is collected or electricity is generated, it is transported away, stored, and utilized by means that have been long used in conventional energy systems. Pipes, ducts, valves, storage tanks, controls, wiring, storage batteries, etc. are already provided for, even though their use may be novel.

Alternative materials, equipment, or methods must be shown, to the building officials' satisfaction, to be at least the equivalent of that prescribed in the codes in strength, fire resistance, durability, quality, effectiveness, and safety. Demonstrating that these six criteria have been met may be difficult. Building officials may require testing, at the applicants' expense, by approved testing facilities (chosen by the building officials) using approved test methods (chosen by the building officials unless methods are specifically provided for in the code). For many kinds of nonsolar equipment, nationally recognized standards, test methods, and listing agencies (Underwriters Laboratories for electrical equipment and the American Gas Association for gas equipment, for example) are specified in the codes.

The effect of these administrative provisions is to give building officials discretion that may result in market fragmentation, delay, additional expense, and uncertainty in the processing of permit applications.

General impediments to solar systems   While the lack of standards may be the most important building-code problem for solar systems, there are other possible impediments. Unrealistically high demands on solar heating systems for buildings may result from high minimum temperature requirements (72° F under the ICBO code's standards) and from large minimum ventilation and light requirements for windows in the model codes. Limitations on the overhang of roofs and awnings may thwart some passive solar designs. Other problems suggested by the literature on solar systems, but not found in the model codes, are limitations on building height that prevent collectors on roofs, chimney and plumbing clearances, retroactive application of current standards to old buildings in retrofit situations, roof slope requirements, etc. [77]

Possibility of realization of these impediments   Impediments due to testing and approval procedures and other various requirements have not been reported as a practical problem for solar equipment users so far. It is possible that building officials will continue to look upon solar systems in a friendly way, and never apply the codes to them in the manner that the codes call for. It seems more likely that as solar systems gain popularity there will be some labor unions and manufacturers of particular types of systems who would benefit if current codes are enforced, and who would thus lobby for their strict application. If opposing forces organize before building codes are modified to accommodate solar equipment, it may be much more difficult to change the codes.

Revisions to Codes

Varying local requirements, if applied strictly, could make solar systems less competitive than conventional systems because of the uncertainty, delay, and expense in processing permit applications. They could easily fragment a potential national market into hundreds or thousands of small markets, or result in unnecessarily expensive products designed to meet the strictest standards found anywhere.

The ideal solution would be nationally recognized standards and testing procedures for the various kinds of solar energy systems, and nationally recognized accreditation agencies to certify compliance with these standards and to grant listings. These standards for materials, equipment, and installation would be adopted by reference in all local and state building codes, and listing would be accepted as sufficient proof of code approval if the equipment were installed in compliance with the conditions given in the listing. Equipment standards must include both performance standards and test methods for collectors, storage, and whole systems. The federal government has developed standards for both commercial and residential applications for use in the solar heating and cooling demonstration program. These standards are being used as the basis for private-sector standards that will eventually become nationally recognized standards appropriate for adoption by building codes.

Certification of compliance can be handled in a number of ways. Manufacturers could be permitted or required to certify compliance with the standards themselves, with either civil or criminal sanctions for false certification, or a trade association could certify compliance based on submissions from manufacturers, with periodic testing by an independent laboratory of production samples.

While these kinds of certification may be useful, especially as a stopgap measure and for building consumer confidence, they would not satisfy the usual building code requirements for listing by an approved testing agency. Under the three model codes, the certification must be done by a testing agency qualified and equipped for experimental testing, and for maintaining an adequate periodic inspection of current production; listings of such organizations are not granted on the producer's representations alone, but on the basis of the agency's own testing, including safety tests.

Federal and State Action

Although a federal solar building code or federally mandated standards would probably be constitutional, they are unnecessary. Federal assistance in coordinating and certifying state solar building-code programs would make sense, however.

A variety of state approaches should be acceptable so long as the variations in code requirements are based on state and local conditions and do not unreasonably burden interstate commerce in solar equipment. It would be best if states adopted statewide mandatory requirements for acceptance of solar energy systems that complied with the latest version of the federal standards until the private-sector standards have been developed. A recently enacted Minnesota law in essence does this.[78]

There might be private, state, or federal certification of solar equipment. States will of course want to test for matters bearing on safety, health, and structural strength. If the standards and test methods used are reasonably consistent, predictable, and not unreasonably costly, the federal government should permit the states to require them. State certification for ratings based on thermal performance of collectors is being done in Florida, California, and other states.

In this country, certification has traditionally been a private-sector function of test labs such as Underwriters Laboratories. But federal certification of solar equipment for heating and cooling of buildings has been recommended by some.[79] Indeed, the federal government, which developed the standards in the first place, may be best able to apply the standards for certification. Early experience with federal certification of systems for the HUD demonstration program proved unsatisfactory because of lengthy delays, however, and that experience may lead to opposition to any federal certification program. Self-certification by manufacturers, provided there is some mechanism for assuring that it is done honestly, should be adequate for removing the code burden, provided the state will accept that as equivalent to code approval.

Various code impediments States could write statewide mandatory codes that would revise building codes to remove the problem provisions or exempt solar systems from application of particular provisions so long as the requirements of safety, health, and structural strength are met. Or the federal government may want to support an effort to revise the three major model codes to remove impediments to the effective use of solar energy. This could be done by the major model code groups. Federal legislation could make such a program mandatory nationwide, or could make adoption by the states voluntary, but with incentives to make state adoption likely. The great advantage of the first option is that it would quickly result in nearly uniform nationwide standards. The federal enactment of even a national solar building code, although unprecedented, would certainly be sustained by the courts as within the broad powers of Congress to regulate commerce. Opposition to such precedent-setting legislation might be overwhelming.

An alternative would be to pass legislation requiring states to set their own standards and revise their building codes by a specified date, or have federally determined standards and codes set for them. This could be a fast solution, but may result in enough local variation in solar energy systems to prove a burden on interstate commerce.

UTILITIES

The role of utilities vis-à-vis solar energy is controversial and complex. Public utilities, which currently provide a substantial portion of the energy used to heat buildings, could lose some potential customers if solar-powered heating systems become widespread. Moreover, solar building owners who use electricity as a backup source of energy could cost utilities far more to serve than other residential customers. Since even idle capacity must be paid for, the costs of serving the occasional user may be higher than those for a customer who uses the same amount of electricity, but has a steady demand. The owner of a solar system using electrical heating as a backup may in fact impose a demand at times of utility peak demand. Although the battle has hardly begun, one utility has already tried to retaliate by imposing a rate structure that reflects the *potentially* higher costs of serving solar customers.[80]

On the other hand, some utilities will see an opportunity to profit from participation in the solar energy market. Natural gas companies may soon have to locate alternative sources of energy because proven gas reserves are steadily declining,[81] and at least one gas company has begun experiments with solar-assisted gas heating systems.[82]

A General Overview

Over 75 percent of the electrical generating capacity in the United States is the property of private power companies.[83] Although these companies are privately owned, their operations are regulated by state public utility commissions (PUCs) because of the "natural monopoly" nature of the utility business.

The Federal Energy Regulatory Commission (FERC) sets accounting standards and reporting requirements. Recent regulations issued by the FERC offer some hope for a more active energy conservation program.[84] The commission announced recognition of the shift in public concern for the "proper utilization and conservation of our natural resources including fuels and raw materials as well as air, water and land." Although action reflecting these concerns was left for a later date,[85] utilities were asked to submit more detailed rate reports.

Recent congressional proposals would expand federal regulation of utilities.[86] One proposal, for example, would dictate permissible rates and other essential utility policies.[87] While Congress almost certainly has the power to regulate utilities under the interstate commerce clause, or on grounds of national security,[88] it seems likely that states will continue to exercise primary responsibility for utility practices.

State public utility commissions have generally acted as overseers rather than initiators of policy, although this may be changing in some states.[89] As a practical matter, utility commissions have also lacked the resources and staff to take an aggressive posture. From the standpoint of solar energy use, the crucial regulatory function is rate approval. Typically, state regulatory agencies first decide how much a utility will be allowed to earn, and then approve rate schedules designed to produce the approved profit margin.[90] The rate of return is a function of the rate base (those investments on which the utility may make a profit). A utility decision to market or lease solar collectors would have to be approved by the utility commission before these expenses could be added to the rate base. It should also be noted that the utility profits only if it makes an investment in capital. Therefore, a utility might finance the purchase of solar collectors by homeowners, but it would stand to profit less than from an investment in generating facilities.

Rate structures are also designed to reflect different costs of service. For example, residential consumers have traditionally paid higher rates than large industrial customers because of the lower costs of billing and metering a single large user. Industries willing to accept interruptible service, that is, the possibility of service cutoffs during peak periods, also receive a lower rate.

There is one major exception to the scope of utility commission jurisdiction: publicly owned utilities are usually exempt from state jurisdiction because they are already publicly controlled. Some utility critics view locally owned utilities as one alternative to the unresponsiveness of privately owned systems.[91] Whether or not this argument is valid, in the short run municipal utilities are too small to play a major role in national energy issues. They accounted for only 10 percent of total installed capacity in 1972.[92]

Rate and Service Discrimination [93]

One crucial question in utility regulation is whether utilities may adopt rates or service policies that either favor or hinder the development of solar heating and cooling. At one extreme, public utilities could refuse to provide any backup service on cloudy days. At the other extreme, the utilities could refuse certain services to customers who did not install solar equipment.

State antidiscrimination laws   One of the major purposes for public regulation of electric utilities is the prevention of unreasonable discrimination or undue preferences.[94] Nearly every state has a statute prohibiting conduct that favors one class of customer while harming another. Such antidiscrimination statutes only proscribe policies that are "unreasonable," "unjust," "undue," or "unlawful." [95] Whether a particular utility rate or service unlawfully discriminates is a question of fact to be determined on a case-by-case basis by the state utility commission. [96] It is, therefore, very difficult to predict how any given discriminatory practice will be dealt with.

Two general principles can be culled from the reported decisions. The first is that preferential treatment is more likely to be found reasonable if it produced indirect benefits to all customers. [97] This principle would favor discrimination that benefits solar systems if that discrimination would reduce rates for all customers by reducing the utility's needs for capital equipment and fuel. Some decisions have gone even further by suggesting that for a practice to be unreasonable or unjust it must not only benefit one class of customers, but must also burden another class. [98] A second principle that emerges from the cases is that utilities may treat different classes of customers in different ways if there is a reasonable economic basis for distinguishing them.[99] Thus, if solar customers cost more or less to serve than do other customers, they may validly be charged different rates and receive different services.

State antidiscrimination statutes are not the only bar to discriminatory practices by utilities. The federal antitrust laws may also outlaw rates or services that single out the owners of solar energy systems for special treatment. It is now clear that the antitrust exemption for state action will not totally immunize public utilities from antitrust liability. A recent decision handed down by the Supreme Court said that a privately owned public utility is not exempt from possible antitrust liability when it furnishes its customers light bulbs without charge.[100] The state action antitrust exemption was found not to apply, although the light bulb promotional practice had been approved (as part of the utility's rate structure) by the state public utility commission, and could be discontinued only with the PUC's permission. In reaching its holding the court noted ". . . state authorization, approval, encouragement, or participation in restrictive private conduct confers no antitrust immunity." [101]

If, to protect its monopoly position, a utility charges a very high price or even refuses to provide backup service to solar customers, such refusals may violate the Sherman Act's prohibition against monopolization.[102] An antitrust violation might also be found if a utility subsidizes its entrance into the solar heating and cooling market by distributing its losses across all utility customers. This could give it an overwhelming advantage.[103]

Service discrimination   An extremely important service discrimination issue is whether a public utility can refuse to provide backup electricity for structures with solar hearing or cooling systems. The short answer is that it appears a utility may not — unless it can demonstrate a compelling case that backup service would cause substantial harm to the utility's existing customers. Refusal to provide service would not only transgress the federal antitrust laws and the antidiscrimination statutes discussed above, but would violate the utility's common law and statutory duty to provide utility service.

The major argument against providing backup service is that it requires the utility to build and maintain expensive peaking equipment that would only be used infrequently, i.e., when cloudy periods have drained the storage facilities of solar structures. This argument is of little consequence, as utilities can condition the receipt of solar backup power on the installation of equipment that will draw power from the utility only during nonpeak periods. Even if such a condition did not eliminate the peak demand induced by solar customers, the public interest in fuel conservation might justify the enforcement of the duty to serve.

How will the duty to serve affect discrimination in favor of solar customers? In particular, can a gas company refuse to provide gas connections to new residences that do not install solar heating and cooling equipment? All indications are that a discriminatory practice would be viewed as reasonable. Present natural gas shortages argue strongly for conditioning the receipt of gas on the implementation of various conservation measures. Some states have taken measures to restrict gas to certain customers or to eliminate its availability for some uses. For example,

New York banned the use of gas in swimming pools and in buildings without adequate insulation.[104]

Rate discrimination   The declining block rate schedule, which imposes a higher fee for the first block purchased, is the most common rate structure for residential customers, and was designed to encourage long-run growth in demand.[105] Another common utility rate structure provides a lower overall price to all-electric customers. At the time such rates were adopted, growth was a source of declining costs, and therefore thought to benefit all of a utility's customers. Although this situation does not currently exist, the all-electric rate continues in many places. [106] The current justification by utilities is that the demand imposed by all-electric users is largely off-peak, that is, when the demands on the utility's capacity are low.[107] Since arguably, the all-electric rate is only for those using electricity for heating as well as other needs, solar users might not qualify.[108]

Conflict between solar energy advocates and utilities surfaced in Colorado in the mid-1970s. Public Service Company of Colorado requested a rate schedule for new residential customers designed in part to capture the extra costs imposed by solar-heated dwellings.[109] The rate schedule the company proposed was a demand/energy rate. It has two components: an energy charge, reflecting the total kilowatt-hours used, and a demand charge, based on the maximum kilowatt demand during any 15-min period. The theory underlying this division is that the demand charge reflects the cost of generating capacity, as opposed to the cost of fuel used to serve the customer. Solar energy advocates were extremely critical of the demand charge concept.[110] The impact on solar users could be devastating since the occasional user would pay a relatively high charge for any occasional demand, despite very low amounts of total energy consumption.

Solar advocates questioned whether the demand from solar-heated buildings was likely to coincide with the utility's peak period; if not, no capacity charge was justified. Several studies have attempted to answer this question by simulating the performance of solar-heated buildings, and comparing their needs for backup energy with utility load curves.[111] One recent study examined six different utilities and concluded:

No general statement can be made. . . . This analysis must be performed on an individual basis, since variations in the ambient weather conditions, load curves, and generation mixes of utilities will be the prime determinants in the magnitude of the impact.[112]

The same studies have also noted the importance of thermal energy storage systems as a potentially significant factor. An appropriately designed building with an adequate storage system could always be served off-peak.[113]

The Colorado Public Utility Commission initially granted the utility's request,[114] but following a rehearing decided that there were numerous general questions that should be addressed in a generic rate hearing.[115] During the interim, the demand charge was left as an option.

An alternative approach fair to both homeowners and the utility is time-of-day pricing. Time-of-day structures charge more for power consumed during peak periods and less during other hours, such as late at night. A homeowner with an energy storage system (whether or not it is part of a solar unit) could buy energy during off-peak times, but use it to provide heat during peak periods. The argument is complicated by questions about the added cost of time-of-day meters and utility claims that present off-peak periods are needed to allow for maintenance.

Several other rate structures have been proposed that have different implications for solar users. A few utilities have flat rates for residential customers. Flat rates are simply a set amount per unit of energy, regardless of the amount purchased.[116] This rate structure is neutral with regard to energy savings. Lifeline rates have been adopted in a few states.[117] Under this system, less is charged for the first units of energy. Its goal is to ease the burden on low-income consumers.[118] This rate may incidentally benefit solar users whose needs for supplemental sources of energy are small enough to fall within the "lifeline" amount.

A rate structure that adversely affects solar energy users may be difficult to challenge under current case law. Several cases have upheld the legality of rate structures that subsidize all-electric customers, despite antidiscrimination laws.[119] If a rate structure that provides a *direct* subsidy for the use of one source of energy is legal, then a rate structure that *incidentally* burdens a competing source of energy is, presumably, also valid. Such facile judicial acceptance of promotional rate structures should not be expected in the future. For instance, a New York court recognized the common impact of rising fuel prices in a recent decision overturning a subsidy for all-electric homeowners.[120] The subsidy, which was to run for a year, was intended to lessen the impact of higher electric rates on residential customers who had previously been induced to buy all-electric homes by favorable rates. The court held that the subsidy "constituted undue preference and advantage" in violation of the state antidiscrimination law.[121]

As a result of this change in financial realities, it may be more defensible for public utility commissions to grant subsidies for *conservation* than for promotion of energy consumption. Some states have adopted legislation specifically authorizing conservation programs, eliminating any doubt about their validity. [122]

Regulatory Burdens on Multiuser Solar Systems

In certain areas it may be difficult to retrofit individual existing buildings with solar collectors. The roofs of some buildings may be ill-suited to accept collectors, and others may be shaded by existing structures. Still others with large flat roofs may have excess room for collectors. In such situations there may be substantial advantages to the development of joint or multiuser solar systems —systems that share the available space that is suited for accepting collectors. There may also exist similar advantages to the use of communal heat-storage systems.

Possible advantages of shared solar systems include not only more efficient use of existing space but reduced construction and maintenance costs and increased efficiency. Already, a number of shared solar facilities have begun operation. A 230-unit apartment building in Brookline, Massachusetts is generating hot water from rooftop collectors.[123] And joint systems are being used for the Oakmead Industrial Park in Santa Clara, California,[124] a Denver office complex,[125] and a luxury hotel in the Virgin Islands.[126]

Scope of PUC jurisdiction   In most states, solar systems would not fall within the jurisdiction of state public utility commissions (PUCs), where those systems are operated and owned by a *single* entity on its own property for its own use (as may be the case with a university heating plant that services several dormitories and classroom buildings). And, to the extent that joint systems are operated by municipal utilities within the bounds of the franchising municipality, there should be no PUC jurisdiction in most states.[127]

Any regulatory jurisdiction that does exist over multiuser systems will be at the state level. Neither the Federal Power Commission nor any other federal agency has authority to regulate the production, sale, or shipment of heated or cooled water.[128]

At the state level, PUC jurisdiction over multiuser solar systems will turn on the interpretation of utility commission statutes. While electric utilities are almost universally regulated, regulation of utilities supplying heat or cold is not nearly so pervasive. Nevertheless, some states do have statutes granting the PUC jurisdiction over entities that provide heat or cold to the public.[129]

In these states the key legal issue on which jurisdiction turns is whether the heating or cooling entity is providing its services to the public. In short, a shared solar system will not be found to be a public utility if its energy is not provided "to the public." The majority rule appears to be that a company is serving the public if it has "dedicated its property to public use."[130] Such dedication exists if the entity is serving, or has evidenced a readiness to serve, an "indefinite public" which has a legal right to receive service.[131] It is difficult, therefore, to predict whether multiuser systems will be subject to the burdens of PUC regulation.

Consequences of PUC jurisdiction   There are several reasons why the owners of a shared solar system should fear PUC jurisdiction. If a shared solar system is found to be a public utility, it must file reports and accounts,[132] serve all customers who demand service within a given area,[133] submit its rate schedules to the PUC for approval,[134] continue providing service until given permission to discontinue,[135] provide safe and adequate service,[136] and comply with limitations on the issuance of securities.[137]

But perhaps the most significant burden that PUC jurisdiction would place on shared solar systems would be the duty to apply for certificates of public convenience and necessity. State utility regulatory statutes universally require that every public utility must obtain a certificate before beginning operation or even construction of its equipment.[138] Not only are certification proceedings often long and expensive, but the PUCs use the certification process to protect the monopoly of existing utilities. The general rule is that an existing utility shall be given a monopoly in its area unless the public convenience and necessity require otherwise. In practice this means that even where the existing utility is providing woefully inadequate and inefficient service, it will be permitted to exercise monopoly control over its service area if it promises to correct its shortcomings.[139]

Would a multiuser solar system found to be a public utility be certified to provide heat and cold to areas being served by existing utilities? If the existing utility already provides heat and cold, the existing utility will probably be permitted to retain its monopoly in the absence of some overwhelming reason to the contrary. The question is more problematical where the existing utility is providing heat and cold indirectly (by selling gas or electricity). A study completed for the Energy Research and Development Administration concludes that there is substantial case precedent for certifying solar energy systems despite the fact that conventional facilities exist for providing heat and cold.[140] The precedents cited in that study suggest that new energy forms should be permitted to compete with existing energy forms if the new form is cheaper, cleaner, or in some way more efficient.[141]

Nevertheless, it is clearly within the authority of some PUCs to deny certification. The mere possibility of PUC jurisdiction over shared solar facilities, and the threat that such jurisdiction may be used to prevent operation of the facilities, is a substantial barrier to the development of joint solar systems.

Legislative action may be the only feasible approach, and could take several forms. A law might simply state that the public interest demands that shared solar facilities be permitted to compete with existing utilities. Such an approach would not preclude PUC regulation of other aspects of joint solar heating and cooling plants. A more drastic approach would be for the legislature to completely exempt solar facilities from PUC jurisdiction. A related proposal, dealing with electrical generators, has been put before the California legislature.[142] It could easily be broadened to encompass suppliers of heat and cold.

Public Utilities and Solar Commercialization

Utility participation in the solar energy market might come in a variety of forms, from simply financing homeowners' purchases of collectors, an approach used in some states to help homeowners install insulation,[143] to the actual provision of solar collectors by utilities.

Many utilities are already considering such programs. The Southern California Gas Company, for example, is testing the use of solar-assisted gas heating for apartment buildings.[144] Other utilities also experiment with solar energy,[145] and the utility-funded Electric Power Research Institute has a division devoted exclusively to solar energy projects.[146]

The merit of utility participation in the solar market is a hotly contested issue.[147] Utility advocates point to several possible advantages:

Since the construction industry is highly "first-cost sensitive," we expect that solar energy will have some difficulty finding early, rapid acceptance. A utility company is used to high first-cost (capital intensive) business ventures. Utility company sponsorship in the "lease to the user" mode will do a lot to reduce this barrier . . .

Second, the sponsorship of a utility company may help overcome market "fragmentation." If the utility company buys the equipment and leases it in a large-scale fashion, the solar industry will face at least one aggregated market (to the gas company). This may provide a large enough incentive to actively stimulate a solar energy system fabrication industry.

Third, a utility company already has a sales/distribution/service network . . .

Finally, . . . utility company sponsorship will help overcome some of the traditional "institutional-cultural biases" against solar energy which exist within the housing industry. [148]

On the other hand, utility critics have been quick to raise the spectre of utility "ownership of the sun," with the attendant evils of "excessive profit-taking and monopolistic favoritism in equipment purchases."[149] Specialists in utility economics have also raised serious questions about the desirability of using utilities to promote solar energy. Roger Noll, although he ultimately concludes that a limited form of utility involvement may be desirable, notes two dangers:

. . . a regulated utility has an incentive to invest in solar technology that is too durable, that is excessively efficient in converting sunlight to usable energy, and that requires inefficiently little maintenance. If permitted this would lead to excessive costs and prices for solar energy, and inefficiently slow adoption of the technology.

Second, regulated utilities can use solar technology strategically to recapture some of the monopoly profits that regulation takes away and to foreclose competition in the solar energy business.[150]

As some form of utility participation in the solar market may occur, it is appropriate to examine the legal framework in which utility participation in the solar market will be regulated. Several alternative regulatory policies will be discussed, and their legal consequences distinguished.[51] First, utilities might ask for a monopoly on the distribution of solar systems. They might hope to do this by denying backup energy to persons not using utility-supplied solar equipment. The utility could either rent or sell the equipment to the customer, but no other business could market competitive systems. Such a program would be extremely controversial;

the necessary regulatory approval is unlikely. It is difficult to imagine any justification for the creation of a monopoly in solar equipment sales.

Exclusive marketing rights would also probably run afoul of federal antitrust laws or state policies against anticompetitive practices. The Supreme Court decision *Cantor* v. *Detroit Edison*, discussed previously, limited state activity to provide exemptions if the challenged activity is central to the purposes of a state's regulatory program.[152] The light bulb exchange program under attack failed to meet the test since "there is no reason to believe that [without the program] Michigan's regulation of its electric utilities will no longer be able to function effectively."[153] A regulatory authority would have to offer a more convincing rationale for a program that even more clearly contravened federal antitrust principles.

A utility is likely to view a regulated mode as desirable because of the opportunities for cost sharing and risk spreading. But whether or not the solar business is regulated, close scrutiny by the public utilities commission will be desirable. The regulatory process is certainly accustomed to the notion of cost sharing; lifeline rates, for example, diverge from simple cost-of-service principles, but are justified by other social objectives. The conservation of nonrenewable resources could easily be recognized as a benefit to all consumers of the utility, and therefore warrant some sharing of expenses from the solar business.[154] In any case, such cost sharing is likely to be tightly constrained by federal antitrust laws. Utility practices that are not explicitly authorized by utility commissions *and* perhaps by legislative action as well may be vulnerable to treble damage suits, a very effective weapon. Any effort to destroy competition by selling below market rates would very likely be challenged, whether or not the lower price is attributable to lower costs or profits.[155]

An alternative to utility participation in the solar energy market is to restrict such activities as much as possible. The extent to which such prohibitions could be imposed depends on the ability of a PUC to assert jurisdiction over the offending activity. There is no legal basis for seeking jurisdiction unless the challenged activity affects the utility's regulated business.[156] However, it seems likely that a sufficient nexus between solar energy and other energy services exists to justify jurisdiction *if* the PUC chooses to exercise it. As a mixed question of fact and law, the agency's judgment is likely to receive only limited deference. On the other hand, there are instances where utility companies were essentially forced to accept limitations on their outside activities.[157]

As an alternative to the distribution of solar equipment, utilities might undertake to act simply as financiers or insurers. This may be an undesirable role from the standpoint of the utility since the profit allowed on its loans is likely to be less than the utility's usual rate of return on investments. Borrowing for solar purposes would also compete with more profitable utility programs, increasing their tremendous capital needs. There are precedents in the insulation financing programs discussed earlier, but the amount of money that would be involved in solar systems is substantially greater—insulation is usually a matter of a few hundred dollars; a solar system costs several thousand and up. An assertive PUC might try to force a utility to finance solar purchases. The utility's certificate of operation is a license subject to conditions on whatever terms the regulatory agency believes necessary.[158]

Conclusions

In each of the three areas discussed, service and rate discrimination, regulatory jurisdiction, and utility participation in the solar market, much uncertainty exists over appropriate regulatory policies and the impact of current law. From the perspective of conservationists and solar energy advocates, this problem is compounded by differences among utilities and states. Very few generalizations are possible. Examining these issues for each utility could be slow, complex, and expensive. This situation could very well impede the commercialization and acceptance of solar energy systems. Utilities are not likely to risk significant sums without some assurance of protection from the antitrust laws. Other distributors may be reluctant to start their own retail businesses if utilities are expected to enter the market. Homeowners will want to know the net cost and savings of their solar systems, a calculation that depends on expectations about future rate structures and available sources of auxiliary energy. Builders of multifamily dwellings may think twice about installing a solar system if they may be subject to PUC regulation.

The time is clearly ripe for legislation and administrative action. The federal government should address those technical issues, such as appropriate methodologies for evaluating the impact of solar systems on utility load patterns, that are common to every state and utility. This is already being done to some extent; several of the studies cited earlier were funded by the federal government. But a larger, more systematic effort in cooperation with utility regulators and utility representatives is appropriate. The federal government should also offer a clearinghouse for technical information and assist states in the formulation of policy agendas.

State legislatures must decide the broad policy issues involved in solar utility relationships. For example, a decision to subsidize the use of solar collectors can be clarified by technical studies about the effect of direct incentives for energy conservation. Political judgments must also be made about the importance of the broad public interest in conservation of nonrenewable fuels, reduced dependence on foreign oils, etc. Moreover, a political decision must be made as to the relative merits of different forms of incentives—tax credits or loan subsidies may be a more equitable and efficient approach than the use of utility rate structures. Since these alternatives are not available to PUCs, state legislatures must make these choices. The federal government may also play some role.

Within the broad policy established by the state, considerable discretion must still be exercised by public utility commissions. The specifics of rate structures, scope of regulatory jurisdiction, and particular utility programs are too technical to be decided by legislative bodies. State legislation should require PUCs to investigate and recommend appropriate policies, subject to legislative review.

POCKETBOOK PROBLEMS: HOME FINANCING, PROPERTY TAXATION, AND MANDATORY INSTALLATION

Home Financing

Because solar homes cost more than conventional ones, persons seeking to finance or retrofit homes with solar energy systems may encounter legal-economic barriers.

Federally chartered savings and loan companies (representing three-fifths of the assets of all savings and loans) are regulated by the Home Owners' Loan Act of 1933.[159] This act says that if a savings and loan makes a home loan of over $55,000, the entire amount of the loan must be put into a "basket" that can never hold more than 20 percent of the corporation's assets. This is important because these imaginary baskets fill quickly. This limit may make it impossible for average-income families to finance average-priced solar homes. The law could be amended to raise the $55,000 limit; or to state that only dollars in excess of $55,000 (and not the entire amount of a loan) must go into the basket; or the law could make exceptions on public policy grounds for energy-conserving homes. For the few homes participating in HUD's demonstration program, the maximum dollar limit is raised by the amount a solar heating or cooling system exceeds the cost of a conventional system.[160] Legislation has already been proposed that would increase (by 20 percent) the loan amount that may be made, insured, or purchased by the Farmers Home Administration, the Federal Housing Administration, and the Government National Mortgage Association, if the increased purchase price is a result of solar heating or cooling equipment.[161] An even better approach, however, is to raise the ceiling, but to also require the use of life-cycle costing when a property is assessed.

Lenders are concerned about the reliability of solar systems, and about the need to finance two separate heating systems, because building officials or lenders often require a 100 percent backup conventional heating system.[162] When appraising a home, some lenders would exclude part or all of the costs of a solar system, although the Federal Home Loan Mortgage Corporation will finance 90 percent of the extra solar cost.[163]

Another type of problem is that many lenders use borrower underwriting criteria that exclude consideration of the cost of heating and cooling a home when they assess an applicant's ability to pay. Nevertheless, all lenders surveyed for a recent study said that energy costs would become increasingly important to their lending decisions, and the study recommends the consideration of home energy costs in lender underwriting procedures. They found that excluding energy costs when figuring an applicant's ability to pay "may be an important constraint on the availability of financing for solar homes."[164]

Life-cycle costing is essential to an appreciation of solar energy. Since nearly all the expense of a solar energy system is the initial cost, its financial advantages over a conventional system are apparent only over a period of use. Industry representatives say that savings-and-loans are open-minded about life-cycle costing, and that perhaps one-third are already using this tool in their evaluations of projects. [165]

This statement is only partly corroborated by the Regional and Urban Planning Implementation study, which found that a "sizable percentage" of lenders surveyed were probably concerned with information on both payback periods and life-cycle costs. Unfortunately, many lenders were unfamiliar with even the concept of life-cycle costing, and very few would regard it as a more useful tool than payback period or capitalized value.[166]

The two main public entities in the secondary market are the Federal Home Loan Mortgage Corporation and the Federal National Mortgage Association. FHLMC encourages low-and moderate-income housing. Partly for this reason they have adopted the same $55,000 figure as the Home Owners' Loan Act, and will not buy mortgages over this amount. Private entities in the secondary market are not limited by this figure.[167] It is the position of both these institutions that the expenses associated with solar equipment should not be mortgagable until they achieve more "market acceptance."[168]

Financing Solar Retrofits

A different set of legal problems is encountered when loans are sought to retrofit homes with solar equipment. Once again the Home Owners' Loan Act of 1933 allows federally chartered savings-and-loans to make only *first* liens on residential properties. Similar restrictions are generally included in state charters.[169]

Since personal installment loans or homeowners' improvement loans generally involve high interest rates, homeowners would prefer lower-interest long-term second mortgages.

Property Taxation

Opinion seems unanimous that solar equipment will add to a structure's assessed value. But to include this addition in assessments made to determine the amount of tax to be levied on a property may be unfair. Many states have passed or are considering legislation that addresses this problem. Enacted laws typically say that solar equipment shall not cause an increase in the valuation of a building.[170]

There may be general legal problems with exempting solar equipment from property tax assessments, since the majority of states have what are known as "uniformity clauses" in their constitutions and/or in their tax laws. Uniformity clauses say that all "similarly situated" property or the "same class of subjects" must be taxed at the same rates. The language and the interpretation of such clauses vary widely from state to state, making it very difficult to predict the success of an effort to exempt solar heating and cooling equipment. In some states, constitutional changes may be needed.

Furthermore, legislation is usually vague on how backup heating systems should be assessed. In most climates, building and health codes require structures with residential occupancy to be equipped with heaters capable of warming habitable rooms to specified temperatures. Requirements vary greatly, but as massive solar storage systems capable of outlasting weeks of cloudy weather are not now cost-effective, most solar homes will require backup systems.

One popular type of state legislation says solar homes shall be assessed as if equipped with a conventional system[171] or "at no more than" the value of a conventional system.[172] Such laws could be interpreted as requiring a "double assessment," i.e., the value of the backup system plus the adjusted value of the solar system.

A similar precision should be sought in statutory definitions of "solar energy system" and like terms. Not only should the exclusion (or inclusion) of legally required backup systems be specified, but it should be made clear whether passive solar systems are also to receive preferential tax treatment.

State legislation should also be cognizant of the fact that some assessments are made on the basis of a property's income production.

Mandatory Installation of Solar Equipment

Some proposals for expediting use of solar energy would require use of the technology in specified circumstances.[173] Less draconian measures have also been suggested (and adopted in Florida) to provide that buildings are constructed so as to make the later addition of solar devices a little easier.[174]

The principal legal questions governing the legality of such requirements is whether they constitute a "taking without compensation" in violation of the Fifth Amendment to the U.S. Constitution. Although this question is difficult to answer with any certainty, there are numerous precedents upholding requirements for building design and construction methods. However, the application of these ordinances to *specific buildings* may be held invalid where the property owner can demonstrate that the costs clearly exceed the benefit.[175] Thus, while requirements for provision of solar energy are likely to be upheld in general, they may be invalidated in the case of specific properties. This is likely to be a much greater barrier to retrofit requirements than to similar demands on new buildings, since the latter will generally be affected equally.

The federal government might become involved in mandatory installation requirements through the Minimum Property Standards of the Department of Housing and Urban Development. No additional legal barriers exist to prevent federal standards.

REFERENCES

1. James D. Phillips, *Assessment of a Single Family Residence Solar Heating System in a Suburban Development Setting,* Solar Heated Residence Annual Research Report, prepared with the support of the National Science Foundation (NSF-RA-N75078, July 1975), p. 92, n. 77; scrutiny of aerial photographs of Long Island, New York, revealed that over half of all existing buildings were suitable for solar water heating and that all new buildings could be equipped with solar water heaters. Fred S. Dubin, *Analysis of Energy Usage on Long Island from 1975 to 1995: The Opportunities to Reduce Peak Electrical Demands and Energy Consumption by Energy Conservation, Solar Energy, Wind Energy and Total Energy Systems,* sponsored by the Suffolk County Department of Environmental Control (Suffolk County, NY, 1975).
2. In the June 1976 issue of *Professional Builder,* on pp. 105–106, Ralph Johnson (vice president of NAHB Research Foundation), Fred Dubin, and D. Elliot Wilbur, Jr. all agree that energy-saving designs are as important as active solar equipment.
3. Such as the $4000 A-frame solar furnace offered by Waverly Homes in Denver.
4. *Solar Utilization News,* October 1976, p. 7.
5. "Solar," *Sunset,* November 1976, pp. 78, 86.
6. Right to Light Act, 1959, 7 & 8 Eliz. 2, c. 56, §§2, 3.
7. 114 So.2d 357, 181 Fla. Supp. 74 (1959).
8. Melvin M. Eisenstadt and Albert E. Utton. "Solar Rights and Their Effect on Solar Heating and Cooling," *Natural Resources Journal,* 16 (1976): 368. They note, however, that the doctrine was generally upheld in the United States in the first half of the 19th Century. Ibid., p. 367.
9. *See Cohen* v. *Perrino,* 355 Pa. 455, 50 A.2d 348 (1947); and cases cited in Annot., 133 A.L.R. 692 (1941).
10. 287 N.E.2d 677, 52 Ill. 2d 301, *cert. denied,* 409 U.S. 1001 (1972).
11. Ralph E. Becker, Jr., "The Common Law—An Obstacle to Solar Heating and Cooling?" *Journal of Contemporary Law* (Winter 1976–77 issue; page references below refer to May 1976 draft).
12. Ibid., p. 12.
13. Karin H. Hillhouse, *Solar Energy and Land Use in Colorado: Legal, Institutional, and Policy Perspectives,* Interim Report of the Solar Energy Project to the National Science Foundation (Washington, DC: Environmental Law Institute, April 1976), p. 33. *See also* Phillips, *Assessment of Single Family Residence Solar Heating System,* pp. 127–129.
14. Donald N. Zillman and Raymond Deeny, "Legal Aspects of Solar Energy Development," *Arizona State Law Journal,* 1976: 25.
15. Becker, "The Common Law," p. 11.
16. Ibid., p. 16.
17. *See* David Bersohn, "Securing Insolation Rights: Ancient Lights, Nuisance, or Zoning?" *Columbia Journal of Environmental Law:* 25.
18. Zillman and Deeny, "Legal Aspects of Solar Energy Development," p. 54.
19. William L. Prosser, *Law of Torts,* 4th ed. (St. Paul: West Publishing Co., 1971), §86, p. 573.
20. Ibid., §88, p. 586.
21. Ibid., §91, p. 611.
22. Ibid., §87, p. 577.
23. *Proceedings of the American Bar Foundation Workshop on Solar Energy and the Law,* Interim Report to the National Science Foundation (Chicago: American Bar Foundation, 1975), p. 20.
24. For an excellent discussion of this very complicated body of law, *see* Phillip Soper, "The Constitutional Framework of Environmental Law," in *Federal Environmental Law,* eds. Erica L. Dolgin and Thomas G. P. Guilbert (St. Paul: West Publishing Co., 1974).
25. Hillhouse, *Solar Energy and Land Use in Colorado,* p. 33.
26. For instance, the easement is extinguished if the two lots become owned by the same person; if both parties sign a written release; if the owner of the privilege does a physical act that shows intention to *never* make use of the easement; if the owner of the land subject to the restriction uses the land for a long time in a manner inconsistent with the easement; or if the owner of the restricted land changes position in reasonable reliance on the statements or acts of the easement holder.
27. Richard L. Robbins, "Building Codes, Land Use Controls and Other Regulations to Encourage Solar Energy Use" (paper presented at the Consumer Conference on Solar Energy Development, Albuquerque, NM, Oct. 2–5, 1976), p. 9. *See Ayres* v. *City Council of Los Angeles,* 34 Cal.2d 31, 207 P.2d 1 (1949).
28. *See,* e.g., *Wing* v. *Forest Lawn Cemetery Assn.,* 15 Cal.2d 472, 480, 101 P.2d 1099, 1103 (1940); *Rogers* v. *Reimann,* 277 Or. 62, 69, 361 P.2d 101, 105 (1961).

29. Norman Williams, Jr., *American Land Planning Law*, 5 vols. (Chicago: Callaghan & Co., 1974–75), 5: §154.02.
30. *See generally* Note, "Legal and Policy Conflicts Between Deed Covenants and Subsequently Enacted Zoning Ordinances," *Vanderbilt Law Review*, 24 (1971): 1031.
31. *Schulman* v. *Sherrill*, 432 Pa. 206, 246 A.2d 643 (1968).
32. *Hysinger* v. *Mullinax*, 204 Tenn. 181, 319 S.W.2d 79 (1958).
33. *See generally* Note, "Beyond the Eye of the Beholder: Aesthetics and Objectivity," *Michigan Law Review*, 71 (1973): 1438; Basile J. Uddo, "Land Use Controls: Aesthetics, Past and Future," *Loyola Law Review*, 21 (1975): 851; Comment, "Planning and Aesthetic Zoning—Getting More Out of What We've Got," *Journal of Urban Law*, 52 (1975): 835; Williams, *American Land Planning Law*, 1: §11, 3: §71; Comment, "Architecture, Aesthetic Zoning, and the First Amendment," *Stanford Law Review*, 28 (1975): 179; and Comment, "Aesthetics Off the Pedestal: Massachusetts Supreme Judicial Court Upholds Aesthetics as Basis for Exercise of the Police Power," *Environmental Law Reporter*, 6 (1976): 10036.
34. *See* cases cited in Williams, *American Land Planning Law*, 1: §11.14, and ordinances summarized ibid., 3: §71.14.
35. Ibid., 1: §11.15.
36. Ibid., §11.19–11.21.
37. 339 N.E.2d 709, *Environmental Law Reporter*, 6 (1976): 20123 (1975).
38. 339 N.E.2d at 717, *Environmental Law Reporter*, 6 (1976): 20123, 20125 (1975).
39. *See, e.g.*, *Washington Post*, June 25, 1976, p. Cl, and ibid., Sept. 1, 1976, p. A14.
40. One of the grass-cutting cases was decided in Wisconsin, one of the states that fully supports esthetic regulation.
41. George Lof, quoted in "Professional Builder's Report on Solar Energy," *Professional Builder*, June 1976, pp. 101, 106.
42. W. Thomas, A. Miller, and R. Robbins, *Overcoming Legal Uncertainties About the Use of Solar Energy Systems*, American Bar Foundation, Chicago, 1978.
43. Mary D. White, "The Allocation of Sunlight: Solar Rights and the Prior Appropriation Doctrine," *University of Colorado Law Review*, (1976): 423.
44. Thomas, Miller, and Robbins, *Overcoming Legal Uncertainties About the Use of Solar Energy Systems*.
45. Ibid.
46. Ibid.
47. Ibid.
48. Ibid.
49. Ibid.
50. Ibid.
51. Ibid.
52. White, "Allocation of Sunlight."
53. Ibid., p. 441.
54. Peter Wolf, *The Future of the City: New Directions in Urban Planning* (New York: Watson-Guptill Publications, Whitney Library of Design, 1974), p. 149.
55. Robbins, "Building Codes, Land Use Controls and Other Regulations to Encourage Solar Energy Use," p. 17.
56. Thomas, Miller, and Robbins, "Legal Issues Related to Use of Solar Energy Systems."
57. 1975 Or. Laws ch. 153.
58. Thomas, Miller, and Robbins, "Legal Issues Related to Use of Solar Energy Systems."
59. Ibid.
60. Ibid.
61. Eisenstadt and Utton, "Solar Rights," p. 413.
62. Ibid., p. 386.
63. *See* National Conference of State Legislatures Renewable Energy Project, *Turning Toward the Sun*, 2 vols. (Denver, RANN Document No. NSF-RA-G-75-052, n.d.), 1:27–29.
64. Or. H.J.R. 3 (1975).
65. S. 3105, 94th Cong., 2d Sess. (1976) (Energy Extension Service Bill). *See also* Alan Hirshberg, "Public Policy for Solar Heating and Cooling," *Bulletin of the Atomic Scientist*, October 1976, pp. 37, 42.
66. Ordinance No. 784, adopted by the City Council of Davis, CA. (Oct. 15, 1975).
67. *See generally* Donald Hagman, *Urban Planning and Land Development Control Law* (St. Paul: West Publishing Co., 1971).
68. *See* Corbin Harwood, *Using Land to Save Energy* (Cambridge, MA: Ballinger Publishing Co., 1977).
69. Oregon Land Conservation and Development Commission, "Statewide Planning Goals and Guidelines" (1 Jan. 1975), Goal 13.
70. Thomas, Miller, and Robbins, "Legal Issues Related to Use of Solar Energy Systems."
71. *See generally* Kenneth Pearlman, "State Environmental Policy Acts: Local Decision Making and Land Use Planning," *Journal of the American Institute of Planners*, 43 (1977): 42.
72. Harwood, *Using Land to Save Energy*.
73. Colo. H.B. 1166 (1976).
74. *See generally* Michael Meshenberg, *The Administration of Flexible Zoning Techniques*, American Society of Planning Officials Advisory Service No. 318 (Chicago, 1976).

75. Ibid., p. 43.
76. Charles G. Field and Steven R. Rivkin, *The Building Code Burden* (Lexington, MA: D. C. Heath and Co., Lexington Books, 1975), p. 43.
77. *See*, e.g., Thomas, Miller, and Robbins, "Legal Issues Related to Use of Solar Energy Systems."
78. Minn. Stat. §116 H.127 (1976).
79. E.g., Elizabeth C. Moore, "No News is Bad News," *Solar Age*, 1 (December 1976), p. 10, 13.
80. The Public Service Company of Colorado cited the increasing use of solar heating as one justification for a controversial new rate structure. *See* Ref. 109 below.
81. *See generally* William Rosenberg, "Conservation by Gas Utilities as a Gas Supply Option," *Public Utilities Fortnightly*, Jan. 20, 1977, p. 13. The validity of claims that proved gas reserves are declining is currently the subject of intense controversy. *See*, e.g., James Miller, "Natural Gas: The Hidden Reserves," *Washington Post*, Feb. 13, 1977, p. Cl. Whatever the outcome of this statistical debate, it remains likely that natural gas will never be as cheap or taken for granted as it once was.
82. Solar-assisted gas energy, or SAGE, is the name of a project of the Southern California Gas Company. *See* Stephen L. Feldman and Bruce Anderson, *The Public Utility and Solar Energy Interface: An Assessment of Policy Options*, prepared for the Energy Research and Development Administration [ERDA Contract No. E(49-18)-2523, 1976], p. 22; and Alan Hirshberg, "Public Policy for Solar Heating and Cooling," *Bulletin of the Atomic Scientist*, October 1976, pp. 37–40. The company recently requested a rate increase to pay for a solar demonstration project. *Solar Energy Intelligence Report*, Feb. 14, 1977, p. 31. A recent survey found over 100 utility projects involving solar energy, most of them in the area of heating and cooling buildings.
83. *Public Power*, January–February 1975, p. 28.
84. Chapter 1—Federal Power Commission, Part 35—Filing of Rate Schedules, 40 Fed. Reg. 48673 (1975).
85. Ibid., p. 48674.
86. E.g., H.R. 12461, 94th Cong., 2d Sess. (1976). *See* Robert Samuelson, "Battle Lines Are Being Generated for Reform of Electric Utility Rates," *National Journal*, Oct. 16, 1976, p. 1474; Hal Willard, "Electric Power: The Struggle Over Controls," *Washington Post*, Aug. 18, 1976, p. C3; The Public Utilities Policies Act of 1978 resolved some of these issues.
87. Willard, "Electric Power," p. C3.
88. A possible limitation is suggested by the Supreme Court decision in *National League of Cities* v. *Usery*, 426 U.S. 833 (1976). The Court held that federal regulations of the wages paid by state governments to their employees constituted an unconstitutional infringement on state sovereignty. The Court also granted certiorari in a Clear Air Act case testing the limits of federal authority to order indirect source controls such as parking bans and automobile inspection programs. *Brown* v. *EPA*, 521 F.2d 827 (9th Cir. 1975), *cert. granted*, 426 U.S. 904 (1976). Direct federal regulation of utilities would not be affected by these cases, but proposals to require state programs could be.
89. Samuelson, "Reform of Electric Utility Rates." *See also* Richard Morgan and Sandra Jarabek, *How to Challenge Your Local Electric Utility* (Washington, DC: Environmental Action Foundation, 1974).
90. Edward Berlin, Charles J. Cicchetti, and William J. Gillen, *Perspective on Power* (Cambridge, MA: Ballinger Publishing Co., 1974); Thomas B. Stoel, Jr., "Energy," in *Federal Environmental Law*, eds. Erica L. Dolgin and Thomas G. P. Guilbert (St. Paul: West Publishing Co., 1974), p. 59.
91. "Utilities and Solar Energy: Will They Own the Sun?" *People and Energy*, October 1976, p. 2; Mark Northcross, "Who Will Own the Sun?" *The Progressive*, April 1976, pp. 14, 16; and Richard Morgan, Tom Riesenberg, and Michael Troutman, *Taking Charge: A New Look at Public Power* (Washington, DC: Environmental Action Foundation, 1976).
92. *Public Power*, January–February 1975, p. 28.
93. The analysis in these sections draws heavily from existing treatises on public utility regulation, particularly A. J. G. Priest, *Principles of Public Utility Regulation*, 2 vols. (Charlottesville, VA: Michie Co., 1969); and George E. Turner, *Trends and Topics in Utility Regulation* (Washington, DC: Public Utilities Reports, Inc., 1969). For an extremely lucid discussion of the economic issues raised by utility regulation, the reader is directed to Alfred E. Kahn, *The Economics of Regulation*, 2 vols. (New York: Wiley, 1970–71), 2: 76. For a review of more recent developments, *see* Berlin, Cicchetti, and Gillen, *Perspective on Power;* Charles Cicchetti, William J. Gillen, and Paul Smolensky, *The Marginal Cost and Pricing of Electricity: An Applied Approach* (Springfield, VA: National Technical Information Service, NTIS PB 255 967, 1976); and Samuel Huntington, "The Rapid Emergence of Marginal Cost Pricing in the Regulation of Electric Utility Rates Structures," *Boston University Law Review*, 55 (1975): 689; rate and service discrimination was addressed by the Public Utilities Policies Act of 1978.
94. To economists "price discrimination" is value-neutral and includes any case where the same product is sold at more than one price. For purposes of this discussion, "discrimination" is used in its more general sense to refer to any distinction in favor of or against a person. The economists' definition pinpoints the issue nicely: what is the relevant "product" or service? The way the product is defined will determine a fair price.
95. Priest, *Principles of Public Utility Regulation*, 1: 286–88.
96. Ibid.
97. Belnap, McCarthy, Spencer, Sweeney, and Harkaway, "Memorandum: Legal and Regulatory Analysis of Conservation Proposal for the Federal Energy Administration, Energy Resource Development" (Dec. 8, 1976), p. 14.

98. Priest, *Principles of Public Utility Regulation*, 1:295, 300–02. *California Portland Cement Co.* v. *Union Pac. R.R.*, 12 P.U.R.3d 482, 485–86 (Cal. Pub. Util. Comm. 1955).
99. Priest, *Principles of Public Utility Regulation*, 1:288.
100. *Cantor* v. *Detroit Edison Co.*, 96 S.Ct. 3110 (1976).
101. Ibid., p. 3116.
102. Refusals to deal are a classic violation of section 2 of the Sherman Act, 15 U.S.C. §2 (Supp. IV 1974). *See, e.g., Otter Tail Power Co.* v. *United States*, 410 U.S. 366 (1972), where a public utility was found to have violated section 2 of the Sherman Act by refusing to sell electricity to a municipally operated distribution system.
103. Such conduct could be viewed as temporary price cutting to put rival solar firms out of business. *See Porto Rican American Tobacco Co.* v. *American Tobacco Co.*, 30 F.2d 234 (2d. Cir 1929). Or, it might be viewed as an illegal tying arrangement in situations where a solar customer's receipt of favorable treatment is conditioned on acceptance of the utility service. Tying arrangements are another classic antitrust violation. *See* 15 U.S.C. §14 (1970); *International Business Machine Corp.* v. *United States*, 298 U.S. 131 (1936).
104. New York Pub. Serv. Comm., Case 26286 (April 16, 1974); and *National Swimming Pool Institute* v. *Alfred Kahn*, 364 N.Y.S.2d 747, 9 P.U.R.4th 237 (1974). *See also* "Ban on Heated Pools Leaves Californians Boiling," *The New York Times*, Feb. 5, 1976.
105. *See* Berling, Cicchetti, and Gillen, *Perspective on Power*, Chaps. 1–3.
106. The all-electric customer is given a small advantage. *See, e.g.*, Federal Power Commission, *National Electric Rate Book: Colorado* (Washington, DC: Government Printing Office, August 1975), p. 3. (Rate books are published for each state and updated periodically.)
107. Letter insert in monthly bill from Potomac Electric Power Co. to customers (January 1977).
108. In conversations with officials at the Energy Research and Development Administration, the authors were told that denials of all-electric rates to solar users have already occurred.
109. Testimony of James H. Ranniger, manager of rates and regulation for the Public Service Company of Colorado, Colorado Public Utilities Commission Investigation and Suspension Docket No. 935 (Sept. 22, 1975), p. 14.
110. Testimony of Ernst Habict, Jr. and William Vickrey for the Environmental Defense Fund, Colorado Public Utilities Commission Investigation and Suspension Docket No. 935 (Sept. 25–26, 1975). *See generally* Gary Mills, "Demand for Electric Rates: A New Problem and Challenge for Solar Heating," *ASHRAE Journal*, January 1977, p. 42.
111. Results of these studies are summarized in G. F. Swetnam and D. M. Jardine, *Energy Rate Initiatives Study of the Interface Between Solar and Wind Energy Systems and Public Utilities*, prepared for the Federal Energy Administration (Mitre Corporation, Technical Report 7431, draft, Dec. 20, 1976); and Feldman and Anderson, *Public Utility and Solar Energy Interface*.
112. Stephen L. Feldman and Bruce Anderson, *Utility Pricing and Solar Energy Design* (NSF/RANN Grant No. APR-75-18006, 1976), p. 117.
113. "Present solar buildings do not generally avail themselves to the exclusive use of off-peak electric power but generally will use a portion of their auxiliary energy during peak periods. This situation can be remedied by modifications to the control system, storage and collector size, use patterns or a combination of these factors. The building owner may incur increased capital costs, which may be offset by a decrease in his energy bill." Ibid., p. 28. This analysis only holds true as a description of heating needs. The technology for cooling with solar energy is less advanced; solar energy cooling and storage could be analyzed in the same way but remains more hypothetical than real. Ibid., p. 18. *See also* S. A. Klein, W. A. Beckman, and J. A. Duffie, "A Design Procedure for Solar Heating Systems," *Solar Energy*, 18 (1976):113.
114. In the Matter of Proposed Increased Rates and Charges Contained in Tariff Revisions Filed by Public Service Company of Colorado, Decision No. 87460 (Colo. Pub. Util. Comm., Oct. 21, 1975).
115. *Home Builders Assn. of Metropolitan Denver* v. *Public Service Co. of Colorado*, Decision No. 89573 (Colo. Pub. Util. Comm, Oct. 26, 1976).
116. "Flat rate" is also used to denote a rate in which the total bill is the same no matter how much power is used, as opposed to a rate in which the per-kilowatt-hour charge is the same no matter how much is used.
117. E.g., Miller-Warren Energy Lifeline Act, 1975 Cal. Stats. ch. 1010. For other examples, *see Energy Users Report* (BNA), Dec. 16, 1976, p. A-25.
118. *See* "Lifeline Rates—Are They Useful?" *Energy Conservation Project Report*, No. 4 (January 1976), p. 13 (*ECP Report* is a publication of the Environmental Law Institute, Washington, DC). Some authorities question whether the lifeline concept is an effective method to aid lower income groups since these persons often consume relatively high amounts of energy.
119. N.J. Stat. Ann. §48:3-1, 48:3-4 (West).
120. *Lefkowitz* v. *Public Serv. Comm.*, No. 593 (N.Y. Ct. App. Dec. 28, 1976), *aff'g* 377 N.Y.S.2d 671, 50 App. Div. 2d 338 (Sup. Ct. 1975).
121. 377 N.Y.S.2d at 674.
122. Cal. Pub. Util. Code §§ 25007, 2781–88 (West); N.J. Stat. Ann. §§ 48:2–48:23 (West).
123. *New England Solar Energy Newsletter*, October 1976, p. 2.
124. Terrence M. Green, "Factory to be Heated by Solar Energy," *Los Angeles Times*, Oct. 31, 1976, p. 9.

125. *Solar Utilization News,* November 1976, p. 3.
126. *Energy Digest,* November 1976, p. 12.
127. In many states, municipal utilities are exempt from PUC jurisdiction. E.g., Fla. Stat. Ann. § 266.02 (West). *But see* Wis. Stat. Ann. § 196.58 (5) (West).
128. Generally, the Federal Power Commission has jurisdiction only over hydroelectric plants and the interstate transport and sale of electricity. *See* 16 U.S.C. §§ 797, 824 (1970).
129. Wis. Stat. Ann. § 196.01 (West); Ill. Rev. Stat. ch. 111 2/3, § 10.
130. E.g., *Allen* v. *California R.R. Comm.,* 179 Cal. 68, 175 P. 466 (1918).
131. "The principal determinative characteristic of a public utility is that of service to, or readiness to serve an indefinite public . . . which has a legal right to demand and receive its services or commodities." *Motor Cargo, Inc.* v. *Board of Township Trustees,* 52 Ohio Op. 257, 258, 117 N.E.2d 224, 226 (C.P. Summit County 1953). *See generally* A. J. G. Priest, "Some Bases of Public Utility Regulation," *Mississippi Law Journal,* 36 (1965): 18. *See, e.g., Peoples Gas Light & Coke Co.* v. *Ames,* 359 Ill. 132, 134 N.E. 260 (1935); *Bricker* v. *Industrial Gas Co.,* 58 Ohio App. 101, 16 N.E. 2d 218 (1937); *Cawkes* v. *Meyer,* 147 Wis. 320, 133 N.W. 157 (1911); *Claypool* v. *Lightning Delivery Co.,* 38 Ariz. 262, 299 P. 126 (1931); *Story* v. *Richardson,* 186 Cal. 162, 198 P. 1057 (1921); *Sutton* v. *Hunziker,* 75 Idaho 395, 272 P.2d 1012 (1954); *Missouri* v. *Brown,* 323 Mo. 818, 19 S.W.2d 1048 (1929); Re Nafe, 4 P.U.R.3d 369 (Ohio Pub. Util. Comm. 1953); *Limestone Rural Tel. Co.* v. *Best,* 56 Okla. 85, 155 P. 901 (1916); *Schumacher* v. *Railroad Comm.,* 185 Wis. 303, 201 N.W. 241 (1924).
132. E.g., Fla. Stat. Ann. § 366.05(1) (West).
133. *See* note 104 above and accompanying text.
134. E.g., Cal. Pub. Util. Code § 454 (West).
135. E.g., Wis. Stat. Ann. § 196.81 (West).
136. E.g., Cal. Pub. Util. Code § 761 (West).
137. E.g., Fla. Stat. Ann. § 366.04 (West).
138. E.g., Ill. Rev. Stat. ch. 111 2/3, § 56.
139. *See, e.g., Kentucky Util. Co.* v. *Public Serv. Comm.,* 252 S.W.2d 885 (Ky. Ct. App. 1952); William K. Jones, *Regulated Industries: Cases and Materials* (Brooklyn: Foundation Press, 1967), p. 347, n. 2.
140. *See generally* Wilson, Jones, Morton, and Lynch, *The Sun: A Municipal Utility Energy Source,* prepared under an agreement with the city of Santa Clara, CA, with the support of the Energy Research and Development Administration (Santa Clara, CA, 1976).
141. Re Markham, 1916A P.U.R. 1007, 1012 (Mo. Pub. Serv. Comm. 1915); Re: Gas Fuel Service, 3 P.U.R. (Nn.s) 55, 60 (Cal. R.R. Comm. 1933); *Southern Pacific Co.* v. *San Francisco-Sacramento R.R.,* 1929A P.U.R. 116, (Cal. R.R. Comm. 1928); e.g., *McFayden* v. *Public Util. Consol. Corp.,* 50 Idaho 561, 299 P. 671 (1930).
142. Cal. A.B. 4069 (1976).
143. *See* note 78 in ERDA final project report, Contract no. E(49-18)-2528.
144. *See* Ref. 81 above.
145. A recent survey found more than 100 electric utilities supporting solar energy research. Most of these projects involved the use of solar energy for heating and cooling buildings. Electric Power Research Institute, *Survey of Electric Utility Solar Projects* (Palo Alto, CA, ER 321-SR, 1977).
146. For a description of EPRI solar research projects, *see* Electric Power Research Institute, *Electric Power Research Institute: Solar Energy Program, Fall of 1976* (Palo Alto, CA, EPRI RP549, 1976). A summary is also provided in Feldman and Anderson, *Public Utility and Solar Energy Interface,* pp. 27–32.
147. *See generally* Feldman and Anderson, *Public Utility and Solar Energy Interface;* and Sweetnam and Jardine, *Energy Rate Initiatives Study.*
148. Alan Hirshberg and Richard Schoen, "Barriers to the Widespread Utilization of Residential Solar Energy: The Prospects for Solar Energy in the U.S. Housing Industry," *Policy Sciences,* 5 (1974):453, 468.
149. "Utilities and Solar Energy: Will They Own the Sun?"; and Northcross, "Who Will Own the Sun?"
150. Roger Noll, "Public Utilities and Solar Energy Development" (unpublished paper on file with the authors, 1976). This paper appears as a section of Feldman and Anderson, *Public Utility and Solar Energy Interface,* pp. 176–198; citations will be provided to this latter version, in this instance p. 183.
151. This delineation follows that used in Feldman and Anderson, *Public Utility and Solar Energy Interface,* pp. 178–181.
152. 96 S. Ct. 3110 (1976).
153. Ibid., p. 3118. *Compare Gas Light Co. of Columbus* v. *Georgia Power Co.,* 440 F.2d 1135 (5th Cir. 1971) (electric utility rates and practices immune from private antitrust suit where PUC gave lengthy consideration to challenged activities); *Washington Gas Light Co.* v. *Virginia Electric & Power Co.,* 438 F.2d 248 (4th Cir. 1971) (electric utility rate preference for underground transmission lines immune from private antitrust even though PUC did not specifically approve).
154. Jay Lake, "Legal Aspects of the Use of Solar Energy for Water and Space Heating" (unpublished paper on file with the authors, 1976), pp. 17–18.
155. Turner, *Trends and Topics in Utility Regulation,* pp. 407–409.
156. *See* Turner, *Trends and Topics in Utility Regulation,* p. 20.

157. *United States* v. *Western Electric and AT&T*, 13 Rad. Reg. (P-H) ¶ 2143, 1956 Trade Reg. Rep. (CCH) ¶ 71, 134 (D.N.J. 1956) (consent judgment); P. M. Meier and T. H. McCoy, *Solid Waste as an Energy Source for the Northeast* prepared for the Energy Research and Development Administration (Upton, NY: Brookhaven National Laboratory, No. 50550, 1976), p. 96.

158. *See*, e.g., Colo. Rev. Stat. § 40-3-111, 40-4-102. This power is frequently exercised in the context of environmental controls; the PUC may license a new facility subject to the condition that it meet all air pollution standards, which require the use of expensive sulfur-removal technology for coal-fired plants. *See* "Pawnee Plant for Morgan Stirs Up Verbal Dust," *Denver Post*, May 9, 1976, p. 18.

159. 12 U.S.C. § 1461 *et seq.* (1970).

160. 42 U.S.C. § 5511 (Supp. IV 1974).

161. H.R. 15015, 94th Cong., 2d Sess. (1976).

162. "Solar," *Sunset*, November 1976, p. 83.

163. David Barrett, Peter Epstein, and Charles M. Haar, *Financing the Solar Home: Understanding and Improving Mortgage Market Receptivity to Energy Conservation and Housing Innovation* (Cambridge, MA: Regional and Urban Planning Implementation, Inc., 1976), pp. 111, 108.

164. Barrett, Epstein, and Haar, *Financing the Solar Home*, p. iii.

165. Comments of Harold Olin, director of architectural and construction research for the U.S. League of Savings Associations, to a Consumer Action Now conference in October 1976.

166. Barrett, Epstein, and Haar, *Financing the Solar Home*, p. 80.

167. Material in the preceding three sentences is based on a conversation with Phil Gasteyer, associate director of the U.S. League of Savings Association.

168. Barrett, Epstein, and Haar, *Financing the Solar Home*, pp. 138, 140.

169. Conversation with Phil Gasteyer.

170. H. Craig Peterson, *The Impact of Tax Incentives and Auxiliary Fuel Prices on the Utilization Rate of Solar Energy Space Conditioning* (Logan, Utah: Utah State University, 1976), p. 10.

171. Ill. H.B. 164 (1975).

172. MD. H.B. 1604, 1975 Md. Laws ch. 509.

173. Thomas, Miller, and Robbins, "Legal Issues Related to Use of Solar Energy Systems."

174. Fla. Stat. Ann. § 553.87 (West Supp. 1975).

175. *City of St. Louis* v. *Brune*, 515 S.W.2d 471 (1974).

Chapter 27

Macroeconomic and Social Impacts of Solar Energy Development

W. SCOTT NAINIS

and

JOAN BERKOWITZ
Arthur D. Little, Inc., Cambridge, MA

INTRODUCTION

The use of solar energy in residential, commercial, and industrial applications impacts upon many aspects of society—social, legal, institutional, and economic. The potential impacts upon society are an important consideration for government policymakers who must decide questions of whether and how to support solar energy through research and development efforts, government subsidies, and other forms of incentives. The potential impacts are equally important for the private sector in planning near- and far-term corporate strategies to meet the needs of anticipated solar technology applications.

Economic impacts are quantifiable and are amenable to impact modeling. One particular impact model is described and illustrated in the main body of this chapter. Before this specific impact model is discussed, it is important to understand in general terms what is involved in an impact assessment and the flow of analysis activities which could and should be involved.

Examples of specific economic impact categories related to solar technology development include:

- Energy Use (by type, amount, and price)
- Materials Use (by type, amount, and price)
- Industrial Activity (net employment change)—total and by sector; net industrial output change—total and by sector
- Environmental Impacts (change in pollutant discharge levels, change in ambient environmental quality)
- Capital Investment (in solar industry, in solar devices, in related industry)
- Total Solar Device Expenditures (out-of-pocket costs, investment in solar equipment)
- Total Solar Device Cost Savings (net energy saving at site, related to fuel prices)

An impact model may or may not include a method for assessing the level of likelihood of solar device penetration.* In a situation where the market development is to also be forecasted, a market response function of some form is required and the economics of solar energy must be projected through the development of a cost performance model. In addition, a model representing the supply response of industry must be compared against the demand-related model in order to balance supply and demand over time.

If the growth projection for solar energy is exogenously provided, then the impact analysis need not consider the impacts of supply-demand balance. However, for this reason, having a projected scenario of solar equipment installation should imply the existence of a rationalized forecast of competitive fuel prices (gas, oil, electricity).

The detail of the impact analysis is determined by the degree to which factors such as geographic differences, economic sector variations, and diversity of solar markets and equipment types are addressed. Figure 27.1 shows the basic logic flow that should be considered in impact model development. Many of the outcome variables such as employment impacts and financial impacts have ripple effects themselves which will further effect the precursor variables. If additional feedback loops are required, it may be advantageous to formulate the impact model as an input-output model and use existing I-O data for the relevant industrial sector; additional I-O coefficients would be required for the noneconomic impacts.

EXAMPLE OF A QUANTITATIVE IMPACT MODEL

The impact analysis model† discussed here has been developed in order to gauge the environmental and resource utilization impacts of various scenarios of growth for the application

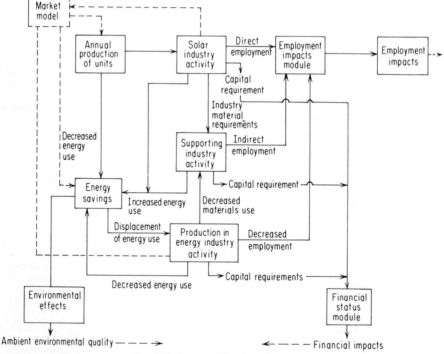

Fig. 27.1   Impact analysis logic diagram.

*See *Solar Heating and Cooling of Buildings; Commercialization*, report Part B Analysis of Market Development, Federal Energy Administration, September 1977.

†A "model" in this sense is a set of mathematical equations that express relationships among sectors of the economy in terms of dollars, quantities of material, energy, or labor. The results of model runs were generated in English units, which are retained herein.

of solar technology in the heating and cooling of buildings. The model may be readily extended to other solar technology applications, provided that appropriate input data are available. The model is computer-based to provide for calculational speed and accuracy in repeated analyses with different input parameters. At the same time, it retains a high degree of flexibility for change in both the type of input data and the logic. In its present form, the model permits a quantitative examination of the following major potential impacts of the expansion of the solar energy industry:

- Fuel savings, by type
- Total capital costs of solar energy devices to consumers
- Net changes in employment
- Capital requirements for the solar energy industry and related industries
- Net changes in air pollution [sulfur oxide (SO_x) and particulate emissions]
- Raw materials requirements for solar system construction

These are not the only impacts that could result from the growth of the solar energy industry. However, it is felt that these variables represent a fundamental set of impacts which are of major importance to some of the principal parties at interest.

A number of possible technical options (i.e., device types and applications) can be analyzed and effects of particular policy decisions can be explored insofar as they are aimed at influencing the level of numerous parameters within the model. For instance, since the interest rate and economic lifetime of the various devices are specified as inputs, the effect of mortgage incentives can be calculated.

Many of the input parameters are not known with any degree of certainty at this time. For this reason, many of the parameters are assigned a range of values to illustrate the types of results obtainable through use of the model. In general, low-, medium-, and high-parameter values can be used to test the sensitivity of the impacts to the levels of important parameters.

Input Data for the Model

The various classes of input data required for operation of the model are discussed below.

Solar energy growth data   The capacity of the solar energy industry for a particular technical option is fed into the model for six milestone planning years (1978, 1980, 1990, 2000, 2010, and 2020 for the example presented); the initial number of installed units is also used. The capacity of the industry is assumed equal to the number of solar units installed in homes or commercial buildings. Since the devices have a finite physical lifetime, the projected number of installations also includes replacement of worn-out units. The annual growth of the solar energy industry is calculated over the planning horizon. The number of total installed units as a function of time is also calculated.

Solar energy device information   Within the model there is the option to simultaneously consider two versions of a particular solar energy device—for example, a medium- and a high-efficiency collector unit. The percentage split between the two versions for the milestone years is input to the model (use of 100 percent for one version allows a single version to be considered).

For each version, information about the device—its specifications of cost, efficiency, mortgage rate, physical life, utilization of materials (glass, steel, aluminum), and manpower requirements—is fed into the model.

Energy market data   The impact model requires the amount of energy (in Btu's per unit) that would be required to supply the energy demand if the solar energy device were not installed. The costs of the various competing fuels—fuel oil, gas, and electricity—is supplied for the milestone years.

Additionally, the fraction of each fuel type that is displaced by installing the solar energy device is required, along with the efficiency of each fuel type.

Air pollution data   The number of pounds of sulfur dioxides and particulates per million Btu's of heat input in home oil burners and large power boilers is input into the model. These data should reflect the expected levels of reasonably attainable pollution control. The total tonnage of sulfur dioxides and particulates expected to be emitted during the target years is utilized to yield the percent change in the amount of pollutant emissions due to the solar energy device utilization.

Support industry data   The capital, man-hour, and kilowatthour requirements for the solar energy support industries (glass, aluminum, steel, and electrical power) are required for the model.

Calculation of the Impacts

Summary of number of installed devices   The total number of solar energy devices installed in each milestone year is calculated and is also broken down by type of device (considered as medium- and high-efficiency devices in the example presented).

Direct energy savings   The total direct energy savings of installed devices is calculated in Btu per year. The savings in barrels of oil and millions of cubic feet (MCF) of natural gas per year are also estimated for each milestone year. Electricity directly saved as a result of installing solar energy devices is calculated in kWh/yr. These direct energy savings are calculated assuming a given ratio of fuels replaced by solar energy and a given efficiency of each energy source.

Materials used for solar energy devices   The amount of glass, aluminum, and steel (tons/yr) required in solar energy device manufacture are derived for the milestone years. The net change in electrical energy requirements due to the combined effects of increased materials production and decreased electrical energy utilization due to the installed solar units is also calculated.

Effects on employment   The employment in the solar energy industry is calculated by multiplying hours of labor needed to produce an installed solar device by the annual installation rate. The additional employment in the support industries is also calculated by considering man-hours of labor required for each ton of material used in the materials industries and each kWh of electricity used in the electrical power industry.

Financial effects—general and for device owner   The annual cost of the solar energy devices ($ sales) is calculated. The solar energy capital requirements based upon the growth of the solar energy industry from the previous year are calculated assuming that capital requirements are proportional to industry expansion and neglecting capital depreciation. The total capital requirements including the device investment ($ sales), solar energy industry capital, and supporting industry capital are also calculated using the above assumptions. The total capital requirements include both the industry investment required and the investment to be made by homeowner and building owners. It is felt that this definition of total capital indicates the burden placed upon capital markets to finance the growth and maintenance of the solar energy heating and cooling industry.

The individual savings are also estimated for each solar energy device based upon an average mix of fuels replaced with the device. Thus, the savings represent an average per unit over all units nationally. The savings are calculated for both versions of the device. The total dollar savings, referred to as "aggregate savings," are calculated by summing over all the devices installed. Aggregate savings is the national total of the average dollar value in annual fuel savings less the annual carrying cost of the solar energy device at a given interest rate and mortgage period of economic lifetime. Aggregate savings represents the national cost to the owners of the solar energy device. A negative aggregate savings represents an average cost to the owners.

Environmental effects   Air pollution effects are indicated by calculating the total tons of sulfur oxides and particulates not emitted due to savings in home fuel and decreased electrical power utilization. These estimates are then compared against the reasonable control technology predictions of total national emissions during the milestone years.

Finally, the grand total of the Btu's saved is calculated by converting all energy data to thermal Btu's.

Sensitivity Analysis and Stochastic Simulation

The use of low, medium, and high values for many of the crucial parameters allows the effects of one particular parameter to be evaluated. The three levels of parameters and their *a priori* 5, 50, and 95 percentage levels also allow the impact to be run iteratively in order to simulate the effects of uncertainty and produce probability distributions on the level of the important impacts.

QUANTITATIVE ESTIMATES OF SOLAR IMPACTS

For purposes of illustration, estimates of the impacts of both residential, single-family solar energy devices and large commercial/multiple dwelling unit devices have been made. The analyses have been made from the macroeconomic viewpoint assuming national average installation with average design specifications.

The markets projected for all devices have been made assuming a leveling off of the installation rate by 2000. Year 2000 installation rates are assumed to be 90 percent of the 2020 rate. For the illustrative input parameters used, significant capital development and industrial expansion begins about 1990. The demand for solar heating units is comprised of those incorporated in new buildings and those retrofitted into existing structures. The fraction of new buildings which utilize solar heating increases from 1980 and reaches a maximum during 2000–2010. Retrofits are calculated for the existing stock of buildings, assuming that a growing fraction of those heating systems needing replacement incorporate solar heating devices.

The market projections to illustrate solar energy heating and cooling device impacts were made assuming that only hot-water and hot-water/heating devices were available for each application with respect to the different building types: residential (single-family and multifamily low- and high-rise), small commercial, and large commercial. The more sophisticated solar cooling was assumed to have an insignificant market penetration.

Table 27.1 summarizes the inventory of existing buildings and new construction in 1972 and gives the assumed levels of penetration of solar hot-water and heating devices in the year 2000, based on the 1972 inventory data. The 30 percent high penetration rate for solar in the single-family market is approximately equal to the present penetration of electric resistance heating in that market. For illustrative purposes, it was assumed arbitrarily that one-third of the solar devices installed would provide hot water alone and two-thirds would provide both hot water and space heating.

Two important general effects can be observed from the impact analysis results. The first relates to the energy pay-back period for solar heating. The second relates to the fact that total net energy savings can be higher than the direct energy savings.

Generally, the total net energy required to produce a solar energy heating and cooling device is greater than the energy it can save within one year. For this reason in the initial years where production is high but the number of installed units is low, the energy consumed in manufacturing can exceed that saved by existing devices. For example, the hot-water-only residential solar energy device saves on the average 20 million Btu per year (including fuel efficiency). Production of the solar energy device requires about 9.2 million Btu (2700 kWh) direct electric energy or 30.6 million Btu of thermal energy. Thus, the ratio of required unit production energy is about 1.5 times the annual energy savings possible with one unit. The break-even energy point should occur before the time when the number of installed units is 1.5 times the annual production rate. For the hot-water-only residential case, this occurs about 1980, given the assumed rate of penetration.

Net total energy saved can exceed the direct energy savings in cases where the electricity saved in any particular year for hot water and heating through use of solar devices is greater than the electricity required to manufacture the solar units installed in the year. The reason is that direct energy savings refer to savings in energy delivered to the building. In the case of electricity resistance heating, 3.3 kWh must be generated at the power plant for every kWh

TABLE 27.1 Assumed Penetration Rate of Solar Devices in Various Building Types in the Year 2000

Type of building	Number of buildings in 1972 (thousand units)		Percent penetration					
	New	Existing	New construction			Retrofit		
			Low	Medium	High	Low	Medium	High
Single-family residential	1,300	48,200	10	20	30	0.10	0.25	0.50
Multifamily low-rise	200	4,140						
Multifamily high-rise	2	40	5	10	15	0.1	0.25	0.5
Small commercial	40	870						
Large commercial	5	125						

delivered to the building. Net savings of electrical energy would thus be 3.3 times the direct energy savings minus the electrical energy used in manufacture.

The device costs are estimated by assembling the cost of collector panels at a given cost per square foot. Additional equipment costs are added where necessary and the labor costs in manufacturing are also included. Each device type has the per square foot additional and labor inputs presented in its basic information table.

Quantified Impact Analysis for Residential Heating with Solar Energy Devices

Single-family/residential—hot-water heating only

Devices of 20-y physical life were considered and both high- and medium-efficiency units were assumed. The device market was projected over the 1978–2010 time horizon. Nearly all data parameters included low, medium, and high estimated values. Table 27.2 indicates the hot-water-only input data.

Table 27.3 indicates the residential fuel prices used as input parameters. It also shows the ratio of fuel types used in residential buildings, the efficiency of the heating systems, and the

TABLE 27.2 Single-Family Residential Hot Water (Aluminum Construction) Input Data

Run condition data					
Six milestone years	1978	1980	1990	2000	2010
Number of years in run	38				

Solar energy growth data				
		Low	Medium	High
Number of installed devices, first milestone year		40	100	200
Annual demand for new units	1978	40	100	200
Annual demand for new units	1980	600	2,200	5,100
Annual demand for new units	1990	6,000	16,000	32,000
Annual demand for new units	2000	60,000	130,000	211,000
Annual demand for new units	2010	62,000	132,000	219,000

Solar energy device specifications				
Cost of devices ($) (78 ft²)	Device 1	600	750	1,000
$8/ft² (100 ft²)	Device 2	750	1,000	1,300
Life of devices (years)	Device 1	15	20	25
	Device 2	15	20	25
Mortgage rate for devices (%)	Device 1	4	9	12
	Device 2	4	9	12
Percent provided by S.E. of demand	Device 1	50	70	90
	Device 2	50	90	98
Aluminum consumption per device	Device 1	180	200	210
(lb), 115 lb/45 ft²	Device 2	230	250	270
Physical life (years) 20				
Glass consumption per device	Device 1	250	275	300
(lb), 155 lb/45 ft²	Device 2	300	350	400
Steel consumption per device (lb)	Device 1	180	200	210
tank size none lb	Device 2	180	200	210
Manpower (h/unit),	Device 1	20	26	35
6 h/45 ft²,	Device 2	25	29	45
16-h installation				
Energy demand (million Btu's)		17	18	19
Split between device 1 and device 2	1978	40	60	(e.g., 100
Split between device 1 and device 2	1980	40	60	represents
Split between device 1 and device 2	1990	35	65	use of
Split between device 1 and device 2	2000	35	65	device 1
Split between device 1 and device 2	2010	30	70	only)

TABLE 27.3   Assumed Residential Energy Market Prices and Other Common Parameters (Constant 1974 $)

		Low	Medium	High
Fuel oil price (¢/gal)	1978	28	37	48
	1980	31	41	54
	1990	38	50	66
	2000	46	61	80
	2010	50	67	88
Natural gas price ($/MCF)	1978	0.91	1.69	2.54
	1980	1.37	1.96	2.94
	1990	1.93	2.76	4.14
	2000	2.72	3.89	5.84
	2010	3.00	4.28	6.42
Electricity price (¢/kWh)	1978	1.78	2.29	3.11
	1980	1.74	2.32	3.48
	1990	1.92	2.57	3.85
	2000	2.05	2.74	4.11
	2010	2.26	3.01	4.52

Ratio of each fuel used

Electricity	Gas	Fuel oil
0.29	0.62	0.087

Efficiency of each fuel type, %

	Fuel oil	Gas	Electricity
Hot-water heating	50	64	92
Space and hot-water heating	63	75	95

Pollution coefficients

Lbs SO$_x$/MMBtu:		
Home heating oil	0.325	
Boilers	1.0	
Lbs Particulate/MMBtu:		
Home heating oil		0.1
Boilers		0.1

amount of pollutants (pounds of SO$_x$ and particulates) generated per million Btu's of energy consumed.

Tables 27.4 to 27.10 indicate the major impacts calculated for the assumed medium-parameter situation. The total installed units would reach about two million in 2010. Net employment gain would peak in the year 2000 at around 1900 employees. Direct energy savings would rise to about 0.04 quadrillion Btu's in 2020. The net energy savings would be negligible until 1990. By 2010 about 0.05 quadrillion Btu's would be saved annually.

Aggregate savings would be slightly negative between 1978 and 1990 but climb to about $60 million annually by 2020. Total annual capital requirements would peak at about $8 million around the year 2000 and would be about $6 million by 2010.

Air pollution impacts would be relatively minor, with reductions of less than 1 percent nationwide estimated both for sulfur oxides and particulates. However, in localized urban residential areas, the reduction in emissions could be very significant. Figure 27.2 indicates that in 2000, from 13,000 to 84,400 tons of SO$_x$ would not be emitted due to installation of solar heating devices of all types.

TABLE 27.4 Impact—Installed Units (Thousands of Units)

Building type	1980 L	1980 M	1980 H	1990 L	1990 M	1990 H	2000 L	2000 M	2000 H	2010 L	2010 M	2010 H
Single-family:												
Hot water	0.4	1.0	3.0	30	90	180	330	750	1,300	900	2,000	3,400
Hot water & heating	0.6	2.0	5.0	60	170	350	660	1,500	2,600	1,800	4,000	6,900
Subtotal	1.0	3.0	8.0	90	260	530	1,000	2,250	4,900	2,700	6,000	10,300
Multifamily/low-rise:												
Hot water				0.5	1	2	24	49	82	78	155	255
Hot water & heating				1	2.5	5	48	99	162	153	310	510
Subtotal				1.5	3.5	7	72	148	245	231	465	765
Multifamily/high-rise:												
Hot water						0.047	0.21	0.47	0.80	0.65	1.42	2.38
Hot water & heating						0.093	0.42	0.95	1.59	1.29	2.84	4.76
Subtotal						0.14	0.63	1.42	2.39	1.94	4.26	7.14
Small commercial	0	0	0	0.85	3.0	9.0	14.0	33.0	64.0	45.0	96.0	170.0
Large commercial	0	0	0	0.17	0.4	1.0	2.0	5.0	9.0	6.0	14.0	24.0
Grand total	1.0	3.0	8.0	90	270	550	1,090	2,440	5,220	2,980	6,680	11,270

TABLE 27.5 Impact—Direct Energy Savings (Quadrillion Btu)

Building type	1980			1990			2000			2010		
	L	M	H	L	M	H	L	M	H	L	M	H
Single-family:												
Hot water	0.000007	0.00002	0.00006	0.0005	0.0016	0.0033	0.006	0.013	0.024	0.016	0.037	0.065
Hot water & heating	0.00004	0.00015	0.00038	0.004	0.011	0.026	0.04	0.10	0.19	0.11	0.28	0.50
Subtotal	0.000047	0.00017	0.00042	0.0045	0.013	0.029	0.05	0.11	0.21	0.13	0.32	0.57
Multifamily/low-rise:												
Hot water	—	—	—	0.00005	0.0013	0.0003	0.0025	0.005	0.009	0.008	0.017	0.029
Hot water & heating	—	—	—	0.0002	0.0005	0.001	0.010	0.022	0.04	0.031	0.07	0.12
Subtotal	—	—	—	0.0002	0.0019	0.0015	0.013	0.03	0.05	0.04	0.09	0.16
Multifamily/high-rise:												
Hot water	—	—	—	0.00001	0.00005	0.00009	0.0004	0.0009	0.0015	0.0011	0.0026	0.0045
Hot water & heating	—	—	—	0.00006	0.0002	0.0004	0.0014	0.0036	0.007	0.0004	0.011	0.02
Subtotal	—	—	—	0.00007	0.0003	0.0005	0.0018	0.0044	0.0082	0.0055	0.013	0.025
Small commercial	6.8^{-6}	3.7^{-5}	0.0003	0.0008	0.0033	0.018	0.013	0.043	0.13	0.04	0.12	0.34
Large commercial	9.0^{-6}	0.00003	0.0002	0.0005	0.002	0.011	0.0064	0.023	0.089	0.019	0.068	0.24
Grand total	0.00006	0.0005	0.0009	0.006	0.02	0.060	0.08	0.21	0.49	0.24	0.61	1.34

TABLE 27.6 Impact—Net Energy Savings (Quadrillion Btu's/Years)

Building type	1980			1990			2000			2010		
	L	M	H	L	M	H	L	M	H	L	M	H
Single-family:												
Hot water	−0.000002	−0.00006	−0.00001	0.00027	0.0018	0.0044	0.003	0.016	0.033	0.0096	0.046	0.094
Hot water & heating	−0.0002	−0.0008	−0.0035	0.0005	0.004	0.0042	0.007	0.042	0.052	0.06	0.21	0.40
Subtotal	−0.0002	−0.0008	−0.0035	0.0008	0.006	0.009	0.010	0.058	0.085	0.07	0.26	0.50
Multifamily/low-rise:												
Hot water	—	—	—	0.00002	0.00012	0.0004	0.0012	0.005	0.012	0.005	0.017	0.04
Hot water & heating	—	—	—	0.0001	0.0003	0.0006	0.003	0.013	0.021	0.021	0.06	0.12
Subtotal	—	—	—	0.0001	0.0004	0.0010	0.0004	0.018	0.033	0.025	0.08	0.16
Multifamily/high-rise:												
Hot water	—	—	—	0.000009	0.00005	0.00012	0.00022	0.0009	0.002	0.0008	0.003	0.0065
Hot water & heating	—	—	—	0.00003	0.0002	0.0003	0.0007	0.003	0.005	0.0034	0.011	0.023
Subtotal	—	—	—	0.0004	0.0002	0.0004	0.0009	0.0004	0.007	0.042	0.013	0.03
Small commercial	−6.2^{-5}	−0.00021	−0.0025	−0.0003	0.0008	0.0086	0.004	0.013	0.078	0.019	0.11	0.39
Large commercial	−2.5^{-5}	−4.6^{-5}	−0.0003	0.00011	0.0015	0.013	0.0017	0.019	0.11	0.013	0.076	0.35
Grand total	−0.00029	−0.00076	−0.0063	0.0008	0.009	0.032	0.009	0.11	0.31	0.17	0.54	1.43

TABLE 27.7 Impact—Net Employment (Thousands)

	1980			1990			2000			2010		
Building type	L	M	H	L	M	H	L	M	H	L	M	H
Single-family:												
Hot water	0.008	0.035	0.13	0.08	0.25	0.72	0.8	0.8	4.6	0.8	1.6	3.7
Hot water & heating	0.09	0.44	1.4	0.9	3.18	8.7	9.2	24.7	57.1	9.1	23.9	53.1
Subtotal	0.10	0.48	1.5	1.0	3.43	9.4	10.0	26.6	61.7	9.9	25.5	56.8
Multifamily/low-rise:												
Hot water	0.0002	0.0005	0.001	0.007	0.02	0.06	0.3	0.9	1.8	0.4	0.9	1.4
Hot water & heating	0.001	0.005	0.014	0.06	0.20	0.54	3.0	7.7	16.9	3.0	7.6	15.8
Subtotal	0.001	0.006	0.015	0.07	0.22	0.60	3.3	8.8	18.7	3.4	8.5	17.2
Multifamily/high-rise:												
Hot water	0.00008	0.0005	0.0014	0.0020	0.0079	0.018	0.049	0.132	0.293	0.048	0.117	0.221
Hot water & heating	0.00074	0.0042	0.0114	0.017	0.068	0.156	0.414	1.135	2.537	0.414	1.103	2.308
Subtotal	0.00018	0.0047	0.0128	0.019	0.076	0.174	0.463	1.1267	2.830	0.462	1.220	2.529
Small commercial	0.018	0.079	0.78	0.26	0.93	4.3	4.4	11.5	27.8	4.5	10.8	22.4
Large commercial	0.009	0.025	0.16	0.093	0.27	0.89	1.1	2.9	6.2	1.1	2.6	2.8
Grand total	0.13	0.59	2.46	1.42	4.93	14.5	19.3	51.1	117.2	19.3	48.6	101.7

TABLE 27.8 Impact—SO_x (Thousand Tons/Yr)*

Building type	1980			1990			2000			2010		
	L	M	H	L	M	H	L	M	H	L	M	H
Single-family:												
Hot water	0.991	0.003	0.004	0.1	0.3	0.65	1.1	2.7	4.9	3.3	7.6	13.5
Hot water & heating	-0.013	-0.059	-0.0004	+0.56	1.80	3.4	6.3	16.2	26.7	21.6	53.2	92.5
Subtotal	-0.012	-0.056	-0.24	0.66	2.10	4.05	7.4	18.9	31.6	25.9	60.8	116.0
Multifamily/low-rise:												
Hot water	—	—	—	0.01	0.03	0.06	0.5	1.1	1.8	1.6	3.5	6.0
Hot water & heating	—	—	—	0.04	0.11	0.24	1.9	4.3	7.7	6.3	14.3	25.9
Subtotal	—	—	—	0.05	0.14	0.30	2.4	5.4	9.5	7.9	17.8	31.9
Multifamily/high-rise:												
Hot water	—	—	—	0.003	0.01	0.002	0.007	0.17	0.3	0.23	0.53	0.94
Hot water & heating	—	—	—	0.01	0.04	0.07	0.28	0.70	1.27	0.91	2.22	4.07
Subtotal	—	—	—	0.01	0.05	0.09	0.36	0.88	1.57	1.14	2.76	5.01
Small commercial	-0.005	-0.017	-0.018	0.072	0.38	2.2	1.3	5.3	17.7	6.9	22.5	62.0
Large commercial	-0.24	0.001	0.009	0.12	0.51	2.9	1.5	6.0	24.0	5.2	19.3	69.5
Grand total	-0.02	-0.06	-0.41	0.91	3.2	9.54	13.0	36.5	84.4	47.0	123.2	284.4

*Net reduction in pollutant production.

TABLE 27.9 Impact—Particulates (Thousand Tons/Year)

Building type	Year											
	1980			1990			2000			2010		
	L	M	H	L	M	H	L	M	H	L	M	H
Single-family:												
Hot water	0.0002	0.0007	0.001	0.018	0.056	0.12	0.20	0.48	0.87	0.58	1.3	2.4
Hot water & heating	-0.0007	-0.004	-0.018	0.11	0.36	0.73	1.2	3.2	5.6	3.9	9.6	17.1
Subtotal	-0.0005	-0.003	-0.017	0.13	0.42	0.85	1.4	3.7	6.5	4.5	10.9	19.5
Multifamily/low-rise:												
Hot water	—	—	—	0.002	0.005	0.01	0.01	0.19	0.33	0.3	0.6	1.1
Hot water & heating	—	—	—	0.001	0.019	0.04	0.03	0.08	1.4	1.1	2.5	4.6
Subtotal	—	—	—	0.009	0.024	0.05	0.05	0.27	1.7	1.4	3.1	5.7
Multifamily/high-rise:												
Hot water	—	—	—	0.0005	0.002	0.003	0.013	0.031	0.54	0.04	0.093	0.16
Hot water & heating	—	—	—	0.002	0.007	0.013	0.05	0.126	0.23	0.16	0.39	0.72
Subtotal	—	—	—	0.002	0.0088	0.016	0.0025	0.0088	0.016	0.199	0.483	0.882
Small commercial	-0.0004	-0.001	-0.014	0.019	0.088	0.50	0.34	1.2	3.8	1.3	4.2	11.5
Large commercial	0.0002	0.001	0.007	0.027	0.11	0.61	0.34	1.3	5.0	1.1	4.0	14.1
Grand total	-0.0007	-0.003	-0.024	0.19	0.65	2.03	2.13	6.48	17.0	8.5	22.7	51.7

TABLE 27.10 Impact—Net Capital Requirements per Annum ($ Billion)

Building type	1980			1990			2000			2010		
	L	M	H	L	M	H	L	M	H	L	M	H
Single-family:												
Hot water	0.0005	0.0024	0.008	0.004	0.014	0.035	0.03	0.08	0.18	0.026	0.07	0.15
Hot water & heating	0.006	0.029	0.11	0.04	0.15	0.40	0.32	0.89	2.07	0.26	0.72	1.6
Subtotal	0.0065	0.031	0.11	0.044	0.16	0.44	0.35	0.97	2.25	0.29	0.79	1.75
Multifamily/low-rise:												
Hot water	0.00001	0.00004	0.0001	0.0004	0.0011	0.003	0.014	0.03	0.07	0.012	0.03	0.06
Hot water & heating	0.0001	0.0008	0.002	0.003	0.009	0.03	0.10	0.3	0.64	0.09	0.23	0.52
Subtotal	0.0001	0.0008	0.002	0.003	0.010	0.003	0.11	0.33	0.71	0.10	0.26	0.58
Multifamily/high-rise:												
Hot water	0.000005	0.000031	0.00009	0.0001	0.00038	0.00088	0.0019	0.0049	0.011	0.0016	0.0042	0.009
Hot water & heating	0.000045	0.00026	0.00081	0.00076	0.0031	0.0076	0.014	0.041	0.096	0.012	0.034	0.078
Subtotal	0.00005	0.00029	0.0009	0.0009	0.0035	0.0085	0.016	0.046	0.011	0.014	0.0056	0.012
Small commercial	0.0019	0.0087	0.081	0.019	0.072	0.30	0.25	0.70	1.52	0.21	0.56	1.17
Large commercial	0.0008	0.0025	0.015	0.0068	0.021	0.068	0.062	0.18	0.38	0.052	0.15	0.29
Grand total	0.01	0.04	0.21	0.07	0.27	0.82	0.79	2.23	4.87	0.67	1.77	3.80

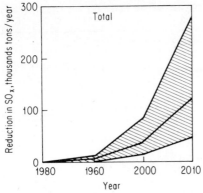

Fig. 27.2 Impact-reduction in SO_x, thousands of tons/year, emissions from using solar heating and hot water in residential and commercial buildings instead of conventional systems. (See Table 27.8.)

The average monthly owner costs are calculated to be less than $5.00 in 1978 with a 9 percent, 20-y (medium parameter) mortgage repayment schedule. By 2010 a slight average savings of about $1.00 per month is realized, based on the illustrative impacts.

The materials utilization in the high market projection are calculated to reach 12,700 tons of steel, 20,600 tons of glass, and 14,800 tons of aluminum per year by 2000.

Single-Family Residential Hot Water and Heating

The larger devices for both hot water and space heating are considered from the point of view of their potential impacts. Again, a medium-efficiency (nominally 60 percent) and a high-efficiency (nominally 70 percent) device were considered. The hot-water and heating input parameters are indicated in Table 27.11.

The major results for the medium-level parameters are indicated in Tables 27.4 to 27.10. The total number of installed units is estimated to climb to about 1.5 million by 2000 and 4.1 million by 2010. The net employment would rise sharply over the 1990–2000 decade to peak at around 25,000 during the year 2000 and level off slightly by 2010.

Direct energy savings are calculated to increase to about 0.1 quadrillion Btu's annually by 2000 and 0.3 quadrillion Btu's by 2010. The total net energy saved is slightly negative during 1978 and 1980, but becomes positive around 1983. After 2000 the net energy saved would rise rapidly and reach 0.20 quadrillion Btu's by 2010. The aggregate savings are calculated to be negative during 1978–2010 reaching −0.2 billion dollars by 2010.

The assumed average efficiency range of the hot-water and heating solar energy residential device produces a total direct energy saving of 0.1 quadrillion Btu's per year in 2000. The average solar energy device is calculated to cost the homeowner more than conventional heating devices.

Annual material usages of 420,000 tons of steel, 200,000 tons of glass, and 150,000 tons of aluminum are projected for 2000. The high market demand impacts for single-family residential devices (both hot water only and hot water and heating) are shown in Tables 27.4 to 27.10. In the year 2000, the number of installed units is projected to be about 4.9 million. Employment is calculated to peak at about 62,000 during the year 2000. The medium market demand trends are roughly parallel with about a 2.5:1 ratio in magnitudes. Total net energy saved would reach 0.09 quadrillion Btu's by 2000 and 0.50 quadrillion Btu's by 2010 based on the assumed input data.

The projected low market demand indicates total installed unit levels of about 1.0 million by 2000. Net employment is calculated to reach nearly 10,000 during 2000. Energy savings in 2000 of 0.01 quadrillion Btu's are insignificant with respect to 1969 projected residential heating energy budgets. Aggregate savings are negative and annual capital requirements are about $0.35 billion by 2000.

Multifamily/low-rise residential—hot-water-only solar units Low-rise residential units were assumed to average four living units per dwelling. Table 27.12 shows the input data used for the national average multifamily/low-rise solar heating devices.

TABLE 27.11   Single-Family Residential Hot-Water and Heating (Aluminum Construction) Input Data

Run condition data					
Six milestone years	1978	1980	1990	2000	2010
Number of years in run	38				

Solar energy growth data				
		Low	Medium	High
Number of installed devices, first milestone year		10	20	30
Annual demand for new units	1978	10	20	30
Annual demand for new units	1980	1,200	4,300	10,200
Annual demand for new units	1990	12,000	33,000	64,000
Annual demand for new units	2000	120,000	255,000	422,000
Annual demand for new units	2010	124,000	264,000	438,000

Solar energy device specifications				
Cost of devices ($) (400 ft^2),	Device 1	3,000	4,000	5,500
$10/ft^2 (500 ft^2)	Device 2	3,750	5,000	6,750
Life of devices (years)	Device 1	15	20	25
	Device 2	15	20	25
Mortgage rate for devices (%)	Device 1	4	9	12
	Device 2	4	9	12
Percent provided by S.E. of demand	Device 1	55	60	67
	Device 2	60	70	80
Aluminum consumption per device	Device 1	950	1,020	1,250
(lb), 115 lb/45 ft^2	Device 2	1,200	1,275	1,500
Physical life (years)	20			
Class consumption per device	Device 1	1,250	1,375	1,500
(lb), 155 lb/45 ft^2	Device 2	1,500	1,725	2,000
Steel consumption per device (lb),	Device 1	2,800	3,000	3,200
tank size 2000 lb	Device 2	3,000	3,500	4,000
Manpower (h/unit),	Device 1	100	135	160
6 h/45 ft^2,	Device 2	110	145	170
80-h installation				
Energy demand (million Btu's)		40	50	55
Split between device 1 and device 2	1978	40	60	
Split between device 1 and device 2	1980	39	61	
Split between device 1 and device 2	1990	38	62	
Split between device 1 and device 2	2000	36	64	
Split between device 1 and device 2	2010	35	65	
Capital requirement for S.E. industry ($ cap/$ output)		0.35	0.40	0.60

The medium market assumed projects about 1300 units by 1990 and 50,000 units installed by 2000. The number of installed units increases to about 150,000 by 2010. Direct energy savings are about 0.001 quadrillion Btu's in 1990, increasing to 0.005 quadrillion Btu's by 2000, and 0.016 by 2010. Net energy savings are slightly lower.

Net employment is calculated to increase by 20 during 1990 and 900 during 2000. Net employment tapers off slightly by 2010. Net capital requirements are calculated as $1.1 million during 1990 and reach $32 million by 2000. SO$_x$ emissions are reduced by 1100 tons per year in 2000, but during 1990 they are reduced by less than 30 tons per year. Particulate emissions are reduced 40 tons during 1990 and 800 tons during 2000.

Materials requirements would reach 6000 tons of aluminum, 8000 tons of glass products, and 6000 tons of steel per year by the year 2000, for the input parameters used.

Table 27.12   Multifamily/Low-Rise – Hot-Water-Only (Aluminum Construction) Input Data

Run condition data					
Six milestone years 1978		1980	1990	2000	2010
Number of years in run 38					

Solar energy growth data					
			Low	Medium	High
Number of installed devices, first milestone one year			0	0	0
Annual demand for new building units	1978		0	0	1
Annual demand for new building units	1980		2	6	9
Annual demand for new building units	1990		100	290	550
Annual demand for new building units	2000		5,100	10,000	17,000
Annual demand for new building units	2010		5,300	11,000	18,000
Split between device 1 and device 2	1978		100	0	
Split between device 1 and device 2	1980		100	0	
Split between device 1 and device 2	1990		100	0	
Split between device 1 and device 2	2000		100	0	
Split between device 1 and device 2	2010		100	0	
Capital requirement for S.E. industry ($ cap/$ output)			0.35	0.40	0.60
Physical life (years)				25	

Solar energy device specifications					
Cost of devices ($), $13/ft²	Device 1		3,600	4,500	6,000
Life of devices (years)	Device 1		15	20	25
Mortgage rate for devices (%)	Device 1		4	9	12
Percent provided by S.E. of demand	Device 1		50	70	90
Aluminum consumption per device (lb), 115 lb/ft²	Device 1		1,080	1,200	1,260
Glass consumption per device (lb), 155 lb/45 ft²	Device 1		1,500	1,650	1,800
Steel consumption per device (lb), tank size 3000 lb	Device 1		1,080	1,200	1,260
Manpower (h/unit), 6 h/45 ft²	Device 1		120	156	210
Energy demand (million Btu's)			102	108	114

Under the high market conditions the number of installed units would reach 80,000 units by 2000 and as many as 250,000 units by 2010. The net energy savings would be 0.01 quadrillion Btu's per year during 2000 and 0.04 million Btu's per year during 2010. Net employment increase would reach 1800 employees during 2000.

Multifamily/low-rise—residential—hot water and heating   Based on input data summarized in Table 27.13, the number of multifamily/low-rise hot-water and space heating units would reach about 2500 by 1990; the number would climb to 100,000 by 2000 and 150,000 by 2010. The units would have a direct energy savings of 0.02 quadrillion Btu's by 2000 resulting in a net energy savings of 0.01 quadrillion Btu's.

Net employment would reach 8,000 by the year 2000. Net capital requirements would reach $275 million by 2000. Environmental impacts would include a reduction in SO_x emissions of 4000 tons per year and particulate emissions of 800 tons per year.

Multifamily/high-rise residential—hot water only   Assumed input data for multifamily/high-rise residential hot-water-only units are summarized in Table 27.14. By the year 2000, only 470 solar hot-water devices would be installed in this example. The direct savings would amount to 0.001 quadrillion Btu's, increasing to 0.003 quadrillion Btu's as the number of installed units increased to 1400 in the year 2010.

Net employment is calculated to increase by 130 during 2000; net capital requirements would

TABLE 27.13 Multifamily/Low-Rise Hot-Water and Heating (Aluminum Construction) Input Data

Run condition data					
Six milestone years	1978	1980	1990	2000	2010
Number of years in run	38				

Solar energy growth data				
		Low	Medium	High
Number of installed devices, first milestone year		0	0	0
Annual demand for new units	1978	0	0	1
Annual demand for new units	1980	4	10	25
Annual demand for new units	1990	200	500	1,100
Annual demand for new units	2000	1,000	2,200	34,000
Annual demand for new units	2010	11,000	22,000	36,000
Capital requirement for S.E. industry ($ cap/$ output)		0.35	0.40	0.60
Physical life (years)			25	

Solar energy device specifications				
Cost of devices ($), $13/ft²		13,500	18,000	25,000
Life of devices (years)		15	20	25
Mortgage rate for devices (%)		4	9	12
Percent provided by S.E. of demand		60	70	80
Aluminum consumption per device, (lb), 115 lb/ft²		4,300	4,600	5,600
Glass consumption per device (lb), 155 lb/45 ft²	Device 1	5,600	6,200	6,800
Steel consumption per device (lb), tank size 6000 lb	Device 1	12,600	13,500	14,400
Manpower (h/unit), 6 h/45 ft²	Device 1	450	610	720
Energy demand (million Btu's)		200	225	250

reach nearly $5 million by that time. In the year 2000 SO_x emissions are reduced 170 tons per year and particulates are reduced by 30 tons per year.

Multifamily/high-rise residential—hot water and heating Assumed input data for multifamily/high-rise residential hot-water and space heating units are given in Table 27.15. By 2000 the medium projection results in 950 units having hot-water and solar devices. The direct energy savings would reach 0.004 quadrillion Btu's during 2000. The 2,800 units projected in 2010 would save about 0.01 quadrillion Btu's directly. The net energy savings in 2000 would be about 0.003 quadrillion Btu's.

Net employment is calculated to increase by about 1000 during the year 2000. Year 2000 net capital requirements would exceed $40 million. Environmental impacts would include a year 2000 reduction in SO_x emission of 700 tons and a particulates emissions reduction of 130 tons.

Commercial hot-water and heating solar energy devices Two different commercially oriented hot-water and heating systems were analyzed; specifically, a 15,000-ft² device that might be suitable for small stores and office buildings and a 30,000-ft² solar energy device that might be used in hospitals, schools, and industrial buildings. Tables 27.16 and 27.17 indicate the base device input data. Table 27.18 provides other required data on commercial fuel prices.

Small commercial hot-water and heating devices. The major impacts of the small (15,000-ft² collector) commercial solar energy device are shown in Tables 27.4 through 27.10. The low market level yields only a very small number of installed units (less than 14,000 by 2000).

The high projection calls for annual production rates reaching 10,000 by 2000 and about

TABLE 27.14   Multifamily/High-Rise Hot-Water-Only (Aluminum Construction) Input Data

Run condition data					
Six milestone years	1978	1980	1990	2000	2010
Number of years in run	38				

Solar energy growth data				
		Low	Medium	High
Number of installed devices, first milestone year		0	0	0
Annual demand for new units	1978	0	0	0
Annual demand for new units	1980	0	0	1
Annual demand for new units	1990	2	6	10
Annual demand for new units	2000	40	90	155
Annual demand for new units	2010	45	100	160
Capital requirement for S.E. Industry ($ cap/$ output)		0.35	0.40	0.60
Physical life (years)			25	

Solar energy device specification				
Cost of devices ($), $13/ft^2	Device 1	60,000	75,000	100,000
Life of devices (years)	Device 1	15	20	25
Mortgage rate for devices (%)	Device 1	4	9	12
Percent provided by S.E. of demand	Device 1	55	75	95
Aluminum consumption per device (lb), 115 lb/ft^2	Device 1	18,000	20,000	21,000
Glass consumption per device (lb), 155 lb/45 ft^2	Device 1	25,000	27,500	30,000
Steel consumption per device (lb), tank size 8000 lb	Device 1	18,000	20,000	21,000
Manpower (h/unit) 6 h/45 ft^2	Device 1	2,000	2,600	3,500
Energy demand (million Btu's)		1,700	1,800	1,900

64,000 installed devices by 2000. The medium projection involves a market of about 50 percent the size of the high market penetration.

Net employment would rise sharply from 780 in 1980 for the high market case to a peak of about 28,000 during the year 2020. In the medium market projection, employment resulting from production and use of the small commercial hot-water and heating device would reach a peak of about 11,500 in 2000. These figures are small compared with the 1969 total manufacturing employment of around 20 million.

The energy savings would be quite significant for the high market projection. Total net savings of energy would reach about 0.4 quadrillion Btu's per year. This amount can be as high as 0.3 percent of the projected 2000 energy budget. Initially the net energy savings would be negative as the solar heating industry expands. Total energy loss during 1978–1983 would easily be compensated for by one year or less of energy savings any time after 1990. The energy savings for the medium market are about 25 percent of those for the high market and would reach about 0.11 quadrillion Btu's per year in 2010.

Aggregate savings are all projected to be negative for the small commercial device; net costs would approach $0.24 billion by 2000. This fact indicates that the small commercial unit with assumed costs, efficiency, and fuel prices would not be competitive for all areas of the country. The annual capital requirements would reach about $0.7 billion per year by the year 2000.

The 1990 capital requirements would be about $0.07 billion. Again, the calculated medium market capital requirements are about 25 percent of the high projection values.

By 2000 about 29,000 tons of steel would be required annually. Glass and aluminum re-

TABLE 27.15 Multifamily/High-Rise Hot-Water and Heating (Aluminum Construction) Input Data

Run condition data					
Six milestone years	1978	1980	1990	2000	2010
Number of years in run	38				

Solar energy growth data				
		Low	Medium	High
Number of installed devices, first milestone year		0	0	0
Annual demand for new units	1978	0	0	0
Annual demand for new units	1980	0	1	1
Annual demand for new units	1990	4	10	20
Annual demand for new units	2000	85	185	310
Annual demand for new units	2010	90	195	325
Annual demand for new units	2020			
Capital requirement for S.E. Industry ($ cap/$ output)		0.35	0.40	0.60
Physical life (years)			25	

Solar energy device specifications			
	Low	Medium	High
Cost of Devices ($) (500 ft²), $13/ft² (900 ft²)	225,000	300,000	412,000
Life of devices (years)	15	20	25
Mortgage rate for devices (%)	4	9	12
Percent provided by S.E. of demand	55	60	70
Aluminum consumption per device	71,000	77,000	94,000
Glass consumption per device (lb), 155 lb/45 ft²	94,000	103,000	113,000
Steel consumption per device (lb), tank size 45,000 lb	210,000	225,000	240,000
Manpower (h/unit), 6 h/45 ft²	7,500	10,100	12,000
Energy demand (million Btu's)	3,400	3,800	4,200

quirements would level off at about 11,000 and 14,000 tons per year, respectively, for the assumed medium market level.

Air pollution effects would be significant. The number of tons SO_x and particulates removed per year by 2000 is around 62,000 and 12,000 tons, respectively, in the high market demand case.

Average individual owner monthly costs for the nation as a whole are calculated to be about $1200 per month in 1978, dropping to about $400 per month in 2020 (constant dollars). Low device costs, low mortgage rate, high device efficiency, or longer economic (and therefore physical) life would reduce the monthly cost. It would take a combination of favorable parameters to bring the 2000 costs to zero or to return a positive savings.

Large commercial hot-water and heating devices. Tables 27.4 through 27.10 indicate the major impacts for the large commercial hot-water and heating devices. The high and medium market projections are indicated. The number of installed large hot-water and heating devices is estimated to be less than for the small unit; the high market projection would reach only 9000 by 2000 and 24,000 by 2010. The net employment would rise slowly with the market demand to about 6,000 employees over the 2000–2010 period under high market demand assumptions.

Under medium market demand assumptions, direct annual energy savings would reach over 0.02 quadrillion Btu's by 2000. The total net annual energy savings would reach about 0.08 quadrillion Btu's by 2010.

TABLE 27.16   Small Commercial Hot-Water and Heating (Aluminum Construction) Input Data

Run condition data						
Six milestone years	1978	1980	1990	2000	2010	2020
Number of years in run	48					

Solar energy growth data					
			Low	Medium	High
Number of installed devices, first milestone year			0	0	0
Annual demand for new units	1978		1	2	4
Annual demand for new units	1980		12	41	260
Annual demand for new units	1990		172	500	1,600
Annual demand for new units	2000		2,900	6,200	10,400
Annual demand for new units	2010		3,100	6,400	10,700
Capital requirement for S.E. industry ($ cap/$ output)			0.35	0.40	0.60
Physical life (years)				25	

Solar energy device specifications			
	Low	Medium	High
Cost of devices ($) (15,000 ft^2)	112,000	150,000	190,000
Life of devices (years)	15	20	25
Mortgage rate for devices (%)	4	9	12
Percent provided by S.E. of demand	65	80	95
Aluminum consumption per device (lb), 115 lb/45 ft^2	32,500	34,500	40,500
Glass consumption per device (lb), 155 lb/45 ft^2	40,000	46,500	53,000
Steel consumption per device (lb), Tank size none lbs	81,000	94,500	108,000
Manpower (h/unit), 600 h-installation	1,720	2,266	2,650
Energy demand (million Btu's)	900	1,300	2,000

Annual aggregate savings would be negative until about the year 2000. By 2010 the aggregate annual savings would reach $47 million. Annual net total capital requirements would reach $0.02 billion by 1990 and rise to $0.18 billion by 2000.

The materials usages would be significant. Use of steel, aluminum, and glass would rise steadily so that by 2000, 82,000 tons of steel, 40,000 tons of glass, and 30,000 tons of aluminum would be required annually.

The year 2000 air pollution reduction for the high market would be 156,000 and 51,000 tons per year for SO_x and particulates, respectively.

The average individual installation would cost about $1800 per month; by 1980 assumed average fuel price increases would reduce the costs to $60 per month during 2000. The effect of favorable levels of average device efficiency, costs or conventional fuel prices would be to provide an average year 2000 monthly savings of as much as $125 per month.

COMBINED IMPACT ANALYSIS

Assuming that each of the solar heating and cooling options is available for development in 1978–2010, a composite quantified impact analysis was performed for all the previously described and analyzed devices, both single-family residential and multifamily units, and the large and small commercial buildings. Tables 27.19 through 27.21 and Figs. 27.3 through 27.7 indicate the impacts for all three market cases. These impacts should be compared against the value of total materials and energy consumption indicated in Table 27.22. Figure 27.8 shows the net capital requirement for solar implementation to the year 2010.

TABLE 27.17   Large Commercial Hot-Water and Heating (Aluminum Construction) Input Data

Run condition data					
Six milestone years	1978	1980	1990	2000	2010
Number of years in run	38				

Solar energy growth data				
		Low	Medium	High
Number of installed devices, first milestone year		0	0	0
Annual demand for new units	1978	1	2	4
Annual demand for new units	1980	3	7	28
Annual demand for new units	1990	34	80	205
Annual demand for new units	2000	395	865	1,485
Annual demand for new units	2010	430	925	1,570
Capital requirement for S.E. industry ($ cap/$ output)		0.35	0.40	0.60
Physical life (years)			25	

Solar energy device specifications			
	Low	Medium	High
Cost of devices ($) (30,000 ft^2)	205,000	275,000	348,000
Life of devices (years)	15	20	25
Mortgage rate for devices (%)	4	9	12
Percent provided by S.E. of demand	65	80	95
Aluminum consumption per device (lb), 115 lb/45 ft^2	65,000	69,000	81,000
Glass consumption per device (lb), 155 lb/45 ft^2	80,000	93,000	106,000
Steel consumption per device (lb), tank size none lbs.	162,000	189,000	216,000
Manpower (h/unit), 1100-h installation	3,000	4,000	4,700
Energy demand (million Btu's)	3,000	5,000	10,000

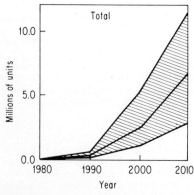

Fig. 27.3 Impact—installed solar hot-water and heating units, millions, in single-family residential and multifamily/low-rise buildings.

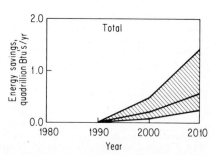

Fig. 27.4 Impact—direct energy savings, quadrillion Btu's/year, for solar hot-water devices in residential and commercial buildings.

TABLE 27.18   Energy Market Prices—Commercial (Constant 1974 $)

	Low	Medium	High
Fuel oil price (¢/gal)	22	29	37
	24	32	46
	29	39	56
	36	48	68
	40	53	75
	44	58	83
Natural gas price ($/MCF)	0.99	1.41	2.12
	1.14	1.63	2.46
	1.61	2.30	3.45
	2.27	3.24	4.86
	2.49	3.56	5.34
	2.74	3.92	5.88
Electricity price (¢/kWh)	1.72	2.29	3.44
	1.74	2.32	3.48
	1.92	2.57	3.85
	2.05	2.34	4.11
	2.26	3.01	4.52
	2.48	3.31	4.97

Efficiency of each fuel type, %

	Fuel oil	Gas	Electricity
Hot-water heating	50	64	92
Space and hot-water heating	76	77	95
Air conditioning			3.60

Distribution, %

	Fuel oil	Gas	Electricity
Hot-water heating	0	60–70	30–40
Hot-water heating and heating	50–60	30–40	5–10
Air conditioning (small)	0	10–15	85–90
Air conditioning (large)	0	50	50

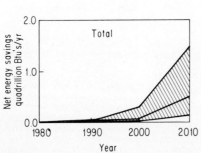

Fig. 27.5   Impact—net energy savings, quadrillion Btu's/year, for solar hot water and heating in residential and commercial buildings.

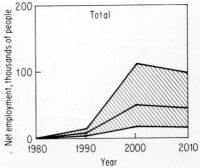

Fig. 27.6   Impact—net employment in the solar hot-water and heating industry.

TABLE 27.19   Total Projected Number of Installed Solar Units (Medium Market Projection)

Building type	Construction inventory, 1972	Hot-water-only		Hot-water and space heating	
		1980	2000	1980	2000
Single-family residential	48×10^6	1,300	750,000	2,200	1,500,000
Multifamily low-rise	4×10^6	4	50,000	8	100,000
Multifamily high-rise	4×10^4	0	470	0	950
Small commercial	870,000	—	—	29	33,000
Large commercial	125,000	—	—	6	4,800

TABLE 27.20   Direct Energy Savings Resulting from Solar Energy Development (Medium Market Projection) (10^{12} Btu/Yr)

Building type	Hot-water-only		Hot-water and heating	
	1980	2000	1980	2000
Single-family residential	0.02	13.5	15.1	1,019
Multifamily low-rise	0.0005	5.3	0.0005	9.7
Multifamily high-rise	0.0004	0.9	0.001	3.6
Small commercial	—	—	0.04	43.1
Large commercial	—	—	0.03	23.7
Total direct savings	0.02	20	15.2	1,099
Total residential heating demand (1969)				8,400

TABLE 27.21   Direct Requirements Resulting from Solar Energy Development (Medium Market Projection) (Tons/Yr)

Building type	Aluminum		Glass		Steel	
	1980	2000	1980	2000	1980	2000
Single-family residential:						
Hot water	250	14,800	350	20,600	220	12,700
Hot water & heating	2,500	150,000	3,400	205,000	7,200	425,000
Multifamily low-rise:						
Hot water	16	24,800	20	34,100	16	24,800
Hot water & Heating	120	190,000	165	260,000	360	558,000
Multifamily high-rise:						
Hot water	330	92,700	450	127,000	330	92,700
Hot water & Heating	2,500	714,000	3,400	955,000	7,500	2,085,000
Small commercial	700	11,000	950	14,400	1,940	29,000
Large commercial	250	29,800	325	40,200	660	81,700
Total	6,700	1,227,000	9,060	1,656,000	18,200	3,309,000
1969 Production	3,900,000		2,200,000		110,000,000	

TABLE 27.22   1974 Values of Impact-Related Variables

Single-family residences	49,200,000*
Employment:	
Total U.S. employment	87,000,000
Total manufacturing employment	20,000,000
Heating equipment employment (except electrical heating equipment)	43,000
Energy demand:	
Residential heating	9.3 quadrillion Btu (Btu $\times$ 10^{15}
Total energy	73 quadrillion Btu (Btu $\times$ 10^{15})
Financial:	
Personal income	$5,450 \times 10^9$
Materials production:	
Steel	125,000,000 tons/yr
Glass	2,600,000 tons/yr
Aluminum	4,500,000 tons/yr

*F.W. Dodge estimate.

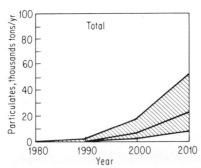

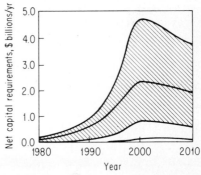

Fig. 27.7   Impact—reduction in particulate, thousand tons/year, emissions from using solar heating and hot water in residential and commercial buildings instead of conventional systems.

Fig. 27.8   Impact—net capital requirements per year, billions of dollars, for implementing the projected solar heating and hot-water market.

During 1980 only 3000 solar heating units are projected under the illustrative input conditions. By 1990 this total number would be expected to reach 290,000. By 2000 about 3 million solar heating units are calculated to be in place across the country. These 3 million units would directly save about 0.7 quadrillion Btu's of energy. The net energy savings in 2000 would be about 0.5 quadrillion Btu's annually. By 2010 if 8.5 million units were in place, the total net energy savings would amount to 2.1 quadrillion Btu's annually.

In 2000 the total net increase in employment is calculated to be over 200,000. The annual net capital requirements would total to over $7.6 billion. Annual SO_x and particulate emissions would be reduced by 140,000 and 26,000 tons, respectively.

SUMMARY

A macroeconomic input model for solar technology development has been described, and illustrative results have been presented for three developmental scenarios. Information on the availability of the computer program and possibilities for modifying it to handle other types of solar applications may be obtained from the authors.

Chapter 28

The Microeconomics of Solar Energy*

ROSALIE T. RUEGG

and

G. THOMAS SAV

Applied Economics Program,
Center for Building Technology,
National Engineering Laboratory,
National Bureau of Standards, Washington, DC

INTRODUCTION

A primary question about solar energy is how it compares economically with alternative sources of energy. Historically, new technologies have seldom emerged in the marketplace unless they were economically competitive with existing technologies, and have usually not obtained a major market share unless they were economically superior to existing technologies. Apart from severe curtailment of conventional energy supplies, it is unlikely that solar energy will be widely used unless it is economically competitive with alternative energy sources, either under natural market conditions or under a system of governmental incentives.

In addressing the question, "Will solar energy pay in a given application?", it is important to consider systems which are optimal in their design and size relative to other components of buildings and facilities in which they are used. Techniques of economic evaluation are required both to design and size solar energy systems for maximum economic benefit and to estimate what the expected savings or losses arising from use of solar energy systems will be.

The purpose of this chapter is to explain and illustrate how the techniques of microeconomic analysis can be used in the design, sizing, and evaluation of solar energy systems. For the purpose of exposition, the focus is on solar hot-water and space-heating systems for residential and commercial buildings. However, the basic concepts and procedures will generally apply to the analysis of solar energy in diverse applications, e.g., industrial process heat systems, power production, and total energy systems.

*Adapted from Ref. 11.

A basic premise of this chapter is that building owners consider the lifetime costs of owning and operating the energy-related components of their buildings or facilities. If owners' behavior is consistent with this basic premise, then it is reasonable to believe that they are willing to incur relatively higher first costs to acquire solar energy systems if they expect to save an equivalent or greater amount in future fuel costs plus an acceptable return on their investment.

It may be argued that owners—and if not owners, certainly speculative builders—focus on first costs rather than on lifetime costs. While this appears to have been true to some extent in the past, there are indications that owners are giving increasing attention to energy costs and to the potential savings from solar energy and energy conservation, and are also beginning to be willing to pay more for buildings and facilities that use relatively less nonrenewable energy.[3]

In the microeconomic analysis of solar energy, we seek to minimize net costs of the energy component of a building or facility, subject to thermal comfort requirements and other constraints. This is done by optimizing the dollar tradeoffs between (1) the owning and operating costs of a solar energy system, (2) the owning and operating costs of a conventional energy system, and (3) the energy conservation investments which can reduce the building's energy requirements.

The first part of this chapter presents basic concepts and methods of economic analysis, and illustrates their use in sample problems. An overview is given of the basic steps in life-cycle cost analysis, discounting cash flows, considerations important to making assumptions, and methods of dealing with uncertainty and inflation. The second part describes the major components of costs and savings associated with solar energy systems, including various types of system costs, energy costs and savings, taxes, and government incentives. The third part describes a methodology for optimizing a solar energy system for maximum net savings. The basic concepts of optimization are set forth—an optimization analysis and an approach to optimization are described in detail. An example optimization problem is solved.

METHODS OF ECONOMIC EVALUATION

When investment costs are high and resulting dollar benefits (savings) are distributed and changing in amount over time—as they are for most applications of solar energy—it is reasonable to make thorough comparisons of lifetime costs and savings before undertaking an investment. "Life-cycle costing" is a method of economic evaluation that is generally appropriate for evaluating an investment whose principal benefits occur in the form of cost savings.

Although the term life-cycle costing is used to refer to a specific method of measuring economic performance, it is also often used in a broader sense to refer to any of several widely used, comprehensive methods that evaluate all significant costs and benefits (savings) over the relevant period of time, taking into account the time value of money. These include the net benefits (or savings) method, the benefit/cost (or savings-to-investment) ratio method, and the internal rate of return method.[9]

Another widely used method of evaluation, the payback method, does not take into account costs and savings over the life of an investment, and, therefore, is not in a strict sense a life-cycle costing method. Even though the payback method gives an incomplete evaluation, it is included here because of its widespread use.

The evaluation methods listed above differ essentially in the manner in which they relate costs and savings. For example, the life-cycle cost method, used to refer to a specific evaluation method, means the summing of acquisition, maintenance, repair, replacement, and energy costs over the life of the investment. The investment that has the lowest total life-cycle cost while meeting the investor's objectives and constraints is the preferred investment.

The net benefits method, as applied to solar energy, finds the difference between the lifetime dollar energy savings from an investment in solar energy and its lifetime dollar costs. A positive dollar value means the investment is profitable, and a negative value indicates losses.

The benefit/cost (or savings-to-investment) ratio method expresses savings, net of recurring costs, from solar energy as a numerical ratio to investment costs. The higher the ratio, the more dollar savings that are realized per dollar of investment cost.

The internal rate of return method finds the interest rate for which lifetime dollar savings are just equal to lifetime dollar costs. This interest rate indicates the rate of return on the investment. The rate of return is then compared to the investor's minimum acceptable rate of return to determine if the investment is desirable.

The payback method finds the period of time that is expected to elapse before cumulative dollar savings from an investment in solar energy will offset the investment costs. If the years and months to pay back are calculated to reflect the time value of money, the method is called discounted payback; if the time value of money is not considered, it is called simple payback.

Although these evaluation methods are closely related to one another, they are not necessarily substitutable methods for dealing with different types of investment decisions. For some types of decisions, the choice of a method is more important than for others. The choice of method is not usually critical, for example, in simple "accept-reject" investment decisions, in that any of the above life-cycle cost methods, used correctly, will indicate if an investment in solar energy will save more than it costs.

The choice of method *is* important for determining the economically optimal size of a solar energy system. For this type of decision, the life-cycle cost method or the net benefits method is recommended over the other methods because either of these two methods will indicate very directly the optimal system size. As long as the total cost of a building is reduced or the net savings from an investment is increased by adding to the capacity of the solar energy system, it pays to do so.

For allocating a limited budget among available investment opportunities, the benefit/cost (savings-to-investment) ratio method or the internal rate of return method is recommended, because the use of either of these methods will result in the selection of investment projects with the largest total return for a given investment budget.

In the case where a fast turnaround on an investment is required (for example, to repay a short-term loan), the payback method may provide useful information.

Now let us define each of these economic evaluation methods algebraically, illustrate their use in sample solar energy problems, and examine their advantages and disadvantages in more detail. (To do this, it is necessary to use concepts of discounted cash-flow analysis. Readers unfamiliar with this type of analysis are referred to the section of this chapter entitled *Discounting Cash Flows*.)

Life-Cycle Costing Method

This evaluation method calculates either the total present value or the total annual value of the lifetime dollar costs associated with each alternative energy system under consideration. The alternative with the lowest cost is more economical, provided it meets other requirements of the investor.

Present value of life-cycle costs   Following is a formula for calculating, in present-value dollars, the total life-cycle costs PV associated with owning and operating any energy system:

$$PV = I - (V_n a^n) + \sum_{j=1}^{n} a^j(M_j + R_j) + \sum_{k=1}^{H} \sum_{j=1}^{n} P_k Q_k b^j \qquad (28.1)$$

where PV = total present-value, life-cycle costs , before taxes, associated with a given energy system

I = total first costs associated with the energy system, including design, purchase, installation, building modification, and the value of useful building space lost

V_n = residual or salvage value at year n, the last year in the evaluation

a = the single-present-value formula computed for a designated year from $j=1$ to n, and discount rate d; i.e., $a^j = (1 + d)^{-j}$

M_j = maintenance costs in year j

R_j = repair and replacement costs in year j

P_k = the initial price of the kth type of conventional energy for energy types $k=1$ to H

Q_k = the quantity required of the kth type of conventional energy (equipment efficiencies should be taken into account in calculating Q_k)

b^j = a formula for finding the present value of an amount in the jth year, escalated at a rate e_k, where k denotes the kth type of energy, and discounted at a rate d; i.e., $b^j = [(1 + e_k)/(1 + d)]^j$

Table 28.1 illustrates in a simple example the use of the life-cycle costing method to compare two alternative energy systems: (1) a combined solar and conventional energy system, and (2) a conventional energy system used alone. The total costs were calculated from Eq. (28.1).

The combined solar/conventional energy system is assumed to cost $20,000, or $8000 more than the conventional system alone (col. 3). The major components of the combined system

TABLE 28.1 Illustration of the Life-Cycle Costing Method in Present-Value Dollars[a]

Type of energy system (1)	Period of analysis, years (2)	Initial investment cost, dollars (3)	Base-year energy costs, dollars (4)	Present value of energy costs,[b] dollars (5)	Replacement costs, dollars (6)	Present value of replacement costs,[c] dollars (7)	Salvage, dollars (8)	Present value of salvage,[d] dollars (9)	Present value of total costs,[e] dollars (10)
Solar/conventional auxiliary	20	20,000	500	6359	0	0	0	0	26,359
Conventional	20	12,000	2000	25,435	6000/15th year	1436	4000/20th year	594	38,277

[a]All costs are evaluated with a discount rate of 10%, a fuel price escalation rate of 5%, an investment horizon of 20 years, and the assumption that the solar energy system has no value remaining after 20 years, and the conventional system has a salvage value of two-thirds of the cost of the replacement.

[b]Derived using the uniform-present-value formula including a factor for fuel price escalation, for a discount rate of 0.05, and 20 years; i.e.,

$$\$6359 = \$500 \sum_{j=1}^{20} \left(\frac{1 + 0.05}{1 + 0.10} \right)^j$$

$$\$25,435 = \$2000 \sum_{j=1}^{20} \left(\frac{1 + 0.05}{1 + 0.10} \right)^j$$

[c]Derived using the single-present-value factor for a discount rate of 0.10 and 15 years; i.e., \$1436 = \$6000 (0.2394).

[d]Derived using the single-present-value factor for a discount rate of 0.10 and 20 years; i.e., \$595 = \$4000 (0.1486).

[e]Derived by summing the present value of investment costs, energy costs, and replacement costs less salvage value; i.e., col. 10 = col. 3 + col. 5 + col. 7 − col. 9.

are expected to last 20 years, as compared with 15 years for the conventional system alone. Setting the period of analysis at 20 years for both systems (col. 2) means that the conventional system will be due for a major replacement (assumed to cost $6000) at the end of the 15th year (col. 6), of which two-thirds of the value is assumed to remain at the end of the 20th year (col. 8). The combined system will require no replacements and is assumed to result in substantial fuel savings, reducing annual energy costs from $2000 to $500, valued in base-year dollars (col. 4). On the basis of an expected fuel price rise of 5 percent per year and a discount rate of 10 percent, the present value of energy costs for the combined system equals $6359 (col. 5), and its total life-cycle cost equals $26,359 (col. 10). For the conventional system alone, the present value of energy costs is $25,435 (col. 5), the present value of replacement cost equals $1436 (col. 7); the present value of the salvage value equals $594 (col. 9), and total life-cycle cost equals $38,277 (col. 10). Hence, over 20 years the total life-cycle cost of the combined system is $11,918 less than the cost of the conventional system, and is, therefore, the preferred investment.

Annual value of life-cycle costs An alternative to expressing life-cycle costs in present-value terms is to express them in annual-value terms, that is, as a stream of level annual costs extending over a period equal in length to the period of analysis. The measure of annual value may be derived directly from the measure of present value by applying the uniform capital recovery factor. Expressed in annual-value terms, the life-cycle costing method is used exactly as it was in the preceding case to find the lowest-cost alternative for accomplishing a given objective. A formula for finding the total annual value AV of life-cycle costs of an energy system is

$$AV = c \times PV \qquad (28.2)$$

where AV = total annual-value life-cycle costs associated with a given energy system
c = the uniform capital recovery formula $i/[1 - (1 + i)^{-n}]$

and other symbols are as previously defined.

Table 28.2 illustrates the evaluation of the life-cycle cost of an investment in annual-value dollars. The present value of the total life-cycle costs of the two energy systems derived in Table 28.1 are here converted to annual-value equivalents using Eq. (28.2). In annual-value dollars, the combined solar/conventional energy system costs $3097 (as compared with the present-value equivalent of $26,359); the conventional system alone costs $4498 in annual-value dollars (as compared with the present-value equivalent of $38,277); and the combined system is shown again to be the least-cost system.

Life-cycle cost per unit of energy A slightly different way of using the life-cycle costing method to evaluate an investment in solar energy is to calculate and compare in annual-value dollars the *life-cycle cost per unit of energy* delivered to a building by alternative energy systems.[4] If the assumptions are made that the conventional heating and cooling system is identical whether or not a solar energy system is added, and that the price of conventional energy per unit purchased is constant with respect to the quantity purchased, the calculations are simplified. In this case, the cost comparison can be made by first finding the life-cycle costs in annual-value dollars of the solar energy system, and dividing this by the amount of energy expected to be supplied by the solar energy system each year. That is, the life-cycle cost in annual-value dollars of a unit of energy provided by solar, US, may be calculated as follows:

TABLE 28.2 Illustration of the Life-Cycle Costing Method in Annual-Value Dollars*

Type of energy system (1)	Period of analysis (2)	Present value of total costs, dollars (3)	Annual value equivalent of total costs,† dollars (4)
Solar/conventional auxiliary	20	26,359	3097
Conventional	20	38,277	4498

*Based on the hypothetical cost data and assumptions given in Table 28.1.
†Derived by Eq. (28.1); i.e., $3097 = $26,359 (0.1175); $4498 = $38,277 (0.1175), where 0.1175 is the uniform capital recovery factor for a discount rate of 10% and 20 years.

$$US = \frac{AV_s}{SES} \qquad (28.3)$$

where US = the unit life-cycle cost in annual-value dollars of solar energy
AV_s = total life-cycle costs in annual-value dollars of the solar energy system [Eq.(28.2)]
SES = amount of solar energy supplied annually, expressed in some unit of measure such as joules/year or Btu/year.

The next step is to derive the per-unit cost of supplying a unit of heating or cooling by conventional energy, UC. This may be found by the following equation:

$$UC = \left[\left(\frac{SP_C}{COP} \sum_{j=1}^{n} b_c^j \right) + \left(\frac{HP_h}{F} \sum_{j=1}^{n} b_h^j \right) \right] c \qquad (28.4)$$

where UC = per-unit cost of supplying conventional energy, after the heating and/or cooling plant is in place
S = energy for cooling as a *fraction* of total annual energy requirement
P_c = purchase price per unit of energy for cooling (units should be consistent with the units in which solar energy costs are measured); COP= coefficient of performance of cooling equipment
b_c^j = a factor for finding the present value of the price per unit of cooling energy in the jth year when it is escalating at a rate e_c; $b_c^j = [(1 + e_c)/(1 + i)]^j$
H = energy for heating as a *fraction* of total annual energy requirements
P_h = purchase price per unit of energy for heating
F = efficiency of the heating furnace expressed as a fraction
b_h^j = a factor for finding the present value of the price per unit of heating energy in the jth year when it is escalating at a rate e_h; $b_h^j = [(1 + e_h)/(1 + i)]^j$
c = the uniform capital recovery factor; $c = i/[1 - (1 + i)^{-n}]$

By comparing the life-cycle cost in annual-value dollars of a unit of energy supplied by the solar energy system, US, with the unit cost of conventional energy, UC, the type of energy with the lowest life-cycle cost can be identified.

Table 28.3 illustrates the use of this method for evaluating the cost-effectiveness of the same solar energy system used in preceding examples. If it is assumed that 210 million Btu (222 GJ) are supplied each year by solar energy (col. 6) and that the life-cycle cost of the solar energy system in annual-value dollars is $841 (col. 5), Eq. (28.3) can be used to derive the unit cost of solar energy. The life-cycle cost per therm (100,000 Btu, 105,500 kJ) of solar energy supplied is found to be $0.40 (col. 7). This is lower than the cost per therm of $1.067 for conventional energy as derived by Eq. (28.4) (col. 8).

Advantages, disadvantages, and recommended applications of the life-cycle costing method

This method is effective for determining if a solar energy system is expected to reduce the total life-cycle costs of the energy components of a building. It is also useful for determining the optimal sizing of a solar energy system, by showing the impact of changes in investment size on lifetime costs. If total life-cycle costs are lower with a solar energy system than with a conventional system alone, the solar energy system is cost-effective. As long as life-cycle costs continue to decline as the size of the solar energy system is increased, it pays to increase the size of the system, other things being equal.

Theoretically the size of a project should be increased only as long as the return at the margin on that investment is equal to or greater than the return at the margin on competing investments. However, as a matter of practice, individual projects are often sized (scaled) apart from consideration of other projects. Then projects of a predetermined design and size are compared with one another.

The life-cycle costing method is often used for overall comparisons of alternative building designs, for example, an energy-conserving design vs. a more conventional design. This is the method that is used in a later section (*Economic Optimization*) to determine the economically efficient size of a solar energy system in conjunction with a conventional energy system and a building envelope.

The life-cycle cost method, however, is not always effective for evaluating investments in solar energy relative to competing investments, because the method does not lend itself to ranking competing investments in terms of their economic efficiency. Although the life-cycle cost

TABLE 28.3 Illustration of the Life-Cycle Cost Method to Compare the Unit Cost of Energy Supplied by Alternative Energy Systems[a]

Differential investment cost of solar, dollars (1)	Annual value of solar investment cost,[b] dollars (2)	Savings in replacement costs salvage,[c] dollars (3)	Annual value of net replacement cost savings,[c] dollars (4)	Annual value of solar costs,[d] dollars (5)	Amount of solar energy supplied each year, Btu (6)	Annual cost per unit of solar energy supplied,[e] dollars (7)	Levelized unit cost of conventional energy,[f] dollars (8)
8000	940	6000/15 year 4000/20 year	99	841	21×10^7	$0.40/10^5$ Btu 4.00/MM Btu	$1.067/10^5$ Btu 10.67/MM Btu

[a] Based on hypothetical cost data presented in Table 28.1, and derived by Eqs. (28.3) and (28.4). Additional assumptions are that the initial cost of the solar auxiliary system is equal to the cost of the conventional system used alone, that the conventional system lasts 5 years longer when used as an auxiliary, that the energy costs of the solar auxiliary system are \$500, that the conventional system is powered by no. 2 fuel oil priced at \$0.50/gal, and that the furnace efficiency is 0.5.
[b] Derived using the uniform capital recovery factor for a discount rate of 10% and 20 years; i.e., \$940 = \$8000(0.1175).
[c] Derived using the single-present-value factors for a discount rate of 10% and 15 years and 20 years and the uniform capital recovery factor for 20 years; i.e., \$99 = [\$6000(0.2394) − \$4000(0.1486)] (0.1175).
[d] Derived by Eq. (28.2), combining the annual value of investment cost and the annual value of net replacement cost savings; i.e., \$841 = \$940 − \$99.
[e] Derived by Eq. (28.3); i.e., \$0.40/10^5 Btu = \$841/21 × 10^7 Btu.
[f] Derived by Eq. (28.4); i.e.,

$$1.067/10^5 = (0.50/\text{gal}/0.5) \left[\sum_{j=1}^{20} \left(\frac{1+0.05}{1+0.10} \right)^j \right] (0.1175)/(1.4 \text{ therm/gal})$$

measure indicates whether total dollar owning and operating costs of a building are increased or decreased by an investment in solar energy, it does not indicate the return on the investment dollar.

Net Benefits (Savings) Method

This method is useful for converting the analysis of an investment such as solar energy, which largely involves costs, to a standard benefit-cost format involving both costs and savings. Lifetime costs are subtracted from lifetime savings to derive net savings from the investment. As in the case of computing total life-cycle costs, net savings may be expressed in either present-value or annual-value dollars.

The method involves the same cost elements and arrives at the same conclusion as the life-cycle costing method, but it is formulated somewhat differently.

Present value of net savings   A formula derived from Eq (28.1) by differencing for calculating net savings in present-value dollars, PNS, from a solar energy investment is

$$\text{PNS} = \sum_{k=1}^{H} \sum_{j=1}^{n} [(P_k \, \Delta Q_k \, b^j)(-1)] - [(\Delta I - \Delta V \,_n a^n) - \sum_{j=1}^{n} a^j (\Delta M_j + \Delta R_j)] \qquad (28.5)$$

where PNS = present value of net dollar savings
Δ = change attributable to solar energy

and other symbols are as previously defined.

This formula allows for the evaluation of a solar energy plus auxiliary system in direct comparison with a conventional energy system. It calculates the net difference between the present value of energy savings due to the solar energy system and the differential investment, maintenance, repair, and replacement costs attributable to the solar energy system.

Table 28.4 illustrates this method in a simple example, based on the same costs and assumptions used in the preceding example. The solar energy system is assumed to cost $8000 more than the conventional system, to save $1500 annually in energy costs (valued in base-year dollars), and to last 20 years—5 years longer than the conventional system—without a major replacement. The present value of energy savings over the 20 years equals $19,077 (col. 4). The present value of the savings in net replacement costs owing to the expected 5-year longer life of the conventional system when paired with a solar system equals $841 (col. 5). Net savings from the investment in solar energy are, therefore, $11,919 (col. 6).

Annual value of net savings   Measured in annual-value dollars, the net benefits (savings) method finds the difference between the additional annual costs associated with a solar energy system and the annual savings expected to result from it. This can be done by applying the appropriate discount formula to the net present value of savings as derived by Eq. (28.3). The resulting formula is

TABLE 28.4   Illustration of the Net Benefits (Savings) Method in Present-Value Dollars*

Period of analysis, years (1)	Differential solar investment costs, dollars (2)	Base-year energy cost savings, dollars (3)	Present-value energy savings,† dollars (4)	Present value of net replacement cost savings,‡ dollars (5)	Net present-value savings,§ dollars (6)
20	8000	1500	19,077	841	11,918

*Based on the hypothetical cost data and assumptions given in Table 28.1.
†Derived using the uniform-present-value formula including a factor for fuel price escalation; i.e.,

$$\$19,077 = 1500 \, \Sigma \left(\frac{1 + 0.05}{1 + 0.10} \right)^{20}$$

‡$841 = 6000(0.2394) - \$4000(0.1486)$.
§Derived by Eq. (28.5); i.e., $\$11,918 = \$19,077 - \$8000 + \$841$.

$$ANS = c \times PNS \tag{28.6}$$

where ANS = net annual savings and c = the uniform capital recovery factor.

Converted to an annual-value basis ($c = 0.1175$), the present-value savings of $11,918, from the preceding sample problem, equals $1400. This means that a savings of $1400 per year for 20 years (in constant dollars) is equivalent in value to a savings of $11,918 realized now.

Advantages, disadvantages, and recommended application of the net benefits (savings) method   This method has essentially the same advantages and disadvantages as the life-cycle costing method; in fact, the two methods are generally interchangeable.

Benefit/Cost (Savings-to-Investment Ratio) Method

Because the benefits of investing in solar energy are in terms of cost savings, a version of the benefit/cost ratio (B/C) method known as the savings-to-investment ratio (SIR) method is often used to evaluate solar energy investment. Like the two preceding methods, this method is based on discounted cash flows. The SIR method, however, expresses savings and investment costs as a ratio rather than as a dollar amount. The formula for computing SIR for an investment in solar energy based on costs and savings expressed in present-value terms (alternatively, annual-value costs may be used) is

$$SIR = [-\sum_{k=1}^{H} \sum_{j=1}^{n} (P_k \, \Delta Q_k \, b^j) - \sum_{j=1}^{n} a^j (\Delta M_j + \Delta R_j)]/[\Delta I - (\Delta V_n \, a^n)] \tag{28.7}$$

where SIR = net discounted savings attributable to solar energy, as a ratio to solar investment costs, and all other terms are as previously defined.

The placement of certain costs in the numerator rather than in the denominator of the SIR is largely an arbitrary decision. Equation (28.7) shows all recurring costs, including replacement costs in the numerator, and investment or first costs less salvage value in the denominator. An alternative is to add the present value of replacement costs to first costs in the denominator. A rationale for the approach taken here—that of the including replacement costs with other recurring costs in the numerator—is that it avoids possible confusion over what are repair costs and what are renewal costs of the investment. While this distinction is not important in evaluation of single solar energy installations, it may cause inconsistent measures when the SIR is used to evaluate and compare multiple installations, unless a consistent treatment is followed.

The acceptance criteria for an investment using the SIR method are (1) that the savings-to-investment ratio be equal to or greater than 1, *both for the total investment and for the last increment in investment costs*, and (2) that the system be expanded in size as long as the ratio for the last increment in the investment is greater than 1.0, and is equal to or greater than the ratio at the margin for the next best investment opportunity.

If there is no budget constraint, it pays to expand the solar energy system to the point that the ratio for the last increment of the investment equals 1.0. Because the SIR ratio tends to fall as an investment in solar energy is expanded, a larger, more efficiently sized version of a given project may have a lower SIR for the total investment than a smaller, less efficient version of the same project.

Using the same hypothetical cost data that was presented in the other illustrations, Table 28.5 illustrates the SIR method of evaluating an investment in solar energy. The initial yearly

TABLE 28.5   Illustration of the SIR Method*

Solar energy savings in base-year dollars (1)	Present value of energy savings,† dollars (2)	Present value of net replacement cost savings,‡ dollars (3)	SIR numerator [(2) + (3)], dollars (4)	SIR denominator (differential investment cost), dollars (5)	SIR ratio (6)
1500	19,077	841	19,918	8000	2.5

*Based on hypothetical cost data presented in Table 28.1, and derived by Eq. (28.7).
†Derived in Table 28.4, col. 4.
‡Derived in Table 28.4, col. 5.

energy savings from the solar energy system of $1500 amounts to $19,077 over the 20 years when expressed in present-value dollars. To this amount is added the net present-value savings in replacement costs realized because of the assumed longer life of the conventional auxiliary system when it is paired with the solar energy system. [If maintenance, repair, and replacement costs had been assumed to be higher with the solar energy system than without it (e.g., $\Delta M_j > 0$), the present-value increase in these costs would have been subtracted from the energy savings in the numerator of the SIR.]

The net replacement cost saved is equal to the present value of the $6000 of replacement cost that would otherwise be incurred in 15 years (i.e., $1436), less the present value of the salvage value that would have been realized from the replacement at the end of the 20-year time horizon (i.e., $595). Thus the numerator of the SIR is equal to $19,918, the denominator is equal to $8000, and the SIR is 2.5. This indicates on the average a $2.50 return per investment dollar over the life cycle.

Advantages, disadvantages, and recommended applications of the SIR method   Like the life-cycle cost method and the net savings method, this method offers the advantage of providing a comprehensive measure of profitability of an investment over its expected life. It offers an advantage over these other two methods by providing a measure which can be compared with comparable measures for competing independent projects to determine the most profitable group of investments available to the investor. That is, *it is a useful method for ranking an investment's profitability relative to other projects.* At the same time, it is generally less suitable than the other two methods for determining the optimal size of a project, because the SIR tends to fall as investment size increases. To use the measure for sizing projects, it is necessary to compute and compare the SIR for increments in investment size.

Internal Rate of Return Method

This method calculates the rate of return an investment is expected to yield. Unlike the other methods, the internal rate of return method does not call for the discounting of cash flows on the basis of a prespecified discount rate. Rather, the method solves for that rate of interest which when used to discount both costs and savings will cause the two to be equal, resulting in a net savings of zero.

The rate of return is generally calculated by a trial-and-error process by which various rates of interest are used to discount cash flows until a rate is found for which the net value of the investment is zero. The method may be described algebraically as follows: given values of other parameters, find the value of i that satisfies the equation

$$-\left[\sum_{k=1}^{H}\sum_{j=1}^{n}(P_k\,\Delta Q_k)b^j\right] = \left[(\Delta I - (\Delta V_n\,a^n) + \sum_{j=1}^{n}(\Delta M_j + \Delta R_j)a^j\right] \qquad (28.8)$$

where i = the compound rate of interest which when inserted in the expressions b^j and a^j solves the equation; this value of i is the internal rate of return on the investment.

The economic efficiency criterion for accepting an investment based on the internal rate of return method is that the calculated rate of return be equal to or larger than the investor's minimum attractive rate of return. The method can be used to compare the return on an investment in solar energy with the return on investment alternatives, in order to maximize the return from a given budget. The selection criterion is to choose individual investments in descending order of their expected rates of return, as long as the rates of return are equal to or exceed the minimum attractive rate of return, until the total budget is exhausted. The criterion for using this method to determine the most economical size of a solar energy system is to increase the size of the system as long as the rate of return on each investment increment is greater than the investor's minimum acceptable rate of return.[9]

Table 28.6 shows the calculation of the internal rate of return method for the sample hypothetical cost data that was used to illustrate the other evaluation methods. By trial and error it is found that the investment saves $2058 when evaluated with a 20 percent interest rate, but loses $201 when evaluated with a 25 percent interest rate. It therefore may be concluded that the rate of interest which will equate the total of savings and costs to zero lies between 20 and 25 percent. As shown in the table, interpolation can be used to determine that the internal rate of return is 24.6 percent. If this rate is greater than the investor's minimum attractive rate of return, the investment is economically attractive.

Advantages, disadvantages, and recommended application of the internal rate of return method   This method shares with the preceding three methods the advantage of providing a comprehensive evaluation of an investment in solar energy. A unique characteristic

TABLE 28.6 Illustration of the Internal Rate of Return Method*

Trial interest rates (1)	Solar differential investment cost, dollars (2)	Present value of energy savings,† dollars (3)	Present value of net replacement cost savings,‡ dollars (4)	Net present value based on the trial interest rates,§ dollars (5)
20	8000	9773	285	2058
25	8000	7634	165	−201

By interpolation, $i = 0.20 + 0.05[\$2058/(2058 + 201)]$
$i = 0.246$, the internal rate of return on the investment for NPV $= 0$

---

*Based on hypothetical cost data presented in Table 28.1, and derived by Eq. (28.8).
†Based on base-year savings of $1500, and derived using the uniform-present-value formula with a factor for fuel escalation, for an interest rate first of 20% and then of 25%.
‡Derived by applying to the replacement cost and salvage value the appropriate single present-value factors for an interest rate first of 20% and then of 25%; i.e., for $i = 0.20$, $285 = \$6000 (0.0649) − \$4000 (0.0261)$; for $i = 0.25$, $165 = \$6000 (0.0352) − \$4000 (0.0115)$.
§$2058 = \$9773 − (\$8000 − \$285); − \$201 = \$7634 − (\$8000 − \$165)$.

of this method, which might sometimes be an advantage, is the lack of necessity to specify the discount rate. However, it is necessary to have an estimate of the minimally attractive rate of return against which the calculated internal rate of return can be compared to decide the desirability of the investment.

The method has several possible disadvantages. Under certain circumstances there may be either no determinable solution or multiple solutions; however, this problem will probably be rare in evaluating solar energy systems.[9] As in the case of the savings-to-investment ratio method, a problem may arise in using the internal rate of return method to determine the economically efficient size of a solar energy system. As an investment is expanded, the rate of return on the overall investment may fall, but the rate of return of the additional investment may nevertheless be above the minimum attractive rate of return. As in the other case, however, this problem can be overcome by using the internal rate of return method to analyze incremental changes in the investment rather than the total investment. Like the savings-to-investment ratio method, this method has the advantage of indicating the relative efficiencies of alternative investments. It therefore may be useful in comparing solar energy with other investment options.

Payback Method

This evaluation method measures the elapsed time between the point of an initial investment and the point at which accumulated savings, net of post-investment accumulated costs, are sufficient to offset the initial investment. As for the other evaluation methods, costs and savings should be discounted to take into account the opportunity cost of capital. The algebraic expression for determining *discounted payback* is the following:

$$\Delta I + [(\sum_{k=1}^{H} \sum_{j=1}^{Y} (P_k \, \Delta Q_k \, b^j) + \sum_{j=1}^{Y} a^j(\Delta M_j + \Delta R_j)] = 0 \qquad (28.9)$$

where Y = number of years to pay back when cash flows are *discounted*, and other variables are as previously defined. If the net savings are constant S, $Y = - \ln (1 - iI/S)/\ln (1 + i)$.

Table 28.7 shows the payback period calculated for the hypothetical investment problem used to illustrate the other methods. Cumulative discounted savings and recurring costs are compared in each year with the initial investment cost, until net savings become positive. With energy prices escalated at a rate of 5 percent and discounted at a rate of 10 percent, energy savings are shown to offset investment costs in the seventh year.

Advantages, disadvantages, and recommended application of the payback method
The payback method has the principal advantages of being easy to understand and familiar to a wide audience. Its appeal also lies in the fact that it allows emphasis to be given to the rapid recovery of investment funds, an objective important to many organizations. Rapid payback is often of critical importance to speculative investors. It can also be important to other investors if financial resources are available for only a short period of time or if there is considerable

TABLE 28.7   Illustration of Discounted Payback Method*

Years into the investment (1)	Amount of initial investment cost, dollars (2)	Present value of cumulative energy savings,† dollars (3)	Present value of other cumulative costs (4)	Present value of net savings,‡ dollars (5)
1	8000	1432	0	−6568
2		2799	0	−5201
3		3103	0	−4897
4		5348	0	−2652
5		6537	0	−1463
6		7672	0	− 328
7		8755	0	+ 755

*Based on hypothetical cost data and assumptions presented in Table 28.1, and derived by Eq. (28.9).
†Derived by applying to the base-year energy savings of $1500 the uniform-present-value factor for each successive year, including a factor for energy price escalation.
‡Net savings is the difference between the initial investment cost and the cumulative present value of savings less recurring costs.

doubt as to the expected life or resale value of the major components of the investment. Often, however, there is too much emphasis on the length of the payback period, and not enough emphasis on overall expected profitability of the investment.

A principal disadvantage of the payback method is that even if based on discounted cash flows, it does not provide a comprehensive evaluation of an investment's profitability. It is not comprehensive because it does not include those cash flows that occur after the point payback is reached. The case is sometimes made that this shortcoming can be overcome by comparing the expected life of the major components of an investment with the estimated payback period to determine how long savings are expected to continue beyond the point that costs are recovered. And, with very simple investment problems involving uncomplicated patterns of costs and savings, this comparison can usually be made. However, even in the simple case the payback method provides no clear, reliable measure of overall economic performance for comparing alternative investments. Comparison of alternatives strictly on the basis of the payback period may lead to inefficient investment decisions, because an investment with a longer payback may be more profitable than an investment with a shorter payback.

CONVERTING COSTS AND SAVINGS TO A COMMON TIME AND COMMON DOLLAR MEASURE

A necessary step in each of the comprehensive economic evaluation methods is the adjustment of the various cost items to an equivalent time basis. This adjustment is necessary because an investment in solar energy results in expenditures and savings that occur both in the present and in the future, and there is a difference between the value of a dollar today and its value at some future time. This time dependency of value reflects not only inflation, which may reduce the purchasing power of a currency, but also reflects the real earning potential of money.

The preceding section included the appropriate adjustments for the timing of cash flows in the formulations of the selected methods of economic evaluation; this section describes in more detail the whys and hows of converting costs and savings to a common time and common dollar measure. First, the treatment of inflation in economic evaluations is described in general. Then discounting of cash flows is described in detail.

Treatment of Inflation

Removal of inflation from cash flows is essential for a valid economic evaluation of an investment. Otherwise, the evaluation is made in variable-value dollars, and makes no economic sense. Removing inflation means measuring cash amounts in terms of the value of the dollar in a base year, usually the time at which the investment decision is made. There are several alternative ways of removing inflation from cash flows. Briefly, they are (1) exclude inflation from the analysis at the outset by assuming that all cash flows are fully and evenly responsive to inflation and, therefore, remain constant in terms of base-year dollars, (2) include expected price changes in cash-flow estimates and then remove inflation prior to discounting by the use of a constant dollar price deflator based on past inflation rates or predicted ones, and (3) include expected price changes in cash-flow estimates (i.e., use *current* dollars) and discount the cash flows with a discount rate that includes the expected rate of inflation, in addition to the investor's real potential earning rate.

The choice of approaches depends in part on the expected pattern of price change. The simplest approach is the first, that is, to omit inflation from analysis by assuming that inflationary effects cancel out, leaving the outcome of the analysis unchanged. For example, other things being equal, if the cost of labor to perform a given maintenance service for a solar energy system is expected to change at about the same rate as prices in general, the cost of that service in the future in constant dollars can be assumed to be the same as it is today. If there is not a compelling reason to assume that the prices of particular goods and services will inflate differently from prices in general, this convention of assuming base-year prices to hold for future constant dollar prices is a widely accepted practice of analysis.

In the absence of income tax effects, inflation generally will not affect the outcome of an investment whose cash flows are fully and evenly responsive to inflation. However, when tax effects are considered, inflation may significantly affect the investment's profitability. For example, under existing tax practices, the tax benefit from depreciation is *received in future dollars, but is measured in the dollars of the initial investment*. This results in a decline in the value of the tax benefit, other things being equal.

Future prices of goods and services relevant to the investment may not, however, respond fully and evenly to inflation in such a way that they remain level in constant dollars. For example, mortgage loans are fixed over time and fall in value when measured in constant dollars. Alternatively, the future value of an item may rise at a rate faster than general price inflation. For example, forces of demand and supply are widely expected to increase the price of non-renewable energy sources faster than most other prices.

If future prices are fixed at current levels or for some other reason are not expected to respond fully to price inflation, it is important that these future values be converted to constant dollars. This can be done by following the second approach above and applying a price deflator to each future annual payment prior to discounting, or by following the third approach and discounting the future annual payments with a discount rate that includes inflation.

If future prices are expected to rise faster than general price inflation, the adjustment to constant dollars is usually made in one of two ways: (1) a variation of the first approach described above, and (2) the third approach described above. In the first approach, inflation *per se* is excluded from the estimates of future cash flows. However, the *differential* price increase, that is, the expected price change over and above the general rate of inflation, is included in the future cash flows. In accordance with this procedure, a real discount rate (exclusive of inflation) is used to further adjust for time preference.

Discounting Cash Flows

As indicated above, a procedure for adjusting for the time value of money is a technique often called "discounting." Discounting refers to the use of interest formulas to convert cash flows occurring at different times to equivalent amounts at a common point in time.

Discount formulas   To discount cash flows to equivalent values at a common time, an appropriate interest formula (or "discount formula") is applied to each cash amount. Alternatively, to simplify the calculation, interest factors (or "discount factors") precalculated from the discount formulas can be multiplied by the cash amounts. The discount formulas most frequently used in investment analysis are the following: (1) the uniform capital recovery formula, which when applied to a present amount converts it to an equivalent series of level annual payments; (2) the single compound amount formula, which when applied to a present amount converts it to an equivalent value at a future time; (3) the single-present-value formula, which when applied to a future amount converts it to an equivalent present value; (4) the uniform compound amount formula, which when applied to a recurring annual amount converts it to an equivalent lump-sum amount at a future time; (5) the uniform sinking fund formula, which when applied to a given future total value indicates the annual amount required to achieve the future total value; and (6) the uniform-present-value formula, used to find the equivalent present value of a series of level annual amounts.

Table 28.A1 in the appendix to this chapter gives the nomenclature and algebraic expression for each of these discount formulas and Tables 28.A2 through 28.A7, give the counterpart discount factors for selected time periods and discount rates.

Discount rates   To apply the discount formulas or factors, it is necessary to select a discount rate. The discount rate should reflect the investing person's or firm's time preference for money (or, more accurately, the resources money can buy). Apart from inflation, time preference reflects the fact that (1) money in hand can be invested to earn a return and (2) money borrowed requires interest to be paid. Of these two factors, the former, often called "the opportunity cost," is generally predominant in establishing a discount rate. If there is a limit on the total funds available for investment from all sources including borrowing, the most important element in determining the discount rate is the rate of return which would be forgone on the next-best investment if a given investment is undertaken. This rate can be higher than the borrowing rate, and needs to be accounted for in an investment analysis even if funds are not borrowed. The discount rate, then, stipulates a minimum rate of return which must be recovered on an investment over and above other investment costs.

A discount rate may be either "nominal" or "real." A nominal discount rate reflects both the effects of inflation and the real earning power of money invested over time. A nominal rate is appropriate to use only in combination with cash flows which also include inflation. Market rates of interest, such as the mortgage rate on building loans, are nominal rates.

A real discount rate reflects only the real earning power of money and not inflation. It is appropriate for evaluating investments if inflation is removed from the cash flows prior to discounting, i.e., if they are already stated in constant dollars. (See *Treatment of Inflation*.)

The relationship between a nominal rate of interest and a real rate of interest may be described as follows:

$$i = d + I + dI \tag{28.10}$$

where i = a nominal rate of interest
d = a real rate of interest
I = the expected rate of inflation

Note that the nominal rate is not equal simply to the real rate plus the rate of inflation. The rate of inflation is reflected in the nominal rate in two terms: I, the rate of inflation, and dI, the product of the real rate and the rate of inflation.[2]

There is no single discount rate—whether calculations are in real or nominal terms—that is appropriate for discounting all cash flows. A wide range of rates is used in practice, and there is a subjective element in the choice of rates.[6] The important thing is that the discount rate reflect the investor's time preference.

The choice of rates can significantly affect the outcome of an evaluation. The higher the rate, the lower the value of future cash flows; the lower the rate, the smaller the effect of discounting on future cash flows.

Timing of cash flows   In the case of most investments, expenditures and receipts or savings occur throughout the year. Some occur annually, others semiannually, quarterly, monthly, weekly, daily, or even nearly continuously as in the case of cash receipts from sales. Seldom, however, are cash flows modeled exactly as they occur within a year. To simplify the analysis, they are commonly assumed to occur in lump sums either at the beginning, middle, or end of each year, or continuously throughout the year.

The discount factors contained in Tables 28.A2 through 28.A7 in the appendix of this chapter can be used to discount cash flows on either a beginning-of-period or an end-of-period basis simply by designating the initial period index as 0 (beginning) or 1 (end). By averaging the discount factors for two consecutive periods, discount rates reflective of middle-of-the-period payments can be developed. For example, the single-present-value factor for midway between the first and second years is 0.9545, for a discount rate of 10 percent. This is derived by averaging 1.000, the discount factor for a cash flow occurring at the beginning of the first year, and 0.9091, the discount factor for the cash flow occurring at the end of the first year.

If a continuous flow model is desired, conversion factors in Table 28.A1 can be used.

Escalation of nonrenewable energy prices   The fact that energy prices are widely expected to increase (escalate) faster than the rate of increase in the general price level is a major force behind the growing interest in solar energy. To reflect this expectation, energy prices are usually escalated in evaluating the economic performance of solar energy. As in the case of other cash flows, however, it is important to eliminate the effects of changes in the purchasing power of the dollar and to express future energy costs in constant dollars (see *Treatment of Inflation*).

The economic performance of the solar energy system tends to be quite sensitive to the choice of a fuel price escalation rate. Selecting a high rate raises savings relative to costs and can easily change the evaluation of a system from economically unprofitable to economically profitable. For example, Table 28.8 shows the effect of several different price escalation rates on the present value of a yearly savings of 100 million Btu of electricity priced initially at $0.035/kWh, and discounted at a rate of 10 percent.

Over a 25-year period, the present-value savings of 100 million Btu of electricity ranges from a low of $9308 with no price escalation and a discount rate of 10 percent to a high of $48,060 with a price escalation rate of 15 percent and a discount rate of 10 percent. When the fuel price escalation rate is exactly equal to the discount rate, they offset one another, and the net present value is equal simply to the yearly savings multiplied by the number of years, or $25,635.

Without escalation of future cash flows, discounting tends to reduce them eventually to the point at which they are so small in present-value terms as to be of little consequence to the outcome of the evaluation. For example, with a 10 percent discount rate, a dollar to be received in five years is worth $0.62 today, in 10 years, $0.38, and in 15 years, only $0.24. However, including a relatively high price escalation rate can cause cash flows expected in the distant future to weigh significantly in the outcome.

Because of the considerable uncertainty about energy prices in the distant future, caution is advisable in specifying a long-term rate of price escalation. One approach that may be taken to reduce the impact of ever-compounding fuel prices in distant years is to limit the period over which the investment is to be evaluated. Another approach is to use a short-term escalation rate for the period in which rising prices seem assured, and a somewhat lower long-term rate for the period for which there is greater uncertainty.[1]

Economic life of principal assets and period of analysis   A life-cycle analysis requires the designation of economic life expectancies for an investment's principal assets. The economic life of an asset is that period during which the asset is expected to be retained in use for its intended purpose at the minimum cost for achieving that purpose. The useful life is that during which the asset is expected to serve a useful purpose.

The economic or useful life expectancy of solar energy systems is difficult to estimate with a high level of confidence. The relatively short period in use of most solar energy systems limits the possibilities for basing life expectancies on actual data samples, although some insight may be gained by examining historical durability data for individual components of solar energy systems that have been used in other, related capacities. For the most part, analysts must rely upon durability information from manufacturers, trade and professional journals, and research reports.

The comprehensive methods of evaluation require the analyst to specify the length of time over which an investment is to be evaluated. In establishing the time-frame of analysis, the following requirements should be met: (1) in comparing solar energy systems with conventional energy systems, all alternatives should be evaluated on the basis of the same time frame; (2) in evaluating an investment for a period shorter or longer than the expected physical lives of the principal components, any significant residual salvage values or replacement costs should be taken into account; (3) the period over which an energy system is to be analyzed should not exceed the life expectancy of the facility in which it is used; e.g., building life normally imposes a constraint on the life of the energy system.

In using a present-value measure to compare alternative energy systems, it is important that the alternatives be evaluated for an equal time period. Often a time cycle is chosen that will

TABLE 28.8   Impact of Alternative Energy Price Escalation Rates on Present-Value Energy Savings in a Hypothetical Case

Energy price escalation rate, percent	Present value of 10×10^7 Btu energy savings over 25 years,* dollars
0	9,308
5	14,858
10	25,635
15	48,060

*Discounted at a rate of 10 %, and priced initially at $0.035/kWh.
NOTE: This example examines only savings, not costs, of a purely hypothetical solar energy system.

permit enough renewals of each alternative that the cost of their economic lives coincides with no remaining salvage value. For example, in comparing a system with a 20-year life with one having a 30-year life, the present value of each could be examined over 60 years. This would require two renewals of the 20-year system and one renewal of the 30-year system.

Alternatively, if the length of time over which the energy system is needed is limited, e.g., 10 years, both systems would be evaluated for 10 years, with any renewal costs on salvage values remaining for either at the end of the 10 years taken into account.

If it is reasonable to assume that the system will be needed indefinitely, the annual cost measure may be calculated on the basis of the economic life of each system. This eliminates the necessity of evaluating both systems for the same length of time and including renewal costs and salvage values. With indefinite need for a system, for example, the annual cost of the 20-year-life system could be calculated on the basis of 20 years and compared with the annual cost of the 30-year system calculated for 30 years. Evaluating alternatives on the basis of their respective lives without the need to include renewals and salvage can in some cases considerably simplify the calculations. The annual-cost measure is sometimes preferred over the present-value measure for this reason.

Examples of discounting   The different patterns of cash flows generated by investments in solar energy systems require the use of the various discounting formulas (see Table 28.A1). To select the appropriate discounting formula, it is necessary to decide whether cash flows are to be expressed in terms of (1) a present value or (2) an annual value, i.e., a series of level values over time. The time basis is largely a matter of preference; either will serve as well as the other for the purpose of an economic comparison, as long as a consistent approach is followed for the cash flows of a given investment.

Table 28.9 illustrates the conversion of various costs and savings, typical of those that might result from an investment in solar energy, to present-value and annual-value equivalents. The examples are based on the specific assumptions given in the table.

The first column describes the type of cash flow. The second column uses a "cash-flow diagram" to show the timing of that particular cash flow. The horizontal line of the cash-flow diagram is a time scale, where P indicates the present; the progression of time moves from left to right, and the numbers between the points represent years. The downward-pointing arrows represent expenditures (cash outflows), and upward-pointing arrows represent receipts or savings (cash inflows).

The first row shows the discounting of planning and design costs assumed to have been incurred at the beginning of the past year. To convert this past cost to a present value, the single compound amount factor for 1 year and the assumed discount rate of 8 percent is used. To convert the past amount to an annual value, it is first converted to a present value and then to an annual value by applying the uniform capital recovery factor. Thus, in cols. 3 and 4 of Table 28.9, it may be seen that paying $500 a year ago is equivalent to paying $540 now (apart from inflation), or to paying $63 each year over the next 15 years, under the stated conditions.

The second and third rows of Table 28.9 show the discounting of current expenditures associated with acquiring and installing the system on the building. These values are already stated as present values and need only be converted to annual values by application of the uniform capital recovery factor. (Tax and financing effects are treated later.)

The fourth row of Table 28.9 shows the discounting of a series of yearly maintenance costs that are level in constant dollars. The uniform-present-value factor is used to convert the series of yearly expenditures of $60 to a present-value equivalent of $514. There is no need for any adjustment to derive the equivalent annual value because the cash flow is already expressed in annual terms.

The fifth row shows the discounting of repair and replacement costs. These costs consist of future payments that occur less often than every year. The single-present-value factor is used to convert the expected amounts of $100 in the fifth year and $100 in the tenth year to the total present-value equivalent of $114. The uniform capital recovery factor is used to convert the present value of $114 to an equivalent annual value of $13.

The sixth row shows the discounting of a single future receipt, the net salvage value of $3000 that the solar energy system is assumed to yield at the end of the period of use. Again, the single-present-value factor is used to convert the future value to a present-value equivalent of $946. The annual-value equivalent of $110 is found by applying the uniform sinking fund factor to see how much money would be necessary to invest annually in order to accumulate the specified amount by the designated time.

The seventh row of Table 28.9 shows the discounting of expected energy savings. The yearly increase in the amount of savings reflects the assumption that energy prices will rise substantially faster than prices in general. At the outset, yearly savings are estimated at $1000. To find the present value of the life-cycle savings, it is necessary both to escalate the yearly savings and to discount them. This is done by using the variation of the uniform-present-value formula from Table 28.A1. The present-value equivalent of yearly energy savings is $12,959, under the stated conditions. The annual-value equivalent is $1514, found by applying the uniform capital recovery factor to the present-value amount.

The discounted cash flows may now be used to evaluate the profitability of the investment by computing, for example, the total life-cycle cost of the investment, the net savings from the investment, or the savings-to-investment ratio. In this simplified example, the net present-value savings from the investment, found by subtracting the total present value of costs, less salvage value, from the total present value of fuel savings, is $3137 (i.e., $12,959 − [$540 + ($9000 − $946) + $600 + $514 + $114] = $3137). In annual-value terms, total costs amount to $1147 and total savings to $1514, resulting in a net annual savings of $367. Thus, under the stated conditions, saving $3137 in present-value terms over 15 years is equivalent to saving $367 each year for the 15 years.

Methods of Treating Uncertainty in Economic Evaluations

In evaluating the economic performance of a solar energy system there is little past experience upon which to base estimates of an uncertain future, and the appropriateness of key assumptions is often questionable. Yet the results of an economic evaluation are dependent on the particular parametric values and assumptions specified by the analyst. This situation—common to many investment decisions, but particularly to a new technology for which there is little historical data base—leads to uncertainty about the accuracy of evaluation results. There is uncertainty in the economic evaluation of solar energy systems in the sense that there may be a disparity between predicted values in the analysis and actual values.

In the examples shown heretofore, and in many economic studies that have been made of solar energy, the simplifying assumption is made that the timing and values of most or all of the cash flows are known with certainty. That is, deterministic life-cycle cost models are used and decisions are based on single-value estimates of the parameters.

In some cases and for certain parameters, it appears reasonable to make this assumption of certainty, either because an error in the estimate will not significantly affect the outcome of the analysis or because the cost of attempting to estimate the uncertainty will exceed the benefits of doing so. Often the approach taken in using deterministic models is to base them on conservative estimates of key parameters in the effort to reduce the possibility of incurring large losses from decisions based on the model results. The user of a deterministic economic evaluation model should, however, be aware of the possible risks inherent in basing a decision on conclusions derived from a model whose structure or parametric values may be questionable.

If practicable, it is advisable to consider the effects of possible variation in those parameters which would significantly affect the economic performance of the investment. There is, for example, considerable uncertainty as to how fast fossil fuel prices will escalate, and this is a critical factor in the economy of a solar energy system. There may also be substantive questions about the life of the system, the costs of maintenance, repairs, and replacements over the life, and the appropriate discount rate. Another important area of uncertainty is system performance as affected by weather conditions. Weather conditions cannot be known precisely in advance and may not in a given year be well described by the data from the "typical year" upon which technical performance evaluations of solar energy systems are usually based. Hence, the economic performance of the system in a given year or period of years may deviate significantly from the predicted performance. In some cases, the building or facility owner will not be particularly sensitive to short-run deviations of the system from predicted performance levels. However, there may be cases in which the owner will be extremely sensitive to deviations from predicted performance. This may occur, for example, with owners of commercial buildings who have low priority when shortages of fuel oil or natural gas occur. Their reason for acquiring a solar energy system may be to provide a fraction of the monthly energy requirements sufficiently large to guard against the threat of an operational shutdown during times of fuel shortages. In this case, it might be critical to the investment decision to know the reliability of differently designed and sized solar energy systems under the weather conditions that might actually be experienced.

In some cases the analyst may be able to estimate the probability of alternative events occurring. Where the probability of alternative outcomes can be predicted, the technique of probability analysis can be used to develop the "expected value" of an investment's profitability.

TABLE 28.9 Examples of Discounting Cash Flows for a Solar Energy Investment*

Type of cash flow (1)	Cash-flow diagram (2)

(1) Engineering, design, and planning

P
−1 1 2 3· 4 5 6 7 8 9 10 11· 12 13 14 15

$500

(2) Purchase and installation

P
1 2 3 4 5 6 7 8 9 10 11 12 13 14 15

$9000

(3) Building modifications

P
1 2 3 4 5 6 7 8 9 10 11 12 13 14 15

$600

(4) Maintenance and operation

P
1 2 3 4 5 6 7 8 9 10 11 12 13 14 15

60 60 60 60 60 60 60 60 60 60 60 60 60 60 60

(5) Repair and replacement

P
1 2 3 4 5 6 7 8 9 10 11 12 13 14 15

$100 $100

(6) Salvage of system, end of 15 years

$3000
P
1 2 3 4 5 6 7 8 9 10 11 12 13 14 15

(7) Fuel savings

P $1060 $1124   1191   1263   1338   1419   1504   1594
1   2   3   4   5   6   7   8

1690   1791   1898   2012   2133   2261   2397
9   10   11   12   13   14   15

*Based on the assumptions that all cash flows are in constant dollars, a real discount rate of 8% is used, the system has already been designed and is ready for purchase and installation, the solar energy system has an expected life of 25 years, conventional energy prices are expected to escalate at a real rate of 6%, the investor has a time horizon of 15 years, and the system has a salvage value of $3000.

Present-value equivalent (3)	Annual-value equivalent (4)

$P = \$500$ (SCA, 8%, 1 year)
  $= \$500$ (1.0800)
  $= \$540$

$A = \$500$ (SCA, 8%, 1 year) (UCR, 8%, 15 years)
  $= \$500$ (1.0800) (0.11683)
  $= \$63$

$P = \$9000$ (1)
  $= \$9000$

$A = \$9000$ (UCR, 8%, 15 years)
  $= \$9000$ (0.11683)
  $= \$1051$

$P = \$600$ (1)
  $= \$600$

$A = \$600$ (UCR, 8%, 15 years)
  $= \$600$ (0.11683)
  $= \$70$

$P = \$60$ (UPV, 8%, 15 years)
  $= \$60$ (8.559)
  $= \$514$

$A = \$60$ (1)
  $= \$60$

$P = \$100$ (SPV, 8%, 5 years) +
  $\$100$ (SPV, 8%, 10 years)
  $= \$100$ (0.6806) + $\$100$ (0.4632)
  $= \$68 + \$46$
  $= \$114$

$A = [\$100$ (SPV, 8%, 5 years) + $\$100$ (SPV, 8%, 10 years)]
  (UCR, 8 %, 15 years)
  $= [\$100$ (0.6806) + $\$100$ (0.46323)] (0.11583)
  $= (68 + 46)$ (0.11583)
  $= \$13$

$P = \$3000$ (SPV, 8%, 15 years)
  $= \$3000$ (0.3152)
  $= \$946$

$A = \$3000$ (USF, 8%, 15 years)
  $= \$3000$ (0.03683)
  $= \$110$

$P = \$1000 \sum_{j=1}^{15} \left(\dfrac{1 + 0.06}{1 + 0.08}\right)^{j} = \$1000$ (UCR, 1.89%, 15 years)

  $= \$1000$ (12.959) $\left(1.89\% = \dfrac{0.08 - 0.06}{1.06} \times 100\%\right)$

  $= \$12,959$

$A = \$1000 \sum_{j=1}^{15} \left(\dfrac{1 + 0.06}{1 + 0.08}\right)^{j}$ (UCR, 8%, 15 years)

  $= \$12,959$ (0.11683)

  $= \$1,514$

NOTE: SCA = single compound amount factor; (SCA, 8%, 1 year) denotes the factor for 1 year at 8% interest
  UPV = uniform-present-value factor
  SPV = single-present-value factor
  UCR = uniform capital recovery factor
  USF = uniform sinking fund factor

Statistical analysis of the variability of predicted values can be used to assess further an investment's risk. Where probabilities cannot be predicted, sensitivity analysis can be used to test the impact on the outcome of alternative values of key parameters, or breakeven analysis can be used to determine the minimum or maximum required values of selected parameters.

SOLAR ENERGY SYSTEM CASH-FLOW COMPONENTS

An Overview of Major Kinds of Costs and Savings

A comprehensive evaluation of solar and conventional heating and/or cooling systems requires the assessment of the following kinds of costs over the life cycle: (1) system acquisition costs, including design, engineering, and "search costs" if they are important, purchase prices, delivery costs, and installation costs; (2) costs or savings due to changes in the building envelope or in other building components that must be modified to accommodate the solar energy system; (3) system repair and replacement costs; (4) maintenance costs; (5) energy costs; (6) salvage values, net of removal and disposal costs; (7) insurance; (8) taxes; and (9) governmental incentives.

These various kinds of costs are required for all parts of the systems being compared. For a solar energy system for a building, the principal parts may comprise the following: (1) solar collector, (2) thermal storage, (3) domestic hot-water system, (4) air-conditioning components, (5) auxiliary energy system, (6) heating and cooling distribution systems, (7) the system control, and (8) any motors, pumps, fans, blowers, wiring, and tubing included in these parts.

If the energy systems that are to be compared have parts which are identical in costs, the costs of these parts may be omitted for the purpose of comparing the economics of the systems. For example, often it is assumed that the acquisition costs, as well as the reliability, durability, maintenance, and repair costs, of the conventional auxiliary system are the same as those of the conventional system used alone. This assumption reduces the number of cost elements to consider and simplifies the analysis.

Where the costs of alternative systems differ, the costs of each may be assessed in full and then compared; or, alternatively, the costs attributable to the solar energy system can be calculated by subtracting the cost of the conventional system from the cost of the combined solar/auxiliary system for each item of cost.

System costs (other than insurance, taxes, and incentives) may be further divided into (1) engineering, (2) materials, (3) labor, and (4) marketing costs. This division of costs may be useful for the purpose of current cost control and for predicting future system costs. For example, it has been predicted that system design, engineering, and search costs will tend to decrease as the number of solar installations increases and the knowledge of system size requirements for alternative locations and standardized designs expands.[5] As another example, there may be cost tradeoffs between materials and labor that could reduce future total system costs. Factory preassembly of solar collectors in modular form, for instance, tends to increase materials costs (as compared with assembly at the job site), but may lower labor costs of assembly through techniques of mass production in the factory.

Some kinds of costs may be difficult to lower. Labor costs for collector installation, for example, may be more difficult to reduce than labor costs for system assembly. Similarly, materials costs may be difficult to reduce apart from a switch to innovative designs that use fewer and/or less expensive materials.

The relative proportions of the various kinds of costs may differ considerably, depending on the type of application. For example, the relative proportions of materials and labor costs may differ for new construction and for retrofit installations. Installation of a collector as an integral part of the building during construction tends to displace a portion of roofing labor costs, whereas retrofit of a collector to an existing building may require additional materials costs for structural support to the roof.[5]

Fixed Costs and Variable Costs

Another cost concept—one that is particularly useful in determining the optimal system size—is the division of costs into fixed and variable components. Fixed, or constant, costs are those that do not change as the size of the solar energy system is increased. (Few costs are truly unchanging as the capacity of the system is changed; but some costs do tend to be somewhat fixed over a restricted range of system sizes.) For example, the need for a reliable energy source may mean that the capacity of the conventional backup system will be held constant over a wide range of solar energy system sizes. Additionally, most of the other components of a solar energy system will have some portion of their costs fixed for alternative total system sizes. For example, a solar energy system of any size will require a minimum of controls, pumps, tanks, heat exchangers, valves, piping, fittings, etc. that will tend to remain about constant over some range as the size of the collector area is expanded. The cost of this minimum set of components is a fixed cost. After a point of increasing collector size, these costs will begin to increase. The increases are variable costs. The cost of the collector panels, storage, and heat exchangers, usually comprises the major item of variable costs. Collector installation generally has both a fixed and a variable cost component.

Figure 28.1 illustrates the fixed and variable elements typical of the major items of costs for a solar energy system. The vertical axis measures costs, and the horizontal axis measures collector area. The upper quadrant of the figure shows items that raise the cost of solar; the lower quadrant shows items that lower the cost. Cost curves that originate at the origin or at the point of origin of the cost curve just below it, have no fixed element. If they maintain a constant distance from the curve just below, they have no variable element. If they diverge as collector area increases, they encompass a variable element.

To estimate the costs of alternative systems, it is useful to identify the fixed and variable portions of system costs, and to calculate the variable portion of costs as a function of collector area. For example, the cost of a component such as storage may be estimated as follows:

$$S = F + sAP \tag{28.11}$$

where S = total storage costs
F = fixed costs for tanks, valves, etc.
s = units of storage medium per unit of collector area
A = collector area
P = price per unit of storage medium

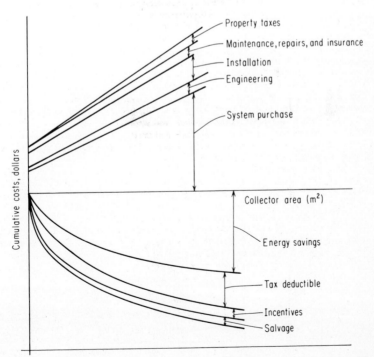

Fig. 28.1   Fixed and variable cost components. (Fixed costs are the differences in the intercepts of the upper arrows.)

By dividing all system costs into fixed and variable components, the cost-estimating equation for the system can be simplified. Because fixed costs are incurred regardless of the size of the system, they influence only whether or not net savings are positive or negative, but do not otherwise influence the optimal size. This is because it is the *changes* in costs and savings as system size is expanded that determine the optimal system size.

Total, Average, and Marginal Costs

The changes in costs and savings that result from an increment in the size of a system are known in economics as the "marginal cost" and the "marginal savings," respectively. The marginal costs and savings are the derivatives of the total costs and savings, and when equated, result in the maximum level of net savings, measured as the difference between total cost and total savings. (See *Economic Optimization* in this chapter.) Figure 28.2 illustrates the typical relationship between total costs, total savings, marginal costs, and average costs. These concepts apply to costs and savings stated in life-cycle cost terms.

Treatment of Acquisition Costs

Costs incurred initially (first costs) are already in present-value terms and require no further adjustment if a present-value model is used. If an annual-value model is used, it is necessary only to multiply the total of first costs by the uniform capital recovery factor UCR.

Acquisition costs may be treated entirely as a first cost if it is assumed that the system is purchased outright, or if no difference is assumed in the discount and the mortgage loan rates. Otherwise, only the down-payment portion of acquisition costs is treated as a first cost, and the present value of the remaining costs is found by first applying the UCR based on the mortgage loan rate of interest, and then applying to the level, periodic mortgage payment the uniform present value factor UPV based on the discount rate. Note that the mortgage payment is fixed and declines in constant dollars during inflationary periods.

Treatment of Maintenance and Insurance Costs

In a present-value model, level yearly recurring costs are brought to a present value by multiplying the amount in the base year by the UPV, provided the analysis is in constant dollars and uses a real discount rate. If the analysis is in nominal dollars and uses a nominal discount

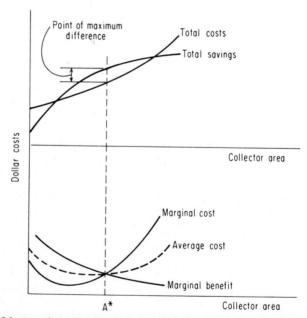

Fig. 28.2   The relationship of total, marginal, and average costs of a solar energy system.

rate, it will be necessary to multiply base-year costs by the modified version of the UPV, incorporating the projected inflation rate.

No adjustment to the base-year costs is needed for an annual-value model. The base-year values can be interpreted as uniform annual values in constant dollars.

Maintenance and insurance costs are often treated as yearly recurring costs, remaining unchanged in constant dollars. Alternatively, these costs, particularly maintenance costs, may be projected to change in real terms, because of deterioration over the life of the system. If the projected change is at a constant rate, the modified UPV can be used to convert to a present value. If the rate of change is not constant, these costs are usually treated as nonrecurring costs.

If additional insurance costs are incurred on the solar energy system, they should be considered in the life-cycle cost analysis. In so doing, it is important to remember that insurance costs represent a tradeoff to the homeowner for incurring damage costs. It is the net cost to the homeowner of damage, I_n, which is relevant, that is, the cost of insurance (insurance premiums) plus damage losses, net of insurance reimbursements collected. The net cost in annual-cost terms is as follows[8]:

$$I_n = I + L - C \qquad (28.12)$$

where I = annual insurance premiums
L = annual damage loss
C = annual insurance reimbursements

Treatment of Energy Savings

If a yearly recurring cash flow, e.g., energy savings, is projected to change at a constant rate over some period of time, the modified UPV for the appropriate period of time can be multiplied by the base-year amount in order to determine the equivalent present value. To determine the equivalent annual value, it is necessary first to find the present value incorporating the price escalation, and then to apply the UCR to the present-value amount. The projected rate of future price escalation can also be changed over time.

Treatment of Repair and Replacement Costs and Salvage Values

Cash flows that do not recur annually, such as repair and replacement costs and salvage values, can be converted to a present-value by applying to each item the single-present-value factor SPV appropriate to the year in which the cash flow occurs. Nonrecurring cash flows can be converted to an annual value by applying the UCR to the present-value amount.

Treatment of Property Taxes and Income Taxes[8]

The effect of taxation on the costs of a solar system to the owner may be considered for two main cases: (1) for the owner-occupied solar residence, and (2) for the rental solar residence or other solar commercial buildings. For both cases, taxes impact on costs in several ways, in some instances raising and in other instances lowering life-cycle costs of the solar heating, ventilation, and air-conditioning (HVAC) system relative to its conventional counterpart. Let us examine the two cases in turn for tax implications.

Owner-occupied residence   For the owner-occupied residence, the primary tax effects are from the property tax and, indirectly, from the income tax. The particular effect of either of these taxes could be expected to vary among individual solar residences, depending upon local property-tax rates, property assessment practices, and the income-tax bracket of the homeowner. The focus here is on the nature of the effects and on the method of including them in the life-cycle cost analysis.

The *property tax*, which is levied as a percentage of a fraction of the market value of a building, would tend to raise the life-cycle costs of a solar HVAC system relative to a conventional system. Life-cycle costs would be raised because, other things equal, the greater first cost of solar HVAC equipment would be reflected in a higher market value for the residence, and hence in a larger assessed value for the solar residence than for a conventional residence with its lower first-cost HVAC system. Thus, the capital intensiveness of an HVAC system influences the amount of property taxes levied on a residence, and thereby alters the life-cycle cost of the HVAC system.

As a simple example, let us compare the property tax on a $60,000 solar residence, of which $8000 is attributable to additional cost of the solar HVAC system, with a counterpart conventionally heated and cooled home valued at $52,000 (i.e., $60,000 − $8000). Given a typical tax rate of 4.50 percent of 50 percent of market value, the $60,000 solar residence would be assessed at $30,000 and taxed $1350. The counterpart conventional home would be assessed $26,000 (i.e., 50 percent of $60,000 − $8000) and taxed $1170. For purpose of illustration, further assume a real discount rate of 2 percent, a constant real property value (including a constant real value for the solar system with replacements made as needed), and a constant property tax rate over a 20-year period of evaluation. (The assumption of a constant real property value and a constant property tax rate means that even though the nominal, or market, property value changes, the yearly property tax remains constant in terms of present prices.) Over 20 years, the property tax on the solar residence would amount in present-value terms to $22,074—i.e., $1350 [(1 + 0.02)^{20} −1]/ [0.02 (1 + 0.02)^{20}] = $22,074; the present value of the property tax on the conventional residence would amount to $19,131—i.e., $1170 [(1 + 0.02)^{20} −1]/[0.02 (1 + 0.02)^{20}] = $19,131. Thus, the effect of the property tax in the above example is to raise the life-cycle cost of the solar residence relative to the conventional system by nearly $3000. This simple illustration indicates that the property tax provides a disincentive for choosing solar HVAC systems.

In the example a constant real property value was assumed for ease of illustration. This assumption implies no real depreciation of the system over time during which necessary replacement of parts is made; i.e., the salvage value, in real terms, is assumed equal to the original first cost. In some cases this assumption might be reasonable. An alternative assumption is that the HVAC system (with parts replacements) depreciates in real terms from the time of purchase, so that little or no real value remains after, say, 20 years. Yet another possible assumption is that the assessed value of the system simultaneously reflects inflation and deterioration. This third assumption is described by the following equation:

$$PV_t = \sum_{j=1}^{n} \frac{tG_j}{(1 + i)^j} (1 - \bar{t}) \tag{28.13}$$

where t = the property tax rate
$\bar{t}$ = the income tax rate
G_j = the assessed value of the HVAC system in year j in present dollars

The present value, or, depending on the model, the annual value, of the property taxes would be computed and added to other costs. This formula would cover both the case of a constant real assessed value for the HVAC system and the case of changing real assessed values over time.

The *income tax*, in contrast to the property tax, would tend to reduce the life-cycle cost of a solar vis-à-vis a conventional residence in two ways. For one thing, the homeowner is able to deduct solar mortgage interest payments from taxable income. The higher first cost of the solar HVAC system, by increasing the size of the mortgage to be amortized, raises interest payments; the higher interest payments can then be deducted from income for purpose of computing income tax. The value of the tax deduction depends on the homeowner's personal income-tax bracket. Consider, for purpose of illustration, the case of a solar HVAC system whose first cost of, say, $8000 comprises part of the homeowner's mortgage. With a 10 percent market rate of interest on the residential mortgage, the $8000 amortized over 20 years would add approximately $940 per year to the mortgage payment. (For simplicity let us assume yearly mortgage payments rather than monthly payments.) The addition to the yearly payment is fixed at $940 over the 20 years, and interest comprises a declining portion of the payment over time. In the first year, interest amounts to $800 (i.e., $8000 × 0.10 = $800), and the principal is reduced by $140 (i.e., $940 − $800 = $140). In the second year, interest is $786 [i.e., ($8000 − $140) 0.10 = $786], and the principal is reduced by $154 (i.e., $940 − $786 = $154). Thus in the first year, the solar system would result in additional interest deductions from taxable income of $800, and, if the homeowner is in a 25 percent income-tax bracket, the end-of-year value of the deduction would be $200 (i.e., $800 × 0.25 = $200). At a 2 percent real discount rate, the present value of this savings would be $196 [i.e., $200/(1 + 0.02)^1 = $196]. In the second year, the present-value return from the income-tax deduction would be $189 [i.e., ($786 × 0.25)/(1 + 0.02)^2 = $189]. Over the full 20 years, the present value of the interest deductions, PV$_I$, would be calculated generally as

$$PV_I = \sum_{j=1}^{n} \frac{\bar{t}\,(L_j m)}{(1+i)^j} \tag{28.14}$$

where $\bar{t}$ = personal income-tax rate
$\quad L_j$ = the additional mortgage loan principal outstanding in period j, i.e., that part associated with the HVAC system
$\quad m$ = the market rate of interest on the mortgage

(During a period of price change, it would be necessary to apply a price index to the amount of the tax deductions to convert them to constant prices or to use a nominal discount rate. This conversion of current dollars to real terms would be necessary because the *tax deductions are fixed, and do not reflect changing prices*.) To account for the effect of tax deductions of interest, the present value is subtracted from the homeowner's life-cycle costs.

In conclusion, property taxes tend to increase the homeowner's cost for a solar system relative to a counterpart conventional system because of the greater capital-intensiveness of the typical solar system. Income-tax effects, on the other hand, tend to reduce the relative costs of the typical solar system, principally because of deduction from the homeowner's taxable income of interest payments which are larger for more capital-intensive systems.

Commercial and industrial buildings   Let us now consider the tax effects on the life-cycle cost of a commercial building equipped with a solar HVAC system, as compared with a commercial building equipped with a conventional HVAC system. In the case of commercial use of solar energy systems, the previously described property tax and income tax effects would also apply. The institutional treatment of property tax and interest charges would be somewhat different for commercial buildings than for owner-occupied houses in that these items of cost would be deductible as business expenses. The effect on costs, however, would be described by the same mathematical expressions developed above. There are, in addition, other income-tax-deductible expenses to consider in evaluating the commercially used system, such as depreciation deductions and deductions of *operating* (including energy) and maintenance expenses. Also, after-tax rental income of commercial buildings may be influenced by the choice between solar and conventional HVAC systems, and therefore, may need to be considered.

The larger capitalized value of a solar energy system would result in increased deductions of depreciation from taxable income. For example, for a straight-line method of depreciation, a first cost of $8000 for the solar system, a 20-year life, and no salvage value, the annual depreciation deduction would be $400. The present value to the building owner of the $400 depreciation in a given year may be found by applying the owner's income tax rate to the $400, and discounting that amount to the present. If inflation is present, the $400 must first be converted to constant dollars before discounting. Alternatively, a depreciation method might be used which does not yield equal yearly amounts in either real or nominal terms (e.g., a declining balance method). A general expression of the present value of the tax deduction resulting from depreciation, PV_D, is

$$PV_D = \sum_{j=1}^{n} \frac{D_j \bar{t}}{(1+i)^j} \tag{28.15}$$

where D_j = depreciation in year j, in present dollars, and $\bar{t}$ = building owner's income-tax rate.

On the other hand, a solar energy system might mean lower operating (fuel) costs than its conventional counterpart. Because of the attendant decline in tax-deductible expenses, tax deductions for operating costs would tend to be lower for a solar system than for its conventional counterpart.

Because of the time value of money, the present value of depreciation expenses is less than the present value of the capital expenses upon which it is based. In contrast, the present value of the deductible operating expenses is approximately equal to the corresponding operating expenses incurred. Consequently, if present-value capital costs are substituted (traded off) for present-value operating costs on a dollar-for-dollar before-tax basis, there will not be a corresponding dollar-for-dollar tradeoff on an after-tax basis. Rather, the present value of after-tax capital costs will increase relatively more than operating costs decline, and after-tax total costs will, therefore, rise as a result of the more capital-intensive system. Hence, the fact that operating costs are fully deductible as a current business expense, while capital costs are deductible only as a depreciation expense, may in some cases bias building owners towards relatively less capital-intensive conventional HVAC systems over solar energy systems.

In most cases, it will probably be reasonable to assume that there are equal private benefits for the solar energy system and the conventional counterpart to which it is compared. However, in the case of some rental properties, particularly low-density rental residences, it may be necessary to take into account possible differences in rental revenue. Where conventionally provided utilities are paid by the tenant, rental revenue should be expected, other things equal, to be higher on a solar residence than on a comparable conventionally equipped residence. That is, the owner of a solar rental residence would incur the costs of solar equipment that would be reflected in higher rent but lower utility bills to the tenant. The owner of the rental solar residence would require higher rental payments to cover higher capital costs. If other things are equal and the market is functioning well, tenants should be willing to pay an additional amount of rent up to the amount of the additional utilities outlay which they would incur in a counterpart conventionally equipped residence (i.e., an amount sufficient to equalize the life-cycle costs to the tenants of counterpart solar and conventional rental units).

Different amounts of benefits (i.e., rental income) for buildings equipped with different HVAC systems mean that benefits of the alternative systems are unequal. To compare the alternative systems, differences in their benefits as well as in their costs should be evaluated. Inequalities in benefits can be treated in the present-value and annual-cost equations as negative costs, by entering, in this case as a negative cost, any additional after-tax rental income generated by the rental solar residence over the conventional counterpart. Annual after-tax rental income Y_T would be expressed as

$$Y_T = (1 - \bar{t})\,Y \qquad (28.16)$$

where Y = additional annual gross rental revenue for a solar residence over its counterpart conventional residence, and $1-\bar{t}$ = the factor applied to obtain after-tax income. This additional amount of annual income, i.e., $(1-\bar{t})Y$, would be converted to present value and subtracted from life-cycle costs.

Treatment of Governmental Incentives[7]

Effective life-cycle costs to owners and users of solar-equipped buildings may be further changed by governmental programs designed to encourage adoption of solar HVAC systems. These programs might offer special incentives for solar systems in the form of tax credits, low-interest loans, or direct grants or subsidies to manufacturers and/or buyers of solar HVAC systems. Alternatively, incentives for solar energy systems might be provided in the form of penalties applied to conventional HVAC systems such as taxing conventional HVAC systems more severely than solar systems. In either case, if the comparative cost to the homeowner is altered by special programs, the cost evaluation should reflect the induced changes.

The method of treating the cost effects of such programs would vary. Subsidies to producers of solar systems, for example, might be reflected in the lower purchase price of the systems, and no additional expression need be introduced into the life-cycle cost model in order to assess this effect. On the other hand, a subsidy to the purchaser of a solar energy system, say, in the form of a low-interest loan for the purchase of a solar home, might require specific evaluation of the interest subsidy, including income tax effects.

Some programs intended to provide incentives for purchase of solar energy systems may do this by reducing previously existing disincentives for solar energy. For example, some states and localities are exempting some part of the first cost of a solar energy system from the property tax (e.g., in Colorado, the law requires that solar systems be assessed at 5 percent of their market values).

Following are seven types of financial incentive policies that appear to be the principal types of incentives under consideration both at the state and federal levels.

1. Direct grant
2. Income-tax credit
3. Property-tax reduction
4. Sales-tax reduction
5. Income-tax deduction for depreciation
6. Loan-interest subsidy
7. Tax on conventional energy

Each of these is briefly described and discussed in turn.

The first type of incentive listed, i.e., a direct grant to the purchaser of a solar energy system, is being used in conjunction with federal programs as well as by a number of states. The Solar Demonstration Program (1974–79), administered by the U.S. Department of Housing and Urban Development (HUD), provided grants for the installation of solar units in new and existing dwellings. Grants can be treated in the evaluation model as a negative first cost.

The second type of incentive listed, i.e., the income-tax credit, involves the reduction of the recipient's income-tax liability by a specified amount. Aside from slight differences in timing, the income tax credit is essentially the same as a direct grant, as long as the credit is reimbursable (i.e., the recipient receives any excess of the tax credit over the amount of income-tax liability). For example, suppose a tax credit of $2000 is allowed to the purchaser of a solar energy system whose personal income-tax liability is only $1500. If the purchaser receives a check for $500, in addition to the waiver of income taxes owed, the purchaser will have the equivalent of a cash payment of $2000 received at that time. Extension of the investment-tax credit to cover solar energy equipment installed in commercial buildings is also a type of tax-credit incentive.

The third type of incentive, i.e., the reduction in property taxes, appears to be the most prevalent form of direct financial incentive now being enacted at the state level. The fourth type of incentive, i.e., the reduction in the sales tax on solar energy equipment, is less frequently used.

The fifth type of incentive, i.e., the allowance of an income-tax reduction for depreciation on the capital costs of solar energy systems, can take several forms. One approach is to expand the current eligibility for capital depreciation deductions from businesses to include home-owners. Another approach is to increase the value of the depreciation, either by shortening the length of time over which the depreciation is written off against yearly tax liability, or by otherwise allowing a more liberal depreciation method. The value of the write-off is increased by shortening the defined life of the system or by using a depreciation method which results in larger deductions initially because the tax savings are thereby obtained more quickly and can be put to profitable use.

The sixth type of incentive is a subsidy to reduce the interest rate charged on loans to purchase solar energy systems. The after-tax value of this incentive may be substantially less than the before-tax value.

The seventh incentive is the imposition of a special tax on conventional energy sources. Because solar energy systems derive their economic value from the cost of alternative sources of energy, raising the price of the alternative sources (e.g., by imposing a new tax or by raising existing taxes) will increase the value of solar energy.

ECONOMIC OPTIMIZATION

The purpose of this section is to explain and illustrate a method for determining the economical size of a solar energy system in a given application. The approach taken allows for the sizing of the solar energy system in conjunction with both the conventional energy system which it may supplement or displace, and energy conservation measures which may be instituted to lower energy requirements. The resulting optimization method incorporates the techniques of discounting and the effects of taxes and incentives that were described earlier in this chapter.

Optimizing the size of a solar energy system, with consideration given to a conventional energy system as well as to building load reduction, is a complex procedure that requires the use of mathematical models and is facilitated by the use of computer programs. The emphasis here is on explaining the fundamental concepts of optimization while describing verbally and algebraically the structure that a mathematical optimization model for solar energy might take. A flowchart is provided to indicate the nature and sequence of operations that would be followed in a computerized version of the optimization model.

The use of an optimization model requires data inputs to calculate the building's energy requirements for heating, cooling, or domestic hot water (whichever is applicable). These data inputs may be obtained from one of the many computer programs for energy analysis, or less precise hand-calculating techniques.

Economic optimization also requires the designation of a number of solar energy system design parameters upon which performance of the system is dependent. Included among these design parameters are collector size, tilt and orientation, storage size, collector surface absorptance, collector surface emittance, number and transmissivity of collector glazings, collector coolant flowrate, and collector heat-transfer coefficient. (See Chaps. 7 through 16, 20, and 21.)

Variation in these design parameters will affect the performance of a solar energy system and change the economics of the system. As the number of design parameters is allowed to vary, the number of variables in the computer simulation increases. As a result, the number of necessary runs of the program expands rapidly and the optimization becomes more difficult. To simplify the analysis, engineering and economic studies have been conducted on the effects of variation in key design parameters. [10] These studies have established economically optimal ranges around the most critical design parameters and have led to the use of values of design parameters for predicting the performance of solar energy systems. (See Chap. 14 for recommended values of design parameters for solar heating systems, for example.)

If the system design parameters, other than collector size, are held constant, solar energy system performance can be determined as a function of collector size. By determining the life-cycle costs of alternative system sizes and the life-cycle savings associated with the performance of the system at each size, the economically optimal collector size can be determined.

The optimization method presented here allows for variations in design parameters. The case examples, however, are based on the assumption that design parameters other than collector area are at their economic optimum and optimization is shown as a function of collector area alone.

Like the other economic evaluation methods described in this chapter, the optimization method is adaptable to diverse applications of solar energy systems; however, the descriptions of the method given here are specific to heating and domestic hot-water applications in buildings. The case examples are further specific to commercial (income-producing) buildings; however the optimization method is applicable to residential, commercial, institutional, and public buildings.

Three topical discussions of optimization follow: first, the basic concepts of economic optimization are presented graphically; second, a method for optimizing a solar energy system in conjunction with a conventional energy system is described; and third, a method for optimizing a solar energy system in conjunction with both a conventional energy system and energy conservation alternatives is described.

General Concepts of Economic Optimization

Economic optimization as applied to solar energy systems means finding the system type and size that will yield the largest possible net savings, or, what is the same thing, that will meet the goal of a desired level of thermal comfort at the lowest life-cycle cost.

There are, however, alternative ways to achieve a desired level of thermal comfort in a building. Alternatives include the use of different heating and cooling equipment (e.g., conventional plants fueled by oil, gas, or electricity; heat pumps; solar energy equipment; or combined solar and conventional systems), different mechanical distribution systems (e.g., a multizone fan system with single-set-point thermostats, or a variable-volume system with dual-set-point thermostats), and different building envelope designs (including numerous energy conservation options such as insulation, double- and triple-glazed windows, and solar shading). To find the least-cost way of meeting the comfort objective, it is important to evaluate solar energy in conjunction with these alternatives. Among the available combinations of alternatives there exist not only possible tradeoffs in technical performance, but also economic tradeoffs, in terms of both relative first costs and future costs and savings. Hence, the economic objective is more specifically to choose that combination of alternatives—which may include solar—that will minimize the total life-cycle costs of the energy-related components of a building, while maintaining the desired level of thermal comfort and meeting any noncomfort constraints.

Preparatory to a mathematical approach, let us examine graphically the basic concepts in optimization and the several potential solutions to an optimization problem in solar energy. For simplicity, let us begin by considering only tradeoffs between a solar energy system and a conventional energy system. Beginning with the simple case, Figs. 28.3 through 28.5 describe graphically three possible outcomes of the tradeoff between solar energy and conventional energy: (1) a 100 percent conventional energy system, (2) a 100 percent solar energy system, and (3) some combination of solar energy and conventional energy systems. Assuming that either system is capable of satisfying the comfort requirements and the noncomfort constraints, the choice to minimize costs will depend on the relative costs of the two systems.

In each figure, the percent of the building's heating requirements that is supplied by the solar energy system, f_s, is measured along the abscissa. The life-cycle costs (LCC) for the solar energy system, the conventional energy system and the total energy system (i.e., solar plus conventional) are measured along the ordinate. Because the increases in system size are necessary to raise the contribution of solar energy and larger sizes mean higher costs, the LCC of the solar energy system increases as the percent of the heating requirements supplied by

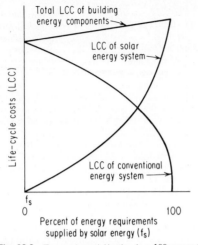

Fig. 28.3   Economic optimization in a 100 percent convention energy system (i.e., $f_s = 0.0$).

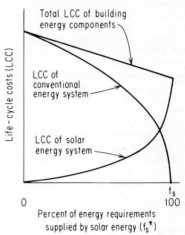

Fig. 28.4   Economic optimization resulting in a 100 percent solar energy system (i.e., $f_s = 100$ percent).

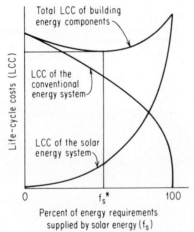

Fig. 28.5   Economic optimization resulting in a combined conventional and solar energy system for least cost.

solar, i.e., f_s, increases. As f_s increases, the conventional energy system's contribution to heating load (100 percent $-f_s$) decreases. Therefore, the LCC of the conventional energy system decreases. The total LCC curves are derived by adding vertically the solar energy LCC and the conventional energy LCC at each point along the horizontal axis. The economic objective is to find that particular combination of conventional energy and solar energy which minimizes total LCC, as indicated by the lowest point on the total LCC curve.

In *Figs. 28.3 through 28.5 it is assumed that as f_s increases, both the size of and the energy consumed by the conventional energy system decrease.* When f_s reaches 100 percent, it is assumed that the conventional energy system can be eliminated. This results in the bowed shape of the total LCC curve of the conventional energy system. However, *if a full conventional energy system is assumed to be required as auxiliary to the solar energy system (i.e., if there are no size tradeoffs for the conventional system), then only conventional fuel costs would* *decrease as f_s increases, and the conventional LCC curve would be a straight line* passing through the end points of the bowed curves shown (if fuel prices are not subject to step rates such as those used for electricity).

Figure 28.3 depicts the case in which the minimum point of the total LCC curve occurs at zero percent solar, indicating that the economically optimal choice is a 100 percent conventional energy system and nonsolar, f_s^*. In this case, solar energy is not cost-effective, because it does not reduce total LCC.

Figure 28.4, in contrast, depicts the case in which the minimum point on the total LCC curve occurs at $f_s^* = 100$ percent solar and zero percent conventional energy. The solar energy system is sufficiently cost-effective that it pays to eliminate entirely the conventional energy system.

In both figures, the total LCC curve either increases or decreases monotonically throughout the range of solar contribution. It is also possible that total LCC may first decrease with increasing solar contribution, reach a minimum point, and then begin to increase. Figure 28.5 depicts the case where the minimum point on the total LCC curve occurs between zero and 100 percent solar. Then the minimum-cost choice is a combination of solar energy providing f_s^* percent of the building's heating requirements and conventional energy providing the remaining $100-f_s^*$ percent of the building's heating requirements.

An energy system that is totally solar will usually not be cost-effective, primarily because of the disportionately high costs of using solar to meet those energy requirements that do not often occur. To meet the peak requirements with solar means a fixed, capital-intensive investment in capacity that is inefficiently utilized during nonpeak requirements. As the total fraction of energy requirements supplied by solar becomes large, increasingly large investments in solar energy equipment are necessary to continue to raise the total fraction supplied by solar by a given increment.

This condition of diminishing returns is depicted in Fig. 28.6, which shows that the fraction of a building's energy requirements supplied by solar energy typically increases at a decreasing rate as the area of the solar collector is increased. In contrast, the costs of supplying a building's energy requirements with a conventional energy system, e.g., an oil furnace, tend to rise in a linear fashion as the fraction supplied increases.

Figure 28.7 shows the typical nature of the technical tradeoffs that can be made between solar collector size and conventional fuel consumption while meeting a building's energy requirements. The ordinate measures the quantity of conventional fuel consumed, and the abscissa measures the size of solar collector. The curve Q_o (the isoquant) gives all the combinations of solar and conventional fuel that will meet the energy requirement Q_o for a given level of comfort. For example, Q_o can be produced by combining $\overline{OP}$ of solar collector area and $\overline{OJ}$ of conventional fuel. Hence, the curve shows the technical tradeoffs between the two inputs. The shape of the curve indicates that the solar collector area required to displace a given amount of conventional energy consumption tends to increase as more of the energy requirements are met by solar.

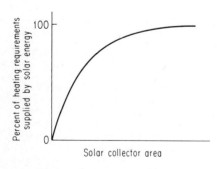

Fig. 28.6 Typical diminishing return relationship between the fraction of heating requirements provided by solar energy and the area of solar collectors.

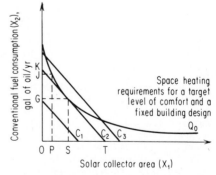

Fig. 28.7 Tradeoff between conventional energy consumption and solar collector area to find least-cost combination of inputs to produce a desired level of comfort Q_o.

It is because of the nature of the technical tradeoffs between solar energy and conventional energy, and the relative prices associated with the technical tradeoffs, that it is usually more feasible economically to use solar in combination with some type of conventional system rather than alone. The problem is to find from Fig. 28.7 the combination of inputs for which total life-cycle costs are minimal.

The economically efficient combination of inputs depends both on the productivity of those inputs and on their relative prices. A necessary condition for arriving at the minimum cost of providing a given level of space heating requirements is that marginal productivity per dollar spent is equal for all inputs. This means that each input will be used up to that level at which its additional contribution to the objective per extra dollar spent is just equal to that for all other inputs. Assuming continuous and smooth functions, this necessary condition can be expressed mathematically for two inputs as

$$\frac{P_1}{\dfrac{\partial Q}{\partial X_1}} = \frac{P_2}{\dfrac{\partial Q}{\partial X_2}} \tag{28.17}$$

where Q = units of space-heating requirements to meet a given comfort level

X_1, X_2 = units of inputs 1 and 2

P_1, P_2 = ∂ cost$_i/\partial Q_i$, the cost per unit of inputs 1 and 2 in present-value, life-cycle terms. [see Eqs. (28.1) and (28.4) for solar and fuel costs]

This expression could be expanded in a different form to accommodate as many inputs as are relevant. The numerators are marginal productivities and the denominators are marginal costs.

This necessary condition is illustrated graphically in Fig. 28.7. The lines C_1, C_2, and C_3 illustrate three of a family of cost functions. Each function indicates a specific total cost, and shows the combinations of inputs 1 and 2 which may be purchased for that total cost. With the expenditure equal to C_2, for example, one could buy either $\overline{OT}$ of X_1 or $\overline{OK}$ of X_2 or any combination of X_1 and X_2 that lies on the cost line C_2. The slope of the cost line indicates the relative costs of the two inputs, i.e., P_1/P_2. Total costs rise as larger quantities of both inputs are purchased, that is, $C_3 > C_2 > C_1$.

Given the relative costs of X_1 and X_2 indicated by the slopes of the cost curves, the lowest total cost at which Q_o comfort level can be produced is C_2, using $\overline{OS}$ of X_1 in combination with $\overline{OG}$ of X_2. For any cost less than C_2 (e.g., C_1), Q_o could not be achieved. To achieve Q_o at any cost greater than C_2 (e.g., C_3) would be inefficient since Q_o can be achieved at the lower cost C_2. Thus the least-cost combination of factors is determined by the point of tangency between the space heating requirements curve Q_o and the cost curve C_2. At the point of tangency,

$$\frac{\dfrac{\partial Q}{\partial X_1}}{P_1} = \frac{\dfrac{\partial Q}{\partial X_2}}{P_2}$$

and the basic optimality rule is met. This point of tangency of the space-heating requirements curve and the cost curve corresponds to the minimum point on the total LCC curve of Fig. 28.5.

Figures 28.3 through 28.5 express LCC as a function of the fraction of energy requirement supplied by a solar energy system, measured along the horizontal axis, and Fig. 28.6 shows the relationship between the fraction of the load supplied and collector area. Figure 28.8 combines the two and expresses LCC as a function of collector area. The minimum total LCC (excluding energy conservation) occurs at a collector area A^* corresponding to a solar contribution of f^*.

Although the choice may sometimes be limited to solar energy vs. conventional energy, in most cases—particularly, for new building design—it will be feasible to consider, in addition, energy conservation investments in the building envelope. If both energy conservation options and energy system options are considered for the same building, it is necessary to compare the engineering performance and economic performance of the various options. For economic optimization, it is necessary to size the options simultaneously, not separately, using a three-dimensional analog of Fig. 28.7 and the equations which describe it.

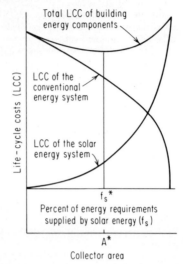

Fig. 28.8 Economic optimization of a solar energy system in relation to collector area. Note that the f_s and A_c scales cannot both be linear.

Example

The following example illustrates how the economically optimal size of a solar energy system of a given design can be determined.

The example is based on a solar space heating system applied to an existing office building in Bismarck, ND. It is assumed that the existing oil-fired boiler will serve as the auxiliary system if a solar energy system is added. The alternative to the combined solar/oil furnace system is the existing oil furnace used alone. For simplicity the evaluation is on a before-tax basis as would be the case for a public or other tax-exempt building. Table 28.10 summarizes other key assumptions.

TABLE 28.10   Optimal Sizing Example—Key Assumptions

Building type	Office, tax-exempt status
Location	Bismarck, ND
Space heating load	8×10^8 Btu/y
Oil furnace efficiency	0.6
Oil furnace useful life	20 y
Oil furnace O & M costs	Identical for furnace operated alone or as solar backup
Solar energy system useful life	25 y
Building life	Indefinite
Study period	20 y
Solar energy system salvage value in 20 y	20% of initial investment cost
Current price of distillate oil	$0.541/gal
Projected differential escalation rates (Department of Energy Mid-Term Energy Forecasting System Rate Projections*)	4.99% first 5 y, 2.16% second 5 y, 5.39% thereafter
Fixed cost of solar energy system	$18,000
Variable cost of solar energy system	$14/ft²
Annual solar energy system O & M cost	1% of initial investment cost
Financing terms	Equity funded
Discount rate	10%, real

*Federal Register, 44, No. 84, Apr. 30, 1979.

We can begin by calculating the life-cycle cost associated with keeping the existing system, the alternative to which the combined solar/auxiliary system is being compared. Since it is assumed that the existing oil furnace is the same system that will serve as the solar auxiliary, and that the same nonfuel O & M costs will be incurred for it regardless, the capital costs and nonfuel O & M costs for the oil furnace can be omitted from the evaluation. Thus, only the energy costs need be evaluated for the oil furnace used alone. These costs (PV_c) may be calculated according to Eq. (28.1) as follows:

$$PV_c = \$0.541 \times 8 \times 10^8/(0.6 \times 140{,}000) \times 11.904 = \$61{,}334 \qquad (28.17)$$

where the terms are, in order: Oil price per gallon, Annual heating load in Btu, Furnace efficiency, Btu content per gallon, UPV factor.

The life-cycle costs of the combined solar/auxiliary system consist of solar investment costs, less salvage value, nonfuel O & M costs, and energy costs. (Again the capital costs and the O & M costs of the auxiliary system can be excluded because they are identical to those of the alternative system used alone.)

To calculate the fuel costs with the solar energy system, system performance data are needed to estimate the fraction of the heating load to be supplied by the solar energy contributed by various sizes of this particular system. A solar performance model such as the f-chart method could be used to estimate solar fractions. For the purpose of this example, the following annual solar fractions apply for four selected collector sizes in increments of 550 ft.^{2}

Collector size, ft^2	Estimated annual solar fraction f_s, %
550	35.7
1100	62.5
1650	75.0
2200	82.5

The estimate of the solar fraction associated with each collector size can be used in Eq. (28.1) to calculate total life-cycle costs of a system of that size.

For a collector size of 550 ft^2, for example, total life-cycle costs can be calculated as follows:

$$PV_{s,c} = [\$18{,}000 + (\$14 \times 550)] + [0.01\,(\$25{,}700)\,8.514] + [\$0.541$$

where the terms in the first bracket are System fixed costs and System variable costs (Collector size), the second bracket is the O & M fraction of Total system costs times UPV factor, and the third bracket begins with Unit oil price.

$$\times 9524 \text{ gal } (1-0.357) \times 11.904] = \$67{,}327 \qquad (28.18)$$

where the terms are Quantity of oil without solar, Solar fraction, and UPV factor.

Total life-cycle costs for the other collector sizes can be calculated in similar fashion. Table 28.11 summarizes the evaluation results for the oil-fired furnace alone and for the four combinations of solar and auxiliary. The table shows that, of the sizes examined, the 1,100-ft^2 system

TABLE 28.11   Example Total Life-Cycle Cost Components and Totals

Collector area ft²	Fuel costs	Solar faction, %	Solar investment + other costs, $	Total LCC, $	Solar net savings, $
0	$61,335	0	0	$61,334	0
550	39,439	35.7	27,888	67,327	−5993
1100	23,001	62.5	36,244	59,244	2090
1650	15,334	75.0	44,599	59,933	1401
2200	10,734	82.5	52,955	63,689	−2355

is the most economical, resulting in an estimated net savings of $2090 in present-value dollars over the 20-year evaluation period.

It should be recognized that this example has illustrated only one aspect of economic optimization: optimal sizing, given a system of particular design and values of key parameters. A more thorough evaluation would consider solar energy systems of alternative design—as well as size, the use of alternative backup systems, and the sensitivity of results to key assumptions.

SUMMARY

This chapter has presented and illustrated basic concepts, procedures, and techniques required to evaluate the economic feasibility of solar energy systems. It has outlined five of the principal techniques of economic evaluation and has demonstrated each in an example problem. The advantages and disadvantages, as well as the recommended applications, of each technique have been explained.

The chapter moved from a general description of the techniques, to a guide for treating inflation and the real earning potential of resources. Key assumptions regarding the selection of discount rate, the timing of cash flows, energy price escalation, economic life, and the time horizon of the investor were discussed.

A section on costs and savings divided the various major items into their components. The significance of fixed and variable costs and marginal and total costs was explained.

To summarize briefly, the main steps that would be taken to evaluate the cost-effectiveness of a solar energy system are the following:

1. Identify the alternatives in solar and conventional energy equipment (and in energy conservation measures) which are to be considered, as well as any constraints that must be met.

2. Select a method of economic evaluation, taking into account the nature of the choice and any special investment considerations.

3. Decide on the basic assumptions and select values for key economic parameters such as the economic lifetime, loan interest rates, discount rate, and present and future energy prices.

4. Specify values for all relevant, significant cash flows associated with each alternative over the designated time horizon.

5. Use discounting techniques appropriate to the selected evaluation method to adjust differently timed cash flows to a comparable basis.

6. Carry out the evaluation method for each alternative to obtain comparative measures of economic performance for each alternative.

7. Select that alternative with the highest economic performance.

REFERENCES

1. Recommended energy price projections by fuel type, by region of the country, and by year are published by the Department of Energy. (See Energy Audit Procedures, *Federal Register*, vol. 42, No. 73, Friday, April 15, 1977, part IV.)

2. Steve H. Hanke, Philip H. Carver, and Paul Bugg, Project Evaluation During Inflation, *Water Resources Research*, vol. II, no. 4, August 1975, pp. 511–514.
3. Harold E. Marshall and Rosalie T. Ruegg, Life-Cycle Costing Guide for Energy Conservation in Buildings, "Energy Conservation Through Building Design," (McGraw-Hill Architectural Record Books, New York, December 1977.)
4. Arthur E. McGarity, Solar Heating and Cooling, National Science Foundation Report 76–37, undated, pp. 14–15.
5. Arthur E. McGarity, Solar Heating and Cooling: An Economic Assessment, National Science Foundation, GPO No. 038–000–00300–3, Washington, DC.
6. OMB Circular A–94, March 27, 1972, requires that a real rate of 10 percent be used in the evaluation of most federal investment decisions. OMB Circular A–104, June 14, 1972, requires that a real rate of 7 percent be used in lease-buy decisions involving real estate.
7. Rosalie T. Ruegg, Evaluating Incentives for Solar Heating, National Bureau of Standards, NBSIR, Washington, DC.
8. Rosalie T. Ruegg, Solar Energy: Methods of Economic Evaluation, National Bureau of Standards NBSIR 1127, Washington, DC, 1975.
9. Gerald W. Smith, Engineering Economy: Analysis of Capital Expenditures, pp. 133–140.
10. Richard A. Tybout and George O. G. Löf, Solar House Heating, *Natural Resources Journal*, vol. 23, April 1970, pp. 269–326.
11. R. T. Ruegg, and G. T. Fav, The Microeconomics of Solar Energy, The National Bureau of Standards, Washington, DC, 1979.

APPENDIX: COMPOUND INTEREST FACTOR EQUATIONS AND TABLES

TABLE 28.A1 Summary of Interest Factors for Cash Flows with End of Period Compounding*

	Interest factor	Quantity known	Quantity to be found	Discrete compounding of a discrete flow expression	Continuous compounding of a discrete flow expression
Single compound amount (SCA)	$(F/P, i, N)$	P	F	$(1+i)^N$	e^{rN}
Single present value (SPV)	$(P/F, i, N)$	F	P	$(1+i)^{-N}$	e^{-rN}
Uniform sinking fund (USF)	$(A/F, i, N)$	F	A	$\dfrac{i}{(1+i)^N - 1}$	$\dfrac{e^r - 1}{e^{rN} - 1}$
Uniform compound amount (UCA)	$(F/A, i, N)$	A	F	$\dfrac{(1+i)^N - 1}{i}$	$\dfrac{e^{rN} - 1}{e^r - 1}$
Capital recovery factor (CRF)	$(A/P, i, N)$	P	A	$\dfrac{i}{1 - (1+i)^{-N}}$	$\dfrac{e^r - 1}{1 - e^{-rN}}$
Uniform present value (UPV)	$(P/A, i, N)$	A	P	$\dfrac{1 - (1+i)^{-N}}{i}$	$\dfrac{1 - e^{-rN}}{e^r - 1}$
Gradient factor (GF)	$(A/G, i, N)$	G	A	$[1/i - N/i(A/F, i, N)]$	—†

* P is a discrete present amount, F a future amount, A a uniform end of period payment, and G the uniform increase in an amount; i is the effective interest rate and r is the nominal rate; N is the number of payment periods.

† Substitute $i = e^r - 1$ into the previous column expression for the uniform gradient factor.

TABLE 28. A2 8% Interest Factors for Discrete Compounding Periods

| | Single payment | | Uniform series | | | | | |
| | Compound amount factor | Present worth factor | Capital, recovery factor | Present worth factor | Sinking fund factor | Compound amount factor | Gradient factor | |
N	$(F/P, 8, N)$	$(P/F, 8, N)$	$(A/P, 8, N)$	$(P/A, 8, N)$	$(A/F, 8, N)$	$(F/A, 8, N)$	$(A/G, 8, N)$	N
1	1.0800	.92593	1.0800	.9259	1.0000	1.0000	.0000	1
2	1.1664	.85734	.56077	1.7832	.48077	2.0799	.4807	2
3	1.2597	.79383	.38803	2.5770	.30804	3.2463	.9487	3
4	1.3604	.73503	.30192	3.3121	.22192	4.5060	1.4038	4
5	1.4693	.68059	.25046	3.9926	.17046	5.8665	1.8463	5
6	1.5868	.63017	.21632	4.6228	.13632	7.3358	2.2762	6
7	1.7138	.58349	.19207	5.2063	.11207	8.9227	2.6935	7
8	1.8509	.54027	.17402	5.7466	.09402	10.636	3.0984	8
9	1.9989	.50025	.16008	6.2468	.08008	12.487	3.4909	9
10	2.1589	.46320	.14903	6.7100	.06903	14.486	3.8712	10
11	2.3316	.42889	.14008	7.1389	.06008	16.645	4.2394	11
12	2.5181	.39712	.13270	7.5360	.05270	18.976	4.5956	12
13	2.7196	.36770	.12642	7.9037	.04652	21.495	4.9401	13
14	2.9371	.34046	.12130	8.2442	.04130	24.214	5.2729	14
15	3.1721	.31524	.11683	8.5594	.03683	27.151	5.5943	15
16	3.4259	.29189	.11298	8.8513	.03298	30.323	5.9045	16
17	3.6999	.27027	.10963	9.1216	.02963	33.749	6.2036	17
18	3.9959	.25025	.10670	9.3718	.02670	37.449	6.4919	18
19	4.3156	.23171	.10413	9.6035	.02413	41.445	6.7696	19
20	4.6609	.21455	.10185	9.8181	.02185	45.761	7.0368	20
21	5.0337	.19866	.09983	10.016	.01983	50.422	7.2939	21
22	5.4364	.18394	.09803	10.200	.01803	55.455	7.5411	22
23	5.8713	.17032	.09642	10.371	.01642	60.892	7.7785	23
24	6.3410	.15770	.09498	10.528	.01498	66.763	8.0065	24
25	6.8483	.14602	.09368	10.674	.01368	73.104	8.2253	25
26	7.3962	.13520	.09251	10.809	.01251	79.953	8.4351	26
27	7.9879	.12519	.09145	10.935	.01145	87.349	8.6362	27
28	8.6269	.11592	.09049	11.051	.01049	95.337	8.8288	28
29	9.3171	.10733	.08962	11.158	.00962	103.96	9.0132	29
30	10.062	.09938	.08883	11.257	.00883	113.28	9.1896	30
31	10.867	.09202	.08811	11.349	.00811	123.34	9.3583	31
32	11.736	.08520	.08745	11.434	.00745	134.21	9.5196	32
33	12.675	.07889	.08685	11.513	.00685	145.94	9.6736	33
34	13.689	.07305	.08630	11.586	.00630	158.62	9.8207	34
35	14.785	.06764	.08580	11.654	.00580	172.31	9.9610	35
40	21.724	.04603	.08386	11.924	.00386	259.05	10.569	40
45	31.919	.03133	.08259	12.108	.00259	386.49	11.044	45
50	46.900	.02132	.08174	12.233	.00174	573.75	11.410	50
55	68.911	.01451	.08118	12.318	.00118	848.89	11.690	55
60	101.25	.00988	.08080	12.376	.00080	1253.1	11.901	60
65	148.77	.00672	.08054	12.416	.00054	1847.1	12.060	65
70	218.59	.00457	.08037	12.442	.00037	2719.9	12.178	70
75	321.19	.00311	.08025	12.461	.00025	4002.3	12.265	75
80	471.93	.00212	.08017	12.473	.00017	5886.6	12.330	80
85	693.42	.00144	.08012	12.481	.00012	8655.2	12.377	85
90	1018.8	.00098	.08008	12.487	.00008	12,723.9	12.411	90
95	1497.0	.00067	.08005	12.491	.00005	18,701.5	12.436	95
100	2199.6	.00045	.08004	12.494	.00004	27,484.5	12.454	100

TABLE 28.A3 10% Interest Factors for Discrete Compounding Periods

	Single payment		Uniform series					
	Compound amount factor	Present worth factor	Capital recovery factor	Present worth factor	Sinking fund factor	Compound amount factor	Gradient factor	
N	(F/P, 10, N)	(P/F, 10, N)	(A/P, 10, N)	(P/A, 10, N)	(A/F, 10, N)	(F/A, 10, N)	(A/G, 10, N)	N
1	1.1000	.90909	1.1000	.9091	1.0000	1.000	.0000	1
2	1.2100	.82645	.57619	1.7355	.47619	2.0999	.4761	2
3	1.3310	.75132	.40212	2.4868	.30212	3.3099	.9365	3
4	1.4641	.68302	.31547	3.1698	.21547	4.6409	1.3810	4
5	1.6105	.62092	.26380	3.7907	.16380	6.1050	1.8100	5
6	1.7715	.56448	.22961	4.3552	.12961	7.7155	2.2234	6
7	1.9487	.51316	.20541	4.8683	.10541	9.4870	2.6215	7
8	2.1435	.46651	.18745	5.3349	.08745	11.435	3.0043	8
9	2.3579	.42410	.17364	5.7589	.07364	13.579	3.3722	9
10	2.5937	.38555	.16275	6.1445	.06275	15.937	3.7253	10
11	2.8530	.35050	.15396	6.4950	.05396	18.530	4.0639	11
12	3.1384	.31863	.14676	6.8136	.04676	21.383	4.3883	12
13	3.4522	.28967	.14078	7.1033	.04078	24.522	4.6987	13
14	3.7974	.26333	.13575	7.3666	.03575	27.974	4.9954	14
15	4.1771	.23940	.13147	7.6060	.03147	31.771	5.2788	15
16	4.5949	.21763	.12782	7.8236	.02782	35.949	5.5492	16
17	5.0544	.19785	.12466	8.0215	.02466	40.543	5.8070	17
18	5.5598	.17986	.12193	8.2013	.02193	45.598	6.0524	18
19	6.1158	.16351	.11955	8.3649	.01955	51.158	6.2860	19
20	6.7273	.14865	.11746	8.5135	.01746	57.273	6.5080	20
21	7.4001	.13513	.11562	8.6486	.01562	64.001	6.7188	21
22	8.1401	.12285	.11401	8.7715	.01401	71.401	6.9188	22
23	8.9541	.11168	.11257	8.8832	.01257	79.541	7.1084	23
24	9.8495	.10153	.11130	8.9847	.01130	88.495	7.2879	24
25	10.834	.09230	.11017	9.0770	.01017	98.344	7.4579	25
26	11.917	.08391	.10916	9.1609	.00916	109.17	7.6185	26
27	13.109	.07628	.10826	9.2372	.00826	121.09	7.7703	27
28	14.420	.06935	.10745	9.3065	.00745	134.20	7.9136	28
29	15.862	.06304	.10673	9.3696	.00673	148.62	8.0488	29
30	17.448	.05731	.10608	9.4269	.00608	164.48	8.1761	30
31	19.193	.05210	.10550	9.4790	.00550	181.93	8.2961	31
32	21.113	.04736	.10497	9.5263	.00497	201.13	8.4090	32
33	23.224	.04306	.10450	9.5694	.00450	222.24	8.5151	33
34	25.546	.03914	.10407	9.6085	.00407	245.46	8.6149	34
35	28.101	.03559	.10369	9.6441	.00369	271.01	8.7085	35
40	45.257	.02210	.10226	9.7790	.00226	442.57	9.0962	40
45	72.887	.01372	.10139	9.8628	.00139	718.87	9.3740	45
50	117.38	.00852	.10086	9.9148	.00086	1163.8	9.5704	50
55	189.04	.00529	.10053	9.9471	.00053	1880.4	9.7075	55
60	304.46	.00328	.10033	9.9671	.00033	3034.6	9.8022	60
65	490.34	.00204	.10020	9.9796	.00020	4893.4	9.8671	65
70	789.69	.00127	.10013	9.9873	.00013	7886.9	9.9112	70
75	1271.8	.00079	.10008	9.9921	.00008	12,709.0	9.9409	75
80	2048.2	.00049	.10005	9.9951	.00005	20,474.0	9.9609	80
85	3298.7	.00030	.10003	9.9969	.00003	32,979.7	9.9742	85
90	5312.5	.00019	.10002	9.9981	.00002	53,120.2	9.9830	90
95	8555.9	.00012	.10001	9.9988	.00001	85,556.8	9.9889	95
100	13,780.6	.00007	.10001	9.9992	.00001	137,796.1	9.9927	100

Table 28.A4　12% Interest Factors for Discrete Compounding Periods

| | Single payment | | Uniform series | | | | | |
| | Compound amount factor | Present worth factor | Capital recovery factor | Present worth factor | Sinking fund factor | Compound amount factor | Gradient factor | |
N	$(F/P, 12, N)$	$(P/F, 12, N)$	$(A/P, 12, N)$	$(P/A, 12, N)$	$(A/F, 12, N)$	$(F/A, 12, N)$	$(A/G, 12, N)$	N
1	1.1200	.89286	1.1200	.8929	1.0000	1.0000	.0000	1
2	1.2544	.79719	.59170	1.6900	.47170	2.1200	.4717	2
3	1.4049	.71178	.41635	2.4018	.29635	3.3743	.9246	3
4	1.5735	.63552	.32924	3.0373	.20924	4.7793	1.3588	4
5	1.7623	.56743	.27741	3.6047	.15741	6.3528	1.7745	5
6	1.9738	.50663	.24323	4.1114	.12323	8.115	2.1720	6
7	2.2106	.45235	.21912	4.5637	.09912	10.088	2.5514	7
8	2.4759	.40388	.20130	4.9676	.08130	12.299	2.9131	8
9	2.7730	.36061	.18768	5.3282	.06768	14.775	3.2573	9
10	3.1058	.32197	.17698	5.6502	.05698	17.548	3.5846	10
11	3.4785	.28748	.16842	5.9376	.04842	20.654	3.8952	11
12	3.8959	.25668	.16144	6.1943	.04144	24.132	4.1896	12
13	4.3634	.22918	.15568	6.4235	.03568	28.028	4.4682	13
14	4.8870	.20462	.15087	6.6281	.03087	32.392	4.7316	14
15	5.4735	.18270	.14682	6.8108	.02682	37.279	4.9802	15
16	6.1303	.16312	.14339	6.9739	.02339	42.752	5.2146	16
17	6.8659	.14565	.14046	7.1196	.02046	48.883	5.4352	17
18	7.6899	.13004	.13794	7.2496	.01794	55.749	5.6427	18
19	8.6126	.11611	.13576	7.3657	.01576	63.439	5.8375	19
20	9.6462	.10367	.13388	7.4694	.01388	72.051	6.0201	20
21	10.803	.09256	.13224	7.5620	.01224	81.698	6.1913	21
22	12.100	.08264	.13081	7.6446	.01081	92.501	6.3513	22
23	13.552	.07379	.12956	7.7184	.00956	104.60	6.5009	23
24	15.178	.06588	.12846	7.7843	.00846	118.15	6.6406	24
25	16.999	.05882	.12750	7.8431	.00750	133.33	6.7708	25
26	19.039	.05252	.12665	7.8956	.00665	150.33	6.8920	26
27	21.324	.04689	.12590	7.9425	.00590	169.37	7.0049	27
28	23.883	.04187	.12524	7.9844	.00524	190.69	7.1097	28
29	26.749	.03738	.12466	8.0218	.00466	214.58	7.2071	29
30	29.959	.03338	.12414	8.0551	.00414	241.32	7.2974	30
31	33.554	.02980	.12369	8.0849	.00369	271.28	7.3810	31
32	37.581	.02661	.12328	8.1116	.00328	304.84	7.4585	32
33	42.090	.02376	.12292	8.1353	.00292	342.42	7.5302	33
34	47.141	.02121	.12260	8.1565	.00260	384.51	7.5964	34
35	52.798	.01894	.12232	8.1755	.00232	431.65	7.6576	35
40	93.049	.01075	.12130	8.2437	.00130	767.07	7.8987	40
45	163.98	.00610	.12074	8.2825	.00074	1358.2	8.0572	45
50	288.99	.00346	.12042	8.3045	.00042	2399.9	8.1597	50

TABLE 28.A5 15% Interest Factors for Discrete Compounding Periods

	Single payment		Uniform series					
	Compound amount factor	Present worth factor	Capital recovery factor	Present worth factor	Sinking fund factor	Compound amount factor	Gradient factor	
N	(F/P, 15, N)	(P/F, 15, N)	(A/P, 15, N)	(P/A, 15, N)	(A/F, 15, N)	(F/A, 15, N)	(A/G, 15 N)	N
1	1.1500	.86957	1.1500	.8696	1.0000	1.000	.0000	1
2	1.3225	.75614	.61512	1.6257	.46512	2.1499	.4651	2
3	1.5208	.65752	.43798	2.2832	.28798	3.4724	.9071	3
4	1.7490	.57175	.35027	2.8549	.20027	4.9933	1.3262	4
5	2.0113	.49718	.29832	3.3521	.14832	6.7423	1.7227	5
6	2.3130	.43233	.26424	3.7844	.11424	8.7536	2.0971	6
7	2.6600	.37594	.24036	4.1604	.09036	11.066	2.4498	7
8	3.0590	.32690	.22285	4.4873	.07285	13.726	2.7813	8
9	3.5178	.28426	.20957	4.7715	.05957	16.785	3.0922	9
10	4.0455	.24719	.19925	5.0187	.04925	20.303	3.3831	10
11	4.6523	.21494	.19107	5.2337	.04107	24.349	3.6549	11
12	5.3502	.18691	.18448	5.4206	.03448	29.001	3.9081	12
13	6.1527	.16253	.17911	5.5831	.02911	34.351	4.1437	13
14	7.0756	.14133	.17469	5.7244	.02469	40.504	4.3623	14
15	8.1369	.12290	.17102	5.8473	.02102	47.579	4.5649	15
16	9.3575	.10687	.16795	5.9542	.01795	55.716	4.7522	16
17	10.761	.09293	.16537	6.0471	.01537	65.074	4.9250	17
18	12.375	.08081	.16319	6.1279	.01319	75.835	5.0842	18
19	14.231	.07027	.16134	6.1982	.01134	88.210	5.2307	19
20	16.366	.06110	.15976	6.2593	.00976	102.44	5.3651	20
21	18.821	.05313	.15842	6.3124	.00842	118.80	5.4883	21
22	21.644	.04620	.15727	6.3586	.00727	137.62	5.6010	22
23	24.891	.04018	.15628	6.3988	.00628	159.27	5.7039	23
24	28.624	.03493	.15543	6.4337	.00543	184.16	5.7978	24
25	32.918	.03038	.15470	6.4641	.00470	212.78	5.8834	25
26	37.856	.02642	.15407	6.4905	.00407	245.70	5.9612	26
27	43.534	.02297	.15353	6.5135	.00353	283.56	6.0318	27
28	50.064	.01997	.15306	6.5335	.00306	327.09	6.0959	28
29	57.574	.01737	.15265	6.5508	.00265	377.16	6.1540	29
30	66.210	.01510	.15230	6.5659	.00230	434.73	6.2066	30
31	76.141	.01313	.15200	6.5791	.00200	500.94	6.2541	31
32	87.563	.01142	.15173	6.5905	.00173	577.08	6.2970	32
33	100.69	.00993	.15150	6.6004	.00150	664.65	6.3356	33
34	115.80	.00864	.15131	6.6091	.00131	765.34	6.3705	34
35	133.17	.00751	.15113	6.6166	.00113	881.14	6.4018	35
40	267.85	.00373	.15056	6.6417	.00056	1779.0	6.5167	40
45	538.75	.00186	.15028	6.6543	.00028	3585.0	6.5829	45
50	1083.6	.00092	.15014	6.6605	.00014	7217.4	6.8204	50

Table 28.A6   20% Interest Factors for Discrete Compounding Periods

| | Single payment | | Uniform series | | | | | |
| | Compound amount factor | Present worth factor | Capital recovery factor | Present worth factor | Sinking fund factor | Compound amount factor | Gradient factor | |
N	$(F/P, 20, N)$	$(P/F, 20, N)$	$(A/P, 20, N)$	$(P/A, 20, N)$	$(A/F, 20, N)$	$(F/A, 20, N)$	$(A/G, 20, N)$	N
1	1.2000	.83333	1.2000	.8333	1.000	1.0000	.0000	1
2	1.4400	.69445	.65455	1.5277	.45455	2.1999	.4545	2
3	1.7280	.57870	.47473	2.1064	.27473	3.6399	.8791	3
4	2.0736	.48225	.38629	2.5887	.18629	5.3679	1.2742	4
5	2.4883	.40188	.33438	2.9906	.13438	7.4415	1.6405	5
6	2.9859	.33490	.30071	3.3255	.10071	9.9298	1.9788	6
7	3.5831	.27908	.27742	3.6045	.07742	12.915	2.2901	7
8	4.2998	.23257	.26061	3.8371	.06061	16.498	2.5756	8
9	5.1597	.19381	.24808	4.0309	.04808	20.798	2.8364	9
10	6.1917	.16151	.23852	4.1924	.03852	25.958	3.0738	10
11	7.4300	.13459	.23110	4.3270	.03110	32.150	3.2892	11
12	8.9160	.11216	.22527	4.4392	.02527	39.580	3.4840	12
13	10.699	.09346	.22062	4.5326	.02062	48.496	3.6596	13
14	12.839	.07789	.21689	4.6105	.01689	59.195	3.8174	14
15	15.406	.06491	.21388	4.6754	.01388	72.034	3.9588	15
16	18.488	.05409	.21144	4.7295	.01144	87.441	4.0851	16
17	22.185	.04507	.20944	4.7746	.00944	105.92	4.1975	17
18	26.623	.03756	.20781	4.8121	.00781	128.11	4.2975	18
19	31.947	.03130	.20646	4.8435	.00646	154.73	4.3860	19
20	38.337	.02608	.20536	4.8695	.00536	186.68	4.4643	20
21	46.004	.02174	.20444	4.8913	.00444	225.02	4.5333	21
22	55.205	.01811	.20369	4.0094	.00369	271.02	4.5941	22
23	66.246	.01510	.20307	4.9245	.00307	326.23	4.6474	23
24	79.495	.01258	.20255	4.9371	.00255	392.47	4.6942	24
25	95.394	.01048	.20212	4.9475	.00212	471.97	4.7351	25
26	114.47	.00874	.20176	4.9563	.00176	567.36	4.7708	26
27	137.36	.00728	.20147	4.9636	.00147	681.84	4.8020	27
28	164.84	.00607	.20122	4.9696	.00122	819.21	4.8291	28
29	197.81	.00506	.20102	4.9747	.00102	984.05	4.8526	29
30	237.37	.00421	.20085	4.9789	.00085	1181.8	4.8730	30
31	284.84	.00351	.20070	4.9824	.00070	1419.2	4.8907	31
32	341.81	.00293	.20059	4.9853	.00059	1704.0	4.9061	32
33	410.17	.00244	.20049	4.9878	.00049	2045.8	4.9193	33
34	492.21	.00203	.20041	4.9898	.00041	2456.0	4.9307	34
35	590.65	.00169	.20034	4.9915	.00034	2948.2	4.9406	35
40	1469.7	.00068	.20014	4.9966	.00014	7343.6	4.9727	40
45	3657.1	.00027	.20005	4.9986	.00005	18,281.3	4.9876	45
50	9100.1	.00011	.20002	4.9994	.00002	45,497.2	4.9945	50

TABLE 28.A7 25% Interest Factors for Discrete Compounding Periods

	Single payment		Uniform series					
	Compound amount factor	Present worth factor	Capital recovery factor	Present worth factor	Sinking fund factor	Compound amount factor	Gradient factor	
N	(F/P, 25, N)	(P/F, 25, N)	(A/P, 25, N)	(P/A, 25, N)	(A/F, 25, N)	(F/A, 25, N)	(A/G, 25, N)	N
1	1.2500	.80000	1.2500	.8000	1.0000	1.0000	.00000	1
2	1.5625	.64000	.69444	1.4400	.44444	2.2500	.44444	2
3	1.9531	.51200	.51230	1.9520	.26230	3.8125	.85246	3
4	2.4414	.40960	.42344	2.3616	.17344	5.7656	1.2249	4
5	3.0518	.32768	.37185	2.6693	.12185	8.2070	1.5631	5
6	3.8147	.26214	.33882	2.9514	.08882	11.259	1.8683	6
7	4.7684	.20972	.31634	3.1661	.06634	15.073	2.1424	7
8	5.9605	.16777	.30040	3.3289	.05040	19.842	2.3872	8
9	7.4506	.13422	.28876	3.4631	.03876	25.802	2.6048	9
10	9.3132	.10737	.28007	3.5705	.03007	33.253	2.7971	10
11	11.642	.08590	.27349	3.6564	.02349	42.566	2.9663	11
12	14.552	.06872	.26845	3.7251	.01845	54.208	3.1145	12
13	18.190	.05498	.26454	3.7801	.01454	68.760	3.2437	13
14	22.737	.04398	.26150	3.8241	.01150	86.949	3.3559	14
15	28.422	.03518	.25912	3.8593	.00912	109.687	3.4530	15
16	35.527	.02815	.25724	3.8874	.00724	138.109	3.5366	16
17	44.409	.02252	.25576	3.9099	.00576	173.636	3.6084	17
18	55.511	.01801	.25459	3.9279	.00459	218.045	3.6698	18
19	69.389	.01441	.25366	3.9424	.00366	273.556	3.7222	19
20	86.736	.01153	.25292	3.9539	.00292	342.945	3.7667	20
21	108.420	.00922	.25233	3.9631	.00233	429.681	3.8045	21
22	135.525	.00738	.25186	3.9705	.00186	538.101	3.8365	22
23	169.407	.00590	.25148	3.9764	.00148	673.626	3.8634	23
24	211.758	.00472	.25119	3.9811	.00119	843.033	3.8861	24
25	264.698	.00378	.25095	3.9849	.00095	1054.791	3.9052	25
26	330.872	.00302	.25076	3.9879	.00076	1319.489	3.9212	26
27	413.590	.00242	.25061	3.9903	.00061	1650.361	3.9346	27
28	516.988	.00193	.25048	3.9923	.00048	2063.952	3.9457	28
29	646.235	.00155	.25039	3.9938	.00039	2580.939	3.9551	39
30	807.794	.00124	.25031	3.9950	.00031	3227.174	3.9628	30
31	1009.742	.00099	.25025	3.9960	.00025	4034.968	3.9693	31
32	1262.177	.00079	.25020	3.9968	.00020	5044.710	3.9746	32
33	1577.722	.00063	.25016	3.9975	.00016	6306.887	3.9791	33
34	1972.152	.00051	.25013	3.9980	.00012	7884.609	3.9828	34
35	2465.190	.00041	.25010	3.9984	.00010	9856.761	3.9858	35

Chapter 29

Energy Conservation in Buildings

GÜNTER BERGMANN
RICHARD BRUNO

and

HORST HÖRSTER
Philips Research Laboratory, Aachen, FRG.

INTRODUCTION

Energy Utilization in Buildings

In 1975 the residential energy consumption in the member countries of the Organization for Economic Cooperation and Development (OECD)* represented 26 percent of their total final consumption.[42] Moreover, energy consumption in buildings also formed a part of each of two other consumer sectors: industry and the sum of the commercial and public service and agricultural subsectors. This amount can only be roughly estimated. However, the 26 percent already indicates that energy utilization in buildings forms one of the major portions of the total energy consumption of industrialized countries.

Most of the energy used in residential and commercial buildings is for heating. The second important part of domestic energy consumption is production of service hot water. In all countries mentioned, only the United States uses a few percent of its energy on air conditioning, i.e., cooling of buildings. Lighting and the operating of various household appliances and similar equipment rank third in energy consumption.

The list below indicates the priority of the efforts that have to be made to reduce the energy consumption in buildings:

---

*The 24 OECD member countries are: Australia, Austria, Belgium, Canada, Denmark, Finland, France, the Federal Republic of Germany, Greece, Iceland, Ireland, Italy, Japan, Luxemburg, the Netherlands, New Zealand, Norway, Portugal, Spain, Sweden, Switzerland, Turkey, the United Kingdom, and the United States.

- More efficient insulation.
- Reduction of the losses of HVAC equipment.
- Higher efficiencies of burners, lamps, machines, electrical appliances, etc.
- Better regulation of heating, cooling, and lighting equipment.
- Recovery and re-use of waste heat.
- Use of alternative energy sources.
- Use of passive measures.

The topics of this list indicate a certain amount of overlapping of the subjects in this and some of the other chapters of this handbook.

To begin with, let us summarize the results of detailed investigations into the potential of energy consumption and conservation in residential and commercial buildings. The figures given are based on statistical data of past energy consumption. Generally, we shall use OECD energy statistics to obtain reliable comparisons of the consumption data of different countries. Quantitative differences in other statistics are quite possible. They are mainly due to different definitions of the terminology used. In some statistics, for instance, the transformation and distribution losses are given as separate consumer sectors in the final consumption, whereas they form part of the difference between "Total Energy Requirement" and "Total Final Consumption" in the OECD statistics.

Table 29.1 shows detailed data for the energy consumption in 1975 and the corresponding mean rates of growth from 1965 through 1972 (a period of undisturbed growth) in the three major economic blocks of OECD: the United States, the European Community (EC), and Japan. It indicates that industry, transportation, and residences form the major consumer sectors. Annual growth rates differ according to different economic situations and developments. An increasing number of dwelling units and the growth in the standard of living raised the consumption rate in the residential and commercial sector more than the average. However, it is mainly due to the low level of energy prices that energy in this sector has been used uneconomically.

The percentages of "residential" with respect to "total" consumption differ widely for different countries. In 1975 they were for

Spain, 11.9	Sweden, 35.5
Japan, 22.5	United Kingdon, 35.8
United States, 23.1	Federal Republic of Germany, 36.6
France, 29.8	Denmark, 44.9

The great differences are mainly due to the fact that heating of buildings is the major component of this sector. As will be shown in detail, it amounts to 70 or 80 percent in some countries.[37,43,53] Consequently, the quantity of energy consumption in the "residential" sector is primarily determined by the climatic conditions.

Let us consider the structure of the energy demand in the residential and commercial sectors

TABLE 29.1   Energy Consumption, 1975, Million Tons of Oil Equivalent, and Average Annual Percent Growth Rate $\bar{a}$ (1965-72)

	United States	European Community	Japan
Total energy requirement, 1975,	1,690.15	871.40	331.88
$\bar{a}$ (1965–72)	4.72	4.51	10.95
Total final consumption, 1975,	1,245.72	660.54	232.35
$\bar{a}$ (1965–72)	4.63	5.03	12.44
of which:			
1. Industry	420.07	256.07	124.61
$\bar{a}$ (1965–72)	3.52	3.33	11.61
2. Transportation	409.49	127.09	41.44
$\bar{a}$	4.88	5.92	10.28
3. Residential	288.24	221.12	52.30
$\bar{a}$ (1965–72)	5.01	5.21	14.34
4. Rest for commercial and public service, agricultural, and nonenergy use			

of some countries by way of illustration. Detailed investigations for the Federal Republic of Germany in this field were made by Reents. [46] This author's study shows that about 60 percent of this sector's consumption is due to private households. The rest is the consumption of "small-scale consumers," including institutional households, public institutions, small businesses, trading companies, farms, etc. Fuel for transportation is not included because it is specified separately. The division into heating, hot-water supply, and other purposes (lighting, power, etc.) is given in Fig. 29.1. It demonstrates that the largest portion in each subsector is for heating of buildings. Figure 29.2 shows the structure of the consumption in households (Federal Republic of Germany, 1974) in relation to the primary energy sources.

A similar analysis covering the "residential and commercial" consumer sectors of the United States in 1960 and 1968 is shown in Table 29.2. [49] About 18 percent of the total, national consumption was for heating of buildings in 1968, 2.3 percent for cooling, and 4.0 percent for domestic hot-water supply. For all purposes other than residential cooling, which shows an exceptionally high growth rate, these figures have remained essentially unchanged since then.

Figure 29.3 illustrates U.S. energy consumption in this sector in relation to the primary energy sources. [29] Statistics of a number of German economic research institutions arrive at similar results for the percentage distribution of the FRG domestic and commercial consumption. [62]

Any improvements can be distinguished by technical improvements of appliances and by changing over to forms of energy for which appliances are more efficient. With due allowance made, these effects can be estimated quantitatively. Table 29.3 shows the results of an evaluation of the mean efficiencies for the U.K. 1970 household use of energy, in percentage efficiencies for conversion from primary to net and from net to useful energy, indicating that the overall mean efficiency increases according to the sequence electricity-solid fuel-gas-oil fuel.

Conservation Potential

Quantitative estimates of the conservation potentials are beset with uncertainties. This has to be kept in mind when the following investigations and their results are considered. In a detailed study of the possibilities of energy conservation on the basis of the existing stock of buildings and probable future development and fluctuations, a Battelle study by Dittert reaches the conclusion that in the Federal Republic of Germany the annual energy conservation achievable up to 1985 by improvements in building construction, heating equipment, and regulatory

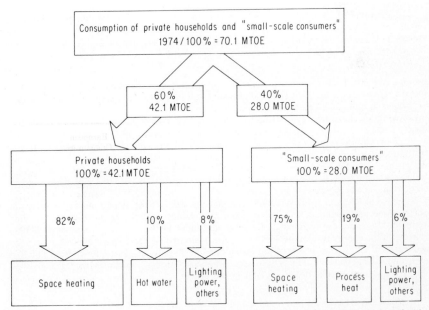

Fig. 29.1   Energy demand ot private households and "small-scale consumers" (FRG, 1974). (From Ref. 49.)

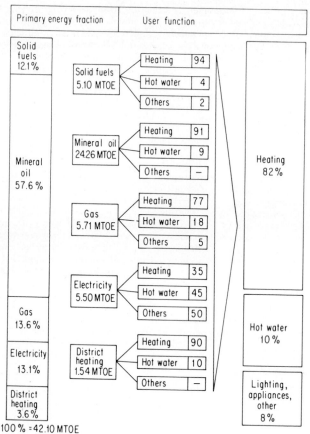

Fig. 29.2 Distribution by function of separate energy carriers. FRG. Private households 1974. (From Ref. 49.)

systems amounts to 20.2 million tons of oil equivalent,* corresponding to 49 percent of the sector's requirements for 1974.[19] The decrease in the percentage of energy consumption due to changes in building envelope is given in detail in Table 29.4.

Conservation potentials of between 40 and 50 percent merely by modifications of the construction standard are estimated by the National Bureau of Standards for the United States,[36] a result which also comes out of a study by the Ford Foundation[42] and the Massachusetts Institute of Technology's Workshop on Alternative Energy Strategies (WAES).[23] Allowing for the generally lighter constructional design in the United States as compared with Europe, a still larger technologically and economically realizable conservation potential should be expected on the basis of the Battelle studies.

To arrive at 50 percent conservation potential the study of the Ford Foundation does not only assume improvements in insulation but also changes in lighting, heating, and cooling systems. However, the technologically achievable limit for heating requirements, including changes in the energy system (heat recovery, use of alternative energy sources), has to be rated at a substantially higher level, of the order of 85 to 90 percent. This follows from detailed experimental and theoretical investigations in Denmark, Sweden, and Germany.[44] Particularly, the German study will be described later in this article.

---

*1 million tons of oil equivalent = 1×10^{13} kcal = 3.968×10^{13} Btu = 11.63 TWh.

TABLE 29.2   Energy Consumption in the United States by End Use, 1960–68, and Average Annual Growth Rate[a]

Sector and end use[a]	Consumption, MTOE[b] 1960	Consumption, MTOE[b] 1968	$\bar{a}$, %	Percentage of national total 1960	Percentage of national total 1968
Residential					
Space heating	122.2	168.2	4.1	11.3	11.0
Water heating	29.2	43.8	5.2	2.7	2.9
Cooking	14.0	16.1	1.7	1.3	1.1
Clothes drying	2.3	5.2	10.6	0.2	0.3
Refrigeration	9.3	17.4	8.2	0.9	1.1
Air conditioning	3.4	10.8	15.6	0.3	0.7
Other[c]	20.4	31.3	5.5	1.9	2.1
Total	200.8	292.7	4.8	18.6	19.2
Commercial					
Space heating	78.4	105.4	3.8	7.2	6.9
Water heating	13.7	16.5	2.3	1.3	1.1
Cooking	2.5	3.5	4.5	0.2	0.2
Clothes drying	13.5	16.9	2.9	1.2	1.1
Refrigeration[d]	14.5	28.0	8.6	1.3	1.8
Air conditioning	18.5	24.8	3.7	1.7	1.6
Other[e]	3.7	25.8	28.0	0.3	1.7
Total	144.7	220.9	5.4	13.2	14.4
Industrial total	462.2	629.0	3.9	42.7	41.2
Transportation total	277.6	382.7	4.1	25.5	25.2
National total	1085.3	1525.3	4.3	100.0	100.0

[a] Consumption by electric utilities has been allocated to each end use.
[b] Million tons of oil equivalent.
[c] Other in residential sector includes lighting, large and small appliances, television, food freezers, etc.
[d] Includes energy consumed for food freezing.
[e] Other in commercial sector is primarily electricity used for lighting and mechanical drives (for computers, elevators, escalators, office machinery, etc.).
From Ref.49, Table 13–2, using results from Stanford Research Institute, Bureau of Mines and others.

The WAES studies of the residential and commercial sectors show wide variations from country to country in the projected improvements. In their scenarios, in some European countries and in Japan the increasing size of dwellings, even if they are well-insulated, tends to offset the gains from improved design and better efficiency. In the United States, on the other hand, better residential dwelling insulation on new and existing structures is projected to improve efficiency by as much as 40 percent by the year 2000, bringing this efficiency measure to a value just slightly better than today's average European standard.

According to the OECD statistics the 1975 U.S. energy consumption in the "residential" sector was 288 million tons of oil equivalent, corresponding to 23 percent of the total consumption. Assuming a conservation potential of 50 percent in this sector, the 1975 total final consumption would have been lower by about 150 million tons of oil equivalent. The corresponding rate of saving of the European Community of "residential" consumption alone (assuming the same potential) would be 110 million tons of oil equivalent and for the total OECD 325 million tons of oil equivalent, which is about 40 percent more than the total final consumption of Japan, the world's second largest national energy consumer.

Recalculation of the corresponding primary energy equivalents on the basis of the saving potentials is obviously very uncertain. Nevertheless, a rough estimate gives an impression of the possibilities that are offered: it shows that the U.S. primary energy conservation potential from "residential" savings can be two-thirds the U.S. crude oil imports of 1975.

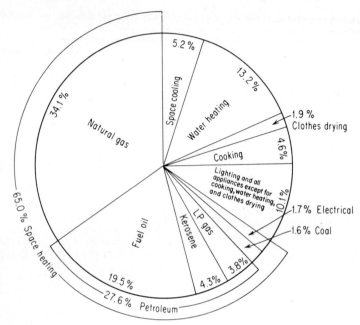

Fig. 29.3 U.S. — Residential energy usage, 1970. (From "The U.S.-Energy Problem," Vol. II Appendices—Part A, Intertechnology Corporation, Nov. 1972, reprinted in "Perspectives on the Energy Crisis," Vol. I, Ann Arbor Science, Ann Arbor, MI, 1977.)

TABLE 29.3   Mean Efficiencies for the U.K. Household Use, 1970, Percentages

	Percentage efficiencies for conversion		
	Primary to net energy	Net to useful energy	Overall mean efficiency*
Electricity	27	82	22
Solid fuel	98	37	36
Gas	84†	69	58†
Oil fuel	92	69	63

*These efficiencies refer to the average total use of each form of energy in households. For central heating alone the percentage for overall mean efficiencies would be approximately given by: electricity, 25 percent; solid fuel, 59 percent; gas, 59 percent, but natural gas 66 percent and oil fuel 64 percent. These are averages and modern appliances (well maintained) may give better figures.
†The use of natural gas only (instead of town gas) would raise the figures to about 94 and 64, respectively. From Ref. 40.

Conservation Measures

Which ways lead to the best exploitation of conservation potentials? Of the following four possibilities:

- Free market regulation (mainly: free price of oil and gas)
- Request and advice regarding rational use
- Development, promulgation, and enforcement of new technologies
- Legislative measures: codes and incentives

we shall focus attention primarily on item 3 in the following paragraphs and consider in passing some aspects of item 4. A "Checklist of Energy Conservation Opportunities, Ranked in Order of Priority According to Climatic Conditions" (Table 29.5) is taken from Ref. 30.

TABLE 29.4 Energy Saving Measures (Percentages Cannot Be Added)

Constructional measures	Savings, %
Additional insulation of outer walls	50–60 (reduction of transmission losses)
Additional insulation of the top ceiling	50–60 (reduction of transmission losses)
Additional insulation of the cellar ceiling	50–60 (reduction of transmission losses)
Replacing one pane by double-pane windows	30–40 (reduction of transmission losses)
Tightening of window and door frames	50–70 (reduction of ventilation losses)
Higher wall temperatures due to improved insulation allow reduction of air temperature by 1 to 2 K	6–12
Domestic hot-water supply by decentralized instead of central production	Up to 35

From: Ref. 19.

Due to the long service life of a house it is most important that the long-term, but most effective, measures to limit energy waste in buildings be accomplished immediately. These methods are improved insulation, heat recovery, and use of alternative energy sources in all new construction. It is less cost-effective to retrofit these measures than to install them in new construction. Nevertheless, savings of 20 to 25 percent are also estimated from retrofit improvements with pay-off times of about 10 years.

Contrary to many conventional assumptions, an evaluation that Arthur D. Little performed for the U.S. Federal Energy Administration of the impact of the new energy conservation building code developed by the American Society of Heating, Refrigerating and Air Conditioning Engineers (ASHRAE), arrived at the result that new buildings can be designed for energy conservation with little or no first-cost penalty.[38,64] The analysis showed that with savings of between 40 and 60 percent as compared with previous building standards the cost of construction of these buildings would be equal to or less than that of traditional buildings. This is also in agreement with conclusions which were drawn from the investigations into "Low Energy Houses" in Denmark, Germany, and Sweden.[44]

The urgency with which saving measures must be introduced becomes evident from the fact that a period of several years usually passes between enacting legislation or technical standards and the recognition of a perceptible economical effect. Table 29.6 gives an estimate of the time lags associated with the various technical measures.

These time lags are comparable to periods for legislative or administrative measures:
- Decreed rationings have immediately observable effects (time lags less than 2 years),
- Financial incentives for conservation measures in existing buildings have medium delay,

and
- Legislative regulations concerning constructions of new buildings (the most effective measures) are impeded by the longest time lags (more than 5 or 10 years).

Figure 29.4 shows as an example the effect of various conservation measures on the temporal development of the energy consumption, computed for one-family houses in Sweden.[27] The beginning is fixed by the date of enacting the new Swedish building codes in 1977, and the full effect is calculated to be reached in 1988.

The economic effects that are due to technological and statutory measures gradually increase over the years to a saturation limit, which is the balancing of technical possibilities by economic considerations. It is consequently lower than the purely technical limit which itself is of course lower than the physical limit.

There are good reasons to estimate the technological limit of the conservation potential for the "operation" of buildings in the range 85 to 90 percent, as compared to buildings of the seventies. It seems also reasonable to assume the economically acceptable limit to be close to 50 percent, perhaps even higher in some areas. It is, however, doubtful, whether this goal can be achieved before the year 2000.

ENERGY DEMAND OF BUILDINGS

A Qualitative Discussion

As demonstrated in the previous section the energy demand of buildings, especially for space heating, corresponds to 25 to 40 percent of the used energy in the different industrialized

TABLE 29.5 A Checklist of Energy Conservation Opportunities, Ranked in Priority According to Climatic Conditions*

This checklist of energy-saving opportunities, appended to the guidelines, includes some items that subsume others. Some seem to border on the obvious, yet many contemporary buildings are testimony to the need for even seemingly obvious measures.

The items are ranked in priority and coded to the following climatic features: For winter, *A* indicates a heating season of 6,000 degree-days or more; *B* a heating season of 4,000 to 6,000 degree-days, and *C* 4,000 degree-days or less. The numeral *1* following these letters indicates sun 60 per cent of daylight time or more and wind nine miles per hour or more; *2* indicates the sun condition but not the wind condition; *3* indicates the wind condition without the sun condition, and *4* the absence of either condition.

For summer, the letter *D* indicates a cooling season or more than 1,500 hours at 80 degrees Fahrenheit; *E* 600 to 1,500 hours at the same temperature and *F* less than 600 hours. The numeral *1* indicates a dry climate of 60 per cent relative humidity or less and *2* indicates 60 per cent or more humidity.

Guidelines that are independent of climate are not rated in priority columns and are marked '*'.

SITE

	Priority			
	1	2	3	N/A
1. Use deciduous trees for their summer sun shading effects and wind break for buildings up to three stories.	A1	A2	A4	C4
	D1	D2	E1	F
2. Use conifer trees for summer and winter sun shading and wind breaks.	C4	C1	C2	A2
	D1	D2	E1	F
3. Cover exterior walls and/or roof with earth and planting to reduce heat transmission and solar gain.	A1	A2	A4	C4
	D1	D2	E1	F
4. Shade walls and paved areas adjacent to building to reduce indoor/outdoor temperature differential.	C2	C1	C3	A2
	D1	D2	E1	F
5. Reduce paved areas and use grass or other vegetation to reduce outdoor temperature buildup.	C2	C1	C3	A2
	D1	D2	E1	F
6. Use ponds, water fountains, to reduce ambient outdoor air temperature around building.	C2	C1	C3	A4
	D1	E1	D2	F
7. Collect rain water for use in building.	*			
8. Locate building on site to induce air flow effects for natural ventilation and cooling.	C2	C1	C3	A4
	F	E1	E2	D2
9. Locate buildings to minimize wind effects on exterior surfaces.	A4	A1	B4	C2
	F	E2	E1	D1
10. Select site with high air quality (least contaminated) to enhance natural ventilation.	C2	C1	C3	A4
	F	E1	E2	D2
11. Select a site which has year-round ambient wet and dry bulb temperatures close to and somewhat lower than those desired within the occupied spaces.	*			
12. Select a site that has topographical features and adjacent structures that provide breaks.	A4	A1	B4	C3
	F	E2	E1	D2
13. Select a site that has topographical features and adjacent structures that provide desirable shading.	C2	C1	B2	A1
	D2	D1	E2	F
14. Select site that allows optimum orientation and configuration to minimize yearly energy consumption.	*			
15. Select site to reduce specular heat reflections from water.	C2	C1	B2	A4
	D2	D1	E2	F
16. Use sloping site to bury building partially or use earth berms to reduce heat transmission and solar radiation.	A4	A1	A3	C2
	D1	D2	E1	F
17. Select site that allows occupants to use public transport systems.	*			

Table 29.5 Energy conservation checklist. (Reproduced from the *AIA Journal* by permission of the publishers.)

BUILDING

	Priority			
	1	2	3	N/A
1. Construct building with minimum exposed surface area to minimize heat transmission for a given enclosed volume.	A4 D1	A1 D2	A3 E1	C2 F
2. Select building configuration to give minimum north wall to reduce heat losses.	A4	A1	A3	C2
3. Select building configuration to give minimum south wall to reduce cooling load.	D1	D2	E1	F
4. Use building configuration and wall arrangement (horizontal and vertical sloping walls) to provide self shading and wind breaks.	A4 D1	A1 D2	B4 E1	C3
5. Locate insulation for walls and roofs and floors over garages at the exterior surface.	A4 D1	A3 D2	A1 E1	C2 F
6. Construct exterior walls, roof and floors with high thermal mass with a goal of 100 pounds per cubic foot.	A4 D1	A1 D2	A3 E1	C3 F
7. Select insulation to give a composite U factor from 0.06 when outdoor winter design temperatures are less than 10 degrees F. to 0.15 when outdoor design conditions are above 40 degrees F.				
8. Select U factors from 0.06 where sol-air temperatures are above 144 degrees F. up to a U volume of 0.3 with sol-air temperatures below 85 degrees F.				
9. Provide vapor barrier on the interior surface of exterior walls and roof of sufficient impermeability to provide condensation.				
10. Use concrete slab-on-grade for ground floors.	A4 D1	A1 D2	A3 E1	C2 F
11. Avoid cracks and joints in building construction to reduce infiltration.	A4 D2	A1 E2	A3 D1	
12. Avoid thermal bridges through the exterior surfaces.	A4 D2	A1 D1	A3 E2	C3 F
13. Provide textured finish to external surfaces to increase film coefficiency.	A4	A1	B4	C2
14. Provide solar control for the walls and roof in the same areas where similar solar control is desirable for glazing.	D2	D1	E2	A
15. Consider length and width aspects for rectangular buildings as well as other geometric forms in relationship to building height and interior and exterior floor areas to optimize energy conservation.	A4 D1	A1 D2	A3 E1	C2 F
16. To minimize heat gain in summer due to solar radiation, finish walls and roofs with a light-colored surface having a high emissivity.	D1	D2	E1	
17. To increase heat gain due to solar radiation on walls and roofs, use a dark-colored finish having a high absorptivity.	A1	A2	A4	C2
18. Reduce heat transmissions through roof by one or more of the following items:				
a. Insulation.	A4 D1	A1 D2	A3 E1	C3 F
b. Reflective surfaces.	C2 D1	C1 D2	C3 E1	A4
c. Roof spray.	D1	E1	F	
d. Roof pond.	D1	E1	F	
e. Sod and planning.	A4 D1	A1 D2	A3 E1	C2 F
f. Equipment and equipment rooms located on the roof.	A4 D1	A1 D2	A3 E1	
g. Provide double roof and ventilate space between.	D1	D2	E1	F
19. Increase roof heat gain when reduction of heat loss in winter exceeds heat gain increase in summer:				
a. Use dark-colored surfaces.	A2	A1	B2	B1
b. Avoid shadows.	A2	A1	B2	B1
20. Insulate slab on grade with both vertical and horizontal perimeter insulation under slab.	A	B	C	
21. Reduce infiltration quantities by one or more of the following measures:				
a. Reduce building height.				

(*continued*)

b. Use impermeable exterior surface materials.

c. Reduce crackage area around doors, windows, etc., to a minimum.

| A4 | A1 | A3 | C4 |
| D2 | E2 | D1 | F |

d. Provide all external doors with weather stripping.

e. Where operable windows are used, provide them with sealing gaskets and cam latches.

f. Locate building entrances on downwind side and provide wind break.

g. Provide all entrances with vestibules; where vestibules are not used, provide revolving doors.

h. Provide vestibules with self-closing weather-stripped doors to isolate them from the stairwells and elevator shafts.

| A4 | A1 | A3 | C4 |
| D2 | E2 | D1 | F |

i. Seal all vertical shafts.

j. Locate ventilation louvers on downwind side of building and provide wind breaks.

k. Provide break at intermediate points of elevator shafts and stairwells for tall buildings.

22. Provide wind protection by using fins, recesses, etc., for any exposed surface having a U value greater than 0.5.

| A4 | A1 | B4 | C2 |

23. Do not heat parking garages.   •

24. Consider the amount of energy required for the protection of materials and their transport on a life-cycle energy usage.   •

25. Consider the use of the insulation type which can be most efficiently applied to optimize the thermal resistance of the wall or roof; for example, some types of insulation are difficult to install without voids or shrinkage.   •

26. Protect insulation from moisture originating outdoors, since volume decreases when wet. Use insulation with low water absorption and one which dries out quickly and regains its original thermal performance after being wet.   •

27. Where sloping roofs are used, face them to south for greatest heat gain benefit in the wintertime.

| A1 | A2 | B1 | C4 |

28. To reduce heat loss from windows, consider one or more of the following:

a. Use minimum ratio of window area to wall area.

b. Use double glazing.

c. Use triple glazing.

d. Use double reflective glazing.

e. Use minimum percentage of the double glazing on the north wall.

A4	B4	C4
A1	B1	C1
A2	B2	C2
A3	B3	C3

f. Manipulate east and west walls so that windows face south.

g. Allow direct sun on windows November through March.

h. Avoid window frames that form a thermal bridge.

i. Use operable thermal shutters which decrease the composite U value to 0.1.

29. To reduce heat gains through windows, consider the following:

a. Use minimum ratio of window area to wall area.

b. Use double glazing.

c. Use triple glazing.

d. Use double reflective glazing.

| D1 | E1 | F |
| D2 | E2 | F |

e. Use minimum percentage of double glazing on the south wall.

f. Shade windows from direct sun April through October.

30. To take advantage of natural daylight within the building and reduce electrical energy consumption, consider the following:

a. Increase window size but do not exceed the point where yearly energy consumption, due to heat gains and losses, exceeds the saving made by using natural light.

b. Locate windows high in wall to increase reflection from ceiling, but reduce glare effect on occupants.

c. Control glare with translucent draperies operated by photo cells.

C2	B2	A2
C1	B1	A2
C3	B3	A3
C4	B4	A4
F	E	D

d. Provide exterior shades that eliminate direct sunlight, but reflect light into occupied spaces.

e. Slope vertical wall surfaces so that windows are self-shading and walls below act as light reflectors.

(continued)

f. Use clear glazing. Reflective or heat absorbing films reduce the quantity of natural light transmitted through the window.

31. To allow the use of natural light in cold zones where heat losses are high energy users, consider operable thermal barriers.

| A4 | A1 | B4 | C3 |

32. Use permanently sealed windows to reduce infiltration in climatic zones where this is a large energy user.

| A1 | A4 | B1 | C3 |
| D1 | D2 | E1 | F |

33. Where codes of regulations require operable windows and infiltration is undesirable, use windows that close against a sealing gasket.

| A1 | A4 | B1 | C3 |
| D1 | D2 | E1 | F |

34. In climatic zones where outdoor air conditions are suitable for natural ventilation for a major part of the year, provide operable windows.

| C2 | C3 | C1 | A4 |
| F | E1 | E2 | D2 |

35. In climate zones where outdoor air conditions are close to desired indoor conditions for a major portion of the year, consider the following:
   a. Adjust building orientation and configuration to take advantage of prevailing winds.
   b. Use operable windows to control ingress and egress of air through the building.
   c. Adjust the configuration of the building to allow natural cross ventilation through occupied spaces.

| F | E1 | E2 | D2 |

   d. Use stack effect in vertical shafts, stairwells, etc., to promote natural air flow through the building.

PLANNING

Priority			
1	2	3	N/A

1. Group services rooms as a buffer and locate at the north wall to reduce heat loss or the south wall to reduce heat gain, whichever is the greatest yearly energy user.

| A4 | A1 | A3 | C2 |
| D1 | D2 | E1 | F |

2. Use corridors as heat transfer buffers and locate against external walls.

| A4 | A1 | A3 | C2 |
| D1 | D2 | E1 | F |

3. Locate rooms with high process heat gain (computer rooms) against outside surfaces that have the highest exposure loss.

| A4 | A1 | A3 | C2 |

4. Landscaped open planning allows excess heat from interior spaces to transfer to perimeter spaces which have a heat loss.

| • | | | |

5. Rooms can be grouped in such a manner that the same ventilating air can be used more than once, by operating in cascade through spaces in decreasing order of priority, i.e., office-corridor-toilet.

| • | | | |

6. Reduced ceiling heights reduce the exposed surface area and the enclosed volume. They also increase illumination effectiveness.

| • | | | |

7. Increased density of occupants (less gross floor area per person) reduces the overall size of the building and yearly energy consumption per capita.

| • | | | |

8. Spaces of similar function located adjacent to each other on the same floor reduce the use of elevators.

| • | | | |

9. Offices frequented by the general public located on the ground floor reduce elevator use.

| • | | | |

10. Equipment rooms located on the roof reduce unwanted heat gain and heat loss through the surface. They can also allow more direct duct and pipe runs reducing power requirements.

| A4 | A3 | B4 | C2 |
| D1 | D2 | E1 | |

11. Windows planned to make beneficial use of winter sunshine should be positioned to allow occupants the opportunity of moving out of the direct sun radiation.

| • | | | |

12. Deep ceiling voids allow the use of larger duct sizes with low pressure drop and reduce HVAC requirements.

| • | | | |

13. Processes that have temperature and humidity requirements different from normal physiological needs should be grouped together and served by one common system.

| • | | | |

14. Open planning allows more effective use of lighting fixtures. The reduced area of partitioned walls decreases the light absorption.

| • | | | |

(*continued*)

15. Judicious use of reflective surfaces such as sloping white ceilings can enhance the effect of natural lighting and increase the yearly energy saved.

*			

VENTILATION AND INFILTRATION

	Priority		
1	2	3	N/A

1. To minimize infiltration, balance mechanical ventilation so that supply air quantity equals or exceeds exhaust air quantity.

1	2	3	N/A
A	B	C	
D	E	F	

2. Take credit for infiltration as part of the outdoor air requirements for the building occupants and reduce mechanical ventilation accordingly.

A	B	C	
D	E	F	

3. Reduce C.F.M./occupant outdoor air requirements to the minimum considering the task they are performing, room volume and periods of occupancy.

A	B	C	
D	E	F	

4. If odor removal requires more than 2,000 cubic feet per minute exhaust and a corresponding introduction of outdoor air, consider recirculating through activated carbon filter.

C	B	C	
D	E	F	

5. Where outdoor conditions are close to but less than indoor conditions for major periods of the year, and the air is clean and free from offensive odors, consider the use of natural ventilation when yearly energy trade-offs with other systems are favorable.

C2	C1	C3	A2
F	E1	E2	D2

6. Exchange heat between outdoor air, intake and exhaust air by using heat pipes, thermal wheels, run-around systems, etc.

A	B	C	
D	E	F	

7. In areas subjected to high humidities, consider latent heat exchange in addition to sensible.

D2	E2	F	

8. Provide selective ventilation as needed; i.e., 5 cubic feet per minute/occupant for general areas and increased volumes for areas of heavy smoking or odor control.

*			

9. Transfer air from "clean" areas to more contaminated areas (toilet rooms, heavy smoking areas) rather than supply fresh air to all areas regardless of function.

*			

10. Provide controls to shut down all air systems at night and weekends except when used for economizer cycle cooling.

A	B	C	
D	E	F	

11. Reduce the energy required to heat or cool ventilation air from outdoor conditions to interior design conditions by considering the following:

   a. Reduce indoor air temperature setting in winter and increase in summer.

A	B	C	
D	E	F	

   b. Provide outdoor air direct to perimeter of exhaust hoods in kitchens, laboratories, etc. Do not cool this air in summer or heat over 50 degrees F. in winter.

A	B	C	
D	E	F	

HEATING, VENTILATION AND AIRCONDITIONING

	Priority		
1	2	3	N/A

1. Use outdoor air for sensible cooling whenever conditions permit and when recaptured heat cannot be stored.

*			

2. Use adiabatic saturation to reduce temperature of hot, dry air to extend the period of time when "free cooling" can be used.

D1	E1	F	

3. In the summer when the outdoor air temperature at night is lower than indoor temperature, use full outdoor air ventilation to remove excess heat and precool structure.

D1	E1	F	

4. In principle, select the air handling system which operates at the lowest possible air velocity and static pressure. Consider high pressure systems only when other trade-offs such as reduced building size are an overriding factor.

*			

5. To enhance the possibility of using waste heat from other systems, design air handling systems to circulate sufficient air to enable cooling loads to be met by a 60 degree F. air supply temperature and heating loads to be met by a 90 degree F. air temperature.

*			

(continued)

6. Design HVAC systems so that the maximum possible proportion of heat gain to a space can be treated as an equipment load, not as a room load.

7. Schedule air delivery so that exhaust from primary spaces (offices) can be used to heat or cool secondary spaces (corridors).

8. Exhaust air from center zone through the lighting fixtures and use this warmed exhaust air to heat perimeter zones.

9. Design HVAC systems so that they do not heat and cool air simultaneously.

10. To reduce fan horsepower, consider the following:
    a. Design duct systems for low pressure loss.
    b. Use high efficiency fans.
    c. Use low pressure loss filters concommitant with contaminant removal.
    d. Use one common air coil for both heating and cooling.

11. Reduce or eliminate air leakage from duct work.

12. Limit the use of re-heat to a maximum of 10 percent of gross floor area and then only consider its use for areas that have atypical fluctuating internal loads, such as conference rooms.

13. Design chilled water systems to operate with as high a supply temperature as possible—suggested goal: 50 degrees F. (This allows higher suction temperatures at the chiller with increased operating efficiency.)

14. Use modular pumps to give varying flows that can match varying loads.

15. Select high efficiency pumps that match load. Do not oversize.

16. Design piping systems for low pressure loss and select routes and locate equipment to give shortest pipe runs.

17. Adopt as large a temperature differential as possible for chilled water systems and hot water heating systems.

18. Consider operating chillers in series to increase efficiency.

19. Select chillers that can operate over a wide range of condensing temperatures and then consider the following:
    a. Use double bundle condensers to capture waste heat at high condensing temperatures and use directly for heating or store for later use.
    b. When waste heat cannot be either used directly or stored, then operate chiller at lowest condensing temperature compatible with ambient outdoor conditions.

20. Consider chilled water storage systems to allow chillers to operate at night when condensing temperatures are lowest.

21. Consider the use of double bundle evaporators so that chillers can be used as heat pumps to upgrade stored heat for use in unoccupied periods.

22. Consider the use of gas or diesel engine drive for chillers and large items of ancilliary equipment and collect and use waste heat for absorption cooling, heating, and/or domestic hot water.

23. Locate cooling towers or evaporative coolers so that induced air movement can be used to provide or supplement garage exhaust ventilation.

24. Use modular boilers for heating and select units so that each module operates at optimum efficiency.

25. Extract waste heat from boiler flue gas by extending surface coils or heat pipes.

26. Select boilers that operate at the lowest practicable supply temperature while avoiding condensation within the furnaces.

27. Use unitary water/air heat pumps that transport heat energy from zone to zone via a common hydronic loop.

28. Consider the use of thermal storage in combination with unit heat pumps and a hydronic loop so that excess heat during the day can be captured and stored for use at night.

29. Consider the use of heat pumps both water/air and air/air if a continuing source of low-grade heat exists near the building, such as lake, river, etc.

(continued)

30. Consider the direct use of solar energy via a system of collectors for heating in winter and absorption cooling in summer.

31. Minimize requirements for snow melting to those that are absolutely necessary and, where possible, use waste heat for this service.

32. Provide all outside air dampers with accurate position indicators and insure that dampers are air-tight when closed.

33. If electric heating is contemplated, consider the use of heat pumps in place of direct resistance heating; by comparison they consume one-third of the energy per unit output.

34. Consider the use of spot heating and/or cooling in spaces having large volume and low occupancy.

35. Use electric ignition in place of gas pilots for gas burners.

36. Consider the use of a total energy system if the life-cycle costs are favorable.

LIGHTING AND POWER

	Priority			
	1 C F	2 B E	3 A D	N/A

1. a. Use natural illumination in areas where effective when a net energy conservation gain is possible vis-a-vis heating and cooling loads.
   b. Provide exterior reflectors at windows for more effective internal illumination.

2. Consider a selective lighting system in regard to the following:
   a. Reduce the wattage required for each specific task by review of user needs and method of providing illumination.
   b. Consider only the amount of illumination required for the specific task considering the duration and character and user performance required as per design criteria.
   c. Group similar task together for optimum conservation of energy per floor.
   d. Design switch circuits to permit turning off unused and unnecessary light.
   e. Illuminate tasks with fixtures built into furniture and maintain low intensity lighting elsewhere.
   f. Consider the use of polarized lenses to improve quality of lighting at tasks.
   g. Provide timers to automatically turn off lights in remote or little-used areas.
   h. Use multilevel ballasts to permit varying the lumen output for fixtures by adding or removing lamps when tasks are changed in location or requirements.
   i. Arrange electrical systems to accommodate relocatable luminaires which can be removed to suit changing furniture layouts.
   j. Consider the use of ballasts which can accommodate sodium metalhalide bulbs interchangeably with other lamps.

3. Consider the use of high frequency lighting to reduce wattage per lumen output. Additional benefits are reduced ballast heat loss into the room and longer lamp life.

4. Consider the use of landscape office planning to improve lighting efficiency. Approximately 25 percent less wattage per foot-candles on task for open planning versus partitions.

5. Consider the use of light colors for walls, floors and ceilings to increase reflectance, but avoid specular reflections.

6. Lower the ceilings or mounting height of luminaires to increase level of illumination with less wattage.

7. Consider dry heat-of-light systems to improve lamp performance and reduce heat gain to space.

8. Consider wet heat-of-light system to improve lamp performance and reduce heat gain to space and refrigeration load.

9. Use fixtures that give high contrast rendition factor at task.

(continued)

10. Provide suggestions to GSA for analysis of tasks to increase use of high contrast material which requires less illumination. *

11. Select furniture and interior appointments that do not have glossy surfaces or give specular reflections. *

12. Use light spills from characteristic areas to illuminate non-characteristic areas. *

13. Consider use of greater contrast between tasks and background lighting, such as 8 to 1 and 10 to 1. *

14. Consider washers and special illumination for features such as plants, murals, etc., in place of overhead space lighting to maintain proper contrast ratios. *

15. For horizontal tasks or duties, consider fixtures whose main light component is oblique and then locate for maximum ESI footcandles on task. *

16. Consider using 250 watt mercury vapor lamps and metal-halide lamps in place of 500 watt incandescent lamps for special applications. *

17. Use lamps with higher lumens per watt input, such as:
    a. One 8-foot fluorescent lamp versus two 4-foot lamps. *
    b. One 4-foot fluorescent lamp versus two 2-foot lamps. *
    c. U-tube lamps versus two individual lamps. *
    d. Fluorescent lamps in place of all incandescent lamps except for very close task lighting, such as at a typewriter paper holder. *

18. Use high utilization and maintenance factors in design calculations and instruct users to keep fixtures clean and change lamps earlier. *

19. Avoid decorative flood-lighting and display lighting. *

20. Direct exterior security lighting at entrances and avoid illuminating large areas adjacent to building. *

21. Consider switches activated by intruder devices rather than permanently lit security lighting. *

22. If already available, use street lighting for security purposes.

23. Reduce lighting requirements for hazards by:
    a. Using light fixtures close to and focused on hazard. *
    b. Increasing contrast of hazard; i.e., paint stair treads and risers white with black nosing. *

24. Consider the following methods of coping with code requirements:
    a. Obtaining variance from existing codes. *
    b. Changing codes to just fulfill health and safety functions of lighting by varying the qualitative and quantitative requirements to specific application. *

25. Consider the use of a total energy system integrated with all other systems. *

26. Where steam is available, use turbine drive for large items of equipment. *

27. Use heat pumps in place of electric resistance heating and take advantage of the favorable coefficient of performance. *

28. Match motor sizes to equipment shaft power requirements and select to operate at the most efficient point. *

29. Maintain power factor as close to unity as possible. *

30. Minimize power losses in distribution system by:
    a. Reducing length of cable runs. *
    b. Increasing conductor size within limits indicated by life-cycle costing. *
    c. Use high voltage distribution within the building. *

31. Match characteristics of electric motors to the characteristics of the driven machine. *

32. Design and select machinery to start in an unloaded condition to reduce starting torque requirements. (For example, start pumps against closed valves.) *

33. Use direct drive whenever possible to eliminate drive train losses. *

34. Use high efficiency transformers (these are good candidates for life-cycle costing). *

(continued)

35. Use liquid-cooled transformers and captive waste heat for beneficial use in other systems.	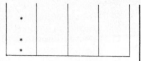
36. In canteen kitchens, use gas for cooking rather than electricity.	
37. Use conventional ovens rather than self-cleaning type.	

*From Ref. 30.

TABLE 29.6   Time Lags between Starting Technical Rationalization Measures and Recognizing the Economic Effect

Measure	Time lag, years
Economical operation	< 2
Technical improvement	1–5
Introduction of new technologies	5–10
Development of new technologies	10–25
Basic research into new technologies	< 25

From Ref. 40.

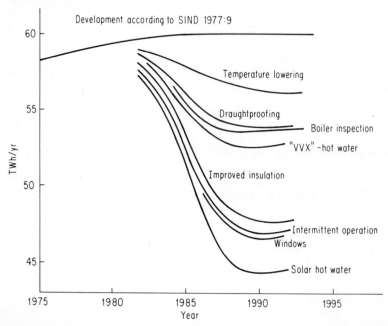

Fig. 29.4 Future development of the energy requirement for one-family houses in Sweden. (From Ref. 27, Fig. 19B/12.)

countries. There is a large potential for energy savings in this sector of the order of 50 percent or more.

The purpose of a house is to protect its inhabitants against the influence of the environment such as rain, cold, heat, the sun, wind, and noise. (See Fig. 29.5.) In a simple sense this is physically accomplished by the outer shell of a building which must fulfill some conditions like stability, tightness, noise protection, etc. Examples of various outer shell or envelope constructions are given below.

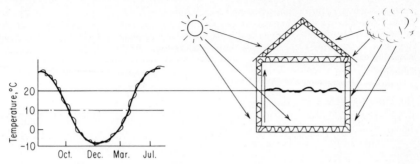

Fig. 29.5 Schematic representation of a building and its environment.

The thermal comfort level requires an adequate indoor temperature, e.g., 20°C, a relative humidity level between 30 and 60 percent, and a regular fresh air input, which corresponds to an air exchange rate of 0.5 to 1.5 of the living space per hour. In modern houses these comfort requirements are achieved by heating and air-conditioning equipment which is sized to the requirements of a specific building.

During the past 50 years the energy demand for space heating of buildings has increased rapidly even in countries with constant population, e.g., Germany. This was caused by the enormous increase in heated and air-conditioned building area per capita. In former years there was only one heated room per family, where an open fireplace acted as a cooking range and heating system. The internal load, that is, the energy dissipated by persons and equipment, was relatively high in the heated room. As a consequence, the extra heating energy was small. Today, the heated and air-conditioned building area per capita in the residential sector has increased by more than 300 percent, compared to the situation 50 years ago. In addition there was a comparable increase in commercial and office buildings. On the other hand, there was little change in building construction practice with respect to the thermal behavior of buildings. Moreover, the tendency to larger window areas may have increased the living standard but it decreased the thermal quality of buildings. Even a double-pane window has in most northern climates a higher net loss than a wall of standard older houses.

The existing and oncoming energy crisis demands a rapid change in the thermal construction of buildings by taking into account the progress in materials, technologies, and analytical techniques. An application of these topics may lead to an energy demand for space heating in new buildings which is about 20 or 10 percent, compared with an average building of the same size today. Energy experimental houses in Europe have demonstrated that this goal can be achieved without losses in thermal comfort and living habits. These buildings have the same amount of living area, window area, etc. as today's buildings.

Limits of Wall Insulation

Before going into detail it may be helpful to discuss the limits of a well-insulated house. Human metabolism requires a certain energy flow from the body which corresponds to about 100 W for an adult, having a body surface of about 2 m². Assuming a living room of ambient temperature of 20°C, the average heat loss coefficient through our clothes is 100 W/(37°C-20°C) 2 m² = 2.94 W/m²C which is about the same as a double-pane window or the heat flow resistance of about 0.6 cm of stagnant air. The average insulation of the outside walls of the living room should transport this internal load in such a way that under average winter outside temperatures the room temperature does not exceed the upper comfort level, i.e., 20°C. Assuming a room of 20-m² living space with an outside wall area of about 50 m² and an average outside temperature during January of about +2°C, the average heat loss coefficient of the outer walls should be greater than

$$U = 2.94 \ \text{W/m}^2\text{C} \cdot \frac{37°\text{C} - 20°\text{C}}{20°\text{C} - 2°\text{C}} \cdot \frac{2 \ \text{m}^2}{50 \ \text{m}^2} = 0.11 \ \text{W/m}^2\text{C}$$

This rough estimation must be corrected by the average number of persons living in this room + the internal load of appliances + light sources, etc. A more precise analysis (see the next section) gives an average internal load of about 3.5 W/m² for a single-family house of four persons which gives for the above example a minimum U-value of about 0.2 W/m²C. The U-values of the Energy-Experimental-Houses mentioned below are: walls—about 0.17 W/m²C and windows—1.9 W/m²C during the day and 1.3 W/m²C at night. The window area is about 20 percent of the living space area. So, the average U-value for the outside walls is about 0.4 W/m²C which is twice the minimum value stated above. This example indicates clearly that with a very excessive wall insulation and improved windows, even in a moderate climate, there is no danger of exceeding the comfort level by only conservation means.

The Energy Balance of a Building

To keep the house within a comfortable temperature range, independent of the weather conditions, there is a need for heating or cooling the house. Under steady-state conditions this energy requirement is the difference between the thermal losses and the thermal gain of the building construction:

$$\text{Heating requirement} = \text{Thermal losses} - \text{Thermal gain}$$

There are several thermal loss mechanisms, which will be discussed briefly.

Thermal losses of walls   According to the second law of the thermodynamics there is always a thermal energy flow from higher temperature levels to lower ones. This energy flow is proportional to the heat conductivity of the wall material.

$$E_{th} = kA \frac{\Delta T}{d}$$

where k = heat conductivity, W/m°C
ΔT = temperature difference between inside and outside wall temperature
d = thickness of the wall, m
A = surface of the wall, m²

For typical building materials k varies from about 0.02 to 1.2 for:

Concrete	1.15 W/m°C
Glass	0.9 W/m°C
Normal cinder Brick stone	0.6 W/m°C
Porous cinder Brick stone	0.17 W/m°C
Glass wool Rock wool	≈ 0.034 W/m°C
Polyurethane Foam	≈ 0.02 W/m°C
Air	≈ 0.02 W/m°C

This table indicates that solid materials have k-values around 0.5 to 1 W/m°C, special porous stones around 0.2 which is an order of magnitude above that of insulating materials like rock wool or foam whose values vary from 0.034 to 0.02. As the value of 0.02 W/m°C for air indicates, stagnant air would be an excellent insulating material. One function of matrix-type insulating materials like rock wool is to create a stagnant layer of air which is as thick as the material thickness and therefore much larger than the boundary layer of air under free convection (< 1 cm). Another function of insulation is to reduce the thermal radiation to nearly zero by multiple scattering. The thermal flow of energy from inside a building to the outside through a wall is governed by the heat conduction of the wall material but is also somewhat influenced by the stagnant air films and thermal radiation on the inner and outer side of the wall. The heat coefficient of the inner wall is assumed to be h_{in} = 8 W/m²°C and for the outer wall is to be h_{out} = 23 W/m²°C, taking into account a certain average wind velocity outside which controls the thickness of the air film. The total heat transfer coefficient is one over the sum of the different thermal resistances which are the inverse of the subsequent thermal flows per square meter and ΔT = 1°C. (See Chap. 4.)

$$U = \frac{1}{[(1/h_{out}) + (d/k) + (1/h_{in})]}, \text{ W/m}^2 \text{ °C}$$

The heat transfer coefficient of a 25-cm thick wall is:

U-values of a wall of different materials,
Thickness 25 cm

Concrete	2.72 W/m² °C
Normal cinder Brick stone	1.76 W/m² °C
Porous cinder Brick stone	0.62 W/m² °C
Glass wool Rock wool	0.13 W/m² °C
Polyurethane Foam	0.08 W/m² °C

The results indicate that for a wall thickness around 25 cm only a wall containing insulating material in a sandwich structure can reach U-values below 0.5 W/m²°C.

Thermal losses of windows   For a single-pane window (Fig. 29.6a) the thermal losses (neglecting edge losses) are found from

$$U_{\text{window}} = \frac{1}{[(1/h_{\text{in}}) + (d/k) + (1/h_{\text{out}})]} = \frac{1}{[(1/8) + (0.004/0.9) + (1/23)]} = 5.8 \text{ W/m}^2\,°\text{C}$$

for a glass thickness of 4 mm. From the formula it can be seen that the thermal resistance of the glass $(0.004/0.9 = 0.0044)$ is very small, compared to the effect of the heat transfer of the inner surface. The glass U-value results in only 4 percent of the total. To understand further improvements of windows it may be helpful to discuss the composition of the h_{in} and h_{out} by thermal radiation and thermal conduction through an air film. The thermal radiation can be linearized with an h_{rad} which is (20°C) about 5.2 W/m² °C for a black body. Glass has an emittance of about 0.88, so the glass surface radiates 4.6 W/m² °C.
With $h_{\text{in}} = h_{\text{i, rad}} + h_{\text{i, cond}}$ for

$$h_{\text{in}} = 4.6 + 3.4 \text{ W/m}^2\,°\text{C} = 8 \text{ W/m}^2\,°\text{C}$$

The corresponding air film would be 5.9 mm thick.
For h_{out} we find $h_{\text{o, cond}}$ to 18.4 W/m² °C with an equivalent air film thickness of about 1 mm.
For a double-pane window (Fig. 29.6b), assuming no convection within the air space ($\approx$ 10 mm) between the inner and outer pane, the U-value of a double-pane window is

$$U = \frac{1}{(1/8) + [1/(4.6 + 2.0)] + (1/23)} = 3.1 \text{ W/m}^2\,°\text{C}$$

which is about one-half of the single-pane window.
For a multipane window with n-panes the U-value is

$$U_{\text{n-pane}} = \frac{1}{(1/8) + [(n-1)/(4.6 + 2.0)] + (1/23)} \text{ W/m}^2\,°\text{C}$$

$$U_{\text{3-pane}} = 2.1 \text{ W/m}^2\,°\text{C}$$
$$U_{\text{4-pane}} = 1.6 \text{ W/m}^2\,°\text{C}$$

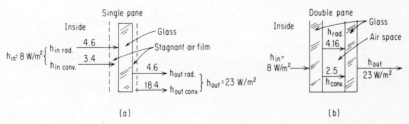

Fig. 29.6a   Single-pane window heat loss factors.   **Fig. 29.6b**   Double-pane window heat loss factors.

Improvements of double-pane windows The derivation of the heat transfer coefficient of a double-pane window indicates that the heat transfer from the inner air space is caused about two-thirds by thermal radiation. Improvements in reducing the heat transfer should reduce thermal radiation. This can be achieved by the application of an infrared reflecting coating which has high infrared reflection ρ for wavelengths greater than 2 μm and good transmission for solar radiation (0.4 < λ < 2 μm).

The transmission for thermal radiation through glass is very low, therefore the infrared emission coefficient ϵ is:

$$\epsilon \approx 1 - \rho$$

For a glass surface with a coating having an average infrared reflectance for thermal radiation of about 0.85, the ϵ-value of glass can be reduced from 0.88 to 0.15.

There are a few substances having such selective properties, e.g., semiconducting materials as SnO_2 or In_2O_3. Figure 29.7 shows the transmittance and reflectance of an In_2O_3 -coating on glass as a function of the wavelength, compared with the wavelength distribution of the solar radiation and thermal radiation of two glass panes with a temperature of 350 and 300 K. Similar optical properties but with a smaller transmission for solar radiation can be obtained from a combination of silver coatings[24] with an antireflection film like ZnS.

By applying an In_2O_3 coating on one inner pane of a double-pane window[26] the infrared emission ϵ of the surface is

$$\epsilon = \frac{(1 - \rho_1)\,(1 - \rho_2)}{1 - \rho_1\rho_2} = 0.14$$

where $\rho_1 = 0.2$ is the reflection of the glass surface and
$\rho_2 = 0.85$ is the reflection of the coating.

The energy flow due to radiation within the space between the two panes is reduced to

$$h_{\text{rad. coatg.}} = 0.14 \times 4.6 \text{ W/m}^2\text{°C} = 0.65 \text{ W/m}^2\text{°C}$$

The corresponding U-value is

$$U_{\text{double pane+ coatg.}} = \frac{1}{(1/8) + [1/(0.65 + 2.0)] + (1/23)} = 1.85 \text{ W/m}^2\,\text{°C}$$

which is about the same as the U-value of a triple pane window. Further improvements can be made by using gases within the space between the panes with a lower heat conductivity than that of air. Best values can be obtained with heavy inert gases like krypton.

Ventilation losses An air exchange of about n = 0.5/h of the volume of the living area is sufficient to fulfill the comfort and health conditions. The power which is needed to heat fresh air to the indoor temperatures is:

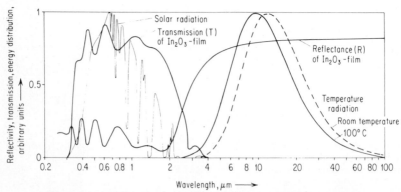

Fig. 29.7 Spectral reflectance of In_2O_3 for radiation loss control from windows.

$$Q_{ventil} = C\, n\, (T_i - T_{out})\ W/m^3$$
$$= 0.35 \times 0.5\, (20 - 2) = 3.15\ W/m^3$$

For a single-family house with a living volume of 300 m³ the corresponding value is

$$Q_{ventil,300}\ m^3 = 945\ W$$

or 22.7 kWh/day.

This minimum air exchange is due to leakage via doors and windows. In older houses uncontrolled air exchange rates of n ≈ 5/h have been reported. This rate is 10 times the minimum value required. The 227-kWh/day loss is equivalent to a heat transmission loss of the outer walls (300 m²) of the above mentioned single-family house of about 31 W/m² or a U-value of 1.75 W/m² °C. This is about the average U-value of a standard house. This figure demonstrates clearly the importance of this loss mechanism.

Air flow via leakage per meter of length of windows and doors is

$$Crack\ air\ flow = a\, (\Delta p)^{2/3}\ [m^3/(h)(m)]$$

where Δp is the difference in pressure, mm H_2O. Normal windows have a-values of about 1.5–3.0 and the best windows have values down to 0.1. The air flow per meter length is for a pressure difference of about 1.5 mm H_2O, which corresponds to average air velocity of about 5 m/s or 18 km/h:

$$Air\ flow = 3 \times 1.5^{2/3} \approx 4\ m^3/(h)(m)$$

For a 20-m window length the total air flow is 80 m³/h or 0.27 air exchanges per hour for a single-family house. This example indicates that very large gaps and poor construction are needed to explain air exchanges of n = 5. On the other hand, a very tight house with a-values of 0.1 needs a controlled air exchange to fulfill the minimum fresh air requirements.

Internally generated loads   Energy is dissipated by the activities of the inhabitants within a building. The internal load comes from persons living in the building and from the energy demand of the household appliances. The internal load depends strongly on the behavior of the inhabitants which is influenced by the standard of living and lifestyle of the occupants. Only few precise data are available on this point.

The "external load" is caused by solar radiation through windows. The solar energy transmitted through a window is mostly absorbed within the rooms. The transmission is influenced by the reflection and absorption losses of the panes. For a single-pane window the reflection losses are about 8 percent. Absorption losses are caused by impurities in the glass, mostly iron oxide, and correspond to about 5–10 percent for normal window glass. The transmission for a single pane is therefore about 87 percent and for a double-pane window about 75 percent. The effect of solar radiation through windows on the energy demand for space heating depends very much on the weather structure and the architectural design of the building. The effect is analyzed in more detail below and in Chap. 16 on passive solar systems. For a single-family house having 20 percent of the outer building surface as window area the effect of this internal load can represent about 30 percent of the space heating requirement in sunny climates.

Absorption of insolation on outer walls   The absorption of solar energy on the outer walls of a building increases the average wall temperature and so reduces the heating demand of a building. Assuming an absorption coefficient α between 0.2 (white) and 0.8 (dark) and an average daily insolation I of 500 W/m² the average temperature increase relative to a wall with $\alpha = 0$ will be

$$\Delta T_s = \frac{\alpha I}{h_0 + U}$$

where h_0 is the external heat transfer coefficient (10 W/m²C) for low wind and U is the U-value of the house wall (1 W/m² K).

The maximum average temperature rise for a dark wall is

$$\Delta T_s = \frac{0.8 \times 500}{11} \approx 36°C$$

For a white wall it is

$$\Delta T_s = \frac{0.2 \times 500}{11} \approx 9°C$$

As a consequence of this temperature increase of the outer walls of a building an additional energy flow to the interior is achieved. This is

$$C_i = U \cdot \Delta T_s$$

or at a maximum 36 W/m². For a single-family house a flat roof alone of 100 m² can therefore result in an additional load on a sunny day of about

$$36 \text{ W/m}^2 \times 100 \text{ m}^2 \times 10 \text{ h} \approx 36 \text{ kWh}$$

As long as the outside wall temperature is lower than the room temperature the so-defined external load reduces the thermal losses which are normally calculated on an air temperature basis. For a cooling situation, where the air temperature is above the room temperature this external load represents cooling requirement. For this example the maximum cooling load could be reduced from 36 to 9 kWh by using a white color. In summary, the example indicates that insolation on external walls may have a considerable influence on the heating and cooling requirement of buildings. More detailed information is given below.

Heating Requirement of a Single-Family House

The influence of different parameters on heating requirements is discussed for the example of a single-family house having a floor area of about 116 m² and a living volume of 290 m³. The window area is 20 m² which corresponds to 10 percent of the outer wall area. These dimensions are for the Philips Experimental House.

In the first case to be analyzed this house is built as an average house in Germany, which is called "Average House." The second case represents the same building with respect to the design but with thermal measures according to a well-insulated house. The third example represents the insulation and heat recovery measures as used in the Philips Experimental House in Aachen/Germany. This house is called "Experimental House." It nearly represents the economic limit for buildings in moderate climates like Central Europe.

Table 29.7 gives the results of the heating requirement of the different versions of the single-family house. The results were obtained from the heating degree method for 3670 degree C-days. A more detailed method based on a dynamic model using hourly weather data is given below. The results indicate that the heating requirement can be reduced from 49,600 kWh for the Average House to 25,300 kWh for the well-insulated house to 8300 kWh for the Experimental House.

CALCULATION OF ENERGY REQUIREMENTS OF BUILDINGS

Predicting the Energy Requirements of Buildings

The simulation of the thermal behavior and the energy requirements of buildings can be performed by either empirically based, simple design or analysis procedures such as the "degree-day" or "bin" methods or by more complex digital methods. The simple procedures currently used in most engineering practice are acceptable for some needs but, as energy conservation measures become more widespread, the simple empirical procedures will not be a good basis for essential energy calculations in buildings.

Digital simulations of two types have commonly been used to obtain information on the thermal performance of building structures. These are: (1) "First principle" methods[31,58,61] and (2) simplified methods.[7,33,51,52] The former approach is based on a variety of solution procedures such as finite element methods[61] or the use of heat balance equations along with conduction transfer functions.[32,39] These "first principle" methods are "exact" in both their description of the details of a building in terms of input construction data and in the results which one may obtain from their heat transfer theory algorithms. Computer programs based on these methods have been increasingly successful in determining the energy requirements of various buildings in different locations. However, such programs are essentially deterministic and once all the

TABLE 29.7 Annual Heat Requirements for a Single-Family House Calculated by the Degree-Day Method

Losses	Average house			Well-insulated house			Experimental house		
Type	U-value*	kWh/year	%	U-value*	kWh/year	%	U-value*	kWh/year	%
Walls, floor, ceiling	1.23	32,630	65.8	0.48	12,600	49.8	0.14	3630	43.7
Windows	5.8	9970	20.1	3.3	5700	22.5	1.5	2570	31.0
Leakage†		7000	14.1		7000	27.7		700	8.4
Controlled ventilation‡								1400	16.8
Total, kWh/year		49,600			25,300			8300	
Oil equivalent, L		6450			3300			1100	

*U-value, W/m^2 K.
†For leakage one air change per hour is taken.
‡For controlled ventilation one air change per hour with 80% heat recovery.
Note: Floor area, 116 m^2; window area, 23.5 m^2; living area volume, 290 m^3. Dimensions are for the Philips Experimental House.

relevant input data, concerning the material properties, construction details, weather and operating schedules (including internal loads), are defined the predicted performance is fixed. If the predicted performance does not agree with measured data[33] there is no obvious way to accommodate for the differences. To further demonstrate this point, it has been found[50] that identical houses under the same weather conditions can show a heat demand variation of a factor of 2. Only part of this difference can be related to differences in occupant behavior, i.e., internal loads and house usage patterns.

Apart from such problems as compensating for "real life" differences, if such first principle programs are used as predictive or design tools they have two other drawbacks:

1. They are time intensive in defining, entering, and checking the detailed input data as well as sorting out and checking the relevant results.[6]

2. They require large computer facilities and can have long run times.

Of the latter approach, most of the simplified methods are easy to program, fast to use, and require relatively little computer run time. Some of the methods utilized to date involve the use of equivalent thermal parameters[33,52] in a single or several node heat capacity model of a building. A similar approach although more exact in its physical nature and solution method is a network approach[7] containing all the basic loss and gain mechanisms and essential physical capacitive effects (see Fig. 29.8). The Fourier transformation methodology[51] is a third simplified method which has found widespread use.[16] However, it is difficult to extend this approach to consider the effects of all heat and mass transfer processes, such as occupation, internal loads, ventilation, and gains-losses of windows occurring within the building shell.

Simplified methods can be used in three ways. Firstly, by using measured data on the performance of a building over short time intervals, such as one or several days, one can define the effective thermal parameters.[33,52] These can then be used to simulate the performance of that building under various weather conditions and over longer time periods in order to identify its requirements and thermal behavior in normal and under critical situations. Furthermore, these parameters, found in this way, can be used to define the thermal quality of any particular building. They thus give direct insight to the thermal weak points of a building and indicate how to improve its thermal performance. The second way of using simplified methods is to apply the results of a "First Principles" approach as the above mentioned "measured data" over short time periods. Finally, in the case of a network model, one can combine the effect of known material properties together in a "static sense" and directly define the effective thermal parameters. These parameters, which relate to losses, gains or capacitive effects, are found from the material properties and construction details of a building and are then directly usable in the simplified model. In either case of defining the constants for a simplified method, the work involved is usually far less than that of setting up the input data for "first principle" models. In the next section these two approaches are compared.

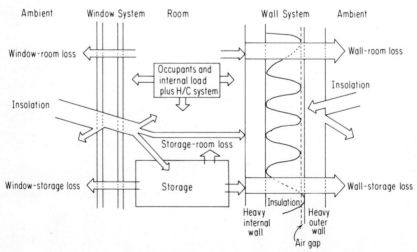

Fig. 29.8 Building thermal network model.

Comparison of a First Principles and Simplified Model

There are several computer packages[6,32,61] which can be used as a "first principle" model to compare with any of the simplified methods discussed above. Here we will restrict ourselves to considering the National Bureau of Standards model NBSLD[32] as the reference "first principles" approach and a network type model[7,13] for the simplified approach. The latter choice was made as it is felt that future simple models need not only be models which consider dynamic effects but also have the possibility to be extended to calculations of buildings with several rooms and /or floors. A network approach is an adequate way of doing this. The simplified model which will be considered here is an extension of the model used by Balcomb et al.[7] with the inclusion of internal loads, occupancy schedules, and maximum power output for the heating and cooling system (see Fig. 29.8). This model can be used in a single internal capacity and multiwall approach (i.e., if the building can be considered as one effective room) or as a multi-internal capacity and wall arrangement for a building having a distinct (e.g., individual temperature settings) set of rooms.

The building design and detailed layout which is taken as a basis for this comparison is a standard building type defined by the NBS for the International Energy Agency (IEA) study group.[35] In simple terms the building is characterized by a single zone house with the dimensions and static physical properties given in Table 29.8. The comparisons given here were calculated for Hamburg (1973), Federal Republic of Germany. The reason for choosing Hamburg as a location was twofold. Firstly, Hamburg has a cold enough climate over a good period of the year and so a reasonably high heating requirement for Central Europe. Secondly, during the heating season on a sunny day the beam component is almost perpendicular to the south-facing windows at noon: This results in a very sensitive energy requirement as a function of time and thus a good test of the programs. Normally, no cooling is required in Hamburg as the typical German house has a high internal heat capacity.

However, the IEA house is a typical light American house with little internal heat capacity but rather good insulation. Thus, as the cooling set temperature is 23.9°C, a small and sensitive cooling requirement is also experienced in Hamburg. This also provides a good test for these programs.

Before considering the differences in the results obtained by these programs it is useful to get a feeling of the size of the dominant weather parameters. Figure 29.9 gives the insolation incident on a 1-m² surface area facing S, W, N, and E, respectively in Hamburg (latitude 53° N) 1973. These results indicate the effect of possible insolation absorption of walls and transmission through windows. Figure 29.10, on the other hand, gives the average monthly temperatures for Hamburg 1973 which drive the heat loss mechanism. The bars in Fig. 29.10 indicate the root mean square (± 1 standard deviation) variation of the daily temperatures which occur within any month. This result gives a measure of how valid it is to use the average as indicative of the loss effects in a month.

Using the detailed IEA-house data, given in Ref. 13 and summarized in Table 29.8, both methods were used to calculate the heating and cooling requirements.* NBSLD did this by calculating the load over each hour for the modified network model by calculating the necessary energy input to the house via a heating/cooling system sized for the coldest/hottest period and coupled to a thermostat with a ±0.5°C hysteresis. The results of these calculations are presented in Fig. 29.11 for heating and Fig. 29.12 for cooling. It is observed *in both cases that the difference in the yearly heating or cooling requirement found by the first principles or simplified method differed by less than 1 percent.* However, on a month-by-month basis the variations can reach almost 8 percent. These larger mean deviations tend to occur mainly for months having lower heating and cooling requirements. Thus, the total energy effect of these errors over the year is negligible.

In order to get a feeling as to what the typical distribution of errors looks like in comparing these models it is necessary to consider the next highest moment, i.e., the root-mean-square deviation between the results found by either of these methods. This number gives an approximate value of the spread of the day-to-day error distribution function from the mean error found. Thus, the ± 1 standard deviation errors about the mean indicates the errors found for about 70 percent of the cases observed.

---

*Requirement is used here to define the net energy delivered or taken away from the house by the HVAC system without distribution or "furnace" losses.

TABLE 29.8 Dimensions and Static Physical Properties of the Reference House

Surface	Length, m	Width, m	Total heat transmission coefficient, W/m²K	Surface area, m²	Wall absorptance, α	U-value glass, W/m²K	Window area, m²	Transmittance glass, τ	Door heat transmission coefficient, W/m²K	Door area, m²	Door absorptance, α
South	12.53	4.27	0.374	51.8	0.8	2.62	1.67	0.8			
West	5.87	4.27	0.374	16.9	0.8	2.62	8.18	0.8			
North	12.53	4.27	0.374	51.1	0.8	2.62	2.42	0.8	1.99	1.12	0.8
East	5.87	4.27	0.374	19	0.8	2.62	4.95	0.8			
Roof	12.53	5.87	0.258	73.6	0.87						
Floor	12.53	5.87	0.158	73.6							

Note: Temperature setting heating: 21.1°C, temperature setting cooling: 23.9°C, maximum lighting load: 6 W/m² maximum equipment load: 18.9 W/m², maximum number of occupants: 5.5 persons, total internal load: 24.8 kWh/day (exclusive of solar gain).

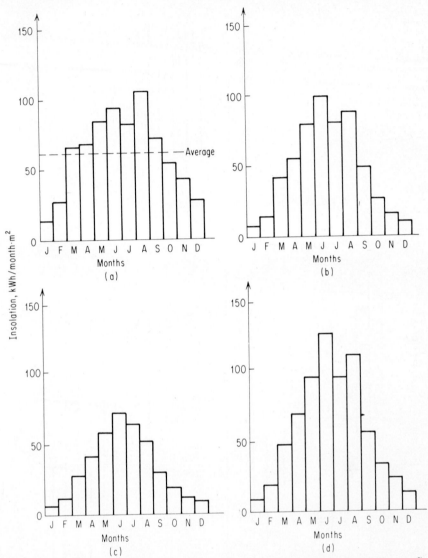

Insolation, kWh/month·m²

Months
(a)

Months
(b)

Months
(c)

Months
(d)

Fig. 29.9 (a) Monthly insolation on a 1-m² area facing south in Hamburg 1973 (total = kWh/year · m²); (b) monthly insolation on 1-m² area facing west in Hamburg 1973 (total = 565 kWh/year · m²); (c) monthly insolation on 1-m² area facing north in Hamburg 1973 (total = 402 kWh/year · m²); (d) monthly insolation on 1-m² area facing east in Hamburg 1973 (total = 689 kWh/year · m²).

The root-mean-square deviation is given in energy terms in Figs. 29.13 and 29.14 for both heating and cooling, respectively. With the exception of first and last months in the heating/cooling season, where not all days are heating/cooling days and so have a wide range at low ·demand, it is seen that the root-mean-square deviation is never more than 10 percent of the average daily heating and 15 percent of the average daily cooling requirement (i.e., 70 percent of all cases lie in this range). The errors between these two models which give rise to this random day-to-day variation in results can be attributed to differences in the algorithms which calculate the effect of:

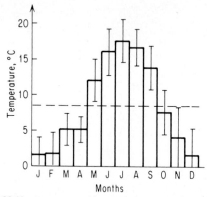

Fig. 29.10 Average monthly temperatures in Hamburg 1973.

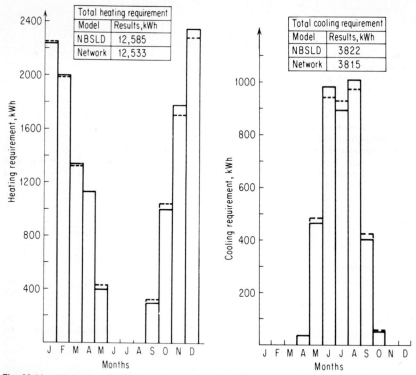

Fig. 29.11 Comparison of NBSLD (−) with network model (−−−) heating Hamburg 1973.

Fig. 29.12 Comparison of NBSLD (−) with network model (−−−) cooling Hamburg 1973.

a. The on/off switching of the heating system (not present in NBSLD).
b. Uncontrolled ventilation losses.
c. Solar gains through windows.

The priority of these reasons is a, b, c for heating and a, c, b for cooling. In both cases the principle effect comes from a. Summed over several days, this results in a negligible error, but when looking on a day-to-day basis a positive or negative shift of the heating/cooling requirement can result at the end of one day to the next day because of the two-point thermostat and

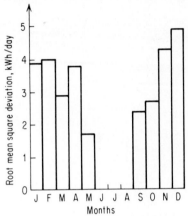

Fig. 29.13 Root-mean-square deviation for network model with respect to NBSLD heating Hamburg 1973.

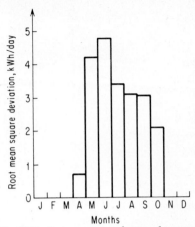

Fig. 29.14 Root-mean-square deviation for network model with respect to NBSLD cooling Hamburg 1973.

maximum power rating heater used. This daily error can be sizable. For uncontrolled ventilation losses the network approach assumes a constant wind velocity of 4.2 m/s. The actual velocity distribution is such for Hamburg 1973 that a standard deviation of 0.85 m/s exists about the mean of 4.2 m/s. This 20 percent possible difference, which is randomly distributed, results in at most a ±7 percent effect[58] on the losses. A rough estimate for these effects indicates that the day-to-day standard deviation may be decreased by almost a factor of 3 while the mean deviation is almost unchanged.

This comparison indicates that the network model results give a detailed account of the *dynamic* behavior of a building which differs by only a few percent of that found from a first principles model. Furthermore, as already shown[33] when comparing such a simple model or first principle model with experiment, the least-squares evaluation of the simple model's input parameters affords a better experiment-theory match than found by using the detailed model. Finally, a detailed model such as the NBSLD requires almost two orders of magnitude more computer time for a heating/cooling calculation than the modified network model considered here.

Because of the good agreement which has been found between these two models all the results used in the discussion below were found using the modified network model.

Effect of Climate Types

The IEA house is representative of a typical American single-family dwelling. After seeing that the two load models give almost identical results for both the heating and cooling requirement under the sensitive conditions prevalent in Central Europe, it is instructive to see how this house reacts under different climate conditions. The climate types chosen for this comparison are those given by the IEA.[35] They are:

Climate type	City
Continental cold	Madison, WI
Desert	Santa Maria, CA
Semitropical	Tokyo, Japan

The yearly results for both heating and cooling in these climates are given in Fig. 29.15 for the IEA house. The general tendency observed here is that in going from latitude 53.6° (Hamburg) to 34.6° (Santa Maria) the heating requirement decreases sharply and the cooling increases correspondingly such that heating dominates cooling at the one end and vice versa. It is also noted that the Central European climate has a milder winter than the more southerly (latitude 43.1°) located Madison with a continental cold climate. However, the continental climate experienced in Madison results, in the summer, in about the same cooling requirement as the warmer desert and subtropical climates of Santa Maria and Tokyo.

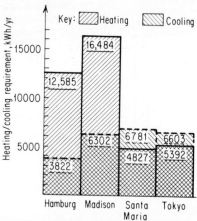

Fig. 29.15 Yearly heating and cooling requirement of IEA house in various climatic regions.

Detailed month-by-month heating and cooling requirements have periods which are almost distinct. The one exception is Santa Maria (Fig. 29.16). Here, each month has both a heating and cooling demand. This can principally be related to the fact, already mentioned above in connection with the cooling requirement for Hamburg (Fig. 29.11 and 29.12), that the IEA house has a low effective internal heat capacity. For a climate like that of Santa Maria a leveling of the ambient day-night temperature fluctuations and the high solar gains via windows during the day does not occur. An increase of the effective internal heat capacity would, for Santa Maria, produce a strong decrease of both the total heating and total cooling requirement. A similar capacity increase for a climate with district heating and cooling systems would only have a negligible effect.

A Central European House: Heating Requirement under Different Building Codes

Before looking in detail at the effect of all the measures one can introduce for energy conservation in buildings, a brief discussion is given on the heating requirement of one house built under three different building codes in Central Europe. The physical properties of these houses as well as other thermal characteristics such as internal loads are summarized in Table 29.9. The outer dimensions of this single-family dwelling are those of the IEA house while the internal

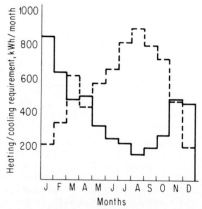

Fig. 29.16 Monthly heating (—) and cooling (−−−) requirement of IEA house in Santa Maria, CA.

TABLE 29.9   Building and Operations Parameters for Three Basic House Types

	Parameter values for basic house types		
Building parameters	Normal house	Swedish standards house	Experimental house
House dimensions L/W/H		12.50 m/5.88 m/4.27 m	
Window areas N/E/S/W		2m²/5m²/8m²/5m²	
Door area		1.11 m²	
Heat loss values, W/m²°C			
U-walls	1.12	0.37	0.17
U-roof	0.778	0.257	0.17
U-cellar	0.789	0.187	0.17
U-door		2.0	
U-windows (day-night)	5.8/5.8	2.242/1.5	1.9/1.2
Transmission coefficient	0.9	0.74	0.7
(Internal) capacity, C_H	40 kWh/°C	2.424 kWh/°C	7 kWh/°C
Effective air exchange per hour	1.58	0.87	0.3
Total heat transmission coeficient, W/°C	549.2	227.7/212.9	119.4/105.4
Relaxation time, h	72.8	10.7/11.4	58.6/66.4

Operations parameters	Normal house	Swedish standards house	Experimental house
Heating temperature		20°C	
Effective hysteresis		0.5°C	
Hysteresis used in program	0.14°C	0.5°C	0.14°C
Cooling temperature		26°C	
Internal load, (Wh, 0h—24 h)	646, 571, 565,	536, 538, 483,	749, 1566,
	1744, 1563, 1263,	1227, 2128, 465,	465, 465,
	813, 1305, 1552,	1773, 1465, 1364,	1422, 677
Cellar temperature, °C, Jan.-Dec.	12, 13, 14,	15, 16, 17, 18,	17, 16, 15, 14, 13,

loads are a slightly modified version of the IEA internal loads. The three codes taken for these houses as a basis for comparison typify:
1. The normal heavy German house construction given in DIN 4108 (Normal House).
2. The new 1978 Swedish building standards (Swedish Standards House).
3. The building standards used in the Philips Experimental House as given in Ref. 12 (Experimental House) which is seen as the near economic limit.

The yearly heating demand of the three house types is given for 3 years in Hamburg as well as three locations in the Federal Republic of Germany in Fig. 29.17. It is found that the heating requirement of the Normal house exceeds the heating requirement of the Swedish standards house by a factor of ≈ 4, while the heating requirement of the Experimental house is between a factor of 35 and 50 less than that of the Normal house and depends greatly on location and year. Clearly the differences in the thermal behavior of the three house types are related to their different insulating standards and the effect of energy sources such as the internal loads and insolation through windows. Taking the ratio of the average heat transfer coefficients (including ventilation effects) given by Table 29.9 one obtains the numbers 5:2:1 for the Normal to Swedish to Experimental house, respectively. If one neglects the error made by taking the arithmetic mean of the day and night values for these heat transfer coefficients then the ratio presented above is an exact measure of the ratio of the gross energy requirement of these houses. This point is observed by looking at the results presented in row 1 of Table 29.10. It should be noted that it is assumed here that the indoor and outdoor temperatures are spatially constant for any of the surfaces considered. The heating requirement, in the form of the energy supplied by the heating system to an actual house, only accounts for part of the total energy flux. Therefore, the heating requirements provided by fuel must form a larger ratio (see Table 29.10) when compared to one another than the gross energy requirements or total heat transfer coefficients.

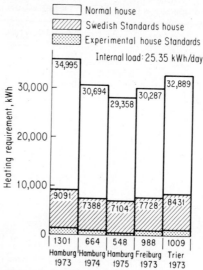

Fig. 29.17   Yearly heating requirement for various houses.

TABLE 29.10   Yearly Heating Requirement and its Dependence on Heat Sources for the Basic House Types (Hamburg, 1973)

	Solar radiation* (via windows)	Internal load*	Open windows*	Heating requirement, kWh/year		
				Normal house	Swedish Standards house	Experimental house
1	−	−	−	51,767	20,517	10,375
2	+	−	−	42,556	15,232	5539
3	−	+	−	42,919	12,409	3152
4	−	−	+	50,734	20,374	10,051
5	+	+	−	35,034	9091	1301
6	+	−	+	42,332	15,232	5539
7	−	+	+	42,375	12,406	3151
8	+	+	+	34,995	9091	1301

*+ : "on" and − : "off".

The effect of changing weather conditions on the heating requirement can be estimated by comparing the yearly heating requirements for five reference years. As Fig. 29.17 shows the maximum variation in the yearly heating requirement in Central Europe is 6000 kWh for the Normal house, 1900kWh for the Swedish Standards house, and less than 800 kWh for the Experimental house. As is to be expected on physical grounds these differences show a closer relation to the total heat transfer coefficients than the heating requirements themselves. This may be a good starting point for a simplified computational approach.

For a more detailed analysis it is necessary to consider the monthly results and to examine the gains and losses of these houses in terms of an energy balance. First, the effect of insolation through windows, internal loads, and contact with ambient air via window opening are considered. The total effect of these sources on the heating requirement can be obtained by switching off all of these in the model and comparing the results obtained with the original ones. The individual contribution of any source can be estimated by computing the heating requirement with a particular source "switched off" or "on" when all others are on or off, respectively. Table 29.10 shows the results of these calculations for the cloudy climate of Hamburg.

The first row of Table 29.10 gives the gross energy requirements of the three basic house types with all sources "off". Since the cooling losses for this case are zero, these energy requirements must be equal to the gross heating requirements. Comparison of row 1 of Table 29.10 with the real physical values (row 8) indicates the total effect of the additional sources. This shows that the Normal house, due to its high basic needs and long heating season, makes the best use of solar and internal gains. From the given 20,000 kWh (see Table 29.11) more than 80 percent is used to cut down on the gross requirement for heating. The Swedish Standards house utilizes 63 percent of the 18,000 kWh available (see Table 29.12) and the Experimental house uses only 55 percent of the available 17,500 kWh (see Table 29.13) for reducing the gross energy requirement for heating. As row 3 of Table 29.10 shows, the internal load is used more effectively for both the Swedish and Experimental house than the radiation through windows (row 2), whereas in the Normal house both of these sources give almost equal contributions. In almost all cases the contribution of the ambient air is less than 2 percent.

Tables 29.11–29.13 show that the seasonal variations of insolation are responsible for the relatively small effect of insolation as a year-long source especially in well-insulated houses which have a shorter heating season. Tables 29.12 and 29.13 indicate that at the times where the Swedish Standards house and Experimental house have a substantial heating requirement, the internal load is comparable or higher than the amount of radiation entering through the windows. On the other hand, at times where the heating demand is low, neither insolation nor internal loads can be fully made use of. This is evident from the column titled "heat not used". It should be noted that this column does not contain the additional conductive losses due to an increased $(T_H > \overline{T}_h)$ indoor temperature.

Most of these additional energy losses are found in the first and third columns of the tables under the title "Losses".

The heating period is an important characteristic of a house especially if considerations are made in using alternative energy for the heating system. If the heating periods of the three basic house types are compared, it is evident that the heating period of the Swedish Standards house is longer than one may expect on grounds of its good insulation. The reason for this behavior lies in the fact that the specific heat capacity of 2.4 kWh/°C is too small to store enough energy for a cool night even in the summer. In other words, the Swedish Standards house is thermally too sensitive to day-night variations in temperature. This is typical of most single-family dwellings in the United States also.

A direct measure for the thermal sensitivity of a building is given by the quotient of its "internal heat capacity" to its total heat transfer coefficient. This quotient assumes the role of a first-order effective relaxation time τ in Eq. (29.1) which governs the decay of the indoor temperature if the outdoor temperature is kept constant and if all internal heating is turned "off":

$$T_{\text{indoor}}(t) \approx (T_{\text{indoor}}(0) - T_{\text{amb}})\, e^{-t/\tau} + T_{\text{amb}} \tag{29.1}$$

In other terms, τ is the time necessary for a $1/e$ reduction of the initial temperature difference. The relaxation time τ for the Swedish Standards house is 11 h (Table 29.9) which confirms the qualitative explanation given above. Two other quantities are of interest for the heating system of any house type. The first is the peak design heating load of the house type. The values of these design peak heating power loads are 16 kW for the Normal, 6 kW for the Swedish Standards and 2 kW for the Experimental house for Hamburg. These also reflect typical values in Central Europe. The second quantity of interest is the average heating power demand necessary for each house type. This value defines the lower bound for a heating system size. Average demands over the heating hours are 6 kW, 1.9 kW, and 0.5 kW for the Normal, Swedish Standards, and Experimental house. Thus the upper and lower bounds of a heating system may vary by a factor of more than 2. A good design may approach the lower bound.

In the above discussion the effects of different energy conservation codes have been shown in some detail. Below the discussion will be continued by looking at the effect of individual measures.

THE MEASURES AND THEIR INFLUENCE ON THE ENERGY DEMAND OF BUILDINGS

In this section the influence of operating conditions and design parameters on the heating requirement of the three building types discussed above will be examined in detail.

TABLE 29.11 Energy Balance of Normal House (Hamburg 1973)

	Gains								Losses			
MO	DIRRAD	DIFRAD	INTLD	CND/AIR	CND/WDW	OPEN WDW	HEATING	TOTGAIN	CND/AIR	CND/WDW	LS/CLR	CLG/URG
1	78	114	786	0	0	0	6104	7081	5153	1587	346	0
2	168	213	710	0	0	0	5259	6349	4637	1428	273	0
3	414	521	786	0	0	0	3975	5696	4155	1280	261	0
4	303	757	760	0	0	0	3641	5460	4016	1237	210	0
5	378	937	786	9	3	45	1211	3370	2385	735	192	0
6	442	1012	760	33	10	77	178	2513	1599	493	204	0
7	277	1054	786	42	13	112	34	2319	1542	475	212	103
8	711	809	786	45	14	213	60	2637	1831	564	261	164
9	506	505	760	17	5	87	778	2658	1921	592	205	18
10	406	322	786	0	0	0	3248	4762	3479	1072	217	0
11	362	180	760	0	0	0	4588	5890	4313	1329	251	0
12	230	111	786	0	0	0	5919	7046	5150	1586	303	0
	4275	6534	9251	146	45	535	34,995	55,780	40,182	12,378	2934	285

Note: MO = Month, DIRRAD = Direct Radiation through Windows, DIFRAD = Diffuse Radiation through Windows, INTLD = Internal Load, CND/AIR = Conduction through Walls, Air Exchange, CND/WDW = Conduction through Windows, OPEN WDW = Gains through open Windows, TOTGAIN = Total Gain (= Total Loss), LS/ CLR = Conductive Losses via Cellar (or Ground), CLG/URG = Heat not Used (Cooling, Shading).

TABLE 29.12 Energy Balance of Swedish Standards House (Hamburg, 1973)

	Gains								Losses			
MO	DIRRAD	DIFRAD	INTLD	CND/AIR	CND/WDW	OPEN WDW	HEATING	TOTGAIN	CND/AIR	CND/WDW	LS/CLR	CLG/URG
1	64	93	786	0	0	0	1834	2777	2217	477	83	0
2	138	175	710	0	0	0	1495	2518	2006	446	66	0
3	341	428	786	0	0	0	944	2498	1897	431	72	99
4	249	622	760	0	0	0	799	2430	1881	445	63	32
5	311	771	786	1	0	0	210	2078	1344	323	73	333
6	363	832	760	2	1	0	37	1994	948	229	72	745
7	228	867	786	1	0	0	3	1886	867	209	69	743
8	585	665	786	0	0	0	31	2068	918	211	75	867
9	416	415	760	0	0	0	150	1742	1054	239	69	389
10	334	265	786	0	0	0	713	2097	1610	359	62	67
11	298	148	760	0	0	0	1174	2380	1894	411	63	9
12	189	92	786	0	1	0	1703	2769	2223	475	73	0
	3515	5372	9251	4	1	0	9091	27,235	18,859	4254	840	3283

Definition of Headings: Cf. Table 29.11.

TABLE 29.13 Energy Balance of Experimental House (Hamburg, 1973)

	Gains								Losses			
MO	DIRRAD	DIFRAD	INTLD	CND/AIR	CND/WDW	OPEN WDW	HEATING	TOTGAIN	CND/AIR	CND/WDW	LS/CLR	CLG/URG
1	61	88	786	0	0	0	478	1412	946	391	75	0
2	130	166	710	0	0	0	292	1298	865	371	62	0
3	322	405	786	0	0	0	43	1556	909	399	84	127
4	235	588	760	0	0	0	0	1584	950	435	84	111
5	294	729	786	1	1	0	0	1809	698	327	89	694
6	344	787	760	1	1	0	0	1893	487	230	80	1096
7	216	820	786	2	1	0	0	1825	431	203	74	1117
8	553	629	786	1	0	0	0	1969	472	211	82	1203
9	394	392	760	0	0	0	0	1547	583	256	87	626
10	316	251	786	0	0	0	6	1358	825	353	81	98
11	282	140	760	0	0	0	109	1291	870	362	69	20
12	179	87	786	0	0	0	373	1423	956	392	68	0
	3325	5082	9251	5	3	0	1301	18,966	8993	3931	936	5091

Definition of Headings: Cf. Table 29.11.

Internal Load Profile and Magnitude

The basic internal load profile (Table 29.9) is assumed for comparative purposes to have the same set of hourly values for each day of the year. For this reason three classes of variations should be studied:

    a. Variations of the daily profile shape.
    b. Variations of the weekly profile shape.
    c. Variations of the average internal load intensity.

An idea of the importance of profile shape is obtained by comparing the heating requirement for a *constant profile* with that resulting from the original profile. As Table 29.14 shows a profile distortion of this kind results in a 0.1 percent (46 kWh/year) change in the heating requirement of the Normal house, a 1 percent (101 kWh/year) change for the Swedish Standards house, and a 2 percent (22 kWh/year) change for the Experimental house. As is to be expected the absolute effect of profile changes is greatest for the Swedish Standards house due to its low effective internal heat capacity and small relaxation time. For this type of house more pronounced distortions of the profile shape can lead to considerably larger changes in the heating requirement especially if the hourly load exceeds the specific heat storage capacity. It can also be shown that profile changes on a weekly time scale (different weekday and weekend profiles) have only small effects if the total annual internal load is fixed. The weekly profiles always give a higher heating requirement than the daily normal profile.

TABLE 29.14 Yearly Heating Requirement as a Function of Internal Load Profile Constant Hourly Profiles—Variation of Profile Intensity

	Heating requirement, kWh/year		
Internal load per hour	Normal house	Swedish Standards' house	Experimental house
0 W (0 kWh/year)	42,332	15,232	5539
528 W (4625 kWh/year)	38,501	11,551	3067
1056 W (9251 kWh/year)	34,949	8890	1279
2112 W (18,501 kWh/year)	28,390	4044	6
Basic profile (9251 kWh/year) (see Table 29.9)	34,995	9091	1301

Heating Temperature Set Point

The thermal performance of a building is mainly determined by its conductive and ventilation losses, which are proportional to the difference between ambient air and indoor temperatures; therefore, the heating set point has a pronounced effect on the heating requirement. It turns out that at a heating set point temperature (see Fig. 29.18) of 26°C the yearly heating requirement of the Normal house, Swedish Standards house, and Experimental house increases by 3700 kWh/°C year (11 percent of total), 1200 kWh/°C year (13 percent of total), and 300 kWh/°C year (24 percent of total), respectively. Thus, from an energy conservation point of view it is desirable to keep as low as possible a yearly heating set point temperature. However, a reduction of the daily heating set point temperature may affect the desired level of comfort.[60] Thus, one may consider the effect of only day-night variations in the set point which do not affect the comfort level. The effects of such a day-night setting are shown in Table 29.15. It is observed that only small gains (<2.5 percent) can be made for both the Normal and Experimental house as they both have comparatively large relaxation times. However, for the Swedish Standards house with its small relaxation time a gain of over 12 percent is attained.*

Although, for comparison in most cases considered here the indoor air temperature set point was considered as 20°C, as far as comfort requirements are concerned, it should be noted that the Normal house due to its poor insulation requires a higher set point temperature to coun-

---

*In stating this, we have not included the effects such as parasitic losses of the heating system should it be on at night. These losses occur due to the on-off heating of the boiler, the capacitive losses of the pipes, and the increased conduction about the wall where the radiators are. All these effects easily increase the effect of a day-night setting for these houses.

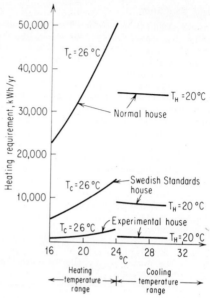

Fig. 29.18 Yearly heating requirements as function of heating and cooling set point temperatures, Hamburg, 1973. The effect of cooling set point is negligible.

TABLE 29.15   Effect of Night Setback on Yearly Heating Requirement of Basic House Types (Hamburg, 1973)

	Heating requirement, kWh/year		
Heating temperature, 22:00–7:00 h	Normal house	Swedish Standards house	Experimental house
20°C	34,995	9091	1301
16°C	34,140	7991	1286

terbalance the lower radiation temperature of the walls and windows. To reach the comfort conditions prevailing in the Swedish Standards and Experimental house at a heating temperature of 20°C, the heating temperature of the Normal house should be raised by about 1°C* on an average. This is equivalent to an additional heating requirement of about 4000 kWh/year.

Cooling Temperature Set Point

The cooling set point gives an upper limit for the indoor temperatures. Therefore, for the use of the specific heat as a storage capacity it may well be that this set point also has an effect on the heating requirement of a house. From Fig. 29.18, however, it is seen that in general for the temperature range 24 to 30°C which is equivalent to a variation of usable storage capacity by a factor of 2, the impact of this set point is negligible. Again, only the yearly heating requirement of the Swedish Standards house is noticeably affected with a rate of change of -100 kWh/°C year at $T_c = 26$°C.

These different responses of the three basic house types may be easily understood in terms of temperature relaxations during the night. Putting $T_{indoor}(o) = T_c = 24$°C at the beginning of the night, Eq. (29.1) indicates that for the Swedish Standards house $T_{indoor}(t)$ remains above the set point of the heating system (19.5°C) for a time span of 12 h only if $T_{amb} \geq 16.5$°C ($T_{amb} \geq 16.5$°C for 1400 h/year in Hamburg, 1973); whereas, an ambient temperature above -3 and

---

*This is based on a static calculation with an equal distribution of external and internal wall areas.

$-5°C$ ($\sim$8600 h/year in Hamburg, 1973) is sufficient for fulfilling this requirement for the Experimental house and the Normal house, respectively.

For $T_{indoor}(0) = T_c = 30°C$, $\overline{T}_{amb}$ may take the value of 12°C ($\sim$2900 h/year in Hamburg, 1973) for the Swedish Standards house, whereas for the Experimental house and the Normal house $\overline{T}_{amb}$ is allowed to fall below $-30°C$ ($T_{amb} \gtrsim -30°C$ for $\sim$8760 h/year) without infringing on the comfort requirements within a time span of 12 h.

This shows that by raising the set point of the cooling temperature, the Swedish Standards house may bridge over more nights without drawing energy from the heating system, whereas the night performance of the Experimental house and the Normal house is hardly affected at all. Variations of the building parameters and their effect on energy demand are given below.

Area and Type of Window

A parameter which is often expected to reduce the heat requirement is increased south-facing window area. The effect of this parameter will be considered for the three house types. A similar study has already been done for the United States by Dean and Rosenfeld as well as others.[3,15,17]

To get a feeling for the thermal effect of other windows, it is of interest to start off with a building having no windows at all and then adding, one by one, the windows to the north, east, and west facing surfaces up to an area of 2 m², 5 m² and 5 m², respectively. The area of the south facing window is then increased from 0 until the whole south wall is covered. The results of this parameter variation are shown in Fig. 29.19. Here, the results are presented for the three house types for Hamburg, 1973 and their normally fitted window and shutter types. For the Normal house, moreover, a variation of window and shutter quality is also made.

From Fig. 29.19 it is seen that for the Normal house with its normal windows (i.e., single glazing, no shutters) there is only a net loss with increasing window area. However, as soon as good shutters are introduced which decrease the night-time losses without affecting the night time wall U-value only energy gains are obtained with increasing window area. It is seen that the effect of these shutters on the Normal house with the normal window dimensions (Table 29.9) is a 4000 kWh/year decrease in the heat requirement. If one chooses to use better windows such as thermopane and/or special insulating shutters then a saving of up to 14,000 kWh/year may be realized if the whole south facing area is filled with windows. However, such a measure is not a practical one as it would, even for the Normal house, drastically affect the comfort level of the occupants. This is so since the temperature felt by the occupants is the mean of the radiative and air temperature. As the window area is greatly increased then under no sun, clear sky, and cold ambient temperatures the occupants will have the sensation of being too cold, whereas under the same conditions with direct sunlight they will have the sensation of being too hot.

The result on the two well-insulated house types to an increase of the south-facing window area is quite different. One, the Swedish Standards house, always has a thermal optimum whereas the other, the Experimental house, does not (excluding 0 m²). The reason for this behavior in the Experimental house is that the walls are so good that even the best windows with their reduced U-values but lower transmittance cannot substitute during the day for the losses incurred at night in this house over its heating season relative to the displaced insulated wall area. The Swedish Standards house operates over almost a double-length heating season vs. the Experimental house and so there are a set of days in its heating year where for each unit area the gained energy over the day can be stored and transferred to compensate for losses at night. There is, however, another set of days where no gain is made over the day or day-night. For the former set as the window area is increased a saturation occurs in that the house has only a finite (small) heat storing capacity; at this point each additional square meter of window put on the south wall just adds to the night-time losses without transferring any net gain. This flat thermal minimum for the Swedish Standards house indicates a net possible saving of the order of only a few hundred kilowatt-hours.

In all cases considered for the Swedish Standards house in the Federal Republic of Germany it was found that the minimum lies between about 8–12 m² of window area which, in fact, is very close to the DIN 4108 window area regulations for this house size.

Insulation

A comparison between the Normal house, Swedish Standards house, and Experimental house has shown the considerable effect that well-insulated walls can have on the heating requirement. In order to get a more precise picture calculations were made for the heating requirement of the Normal house as a function of its additional insulation thickness for an insulation conductivity of 0.04 W/m²°C. Figure 29.20 shows the results for the Normal house and indicates

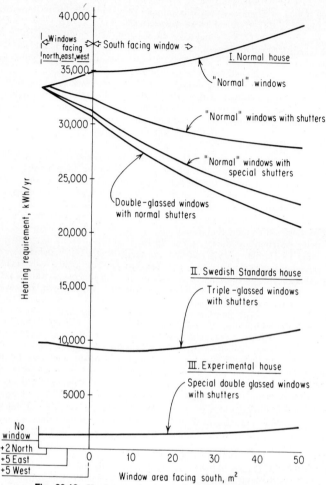

Fig. 29.19 Heating requirement as function of window area.

that a substantial reduction in the heating requirement is achieved by the first few centimeters. For example, 1 cm additional insulation results in a 4000-kWh/year saving, whereas 5 cm gives a saving of 11,000 kWh/year. This is due to the reciprocal dependence of the losses on the insulation thickness. It is seen that the asymptote for insulation effects lies at about 16,000 kWh/year or 19,000 kWh/year. This is a result of the poor windows and infiltration losses as well as cellar losses which are not affected by added insulation. The effect of increasing insulation is much less for the Swedish Standards house and Experimental house. A 5-cm increase in insulation has a ~2300-kWh/year effect on the heating requirement of the Swedish Standards and a ~700-kWh/year effect on the Experimental house.

It should be noted that variation of insulation in the above model is also equivalent to variation of tightness as both directly affect the total heat transfer coefficient. As far as economic aspects are concerned more detailed economic evaluation procedures have been developed by Fisk[25] and Ruegg.[48]

Absorbing Walls

As another heat-saving measure one may consider the effect of different color paints—covering the east, south, and west walls—on absorbing solar radiation and thus reducing the heat requirement by increasing the outside wall temperature. Figure 29.21 shows the results found

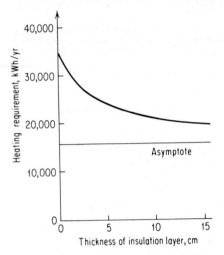

Fig. 29.20 Yearly heating requirement as function of wall insulation, normal house, Hamburg, 1973.

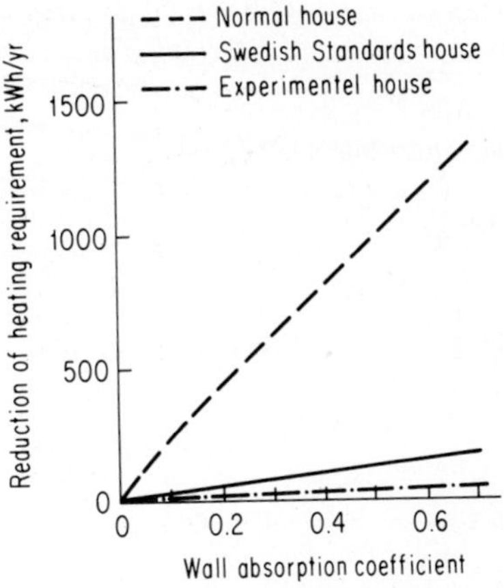

Fig. 29.21 Reduction of yearly heating requirement as function of solar absorbtion of east, south, and west facing walls, Hamburg, 1973.

for a wall with an absorption coefficient between 0 and .7 in terms of the energy reduction for heating. Because of the good insulating standards of the Swedish Standards and Experimental house and since h_{air} is so high (19-25 W/m^{2}K), a negligible effect is observed. However, for the Normal house up to 1500 kWh/year ($\sim$4 percent) can be saved. The only way such results can be obtained for the Swedish Standards and Experimental house under Central European climatic conditions is by a Trombe wall approach [2,7,59] where the h_{air} is effectively decreased by putting a glass layer in front of the painted wall area and combining it with a storage. In doing so a several percent decrease may be realized; however, a penalty of higher mean temperatures need be paid in the summer months.

Specific Heat as Storage in Walls

Due to its storage function the specific heat capacity influences the heating requirement.[51] Here only the Swedish Standards house is considered. As Fig. 29.22 shows, the heating requirement of the Swedish Standards house is not too strongly dependent on the capacity size. A reduction of about 1000 kWh (11 percent) is achieved by increasing the capacity of the Swedish Standards house from its normal value of 2.4 to 10 kWh/°C or from a ⅓-day capacity to a daily storage capacity effect; there is only a negligible change observed until one reaches the level of a week's storage capacity (i.e., about 50 kWh/°C for Swedish Standards house). There again an effect is observed as now one can bridge over short bad weather periods. As already mentioned the other house types are affected even less by changes in the internal heat capacity.

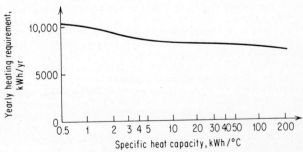

Fig. 29.22 Yearly heating requirement as function of specific heat capacity of Swedish Standard's house ($C_{normal} = 2.424$ kWh/°C), Hamburg, 1973.

Infiltration

The case of air infiltration takes on two forms: first, uncontrolled ventilation [1,55,56,57] and second, controlled ventilation. [18,20] The uncontrolled/losses of a building can be defined as:

$$Q_{L_U}^A = n_U \, V_H \, \rho_A C_A \, (T_H - T_{amb})$$ (29.2)

where n_U = the number of uncontrolled air changes per hour
V_H = the interior volume of the building
ρ_A, C_A = the relevant air density and heat capacity, respectively
T_H = the relevant building temperature and
T_{amb} = the ambient temperature

In the case of a building with only controlled ventilation losses these losses may be written as

$$Q_{L_C}^A = n_C V_H \rho_A C_A \, (1 - \eta_{CV}) \, (T_H - T_{amb})$$ (29.3)

where n_C = the number of controlled air changes per hour and
η_{CV} = the efficiency (based on the total enthalpy) of the heat recovery device

It is easily seen that the total ventilation loss, controlled and uncontrolled may, to first order, be written as

$$Q_L^A = \left[(n_U + n_C)\left(1 - \frac{n_C \, \eta_{CV}}{(n_U + n_C)} \right) \right] V_H \rho_A \, C_A \, (T_H - T_{amb})$$ (29.4)

where the factor in square brackets takes on the role of an effective air change factor per hour n_e. This factor is naturally a measure of the average air diffusion and so is implicitly dependent on wind velocity, pressure differences, relative humidity, and temperature. [1,18,20,32,55] We define an average $\bar{n}_e$ in terms of the average climatic properties of any location and for each of the three house types given above. Thus $\bar{n}_e$ is a normalization parameter giving a measure of the effective tightness of a building. Varying this yearly average effective air change number for each house type results in similar changes in the heating requirement in each case. Specifically, changing $\bar{n}_e$ from 1.0 to 0.0 results in a saving of ~6200 kWh/yr for the experimental house, ~7900 kWh/yr for the Swedish Standards house, and ~9500 kWh/yr for the normal house.

Thus, tightening a house up and installing a heat recovery device has the same energy effect on each of the above houses for the same change in $\bar{n}_e$. However, the percent change in the heating requirement is far greater for the Experimental house than for the Normal house. So, the most important factor determining the heating requirement in the Experimental house where all other conservation measures have been adopted, is the air infiltration. Only a small change of 0.1/h results in a 620-kWh/year increase of the heating requirement. In this case heat recovery and controlled ventilation are imperative. The case of control application to energy conservation is not only limited to this sort of heat recovery mentioned above. In fact, for multizone building structures it takes the form of air handling or ventilation management. [5] Here it is necessary not only to minimize the uncontrolled ventilation but also to use an optimum control strategy. A detailed account of such approaches is given in the HVAC literature for multistory buildings. [5]

A unique method of decreasing infiltration losses was used in the Philips Experimental House. This method involved coupling the controlled air inlet of a tight house to a hollow cinderbrick wall (Fig. 29.23) normally used for water drainage about the cellar walls. Coupling this wall thermally to the earth by burial makes the wall an effective heat exchanger. Since the earth at the depth of 3-5 m (depending on the location) has an almost constant temperature which is equal to the average yearly ambient temperature of the location, such a wall system can be used for cooling the incoming air to the house in the summer and preheating it in the winter. Figure 29.24 shows the results of such a system. Here it is observed that although over the month of December the average daily ambient temperature varied from about 7 to −4°C the outlet temperature of the hollow cinderbrick wall was almost constant at 12°C. Furthermore, it is observed that about 500 kWh of preheating were obtained from this wall. During summer, ambient temperature and humidity fluctuations are damped out by this wall. It was found that

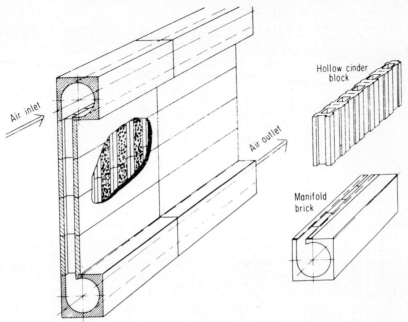

Fig. 29.23 Porous block wall section and details.

a constant one ton (12,000 Btu/hr, 3.5 kW) cooling effect could be obtained by this wall which acts as a half-year phase shifter. (See Fig. 29.24b.)

A simple rule of thumb which may be used to estimate the pre-heating potential of such a wall at steady state is

$$\frac{k}{2}\,[\overline{T}_{earth}\,(d_{min}) - \overline{T}_{amb}],\ \text{W/m}^2 \tag{29.5}$$

where k is the effective earth conductivity, W/mK

$\overline{T}_{amb}$ is the mean monthly ambient temperature and

$\overline{T}_{earth}\,(d_{min})$ is the unperturbed temperature of the earth at the bottom, d_{min}, of the wall.

Under nonsteady-state conditions a much higher exploitation rate may be achieved over short time periods (i.e., until the "neighboring" heat capacity is drained).

Energy Consumption due to Different Heating System Types

Most calculations which are made, including the ones presented above, do not take into account the effect of different types of *real* heating systems along with their resulting associated real air circulation. Typically, each type of heating system (e.g., floor heating, warm water or electrical radiator heating, or forced warm air convective heating) results in distinct radiation exchanges between walls and/or convective exchanges determined by the inside air circulation. These, in turn, influence the comfort of the inhabitant, the temperature setting he chooses, the distribution of temperatures at the walls and in the air and so, finally, the total energy requirement of a building. Several interesting studies have been reported on these points. [14,34,60] It has been found [34] that large area radiative heating such as floor and ceiling heating requires less energy and tends to have a better "comfort" effect than hot-water radiator and warm air forced convection. Moreover, forced convection tends to dominate over the use of wall-located hot-water radiators. These points should be considered when an optimization in terms of energy conservation measures is considered as they can account for up to a 10 percent effect in increasing energy use for heating. [14]

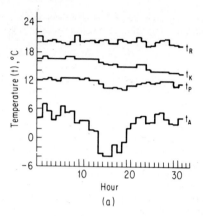

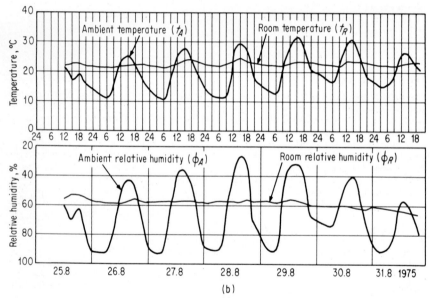

Fig. 29.24   (a) Room t_R, celler t_K, porous wall t_p, and ambient temperatures t_A for the Philips house, December 1975; (b) temperatures and humidity ϕ for August, 1975.

Heat Recovery from Hot Water

An important energy loss which occurs in buildings but is not associated with the building envelope or HVAC system is the heat lost from waste hot water. Typically, in the residential sector a hot-water usage load of 50-150 L/day per person may be expected on an average in western countries. This broad band in average usage reflects effects such as age, social level, and the particular country considered. In Central Europe one can say that the typical average cold water inlet temperature is about 12°C and the hot-water usage temperature about 45°C[10] which for about 75 L/day average per person gives an average energy consumption of about 1120 kWh/person per year. In the United States a typical average inlet temperature of 10°C may be used (for the northern states) and a hot-water usage temperature of 50°C[35] for an average of 100 L/day per person. This gives an average energy consumption for hot-water production of about 1700 kWh/year per person. Comparing these figures with the heating requirements or heating and cooling requirements mentioned above it is noted that *new house codes such as the Swedish Standards imply a yearly heating or cooling load which is of the same order as the yearly hot-water load.* How then can this waste energy be made useful?

There are many systems which have been proposed and put to use to recover the heat contained in service hot water. One such system is a delayed action storage pipe contained in the floor of a house.[47] Such a system is capable of storing a bit over a day's hot-water supply and allows the heat to diffuse back into the living area, thus reducing the heating requirement during the heating season. Typically, for a Swedish Standards house in Central Europe such a system would allow for about a 35 percent heat recovery over the year with no additional energy input. Another system which may be used is presented in Fig. 29.25. It is a system which uses a cold-water pre-heating circuit plus heat recovery via a heat pump and a waste tank to the hot-water user tank.[11] Although new problems arise here insofar as the forming of scum inside of the waste water tank, which has as one consequence the lowering of the performance of the heat exchangers, it is, nevertheless, an attractive alternative.

Using a simple calculation method without the pre-heating described in Fig. 29.25 it can easily be estimated (for tank volume $V = 100$ L/person, pipe heat loss ≈ 25 kWh/year per person, waste water tank heat loss ≈ 50 kWh/year per person, average waste water tank temperature $\approx 30°C$) that the total amount of auxiliary required is $Q_{aux} \approx 260$ kWh/year per person, using a ~ 200-W/person heat pump. A further boost in saving of a few percent may be achieved[11] if a pre-heating circuit is used. However, it is doubtful that it would pay its way if the heat pump is already in use.

BUILDING CODES

Legislative Activities

The rational use of energy has become an important goal in modern building codes. All aspects under which regulations are developed, namely, technical, economic, environmental, hygienic, aesthetic as well as safety and security ones, must be attuned to the economic optimization of the total energy consumption in buildings.

At the present time the codes and regulations in many countries are in the course of legislation or discussion and will be enacted in the near future. Therefore it is too early to present a general review of such regulations. Here we shall consider some examples that are already known in order to demonstrate the trends and some possible legislative initiatives.

Generally the official codes rely on technical standards which have been developed from experience and a knowledge of the thermophysical properties of building materials, taking into account the climatic conditions in the particular country and different user purposes. These standards may be more or less performance-based, i.e., they may be setting goals without specifying the methods, materials, or processes to be used in achieving them, or they may to some extent be prescriptive, which means achieving a given goal by specifying the materials and construction of the components.

Energy conservation activities in the United States result from federal as well as state legislation. A large number of energy conservation measures came from state legislation stimulated by public concern or by federal requirement.

During the early stages of these activities the approaches to energy conservation at state level differed widely, primarily because at least half the states did not have any statewide building codes. Consequently, the preparation of conservation measures included such topics as

- Setting up study committees.
- Enacting new codes.
- Adding conservation standards to existing building codes.

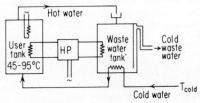

Fig. 29.25   Heat recovery from waste warm water by use of a heat pump and a heat exchanger.

- Implementing regulations concerning the quality and installation of energy conservation materials.
- Funding conservation programs, giving financial incentives.

Every state has now at least an energy conservation plan, owing primarily to the Federal Energy Policy and Conservation Act (EPCA), signed into law in December 1975. That act approved a 3-year, $150 million program for state energy conservation plans designed to reduce energy consumption by at least 5 percent below the expected energy consumption level for 1980.

To qualify for EPCA funds, the state programs had to include:
- Mandatory thermal efficiency and insulation standards for new and renovated buildings.
- Programs to promote carpools, vanpools, and public transportation.
- A right-turn-on-red law or regulation.
- Mandatory lighting efficiency standards for public buildings.
- Mandatory energy efficiency standards for procurement practices of the state and its political subdivisions.

The state programs prompted by EPCA often included innovative conservation plans in addition to those prescribed by the act, and some programs anticipated even more than a 5 percent saving of energy before the next decade. Oregon passed an energy conservation bill with a goal of 20 percent energy reduction by 1980.

State legislatures that attempted partial or total compliance with EPCA's building energy efficiency requirements took two general approaches—authorization for development of a standard by a designated body, or specification in a bill of a particular code to be adopted, frequently one recommended by Congress as a minimum standard. That standard is ASHRAE 90-75.

More weight was given to state energy conservation measures in 1976 with the enactment of the federal Energy Conservation and Production Act (ECPA). ECPA extended the life of the Federal Energy Administration by 18 months and directed HUD to develop, by 1979, mandatory performance standards for energy efficiency in all new commercial and residential buildings. The act also set aside $5 million in the 1977 fiscal year to help states and cities implement the new standards or certify that their building codes met them.

ECPA denies federal financial assistance for construction of any new commercial or residential building in any state which does not adopt the standard or for any building that does not comply with the new regulations. For further information on state legislation activities see Ref. 54.

EPCA and ECPA, together with the President's Energy Program, give the guidelines for all efforts in the field of residential energy conservation. The key provisions of these programs concerning household energy use are summarized in Table 29.16.

An economic analysis and a comparison of the conservation programs using a model developed for the Federal Energy Administration and Energy Research and Development Administration, which projects consumption up to the year 2000, showed that the programs could save both energy for the United States and money for individual households.[28]

Standards in the United States

An example standard is ASHRAE Standard 90-75 which provides design requirements to improve utilization of energy in new buildings. The requirements are directed toward the design of building envelopes with adequate thermal resistance and low air leakage and toward the design and selection of mechanical, electrical, service, and illumination systems and equipment which will make possible the effective use of energy in new buildings. It sets forth requirements for the design of new buildings, covering their exterior envelopes and selection of their HVAC, service water heating, electrical distribution and illuminating systems, and equipment for the effective use of energy. It also covers new buildings that provide facilities or shelter for public assembly, educational, business, mercantile, institutional, warehousing, and residential occupancies, as well as those portions of factory and industrial occupancies which are used primarily for human occupancy, such as office space.

The standard contains detailed instructions and, where necessary, calculation procedures on
- exterior envelopes
- HVAC systems
- HVAC equipment
- service water heating
- electrical distribution systems
- lighting power budget determination procedures
- energy requirements for building designs based on systems analysis

TABLE 29.16   Recent Federal Legislation and Proposals Affecting Residential Energy Use

Energy Policy and Conservation Act (PL 94-163, December 22, 1975)
Residential equipment and appliance labeling (FEA, FTC)
Residential equipment and appliance efficiency targets (FEA)
State energy conservation plans (FEA):
  Thermal efficiency standards for new and renovated buildings
Energy Conservation and Production Act (PL 94-385, August 14, 1976)
Thermal standards for new buildings (HUD)
Financial assistance to weatherproof existing buildings (FEA)
State conservation plans (FEA):
  Public education programs
  Energy audits
Conservation assistance for existing buildings (HUD):
  Demonstration programs
  Financial assistance
Energy conservation obligation guarantees (FEA)
U.S. President's Energy Program (April 20, 1977)
Replacing voluntary appliance efficiency targets by mandatory standards
Thermal standards for new buildings (HUD) by 1980
Retrofit 90 percent of existing residences:
  Tax credits
  Utility conservation programs
  Provision of capital at low-interest rates
  Rural home weatherproofing program
  Increased funding for low-income weatherproofing program

From Ref. 28.

■ requirements for buildings utilizing solar, wind, or nondepleting energy sources.
"Buildings" are divided, according to their user purpose, into two main groups:
  A. Residential with
        A1: detached one- and two-family dwellings,
        A2: all other residential buildings, three stories or less, hotels, and motels.
  B. All other buildings not covered by A1 and A2.
The most important regulations with respect to "energy conservation in buildings" are contained in the chapter on "Exterior Envelope Requirements": the minimum requirements for thermal design of the exterior envelope. The most essential regulation is shown in Fig. 29.26: the minimum average thermal transmittance U_o of the gross wall area of residential houses as a function of annual heating degree days. (See Chapter 13.)

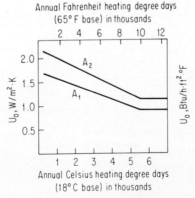

Fig. 29.26 Maximum gross wall area heat transmission loss factor U_o for residential buildings according to ASHRAE Standard 90–75.

The average thermal transmittance of the gross wall area is defined as follows:

$$U_o = \frac{U_{\text{wall}} A_{\text{wall}} + U_{\text{window}} A_{\text{window}} + U_{\text{door}} A_{\text{door}}}{A_o}$$

where U_o = the average thermal transmittance of the gross wall area, Btu/h · ft² · F (W/m² · K)

A_o = the gross area of exterior walls, ft² (m²)

U_{wall} = the thermal transmittance of all elements of the opaque wall area, Btu/h · ft² · F (W/m² · K)

A_{wall} = opaque wall area, ft² (m²)

U_{window} = the thermal transmittance of the window area, Btu/h · ft² · F (W/m² · K)

A_{window} = window area (including sash), ft² (m²)

U_{door} = the thermal transmittance of the door area, Btu/h · ft² · F (W/m² · K) and

A_{door} = door area, ft² (m²)

Some examples for U_o values of some U.S. cities are given in Table 29.17.

In addition to the regulations concerning heat transmission by conduction and radiation, the code also includes minimum performance standards for air leakage rates for all buildings. Standards more stringent than ASHRAE 90-75 are under development.

Legislation and Standards in the Federal Republic of Germany

The new building code of the Federal Republic of Germany is an example of another possibility to consider different building types.[63] It is based on the old technical standard DIN 4108, which was ultimately revised in November 1975.[8]

In this standard, buildings are differentiated according to their inside temperatures:
1. normal (T > 19°C) and
2. low (12°C < T < 19°C),

and their purpose:
3. sports halls and assembly buildings.

In order to maintain some flexibility in the structural design, the limitation of the heat transmission losses of "normal" residential houses may be proved alternatively: either by indicating that minimum requirements applicable to all structural elements of the outer shell are fulfilled (see Fig. 29.22) or by demonstrating that a maximum overall thermal transmission loss factor can be guaranteed. This factor U_{max} depends on the ratio A/V, with A the heat transmitting envelope area and V the volume of the building, as given in Fig. 29.27. The U-value for windows and doors has to be lower than 3.5 W/m²K. The air infiltration rates have to be limited to a

TABLE 29.17 Recommended Heating Outdoor Design Temperature, Heating Degree-Days, and Prescribed Maximum Average Thermal Transmittance for Type "A1"—Buildings According to ASHRAE 90-75 for Some U.S. Cities

City	Design temperature		Heating degree-days		U_o	
	°C	°F	°C	°F	W/m² · K	Btu/h · ft² · F
Fairbanks, AK	−51	−60	7933	14,279	0.91	0.160
Milwaukee, WI	−26	−15	4242	7635	1.11	0.195
Chicago, IL	−23	−10	3688	6639	1.19	0.209
Denver, CO	−23	−10	3491	6283	1.22	0.215
New York, NY	−15	5	2899	5219	1.31	0.230
Albuquerque, NM	−12	10	2416	4348	1.38	0.243
Washington, DC	−12	10	2347	4224	1.39	0.245
San Francisco, CA	2	35	1675	3015	1.49	0.263
Phoenix, AZ	2	35	981	1765	1.60	0.281
Miami, FL	7	45	119	214	1.73	0.304
Honolulu, HI	16	60	0	0		

value which is lower than $a = 2.0$ m³/h and m of sash crack for one-family houses with a presumed air pressure difference of 0.01 kN/m². For comparison, the ASHRAE 90-75 recommended a-value is 2.8 m³/h · m (0.5 ft³/min · ft).

The relation between U_{max} and A/V, given in Fig. 29.27, allows for the fact that heat transmission losses related to volume or living area increase with increasing A/V for buildings with comparable thermal insulation. The A/V-values for one-family houses, for example, lie between 0.6 and 1.2 m⁻¹ (applicable to, respectively, two stories with 45° inclined roof and one-story bungalow type) and in the case of 10-story department houses in the range of 0.3 to 0.4 m⁻¹.

According to the more or less homogeneous weather situation in Germany it seemed admissible to neglect the influence of different climatic conditions in the new codes. Heating degree-days in Germany are about 3300°C-days.

In accordance with the new thermal performance codes a conservation potential of 40 percent for the total heating demand of residential buildings is predicted. For comparison: calculation of the maximum conservation potential due only to insulation (using the same conditions for the computational model) yields savings of 60 percent.[46]

Regulations or at least recommendations concerning the utilization of environmental conditions by passive means are lacking in the German codes, as they are also in the United States or other countries. The thermal performance of windows is taken into consideration, but only insofar as their thermal insulating properties are concerned. Their influence on the dynamic energy balance of buildings and its control by appropriate location, size, and degree of window shading is not taken into account.

Codes and Standards in Sweden

Sweden seems to be the most progressive country in the field of energy conservation in buildings. Since July 1, 1977 codes have been enforced that exceed the scope of others by far. One reason for this may be the geographic and climatic situation of that country, which puts more emphasis on the energy problem.

The general requirements of the conservation code are described as follows: "Buildings with associated installations are so arranged that the energy consumption is limited, having regard to the requirement for good energy conservation. Heat-emission and air-leakage through the enclosing components of a building shall be restricted. In addition, attention must be paid to the possibility of using solar radiation during the cold period of the year to limit energy consumption and also to the output and energy-based consequences of solar radiation during the warm season."

The codes prescribe in great detail the outer and inner construction of buildings for various user purposes, the amount of thermal insulation and air tightness, the thermal indoor climate and air quality, the design of distribution and control systems, and the installation of heat recovery equipment in certain commercial buildings. The codes also contain crisis regulations concerning contingency measures for the case of reduction or cessation of the supply of imported fuel.

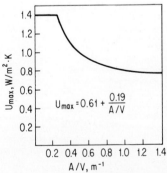

Fig. 29.27 Maximum gross wall area heat transmission loss factor U_{max} for residential buildings in Germany according to DIN 4108.

The maximum permitted heat transmission factor for outer walls of residential buildings ($T > 18°C$) near Stockholm (4652°C-heating degree-days according to ASHRAE definition) is 0.30 W/m² K; in more northerly regions 0.25 W/m² K, the U-values for roofs are, respectively, 0.20 and 0.17 W/m² K. In view of the importance and rigidity of the thermal insulation codes these are given here in detail in Table 29.18. The paragraphs §211 and §212 referred to in this table read as follows:

§ 2   *Heat insulation*
§ 21   *Premises intended to be heated to a temperature higher than* 18°C
§211   "Components for rooms intended to be heated to more than +18°C, such as habitable rooms, are to be arranged so that their heat-transfer coefficients do not exceed the values given in Table 33.21, columns 3 and 4. The window area is determined in relation to the requirement for good energy conservation, but still observing the regulation concerning daylight in Chapter 38.

"Normally approval is given to a window area of a building amounting to 15 percent of the external floor area, with the addition of 13 percent of the internal floor area. The term 'external floor area' refers to that part of the floor area which is bounded by the outer sides of the external walls in which windows are permitted and a line 5.0 m from the external side of these walls."

§212   "The heat-transfer (thermal conductivity) coefficient for building components in Table 33.21, columns 3 and 4, may be exceeded, and deviations from the regulation in §211

TABLE 29.18   Detailed Compilation of Prescribed Maximum Heat Transmission Factors from Swedish Thermal Insulation Regulations

		Requirement according to §211 in temperature zones according to Fig. 33.21		Limiting values according to §212 in temperature zones according to Fig. 33.21	
	Building component	I + II	III + IV	I + II	III + IV
Group	Description				
1	2	3	4	5	6
1	Walls directly exposed to the open air or through earth to the open air	0.25	0.30	0.50	0.60
2	Roofs, or attic floors with roof above, exposed to the open air	0.17	0.20	0.50	0.60
3	Floor structures against the open air	0.17	0.20	0.35	0.40
4	Floor structures over closed space with ventilation openings which should not exceed 0.20 m² per 100 m² surface area of the joist layer	0.30	0.30	0.40	0.45
5	Floor structure on earth (see §247)	0.30	0.30	0.40	0.40
6	Windows and doors facing the open air				
6.1	Nonglazed portions of doors (including frame)	1.00	1.00	1.50	1.50
6.2	Windows and windows in doors (including lintels and frames)*	2.00	2.00	3.00†	3.00†
7	Walls and floors against storage cellars, or other spaces heated to some degree, with temperatures which can drop below 10°C but not below 0°C	0.50	0.50	—	—
8	Walls and floors against staircases, rooms in cellars or other heated spaces with temperatures which can drop below +18°C but not below +10°C	1.00	1.00	—	—

*Frames and lintels shall be so designed that unsuitable cold-bridges do not arise.
†This concerns the U-value for glass components. In rooms where the area of windows and glazed doors within the external measurements of the frame, A_f, amounts to 60% or more of the internal wall area, the value 2.50 applies. A_f does not include the area of nonglazed ventilating panels and nonglazed portions of glazed doors.
Table 33.21: Maximum permitted thermal conductivity coefficient (U-values), W/m²°C for building components for rooms intended to be heated to temperatures higher than +18°C.

may be approved with respect to limitation of window area if the following conditions are shown to be fulfilled:

a. The total heat loss through all the components enclosing the room shall not be greater than if the requirements for the U-value of walls, ceilings, floors, windows, and doors, had been observed in accordance with columns 3 and 4, while at the same time the requirement for limitation of the window area in accordance with §211 has been met.

b. The heat-transfer coefficient of the various components shall not exceed the limiting values given in Table 33.21, columns 5 and 6.

"In calculating the total heat loss it is assumed that the window area amounts to the value approved in §211, i.e. that it is 15 percent of the external floor surface with the addition of 13 percent of the inner floor surface.

"If higher values for the heat-transfer coefficient are chosen for certain building components than those given in Table 33.21, columns 3 and 4, condition (a) can be applied in the following way: The difference between the chosen value and the tabulated value, multiplied by the area of the component, is compensated by a corresponding improvement in the U-value, compared with the actual tabulated value, of another component of the same area.

"On increasing the window area above the approved value, a corresponding procedure is adopted.

"Calculation rules are given in a separate publication of the National Board of Urban Planning."

The climatic regions I and II refer to the colder temperature zones in northern Sweden, III and IV to the central and southern part including Stockholm. The thermal performance of windows, given in Table 29.18, is equivalent to that of three-pane windows.

The new regulations aim at an improvement of 50 to 100 percent in the insulation of residential buildings as compared with the standard of 1977. The Swedish energy commission is expecting savings of 22 to 51 TWh/year (1.9 to 4.4 million tons of oil equivalent/year) for 1985, depending on the future development of energy prices. This corresponds to a conservation potential of 15 to 35 percent in the residential sector.

The aim of energy management in Sweden is to slow down the rate of increase in the country's total energy consumption to 2 percent per annum up to 1985. The energy policy means that energy consumption in the housing sector which can be attributed to heating, ventilation, hot water supply, and other domestic usage will have to be reduced over the above period despite the fact that the housing stock is expected to continue growing both in terms of number of dwellings and volume. The savings scheme assumes that a consumption cut of 0.9 percent per annum will be necessary.*

INCENTIVES

Definition

A great many barriers slow down and/or impede the introduction of measures for the rational use of energy. These are not only the cost of investments for insulation material and devices such as heat recovery equipment, solar collectors, heat pump systems, double glazing, shutters, or the costs of their installation, but also a series of sociopolitical constraints, which may perhaps be less evident but probably more significant. One example is the unequal competition between conventional fuels and solar; nuclear electricity, for example, is governmentally subsidized, coal attracts tax credits, depletion allowances are given for oil, and oil and gas prices are regulated to keep them low.[45] To solve these basic conflicts of energy politics is quite a complex matter and includes a great number of political, social, and economic aspects.

The main purpose of incentives is to diminish the first costs of purchase and installation, because a high capital content is the characteristic feature of all new and unconventional techniques of energy conservation.

The incentives must be adapted to the requirements of the two target groups:

- Producers —manufacturers
  —architects
  —property developers
  —building contractors, etc.

---

*Private communication by J. Höglund, member of the Swedish Energy Commission.

- Consumers   —private home owners (single-family residences)
              —rental housing owners
              —commercial consumers
              —institutional consumers
              —government

Incentives for Producers

Manufacturers of new and unproved technical equipment bear a higher business risk, because during the initial stages of production the actual production costs and, consequently, the costs to the consumer are high due to the diseconomies of small-scale production and the costs of learning the production process. The under-capitalization of a great proportion of building trade companies is a reason for their unwillingness to take the risks involved in innovations.

On the other hand, acceptance of the principles of a rational use of energy and the active and passive utilization of solar energy by architects, property developers, and building contractors is an important factor in the realization of low energy demand programs. According to the new guidelines, alterations in current building designs in order to incorporate collectors, photovoltaic cells, special air ducts, heat exchangers, storage beds, better insulation of windows, etc., and in the choice of building materials will have to be made, and so will adaptations of thermally optimized buildings to the structure of the surroundings. New activities cause delays and consequently costs that can only partially be compensated by the promise of lower energy consumption to the house buyer.

Incentives to producers are:

- Funds for research and development.
- Technical assistance, education, and information.
- Compensation for losses and delays.
- Sales tax relief.
- Rapid depreciation allowances.
- Governmental loans.
- Investment credits.

Large amounts of government funds for research and development have already been given to industry, universities, and public institutions. The high quality standard of today's solar systems and components—admitting that further improvements are still necessary—is due primarily to the strong thrust resulting from extensive governmental funding in that direction.

Special incentives should be considered to promote production of standardized prefabricated components, factory made heating and cooling systems, dwelling units, and prefabricated houses. Tax reliefs, local government low interest loans, additional investment credits, or high depreciation allowances reduce the initial production costs.

Provision of technical assistance, education and information, and compensation for delays are incentives to encourage architects and builders to choose new ways, beyond those given by the new codes. Technical assistance reduces delays in installing equipment and provides greater assurance of the quality of the installed systems.

Incentives for Consumers

High costs of purchase and installation are the major barriers for the private home owner to insulate his house by retrofitting or to use alternative forms of energy. Moreover, with home-buyers not usually accustomed to thinking in terms of lifecycle costing and long-range fuel savings, the desirability of investment in low energy demand equipment is normally not apparent. Therefore, financial assistance should be forthcoming to enhance the desirability of the investment where the benefits resulting from the introduction of low energy equipment would prove profitable.

Owners of rented houses, apartments, and commercial buildings are, on the other hand, concerned primarily with the profitability of solar heating, cooling, and power generating systems and extra insulation, as compared with conventional systems, as well as with the effects on cash flow and return on investment.

Incentives to these consumer groups are:

- Income tax relief by deduction or credit.
- Accelerated depreciation.
- Investment credits.
- Direct subsidies.

- Governmental loans.
- Assistance with availability of mortgages for loans and reduced interest rates.

Income tax deductions can be granted for part of the costs of solar or unconventional equipment utilizing renewable energy sources as well as for installing highly insulating walls or windows. The homeowner (or consumer) deducts this amount from his adjusted gross income. One weakness of this approach is that, because of the graduated rates of income tax in most countries, the actual tax savings from a deduction become greater at higher tax brackets of the taxpayer, thus favoring high income groups.

With the tax credit approach the taxpayer is permitted to reduce his actual tax liability by an amount given by a certain percentage of his investments. The advantage of a tax credit is that its value is independent of the income level of the consumer.

Accelerated depreciation or investment credits are incentive measures which appeal particularly to profit-oriented consumers.

An alternative to direct financial assistance which avoids using the tax structure is to help the homeowner by means of a direct subsidy, whereby the government simply pays part of the initial cost of the equipment or part of the annual mortgage cost.

A government can play a role in assuring the availability of mortgage funds to finance the incremental costs of solar and other energy saving equipment. To overcome difficulties of mortgage availability that are due to the careful attitude of lending institutions toward new product systems and unproven designs it can guarantee or insure bank loans. As an alternative to this, the government can lend homebuyers an amount sufficient to cover all or part of the equipment costs. The loan could be at a rate lower than prevailing bank rates. Such a loan, either alone or in conjunction with a direct subsidy or tax credit, could be an incentive for the application of conservation measures by reducing the monthly mortgage payments.

The application of energy saving measures can be made a condition for the provision of special government-guaranteed mortgages, such as Federal Housing Administration (FHA) or Veteran's Administration (VA) mortgages. These programs, for instance, already contain certain constructional requirements and regulations as a condition for an approved mortgage. Active and passive solar as well as energy conservation equipment can be made an additional requirement when these government-supported mortgages are applied for.

Nonprofit and public institutional and government buildings are easy to use for opening a solar and conservation consumer market and for demonstration purposes in this field. In this case solar equipped and low energy demand buildings can be built under government contract. The federal government has an extensive program of direct construction grants to nonprofit and public institutions. It can effectively use these grant programs to encourage or require energy demand reduction by installation of low energy demand equipment in new institutional constructions.

Actual Efforts by Governments

During recent years there has been considerable activity in the state legislatures of the United States to promote legislation that would support energy conservation efforts in buildings. The results of these activities i.e., codes, regulations, bills, resolutions, files, etc. are regularly collected and reviewed by the Office of Building Standards and Codes Services, Center for Building Technology, Institute for Applied Technology, National Bureau of Standards, Washington, D.C.

Recent summaries of state legislative endeavor[21,22,31,54] show that tax credits or deductions for conservation improvements were the most popular building efficiency incentives. Of about 70 bills, enacted by 30 state legislatures in 1977, some 30 dealt with financial support to consumers or houseowners. Seventeen of them provided for tax exemptions, deductions, or rebates, seven for tax credits, and the rest for direct subsidies, loans, and exemptions of solar equipment from property taxation. Nearly all of these financial reliefs relate to the utilization of renewable energy sources, mainly solar, but only six provide support for energy conservation measures explicitly or in combination with solar installations.* This distribution reflects the characteristic feature of the activities and goals, perceptible up to now in this field: an over-emphasis on the use of unconventional energy sources at the expense of energy conservation.

---

*These statistical considerations are based on Ref. 54 and on a compilation of 1977 state legislation acts, published in "Solar Energy Intelligence Report", December 1977.

Three examples of tax reliefs will be described in detail:

1. California allows an income tax credit of 55 percent of the cost incurred by any taxpayer for purchase and installation of a solar energy system on premises owned and controlled by the taxpayer, and it limits credit to $3000. It allows owners of premises, other than single-family dwellings, for which installations exceed $6000, the choice of whichever is the greater: either $3000, or 25 percent of the cost of purchase and installation. This bill stipulates that if federal income tax credits are enacted for purchase and installations, the allowable state credit will be reduced so that the combined credit will not exceed the 55 percent. The bill considers conservation measures installed with the system to improve efficiency or reduce backup costs.

2. Another tax deduction system is used by Montana:

It allows income and corporate license tax deductions for energy conservation and nonfossil energy generating system expenses according to the following schedule:

For residential buildings: 100 percent of the first $1000 spent, 50 percent of the next $1000 spent, 20 percent of the next $1000 spent, and 10 percent of the next $1000 spent. Maximum allowable deduction would be $1800. For nonresidential buildings the figures for consumption are doubled.

3. This latter schedule resembles the federal tax incentives in the National Energy Act: 30 percent on the first $2000 and 20 percent on the next $8000, for a total maximum tax credit of $2200 on active solar systems and 15 percent of conservation cost up to $300.

Parallels to the development of governmental funding and incentive programs can be found in a number of countries that attempt to promote solar and energy conservation technologies. The government of the Federal Republic of Germany, for instance, tried to enact an incentive program in September 1977 that provided for 25 percent subsidies for the installation of energy conservation equipment, including solar, up to about $1500. The first attempt to pass that program failed, mainly for political reasons; it was enacted a few months later in 1978 with only minor modifications of the original version. This federal-based incentive program of the Federal Republic of Germany is flanked by acts of the constituent states of the Federal Republic of Germany and a number of additional federal incentive bills that form part of structural development, labor provision, and clearing and reconstruction programs.

REFERENCES

1. E.A. Arens and P.B. Williams: "The Effect of Wind on Energy Consumption in Buildings," *Energy and Buildings*, 1, p. 77, 1977.
2. F. Arumi, and M. Hourmanesh: "Energy Performance of Solar Walls: A Computer Analysis," *Energy and Buildings*, 1, p. 167, 1977.
3. F. Arumi: "Daylighting as a Factor in Optimizing the Energy Performance of Buildings," *Energy and Buildings*, 1, p. 175, 1977.
4. F. Arumi and S. Jaeger: "A Comparison of Thermal Requirements of Buildings," *Energy and Buildings*, 1, p. 159, 1977.
5. ASHRAE, "Control Applications for Energy Conservation," *Proceedings*, ASHRAE Semiannual Meeting, Feb. 3–7, Los Angeles, *LA-74-9*, 1974.
6. J.M. Ayres: "Predicting Building Energy Requirements," *Energy and Buildings*, 1, p. 11, 1977.
7. J.D. Balcomb, J.C. Hedstrom, and R.D. McFarland: "Simulation as a Design Tool," *Proceedings*, Conference on Passive Solar Heating and Cooling, Albuquerque NM, p. 238, 1976.
8. Beiblatt DIN 4098, Nov. 1975, Beuth-Verlag BmbH, Berlin 30.
9. R.H. Bezdek and P.D. Maycock: *Proceedings*, Joint Conference 1976 of the American Section of the ISES and the Solar Energy Society of Canada Inc., Winnipeg, vol. 9, p. 64.
10. R. Bruno, W. Hermann, H. Hörster, R. Kersten, and F. Mahdjuri: "Analysis and Optimization of Solar Hot Water System," *Energy Research*, 1, p. 329, 1977.
11. R. Bruno, W. Hermann, H. Hörster, R. Kersten, and F. Mahdjuri: "The Philips Experimental House," *Proceedings*, ISES London on European Solar Houses, p. 1, April 1976.
12. R. Bruno, W. Hermann, H. Hörster, R. Kersten, and K. Klinkenberg: "The Philips Experimental House: A System's Performance Study," *Proceedings*, CCMS and German ISES-Conference, Düsseldorf, April 1978.
13. R. Bruno, and B. Steinmüller: "The Energy Requirements of Buildings," to be published, 1980.
14. Centre Scientifique et Technique de la Construction Syndicat d'etudes inter industries, "Recherches sur les Performances dans le Batiment," Rue de Lombard 26, Bruxelles, Belgium, June 1977.
15. D. Claridge: "Window Management and Energy Savings," *Energy and Buildings*, 1, p. 57, 1977.
16. M.G. Davies: "The Thermal Admittance of Layered Walls," *Building Sciences*, 8, p. 207, 1973.
17. E. Dean and A.H. Rosenfeld: "Modeling Natural Energy Flow in Houses," *Energy and Buildings*, 1, p. 19, 1977.
18. F. Dehli, and H. Vicktor: "Regenerative Wärmetauscher zur Wärmerückgewinnung in Lüftungs- und Klimatechnischen Anlagen," *Elektrowärme International 30*, p. 187, 1972.

19. B. Dittert: "Möglichkeiten der Energieeinsparung im Gebäudebestand," Battelle Institut, Frankfurt/Main, 1977.
20. E. Dreher: "Wärmerückgewinnung in der Luft und Klimatechnik," Schweizerische Blätter für Heizung und Lüftung, p. 2, 1972.
21. Robert M. Eisenhard: "State Solar Energy Legislation of 1976, Review of Statutes Relating to Buildings," NBSIR 77-1297, available from NTIS, Springfield, VA.
22. Robert M. Eisenhard: "Building Energy Authority and Regulations Survey: State Activity," NBSIR 76-986, 1976.
23. "Energy, Global Prospects 1985-2000," Report of the Workshop of Alternative Energy Strategies, McGraw-Hill Book Co. New York, 1977.
24. J. C. Fan et al.: Applied Physics Letters 25, 693, 1974.
25. D.J. Fisk: "Microeconomics and the Demand for Space Heating," Energy, 2, p. 391, 1977.
26. R. Groth and E. Kauer: Philips Technische Rundschau, 25, 352, 1963; F. Mahdjuri, Energy Research 1, 135, 1977.
27. J. Höglund, F. Peterson, S. Sandesten, and S. Stillesjö: "Energibehov för bebyggelse, hushallnings-möjligheter," Report of an expert group for the Swedish Energy Commission, Stockholm, Dec. 1977.
28. E. Hirst, and J. Carney: "Residential Energy Conservation," Energy Policy, p. 211–222, Sept. 1977.
29. Intertechnology Corporation, The US-Energy Problem, Vol. II, Appendices Part A, Nov. 1972, reprinted in Perspectives on the Energy Crisis, Vol. 1, Ann Arbor Science, Ann Arbor, MI, 1977.
30. Jan F. Kreider and Frank Kreith: Solar Heating and Cooling, McGraw-Hill Book Co., New York, 1977.
31. Molly Kuntz: "1977 State Energy Conservation Legislation," Report published by the National Conference of State Legislatures, NCSL, 1978.
32. T. Kusuda: "NBSLD, The Computer Program for Heating and Cooling Loads in Buildings," NBS Building Science Series 69, 1976.
33. T. Kusuda, T. Tsuchiya, and F.J. Powell: "Prediction of Indoor Temperature by Using Equivalent Thermal Mass Response Factors," Proceedings, 5th Symposium on Temperature, NBS, p. 1345, 1971.
34. J. Lebrun, and D. Marret: "Heat Losses of Buildings with Different Heating Systems," Proceedings, Energy Use Management Conference, p. 471, Tucson, AZ, Oct. 1977.
35. J. Lemming, and S. Svendson, eds.: IEA Solar Heating and Cooling Program Task 1 Report, "Investigation of the Performance of Solar Heating and Cooling Systems," 1978.
36. G.A. Lincoln: "Energy Conservation," in Energy, Use, Conservation and Supply, AAAS, New York, 1974.
37. G.A. Lincoln: "Energy Conservation," in Energy, Use, Conservation and Supply, AAAS, New York, 1974.
38. Arthur D. Little Inc., "An Impact Assessment of ASHRAE Standard 90-75, Energy Conservation in New Building Design," Dec. 1975.
39. G.P. Mitalas, and D.G. Stephenson: "Room Thermal Response Factors." ASHRAE Transactions, II, p. III 2.1, 1967.
40. National Economic Development Office, "Energy Conservation in the United Kingdom," London, 1974.
41. Nuclear Energy Policy Study Group, "Nuclear Power, Issues and Choices," Ballinger Publishing Co., Cambridge, MA, 1977.
42. OECD, Energy Balances of OECD-Countries 1973–75, Paris 1977/1960–74, Paris, 1976.
43. OECD, Energy Prospects to 1985, Vol. I, OECD Paris, 1974.
44. "Performance of Solar Heating and Cooling Systems," Proceedings, Joint Conference CCMS and German Section of ISES, Düsseldorf, April 1978.
45. Ron Peterson, President of Grumman Energy Systems, in an address "The Solar Choice," 5th Energy Technology Conference and Exhibition, Feb. 27–March 1, Washington, DC (see "Solar Energy Report," vol. II, no. VI, 16 March, 1978).
46. H. Reents: "Die Entwicklung des sektoralen End- und Nutzenergiebedarfs in der Bundesrepublik Deutschland," Bericht 1452, KFA Jülich, Aug. 1977.
47. B. Rosengren, and E. Morawetz: "The Termoroc House, an Experimental Low Energy House in Sweden," Proceedings, ISES London on European Solar Houses, p. 11, April 1976.
48. R.T. Ruegg: "Solar Heating and Cooling in Buildings: Methods of Economic Evaluation," National Bureau of Standards, Building Economics, NBSIR 75-712, COM-75-11070, July 1975.
49. Science and Public Policy Program, "Energy Alternatives: A Comparative Analysis," University of Oklahoma, Norman, OK, May 1975
50. R.H. Socolow and R.C. Sonderegger: "The Twin River Program on Energy Conservation in Housing: Four Year Summary Report," Center for Environmental Studies Report No. 32, Princeton University, 1976.
51. R.C. Sonderegger: "Harmonic Analysis of Building Thermal Response Applied to the Optimal Location of Insulation within the Walls," Energy and Buildings, 1, p. 131, 1977.
52. R.C. Sonderegger: "Modeling Residential Heat Load from Experimental Data: The Equivalent Thermal Parameters of a House," Proceedings, Energy Use Management Conference, Tucson AZ, p. 183, 1977.
53. Stanford Research Institute, "Patterns of Energy Consumption in the United States," Menlo Park, CA, 1972.
54. "State Conservation Measures in 1977," Energy Report to the States, vol. 4, no. 2, Feb. 1978.
55. G.T. Tamura and A.G. Wilson: "Air Leakage and Pressure Measurements on Two Occupied Houses," ASHRAE Journal, 5, p. 65, 1963.

56. G.T. Tamura: "Predicting Air Leakage for Building Design," *Proceedings*, 6th CIB Congress, *II/2*, p. 368, 1974.

57. G.T. Tamura: "Measurement of Air Leakage Characteristics of House Enclosures," *ASHRAE Transactions*, 2339, 1975.

58. Task Group on Energy Requirements for Heating and Cooling, ASHRAE, "Subroutine Algorithms for Heating and Cooling Loads to Determine Building Energy Requirements," New York, 1975.

59. F. Trombe, J.F. Robert, M. Cabanat, and B. Sesolis: "Some Performance Characteristics of the CNRS Solar House Collectors," *Proceedings*, Passive Solar Heating and Cooling, p. 201, Albuquerque, NM, May 1976.

60. J. Hamay, L. Laret, J. Lebrun, D. Marret, and P. Nusgens: "Thermal Comfort and Energy Consumption in Winter Conditions—A New Experimental Approach," *ASHRAE Transactions*, 1978.

61. R.J.A. van der Bruggen: "Digitale Berekening van de Dynamische Warmtehuishouding van Gebouwen," Technische Hogeschool Eindhoven report, BIB-2771, 7828, 1975.

62. VDI, "Entwicklung des Energiebedarfs und Möglichkeiten der Bedarfsdeckung," Berlin, Dec. 1977, VDI-Bericht Nr. 300.

63. "Verordnung, über einen energiesparenden Wärmeschutz in Gebäuden," Federal Law 1977, part I, p. 1554 et seq., FRG.

64. D. Elliott Wilbur, Jr.: "The Potential for Energy Conservation in Buildings," ADL Impact Services, Sept. 1977.

The International System of Units and Conversion Factors

TABLE A1.1 The Seven SI Base Units

Quantity	Name of unit	Symbol
Length	Meter	m
Mass	Kilogram	kg
Time	Second	sec
Electric current	Ampere	A
Thermodynamic temperature	Kelvin	K
Luminous intensity	Candela	cd
Amount of a substance	Mole	mol

TABLE A1.2 SI Derived Units

Quantity	Name of unit	Symbol
Acceleration	Meters per second squared	m/sec^2
Area	Square meters	m^2
Density	Kilogram per cubic meter	kg/m^3
Dynamic viscosity	Newton-second per square meter	$N \cdot sec/m^2$
Force	Newton ($= 1 \, kg \cdot m/sec^2$)	N
Frequency	Hertz	Hz
Kinematic viscosity	Square meter per second	m^2/sec
Plane angle	Radian	rad
Potential difference	Volt	V
Power	Watt ($= 1 \, J/s$)	W
Pressure	Pascal ($= 1 \, N/m^2$)	Pa
Radiant intensity	Watts per steradian	W/sr
Solid angle	Steradian	sr
Specific heat	Joules per kilogram-Kelvin	$J/kg \cdot K$
Thermal conductivity	Watts per meter-Kelvin	$W/m \cdot K$
Velocity	Meters per second	m/sec
Volume	Cubic meter	m^3
Work, energy, heat	Joule ($= 1 \, N \cdot m$)	J

TABLE A1.3 SI Prefixes

Multiplier	Symbol	Prefix
10^{12}	T	Tera
10^{9}	G	Giga
10^{6}	m	Mega
10^{3}	k	Kilo
10^{2}	h	Hecto
10^{1}	da	Deka
10^{-1}	d	Deci
10^{-2}	c	Centi
10^{-3}	m	Milli
10^{-6}	μ	Micro
10^{-9}	n	Nano
10^{-12}	p	Pico
10^{-15}	f	Femto
10^{-18}	a	Atto

TABLE A1.4 Physical Constants in SI Units

Quantity	Symbol	Value
Avogadro constant	N	6.022169×10^{26} kmol^{-1}
Boltzmann constant	k	1.380622×10^{-23} J/K^{1}
First radiation constant	$C_1 = 2\pi h c^2$	3.741844×10^{-16} W $\cdot$ m^2
Gas constant	R	8.31434×10^{3} J/kmol $\cdot$ K
Planck constant	h	6.626196×10^{-34} J $\cdot$ sec
Second radiation constant	$C_2 = hc/k$	1.438833×10^{-2} m $\cdot$ K
Speed of light in a vacuum	c	2.997925×10^{8} m/sec^{1}
Stefan-Boltzmann constant	σ	5.66961×10^{-8} W/m^2 $\cdot$ K^4

TABLE A1.5 Conversion Factors

Physical quantity	Symbol	Conversion factor
Area	A	1 ft² = 0.0929 m² 1 in² = 6.452 × 10⁻⁴ m²
Density	ρ	1 lb$_m$/ft³ = 16.018 kg/m³ 1 slug/ft³ = 515.379 kg/m³
Heat, energy, or work	Q or W	1 Btu = 1055.1 J 1 cal = 4.186 J 1 ft · lb$_f$ = 1.3558 J 1 hp · hr = 2.685 × 10⁶ J
Force	F	1 lb$_f$ = 4.448 N
Heat flow rate	q	1 Btu/hr = 0.2931 W 1 Btu/sec = 1055.1 W
Heat flux	q/A	1 Btu/hr · ft² = 3.1525 W/m²
Heat-transfer coefficient	h	1 Btu/hr · ft² · F = 5.678 W/m² · K
Length	L	1 ft = 0.3048 m 1 in = 2.54 cm 1 mi = 1.6093 km
Mass	m	1 lb$_m$ = 0.4536 kg 1 slug = 14.594 kg
Mass flow rate	$\dot{m}$	1 lb$_m$/hr = 0.000126 kg/sec 1 lb$_m$/sec = 0.4536 kg/sec
Power	$\dot{W}$	1 hp = 745.7 W 1 ft · lb$_f$/sec = 1.3558 W 1 Btu/sec = 1055.1 W 1 Btu/hr = 0.293 W
Pressure	p	1 lb$_f$/in² = 6894.8 Pa (N/m²) 1 lb$_f$/ft² = 47.88 Pa (N/m²) 1 atm = 101,325 Pa (N/m²)
Radiation	I	1 langley = 41,860 J/m²
Specific heat capacity	c	1 Btu/lb$_m$ · °F = 4187 J/kg · K
Internal energy or enthalpy	e or h	1 Btu/lb$_m$ = 2326.0 J/kg 1 cal/g = 4184 J/kg
Temperature	T	$T(°R) = (9/5)T(K)$ $T(°F) = [T(°C)](9/5) + 32$ $T(°F) = [T(K) - 273.15](9/5) + 32$
Thermal conductivity	k	1 Btu/hr · ft · °F = 1.731 W/m · K
Thermal resistance	R_{th}	1 hr · °F/Btu = 1.8958 K/W
Velocity	V	1 ft/sec = 0.3048 m/sec 1 mi/hr = 0.44703 m/sec
Viscosity, dynamic	μ	1 lb$_m$/ft · sec = 1.488 N · sec/m² 1 cP = 0.00100 N · sec/m²
Viscosity, kinematic	ν	1 ft²/sec = 0.09029 m²/sec 1 ft²/hr = 2.581 × 10⁻⁵ m²/sec
Volume	V	1 ft³ = 0.02832 m³ 1 in³ = 1.6387 × 10⁻⁵ m³ 1 gal (U.S. liq.) = 0.003785 m³
Volumetric flow rate	$\dot{Q}$	1 ft³/min = 0.000472 m³/sec

Nomenclature*

a A dimension; thermal admittance

a_s Solar-azimuth angle (positive east of south), rad (deg)

a_w Wall-azimuth angle (positive east of south), rad (deg)

A Area, m² (ft²); empirical constant used in exponent of diode equation; ideally equal 1, practically up to 2 or even greater.

Å   Angstrom unit

AF   Additional softening factor photorall IV-characteristic

B Daily total beam radiation, kWh/(m²) (day) [Btu/(day) (ft²)]

$\overline{B}$ Monthly-averaged, daily beam radiation, kWh/(m²) (day) [Btu/(day) (ft²)]

c Speed of light, m/s (ft/s)

c_p Specific heat at constant pressure, kJ/(kg) (°C), [Btu/(lb) (°F)]

c_r Specific heat at constant volume, kJ(kg) (°C), [Btu/(lb) (°F)]

C Cost, $, D.M., Fr., £, etc.

CC   Opaque cloud cover expressed as tenth of sky covered: CC = 0 indicates a clear sky; CC = 10 indicates the sky is fully covered with clouds

CF   Curve factor of IV-characteristic

COP   Coefficient of performance

CR   Concentration ratio

CRF   Capital recovery factor

C_{se} Average annual cost of delivered solar energy, $/kJ ($/MMBtu)

d,D A dimension, thickness or diameter, m (ft)

D Daily total diffuse (scattered) radiation, kWh/(m²) (day) [Btu/(day) (ft²)]; diffusion constant (cm²/s)

$\overline{D}$ Monthly-average, daily diffuse (scattered) radiation, kWh/(m²) (day) [Btu/(day) (ft²)]

e Internal energy; eccentricity of earth orbit; electron charge

E Energy, kWh (Btu); electric field strength, V/cm

$E_{b\lambda}$ Spectral emissive power of a black body, i.e., radiative heat flux in a small bandwidth interval centered at λ, W/(m²) (μm) [Btu/(h) (ft²) (μm)]

*Symbols not in this list are used in only one chapter and are defined in that chapter.

E_B Black body emissive power, W/m² [Btu/(h) (ft²)]

f Focal length, m (ft); friction factor; f-ratio of a lens or mirror

f_s Fraction of energy demand delivered by solar system

F Fin efficiency; fuel

F' Plate efficiency, HWB collector model

FF   Fill factor of photocell IV-characteristic

F_{ij} Radiation shape factor

F_R Heat removal factor, HWB collector model

g Gravitational acceleration, m/s² (ft/s²); beam spread parameter

G Mass flow per unit area, kg/(m²) (h) [lb/(ft²) (h)]

Gr   Grashof number

h Enthalpy, kJ/kg (Btu/lb); Planck's constant; convective heat transfer coefficient, W/(m²) (°K) [Btu/(h) (ft²) (°F)]

h_s Local solar-hour angle (measured from local solar noon; 1 h = 15°; positive before noon)

h_{ss}, h_{sr} Hour angle between sunset (or sunrise) and local solar noon

H Daily total horizontal radiation, kWh/(m²) (day) [Btu/(ft²) (day)]

$\overline{H}$ Monthly-averaged, daily horizontal (global) total radiation, kWh/(m²) (day) [Btu/(ft²) (day)]

i Incidence angle, rad (deg); interest rate, decimal percent

I Current, amp; insolation, i.e., the instantaneous hourly solar radiation on a surface, W/m² [Btu/(h) (ft²)]; use subscripts to denote beam, diffuse, reflected, horizontal, tilted, direct normal, etc.

I_o Solar constant, i.e., the amount of solar energy received by a unit area of surface placed perpendicular to the sun's rays outside the earth's atmosphere at the earth's mean distance from the sun, W/m² [Btu/(h) (ft²)]

j Current density, A/m²

k Thermal conductivity, W/m°C [Btu/(h) (ft) (°F)] Boltzmann's constant; mass diffusivity

k_T Instantaneous or hourly ratio of horizontal total radiation on a terrestrial surface to that on the corresponding extraterrestrial surface, i.e., the instantaneous or hourly clearness index; sometimes called percent of possible radiation

K Extinction coefficient, m^{-1} (ft^{-1})

K_T Daily ratio of horizontal total radiation on a terrestrial surface to that on the corresponding extraterrestrial surface, i.e., the daily clearness index

$\overline{K}_T$ Ratio of monthly-averaged, daily horizontal total radiation on a terrestrial surface to that on the corresponding extraterrestrial surface, i.e., the monthly clearness index

l A length, m(ft)

L Latitude, rad (deg); length; thermal load or demand; minority carrier diffusion length, m (ft)

m Air mass, i.e., the distance through which radiation travels from the outer edge of the earth's atmosphere to a recovery point on the earth divided by the distance radiation travels to a point on the equator at sea level when the sun is overhead

m, M Mass, kg (lb$_m$)

$\dot{m}$ Mass flowrate, kg/s or kg/(s) (m²) [lb$_m$/h or lb/(h) (ft²)]; see G

n Index of refraction; number

$\hat{n}_c$ Unit vector normal to collector

$\hat{n}_s$ Unit vector from point on earth's surface to sun

N Density of particles, such as photons, cm^{-2}

NTU   Number of transfer units

Nu   Nusselt number

P Pressure, N/m² (lb/in²); period of sinusoidal function; power

PP   Percent of possible sunshine

P_f Packing factor

Pr   Prandtl number

PV   Present value

PWF   Present worth factor

q Heat flux, W/m^2 [Btu/(h) (ft^2)] (with appropriate subscripts, e.g., q_R = total radiation emitted by R); electronic charge (1.6×10^{-13} C)

Q Quantity of energy or heat, kJ or kWh (Btu)

Q_u Useful thermal energy gain, kW or kWh (Btu/h or Btu)

r Radius, m (ft); escalation rate; reflectance

R Thermal resistance, (°C) (m^2)/W [(°F) (ft^2) (h)/Btu]; tilt factor—denote radiation type by subscript b, d, r, or t; electrical resistance, Ω; ratio; reflectance

Ra   Rayleigh number (= GrPr)

Re   Reynolds number = $\rho v D / \mu$

s Salinity; surface recombination velocity, cm/s

t Time, s (h); thickness, m (ft)

T Temperature, K or °C (°F or °R); tax or tax rate

U Overall heat transfer coefficient, $W/(m^2)$ (°C) [Btu/(h) (ft^2) (°F)]

U_c, U_L Collector loss coefficient, $W/(m^2)$ (°C) [Btu/(h) (ft^2) (°F)]

v Velocity, m/s (ft/s)

V Volume, m^3 (ft^3); voltage, V

VF   Voltage factor of photocell IV-characteristic

W Humidity ratio, kg water/kg dry air (lb water/lb dry air); width

x A dimension or coordinate

z Zenith angle, rad (deg); altitude above mean sea level, m (ft)

GREEK SYMBOLS (MAY BE SUBSCRIPTS)

α Absorptance, solar altitude angle, rad (deg)

$(\alpha\tau)_e$ Effective $\alpha\tau$ product

β Collector tilt angle from horizontal plane, rad (deg); thermal expansivity

β_0 Collector tilt from equatorial plane, rad (deg)

Γ Fraction of radiation falling on certain region of absorber

γ Solar profile angle; optical intercept factor

δ Declination, rad (deg); Kronecker delta

Δ Difference, change in

Δ_m Error in mirror surface and alignment (one-sided deviation from perfect)

Δ_s Angular half width of sun = 4.7 mrad = 1/4°

ϵ Emittance (with appropriate subscripts); heat-exchanger effectiveness

η Efficiency; effectiveness

η_o Optical efficiency

θ, ϕ Angles; latitude, rad (deg)

λ Wavelength; Lagrange multiplier

μ Dynamic viscosity, kg/m·s (lb/ft = s); mean value; charge carrier mobility (cm^2 V^{-1} ϵ^{-1})

ν Frequency; kinematic viscosity, m^2/s (ft^2/s)

ρ Density, kg/m^3 (lb/ft^3); reflectance

σ Stefan-Boltzman constant, W/m^2 K^4 [Btu/(h) (ft^2) (R^4)]; standard deviation; stress solidity ratio

σ_m Standard deviation of mirror errors (one-sided deviation from perfect)

σ_s Standard deviation of sun

τ Transmittance; shear stress; torque; minor carrier lifetime, s

ψ Ground cover factor

ω Solid angle

SUBSCRIPTS

a Air; ambient conditions; aperture; absorber

ab   Absorbent

abs   Absorbed

acc   Accepted

ann   Annual

Ar   Array

atm   Atmospheric

aux   Auxiliary

b, B Beam; particle bed; base region of solar cell

c Collector; convection; condenser; concentrator

coll   Collector

C Carnot cycle

d Diffuse (scattered); delivery conditions; dust; diode

day   Day

dn   Direct normal

D Diffused region of solar cell

e Envelope

es   Envelope to surroundings

eff   Effective

exp   Expansion

f Fluid; fin; fuel

g Glass; global

G Indicates energy gap

h,H Horizontal surface; hydraulic

hp   Heat pump

hw   Hot water

hx   Heat exchanger

i Infiltration; inside; intrinsic

in   Inlet or inside

ir   Infrared

ins   Due to photon of insufficient energy

j Junction (ideal)

k Conduction

l Loss

L Attached to I or j designates light-generated current or current density

m Mirror; air mass; monthly totals; maintenance

max   Maximum

min   Minimum

mp   Maximum power

M Mirror

n Normal

o Reference or standard; extra-terrestrial

oc   Open current

opt   Optimum
out   Outlet or outside
  p Photovoltaic; holes
ph   Photon(s)
  $\perp$ Perpendicular
  $\|$ Parallel
  r Reflected; roof; refracted; receiver
rad   Radiation or radiative
ref   Refrigerant
refl   Reflected
  R Rankine cycle; receiver; rear boundary plane
  s Solar; surface
sc   Short circuit
sky   sky
  sr   Sunrise; series
  ss   sunset
  t Tilted surface; top, total
terr   Terrestrial
tot   Total
  T Total; transition region of solar cell
  u Useful
  v Ventilation
  w Wind; wall; water
  y Yearly
  z Zenith
  ∞ Sink or environmental conditions

BIOGRAPHIES

J. DOUGLAS BALCOMB is assistant energy division leader for solar programs, Los Alamos Scientific Laboratory. In 1979 he was chairman, International Solar Energy Society/American Section and a founder of the passive systems division of ISES. He was also chairman (two terms) of the New Mexico Solar Energy Association and is a member of several solar energy advisory committees. He is the author of numerous technical papers and articles on passive solar heating and other solar heating and cooling methods. He received a PhD in engineering from Massachusetts Institute of Technology (1961) and a BS in electrical engineering from the University of New Mexico.

ROBERT BARBER is a principal of Barber-Nichols Engineering Co., a research and development firm specializing in turbomachinery and thermodynamic applications. Prior to the formation of Barber-Nichols in 1966, he was a principal engineer at Sundstrand Aviation and research engineer at United Aircraft Corporation. He spent 20 years in the field of energy conversion, during which he was widely published. He holds patents in the area of turbomachinery and thermodynamics, and is listed in *Who's Who in the West*. Mr. Barber received his BS in mechanical engineering from Oregon State College in 1957 and his MS in mechanical engineering from Rensselaer Polytechnic Institute in 1960. He is registered to practice engineering in Colorado and California.

WILLIAM A. BECKMAN is professor of mechanical engineering at the University of Wisconsin—Madison. He received his PhD from the University of Michigan in 1964. He was a visiting senior research scientist at CSIRO, Australia in 1968–69 and in 1977–78. Dr. Beckman served as chairman of the Solar Energy Division, American Society of Mechanical Engineers, 1973–74. He co-authored *Solar Energy Thermal Processes* with Dr. John A. Duffie (New York: John Wiley and Sons, 1974). He is a frequent contributor to the growing body of literature concerning the transfer of heat and solar energy.

GÜNTER BERGMANN is a member of the Energy Systems Project at Philips Research Laboratories in Aachen, West Germany. He obtained his diploma in physics from the Technical University of Braunschweig in 1957 and his Dr.rer.nat. from the University of Erlangen in 1962. From 1958 to 1964 he worked for AEG Research Laboratories in Frankfurt/Main on transport phenomena in compound semiconductors. Since 1964 he has been working with Philips Research Laboratories in Aachen on a variety of problems in the field of solid state and gas discharge physics. In 1976 he joined the Energy Systems Project as a coordinator for projects in energy use management and energy conservation in buildings. He is a member of EPS and the German ISES Section.

JOAN BERKOWITZ, senior supervisory staff member in Arthur D. Little, Inc.'s Chemical Systems Section, has been actively concerned with technical and experimental programs in environmental management and materials development for many years. She joined Arthur D. Little in 1957, after a two-year post-doctoral appointment at Yale University. She received her BA from Swarthmore College and her PhD in physical chemistry from the University of Illinois. She is a vice-president of The Electrochemical Society.

SELWYN BLOOME joined Fred Dubin & Associates in 1956, having previously been a designer of mechanical systems in other consulting engineering offices and for power plant constructors in New York City. He is currently executive vice-president and secretary of Dubin-Bloome Associates, P.C. and partner, Fred S. Dubin Associates International. He has been a consulting engineer since 1948 and has served as a design critic at the Schools of Architecture at Pratt Institute; he is a member of many professional organizations. Mr. Bloome has a BS in mechanical engineering from Pratt Institute (1948) and has education in advanced engineering from Cornell University and Ohio State University.

ELDON C. BOES received his PhD in mathematics from Purdue University in 1968. After spending eight years as a mathematics professor, he joined the solar energy department of Sandia Laboratories in 1974. His primary areas of interest in solar energy are solar radiation resource assessment, solar data modeling, photovoltaic concentrator development, and photovoltaic system design. He has authored numerous articles and reports on solar data analysis.

RICHARD BRUNO is a member of ISES and an energy systems analyst at Philips Research Laboratories in Aachen, West Germany. He did his honors bachelor's degree in mathematics and physics at McGill University in Montreal and obtained his PhD in theoretical physics at McMaster University in 1971. After a short post-doctoral he moved to the Technical University of Munich as an assistant professor in solid state physics. He spent the year of 1973–74 as head of the biometrics group at the A.K.L. in Klinik Höhenried before becoming a member of the Energy Systems Project at Philips Research Laboratory in Aachen. During the last few years he has done extensive work on the heat and mass transfer aspects of alternative energy systems as well as on methods for energy conservation.

KEN BUTTI (*left*) and **JOHN PERLIN** (*right*) are Los Angeles based solar energy consultants. They have spent four years extensively researching the worldwide historical use of solar energy. Their research has culminated in museum displays, audio-visual materials, various publications, and their complete history of solar energy: *A Golden Thread: 2500 Years of Solar Architecture and Technology* (New York: Van Nostrand Reinhold Co., 1980).

RICHARD C. CAPUTO is a senior analyst at SERI where he contributed to a number of activities such as the formulation of a solar initiative, contributed to the future energy scenarios developed for the domestic council review (DPR), and developed parity policy scenarios for the national accelerated commercialization plan for solar energy. Mr. Caputo also developed models of future solar energy use in the United States for environmental and social impact evaluation of solar energy. He has an ME in mechanical engineering from Carnegie Institute of Technology (1965) and a BE from Manhattan College, New York City (1961). Graduate study at Penn State University (1966–1967) was in fluid mechanics and thermodynamics.

NORMAN L. DEAN is an Attorney at the law office of Bruce J. Terris and his practice principally involves public interest cases with an emphasis on environmental litigation. Also he is an associate at Johns Hopkins University School of Hygiene and Public Health, teaching a course on environmental law and policy. Mr. Dean was formerly project director and staff attorney at the Environmental Law Institute. He is the author of *Energy Efficiency in Industry* (published by Ballinger, 1979) and is vice co-chairman of a Task Force on Fuel Utilization and Conservation of the National Coal Policy Project. He has numerous publications and has made many speeches in the fields of toxic substances, industrial cogeneration, federal coal leasing policy and energy conservation.

ALAN S. MILLER is a project attorney for the Natural Resources Defense Council, Washington, D. C., and visiting professor of law at Iowa State University. He was formerly a Fulbright Scholar at Macquarie University of Australia, and staff attorney at the Environmental Law Institute. Mr. Miller has a JD and Master of Public Policy from the University of Michigan and is co-author with William A. Thomas and Richard L. Robbins of *Overcoming Legal Uncertainties About Use of Solar Energy Systems* (Chicago: American Bar Foundation, 1978). He has numerous other publications in the areas of solar law, utility regulation, land use, and environmental policy.

GAIL BOYER HAYES is director of energy research at the Environmental Law Institute, Washington, D. C. She was formerly with the President's Council on Environmental Quality and the Illinois Environmental Protection Agency. Ms. Hayes was a television and magazine journalist before turning to environmental law and is the author of *Solar Access Law* (a project for the U. S. Department of Housing and Urban Development, published in 1979). She is director of research projects in fields of solar law, energy conservation, helium conservation, and federal coal leasing policy and has a JD from the University of California (Hastings School of Law).

FRANK O. MEEKER is a research attorney at the Environmental Law Institute, Washington, D. C. He was formerly an attorney for the Domestic Public Land Mobile Radio Service in the Common Carrier Bureau of the Federal Communications Commission and a translator of Japanese legal documents into English. Mr. Meeker is co-author of *Solar Access Law* (a project for the U. S. Department of Housing and Urban Development, published in 1979). His areas of expertise include energy conservation, solar energy, building codes, government patent policy, and state lighting standards, and he has a JD from University of Washington Law School and a BS in electrical engineering from Stanford University.

GRANT P. THOMPSON is a senior associate at the Conservation Foundation, Washington, D. C. He was formerly institute fellow and director of energy research for the Environmental Law Institute, and an associate with Jones, Day, Cockley & Reavis. Mr. Thompson is environmental co-chairman of the Energy Pricing Task Force, National Coal Policy Project, chairman of Consumer Action Now Solar Incentives Workshop, and author of *Building to Save Energy— Legal & Regulatory Approaches* (published by Ballinger in 1979). He has an LLB from Yale Law School and an MA from Oxford University.

FRED DUBIN, PE, is the president of Dubin-Bloome Associates, P. C. He holds a BS in mechanical engineering from Carnegie Tech and an MS in architecture from Pratt Institute. He is a fellow of ASHRAE and ACEC. Mr. Dubin has served as a professor at Columbia and U. S. C. Schools of Architecture. He has been the partner in charge of projects such as the Salk Institute for Biological Studies, the State University at Buffalo, Amherst Campus, and the solar installation for Shiraz Technical Institute.

WILLIAM DUFF is currently an associate professor of mechanical engineering at Colorado State University. He was formerly a senior systems analyst with Stanford Research Institute. He received a BS in mechanical engineering from Cornell University, an MBA from the University of Pennsylvania's Wharton School, an MS in statistics, and a PhD in industrial engineering from Stanford University. Professor Duff has worked with the Solar Energy Applications Laboratory at Colorado State University since 1973. While there, he conducted research in solar heating and cooling of buildings, solar thermal electric power generation, and photovoltaics.

JOHN A. DUFFIE is professor of chemical engineering at the University of Wisconsin—Madison. He received his PhD from the University of Wisconsin in 1951, and has served as instructor at Rensselaer Polytechnic Institute, research engineer for du Pont, and scientific liaison officer for the Office of Naval Research. He has been engaged in solar energy work since 1954, is a past president of the International Solar Energy Society, and has been a Fulbright fellow in Australia in 1964 and 1976. He is a fellow of AIChE and is the author of books and papers in the solar energy field.

GORDON L. DUGGER is supervisor of the aeronautics division and deputy director, ocean energy programs, at the Applied Physics Laboratory of The Johns Hopkins University. He holds BChE and MSE degrees from the University of Florida and a PhD from Case Tech. He did fundamental combustion research at the NACA (now NASA) Lewis Laboratory in Cleveland (1947–54) and was supervisor of chemical process development at the International Minerals and Chemical Laboratory's Research Station (1954–57). Dr. Dugger currently directs APL's work on OTEC heat exchangers, platform design, mission investigations, and system integration. Dr. Dugger has written or co-authored some 100 major reports, papers, and book chapters, including 15 on OTEC.

FRANK EDLIN, PE, has a BS in industrial and chemical engineering from Kansas State University (1931). He worked for the Du Pont Co. from 1937 to 1965 and was executive secretary of the International Solar Energy Society from 1965 to 1967. From 1966 to 1975 he was vice-president of International Plastics, Inc. He received the Manhattan District, Silver Medal in 1945 and is a member of RESA, Sigma Xi, American Association for the Advancement of Science, American Chemical Society, American Institute of Chemical Engineers, American Society of Mechanical Engineers, and the International Solar Energy Society. He has 98 U. S. and foreign patents and has had articles in 17 publications.

DON L. EVANS received a BS in mechanical engineering from the University of Cincinnati in 1962 and a PhD in mechanical engineering and astronautical sciences from Northwestern University in 1967. Since 1966 he has been a member of the mechanical engineering faculty at Arizona State University. His main fields of interest are solar energy technology and utilization and optical diagnostics of high temperature fluid flows.

ANNE FEGE is currently a consultant in bioconversion technology. Formerly she worked as the terrestrial biomass production program manager in the Fuels from Biomass Systems Branch, Department of Energy. She received her Masters of Forest Science degree from Yale School of Forestry, and completed undergraduate biology studies at Kalamazoo College in Michigan. Ms. Fege has previously worked for the Council on Environmental Quality and the Energy Research and Development Administration in managing research programs on the environmental impacts of energy technologies.

MALCOLM D. FRASER is a senior engineer and program manager with InterTechnology/Solar Corporation located in Warrenton, Va., and has been associated with ITC/Solar since 1973. He holds BS, MS, and DSc degrees in chemical engineering from Massachusetts Institute of Technology. As manager of the Alternate Energy Engineering Division at ITC/Solar, he has been involved in research and development on alternate sources of energy, including the production of biomass for its fuel value, the use of solar thermal energy for industrial process heat, and proprietary systems for generating power from waste heat and solar energy. He was previously associated with Gulf Research and Development Company where he was engaged in process development. He holds several patents dealing with petrochemical processes.

HORST HÖRSTER obtained his diploma and doctorate degrees in physics from the Technical University of Aachen. He has been active in research with the Philips Research Laboratories in Aachen in the fields of optical properties of materials, heat and mass transfer, energy conservation, and utilization of solar energy in buildings. He is the director of the department of applied physics and supervises the research programs for the Energy Systems Project. He has contributed several papers to ISES meetings and to European energy meetings. He is chairman of the German Section of ISES.

WILLIAM HUGHES is director of the Engineering Energy Laboratory and Clark A. Dunn professor of engineering at Oklahoma State University. He has been active in alternative energy systems research for almost 20 years with primary concentration in such subjects as energy storage, fuel cells and electrolysis, wind and solar, special electrical energy devices, etc. He is the author and co-author of numerous papers and has several patents in the field. He is a Fellow of IEEE.

SANFORD A. KLEIN is assistant professor of mechanical engineering at the University of Wisconsin—Madison. He received his PhD in chemical engineering from Wisconsin in 1976, and has been an NSF post-doctoral Fellow and lecturer in chemical engineering at Wisconsin. He has authored and co-authored many papers dealing with computer simulation, analysis, and design of solar energy systems. He is co-author with William A. Beckman and John A. Duffie of the book *Solar Heating Design* (New York: Wiley-Interscience, 1977).

JAN F. KREIDER is a consulting engineer specializing in the design and economic analysis of solar energy and energy conservation systems. He is also a visiting professor at the University of Colorado (Boulder) and has written five books and several dozen technical papers on various solar technologies. Dr. Kreider organized the most widely taught solar energy seminars for practicing professionals in the United States—seminars also presented in Europe, the West Indies, and Central America. Dr. Kreider was the solar consultant for the largest and the second-largest solar heating systems in the world, as well as many other very large solar projects. He has also assisted in the design of many solar

systems for residences and has served as consultant to many U.S. and foreign firms, government agencies, and private clients. Dr. Kreider received his BS degree from Case Institute of Technology, and his MS and PhD degrees in engineering from the University of Colorado; he is a registered professional engineer and member of several professional societies.

FRANK KREITH is branch chief of the Thermal Conversion Branch of the Solar Energy Research Institute. He has authored over 100 publications and eight books, some translated into foreign languages, on heat transfer, radiation, and solar energy. Before joining SERI, Dr. Kreith was professor of engineering at the University of Colorado for 19 years. He received a BS degree in mechanical engineering from the University of California at Berkeley in 1945, an MS in engineering from UCLA in 1949, and a Doctorate of the University of Paris (Sorbonne) in 1964. Dr. Kreith is a fellow of the American Society of Mechanical Engineers and a member of the American Association for the Advancement of Science. He has been the recipient of Guggenheim and Fulbright Fellowships and has been a consultant to industry and government and served with the National Science Foundation and the U.S. Academy of Sciences.

THOMAS A. LAWAND did his BS degree in chemical engineering at McGill University and earned his MS in agricultural engineering at Macdonald College of McGill University. He has been employed with the Brace Research Institute since November 1959 and is presently director of field operations, where he has lead in the development of natural energy sources for arid areas of the Third World.

HENRY S. LIERS has been manager of the solar engineering division at InterTechnology/Solar Corporation since 1975. During this time he was the senior project engineer on ITC/Solar's government study to develop the National Plan for Solar Heating and Cooling of Commercial Buildings, as well as on numerous solar heating and cooling design and construction projects. Dr. Liers was involved in studying solar systems performance and analysis on ITC/Solar's Industrial Process Heat Survey. Also, Dr. Liers has been program manager on several ITC/Solar photovoltaic projects. In 1969 Dr. Liers received his PhD in physics at the University of Minnesota. From 1971–1975 he was employed by the Naval Research Laboratory where, among his duties, he was a member of the Laboratory's Energy Advisory Committee.

GEORGE O.G. LÖF is technical director and chairman of the Board of the Solaron Corporation and technical advisor at the Solar Energy Applications Laboratory at Colorado State University. He is a former director of that Laboratory, past president of the International Solar Energy Society, and has written over 150 articles, papers, and books on solar energy over the past 35 years. He is the recipient of numerous awards and prizes, including the Lyndon Baines Johnson Foundation Award in 1976. His training was in chemical engineering, and he holds degrees from the University of Denver and Massachusetts Institute of Technology.

DENNIS K. McLAUGHLIN received his undergraduate education at the University of Manitoba. He attended Massachussetts Institute of Technology for graduate work in the Department of Aeronautics and Astronautics, receiving the SM, EAA, and PhD degrees (the latter in 1970). He joined the faculty at Oklahoma State University in 1970 and is currently an associate professor there. His research interests include experimental fluid dynamics, aerodynamic noise, and wind turbine aerodynamics.

FREDERICK E. NAEF is regional marketing manager for Lockheed Missiles and Space Company. Operating from the Washington, D.C. area, his portfolio includes programs conducted by the Ocean Systems group in Sunnyvale, Calif., the Research Laboratories in Palo Alto, Calif., and the Oceanography Laboratory in San Diego, Calif. Mr. Naef has a BS degree from the U.S. Naval Academy and an MBA degree from the University of Delaware.

ALWIN NEWTON is a 1930 BS graduate of Syracuse University and a 1932 MS graduate of Massachusetts Institute of Technology. He has managed engineering and research in heating and air conditioning for firms such as Honeywell, Chrysler, Coleman, and Borg-Warner. He has done pioneering work in Solar HVAC since 1932, and designed two successful solar-heated homes in 1937. Mr. Newton retired from York Division, Borg-Warner as vice-president and director of research in 1972. He has since consulted in the solar HVAC field for government agencies such as ERDA, DoE, NBS and NASA, and consults for manufacturers and engineering firms in the same fields. He has a total of 209 patents, many of which have been the basis for equipment design.

DARYL PRIGMORE is the office manager at Barber-Nichols Engineering Company. Since joining Barber-Nichols in 1972, his activities have focused on turbomachinery and thermodynamic applications. He has published numerous papers in the energy conversion field and holds a patent on a unique thermodynamic approach for energy utilization in a geothermal energy system. Mr. Prigmore received his BS and MS (with honors) degrees in mechanical engineering from Colorado State University in 1971 and 1972, respectively. He is a registered professional engineer in Colorado.

ARI RABL is currently senior scientist at the Solar Energy Research Institute in Golden, Colo. Recipient of a Fulbright grant, he obtained his PhD in physics from the University of California at Berkeley in 1969, and had postdoctoral appointments at the Weizmann Institute and at Ohio State University. For several years he worked on design, analysis, and testing of solar collectors at Argonne National Laboratory and at the University of Chicago. He has published many articles in the fields of high energy physics, applied optics, heat transfer and solar energy, and has five patents on solar collectors.

R. RAMAKUMAR received his PhD in electrical engineering from Cornell. After serving for several years in Coimbatore Institute of Technology, India, he came to Oklahoma State University where he is currently a professor of electrical engineering. He has published extensively on various aspects of energy and is keenly interested in the harnessing of renewable energy sources in developing countries. His professional affiliations include the International Solar Energy Society and Senior Membership in the IEEE.

ROSALIE T. RUEGG is program leader for solar energy economics in the Building Economics and Regulatory Technology Division of the National Bureau of Standards. The author of many publications on the economics of active and passive solar energy and energy conservation, she has taught energy design economics through the University of California Extension Program and has conducted workshops and lectured widely on these and related topics. A former Woodrow Wilson Fellow and member of Phi Beta Kappa, she holds an MA degree in economics from the University of Maryland.

G. THOMAS SAV, an economist in the Building Economics and Regulatory Technology Division of the National Bureau of Standards, has recently completed several research studies in the microeconomics of solar energy. Previously with the U.S. Nuclear Regulatory Commission, he has conducted econometric studies of electricity demand, cost-benefit studies of alternative pricing policies for energy, and cost analyses of nuclear power generators. Mr. Sav holds an MA degree from the George Washington University and is currently a PhD candidate there.

GARY SKARTVEDT has been technical director of American Heliothermal Corporation since 1975 where he has designed solar space heating and domestic hot-water systems using flat-plate solar collectors. Prior to 1975 Mr. Skartvedt spent 17 years as an aerospace engineer in Martin Marietta's Denver Division. As manager of the propulsion and mechanical engineering department he was responsible for the design of spacecraft propulsion and thermal control systems. Mr. Skartvedt was educated at Stanford University, receiving a BS in mechanical engineering in 1956 and an MS in mechanical engineering in 1975.

J. EDWARD SNYDER, III received his BS in mechanical engineering from Massachusetts Institute of Technology in 1967, his MS from Catholic University, and his PhD from Massachusetts Institute of Technology in 1975, specializing in dynamic systems and control. After undergraduate studies, he worked as an engineering officer in the U.S. Navy. Dr. Snyder is currently a senior project engineer at TRW Systems and Energy and has been involved in the design of ocean thermal energy conversion systems for the past several years. He has published material in several technical journals, is a member of the ASME, and is a registered professional mechanical engineer in California.

CHARLES J. SWET received his BS in naval architecture and marine engineering from the Massachusetts Institute of Technology in 1946. He has held various technical and management positions in the ship-building, textile, steam power and aerospace industries, and has been active since the late 1960s in the development of terrestrial solar thermal systems. He fomerly managed the thermal energy storage program of the U.S. Department of Energy and is currently an energy conversion consultant.

HARRY TABOR is an applied physicist, and for many years was director of the National Physical Laboratory of Israel. He is currently scientific director of the Scientific Research Foundation (SRF) in Jerusalem. Also, he has received international recognition—the 1975 Energy Award of the Royal Society of London and the 1977 Gold Medal of the Diesel Foundation—for his pioneering work in solar energy use related to selective surfaces, collector design, mirror systems, heat engines, and solar ponds. His broader interest in energy problems has led him to initiate, within SRF, an important program in electric vehicles.

ZVI WEINBERGER is a theoretical physicist associated with Solmat Systems Ltd., Jerusalem, and is presently devoting research efforts towards the fuller understanding of the broad spectrum of solar pond problems which must be solved before useful exploitation of solar ponds can be achieved. He was formerly head of the Optics Group of the National Physical Laboratory of Israel where he developed the theory of multilayer selective coatings for metal substrates. He is also academically affiliated with the Jerusalem College of Technology where he directs the optics research program.

WALTER T. WELFORD is professor of physics at Imperial College, London, a School of the University of London. He worked in the optical industry before joining the academic staff of Imperial College as a lecturer in 1951 and has since worked in and taught all aspects of applied optics. He has about 100 publications, including five books. Mr. Welford is a vice-president of the ICO and chairman of the Optical Group of the Institute of Physics. He has PhD and DSc degrees from the University of London and was awarded the Thomas Young Medal in 1973.

BYRON WINN has been actively involved in solar energy since 1973 and is founder and president of Solar Environmental Engineering Co., a Colorado corporation founded early in 1974. He is also a professor of mechanical engineering and member of the Solar Energy Applications Laboratory at Colorado State University. He has directed workshops in the design of solar systems and has written many papers and chapters in books relating to solar energy systems. He serves as a consultant to several industrial and governmental organizations. His graduate study was at Stanford University, and his undergraduate study was at the University of Illinois.

ROLAND WINSTON received undergraduate and graduate training in physics at the University of Chicago (PhD in 1963). After teaching at the University of Pennsylvania, he joined the faculty of the University of Chicago, where he is professor of physics. His primary research interests are high energy physics, infrared astronomy, and solar energy. Since the mid-1960s, he has been developing the optics of nonimaging concentrators and has applied the principles to high energy physics particle detectors, visual receptors, and solar energy concentrators. He was a Sloan Fellow in 1972 and a Guggenheim Fellow in 1978.

MARTIN WOLF is professor of electrical energineering and science at the University of Pennsylvania. Educated at the University of Göttingen, Germany (Dipl. Phys. 1952), his first activities were in microwave measurements and UHF component development. (Admiral Corp. 1952–55). After switching into semiconductor device research and development, he made many contributions to solar cell development and the understanding, performance improvement, and cost reduction of these devices (section manager, Hoffman Semiconductor Division, 1955–61; general manager, Heliotek, 1961–65; and manager of physical research, RCA Astro-Electronics Division, 1965–70).

Index